Advances in Atomic Molecular Collisions

Lokesh C. Tribedi
Editor

Advances in Atomic Molecular Collisions

Springer

Editor
Lokesh C. Tribedi
Department of Nuclear and Atomic Physics
Tata Institute of Fundamental Research
Mumbai, Maharashtra, India

ISBN 978-981-97-7062-5 ISBN 978-981-97-7063-2 (eBook)
https://doi.org/10.1007/978-981-97-7063-2

This Springer imprint is published by the registered company Springer Nature Singapore Pte Ltd.
The registered company address is: 152 Beach Road, #21-01/04 Gateway East, Singapore 189721, Singapore

Preface

This volume deals with a number of topics in the broad field of atomic molecular collision physics. The atomic and molecular physics has undergone extraordinary transformations in last few decades which is driven by a convergence of breakthroughs in sophisticated instruments, theoretical advancements, and computational capabilities. Rapid advancement and availability of a number of sophisticated tools such as, accelerators, ion sources, lasers, synchrotrons, free electron lasers, molecular storage rings, etc., have led to this remarkable transformation. On the other hand, there has been a phenomenal growth in the theoretical front which has not only complemented the experimental data, but also have predicted new mechanisms guiding the experimentalists to venture into the new and challenging experiments. This comprehensive exploration of the intriguing field of molecular collisions aims to capture the essence of these exciting developments, offering a comprehensive overview of the latest insights into the field. Due to the availability of increasingly powerful computers quantum simulations are able to unravel the behavior of complex molecular systems with unprecedented accuracy. This present volume aims to provide experts, researchers, the beginners, and students with a subtle understanding of the frontiers in the field.

The thirteen articles are prepared such that the newcomers in the field will get a glimpse into the field. It will also offer a useful review for the expert researchers. A judicial balance has been maintained between the experiment and theory. All the 30 authors are chosen from different countries, such as, France, Spain, Germany, Greece, Argentina, USA, Canada, Japan, and India, based on their wide experience and expertise in the fields. The articles deal with relatively new discoveries in atomic and molecular physics namely the ICD (inter Coulombic decay) and the Young type electron interferences in a nanoscale molecular double slits as well as the latest development in the high precision atom-clock using highly charged ions. The experimental techniques and the related scientific achievements using an electron spectrometer, the EUV/VUV spectrometer, and high-resolution COLTRIMS have been dealt in detail. The physics associated with the post-collision interaction, the atomic and molecular ICD, spatial coherence in molecular double slit scattering, eikonal impulse approximation in electron transfer, electron capture, and Shannon entropy, fast and ultra-fast irradiation in clusters, atomic and molecular dimers, ionization and electron emission are delved in various chapters. Three chapters are devoted to explore the modeling of various collision dynamics, namely, ICD and ionization involving water targets.

Each chapter, authored by leading scientists in their respective fields, explores into specific facets of atomic collisions, exploring not only the current state of knowledge but also indicating occasionally the pathways forward. The volume thus delves into various aspects, ranging from theoretical frameworks to cutting-edge experimental techniques By bridging theory with experiment in the collection of articles book seeks to inspire curiosity and foster a deeper appreciation for the intricacies of the molecular sciences. Readers including the beginners will have the opportunity to appreciate the complexities of the collision dynamics, explore novel theoretical and experimental techniques, learn new discoveries, and gain insights into the practical applications of atomic collisions in diverse scientific disciplines, such as, science of plasmas, astrochemistry, astrophysics, radiobiology, radiation damage, laser and accelerator technology, etc. It is the hope of the editor and the authors that readers will find inspiration in these articles in order to embark on their own journey of discovery in this fascinating field of research.

I would like to extend my gratitude to all the authors who have spent their valuable time, and shared their expertise and knowledge to make this volume a reality. My appreciation is due to the reviewers and colleagues for providing various crucial inputs throughout this journey. In particular, my special appreciation goes to Roberto Rivarola, Amine Cassimi, Laszlo Gulyas, Karoly Tokesi, Hajime Tanuma and Peter Beiersdorfer for their reviews, suggestions, and feedback which helped to enhance the quality of the volume. It is my pleasure to thank my colleagues, ex-students/post-docs, Deepankar Misra, Ajay Kumar Siddharth Kasthurirangan Sanjeev Maurya, and Madhusree Roy Chowdhury for their contributions towards reviewing, editing and proofreading of some of the articles. I am grateful to my student Debasmita Chakraborty for her diligent overall effort and for spending quality time for refining the articles besides assisting me for solving several technical issues. Finally, I must express my sincere thanks to the 'Springer Nature' team for continuous assistance, help, and patience throughout the entire project.

As an editor my main goal is to assemble a collection that reflects the dynamism of contemporary research in atomic collisions and related physics. It is the hope that this compilation serves as a resource of existing knowledge and a catalyst for future explorations through the fascinating realms of atomic-molecular based interdisciplinary sciences where every discovery uncovers new layers of complexity. Hope the effort will be fruitful if it serves as a valuable resource for researchers, students, and enthusiasts alike, pushing the boundaries of our understanding of atomic physics and its applications.

Mumbai, India
July 2024

Prof. Lokesh C. Tribedi

An introduction to the Editor and the book

The Editor of this book, Prof. Dr. Lokesh C Tribedi, is an internationally renowned researcher in the area of Atomic and Molecular Collision physics. He is a Senior Professor at the TATA Institute of Fundamental Research in Mumbai, India. His expertise in experimental Atomic Physics involves among many other research subjects: ion-molecule or ion-solid interactions, disentangling different mechanisms, such as, Young type electron interference at molecular double slits, collective plasmon resonances and wake field effect, ionization and fragmentation, radiative electron capture, innershell processes and relativistic effects in atomic collisions-just to mention a few. He is well known for his contributions in the field of collision dynamics involving simplest atom to nanoparticles and large molecules having radio-biological and astrophysical applications. He is the leader of an internationally recognized highly vibrant research group in India and has been playing an active role in many international advisory committees and collaborations.

This book displays recent advances in Atomic and Molecular Collisions with chapters including the collaboration of well-recognized specialists, giving at same time a contemporary vision on different related aspects and opening the possibility of new lines to explore. Thus, it is useful for expert researchers as well to guide new interested students.

Rosario, Argentina
October 2024

Prof. Roberto Rivarola
(an eminent scientist
and internationally reputed theorist
in atomic physics)

A Brief Overview of the Scientific Contents

The present volume is a collection of articles on recent advancements in atomic collision physics and related phenomena including experiments as well as theoretical techniques.

The article 'COLTRIMS in Collision Physics' by Amine Cassimi, Xavier Fléchard, Benoit Gervais, Alain Méry, and Jimmy Rangama provides an excellent review about the experimental techniques and the kinematics involved in COLd Target Recoil Ion Momentum Spectroscopy. This technique has been applied not only in ion-atom and molecule collisions but also to photo-ionization as well as in strong field interaction with matter. The COLTRIMS allows one to conduct the kinematically complete collision experiments to investigate the dissociation dynamics and thus revealing the full dynamics of quantum systems. Numerous examples have been provided in regard to its application not only for atoms and molecules but also for the collision dynamics involved in atomic and molecular dimers, which is a step forward towards clusters.

The article by M. Schultz, titled, 'The post-collision interaction in ionization by proton impact' deals with few-body problem (FBP) which remains as one of the most fundamentally important and yet unsolved problems in physics. The models describing the FBP mainly depend on numerical approaches besides a few approximations which have to be tested by experiments. The ionization, which involves three independently moving particles, can be an ideal testing ground for the FBP. In perturbation theory, one higher-order process, i.e., post-collision interaction (PCI), has been studied extensively in-ion-atom ionization which involves at least two interactions between the projectile and target electron. The signature of PCI was revealed in many experiments involving electron spectroscopy and recoil-ion momentum spectroscopy. In this chapter the author presents a comprehensive review and some of the most significant insights obtained from the literature.

The article "Ionization of water, ammonia, and methane by proton collision: experimental and electronic configuration studies" by Wania Wolf deals with the proton induced ionization and fragmentation water, ammonia, and methane which are omnipresent molecules. The fragmentation of these molecules may involve the breaking of bonds, formation of highly reactive fragment species, and the formation of complex molecules in the environment. These products enrich the understanding of the chemistry of planetary atmospheres, ecosystems, or radiation therapy. The author reviews a large set of data from the literature in order to present a relatively simple approach to obtain a broader picture of the fragmentation process which can be extended to complex molecules.

The zero-degree Auger electron spectroscopy has been tremendously successful in identifying the underlying mechanisms of ion-atom ionization. A series of pioneering experiments at HMI, Berlin, and Kansas State University by N. Stolterfoht, Theo Zouros, P. Richard have led these new directions which have been extended by L. C. Tribedi (Mumbai), J. Tanis (Kalamazoo), Freemont (Caen) and A. Itoh (Kyoto). The latest development in this field is well reviewed by Emmanouil P. Benis in the article titled: "Fast ion-atom collisions: Electron spectroscopy of mixed-state beams". The author reviewed the experimental and theoretical findings from fast ion-atom collisions involving mixed-state ion beams based on the zero-degree Auger projectile spectroscopy technique. The author summarizes the results on electron capture as well as transfer excitation and indicates about the ongoing experiments and future direction.

The spectroscopic studies of highly charged ions in the EUV range are an important aspect of atomic physics research as well as for various applications. The experimental studies on this topic have been introduced in a comprehensive article by N. Nakamura, titled: "EUV spectroscopy of highly charged ions with an electron beam ion trap". The study mainly focuses on experiments done with EBITs in Japan. The article indicates the importance of the EUV technique for the study of multipole transitions: E2, M2, E3, and M3 due to the strong dependence of the transition probability on the transition energy as well as the loss of resonances in case of certain HCIs including W-ions important for the ITER related science. The various applications of the EUV techniques towards lithography, astrophysics, ITER-plasma physics, fusion-plasma, etc., are introduced.

The response of clusters and molecules to electromagnetic perturbations has been a topic of numerous investigations. The dynamics of irradiation-involving ionization, electron transport, and dissipative effects leading to thermalization have been reviewed in the article "On the dynamics of fast and ultra-fast irradiation in clusters and molecules" by P.-G. Reinhard and P. M. Dinh and Eric Suraud. They show theoretically, that the clusters and molecules offer a remarkable field of investigation for understanding localized processes in the early phases of the system's response using ultra-fast laser pulses, starting from femtosecond to the attosecond domain. A major physical consequence of the step from simple to more complex molecular systems is the onset of collective effects which are known to play a considerable role in irradiation.

The article-"Inspecting state-selective distributions due to the charge exchange collisions of bare ions with hydrogen"-by Maria S. Gravielle deals with state-selective electron capture process. Within the framework of the eikonal impulse (EI) approximation the author derived state-selective electron capture cross sections from H(1s) by energetic bare HCIs, such as, Li to O-ions. A good agreement of the EI model with the experiments and a realistic description of the process for all the considered collision systems which is a good basis for a derivation of the scaling laws based on the analytical properties of the EI model which is of importance for many fields of science, fusion and astrophysical plasmas etc.

In the article "Inspecting the information quantity in ion-hydrogen electron capture process with the Shannon entropy", Jorg E. Miraglia discusses the information entropy taken place in a single electron capture process. The Shannon entropy is an important quantity in the information theory built up on the information inherent to the possible outcome of the variables. The concept of information entropy was proposed by C. Shannon in his seminal paper of 1948. Since then the information entropy has been used in several fields including Chemistry, heavy ion collision, and even in inelastic collisions as a source of entropy. The quantum numbers of the captured electron, i.e., *n*, *l*, and *m* are considered as all the possible events taking place, like the faces of a "dice". The data sets obtained by M. S. Gravielle in this book are used for the analysis.

The authors M. A. Quinto, N. Esponda, A. Larouze, C. Champion, R. D. Rivarola, and J. M. Monti deal with this topic in the article titled: "Target ionization and electron loss processes induced by neutral and charged hydrogen and helium projectiles in water molecule". One important application of ionization is its use as input to model the radiobiology in hadron therapy and related radiation damage. Such ionizing processes, electron capture, and fragmentation are responsible for the slowing-down of the fast ions in the biological matter which is commonly modeled by water. The neutral-hydrogen or helium induced interactions with molecules have been rarely experimentally investigated. This article deals with the ionization in neutral atom collision with water under CDW-EIS approximation which also includes two major corrections, such as, the energy and the momentum-transfer dependent dynamic projectile charge-state. The theoretical description along with the comparison with the experimental data is the characteristic of this article.

Apart from the state-of-the-art quantum mechanical models for atomic scattering, often the CTMC calculations are used, which are easier to handle. More recently this technique has been successfully applied for collisions involving molecular targets. The article "Classical-trajectory time-dependent mean-field theory for ion-molecule collision problems" by Alba Jorge, Marko Horbatsch, and Tom Kirchner deals with collisions with molecules, such as, water and ammonia. The radiobiological application of collisional aspects involving water is of contemporary interest. They explore the influence of the multicenter model potential for the molecular structure as well as the dynamical charge screening for the projectile-target interaction.

A few key examples where the HCIs could provide a better probe include the parity and time reversal violations, atomic clocks, variation in the fundamental constants, search for new physics beyond the Standard model (BSM) of particle physics. The highly charged ions are reported to have a fractional frequency shift $\sim 10^{-20}$ or smaller, and therefore, are the promising candidates for optical atomic clocks. The article "Perturbed relativistic coupled-cluster calculations of the properties of Ar13+" by B. K. Mani, Palki, Suraj Pandey, and D. Angom reports the development of a perturbed relativistic coupled-cluster theory for the study of the parity nonconservation in atomic systems. They have computed the electric dipole polarizability of ground and the first excited state of Ar^{13+} ions a promising candidate for the modern atomic clock.

Recent times has seen discovery of new mechanisms in atomic collisions. One of such fundamental process is known as intermolecular Coulombic decay (ICD). This is considered as an unconventional nonlocal electronic decay mechanism, featuring ultra-fast non-radiative decay process and ultralong-range for energy transfer between neighboring atoms, in a dimer or cluster. Since then liquid water has been a topic of very active research in the study of ICD which has triggered a remarkable concerted experimental and theoretical research effort. This stems in part from the importance of ICD as a source of low-energy electrons which is known to play a major role in radiation damage in biological matter in aqueous environments. The authors, Philippe F. Weck and Eunja Kim, have unfolded the theoretical methods to describe the ICD in water in the article: 'Intermolecular Coulombic Decay: Geometric and Electronic Structures of Ionized Water'

Similarly another new mechanism influencing ejected electron spectrum is known as Young type electron interference in a molecular double slit which is discovered about two decades back. Although this mechanism was predicted in mid-sixties but detail experimental proof and a clear understanding of the process was unveiled nearly three-to-four decades after its proposition. Since then a plethora of high quality data was published on photon-, ion- and e-impact ionization di-atomic molecules, which have provided rich, new, and insightful information. The implications of these studies towards fundamental quantum mechanical issues are also indicated. This topic is touched upon in the article "A brief review of Young type electron interference" by myself.

Mumbai, India
July 2024

Prof. Lokesh C. Tribedi

Contents

COLTRIMS in Collision Physics

Amine Cassimi, Xavier Fléchard, Benoît Gervais, Alain Méry, and Jimmy Rangama

Abstract The COLd Target Recoil Ion Momentum Spectroscopy (COLTRIMS) has revolutionized Collision Physics in the 90s. By introducing coincidence and imaging techniques in their experimental setup, three pioneering groups (KSU, IKF, and CIMAP) have given a direct access to collision dynamics and correlations between the reaction products. This technique has been first applied to ion–atom and ion–molecule collisions for both high and low projectile energies, from keV to GeV, and soon after to photo-ionization by synchrotron radiation. With the recent advent of short pulse-high power lasers, COLTRIMS technique has been naturally and successfully introduced leading to world-class results on strong field's interaction with matter.

1 Introduction

The COLd Target Recoil Ion Momentum Spectroscopy (COLTRIMS) is the first technique allowing kinematically complete collision experiments in atomic and molecular physics and thus revealing the full dynamics of quantum systems. Born in the field of atomic collisions in the 90s, it has been adopted more recently in photo-ionization experiments, first using synchrotron radiation, then with high power femto- and attosecond lasers and today in XFEL facilities. COLTRIMS is nowadays the first-choice method to investigate the dissociation dynamics resulting from ionization of atoms and molecules.

The beginning of the interest in recoil ions in atomic collisions could be dated in the late 70s, starting by the indirect discovery by X Ray spectroscopy [1] and electron spectroscopy [2] that multiply charged ions can be produced in a single fast ion–atom collision. At that time, the interest in high-energy atomic collisions was increasing using nuclear physics accelerators (KSU, GSI UNILAC, LBNL SuperHILAC, etc.).

A. Cassimi (✉) · B. Gervais · A. Méry · J. Rangama
CIMAP Laboratory, CEA/CNRS/ENSICAEN/UNICAEN, Caen, France
e-mail: amine.cassimi@ganil.fr

X. Fléchard
LPC, CNRS/ENSICAEN/UNICAEN, Caen, France

L. C. Tribedi (ed.), *Advances in Atomic Molecular Collisions*,
https://doi.org/10.1007/978-981-97-7063-2_1

The first evidence that recoil ions were not only multiply charged but also slowly moving appeared in 1979 [3]. In a single collision between 25–45 MeV Cl^{12+} and Ar or Ne, recoil ions with a charge state 11 + for Ar and 8 + for Ne have been observed with a kinetic energy estimated to be lower than 30 eV. These characteristics quickly increased their interest for ion spectroscopy free from Doppler broadening [4], compared to beam foil spectroscopy. Another application foreseen for these newly discovered highly charged slow ions was their use as a secondary ion source for low-energy collisions. Such a recoil ion source has been constructed and exploited for 250–600 eV per charge Ar^{1-10+} collisions with different noble gases (He, Ne, Ar, and Xe) [5]. The main achievement of these experiments was to evidence autoionization in multiple electron capture (named transfer ionization (TI) at that time). With the same idea, to take advantage of the low energy of the recoil ions, and with the same goal, the study of electron capture in low-energy collisions, C. R. Vane et al. injected Ne^{10+} recoil ions produced in high-energy collisions into an electrostatic ion trap filled with a target gas [6]. These series of experiments measured the projectile and target charge state coincidences to determine electron capture cross sections at low collision energy. The first review articles on recoil ion physics appeared in 1983 [7] and in 1991 [8]. Many of these recoil ion sources have been constructed at the end of the 80s [9, 10] based on high-energy ion accelerators. It was even proposed that all ion accelerators should have such an ion source [11]. Such a "very cold ion source" has also been proposed using synchrotron radiation inner-shell ionization [12] for which room temperature kinetic energies have been measured for recoil ions up to Kr^{9+} using a time-focusing time-of-flight spectrometer. It was the time of the race toward low-energy highly charged ion production between the "hammer" (ion impact) and the "scalpel" (photons) methods!

In the late 80s came the time of experiments dedicated to the recoil ion energy measurements in ion atom collisions. These experiments used either a time-focusing time-of-flight spectrometer [13] or magnetic spectrometers without any extraction electric field. The discussion on an expected role of the electrons in the collision dynamics started but no clear effect could be seen [14], and the reaction Q-value was used to explain the recoil ion angular distribution [15]. The main output of this research period was that the resolution on the recoil energy measurement was limited by the target temperature.

The main motivation in improving the energy resolution was to access the atomic collision impact parameters. At that time, the interest in high-energy atomic collisions was increasing, following nuclear physics research topics in which newly constructed facilities delivered their first ion beams (GANIL in 1983, followed by GSI/ SIS18 in 1990). Such high energies were making access to scattering angles nearly impossible in ion–atom collisions for which the impact parameters are large. The goal was to reach a sufficient resolution on the measurement of the recoil ion energy due to momentum transfer, which contains the scattering angle information.

In 1988, the three groups involved in the search of a method to reduce the initial velocity distribution proposed different technical solutions. The Manhattan group (KSU) proposed to use a collimated gas jet based on multicapillaries, while the

Frankfurt group (IKF) started the construction of a 30 K cooled gas target using a cold head [16]. The Caen group (CIMAP) planned to develop a supersonic gas target.

The first official Recoil Ion Momentum Spectroscopy (RIMS) experiment was published in 1992 [17] with the full momentum vector determination through position and time-of-flight techniques, leading to the first low-resolution Q-value measurement in electron capture experiments. At that time, the standard and widely used projectile energy gain technique was limited to low collision energies due to the projectile energy spread limiting the Q-value resolution. The new RIMS technique allowed Q-value determination for a much wider range of collision energy. One year later, RIMS was applied to multiple ionization at high collision energy, addressing already the role of electron momentum in the collision dynamics [18].

The same year the Caen group and the Frankfurt group (who gave up the cold gas target and adopted the supersonic jet solution) presented in conferences their setup based on a supersonic target, which would be the final configuration for the Cold Target Recoil Ion Momentum Spectroscopy (COLTRIMS). The supersonic gas jet was the ideal target for this technique as it combines the two unavoidable assets of low temperature (low velocity spread) and reduced size (1 mm in diameter) that are required to achieve the needed momentum and position resolutions.

It may be interesting to have a look into the pioneer groups who developed the COLTRIMS. These are the Kansas State group, the Frankfurt group, and the Caen group. All three of them are atomic physics groups close to the nuclear physics community. This means that they were familiar with particle detectors (including position-sensitive detectors), fast electronics, and multi-parameter data acquisition, an expertise which at that time was not commonly found in atomic physics laboratories.

The first COLTRIMS experiments were performed in 1994 and from then on have been successfully applied to a wide range of processes in atomic collisions. Low-energy projectile ionization has been the first process studied using the newly developed COLTRIMS technique [19, 20]. COLTRIMS was able to clearly separate electron–electron and electron–nucleus interaction contributions. At the same time, in a high collision energy experiment performed by the Frankfurt group, a COLTRIMS setup equipped with electron detection showed the important role of the ionized electron in the collision dynamics of single ionization [21]. This was the first kinematically complete experiment which established the COLTRIMS technique as a breakthrough for atomic and molecular collisions. These first successes of the COLTRIMS have rapidly been followed by many others. Electron capture at high energy [22] in which a projectile scattering angle uncertainty of 1 μrad was achieved, multiple ionization [21, 23–25] evidencing the role of the electrons in the momentum balance, state-resolved low-energy electron capture [26] determining the final n-state of electrons captured by Ne^{10+} and Ar^{18+} bare ions, electron correlation in double ionization [27], low-energy ionization [28] discussing the saddle point ionization mechanism, and relativistic energy ionization [29] showing the equivalence between collision ionization and photo-ionization were the next processes to be thoroughly analyzed and revealed by COLTRIMS. Two review articles compile the successes of the first decade of COLTRIMS experiments [30, 31].

It is worth noting that from the very beginning, COLTRIMS has been introduced in photo-ionization studies with synchrotron radiation. The first achievement in this field was the clear separation between photo-absorption and Compton effects [32]. The COLTRIMS technique, invented for atomic collision physics, has been adopted later (initiated by the Frankfurt group around the year 2000 [33–35]) by many "photo-ionization" teams and has revolutionized the field of atto-second science.

In 1999, thanks to its many particle coincidence capabilities and its 4π solid angle besides its momentum resolution, COLTRIMS entered the field of electron impact ionization with the first e-3e experiment [36]. The technique was proven to be so efficient that experimental fivefold differential cross sections have been obtained.

Meanwhile, molecular fragmentation has been studied in the pioneering work of Prof. H. Tawara [37] and the first experiment measuring molecular fragments momentum in coincidence has been performed 10 years later [38]. However, the first COLTRIMS experiment devoted to ion-induced molecular fragmentation has been performed in 2000 [39] showing that populated excited states of the molecular ion produced during the collision can be clearly separated on the Kinetic Energy Release (KER) spectra, with a resolution of the order of 300 meV. The first experimental measurement of the electron emission angular distribution in the molecular frame appeared in 2001 [40]. In this experiment using the COLTRIMS technique, the photoelectron and molecular fragments momentum vectors were measured simultaneously, evidencing complex diffraction patterns in the photo-ionization process. The fourth review article on COLTRIMS technique was published at that time [41].

For more than 30 years, the Caen group has been working in the field of recoil ion physics. It is one of the three pioneering groups who invented and developed the COLTRIMS technique. In the 90s, the group has been involved in recoil ion photon spectroscopy [42] and constructed a recoil ion source for very low-energy ion–atom collisions [10, 43]. Meanwhile, the Caen group proposed to use a supersonic gas jet as the cold target and constructed a first-generation COLTRIMS setup around it [44]. It finally appeared that this was the technical breakthrough which made the COLTRIMS possible and has been adopted worldwide. This spectrometer has led to several highlights on high-velocity ionization process [23] as well as on low-energy capture process [26, 45, 46]. Electron momentum measurement has also been added in order to combine COLTRIMS with coincident Auger electron spectroscopy and apply it to low-energy double capture process study [47]. Then the Caen group moved to more complex targets such as molecules and demonstrated that COLTRIMS is a high-resolution spectrometer for dissociating excited state identification of molecular ions [39]. Again, electron detection added to molecular fragments detection allowed to access the electron emission distribution in the molecular frame and search for interference effects in H_2 ionization by fast ions [48, 49]. The Caen group went further by developing a magneto-optical trap for MOTRIMS experiments [50]. This led to the precise measurement of matter-wave diffraction patterns in ion–atom collisions [51] and allowed the experimental study of collisions with oriented targets [52]. Using the COLTRIMS philosophy, the Caen group also developed a setup around a Paul trap to study the β-decay of trapped radioactive species in order to test the standard model. This setup permitted to perform a perfect physics textbook experiment on pure shake-off process [53]. In the recent years, the Caen group has concentrated its efforts on the study of atomic and molecular dimers with several breakthroughs advancing our knowledge on the effect of close surrounding on atomic collision processes [54–60].

This chapter starts with a description of a typical COLTRIMS experimental setup recalling the main features and elements composing such a spectrometer. Then, few equations will illustrate which physics information is extracted from the recoil ion momentum vector. The special case of molecular targets will be presented with a focus on the use of momentum conservation to improve the spectrometer resolution. The next four paragraphs review a selection of several highlights from the group of Caen related to ion collisions with several different targets: atoms, molecules, atomic dimers, and finally molecular dimers.

2 The COLTRIMS Technique

2.1 Experimental Technique

The basic principle of the spectrometers developed by the pioneering groups (Kansas State University, Frankfurt University, and Caen group) is very simple and based on the projection of the ionized target atom, the so-called recoil ion, on a position-sensitive detector by a uniform electric field (0.3 to 5 V/cm) surrounding the collision region (Fig. 1). The recoil ion momentum is deduced from both the position measurement and the time of flight between the collision region and the detector. Thus, a high resolution is only achievable if the initial momentum spread of the target atoms is narrow and if the target size is small. These two conditions are perfectly fulfilled with a supersonic expansion gas jet. Regarding the spectrometer, it has to have a high collection efficiency close to 4π solid angle, and a high resolution for both impact position and time-of-flight measurements. Furthermore, electric (and possibly magnetic) fields have to be precisely controlled for an accurate trajectory reconstruction.

The resolution is further improved by using additional field free regions to fulfill time-focusing conditions (known as Wiley–McLaren condition [62]), or inhomogeneous extracting fields introducing lens effects leading to a better position resolution than the collision area size [46].

The target is the crucial point and the COLTRIMS was born the day a satisfying solution has been found. The Caen group proposed to use a supersonic gas jet as the target in 1992 and rapidly the other groups adopted this device and built spectrometers around it. The adiabatic expansion of the target gas through a small nozzle (10 to 75 μm) at high pressure (1 to 30 bars) forming the supersonic jet provides highly mono-kinetic atoms. While their mean kinetic energy is around 66% higher than the thermal Boltzmann mean value, the momentum spread in the jet is typically below $\pm$ 0.05 a.u. in the parallel direction and $\pm$0.035 a.u. in the transverse direction [61]

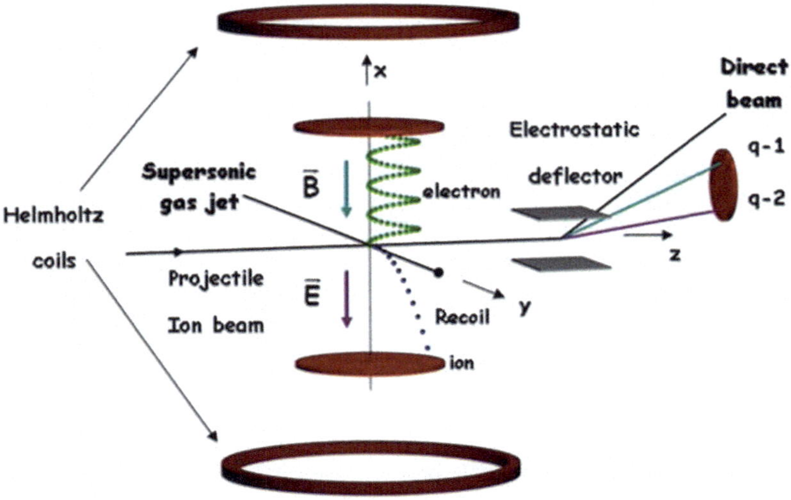

Fig. 1 Typical COLTRIMS spectrometer configuration including a magnetic field for electron detection and the projectile charge state analysis by an electrostatic deflector (from [47])

(Fig. 2). A second advantage of such a gas jet is its small transverse dimension. The typical diameter is of the order of 1 mm and directly benefits to the position measurement resolution. The jet diameter may even be reduced down to 100 μm by collimation when the projectile beam has this size too, as it is often the case in high power laser experiments. A final advantage is its density which can be as high as 10^{12} atoms/cm^3 in the collision region while keeping high vacuum in the collision chamber.

The nearly 4π detection solid angle of such spectrometer configuration makes it compatible with coincident measurement of other particles such as scattered projectiles, electrons, or fragments in case of an exploding molecular target. Note that for an efficient collection efficiency of the electrons, a confining magnetic field of a few Gauss may be superimposed to the extraction electric field. Concerning the projectile, it has been shown that the addition of a charge and momentum analysis in coincidence with the COLTRIMS technique is a powerful way to separate the different processes occurring during an ion–atom collision such as single electron capture, double electron capture (whether it is auto-ionizing or not), and ionization.

COLTRIMS has been applied successfully also to molecular targets as soon as position-sensitive detectors with multi-hit capabilities appeared. The use of such detectors, the next innovation for the spectrometer, allows momentum reconstruction of several fragments detected almost simultaneously. Access was thus given to high-resolution Kinetic Energy Release (KER) measurements (resolution of the order of the vibrational states separation 300 meV) and to the identification of dissociating molecular ion excited states [39]. Such a resolution results from the smart use of the momentum conservation among the fragments coming from a single dissociation.

state selective capture cross sections can be experimentally determined for single capture [26] as well as for multiple capture [46].

2.3 3D Focusing for Molecules: Enhanced Resolution and Selectivity

In the case of molecules, one can proceed as for atoms to determine the fragment momenta. However, for the particular fragmentation channels for which all fragments are detected, meaning all fragments are charged, momentum conservation may be used. Indeed, in the center of mass frame, the momentum sum on all fragments is equal to zero. For two-body fragmentation, this means that fragments are ejected back to back. This leads to the two following important and helpful properties.

For a simple spectrometer in which the fragments are extracted by a single uniform electric field region (no free flight region), the momentum components P_{ix}, P_{iy}, and P_{iz} can be calculated from the initial coordinates (x_0, y_0, z_0); the impact coordinates on the detector (x_i, y_i, z_i); the fragment charge state q_i; and the time of flight T_i, for each fragment denoted as i. The quantities V_{jet} and E are the molecular jet velocity, aligned along the y-axis, and the electric field, aligned along the x-axis, respectively. The z-axis is the ion beam axis (see Fig. 1).

$$\begin{cases} P_{iz} = m_i(z_i - z_0)/T_i \\ P_{iy} = m_i[(y_i - y_0)/T_i - V_{jet}] \\ P_{ix} = \frac{m_i}{T_i}(x_i - x_0) - \frac{1}{2}q_i T_i E \end{cases} \tag{7}$$

For clarity, let's take the simple case of a two-body dissociation (the many-body case may be derived easily). By using the fact that the momentum components are equal, one can eliminate the initial coordinates and calculate the momentum of the fragments by using only the impact coordinates on the detector. This means that the collision region doesn't have to be known and can be large without harming the momentum resolution. Momentum conservation acts as a 3D focusing of the collision region and eliminates the need for a well-defined, point-like, collision region!

$$\begin{cases} P_{1z} = P_{2z} = \frac{m_1 m_2 (z_1 - z_2)}{m_1 T_2 + m_2 T_1} \\ P_{1y} = P_{2y} = \frac{m_1 m_2}{m_1 T_2 + m_2 T_1}\{y_1 - y_2 - (T_1 - T_2)V_{jet}\} \\ P_{1x} = P_{2x} = \frac{m_1}{T_1} L_0 - \frac{1}{2} q_1 T_1 E \end{cases} \tag{8}$$

where L_0 is the spectrometer length. From the above equations, one immediately sees that the momentum resolution depends solely on the position-sensitive detector resolution and on the time measurement resolution. The collision region extension does not play any role anymore.

But one can go further. The two sets of Eqs. (7) and (8) may be rewritten in a way to extract the initial coordinates (x_0, y_0, z_0) of the molecule prior to its fragmentation.

Equation (9) show that it is then possible to calculate back the collision point within the collision region!

$$\begin{cases} z_0 = \frac{m_1 T_2 z_1 + m_2 T_1 z_2}{m_1 T_2 + m_2 T_1} \\ y_0 = \frac{1}{m_1 T_2 + m_2 T_1}\left\{m_1 T_2 y_1 - m_2 T_1 y_2 - T_1 T_2 (m_1 + m_2) V_{jet}\right\} \\ L_0 = \frac{1}{2}\left(\frac{T_1 T_2}{m_1 T_2 + m_2 T_1}\right)(q_1 T_1 + q_2 T_2) E \end{cases} \quad (9)$$

The distribution for each of the three initial coordinates will be used to filter the data and isolate a particular fragmentation channel by applying filters. Figure 3 shows a coincidence spectrum obtained in the case of a mixture of H_2O, D_2O, and HDO. Four fragmentation channels are possible and visible on the spectrum:

- $H^+ + OH^+$
- $H^+ + OD^+$
- $D^+ + OH^+$
- $D^+ + OD^+$

Furthermore, these fragmentation patterns are superimposed on three-body fragmentation patterns and a background of random coincidences. The challenge is to isolate one single fragmentation channel and clean it from random coincidences.

From the set of Eq. (9), the initial coordinate distributions are calculated from all events of Fig. 3 and assuming H^+ and OD^+ fragments. Figure 4 shows the consequence of the sequential application of filters on those distributions as a progressive cleaning from other channels and background.

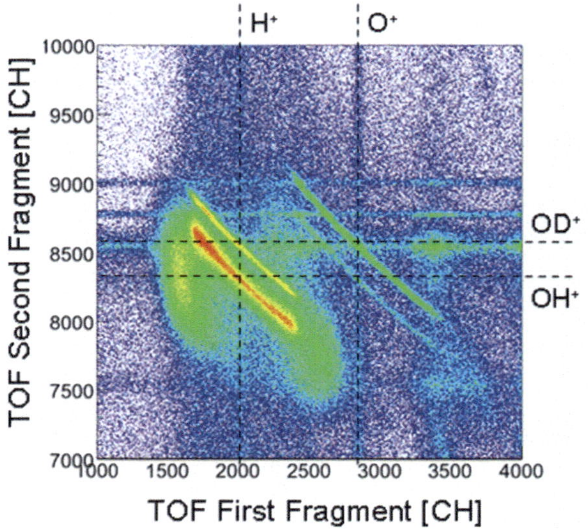

Fig. 3 Coincidence map showing four fragmentation channels for a mixture of H_2O, D_2O, and HDO molecules

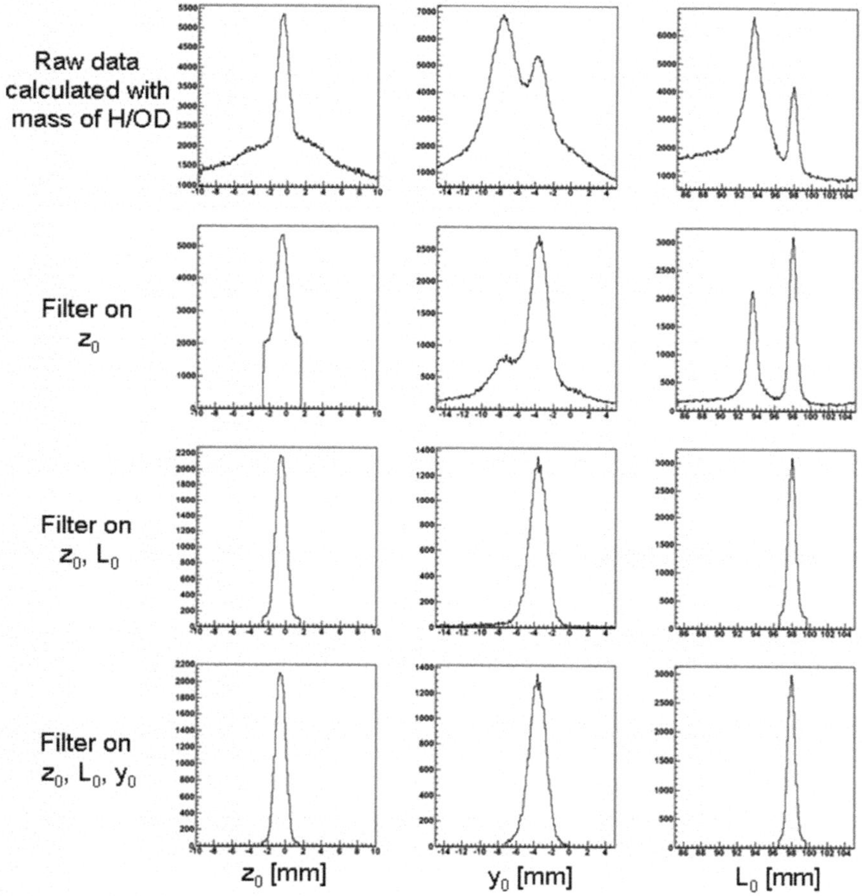

Fig. 4. 3D filtering on initial coordinates distributions calculated for the $H^+ + OD^+$ fragmentation channel

Finally, the filtered coincidence map contains only the events coming from the selected channel, here the channel $H^+ + OD^+$, and cleaned out from the random coincidence background (Fig. 5).

As a conclusion, making use of momentum conservation among the fragments, when it is possible, leads to an improved momentum resolution (3D focusing), a high selectivity between the different fragmentation channels and a strong rejection of random coincidence background.

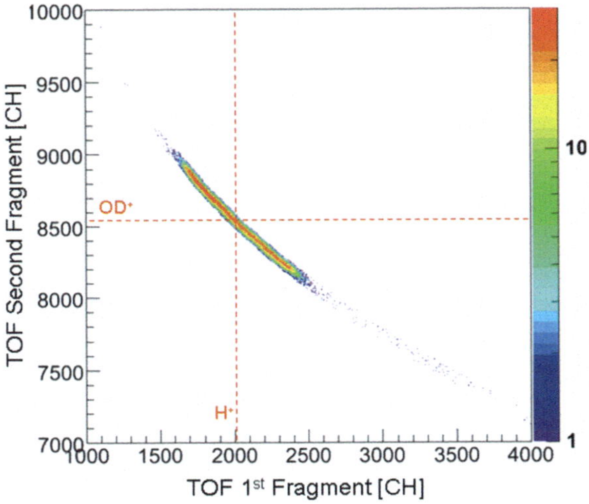

Fig. 5 Events corresponding to the single $H^+ + OD^+$ fragmentation channel after filtering

3 Ion–Atom Collisions

The initial and strong motivation to develop the COLTRIMS technique was certainly the study of the ionization of atoms in collisions with high-energy ions. The idea was to access the impact parameter-dependent cross sections, inferred from the projectile scattering angle. However, it has been clear very soon that electrons play a significant role in the collision momentum balance [24] and their momentum measurement was required for impact parameter determination.

On the contrary, for low collision energy (v < 1 a.u.) where the electron capture process dominates, electrons are bound either to the target, before the collision, or to the projectile, after the collision. In that case, reaction Q-value and scattering angle can be deduced from the recoil ion momentum solely. The very first experiment has been performed with Ar^{18+} and Ne^{10+} projectile ions. At that time, such ion beams were quite exotic because these ions were extremely difficult to produce and extract from the existing ion sources (in that case an ECR ion source). Even if beam intensities were not high enough to be measured with Faraday cups, COLTRIMS is so sensitive that it has been possible to confirm experimentally the $q^{0.75}$ law predicted by CTMC calculations for the capture state principal quantum number [26] up to such high projectile charge states. Later on, doubly differential cross sections for C^{5+} + He have shown that COLTRIMS high resolution gives access to quantum interference effects known as Stueckelberg oscillations [63].

But the strength of COLTRIMS has been clearly proven in double capture process studies. Before this technique was available, two techniques were widely used: energy gain and electron spectroscopies. The energy gain technique was based on the projectile energy change during the collision and the coincident measurement of both projectile and target charge states. True double capture (TDC) and auto-ionizing double capture (ADC) processes were separated and Q-values measured.

The final projectile doubly excited electronic state, in both TDC and ADC cases, was deduced from the Q-value. The only limitation of this technique comes from its energy resolution, directly linked to the projectile beam energy spread, limiting its use to low-energy collisions. Electron spectroscopy doesn't have this limitation and could determine the final projectile state from Auger electron energy measurement. However, this technique is only sensitive to ADC and no experimental information on TDC was accessible. With COLTRIMS, it has been possible to extend the projectile energy range of these measurements and gain access to the energy dependence of the double electron capture process. It was the perfect tool to enter the debate and try to answer the question: does the electron–electron interaction play a role in the double electron capture process? A rough criterion to estimate this electron correlation is to determine whether the two electrons are captured on quasi-equivalent (QE) or non-equivalent (NE) configurations on the projectile, which is made possible thanks to the Q-value resolution of COLTRIMS. The idea is that QE configurations are populated by two independent electron–nucleus interactions while NE configurations rely on electron–electron interaction during the collision. COLTRIMS gave access to the measurement of the collision velocity dependence of the weight of these two contributions on a large range and confirmed that NE configuration population decrease as the energy increase. However, this simple picture was corrected by a combined experimental and theoretical study [46] which showed that NE configurations can also be populated by a transfer from QE configurations due to post-collision effects.

In some cases, QE and NE configurations may be degenerated and correspond to the same Q-value. A full COLTRIMS setup, including electron momentum measurement, is able to remove the degeneracy and separate these configurations taking advantage of their different branching ratio between radiative decay and Auger effect (Fig. 6).

In the 90s, magneto-optical traps (MOT) became almost standard tools in many atomic physics laboratories. Soon after, with the idea that the trapped atomic cloud of a MOT could be used in place of a supersonic jet, three groups (in Denmark, Netherlands, and Kansas) extended the COLTRIMS technique to the so-called MOTRIMS (for magneto-optical trap recoil ion spectroscopy) technique. This technique complements COLTRIMS in the field of ion–atom collisions by providing new targets, such as alkali metal targets, at very low temperature (a fraction of mK) and with a cloud diameter smaller than 1 mm. Moreover, the nature of the target makes it quite easily prepared in an excited state or polarized with a chosen orientation using dedicated lasers for optical pumping. In Caen, we considered the small size and low temperature of the target as an opportunity to push further the resolution on the recoil momentum measurement and scrutinize ion–atom collision dynamics with even more details. The resolution finally achieved on the recoil ion momentum allowed the observation of diffraction-like oscillations in projectile scattering angle distributions. These oscillations were observed for the single electron capture process in low-energy collisions (from 1 to 5 keV) between Na^+ projectiles and Rb atomic targets. They result from the limited range of impact parameters involved in electron charge transfer toward given states and leading to the diffraction of the matter wave of the projectile. These features were evidenced thanks to a resolution in scattering angle better than

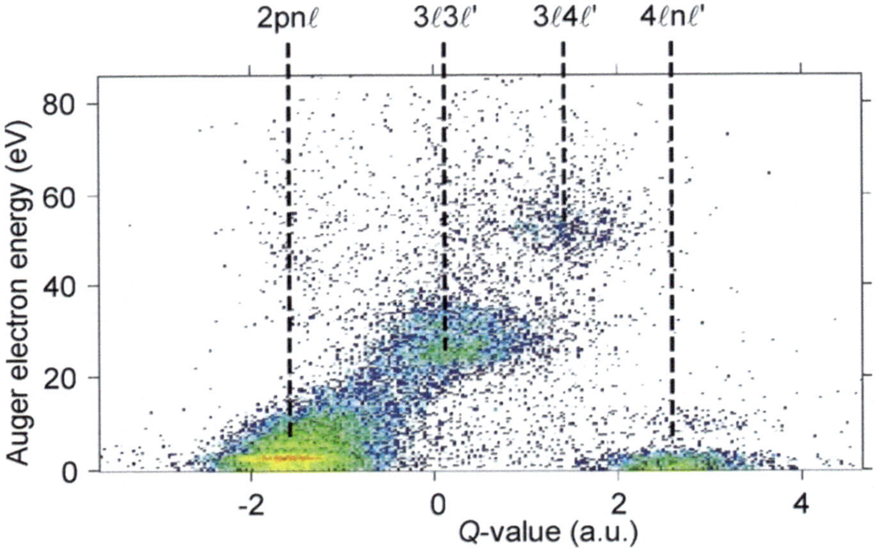

Fig. 6 Two-dimensional intensity spectrum with respect to Auger electron energy and reaction Q-value, for auto-ionizing double capture in 138 keV O^{6+} + He collisions (from [47])

30 μrad. It was thus possible to test semi-classical molecular orbital close-coupling calculations with unprecedented experimental details [51].

In a second step, the ability to polarize the ^{87}Rb atomic target was exploited to explore ion collisions with targets prepared into well-defined oriented states. With such oriented targets, a left-right asymmetry in the transverse momentum components can be expected as, in a classical picture, the relative velocity between the projectile and the active electron depends on the impact parameter position with respect to the electron angular momentum (Fig. 7). Such an asymmetry was indeed observed and measured (Fig. 8), providing a detailed picture of the underlying dynamics, beyond our basically classical intuition [52].

4 Ion–Molecule Collisions

Molecular fragmentation, induced either by ion or photon impact, has been studied for several years using coincident time-of-flight mass spectroscopy. In photo-ionization experiments, such setups are known as the PIPICO technique (**P**hoto**I**on–**P**hoto**I**on **CO**incidence). These experiments could efficiently separate the different fragmentation channels and obtained low-resolution KER distributions allowing discussing their mean values and their width. The main model used at that time to interpret the data was the Coulomb explosion model. This simple model assumes, i.e., in the case of a two-body fragmentation, that the final Kinetic Energy Release (KER) is due

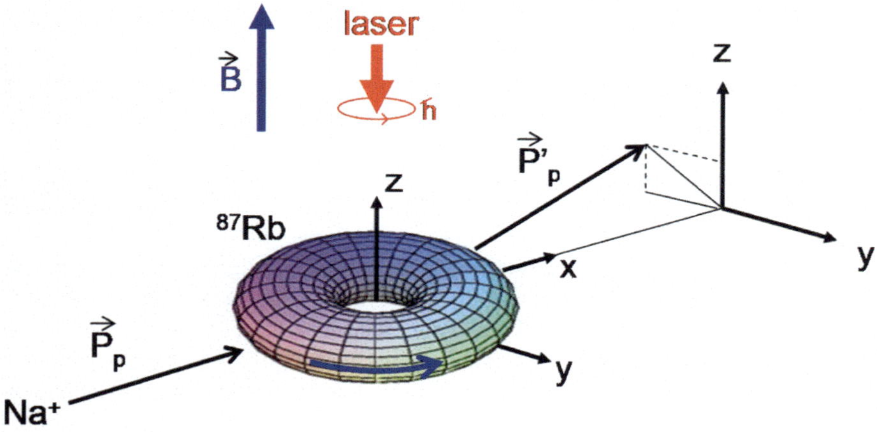

Fig. 7 Schematic drawing illustrating the principle of the experiment with oriented targets. The circularly polarized light of a laser is used to populate sublevels $m_{F'} = F'$ or $m_{F'} = -F'$ of the target

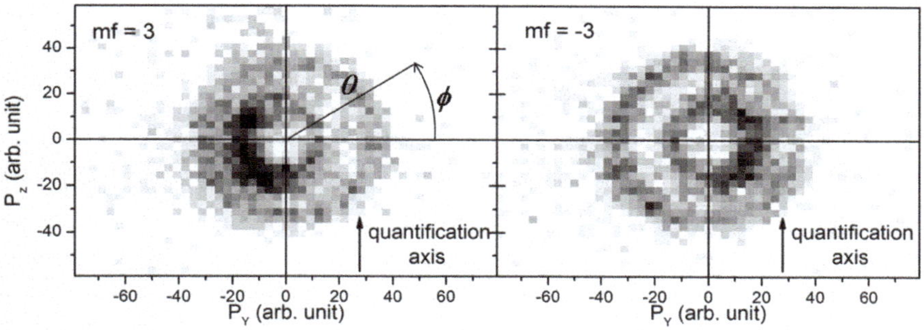

Fig. 8 Recoil momentum distribution in the transverse plane for a Rb target prepared in a $5^2p_{3/2}(F' = 3)$ hyperfine state with angular momentum projection $m_{F'} = 3$ (left panel) and $m_{F'} = -3$ (right panel). The results were obtained in Na^+ + Rb collisions at 1 keV. Both the diffraction pattern due to the matter-wave diffraction and the asymmetry due to the target orientation are clearly visible

to the Coulomb repulsion of two point charges, holding the final charge of the two fragments and standing at the initial position of the fragment atoms in the neutral molecule. This model is able to predict a mean KER value and a KER distribution width by converting the neutral molecule nuclear wave function using the Coulomb potential energy curve of the two charged fragments. Most of the experiments showed that the experimental KER mean values as well as the KER distribution widths were smaller than the model predictions.

As soon as multi-hit position-sensitive detectors appeared, COLTRIMS has been applied to ion impact induced molecular fragmentation. The strong improvement of the resolution obtained with this technique immediately explained the limitation of the pure Coulomb explosion model. Indeed, the first experiments have readily shown that KER is determined by the potential energy surfaces of the fragmenting molecular ion populated by the collision [39, 64]. COLTRIMS provides a new spectroscopy of dissociating states which would not have been accessible with photon spectroscopy.

Later, COLTRIMS has been used to study water molecule fragmentation. More precisely, we measured the ratio between H^+/OD^+ and D^+/OH^+ fragmentation of HDO molecule induced by ion impact. The former channel, H^+/OD^+, has been found to be the most favored, dominating by a factor of 6.5 over the D^+/OH^+ channel [65]. This has been qualitatively explained by the fact that the lighter, and thus faster, H^+ fragment escapes more easily the hydroxyl ion binding potential than the D^+ fragment. This simple picture has been more precisely and quantitatively explained by theoretical calculations of the potential energy surfaces of the water dication [66].

One of the very first complete pictures of the fragmentation of a molecule has been obtained by adding electron momentum measurement (Fig. 2). With such a setup, D_2 molecule fragmentation induced by 13.6 MeV/u S^{15+} ion impact, all the involved and competing processes leading to dissociative ionization have been identified, namely, direct single ionization (DSI), auto-ionization of doubly excited states (AI), ionization excitation (IE), and double ionization (DI) [48]. These results are supported by theoretical calculations (Fig. 9).

5 Ion–Atomic Dimer Collisions

Contrary to covalent molecules, van der Waals atomic dimers, bound by weak dispersion forces, are characterized by much larger internuclear distances (typically more than four times larger). Such distances ask new questions. The two main questions concerned are, on one hand, the existing charge exchange and de-excitation processes between the two atomic centers and, on the other hand, the behavior of the dimer compared to two independent atoms sitting next to each other. We tried to answer these questions through a series of experiments [54–56, 67–69] in which projectiles collide with Ar_2 and Ne_2 dimers produced by a supersonic gas jet. These two questions are interdependent as they could be formulated in the following single question: how collision processes in ion–atom collisions are affected by the atomic target environment? In the present case, the environment, consisting of a single neighbor atom without any shared electron (no covalent bond), is the simplest we can simulate experimentally, but allows a detailed microscopic analysis of its effect on the ion–atom collision.

Indeed, many processes appear following dimer ionization which range from the simple Coulomb explosion due to the simultaneous ionization of the two atoms, to the more exotic Interatomic Coulombic Decay (ICD) which has attracted a huge interest in photo-ionization experiments. Due to the large distance between the two atoms in the dimer, Coulomb explosion corresponds to the dissociation of the dimer, in which both atoms are ionized, following a pure Coulombic potential energy surface. Thus,

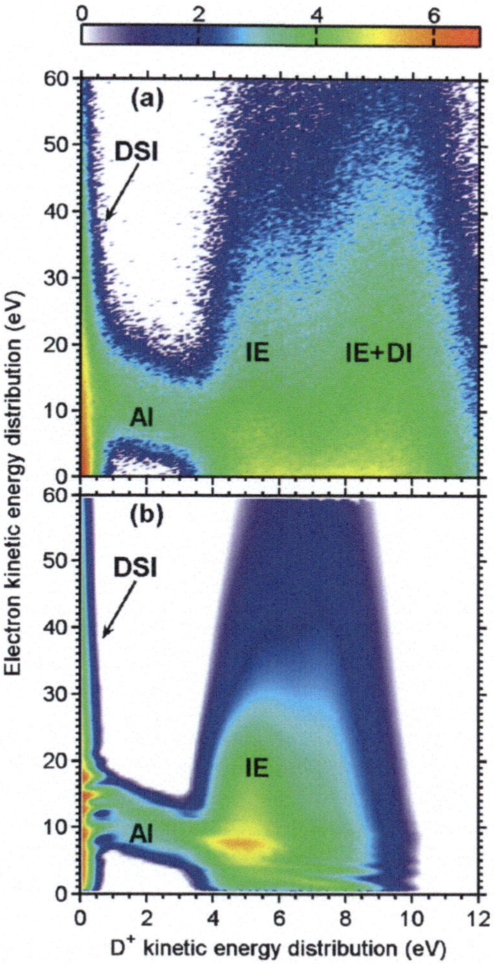

Fig. 9 Electron kinetic energy versus D-+-$^+$ fragment kinetic energy for 13.6 MeV/u $\mathbf{S}^{15+}$-D_2 collisions. **a** Experiment. **b** Theoretical calculation. DSI = direct single ionization, AI = auto-ionizing doubly excited states, IE = ionization + excitation, DI = double ionization (taken from [48])

the mean KER values as well as the KER distribution width are perfectly reproduced by the Coulomb explosion model [54].

In the case of an asymmetric double capture in which two electrons are captured on one of the two atoms, the Radiative Charge Transfer (RCT) process takes place. Indeed, because of the electron cloud polarization of the neutral atom by the doubly charged neighbor, the potential energy surface of the pair is attractive. At a certain interatomic distance, a radiative transition occurs from the initial potential energy surface (PES) to the Coulombic one. It corresponds to the transfer of one electron from the neutral partner to the doubly charged one [54] (Fig. 10). Note that this process is specific to homonuclear atomic pair A_2, for which A^{2+}-A dissociation limit is higher in energy than the A^+-A^+ dissociation limit.

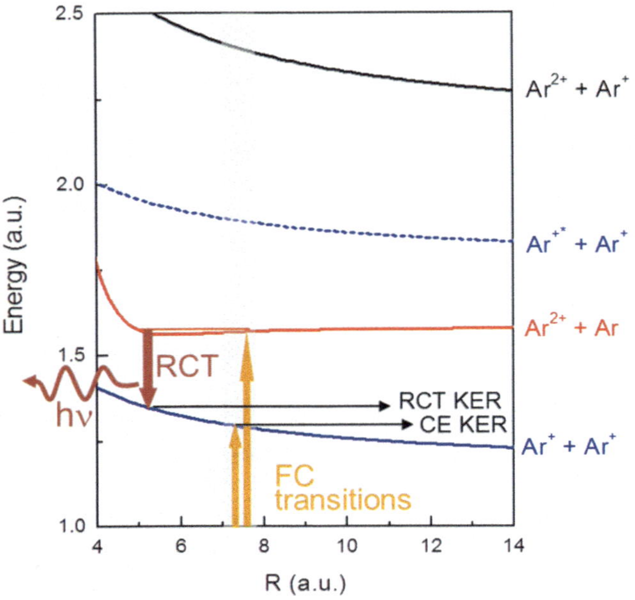

Fig. 10 Calculated potential energy curves for different fragmentation channels of $Ar_2{}^{2+}$ and $Ar_2{}^{3+}$. The gray band corresponds to the Franck–Condon region

The last process has been evidenced 20 years ago in a photo-ionization experiment [70] after being predicted theoretically and widely studied since then. This is the Interatomic Coulombic Decay (ICD) which may occur, i.e., in Ne_2 dimer, after an inner-valence shell ionization of one of the two atoms of the dimer. This vacancy can then be filled by an electron transfer from the neutral partner via a transition whose energy leads to the ejection of one electron, ending with two singly charged atoms. In collision physics, ICD has first been observed induced in high-energy projectile ions collisions [68, 69]. It is only recently that unexpectedly, ICD has been evidenced in low-energy ion–dimer collisions and only for the lower projectile charge state. This was surprising as multiply charged ions are known to capture preferentially outer shell electrons. Indeed, inner-valence electron capture is possible with low charge state projectile ions such as 2.8 keV/u O^{3+} [56]. This is an important finding as ICD is the only process producing low-energy electrons at such low collision energies which do correspond to the region where ions stop in matter. In hadron therapy, this region is where the tumor is located and these electrons correspond to an additional deposited energy dose.

But how "independent" are the two atoms of a dimer? To try to answer this question, we have measured the multiple capture cross sections, either total or differential in dimer angle (relative to the beam axis), and compared them to the predictions of a purely atomic model. In a Monte Carlo simulation, using the classical over the barrier model (COBM), widely used in low-energy electron capture process, we assumed that the dimer is composed of completely independent atoms [58]. If strong molecular effects are present, deviations from the simulation should appear when compared to experimental data. This model has been able to reproduce relative yields for the different electron capture and fragmentation channels as well as the angular distributions for the different fragmentation channels [55]. Furthermore, the use of model calculations combined with the measurements have shown that electrons are preferentially captured from the atomic site standing closer to the projectile, the "near site", during the collision.

6 Ion–Molecular Dimer Collisions

The question of the target surrounding is of importance whenever one wishes to extrapolate gas phase results to condensed matter such as liquids or solids. It is the case for biological molecules and their fragmentation measured in the gas phase. Are these results pertinent and useful to help understanding processes involved for example in a solvent (i.e., in hadron therapy)? Molecular dimers are good candidates to try to answer this question. They provide a mean to see how fragmentation of a molecule depends on whether the molecule is isolated or close to a molecule nearby, thus mimicking in a simple way the environment of a medium denser than the gas phase.

Once again, COLTRIMS is the perfect method to study the dynamics of such systems by separating unambiguously the different fragmentation channels as well as giving the complete kinematic picture of the dissociation process dynamics involving several fragments. In a recent study, we measured the KER spectrum of N_2 molecule dication for $N^+ + N^+$ and $N^+ + N^{2+}$ fragmentation channels after 90 keV Ar^{9+} ion collision, for isolated molecules and for $(N_2)_2$ dimers [57]. Unexpectedly, and apart from an energy shift due to the Coulomb repulsion caused by the presence of the charged neighbor $N_2{}^+$, the KER measured for the molecule coming from a dimer is very similar to the one for the isolated N_2 molecule (Fig. 11). The presence of neighbor molecules does not seem to modify the first steps of the dissociation, which is usually called the chemical-physics step in radiation chemistry. In both cases, the same excited states of the molecular ion $N_2{}^{2+}$ are populated during the collision with similar cross section.

The same conclusion has been derived for $(CO)_2$ dimers very recently [60] reinforcing the first observation. Furthermore, in the case of CO dimers, it has been possible to separate the metastable excited states of CO^{2+} molecular ion from the dissociating ones as they lead to different fragmentation dynamics of the $(CO)_2$ dimers: the former experiences a sequential dissociation while the later presents a concerted fragmentation dynamics.

COLTRIMS provides also a powerful tool to study the geometry of molecular dimers as a Coulomb imaging technique when dimers are multiply ionized by ion impact. A first experiment devoted to $(CO)_2$ dimers has been performed with slow

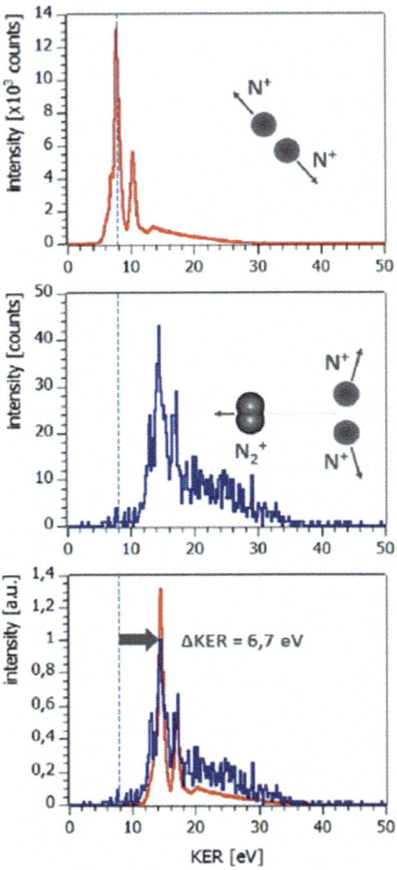

Fig. 11 (from top to bottom) 1-KER distribution of ($N^+ + N^+$) two-body fragmentation channel of an isolated molecule. 2-KER distribution of the ($N^+ + N^+ + N_2^+$) three-body fragmentation channel of an $(N_2)_2$ dimer. 3-Superposition of the two KER distributions. The two-body KER distribution has been shifted by 6.7 eV for comparison. 6.7 eV corresponds to the potential energy of 2 charges (1+ and 2+) standing at a distance corresponding to the van der Waals bond length in $(N_2)_2$ dimer: 4.3 Å

highly charged Ar ion projectiles with the aim to reveal the relative orientation of the two CO molecules within the neutral dimer [59].

7 Conclusion

COLTRIMS, invented by three pioneer groups (KSU, IKF, CIMAP) from three different countries (USA, Germany, France), has revolutionized atomic and molecular physics by introducing in the field of atomic collisions multi-parameter techniques inherited from nuclear physics. This is the first 4π multi-particle detector ever designed for atomic physics. The scope of this review was to show a few highlights obtained in this field but many more have been obtained in photo-ionization either by synchrotron radiation or by high power lasers and today by Free Electron

Lasers (XFEL). This technique reveals the tiny secrets of the dynamics of small quantum systems with an incredible level of details. Its extraordinary luminosity makes measurements of multiply differential cross sections feasible with complex target systems such as molecular clusters. A large number of laboratories have adopted the COLTRIMS and today such spectrometers are spread all over the world. We can be confident that this 30-year-old spectroscopy will continue in the future to reveal the amazing dynamics of multi-particle quantum systems. The story continues.

Acknowledgements We would like to thank our colleagues who contributed all along the years to the great COLTRIMS adventure: D. Hennecart, D. Lelièvre, J.P. Grandin, P. Jardin, J.P. Lemoigne, L. Guillaume, S. Duponchel, C. Caraby, M. Tarisien, G. Laurent, T. Muranaka, J. Blieck, A. Leredde, C. Couratin, W. Iskandar, and many others, among which are the technical staff of CIMAP and GANIL facility.

References

1. R.L. Kaufmann, F. Hopkins, C.W. Woods, P. Richard, Phys. Rev. Lett. **31**, 621 (1973)
2. R. Mann, F. Folkmann, K.O. Groeneveld, Phys. Rev. Lett. **37**, 1674 (1976)
3. C.L. Cocke, Phys. Rev. A **20**, 749 (1979)
4. H.F. Beyer, F. Folkmann, K.H. Schartner, J. Phys. B At. Mol. Opt. Phys. **13**, 2459 (1980)
5. C.L. Cocke, R. Dubois, T.J. Gray, E. Justiniano, C. Can, Phys. Rev. Lett. **46**, 1671 (1981)
6. C.R. Vane, M.H. Prior, R. Marrus, Phys. Rev. Lett. **46**, 107 (1981)
7. T.J. Gray, C.L. Cocke, IEEE Trans. Nuc. Sci. **30**, 937 (1983)
8. C.L. Cocke, R.E. Olson, Phys. Rep. **205**, 153 (1991)
9. E.H. Nielsen, L.H. Andersen, A. Barany, H. Cederquist, J. Heinemeier, P. Hvelplund, H. Knudsen, K.B. MacAdam, J. Sorensen, J. Phys. B At. Mol. Opt. Phys. **18**, 1789 (1985)
10. A. Gosselin, D. Hennecart, S. Hicham, X. Husson, H. Kucal, D. Lecler, A. Lepoutre, A. Cassimi, J.P. Grandin, J. Phys. III **4**, 1765 (1994)
11. J. Ullrich, C.L. Cocke, S. Kelbch, R. Mann, P. Richard, H. Schmidt-Bocking, J. Phys. B At. Mol. Opt. Phys. **17**, L785 (1984)
12. R.T. Short, C.S. O, J.C. Levin, I.A. Sellin, L. Liljeby, S. Huldt, S.E. Johansson, E. Nilsson, D.A. Church, Phys. Rev. Lett. **56**, 2614 (1986)
13. J.P. Grandin, D. Hennecart, X. Husson, D. Lecler, I. Lesteven-Vaisse, D. Lisfi, Europhys. Lett. **6**, 683 (1988)
14. J. Ullrich, H. Schmidt-Böcking, Phys. Lett. A **125**, 193 (1987)
15. C.E. Gonzales Lepera, M. Breinig, J. Burdorfer, R. Deserio, S.B. Elston, J.P. Gibbons, H.P. Hulskotter, L. Liljeby, R.T. Short, C.R. Vane, Nucl. Instrum. Methods Phys. Res. B **24/25**, 316 (1987)
16. R. Dörner, J. Ullrich, H. Schmidt-Bocking, R.E. Olson, Phys. Rev. Lett. **63**, 147 (1989)
17. R. Ali, V. Frohne, C.L. Cocke, M. Stockli, S. Cheng, M.L.A. Raphaelian, Phys. Rev. Lett. **69**, 2491 (1992)
18. V. Frohne, S. Cheng, R. Ali, M. Raphaelian, C.L. Cocke, R.E. Olson, Phys. Rev. Lett. **71**, 696 (1993)
19. R. Dörner, V. Mergel, R. Ali, U. Buck, C.L. Cocke, K. Froschauer, O. Jagutzki, S. Lencinas, W.E. Meyerhof, S. Nuttgens, R.E. Olson, H. Schmidt-Bocking, L. Spielberger, K. Tokesi, J. Ullrich, M. Unverzagt, W. Wu, Phys. Rev. Lett. **72**, 3166 (1994)
20. W. Wu, K.L. Wong, R. Ali, C.Y. Chen, C.L. Cocke, V. Frohne, J.P. Giese, M. Raphaelian, B. Walch, R. Dorner, V. Mergel, H. Schmidt-Bocking, W.E. Meyerhof, Phys. Rev. Lett. **72**, 3170 (1994)

21. R. Moshammer, J. Ullrich, M. Unverzagt, W. Schmidt, P. Jardin, R.E. Olson, R. Mann, R. Dorner, V. Mergel, U. Buck, H. Schmidt-Bocking, Phys. Rev. Lett. **73**, 3371 (1994)
22. V. Mergel, R. Dorner, J. Ullrich, O. Jagutzki, S. Lencinas, S. Nuttgens, L. Spielberger, M. Unverzagt, C.L. Cocke, R.E. Olson, M. Schulz, U. Buck, E. Zanger, W. Theisinger, M. Isser, S. Geis, H. Schmidt-Bocking, Phys. Rev. Lett. **74**, 2200 (1995)
23. P. Jardin, A. Cassimi, J.P. Grandin, H. Rothard, J.P. Lemoigne, D. Hennecart, X. Husson, A. Lepoutre, Nucl. Instrum. Methods Phys. Res. B **98**, 363 (1995)
24. P. Jardin, A. Cassimi, J.P. Grandin, D. Hennecart, J.P. Lemoigne, Nucl. Instrum. Methods Phys. Res. B **107**, 41 (1996)
25. M. Unverzagt, R. Moshammer, W. Schmitt, R.E. Olson, P. Jardin, V. Mergel, J. Ullrich, H. Schmidt-Bocking, Phys. Rev. Lett. **76**, 1043 (1996)
26. A. Cassimi, S. Duponchel, X. Fléchard, P. Jardin, P. Sortais, R.E. Olson, D. Hennecart, Phys. Rev. Lett. **76**, 3679 (1996)
27. R. Moshammer, J. Ullrich, H. Kollmus, W. Schmitt, M. Unverzagt, O. Jagutzki, V. Mergel, H. Schmidt-Bocking, R. Mann, C.J. Wood, R.E. Olson, Phys. Rev. Lett. **77**, 1242 (1996)
28. R. Dörner, H. Khemliche, M.H. Prior, C.L. Cocke, J.A. Gary, R.E. Olson, V. Mergel, J. Ullrich, H. Schmidt-Bocking, Phys. Rev. Lett. **77**, 4520 (1996)
29. R. Moshammer, W. Schmitt, J. Ullrich, H. Kollmus, A. Cassimi, R. Dorner, O. Jagutzki, R. Mann, R.E. Olson, H.T. Prinz, H. Schmidt-Bocking, L. Spielberger, Phys. Rev. Lett. **79**, 3621 (1997)
30. J. Ullrich, R. Moshammer, R. Dorner, O. Jagutzki, V. Mergel, H. Schmidt-Bocking, L. Spielberger, J. Phys. B At. Mol. Opt. Phys. **30**, 2917 (1997)
31. A. Cassimi, Phys. Scr. **T80**, 98 (1999)
32. L. Spielberger, O. Jagutzki, R. Dorner, J. Ullrich, U. Meyer, V. Mergel, M. Unverzagt, M. Damrau, T. Vogt, I. Ali, K. Khayyat, D. Bahr, H.G. Schmidt, R. Frahm, H. Schmidt-Bocking, Phys. Rev. Lett. **74**, 4615 (1995)
33. T. Weber, M. Weckenbrock, A. Staudte, L. Spielberger, O. Jagutzki, V. Mergel, F. Afaneh, G. Urbasch, M. Vollmer, H. Giessen, R. Dorner, Phys. Rev. Lett. **84**, 443 (2000)
34. R. Moshammer, B. Feuerstein, W. Scmitt, A. Dorn, C.D. Schroter, J. Ullrich, H. Rottke, C. Trump, M. Wittmann, G. Korn, K. Hoffmann, W. Sandner, Phys. Rev. Lett. **84**, 447 (2000)
35. T. Weber, H. Giessen, M. Weckenbrock, G. Urbasch, A. Staudte, L. Spielberger, O. Jagutzki, V. Mergel, M. Vollmer, R. Dorner, Nature **405**(6787), 658 (2000)
36. A. Dorn, R. Moshammer, C.D. Schroter, T.J.M. Zouros, W. Schmitt, H. Kollmus, R. Mann, J. Ullrich, Phys. Rev. Lett. **82**, 2496 (1999)
37. H. Tawara, T. Tonuma, H. Shibata, S.H. Be, H. Kumagai, M. Kase, T. Kambara, I. Kohno, Z. Phys. D **125**, 239 (1986)
38. U. Werner, K. Beckord, J. Becker, H.O. Lutz, Phys. Rev. Lett. **74**, 1962 (1995)
39. M. Tarisien M. Tarisien, L. Adoui, F. Frémont, D. Lelièvre, L. Guillaume, J.Y. Chesnel, H. Zhang, A. Dubois, D. Mathur, S. Kumar, M. Krishnamurthy, A. Cassimi, J. Phys. B: At. Mol. Opt. Phys. **33**, L11 (2000)
40. A. Landers, T. Weber, I. Ali, A. Cassimi, M. Hattass, O. Jagutzki, A. Nauert, T. Osipov, A. Staudte, M.H. Prior, H. Schmidt-Bocking, C.L. Cocke, R. Dorner, Phys. Rev. Lett. **87**, 013002 (2001)
41. R. Dörner, V. Mergel, O. Jagutzki, L. Spielberger, J. Ullrich, R. Moshammer, H. Schmidt-Bocking, Phys. Rep. **330**, 95 (2000)
42. S. Hicham, P. Boduch, A. Cassimi, M. Chantepie, A. Gosselin, J.P. Grandin, D. Hennecart, X. Husson, D. Lecler, A. Lepoutre, L.H. Zhang, J. Phys. B At. Mol. Opt. Phys. **26**, 1935 (1993)
43. A. Gosselin, P. Boduch, D. Hennecart, S. Hicham, X. Husson, D. Lecler, A. Lepoutre, A. Cassimi, J.P. Grandin, P. Jardin, J. Phys. B At. Mol. Opt. Phys. **28**, 445 (1995)
44. P. Jardin, J.P. Grandin, A. Cassimi, J.P. Lemoigne, A. Gosselin, X. Husson, D. Hennecart, A. Lepoutre, AIP Conf. Proc. **274**, 291 (1993)
45. X. Fléchard, S. Duponchel, L. Adoui, A. Cassimi, P. Roncin, D. Hennecart, J. Phys. B At. Mol. Opt. Phys. **30**, 3697 (1997)

46. X. Fléchard, C. Harel, H. Jouin, B. Pons, L. Adoui, F. Frémont, A. Cassimi, D. Hennecart, J. Phys. B At. Mol. Opt. Phys. **34**, 2759 (2001)
47. G. Laurent, M. Tarisien, X. Fléchard, P. Jardin, L. Guillaume, P. Sobocinski, L. Adoui, A. Bordenave-Montesquieu, D. Bordenave-Montesquieu, J.Y. Chesnel, F. Frémont, D. Hennecart, E. Liénard, L. Maunoury, P. Moretto-Capelle, A. Cassimi, Nucl. Instrum. Meth. Phys. Res. B **205**, 546 (2003)
48. G. Laurent, J. Fernandez, S. Legendre, M. Tarisien, L. Adoui, A. Cassimi, X. Fléchard, F. Frémont, B. Gervais, E. Giglio, J.P. Grandin, F. Martin, Phys. Rev. Lett. **96**, 173201 (2006)
49. G. Laurent, P.D. Fainstein, M.E. Galassi, R.D. Rivarola, L. Adoui, A. Cassimi, J. Phys. B At. Mol. Opt. Phys. **35**, L495–L501 (2002)
50. J. Blieck, X. Fléchard, A. Cassimi, H. Gilles, S. Girard, D. Hennecart, Rev. Sc. Instrum. **79**, 103102 (2008)
51. A. Leredde, A. Cassimi, X. Fléchard, D. Hennecart, H. Jouin, B. Pons, Phys. Rev. A **85**, 032710 (2012)
52. A. Leredde, X. Fléchard, A. Cassimi, D. Hennecart, B. Pons, Phys. Rev. Lett. **111**, 133201 (2013)
53. C. Couratin, P. Velten, X .Fléchard, E. Liénard, G. Ban, A. Cassimi, P. Delahaye, D. Durand, D. Hennecart, F. Mauger, A. Mery, O. Navillat-Cuncic, Z. Patyk, D. Rodriguez, J.C. Thomas, Phys. Rev. Lett. **108**, 243201 (2012)
54. J. Matsumoto, A. Leredde, X. Fléchard, K. Hayakawa, H. Shiromaru, J. Rangama, C.L. Zhou, S. Guillous, D. Hennecart, T. Muranaka, A. Mery, B. Gervais, A. Cassimi, Phys. Rev. Lett. **105**, 263202 (2010)
55. W. Iskandar, J. Matsumoto, A. Leredde, X. Fléchard, B. Gervais, S. Guillous, D. Hennecart, A. Mery, J. Rangama, C.L. Zhou, H. Shiromaru, A. Cassimi, Phys. Rev. Lett. **113**, 143201 (2014)
56. W. Iskandar, J. Matsumoto, A. Leredde, X. Fléchard, B. Gervais, S. Guillous, D. Hennecart, A. Mery, J. Rangama, C.L. Zhou, H. Shiromaru, Phys. Rev. Lett. **114**, 033201 (2015)
57. A. Méry, A.N. Agnihotri, J. Douady, X. Fléchard, B. Gervais, S. Guillous, W. Iskandar, E. Jacquet, J. Matsumoto, J. Rangama, F. Ropars, C.P. Safvan, H. Shiromaru, D. Zanuttini, A. Cassimi, Phys. Rev. Lett. **118**, 233402 (2017)
58. W. Iskandar, X. Fléchard, J. Matsumoto, A. Leredde, S. Guillous, D. Hennecart, J. Rangama, A. Mery, B. Gervais, H. Shiromaru, A. Cassimi, Phys. Rev. A **98**, 012701 (2018)
59. A. Méry, V. Kumar, X. Fléchard, B. Gervais, S. Guillous, M. Lalande, J. Rangama, W. Wolff, A. Cassimi, Phys. Rev. A **103**, 042813 (2021)
60. A. Méry, X. Fléchard, S. Guillous, V. Kumar, M. Lalande, J. Rangama, W. Wolff, A. Cassimi, Phys. Rev. A **104**, 04281 (2021)
61. P. Jardin, J.P. Grandin, H. Rothard, A. Cassimi, J.P. Lemoigne, A. Gosselin, X. Husson, D. Hennecart, A. Lepoutre, In: *XVIIIth International Conference on the Physics of Electronic and Atomic Collisions* (University of Aarhus, Denmark, 1993), 21–27 July 1993
62. W.C. Wiley, I.H. McLaren, Rev. Sci. Instrum. **26**, 1150 (1955)
63. H. Zhang, X. Fléchard, A. Cassimi, L. Adoui, F. Frémont, D. Lecler, G. Cremer, L. Guillaume, D. Lelièvre, A. Lepoutre, D. Hennecart, Phys. Rev. A **60**, 3694 (1999)
64. A. Cassimi, M. Tarisien, G. Laurent, P. Sobocinski, L. Adoui, J.Y. Chesnel, F. Frémont, G. Gervais, D. Hennecart, in *Many-particle quantum dynamics in atomic and molecular fragmentation*, ed. by J. Ullrich, V.P. Shevelko, Springer series on "Atomic and Optical Plasma Physics" (Springer, Berlin, Heidelberg, New-York, 2003). ISSN 1615–5653
65. S. Legendre, E. Giglio, M. Tarisien, A. Cassimi, B. Gervais, L. Adoui, J. Phys. B At. Mol. Opt. Phys. **38**, L233 (2005)
66. B. Gervais, E. Giglio, L. Adoui, A. Cassimi, D. Duflot, M.E. Galassi, J. Chem. Phys. **31**, 024302 (2009)
67. J. Matsumoto, A. Leredde, X. Fléchard, K. Hayakawa, H. Shiromaru, J. Rangama, C.L. Zhou, D. Hennecart, T. Muranaka, A. Mery, B. Gervais, A. Cassimi, Phys. Scr. **T144**, 01401 (2011)
68. H.K. Kim, H. Gassert, M.S. Schoffler, J.N. Titze, M. Waitz, J. Voigtsberger, F. Trinter, J. Becht, A. Kalinin, N. Neumann, LPh.H. Schmidt, O. JagutzkiI, A. Czascht, H. Merabet, H. Schmidt-Bocking, T. Jahnke, A. Cassimi, R. Dorner, Phys. Rev. A **88**, 042707 (2013)

69. H.K. Kim, H. Gassert, J.N. Titze, M. Waitz, J. Voigtsberger, F. Trinter, J. Becht, A. Kalinin, N. Neumann, C.L. Zhou, L.P.H. Schmidt, O. JagutzkiI, A. Czascht, M.S. Schoffler, H. Merabet, H. Schmidt-Bocking, T. Jahnke, H.J. Ludde, A. Cassimi, R. Dorner, Phys. Rev. A **89**, 02270 (2014)
70. T. Jahnke, A. Czascht, M.S. Schoffler, S. Schosser, A. Knapp, M. Kasz, J. Titze, C. Wimmer, K. Kreidi, R.E. Grisenti, A. Staudte, O. Jagutzki, U. Hergenhahn, H. Schmidt-Bocking, R. Dorner, Phys. Rev. Lett. **93**, 163401 (2004)

The Post-collision Interaction in Ionization by Ion Impact

M. Schulz

The most important goal of research in ion–atom collisions is to advance our understanding of the few-body problem. In perturbation theory understanding, the few-body dynamics means accurately describing the relative importance of the contributions from the various first- and higher order contributions to the process of interest. One higher order mechanism which often plays a prominent role is known as post-collision interaction (PCI). Since its signatures were discovered in the ejected electron spectra, they were also found in cross sections differential in kinematic parameters of the scattered projectile and the recoiling target ion. More recently, numerous fully differential measurements were reported. In this chapter, the progress achieved by these experiments and theoretical calculations are reviewed.

1 Introduction

Ever since the famous Rutherford experiment [1], collision studies played a crucial role in advancing our understanding of the properties of atomic as well as sub-atomic systems. Based on the results of his measurements, Rutherford was able to develop the first viable atomic model. However, it could explain neither the discrete energy spectrum of light emitted by atoms nor the stability of atoms. With the advent of quantum mechanics, both of these apparent puzzles could be solved. Since the Coulomb force acting between two charged point particles is practically completely understood, the Schrödinger equation can be solved exactly at least for the hydrogen atom and H-like ions [e.g., 2], which represent two-particle systems.

M. Schulz (✉)
Physics Department and LAMOR, Missouri University of Science and Technology, Rolla, MO 65409, USA
e-mail: schulz@mst.edu

L. C. Tribedi (ed.), *Advances in Atomic Molecular Collisions*,
https://doi.org/10.1007/978-981-97-7063-2_2

For more complex many-particle atoms or ions, theory is confronted with a major complication, which does not stem from any shortcoming of the existing atomic model. Rather, it is due to the fact that the Schrödinger equation is not analytically solvable for more than two mutually interacting particles even when applied to a flawless atomic model and when the underlying forces acting within and on the system are precisely known. This is known as the few-body problem (FBP), which remains one of the most fundamentally important and yet unsolved problems in physics. Because of the FBP theory has to resort to numerical approaches and the assumptions and approximations entering in these methods have to be tested by detailed experimental data.

For the calculation of properties of stationary atoms, such as energy levels or wave functions, powerful numeric methods, such as the multi-configuration Hartree–Fock approach [e.g., 3], were developed, which often provide very accurate results. However, the modeling of dynamic systems, like atomic collisions leading to electronic transitions, is much more challenging. As a result, calculations of cross sections for such transitions tend to yield significantly less satisfactory agreement with experiment than atomic structure calculations [e.g., 4]. Regarding the FBP, ionization takes a prominent role among the inelastic processes that can occur in the collision because the final state involves three independently moving particles (the scattered projectile, the recoiling target ion, and the ejected electron). In contrast, in excitation and capture, the active electron remains bound to one of the collision partners. While the dynamics of such transitions still represent a three-body problem, at least kinematically excitation and capture can be regarded as two-body scattering processes, in contrast to ionization. Therefore, the latter process is particularly suitable to study the FBP.

The theoretical modeling of inelastic processes in atomic collisions can be grouped in two categories: perturbative [e.g., 5–14] and non-perturbative approaches [e.g., 15–22]. Non-perturbative models tend to be more complete in that they employ basis sets consisting of a large number of states so that couplings between the process of interest and other reaction channels can be accounted for. However, non-perturbative calculations tend to be numerically considerably more involved than perturbative approaches. Nevertheless, with rapidly increasing computational resources non-perturbative calculations became increasingly common over the last 2 to 3 decades for electron impact [15–17]. For ion impact, such approaches are even more challenging because of the much larger projectile mass necessitating a much larger number of angular momentum states to describe the scattered projectiles. As a result, for ion impact, non-perturbative calculations became common [18–22] with some delay relative to electron impact and perturbative methods still play an important role.

Notwithstanding the aforementioned disadvantages of perturbative compared to non-perturbative models one important strength is their transparency. For example, in the Born series, each expansion term corresponds to a well-defined reaction mechanism, characterized by the interaction or sequence of interactions between the projectile and the target atom (i.e., the perturbation) contained in the operator of that term. Here, the term containing only a single projectile target interaction is referred to as the first-order term and those containing multiple interactions as higher order terms. In perturbation theory, understanding the few-body dynamics in a collision basically means accurately describing the relative importance of the various higher order and the first-order contributions.

One higher order process that has been studied extensively for target ionization by ion impact is known as post-collision interaction (PCI) [e.g., 23–44]. This process involves at least two interactions between the projectile and the active target electron. In the primary interaction, sufficient energy is transferred by the projectile to the electron for the latter to be lifted to the continuum. In the second interaction, the projectile and the electron attract each other in the outgoing part of the collision so that they are both focused toward the initial projectile beam axis. PCI was first observed in ejected electron spectra recorded in the pioneering work of Crooks and Rudd [23]. Since then, signatures of PCI have been observed in electron spectra taken in many further experiments [e.g., 24–30], in recoil-ion momentum spectra [33], in double differential scattered projectile spectra [34–36], and in fully differential cross sections [e.g., 37–41]. In this chapter, these studies are reviewed and some of the most significant insights obtained from them are analyzed.

2 Double Differential Ejected Electron Spectra

The first differential measurements on ionization by ion impact were performed by recording ejected electron energy spectra for fixed ejection angles [45–47]. At relatively large angles (measured relative to the initial projectile beam direction), above about 10 eV the double differential cross sections (DDCS) (in energy and electron solid angle) tend to fall off monotonically with increasing energy. This behavior is qualitatively reproduced by the first-Born approximation (FBA). Initially, no data were taken at electron angles smaller than about 20° [45, 46]. However, when later the measurements were extended to angles as small as 10° [47, 48], shown in Fig. 1, two relatively broad structures were observed in the energy dependence of the DDCS. The second structure, now known as the "binary encounter peak" (BEP), occurs approximately at four times the projectile energy divided by the projectile mass (≅400 eV) and thus corresponds to an electron speed twice as large as the projectile speed. It is qualitatively reproduced by the FBA and can thus be regarded as a signature of the first-order process. The first structure occurs near the projectile energy divided by its mass (≅100 eV) corresponding to an electron speed equal to the projectile speed (velocity matching).

These structures at 100 eV were initially believed by the authors to be due to experimental artifacts [47] and were thus not further discussed in their publication. However, they did trigger theoretical studies, which incorporated higher order contributions [48, 50]. They predicted that at 0° the DDCS should exhibit a peak with an intensity proportional to $|v_p—v_{el}|^{-2}$, where v_p and v_{el} are the projectile and electron speeds, respectively. The discontinuity at the velocity matching $v_p = v_{el}$ results in a cusp shape of the peak and the term "cusp peak" is now well established in

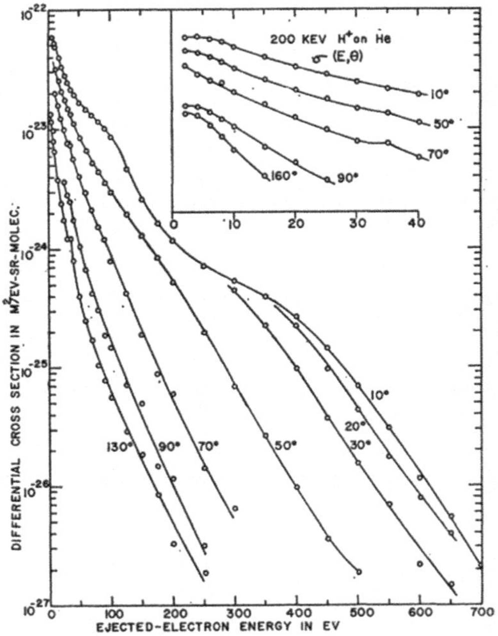

Fig. 1 Double differential ejected electron spectra for 200 keV p + He collisions [47]. The polar electron angle is fixed at various values as indicated in the plot and the data are shown as a function of electron energy. The inset shows the cross sections zoomed in on small energies

the literature. Theory also predicted that the cusp peak should only occur in a very narrow angular range in the forward direction. The "hump" observed by Rudd and co-workers at 10° can then be interpreted as a residue of the cusp peak. Indeed, at this angle the calculation of Salin for ionization of H also found a "hump" at an electron energy equal to the projectile energy divided by its mass [49]. These theoretical results prompted Crooks and Rudd to extend the experiments on ionization in p + He collisions to an electron ejection angle of 0° [23]. Their results are shown in Fig. 2 for three projectile energies. In each case, a very pronounced cusp peak can be seen at the right energy reflecting the $|v_p—v_{el}|^{-2}$-dependence predicted by Salin. The shape of these cusp peaks observed in the experiment was reproduced by the calculations of Macek as well.

The mechanism leading to the cusp peak was adequately termed "charge transfer into continuum states" by Crooks and Rudd, referring to unbound low-energy states in the projectile frame. Later, the term "electron capture to the continuum (ECC)" became common. This term is somewhat misleading and even contradictory. Capture implies that a particle (typically an electron) makes a transition from being bound to another particle (typically the target atom) to being bound to a third particle (typically the projectile). At the same time, the term continuum state implies an unbound state. What is actually happening in this process is that an electron initially bound to the target atom makes a transition to an unbound state with an energy barely above zero in the projectile frame, i.e., it is a target ionization rather than a capture process. This transition proceeds through at least two interactions between the projectile and

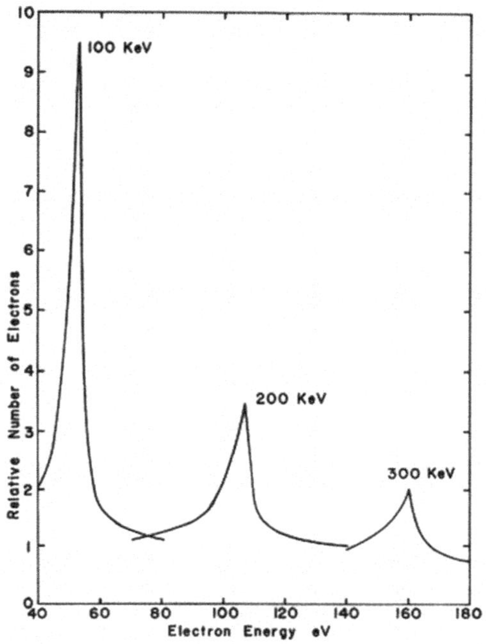

Fig. 2 Double differential ejected electron spectra for p + He collisions at three different projectile energies [23]. The polar electron angle is fixed at 0° and the data are shown as a function of electron energy. In each case, a sharp cusp-shaped peak is observed at an energy corresponding to the electron speed matching the projectile speed

the active target electron. In the first interaction, the projectile transfers sufficient energy for the electron to be released from the target atom. In the second interaction, the projectile and the ejected electron attract each other in the outgoing part of the collision. Thereby, the electron is accelerated toward the projectile velocity, both in magnitude and direction. This focusing effect explains the strongly enhanced cross sections at the velocity matching for electrons ejected in the forward direction and the narrow angular range in which the cusp peak is observed.

The existence of the cusp peak was confirmed in many follow-up experiments [e.g., 25–32]. In Fig. 3, DDCS are shown for 28.5 MeV F^{9+} + H_2 [28–30] for electrons emitted at 0° as a function of electron energy. The cusp peak at 800 eV and the BEP at about 3 keV are clearly visible. The dotted curve represents an FBA calculation and the dashed curve a strong potential Born (SPB) model, both by Brauner and Macek [51]. Once again, as expected, the FBA results reproduce the BEP rather well, but the cusp peak is completely absent. In the SPB, the transition amplitude is expanded in powers of the target to projectile charge ratio up to the second term (i.e., it represents a higher order calculation). As a result, both the BEP and the cusp peak are well described. However, outside these peak structures, there are significant discrepancies to experiment. The two solid lines represent two different versions of the continuum distorted wave—eikonal initial state (CDW-EIS) model [52], in which higher order contributions are accounted for in the wave functions of the system. While one of these calculations (thin solid curve) yields very similar results to the SPB, somewhat improved agreement is achieved with the other (thick solid curve), especially in the

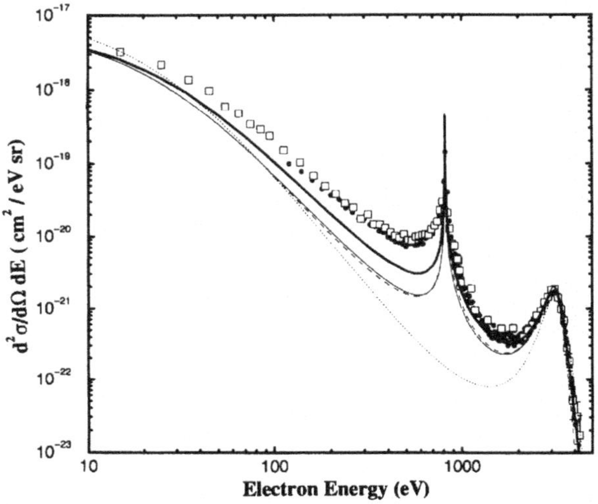

Fig. 3 Double differential ejected electron spectra for 28.5 MeV F^{9+} + H_2 collisions [28]. The polar electron angle is fixed at 0° and the data are shown as a function of electron energy. For curves, see text

region between the cusp and the BEP. However, considerable discrepancies remain below the cusp energy. Remarkably, the best agreement is obtained with a classical simulation [52], shown as open squares.

As a crude criterion for the range of validity of the FBA the condition $\eta \ll 1$ is often used, where η (the perturbation parameter) is the projectile charge to speed ratio. 28.5 MeV F^{9+} corresponds to $\eta = 1.16$ and it is thus not at all surprising that the FBA does not even yield qualitative agreement with experiment. Because of the higher order contributions accounted for in CDW-EIS, it was suggested that the condition for the range of validity should be modified to $\eta' = Q_p/{v_p}^2 \ll 1$ [53]. For 28.5 MeV F^{9+} η' = 0.15, which satisfies the condition $\eta' \ll 1$ quite well. Although with the CDW-EIS calculation the agreement with experiment is dramatically improved compared to the FBA, the considerable residual discrepancies nevertheless seem to suggest that $\eta' =$ 0.15 is still too large. On the other hand, for 3.6 MeV/A Au^{53+} + He, corresponding to $\eta' = 0.37$, measured double differential (in the longitudinal and transverse electron momentum components) electron spectra were surprisingly well reproduced by a CDW-EIS calculation [54]; the agreement was at least not worse than in the case of 28.5 MeV F^{9+} + He. One possible explanation is that the condition for the validity of the CDW-EIS approach could change for very large projectile charges. However, the discrepancies for the Au^{53+} projectiles become severe when fully differential cross sections (FDCS) were analyzed [37].

3 Double Differential Scattered Projectile Spectra

The cusp peak can be regarded as a manifestation of a mutual focusing of the projectile and the electron in the outgoing part of the collision. One might therefore expect that PCI effects on the electron are reflected by corresponding features in cross sections differential in projectile parameters. For example, since the two particles attract each other toward the beam axis, one would expect the average scattering angle of the projectile to be smaller compared to first-order ionization events, where PCI is absent. Such an effect was indeed observed by Vajnai et al. [34] for ionization in p + He collisions for projectile energies ranging from 50 to 150 keV. In Fig. 4, the width of the double differential scattered projectile angular distribution is shown for a collision energy of 75 keV as a function of the electron to projectile speed ratio v_e/v_p. At $v_e/v_p = 1$ a pronounced change of slope, and perhaps even a slight minimum, is found. While for $v_e/v_p < 1$ the width increases rather slowly with increasing electron speed, above $v_e/v_p = 1$ an abrupt strong increase of the width is observed.

The dashed curve in Fig. 4 represents an FBA calculation. It is in poor agreement with the experimental data: it generally drastically overestimates the width and it does not show any abrupt change of slope near $v_e/v_p = 1$ at all. On the other hand, a calculation incorporating PCI following the method of Salin [49] does reproduce the abrupt change of slope and for $v_e/v_p < 1$ it is even in nice quantitative agreement with experiment. This shows that indeed the change of slope is due to PCI. Later, similar data were obtained for ionization of molecular and atomic hydrogen by proton impact [35]. The widths for these collision systems are shown as a function of v_e/v_p in Fig. 5. Here, the effect is even more pronounced than for ionization of He in that a pronounced minimum is observed for H_2 at $v_e/v_p = 1$. For H the width also strongly

Fig. 4 Widths of the double differential scattered projectile angular distribution as a function of the electron to projectile speed ratio in 75 keV p + He collisions [34]. For curves, see text

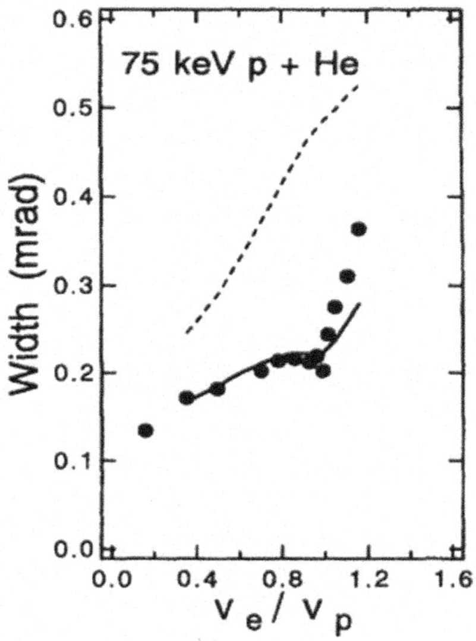

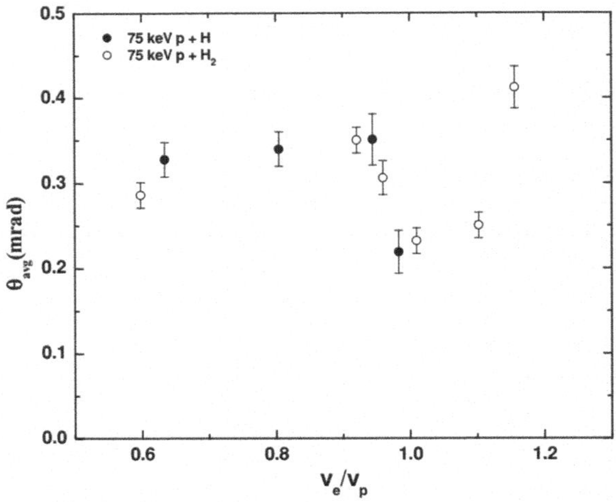

Fig. 5 Average projectile scattering angle in 75 keV protons colliding with H (solid data points) and H_2 (open data points) as a function of the electron to projectile speed ratio [35]

decreases between the two data points just below and at $v_e/v_p = 1$, but a minimum was not found because no data were taken for $v_e/v_p > 1$.

So far, we described PCI in terms of a process involving at least two projectile–target electron interactions. However, as pointed out in [35], because of momentum conservation, momentum must be transferred to the recoil-ion as well in an interaction with either the electron or the projectile. This can be illustrated by using a classical analogy: if two billiard balls collide, they will subsequently move apart. A second collision between the same balls is only possible if at least one of them gets redirected in a collision with a third body. The third body can either be a third billiard ball or the wall of the table. In PCI the third body is the recoil ion. PCI can thus proceed by the interaction sequences PE-ET-PE or PE-PT-PE (or higher order processes), where P, E, and T stand for projectile, electron, and target ion, respectively. The first sequence corresponds to a third-order process in perturbation theory. However, the second sequence represents a second-order process because the ET interaction is contained in the target wave function and is not part of the perturbation. The requirement of the recoil ion participating in PCI makes it a two-center process, as pointed out by Sarkadi et al. [42].

4 Recoil-Ion Momentum Spectra

Since the recoil ion is necessarily involved in PCI one might expect features characteristic to this mechanism as well. Indeed, the following kinematic analysis shows that the longitudinal recoil-ion momentum component assumes a minimum value when $v_{el} = v_p$: The projectile energy loss ε is given by

$$\varepsilon = K_i - K_f = \left(p_0 - p_f\right)\left(p_0 + p_f\right)/2M \tag{1}$$

Because of the extremely small projectile scattering angles (typically less than 1 mrad even for light ions), the final kinetic projectile energy is almost exclusively contained in the longitudinal motion of the projectile. Therefore, the final projectile momentum p_f can to a very good approximation be replaced by its longitudinal component p_{fz}. The term (p_0—p_f) in Eq. (1) then becomes the longitudinal component q_z of the momentum transfer $\boldsymbol{q} = \boldsymbol{p}_0 — \boldsymbol{p}_f$. Likewise, the term $(p_0 + p_f)/2\,M$ is to a very good approximation equal to the initial projectile speed v_p. Therefore, ε and q_z are related to each other by

$$\varepsilon = q_z v_p \tag{2}$$

Since $\boldsymbol{q}$ is equal to the sum of the ejected electron and recoil-ion momenta, momentum conservation yields

$$p_{recz} = \varepsilon / v_p - v_{elz} \tag{3}$$

The energy loss is equal to the sum of the kinetic energy of the electron and the target ionization potential I. Breaking the kinetic energy into fractions contained in the longitudinal and transverse motion of the electron we get (in atomic units)

$$p_{recz} = \left(v_{elz}{}^2 + v_{elt}{}^2 + 2I\right) / 2v_p - v_{elz} \tag{4}$$

Finally, taking the derivative of p_{recz} in v_{elz} and setting it to 0 yields $v_{elz} = v_p$, i.e., the longitudinal recoil-ion momentum assumes a minimum value when the electron speed matches the projectile speed. By substituting $v_{elz} = v_p$ and $v_{elt}{}^2 = 0$ in Eq. (4) we obtain for the minimum longitudinal recoil-ion momentum p_{min}

$$p_{min} = I / v_p - 1 / 2v_p \tag{5}$$

Recoil-ion momentum spectra were recorded by Weber et al. [33] for d + He collisions. These spectra are shown for collision energies ranging from 200 to 400 keV in Fig. 6. The vertical dash-dotted lines indicate the longitudinal recoil momenta for electrons ejected with zero kinetic energy (right line) and p_{min} calculated with Eq. (5) (left line). Indeed, a sharp cut-off is observed at p_{min} for all projectile energies, and this represents a characteristic manifestation of PCI.

Another manifestation of PCI was found in the longitudinal recoil-ion momentum spectra for fast highly charged ion impact by Moshammer et al. [55]. In Fig. 7, the longitudinal electron, recoil-ion, and projectile momentum spectra, taken for 3.6 MeV/A Ni^{24+} + He collisions, are plotted. For these large projectile energies, the cusp energy is in the keV regime, for which the electron detection efficiency in that experiment was practically zero. Therefore, these spectra are unaffected by cusp electrons. Nevertheless, signatures of PCI can be seen in these data. In a first-order process, the residual target ion remains essentially passive and one would expect the recoil-ion momentum spectrum to be centered at 0. The shift of the recoil ions to negative momenta and of the electrons to positive momenta seen in Fig. 7 was

Fig. 6 Double differential recoil-ion momentum spectrum as a function of the longitudinal and transverse momentum components for d + He collisions at three different projectile energies [33]

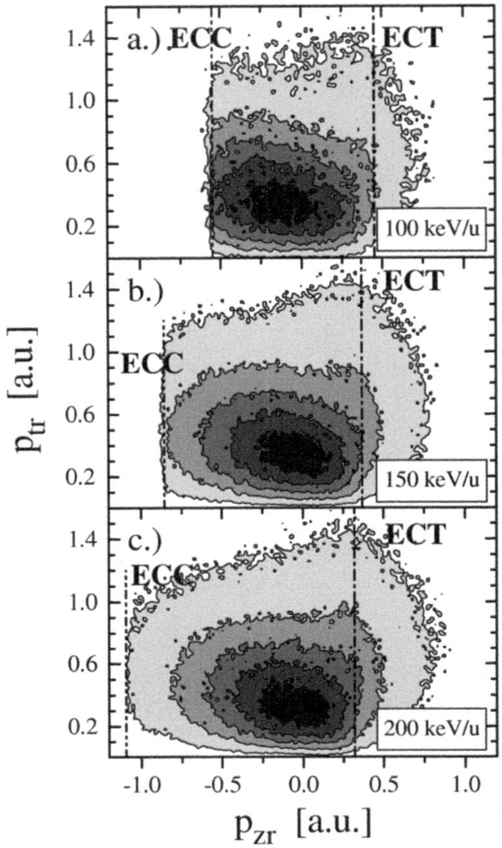

interpreted by the authors as due to PCI. Because of the long-range nature of the Coulomb force, the secondary projectile–target atom interaction in the outgoing part of the collision attracts the electron in the forward direction and repels the target ion in the backward direction. This shows that PCI can even be important far away from the matching velocity.

5 Fully Differential Cross Sections

In the previous sections, we showed that signatures of PCI can be observed in all collision fragments of an ionization process. However, in the experiments that were reviewed so far only one of these particles was detected (except for [55]). If the momenta of two particles are measured simultaneously even more detailed information about the collision dynamics can be obtained. The momentum of the third

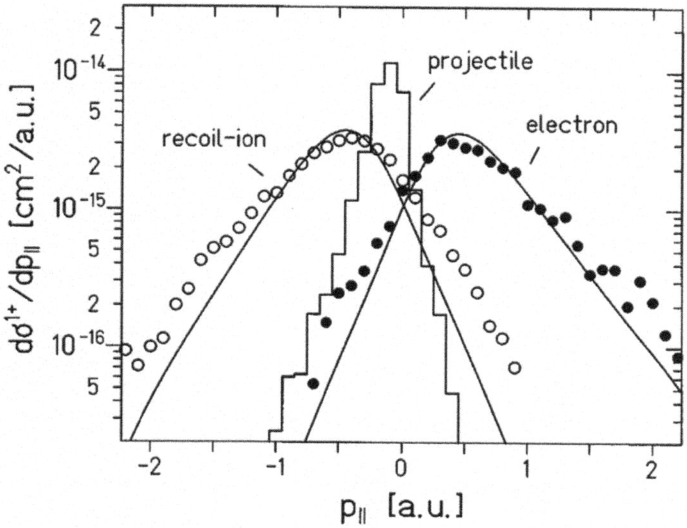

Fig. 7 Longitudinal momentum spectra for the ejected electrons, the recoil ions, and the scattered projectiles in 3.6 MeV/A Ni^{24+} + He collisions [55]

(undetected) particle is then determined by the kinematic conservation laws. In the experiments performed so far for ion impact, the recoil ion was detected in coincidence with either the electron [e.g., 37, 38, 56, 57] or the scattered projectile [e.g., 38–40]. Such kinematically complete experiments can provide detailed information about correlations between the collision products and fully differential cross sections (FDCS) can be extracted from them.

The first experimental FDCS on ionization by ion impact were reported for 1.2 GeV C^{6+} + He collisions [56, 57]. However, for this collision system, PCI effects are negligible because of the very small perturbation parameter ($\eta = 0.1$). Furthermore, the speed of the detected electron was only about 1% of the projectile speed. However, later FDCS were measured for a similar projectile (O^{7+}), but at a much smaller energy of 16 MeV corresponding to $\eta = 1.1$ [38]. Here, too, only electrons with an energy small compared to the cusp energy were detected. Nevertheless, another signature of PCI, apart from the cusp peak, could be identified.

In Fig. 8, FDCS are plotted for electrons with an energy of 8 eV ejected into the scattering plane, spanned by the initial and final projectile momenta, and fixed q of (from left to right) 0.5, 1.5, and 4 a.u. as a function of the polar electron angle θ_{el}. In each case, a pronounced peak structure, known as the binary peak, in a direction varying with q, is seen. In a first-order process, the binary peak is known to be centered on the direction of $\boldsymbol{q}$, which is indicated by the arrow in each case. In the data, only the FDCS for q = 1.5 a.u. maximize close to the direction of $\boldsymbol{q}$. For q = 0.5 a.u. and especially for q = 4 a.u., the binary peak is strongly shifted in the forward direction. It is thus clear that at least for these latter two cases higher order mechanisms strongly contribute to ionization.

Based on the discussion of the data shown in Fig. 7 by Moshammer et al. [55] one might suspect that the forward shift of the binary peak is due to PCI. However, the authors of [38] demonstrated that there are other higher order mechanisms which can also lead to a forward shift. In Fig. 9, the shift θ_q—θ_{BP} is plotted as a function of the transverse component of $\boldsymbol{q}$. A pronounced minimum can be seen at about q = 1.5 a.u. and a large forward shift both below and above that minimum. The dashed curve in Fig. 9 represents a CDW-EIS calculation not incorporating the interaction between the projectile and the residual target ion (PT interaction). It shows a monotonically increasing shift with increasing q and reasonably well reproduces the data above the minimum. Since this calculation does not account for the PE-PT-PE interaction sequence leading to PCI, this strongly suggests that the PE-ET-PT sequence is the dominant mechanism and that indeed PCI is mostly responsible for the shift at least at large q.

Based on the observation that the CDW-EIS calculation does not reproduce the increasing shift with decreasing q below the minimum one could be tempted to conclude that in this region the shift is mostly due to the PE-PT-PE sequence. Indeed, the data are qualitatively reproduced in this region by a CDW-EIS calculation including the PT interaction, shown as the solid curve in Fig. 9. However, an even stronger increase of the shift with decreasing q (and better agreement with the data at small q) is obtained with a modified FBA which accounts for the PT interaction in terms of the eikonal approximation and for the electron target–ion interaction in the final state by a Coulomb wave. This calculation treats the PE interaction only to first order and thus does not account for PCI at all. If, on the other hand, the Coulomb wave is replaced by a plane wave (dash-dotted curve in Fig. 9) the shift decreases with decreasing q and even becomes negative for q < 0.7 a.u. The authors of [38] therefore considered the possibility that at small q the forward shift of the binary peak observed in the data may be mostly due to an interaction sequence PE-PT-ET. Nevertheless, this work showed that, apart from the cusp peak, another signature of

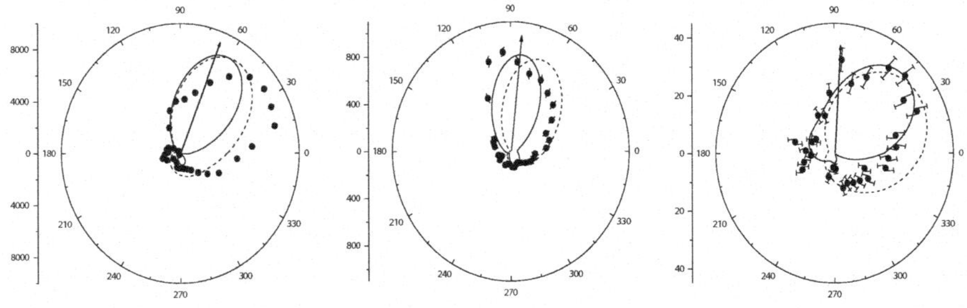

Fig. 8 Fully differential cross sections for electrons with an energy of 8 eV ejected in 16 MeV O^{7+} + He collisions into the scattering plane [38]. The momentum transfer is fixed at (from left to right) 0.5 a.u., 1.5 a.u., and 4 a.u. The arrows indicate the direction of the momentum transfer. For curves see text

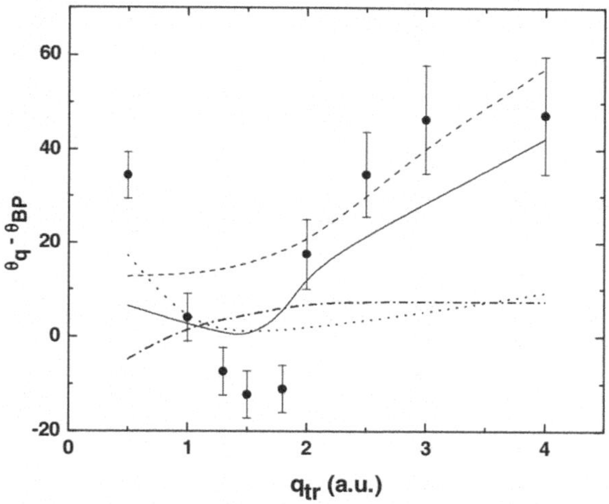

Fig. 9 Shift of the binary peak in the fully differential cross sections for 16 MeV O^{7+} + He collisions relative to **q** (θ_q—θ_{BP}) as a function of q [38]. The ejected electron energy is 8 eV. For curves, see text

PCI in electron spectra is a forward shift of the binary peak (relative to **q**), at least at relatively large q.

For projectile energies in the MeV/A regime, fully differential measurements for electrons ejected with a speed close to the projectile speed are very difficult because they have an energy in the keV regime. However, recently several studies were reported for 75 keV protons colliding with He and H_2 in the velocity-matching regime corresponding to electron energies of only around 40 eV [39–41]. For ionization of molecular [43] and atomic hydrogen [44] by this projectile species and energy theory had predicted a sharp and strong forward peak in the fully differential angular electron distribution near the velocity matching. In a first experiment for ionization of H_2 no structure at all was observed in the forward direction [58]. However, this was later blamed on an unnoticed hair crack in the anode of the projectile detector after the experiment was repeated [39].

The FDCS for electrons with energies ranging from 34.6 to 44.6 eV (corresponding to projectile energy losses between 50 and 60 eV) ejected into the scattering plane in 75 keV p + H_2 collisions are shown for a scattering angle of $\theta_p = 0.55$ mrad in Fig. 10 as a function of θ_{el}. The matching velocity corresponds to an energy loss of 57 eV. This time, a pronounced and strong forward peak is observed which is at least partly resolved from the binary peak. As expected, the ratio between the intensity in the forward peak relative to the binary peak maximizes at the matching velocity. The dashed and solid curves in Fig. 10 represent CDW-EIS and 3-body distorted wave (3DW) calculations, respectively. These models are conceptually very similar and should basically catch the same physics. Although the experimental results confirmed the theoretical prediction of a strong forward peak, it was thus very surprising that both theories make very different predictions on the FDCS and are not even in qualitative agreement with experiment. In contrast, well below the matching velocity regime, at least the shape of the angular dependence of the FDCS was reasonably well reproduced by the 3DW model, even outside the scattering plane [59]. Later, similar results were obtained for ionization of He [40]. These studies suggest that the velocity-matching regime is particularly challenging for perturbative methods. This is not really surprising because at this very small relative velocity between the projectile and the electron these particles can interact for a long time with each other and the projectile can no longer be viewed as a small perturbation to the target. Therefore, non-perturbative calculations are called for.

Instead of fixing the electron energy E_{el} and plotting the FDCS as a function of θ_{el} of course one can also plot the FDCS as a function of E_{el} for fixed θ_{el}. Not surprisingly, these plots for $\theta_{el} = 0°$ showed a very strong cusp peak both for ionization of H_2 and He [41]. What was surprising is the behavior of these spectra when θ_{el} was fixed at a relatively large angle, where the DDCS usually don't even show a trace of a cusp peak. In Fig. 11, a fully differential energy spectrum is shown for ionization of He where θ_{el} is fixed at 30°. The projectile scattering angle ranges from 0.1 to 0.5 mrad, as indicated in Fig. 11. A noticeable, although compared to $\theta_{el} = 0°$ strongly suppressed, cusp peak was observed, which became increasingly pronounced with increasing θ_p. The solid blue and red curves in Fig. 11 represent calculations based

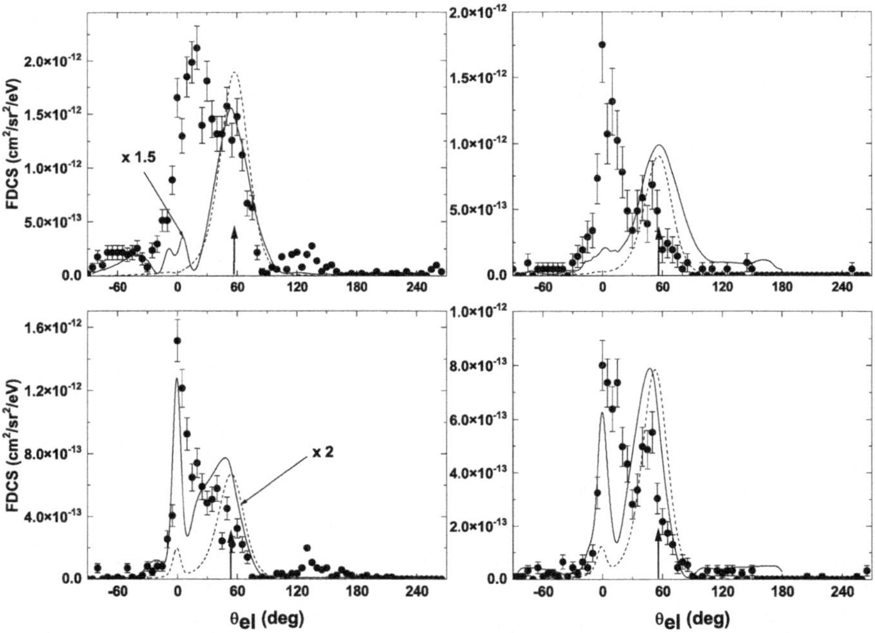

Fig. 10 Fully differential cross sections for electrons ejected in 75 keV p + H_2 collisions into the scattering plane as a function of the polar electron angle [39]. The projectile scattering angle is fixed at 0.55 mrad and the projectile energy losses at values ranging from 50 to 60 eV, where 57 eV is closest to the projectile–electron velocity matching. The arrows indicate the direction of the momentum transfer. For curves, see text

on the same CDW-EIS and 3DW models as in Fig. 10. Although here, too, there are considerable discrepancies between both calculations and the measurement, the 3DW model does agree with experiment on the presence of a cusp peak, which becomes increasingly important with increasing θ_p. The CDW-EIS approach predicts a cusp peak only at the largest θ_p. Furthermore, in both experiment and theory a small shift to smaller E_{el} compared to the cusp energy (indicated by the arrows) is seen. The importance of the cusp peak especially at large θ_p is reminiscent of a similar trend found in the shift of the binary peak relative to the direction of **q** in 16 MeV O^{7+} + He collisions (Fig. 9). This similarity may be taken as an indication that at least at large θ_p the cusp peak is, like the shift of the binary peak, mostly due to the PE-ET-PE interaction sequence leading to PCI.

Returning to the angular dependence of the FDCS, the double-peak structure may not seem to be particularly remarkable given that in first order the binary peak must occur in the direction of **q** and a peak at 0° appears to be plausible because in PCI the projectile and the electron focus each other toward the beam axis. However, it should be noted that the binary peak cannot be completely described within first order, but rather its shift relative to **q** is caused by higher order contributions, mostly by PCI. Therefore, the forward and the shifted binary peaks are due to the same mechanism. In a classical picture, it is then very difficult to understand why a small

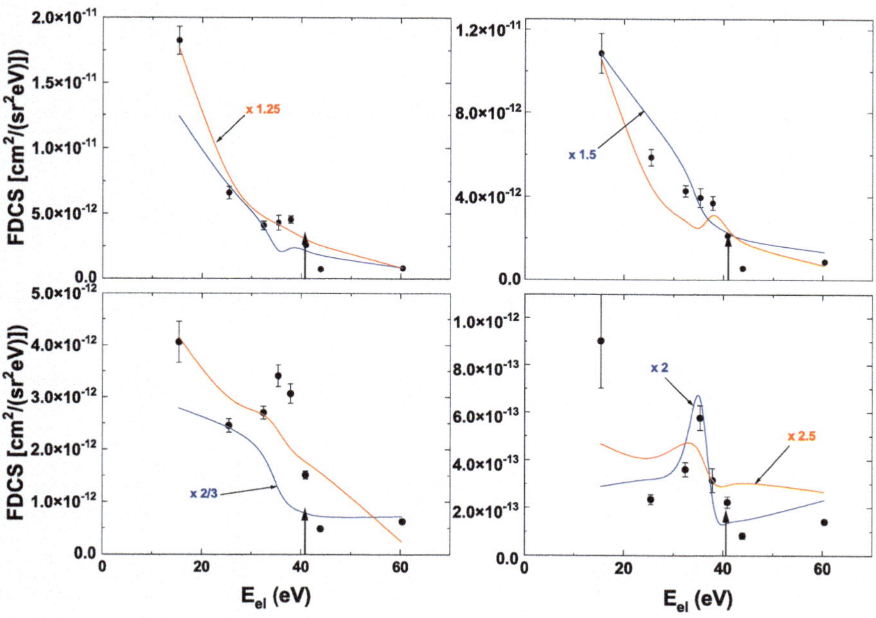

Fig. 11 Fully differential cross sections for electrons ejected in 75 keV p + He collisions into the scattering plane as a function of ejected electron energy [39]. The polar electron angle is fixed at 30° and the projectile scattering angle at values ranging from 0.1 to 0.5 mrad. The arrows indicate the cusp energy (velocity matching). For curves, see text

shift caused by PCI (location of binary peak) and a large shift (forward peak) are likely, but an intermediate shift (minimum separating both peaks) is unlikely. Classically, one would expect a single peak with an enhanced small-angle wing extending well into the forward region. In quantum mechanics, a possible explanation for the double-peak structure is interference between the first-order and higher order amplitudes. This hypothesis was tested by Bastola et al. [60] by measuring the FDCS for ionization in 75 keV p + He collisions near the velocity matching for an incoherent projectile beam. The data were then compared to FDCS measured earlier for a coherent beam [40]. The ratios between the coherent and incoherent FDCS, which, like in classical optics, represent the interference term, are plotted in Fig. 12 for $\theta_p = 0.3$ mrad (left panel) and 0.5 mrad (right panel) as a function of θ_{el}. Indeed, pronounced structures were observed signifying interference effects. This interpretation was qualitatively (despite significant quantitative discrepancies to experiment) supported by calculations presented in [60].

The collision system with the largest perturbation parameter ($\eta = 4.4$) for which FDCS were measured so far is 3.6 MeV/A Au^{53+} + He [37]. However, only ejected electrons with energies of less than 100 eV were detected, which is very far from the cusp energy (1.97 keV). In Fig. 13, these are shown for electrons with an energy of 17.5 eV ejected into the scattering plane and with q fixed at 0.65 a.u. A somewhat similar shape as in the velocity-matching regime for 75 keV p + He collisions, with a strong forward peak, is found. However, there is no separate binary peak, but rather only a "shoulder" on the large-angle wing of the forward peak which is consistent with an unresolved binary peak once again significantly shifted in the forward direction. Considering the large relative velocity between the ejected electron and the projectile a forward peak was not necessarily expected. On the other hand, it is not too surprising that the very large projectile charge compensates for the larger average distance

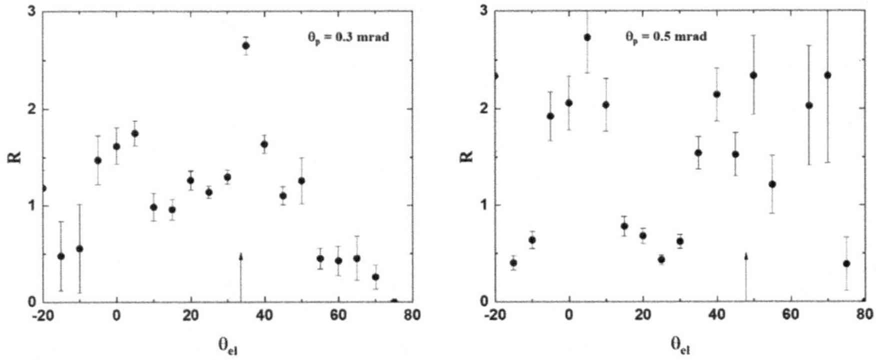

Fig. 12 Ratio between fully differential cross sections for electron emission into the scattering plane in 75 keV p + He collisions recorded for coherent and incoherent projectiles as a function of polar electron emission angle [60]. The projectile energy loss was fixed at 68.5 eV and the scattering angle at 0.3 mrad (left panel) and 0.5 mrad (right panel). The arrows indicate the direction of the momentum transfer

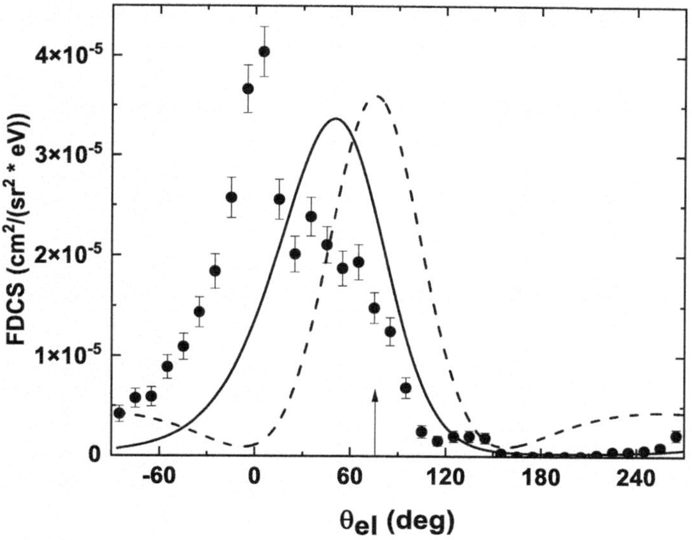

Fig. 13 Fully differential cross sections for electrons with an energy of 17.5 eV ejected in 3.6 MeV/A Au^{53+} + He collisions into the scattering plane [37]. The momentum transfer is fixed at 0.65 a.u. The arrows indicate the direction of the momentum transfer. For curves see text

between both particles. Therefore, the attractive force that the projectile exerts on the electron could well be similar to the one in the velocity-matching regime for proton impact both in magnitude and in direction.

The curves in Fig. 13 represent an FBA calculation (dashed curve) and a CDW-EIS calculation including the interaction between the projectile and the residual target ion. For ionization from a 1 s state, the FBA must be strictly symmetric about **q** and therefore cannot yield a shift of the binary peak. In contrast, the CDW-EIS calculation confirms the forward shift of the binary peak observed in the experiment. However, neither calculation shows even the slightest indication of a forward peak. Later, a small forward peak was found in a 3DW calculation for an electron energy of 20 eV [61]. Overall, like in the velocity-matching regime for p + He collisions, perturbative calculations are in poor agreement with the experimental data.

6 Conclusions

In perturbation theory, advancing our understanding of the few-body dynamics in atomic collisions essentially means understanding the various higher order mechanisms contributing to a specific process and their relative contributions to the cross sections. One higher order process that has attracted strong interest since the pioneering work of Crooks and Rudd [23] is known as post-collision interaction

(PCI). This process involves at least two interactions between the projectile and the active electron and at least one interaction between the residual target ion and the projectile or the electron. Since then, experimental advancements have resulted in a rich literature on PCI. Its signatures have been observed in spectra of all final-state particles of ionization processes, i.e., the ejected electron, the scattered projectile, and the residual target ion. Even numerous fully differential measurements were performed, and they provide the most sensitive tests of theoretical models.

The experimental observations and comparison to theory led to a significantly improved qualitative understanding of the few-body dynamics underlying PCI. To just mention a few examples, we now know the leading-order interaction sequences which mainly contribute to PCI. We understand that the projectile electron–electron target–projectile electron sequence becomes increasingly important with increasing projectile scattering angle, at least as far as the shift of the binary peak is concerned (previously it was assumed that PCI generally becomes more important with decreasing scattering angle). And we have strong indications that the double-peak structure in the FDCS in the velocity-matching regime is due to the interference between various transition amplitudes. At the same time, perturbative approaches, which often reproduce measured FDCS for small electron energies reasonably well, still have major difficulties in describing experimental data in the velocity-matching regime. This is probably due to the long interaction time between the projectile and the electron, resulting from the small relative velocity, so that the projectile cannot be treated as a small perturbation to the target atom. For this regime, it appears to be particularly important to develop non-perturbative approaches.

It is clear that further work on PCI, especially theoretical studies, is needed. However, it is equally important to investigate the few-body dynamics of other higher order mechanisms. For example, our understanding of processes involving the projectile–residual target ion interaction is still rather incomplete. In the velocity-matching regime and for large projectile charges such studies are not feasible because there PCI is the dominant higher order process, and it will thus mask up other higher order contributions. Numerous FDCS measurements have been performed for ejected electron energies well below the cusp energy, but no data have been reported yet for electron energies significantly above the cusp energy. Such measurements are currently in progress.

Acknowledgements This work was supported by the US National Science Foundation under grant no. PHY-2011307.

References

1. E. Rutherford, Philos. Mag. **21**, 669 (1911)
2. B.H. Bransden, C.J. Joachain, *Physics of Atoms and Molecules* (Prentice Hall, Harlow, England, 2003)

3. C. Froese-Fischer, T. Brage, P. Johnson, *Computational Atomic Structure: An MCHF Approach*, (CRC Press, 1997)
4. M. Schulz ed., *Ion-Atom Collisions. The Few-Body Problem in Dynamic Systems* (De Gruyter, Berlin/Boston, 2019)
5. R.T. Pedlow, S.F.C. O'Rourke, D.S.F. Crothers, Phys. Rev. A **72**, 062719 (2005)
6. P.D. Fainstein, V.H. Ponce, R.D. Rivarola, J. Phys. B **24**, 3091 (1991)
7. M.D. Sanchez, W.R. Cravero, C.R. Garibotti, Phys. Rev. A 61, (2000)
8. V.D. Rodriguez, R.O. Barrachina, Phys. Rev. A **57**, 215 (1998)
9. L. Gulyás, S. Egri, A. Igarashi, Phys. Rev. A **99**, 032704 (2019)
10. M.F. Ciappina, W.R. Cravero, M. Schulz, R. Moshammer, J. Ullrich, Phys. Rev. A **74**, 042702 (2006)
11. A.B. Voitkiv, B. Najjari, J. Ullrich, J. Phys. B **36**, 2591 (2003)
12. D.H. Madison, D. Fischer, M. Foster, M. Schulz, R. Moshammer, S. Jones, J. Ullrich, Phys. Rev. Lett. **91**, 253201 (2003)
13. O. Chuluunbaatar, S.A. Zaytsev, K.A. Kouzakov, A. Galstyan, V.L. Shablov, Yu.V. Popov, Phys. Rev. A **96**, 042716 (2017)
14. A.L. Godunov, V. Schipakov, M. Schulz, J. Phys. B **31**, 4943 (1998)
15. T.N. Rescigno, M. Baertschy, W.A. Isaacs, C.W. McCurdy, Science **286**, 2474 (1999)
16. J. Colgan, M. Foster, M.S. Pindzola, I. Bray, A.T. Stelbovics, D.V. Fursa, J. Phys. B **42**, 145002 (2009)
17. M. McGovern, D. Assafrao, J.R. Mohallem, C.T. Whelan, H.R.J. Walters, Phys. Rev. A **81**, 042704 (2010)
18. I. Bray, D.V. Fursa, Phys. Rev. A **54**, 2991 (1996)
19. A.T. Stelbovics, I. Bray, D.V. Fursa, K. Bartschat, Phys. Rev. A **71**, 052716 (2005)
20. I.B. Abdurakhmanov, J.J. Bailey, A.S. Kadyrov, I. Bray, Phys. Rev. A **97**, 032707 (2018)
21. O.J. Kroneisen, H.J. Lüdde, T. Kirchner, R.M. Dreizler, J. Phys. A **32**, 2141 (1999)
22. S. Borbély, J. Feist, K. Tőkési, S. Nagele, L. Nagy, J. Burgdörfer, Phys. Rev. A **90**, 052706 (2014)
23. G.B. Crooks, M.E. Rudd, Phys. Rev. Lett. **25**, 1599 (1970)
24. N. Stolterfoht, D. Schneider, J. Tanis, H. Altevogt, A. Salin, P.D. Fainstein, R. Rivarola, J.P. Grandin, J.N. Scheurer, S. Andriamonje, D. Bertault, J.F. Chemin, Europhys. Lett. **4**, 899 (1987)
25. M. Breinig, S.B. Elston, S. Huldt, L. Liljeby, C.R. Vane, S.D. Berry, G.A. Glass, M. Schauer, I.A. Sellin, Phys. Rev. A **25**, 3015 (1982)
26. W. Meckbach, P.J. Focke, A.R. Goni, S. Suares, J. Macek, M.G. Menendez, Phys. Rev. Lett. **57**, 1587 (1986)
27. L. Sarkadi, J. Pálinkás, A. Kövér, D. Berényi, T. Vajnai, Phys. Rev. Lett. **62**, 527 (1989)
28. D.H. Lee, P. Richard, T.J.M. Zouros, J.M. Sanders, J.L. Shinpaugh, H. Hidmi, Phys. Rev. A **41**, 4816 (1990)
29. L.C. Tribedi, P. Richard, Y.D. Wang, C.D. Lin, R.E. Olson, Phys. Rev. Lett. **77**, 3767 (1996)
30. Y.D. Wang, L.C. Tribedi, P. Richard, C.L. Cocke, V.D. Rodríguez, C.D. Lin, J. Phys. B **29**, L203 (1996)
31. O. Jagutzki, R. Koch, A. Skutlartz, C. Kelbch, H. Schmidt-Böcking, J. Phys. B **24**, 993 (1991)
32. G. Bernardi, P. Focke, W. Meckbach, Phys. Rev. A **55**, R3983 (1997)
33. Th. Weber, Kh. Khayyat, R. Dörner, V.D. Rodríguez, V. Mergel, O. Jagutzki, L. Schmidt, K.A. Müller, F. Afaneh, A. Gonzalez, H. Schmidt-Böcking, Phys. Rev. Lett. **86**, 224 (2001)
34. T. Vajnai, A.D. Gaus, J.A. Brand, W. Htwe, D.H. Madison, R.E. Olson, J.L. Peacher, M. Schulz, Phys. Rev. Lett. **74**, 3588 (1995)
35. M. Schulz, A.C. Laforge, K.N. Egodapitiya, J.S. Alexander, A. Hasan, M.F. Ciappina, A.C. Roy, R. Dey, A. Samolov, A.L. Godunov, Phys. Rev. A **81**, 052705 (2010)
36. A. Silvus, M. Dhital, S. Bastola, J. Buxton, Z. Klok, E. Ali, M.F. Ciappina, B. Boggs, D. Cikota, D.H. Madison, M. Schulz, J. Phys. B **52**, 125202 (2019)
37. M. Schulz, R. Moshammer, A.N. Perumal, J. Ullrich, J. Phys. B **35**, L161 (2002)
38. M. Schulz, B. Najjari, A.B. Voitkiv, K. Schneider, X. Wang, A.C. Laforge, R. Hubele, J. Goullon, N. Ferreira, A. Kelkar, M. Grieser, R. Moshammer, J. Ullrich, D. Fischer, Phys. Rev. A **88**, 022704 (2013)

39. M. Dhital, S. Bastola, A. Silvus, A. Hasan, B.R. Lamichhane, E. Ali, M.F. Ciappina, R.A. Lomsadze, D. Cikota, B. Boggs, D.H. Madison, M. Schulz, Phys. Rev. A **99**, 062710 (2019)
40. M. Dhital, S. Bastola, A. Silvus, B.R. Lamichhane, E. Ali, M.F. Ciappina, R. Lomsadze, A. Hasan, D.H. Madison, M. Schulz, Phys. Rev. A **100**, 032707 (2019)
41. M. Dhital, S. Bastola, A. Silvus, J. Davis, B.R. Lamichhane, E. Ali, M.F. Ciappina, R. Lomsadze, A. Hasan, D.H. Madison, M. Schulz, Phys. Rev. A **102**, 032818 (2020)
42. L. Sarkadi, L. Gulyas, L. Lugosi, Phys. Rev. A **65**, 052715 (2002)
43. U. Chowdhury, M. Schulz, D.H. Madison, Phys. Rev. A **83**, 032712 (2011)
44. H.R. Walters, C.T. Whelan, Phys. Rev. A **92**, 062712 (2015)
45. E. Blauth, Z. Physik **147**, 228 (1957)
46. C.E. Kuyatt, T. Jorgensen, Phys. Rev. **130**, 1444 (1963)
47. M.E. Rudd, T. Jorgensen Jr., Phys. Rev. **131**, 666 (1963)
48. M.E. Rudd, C.A. Sautter, C.L. Bailey, Phys. Rev. **151**, 20 (1966)
49. A. Salin, J. Phys. B **2**, 631 (1969)
50. J. Macek, Phys. Rev. A **1**, 235 (1970)
51. M. Brauner, J.H. Macek, Phys. Rev. A **46**, 2519 (1992)
52. D.R. Schultz, C.O. Reinhold, Phys. Rev. A **50**, 2390 (1994)
53. D.S.F. Crothers, J.F. McCann, J. Phys. B **16**, 3229 (1983)
54. W. Schmitt, R. Moshammer, F.S.C. O'Rourke, H. Kollmus, L. Sarkadi, R. Mann, S. Hagmann, R.E. Olson, J. Ullrich, Phys. Rev. Lett. **81**, 4337 (1998)
55. R. Moshammer, J. Ullrich, H. Kollmus, W. Schmitt, M. Unverzagt, H. Schmidt-Böcking, C.J. Wood, R.E. Olson, Phys. Rev. A **56**, 1351 (1997)
56. M. Schulz, R. Moshammer, D.H. Madison, R.E. Olson, P. Marchalant, C.T. Whelan, S. Jones, M. Foster, H. Kollmus, A. Cassimi, J. Ullrich, J. Phys. B **34**, L305 (2001)
57. M. Schulz, R. Moshammer, D. Fischer, H. Kollmus, D.H. Madison, S. Jones, J. Ullrich, Nature **422**, 48 (2003)
58. A. Hasan, T. Arthanayaka, B.R. Lamichhane, S. Sharma, S. Gurung, J. Remolina, S. Akula, D.H. Madison, M.F. Ciappina, R.D. Rivarola, M. Schulz, J. Phys. B **49**, 04LT01 (2016)
59. A. Hasan, S. Sharma, T.P. Arthanayaka, B.R. Lamichhane, J. Remolina, S. Akula, D.H. Madison, M. Schulz, J. Phys. B **47**, 215201 (2014)
60. S. Bastola, M. Dhital, S. Majumdar, A. Hasan, R. Lomsadze, J. Davis, B. Lamichhane, S. Borbély, L. Nagy, M. Schulz, Atoms **10**, 119 (2022)
61. M. Foster, D.H. Madison, J.L. Peacher, J. Ullrich, J. Phys. B **37**, 3797 (2004)

Ionization of Water, Ammonia and Methane by Proton Collision: Experimental and Electronic Configuration Studies

Wania Wolff

Abstract The isoelectronic species water, ammonia and methane play an important role in the effects of ionizing radiation with profound implications in surrounding environments. The interest on these molecules remains unabated in diverse fields ranging from atmospheric and astro-chemistry to biomedical applications. The ionized molecules, their radicals and charged fragments are prolific sources for the formation and damage of organic compounds. In this study we present a short review of single ionization and fragmentation of water, ammonia and methane induced by swift proton collision. The formation of ionic species depends on the selection of which orbital electron(s) from the molecules are removed by the proton. The molecule is initially destabilized, then a stable, metastable or dissociative configuration state is reached. An ionic species is formed dependent on the excited state the molecule is left by the electron removal mechanism and its production can be attributed to several final ionic states. It is our goal to understand the mechanism of water, ammonia and methane dissociation based on the knowledge of the distribution of single primary vacancies produced by the swift proton collision. The electronic configuration and orbital symmetry of the molecule are playing a more dominant role in controlling the ionization and fragmentation processes.

1 Introduction

Located in the first-row hydrides, water (H_2O), ammonia (NH_3) and methane (CH_4) are among the most important and ubiquitous chemical species. Various experimental investigations have been reported to date on the interaction of swift protons with these molecules [1–21]. The fragmentation of these molecules by ionizing particles may involve the simultaneous breaking of one or several bonds, rearrangement and attachment of the radicals with the consequent formation of highly reactive fragment species, ions and neutrals. The subsequent reactions with other molecules

W. Wolff (✉)
Physics Institute, Federal University of Rio de Janeiro, Rio de Janeiro, Brazil
e-mail: wania@if.ufrj.br

L. C. Tribedi (ed.), *Advances in Atomic Molecular Collisions*,
https://doi.org/10.1007/978-981-97-7063-2_3

in the environment lead to the formation of complex molecules. These products enrich the chemistry and impact the inventory of many physicochemical environments subjected to radiation, such as planetary atmospheres, ecosystems, cosmic objects or the human body exposed to radiation in medical diagnosis or radiation therapy [22–30].

The spectroscopic (i.e. FTIR, Raman, fluorescence, absorption, electron emission spectra) properties of water, ammonia and methane have been widely studied. Water, ammonia and methane are isoelectronic molecules and have an approximately tetrahedral arrangement of four electron pairs. In methane there are no lone pairs, ammonia and water molecules contain one and two lone pairs of electrons exhibiting a distorted geometry, ammonia, pyramidal angle of 107, and water v shape or angular angle of 104.5 in contrast to methane—tetrahedral angle of 109.5.

They all have sp3 hybridization. The bonds in a **methane** (CH_4) molecule are formed by four separate but equivalent **orbitals**; a single 2*s* and three 2*p* **orbitals** of the carbon hybridize into four sp^3 **orbitals.** Each of the four sp^3 hybrid orbitals overlaps with the 1*s* orbital of a hydrogen atom. In the **ammonia** molecule (NH_3), 2*s* and 2*p* **orbitals** create four sp^3 hybrid **orbitals**, one of which is occupied by a lone pair of electrons meaning that three sp^3 orbitals are bonded to a hydrogen. In the water molecule one 2*s* and three 2*p* orbitals of oxygen are hybridized with two pairs of bonding electrons and two lone pairs and the hybridization of water is also *sp3* and only two *sp3* are bonded to a hydrogen, (Fig. 1).

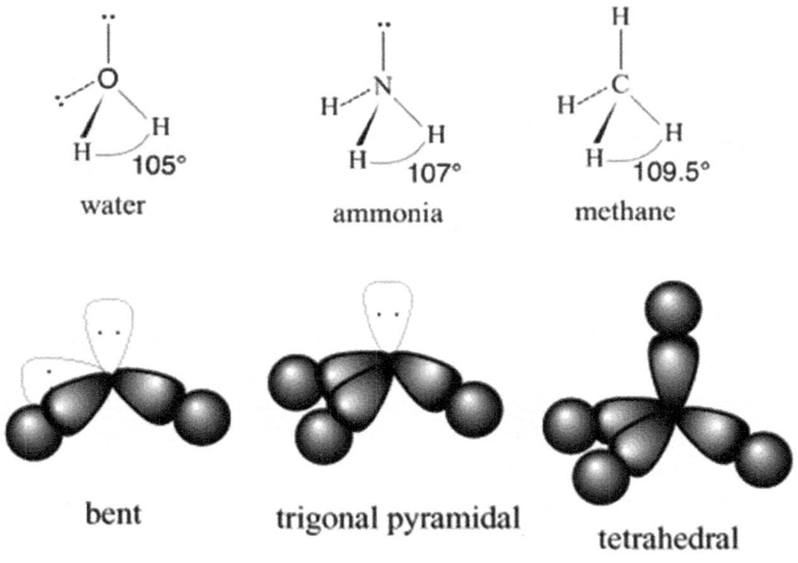

Fig. 1 Structures of water, ammonia and methane

The ground state molecular configuration of water is $(\text{core})^2$, $(1b_1)^2$, $(3a_1)^2$, $(1b_2)^2$ and $(2a_1)^2$, [13] of ammonia $(\text{core})^2$, $(3a_1)^2$, $(1e_1)^4$ and $(2a_1)^2$ [14] and of methane is $(\text{core})^2$, $(2a_1)^2$ and $(1t_2)^6$ [15] (Fig. 2).

Experimental techniques like electron momentum spectroscopy (EMS), high resolution (*e*, 2*e*) and (*e*, 2*e*+ ion) experiments as well as theoretical approaches such as multi-reference theories and symmetry-adapted-cluster configuration interaction calculations improved significantly the characterization of the electronic configuration of the ionized states of water [31–35], ammonia [36–43] and methane [44–50]. All the reported theoretical results and experimental data indicate a partial breakdown of the orbital picture describing the ionization process for the innermost orbital of water, ammonia and methane [51, 52]. The strong influence of electron–electron correlation during relaxation following the production of a vacancy in an inner valence was included in the picture. In the ionization of the inner valence orbitals, many body structures were observed in the ionization spectra in the form of a dispersion of the innermost valence ionization intensity.

In the binding energy spectra measured using electron momentum spectroscopy of water, [31, 34, 35], ammonia [36, 38–40] and methane [45–50], the outermost one-electron ionization states were resolved and appeared as sharp peaks. They are related to single hole states which were assigned to vertical transitions. In contrast, the main peak of the ionization state in the inner-valence region is quite broad with a tail extending to high energy accompanied by shallow peaks on the high energy side spreading over 10–20 eV, depending on the molecule. The main inner valence state shares some of its intensity (pole strength) with these nearby bands—satellite bands. These ionized states are 2-hole l-particle (2 *h*-1*p*) states originating from the excitation-ionization two-electron processes and shake-up processes. This is in qualitative agreement with the prediction that ionization of inner valence electrons of saturated hydrocarbons can be properly described as the creation of quasi-holes [52]. These satellite states are described by calculations using a linear combination of several states with considerable configuration mixing [35, 38, 50]. The density of (2 *h*-1*p*) configurations is high and can no longer be expected to correspond to only a single hole configuration.

Figure 3 represents two processes that lead to ionic final states, one originating from the vertical transition of the $2a_1$ inner valence state (*ion*) and the other from the satellite shake-up state (*ion* and *exc* in Fig. 3a,) for water, ammonia and methane. The satellite states indicated in the figure are $(3a_1^{-1}\ 1\text{b}_2^{-1}\ 2b_2)$ for water [35], $(3a_1^{-1}\ 1e^{-1}\ 9e)$ for ammonia [38] and $(1t_1^{-2}\ 2t_2)$ for methane [50].

Electron–electron-ion triple coincidence spectroscopy (*e*, 2*e* + ion) is a powerful orbital imaging technique and offers a kinematically complete study of electron impact ionization events [47]. By measuring the kinetic energies of two ejected electrons, (E_1, E_2) the binding energy (BE) spectra can be obtained (BE $= E_0 - E_1 - E_2$). By measuring the kinetic energies of the reaction products, of the outgoing electrons in coincidence with the ionic fragments, the ionic species can be identified with their kinetic energies, binding energies (BE) and ionization channels. Recently, an improved experimental setup based on a reaction microscope with high mass resolution, detection efficiency of the fragment ions and binding energy resolution

Fig.2 Ground state electronic configuration of water (**a**), ammonia (**b**) and methane (**c**)

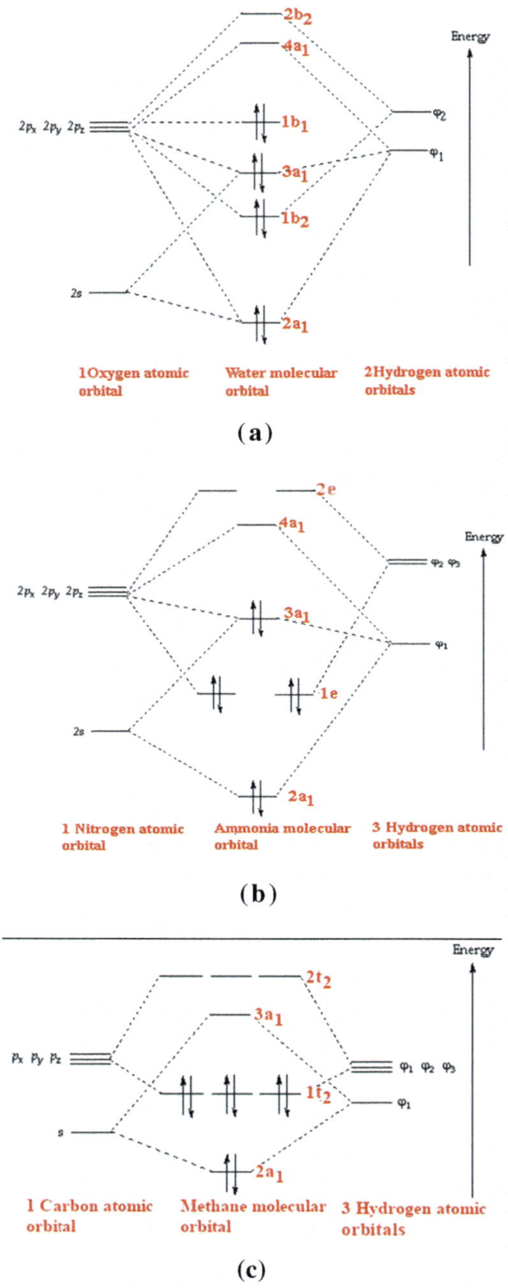

Fig. 3 Final states of the inner valence orbital. Vertical transitions and shake-up process are indicated by arrows for water (**a**) ammonia, (**b**) and methane (**c**)

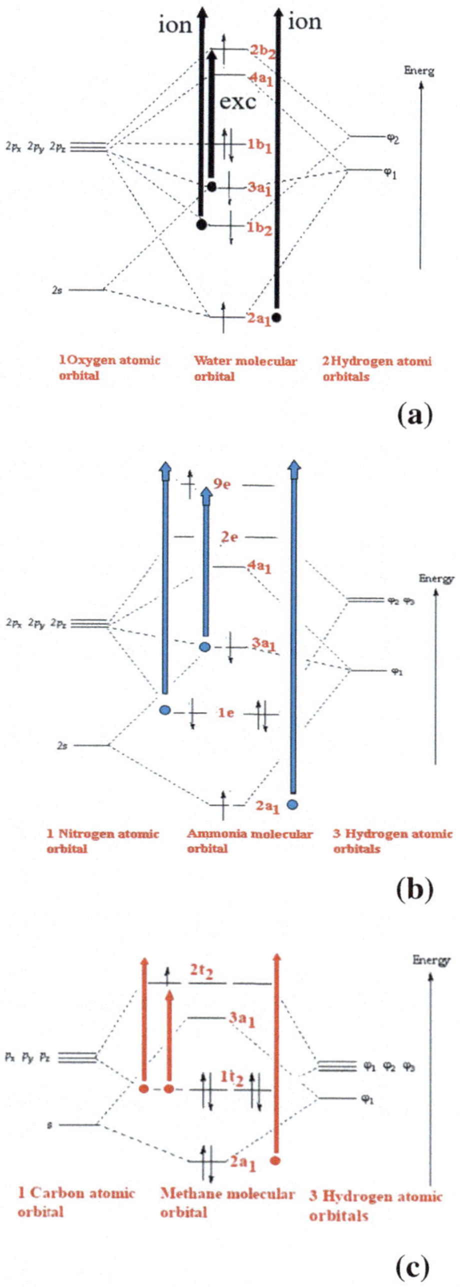

measured each of the reaction products of methane [47]. The complex experiment was able to discriminate between the levels of the ionized valence states including the ionization/excitation satellites states. Their relative abundances were responsible for the formation of a specific fragment and its kinetic energy distribution.

The purpose of this review is to present a relatively simple experimental-theoretical model approach to obtain a broader picture of the fragmentation process of water, ammonia and methane induced by collision with protons at intermediate velocities [11, 13–15]. This approach was already extended to more complex molecules [53, 54]. In multielectron molecules, several final electronic states are associated with each of the fragment ions produced and it is a major challenge to predict the dissociation of a molecule from the knowledge of the distribution of primary vacancies of the molecular orbitals produced by the projectile. In the case of water, ammonia and methane vertical transitions from the valence molecular orbitals are responsible for most of the ionic states [13–15, 47], but as outlined there are additionally several satellite bands assignable to transitions converging to an excited ionic state prepared by the shake-up process.

In ion collisions, electron removal from orbitals can be due to capture and ionization events and most of these processes result in some dissociating states which undergo fragmentation [11]. Protons at an intermediate velocity range can remove any electron from a molecule, from the outermost to the innermost molecular orbitals. The removal of a specific electron leads to a new electronic configuration of the molecule that is stabilized as the parent molecule ion or break-up dissociating into channels with one or more ionic species and the corresponding neutral fragments. One electron ejection dominates the production of the parent ion and of cation and the corresponding neutral fragment. Only a small fraction of molecules is doubly ionized and dissociates into ion pairs [11, 13–15].

For swift projectiles, the collision time of the order of 10^{-17} s is much shorter than the relaxation or dissociation time, which is typically of the order of the vibration time (10^{-14}–10^{-12} s). Then the intensity of the ejected fragments associated with the various molecular orbitals can be considered as characteristics of the molecule that does not depend on the velocity of the projectile. The velocity and type of the projectile determine the population of primary electron vacancies produced in the molecular orbitals in a first step and the yields of the molecular parent ion and fragments produced by the relaxation of the charged molecule with its subsequent fragmentation in a second step.

This review is presented as follows. The introduction presented the structure and orbitals of water, ammonia and methane, outlined key features of the (*e*, 2*e*) and (*e*, 2*e*+ ion) techniques. In the experimental section details of the setup of recent experiments with protons on water, ammonia and methane [13–15] are presented and procedures to efficiently discriminate single- from multiple-ionization events. Absolute fragmentation cross sections over a wide range of proton energies were obtained based on standard time-of-flight spectrometry and coincidence techniques. Relevant experimental procedures, such as optimizing beam transmission, sample handling and species detection are described in this section.

Few experimental studies were dedicated in measuring single charged and multiple charged ions or ion pairs in coincidence of water, [8, 13] ammonia [14] and methane [16] from several keV to MeV proton energies. Here we present data from our measurements of the ionization and subsequent dissociation of water, ammonia and methane molecules in a gas phase induced by protons with energies varying from 125 to 2700 keV [13, 14, 16]. They are compared with published data from other investigations colliding keV to MeV protons onto the three molecules, water [8–13], ammonia, [14, 15] methane [16–21].

The discussion of the cross sections data addresses the total and partial ionization cross sections, i.e., the non-dissociative and dissociative cross sections. In the intermediate-velocity regime single ionization predominates over multiple ionization, while charge exchange, which prevails at lower velocities, is no longer important [11]. Therefore, the focus here is on the single ionization process. In the data reduction, the discrimination of the singles events from doubles and its impact on the cross-section's values are evaluated. Each partial ionization cross section quantifies the contribution of one or more molecular states ionization for forming a specific ion. To estimate the contribution of each molecular orbital in the ionic species formation a semi-classical fragmentation model was applied to estimate the contribution of each molecular state in the ionic species formation by matching calculated ionization with measured fragmentation cross sections. The model used to generate the predicted (experimental) cross sections is described in detail based on the principles of the Born approximation and the ionic states of the molecules. Results from experiment and model suggest the contribution of the ionic states in the ionic species formation. Conclusions from this review are briefly outlined in the last section.

2 Experimental Procedures

The 125–2700 keV proton experiments [13–15] were carried out at the Pelletron accelerator facility at the Physics Institute of the Rio de Janeiro Federal University. Negative hydrogen ions (H^-) are extracted from the SNICS-II ion source at an energy of 7 kV, accelerated to an energy of about 27 keV, selected in mass by a Wien filter and injected into a two-stage electrostatic accelerator. In the first stage, the beam is accelerated towards the high voltage terminal located at the centre of the accelerator, passing through a cell filled with molecular nitrogen tripper gas. The removal of two electrons by collision with the N_2 produces a proton beam. Applying a terminal voltage V_T (in kV) the proton beam follows a second acceleration towards the exit of the accelerator reaching a final beam energy of $[27 + (1 + q)\ V_T]$ keV. The proton beam is mass, energy and charge selected by a switching magnet and directed towards the ion-atom collision beam line located at 15∘ angle relative to the main accelerator beam line. The beam optics and two sets of adjustable 4-jaw slits ensure a small parallel clean from neutral and charge-state impurities. Parallel plates and a position sensitive detector positioned at the end of the beam line allow the alignment of the beam with the time-of-flight spectrometer and the flux of injected molecules.

The proton beam crosses the effusive jet of the molecules at right angles on the focus of the time-of-flight mass spectrometer. Protons were made to interact under single collision conditions with the neutral molecules effusing from a grounded needle. The flow of the molecules is regulated by an ultrahigh leak valve. The pressures of the experimental chamber with and without gas load were typically 1–3 10^{-6} Torr and 8–10 10^{-8} Torr, respectively. The position of the gas jet with respect to the collimated projectile beam was adjusted to maximize their overlap in the focus of the time-of-flight spectrometer. A detailed description of the transmission of the proton to the interaction region is presented in Ref. [54]. It is worth mentioning that the same experimental setup and procedures were applied to the proton-water, proton-ammonia and proton-methane measurements.

The liquid water sample was kept in a water bath at fixed temperature to maintain the sublimation flow constant. In the case of several liquid organic molecules, it is necessary to keep the liquid stored in vial in a bath of an ice-salt-alcohol solution, (which temperature is monitored by a temperature sensor) to maintain the sublimation flow almost constant. Methane is easier to handle as it is delivered in its natural gaseous form, methane gas is colourless, odourless and tasteless. Ammonia is gaseous at room temperature, ammonia gas is a colourless gas with a distinct pungent smell, liquefied to its own vapor pressure. Ammonia and methane gases make measurements more straightforward to conduct in comparison to liquid samples. Possible contamination of the system by critical impurities such as air and H_2O were monitored. The injection line was heated before injection of gaseous ammonia and methane and dried with gaseous N_2, while during experiments the injection line was kept at room temperature. No water fragment ions were detected in the ammonia and ammonia experiments. A small amount of N_2^+ was detected. The literature values of the dissociative cross sections of N_2 [20] were used to determine the fraction of N_2^+ molecules that dissociate giving rise to a N^+ ion, and the contribution was accordingly subtracted.

For molecules in liquid phase, it is essential to apply degassing procedures. Multiple freeze pump thaw cycles are a common and effective method for small-scale degassing of liquids, by which dissolved gases are removed from a liquid. The presence of dissolved gases such as nitrogen, oxygen or carbon dioxide can interfere with spectroscopic measurements or can induce unwanted bubble formation. The liquid is frozen by liquid nitrogen, while air and other gases are pumped away. The pump was valved off and the frozen liquid warmed up to drive out dissolved gas left. The process is repeated 3–6 times until no bubbles of gas are seen in the warming up process.

Singles and ion-pairs spectra were obtained using time-of-flight mass spectroscopy combined with multi-hit coincidence technique. Detection of the first electron collected by the system triggers the timing measurement of recoil ions. After ionization, the ejected electrons and ionic fragments are extracted in opposite directions by a homogeneous electric field of 700 V/cm, being guided throughout the TOF. The electrons are focused by an Einzel type lens and detected by a channeltron (electron multiplier detector). The ionic fragments are pre-accelerated, focused and directed towards a field-free drift tube and detected by a dual microchannel-plate detector (MCPs) in a chevron stack. Standard procedures for eliminating mass discrimination effects in time-of-flight mass spectra by varying experimental parameters such as the MCP voltage and the CFD discrimination threshold were done.

A coincidence setup based on a fast time to digital converter (TDC), sets the emitted electron(s) as the start pulse input and the ionic fragments (recoil ions) as the sequential STOP input signals where the light mass, H^+ recoil, provides the first stop and the heavier mass recoil provides the second stop. Two time-of-flight spectra are generated: a single spectrum is produced from events in which only a single ion arrived at the detector. When two ions arrive at the detector following double ionization, ion-pair coincidence spectra are generated. The ion-pair spectra are displayed as two-dimensional histograms with the time of flight of the first detected ion t_1 plotted on the vertical axis, and the time of flight of the second detected ion t_2 plotted on the horizontal axis.

3 Data Reduction

A relevant issue that needs to be addressed is the detection efficiencies of the recoils ions and electrons that are not 100%. The efficiency related to the multi-hit detection conjugated with the Coulombic explosion effect must be considered carefully. The true yields of single-ion and ion-pair fragment production follow a procedure presented in reference [56, 57]. The lack of a sequential detection of one ion of a pair can contribute as a false event to the singles yields producing a false enhancement of single-ionization channels. It is necessary to correct the measured ion yields to the detection efficiencies to obtain the true ion yields.

The procedure to correct the measured yields and obtain the true yields relies on determining the detection efficiencies for the recoil ion ε_R and for the electron ε_e. The detection efficiencies account for the transmission efficiency of the grids, which define the electric fields in the spectrometer as well as the efficiency of the detector(s) and electronics. The values of ε_R and ε_e are determined by carrying out separate measurements under the same experimental conditions as for water, ammonia and methane experiments. For the recoil ion case, we used the data from the literature Ref. [18] to obtain the detection efficiency. Although the MCP is set to work at the saturation mode [48], the efficiencies for detection, for example, CH_n^+, NH_n^+ ions and the H^+ recoils are different. H^+ and H_2^+ fragments are emitted with larger kinetic energies than the other recoil ions. The value of ε_R was assumed to be ε_H for H + and ε_{ion} for the heavier recoil's ions, respectively. The values of ε_H and ε_{ion} varied from 0.17 to 0.13 and from 0.26 to 0.20, respectively along the sequence of water, ammonia and methane experiments.

The electron detection efficiency ε_e was obtained by comparing the yield of single ionization of neon (Ne^+) by proton impact, using two different START-STOP signal arrangements measured simultaneously: one arrangement with the electron giving the START and Ne^+ the STOP signal, the other with the proton as START and Ne^+ as STOP. The proton signal was obtained using a solid-state detector, and an efficiency of 100% was assured if the counting rate was kept below 2000 particles per second. The ratio between these yields provides the value for ε_e as ε_R cancels out. The values obtained along the measurements varied from 0.2 to 0.3.

The true counts N_{true} (Y+) originated from the parent ion (H_2O^+, $NH_3{}^+$ or $CH_4{}^+$) that dissociates into a neutral X and Y^+ ion or into an ion pair X^+ and Y^+ can be obtained once the detection efficiencies involved in the measurement are determined. The false coincidence events were minimized in the offline data reduction analysis and the aliasing of coincidence events into lower-order events [55, 56] were subtracted. The equations which provide the true counts N_{true} (N^t) from the measured counts $N_{measured}$ (N^m) are as follows:

$$N^t(P^+) = \frac{N^m(P^+)}{\varepsilon_e \varepsilon_R}$$

$$N^t(Y^+) = \frac{N^t(Y^+)}{\varepsilon_e \varepsilon_R} - \frac{(1-\varepsilon_H)}{\varepsilon_e \varepsilon_R \varepsilon_H} N^m(X^+, Y^+)$$

$$N^t(X^+) = \frac{N^t(X^+)}{\varepsilon_e \varepsilon_H} - \frac{(1-\varepsilon_R)}{\varepsilon_e \varepsilon_R \varepsilon_H} N^m(X^+, Y^+)$$

$$N^t(X^+, Y^+) = \frac{N^t(X^+)}{\varepsilon_H \varepsilon_R[1-(1-\varepsilon_e)^2]} N^m(X^+, Y^+)$$

$$N^t(P^{2+}) = \frac{N^{tm}(P^{+2})}{\varepsilon_R[1-(1-\varepsilon_e)^2]}$$

$$N^t(Y^{2+}, H^+) = \frac{N^m(Y^{2+}, H^+)}{\varepsilon_H \varepsilon_R[1-(1-\varepsilon_e)^2]}$$

where P, X and Y indicate parent molecule (CH_4, NH_3 and H_2O), hydrogen atom and molecule (H and H_2) and radicals ($CH_{n=0-3}$, $NH_{n=0-2}$, $OH_{n=0-1}$), respectively.

Following this data reduction, the fragment ions that originate from single (s) ionization events were separated from those resulting from the double (d) and triple (t) ionization events. After subtracting the false events, for example for the ammonia case, at the proton impact velocity of 4.7 a.u., counting all true NH^+ and N^+ ions irrespective of the degree of ionization, an enhancement of 21% and 44% relative to the true pure single counts, i.e., true $NH^+(s + d + t) = 1.21 \times$ true $NH^+(s)$ and true $N^+(s + d + t) = 1.44 \times$ true $N^+(s)$ results [14].

Charge separation in doubly or multiply charged molecular ions is responsible for a significant number of energetic fragments due to the Coulomb repulsion between the separating ionic species. The inefficient collection of translationally energetic fragment ions, like H+ cannot be discarded in experiments. Measurements of the kinetic energies of the fragment's ions can estimate truly the detection efficiency of energetic ions species. Fragment ions, particularly those arising from dissociative multiple ionization can be formed with significant kinetic energies, which inhibit their detection. Losses for those fragment ions can be corrected. Peaks representing channels where two identical ions, like hydrogen pairs, are detected in coincidence are commonly required due to dead time discrimination. The pairs missed can be estimated from a time-of-flight difference (t_2–t_1) plot (ion intensity versus (t_2–t_1)). A flat-topped symmetrical peak is observed if all ion pairs were detected, otherwise the peak in the (t_2–t_1) spectrum exhibits a hollowed peak instead of the expected square peak. The missed counts that make the hollowed area can be extrapolated to the limit $t_2 = t_1$ using simple geometry.

The absolute cross sections σ (singles, ion pairs) were obtained by normalizing the true yield of parent ion, fragment ions and ion pairs to the total electron production σ_-cross sections of references available in the literature for water [2, 5] and ammonia [3, 4, 6, 7] and methane. [3, 4] In these references, existing data on the ionization of neutral atoms and molecules by proton impact were reviewed. A mathematical equation was fitted to the weighted experimental data for each molecule, and these were adjusted to be consistent with appropriate theoretical calculations and electron impact and photoionization data. The recommended values of total ion production cross sections for proton-impact ionization of water, ammonia and methane were here considered for normalization. At energies above 200 keV the ion (σ_+) and electron (σ_-) production are the same, the contribution of the capture process in the production of ions is negligible [5]. In methane and water, for example, at 150 keV and 200 protons, respectively, the capture cross section σ_c amounts to 6 and 7% of the σ_- and the tendency of σ_c is of a steep decaying function for higher energies. Total ionization cross sections were mostly insensitive to isoelectronic molecules, as no remarkable structures (onset and shape) in the cross sections are present in the proton intermediate energy range (Fig. 4).

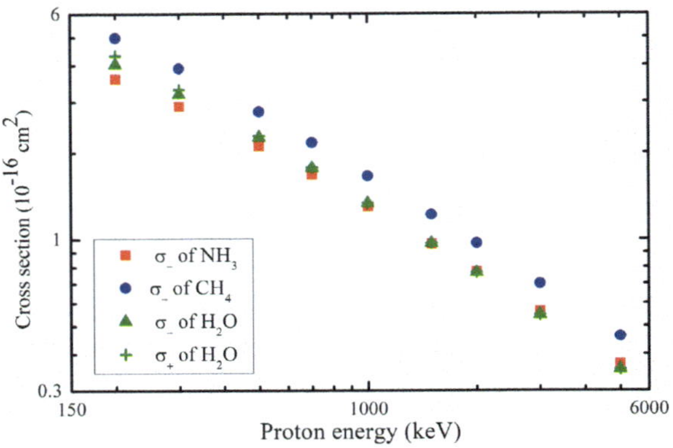

Fig. 4 Total electron and ion production cross sections for water [5] and total electron production for ammonia and methane [3, 4] as function of the proton energy

The sum of all true single and ion-pairs were normalized to the total electron cross section at each proton energy.

$$\sum_{f1}^{fn} \sigma(fn = singles) + \sum_{ion-pair1}^{ion-pair,n} \sigma(ionpair, n = doubles) = \sigma_-$$

$$\sigma(fn) = Yield\{fn\} * \sigma_- = \sum \frac{N\{fn\}}{\sum(N^t\{fn\}) + \sum(N^t\{ionpair, n\})} * \sigma_-$$

Single and ion-pair maps of coincidence spectra were recorded at different ionizing proton energies. Four experimental data on water in a wide energy range of protons with 100–350 keV [8], 20–150 keV [9, 10], 15–100 keV, 500–3500 keV [11] and 300–2000 keV [13] are scattered throughout the literature, and a sample collection is presented in Fig. 5. The data sets of partial cross sections are shown for the parent ion H_2O^+ and OH^+ and O^+ fragments. For the parent ion the data agrees quite well, but a dispersion is observed for the fragment ions, becoming wider for the single charged atoms, O^+.

For ammonia partial ionization cross sections were quantified in one data set in the energy range of 125–2700 keV [14]. Only branching ratios leading to H_2^+ and H_3^+ formation at protons with 4 MeV are available in the literature [15]. On the other hand, in the case of methane there are five data sets ranging from protons of 5–45 keV [17], 150–2700 keV [16], 50–6000 keV [20], 500–3500 keV [19] and 1000–12,000 keV [18] available in the literature (see Fig. 6) Similar as in the water case, the parent ion cross sections agree between the data sets quite well, but a dispersion is observed for the fragment ions, becoming wider for the single charged atoms, C^+. Other experimental results on proton impact, found in the literature for water [1, 2, 5], ammonia [6, 7] and methane [3, 4] concern only the total ion and electron production and do not provide any information on the separation between *n*-fold ionization contributions in the fragment-ion production from dissociative ionization.

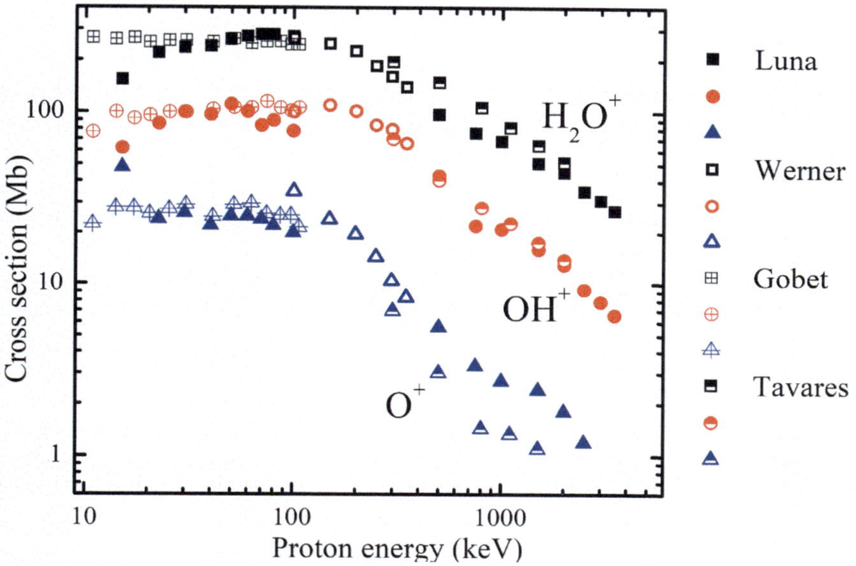

Fig. 5 Cross sections for parent ion H_2O^+ and ionic fragments OH^+ and O^+ of water available in the literature. First authors of the experimental works are indicated as legends: Luna [11], Werner [8], Gobet [9, 10] and Tavares [13]

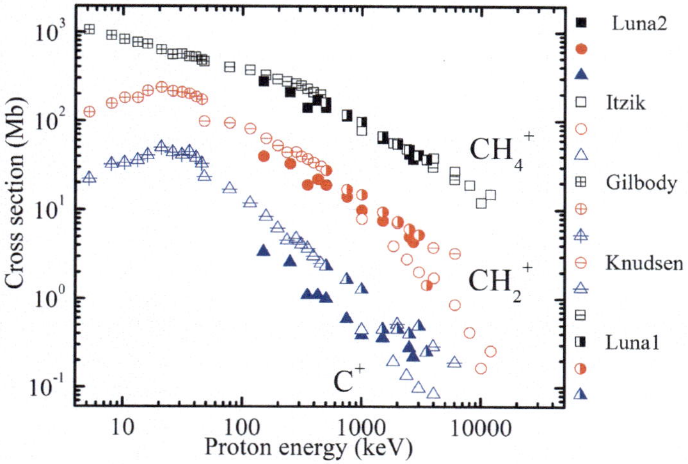

Fig. 6 Cross sections for parent ion CH_4^+ and ionic fragments CH_2^+ and C^+ of methane available in the literature. First authors of the experimental works indicated as legends: Luna2 [19], Itzik [18], Browning [17], Knudsen [20] and Luna1 [16]

4 Analysis and Discussion

In multi-electronic molecules, final electronic states are associated with each one of the ionic species produced. The interaction of the projectile with a molecule in the collisional phase can lead to the ionization process, creating a primary vacancy in the molecule. The molecular ion can populate a repulsive state after ionization and dissociate into neutral and positively charged fragments. If the collisional time is fast enough, this relaxation of the ionized/excited molecule through the dissociative ionization process can be considered as a post-collisional process.

At these collision energies, valence electrons are more likely to be removed in the projectile-molecule interaction. The stable molecular parent ion should originate mostly or mainly from the outermost valence orbital. But the removal of one outer or inner valence electron should also result in the fragmentation of the molecular ion. The removal of electrons from outer and inner valence orbitals produces a characteristic fragmentation pattern of every molecule which is a particular signature of the primary vacancies and final ionized states of the molecule.

From the theoretical view, the cross section of ionization by swift projectiles is to some degree insensitive to the details of the wave functions of the initial and final states of the *s* and *p* active electron. The results of Born approximation calculations showed that the matrix elements are essentially the same for electrons in the 2, 3 and 4 *p*-states, and have similar values for 3 and 4*s* electrons, indifferent to details of the shape of the wave functions [58]. This characteristic allows in a first approach to propose a semi-classical model to estimate the ionization cross sections of some molecules based on the Born approximation and depending essentially on the binding energy of the active electrons. The removed electrons of water, ammonia and methane that contribute to the ionization by swift projectiles are the *s* and *p* atomic orbitals, which compose the outer and inner valence molecular orbitals.

Considering these arguments, a model approach is proposed to describe the ionization cross section using as essential input the ionization energies of the removed electrons as the molecular parameter [13–15, 53, 59]. The ionization cross sections of protons follow a scaling law with constants A, B and C, the projectile energy x in mass E/M in keV/u, the ionization potentials I of a particular molecular orbital (state) j, the parameter Z_j that accounts for the number of electrons active in the molecular orbital j *and* δ_j is a fixed parameter. The functional equation describes well the behaviour at high energies as it is based on the Born approximation, becoming less accurate at lower energies.

$$\frac{\sigma_j}{Z_j \delta_j} I_j^2 = \frac{A ln(1 + Bx)}{x} - \frac{AB}{(1 + Cx)^2}$$

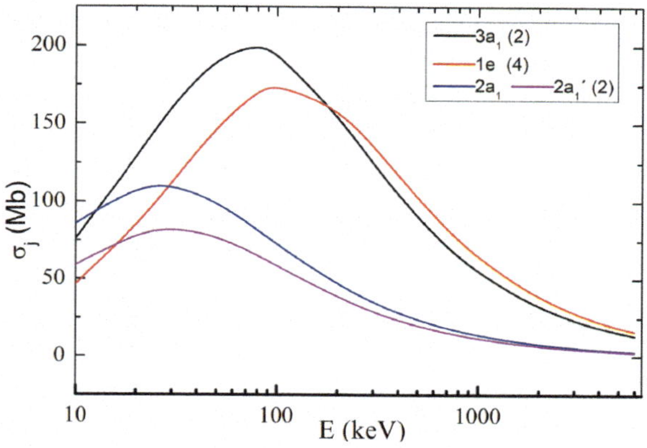

Fig. 7 Model cross sections for the outer and inner valence orbitals, including satellite states as function of the proton energy for ammonia. Occupancy of the orbitals are indicated in parenthesis.

The selections of two sets of parameters *A*, *B* and *C* discriminate the transition processes of the outer valence orbitals and of the inner valence orbitals including the satellite states (see Fig. 7 for ammonia cross sections).

The shape and intensity of the experimental and model single ionization cross sections over a wide range of projectile energies should suggest a correlation between one electron vacancies and ions production. All eight active electrons from the valence orbitals participate in the ion's formation. The model cross sections are distributed among the ion's experimental cross sections. The fragment *i* species cross sections σ^i are obtained from the cross section σ_j for each valence orbital (state) *j* contributing to its formation through an appropriate percentage f_j^i (fragmentation matrix). There are n valence states and m ionic species. The total σ^i is the sum over all valence states of the product of the molecular orbital cross section σ_j weighted by a percentage f_j^i e with the constraint that the sum over the fragment species of the fragment matrix elements f_j^i of each molecular orbital is unity. Initially, the parameters were adjusted to give better agreement with the measured total electron cross sections of References [3–5]. the matrix elements are not unique and some freedom to change the tabulated values exists.

$$\sigma^i = \sum_{j=1}^{n} f_j^i \sigma_j \sum_{i=1}^{m} f_j^i = 1$$

The f_j^i are associated with the water molecular orbitals $1b_1$, $3a_1$, $1b_2$, (outer valence orbitals) $2a_1$ including the $2a_1{'}$ satellites bands (inner valence orbitals) with ionization potentials 12.6 (1), 14.8 (1), 18.7 (1), 32.4 (1), 34.8 (0.27) and 37.5 (0.14) [35] respectively, with the ammonia orbitals $3a_1$, $1e_4$ (outer valence orbitals) and $2a_1$ and $2a_1{'}$ satellite bands (inner valence orbitals) with ionization potential 10.85 (1), 16.4 (1), 27.6 (1), 30.3 (0.16), 33.2 (0.13), 36.5 (0.13) and 41.8 (0.10) [36, 38] respectively, and with methane orbitals $1t_2$, (outer valence orbitals) $2a_1$ and $2a_1{'}$ satellite bands (inner valence orbitals) with ionization potential 14.13 (1), 23.05 (1), 25.4 (0.08), 28.56 (0.06), 31.45 (0.15) and 35.28 (0.06). [48], respectively. The spectroscopic intensities (pole strengths) are indicated in parentheses and are given relative to the main $2a_1$ band.

The inner valence satellite's states are composed of a set of states grouped in bands, the onset and width of the bands extend in an energy range and have a mean intensity. The values considered in the analysis are from the most recent EMS or (*e*, 2*e*) experiments. Experimental data as well as calculations of the configuration mixing states present some differences in the state's energies and pole strengths. In the semi-classical model, only the main states' energies and a mean energy value weighted by the pole strength for the satellite states were considered in the analysis as higher degree of fineness not justified. Bands which lay above the double-ionization threshold were not considered as they more likely lead subsequently to dissociative double-ionization channels. The double ionization threshold for water, ammonia and methane measured by electron impact experiments were 35.1, 34.8 eV 35.1 eV [60–62] For methane, recent high resolution (*e*, 2*e* + ion) spectra showed that three satellite states describing an envelope structure are responsible for the fragment's formation [47].

The contribution of each molecular state in the ionic species formation was obtained by matching calculated ionization with measured fragmentation cross sections (see Figs. 8 and 9 and Tables 1 and 2 for details). Only the recent measurements for the water, [13] ammonia [14] and methane [15] were considered in the analysis. The proton energy range extends from 300 to 2000 keV, 250 to 2700 keV and 150 to 2700 keV, for water, ammonia and methane respectively. The calculated cross section should reproduce better the experimental cross section at high energies.

The parent ions of all three molecules are formed from six active electrons of the outer valence orbital, but in contrast the single charged atoms C^+, N^+ and O^+ are only created from the satellite states. The fragments $CH_3{}^+$ and $NH_2{}^+$ produced with the emission of neutral hydrogen originate from the outer and main inner valence orbitals, while the fragments attached to one hydrogen atom, CH^+, and NH^+ have contributions from the inner valence and satellite states (see Table 3).

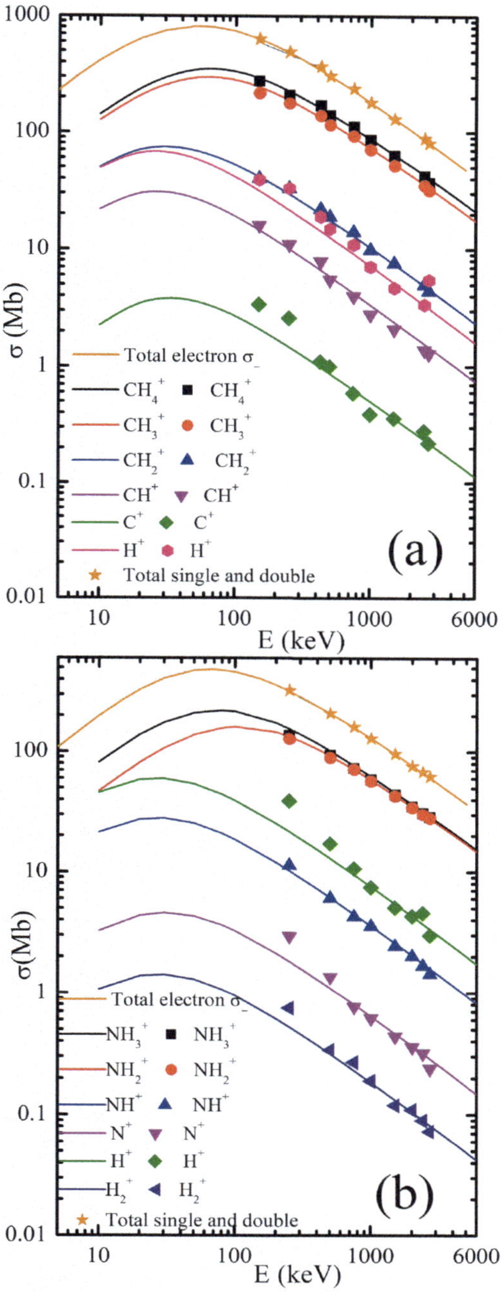

Fig. 8 Experimental cross sections for the ions species and matching model cross sections in Mbarn for **a** methane and **b** ammonia as function of the proton energy

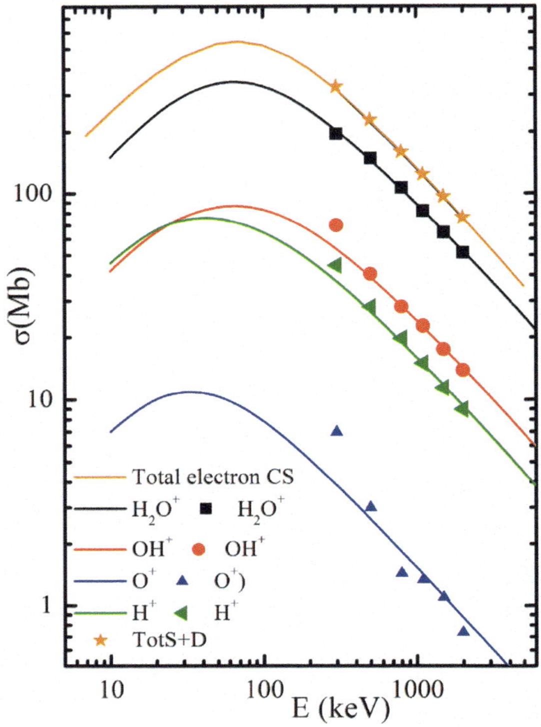

Fig. 9 Experimental cross section for the ion's species and matching model cross sections for water

Table 1 Fragmentation matrix f_j^i for methane and ammonia. Unity corresponds to 100% of contribution of the ion formation to the ionization to the specified molecular states

Methane				Ammonia				
	$1t_2^{-1}$	$2a_1^{-1}$	$2a_1'^{-1}$		$3a_1^{-1}$	$1e^{-1}$	$2a_1^{-1}$	$2a_1'^{-1}$
CH_4^+	0.54			NH_3^+	1	0.12		
CH_3^+	0.44	0.08		NH_2^+		0.88	0.07	
CH_2^+	0.02	0.25	0.12	NH^+			0.21	0.1
CH^+		0.13	0.04	N^+				0.05
C^+			0.03	H^+			0.49	0.06
H^+		0.32	0.03	H_2^+			0.01	0.005

Only one ionization experiment at 66 eV electron energy and sole on methane provide information on the vacancy contributions in the fragment-ion production from dissociative ionization [47]. A comparison of the model estimation for the proton data of the molecular assignments to the fragment ions is in good agreement with those experimental electron data but a direct comparison of the quantitative contribution of the molecular states to the ion's formation is not straightforward.

Table 2 Fragmentation matrix f_j^i for water. Unity corresponds to 100% of contribution of the ion formation to the ionization to the specified states. Table 3 shows that some vacancy patterns can be associated with the ions production

Water					
	$1b_1^{-1}$	$3a_1^{-1}$	$1b_2^{-1}$	$2a_1^{-1}$	$2a_1^{\prime -1}$
H_2O^+	1	1	0.02		
OH^+			0.68	0.27	0.05
O^+					0.13
H^+			0.30	0.45	0.10

For the CH_2 + fragment (reaction channel $CH_2^+ + H_2$), the relative contribution of the $1t_2^{-1}$, $2a_1^{-1}$ and $2a_1^{\prime -1}$ final states was estimated as 25%, 56% and 19%, while the electron experiment measured 33%, 44% and 23%. In contrast the proton experiment model suggests for CH^+ a high contribution of the $2a_1^{-1}$ state, 62%, and the electron experiment only 30%. It is noticed that the electron energy of 66 eV corresponds to 121 keV protons, and at this energy the contribution of double ionization in the formation of C^+ and CH^+ can be quite relevant, as it is for the equivalent proton velocity. The cross sections of single and double ionization of methane for 150 keV protons impact are displayed in Fig. 10 for evaluation. The electron impact experiment does not discriminate single from double ionization events.

The contribution to the ion's formation proceeds through the ionization to vibrational and continuum levels of the orbitals, outer valence satellite states, Rydberg states and other ionization processes, such as dissociative resonant autoionization and dissociation through cascade [35, 37, 38, 47, 48]. The (e, 2e + ion) spectra identified that a shift in energy occurs in the final state's energies of the ions and such finesse is not included in the simple model calculation. CH_4^+ and CH_3^+ fragment is attributed to the ionization to the low and high vibrational levels of the $1t_2^{-1}$ state, respectively. The vibrational continuum of the $1t_2^{-1}$ states, the low vibrational levels of the $2a_1^{-1}$ and ionization-excitation satellite states lead to the CH_2^+ fragment. The CH^+ fragment arises from $2a_1^{-1}$ state and several ionization-excitation satellite states, such as the $1t_2^{-1} - 2t_2$ and $1t_2^{-1} - 3a_1$ states and ionization-excitation satellite states are the major contributions to the C^+ ion. In the proton-model calculation single ionization energies were chosen for the $1t_2$, $2a_1$ and $2a_1$ vertical and $2a_1^{\prime}$ shake-up potentials. In the proton data-model approach only the mechanics is considered, where the projectile interacts with one electron either removing through vertical transition or removing one with additional excitation of another via a shake-up process, promoting the molecule to an ionized configuration mixing state.

Table 3 Ion's species formation attributed to the ionization to the molecular states are indicated by "x" for the methane, ammonia, and water

Methane				Ammonia					Water					
	$1t_2^{-1}$	$2a_1^{-1}$	$2a_{1'}^{-1}$		$3a_1^{-1}$	$1e^{-1}$	$2a_1^{-1}$	$2a_1^{'-1}$		$1b_1^{-1}$	$3a_1^{-1}$	$1b_2^{-1}$	$2a_1^{-1}$	$2a_1^{-1'}$
	6	2			2	4	2			2	2	2	2	
CH_4^+	x			NH_3^+	x	x			H_2O^+	x	x	x		
CH_3^+	x	x		NH_2^+		x	x		OH^+			x	x	x
CH_2^+	x	x	x	NH^+			x	x	O^+					x
CH^+		x	x	N^+				x	H^+			x	x	x
C^+			x	H^+			x	x						
H^+		x	x	H_2^+			x	x						

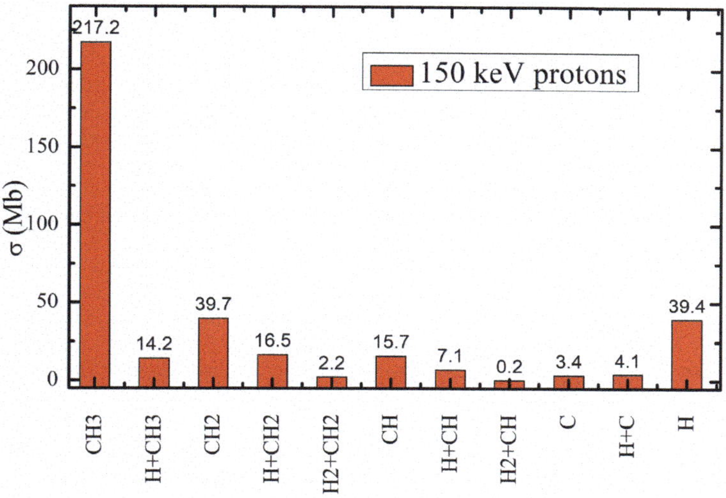

Fig. 10 Cross sections in Mb for the singles and doubles channels indicated in the x-axis. Values in the figure denote the absolute cross section values

5 Conclusion

A review of molecular ion species production by proton impact was presented for water, ammonia and methane. Here, we focus to uncover some patterns among the single vacancy production of the isoelectronic molecules in the view of their electronic configuration. Relevant experimental details in the data collection and reduction were outlined and the data analysis was focused on the proton interaction that leads to a final single ionized state either by forming a single hole state, when the proton removes an electron from the outermost valence orbital or by creating a vacancy in the inner valence orbital by shake-up processes, populating $2h$-$1p$ states of the satellite states. In the case of water, ammonia and methane vertical transitions from the valence molecular orbitals are responsible for most part of the ions production but the formation of atomic ions and ions attached to a single hydrogen are attributed to satellite bands prepared by the shake-up process. The experimental cross sections were quite well described by the model cross sections, which incorporate into the picture of ionization the satellite structures of the molecules indicating the existence of valence excited states and the relevance in the formation.

References

1. L.H. Toburen, M.Y. Nakai, R.A. Langley, Measurement of high-energy charge-transfer cross sections for incident protons and atomic hydrogen in various gases. Phys. Rev. **171**, 114 (1968)
2. L.H. Toburen, W.E. Wilson, Energy and angular distributions of electrons ejected from water vapor by 0. 3--1. 5 MeV protons. J. Chem. Phys. **66**, 5202–5213 (1977)
3. M.E. Rudd, R.D. DuBois, L.H. Toburen, C.A. Ratcliffe, T.V. Goffe, Cross sections for ionization of gases by 5–4000-keV protons and for electron capture by 5–150-keV protons. Phys. Rev. A **28**, 3244 (1983)
4. M.E. Rudd, Y.-K. Kim, D.H. Madison, J.W. Gallagher, Electron production in proton collisions: total cross sections. Rev. Mod. Phys. **57**, 965 (1985)
5. M.E. Rudd, T.V. Goffe, R.D. DuBois, L.H. Toburen, Cross sections for ionization of water vapor by 7–4000 keV protons. Phys. Rev. A **31**, 492 (1985)
6. D.J. Lynch, L.H. Toburen, W.E. Wilson, Electron emission from methane, ammonia, monomethylamine, and dimethylamine by 0.25–2.0 MeV protons. J. Chem. Phys. **64**, 2616–2622 (1976)
7. R.J. McNeal, Production of positive ions and electrons in collisions of 1–25-keV protons and hydrogen atoms with CO, CO_2, CH_4, and NH_3. J. Chem. Phys. **53**, 4308 (1970)
8. U. Werner, K. Beckord, J. Becker, H.O. Lutz, 3D imaging of the collision-induced coulomb fragmentation of water molecules. Phys. Rev. Lett. **74**, 1962 (1995)
9. F. Gobet, B. Farizon, M. Farizon, M.J. Gaillard, M. Carreé, M. Lezius, P. Scheier, T.D. Mark, Total, partial, and electron-capture cross sections for ionization of water vapor by 20–150 keV protons. Phys. Rev. Lett. **86**, 3751 (2001)
10. F. Gobet, S. Eden, B. Coupier, J. Tabet, B. Farizon, M. Farizon, M.J. Gaillard, M. Carreé, S. Ouaskit, T.D. Mark, P. Scheier, Ionization of water by (20–150)-keV protons: separation of direct-ionization and electron-capture processes. Phys. Rev. A **70**, 062716 (2004)
11. H. Luna, A.L.F. de Barros, J.A. Wyer, S.W.J. Scully, J. Lecointre, P.M.Y. Garcia, G.M. Sigaud, A.C.F. Santos, V. Senthil, M.B. Shah, C.J. Latimer, E.C. Montenegro, Water-molecule dissociation by proton and hydrogen impact. Phys. Rev. A **75**, 042711 (2007)
12. E.C. Montenegro, Fragmentation of water by heavy ions. J. Phys.: Conf. Ser. **194**, 012049 (2009)
13. A.C. Tavares, H. Luna, W. Wolff, E.C. Montenegro, Double ionization of water molecules induced by swift protons. Phys. Rev. A **92**, 032714 (2015)
14. W. Wolff, H. Luna, E.C. Montenegro, L.C. Rodrigues Jr., Multiple fragmentation mechanisms in ammonia: collisions with protons in the intermediate velocity regime. Phys. Rev. A **102**, 052821 (2020)
15. B. Jochim, A. Lueking, L. Doshier, S. Carey, E.Wells, E. Parke, M. Leonard, K.D. Carnes, I. Ben-Itzhak, Rapid formation of H_3 + from ammonia and methane following 4 MeV proton impact. J. Phys. B: At., Mol. Opt. Phys. **42**, 091002 (2009).
16. H. Luna, W. Wolff, E.C. Montenegro, CH_4 fragmentation from single and double ionization by proton and electron impact. Phys. Rev. A **99**, 012709 (2019)
17. R. Browning, H.B. Gilbody, Fragmentation of molecular gases by 5–45 keV protons. J. Phys. B At. Mol. Opt. Phys. **1**, 1149 (1968)
18. I. Ben-Itzhak, K.D. Carnes, D.T. Johnson, P.J. Norris, O.L. Weaver, Velocity dependence of ionization and fragmentation of methane caused by fast-proton impact. Phys. Rev. A **49**, 881 (1994)
19. H. Luna, E.G. Cavalcanti, J. Nickles, G.M. Sigaud, E.C. Montenegro, CH4 ionization and dissociation by proton and electron impact. J. Phys. B At. Mol. Opt. Phys. **36**, 4717 (2003)

20. H. Knudsen, U. Mikkelsen, K. Paludan, K. Kirsebom, S.P. Moller, E. Uggerhoj, J. Slevin, M. Charlton, E. Morenzoni, Non-dissociative and dissociative ionization of N_2, CO, CO_2, and CH_4 by impact of 50–6000 keV protons and antiprotons. J. Phys. B: At., Mol. Opt. Phys. **28**, 3569 (1995)
21. I.B.E. Fuentes, J. López-Patiño, F.B. Yousif, H. Martínez, Lonization and fragmentation of CH_4 by proton impact. Int. J. Mass Spectrom. **411**, 21–26 (2016)
22. B.C. Garrett, et al., Role of water in electron-initiated processes and radical chemistry: issues and scientific advances. Chem. Rev. **105**(1), 355–390 (2005)
23. E.J. Hall, *Radiobiology for the Radiologist*, 5th ed. (Lippincott, Williams, & Wilkins, Philadelphia, PA, 2000)
24. V.N. Salinas, M.R. Hogerheijde, E.A. Bergin, L.I. Cleeves, C. Brinch, G.A. Blake, D.C. Lis, G.J. Melnick, O. Pani´c, J.C. Pearson, L. Kristensen, U.A. Yildiz, E.F. van Dishoeck, First detection of gas-phase ammonia in a planet-forming disk. NH_3, N_2H+, and H_2O in the disk around TW hydrae. Astron. Astrophys. **591**, A122 (2016)
25. K.M. Pontoppidan, C. Salyk, A. Banzatti, G.A. Blake, C. Walsh, J.H. Lacy, M.J. Richter, The nitrogen carrier in inner protoplanetary disks. Astrophys. J. Lett. **874**, 92 (2019)
26. K. Altwegg, H. Balsiger, N. Hänni, M. Rubin, M. Schuhmann, I. Schroeder, T. Sémon, S. Wampfler, J.-J. Berthelier, C. Briois, M. Combi, T.I. Gombosi, H. Cottin, J. De Keyser, F. Dhooghe, B. Fiethe, S.A. Fuselier, Evidence of ammonium salts in comet 67p as explanation for the nitrogen depletion in cometary comae. Nat. Astron. **4**, 533 (2020)
27. L.Y. Stein, M.G. Klotz, The nitrogen cycle. Curr. Biol. **26**, R94 (2016)
28. C.H. Cheng, S.A.T. Redfern, Impact of interannual and multidecadal trends on methane-climate feedbacks and sensitivity. Nat Commun. **13**, 3592 (2022)
29. D. Qasim, G. Fedoseev, K.J. Chuang et al., An experimental study of the surface formation of methane in interstellar molecular clouds. Nat. Astron. **4**, 781–785 (2020)
30. E.H. Wilson, S.K. Atreya, Current state of modelling the photochemistry of Titan's mutually dependent atmosphere and ionosphere. J. Geophys. Res. **109**, E06002 (2004)
31. A.O. Bawagan, L.Y. Lee, K.T. Leung, C.E. Brion, An investigation of the valence orbitals of water by high momentum resolution electron momentum spectroscopy. Chem. Phys. **199**, 367–382 (1985)
32. H. Nakatsuji, T. Yonezawa, Cluster expansion of the wavefunction. satellite peaks of the inner-valence ionization of H_2O studied by the SAC and SAC-CI theories. Chem. Phys. Lett. **87**, 426 (1982)
33. H. Ågren, H. Siegbahn, Semi-internal correlation in the auger electron spectrum of H_2O. Chem. Phys. Lett. **69**, 424 (1980)
34. R. Cambi, G. Ciullo, A. Sgamellorti, C.E. Brian. J.P.D. Cook, L.E. McCarthy, E. Weigold, Experimental and theoretical binding energy spectra and momentum distributions for the valence orbitals of H_2O. Chem. Phys. **91**, 373 (1984); Erratum **98**, 166 (1985)
35. C.G. Ning, B. Hajgato, Y.R. Huang, S.F. Zhang, K. Liu A, Z.H. Luo, S. Knippenberg, J.K. Deng, M.S. Deleuze, High resolution electron momentum spectroscopy of the valence orbitals of water. Chem. Phys. **343**, 19–30 (2008)
36. A.O. Bawagan, R. Mijller-Fiedler, C.E. Brion E.R. Davidson, C. Boyle, The valence orbitals of NH_3 by electron momentum spectroscopy: quantitative comparisons using hartree-fock limit and correlated wave functions. Chem. Phys. **120**, 335–357 (1988)
37. L. Ishikawa, T. Odagiri, K. Yachi, T. Nakazato, M. Kurokawa, M. Kitajima, N. Kouchi, Doubly excited states of ammonia produced by photon and electron interactions. J. Phys. B At. Mol. Opt. Phys. **41**, 195204 (2008)
38. M. Ishida, M. Ehara, And hiroshi nakatsuji, outer- and inner-valence ionization spectra of NH_3, PH_3 and AsH_3: symmetry adapted cluster configuration interaction general-R study. J. Chem. Phys. **116**, 1934 (2002)
39. S.T. Hood, A. Hamnett, C.E. Brion, An (*e*, 2*e*) study of ammonia: binding energies and momentum distributions of valence electrons. Chem. Phys. Lett. **39**, 252 (1976)
40. K.L. Nixon, A.J. Murray, H. Chaluvadi, C. Ning, J. Colgan, D.H. Madison, Low energy (*e*, 2*e*) coincidence studies of NH_3: results from experiment and theory. J. Chem. Phys. **138**, 174304 (2013)

41. G. Fronzoni, G. De Alti, P. Decleva, A. Lisini, CI calculations of the inner and outer valence photoelectron spectra of the fourth row hydrides. J. Electron Spectrosc. Relat. Phenom. 58, 375-392 (1992)
42. G.R. Wight, M.J.V. der Wiel, C.E. Brion, Dipole oscillator strengths for excitation, ionisation, and fragmentation of NH_3 in the 5–60 eV region. J. Phys. B: At. Mol. Phys. **10**, 1863 (1977)
43. C.E. Brion, A. Hamnett, G.R. Wight, M.J. Van der Wiel, Branching ratios and partial oscillator strengths for the photoionization of NH_3 in the 15–50 eV region. J. Electron Spectrosc. Relat. Phenom. **12**, 323 (1977)
44. X. Liu, D.E. Shemansky, Analysis of electron impact ionization properties of methane. J. Geophys. Res. **111**, A04303 (2006)
45. S. Xu, X. Ma, X. Ren, A. Senftleben, T. Pflüger, A. Dorn, J. Ullrich, Formation of protons from dissociative ionization of methane induced by 54 eV electrons. Phys. Rev. A **83**, 052702 (2011)
46. S. Xu, X. Ma, X. Ren, A. Senftleben, T. Pflüger, S. Yan, P. Zhang, J. Yang, J. Ullrich, A. Dorn, An (e, $2e$ + ion) investigation of dissociative ionization of methane. J. Chem. Phys. **138**, 134307 (2013)
47. X. Ren, T. Pflüger, M. Weyland, W.Y. Baek, H. Rabus, J. Ullrich, A. Dorn, High-resolution (e, $2e$ + Ion) study of electron-impact ionization and fragmentation of methane. J. Chem. Phys. **142**, 174313 (2015)
48. S.A.C. Clark, T.J. Reddish, C.E. Brion, E.R. Davidson, R.F. Frey, The valence orbital momentum distributions and binding energy spectra of methane by electron momentum spectroscopy: quantitative comparisons using near Hartree-Fock limit and correlated wavefunctions. Chem. Phys. **143**, 1–10 (1990)
49. R. Cambi, G. Ciullo, A. Sgamellotti, F. Tarantelli, R. Fantoni, A. Giardini-Guidon, A. Sergio, Lonization of CH_4 and some fluoromethanes: a green's function study and an (e, $2e$) spectroscopic investigation. Chem. Phys. Lett. **80**, 296–300 (1981)
50. M.C. Göthe, B. Wannberg, L. Karlsson, S. Svensson, P. Baltzer, F.T Chau, M.Y. Adam, Xray, ultraviolet, and synchrotron radiation excited inner valence photoelectron spectra of CH_4. J. Chem. Phys. **94**, 2536 (1991)
51. L.S. Cederbaum, W. Domcke, J. Schirmer, W.V. Niessen, Correlation effects in the ionization of molecules: breakdown of the molecular orbital picture, in *Advances in Chemical Physics* (Wiley, 2007), pp. 115–159
52. L.S. Cederbaum, W. Dorncke, J. Schirmer, W. von Niessen, G.H.F. Diercksen, W.P. Kraemer, Correlation effects in the ionization of hydrocarbons. J. Chem. Phys. **69**, 1591 (1978)
53. E.C. Montenegro, L. Sigaud, W. Wolff, H. Luna, Ferreira, natalia, some dynamical features of molecular fragmentation by electrons and swift ions. Phys. Procedia **66**, 39–45 (2015)
54. W. Wolff, H. Luna, L. Sigaud, A.C. Tavares, E.C. Montenegro, Absolute total and partial dissociative cross sections of pyrimidine at electron and proton intermediate impact velocities. J. Chem. Phys. **140**, 064309 (2014)
55. W. Wolff, I.J. de Souza, A.C. Tavares, G.F.S. de Oliveira, H. Luna, Electron-recoil ion and recoil ion-projectile coincidence techniques applied to obtain absolute partial collision cross sections. Rev. Sci. Instrum. **83**, 123107 (2012)
56. I. Ben-Itzhak, S. G. Ginther, K.D. Carnes, Multiple electron removal and molecular fragmentation of CO by fast F4+ impact. Phys. Rev. A **47**, 2827 (1993)
57. M. Simon, T. LeBrun, P. Morin, M. Lavollée, J.L. Maréchal, A photoelectron-ion multiple coincidence technique applied to core ionization of molecules. Nucl. Instrum. Methods Phys. Res., Sect. B **62**, 167 (1991)
58. H. Bethe, Ann. d. Physik **5**, 325 (1930)
59. E.C. Montenegro, G.M. Sigaud, R.D. DuBois, Projectile, and target scaling of the total ionization cross sections of atoms and molecules. Phys. Rev. A **87**, 012706 (2013)
60. F. Frémont, C. Leclercq, A. Hajaji, A. Naja, P. Lemennais, S. Boulbain, V. Broquin, J.-Y. Chesnel, Fragment emission following multiple ionization in 20–200 eV e^-+H_2O collisions. Phys. Rev. A **72**, 042702 (2005)

61. R. Locht, M. Davister, W. Denzer, H.W. Jochims, H. Baumgärtel, About the double ionization of ammonia and carbon dioxide: a comparison between photoionization and electron impact. Chem. Phys. **138**, 433–440 (1989)
62. M.D. Ward, S.J. King, S.D. Price, Electron ionization of methane: the dissociation of the methane monocation and dication. J. Chem. Phys. **134**, 024308 (2011)

Fast Ion-Atom Collisions: Electron Spectroscopy of Mixed-State Beams

Emmanouil P. Benis

Abstract The problem of the dynamics of many-body quantum systems under intense and ultra-fast perturbations is ideally addressed in energetic collisions of few-electron ions with atomic targets. Collisions with multiunpaired electron-ion cores, offered by mixed-state ion beams, are most challenging for studying electron dynamics, both theoretically and experimentally. Here, we present the recent advances of our experimental and theoretical work in the field of fast ion-atom collisions with mixed-state ion beams. The fundamental processes of single electron capture (SEC) and transfer excitation (TE) are revisited under new experimental results and a theoretical treatment within a three-active electron close-coupling approach (3eAOCC). In SEC, our results invalidated the frozen core approximation and related spin statistics populations, additionally proposing a distinctive screening effect due to the Pauli exclusion principle (Pauli shielding). In TE, 3eAOCC theory fills the gap of a long awaited coherent treatment of the TE process that includes its resonant and nonresonant features, showing excellent agreement with the measurements, and proposing a new nonresonant one-step TE mechanism. The report also includes the description of the basic principles of the experimental methods and the used instrumentation, as well as techniques for the experimental treatment of the mixed-state beams.

1 Introduction

Energetic (MeV/u) collisions of few-electron ions with atomic targets provide ideal laboratories to address the problem of the dynamics of many-body quantum systems under intense, ultra-fast perturbations [1–3]. The interplay between the electron-nucleus (e-n) and electron-electron (e-e) interactions, as well as electron exchange effects, occurring on time scales shorter than 1 fs, is a complex dynamical problem, the theory and detailed modeling of which is still challenging. The level of sophistication of theory is further driven by results from *state-selective* differential cross

E. P. Benis (✉)
Department of Physics, University of Ioannina, 45110 Ioannina, Greece
e-mail: mbenis@uoi.gr

L. C. Tribedi (ed.), *Advances in Atomic Molecular Collisions*,
https://doi.org/10.1007/978-981-97-7063-2_4

section measurements, offered by high-resolution studies of projectile ions [4, 5]. Moreover, from the abundant experimental and theoretical studies of the electron processes occurring in fast ion-atom collisions, the most challenging, both theoretically and experimentally, is the collisions with multiunpaired electron-ion cores [6]. In theory, certain fundamental questions such as the population of states according to spin statistics or the validity of frozen core approximations are revisited. In experiment, the projectiles with multiunpaired electrons are found as admixtures to ground state beams, in cases where the occupied states of the multiunpaired electrons are metastable. Then, the experimental challenge is to separate the contributions from only the metastable parts and obtain the information about the corresponding cross sections.

Metastable beams have been used in investigations of high-resolution projectile electron spectroscopy, as, for example, in single [6–9] and double [10] electron capture, excitation [11, 12], transfer excitation [13–17], the production of triply-excited states [18], and superelastic scattering [19, 20]. In addition, metastable states have also been used in a variety of other atomic physics investigations including the study of photoionization [21], electron impact ionization [22, 23], tokamak high-energy charge exchange [24] and edge impurities [25], electron capture and excitation [26], slow collisions of quasi-symmetric heavy systems [27], beam-two foil spectroscopy [28], and even two-electron quantum entanglement [29].

Here, we review our experimental and theoretical findings from fast ion-atom collisions involving mixed-state ion beams. The paper is structured as follows: In Sect. 2, a description of the zero-degree Auger projectile spectroscopy (ZAPS) technique, developed for high-resolution studies, is presented. It includes the description of the spectrograph in use, the current ZAPS setup, and the treatment of long-lived Auger states in determining the corresponding double differential cross section (DDCS) of a process. In Sect. 3, the mixed-state ion beams are described. Production mechanisms, lifetime considerations, and methods for determining the metastable beam fraction are presented. In Sect. 4, we summarize our results on the studies of the process of single electron capture. In Sect. 5, we summarize our results on the studies of the process of transfer excitation. Finally, in Sect. 6, we report some ongoing projects and discuss about our near-future plans.

2 Zero-Degree Auger Projectile Spectroscopy

ZAPS has been one of the most powerful tools in investigating the dynamics of fast ion-atom collisions ever since its introduction in the early 80s [30]. In ZAPS, the electrons emitted by the projectile are detected at the laboratory angle of zero degrees with respect to the projectile velocity, thus attaining the optimal kinematic conditions for the highest possible resolution. This laid the grounds for LS-resolved Auger measurements (also called state-selective) that offer the most stringent tests for advanced collision theories. A detailed review of the ZAPS technique principles can be found in Refs. [5, 31].

2.1 Kinematic Effects

A brief introduction to the ZAPS kinematics will be given here for completeness purposes. Auger electrons emitted from scattered projectiles are kinematically influenced. For the case of MeV/u collisions projectiles are scattered through very small angles ($\sim$ mrads), resulting in negligible effects both on the energy loss and on the projectile electron trajectories. Thus, the projectile ion scattering angle in swift collisions can be safely assumed to be zero, and thus a simple velocity vector addition model is sufficient for determining all kinematic effects.

The velocity $\mathbf{v}$ of the Auger electron in the laboratory frame is obtained by adding the projectile velocity $\mathbf{V}_p$ to the velocity $\mathbf{v}'$ of the electron in the projectile rest frame as shown in Fig. 1a. Denoting by primed symbols the quantities in the projectile rest frame hereafter, the electron kinetic energy $\epsilon = \frac{1}{2}mv^2$ in the laboratory frame is related to the corresponding rest frame electron kinetic energy $\epsilon' = \frac{1}{2}mv'^2$ as

$$\epsilon = \epsilon' + t + 2\sqrt{\epsilon' t}\ \cos\theta' \tag{1}$$

or equivalently

$$\epsilon' = \epsilon + t - 2\sqrt{\epsilon t}\ \cos\theta, \tag{2}$$

where

$$t = \frac{1}{2}mV_p^2 = \frac{m}{M}\ E_p\ = 548.58\ \frac{E_p(\mathrm{MeV})}{M(\mathrm{u})}\ \ (\mathrm{eV}) \tag{3}$$

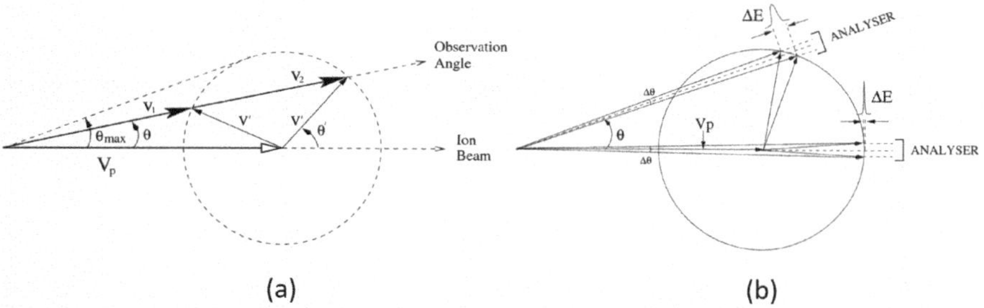

Fig. 1 **a** Velocity addition diagram. The electron velocity $\mathbf{v}'$ in the projectile rest frame is transformed in the laboratory frame according to the vector addition rule as $\mathbf{v} = \mathbf{V_p} + \mathbf{v}'$. **b** Geometrical interpretation of the broadening effect due to the laboratory observation angle. It is clearly seen how observation at zero degrees reduces substantially the line energy broadening

is the reduced projectile energy, also known as the *cusp* energy. E_p and M are the kinetic energy and mass of the projectile, respectively, while m is the electron mass. Introducing the parameter ζ [5]

$$\zeta \equiv \sqrt{\frac{t}{\epsilon'}} = \frac{V_p}{v'} \tag{4}$$

and using simple trigonometric rules, the energy in the laboratory frame can be written as

$$\epsilon_{\pm}(\theta) = \epsilon' \left(\zeta \cos\theta \pm \sqrt{1 - \zeta^2 \sin^2\theta} \right)^2 \qquad (\zeta > 1, \quad 0 \leq \theta \leq \arcsin\frac{1}{\zeta}). \tag{5}$$

For zero-degree observation and $\zeta > 1$, Eq. 5 reads

$$\epsilon_{+} = \epsilon'(1+\zeta)^2 \qquad (\theta' = 0°) \tag{6}$$

$$\epsilon_{-} = \epsilon'(1-\zeta)^2 \qquad (\theta' = 180°). \tag{7}$$

Kinematic effects change the position, energy width, and intensity of an Auger line and are referred to in the literature as *shifting, stretching* and *enhancing*, respectively. Shifting is described by Eqs. 6 and 7. Stretching and enhancing affect the double differential and differential cross sections, which suffer the following transformations between projectile and laboratory rest frames, respectively:

$$\frac{d^2\sigma}{d\Omega d\epsilon} = |1 \pm \zeta| \frac{d^2\sigma}{d\Omega' d\epsilon'} \tag{8}$$

$$\frac{d\sigma}{d\Omega} = (1 \pm \zeta)^2 \frac{d\sigma}{d\Omega'}, \tag{9}$$

where $(+)$ and $(-)$ correspond to $\theta' = 0$ and $\theta' = 180$, respectively.

In addition to the above, experimental parameters such as the finite acceptance angle $\Delta\theta$ of the spectrometer degrade the energy resolution of the spectra. The effect is referred to as kinematic *broadening* [4], and it is geometrically explained in Fig. 1b. The first-order broadening coefficient in terms of θ and ϵ is calculated as

$$\Delta B_{\theta}^{(1)} = -2 \sin\theta \Delta\theta \left(\frac{\sqrt{\epsilon t}}{1 - \sqrt{\frac{t}{\epsilon}} \cos\theta} \right), \tag{10}$$

which for 0° observation vanishes. Thus, zero-degree observation offers the optimum resolution in projectile spectroscopy as the first-order broadening effects are absent. Relativistic effects in the above kinematics considerations can be found in Ref. [32]. For collision energies of up to 2 MeV/u, as discussed here, their effect on the cross section transformations is negligible. However, for the shifting of the electron energies in the laboratory frame these are clearly observable and thus are considerable in the correct identification of the Auger lines.

2.2 *The HDA Spectrograph*

Initially, ZAPS setups exclusively used the tandem parallel-plate analyzer (PPA)[1] electron spectrometer for the measurements [33, 34]. There, the Auger spectrum was obtained after multiple energy scans to reach the desirable statistics. The field was much improved with the replacement of the tandem PPA spectrometer by a large area hemispherical deflector analyzer (HDA) equipped with a pre-retardation entry lens and a two-dimensional position-sensitive detector (PSD), capable of obtaining simultaneously an electron spectrum of about 20% of the tuning energy, avoiding thus the time-consuming scanning process. The new spectrograph was at least 100 times more efficient than the PPA spectrometer, thus offering the option for measurements of very high statistics and/or small cross sections. The HDA spectrograph was initially developed and operated at the J.R. Macdonald Laboratory of Kansas State University at the turning of the twenty-first century. Since 2013 it has been transferred, reinstalled, and operated at the NCSR "Demokritos" 5.5 MV Tandem accelerator facility [35] in Athens, thus initiating the field of Atomic Physics with accelerators to Greece (Atomic Physics with Accelerators: Projectile Electron Spectroscopy—APAPES initiative [36]). A detailed description of the HDA spectrograph can be found in Refs. [37–40]. Here, we will briefly describe some of its energy resolution aspects for completeness purposes.

The unique characteristic of our HDA is its paracentric entry, i.e., an entry that is not located at the center of the distance between the two hemispherical plates but closer to the inner shell. This is for compensating for the large fringing fields, inherent to spectrographs with large areas and openings. As proven initially by SIMION simulations [38] and later on experimentally [41], the HDA paracentric entry combined with a focusing/retardation entry lens can fully compensate for the large fringing fields, as clearly shown in Fig. 2. Thus, energy resolution remains almost unaffected by the large fringing fields, close to the ideal field value [40].

The energy resolution can be further improved by applying deceleration to the incoming electrons, a common technique in electron spectroscopy [42]. For the HDA spectrograph, electron deceleration is achieved at the entry lens, which is appropriately tuned to a deceleration factor F, defined as the ratio of the value of the initial electron energy to the value of the decelerated. In addition, the lens focuses the electrons at the entry of the HDA, thus both improving their energy resolution

[1] Often called as "Nikotron" among the specialists, from the name Dr. Nikolaus Stolterfoht who introduced it [30].

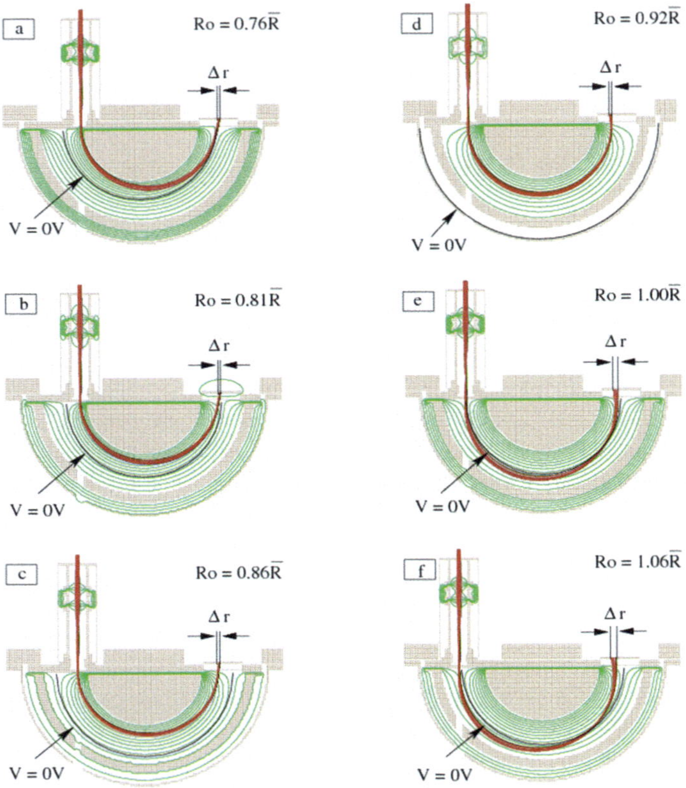

Fig. 2 SIMION simulation for six different analyzer entries R_0. Red lines: Electron trajectories originating from the gas-cell area and focused under the same lens voltages at different HDA entries R_0. Green lines: HDA equipotential lines with the location of the zero value explicitly shown. It is evident that there is an improvement in focusing trace width Δr for the cases of $R_0 < \overline{R} = \frac{R_1+R_2}{2}$ **a–d** compared to the conventional central entrance at $\overline{R}$ (**e**)

and increasing the detection efficiency. In Fig. 3, we show energy resolution studies for various F values performed with the KLL peak of $2p^2\,{}^1D$ obtained in collisions of 21.6 MeV $F^{8+} + H_2$.

2.3 The ZAPS Setup

A cross-sectional view of the ZAPS setup along with a typical Auger spectrum measurement is shown in Fig. 4. The electron spectrograph is the HDA already described. The gas target is a gas cell that consists of two concentric cylinders with 2 mm minimum diameter openings for succeeding double differential pumping conditions and

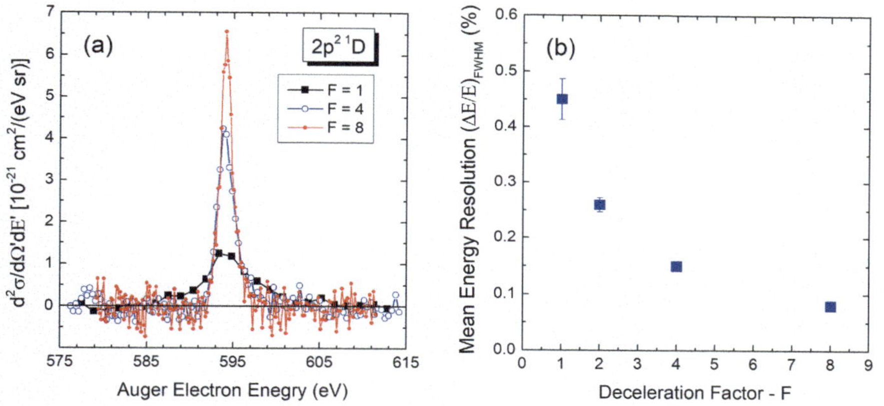

Fig. 3 **a** Comparison of the DDCS Auger spectra for the $2p^2\ ^1D$ peak, obtained in collisions of 21.6 MeV $F^{8+} + H_2$, at deceleration factors of $F = 1, 4$, and 8. **b** The HDA spectrograph mean energy resolution at FWHM, obtained from the corresponding data of (**a**), as a function of the deceleration factor F

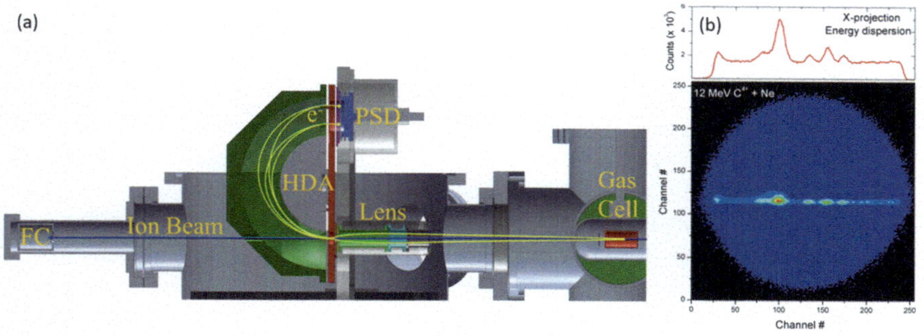

Fig. 4 **a** CAD cross-sectional view of the ZAPS setup. Electrons emitted at zero degrees with respect to the ion beam velocity enter the injection lens to be decelerated and focused onto the entry of the HDA. Subsequently, they are analyzed in energy by the HDA and imaged onto the PSD along the dispersion axis. **b** Typical two-dimensional image of KLL Auger electrons. The projection of the image along the dispersion axis (top graph) corresponds to the Auger electron spectrum

thus prevent the gas from polluting the spectrograph chamber. The pressure of the gas cell is controlled with an accuracy of 0.1 mTorr by a Baratron feedback system, necessary for absolute cross section measurements. The doubly differential pump scheme allows for the spectrograph chamber to maintain a base pressure of about 3×10^{-7} Torr, even when the gas cell is loaded to 20–40 mTorr He.

The ion beam travels through the gas cell where it interacts with the gas target atoms/molecules and upon exiting continues on through the spectrograph to be collected in a Faraday cup for normalization purposes after exiting through a small hole in the back of the HDA. Auger electrons generated from the decay of the populated projectile states and emitted in the forward direction along the ion beam ($\theta = 0°$) are decelerated and focused by the injection lens onto the entry of the HDA to improve their energy resolution and collection efficiency. They are then energetically analyzed by the HDA and imaged onto the PSD along the dispersion axis [39]. The projection of the image onto the dispersion axis results in the Auger electron spectrum, as shown in the inset of Fig. 4b.

2.4 DDCS Measurements

The electron DDCS is determined from the raw data recorded at the PSD area. First an energy calibration of the channel number axis is necessary. This can be done either by the use of an electron gun [43], or by known Auger lines in the recorded Auger spectra. Then, DDCS can be obtained according to the following formula [5, 15]:

$$DDCS_j \equiv \frac{d^2\sigma_j}{d\Omega dE_j} = \frac{N_j^e}{N_I\, L_{eff}\, n\, \overline{\Delta\Omega}\, \Delta E_j\, T\, \eta}, \tag{11}$$

where N_j^e is the number of electrons detected in channel j, L_{eff} is the effective length of the target gas cell, N_I is the number of ions collected in the Faraday cup, n is the target gas density, $\overline{\Delta\Omega}$ is the effective solid angle averaged over the length of the gas cell, ΔE_j is the energy step per channel in the spectrum, and T is the analyzer transmission determined here by three meshes of 90% transmission each. The absolute overall spectrograph efficiency η can be obtained by performing auxiliary *in situ* measurements of either elastically scattered (binary encounter) electrons from bare ion beams, as typically done for increased accuracy in all such ZAPS measurements [5, 44], and/or measurements of the target Ne-KLL Auger production utilizing proton beams [5, 45]. For the works reported here, both methods gave very similar results, i.e., an overall efficiency $\eta = (50 \pm 5)\%$. It should be also noted that our present data acquisition system allows for a maximum count rate of 100 kHz with negligible dead time. The DDCS measured in the laboratory frame is then transformed to the projectile rest frame according to the kinematics transformation of Eq. 8.

2.5 Long-Lived Auger States

Ion-atom collisions may result in the formation of long-lived Auger states, especially in case where the ion beam is delivered in a mixed-state that has a metastable component, as, for example, the He-like ($1s^2\,{}^1S$, $1s2s\,{}^{1,3}S$) beam.[2] Long-lived Auger transitions are characterized by a decay length $z = V_p\tau$, where V_p and τ are the projectile velocity and the lifetime of the state, respectively, that is longer than the

[2] Here, we reserve the term *metastable* for the ion beam excited states and the term *long-lived* for the Auger states, for clarity, although there are no essential physical differences between them.

length of the target volume. Considering typical Auger lifetimes of $\tau < 10^{-3}$ ns, and fast projectiles with typical energies of 1 MeV/u (corresponding to a speed of 13.9 mm/ns), most Auger states are considered prompt, and thus decay well within the target volume. Consequently, for prompt Auger states, the electron detection solid angle depends only on geometrical parameters of the experimental setup and can be straightforwardly estimated. However, long-lived Auger states with typical lifetimes $\tau > 1$ ns decay all along the projectile path from the target area to the spectrograph and even through it. This results in a variable detection solid angle, the estimation of which relies on model calculations that include the lifetimes of the Auger decays. As an example of long-lived Auger state, we may refer to the $1s2s2p\,^4P_J$ state[3] for which the $J = 1/2, 3/2, 5/2$ levels are characterized by inherent long lifetimes [46].

2.5.1 Geometrical Considerations

For our ZAPS setup shown in Fig. 4, the variable detection solid angle, hereafter called *effective* solid angle, averaged over the length of the gas cell is calculated according to the geometrical parameters detailed in Fig. 5a as

$$\overline{\Delta\Omega}(L, V_p\tau, L_c) = \frac{1}{L_c}\int_{z'=0}^{L_c} dz' \int_{z=0}^{L-z'} dz \frac{\exp\left(-\frac{z}{V_p\tau}\right)}{V_p\tau} \Delta\Omega_0(L - z' - z), \qquad (12)$$

where symbols are explained in the caption of Fig. 5. Similarly, the detection solid angle for a prompt state averaged over the length of the gas cell is written as

$$\overline{\Delta\Omega_0}(L_c, s_0) = \frac{1}{L_c}\int_{z'=0}^{L_c} dz' \Delta\Omega_0(L_c/2 + s_0 - z'), \qquad (13)$$

where $\Delta\Omega_0$ is the solid angle for point source emission into the lens entry aperture of radius r at distance s given by

$$\Delta\Omega_0(s) = 2\pi\left(1 - \frac{s}{\sqrt{r^2 + s^2}}\right). \qquad (14)$$

Considering that in Auger spectroscopy the J-levels of a $^{2S+1}L_{J,M_J}$ state are in general not resolved, we introduce the effective solid angle correction factor G_{τ_J} as the ratio of the effective solid angle of a long-lived J-level to the solid angle of a prompt state as

---

[3] Even though the correct spectroscopic description of a state is $^{2S+1}L_{J,M_J}$, it is common to use the word *state* both for the $^{2S+1}L_J$ level or the ^{2S+1}L term, since in Auger spectroscopy the typical energy resolution is limited to the energies of the ^{2S+1}L terms. A more correct terminology for the Auger states would be LS-resolved states. Here, we will occasionally use the word *state* for the $^{2S+1}L_J$ level or the ^{2S+1}L term, for simplicity.

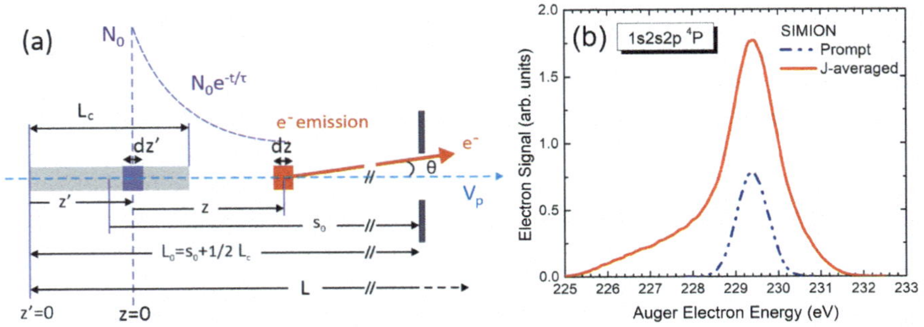

Fig. 5 **a** Schematic of the detection geometry. N_0 ions enter the gas cell with velocity V_p and get excited between z' and $z' \pm dz$. Assuming a long lifetime τ, the excited ionic state travels a further distance z before it decays emitting an Auger electron at angle θ which enters the lens and is eventually energy analyzed by the HDA (not shown). L_c denotes the length of the gas cell, s_0 is the distance from the center of the gas cell to the lens entry, and L is the maximum distance beyond which contributions are negligible ($L = s_0 + 6$ mm). **b** SIMION simulations for the line profiles of the J-averaged $C^{3+}(1s2s2p\,^4P)$ term Auger decaying to the $C^{4+}(1s^2)$ ground state (red line) and its corresponding prompt state (blue dash-dotted line)

$$G_{\tau_J} \equiv \frac{\overline{\Delta\Omega_J}(L, V_p\tau, L_c)}{\overline{\Delta\Omega_0}(L_c, s_0)}. \tag{15}$$

Then, the overall effective solid angle correction factor G_τ of an LS-resolved Auger state is determined by the statistical average of the G_{τ_J} correction factors of the J-levels of the LS-resolved Auger state. Thus, we have

$$G_\tau = \frac{\sum_J (2J+1)\xi_J G_{\tau_J}}{\sum_J (2J+1)\xi_J}, \tag{16}$$

where ξ_J is the Auger yield of the J-level. Finally, the effective solid angle for an LS-resolved Auger state, that is introduced in Eq. 12, is estimated as

$$\overline{\Delta\Omega} = G_\tau\, \overline{\Delta\Omega_0}(L_c, s_0). \tag{17}$$

2.5.2 Monte Carlo-Type Simulations

The above geometrical approach can be straightforwardly applied to spectrometers that do not have focusing elements, such as the tandem parallel-plate analyzer [5]. For spectrometers that have focusing elements, e.g., our hemispherical spectrograph, more sophisticated approaches are required. Thus, for our ZAPS geometry, the overall effective solid angle correction factor G_τ has been determined using Monte Carlo-type simulations within the SIMION ion optics package [32, 47]. For this, the spectrograph layout was modeled in the SIMION design environment and a numerical code, that realized Eqs. 12–16, was developed in the Lua language environment. Then, G_{τ_J} was estimated in the following way: Assuming that a number N_0 of projectiles is initially populated in the J-level, then the number of observed Auger electrons will be

$$N^e_{obs_J} \sim \xi_J N_0 \overline{\Delta\Omega_J}(L, V_p\tau_J, L_c). \tag{18}$$

However, the true number of detected electrons, i.e., those that originate only from the gas cell area, should be

$$N^e_{true_J} \sim \xi_J N_0 \overline{\Delta\Omega_0}(s_0, L_c), \tag{19}$$

and thus

$$\frac{N^e_{obs_J}}{N^e_{true_J}} = \frac{\overline{\Delta\Omega_J}(L, V_p\tau_J, L_c)}{\overline{\Delta\Omega_0}(s_0, L_c)} \equiv G_{\tau_J}. \tag{20}$$

Therefore, the ratio of the number of observed electrons of a long-lived J-level to the number of true electrons, which corresponds to a prompt state of the same initial population N_0, essentially describes the overall effective solid angle correction factor G_{τ_J}. An example of the line profiles of the J-averaged $C^{3+}(1s2s2p\,^4P)$ term Auger decaying to the $C^{4+}(1s^2)$ ground state and its corresponding prompt state generated by SIMION calculations is given in Fig. 5b.

2.5.3 Measurements

Our SIMION results were also confirmed experimentally using Be-like mixed-state $O^{4+}(1s^22s^2\,^1S, 1s^22s2p\,^3P)$ beams [48]. During collisions with H_2 or He targets the *needle ionization*[4] of the 1s electron [49] of the $O^{4+}(1s^22s2p\,^3P)$ state results in the production of the $1s2s2p$ configuration which in the LS-coupling scheme results in the $1s2s2p\,^4P$ (denoted as ^4P) and $1s(2s2p\,^3P)\,^2P$ (denoted as $^2P_-$) terms. We omit the formation of the $1s(2s2p\,^1P)\,^2P$ (denoted as $^2P_+$) term since in the LS-coupling scheme the 2s and 2p electrons interact strongly as parts of the same shell and are negligibly affected by the K-shell configuration. Thus, even after the 1s ionization, the L-shell electrons should maintain their ^3P coupling.

The ^4P and the $^2P_-$ terms are used in the experimental determination of the overall effective solid angle correction factor G_τ. The ^4P_J levels Auger decay to the $1s^2$ ground state having long lifetimes (> 10 ns) [50], while the $^2P_-$ term Auger decays promptly to the $1s^2$ ground state. The production population statistics of the ^4P and $^2P_-$ terms should result in the cross section ratio of $\sigma(^4P) : \sigma(^2P_-) = 2 : 1$, as dictated by the multiplicity of the terms. A direct measurement of this ratio requires also the necessary correction due to the long lifetime of the ^4P term and therefore the determination of G_τ. Thus, we have

[4] The process where the target acts like a needle ionizing just the 1s electron without affecting any of the other electrons.

$$\frac{\sigma(^4P)}{\sigma(^2P_-)} = \frac{Z(^4P)/(\xi_{^4P} G_\tau)}{Z(^2P_-)/\xi_{^2P_-}} = 2 \tag{21}$$

and therefore

$$G_\tau = \frac{1}{2}\frac{Z(^4P)}{Z(^2P_-)}\frac{\xi_{^2P_-}}{\xi_{^4P}}, \tag{22}$$

where Z denotes the measured normalized electron yields obtained from the Auger peak areas.

It should be emphasized that the choice of Be-like mixed-state beams for the determination of G_τ relies on the fact that the formation of the long-lived ^4P_J levels is free from cascading effects, originating from higher lying radiatively decaying ^4L_J levels [51]. Since the long-lived ^4P_J levels are directly produced from the 1s ionization, the population of higher lying ^4L_J levels is a higher order process considered negligible.

In Fig. 6a, we present typical KLL Auger spectra obtained for collisions of 20 MeV $O^{4+}(1s^22s^2\,{}^1S,\, 1s^22s2p\,{}^3P)$ with He targets used for the determination of the correction factor G_τ. The data were recorded with a deceleration factor of $F = 8$ for sufficient energy resolution. Corresponding distributions generated from SIMION simulations for the same experimental conditions were fitted to the experimental data, thus extracting the normalized Auger yields for the ^4P and $^2P_-$ terms, necessary for estimating the correction factor G_τ via Eq. 22. It should be noted that, aside from the geometrical parameters and the Auger lifetimes, the correction factor G_τ also depends on the deceleration factor F and lens voltage scheme applied. In Fig. 6b, we present a SIMION study for the correction factor G_{τ_J} as a function of the parameter $V_p\tau$ for the deceleration factor of $F = 4$, mostly used in our ZAPS measurements.

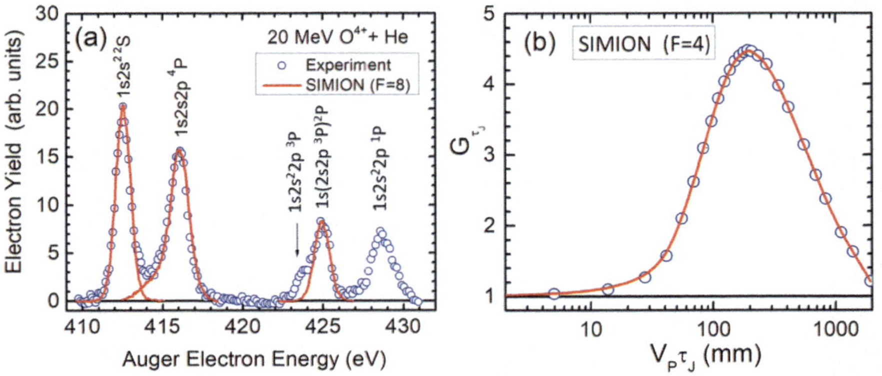

Fig. 6 **a** Open circles: Experimental KLL Auger electron spectra for the collision system of 20 MeV $O^{4+}(1s^22s^2\,{}^1S,\, 1s^22s2p\,{}^3P) + He$. The hemispherical spectrograph deceleration factor was set to $F = 8$ for higher energy resolution. Red solid lines: SIMION generated Auger distributions fitted to the corresponding Auger peaks. **b** SIMION study of the effective solid angle correction factor G_{τ_J} at deceleration factor of $F = 4$

Table 1 The overall effective solid angle correction factor G_τ determined experimentally and with SIMION simulations using Be-like $O^{4+}(1s^2 2s^2\,{}^1S,\, 1s^2 2s2p\,{}^3P)$ beams, for deceleration factors of $F = 4$ and $F = 8$

E_p	G_τ, $F = 8$		G_τ, $F = 4$
(MeV)	(SIMION)	(Exp.)	(SIMION)
8			2.88
10			2.82
12	2.28	2.0±0.2	2.73
14			2.57
16	2.11	1.9±0.2	2.53
20			2.25
24	2.03	1.8±0.2	2.40

In all the above SIMION simulations, the lifetimes for the three J-levels of the ${}^4P_J$ level as well as the corresponding Auger yields were obtained from Ref. [46]. In Table 1, we summarize the results of the G_τ study using oxygen beams.

3 Mixed-State Ion Beams

Highly charged ions are readily produced by passing a lower charge state beam through a thin foil or gas target, where electrons can be stripped from the projectile. In tandem Van de Graaff accelerators these projectile electron strippers are located inside the accelerator terminal. There, the incoming negative ions are stripped off their electrons resulting in a positively charged ion beam that is further accelerated to the exit of the terminal. The resulting positively charged ion beams have a Gaussian-like charge state distribution centered around the mean charge state, depending on the energy of the incoming negative ion beam as well as the stripping medium. The higher the energy of the beam and the density of the medium, the higher the mean charge state attained [52, 53]. The desired charge state and beam energy are then selected by means of analyzing and switching magnets, as well as the necessary magnetic focusing elements and then delivered to the experimental area. In order to produce more intense few electron or even bare ion beams, additional stripping points are provided after the beam exits the accelerator known as *post-strippers*. Various computer codes have been developed to predict the charge state distributions resulting either from stripping or post-stripping processes, such as ETACHA [54], CHARGE [53, 55–57], and TARDIS [58].

Selection of a charge state at a certain energy may result in more than one ionic electron configurations, ground state, and additional metastable components that cannot be magnetically separated. As an example, we may refer to the case of He-like ionic beams that are delivered by tandem Van de Graaff accelerators in mixed

Table 2 Theoretical lifetimes (given in s) of the metastable He-like $1s2s\,^1S$, $1s2s\,^3S$ and Be-like $1s^22s2p\,^3P_1$ states for the atomic numbers $3 \leq Z \leq 10$

Z	$1s2s\,^1S$ $^{(*)}$	$1s2s\,^3S$ $^{(*)}$	$1s^22s2p\,^3P_1$ $^{(\dagger)}$
3	5.1×10^{-4}	4.9×10^{1}	–
4	5.5×10^{-5}	1.8×10^{0}	–
5	1.1×10^{-5}	1.5×10^{-1}	9.8×10^{-2}
6	3.0×10^{-6}	2.1×10^{-2}	9.7×10^{-3}
7	1.1×10^{-6}	3.9×10^{-3}	1.7×10^{-3}
8	4.3×10^{-7}	9.6×10^{-4}	4.4×10^{-4}
9	2.0×10^{-7}	2.8×10^{-4}	1.4×10^{-4}
10	1.0×10^{-7}	9.2×10^{-5}	5.3×10^{-5}

$^{(*)}$ From Refs. [63, 64]
$^{(\dagger)}$ From Refs. [65, 66]

$1s^2\,^1S$ ground and $1s2s\,^{1,3}S$ metastable states. The mixed-state production is a general feature encountered in accelerators such as tandem Van de Graaff [59, 60], storage rings [61, 62], as well as other lower energy highly charged ion sources [22].

Mixed-state beams complicate absolute cross-sectional measurements since the accurate determination of the beam content is also required. In the literature, the term *apparent* cross section is used instead for measurements that use mixed-state beams [9, 21]. However, metastable beams offer access to collision channels that are not readily accessed from the ground state beams. For example, the $1s2s2p\,^4P_J$ state populated in collisions of He-like mixed-state ($1s^2\,^1S$, $1s2s\,^{1,3}S$) beams with gas targets is solely populated from the metastable $1s2s\,^3S$ part by single electron capture.

To fully exploit the metastable component of a beam in collision experiments, the initial metastable fraction content as well as the lifetimes of the metastable states involved should be known. The latter is necessary for metastable beam states with lifetimes short enough to vary during their transportation to the experimental area, in order to estimate the surviving fraction at the target. In Table 2, the lifetimes of the metastable He-like $1s2s\,^1S$, $1s2s\,^3S$ and Be-like $1s^22s2p\,^3P_1$ states, of interest here, are given for the atomic numbers $3 \leq Z \leq 10$. A rapid drop of the lifetimes with increasing atomic number Z is evident.

3.1 Fractions Determination

Several techniques can be found in the literature for determining the metastable content of mixed-state ion beams in various types of facilities (see Ref. [67] and references therein). Here, we shall describe our results for He-like and Be-like ion beams delivered by tandem Van de Graaff accelerators.

3.1.1 $1s2s\,^3S$

Schiebel et al. reported on the first determination of the $1s2s\,^3S$ fraction in Si^{12+} ions based on K X-ray yield measurements [68]. Terasawa et al. used $F^{7+}(1s^2\,^1S, 1s2s\,^{1,3}S)$ mixed-state ion beams to determine the metastable fraction based on target K X-ray data obtained from the beam bombardment on thin Ti layers evaporated on carbon foils [69]. The photon-particle coincidence technique was used in Ref. [70] to determine the metastable $He(1s2s\,^3S)$ fraction produced by electron capture neutralization of slow 25–90 keV He^+ ions in H_2. Electron impact ionization experiments for Li^+ mixed-state beams were reported in Ref. [22]. Similar studies were also reported for other ions as well [71]. Recently, photoionization measurements for mixed-state C^{4+} ions were compared to dielectronic recombination cross sections of $C^{5+} + e^-$ through the principle of detailed balance [72].

Benis et al. [73] proposed a method based on the measurement of relative Auger electron yields from doubly excited $1s2l2l'$ states formed in collisions of He-like $(1s^2\,^1S, 1s2s\,^{1,3}S)$ mixed-state beams with H_2 or He targets [73]. The critical assumption of the method is that the $1s2s2p\,^4P$ state is solely produced by single electron capture to the $1s2s\,^3S$ metastable state, while the $1s2p^2\,^2D$ is primarily produced from the $1s^2\,^1S$ ground state by the process of transfer and excitation. The method requires two independent measurements of the same electron spectrum at the same collision energy, but using mixed-state beams having quite different $1s2s\,^3S$ metastable beam fraction in each. Then, the metastable fraction is determined only by the apparent single differential cross sections (SDCS), $d\sigma[x] \equiv d\sigma[x]/d\Omega'$, of the ^4P and ^2D Auger peaks as

$$f_{^3S}^{[i]} = d\sigma_i[^4P]\,\frac{d\sigma_2[^2D] - d\sigma_1[^2D]}{d\sigma_1[^4P]d\sigma_2[^2D] - d\sigma_2[^4P]d\sigma_1[^2D]}, \quad i = 1, 2, \tag{23}$$

where $i = 1, 2$ refers to the high and low value metastable fractions, respectively. Thus, the method does not suffer from uncertainties arising either from theoretical calculations or long lifetimes and cascading effect considerations. A similar approach has been reported earlier by Zamkov et al. [74, 75], which however required theoretical calculations for the cross sections of transfer excitation and electron capture, as well as modeling calculations for the corrected yield detection of the long-lived $1s2s2p\,^4P$ state.

Typical spectra of He-like mixed-state oxygen beams used for fractions determinations are shown in Fig. 7. The 20 MeV spectra were obtained after stripping the O^- beam in the tandem accelerator terminal either with thin carbon foils (foil terminal stripping, FTS) or with N_2 gas targets (gas terminal stripping, GTS). The 14 MeV spectra were obtained after post-stripping the delivered O^{4+} beam (obtained either with GTS or FTS) into thin carbon foils (foil post-stripping, FPS) or with N_2 gas targets (gas post-stripping, GPS). It is clearly seen that the gas-stripped ion beams have a narrower energy distribution which is evident in the broadening of the observed projectile Auger peaks. This is due to the considerable straggling that the ions suffer in foil stripping [76]. Moreover, it is evident that the amount of metastables may

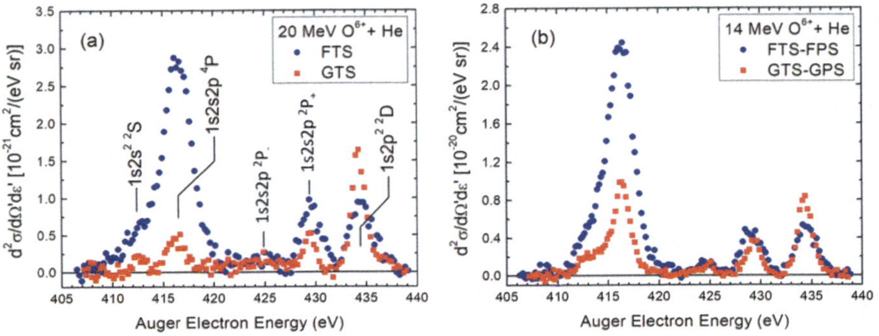

Fig. 7 Typical spectra of He-like mixed-state oxygen beams in collisions with He targets used for fraction determinations. **a** 20 MeV. Blue filled circles: Stripping with FTS. Red filled squares: Stripping with GTS. **b** 14 MeV. Blue filled circles: Post-stripping the O^{4+} beam with FPS. Red filled squares: Post-stripping the O^{4+} beam with GPS. See text for stripping details

vary, and thus be controlled, depending on the stripping location, the energy, and the stripping medium. This is an important feature for the mixed-state beams as it offers the opportunity to exploit all the beam components in collision experiments, as it will be detailed below. For example, it is interesting to notice the negligible $1s2s\,^3S$ fraction in the GTS case corresponding to almost pure ground state beams. Typical high value metastable fractions $f_{^3S}$ were measured to be about 17–23% depending on the stripping process.

3.1.2 $1s^22s2p\,^3P$

As mentioned in Sect. 2.5, Be-like beams are delivered by tandem Van de Graaff accelerators in a mixture of $1s^22s^2\,^1S$ ground and $1s^22s2p\,^3P$ metastable states. During the collisions of these states with H_2 or He targets, the needle ionization of the $1s$ electron of the $1s^22s2p\,^3P$ state results in the production of the ^4P and $^2P_-$ states. Similarly, $1s$ ionization of the $1s^22s^2\,^1S$ ground state results in the $1s2s^2\,^2S$ excited state. The K-vacancy production cross sections from the ground state and the metastable state can be expected to be equal [77], since the $1s$ needle ionization process does not depend strongly on the L-shell configuration, i.e., $\sigma_{1s}(1s^22s^2) = \sigma_{1s}(1s^22s2p\,^3P)$. Moreover, as mentioned in Sect. 2.5, population statistics of the ^4P and $^2P_-$ states should result in the ratio $\sigma(^4P) : \sigma(^2P_-) = 2 : 1$. Therefore, the following ratios for the production cross sections should hold: $\sigma(^2S) : \sigma(^4P) : \sigma(^2P_-) = 3 : 2 : 1$. Then, the metastable fraction $f_{^3P}$ is obtained as [48]

$$f_{^3P} \equiv \left(1 + \frac{d\sigma[^2S]}{d\sigma[^4P] + d\sigma[^2P_-]}\right)^{-1} = \left(1 + \frac{1}{3}\frac{d\sigma[^2S]}{d\sigma[^2P_-]}\right)^{-1}, \qquad (24)$$

Table 3 Measurements of the metastable fraction, $f_{^3P}$ (Eq. 24), for $O^{4+}(1s^22s^2\,{}^1S,\,1s^22s2p\,{}^3P)$ mixed-state beams for various collision energies E_P. The Be-like oxygen beams were obtained after stripping both in N_2 gas targets (GTS) and thin carbon foils (FTS)

E_P (MeV)	Stripping method	$f_{^3P}$ (± 0.05)
12	FTS	0.66
	GTS	0.59
16	FTS	0.65
	GTS	0.57
20	FTS	0.65
	GTS	0.56

where $d\sigma[x]$ denotes the apparent SDCS of the corresponding state x : ${}^2S, {}^2P_-, {}^4P$, in the Auger spectrum. In Table 3, we present measurements of the metastable fraction $f_{^3P}$ for $O^{4+}(1s^22s^2\,{}^1S,\,1s^22s2p\,{}^3P)$ mixed-state beams, obtained from KLL Auger spectra similar to those presented in Fig. 6a. The Be-like oxygen beams were delivered after stripping the O^- in the accelerator terminal both in N_2 gas targets (GTS) and in thin carbon foils (FPS). GTS is seen to result in about 10% smaller fractions than FPS.

3.2 *Single Differential Cross Section Determination*

The two-measurement method for determining the content of a He-like mixed-state beam, proposed by Benis et al. [73], can also be used to determine the SDCS of the $1s2l2l'\,{}^{2,4}L$ doubly excited states from either the ground state or the metastable state. The true SDCS, given in the projectile rest frame, are derived according to the general form [78]:

$$\frac{d\sigma_m[x]}{d\Omega'} = \frac{d\sigma_2[x]\,d\sigma_1[{}^2D] - d\sigma_1[x]\,d\sigma_2[{}^2D]}{d\sigma_1[{}^2D] - d\sigma_2[{}^2D]} \tag{25}$$

$$\frac{d\sigma_g[x]}{d\Omega'} = \frac{d\sigma_2[x]\,d\sigma_1[{}^4P] - d\sigma_1[x]\,d\sigma_2[{}^4P]}{d\sigma_1[{}^4P] - d\sigma_2[{}^4P]}, \tag{26}$$

where x : ${}^2P_{\pm}, {}^4P, {}^2D$. Equations 25–26 are seen to also satisfy the main assumption of the method $\frac{d\sigma_g[{}^4P]}{d\Omega'} = \frac{d\sigma_m[{}^2D]}{d\Omega'} = 0$.

Our two-measurement method can also be used to obtain true SDCS for the heavier than H_2 and He gas targets, where the assumptions of the two-measurement method do not hold. In this case, the following system of equations is always valid independent of any requirements [78]:

$$\left(1 - f_{^3S}^{[1]}\right) \frac{d\sigma_g}{d\Omega'}[x] + f_{^3S}^{[1]} \frac{d\sigma_m}{d\Omega'}[x] = d\sigma_1[x], \tag{27}$$

$$\left(1 - f_{^3S}^{[2]}\right) \frac{d\sigma_g}{d\Omega'}[x] + f_{^3S}^{[2]} \frac{d\sigma_m}{d\Omega'}[x] = d\sigma_2[x]. \tag{28}$$

The fractions $f_{^3S}^{[1]}$ and $f_{^3S}^{[2]}$ are estimated from Eqs. 23 using the apparent SDCS $d\sigma_i[x]$, determined from collisions with the H_2 or He targets. Then, solving the system of two linear equations, Eqs. 27–28, for the two unknowns, $d\sigma_g[x]/d\Omega'$ and $d\sigma_m[x]/d\Omega'$, we obtain

$$\frac{d\sigma_g}{d\Omega'}[x] = \frac{f_{^3S}^{[2]} d\sigma_1[x] - f_{^3S}^{[1]} d\sigma_2[x]}{f_{^3S}^{[2]} - f_{^3S}^{[1]}}, \tag{29}$$

$$\frac{d\sigma_m}{d\Omega'}[x] = \frac{\left(1 - f_{^3S}^{[1]}\right) d\sigma_2[x] - \left(1 - f_{^3S}^{[2]}\right) d\sigma_1[x]}{f_{^3S}^{[2]} - f_{^3S}^{[1]}}. \tag{30}$$

4 Single Electron Capture

The fundamental process of single electron capture (SEC) and the related dynamics of a many-body quantum system is a challenging theoretical problem. Modern theoretical SEC studies include semi-classical close-coupling approaches such as atomic orbital close-coupling (AOCC) [79–81], two-center basis generator method (TC-BGM) [82, 83], molecular orbital close-coupling (MOCC) [84], as well as lattice-based finite-difference calculations [85], classical trajectory Monte Carlo (CTMC) simulations [86], continuum distorted-wave approaches (CDW) at sufficiently high incident energies [87], and other perturbative methods. Numerous SEC experimental studies have reported results for total cross section dependencies [88]. However, only state-selective SEC studies have offered the most stringent tests of theory. These have been obtained with the techniques of energy gain spectroscopy [89, 90], X-ray spectroscopy [27, 91], Auger spectroscopy [7, 8, 92, 93], and COLTRIMS [94–99]. The results of these studies are also essential for the development of a number of fields such as astrophysics [100, 101] and plasma physics [102].

Recently, SEC involving multi-open-shell excited ions, a demanding theoretical problem that inherently bears important dynamic electron correlation effects, has received attention [6, 8, 83, 92, 93]. These SEC studies offered the testing grounds for examining the validity of the single-active electron picture and frozen core approximations, as well as arguments concerning the population of levels following spin statistics [6, 8, 92, 93, 103]. Multi-open-shell excited ions and related SEC channels are offered in collisions of mixed-state ions with gas targets. Especially, He-like mixed-state ions give rise to doubly excited states much more efficiently explored by state-resolved Auger spectroscopy, particularly for low-Z ions [7].

Our multi-open-shell ion core SEC studies were performed for collisions of $C^{4+}(1s^2\,{}^1S,\,1s2s\,{}^3S)$ projectiles with gas targets using the ZAPS technique and applying the two-measurement method to isolate the processes that correspond only to the $C^{4+}(1s2s\,{}^3S)$ projectiles, i.e.:

$$
\begin{aligned}
C^{4+}(1s2s\,{}^3S) + T \rightarrow C^{3+}(1\,s2s2l\,{}^{2,4}L) + T^{+} \\
\hookrightarrow C^{4+}(1\,s^2) + e_A^-(\theta = 0^\circ),
\end{aligned}
$$

where T is the gas target, i.e., H_2, He, Ne, and Ar, and $\theta = 0^\circ$ is the electron observation angle with respect to the projectile beam direction. The emphasis of our study was primarily given to the ratio R of the ${}^4P/{}^2P_\pm$ SEC cross sections, as a measure of the spin statistics population ratio, i.e.:

$$
R = \frac{\sigma[{}^4P]}{\sigma[{}^2P_+] + \sigma[{}^2P_-]}. \tag{31}
$$

Considering only spin multiplicity, the ratio results in $R = 1$ (although this argument should be strictly applied to degenerate states), while in the case of frozen core approximation, where only the ${}^4P$ and ${}^2P_-$ can be produced from the initial $1s2s\,{}^3S$ state, the ratio results in $R = 2$ [92, 103].

For Auger decays, the relation between total cross section and SDCS is described by the well-known Auger electron angular distribution LS-coupling formula [104] written as an expansion of even-order Legendre polynomials, $P_k(\cos\theta')$:

$$
\frac{d\sigma[{}^2L]}{d\Omega'}(\theta) = \overline{\xi}[{}^2L]\,\frac{\sigma_{\text{tot}}[{}^2L]}{4\pi}\sum_{k=0,\text{even}}^{2L} A_k(L)\,P_k(\cos\theta'), \tag{32}
$$

where $\sigma_{\text{tot}}[{}^{2L+1}L] = \sum_{M_L}\sigma(L, |M_L|)$ is the total production cross section for the LS-multiplet, and $\overline{\xi}[{}^{2S+1}L] \equiv \sum_J (2J+1)\xi_J/[(2S+1)(2L+1)]$ is the Auger yield of the LS-multiplet computed as the statistical average over its J components ξ_J. $A_k(L)$ are the angular distribution parameters. For $\theta = 0^\circ$, as in our ZAPS measurements, Eq. 32 results for the ${}^2S$ and ${}^2P_\pm$ states, respectively [9]:

$$
\frac{d\sigma[{}^2S]}{d\Omega'}(\theta = 0^\circ) = \overline{\xi}[{}^2S]\,\frac{\sigma[{}^2S](0,0)}{4\pi}, \tag{33}
$$

$$
\frac{d\sigma[{}^2P_\pm]}{d\Omega'}(\theta = 0^\circ) = 3\,\overline{\xi}[{}^2P_\pm]\,\frac{\sigma[{}^2P_\pm](1,0)}{4\pi}. \tag{34}
$$

The $\theta = 0^\circ$ SDCS are seen to be sensitive only to the $M_L = 0$ component of the partial cross sections. For the $1s2s2p\,{}^4P$ state Eq. 34 was assumed to be a good approximation, since for low-Z ions the spin-orbit interaction has a negligible effect on the production of the resonances. Then, the ratio R can be written for both calculations and measurements equivalently to Eq. 31 as [9]:

$$R = \left[\frac{\sigma[^4P](1,0)}{\sigma[^2P_+](1,0) + \sigma[^2P_-](1,0)} \right]_{\text{Th}} = \left[\frac{\frac{1}{\overline{\xi}[^4P]} \frac{d\sigma[^4P]}{d\Omega'} \frac{1}{G_\tau}}{\frac{1}{\overline{\xi}[^2P_+]} \frac{d\sigma[^2P_+]}{d\Omega'} + \frac{1}{\overline{\xi}[^2P_-]} \frac{d\sigma[^2P_-]}{d\Omega'}} \right]_{\text{Exp}} . \tag{35}$$

In addition, further evidence is provided by examining also the cross section ratio r of $^2P_+$ to $^2P_-$ similarly given by [9]:

$$r = \left[\frac{\sigma[^2P_+](1,0)}{\sigma[^2P_-](1,0)} \right]_{\text{Th}} = \left[\frac{\frac{1}{\overline{\xi}[^4P_+]} \frac{d\sigma[^4P_+]}{d\Omega'}}{\frac{1}{\overline{\xi}[^2P_-]} \frac{d\sigma[^2P_-]}{d\Omega'}} \right]_{\text{Exp}} \tag{36}$$

with the explicit indication that the cross sections are evaluated for $M_L = 0$. For carbon, the ξ_J Auger yields are all very close to 1 and thus we have taken $\overline{\xi}[^{2S+1}L] = 1$ for all $L = 0, 1$.

4.1 3eAOCC Theory and Cascade Feeding

The theoretical treatment of the SEC process was based on *ab initio* dynamical calculations involving *three* active electrons within a full configuration interaction. It relies on a semi-classical atomic orbital close-coupling approach (3eAOCC) with asymptotic descriptions of the atomic collision partners [80, 105, 106]. For this, the time-dependent Schrödinger equation is solved nonperturbatively, including all couplings related to the static and dynamic interelectronic repulsions and effects stemming from the Pauli exclusion principle. It goes beyond frozen core models [83, 92], where only one active electron is considered in the dynamics. The atomic center electronic structures are represented in terms of sets of Gaussian-type orbitals and selected products of these GTOs, in order to obtain fully antisymmetrized electron wave functions. The He target is described by a model potential binding just a single electron. Total cross sections are computed for all processes spanned by the basis sets and then multiplying by a factor of 2 to account for the two electrons on the H_2 and He targets. The ratio R, given in Eq. 35, is evaluated using the $M_L = 0$ partial cross sections for the three P states under consideration, in accordance with the ZAPS sensitivity, as described in Eqs. 33 and 34.

Although the $1s2s2p\ ^{2,4}P$ states correspond to the strongest capture channel, SEC can also populate higher lying quartet and doublet $1s2snl\ ^{2,4}L$ ($n > 2$) states. These states may decay to the initial $1s2s2p$ configuration radiatively, by E1 transitions, thus increasing the initial $n = 2$ population. This mechanism, known as *cascade feeding*, may be concluded in one or more steps and then termed as first order, second order, etc. The alteration of the initial population of the $1s2s2p$ configuration has significant consequences in the determination of the ratio R of the cross sections of the ^4P to ^2P states, and thus contributions from cascade feeding effects should be accounted for.

Specifically, the $1s2snl\,^2L$ doublets do not affect the initial population of the $1s2s2p\,^2P$ states appreciably, since they have much stronger Auger decay rates to the $1s^2$ ground state compared to the corresponding radiative decays, and thus are quickly depleted. On the contrary, the $1s2snl\,^4L$ quartet states have very weak Auger decay rates to the $1s^2$ ground state, and predominantly radiatively decay via cascades to the lowest quartet state, i.e., the $1s2s2p\,^4P$ state [51, 83, 93].

Recently, we have presented a detailed treatment of the selective cascade feeding mechanism [51]. The population at the time t of the n-th level denoted $N_n(t)$ is assumed to be the result of cascade feeding from higher energy levels ($k > n$) by radiative transitions with rates A^r_{kn}, as well as depopulation by radiative transitions to all lower energy levels $f = 1, 2, \ldots i$ ($i < n$), with rates A^r_{nf}. Additional depopulation due to autoionizing transitions from the n level to all final states f', with rate $A^a_{nf'}$, is also considered. An example of the schematic decay of a five-level system is shown in Fig. 8a. The coupled differential equations governing the population of any particular level n are given [107] with the addition of the depopulating autoionization channel [83] by

$$\frac{dN_n}{dt} = \sum_{k=n+1}^{m} A^r_{kn} N_k(t) - \left(\sum_{f=1}^{n-1} A^r_{nf} + \sum_{f'} A^a_{nf'} \right) N_n(t). \tag{37}$$

The above coupled differential equations were solved resulting in the final population of the $1s2s2p\,^4P$ state including up to third-order cascades. Then, we quantified the result for the $C^{3+}(1s2s2p\,^4P)$ state, by calculating the radiative and Auger decay rates of the involved states, corresponding to the principal quantum numbers of

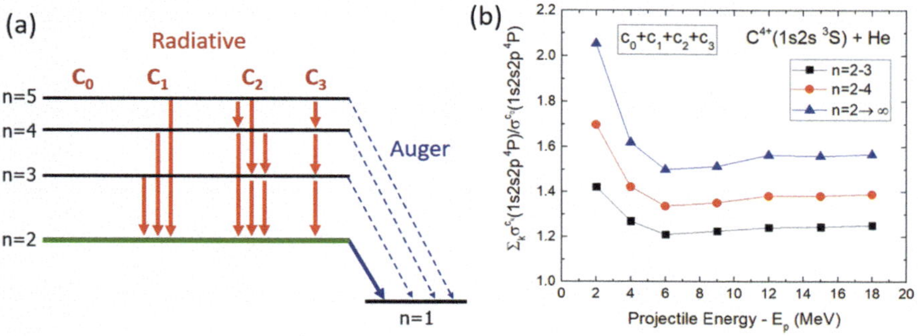

Fig. 8 **a** Schematic decay of a five-level system illustrating radiative cascade feeding to the level 2. The radiative transitions are indicated by red arrows, while the Auger transitions are indicated by blue arrows. The dashed lines correspond to very weak transitions. Cascades are separated into feeding orders c_0, c_1, c_2, and c_3 requiring a total of 0, 1, 2, or 3 sequential transitions, respectively. **b** Ratio of the cumulative cascade contributions up to third order of the $1s2snl\,^4L$ state populations to the $1s2s2p\,^4P$ as a function of the projectile energy E_p, for increasing principal quantum number n. Black filled squares: $n \leq 3$; Red filled circles: $n \leq 4$; Blue filled triangles: Extrapolation to include all $n \to \infty$

$n = 2 - 4$, using the Cowan code. An extrapolation of these results to include all $n \to \infty$, according to the well-known n^{-3} law for the SEC process [83, 108], was also considered. Moreover, the corresponding SEC cross sections were calculated within the 3eAOCC theory with the exception of the $l = 2$ and $l = 3$ that were evaluated within a *one*-electron treatment (1eAOCC).

In Fig. 8b, we present our results about the cascade feeding of the $1s2snl\,^4L$ state as a function of the projectile energy E_p, for increasing principal quantum number n. Third-order cascade contributions were included which were at the convergence limit. A significant increase of the population of the $1s2s2p\,^4P$ state is evident that plays a vital role in the determination of the ratio R of the cross sections of the ^4P to ^2P states.

The existence of the higher lying quartet and doublet $1s2snl\,^{2,4}L$ ($n > 2$) states has been shown experimentally [109]. In Fig. 9, we present Li-like KLn ($n \geq 2$) Auger spectra obtained in collisions of 6 MeV $C^{4+}(1s^2, 1s2s\,^3S)$ with He targets. R-matrix calculations, convoluted with the target Compton profile and the experimental energy resolution, indicate the location of the Auger lines corresponding to the ground state $1s^2$. Auger peaks that are not reproduced by the R-matrix calculations are identified by Cowan code calculations as doubly excited $1s2snl\,^2L$ ($n > 2$) states produced by SEC from the $C^{4+}(1s2s\,^3S)$ part of the beam. Quartet $1s2snl\,^4L$ ($n > 2$) states are not recorded in the Auger spectrum since they have very high radiative decay rates compared to their Auger decay rates. However, since the SEC process proceeds for both doublets and quartets, the existence of quartets states is indirectly evident.

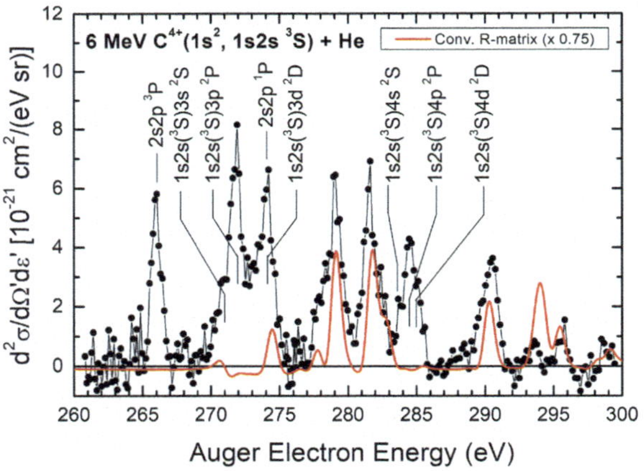

Fig. 9 Li-like KLn ($n \geq 2$) Auger spectra obtained in collisions of 6 MeV $C^{4+}(1s^2, 1s2s\,^3S)$ with He targets. Red line: R-matrix calculations convoluted with the target Compton profile and the experimental energy resolution. The rest of the spectrum contributions are from the $1s2s\,^3S$ beam component. The marked states were identified by Cowan code calculations

4.2 Results

In Fig. 10a, we present a typical set of zero-degree Auger electron spectra obtained in collisions of C^{4+} ($1s^2\,{}^1S$, $1s2s\,{}^3S$) mixed-state beams with He gas targets. For each collision energy two sets of data were recorded corresponding to different mixed-state beam content. Thus, by applying our two-measurement technique we were able to extract the values of the SDCS for each $1s2l2l'\,{}^{2,4}L$ doubly excited state. Then the ratio R was determined according to Eq. 35. The results are presented in Fig. 10b along with the theoretical 3eAOCC predictions.

The experimental value of R is seen to be around the value of two, showing a plausible agreement with the value of the frozen core approximation. The picture though

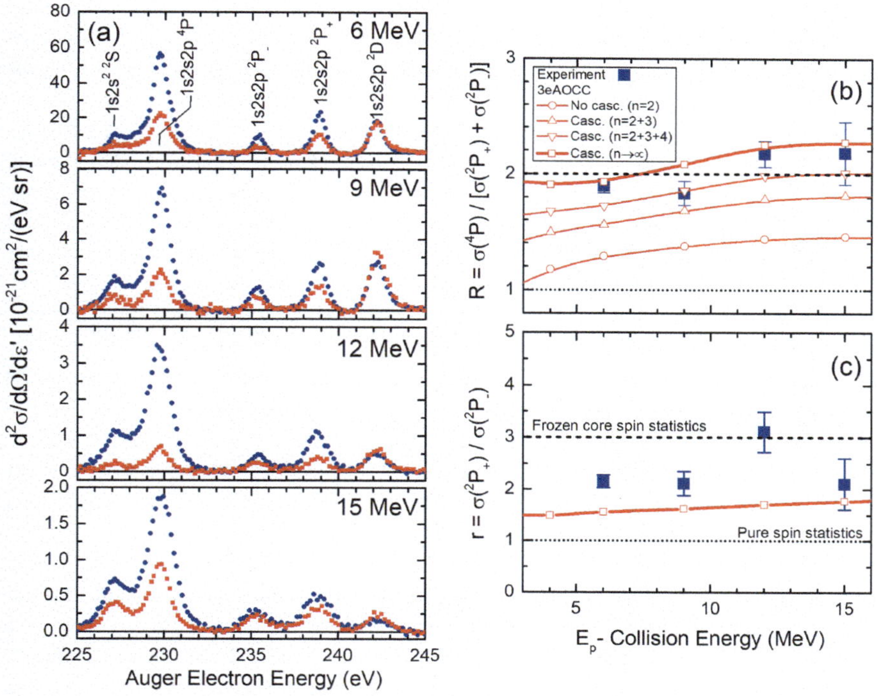

Fig. 10 **a** Projectile rest frame 0° DDCS for collisions of 6, 9, 12, and 15 MeV C^{4+} ($1s^2\,{}^1S$, $1s2s\,{}^3S$) mixed-state beams with He gas targets. Red filled squares: electron spectra obtained from the low f_{3S} beam fraction. Blue filled circles: electron spectra obtained from the high f_{3S} beam fraction. **b** Ratio R of ${}^4P$ to the sum of ${}^2P_+$ and ${}^2P_-$ cross sections for $M_L = 0$, as given by Eq. 35. Red solid lines correspond to the 3eAOCC calculations, with the order of the cascade contributions to the ${}^4P$ separately given. The dashed line corresponds to the value for the frozen core spin statistics, while the dotted line corresponds to the value for the pure spin statistics. Only statistical uncertainties are shown. **c** Same as in (**b**) but for the ratio r of ${}^2P_+$ and ${}^2P_-$ cross sections for $M_L = 0$, as given by Eq. 36

is far from being correct. The theoretical value of R calculated without cascades is found to lie between 0.9 and 1.5. However, after including the cascade feeding of the ^4P state from higher lying quartet states, the value of the ratio R approaches the value of 2 in agreement with the experimental value. In Fig. 10b, we show the order of the cascade contributions to the ^4P separately from the energy shells of $n = 2, n = 2 + 3, n = 2 + 3 + 4$, and $n \to \infty$, according to the n^{-3} law for capture to account for all shells. This unambiguously demonstrates that cascades make an important contribution to the measured ratio R, while masking the spin statistics considerations advocated in the past.

The 3eAOCC theoretical results are seen to depart significantly from the two limits of the spin statistics arguments, indicating that the open-shell levels of the $1s2s2p$ configuration are populated nonstatistically. Thus, even for collision timescales in the attosecond regime, accounted for in our fast ion-atom collisions, the $1s2s$ projectile core electrons cannot be considered frozen. Instead, electronic correlations and channel couplings are seen to play a crucial role in determining the state populations following a SEC process. Therefore, single-active electron models do not describe SEC in its fine details, although they may provide reasonable cross sections.

In a simplified SEC picture, starting from the initial $1s2s\ ^3S$ state ($\equiv\uparrow\uparrow$), a spin-up or spin-down target electron can be directly transferred to the projectile to create the ^4P state or the $1s(2s2p^1P)^2P_+$ state, respectively. However, the creation of the $1s(2s2p^3P)^2P_-$ state requires, in addition to transfer, a spin exchange between the active target electron and one of the projectile electrons. This involves a second-order process, less likely than the direct capture mechanism populating the ^4P and $^2P_+$ states. This simple picture is supported by calculations based on the Oppenheimer-Brinkman-Kramers (OBK) approximation [6] showing that the levels ^4P and $^2P_+$ are dominantly driven by the nucleus-electron attraction matrix element while the $^2P_-$ level is exclusively controlled by the much weaker exchange electron-electron couplings.

In a pure frozen core picture, the capture of a spin-down electron to the lower energy level of the $1s2s$ configuration (i.e., our initial ^3S state) should intuitively lead to the production of the lower energy $1s2s2p\ ^2P$ state, symbolized here by $^2P_-$ (a spin-up electron can produce only the ^4P state). However, the OBK treatment gives rise to a negligibly small cross section for that $^2P_-$, in agreement with our elaborate ab initio 3eAOCC calculations, which is counterintuitive according to the pure frozen core picture. This selectively bars the direct capture to the $^2P_-$ by shielding the Coulomb attraction between the projectile nucleus and the active electron. This mechanism is only active where the Pauli exclusion principle can be advocated, and we therefore refer to it as *Pauli shielding*. It should be noted that this mechanism is different from other Pauli excitation, exchange, and blocking mechanisms reported in the literature [8, 110–112].

In addition, we provide here further evidence for this finding by examining also the ratio r of the cross sections of the $^2P_+$ to $^2P_-$ states, defined in Eq. 36. In Fig. 10c, we compare 3eAOCC calculations of r to experimental 0° SDCS. Considering only the spin statistics of the final $^2P_\pm$ states (pure spin statistics), r should have the value of 1 since both are doublets. However, assuming the frozen core approximation, and

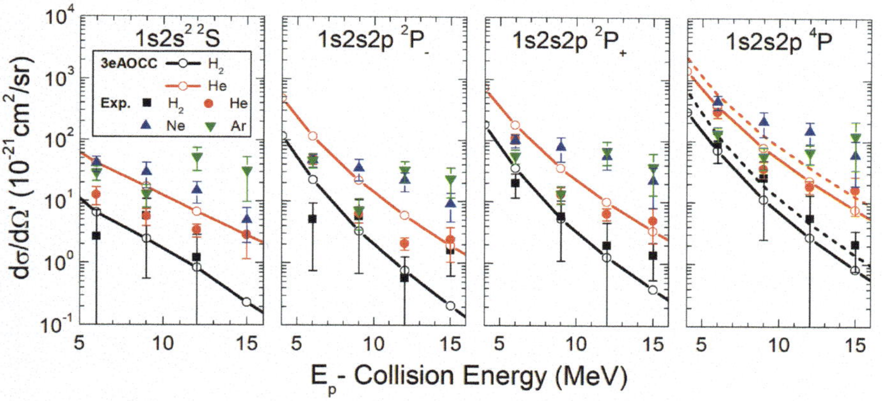

Fig. 11 0° Auger electron SDCS for the $1s2s^2\,{}^2S$, $1s2s2p\,{}^2P_-$, $1s2s2p\,{}^2P_+$ and $1s2s2p\,{}^4P$ states obtained in collisions of $C^{4+}(1s2s\,{}^3S)$ with H_2, He, Ne, and Ar targets. Symbols: Measurements (squares—H_2, circles—He, up triangles—Ne, down triangles—Ar). Only statistical uncertainties are shown. Solid lines with symbols: 3eAOCC calculations (black—H_2, red—He). Dotted lines also include the cascade feeding of the $1s2s2p\,{}^4P$ states from higher lying ($1s2snl\,{}^4L$, $n > 2$) quartet states

the recoupling scheme advocated in Ref. [103], r should have the value of 3. From Fig. 10c, it is evident that the values of r for both measurements and calculations lie in between these two limits, around the value of 2. The deviation of r from its pure statistical value of $r = 1$ can be interpreted considering the Pauli shielding mechanism, which predicts that the $1s2s2p\,{}^2P_-$ SEC cross section is inherently smaller than the corresponding to the ${}^2P_+$ state.

In Fig. 11, we present the experimental and theoretical SDCSs for the $1s2s^2\,{}^2S$, $1s2s2p\,{}^2P_-$, $1s2s2p\,{}^2P_+$, and $1s2s2p\,{}^4P$ states obtained in collisions of the pre-excited $C^{4+}(1s2s\,{}^3S)$ beams with H_2 and He targets, while only experimental results are presented for the heavier Ne and Ar targets [9]. For the H_2 and He targets, a reasonable agreement between the calculations and the measurements is evident. Both experimental and theoretical SDCS are rapidly decreasing with the projectile energy, a dependence expected for the SEC process in fast collisions [88]. For the case of He, the SDCS calculations are seen to be larger than the measurements, close to an average factor of 2, except at the higher projectile energy of 15 MeV, for which the agreement is very good. The strongest deviations are for the ${}^2S$ state, possibly due to the difficulty to extract the Auger yield from the low-energy tail of the ${}^4P$ peak. Even better agreement is seen for the case of H_2, although the statistical error bars are larger than for He, due to the fact that H_2 has smaller SEC cross sections compared to He, and thus suffers from poorer statistics in the measurements.

For Ne targets, the cross section drops monotonically with increasing projectile energy, although not as rapidly as for the H_2 or He targets. This indicates that the SEC process with Ne resembles that of He and could therefore be possibly modeled in a similar way. On the contrary, for Ar targets, the cross section shows a nonmonotonous dependence on the projectile energy. Clearly this is inherently different from the Ne and He cases and may be due to the contribution of inner-valence electrons.

5 Transfer Excitation

Transfer excitation (TE) is a two-electron process involving the excitation of a projectile electron with the simultaneous transfer of a target electron resulting in a doubly excited projectile state. TE is closely related to the electron-ion collision process of dielectronic capture, where a free electron is captured by an ion, with important applications to plasma cooling (e.g., see [113] and references therein). In asymmetric collisions of heavy projectiles with light targets, two distinct peaks are typically observed in the TE cross sections as a function of impact energy [114, 115]: A high-energy peak designated as resonant transfer excitation (RTE) [116], and a low-energy peak designated as nonresonant transfer excitation (NTE) [114, 117, 118]. State-selective TE measurements have been realized by high-resolution Auger projectile spectroscopy [14, 15, 119–126].

The two TE mechanisms, RTE and NTE, have been described incoherently by different approaches. RTE has been modeled by a two-center one-step mechanism mediated by e-e interaction (TCee) through the impulse approximation (IA), as a quasi-free resonant electron scattering, analogous to the inverse Auger process [14, 103, 120, 127]. IA is an electron impact theory adjusted for the initial momentum distribution of the target electron (Compton profile). On the other hand, NTE has been modeled by a sequence of uncorrelated excitation and transfer events, mediated by separate e-n interactions [128, 129]. Such *ad hoc* treatments cannot be considered satisfactory, calling for the necessity of a coherent approach to properly describe TE within an ion-atom collision framework. Attempts to describe TE in a coherent dynamical treatment have been reported in the literature, such as the two-electron atomic orbital close-coupling treatment [128] and the continuous distorted-wave four-body approach [130–132]. However, both attempts involved two-electron collision systems and showed poor agreement with measurements.

Recently, we presented the first non-perturbative treatment of transfer excitation realized within the concept of the 3eAOCC theory [17]. Specifically, we focused on the collisions of He-like $C^{4+}(1s^2)$ ground state beams with He targets, resulting in the doubly excited $C^{3+}(1s2p^2\,{}^2D)$ state (for short ${}^2D$), i.e., on the TE process

$$\begin{aligned} C^{4+}(1\,s^2) + \mathrm{He} \rightarrow\ & C^{3+}(1\,s2p^2\,{}^2D) + \mathrm{He}^+ \\ & \qquad \hookrightarrow C^{4+}(1\,s^2) + e_A^-(\theta = 0°). \end{aligned}$$

As already mentioned, He-like beams are delivered in a $(1s^2,\ 1s2s\ {}^{1,3}S)$ mixed-state, and thus we used either our two-measurement technique or single measurements of known ground state fraction to extract the production SDCS of the ${}^2D$ state.

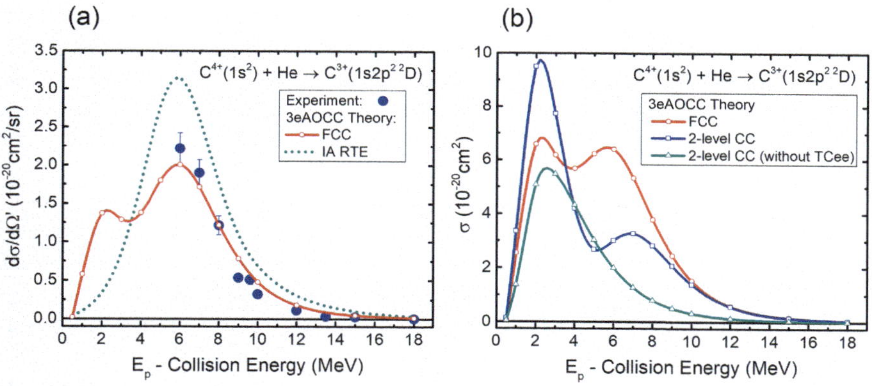

Fig. 12 **a** Projectile energy dependence of absolute 0° Auger SDCS, $d\sigma/d\Omega'$, for the production of ^2D states by TE. Blue filled circles: Experiment. Error bars shown only when larger than symbol. Red line with open circles: Full 3eAOCC theory (FCC). Green dotted line: IA-RTE. **b** Projectile energy dependence of theoretical total cross sections $\sigma_{\rm tot}$ for the production of the $1s2p^2\,^2D$ state. Red line with open circles: Full 3eAOCC theory corresponding to the FCC shown in (**a**). Blue line with open squares: two-level CC calculations. Green lines with open triangles: two-level CC without TCee

In Fig. 12a, the experimental zero-degree Auger SDCSs are presented as a function of the collision energy E_p, together with our 3eAOCC theoretical results, noted as FCC to account for the full close-coupling calculations. The theoretical SDCS were obtained from the calculations of the total cross sections modified by the Auger angular distribution of Eq. 32, which for the ^2D state results

$$\frac{d\sigma_A}{d\Omega'}(\theta = 0°) = \overline{\xi}\,\frac{5\,\sigma(2,0)}{4\pi}, \tag{38}$$

showing that at this observation angle the SDCS depends only on the $M_L = 0$ component of the partial cross section $\sigma(2, 0)$.

The FCC SDCS results show two distinct peaks, the low-energy being around 2 MeV and the high-energy around 6 MeV. The high-energy peak is seen to be in excellent agreement with the measurements, not only in the energy dependence, but also in the absolute scale. The low-energy peak could not be reached experimentally, since the corresponding collision energies are not supported by the tandem Van de Graaff accelerator in use. Model results from the RTE impulse approximation (IA-RTE) [121] are also presented. These are seen to be about 30% larger than experiment, a known IA-RTE deficiency for low-Z ions [16].

To get further insight into the electronic dynamics, we performed 3eAOCC 2-level calculations that used a restricted basis set, including only the initial He target and projectile ground states, and the final ^2D state with its five M_L magnetic states. This model (referred to as two-level CC) blocks any single excitation and single capture channels, and therefore deliberately suppresses any two-step mechanisms involving excitation and capture such as NTE [128]. As in the FCC calculations, they include all interactions and couplings between the states. In addition, we further developed a version of the two-level CC model [referred to as two-level CC (without TCee)], in which the two-center e-e coupling is removed, thus further blocking also processes mediated by TCee such as RTE. The comparative results for the above models are presented in Fig. 12b.

As a first general observation, it is seen that the cross sections from the FCC and the two-level CC calculations show a similar double peak structure. However, quantitative differences clearly demonstrate that couplings to other than the initial and final states, excluded in the two-level CC model, are important.

When the two-center bi-electronic repulsion matrix elements are neglected, the resulting cross section does not show the high-energy peak, contrary to the FCC and two-level CC calculations. This result unambiguously shows that the high-energy peak is mediated by TCee interactions. This finding is supported by collision impact parameter analysis offered by the 3eAOCC calculations, as reported in Ref. [17], where it was shown that the high-energy peak is promoted at large impact parameters, as expected for TCee interactions. Moreover, the finding is further corroborated by the strong $M_L = 0$ contributions. Thus, the underlying mechanism and related dependencies of the high-energy peak, so far modeled within the IA-RTE and the resonant character of inverse Auger mechanism, are validated through our 3eAOCC theory, an ion-atom collision theory that was long awaited.

We next consider the low-energy ($\sim$2 MeV) peak, so far interpreted by a two-step mechanism [118, 128]. Since the two-level calculations still show this peak its origin must necessarily arise from a one-step mechanism. Furthermore, by removing also the TCee interaction the total cross section still shows the low-energy peak, albeit somewhat reduced in magnitude. This indicates that the production of ^2D in the two-level CC is mediated by (e-e) and to some extent by (e-n) interactions in this energy range, although not in a two-step process, as traditionally NTE is modeled, but in one step. Moreover, collision impact parameter analysis showed that the low-energy peak arises from small impact parameters [17]. These unexpected results point to the existence of a nonresonant but correlated (single-step) transfer excitation mechanism.

To qualitatively understand this correlated TE mechanism, we applied an OBK approximation to the two-level CC model [17]. Despite the fact that OBK model calculations are derived from a first-order treatment with known drawbacks, they highlight the two mechanisms responsible for the two peaks shown in the cross sections. Thus, the OBK model showed that the low-energy peak may be promoted by the $1s \rightarrow 2p$ excitation of the projectile by the target nucleus, with the simultaneous transfer of the target electron to the projectile mediated by the two-center overlap of the $2p$-projectile and $1s$-target orbitals. We refer to this new TE mechanism as *nonresonant correlated transfer excitation* (NCTE) [17], which along with the uncorrelated NTE mechanism contributes to the formation of the low-energy TE peak.

6 Ongoing and Near-Future Projects

6.1 Single and Double Electron Capture

Our SEC results and corresponding interpretation, as determined by the ratio R of the cross section of the ^4P to ^2P states, call for a broader validity investigation. The role of the electron correlation contributions to the SEC process fully accounted for in the 3eAOCC theory, and their interplay with terms corresponding to the projectile charge Z will be examined in an isoelectronic sequence study. The investigation will primarily involve the He-like ($1s^2$, $1s2s$) mixed-state projectiles of boron, oxygen, and fluorine, readily available by the tandem Van de Graaff accelerator. The data for oxygen have already been obtained, as partially shown in Fig. 7.

As already mentioned, 3eAOCC theory accounts for three active electrons, and thus necessitates the use of a single-electron approximate model for the He target. In order to avoid such modeling and preserve the fully quantum mechanical treatment of the capture process in the 3eAOCC theory, experiments on capture involving three active electrons are necessary. Thus, we plan to use H-like C^{5+} projectiles in collisions with He targets and study the process of double electron capture (DEC), fully described by the 3eAOCC theory without any target modeling. In addition, this three-active electron collision system does not suffer from additional cascading effects and related calculations. Thus, the expected results will serve as the most stringent test for the 3eAOCC theory.

6.2 He-Like Hollow States

Projectile He-like ($1s^2$, $1s2s\ ^{1,3}S$) mixed-state beams colliding with gas targets provide a route to study the formation and related dynamics of the doubly excited $2s2p\ ^{3,1}P$ hollow states. Hollow states lack a thorough investigation in the literature (e.g., [133]), possibly due to the inherent experimental complexity that the use of mixed-state beams brings in. The production of the $2s2p\ ^{3,1}P$ hollow states proceeds primarily by the $1s \rightarrow 2p$ excitation either from the $1s2s\ ^3S$ for the $2s2p\ ^3P$ or the $1s2s\ ^1S$ for the $2s2p\ ^1P$ states, respectively. Thus, the $1s2s\ ^1S$ metastable component, usually neglected in SEC and TE measurements, plays a non-negligible role here and should be carefully treated. The fundamental process of $1s \rightarrow 2p$ excitation inherently highlights the role of electron-nucleus and electron-electron interactions that has been addressed in the past in similar studies [121, 134]. In Fig. 13, we present typical zero-degree Auger spectra showing the $1s2l2l'$ and $2s2p$ configurations in collisions of 12 MeV $C^{4+}(1s^2\ ^1S, 1s2s\ ^3S)$ mixed-state beams with He gas targets. The systematic study of these hollow states as a function of the projectile energy for various low-Z projectiles is currently under investigation. The results will be examined within the three-active electron concept of the 3eAOCC theory and corresponding calculations.

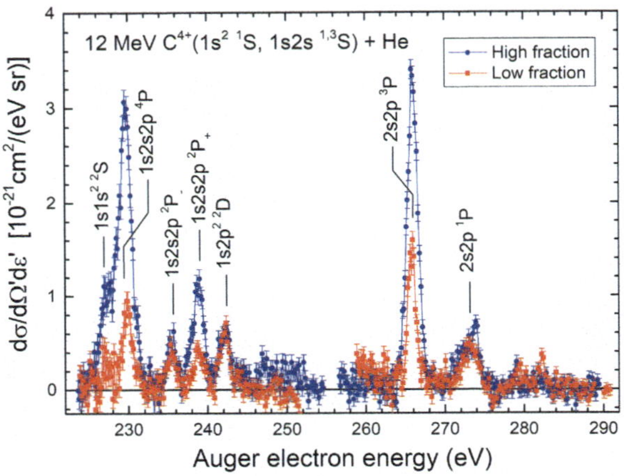

Fig. 13 Projectile rest frame 0° DDCS Auger spectra showing the $1s2l2l'$ and $2s2p$ configurations in collisions of 12 MeV $C^{4+}(1s^2\ ^1S, 1s2s\ ^{1,3}S)$ mixed-state beams with He gas targets. Blue filled circles: electron spectra obtained from the high f_{3S} beam fraction. Red filled squares: electron spectra obtained from the low f_{3S} beam fraction

6.3 Cusp Electrons

The APAPES installation has been primarily developed for ZAPS studies offering high resolution and enhanced efficiency measurements. However, its operation in low-resolution mode has recently been reported in measurements of cusp electrons, i.e., electrons that are emitted with velocities distributed around the projectile velocity [135]. The cusp peak distribution is much wider than the active energy window of the HDA spectrograph, which is about 20% of the tuning electron energy. Thus, several energy windows, obtained at the appropriate tuning energies, are necessary to be overlapped to obtain the cusp peak. An example of piecing together four overlapping energy windows is shown in Fig. 14a for the collision system 1.5 MeV p + He.

He-like mixed-state beams offer the grounds to study the processes that contribute to the formation of the cusp peak in collisions of multi-electron open-shell ions with gas targets. For MeV/u collisions, the cusp contributing processes are: (i) The electron capture to the continuum (ECC) where a target electron is captured to the continuum of the projectile [136–139] and (b) the electron loss to the continuum (ELC) where a target electron is lost to the continuum of the projectile [140–142]. Both ECC and ELC mechanisms are described within the concept of continuum distorted-wave (CDW) theories (see Ref. [143] and references therein), and their application to collisions of multi-electron open-shell ions is expected to be challenging. In Fig. 14b, we show a set of first measurements on cusp electrons resulting from mixed-state $O^{4+}(1s^2, 1s2s)$ beams with different metastable fraction f_{1s2s}, in collisions with He targets. The goal is to apply our two-measurement technique and extract the cusp peaks corresponding only to the $(1s2s)$ metastable part of the ion beam.[5]

[5] While this chapter was finalized, results of our work on this subject have been published in Ref. [144].

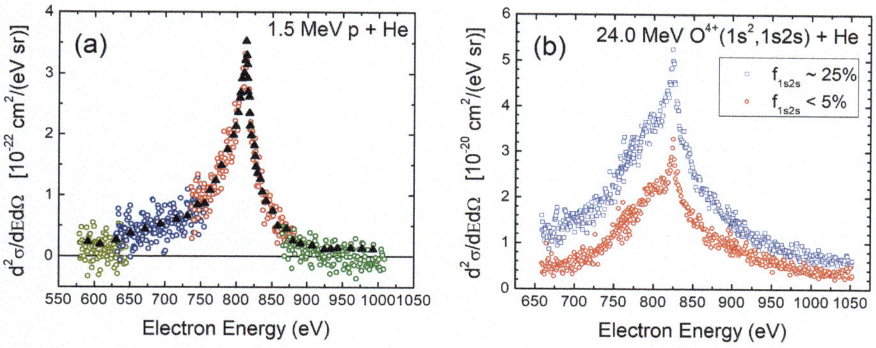

Fig. 14 **a** Zero-degree DDCS electron spectra for collisions of 1.5 MeV protons with He gas targets. Multicolored circles: the four overlapping energy windows corresponding to the tuning energies of W = 585, 685, 800, and 940 eV, covering the cusp peak. Black filled triangles: A weighted average of the data of the four overlapping energy windows. **b** Cusp electron spectra obtained in collisions of 24 MeV $O^{4+}(1s^2, 1s2s)$ mixed-state beam with He targets. Blue open circles: High value metastable fraction, $f_{1s2s} \sim 25\%$. Red open circles: Low value metastable fraction, $f_{1s2s} < 5\%$

6.3.1 Coincidences

Recently, the APAPES setup had been upgraded by including the option of performing coincidence measurements. The beamline was extended beyond the last Faraday cup (see Fig. 4a), to include a charge state selection magnet and related hardware detection components. Thus, the electron spectrum can be measured in coincidence with the change in the projectile charge state. This operation mode is appropriate for studying processes such as ELC, which now can be discriminated from ECC in cusp electron measurements. Such coincidence studies have been reported in the literature [138, 142, 145–147]; however, in our case, we plan to combine the coincidence measurements with the two-measurement technique, thus obtaining only the ELC contribution from the $(1s2s)$ part of the He-like $(1s^2, 1s2s)$ mixed-state ion.

7 Concluding Remarks

In this chapter, we summarized our recent investigations and results obtained in collisions of fast mixed-state ion beams with gas targets. We presented the experimental technology and techniques developed for these studies, as well as the recent theoretical concepts for interpreting the findings. Primarily we focused on the study of the fundamental processes of single electron capture and transfer excitation. Our results in single electron capture invalidated the frozen core approximation and related spin statistics populations, additionally proposing a distinctive screening effect due to the Pauli exclusion principle (Pauli shielding). In transfer excitation, the 3eAOCC theory filled the gap of a long awaited coherent treatment of the TE process that includes its resonant and nonresonant features, showing excellent agreement with the measurements at the resonant part, and proposing a new nonresonant one-step TE mechanism. These advances were made possible not only from the progress in ion-atom collision theories, but also due to improvements of the experimental techniques that exploit mixed-state beams. Our advances show that fast ion-atom collision is a research field that still provides invaluable information about fundamental atomic processes, promoting the understanding of the physics in these conditions. It is worth mentioning that, fast ion-atom technology and related techniques extend their use to closely related fields, as, for example, in the investigations of Young-type experiments [148, 149], water molecules [150, 151], biomolecules [152], fullerenes, and plasmons [153, 154].

Acknowledgements The work reported here has received extended support from the projects: (a) the European Union (European Social Fund (ESF)) and Greek national funds through the Operational Program "Education and Lifelong Learning" of the National Strategic Reference Framework (NSRF) Research Funding Program: THALES. Investing in knowledge society through the European Social Fund, grant number MIS 377289. (b) "Cluster of Accelerator Laboratories for Ion-Beam Research and Applications–CALIBRA" (No. MIS 5002799), which is implemented under the Action "Reinforcement of the Research and Innovation Infrastructure," funded by the Operational Programme "Competitiveness, Entrepreneurship and Innovation" (NSRF 2014–2020) and co-financed by Greece and the European Union (European Regional Development Fund).
I would like to express my gratitude to Prof. Theo Zouros for his continuous support all these years, and mostly for his heroic efforts in developing the ZAPS setup both in the J.R. Macdonald and "Demokritos" laboratories. I am also thankful to Prof. Alain Dubois for valuing our work and providing his 3eAOCC calculations that made the interpretation of our experimental findings possible. I am also indebted to the personnel of J. R. Macdonald and "Demokritos" laboratories for the excessive support all these years. Last but not least, I would like to express my thanks to post-doc Anastasios Dimitriou, PhD students Ioannis Madesis, Angelos Laoutaris, and Stefanos Nanos, as well as MSc student Andreas Biniskos, whose hard work and devotion made the APAPES project a reality.

References

1. M.J.J. Vrakking, F. Lepine, eds., *Attosecond Molecular Dynamics*. Theoretical and Computational Chemistry Series (The Royal Society of Chemistry, 2019)
2. Z. Tao, C. Chen, T. Szilvási, M. Keller, M. Mavrikakis, H. Kapteyn, M. Murnane, Direct time-domain observation of attosecond final-state lifetimes in photoemission from solids. Science **353**(6294), 62–67 (2016)
3. B. Vodungbo, B. Tudu, J. Perron, R. Delaunay, L. Müller, M.H. Berntsen, G. Grübel, G. Malinowski, C. Weier, J. Gautier, G. Lambert, P. Zeitoun, C. Gutt, E. Jal, A.H. Reid, P.W. Granitzka, N. Jaouen, G.L. Dakovski, S. Moeller, M.P. Minitti, A. Mitra, S. Carron, B. Pfau, C. von Korff Schmising, M. Schneider, S. Eisebitt, J. Lüning, Indirect excitation of ultrafast demagnetization. Sci. Rep. **6**, 18970 (2016)
4. N. Stolterfoht, High resolution Auger spectroscopy in energetic ion atom collisions. Phys. Rep. **146**, 315–424 (1987)
5. T.J.M. Zouros, D.H. Lee, Zero degree Auger electron spectroscopy of projectile ions, in *Accelerator-Based Atomic Physics: Techniques and Applications*, ed. by S.M Shafroth, J.C. Austin, Chapter 13 (AIP, Woodbury, NY, 1997), pp. 426–479

6. I. Madesis, A. Laoutaris, T.J.M. Zouros, E.P. Benis, J.W. Gao, A. Dubois, Pauli Shielding and Breakdown of Spin Statistics in Multielectron Multi-Open-Shell Dynamical Atomic Systems. Phys. Rev. Lett. **124**, 113401 (2020) (and Supplemental Material)
7. D.H. Lee, P. Richard, J.M. Sanders, T.J.M. Zouros, J.L. Shinpaugh, S.L. Varghese, Electron capture and excitation studied by state-resolved KLL Auger measurement in 0.25–2 MeV/u $F^{7+}(1s^2\ ^1, 1s2s^3S)$ + H_2/He collisions. Nucl. Instrum. Methods Phys. Res. Sect. B **56–57**, 99–103 (1991)
8. J.A. Tanis, A.L. Landers, D.J. Pole, A.S. Alnaser, S. Hossain, T. Kirchner, Evidence for Pauli exchange leading to excited-state enhancement in electron transfer. Phys. Rev. Lett. **92**(13), 133201 (2004)
9. I. Madesis, A. Laoutaris, S. Nanos, S. Passalidis, A. Dubois, T.J.M. Zouros, E.P. Benis, State-resolved differential cross sections of single electron capture in swift collisions of $C^{4+}(1s2s^3S)$ ions with gas targets. Phys. Rev. A **105**, 062810 (2022)
10. M. Mack, A. Niehaus, Double electron capture by He-like ions: collision energy dependence of the reaction window. Nucl. Instrum. Methods Phys. Res. Sect. B **23**(1–2), 116–119 (1987)
11. M. Mack, A. Niehaus, Radiative and Auger decay channels in K-shell excited Li-like ions (Z = 6–8). Nucl. Instrum. Methods Phys. Res. Sect. B **23**(1–2), 109 – 115 (1987)
12. T.J.M. Zouros, D.H. Lee, P. Richard, Projectile 1s→2p Excitation Due to Electron-Electron Interaction in Collisions of O^{5+} and F^{6+} Ions with He and H_2 Targets. Phys. Rev. Lett. **62**, 2261–2264 (1989)
13. T.J.M. Zouros, D.H. Lee, J.M. Sanders, J.L. Shinpaugh, T.N. Tipping, S.L. Varghese, P. Richard, High resolution studies of electron capture and excitation by 0° projectile electron spectroscopy. Nucl. Instrum. Methods Phys. Res. Sect. B **40/41**, 17 (1989)
14. T.J.M. Zouros, D.H. Lee, P. Richard, J.M. Sanders, J.L. Shinpaugh, S.L. Varghese, K.R. Karim, C.P. Bhalla, State-selective observation of Resonance Transfer Excitation (RTE) in collisions of F^{6+} with He and H_2 targets. Phys. Rev. A **40**, 6246 (1989)
15. D.H. Lee, P. Richard, J.M. Sanders, T.J.M. Zouros, J.L. Shinpaugh, S.L. Varghese, KLL resonant transfer and excitation to $F^{6+}(1s2l2l')$ intermediate states. Phys. Rev. A **44**, 1636–1643 (1991)
16. T.J.M. Zouros, Resonant Transfer and Excitation associated with Auger electron emission, in *Recombination of Atomic Ions*, ed. by W.G. Graham, W. Fritsch, Y. Hahn, J. Tanis, vol. 296, New York, 6–9 Oct. 1991 (NATO Advanced Study Institute Series B: Physics, Plenum Publishing Corporation, 1992), pp. 271–300
17. A. Laoutaris, S. Nanos, I. Madesis, S. Passalidis, E.P. Benis, A. Dubois, T.J.M. Zouros, Coherent treatment of transfer-excitation processes in swift ion-atom collisions. Phys. Rev. A **106**, 022810 (2022)
18. E.P. Benis, T.J.M. Zouros, T.W. Gorczyca, M. Zamkov, P. Richard, Isoelectronic study of triply excited Li-like states. J. Phys. B **36**, L341–L348 (2003)
19. P.A. Závodszky, H. Aliabadi, C.P. Bhalla, P. Richard, G.T.óth, J.A. Tanis, Superelastic scattering of electrons from highly charged ions with inner shell vacancies. Phys. Rev. Lett. **87**, 033202 (2001)
20. A.S. Alnaser, A.L. Landers, D.J. Pole, S. Hossain, O.A. Haija, T.W. Gorczyca, J.A. Tanis, H. Knutson, Supereleastic scattering of electrons from metastable He-like C^{4+} and O^{6+} ions. Phys. Rev. A **65**, 042709 (2002)
21. A. Müller, E. Lindroth, S. Bari, A. Borovik, P.-M. Hillenbrand, K. Holste, P. Indelicato, A.L.D. Kilcoyne, S. Klumpp, M. Martins, J. Viefhaus, P. Wilhelm, S. Schippers, Photoionization of metastable heliumlike $C^{4+}(1s2s\ ^3S_1)$ ions: precision study of intermediate doubly excited states. Phys. Rev. A **98**, 033416 (2018)
22. A. Borovik Jr., A. Müller, S. Schippers, I. Bray, D.V. Fursa, Electron impact ionization of ground-state and metastable Li^+ ions. J. Phys. B **42**(2), 025203 (2009)
23. A.C. Renwick, I. Bray, D.V. Fursa, J. Jacobi, H. Knopp, S. Schippers, A.M.üller, Electron-impact ionization of B^{3+} ions. J. Phys. B **42**(17), 175203 (2009)
24. T. Schlummer, O. Marchuk, D.R. Schultz, G. Bertschinger, W. Biel, D. Reiter, T. Textor-Team, Comparison of effective rate coefficients for high energy charge-exchange with measurements of the Rydberg series of Ar^{16+} at the tokamak textor. J. Phys. B **48**(14), 144033 (2015)

25. Z.Y. Cui, S. Morita, H.Y. Zhou, X.T. Ding, P. Sun, M. Kobayashi, X.W. Cui, Y. Xu, X.L. Huang, Z.B. Shi, J. Cheng, Y.G. Li, B.B. Feng, S.D. Song, L.W. Yan, Q.W. Yang, X.R. Duan, Enhancement of edge impurity transport with ECRH in the HL-2A tokamak. Nucl. Fus. **53**(9), 093001 (2013)
26. L. Liu, D. Jakimovski, J.G. Wang, R.K. Janev, Electron capture and excitation in H^+ - He (1s2s; ^{1,3}S) collisions. J. Phys. B **45**(22), 225203 (2012)
27. M. Trassinelli, C. Prigent, E. Lamour, F. Mezdari, J. Mérot, R. Reuschl, J.-P. Rozet, S. Steydli, D. Vernhet, Investigation of slow collisions for (quasi) symmetric heavy systems: what can be extracted from high resolution x-ray spectra. J. Phys. B **45**(8), 085202 (2012)
28. T. Nandi, M. Oswal, S. Kumar, A. Jhingan, S.R. Abhilash, S. Karmakar, Radiative resonant energy transfer: a new excitation process of beam-foil interaction. J. Quant. Spectrosc. Rad. Tran. **113**(10), 783–788 (2012)
29. Y.-C. Lin, Y.K. Ho, Quantum entanglement for two electrons in the excited states of Helium-like systems **93**(6), 646–653 (2015)
30. A. Itoh, T. Schneider, G. Schiwietz, Z. Roller, H. Platten, G. Nolte, D. Schneider, N. Stolterfoht, Selective production of Auger electrons from fast projectile ions studied by zero-degree Auger spectroscopy. J. Phys. B **16**(21), 3965–3971 (1983)
31. N. Stolterfoht, Zero-degree Auger spectroscopy in energetic, highly charged ion-atom collisions. J. Electron Spectrosc. and Relat. Phenom. **67**(2), 309–339 (1994)
32. S. Doukas, I. Madesis, A. Dimitriou, A. Laoutaris, T.J.M. Zouros, E.P. Benis, Determination of the solid angle and response function of a hemispherical spectrograph with injection lens for Auger electrons emitted from long lived projectile states. Rev. Sci. Instrum. **86**(4), 043111 (2015)
33. A. Itoh, D. Schneider, T. Schneider, T.J.M. Zouros, G. Nolte, G. Schiwietz, W. Zeitz, N. Stolterfoht, Selective production of Li, Be and B-like K vacancy states in fast Ne projectiles studied by zero-degree Auger spectroscopy. Phys. Rev. A **31**, 684 (1985)
34. D.H. Lee, T.J.M. Zouros, J.M. Sanders, J.L. Shinpaugh, T.N. Tipping, S.L. Varghese, B.D. DePaola, P. Richard, A zero-degree tandem electron spectrometer and Rydberg analyzer for projectile electron studies in ion-atom collisions. Nucl. Instrum. Methods Phys. Res. Sect. B **40–41**, 1229 (1989)
35. S. Harissopulos, M. Andrianis, M. Axiotis, A. Lagoyannis, A.G. Karydas, Z. Kotsina, A. Laoutaris, G. Apostolopoulos, A. Theodorou, T.J.M. Zouros, I. Madesis, E.P. Benis, The Tandem Accelerator Laboratory of NCSR Demokritos: current status and perspectives. Eur. Phys. J. Plus **136**, 617 (2021)
36. APAPES Atomic Physics with Accelerators: Projectile Electron Spectroscopy. http://apapes.physics.uoc.gr
37. E.P. Benis, K. Zaharakis, M.M. Voultsidou, T.J.M. Zouros, M. Stöckli, P. Richard, S. Hagmann, A new hemispherical analyser with 2-D PSD and focusing lens for use in 0° electron spectroscopy. Nucl. Instrum. Methods Phys. Res. B **146**, 120–5 (1998)
38. E.P. Benis, T.J.M. Zouros, Improving the energy resolution of a hemispherical spectrograph using a paracentric entry at a non-zero potential. Nucl. Instrum. Methods Phys. Res. A **440**, 462–465 (2000)
39. T.J.M. Zouros, E.P. Benis, The hemispherical deflector analyser revisited. I. motion in the ideal $1/r$ potential, generalized entry conditions, Kepler orbits and spectrometer basic equation. J. Electron Spectrosc. Relat. Phenom. **125**, 221–248 (2002)
40. E.P. Benis, T.J.M. Zouros, The hemispherical deflector analyser revisited. *II. E*lectron-optical properties. J. Electron Spectrosc. Relat. Phenom. **163**, 28–39 (2008)
41. M. Dogan, M. Ulu, G.G. Gennarakis, T.J.M. Zouros, Experimental energy resolution of a paracentric hemispherical deflector analyzer for different entry positions and bias. Rev. Sci. Instrum. **84**(4), 043105 (2013)
42. T.J.M. Zouros, E.P. Benis, Optimal energy resolution of a hemispherical analyzer with virtual entry. Appl. Phys. Lett. **86**, 094105 (2005)
43. O. Sise, G. Martinez, I. Madesis, A. Laoutaris, A. Dimitriou, M. Fernandez-Martin, T.J.M. Zouros, The voltage optimization of a four-element lens used on a hemispherical spectrograph

with virtual entry for highest energy resolution. J. Electron Spectrosc. and Relat. Phenom. **211**, 19–31 (2016)
44. D.H. Lee, P. Richard, T.J.M. Zouros, J.M. Sanders, J.L. Shinpaugh, H. Hidmi, Binary encounter electrons observed at zero degrees in collisions of 1–2 Me V/amu H^+, C^{6+}, N^{7+}, O^{8+} and F^{9+}. Phys. Rev. A **41**, 4816–4823 (1990)
45. C.W. Woods, R.L. Kauffman, K.A. Jamison, N. Stolterfoht, P. Richard, K-shell Auger-electron production cross sections from ion bombardment. Phys. Rev. A **13**(4), 1358–1369 (1976)
46. E.P. Benis, S. Doukas, T.J.M. Zouros, P. Indelicato, F. Parente, C. Martins, J.P. Santos, J.P. Marques, Evaluation of the effective solid angle of a hemispherical deflector analyser with injection lens for metastable Auger projectile states. Nucl. Instrum. Methods Phys. Res. Sect. B **365**, 457 - 461 (2015)
47. http://www.simion.com
48. E.P. Benis, I. Madesis, A. Laoutaris, S. Nanos, T.J.M. Zouros, Experimental determination of the effective solid angle of long-lived projectile states in zero-degree Auger projectile spectroscopy. J. Electron Spectrosc. Relat. Phenom. **222**, 31–39 (2018)
49. N. Stolterfoht, P.D. Miller, H.F. Krause, Y. Yamazaki, J.K. Swenson, R. Bruch, P.F. Dittner, P.L. Pepmiller, S. Datz, Surgery of fast, highly charged ions studied by zero-degree Auger spectroscopy. Nucl. Instrum. Methods Phys. Res. Sect. B **24/25**, 168–172 (1987)
50. E.P. Benis, S. Doukas, T.J.M. Zouros, Evidence for the non-statistical population of the $1s2s2p\,^4P$ metastable state by electron capture in 4 MeV collisions of $B^{3+}(1s2s\,^3S)$ with H_2 targets. Nucl. Instrum. Methods Phys. Res. Sect. B **369**, 83–86 (2016)
51. T.J.M. Zouros, S. Nikolaou, I. Madesis, A. Laoutaris, S. Nanos, A. Dubois, E.P. Benis, Radiative cascade repopulation of $1s2s2p\,^4P$ states formed by single electron capture in 2–18 Me V collisions of $C^{4+}(1s2s\,^3S)$ with He. Atoms **8**(3), 61 (2020)
52. H.D. Betz, Charge states and charge-changing cross sections of fast Heavy ions penetrating through gaseous and solid media. Rev. Mod. Phys. **44**, 465–539 (1972)
53. R.O. Sayer, Semi-empirical formulas for heavy-ion stripping data. Rev. Phys. Appl. (Paris) **12**(10), 1543–1546 (1977)
54. J.P. Rozet, C. Stéphan, D. Vernhet. ETACHA: a program for calculating charge states at GANIL energies. Nucl. Instrum. Methods Phys. Res. Sect. B **107**(1–4), 67–70 (1996)
55. V.S. Nikolaev, I.S. Dmitriev, Semi-empirical method for the calculation of the equilibrium distribution of charges in a fast-ion beam. Soviet Physics J. Exptl. Theoret. Phys. (U.S.S.R) **20**(2), 409–415 (1965)
56. K. Shima, N. Kuno, M. Yamanouchi, H. Tawara, Equilibrium charge fractions of ions of Z = 4–92 emerging from a carbon foil. At. Data Nucl. Data Tables **51**, 173–241 (1992)
57. G. Schiwietz, P.L. Grande, Improved charge-state formulas. Nucl. Instrum. Methods Phys. Res. Sect. B **175–177**, 125–131 (2001)
58. E.-M. Asimakopoulou, TARDIS (Transmitted chARge DIStribution). Technical report, Institute of Nuclear and Particle Physics, Demokritos and Department of Physics, University of Athens (2014)
59. L.H. Andersen, J. Bolbo, P. Kvistgaard, State-selective dielectronic-recombination measurements for He- and Li-like carbon and oxygen ions. Phys. Rev. A **41**, 1293–1302 (1990)
60. L.H. Andersen, G.Y. Pan, H.T. Schmidt, N.R. Badnell, M.S. Pindzola, Absolute measurements and calculations of dielectronic recombination with metastable He- N, F, and Si ions. Phys. Rev. A **45**, 7868–7875 (1992)
61. L.H. Andersen, P. Hvelplund, H. Knudsen, P. Kvistgaard, State-selective dielectronic-recombination measurements for He-like oxygen ions in an electron cooler. Phys. Rev. Lett. **62**, 2656 (1989)
62. G. Kilgus, D. Habs, D. Schwalm, A. Wolf, R. Schuch, N.R. Badnell, Dielectronic recombination from ground state of heliumlike carbon ions. Phys. Rev. A **47**, 4859 (1993)
63. G.W.F. Drake, G.A. Victor, A. Dalgarno, Two-photon decay of the singlet and triplet metastable states of Helium-like ions. Phys. Rev. **180**, 25–32 (1969)

64. G.W.F. Drake, Theory of relativistic magnetic dipole transitions: lifetime of the metastable $2\,^3S$ state of the Helium-like ions. Phys. Rev. A **3**, 908–915 (1971)
65. C. Froese Fischer, G. Gaigalas, Note on the $2s^2\ ^1S_0$–$2s2p^3P_1$ intercombination line of B II and C III. Phys. Scr. **56**(5), 436 (1997)
66. C. Froese Fischer, Multiconfiguration Dirac-Hartree-Fock calculations for Be-like intercombination lines revisited. Phys. Scr. **62**(6), 458 (2000)
67. E.P. Benis, I. Madesis, A. Laoutaris, S. Nanos, T.J.M. Zouros, Mixed-state ionic beams: an effective tool for collision dynamics investigations. Atoms **6**(4), 66 (2018)
68. U. Schiebel, B.L. Doyle, J.R. Macdonald, L.D. Ellsworth, Projectile K x rays from Si^{12+} ions in the $1s2s\ ^3S_1$ metastable state incident on helium gas. Phys. Rev. A **16**, 1089 (1977)
69. M. Terasawa, T.J. Gray, S. Hagmann, J. Hall, J. Newcomb, P. Pepmiller, P. Richard, Electron capture by and electron excitation of two-electron fluorine ions incident on helium. Phys. Rev. A **27**, 2868–2875 (1983)
70. E.H. Pedersen, Metastable-atom population of fast, neutral Helium beams. Phys. Rev. Lett. **42**(7), 440–443 (1979)
71. A. Müller, A. Borovik, K. Huber, S. Schippers, D.V. Fursa, I. Bray, Double − K − vacancy states in electron-impact single ionization of metastable two-electron $N^{5+}(1s2s\ ^3S_1)$ ions. Phys. Rev. A **90**, 010701 (2014)
72. A. Müller, A. Borovik, T. Buhr, J. Hellhund, K. Holste, A.L.D. Kilcoyne, S. Klumpp, M. Martins, S. Ricz, J. Viefhaus, S. Schippers, Near- K-edge single, double, and triple photoionization of C^+ ions. Phys. Rev. A **97**, 013409 (2018)
73. E.P. Benis, M. Zamkov, P. Richard, T.J.M. Zouros, Technique for the determination of the $1s2s\ ^3S$ metastable fraction in two-electron ion beams. Phys. Rev. A **65**, 064701 (2002)
74. M. Zamkov, H. Aliabadi, E.P. Benis, P. Richard, H. Tawara, T.J.M. Zouros, Energy dependence of the metastable fraction in $B^{3+}(1s^2\ ^1S, 1s2s\ ^3S)$ beams produced in collisions with thin-foil and gas targets. Phys. Rev. A **64**, 052702 (2001)
75. M. Zamkov, E.P. Benis, P. Richard, T.J.M. Zouros, Fraction of metastable $1s2s^3$ S ions in fast He-like beams ($Z = 5 - -9$) produced in collisions with carbon foils. Phys. Rev. A **65**, 062706 (2002)
76. P. Hvelplund, Energy loss and straggling of 100–500-ke V ^{90} Th, ^{82} Pb, ^{80} Hg and ^{64} Gd in H_2. Phys. Rev. A **11**, 1921–1927 (1975)
77. D.H. Lee, T.J.M. Zouros, J.M. Sanders, P. Richard, J.M. Anthony, Y.D. Wang, J.H. McGuire, K-shell ionization of O^{4+} and C^{2+} ions in fast collisions with H_2 and He gas targets. Phys. Rev. A **46**, 1374–1387 (1992)
78. E.P. Benis, T.J.M. Zouros, Determination of the $1s2\ell2\ell'$ state production ratios $^4P^o/^2$P, $^2D/^2$ P and $^2P_+/^2P_-$ from fast $(1s^2, 1s2s^3S)$ mixed-state He-like ion beams in collisions with H_2 targets. J. Phys. B **49**(23), 235202 (2016)
79. C.T. Plowman, K.H. Spicer, I.B. Abdurakhmanov, A.S. Kadyrov, I. Bray, Singly differential cross sections for direct scattering, electron capture, and ionization in proton-hydrogen collisions. Phys. Rev. A **102**, 052810 (2020)
80. J.W. Gao, Y. Wu, J.G. Wang, N. Sisourat, A. Dubois, State-selective electron transfer in He^+ + He collisions at intermediate energies. Phys. Rev. A **97**, 052709 (2018)
81. D.L. Guo, J.W. Gao, S.F. Zhang, X.L. Zhu, Y. Gao, D.M. Zhao, R.T. Zhang, Y. Wu, J.G. Wang, A. Dubois, X. Ma, State-selective single-electron capture in intermediate-energy $C^{4+} + He$ collisions. Phys. Rev. A **103**, 032827 (2021)
82. S. Knoop, M. Keim, H.J. Lüdde, T. Kirchner, R. Morgenstern, R. Hoekstra, State selective single-electron capture in $O^{6+}+$ Na collisions. J. Phys. B **38**(17), 3163–3172 (2005)
83. D. Röhrbein, T. Kirchner, S. Fritzsche, Role of cascade and Auger effects in the enhanced population of the $C^{3+}(1s2s2p^4\,P)$ states following single-electron capture in $C^{4+}(1s2s^3S)$-He collisions. Phys. Rev. A **81**(4), 042701 (2010)
84. L.F. Errea, B. Herrero, L. Mendez, A. Riera, Molecular calculation of total and differential charge transfer cross sections in $C^{4+}+$ He collisions. J. Phys. B **28**(4), 693–709 (1995)
85. D.R. Schultz, M.R. Strayer, J.C. Wells, Lattice, time-dependent Schrödinger equation solution for ion-atom collisions. Phys. Rev. Lett. **82**, 3976–3979 (1999)

86. D.L. Guo, X. Ma, R.T. Zhang, S.F. Zhang, X.L. Zhu, W.T. Feng, Y. Gao, B. Hai, M. Zhang, H.B. Wang, Z.K. Huang, State-selective electron capture in 30- and 100-ke V He^{+}+ He collisions. Phys. Rev. A **95**, 012707 (2017)
87. A.B. Voitkiv, J. Ullrich, Three-body Coulomb dynamics in hydrogen ionization by protons and antiprotons at intermediate collision velocities. Phys. Rev. A **67**, 062703 (2003)
88. A.S. Schlachter, J.W. Stearns, W.G. Graham, K.H. Berkner, R.V. Pyle, J.A. Tanis, Electron capture for fast highly charged ions in gas targets: an empirical scaling rule. Phys. Rev. A **27**, 3372–3374 (1983)
89. M. Barat, P Roncin, Multiple electron capture by highly charged ions at keV energies. J. Phys. B **25**(10), 2205–2243 (1992)
90. E.Y. Kamber, O. Abu-Haija, J.A. Wardwell, State-selective electron capture by O^{3+} ions from atomic and molecular targets. Phys. Rev. A **77**, 012701 (2008)
91. D. Dijkkamp, A. Brazuk, A.G. Drentje, F.J. de Heer, H. Winter, State-selective single-electron capture by 80 ke V C^{4+} ions from He, H_2 and Li. J. Phys. B **16**(11), L343 (1983)
92. D. Strohschein, D. Röhrbein, T. Kirchner, S. Fritzsche, J. Baran, J.A. Tanis, Nonstatistical enhancement of the $1s2s2p^4P$ state in electron transfer in 0.5–1.0-MeV/u $C^{4,5+}$ + He and Ne collisions. Phys. Rev. A **77**(2), 022706 (2008)
93. T.J.M. Zouros, B. Sulik, L. Gulyás, K. Tökési, Selective enhancement of $1s2s2p^4P_J$ metastable states populated by cascades in single-electron transfer collisions of $F^{7+}(1s^2/1s2s^3S)$ ions with He and H_2 targets. Phys. Rev. A **77**(5), 050701 (2008)
94. M.A. Abdallah, W. Wolff, H.E. Wolf, E.Y. Kamber, M. Stöckli, C.L. Cocke, Single and double electron capture from He by Ar^{16+} studied using cold-target recoil-ion momentum spectroscopy. Phys. Rev. A **58**, 2911–2919 (1998)
95. E.Y. Kamber, M.A. Abdallah, C.L. Cocke, M. Stöckli, State-selective electron capture in (0.5–50)-keV/u $C^{5+} - He$ collisions studied by cold-target recoil-ion momentum spectroscopy. Phys. Rev. A **60**, 2907–2916 (1999)
96. D. Fischer, B. Feuerstein, R.D. DuBois, R. Moshammer, J.R. Crespo Lopez-Urrutia, I. Draganic, H. Lorch, A.N. Perumal, J. Ullrich, State-resolved measurements of single-electron capture in slow Ne^{7+}- and Ne^{8+}-helium collisions. J. Phys. B **35**, 1369 (2002)
97. S. Knoop, D. Fischer, Y. Xue, M. Zapukhlyak, C.J. Osborne, T. Ergler, T. Ferger, J. Braun, G. Brenner, H. Bruhns, C. Dimopoulou, S.W. Epp, A.J. González Martínez, G. Sikler, R. Soria Orts, H. Tawara, T. Kirchner, J.R. Crespo López-Urrutia, R. Moshammer, J. Ullrich, R. Hoekstra, Single-electron capture in keV $Ar^{15+...18+}$ +He collisions. J. Phys. B **41**(19), 195203 (2008)
98. M.S. Schöffler, J. Titze, LPh.H. Schmidt, T. Jahnke, N. Neumann, O. Jagutzki, H. Schmidt-Böcking, R. Dörner, I. Mančev, State-selective differential cross sections for single and double electron capture in $He^{+,2+} - He$ and $p - He$ collisions. Phys. Rev. A **79**, 064701 (2009)
99. D.L. Guo, X. Ma, S.F. Zhang, X.L. Zhu, W.T. Feng, R.T. Zhang, B. Li, H.P. Liu, S.C. Yan, P.J. Zhang, Q. Wang, Angular- and state-selective differential cross sections for single-electron capture in p-He collisions at intermediate energies. Phys. Rev. A **86**, 052707 (2012)
100. P. Beiersdorfer, J.H. Scofield, A.L. Osterheld, X-ray-line diagnostic of magnetic field strength for high-temperature plasmas. Phys. Rev. Lett. **90**, 235003 (2003)
101. T.R. Kallman, P. Palmeri, Atomic data for x-ray astrophysics. Rev. Mod. Phys. **79**(1), 79 (2007)
102. V.P. Shevelko, H. Tawara (eds.), *Atomic Processes in Basic and Applied Physics* (Springer, Berlin, Heidelberg, New York, 2012)
103. E.P. Benis, T.J.M. Zouros, T.W. Gorczyca, A.D. González, P. Richard, Elastic resonant and nonresonant differential scattering of quasifree electrons from $B^{4+}(1s)$ and $B^{3+}(1s^2)$ ions. Phys. Rev. A **69**, 052718 (2004). Erratum, Phys. Rev. A **73**, 029901 (2006)
104. W. Mehlhorn, K. Taulbjerg, Angular distribution of electrons from autoionising states with unresolved fine structure. J. Phys. B **13**(3), 445 (1980)
105. N. Sisourat, I. Pilskog, A. Dubois, Non perturbative treatment of multielectron processes in ion-molecule scattering: application to He^{2+}- H_2 collisions. Phys. Rev. A **84**, 052722 (2011)

106. N. Sisourat, A. Dubois, Semiclassical close-coupling approaches, in *Ion-Atom Collision-The Few-Body Problem in Dynamic Systems*, ed. by M. Schultz (de Gruyter, Berlin/Boston, 2019), pp. 157–178
107. L.J. Curtis, A diagrammatic mnemonic for calculation of cascading level populations. Am. J. Phys. **36**, 1123–1125 (1968)
108. S.M. Younger, W.L. Wiese, Theoretical simulation of beam-foil decay curves for resonance transitions of heavy ions. Phys. Rev. A **17**, 1944–1955 (1978)
109. E.P. Benis, I. Madesis, A. Laoutaris, S. Nikolaou, A. Dubois, T.W. Gorczyca, T.J.M. Zouros, Population of the $1s1s(^3S)nl\,^2L$ states in collisions of mixed-state $(1s^2\,^1S,\ 1s2s\,^3S)$ B^{3+} and C^{4+} ion beams with He and H_2 targets. X-Ray Spectrom. **49**, 54–59 (2020)
110. W. Brandt, R. Laubert, Pauli excitation of atoms in collision. Phys. Rev. Lett. **24**, 1037–1040 (1970)
111. W. Brandt, R. Laubert, K-shell ionization cross sections in asymmetric heavy-ion-atom collisions. Phys. Rev. A **11**, 1233–1238 (1975)
112. N. Stolterfoht, A. Mattis, D. Schneider, G. Schiwietz, B. Skogvall, B. Sulik, S. Ricz, Time-ordering effects in K-shell excitations of 170-Me V Ne^{7+} colliding with gas atoms: single excitation. Phys. Rev. A **48**, 2986–94 (1993)
113. F.B. Rosmej, V.A. Astapenko, V.S. Lisitsa, L.A. Vainshtein, Dielectronic recombination in non-LTE plasmas. Matter and Radiation at Extremes **5**(6), 064201 (2020)
114. M. Clark, D. Brandt, J.K. Swenson, S.M. Shafroth, Nonresonant electron transfer and projectile K-electron excitation in Ion-Atom Collisions. Phys. Rev. Lett. **54**, 544–546 (1985)
115. J.A. Tanis, E.M. Bernstein, M.W. Clark, W.G. Graham, R.H. McFarland, T.J. Morgan, B.M. Johnson, K.W. Jones, M. Meron, Evidence for uncorrelated electron capture and K-shell excitation in S^{13+} + He collisions. Phys. Rev. A **31**, 4040 (1985)
116. J.A. Tanis, E.M. Bernstein, W.G. Graham, M. Clark, S.M. Shafroth, B.M. Johnson, K.W. Jones, M. Meron, Resonant behavior in the projectile X-ray yield associated with electron capture in S+ Ar Collisions. Phys. Rev. Lett. **49**, 1325 (1982)
117. P.L. Pepmiller, R. Prichard, J. Newcomb, R. Dillingham, J.M. Hall, T.J. Gray, M. Stöckli, A study of transfer excitation in F^{8+} + He, Ne and Ar collisions. IEEE Trans. Nucl. Sci., NS-**30**, 1002 (1983)
118. D. Brandt, A simple classical model for the impact parameter dependence of electron capture. Nucl. Instrum. Methods **214**, 93 (1983)
119. A. Itoh, T.J.M. Zouros, D. Schneider, U. Stettner, W. Zeitz, N. Stolterfoht, Transfer excitation in He^++ He collisions studied by zero-degree electron spectroscopy. J. Phys. B **18**, 4581–4587 (1985)
120. J.K. Swenson, Y. Yamazaki, P.D. Miller, H.F. Krause, P.F. Dittner, P.L. Pepmiller, S. Datz, N. Stolterfoht, Observation of resonant transfer and excitation to specific LS-Coupled states in O^{5+} + He Collisions by high-resolution, 0° Auger-Electron Spectroscopy. Phys. Rev. Lett. **57**, 3042–3045 (1986)
121. T.J.M. Zouros, C.P. Bhalla, D.H. Lee, P. Richard, Effects of alignment and interference in resonant transfer and excitation for F^{6+} and O^{5+} collisions with H_2 in 0° Auger measurements. Phys. Rev. A **42**, 678–681 (1990)
122. B.D. DePaola, R. Parameswaran, W.J. Axmann, High-resolution state-selective study of transfer with excitation in the F^{8+}+ H_2 system. Phys. Rev. A **41**, 6533–6535 (1990)
123. B.D. DePaola, RTE studied in electron emission. Nucl. Instrum. Methods Phys. Res. Sect. B **56–57**, 154–157 (1991)
124. R. Parameswaran, C.P. Bhalla, B.P. Walch, B.D. DePaola, Resonant transfer and excitation in collisions of C^{5+} with H_2 and He targets. Phys. Rev. A **43**, 5929–5932 (1991)
125. J.M. Anthony, S.M. Shafroth, M. Benhenni, E.N. Strait, T.J.M. Zouros, L.D. Hendrik, D.M. Peterson, Production of doubly core-excited configurations in C^{4+} projectiles through resonant transfer excitation. J. de Phys. (Paris) **48**(C9), 301–306 (1987)
126. M. Benhenni, S.M. Shafroth, J.K. Swenson, M. Schulz, J.P. Giese, H. Schöne, C.R. Vane, P.F. Dittner, S. Datz, Angular distribution of Auger electrons emitted through the resonant transfer and excitation process following O^{5+}+ He collisions. Phys. Rev. Lett. **65**, 1849–1852 (1990)

127. D. Brandt, Resonant transfer and excitation in ion-atom collisions. Phys. Rev. A **27**, 1314–1318 (1983)
128. W. Fritsch, C.D. Lin, Analysis of electron correlation in simultaneous electron transfer and excitation in atomic collisions. Phys. Rev. Lett. **61**, 690–693 (1988)
129. A. Jain, R. Shingal, T.J.M. Zouros, State-selective non-resonant transfer excitation (NTE) in 50–500 ke V ^{3} He^{+}+ H_2 and He collisions. Phys. Rev. A **43**, 1621–1624 (1991)
130. H. Bachau, R. Gayet, J. Hanssen, A. Zerarka, Transfer and excitation in ion-atom collisions at high impact velocities: A unified continuum distorted wave treatment of resonant and non-resonant modes in four body approach: II. Application to the collisions S^{15+}(1s) + H(1s). J. Phys. B **25**, 839–852 (1992)
131. R. Gayet, J. Hanssen, L. Jacqui, Transfer and excitation in ion-atom collisions at high impact velocities: III. Application of the CDW- 4B theory to an almost symmetrical system: He^{+}+He. J. Phys. B **28**, 2193–2208 (1995)
132. R. Gayet, J. Hanssen, L. Jacqui, M.A. Ourdane, Transfer and excitation in ion-atom collisions at high impact velocities: IV. Application of the CDW- 4B theory to an almost symmetrical system: He^{+}+H. J. Phys. B **30**, 2209–2219 (1997)
133. M. Zamkov, H. Aliabadi, E.P. Benis, P. Richard, H. Tawara, T.J.M. Zouros, Absolute cross sections and decay rates for the triply excited $B^{2+}(2s2p^2\,{}^2D)$ resonance in electron-metastable-ion collisions. Phys. Rev. A **65**, 032705 (2002)
134. A. Gumberidze, D.B. Thorn, C.J. Fontes, B. Najjari, H.L. Zhang, A. Surzhykov, A. Voitkiv, S. Fritzsche, D. Banaś, H. Beyer, W. Chen, R.D. DuBois, S. Geyer, R.E. Grisenti, S. Hagmann, M. Hegewald, S. Hess, C. Kozhuharov, R. Märtin, I. Orban, N. Petridis, R. Reuschl, A. Simon, U. Spillmann, M. Trassinelli, S. Trotsenko, G. Weber, D.F.A. Winters, N. Winters, D. Yu, T. Stöhlker, Electron- and proton-impact excitation of hydrogenlike uranium in relativistic collisions. Phys. Rev. Lett. **110**, 213201 (2013)
135. N. Stolterfoht, R.D. Dubois, R.D. Rivarola, *Electron Emission in Heavy Ion-Atom Collisions* (Springer Series on Atoms and Plasmas, Berlin, 1997)
136. G.B. Crooks, M.E. Rudd, Experimental evidence for the mechanism of charge transfer into continuum states. Phys. Rev. Lett. **25**, 1599 (1970)
137. P.D. Fainstein, V.H. Ponce, R.D. Rivarola, Two-centre effects in ionization by ion impact. J. Phys. B **24**, 3091–119 (1991)
138. P.-M. Hillenbrand, S. Hagmann, D.H. Jakubassa-Amundsen, J.M. Monti, D. Banaś, K.-H. Blumenhagen, C. Brandau, W. Chen, P.D. Fainstein, E. De Filippo, A. Gumberidze, D.L. Guo, M. Lestinsky, Yu.A. Litvinov, A. Müller, R.D. Rivarola, H. Rothard, S. Schippers, M.S. Schöffler, U. Spillmann, S. Trotsenko, X.L. Zhu, T. Stöhlker, Electron-capture-to-continuum cusp in U^{88+}+N_2 collisions. Phys. Rev. A **91**, 022705 (2015)
139. S. Nanos, M.A. Quinto, I. Madesis, A. Laoutaris, T.J.M. Zouros, R.D. Rivarola, J.M. Monti, E.P. Benis, Subshell contributions to electron capture into the continuum in Me V/u collisions of deuterons with multielectron targets. Phys. Rev. A **105**, 022806 (2022)
140. D. Burch, H. Wieman, W.B. Ingalls, Electron loss in high-energy oxygen-ion collisions. Phys. Rev. Lett. **30**, 823–826 (1973)
141. W.E. Wilson, L.H. Toburen. Electron emission in H_2^{+}-H_2 collisions from 0.6 to 1.5 MeV. Phys. Rev. A **7**, 1535–1544 (1973)
142. P.-M. Hillenbrand, S. Hagmann, A.B. Voitkiv, B. Najjari, D. Banaś, K.-H. Blumenhagen, C. Brandau, W. Chen, E. De Filippo, A. Gumberidze, D.L. Guo, C. Kozhuharov, M. Lestinsky, Y.A. Litvinov, A. Müller, H. Rothard, S. Schippers, M.S. Schöffler, U. Spillmann, S. Trotsenko, X.L. Zhu, Th. Stöhlker, Electron-loss-to-continuum cusp in U^{89+} + N_2 collisions. Phys. Rev. A **90**, 042713 (2014)
143. J.M. Monti, M.A. Quinto, R.D. Rivarola, A Complete CDW theory for the single ionization of multielectronic atoms by bare ion impact. Atoms **9**(1), 3 (2021)
144. S. Nanos, N.J. Esponda, P.-M. Hillenbrand, A. Biniskos, A. Laoutaris, M.A. Quinto, N. Petridis, E. Menz, T.J.M. Zouros, T. Stöhlker, R.D. Rivarola, J.M. Monti, E.P. Benis, Cusp electrons production in collisions of open-shell He-like oxygen ions with atomic targets. Phys. Rev. A **107**, 062815 (2023)

145. P.-M. Hillenbrand, S. Hagmann, D. Atanasov, D. Banaś, K.-H. Blumenhagen, C. Brandau, W. Chen, E. De Filippo, A. Gumberidze, D.L. Guo, D.H. Jakubassa-Amundsen, O. Kovtun, C. Kozhuharov, M. Lestinsky, Y.A. Litvinov, A. Müller, R.A. Müller, H. Rothard, S. Schippers, M.S. Schöffler, U. Spillmann, A. Surzhykov, S. Trotsenko, N. Winckler, X.L. Yan, V.A. Yerokhin, X.L. Zhu, T. Stöhlker, Radiative-electron-capture-to-continuum cusp in $U^{88+}+N_2$ collisions and the high-energy endpoint of electron-nucleus bremsstrahlung. Phys. Rev. A **90**, 022707 (2014)
146. P.-M. Hillenbrand, S. Hagmann, M.E. Groshev, D. Banaś, E.P. Benis, C. Brandau, E. De Filippo, O. Forstner, J. Glorius, R.E. Grisenti, A. Gumberidze, D.L. Guo, B. Hai, M.O. Herdrich, M. Lestinsky, Yu.A. Litvinov, E.V. Pagano, N. Petridis, M.S. Sanjari, D. Schury, U. Spillmann, S. Trotsenko, M. Vockert, G. Weber, V.A. Yerokhin, T. Stöhlker, Radiative electron capture to the continuum in $U^{89+}+N_2$ collisions: experiment and theory. Phys. Rev. A **101**, 022708 (2020)
147. P.-M. Hillenbrand, K.N. Lyashchenko, S. Hagmann, OYu. Andreev, D. Banaś, E.P. Benis, A.I. Bondarev, C. Brandau, E. De Filippo, O. Forstner, J. Glorius, R.E. Grisenti, A. Gumberidze, D.L. Guo, M.O. Herdrich, M. Lestinsky, Y.A. Litvinov, E.V. Pagano, N. Petridis, M.S. Sanjari, D. Schury, U. Spillmann, S. Trotsenko, M. Vockert, A.B. Voitkiv, G. Weber, T. Stöhlker, Electron-loss-to-continuum cusp in collisions of U^{89+} with N_2 and Xe. Phys. Rev. A **104**, 012809 (2021)
148. N. Stolterfoht, B. Sulik, V. Hoffmann, B. Skogvall, J.Y. Chesnel, J. Rangama, F. Frémont, D. Hennecart, A. Cassimi, X. Husson, A.L. Landers, J.A. Tanis, M.E. Galassi, R.D. Rivarola, Evidence for interference effects in electron emission from H_2 colliding with 60 Me V/u Kr^{34+} ions. Phys. Rev. Lett. **87**, 023201 (2001)
149. M.R. Chowdhury, C.R. Stia, C.A. Tachino, O.A. Fojon, R.D. Rivarola, L.C. Tribedi, Ionization of N_2 in collisions with fast electrons: Evidence of an interference effect. Phys. Rev. A **94**, 052703 (2016)
150. H. Luna, W. Wolff, E.C. Montenegro, André C. Tavares, H.J. Lüdde, G. Schenk, M. Horbatsch, T. Kirchner, Ionization and electron-capture cross sections for single- and multiple-electron removal from H_2O by Li^{3+} impact. Phys. Rev. A **93**, 052705 (2016)
151. Abhijeet Bhogale, S. Bhattacharjee, M. Roy Chowdhury, C. Bagdia, M.F. Rojas, J.M. Monti, A. Jorge, M. Horbatsch, T. Kirchner, R.D. Rivarola, L.C. Tribedi, Electron emission from water vapor under the impact of 250-keV protons. Phys. Rev. A **105**, 062822 (2022)
152. A. Mandal, M.R. Chowdhury, C. Bagdia, J.M. Monti, R.D. Rivarola, P.F. Weck, L.C. Tribedi, Radiobiological effectiveness of iodouracil and the influence of atomic giant resonance. Phys. Rev. A **102**, 062811 (2020)
153. C. Bagdia, A. Bhogale, L. Gulyás, L.C. Tribedi, Giant dipole plasmon resonance in fluorene (C_{13} H_{10}) molecules. Phys. Rev. A **104**, L060802 (2021)
154. S. Kasthurirangan, C.-Z. Gao, P.M. Dinh, L. Gulyás, E. Suraud, L.C. Tribedi, Observation of giant quadrupole plasmon resonance in C_{60} in fast ion collisions. Phys. Rev. A **106**, 012820 (2022)

highly charged impurity tungsten ions, which will be introduced from the divertor and may cause serious radiation loss.

The emission of highly charged ions in the EUV range is useful not only in the diagnostics of plasma but also in industrial applications. For example, strong EUV emission called the unresolved transition array (UTA) of highly charged tin plasma at 13.5 nm is bringing next-generation highly integrated devices [9]. Even higher integration is planned and developed using the radiation at around 6–7 nm of highly charged lanthanide ions [10]. Emission in the further shorter 2–4 nm wavelength region is expected to be used for the "water window" source that realizes living cell microscopy [11].

As a unit of wavelength, in addition to the SI unit "nm", the non-SI unit Å (1 Å = 0.1 nm) is also often used for historical reasons, especially in the astrophysics area.

2 Apparatus

2.1 Electron Beam Ion Trap: EBIT

An electron beam ion trap (EBIT) [12] was developed at the Lawrence Livermore National Laboratory in 1986 based on the principle of an electron beam ion source (EBIS) [13]. An EBIT and an EBIS produce highly charged ions with the same principle and thus have practically the same structure consisting of a Penning-like ion trap and a high-energy, high-density electron beam traveling through the trap as shown in Fig. 2. The ion trap consists of three (or more) successive cylindrical electrodes which form a well potential by applying a positive potential to both ends with respect to the middle electrode. The electron beam is compressed by a strong magnetic field generally produced with a superconducting magnet. The well potential confines ions axially whereas the space charge potential of the compressed electron beam and the strong axial magnetic field achieve radial confinement. Highly charged ions are produced through successive electron impact ionization of the trapped ions. The main difference between an EBIS and an EBIT is the type of magnet. A solenoid type is generally employed in an EBIS whereas an EBIT utilizes a Helmholtz-like split type that enables observations of radiation from the trapped ions through observation slits opened in the middle of the trap electrode.

The first several EBITs [12, 14, 15] were designed to have a maximum energy of about 30 keV. After that several "super" types of EBITs having a higher electron energy of more than 100 keV were built [16–19]. Compact-type EBITs with an electron energy of a few keV or less were also developed by several groups [20–23].

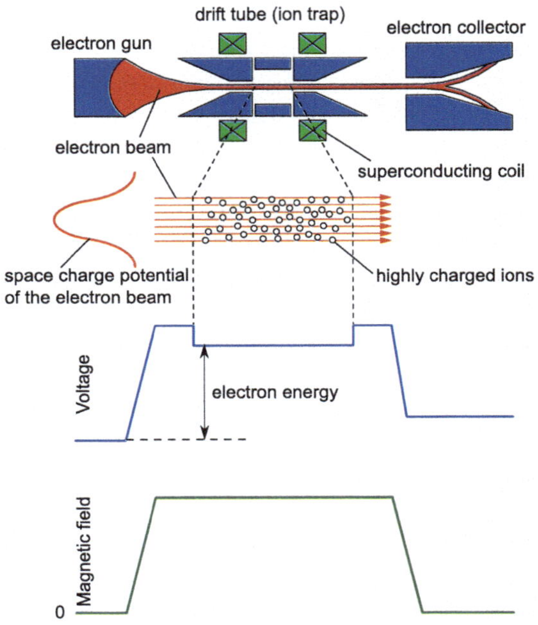

Fig. 2 Principle of an electron beam ion trap

2.2 Spectrometers

Grating spectrometers for the visible range generally have lenses or concave mirrors for light focusing and collimating before and after the grating. However, since there are no effective lenses and normal incidence mirrors in the EUV range, concave gratings that combine the functions of focusing and dispersion are generally used. Such gratings are generally used in the grazing incidence configuration because the diffraction efficiency near the normal incidence is low for the EUV range. The aberration-corrected concave gratings designed by Harada et al. [24–26] are widely used. The variable spacing ruling pattern of their gratings is minutely designed to have a flat focusing plane (flat-field focusing) to use them as a polychromator with an imaging device such as a charge-coupled device (CCD).

Figure 3 [27] shows one of the typical setups for the use of the flat-field grazing incidence concave grating with an EBIT. Generally, wavelength dispersive gratings should be used with an entrance slit. However, for the spectroscopy with an EBIT, gratings can be used without an entrance slit because emission from the trapped highly charged ions is excited by a thin electron beam whose typical diameter of 100 μm or less, which can be regarded as a slit. The slit-less configuration, which is employed for several EBITs [28–31], enables highly efficient observations although configurations with grazing incidence mirrors and a slit are also used [32–34].

As a detector of grating spectrometers, charge-coupled devices (CCD) that have a high sensitivity for the EUV range are frequently used. In ordinary CCD used for

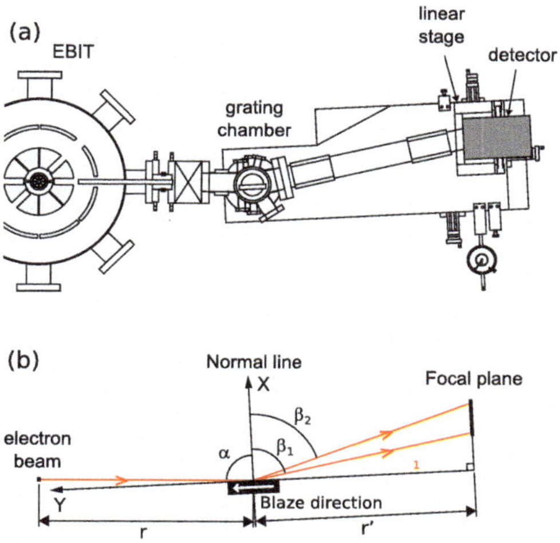

Fig. 3 **a** A cross-sectional diagram of the EUV spectrometer installed at the Tokyo-EBIT. **b** Schematic drawing of the spectrometer arrangement [27]

the visible range, electrodes for collecting and transporting the charge created in the detector are placed on the front (photon incident side) of the sensor. However, the electrodes are generally not transparent for the EUV range; thus back-illuminated types of CCD, in which the electrodes are placed on the back side of the sensor, are generally used for the EUV range. Since there is no effective window for the EUV range, the sensor should be installed in a vacuum. In order to suppress the thermal noise, the detector is generally cooled with a Peltier device or liquid nitrogen. Position-sensitive devices with microchannel plates (MCPs) are also frequently used. Generally, to enhance the sensitivity against EUV photons, the incidence surface of the MCPs is coated with typically alkali halides, such as CsI and KBr.

3 Astrophysics-Related Studies

Iron is the most stable nucleus because it has the largest binding energy per nucleon, i.e., lighter nuclei stabilize by fusion whereas heavier nuclei stabilize by fission. Thus, iron is a highly abundant heavy element in the universe, and emission lines of highly charged iron ions are very important in the spectroscopic diagnostics of astrophysical plasmas. In particular, for the solar corona, whose temperature is in the order of 10^6 K ($\sim 10^2$ eV), EUV emission of iron ions is essential. Since solar observations have a long history, a lot of iron lines have been observed and identified, and their wavelengths are rather well known as compiled in databases [35, 36]. However, the atomic data that determine their line intensity, such as transition probabilities and collisional rates, are not well known in spite of the fact that they are crucial for diagnostics. For example, electron density, one of the most important parameters in the astrophysical plasmas, is generally diagnosed from the density dependence of line intensity ratios, which is determined by the collisional and radiative processes of highly charged ions. The density dependence given by model calculations is used in practical diagnostics of astrophysical plasmas; thus, the model should be examined with an experimental benchmark in order to ensure the accuracy of the diagnostics. An EBIT can provide a good benchmark as it provides a well-defined condition plasma consisting of a quasi-monoenergetic and quasi-unidirectional electron beam and trapped ions with a narrow charge distribution.

The electron density of the electron beam in an EBIT is determined by the electron beam current and the electron beam radius. However, in order to study the electron density dependence of line intensity ratios, the effective density experienced by the trapped ions is important. The effective electron density n_{eff} is determined from the average collision frequency $\overline{\nu}$ as $n_{\mathrm{eff}} = \overline{\nu}/v_e\sigma$, where v_e is the electron velocity and σ is the collision cross section; thus, the overlap between the electron beam and the trapped ion cloud, i.e., the spatial distribution of them, should be known [37, 38]. The electron spatial distribution can be obtained by imaging the spatial distribution of emission in the EUV range as the lifetime for high energy transitions is generally so short ($\sim 10^{-10}$ s) that the transition occurs within the electron beam. On the other hand, the ion spatial distribution can be obtained by imaging the spatial distribution of emission in the visible range as the transitions in the visible range are generally forbidden magnetic dipole transitions between fine structure splitting so that the transition lifetime is so long ($\sim 10^{-3}$ s) that the transition can occur even outside the electron beam.

Figure 4 [39] shows an example of the electron density dependence of iron line intensity ratios obtained with a compact EBIT in Tokyo for the line pairs that are important for density diagnostics of the solar corona. Density sensitivity obtained by collisional radiative model calculations (plotted by the solid and dashed lines in the figure) can be examined by the experimentally obtained ratios. Density sensitivity has also been examined for other line pairs of iron ions with several EBITs [40–43]. The density sensitivity of not only iron ions but also other ions relevant to the solar corona, such as argon, has also been studied [37, 42].

Since an EBIT has a quasi-monoenergetic electron beam, it can also be used for studying electron energy dependence. In particular, resonant processes, such as dielectronic recombination, and resonant excitation, can be studied efficiently by observing the electron energy dependence of emission spectra. Figure 5 shows the energy dependence of iron EUV spectra studied with a compact EBIT in Tokyo [44]. The spectra were obtained by the fast sweeping of electron energy in order for the charge state abundance not to change during the sweeping. As seen in the figure, emission intensity enhancement due to resonance excitation can be found at 400 and 420 eV, which correspond to resonance for Fe^{15+} and Fe^{14+}. Such resonant processes are also important for determining line intensity observed in astrophysical plasmas.

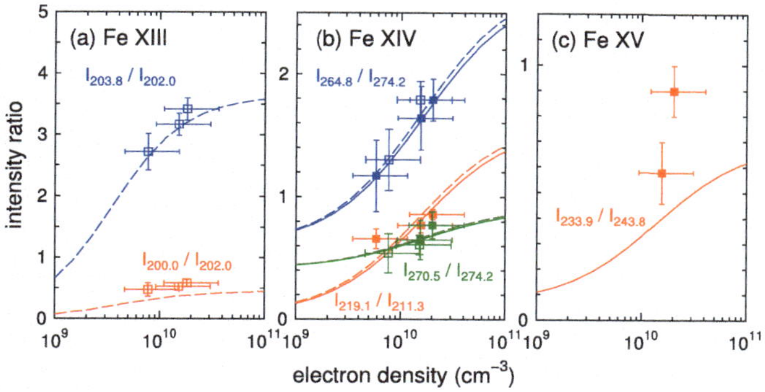

Fig. 4 Intensity ratio of density-sensitive lines of **a** Fe XIII, **b** Fe XIV, and **c** Fe XV [39]. The open (400 eV) and solid (500 eV) squares are the experimental ratios obtained for different electron beam energies. The dashed and solid lines are the theoretical ratios obtained from a collisional radiative model calculation for electron beam energies of 400 and 500 eV, respectively. Subscript of intensity corresponds to the wavelength (in Angstrom) of a particular line

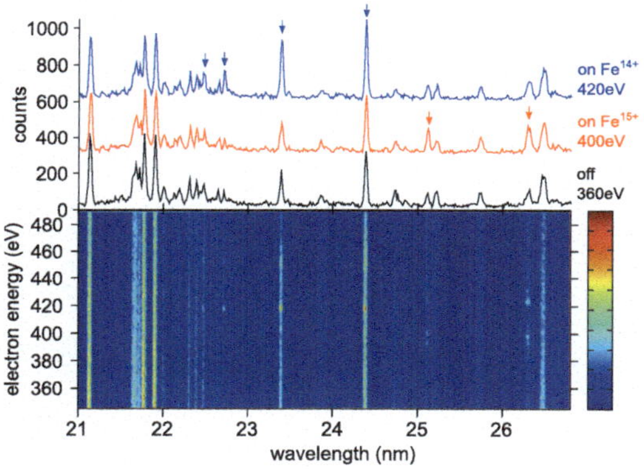

Fig. 5 Lower panel: two-dimensional spectra showing the electron energy dependence of the observed EUV intensity. Upper panel: EUV spectra sliced at electron energies of 360 (off-resonance energy), 400 (resonance energy for Fe^{15+}), and 420 eV (resonance energy for Fe^{14+}). [44] The lines indicated by the blue and red arrows represent the transitions in Fe^{14+} and Fe^{15+}, respectively

4 Fusion-Related Studies

Spectroscopic observation of tungsten ions plays a key role in the diagnostics of the ITER plasma. Since the spectroscopic data of tungsten ions required for the ITER diagnostics span wide ranges of charge states and wavelengths, systematic observations of tungsten spectra have been performed using EBITs over the world (for example, see Ref. [45–49]). Figure 6 shows an example of tungsten spectra [50] obtained with a compact EBIT at NIFS Japan with electron energies of 490–1320 eV. The lines that were not observed at lower energies are indicated by the arrows in each

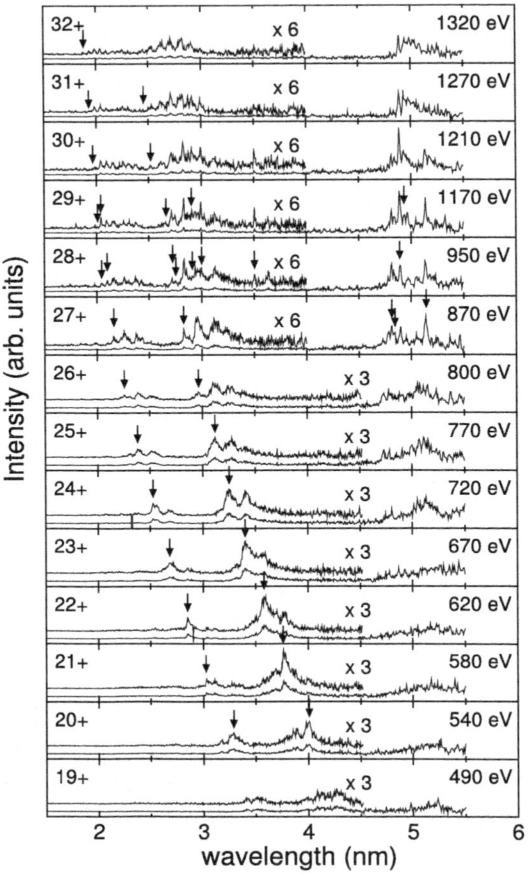

Fig. 6 EUV spectra of highly charged tungsten ions obtained with a compact EBIT at electron energies from 490 to 1320 eV [50]

spectrum. They are considered to be the emission from the highest charge state at each electron energy, which is listed on the left side of the figure. For example, several lines indicated by the arrows in the 870 eV spectrum can be assigned to W^{27+} because they were not observed at 800 eV, which is below the ionization energy of W^{26+} (833.4 eV [51]). Such a method based on energy dependence is useful for charge assignment of previously unreported lines although some care must be paid because the experimental electron energy is difficult to know exactly and also because the ionization cross section near the threshold is not so large that the production rate can be small. Thus, the final assignment should be done with the aid of the comparison with theoretical model calculations. The lines observed in Fig. 6 were identified as the $4f$–$(5g,6g)$, and $4d$–$(5f,5p)$ transitions through comparisons with collisional radiative model calculations [50]. Identified lines can be used to understand complex fusion plasma spectra containing contributions from a wide range of charge states. For example, the identified lines in Fig. 6 were used to understand tungsten spectra obtained with a large helical device (LHD) [52].

5 EUV Source-Related Studies

Radiation in the EUV range is important for industrial applications, such as lithography and water window imaging. In such applications, a strong broad spectral feature from laser-produced plasma (LPP) called unresolved transition arrays (UTA) is expected to be used. The wavelength of UTA depends on the element; thus, the target material should be chosen depending on the required wavelength. For example, tin is selected for an EUV source used in state-of-the-art lithography as tin plasma is known to reveal strong UTA emission at around 13.5 nm, for which highly efficient Mo/Si multilayer mirrors are available for constructing the optical system required for the lithography. For the next-generation lithography called beyond EUV lithography, radiation at around 6.7 nm is planned to be used based on the high reflectivity of available mirrors such as La/B_4C multilayer mirrors. For the source of 6.7 nm emission, UTA emission from lanthanide elements, such as gadolinium, is expected to be efficient. On the other hand, for a "water window" source at the wavelength range between 2.3 and 4.4 nm, UTA emission from bismuth is expected to be used.

In order to actually apply those UTA to light sources, a detailed emission mechanism should be known to maximize efficiency. In this regard, an EBIT is useful because complex LPP spectra emitted from a broad charge state range can be disentangled into charge-separated spectra [9, 53–57]. For example, Fig. 7 shows spectra of highly charged gadolinium ions obtained with a compact EBIT and the Tokyo-EBIT. By comparing the LPP spectra of gadolinium with those EBIT spectra, the optimum conditions for applying its emission to the beyond EUV lithography were deduced [58].

6 Atomic and Plasma Physics-Related Studies

6.1 Multipole Transitions

Multipole transitions are generally strongly forbidden in visible emission spectra. However, due to the strong dependence of the transition probability on transition energy, multipole transitions, such as electric and magnetic quadrupole ($E2$ and $M2$) transitions, become important and often observed in EUV spectra of highly charged ions [60, 61]. Higher multipole transitions, such as electric and magnetic octupole ($E3$ and $M3$) transitions, can also be observed in emission spectra [59, 62]. Figure 8 shows the EUV spectra of tungsten ions observed with a compact EBIT at NIFS, Japan [59]. Three prominent lines that were not observed at an electron energy of 800 eV but observed at 870 eV were assigned to silver-like tungsten W^{27+} from the energy dependence. Based on the comparison with a collisional radiative model calculation shown in the top panel, the two lines at around 9 nm were identified to be the $4f_{7/2,5/2}$–$5s$ $E3$ transitions. Those transitions have much smaller transition probabilities than electric dipole-allowed transitions, but are observable

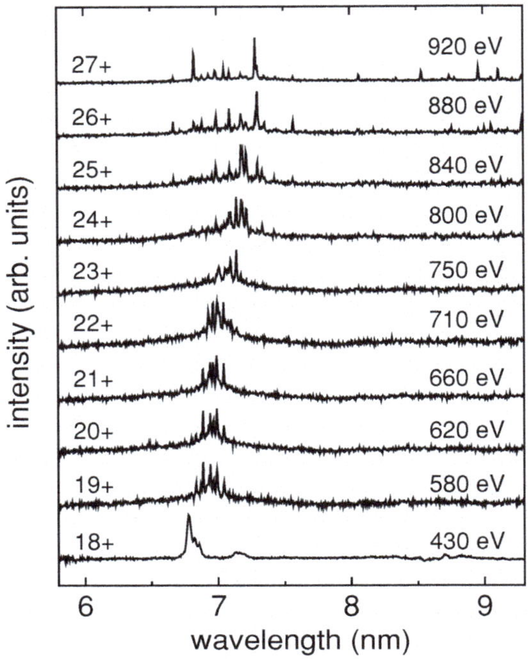

Fig. 7 EUV emission spectra of gadolinium ions. The numbers in the right- and the left-hand sides represent the electron energy at which each spectrum was obtained and the maximum charge state at each electron energy. A compact EBIT (CoBIT) [20] was used for the 430 eV spectrum whereas the Tokyo-EBIT was used for the spectra obtained with the other energies. [58]

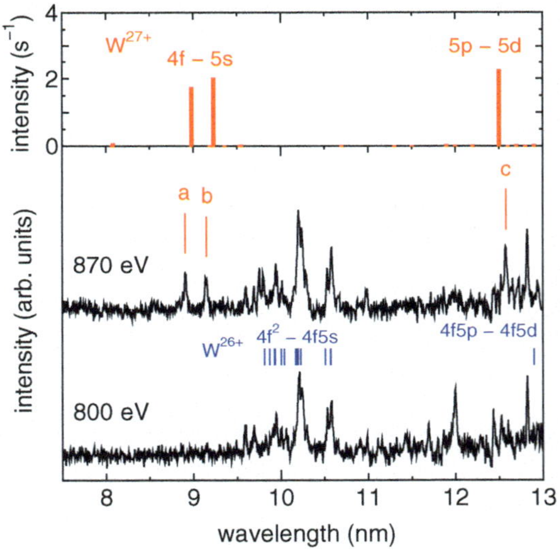

Fig. 8 EUV spectra of tungsten ions obtained with a compact EBIT with electron energies of 800 and 870 eV [59]. The upper panel shows the collisional radiative model spectrum for W^{27+}. The lines a and b correspond to the $4f_{5/2,7/2}$—$5s$ electric octupole transitions

in low-density EBIT plasmas when the upper level is a long-lived metastable state which can have a large population. Because of the strong atomic number dependence of the transition probability, the transition probability becomes much larger for silver-like heavier ions. However, it was found that the $E3$ transitions are hard to observe for higher atomic number elements due to the rapid decrease of the upper-level ($5s$) population. It was found that the $E3$ transitions are specifically prominent in the narrow atomic number region around $Z \sim 74$ by the anomalous behavior of the cascading contribution [59].

6.2 Disappearance of Resonance Lines in Pm-Like Ions

The promethium-like isoelectronic sequence has a simple alkali-metal-like structure with the $4f^{14}5s$ configuration for heavy elements; thus, it has been considered to be one of the most important sequences for plasma diagnostics. In 1980, Curtis and Ellis [64, 65] theoretically predicted the alkali-metal structure and the availability of its $5s$—$5p$ resonance EUV lines in the diagnostics of high-temperature plasmas. Several experiments for identifying the $5s$—$5p$ resonance line have been performed since then [66–73], but all of them failed to identify the resonance line definitely. Three decades after the prediction by Curtis and Ellis, experimental studies with a compact EBIT in Tokyo [63, 74] showed that the reason for the lost resonance line is the population trapping to the $4f^{13}5s^2$ metastable state.

Figure 9 [63] shows the EUV spectra of tungsten, gold, and bismuth obtained with the compact EBIT. The lines experimentally assigned to the promethium-like ions are indicated by the arrows. In the lower panel of Fig. 9, CR model spectra of the promethium-like ions are shown where the $4f^{14}5s$–$4f^{14}5p$ and $4f^{13}5s^2$–$4f^{13}5s5p_{3/2}$ transitions are indicated by the dashed and dotted lines, respectively. As seen in the model spectra, which well reproduce the experimental spectra, the $4f^{13}5s^2$–$4f^{13}5s5p$ transitions are much more prominent compared with the $4f^{14}5s$–$4f^{14}5p$ transitions due to the population trapping to the $4f^{13}5s^2$ metastable state.

7 Summary and Outlook

In this chapter, spectroscopic studies of highly charged ions in the EUV range were briefly introduced mainly focusing on studies done with EBITs in Japan. Almost all the spectra shown in this chapter were obtained with a compact-type EBIT, for which the construction and running costs are relatively low. Thus, such studies are not limited to specific groups and are expected to be widely performed. In addition to passive emission spectroscopic studies, such as those introduced in this chapter, excitation spectroscopy using new high-energy lasers has also been performed recently [75, 76]. Such an active spectroscopic technique is expected to become a trend in the near future with the recent development of new light sources.

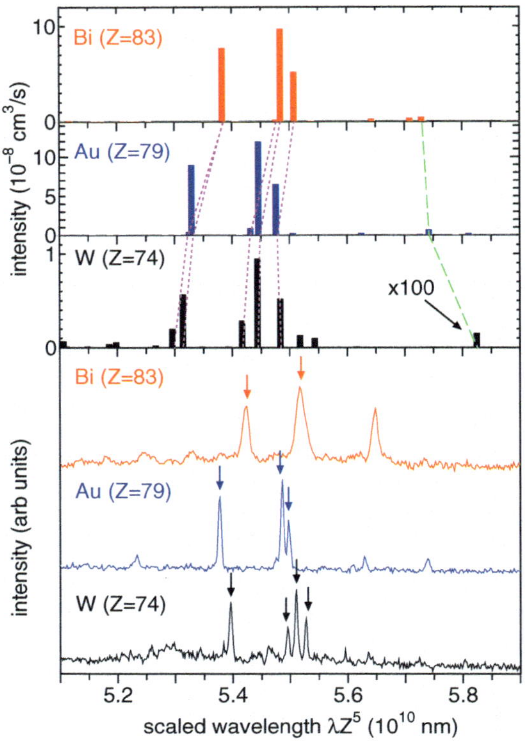

Fig. 9 Experimental (lower) and collisional radiative model (upper) spectra for $5s$—$5p_{3/2}$ transitions of promethium-like tungsten, gold, and bismuth [63]. The horizontal axis is wavelength (λ) scaled by Z^5 ($\lambda \cdot Z^5$ [10^{10} nm])

Acknowledgements The author acknowledges Prof. Lokesh C. Tribedi for proofreading the original manuscript and Drs. Hiroyuki A. Sakaue and Hayato Ohashi for providing the original figures and kind assistance.

References

1. R. Aymar, P. Barabaschi, Y. Shimomura, Plasma Phys. Controlled Fus. **44**(5), 519 (2002). https://doi.org/10.1088/0741-3335/44/5/304
2. P. Bryans, N.R. Badnell, T.W. Gorczyca, J.M. Laming, W. Mitthumsiri, D.W. Savin, Astrophys. J. Supplem. Ser **167**(2), 343 (2006). https://doi.org/10.1086/507629
3. J.L. Culhane, L.K. Harra, A.M. James, K. Al-Janabi, L.J. Bradley, R.A. Chaudry, K. Rees, J.A. Tandy, P. Thomas, M.C.R. Whillock, B. Winter, G.A. Doschek, C.M. Korendyke, C.M. Brown, S. Myers, J. Mariska, J. Seely, J. Lang, B.J. Kent, B.M. Shaughnessy, P.R. Young, G.M. Simnett, C.M. Castelli, S. Mahmoud, H. Mapson-Menard, B.J. Probyn, R.J. Thomas, J. Davila, K. Dere, D. Windt, J. Shea, R. Hagood, R. Moye, H. Hara, T. Watanabe, K. Matsuzaki, T. Kosugi, V. Hansteen, Ø. Wikstol, Solar Phys. **243**(1), 19 (2007). https://doi.org/10.1007/s01007-007-0293-1
4. R.A. Harrison, E.C. Sawyer, M.K. Carter, A.M. Cruise, R.M. Cutler, A. Fludra, R.W. Hayes, B.J. Kent, J. Lang, D.J. Parker, J. Payne, C.D. Pike, S.C. Peskett, A.G. Richards, J.L. Gulhane,

K. Norman, A.A. Breeveld, E.R. Breeveld, K.F. Al Janabi, A.J. Mccalden, J.H. Parkinson, D.G. Self, P.D. Thomas, A.I. Poland, R.J. Thomas, W.T. Thompson, O. Kjeldseth-Moe, P. Brekke, J. Karud, P. Maltby, B. Aschenbach, H. Brauninger, M. Kuhne, J. Hollandt, O.H.W. Siegmund, M.C.E. Huber, A.H. Gabriel, H.E. Mason, B.J.I. Bromage, Solar Phys. **162**(1), 233 (1995). https://doi.org/10.1007/BF00733431
5. W.M. Neupert, G.L. Epstein, R.J. Thomas, W.T. Thompson, Solar Phys. **137**(1), 87 (1992). https://doi.org/10.1007/BF00146577
6. T. Sugie, A. Costley, A. Malaquias, C. Walker, J. Plasma Fus. Res. **79**(10), 1051 (2003). https://doi.org/10.1585/jspf.79.1051. http://www.jstage.jst.go.jp/article/jspf/79/10/79_10_1051/_article
7. J. Clementson, P. Beiersdorfer, E.W. Magee, H.S. McLean, R.D. Wood, J. Phys. B: Atom. Mol. Opt. Phys. **43**(14), 144009 (2010). https://doi.org/10.1088/0953-4075/43/14/144009. https://iopscience.iop.org/article/10.1088/0953-4075/43/14/144009
8. C. Biedermann, R. Radtke, R. Seidel, T. Putterich, Physica Scripta **T134**, 014026 (2009). https://doi.org/10.1088/0031-8949/2009/T134/014026. https://iopscience.iop.org/article/10.1088/0031-8949/2009/T134/014026
9. O.O. Versolato, Plasma Sour. Sci. Technol. **28**(8), 083001 (2019). https://doi.org/10.1088/1361-6595/ab3302. https://iopscience.iop.org/article/10.1088/1361-6595/ab3302
10. N. Mojarad, J. Gobrecht, Y. Ekinci, Sci. Rep. **5**(1), 9235 (2015). https://doi.org/10.1038/srep09235. https://www.nature.com/articles/srep09235
11. J.C. Solem, G.C. Baldwin, Science **218**(4569), 229 (1982). https://doi.org/10.1126/science.218.4569.229. https://www.science.org/doi/10.1126/science.218.4569.229. Publisher: American Association for the Advancement of Science
12. R.E. Marrs, M.A. Levine, D.A. Knapp, J.R. Henderson, Phys. Rev. Lett. **60**(17), 1715 (1988). https://doi.org/10.1103/PhysRevLett.60.1715. https://link.aps.org/doi/10.1103/PhysRevLett.60.1715
13. E.D. Donets, V.P. Ovsyanniko, Soviet Physics? JETP [translation of Zhurnal Eksperimentalnoi i Teoreticheskoi Fiziki] **53**, 466 (1981)
14. J.D. Gillaspy, Physica Scripta T **71**, 99 (1997). https://doi.org/10.1088/0031-8949/1997/T71/017. https://iopscience.iop.org/article/10.1088/0031-8949/1997/T71/017
15. J.D. Silver, A.J. Varney, H.S. Margolis, P.E.G. Baird, I.P. Grant, P.D. Groves, W.A. Hallett, A.T. Handford, P.J. Hirst, A.R. Holmes, D.J.H. Howie, R.A. Hunt, K.A. Nobbs, M. Roberts, W. Studholme, J.S. Wark, M.T. Williams, M.A. Levine, D.D. Dietrich, W.G. Graham, I.D. Williams, R. OfNeil, S.J. Rose, Rev. Sci. Instrum. **65**(4), 1072 (1994). https://doi.org/10.1063/1.1145066. http://aip.scitation.org/doi/10.1063/1.1145066
16. R.E. Marrs, P. Beiersdorfer, S.R. Elliott, D.A. Knapp, T. Stoehlker, Physica Scripta **T59**, 183 (1995). https://doi.org/10.1088/0031-8949/1995/T59/023. https://iopscience.iop.org/article/10.1088/0031-8949/1995/T59/023
17. N. Nakamura, J. Asada, F.J. Currell, T. Fukami, T. Hirayama, K. Motohashi, T. Nagata, E. Nojikawa, S. Ohtani, K. Okazaki, M. Sakurai, H. Shiraishi, S. Tsurubuchi, H. Watanabe, Physica Scripta **T73**, 362 (1997). https://doi.org/10.1088/0031-8949/1997/T73/119. https://iopscience.iop.org/article/10.1088/0031-8949/1997/T73/119
18. J.R.C. Lopez-Urrutia, J. Braun, G. Brenner, H. Bruhns, C. Dimopoulou, I.N. Dragani?, D. Fischer, A.J.G. Martinez, A. Lapierre, V. Mironov, R. Moshammer, R.S. Orts, H. Tawara, M. Trinczek, J. Ullrich, J. Phys.: Conf. Ser. **2**(1), 42 (2004). https://doi.org/10.1088/1742-6596/2/1/006
19. Y. Fu, K. Yao, B. Wei, D. Lu, R. Hutton, Y. Zou, J. Instrum. **5**(08), C08011 (2010). https://doi.org/10.1088/1748-0221/5/08/C08011. https://iopscience.iop.org/article/10.1088/1748-0221/5/08/C08011
20. N. Nakamura, H. Kikuchi, H.A. Sakaue, T. Watanabe, Rev Sci Instrum **79**(6), 063104 (2008). https://doi.org/10.1063/1.2939393. http://aip.scitation.org/doi/10.1063/1.2939393
21. J. Xiao, Z. Fei, Y. Yang, X. Jin, D. Lu, Y. Shen, L. Liljeby, R. Hutton, Y. Zou, Rev. Sci. Instrum. **83**(1), 013303 (2012). https://doi.org/10.1063/1.3675575. http://aip.scitation.org/doi/10.1063/1.3675575

22. S. Liang, Q. Lu, X. Wang, Y. Yang, K. Yao, Y. Shen, B. Wei, J. Xiao, S. Chen, P. Zhou, W. Sun, Y. Zhang, Y. Huang, H. Guan, X. Tong, C. Li, Y. Zou, T. Shi, K. Gao, Rev. Sci. Instrum. **90**(9), 093301 (2019). https://doi.org/10.1063/1.5112154. http://aip.scitation.org/doi/10.1063/1.5112154
23. P. Micke, S. Kuhn, L. Buchauer, J.R. Harries, T.M. Bucking, K. Blaum, A. Cieluch, A. Egl, D. Hollain, S. Kraemer, T. Pfeifer, P.O. Schmidt, R.X. Schussler, C. Schweiger, T. Stohlker, S. Sturm, R.N. Wolf, S. Bernitt, J.R. Crespo Lopez-Urrutia, Rev. Sci. Instrum. **89**(6), 063109 (2018). https://doi.org/10.1063/1.5026961. http://aip.scitation.org/doi/10.1063/1.5026961
24. T. Harada, K. Takahashi, H. Sakuma, A. Osyczka, Appl. Opt. **38**(13), 2743 (1999). https://doi.org/10.1364/AO.38.002743. https://opg.optica.org/abstract.cfm?URI=ao-38-13-2743
25. T. Harada, T. Kita, Appl. Opt. **19**(23), 3987 (1980). https://doi.org/10.1364/AO.19.003987. https://opg.optica.org/abstract.cfm?URI=ao-19-23-3987
26. T. Kita, T. Harada, N. Nakano, H. Kuroda, Appl. Opt. **22**(4), 512 (1983). https://doi.org/10.1364/AO.22.000512. https://opg.optica.org/abstract.cfm?URI=ao-22-4-512
27. H. Ohashi, J. Yatsurugi, H.A. Sakaue, N. Nakamura, Rev. Sci. Instrum. **82**(8), 083103 (2011). https://doi.org/10.1063/1.3618686. http://aip.scitation.org/doi/10.1063/1.3618686
28. H.A. Sakaue, D. Kato, N. Nakamura, E. Watanabe, N. Yamamoto, C. Chen, T. Watanabe, J. Phys.: Conf. Ser. **163**, 012020 (2009). https://doi.org/10.1088/1742-6596/163/1/012020. https://iopscience.iop.org/article/10.1088/1742-6596/163/1/012020
29. A. Lapierre, J.R. Crespo Lopez-Urrutia, T.M. Baumann, S.W. Epp, A. Gonchar, A.J. Gonzalez Martinez, G. Liang, A. Rohr, R. Soria Orts, M.C. Simon, H. Tawara, R. Versteegen, J. Ullrich, Rev. Sci. Instrum. **78**(12), 123105 (2007). https://doi.org/10.1063/1.2818808. http://aip.scitation.org/doi/10.1063/1.2818808
30. Z. Shi, R. Zhao, W. Li, B. Tu, Y. Yang, J. Xiao, S. Huldt, R. Hutton, Y. Zou, Rev. Sci. Instrum. **85**(6), 063110 (2014). https://doi.org/10.1063/1.4876597. http://aip.scitation.org/doi/10.1063/1.4876597
31. P. Beiersdorfer, E.W. Magee, E. Trabert, H. Chen, J.K. Lepson, M.F. Gu, M. Schmidt, Rev. Sci. Instrum. **75**(10), 3723 (2004). https://doi.org/10.1063/1.1779609. http://aip.scitation.org/doi/10.1063/1.1779609
32. B. Blagojevi?, E.O. Le Bigot, K. Fahy, A. Aguilar, K. Makonyi, E. Takacs, J.N. Tan, J.M. Pomeroy, J.H. Burnett, J.D. Gillaspy, J.R. Roberts, Rev. Sci. Instrum. **76**(8), 083102 (2005). https://doi.org/10.1063/1.1988227. http://aip.scitation.org/doi/10.1063/1.1988227
33. P. Beiersdorfer, J.R.C. Lopez-Urrutia, P. Springer, S.B. Utter, K.L. Wong, Rev. Sci. Instrum. **70**(1), 276 (1999). https://doi.org/10.1063/1.1149324. http://aip.scitation.org/doi/10.1063/1.1149324
34. C. Biedermann, R. Radtke, J.L. Schwob, P. Mandelbaum, R. Doron, T. Fuchs, G. Fußmann, Physica Scripta **T92**(1), 85 (2001). https://doi.org/10.1238/Physica.Topical.092a00085. https://iopscience.iop.org/article/10.1238/Physica.Topical.092a00085
35. K.P. Dere, E. Landi, H.E. Mason, B.C. Monsignori Fossi, P.R. Young, Astron. Astrophys. Suppl. Ser. **125**(1), 149 (1997). https://doi.org/10.1051/aas:1997368. http://aas.aanda.org/10.1051/aas:1997368
36. A. Kramida, Yu. Ralchenko, J. Reader, and NIST ASD Team (2020). NIST Atomic Spectra Database (ver. 5.8), [Online]. https://physics.nist.gov/asd [2021, June 14]. National Institute of Standards and Technology, Gaithersburg, MD
37. N. Nakamura, N. Numadate, Y. Kono, I. Murakami, D. Kato, H.A. Sakaue, H. Hara, Astrophys. J. **921**(2), 115 (2021). https://doi.org/10.3847/1538-4357/ac1c6f. https://iopscience.iop.org/article/10.3847/1538-4357/ac1c6f
38. T. Arthanayaka, P. Beiersdorfer, G.V. Brown, M.F. Gu, M. Hahn, N. Hell, T. Lockard, D.W. Savin, Astrophys. J. **890**(1), 77 (2020). https://doi.org/10.3847/1538-4357/ab67b4. Publisher: American Astronomical Society
39. N. Nakamura, E. Watanabe, H.A. Sakaue, D. Kato, I. Murakami, N. Yamamoto, H. Hara, T. Watanabe, Astrophys. J. **739**(1), 17 (2011). https://doi.org/10.1088/0004-637X/739/1/17. https://iopscience.iop.org/article/10.1088/0004-637X/739/1/17

40. E. Shimizu, S. Ali, T. Tsuda, H.A. Sakaue, D. Kato, I. Murakami, H. Hara, T. Watanabe, N. Nakamura, Astron. Astrophys. **601**, A111 (2017). https://doi.org/10.1051/0004-6361/201630199
41. G.Y. Liang, T.M. Baumann, J.R.C. Lopez-Urrutia, S.W. Epp, H. Tawara, A. Gonchar, P.H. Mokler, G. Zhao, J. Ullrich, Astrophys. J. **696**(2), 2275 (2009). https://doi.org/10.1088/0004-637X/696/2/2275. https://iopscience.iop.org/article/10.1088/0004-637X/696/2/2275
42. J.K. Lepson, P. Beiersdorfer, M.F. Gu, P. Desai, M. Bitter, L. Roquemore, M.L. Reinke, AIP Conf. Proc. **1438**(1), 136 (2012). https://doi.org/10.1063/1.4707868
43. N. Yamamoto, T. Kato, H. Funaba, K. Sato, N. Tamura, S. Sudo, P. Beiersdorfer, J.K. Lepson, Astrophys. J. **689**(1), 646 (2008)
44. T. Tsuda, E. Shimizu, S. Ali, H.A. Sakaue, D. Kato, I. Murakami, H. Hara, T. Watanabe, N. Nakamura, Astrophys. J. **851**(2), 82 (2017). https://doi.org/10.3847/1538-4357/aa97e5. https://iopscience.iop.org/article/10.3847/1538-4357/aa97e5
45. R. Radtke, C. Biedermann, P. Mandelbaum, J.L. Schwob, J. Phys.: Conf. Ser. **58**, 113 (2007)
46. J. Clementson, T. Lennartsson, P. Beiersdorfer, Atoms **3**(3), 407 (2015). https://doi.org/10.3390/atoms3030407. http://www.mdpi.com/2218-2004/3/3/407
47. W. Li, Z. Shi, Y. Yang, J. Xiao, T. Brage, R. Hutton, Y. Zou, Phys. Rev. A **91**(6), 062501 (2015). https://doi.org/10.1103/PhysRevA.91.062501. https://link.aps.org/doi/10.1103/PhysRevA.91.062501
48. T. Das, Y.A. Podpaly, J. Reader, J.D. Gillaspy, Y. Ralchenko, European Phys. J. D **72**(7), 124 (2018). https://doi.org/10.1140/epjd/e2018-90118-7. http://link.springer.com/10.1140/epjd/e2018-90118-7
49. M. Mita, H.A. Sakaue, D. Kato, I. Murakami, N. Nakamura, J. Phys.: Conf. Ser. **875**, 012019 (2017). https://doi.org/10.1088/1742-6596/875/2/012019. https://iopscience.iop.org/article/10.1088/1742-6596/875/2/012019
50. H.A. Sakaue, D. Kato, N. Yamamoto, N. Nakamura, I. Murakami, Phys. Rev. A **92**(1), 012504 (2015). https://doi.org/10.1103/PhysRevA.92.012504. https://link.aps.org/doi/10.1103/PhysRevA.92.012504
51. A.E. Kramida, J. Reader, Atomic Data and Nuclear Data Tables **92**(4), 457 (2006). https://doi.org/10.1016/j.adt.2006.03.002. https://www.sciencedirect.com/science/article/pii/S0092640X06000167
52. S. Morita, C.F. Dong, M. Goto, D. Kato, I. Murakami, H.A. Sakaue, M. Hasuo, F. Koike, N. Nakamura, T. Oishi, A. Sasaki, E.H. Wang, (Gaithersburg, Maryland, USA, 2013), pp. 143–152. https://doi.org/10.1063/1.4815848. http://aip.scitation.org/doi/abs/10.1063/1.4815848
53. H. Ohashi, S. Suda, H. Tanuma, S. Fujioka, H. Nishimura, K. Nishihara, T. Kai, A. Sasaki, H.A. Sakaue, N. Nakamura, S. Ohtani, J. Phys.: Conf. Ser. **163**, 012071 (2009). https://doi.org/10.1088/1742-6596/163/1/012071. https://iopscience.iop.org/article/10.1088/1742-6596/163/1/012071
54. Z. Bouza, J. Scheers, A. Ryabtsev, R. Schupp, L. Behnke, C. Shah, J. Sheil, M. Bayraktar, J.R.C. Lopez-Urrutia, W. Ubachs, R. Hoekstra, O.O. Versolato, J. Phys. B: Atom. Mol. Opt. Phys. **53**(19), 195001 (2020). https://doi.org/10.1088/1361-6455/aba3a8. https://iopscience.iop.org/article/10.1088/1361-6455/aba3a8
55. J. Scheers, C. Shah, A. Ryabtsev, H. Bekker, F. Torretti, J. Sheil, D.A. Czapski, J.C. Berengut, W. Ubachs, J.R.C. Lopez-Urrutia, R. Hoekstra, O.O. Versolato, Phys. Rev. A **101**(6), 062511 (2020). https://doi.org/10.1103/PhysRevA.101.062511. https://link.aps.org/doi/10.1103/PhysRevA.101.062511
56. D. Kilbane, G. O'Sullivan, J.D. Gillaspy, Y. Ralchenko, J. Reader, Phys. Rev. A **86**(4), 042503 (2012). Number of pages: 7
57. H. Ohashi, H.A. Sakaue, N. Nakamura, J. Phys.: Conf. Ser. **488**(6), 062017 (2014). https://doi.org/10.1088/1742-6596/488/6/062017. https://iopscience.iop.org/article/10.1088/1742-6596/488/6/062017
58. H. Ohashi, T. Higashiguchi, B. Li, Y. Suzuki, M. Kawasaki, T. Kanehara, Y. Aida, S. Torii, T. Makimura, W. Jiang, P. Dunne, G. O'Sullivan, N. Nakamura, J. Appl. Phys. **115**, 033302 (2014)

59. H.A. Sakaue, D. Kato, I. Murakami, H. Ohashi, N. Nakamura, Phys. Rev. A **100**(5), 052515 (2019). https://doi.org/10.1103/PhysRevA.100.052515. https://link.aps.org/doi/10.1103/PhysRevA.100.052515
60. W.E. Behring, L. Cohen, U. Feldman, G.A. Doschek, Astrophys. J. **203**, 521 (1976). https://doi.org/10.1086/154107. https://ui.adsabs.harvard.edu/abs/1976ApJ...203..521B. ADS Bibcode: 1976ApJ...203..521B
61. M. Klapisch, J.L. Schwob, M. Finkenthal, B.S. Fraenkel, S. Egert, A. Bar-Shalom, C. Breton, C. DeMichelis, M. Mattioli, Phys. Rev. Lett. **41**, 403 (1978)
62. E. Trabert, P. Beiersdorfer, G.V. Brown, K. Boyce, R.L. Kelley, C.A. Kilbourne, F.S. Porter, A. Szymkowiak, Phys. Rev. A **73**(2), 022508 (2006). https://doi.org/10.1103/PhysRevA.73.022508. https://link.aps.org/doi/10.1103/PhysRevA.73.022508
63. Y. Kobayashi, K. Kubota, K. Omote, A. Komatsu, J. Sakoda, M. Minoshima, D. Kato, J. Li, H.A. Sakaue, I. Murakami, N. Nakamura, Phys. Rev A **92**(2), 022510 (2015). https://doi.org/10.1103/PhysRevA.92.022510. https://link.aps.org/doi/10.1103/PhysRevA.92.022510
64. L.J. Curtis, D.G. Ellis, Phys. Rev. Lett. **45**(26), 2099 (1980). https://doi.org/10.1103/PhysRevLett.45.2099. http://link.aps.org/doi/10.1103/PhysRevLett.45.2099. Publisher: American Physical Society
65. D.G. Ellis, L.J. Curtis, Nucl. Instrum. Methods **202**, 339 (1982)
66. B.M. Johnson, K.W. Jones, T.H. Kruse, L.J. Curtis, D.G. Ellis, Nucl. Instrum. Methods **202**, 53 (1982)
67. C.E. Theodosiou, V. Raftopoulos, Phys. Rev. A **28**(2), 1186 (1983). https://doi.org/10.1103/PhysRevA.28.1186. http://link.aps.org/doi/10.1103/PhysRevA.28.1186. Publisher: American Physical Society
68. E. Trabert, P.H. Heckmann, Zeitschrift fur Physik D **1**, 381 (1986)
69. V. Kaufman, E. Trabert, P.H. Heckmann, G. Moller, V. Lodwig, J.H. Blanke, Physica Scripta **42**, 705 (1990)
70. K.B. Fournier, M. Finkenthal, S. Lippmann, C.P. Holmes, H.W. Moos, W.H. Goldstein, A.L. Osterheld, Phys. Rev. A **50**(5), 3727 (1994). https://doi.org/10.1103/PhysRevA.50.3727. http://link.aps.org/doi/10.1103/PhysRevA.50.3727. Publisher: American Physical Society
71. Ö. Andersson, A method for atomic spectroscopy of highly charged ions in the Pm isoelectronic sequence. Ph.D. thesis, Lund University (1995)
72. R. Hutton, Y. Zou, J.R. Almandos, C. Biederman, R. Radtke, A. Greier, R. Neu, Nucl. Instrum. Methods B **205**, 114 (2003)
73. S. Wu, R. Hutton, Canadian J. Phys. **86**, 125 (2008)
74. Y. Kobayashi, D. Kato, H.A. Sakaue, I. Murakami, N. Nakamura, Phys. Rev. A **89**(1), 010501 (2014). https://doi.org/10.1103/PhysRevA.89.010501. https://link.aps.org/doi/10.1103/PhysRevA.89.010501
75. S.W. Epp, J.R.C. Lopez-Urrutia, G. Brenner, V. Mackel, P.H. Mokler, R. Treusch, M. Kuhlmann, M.V. Yurkov, J. Feldhaus, J.R. Schneider, M. Wellhofer, M. Martins, W. Wurth, J. Ullrich, Phys. Rev. Lett. **98**(18), 183001 (2007). https://doi.org/10.1103/PhysRevLett.98.183001. https://link.aps.org/doi/10.1103/PhysRevLett.98.183001
76. S.W. Epp, J.R.C. Lopez-Urrutia, M.C. Simon, T. Baumann, G. Brenner, R. Ginzel, N. Guerassimova, V. Mackel, P.H. Mokler, B.L. Schmitt, H. Tawara, J. Ullrich, J. Phys. B: Atom. Mol. Opt. Phys. **43**(19), 194008 (2010). https://doi.org/10.1088/0953-4075/43/19/194008. https://iopscience.iop.org/article/10.1088/0953-4075/43/19/194008

On the Dynamics of Fast and Ultrafast Irradiation in Clusters and Molecules

P.-G. Reinhard, P. M. Dinh, and E. Suraud

1 Introduction

The response of clusters and molecules to electromagnetic perturbations has focused numerous investigations over the years, in relation to the wider understanding of the physical mechanisms of irradiation [1, 2]. Irradiation of matter, namely, energy deposition by radiation, such as by lasers or charged particles, is a crucial scientific issue where microscopic mechanisms, down to electronic level, directly impacts the macroscopic world. Irradiation takes place essentially $<<$ instantaneously $>>$, within less than the least electronic motion. It furthermore occurs primarily in a very localized manner at the atomic scale. The dynamics of irradiation is by nature far-off equilibrium, involving ionization, electron transport, and large correlation effects, including dissipative effects leading to thermalization. Irradiation covers a multitude of time scales from the microscopic ones up to even human ones. It is a particularly challenging issue which requires to analyze and understand, experimentally and, even more so, theoretically, the interaction of matter (at fully quantum level down to electrons) with possibly strong electromagnetic fields. It covers an impressive range of applications (radiation damage) from biology/medicine [2] to our everyday life invaded by electronic devices [1], miniaturization making them increasingly sensitive to radiation.

There is still a long way to describe irradiation in all its detail. But analyzing the response of clusters and molecules to such situations is certainly an important step forward, in particular, in relation to the originally localized nature of the irradiation process. Clusters and molecules thus offer a remarkable field of investigation for

P.-G. Reinhard
Institut für Theoretische Physik, Universität Erlangen, 91058 Erlangen, Germany

P. M. Dinh · E. Suraud (✉)
Laboratoire de Physique Théorique, Université de Toulouse, CNRS, UPS, 31062 Toulouse Cedex, France
e-mail: suraud@irsamc.ups-tlse.fr

L. C. Tribedi (ed.), *Advances in Atomic Molecular Collisions*,
https://doi.org/10.1007/978-981-97-7063-2_6

understanding such localized processes, at least in the early phases of the system's response. In parallel, the rapid progress in laser technology has opened remarkable new possibilities for the investigation of this response down to electronic time scales [3–6]. This allows the optical manipulation of clusters and molecules, at the level of electrons themselves. Pump-and-probe scenario, in which a pump laser is used to excite the system and a probe laser allows to record the system's response as a function of the delay between the two pulses, is a typical operational setup [6]. Correlatively, the more and more detailed experimental analysis of emitted electrons, in particular, their energy and/or angle-resolved production cross section [7], provides the necessary information for a better understanding of the response of the system [8]. And it also provides stronger and stronger constraints on theoretical investigations, which have experienced a remarkable development over recent years. One of the major directions of research in the field would be to consider more and more complex systems, which might also open the path to various applications. Such a goal can nevertheless only be reached while maintaining the high level of theoretical understanding reached in simple systems (atoms and molecules), where rather well-understood basic mechanisms have been identified.

In this contribution, we shall explore, from a theoretical viewpoint, a few dynamical scenarios using fast and ultrafast laser pulses. Starting from femtosecond (fs) processes we shall explore Ultrafast (UF) processes down to the attosecond (as) domain. Pump-and-probe setups provide a standard tool of investigation of such processes and we shall consider several examples thereof, mixing IR and XUV pulses in various combinations.

A major physical consequence of the step from simple to more complex molecular systems is the onset of collective effects and/or resonances which are known to play a considerable role in irradiation processes with long laser pulses. A basic question in the field is thus to evaluate the potential impact of collective modes on the dynamics of electrons irradiated by shorter and shorter laser pulses. The point might look only marginally relevant in view of the considered time scales as the laser duration in the fs regime at best matches typical collective oscillation periods and/or appears much shorter than the collective oscillations. But decreasing laser duration down to a few fs (as nowadays feasible) enhances uncertainty on laser frequency and leads to a wider coupling between laser and resonance extending over a larger spectral range.

Laser pulses in the fs range furthermore give access to effects connected to the so-called Carrier Envelope Phase (CEP), namely, the phase of the oscillating part of the laser signal with respect to the pulse envelop [8] (Sect. 3.1). While this detail is irrelevant for long laser pulses, it becomes essential when stepping down to pulse envelops of a few fs. And CEP effects are experimentally visible [9], in particular, in the way the electrons are emitted by the system, typically in terms of left/right asymmetry (with respect to laser polarization axis). We are thus directly addressing the actual dynamics of electron emission within such scenarios, at a much more detailed level of investigation than the mere total ionization yields. Measurements have been performed, in particular, in the case of C_{60} with an IR laser field's far-off resonance, thus in a "safe" spectral region, as far as the coupling to collective modes is concerned. Still, the details of the dynamics of irradiation in such UF processes in complex systems, in particular, ionization mechanisms, can be analyzed, to some extent, in terms of approaches developed for simpler objects (Sect. 3.2).

Since a few years laser technology allows to address even shorter time scales down to the attosecond (as) regime, which is expected to fully resolve electron dynamics [4, 5]. Standard experimental setups for reaching such UF pulses lead to the production of a train of a few attosecond pulses, separated by fs intervals. The typical ionization scenario couples an IR pump with an XUV attosecond pulse train built thereof (Sect. 4). While the IR pulse provides the necessary dynamical perturbation of the electron cloud (a soft oscillation with respect to the ionic background), ionization is attained by the probe XUV attosecond pulse train. Most experiments have focused on atoms [10, 11] and simple molecules [12, 13] and evaluated the net ionization signal as a function of delay time between the IR pulse and the attosecond pulse train. A few results study the system's response in terms of the more detailed Photo-Electron Spectra (PES, Sect. 4.2). Typical ionization signals show oscillatory pattern (Sect. 4.1). The modulation of oscillations with half the IR period (the separation between two successive attosecond pulses) in the net ionization signal biases the signal toward laser characteristics rather than toward system's properties (Sect. 4.3). A way out is to consider streaking experiments on the basis of UF IR and XUV pulses, which partially overcomes the problem (Sect. 4.4). Finally, pure UF XUV pulses provide another type of interesting scenario which we shall explore at the end of this contribution.

The paper is organized as follows. Section 2 provides basics of our theoretical approach. Fast processes in the fs regime are addressed in Sect. 3, while UF scenarios are addressed in Sect. 4. We finally give some conclusions and perspectives.

2 Theoretical Modeling

2.1 Realistic TDDFT Simulations

The basic tool used here to describe electron dynamics is Time-Dependent Density-Functional Theory (TDDFT) [14–16]. We mostly work here at the Time-Dependent Local-Density Approximation (TDLDA) level with the Perdew–Wang exchange correlation functional [17]. Pure LDA does not produce a correct Ionization Potential (IP), which amounts to violate Koopmans' theorem [18]. We thus augment it by a Self-Interaction Correction (SIC) [19] at the simplified and most efficient, but robust, level of Average Density SIC (ADSIC) [20]. The latter leads to a proper description of the ionization threshold for a great variety of systems [21]. We also present a few calculations including electronic dynamical correlations at the level of a quantum Relaxation Time Ansatz (RTA) [22, 23].

We represent the state of the system by single-particle (s.p.) wave functions: $\{\varphi_\alpha(\mathbf{r}, t), \alpha = 1, ..., N\}$ where N is the number of active electrons considered in the simulation. The TDLDA time evolution of single-electron wave functions follows the time-dependent Kohn–Sham equations (using atomic units $\hbar$=1, e=1, m_e=0.5)

$$i\partial_t\varphi_\alpha(\mathbf{r},t) = \hat{h}\varphi_\alpha(\mathbf{r},t) = (-\Delta + U_{KS}(\mathbf{r},t) + U_{las}(\mathbf{r},t))\varphi_\alpha(\mathbf{r},t) \tag{1}$$

$$U_{KS}(\mathbf{r},t) = U_{ion}(\mathbf{r}) + \frac{\delta E[\varrho]}{\delta\varrho} \tag{2}$$

where U_{KS} is Kohn–Sham potential at LDA + ADSIC level and U_{laser} the potential delivered by the laser (Sect. 2.3). The TDLDA equation of motion can be recast in matrix form, using the s. p. Hamiltonian $\hat{h}$, as $i\partial_t\hat{\rho} = [\hat{h}, \hat{\rho}]$ where $\hat{\rho}$ is the one-body density matrix. The RTA can then be written as $i\partial_t\hat{\rho} = [\hat{h}, \hat{\rho}] + I_{coll}$ in which the collision term I_{coll} is approximated at RTA level as

$$I_{coll} = \frac{\hat{\rho}_{eq} - \hat{\rho}}{\tau_{RTA}} \tag{3}$$

where $\hat{\rho}_{eq}$ is an instantaneous equilibrium state and τ_{RTA} the relaxation time evaluated semi-classically [22].

The Kohn–Sham potential U_{KS} is composed of the potential of the ionic background U_{ion} complemented by the electronic contribution from Coulomb (Hartree) and exchange–correlation terms. The latter term can be deduced as functional derivative of the energy-density functional $E[\varrho]$ with respect to the key one-electron density ϱ (diagonal part of $\hat{\rho}$) and can be expressed as

$$\varrho(\mathbf{r}) = \sum_\alpha w_\alpha(t)|\varphi_\alpha(\mathbf{r},t)|^2 \tag{4}$$

in terms of the time-dependent s.p. wave functions $\varphi_\alpha(\mathbf{r},t)$ with occupation numbers $w_\alpha(t)$. The summation here runs on all the states taken into account in the simulation. At LDA level, we only consider occupied states which have, by construction, $w_\alpha = 1$ independent of time. At RTA level one has to treat simultaneously a larger set of states, some initially occupied and some unoccupied. Dynamical correlations will depopulate initially occupied states and populate initially unoccupied states. This makes the w_α's fractional and time dependent [22, 24].

The ionic background U_{ion} is provided by pseudopotentials of Goedecker type [25]. For sodium, we use a simpler soft local pseudopotential [26]. We checked that these pseudopotentials reproduce the relevant basic properties (ionization potential (IP), single-electron energies) sufficiently well at ADSIC level. Ions, when moving, follow classical equations of motion at the level of Ehrenfest dynamics [27]

The (real time) TDLDA/RTA equations are solved numerically on a grid in coordinate space with grid spacing optimized to each species, depending on the parameters of the pseudopotential. Box sizes are chosen large enough to avoid any numerical artifact. Static iterations toward the electronic ground state are performed with the damped gradient method [28]. Time evolution employs the time-splitting technique [29], with a step size optimized to each material. Ionization is accounted for via absorbing boundary conditions using a mask function [30] over a large enough absorbing margin. The examples presented below have been computed using our 3D

TDLDA/RTA package QDD [24] or, in some cases, a 2D version thereof. For more details on the numerics, see [24, 31–33].

2.2 *Observables*

We will consider several complementing observables. The first one is the dipole moment

$$D(t) = \int d^3\mathbf{r}\,\mathbf{r}\,\varrho(\mathbf{r}, t). \tag{5}$$

The dipole signal is essential to analyze the dynamical response of any system. When recorded from initializing the dynamics with a small kick, it immediately provides the optical response of the system by spectral analysis [34].

The second observable is the electron content n_α per s.p. wave function:

$$n_\alpha(t) = w_\alpha \int d^3\mathbf{r}\,|\varphi_\alpha(\mathbf{r}, t)|^2. \tag{6}$$

The total number of escaped electrons then reads $N_{esc}(t) = N_0 - \sum_\alpha n_\alpha(t)$, where N_0 is the initial number of electrons.

We also analyze electronic emission dynamics through Photo-Electron Spectra (PES). They are evaluated in the following way. At several measuring points shortly before the absorbing boundaries begin, we record the time evolution of each s.p. wave function. It is then Fourier transformed to the frequency domain, which provides the spectrum of kinetic energies of the escaping electrons. This is the PES [33, 35, 36]. It has to be complemented by an additional phase correction in the case of strong fields [37]. We also consider the Photoelectron Angular Distribution (PAD). At the measuring points used to evaluate the PES, one integrates the signal over time to obtain the number of electrons emitted in a given direction, that is the PAD. Of course PAD and PES can be integrated to recover the total number of emitted electrons. Thus N_{esc} can be obtained by energy integration over the PES and angular integration (with properly computed elementary solid angles) over PAD.

2.3 *Excitation Mechanisms*

In this contribution, we focus on excitations of clusters and molecules using short to ultrashort laser pulses. In the case of ultrafast laser excitations, the frequency selection of the laser is blurred and it then produces an electromagnetic kick rather similar to the one delivered by a swift ion. The tools we are using can be applied to such collisional processes but we shall not address such cases here explicitly.

Our laser pulses are modeled as classical (coherent) linearly polarized (z) photon fields. The corresponding potential reads

$$V_{\text{pulse}} = eE_0 z\, f(t) \cos\left[\omega_{\text{las}}(t - T_{\text{pulse}})\right] \tag{7a}$$

$$f(t) = \begin{cases} \sin^2\left(\pi \dfrac{t}{2T_{\text{pulse}}}\right) & t \in \{0,\ 2T_{\text{pulse}}\} \\ 0 & \text{otherwise.} \end{cases} \tag{7b}$$

The pulse parameters are the frequency ω_{las}, the pulse duration T_{pulse}, and the field strength E_0 (intensity being $I \propto E_0^2$). For a single laser pulse, the laser potential in Eq. (1) $U_{\text{laser}} = V_{\text{pulse}}$. In the case of pump-and-probe scenarios, we consider two successive laser pulses with possibly different parameters and delayed with respect to each other (delay Δt), so that $U_{\text{laser}} = V_{\text{pulse},1} + V_{\text{pulse},2}[\Delta t]$. Note that in the cases considered here we keep the laser polarization the same for the pump-and-probe pulses.

3 Dynamics in the Fs Range

3.1 Ionization Asymmetry and Carrier Phase Envelop Effects

Extremely short laser pulses become sensitive to one specific pulse parameter (negligible for longer pulses), namely, the phase of the oscillating part of the laser pulse relative to the pulse envelope. The latter quantity is called the Carrier Envelope Phase (CEP) and is illustrated in the left panels of Fig. 1. It is obvious that the shorter the pulse (left upper panel) the larger the effect of the CEP: the sequence of forward–backward pushes becomes strongly CEP dependent which changes the impact on the system. Furthermore, such effects typically occur at the time scale of the electronic response. One thus expects the CEP to have a dramatic impact on the direction of emission of electrons and this should be visible on the PAD, the effect fading out progressively for longer pulses. In such cases, the large number of oscillations within the envelop tend to average out a preferred direction.

In order to simply analyze such effects, we compress the PAD to one relevant number and seek for an effect on the direction of emission as a function of CEP. An appropriate observable is the asymmetry [39], which one can express from the emission yield $\mathcal{Y}(\theta)$ along a direction θ, the latter being measured with respect to the laser polarization axis $\theta = 0$

$$\eta = \frac{\int_0^{\theta_0} d\theta\ (\mathcal{Y}(\theta) - \mathcal{Y}(\pi - \theta))}{\int_0^{\theta_0} d\theta\ (\mathcal{Y}(\theta) + \mathcal{Y}(\pi - \theta))}. \tag{8}$$

In Eq. (8) θ_0 defines a cone around laser polarization axis within which one gathers the emission signal. In the right panel of Fig. 1, a value of $\theta_0 = \pi/12$ has been chosen

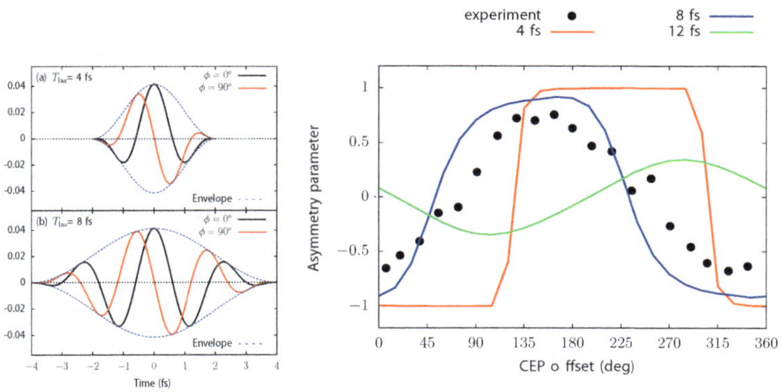

Fig. 1 Impact of the Carrier Envelope Phase (CEP) of the photon pulse. Test case is C_{60}. Left panels: pulse profiles for two values of CEP $\phi = 0^o$ and 90°. Pulse duration of 4 fs pulse (upper) and 8 fs (lower). The pulse envelope is plotted with dashed lines. Right panel: TDLDA asymmetry parameter η Eq. (8), as a function of the CEP for different pulse lengths in comparison to experimental results of [38]. Adapted from [39]

and η is shown as a function of CEP for a few pulse lengths. In the case of the shortest pulse length (4 fs), one observes a sharp transition between forward and backward emission as a function of CEP. The transition smears out with increasing pulse length up to almost disappear already for the longest 12 fs pulse. The actual experimental pulse length is not well known. But the experimental η shows a pronounced forward–backward change with CEP. The comparison with TDLDA values of η points out toward an experimental pulse length around 8 fs.

3.2 *Long and Short Trajectories of Emitted Electrons*

Emission from nanometric tips subject to possibly intense laser fields constitutes a promising source of a well-collimated and coherent beam of ultrafast electrons [41]. These electrons could be used for diffraction experiments [42] as well as electron microscopy [43]. The regime in which emitted electrons recollide with the tip in the course of the laser pulse might be especially interesting, as it could allow to generate extremely short electron pulses [44] and thus pave the way to subfemtosecond and subnanometric probing of matter [45]. For testing purposes, we replace here the tip by a C_{60} and study the response of the system to ultrashort pulses in terms of recollision scenarios.

The results of our analysis are displayed in Fig. 2. Let us first focus on its lower panel where the PES is displayed. One observes there a plateau, modulated by some oscillations, which are the candidate to recollision scenarios. Similar structures have

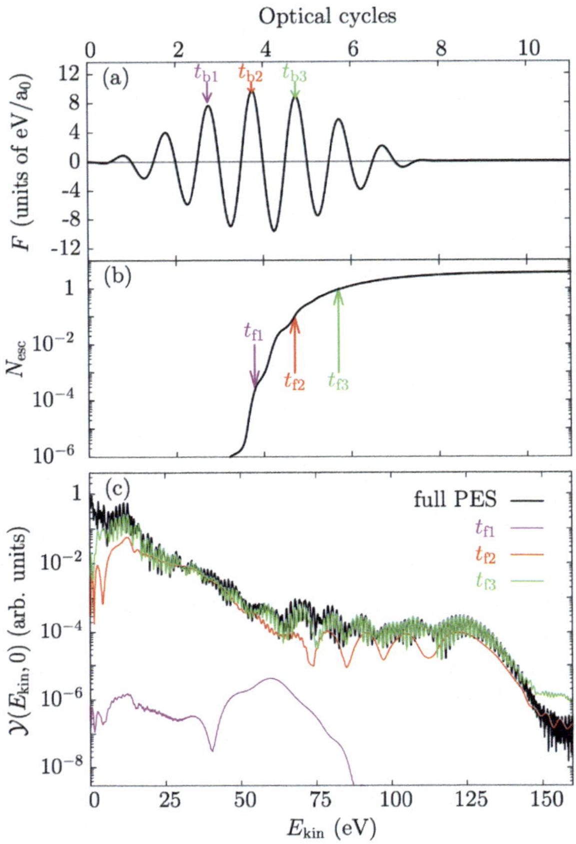

Fig. 2 Time-resolved analysis of photoemission from C60 irradiation. Upper and middle panels, both as a function time: force acting on electrons (including "birth" times) and ionization (including "final" time), respectively. Lower panel: PES. Laser duration is $T_{\text{pulse}} = 24$ fs, intensity 1.6×10^{14} W/cm^{-2} and frequency $\omega_{\text{las}} = 1.36$ eV. From [40]

been experimentally observed in the photoemission spectra of rare gases, for example, in Argon atom [46], following irradiations by strong infrared laser pulses. These structures have been interpreted as stemming from interference effects between several electron trajectories either within the same optical cycle or in subsequent cycles, but leading to the same final states [47]. Two types of trajectories have been identified, labeled as "short" and "long" trajectories, due to their different excursion times. Within the three-step model [48] the electrons are born close to the field maxima, the place where the tunnel ionization probability is maximal.

We focus here on electron emission in the forward direction and thus choose instants based on a classical picture of electron emission. We denote by $t_{b/1/2/3}$ the "birth" times of the electrons taken at the three largest maxima of the electric force (see upper panel of Fig. 2). They correspond to 2.75, 3.75, and 4.75 τ_{IR}, where τ_{IR} = 3 fs is the single optical cycle duration (laser period $\tau_{IR} = 2\pi/\omega_{\text{las}}$). We then correlate them with the associated detection (final) times $t_{f/1/2/3}$. They are separated by a total delay time δt which depends on kinetic energy. It is obtained by adding the return time of the electron and the time to reach an "asymptotic" detection point at 90 a_0 from the cluster. This leads to the estimated "final" times $t_{f/1/2/3} = 3.79$, 4.71 and 5.71 τ_{IR} also indicated in the middle panel of Fig. 2.

The bottom panel of Fig. 2 finally plots the associated PES which allows to understand how the full PES builds up in time. This analysis should make sense for high-energy electrons. Clearly, contributions from t_{f1} are negligible probably because the field at t_{b1} is still too small. Most of the PES, including plateau and modulations, is obtained for t_{f2} corresponding to a t_{b2} at the highest field value of the pulse. Modulations can be associated to interferences from two electron trajectories and can in this case be coined as "intracycle" interferences, i.e., interferences stemming from electron trajectories originating during the same optical period. The PES at t_{f3} only adds high-frequency modulations in the spectra. These ones stem from interferences between electron trajectories born at times separated by the optical period, and can thus be coined as "intercycle" interferences. We thus recover, in this complex case of C_{60} basic mechanisms very similar to the ones observed in simpler systems.

4 Dynamics with Atto Pulses

4.1 Ionization Modulation with Time Delay

One of the striking results of experiments combining a long IR pump pulse to a long XUV atto train is the observed modulation of ionization with delay time Δt between the two pulses. The effect has been observed both in atoms [45, 49, 50] and simple molecules [11–13]. It has been analyzed both in model calculations [51] and in numerous more or less realistic simulations ranging from Time-Dependent Schrödinger Equation (TDSE, [52]) to TDLDA [39]. In terms of total ionization, where the modulation of signal is particularly striking, the agreement between experimental and theoretical results is quite good. For more detailed quantities, such as the PES, less data are available and accordingly less computations have been performed. We look here at the example of the Ar atom.

Results for PES in Ar are shown in Fig. 3 where we compare TDLDA results to experimental ones. The PES yield is plotted as a function of kinetic energy (horizontal) and delay Δt (vertical) between pump-and-probe pulses. The comparison with experimental results has nevertheless to be taken with a grain of salt as we did not have access to details of the post-processing of experimental data. Still, the figure shows similar (theory versus experiments) emission pattern reaching far out to high electron energies.

At high energies, the pronounced fringes have previously been interpreted as a ponderomotive effect in the case of strong IR fields. The idea is then that the momentum of electrons is augmented by the vector potential at the moment of ionization. One then expects to see more electron emission for energies greater than 25 eV.

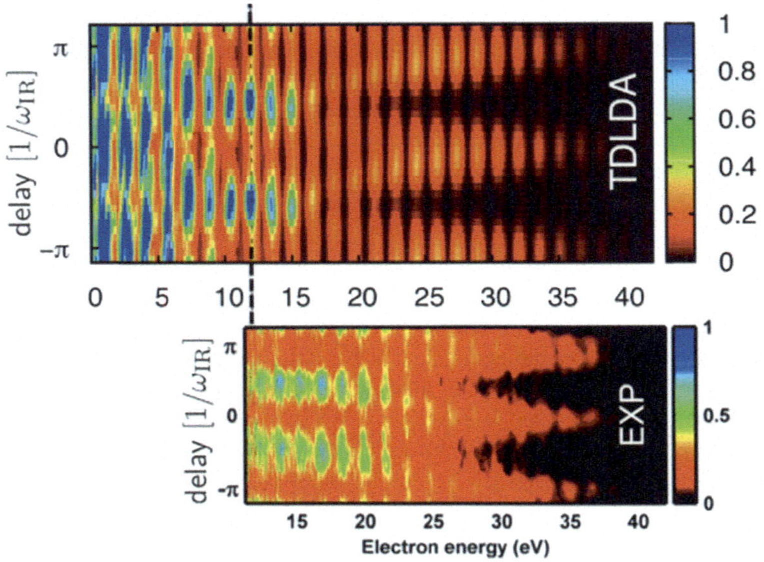

Fig. 3 Plot in an Ar atom of the PES in the plane of kinetic energy and the time delay. Top panel: TDLDA calculations. Bottom panel: Experimental data for kinetic energies between 12 and 42 eV. The vertical dashed line indicates the onset of the energy range from experiments. In each panel, the scale is rescaled to a maximum value of the PES set to 1. Laser parameters are: Atto pulse intensity $I_X = 10^{12}$ W/cm^{-2}, frequency $\omega_X = 30$ eV, duration $T_X = 0.21$ fs; recurrence time of XUV pulses $T_{train} = \tau_{IR}/2 = 1.33$ fs, number of atto pulses $n_{APT} = 7$; IR frequency $\omega_{IR} = 1.55$ eV, intensity $I_{IR} = 3\times10^{13}$ W/cm^{-2} and duration $T_{IR} = 105$ fs

TDLDA results, in this respect, do reproduce the main characteristics of the experiment. In the low-energy regime (≤ 10 eV), the situation is different with a very weak dependence on delay time, but this unfortunately occurs below the experimental energy window.

4.2 *Structure of the PES*

It is interesting to explore a bit more in detail the content of the PES in such a laser setup with an IR pump followed by an atto train probe with variable delay. For a laser pulse characterized by a single frequency ω_{las}, the simple rule for analyzing PES spectra is to consider it as a mapping of the single-particle Density of States $\{\varepsilon_\alpha\}$, namely, to write the PES yield as

$$\mathcal{Y}(E_{kin}) \leftrightarrow \sum_\nu \sum_\alpha \delta(E_{kin} - E_{\alpha,\nu}) \tag{9a}$$

$$E_{\alpha,\nu} = \varepsilon_\alpha + \nu\omega_{las}, \tag{9b}$$

where ν is the number of involved photons to emit electron α from its level of energy ε_α. The situation becomes much more involved when considering the above pump-and-probe setup with an atto train. And the simple peak attribution of Eq. (9b) fails. One has then to consider a much wider perspective and integrate into the analysis of the PES, the other possible mechanisms of emission, as mediated for example by collective motion (dipole, quadrupole, etc.).

This point is illustrated in Fig. 4 in the case of the Helium atom. Because of the more involved frequency content of the laser signal delivered by the IR + XUV setup one is bound to analyze the PES by replacing ω_{las} by the full dipole response of the signal $P_D(\omega = E_{\text{kin}} - \varepsilon_{1s})$, in which the sequence of single-particle levels is reduced to the 1 s state in the case of He. Mind that here $P_D \propto |\tilde{D}(\omega)|^2$ where $\tilde{D}(\omega)$ is the Fourier transform of the electronic dipole in time Eq. (5). As can be seen from Fig. 4, only part of the PES peaks can be understood this way. One has to recur to the quadrupole response P_Q to unambiguously identify PES peaks. Again P_Q is the power spectrum associated to electronic quadrupole, in the same way as the dipole. In the present case, the monopole response (or higher multipolarity contributions) turns out to be strongly suppressed and is thus not included in the figure, but it should in principle also be included for a proper comparison. Note that both P_D and P_Q take similar values after the same x-axis rescaling to make them match with the PES scale. This clearly indicates that both dipole and quadrupole channels do respond with comparable amplitudes. The sum of both P_D and P_Q contributions provides a perfect reproduction of the PES peaks both in terms of positions and, to a large

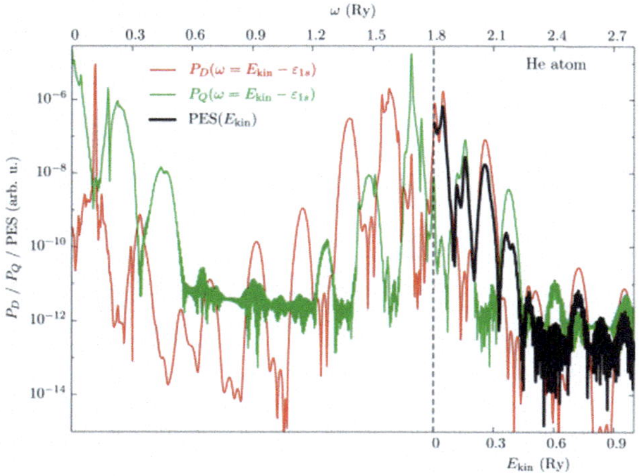

Fig. 4 PES (black) , dipole P_D (red) an quadrupole P_Q (green) power for a He atom excited by a combination of an XUV attotrain with an IR pulse. The vertical solid line indicates the ionization threshold. The P_D and P_Q have been scaled with the same factor to match the height of the PES peaks. Laser parameters are: IR frequency of 1.56 eV, duration 105 fs and intensity 1011 W.cm^{-2}; XUV train of 7 pulses each of 1 fs duration and frequency 23 eV and intensity 1012 W.cm^{-2}. Extracted from [53]

extent, of relative amplitudes. In particular, the quadrupole spectrum P_Q precisely explains the peaks missed by the dipole spectrum.

Altogether, in order to understand the structure of the PES obtained with complex IR + XUV atto train setups, one is thus bound to generalize the simple rule of analysis given by Eq. (9b) to include relative contributions from the electronic response in its dominant channels, namely, monopole, dipole, and quadrupole, as the first moments of the electronic cloud. Altogether this delivers an expression

$$\mathcal{Y}(E_{\text{kin}}) \leftrightarrow \sum_{\alpha}[\eta_M P_M(E_{\text{kin}} - E_{\alpha,\nu}) + \eta_D P_D(E_{\text{kin}} - E_{\alpha,\nu}) + \eta_Q P_Q(E_{\text{kin}} - E_{\alpha,\nu}) + \ldots], \quad (10)$$

where weights η's will depend on the actual situation. Of course this generalized expression contains the rule Eq. (9b) which corresponds to the simple case for which $P_D(\omega) \sim \delta(\omega)$. We have included here only the first multipoles but one could include more terms to attain an even better description of the excitation spectrum of the system, if needed.

4.3 Physical Content

The striking feature of the ionization pattern following irradiation by a long IR pump pulse followed by a rather long train of XUV pulses as probe is the remarkable regularity of the modulation of ionization. It has to be kept in mind that, for technical reasons, the repetition rate of the atto pulses is twice the IR frequency. In other words, atto pulses are separated by half the IR period. A simple way to visualize ionization modulations, with respect to time delay, is thus to switch to frequency domain by Fourier transforming ionization signal from delay time Δt to corresponding frequency. One can then define a Power of Yield (PoY) as

$$\text{PoY}(\omega) = |\int d\Delta t[\mathcal{Y}(\Delta t) - \mathcal{Y}_{\text{lin}}(\Delta t)]|^2, \quad (11)$$

where $\mathcal{Y}_{\text{lin}}$ subtracts the global linear trend of the signal to reduce background in the Fourier spectrum. A typical PoY is shown in the upper panel of Fig. 5. It merges both experimental and theoretical results and clearly points out the dominance of the double IR frequency, which is the frequency corresponding to the repetition rate of XUV pulses in the train.

The analysis can be further extended to include PES and PAD characteristics of the signal. This can be done only by characterizing the latter quantities, each by one number. For the PES one computes the average kinetic energy $\overline{E}_{\text{kin}}$ which is simply the energy weighted integral of the PES over kinetic energy. Note that the total ionization is the integral of the PES over kinetic energy. The PAD is characterized by its shape (or quadrupole) anisotropy β_2, which is the angle-integrated PAD pondered

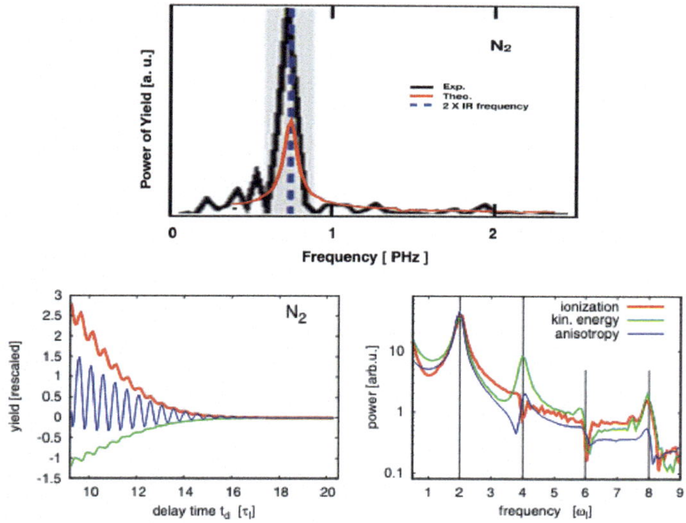

Fig. 5 Upper panel: Comparison of experimental and theoretical PoYs in N_2 following an IR pump and XUV train probe excitation: IR frequency $\omega_{\text{IR}} = 0.4$ eV, intensity 10^{12} W/cm^{-2}, total duration 30 fs, XUV frequency ω_{XUV}= 25 eV, intensity10^9 W/cm^{-2} with 22 atto pulses in the train. The gray band corresponds to the experimental frequency window used to recover ionization oscillations. Lower panels: Analysis of the upper panel setup on a longer time window, considering both PoY and power spectra associated to average kinetic energy and to shape anisotropy. Signals as a function of delay $t_d = \Delta t$ (in units of IR period $\tau_{\text{I}} = \tau_{\text{IR}}$) in the left lower panel, as a function of frequency (in units of IR frequency $\omega_{\text{I}} = \omega_{\text{IR}}$) in the right lower panels. Data have been rescaled to asymptotic zero value to fit the same figure. Adapted from [54]

by the second Legendre polynomial P_2. Once the average kinetic energy and β_2 anisotropy are extracted as a function of delay time they are Fourier transformed to frequency in the same way as the PoY. The bottom panel of Fig. 5 shows the computed results (theory) for the three quantities in delay time (left part) and in frequency domain (right part). The window of delay times is taken in the second half of the pulse as can be seen from the overall decrease (modulo oscillations) of the net ionization, following the IR pulse envelop. The three observables show different global trends as a function of delay time. But the detailed regularly repeated pattern within one IR cycle are the same. As a result, the PoY and power spectra (right panel) show the same pattern for all three observables with peaks of decreasing strength only at even multiples of the IR frequency.

The conclusion is clear and rather obvious: the spectra are fully dominated by the IR frequency and multiples thereof. This means, in particular, that one does not learn anything specific to the system. The ionization signals are dominated by the laser. In order to better understand the origin of these features, we developed a simple, close to analytical, model [54]. The system has two bound levels, one occupied, one empty, and a continuum. The derivation relies on standard methods of strong field physics. For a single atto pulse, the solution is analytical and leads to a total yield $\mathcal{Y}$ as a function of delay time Δt

$$\mathcal{Y} \propto (c_1 - c_2[E_{\text{IR}} + E_{\text{XUV}}(\Delta t)]^2), \tag{12}$$

where E_{IR} and E_{XUV} are the IR and XUV field amplitudes, respectively, c_1 and c_2 are constants depending on the energy of the occupied level, the amplitude of dipole transitions and XUV frequency. The important point in Eq. (12) is the quadratic dependence on the sum of IR and XUV amplitudes which immediately leads to the $2\omega_{\text{IR}}$ dominant frequency content of the ionization signal. This simple model thus fully confirms what any realistic computations obtain, in accordance with experimental results. Even more so, it is interesting to note that the 2 ω_{IR} repetition rate of atto pulses in only enforcing and strengthening the weight of 2 ω_{IR} within the frequency content of the ionization modulation. Indeed with only one atto pulse the dependency is already present. Still, it might be interesting to try to avoid strengthening this 2 ω_{IR} frequency component by considering a different laser setup without the atto train modulated by the IR pulse, the hope being to attain system's specific properties and not only laser characteristics. This is explored in Sect. 4.4.

4.4 A Streaking Example

In order to reduce the role of the IR laser, we now consider a different laser setup. We choose an ultrashort pump IR pulse of duration around its own period τ_{IR}. This implies a huge impact of the CEP, which we thus choose in order to maximize the laser quick. The probe XUV pulse has itself a very short duration, below τ_{IR}. And the XUV pulse is not repeated (no train).

Observables are the same ones as in Sect. 4.3, namely, the total ionization yield N_{esc}, the average kinetic energy $\overline{E}_{\text{kin}}$, and the quadrupole anisotropy β_2. They are complemented here by the dipole anisotropy β_1 (left right asymmetry), which is the angle-integrated PAD pondered by the first Legendre polynomial P_1. Both β_1 and β_2 are evaluated with respect to laser polarization axis.

As a first test case, we chose a simple metal cluster Na_9^+. Of course the laser setup allows to avoid any overwhelming of the IR pump. But all ionization signals become dominated by the dipole plasmon which, due to the very short duration of the IR pulse, is immediately excited, independent of the IR frequency. The net results are thus rather disappointing but might be biased by the fact that Na_9^+ has indeed a very clean and strong dipole plasmon, which thus tends to easily dominate any dynamical response. The next step is thus to consider another system. We take the case of a small organic molecule, namely, C_2H_2. Results of our analysis of ionization are presented in Fig. 6. The figure shows that the IR+XUV analysis is not that straightforward. It is much fragmented even more than the pure dipole response itself (upper panels). The different observables do produce quite different power spectra. This holds true for all signals except for the emission asymmetry β_1.

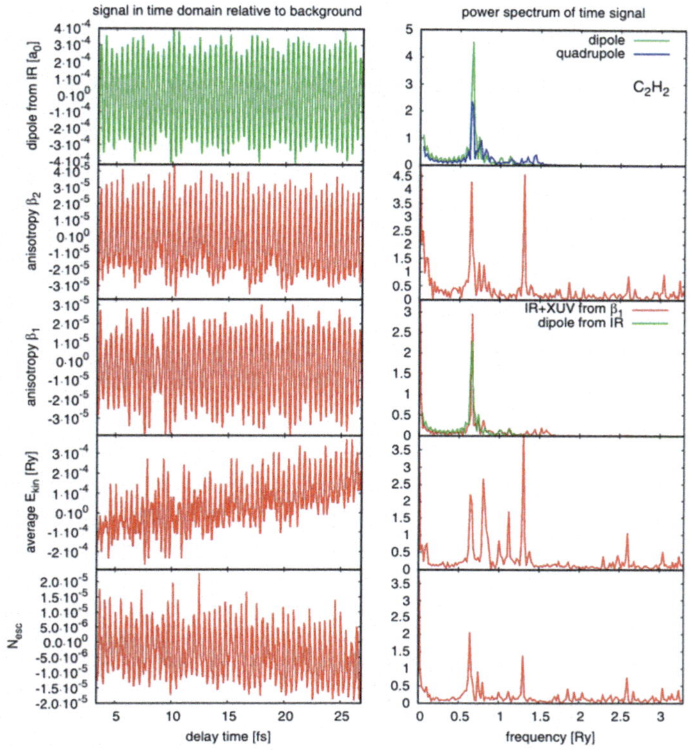

Fig. 6 Signals from different observables in time domain (left panels) and frequency domain (right panels) for C_2H_2. The laser has the following parameters: IR intensity $I_{\text{IR}} = 10^{13}\text{W/cm}^2$, IR frequency ω_{IR} 1.36 eV, IR pulse length T_{IR} = 3.04 fs (= IR period); XUV intensity $I_{\text{X}} = 10^{14}\text{W/cm}^2$, XUV frequency ω_{X}= 40.8 eV; XUV pulse length $T_{\text{X}} = \tau_{\text{IR}}/6 = 0.51$fs which means 5 XUV cycles in the pulse. Observables are ionization yield N_{esc}, average kinetic energy $\overline{E}_{\text{kin}}$, dipole anisotropy β_1 (left right asymmetry along laser polarization axis), and quadrupole anisotropy β_2. The uppermost panels show dipole signal following pure IR excitation. The upper right panel shows in addition the power spectrum of the quadrupole following pure IR excitation. From [55]

We have also considered higher multipole responses than the mere dipole. The quadrupole spectrum is thus also displayed in the right upper panel. It nevertheless does not bring new insight. A striking feature of all spectra, but the dipole asymmetry, is the appearance of a strong peak at twice the frequency of the strongest dipole peak in the spectrum. If one imagines that the whole dipole spectrum, possibly also with contributions of the quadrupole spectrum, is frequency doubled by folding the pure IR spectra with themselves one could explain the the spectra from a superposition of dipole and quadrupole contributions (yield $\mathcal{Y} \propto c_D(D_0 + D(t))^{\gamma_D} + c_Q(Q_0 + Q(t))^{\gamma_Q}$). This is in line with our analysis of PES (Sect. 4.2). This argument, though, remains qualitative and calls for more detailed analysis of such involved spectra.

Contrarily to other signals, the dipole anisotropy β_1 nevertheless displays a very clean signal by itself, directly matching the dipole response. This is particularly encouraging as β_1 is a rather simple and robust quantity to access experimentally speaking and we know that it provides a direct indicator of the CEP in the case of short IR pulses (Sect. 3.1 and [53]). This promising result might call for further studies under varied conditions (systems, pulse parameters) but it has to be noted that the setup to reach such an effect is rather involved and one might question whether it is worth the effort to reach a quantity accessible with other means.

4.5 Surprises with Ultrashort XUV Pulses

After having explored long IR pumps + atto train (Sect. 4.1) and shorter IR pulses plus ultrashort XUV probes (Sect. 4.4), it is interesting to consider another type of setup now starting with an ultrafast XUV pump. One might of course consider pump-and-probe setups there as well, but, as we shall see, such a simple pump setup is already delivering quite surprising results and thus deserves a dedicated investigation by itself.

Ultrafast XUV pulses have been used since some time in experiments for exploring UF dynamics [5]. From a theoretical point of view, a simple way to model such an excitation is to create an instantaneous hole [56, 57] and then follow the ensuing response of the system. This considerably simplifies the picture: depending on the XUV frequency one wants to mock up, one will deplete a more of less deep level, for example, by setting its occupation number to zero at initial time. Such an excitation model is routinely used, for example, to explore charge migration and charge transfer in molecular systems [58–62].

Such an initialization nevertheless overlooks the actual mechanism and rearrangements taking place during the emission of the electron. This is probably harmless when considering processes on "long" times but when considering UF processes down or below 1 fs the "instantaneous" excitation loses its meaning because of an obvious conflict of time scales. It is thus interesting to look for laser setups leading to such UF processes and to follow them explicitly in time. This is what we are considering here, in the case of the simple N_2 molecule.

The dipole response Eq. (5) of an irradiated N_2 is displayed in Fig. 7 as a function of time. The XUV frequency is $\omega_{\text{XUV}} = 63.9$ eV well out of resonance of dominant optical transitions in N_2. One thus expects the dipole signal to die out with the driving laser pulse. This is effectively what one observes. But the surprise comes from the fact that after a few fs one observes a revival of the dipole. The effect is specific to a frequency range around 60 eV and is associated to a clear population inversion. The most deeply bound electron level is then strongly depleted, while higher energy levels remain much more populated.

The result of Fig. 7 is not an accident. We found numerous cases and laser scenarios displaying the same instability in the dipole signal. The key quantity seems to be the deposited excitation energy (resulting from the energy absorbed from the laser minus

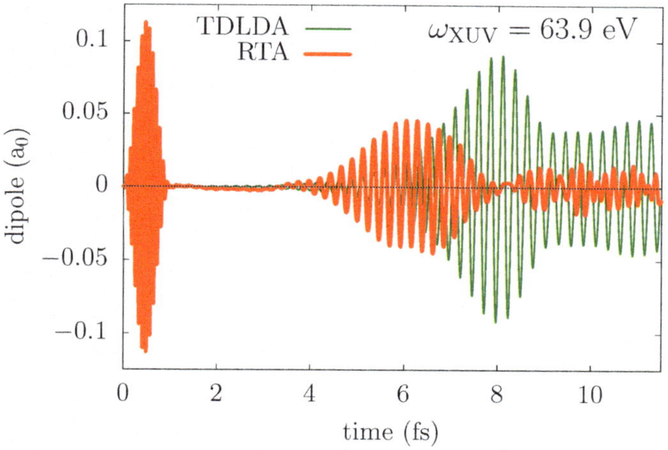

Fig. 7 Time evolution of the dipole moment within TDLDA and RTA. Laser frequency is $\omega_{XUV} =$ 63.9 eV, pulse duration 1 fs and intensity 1.12×10^{16} W/cm^{-2}

the one carried away by electronic emission) or alternatively (both quantities seem strongly correlated) the amount of depletion, especially the one of the deepest bound levels. Another important degree of freedom, namely, ionic motion, does not play any role in that case. This makes sense in view of the extremely short time scales and ion masses. One should nevertheless also consider the level of theory for describing electronic motion. In particular, the potential impact of dynamical electron–electron correlations requires a closer look. We thus also tested electronic dissipation within RTA [22, 24] which accounts for incoherent dynamical correlations. The result is also shown in Fig. 7 in comparison to the TDLDA time evolution of the dipole moment. Dissipation somewhat reduces the dipole instability (at least in amplitude) but sizable dipole oscillations remain. This may not be surprising when considering associated time scales. The dipole amplitude increases one order of magnitude in 2 fs while, at the given excitation energy, the relaxation time for N_2 is about 5 fs. There is thus competition but RTA cannot fully win. In order to observe the instability XUV pulses have thus to be shorter than the RTA relaxation time and shorter than the instability time. In the UF cases considered here, the results tend to validate TDLDA.

Depletion and dipole moment are nevertheless not directly visible from experiments. But one could identify the instability via the PES. Figure 8 shows the PES after an irradiation of N_2 by a 1fs pulse with $\omega_{XUV} = 63.9$ eV and intensity 1.2×10^{16} W/cm^{-2}. One compares different PES: one corresponding to an early time window below 10 fs, basically before the instability becomes visible and one at late times (100 fs). The early PES exhibits the standard peaks associated to a standard one-photon process (Sect. 4.2) with energies at $\varepsilon_\alpha + \omega_{XUV}$ (indicated as vertical dashes in Fig. 8). These peaks are very broad because of the shortness of the pulse and of the strong downshift of s.p. energies (about 8 eV with respect to the ground state s.p. energies) due to the sizable ionization and associated Coulomb charging [33]. The late PES

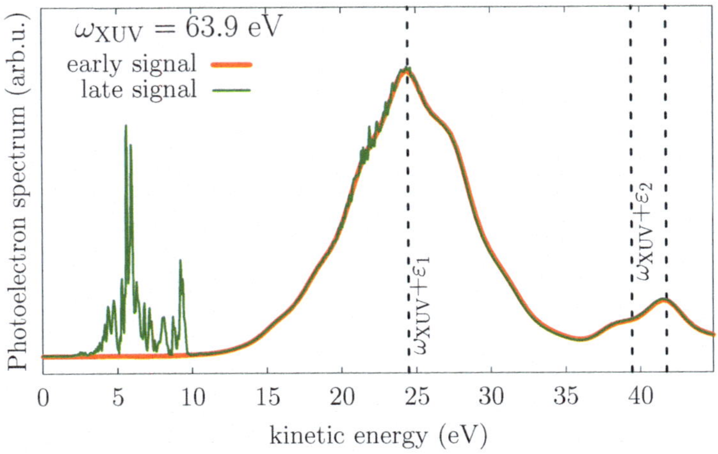

Fig. 8 PES (linear scale), following a 1 fs XUV pulse of frequency $\omega_{XUV} = 63.9$ eV, and intensity 1.2×10^{16} W/cm^{-2}, computed at 10 fs (thick green line) and at 100 fs (thin black line). The faint dashed vertical lines indicate the s.p. energies shifted by the pulse frequency, see text for detail

consists of the early PES plus an additional low-energy structure. The latter does not fit to any combination $\varepsilon_\alpha + n\omega_{XUV}$ (with integer n) and is interpreted as a signature of the dipole instability. Indeed when Fourier transforming the dipole moment signal at late times, one obtains a dominant oscillation frequency of $\omega_{instab} = 16.4$ eV. And it turns out that the low-energy peaks might correspond to the actual s.p. energies (IP$\approx$ 28.6 eV) plus $2\omega_{instab}$.

Investigations on the dipole instability are still going on. Similar instabilities are observed in numerous laser setups. We checked in particular longer pulses (up to 30 fs) at lower intensities (down to some 10^{13} W/cm^{-2}). We also observed similar effects in other molecules. Such investigations are touchy as one is facing unstable situations in which any detail matters. We did not find yet any cases in which RTA could fully suppress the instability observed at TDLDA level, because time scales were not favorable. But there might be cases in which the situation could be different. Still, qualitatively, the instability is robust once a population inversion is attained. And such a population inversion is intrinsic, by construction, to any instantaneous hole excitation, as often modeled to describe UF processes in clusters and molecules. There is thus an interesting question to disentangle here, in particular, at experimental level to see whether one is facing a physical phenomenon or just facing a limitation of the theory.

5 Conclusions

In this paper, we have investigated the theoretical description of fast and Ultrafast (UF) processes in clusters and molecules following irradiation by short laser pulses. Such situations provide crucial insights into the understanding of irradiation processes in matter. We concentrated here on laser excitations but radiation damage by swift particles is equally important and can be analyzed, to a large extent, with similar methods. In the limit of UF processes, involved mechanisms are in addition rather similar when considering collision with fast particles or laser irradiation with UF pulses.

The basic theoretical tool is TDDFT used at TDLDA level complemented by a simple self-interaction correction to ensure proper ionization characteristics. Dynamical equations are solved in real time which allows to consider strongly out of equilibrium situations. In the latter cases, TDLDA can be further complemented by a Relaxation Time Ansatz (RTA) which accounts, in an approximate but realistic way, for incoherent correlations beyond the effective mean level provided by TDLDA. Key observables to analyze the dynamics are the optical response and ionization. The latter can be dealt with at various levels, from total ionization to energy (PES) and/or angle (PAD) resolved signals.

We started by investigating short processes in the fs range. We have seen the key role played by the Carrier Phase Envelop (CEP) on such cases. We have also analyzed in detail ionization mechanisms in complex systems such as C_{60}, following simple models developed in atoms. We have seen that we could recover similar behaviors.

The next step was to consider various pump-and-probe combinations in the attosecond regime. The widely used long IR pump + long XUV atto train probe has been analyzed in various systems. Regular modulations of ionization observed both in total ionization and in PES and PAD signals basically reflect the laser pulse, which blurs any access to specific system's properties. When considering shorter pulses down to the fs total duration for both IR and XUV pulses ionization signals may become more involved and difficult to analyze. We identified the left–right asymmetry of emission (with respect to laser polarization axis) as a promising observable directly reflecting a specific systems property, namely, its optical response. One may nevertheless wonder whether such a complicated pump-and-probe setup is well suited to attain such properties, accessible by other/simpler means.

We finally explored UF XUV pulses leading to the creation of a deeply lying hole. Such setups lead to an instability of the dipole response well after the laser pulse is over. Such an effect could be identified from the low-energy part of the PES on long times, well after the laser pulse is over. Because of unfavorable time scales the instability is only marginally affected by inclusion of incoherent dynamical correlations at RTA level. More investigations, both experimental and theoretical ones, would certainly bring a better understanding of the phenomenon. Work along that line is in progress.

Acknowledgements This work was granted access to the HPC resources of CalMiP (Calcul en Midi-Pyrénées) under the allocation P1238, and of RRZE (Regionales Rechenzentrum Erlangen).

References

1. J.L. Autran, D. Munteanu, *Soft Errors: From Particles to Circuits* (CRC Press, London, 2015)
2. A. Solov'yov (ed.), *Nanoscale Insights into Ion-Beam Cancer Therapies* (Springer, Cham, 2017)
3. U. Keller, Nature **424**, 831 (2003)
4. T. Brabec, F. Krausz, Rev. Mod. Phys. **72**, 545 (2000)
5. F. Krausz, M. Ivanov, Rev. Mod. Phys. **81**, 163 (2009)
6. T. Fennel, K.-H. Meiwes-Broer, J. Tiggesbäumker, P.-G. Reinhard, P.M. Dinh, E. Suraud, Rev. Mod. Phys. **82**, 1793 (2010)
7. C. Bordas, F. Paulig, H. Heln, D.L. Huestis, Rev. Sci. Instr. **67**, 2257 (1996)
8. I. Seideman, Ann. Rev. Phys. Chem. **53**, 41 (2002)
9. G. Paulus et al., prl **91**, 253004 (2003)
10. M. Holler, F. Schapper, L. Gallmann, U. Keller, prl **106**, 123601 (2011)
11. G. Sansone, F. Kelkensberg, J.F. Pérez-Torres, F. Morales, M.F. Kling, W. Siu, O. Ghafur, P. Johnsson, M. Swoboda, E. Benedetti, F. Ferrari, F. Lépine, J.L. Sanz-Vicario, S. Zherebtsov, I. Znakovskaya, A. L'Huillier, MYu. Ivanov, M. Nisoli, F. Martín, M.J.J. Vrakking, Nature **465**, 763 (2010)
12. F. Kelkensberg, W. Siu, J.F. Pérez-Torres, F. Morales, G. Gademann, A. Rouzée, P. Johnsson, M. Lucchini, F. Calegari, J.L. Sanz-Vicario, F. Martín, M.J.J. Vrakking, Phys. Rev. Lett. **107**, 043002 (2011)
13. C. Neidel, J. Klei, C. Yang, A. Rouzée, M.J.J. Vrakking, K. Klünder, M. Miranda, C.L. Arnold, T. Fordell, A. L'Huillier, M. Gisselbrecht, P. Johnsson, M.P. Dinh, E. Suraud, P.-G. Reinhard, V. Despré, M.A.L. Marques, F. Lépine, Phys. Rev. Lett. **111**, 033001 (2013)
14. E. Runge, E.K.U. Gross, Phys. Rev. Lett. **52**, 997 (1984)
15. M.A. Marques, C.A. Ullrich, F. Nogueira, A. Rubio, K. Burke, E.K. Gross, *Time Dependent Density Functional Theory* (Springer, Berlin, 2006)
16. H. Haken, H.C. Wolf, *Time-Dependent Density-Functional Theory: Concepts and Applications* (Oxford University Press, Oxford, 2012)
17. J.P. Perdew, Y. Wang, Phys. Rev. B **45**, 13244 (1992)
18. T. Koopmans, Physica **1**, 104 (1934)
19. J.P. Perdew, A. Zunger, Phys. Rev. B **23**, 5048 (1981)
20. C. Legrand, E. Suraud, P.-G. Reinhard, J. Phys. B **35**, 1115 (2002)
21. P. Klüpfel, P.M. Dinh, P.-G. Reinhard, E. Suraud, Phys. Rev. A **88**, 052501 (2013)
22. P.-G. Reinhard, E. Suraud, Ann. Phys. (N.Y.) **354**, 183 (2015)
23. P.M. Dinh, L. Lacombe, P.-G. Reinhard, E. Suraud, M. Vincendon, Eur. Phys. J. B **91**, 246 (2018)
24. P.M. Dinh, M. Vincendona, F. Coppens, E. Suraud, P.-G. Reinhard, Comput. Phys. Commun. **270**, 108155 (2022)
25. S. Goedecker, M. Teter, J. Hutter. Phys. Rev. B **54**, 1703 (1996)
26. S. Kümmel, J. Akole, M. Manninen, Phys. Rev. Lett. **84**, 3827 (2000)
27. P. Ehrenfest, Zeit. Phys. **45**, 455 (1927)
28. P.-G. Reinhard, R.Y. Cusson, Nucl. Phys. A **378**, 418 (1982)
29. M.D. Feit, J.A. Fleck, A. Steiger, J. Comput. Phys. **47**, 412 (1982)
30. C.A. Ullrich, J. Mol. Struct. (THEOCHEM) **501–502**, 315 (2000)
31. F. Calvayrac, P.-G. Reinhard, E. Suraud, C.A. Ullrich, Phys. Rep. **337**, 493 (2000)
32. P.-G. Reinhard, E. Suraud, *Introduction to Cluster Dynamics* (Wiley, New York, 2004)
33. P. Wopperer, P.M. Dinh, P.-G. Reinhard, E. Suraud, Phys. Rep. **562**, 1 (2015)
34. F. Calvayrac, P.-G. Reinhard, E. Suraud, Ann. Phys. (N.Y.) **255**, 125 (1997)
35. A. Pohl, P.-G. Reinhard, E. Suraud, Phys. Rev. Lett. **84**, 5090 (2000)
36. A. Pohl, P.-G. Reinhard, E. Suraud, J. Phys. B **34**, 4969 (2001)
37. P.M. Dinh, P. Romaniello, P.-G. Reinhard, E. Suraud, Phys. Rev. A **87**, 032514 (2013)
38. H. Li, B. Mignolet, G. Wachter, S. Skruszewicz, S. Zherebtsov, F. Süßmann, A. Kessel, S.A. Trushin, N.G. Kling, M. Kübel, et al., Phys. Rev. Lett. **114**, 123004 (2015)

39. C.-Z. Gao, P.M. Dinh, P.-G. Reinhard, E. Suraud, Phys. Chem. Chem. Phys. **19**, 19784 (2017)
40. C.-Z. Gao, P.-M. Dinh, P. Kluepfel, C. Meier, P.-G. Reinhard, E. Suraud, Phys. Rev. A **93**, 022506 (2016)
41. S. Hilbert, A. Neukirch, C. Uiterwaal, H. Batelaan, JPB. **42**, 141001 (2009)
42. J.C. Williamson, J. Cao, H. Ihee, H. Frey, A.H. Zewail, Nature **386**, 159 (1997)
43. V.A. Lobastov, R. Srinivasan, A.H. Zewail, Proc. Natl. Acad. Sci. USA **102**, 7069 (2005)
44. P.M. Paul, E.S. Toma, P. Breger, G. Mullot, F. Augé, P. Balcou, H.G. Muller, P. Agostini, Science **292**, 1689 (2001)
45. P. Agostini, L.F. DiMauro, Rep. Prog. Phys. **67**, 813 (2004)
46. G.G. Paulus, F. Grasbon, H. Walther, R. Kopold, W. Becker, pra **64**, 021401 (2001)
47. M. Lewenstein, P. Balcou, M.Y. Ivanov, A. L'Huillier, P.B. Corkum, pra **49**, 2117 (1994)
48. P.B. Corkum, prl **71**, 1994 (1993)
49. P. Johnsson, R. López-Martens, S. Kazamias, J. Mauritsson, C. Valentin, T. Remetter, K. Varjú, M.B. Gaarde, Y. Mairesse, H. Wabnitz, P. Salières, Ph. Balcou, K.J. Schafer, A. L'Huillier, Phys. Rev. Lett. **95**, 013001 (2005)
50. P. Johnsson, J. Mauritsson, T. Remetter, A. L'Huillier, K.J. Schafer, Phys. Rev. Lett. **99**, 233001 (2007)
51. P. Rivière, O. Uhden, U. Saalmann, J.M. Rost, New J. Phys. **11**(5), 053011 (2009)
52. B.I. Schneider, J. Feist, S. Nagele, R. Pazourek, S. Hu, L.A. Collins, J. Burgdörfer, Recent advances in computational methods for the solution of the time-dependent Schrödinger equation for the interaction of short, intense radiation with one and two electron systems: application to He and H^2 (Springer, New York, 2011), pp. 149–208
53. C.-Z. Gao, P.M. Dinh, P.-G. Reinhard, E. Suraud, Ann. Phys. **360**, 98–112 (2015)
54. T. Brabec, P.M. Dinh, C.-Z. Gao, C. McDonald, P.-G. Reinhard, E. Suraud, EPJD **73**, 212 (2019)
55. P.-G. Reinhard, E. Suraud, EPJD **74**, 162 (2020)
56. L.S. Cederbaum, W. Domcke, J. Schirmer, W. Von Niessen, Adv. Chem. Phys. **65**, 115 (1986)
57. L.S. Cederbaum, J. Zobeley, Chem. Phys. Lett. **307**, 205 (1999)
58. F. Remacle, R.D. Levine, P. N. A. S. **103**, 6793 (2006)
59. A.I. Kuleff, L.S. Cederbaum, Phys. Rev. Lett. **106**, 053001 (2011)
60. A.I. Kuleff, N.V. Kryzhevoi, M. Pernpointner, L.S. Cederbaum, Phys. Rev. Lett. **117**, 093002 (2016)
61. C.E.M. Gonçalves, R.D. Levine, F. Remacle, Phys. Chem. Chem. Phys. **23**, 12051 (2021)
62. Phys. Chem. Chem. Phys. **23**, 23005 (2021)

Inspecting State-Selective Distributions of Charge Exchange Cross Sections Due to Bare Ion Collisions with Hydrogen

M. S. Gravielle

Abstract State-selective electron capture cross sections due to collisions of multicharged bare ions—Li^{3+}, Be^{4+}, B^{5+}, C^{6+}, N^{7+}, and O^{8+}—with atomic hydrogen in its ground state are investigated within the eikonal impulse (EI) approximation. EI results for total and partial capture to final levels with main quantum numbers n ranging from 1 to 9 are analyzed considering impact energies from 25 to 1000 keV/amu. Furthermore, nl-selective cross sections for Be^{4+} projectiles are presented to illustrate the performance of the subshell electron distributions obtained with the EI method. In all the cases, the EI cross sections show a good agreement with available experimental and theoretical data in the intermediate and high-energy regions. Scaling laws for n-, nl-, and nlm-distributions, derived from the EI approach, are also discussed as useful tools to predict big atomic data sets for plasma and hadron-therapy applications.

1 Introduction

Electron capture by ion impact on atomic targets has been extensively studied over time. In the high and intermediate-energy regimes, that is, for projectile velocities higher than or comparable to the mean electron velocities corresponding to the initial and final bound states, several theoretical methods based on perturbative and distorted-wave theories were proposed [1]. But most of the efforts of these early works were put into the description of total capture cross sections. In contrast, selective probabilities for electron capture to different shells or subshells of the projectile were much less investigated. However, in the last decade, this topic has attracted renewed attention as a consequence of developments in plasma physics [2, 3], hadron therapy [4, 5], and other applications, which require accurate data of selective atomic collision processes as input for numerical simulations.

M. S. Gravielle (✉)
Instituto de Astronomía y Física del Espacio (UBA-CONICET), Ciudad Universitaria, (C1428EGA) Buenos Aires, Argentina
e-mail: msilvia@iafe.uba.ar

L. C. Tribedi (ed.), *Advances in Atomic Molecular Collisions*,
https://doi.org/10.1007/978-981-97-7063-2_7

State-selective charge exchange cross sections due to collisions of multicharged ions with hydrogen atoms are extremely relevant for the diagnostic of confined plasmas [6], where the charge exchange spectroscopy (CXRS) technique, based on the subsequent radiation emission, is usually applied to know the impurity density and the temperature of the plasma [7–9]. Selective nl-electron capture cross sections for such ion-H collisions play also an important role in the study of astrophysical plasmas [10]. Nowadays, the spectroscopic analysis of the X-ray emission produced following the charge exchange between solar wind ions and H atoms of cometary and planetary atmospheres can be used to investigate the characteristics of the solar wind and the astrophysical environment [11–13]. Moreover, precise predictions of charge exchange contributions to the projectile stopping [14] are needed in hadron therapy [15, 16] in order to simulate the radiobiological damage induced by the ion impact.

The goal of this article is to examine state-selective electron capture cross sections produced by high- and intermediate-energy collisions between multicharged bare ions and H($1s$) by using the prior version of the eikonal impulse (EI) approximation [17, 18]. The EI approach is a distorted-wave method that makes use of the impulse wave function, $\psi_f^{(I)}$, in the final channel, and the eikonal function, $\psi_i^{(E)}$, in the initial one. The $\psi_f^{(I)}$ wave function takes into account the Coulomb distortion introduced by the residual charge of the target in the outgoing channel, while $\psi_i^{(E)}$ includes the Coulomb phase associated with the large-distance electron–projectile interaction in the entrance channel, satisfying the Coulomb boundary condition. It should be noticed that although the impulse wave function does not tend to the proper asymptotic limit [19], it provides a good representation of the scattering problem at the small distances involved in the charge exchange process. Furthermore, both wave functions, $\psi_f^{(I)}$ and $\psi_i^{(E)}$, have the proper normalization, i.e., $\left\langle \psi_a^{(I)} \middle| \psi_b^{(I)} \right\rangle = \left\langle \psi_a^{(E)} \middle| \psi_b^{(E)} \right\rangle = \delta_{ab}$, preserving the conservation of the probability. These specific features make the prior EI approximation a reliable method to deal with electron capture at high and intermediate energies, having been successfully applied to a wide variety of collision systems [17, 18, 20–23].

In this work, we consider the charge exchange reaction

$$\mathrm{A}^{Z_P+} + \mathrm{H}(1s) \rightarrow \mathrm{H}^+ + \mathrm{A}^{(Z_P-1)+}(n, l, m), \tag{1}$$

where Z_P is the charge of the fully stripped projectile, Z_T is the nuclear charge of the atomic target (in our case, $Z_T = 1$) initially in the ground state ϕ_i, and n, l, m are the quantum numbers of the final electronic state ϕ_f, bounded to the projectile. Our study will be confined to bare ions with charges Z_P in the range $3 \leq Z_P \leq 8$, colliding with atomic hydrogen with energies varying from 25 to 1000 keV/amu. Since in the reaction of Eq. (1), the mean electronic velocities of the initial and final bound states are, respectively, $Z_T/1$ and Z_P/n, these energies correspond to impact velocities v in the intermediate $Z_T \lesssim v \lesssim Z_P/n$ and high $Z_T, Z_p/n < v$ velocity regimes [1].

The article is organized as follows. The EI approach is summarized in Sect. 2, while EI results for total and partial electron capture to different shells and subshells of the projectile are presented and discussed in Sect. 3. Scaling rules for state-selective electron exchange cross sections are shown in Sect. 4, and in Sect. 5 we outline our conclusions. Atomic units (a.u.) are used unless otherwise stated.

2 EI Approximation

The Hamiltonian H that corresponds to the collision system of Eq. (1) reads

$$H = -\frac{1}{2\nu_T}\nabla^2_{\mathbf{R}_T} - \frac{1}{2\mu_T}\nabla^2_{\mathbf{r}_T} - \frac{Z_T}{r_T} - \frac{Z_P}{r_P} + \frac{Z_T Z_P}{R} = -\frac{1}{2\nu_P}\nabla^2_{\mathbf{R}_P} - \frac{1}{2\mu_P}\nabla^2_{\mathbf{r}_P} - \frac{Z_T}{r_T} - \frac{Z_P}{r_P} + \frac{Z_T Z_P}{R}, \quad (2)$$

where $\mathbf{R}$ is the internuclear vector; $\{\mathbf{R}_T, \mathbf{r}_T\}$ and $\{\mathbf{R}_P, \mathbf{r}_P\}$ are two equivalent coordinate systems (displayed in Fig. 1); and $\nu_T = m_P(m_T+1)/M$, $\mu_T = m_T/(m_T+1)$, $\nu_P = m_T(m_P+1)/M$, and $\mu_P = m_P/(m_P+1)$ are the corresponding reduced masses, with m_T (m_P) the nuclear mass of the target (projectile) and $M = m_T + m_P + 1$. Hence, the charge exchange reaction given by Eq. (1) is associated with the transition

$$\Phi_i = (2\pi)^{-3/2}\exp(i\mathbf{K}_i\cdot\mathbf{R}_T)\phi_i(\mathbf{r}_T) \longrightarrow \Phi_f = (2\pi)^{-3/2}\exp(i\mathbf{K}_f\cdot\mathbf{R}_P)\phi_f(\mathbf{r}_P), \quad (3)$$

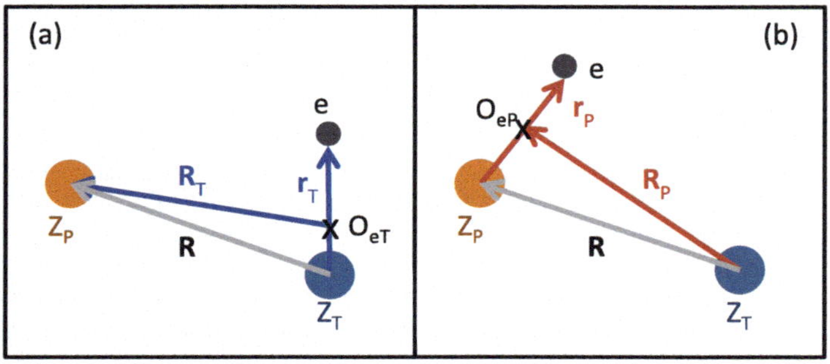

Fig. 1 Depiction of position coordinates for the three-particle collision system: target nucleus (Z_T), projectile nucleus (Z_P), and electron (e). Coordinate systems $\{\mathbf{R}_T, \mathbf{r}_T\}$ and $\{\mathbf{R}_P, \mathbf{r}_P\}$ are, respectively, displayed in the left and right panels. In each panel, O_{ec} indicates the position of the center of mass for the subsystem $e - Z_c$, with $c = T$ or P

where Φ_i and Φ_f are, respectively, the initial and final unperturbed states, with $\mathbf{K}_i$ ($\mathbf{K}_f$) being the initial (final) projectile momentum and ϕ_i (ϕ_f) being the initial (final) electron state, bounded to the target (projectile) with energy $\epsilon_i(\epsilon_f)$.

Within the prior EI approximation, the transition matrix for the above process reads

$$T^{(EI)} = \left\langle \psi_f^{(I)} \left| H - \varepsilon \right| \psi_i^{(E)} \right\rangle, \tag{4}$$

where ε is the total energy, which remains constant during the collision, i.e., $\varepsilon = K_i^2/(2\nu_T) + \epsilon_i = K_f^2/(2\nu_P) + \epsilon_f$. The distorted-wave function $\psi_f^{(I)}$ is derived by means of the impulse approximation as [17]

$$\psi_f^{(I)} = (2\pi)^{-3} \int d\mathbf{q}\, \widetilde{\phi}_f(\mathbf{q}) \exp\left[i\left(\mathbf{K}_f - \mathbf{q}\right) \cdot \mathbf{R}_T\right] \exp\left[i\left(\mathbf{v} + \mathbf{q}\right) \cdot \mathbf{r}_T\right] D^-\left(Z_T, \mathbf{v} + \mathbf{q}, \mathbf{r}_T\right), \tag{5}$$

where $\mathbf{v}$ is the projectile velocity, the tilde denotes the Fourier transform of the bound state, and

$$D^{\pm}\left(Z, \mathbf{k}, \mathbf{r}\right) = C^{\pm}(a)\, {}_1F_1(\pm ia, 1, \pm ikr - i\mathbf{k}\cdot\mathbf{r}) \tag{6}$$

is the Coulomb distortion associated with the interaction of a free electron placed at $\mathbf{r}$ with momentum $\mathbf{k}$ with a charge Z placed at the origin, with the sign + (−) indicating outgoing (incoming) asymptotic conditions. In Eq. (6), ${}_1F_1$ is the confluent hypergeometric function and

$$C^{\pm}(a) = \exp\left[\pi a/2\right] \Gamma(1 \mp ia), \tag{7}$$

with Γ denoting the gamma function and $a = Z/k$.

In turn, the initial eikonal wave function reads

$$\psi_i^{(E)} = \Phi_i E^+\left(Z_P, -\mathbf{v}, \mathbf{r}_P\right), \tag{8}$$

where Φ_i is given in Eq. (3) and

$$E^{\pm}\left(Z, \mathbf{k}, \mathbf{r}\right) = \lim_{r \twoheadrightarrow +\infty} D^{\pm}\left(Z, \mathbf{k}, \mathbf{r}\right) = \exp\left[\mp ia \ln(kr \mp \mathbf{k}\cdot\mathbf{r})\right]. \tag{9}$$

From Eqs. (8) and (5), we observe that both $\psi_i^{(E)}$ and $\psi_f^{(I)}$ take into account the Coulomb distortions introduced, respectively, by the asymptotic projectile and target charges on the initial and final unperturbed scattering states.

As usual within any distorted-wave method, the EI transition matrix is evaluated from Eq. (4) without including the Hamiltonian term corresponding to the internuclear interaction, i.e., $+Z_T Z_P/R$, because it contributes to the transition amplitude with a phase factor that does not affect the state-selective cross section [24]. By using the final residual potential $V_f^{(I)}$, obtained as $V_f^{(I)}\psi_f^{(I)} = (H - \varepsilon)\psi_f^{(I)}$, the transition matrix $T^{(EI)}$ can be expressed as [18]

$$T_f^{(EI)} = \left\langle \psi_f^{(I)} \left| V_f^{(I)\dagger} \right| \psi_i^{(E)} \right\rangle = (2\pi)^{-3/2} \int d\mathbf{q}\, \widetilde{\phi}_f^*(\mathbf{q}) \left(L_T^{D-}(\mathbf{q}+\mathbf{v}, \mathbf{W}_P - \mathbf{v})\right)^* \times \left[(\frac{q^2}{2} - \epsilon_f) I_P^{E+}(-\mathbf{v}, \mathbf{q}+\mathbf{W}_P) + J_P^{E+}(-\mathbf{v}, \mathbf{q}+\mathbf{W}_P) \right], \quad (10)$$

while by using the initial residual potential $V_i^{(E)}$, derived from $V_i^{(E)}\psi_i^{(E)} = (H - \varepsilon)\psi_i^{(E)}$, it reads [18]

$$T_i^{(EI)} = \left\langle \psi_f^{(I)} \left| V_i^{(E)} \right| \psi_i^{(E)} \right\rangle = -(2\pi)^{-3/2} \int d\mathbf{q}\, \widetilde{\phi}_f^*(\mathbf{q}) \times \left[\frac{1}{2} \left(L_T^{D-}(\mathbf{q}+\mathbf{v}, \mathbf{W}_P - \mathbf{v})\right)^* H_P^{E+}(-\mathbf{v}, \mathbf{q}+\mathbf{W}_P) + \left(\mathbf{M}_T^{D-}(\mathbf{q}+\mathbf{v}, \mathbf{W}_P - \mathbf{v})\right)^* \cdot \mathbf{K}_P^{E+}(-\mathbf{v}, \mathbf{q}+\mathbf{W}_P) \right], \quad (11)$$

where

$$\mathbf{W}_P = \mathbf{K}_i - \mu_P \mathbf{K}_f \quad (12)$$

is the transferred momentum vector; the asterisk indicates the complex conjugate; and the functions $H, I, J, \mathbf{K}, L$, and $\mathbf{M}$ correspond to the Nordsieck-type integrals defined in the appendix, which have closed forms. Notice that in our case, since hydrogenic projectile and target are considered, Eqs. (10) and (11) give the same result, that is, $T_f^{(EI)} = T_i^{(EI)}$. However, $T_f^{(EI)}$ differs from $T_i^{(EI)}$ when the use of approximate ϕ_f or ϕ_i wave functions is required to deal with multielectronic particles.

3 EI Cross Sections

The prior EI approach is here used to study state-selective electron exchange produced by impact of Li^{3+}, Be^{4+}, B^{5+}, C^{6+}, N^{7+}, and O^{8+} on atomic hydrogen in its ground state, considering intermediate and high incidence energies, in the range from 25 to 1000 keV/amu. For these collision systems, the EI cross sections corresponding to capture to the different final states ϕ_f, with quantum numbers n, l, m, are expressed as

$$\sigma_{nlm}^{(EI)} = (2\pi)^5 \nu_T \nu_P \frac{K_f}{K_i} \int_0^{\pi} d\theta \sin\theta \, \left|T^{(EI)}\right|^2, \quad (13)$$

where θ is the scattering angle, measured with respect to the $\mathbf{K}_i$-direction. In this work, the transition matrix $T^{(EI)}$ was evaluated from Eq. (11) by numerically solving the three-dimensional integral over $\mathbf{q}$ with a step-adaptive method, with a relative error lower than 1%. The integral on θ involved in Eq. (13) was calculated by using about 60 scattering angles, with a relative error lower than 0.1%.

3.1 Total and n-Selective Cross Sections

We start analyzing total and n-selective electron capture cross sections to the ground and excited levels of the projectile for Z_P ranging from 3 to 8. For all the considered ionic projectiles, we show total and partial capture cross sections to shells n from 1 to 9 in Figs. 2, 3, 4, 5, 6 to 7. Total capture cross sections were derived as

$$\sigma_{total}^{(EI)} = \sum_{n=1}^{9} \sigma_n^{(EI)} + 4.028\, \sigma_9^{(EI)}, \tag{14}$$

where $\sigma_n^{(EI)} = \sum\limits_{l=0}^{n-1} \sum\limits_{m=-l}^{l} \sigma_{nlm}^{(EI)}$ denotes the partial capture cross section to the n shell of the projectile, with $\sigma_{nlm}^{(EI)}$ given by Eq. (13), and the last term corresponds to the contribution of excited levels with $n \geqslant 10$, which was estimated by using the well-known Oppenheimer rule $\sigma_n^{(EI)} \sim n^{-3}$ [24]. This term was found negligible for A^{Z_P+} projectiles with $Z_P = 3$ and 4, but it becomes relevant at the lower impact energies as Z_P increases. For example, for $Z_P = 5$ at $\varepsilon = 25$ keV/amu, the total cross section evaluated without considering this extrapolation rule for higher excited levels, i.e.,

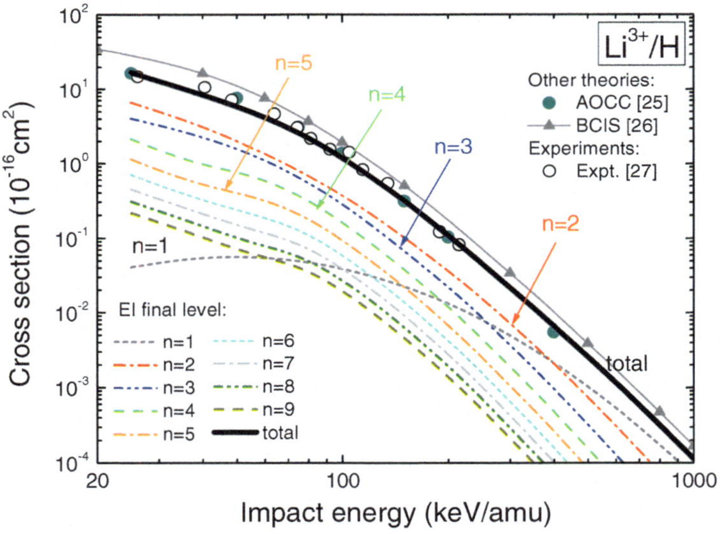

Fig. 2 Electron capture cross sections for Li^{3+} ions impinging on H(1 s) as a function of the incidence energy. EI results: the black solid line shows total capture cross sections, as given by Eq. (14); lines for the partial capture cross sections to the different n levels, with $n = 1, \ldots, 9$, are indicated in the figure. Other calculations: full circles, AOCC from Ref. [25]; full up triangles, BCIS from Ref. [26]. Experiments: empty circles, data from Ref. [27]

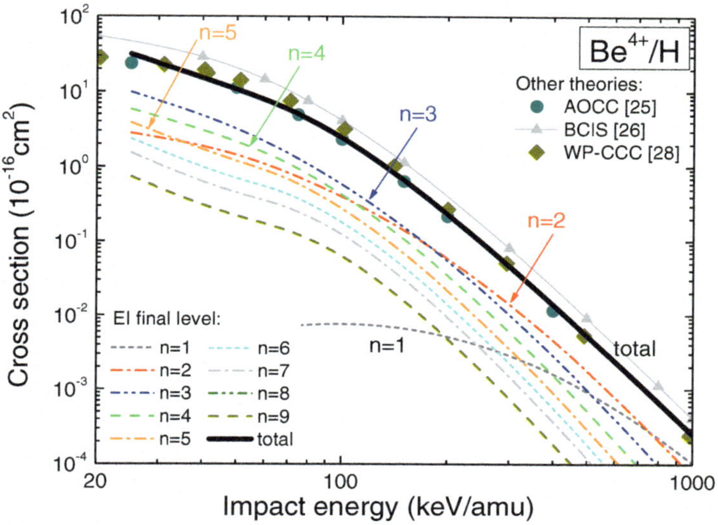

Fig. 3 Analogous to Fig. 2 for Be^{4+} projectiles. Other calculations: full diamonds, WP-CCC from Ref. [28]

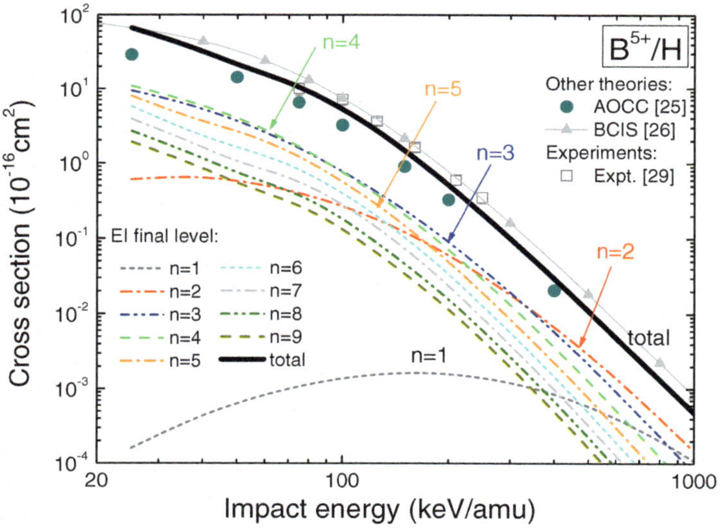

Fig. 4 Analogous to Fig. 2 for B^{5+} projectiles. Experiments: empty squares, data from Ref. [29]

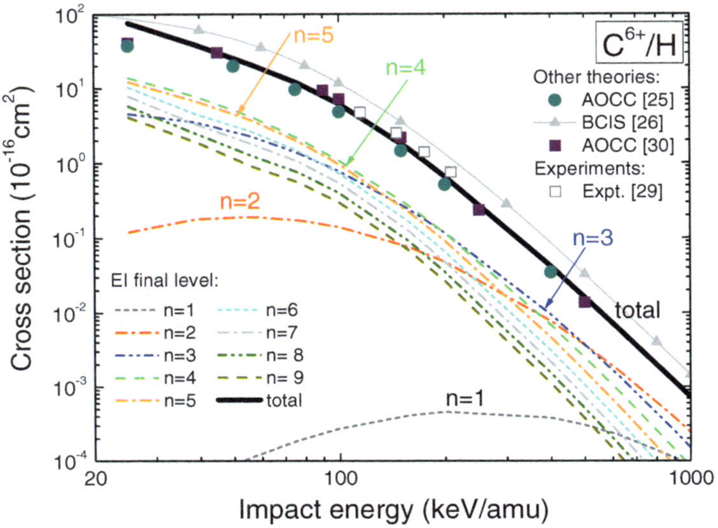

Fig. 5 Analogous to Fig. 2 for C^{6+} projectiles. Other calculations: full squares, AOCC from Ref. [30]. Experiments: empty squares, data from Ref. [29]

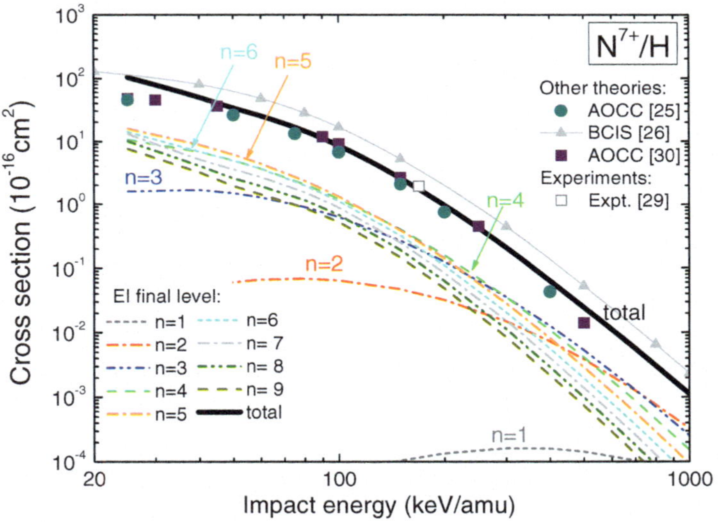

Fig. 6 Analogous to Fig. 2 for N^{7+} projectiles. Other calculations: full squares, AOCC from Ref. [30]. Experiments: empty squares, data from Ref. [29]

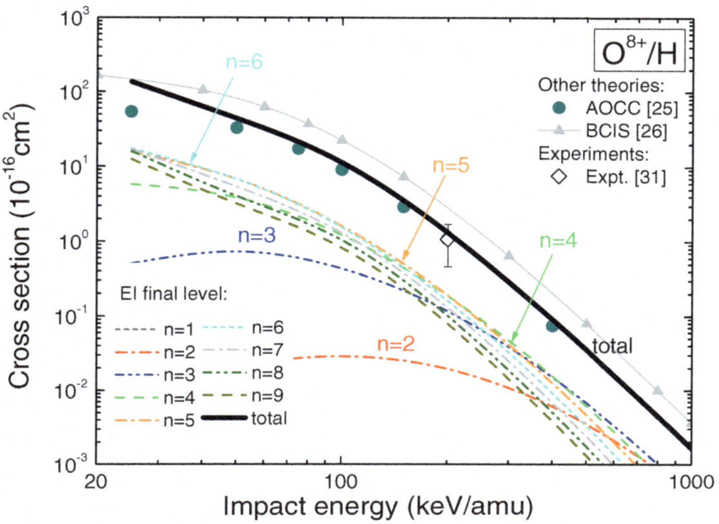

Fig. 7 Analogous to Fig. 2 for O^{8+} projectiles. Experiment: empty diamond, data from Ref. [31]

$\sigma^{(EI)}_{total(n=1-9)} = \sum_{n=1}^{9} \sigma_n^{(EI)}$, differs about 35% from that obtained from Eq. (14), while for $Z_P = 8$ such a difference is about 37%. Moreover, the contribution of highly excited levels, with $n \geqslant 10$, strongly decreases as the impact energy increases.

Total EI cross sections are compared with available experimental data on atomic hydrogen [27, 29, 31] in Figs. 2, 4, 5, 6 to 7. Notice that these experiments were carried out more than 40 years ago. In addition, for some of the studied collision systems there are not experimental values (as it is the case for Be^{4+}/H) or only one value was reported (for N^{7+}/H and O^{8+}/H) in the considered energy range. Therefore, in order to inspect the reliability of the predictions of the prior EI approximation, we also contrast the EI results with total electron capture cross sections obtained from atomic-orbital close-coupling (AOCC) calculations, reported in Refs. [25, 30, 32], as well as with values recently published in Refs. [26, 28], respectively, derived with the boundary-corrected continuum intermediate states (BCIS) and the wave-packet convergent close-coupling (WP-CCC) methods. A comparative study with theoretical values derived within the classical trajectory Monte Carlo (CTMC) approach, which are available in the literature [23], will be presented in Sect.4.

For all the studied collision systems, our total EI cross sections show a remarkably good agreement with the AOCC results by Toshima [25], as observed in Figs. 2, 3, 4, 5, 6 to 7. Furthermore, EI total capture results lay close to the AOCC calculations by Igenbergs et al. [30, 32] for Be^{4+}, C^{6+}, and N^{7+} projectiles and the WP-CCC values by Antonio et al. [28] for Be^{4+}. Instead, the BCIS results by Delibašić et al. [26] run above the EI values, being such a difference equal to factor between 1.4 and 2 approximately, depending on the incident projectile. Notice that for Be^{4+}, the total capture cross sections from Ref. [32] almost perfectly agree with those from Ref. [28] and for the sake of clarity they are not displayed in Fig. 3.

Concerning the comparison with the available experimental data, total EI cross sections are in very good agreement with the experiments by Shah et al. [27] for Li^{3+} ions (Fig. 2) and by Goffe et al. [29] for C^{6+} projectiles (Fig. 5). Also the EI approach predicts very well the value of the unique experimental result for N^{7+} in Fig. 6, extracted from Ref. [29], and that by Meyer et al. [31] for O^{8+} in Fig. 7. In turn, the experimental comparison for the system B^{5+}/H deserves a special mention. For this case, in Fig. 4, we observe that the experimental data [29] tend to the BCIS curve as the energy increases, while our total EI results, albeit quite close to the measured values, run between the AOCC and BCIS cross sections. In addition, for projectiles with $Z_P \geqslant 5$ we found that the total EI cross section at $\varepsilon = 25$ keV/amu (i.e., $v \approx Z_T = 1$) overestimates the corresponding Toshima's value [25], tending to the BCIS result [26] in the lowest limit of the intermediate velocity range.

3.2 *nl*-Selective Cross Sections

With the aim of inspecting thoroughly partial nl-capture cross sections for bare ions impinging on atomic hydrogen, we chose the collision Be^{4+} /H(1 s) as benchmark. Charge exchange for this collision system has been intensively investigated in the last time due to its importance for current fusion projects, like the International Thermonuclear Experimental Reactor (ITER) [2] and the Joint European Torus (JET) [3]. In both fusion experiments, beryllium atoms are released from device components in contact with the plasma, ending up as fully stripped ions that can collide with hydrogen atoms with intermediate and moderate high energies. It produces electron capture to excited projectile levels, whose subsequent radiative decay can be used for the plasma diagnostic by means of the CXRS technique, for which nl-capture cross sections are essential ingredients.

From Fig. 3, we observe that the most important contribution to the total electron capture by Be^{4+} projectiles, in the considered energy range, comes from the final shells $n = 2$, 3, and 4. Therefore, in Fig. 8, we contrast the EI results for these levels with recent WP-CCC data [28]. Also, values for 100 and 500 keV/amu extracted from Ref. [33], which were obtained by numerically solving the time-dependent Schrödinger equation (GTDSE), are shown in Fig. 8. We observe that the partial EI cross sections for the final levels $n = 2, 3$, and 4 are in good agreement with the WP-CCC data, as well as with the GTDSE values, which almost coincide with the WP-CCC ones. As expected, the accord between the n- selective EI cross sections and the corresponding WP-CCC and GTDSE results improves as the impact energy increases. In the opposite limit, at the lowest energy (25 keV/amu) $\sigma_3^{(EI)}$ and $\sigma_4^{(EI)}$ are slightly below the respective WP-CCC curves, while $\sigma_2^{(EI)}$, which dominates the total cross section for $\varepsilon \gtrsim 200$ keV/amu, overestimates the WP-CCC result for $n = 2$ by a factor 2.5.

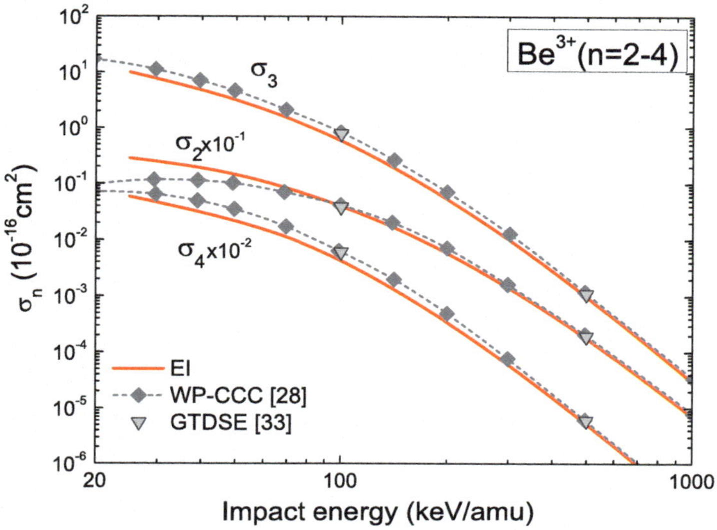

Fig. 8 Partial electron capture cross sections to the final levels $n = 2$, 3, and 4, as a function of the incidence energy, for Be^{4+}/H(1 s) collisions. Red solid lines, EI cross sections; gray dashed lines with full diamonds, WP-CCC results from Ref. [28]; full down triangles, GTDSE results from Ref. [33]. The n-selective cross sections are multiplied by ten powers, as indicated in the figure

Partial EI cross sections for capture to the different sublevels nl of $Be^{3+}(n = 2)$, $Be^{3+}(n = 3)$, and $Be^{3+}(n = 4)$ are respectively shown in Figs. 9, 10, and 11. Analogous to Fig. 8, our EI results are compared with WP-CCC values extracted from [28]. For $n = 2$ Fig. 9, the EI results for capture to the sublevels 2 s and 2p are in a very good accord with the WP-CCC values for $\varepsilon \gtrsim 60$ keV/amu. But the agreement deteriorates as the velocity decreases and for $v \approx Z_T = 1$ a.u., the EI cross sections run above the WP-CCC curve.

A general good agreement between the EI and WP-CCC nl-selective cross sections is also observed for the shell $n = 3$ in Fig. 10. In this case, the EI results for the final subshells 3p and 3d run slightly below the WP-CCC values in the intermediate-energy range, tending to the close-coupling calculations as the energy increases. Instead, the EI curve for the sublevel 3 s departs moderately from the WP-CCC one in the energy range from 100 to 400 keV/amu, being close to the WP-CCC values not only at higher but also at lower energies.

Finally, in Fig. 11, we observe an excellent accord between both approaches, EI and WP-CCC, for capture cross sections to the subshell 4f, in the whole energy range. This sublevel corresponds to the maximum l value, which provides the most important contribution to $\sigma_4^{(EI)}$. But for $l = 2$ and 1 the agreement gradually worsens as the energy decreases, and the corresponding EI results underestimate the WP-CCC curves for $\varepsilon < 100$ keV/amu. Unexpectedly, in such an energy region, the EI cross sections for $l = 0$ almost coincide with the WP-CCC data, although they run

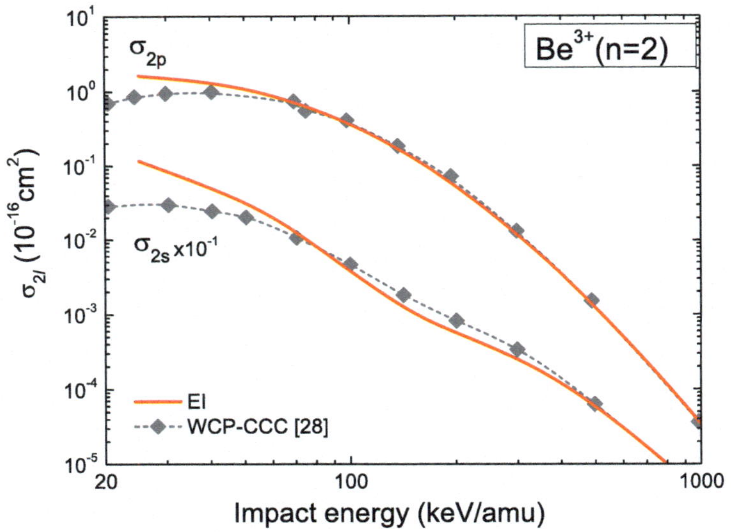

Fig. 9 Analogous to Fig. 8 for selective electron capture cross sections to the different subshells of the final level $n = 2$

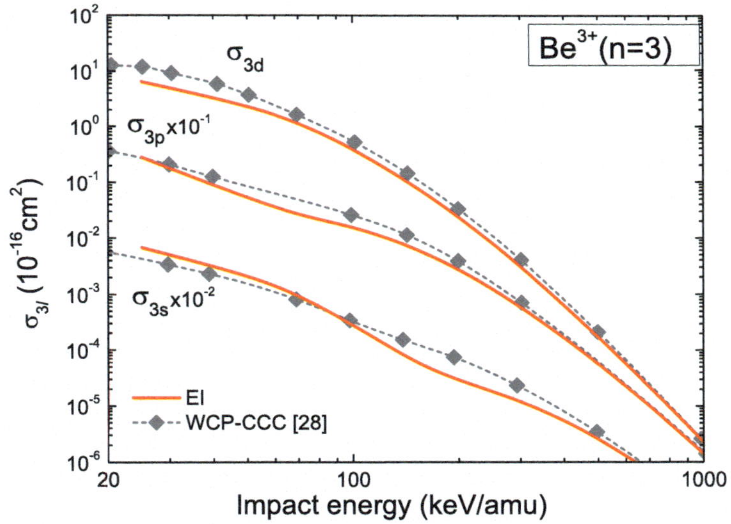

Fig. 10 Analogous to Fig. 8 for selective electron capture cross sections to the different subshells of the final level $n = 3$

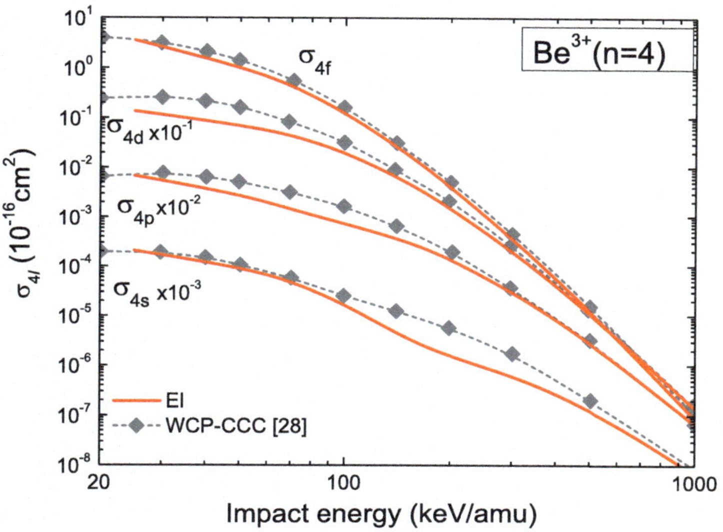

Fig. 11 Analogous to Fig. 8 for selective electron capture cross sections to the different subshells of the final level $n = 4$

below the close-coupling values in the energy range from 100 to 400 keV/amu, as also observed for the state 3 s in Fig. 10.

Therefore, the comparisons of n- and nl-selective EI cross sections with close-coupling results, presented in Figs. 8, 9, 10 and 11, illustrate the satisfactory performance of the prior EI approach to describe the charge exchange reaction of Eq. (1). Furthermore, we found that the agreement between both methods is specially good (even at intermediate impact energies) for the sublevels $l = n - 1$, which dominate $\sigma_n^{(EI)}$.

4 Scaling Rules for State-Selective Cross Sections

Taking into account that most of the applications of collision data in plasma physics and hadron therapy require the use of a large number of atomic cross sections, covering different collision partners, incidence energies, and final states, in this section, we will focus on scaling rules for electron capture cross sections to specific final states. Such state-selective scaling rules are expected to be useful tools to incorporate accurate predictions of charge exchange process involving the impact of fully stripped ions on atomic hydrogen in complex numerical simulations.

For asymmetric collisions with $Z_P > Z_T$, as the ones considered here, it is possible to demonstrate that the EI capture cross section to the (n, l, m) final state nearly satisfies [20]

$$\sigma_{nlm}^{(EI)} \simeq \tilde{z}_P^{-7} \left| C^+(Z_P/v) \right|^2 u_{nlm}(\tilde{W}), \tag{15}$$

where $\tilde{z}_P = Z_P/n$ denotes the mean value of the final electron velocity, the function $C^+(a)$ is given by Eq. (7), and u_{nlm} represents a universal function of the scaled transferred momentum $\tilde{W}$. This parameter is defined as

$$\tilde{W} = \frac{W}{\tilde{z}_P} = \frac{\tilde{v}^2 + \tilde{z}_T^2 - 1}{2\tilde{v}}, \tag{16}$$

with $W = \mathbf{W}_P \cdot \hat{v}$ being the component of the transferred momentum (Eq. (12)) parallel to the incidence velocity. In Eq. (16), the scaled parameters $\tilde{v}$ and $\tilde{z}_T$ are defined as

$$\tilde{v} = \frac{v}{\tilde{z}_P}, \quad \tilde{z}_T = \frac{Z_T}{\tilde{z}_P}, \tag{17}$$

with $Z_T = 1$ for hydrogen targets. It should be noticed that the approximate relation given by Eq. (15) is valid for $\tilde{W} > 0$ values, and, in addition, the condition $\tilde{z}_T < 1$ is required for impact energies lower than 100 keV/amu [20, 23].

The scaling rule for partial cross sections σ_n is straightforwardly derived from Eq. (15) as

$$S_n = \frac{\sigma_n\, \tilde{z}_P^7}{|C^+(Z_P/v)|^2} \simeq U_n(\tilde{W}), \tag{18}$$

where U_n is a roughly universal function of $\tilde{W}$. Although Eq. (18) was obtained from the EI approach, it was found to be valid for a wide variety of experimental and theoretical results [23]. In particular, in Ref. [23] n-selective EI cross sections for final levels $n = 4, ..., 9$ were combined with CTMC results to derive U_n functions with a weak dependence on n, i.e., $U_n(\tilde{W}) \simeq U(\tilde{W})$. In Fig. 12, we show how Eq. (18) allows one to join EI and CTMC n-selective capture cross sections for different bare ions with $Z_P = 4, ..., 8$ on a single curve $U(\tilde{W})$, whose analytical expression (for σ_n in cm^2) is [23]

$$U(\tilde{W}) = \left[\frac{a_1}{\left(b_1 + c_1 \tilde{W}^2\right)^4} - \frac{a_2}{\left(b_2 + c_2 \tilde{W}^2\right)^4} \right] \times 10^{-14}\ \text{cm}^2, \tag{19}$$

where $a_1\,(a_2) = 3.46\,(0.01)$, $b_1\,(b_2) = 2.63\,(2.21)$, and $c_1\,(c_2) = 6.23\,(1.43)\,(\text{a.u.})^{-2}$.

Furthermore, from Eq. (15) we found that the nl-distributions, defined as $P_{nl} = \sigma_{nl}/\sigma_n$, approximately follows the scaling rule:

$$P_{nl} = \frac{\sigma_{nl}}{\sigma_n} \simeq U_{nl}(\tilde{W}), \tag{20}$$

with U_{nl} being again a universal function. Figure 13 displays the U_{nl} functions for the different nl-subshells with $n = 4, ..., 9$ and $l = 0, ..., n-1$. The numerical values of such universal distributions were reported in the appendix of Ref. [23].

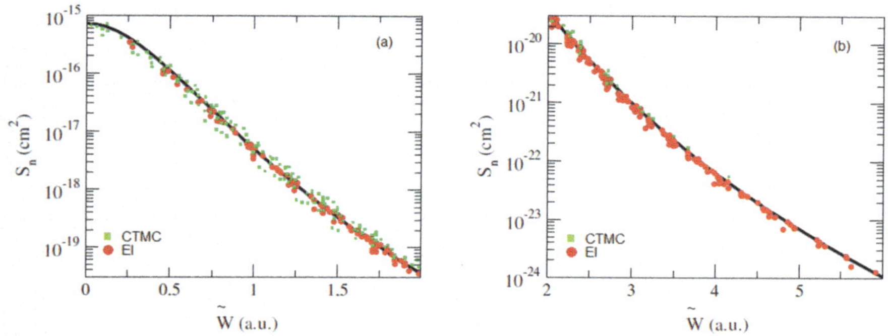

Fig. 12 Scaled cross sections S_n given by Eq. (18), as a function of the scaled momentum $\tilde{W}$ (Eq. (16)), for different n values in the range $n = 4 - 9$. Results derived with the approaches: • EI, ■ CTMC. Solid black line, universal curve [23] given by Eq. (19). Panels (a) and (b) show results for different $\tilde{W}$-ranges. Figure taken from Ref. [23]

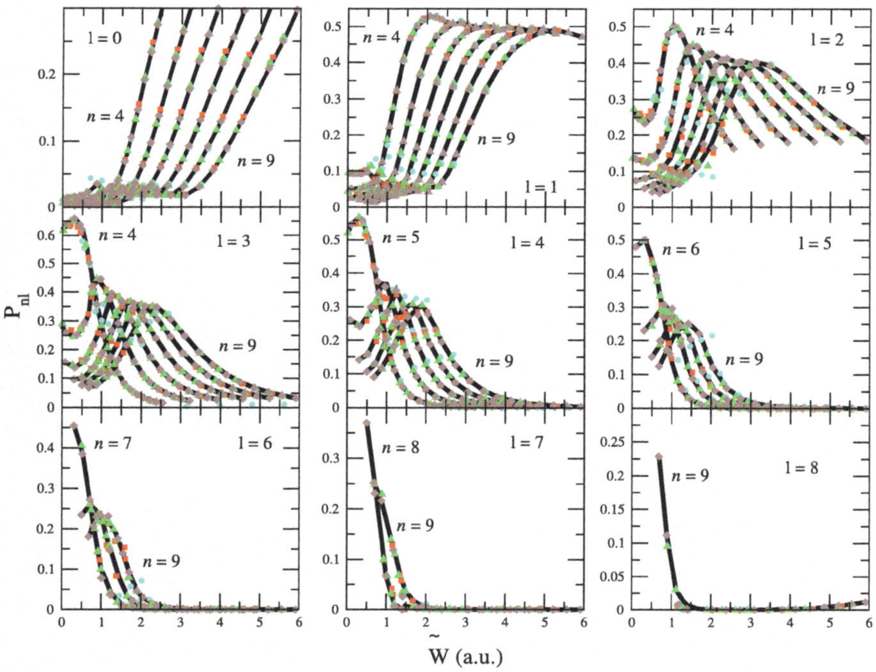

Fig. 13 Joined CTMC and EI P_{nl} distributions as a function of $\tilde{W}$. Solid black line, present scaling, given by Eq. (20); results for different projectiles: • Be^{4+}, ■ C^{6+}, ▲ N^{7+}, ♦ O^{8+}. In each panel, the curves shift from left to right as n increases. Figure taken from Ref. [23]

In addition, the scaling rule given by Eq. (15) can be used to obtain universal nlm-distributions, defined as $P_{nlm} = \sigma_{nlm}/\sigma_{nl} \simeq u_{nlm}(\tilde{W})$. Examples of such universal magnetic distributions can be found in Figs. 7 and 8 of Ref. [23] for the $n = 6$ and $n = 7$ shells, respectively.

5 Conclusions

In this work, we have inspected total and state-selective electron capture cross sections produced by collisions of multicharged bare ions with atomic hydrogen by using the EI approximation. Ionic projectiles with charges Z_P ranging from 3 to 8 and impact energies varying from 25 to 1000 keV/amu were considered. For all the projectiles—Li^{3+}, Be^{4+}, B^{5+}, C^{6+}, N^{7+}, and O^{8+}—we have evaluated partial capture cross sections to the final levels $n = 1, ..., 9$, which provide the most important contributions to total cross sections.

Present EI results for total capture were compared with available experimental and theoretical data, showing a very good agreement in the intermediate and high-energy regimes. In turn, to scrutinize the performance of the n- and nl-partial cross sections derived within the EI approach, we use the Be^{4+}/H(1 s) as example, considering capture to the excited levels $n = 1$, 2, and 3. These n-selective EI cross sections show an overall good accord with recent close-coupling calculations [28] in the whole velocity range. Similar level of agreement is also observed for the corresponding nl-selective cross sections, for which the accord is remarkably better for the highest l values. In addition, n- and nl-scaling rules derived from the EI approximation are discussed, displaying universal curves obtained by combining EI and CTMC state-selective cross sections, as reported in Ref. [23].

In conclusion, we have demonstrated that the EI method provides an appropriate description of total and partial capture cross sections in the intermediate and high-energy ranges, allowing us to derive scaling rules that are useful to predict selective electron distributions for different highly charged ions.

Appendix

The Nordsieck-type integrals involved in Eqs. (10) and (11) are defined as

$$\begin{bmatrix} H_c^{X\pm}(\mathbf{k},\mathbf{q}) \\ I_c^{X\pm}(\mathbf{k},\mathbf{q}) \\ J_c^{X\pm}(\mathbf{k},\mathbf{q}) \\ \mathbf{K}_c^{X\pm}(\mathbf{k},\mathbf{q}) \end{bmatrix} = (2\pi)^{-3/2} \int d\mathbf{r}_c \exp(-i\mathbf{q}\cdot\mathbf{r}_c) \begin{bmatrix} \nabla^2_{\mathbf{r}_c} \\ 1 \\ V_c(r_c) \\ \nabla_{\mathbf{r}_c} \end{bmatrix} X_c^{\pm}(\mathbf{k},\mathbf{r}_c), \qquad (21)$$

and

$$\begin{bmatrix} L_c^{X\pm}(\mathbf{k},\mathbf{q}) \\ \mathbf{M}_c^{X\pm}(\mathbf{k},\mathbf{q}) \end{bmatrix} = (2\pi)^{-3/2} \int d\mathbf{r}_c \exp(-i\mathbf{q}\cdot\mathbf{r}_c)\, X_c^{\pm}(\mathbf{k},\mathbf{r}_c) \begin{bmatrix} 1 \\ \nabla_{\mathbf{r}_c} \end{bmatrix} \phi_c^*(\mathbf{r}_c), \quad (22)$$

for $c = T$ or P, where $X_c^{\pm}(\mathbf{k},\mathbf{r})$ denotes $D^{\pm}(Z_c,\mathbf{k},\mathbf{r})$ or $E^{\pm}(Z_c,\mathbf{k},\mathbf{r})$, as given by Eqs. (6) and (9), respectively, $V_c(r_c) = -Z_c/r_c$ is the Coulomb potential, and $\phi_T = \phi_i$ and $\phi_P = \phi_f$ are the corresponding bound wave functions. Closed forms of these integrals are reported in Ref. [34].

Acknowledgements We acknowledge funding from ANPCYT (PICT-2017-1201; PICT-2017-2945; PICT-2020-1755; PICT-2020-1434) of Argentina.

References

1. D.P. Dewangan, J. Eichler, Charge exchange in energetic ion-atom collisions. Phys. Rep. **247**, 59–219 (1994). (and references therein)
2. International Thermonuclear Experimental Reactor(ITER). https://www.iter.org/
3. Joint European Torus (JET). https://www.euro-fusion.org/devices/jet/
4. M.A. Quinto, J.M. Monti, C.A. Tachino, P.F. Weck, O.A. Fojon, C. Champion, R.D. Rivarola, The physics of irradiation of biological matter by ion beams. Rad. Phys. Chem. **167**, 108337 (2020)
5. S. Rossi, Hadron therapy achievements and challenges: The CNAO experience. Physics **4**, 229–257 (2022)
6. D.G. Whyte, R.C. Isler, M.R. Wade, D.R. Schultz, P.S. Krstic, C.C. Hung, W.P. West, Argon density measurements from charge-exchange spectroscopy. Phys. Plasmas **5**, 3694 (1998)
7. R.C. Isler, An overview of charge-exchange spectroscopy as a plasma diagnostic. Plasma Phys. Control. Fusion **36**, 171 (1994)
8. D.M. Thomas, Beams, brightness, and background: Using active spectroscopy techniques for precision measurements in fusion plasma research. Phys. Plasmas **19**, 056118 (2012)
9. N.D. Cariatore, S. Otranto, R.E. Olson, Classical description of H(1s) and H*(n = 2) for cross-section calculations relevant to charge-exchange diagnostics. Phys. Rev. A **91**, 042709 (2015)
10. D.W. Savin, N.S. Brickhouse, J.J. Cowan, R.P. Drake, S.R. Federman, G.J. Ferland, A. Frank, M.S. Gudipati, W.C. Haxton, E. Herbst, S. Profumo, F. Salama, L.M. Ziurys, E.G. Zweibel, The impact of recent advances in laboratory astrophysics on our understanding of the cosmos. Rep. Prog. Phys. **75**, 036901 (2012)
11. T.E. Cravens, X-ray emission from comets. Science **296**, 1042 (2002)
12. S. Otranto, R.E. Olson, P. Beiersdorfer, X-ray emission cross sections following charge exchange by multiply charged ions of astrophysical interest. Phys. Rev. A **73**, 022723 (2006)
13. D.R. Schultz, S.Y. Ovchinnikov, P.C. Stancil, T. Zaman, Elastic, charge transfer, and related transport cross sections for proton impact of atomic hydrogen for astrophysical and laboratory plasma modeling. J. Phys. B **49**, 084004 (2016)
14. G. Schiwietz, P.L. Grande, Introducing electron capture into the unitary-convolution-approximation energy-loss theory at low velocities. Phys. Rev. A **84**, 052703 (2011)
15. D. Schardt, T. Elsässer, D. Schulz-Ertner, Heavy-ion tumor therapy: physical and radiobiological benefits. Rev. Mod. Phys. **82**, 383–425 (2010)

16. R.D. Rivarola, M.E. Galassi, P.D. Fainstein, C. Champion, Computation of distorted wave cross sections for high-energy inelastic collisions of heavy ions with water molecules. Adv. Quantum Chem. **65**, 231–267 (2013)
17. M.S. Gravielle, J.E. Miraglia, Electron capture in asymmetric collisions. Phys. Rev. A **38**, 5034–5037 (1988)
18. M.S. Gravielle, J.E. Miraglia, Eikonal impulse approximation in electron-capture processes. Phys. Rev. A **44**, 7299–7306 (1991)
19. M.S. Gravielle, J.E. Miraglia, Boundary conditions of the exact impulse wave function. Phys. Rev. A **55**, 1540–1543 (1997)
20. M.S. Gravielle, J.E. Miraglia, State-selective scaling in electron capture by multicharged ions on light atoms. Phys. Rev. A **51**, 2131–2139 (1995)
21. M.S. Gravielle, J.E. Miraglia, State-selective scaling for capture to the level $n = 4$ by multicharged ions on light atoms. Phys. Rev. A **52**, 851–854 (1995)
22. C.C. Montanari, M.S. Gravielle, J.E. Miraglia, Capture from $n = 2$ states of hydrogen by heavy fast ions. Phys. Scrip. **56**, 279–283 (1997)
23. A. Jorge, C. Illescas, J.E. Miraglia, M.S. Gravielle, Scaling for state-selective charge exchange due to collisions of multicharged ions with hydrogen. J. Phys. B **48**, 235201 (2015)
24. Dž. Belkić, R. Gayet, A. Salin, Electron capture in high-energy ion-atom collisions. Phys. Rep. **56**, 279–369 (1979)
25. N. Toshima, Ionization and charge transfer of atomic hydrogen in collision with multiply charged ions. Phys. Rev. A **50**, 3940–3947 (1994)
26. D. Delibašić, N. Milojević, I. Mančev, Dž. Belkić, Electron transfer from atomic hydrogen to multiply-charged nuclei at intermediate and high energies. At. Data. Nucl. Data Tables **139**, 101417 (2021)
27. M.B. Shah, T.V. Goffe, H.B. Gilbody, Electron capture and loss by fast lithium ions in H and H_2. J. Phys. B **11**, L233 (1978)
28. N.W. Antonio, C.T. Plowman, I.B. Abdurakhmanov, I. Bray, A.S. Kadyrov, Integrated total and state-selective cross sections for bare beryllium ion collisions with atomic hydrogen. J. Phys. B **54**, 175201 (2021)
29. T.V. Goffe, M.B. Shah, H.B. Gilbody, One-electron capture and loss by fast multiply charged boron and carbon ions in H and H_2. J. Phys. B **12**, 3763 (1979)
30. K. Igenbergs, J. Schweinzer, A. Veiter, L. Perneczky, E. Frühwirth, M. Wallerberger, R.E. Olson, F. Aumayr, Charge exchange and ionization in N^{7+}-, N^{6+}-, C^{6+}- H(n = 1, 2) collisions studied systematically by theoretical approaches. J. Phys. B **45**, 065203 (2012)
31. F.W. Meyer, R.A. Phaneuf, H.J. Kim, P. Hvelplund, P.H. Stelson, Single-electron-capture cross sections for multiply charged O, Fe, Mo, Ta, W, and Au ions incident on H and H_2 at intermediate velocities. Phys. Rev. A **19**, 515–525 (1979)
32. K. Igenbergs, J. Schweinzer, F. Aumayr, Charge exchange in Be^{4+} - H(n = 1, 2) collisions studied systematically by atomic-orbital close-coupling calculations. J. Phys. B **42**, 235206 (2009)
33. A. Jorge, J. Suárez, C. Illescas, L. Errea, L. Méndez, Application of a grid numerical method to calculate state-selective cross sections for electron capture in $Be^{4+} + H(1s)$ collisions
34. M.S. Gravielle, J.E. Miraglia, Some Nordsieck integrals of interest in radiation and atomic collision theories. Comp. Phys. Commun. **69**, 53–58 (1992)

Inspecting the Information Quantity in Ion-Hydrogen Electron Capture Process with the Shannon Entropy

J. E. Miraglia

Abstract Information Shannon entropy is calculated for the electron capture process from hydrogen atoms by multicharged Coulomb projectiles. The principal, orbital, and magnetic quantum numbers (n, l, and m) of the electron in the final state on the projectile are considered as all the possible events taking place, like the faces of a "dice". The big data set of the theoretical eikonal impulse approximation values published by M.S. Gravielle in this book is used. Comparisons with the simple Brinkman–Kramers approximation are displayed when possible. The Shannon entropy is calculated over the discrete variables considering three different sets of quantum numbers, just on n; on n and l; and on n, l, and m. It is found that the more the quantum numbers, the larger is the entropy, indicating that the "surprise" increases with the incorporation of new variables, or, equivalently, new faces of the "dice". Mathematical limits for large and small velocities are calculated.

PACS numbers: 02.50.-r · 34.70.+e

1 Preliminary

Let us consider an experiment where bare heavy projectiles of Coulomb charge Z impact on hydrogen atoms in the ground state $H(1s)$. The exiting beam can be deflected in an electric field to collect just the rearranged channels, discarding the excitation and ionization outcomes. We can further build up another devise to determine the final principal, orbital and magnetic quantum numbers (n, l and m, respectively) of the electron state on the exiting ion. In another words, we are dealing with the well-known rearrangement collision

J. E. Miraglia (✉)
Instituto de Astronomía y Física del Espacio. Consejo Nacional de Investigaciones Científicas y Técnicas, Departamento de Física. Facultad de Ciencias Exactas y Naturales. Universidad de Buenos Aires, Casilla de Correo 67, Sucursal 28, C1428EGA Buenos Aires, Argentina
e-mail: miraglia@iafe.uba.ar

L. C. Tribedi (ed.), *Advances in Atomic Molecular Collisions*,
https://doi.org/10.1007/978-981-97-7063-2_8

$$P^{Z+} + H(1s) \rightarrow P^{(Z-1)+}(nlm) + H^{+}. \tag{1}$$

This process is very well known and there are in the literature hundreds of experiments dealing with this reaction. Also, there are several well established theories to deal with these kind of collisions. The capture cross section σ_{nlm} is the relevant magnitude to describe its importance. We can see the entire apparatus as a black box (machine as it is mentioned in the information field) in which we have entering projectiles and outgoing variables

$$\begin{cases} n & \in [1, \infty], \\ l & \in [0, n-1], \\ m & \in [-l, l], \end{cases} \tag{2}$$

as they were the faces of a quantum-number "dice". The machine "throws" different nlm−distributions depending on the impact velocity v and on the Coulomb charge Z of the projectile. We now pose the following question: can we determine the information (entropy) of the resulting nlm-quantum-number outputs? This article answers this question by resorting to the Shannon entropy.

The concept of information entropy was proposed by C. Shannon in his seminal paper of 1948 [1]. Since then the information entropy has been used in several fields including Chemistry [2], heavy ion collisions [3], and even in inelastic collisions as a source of entropy [4]. In this article, we simply tackle the information entropy taken place in a single electron capture process given by (1). The article is organized as follows: in Sect. 2, we present and discuss the theoretical values used, and in Sect. 3 we applied the Shannon entropy to capture desegregating three levels of information, i.e., just on n; on n and l; and on n, l, and m. Section 4 presents the results and the discussions.

2 The Electron Capture Data Set

The ultimate data to be scrutinized should be the experimental values. However, there is no complete experiment in the literature. By complete we mean a detailed and thorough nlm-resolution for different projectiles and velocities. Instead, we will resort to the theoretical calculations that have been accumulated in the last 30 years to tackle several experimental challenges and data macro-analysis [5]. In this period, our group has gathered a large amount of capture cross sections (of the order of 20000) calculated with the Eikonal Impulse approximation (EIA) [6]. In a previous chapter of this book, M. S. Gravielle has presented a complete review of the EIA covering dependencies on Z and v [5]. The EIA is an elaborated distorted wave method including high perturbative orders preserving the correct Coulomb asymptotic conditions at large distances. The EIA theory was successfully tested with several experiments [6, 7] and it was proven to be valid at intermediate and high impact velocities, that is, for $v^2/Z > 1$. The EIA was also used to inspect scaling rules

Table 1 Data set of EIA cross sections used: Z is the projectile charge, N_v is the number of velocities, $R_v = [v_i, v_f]$ is the range of velocities of the data in atomic units, $R_n = [1, n_f]$ is the range of principal quantum numbers, n_f is the final principal quantum number, and $N_{\sigma_{nlm}}$ is the total number of cross section values calculated. For details, see Ref. [5]

Z	N_v	R_v	R_n	$N_{\sigma_{nlm}}$
1	13	[1., 7.]	[1, 4]	390
2	13	[1., 7.]	[1, 4]	390
3	13	[1., 7.]	[1, 9]	3705
4	7	[1.,6.3]	[1, 9]	1995
5	11	[2., 6.6]	[1, 9]	3135
6	11	[1., 7.]	[1, 9]	3135
7	10	[1., 4.5]	[1, 9]	2850
8	15	[1., 8.9]	[1, 9]	4275

with Z and v [7–10]. Lately, we have tested that the Benford law validates our data set for the first and second digits [11]. A detail inventory of the results published by M.S. Gravielle is listed in the table. For each impact charge Z, we list in the table the number of velocities N_v, the range of velocities in atomic units $R_v = [v_i, v_f]$, the range of principal quantum numbers $R_n = [1, n_f]$, where n_f is the final principal quantum number calculated, and the total number of cross sections calculated $N_{\sigma_{nlm}}$. In accordance with (2) a total of $N_{\sigma_{nlm}} = n_f(1 + n_f)(1 + 2n_f)/6$ cross sections were considered for each impact velocity (Table 1).

Although our interest is focused on the elaborated EIA, we have at hand the very valuable Brinkman–Kramers approximation (BKA) [12, 13] introduced almost a century ago, the cross section for capture to a principal number n reads

$$\sigma_n^{BKA} = \frac{2^{18}\pi Z^5}{5n^3} \frac{v^8}{D_n^5}, \tag{3}$$

$$D_n = v^4 + 2v^2\left(\frac{Z^2}{n^2} + 1\right) + \left(\frac{Z^2}{n^2} - 1\right)^2. \tag{4}$$

A priori there are certain limitations in our analysis. The first point is that we should inspect all the channels to have a complete picture of all the processes involved, namely, excitation and ionization as well, but the amount of data required is today prohibitive. For this reason, we concentrate just on the capture data available to us. It means that projectiles that do not capture are discarded in our count. The second point is related to the determination of the probability magnitude. The measured capture probability turns out to be the product of the capture cross section σ_{nlm} times the target density times the transit distance corresponding to a specific experimental situation. Our starting data is just the cross section having a dimension of surface, and it is not dimensionless like any probability, as required by the Shannon entropy expression (see Eq. (9) below). Therefore, we resort to the so-called *shape factor* S_{nlm},

$$S_{nlm} = \frac{\sigma_{nlm}}{\sigma}, \tag{5}$$

being σ the total cross section

$$\sigma = \Sigma_n \sigma_n = \Sigma_{nl} \sigma_{nl} = \Sigma_{nlm} \sigma_{nlm}. \tag{6}$$

The shape factor is now suitable to be used in the Shannon entropy expression: it is normalized and dimensionless. Both theories EIA and BKA present different shape factors at intermediate velocities, but one may expect that S_{nlm}^{EIA} tends to be near to S_{nlm}^{BKA} at large velocities and small perturbations. In this region, σ_n^{BKA} behaves as n^{-3} (Oppenheimer rule, see Eq. (3)) which is related to the value of the projectile electron density at the origin. We will largely use this property below. On the other hand, as $v \to 0$, resonant charge exchange holds, i.e., where $D_n \sim 0$,

$$\left(\frac{Z}{n} - 1\right)^2 + v^2 \sim 0, \tag{7}$$

and so σ_n is expected to peak at $n \sim Z$. The resonant capture implies the following reaction:

$$P^{Z+} + H(1s) \to P^{(Z-1)+}(n = Z) + H^+. \tag{8}$$

Note that although the Shape factors S_n^{EIA} and S_n^{BKA} may be rather equal at high velocities, cross-sectional absolute values σ_n^{EIA} and σ_n^{BKA} can be much different each other. For example, one can expect that $\sigma_{100}^{EIA} \sim 0.3\sigma_{100}^{BKA}$ at moderately high (non-relativistic) velocities [14].

3 Shannon Entropy

Shannon entropy H_X is calculated in terms of the dimensionless non-negative probability $P(x_j)$ over a discrete (may be continuum) variable x_j as follows:

$$H_X = -\sum_j P(x_j) \ln P(x_j), \tag{9}$$

where the summation covers all the possible values. We will restrain the logarithm in base e, although bases 2 and 10 are sometimes used. Equation (9) can be also read as the mean value of the "surprise", i.e., $H_X = \langle \ln 1/P(x) \rangle$.

3.1 H_N Entropy

Restricting the dice to the space of the principal quantum number n, the Shannon entropy for capture reads

$$H_N = -\sum_{n=1}^{\infty} S_n \ln S_n, \tag{10}$$

and $S_n = \Sigma_{lm}\sigma_{nlm}/\sigma$, with $\Sigma_n S_n = 1$. Here it is important to recall once again that the use of normalization (5) makes the argument of the logarithm dimensionless as required [2]. Along this work it will be very useful to examine the high velocity limit. By inspecting Eq. (3), we expect there the Oppenheimer behavior to hold, i.e.,

$$S_n^{BKA} \underset{v\to\infty}{\to} S_n^{\infty} = \frac{K}{n^3}, \tag{11}$$

and $K = 0.8319$ is derived by the normalization. The Shannon entropy limit reads in this case

$$H_N^{\infty} = -\sum_{n=1}^{\infty} \frac{K}{n^3} \ln\left(\frac{K}{n^3}\right) = 0.6785, \tag{12}$$

which will be a very important value to compare with along this article. On the other side, at very low velocities one can expect a full resonant capture (see Eq. (8)), $S_n = S_n^{\delta} = \delta_{n,Z}$, and so $H_N^{\delta} = 0$. Therefore, we have now a good idea of the shape of the Shannon entropy H_N,

$$H_N \to \begin{cases} H_N^{\delta} = 0, & v \to 0 \quad \text{(resonant capture)} \\ H_N^{\infty} = 0.6785 & v \to \infty \quad \text{(Oppenheimer rule)} \end{cases}. \tag{13}$$

3.2 H_{NL} Entropy

In similar fashion, we can extend our space to the angular momentum number incorporating the dice face l, then

$$H_{NL} = -\sum_{nl} S_{nl} \ln S_{nl}, \tag{14}$$

$S_{nl} = \Sigma_m \sigma_{nlm}/\sigma$, and $\Sigma_{nl} S_{nl} = 1$. Considering the following structure for S_{nl}

$$S_{nl} = S_n\, f_{nl} \tag{15}$$

so that $\Sigma_l\, f_{nl} = 1$, H_{NL} reads

$$H_{NL} = -\sum_{nl} S_n f_{nl} \ln(S_n f_{nl}) = H_N + \sum_n S_n h_L(n) \tag{16}$$

$$h_L(n) = -\sum_l f_{nl} \ln f_{nl}. \tag{17}$$

Since $f_{nl} \leqslant 1$, we can assert that $H_{NL} \geqslant H_N$, which is expectable: the l-distribution for each n introduces an additional source of entropy. An observation that results interesting is that the term $h_L(n)$ is again a Shannon entropy, and the contribution $\Sigma_n S_n h_L(n)$ is simply an average of these entropies, leading to a non-additive property of the Shannon entropy. We can establish two limits as explained next.

3.2.1 Sharp L-Distribution

If we had a sharp l-distribution that we denote again with the superindex δ, i.e.,

$$f_{nl} = f_{nl}^{\delta} = \delta_{l,l_0}, \tag{18}$$

satisfying, obviously, $\Sigma_{l=0}^{l=n_f-1} \delta_{l,l_0} = 1$. The contribution to the Shannon entropy to H_{NL} is then

$$h_L^{\delta}(n) = -\sum_l \delta_{l,l_0} \ln \delta_{l,l_0} = 0, \tag{19}$$

and so $H_{NL} = H_N$, which is very expected since δ_{l,l_0} implies a "certainty", not any surprise. This assumption is not so bold: capture cross sections at large velocities peak at $l \sim 0$, no matter what is the value of n.

3.2.2 Uniform L-Distribution

This is the opposite case, we have here that all values of l are equally possible, i.e.,

$$f_{nl} = f_{nl}^{U} = \frac{1}{n}, \tag{20}$$

and so $\Sigma_{l=0}^{l=n-1} f_{nl}^{U} = 1$. The Shannon entropy that we denote with the superindex U (uniform) is

$$H_{NL}^{U} = -\sum_{nl} S_n f_{nl}^{U} \ln(S_n f_{nl}^{U}) = H_N + \sum_n S_n h_L(n) \tag{21}$$

$$h_L(n) = -\sum_{l=0}^{n-1} \frac{1}{n} \ln \frac{1}{n} = \ln n. \tag{22}$$

The expression $\ln n$ is known as the Hartley or max-entropy. Therefore, the second term in Eq. (21) $\Sigma_n S_n \ln n$ can be seen as a simple average over S_n of the Hartley entropy. Furthermore, assuming the Oppenheimer distribution $S_n^{\infty} = K/n^3$ one finds

$$\sum_n S_n^{\infty} h_n^U = 0.1648, \tag{23}$$

and so we would expect 0.1648/0.6785 = 24% larger entropy than H_N^{∞}, however this is not the actual case, at large velocities Eq. (18) holds.

3.3 H_{NLM} Entropy

Straightforwardly, we define

$$H_{NLM} = -\sum_{nlm} S_{nlm} \ln S_{nlm}. \tag{24}$$

Considering the same structure: $S_{nlm} = S_{nl}\ f_{nlm}$, so that now $\Sigma_m\ f_{nlm} = 1$ then

$$H_{NLM} = H_{NL} + \sum_{nl} S_{nl} h_M(n, l), \tag{25}$$

$$h_M(n, l) = -\sum_m f_{nlm} \ln f_{nlm}. \tag{26}$$

In similar fashion, the sharp distribution, $f_{nlm}^{\delta} = \delta_{m,m_0}$, introduces no additional entropy, and so $H_{NLM}^{\delta} = H_{NL}$. We could consider an extreme (unrealistic) case: a double Harvey distribution, i.e., a double uniform distribution in l and m,

$$f_{nlm}^{UU} = f_{nl}^{U} \frac{1}{2l+1} = \frac{1}{n} \frac{1}{2l+1}, \tag{27}$$

and assuming the Oppenheimer distribution, it produces

$$\sum_{nl} S_n^{\infty} h_M^{UU}(nl) = \frac{K}{n^3} \frac{1}{n} \frac{1}{2l+1} = 0.2938, \tag{28}$$

producing the maximum possible value at the high velocity limit. Resuming, we have the following hierarchy:

$$H_{NLM} \geqq H_{LM} \geqq H_N. \tag{29}$$

Namely, the more quantum numbers we consider, the larger is the entropy, or equivalently more surprises are activated.

3.4 Limits

The zero of the Shannon entropy corresponds to the point of no surprise at all, this is a dice with just one face, i.e., the full sharp condition,

$$S_{nlm}^{\delta\delta\delta} = \delta_{n,n_0}\delta_{l,l_0}\delta_{m,m_0}. \tag{30}$$

As $v \to 0$, reaction (8) holds, so $S_n \longrightarrow \delta_{n,Z}$, and $H_N \to 0$, no matter the value of Z. But this is not the case for the H_{NL}. Indeed, by setting the low energy limit $S_n \longrightarrow \delta_{n,Z}$, we have

$$H_{NL} \underset{v\to 0}{\to} -\sum_l f_{Zl} \ln f_{Zl}. \tag{31}$$

The only case that $H_{NL} = 0$ corresponds to the sharp l-distribution, i.e., $f_{Zl} = \delta_{n,Z}\delta_{l,l_0}$, which is the case of proton impact ($Z = 1$) at very low velocities (resonant capture). Capture to $n = 1$, forcibly implies $l = 0$, $f_{Zl} = \delta_{1,Z}\delta_{l,0}$ and so $H_{NL} = 0$. For other projectile charges f_{Zl} spans in the range $l \in [0, Z-1]$ and therefore $H_{NL} > 0$. The explanation is again simple: l -distributions different from $\delta_{l,0}$ incorporates new surprises. This behavior extrapolates straightforwardly to H_{NLM}. Again, for proton impact the electron inevitably falls on $n = 1$, $l = 0$, and forcibly $m = 0$, then $f_{nlm} = f_{nlm}^{\delta\delta\delta} = \delta_{n,1}\delta_{l,0}\delta_{m,0}$, which is the zero condition given by Eq. (30), and so $H_{NLM} = 0$. For other projectiles ($Z > 1$), as $v \to 0$, we have a range of probabilities, $m \in [-l, l]$, and so $H_{NLM} > 0$.

4 Results

The calculation of the EIA cross section requires a heavy task: a three-dimensional numerical integration in the momentum space of the initial electronic state and a further one on the angular distribution to obtain a cross section σ_{nlm}^{EIA}. These procedures drag numerical errors, especially for large n and l due to the number of oscillations of the final projectile wave functions and to the oscillations of the continuum wave functions. Although the number n_f ranges between $n_f = 4$ (for $Z = 1$ and 2) to $n_f = 9$ (for $Z > 2$) it is not enough for the Shannon entropy to converge. To make a comparison, consider the simple BKA at large velocities and $Z = 8$, contributions to the cross section σ^{BKA} for $n > n_f = 8$ represent 1%, while for the entropy S_N^{BKA} the contribution for $n > n_f = 8$ is more than 10% larger. In consequence, we are forced to use an extrapolation for large values of n. To extend the cross section for any approximation from n_f to ∞, one can rely on the BKA by using the well-known extrapolation

$$\sigma^{EIA} \simeq \sum_{n=1}^{n_f} \sigma_n^{EIA} + \frac{\sigma_{n_f}^{EIA}}{\sigma_{n_f}^{BKA}} \sum_{n=n_f+1}^{\infty} \sigma_n^{BKA}, \tag{32}$$

for the Shannon entropy we use here the same structure, i.e.,

$$H_N^{EIA} \simeq -\sum_{n=1}^{n_f} S_n^{EIA} \ln S_n^{EIA} - \frac{S_{n_f}^{EIA} \ln S_{n_f}^{EIA}}{S_{n_f}^{BKA} \ln S_{n_f}^{BK}} \sum_{n=n_f+1}^{\infty} S_n^{BKA} \ln S_n^{BKA}, \tag{33}$$

and the same procedure is used to extend H_{NL} and H_{NLM}. In the eight figures of this work, we display H_N, H_{NL} and H_{NLM} for EIA in different colors: red, blue, and green, respectively, for Z = 1 to 8, and H_N for BKA is displayed in black. Dashed lines correspond to the sum up to n_f and solid lines include the extension up to $n = \infty$ as given by Eq. (33).

By inspecting these figures one can extract general behaviors, as resumed next.

i. For the starting case $Z = 1$, the results nicely approach the high velocity limit $H_N^{\infty} = 0.6785$, and further it runs near the BKA indicating the validity of the shape factor of the first perturbative order for this case.
ii. For larger charges ($Z >> 1$), the high velocity tendency starts to departure from H_N^{∞}, which is expected to hold for much larger velocities ($v >> Z$).
iii. H_{NL} and H_{NLM} increase with Z. For larger values of Z, larger n are populated, and correspondingly larger values of l and m are accessible.
iv. For all the cases, $H_{NLM} \geqq H_{NL} \geqq H_N$ is verified, as predicted by Eq. (29), indicating the surprise increases with the incorporation of the new quantum number.
v. For all the cases, one can get a glimpse that Shannon entropy $H_N \to 0$ as $v \to 0$, as observed in Eq. (13).
vi. The behavior of the Shannon entropy is smooth in terms of v contrasting to the dramatic fall of the cross section σ covering several orders of magnitudes. Instead the Shannon entropy ranges within one order of magnitude. We have observed that minor integration errors in the cross sections σ_{nlm} may introduce small wrinkles in the entropy curve. To some extent the smoothness of the Shannon entropy can be used as a tool to diagnose the reliability of the numerical set.

What called our attention is that all the surprises can be cast in one single number given by the Shannon entropy. As we derive the three level, H_N, H_{NL}, and H_{NLM}, we can enhance the surprises including more faces of the quantum-number dice. The implication of these outputs can be very powerful for other theoretical methods. For example, we have in mind the classical trajectory Monte Carlo (CTMC). This method is quite useful to calculate total cross sections, but sometimes it is also used to determine the final quantum numbers n, and l. These numbers are determined when the outgoing atoms landfall in a given box depending on its energy and momentum. A comparison of the CTMC Shannon entropy with the present values would give us a good diagnostic on the quality of the so-designed boxes.

A better calculation would introduce another factor: the Stark mixing. New values of l can be populated through a transition $nlm \rightarrow nl'm$ in a presence of the electric field of the remaining proton. More values of l can be populated and so larger values of H_{NL} are expected.

Another point to be investigated is that at very large velocities, say much larger than 10 MeV/amu for proton, the capture process is dominated by the so-called Thomas mechanism which is included in the EIA and not in the first-order BKA. The limit behaviors are different $\sigma_{100}^{EIA} \propto v^{-11}$ while $\sigma_{100}^{BKA} \propto v^{-12}$. The l-distributions for a given n are also different [14, 15]. This would indicate that H_{nl}^{EIA} may differ a little from H_{nl}^{BKA}.

I am in debt to M. S. Gravielle for valuable discussions and comments. Financial supports from: Consejo Nacional de Investigaciones Científicas y Técnicas (CONICET), Agencia Nacional de Promoción Científica y Técnológica (ANPCyT), and Universidad of Buenos Aires are acknowledged (Figs. 1, 2, 3, 4, 5, 6, 7, and 8).

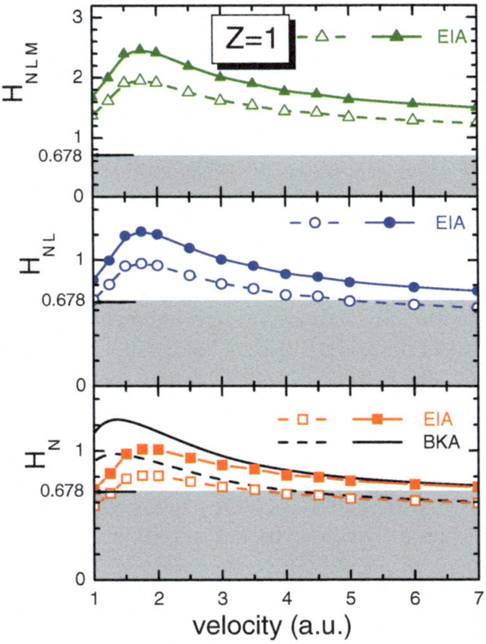

Fig. 1 (Color on line) Shannon entropy H_N^{EIA} (low panel, in red), H_{NL}^{EIA} (medium panel, in blue), and H_{NLM}^{EIA} (in green, upper panel), as a function of the impact velocity v for $Z = 1$ as indicated. Shannon Brinkman–Kramers H_N^{BKA} is displayed in black. Dashed lines denote the entropy calculated up to n_f, while solid lines correspond to the n-extrapolation as indicated in Eq. (33). Symbols are the velocities calculated as indicated in the table. Entropy values less than $H_N^{\infty} = 0.678$ are displayed in light gray

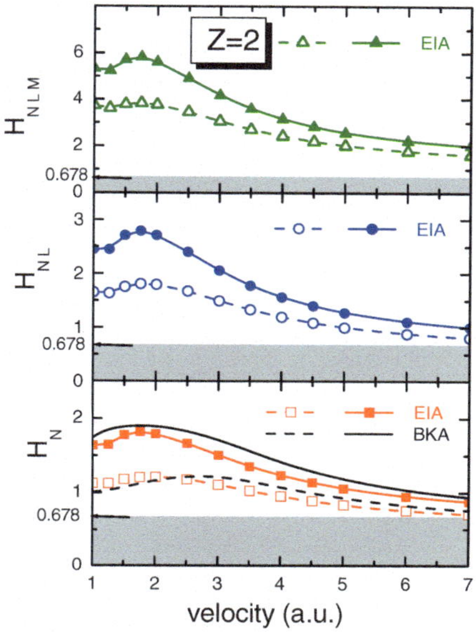

Fig. 2 (Color on line) Similar to Fig. 1 for Z = 2

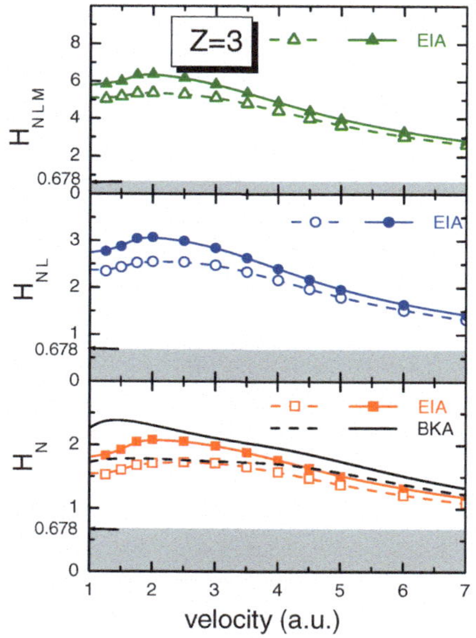

Fig. 3 (Color on line) Similar to Fig. 1 for Z = 3

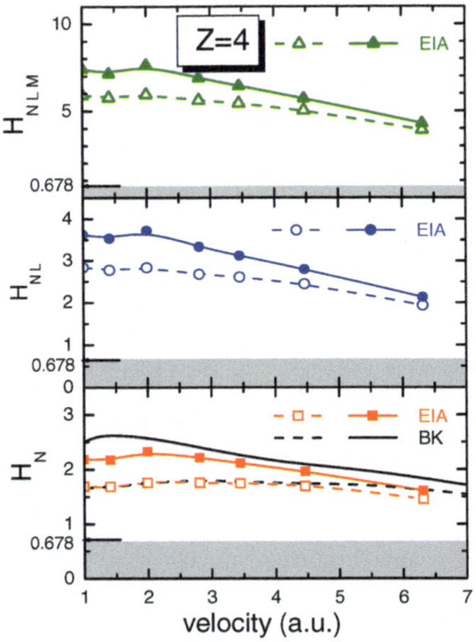

Fig. 4 (Color on line) Similar to Fig. 1 for Z = 4

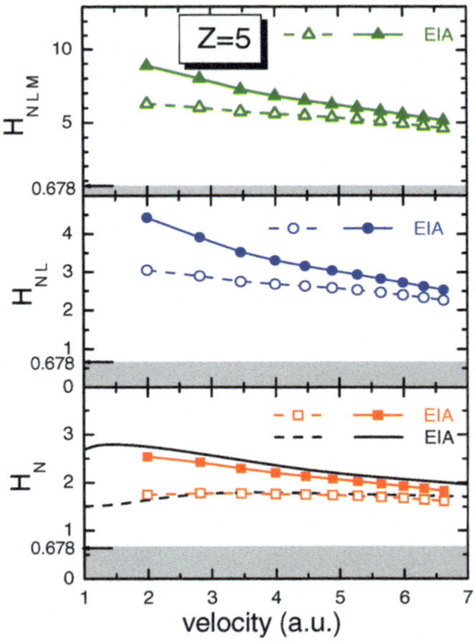

Fig. 5 (Color on line) Similar to Fig. 1 for Z = 5

Fig. 6 (Color on line)
Similar to Fig. 1 for Z = 6

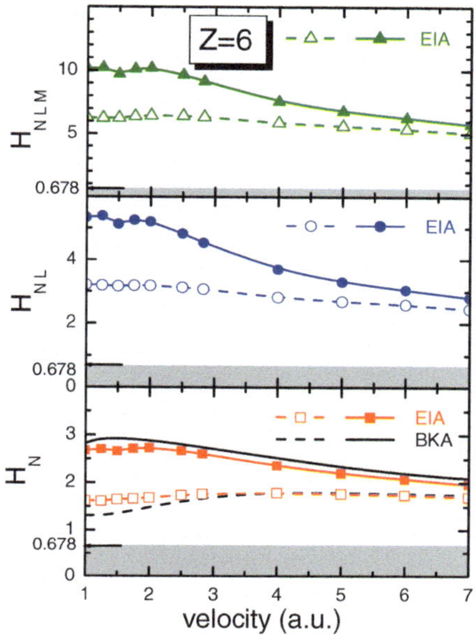

Fig. 7 (Color on line)
Similar to Fig. 1 for Z = 7

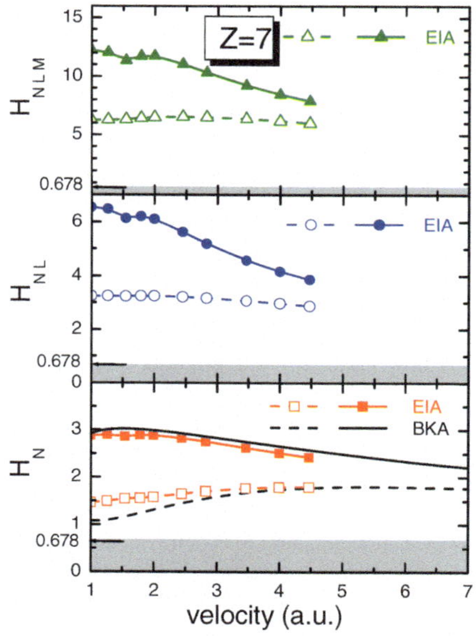

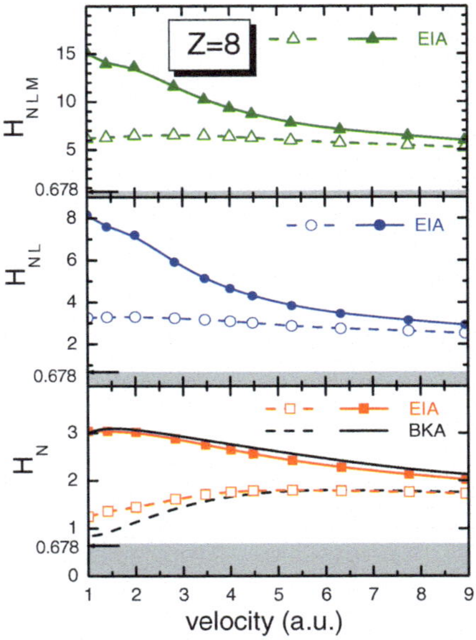

Fig. 8 (Color on line) Similar to Fig. 1 for Z = 8

References

1. C.E. Shannon, *A Mathematical Theory of Communication*. Bell Syst. Tech. J. **27**, 379 (1949), and **27**, 623 (1948)
2. N. Flores-Gallego, A new approach of Shannon entropy in atoms. Chem. Phys. Lett. **650**, 57–49 (2016)
3. C.-W. Maa, Y.-G. Mab, Shannon information entropy in heavy-ion collisions. Prog. Part. Nucl. Phys. **99**, 120–158 (2018)
4. J. Schlitter, Inelastic collisions as a source of entropy? (9 Dec 2011). arXiv:1104.4272v3 [cond-mat.stat-mech]
5. M.S. Gravielle, Previous Chapter of this book (2022)
6. M.S. Gravielle, J.E. Miraglia, Eikonal impulse approximation in electron capture processes. Phys. Rev. A **44**, 7299–7306 (1991)
7. M.S. Gravielle, J.E. Miraglia, State-selective scaling in electron capture by multicharged ions on light atoms. Phys. Rev. A **51**, 2131–2139 (1995)
8. M.S. Gravielle, J.E. Miraglia, State-selective scaling for capture to level n = 4 by multicharged ions on light atoms. Phys. Rev. A **52**, 851–854 (1995)
9. C.C. Montanari, M.S. Gravielle, J.E. Miraglia, Propensity rule for the magnetic substate distribution in electron capture at high impact energies. Phys. Rev. A **56**, 4314–4316 (1997)
10. A. Jorge, C. Illescas, M.S. Gravielle, J.E. Miraglia, Scaling for state-selective charge exchange due to collisions of multicharged ions with hydrogen. J. Phys. B **48**, 235201 (2015)
11. J.E. Miraglia, M.S. Gravielle, Atomic ionization, capture, and stopping cross sections by ion impact examined with the Benford law, in *Advances in Quantum Chemistry*, ed. by J. Oddershede (Elsevier, Amsterdam, 2022), to be published
12. H.C. Brinkman, H.A. Kramers, Proc. Acad. Sci. Amsterdam **33**, 973 (1930)

13. M.R.C. McDowell, J.P. Coleman, *Introduction to the Theory of Ion-Atom Collisions* (Amsterdam North Holland) Chapter 7 (1970)
14. K. Dettmann, G. Liebfried, High energy limit in rearrangement collisions. Z. Phys. **218**, 1 (1969)
15. L.J. Dubé, J.S. Briggs, Systematics of the single and double electron scattering contributions to charge exchange. J. Phys. B: At. Mol. Phys. **14**, 4595–4617 (1981)

Target Ionization and Electron Loss Processes Induced by Neutral and Charged Hydrogen and Helium Projectiles in Water Molecule

M. A. Quinto, N. J. Esponda, A. Larouze, C. Champion, R. D. Rivarola, and J. M. Monti

Abstract This work aims to present cross sections, for electron removal processes in neutral-atom collisions with water molecules, computed within the CDW-EIS (Continuum Distorted Wave-Eikonal Initial State) approximation. For describing these reactions, two corrections are taken into account in the present model: the mean projectile charge as a function of the impact energy and the dynamic projectile charge depending on the momentum transfer during the interaction. The theoretical data, in terms of the double differential and the total cross sections, for target ionization and projectile electron loss processes is compared with existing data.

1 Introduction

Electron removal from molecules induced by ion impact is a process of fundamental importance in many areas of science like astrophysics, physics, and chemistry of planetary atmospheres [1], as well as in radiobiology to quantify the damage induced

M. A. Quinto (✉) · N. J. Esponda
Instituto de Física Rosario (CONICET-UNR), Bv 27 de Febrero 210 bis, 2000 Rosario, Argentina
e-mail: quinto@ifir-conicet.gov.ar

N. J. Esponda
e-mail: esponda@ifir-conicet.gov.ar

A. Larouze · C. Champion
CELIA, Centre Lasers Intenses et Applications, Université de Bordeaux - CNRS - CEA, 33400 Talence, France
e-mail: alexandre.larouze@u-bordeaux.fr

C. Champion
e-mail: christophe.champion@u-bordeaux.fr

R. D. Rivarola · J. M. Monti
Instituto de Física Rosario (CONICET-UNR), Bv 27 de Febrero 210 bis, 2000 Rosario, Argentina
e-mail: rivarola@ifir-conicet.gov.ar

J. M. Monti
e-mail: monti@ifir-conicet.gov.ar

L. C. Tribedi (ed.), *Advances in Atomic Molecular Collisions*,
https://doi.org/10.1007/978-981-97-7063-2_9

by particles over cells and tissues [2]. Indeed, such ionizing processes are predominant in the slowing-down of highly charged ions in biological matter, the latter being commonly modeled by water. In this context, let us note that neutral-hydrogen or neutral-helium collisions with atoms and molecules have been rarely experimentally investigated [3–7].

For proton energy around and below 100 keV, the electron capture becomes of the same order of importance, or even larger, than the ionization leading then to a substantial amount of neutral atoms in the beam. However, in spite of their relevance, cross section measurements for water ionization by neutral-atom impact remain limited. In this context, Toburen et al. reported the first total electron loss cross sections for 100–2500 keV hydrogen atoms in water by using the equilibrium fraction method to analyze the multiple-charge-state beam at the exit of a gas cell [8]. Later on, Bolorizadeh and Rudd measured doubly differential cross sections (DDCSs) for electron production emission in collisions of 20–150 keV neutral hydrogen with water molecules [9]. Also, they reported the total cross sections (TCSs) of the target ionization and the electron loss processes by numerical integration of DDCSs.

More recently, Gobet et al. [10] carried out cross-sectional measurements for the ionization of water molecules impacted by 20–150 keV atomic hydrogen. Event-by-event coincidence analysis of the product ions and the projectile enabled partial cross sections for target ionization and target plus projectile ionization, as well as total cross sections for electron loss reactions. In the recent work of Luna et al. [11] time-of-flight–coincidence-counting arrangement is reported to perform dissociative and non-dissociative measurement reactions of water by 8–100 keV neutral-hydrogen impact. The measurements were then carried out for the three above-mentioned interactions induced by neutral-hydrogen impact, namely, the target ionization, the electron capture, and the projectile electron loss processes.

Under these conditions, and at least in part because of the absence of experimental data to compare with, there has been little theoretical work done on neutral impact ionizing processes. Many semiempirical models were then proposed to overcome the lack of theoretical support and then to predict the occurrence of neutral-hydrogen-induced ionization. Among the existing approaches, we find the one proposed by Dingfelder et al. [12] who used the single differential cross sections SDCSs by proton impact to describe the water ionization induced by neutral hydrogen. The SDCSs for neutral-hydrogen impact are obtained by a slight modification of those by proton impact using a function depending on the incident energy. Thus, on the basis of the data of Bolorizadeh and Rudd [9, 13] as well as those of Toburen [8, 14], Dingfelder et al. suggested a scaling factor between proton and neutral-hydrogen SDCSs, assumed to be higher than unity at low incident energies and lower than unity for higher energies, in view of the nuclear charge screening induced by the bound electron of hydrogen. Similarly, Uehara et al. [15] parameterized the neutral-hydrogen ionization of water molecules by integrating twice the double differential cross sections provided by Bolorizadeh and Rudd [9]. However, as highlighted above, the total cross section includes the contributions of the electrons ejected from both target and projectile, i.e., corresponds to both the target and projectile ionization. Finally, Quinto et al. have studied the target ionization of water molecules by neutral hydrogen within the CDW-EIS approximation. The results obtained in terms of double differential and total cross sections exhibit a good agreement with experimental data [16].

Besides, recent microdosimetric simulations have indicated that the α-emitter radionuclides are promising candidates in targeted radionuclide therapy [2]. Therefore, the series of helium, He^0, He^+, and He^{2+} projectiles, have recently gained great importance in collision with water molecules. In this context, Uehara et al. studied the electron removal processes from water molecules impacted by neutral and charged helium projectiles. A semiempirical approximation was employed to fit the experimental data set [17]. For projectile ions (H^+, He^{2+}, O^{8+}, and Si^{13+}), several theoretical and experiential works have been done to describe the ionization of water molecules [18–24]. For He^+ projectile, it exists in literature both experimental data in terms of total and differential cross sections [13, 25] and the theoretical approaches proposed by Garcia et al. [26] and Kovacs et al. [27]. Whereas the case of the collisions between neutral Helium projectile and water molecule has been little studied.

The current work aims to go beyond any empirical approach and then provide a theoretical description of the target ionization and projectile electron loss in water within the CDW-EIS approximation. In the sequel, we detail the theoretical framework of our model and then compare the theoretical predictions in terms of DDCSs, SDCSs, and TCSs with existing theories and available experimental data.

Atomic units (a.u.) are used throughout unless indicated otherwise.

2 Theory

In this study, we employ the Continuum Distorted Wave Eikonal Initial State (CDW–EIS) approximation to describe the target ionization and projectile electron loss. The CDW-EIS model initially used to calculate the ionization of mono-electronic targets by bare projectile impact has been enhanced over the years. Fainstein et al. [28] extended the CDW-EIS model to the case of multi-electronic targets considering an effective one active-electron target and then assuming the other target electrons as passive ones. This implies that the passive electrons remain in their initial orbitals during the collision. Then, Monti et al. [29] applied this extended approximation to the case of ionization by dressed projectiles. Within the independent electron model and following the procedure given in [30], the multi-electronic Hamiltonian can be reduced to a single-active-electron one given by

$$H_e = -\frac{1}{2}\nabla^2 + V_T(\mathbf{x}) + V_P(\mathbf{s}) + V_S(\mathbf{R}) \tag{1}$$

where $\mathbf{x}$ and $\mathbf{s}$ represent the active-electron position in the reference frame fixed to the target and the projectile, respectively. The potential $V_T(\mathbf{x})$ includes the interaction of the active electron with the rest of the target, while the term $V_P(\mathbf{s})$ describes the interaction between the target-active electron and the projectile, and $V_S(\mathbf{R})$ accounts for the interaction of the projectile with the target nucleus and its passive electrons. It has been shown that, within the straight-line version of the impact parameter approximation, the $V_S(\mathbf{R})$ potential gives rise to a phase factor, which only affects the projectile angular scattering and not the electron dynamics [28, 31, 32]. This term will drop in the following analysis. The first order of the *prior* version of the ionization transition amplitude within the CDW-EIS approximation can be written as

$$\mathcal{A}_{if}^{-}(\boldsymbol{\rho}) = -i \int_{-\infty}^{\infty} dt \left\langle \chi_f^- \right| \left(H_e - i\frac{\partial}{\partial t} \right) \left| \chi_i^+ \right\rangle \tag{2}$$

where the initial and final distorted wave functions (χ_i^+ and χ_f^-, respectively) are proposed as

$$\chi_i^+(\mathbf{x}, \mathbf{s}, t) = \Phi_i(\mathbf{x}, t)\mathcal{L}_i^+(\mathbf{s}) \tag{3}$$

$$\chi_f^-(\mathbf{x}, \mathbf{s}, t) = \Phi_f(\mathbf{x}, t)\mathcal{L}_f^-(\mathbf{s}) \tag{4}$$

with $\Phi_i(\mathbf{x}, t) = \phi_i(\mathbf{x}) \exp(i\epsilon_i t)$ and $\Phi_f(\mathbf{x}, t) = \phi_f(\mathbf{x}) \exp(i\epsilon_f t)$ the initial-bound and final-continuum wave function solutions of the time-dependent Schrödinger equation, and ϵ_i and ϵ_f the corresponding electron energies, respectively. The description of the different initial ground-state molecular orbitals is assumed to be represented by linear combinations of their atomic compound orbitals (LCAOs). The corresponding binding energies are computed employing the Molecular Orbital Linear Combination of Atomic-Orbital Self-Consistent Field method (MO-LCAO-SCF), whereas a complete neglect of differential overlap (CNDO) approximation is employed for the effective occupation electron analysis of the different molecular orbitals [33]. For water molecules, the corresponding atomic effective electron populations and binding energies are reported in Table 1. Thus, the cross section, for a given molecular orbital l, is given as

$$\sigma_l = \sum_{j=1}^{N_l} c_{l,j}\, \sigma_{if,j} \tag{5}$$

Table 1 Population and binding energies of the H_2O molecular orbitals

Molecular orbital	Population	Binding energy ϵ_i (a.u)
1a1	$2.0\ O_{1s}$	−19.84
2a1	$1.48\ O_{2s} + 0.52\ H_{1s}$	−1.18
1b2	$1.18\ O_{2p} + 0.82\ H_{1s}$	−0.67
3a1	$0.22\ O_{2s} + 1.44\ O_{2p} + 0.34\ H_{1s}$	−0.54
1b1	$2.0\ O_{2p}$	−0.46

where N_l refers to the number of atoms for each molecular orbital. $\sigma_{if,j}$ denotes the atomic-orbital cross section involved in its LCAO description, $c_{l,j}$ being the corresponding atomic effective occupation electron numbers. The different ground-state atomic orbitals are described using Roothaan–Hartree–Fock atomic wave functions [34]. The ionization of at least one electron for the water molecule (also known as net ionization [35]) can be obtained by summing the contributions from all the molecular orbitals:

$$\sigma_{H_2O} = \sum_{l=1}^{5} \sigma_l \tag{6}$$

The target final-continuum state $\phi_f(\mathbf{x})$ is chosen as an hydrogenic continuum function defined as

$$\phi_f(\mathbf{x}) = \frac{1}{(2\pi)^{3/2}} \exp(i\,\mathbf{k}.\mathbf{x})\; N^*(\lambda)\; {}_1F_1\,[-i\,\lambda,\, 1,\, -i\;(kx + \mathbf{k}.\mathbf{x})] \tag{7}$$

with $\epsilon_f = k^2/2$ the final electron energy, being $\mathbf{k}$ the ejected electron momentum in the target reference frame. The term $N(a) = \exp(\pi a/2)\Gamma(1 - ia)$ (with Γ being the Euler's Gamma function) is the normalization factor of the ${}_1F_1$ hypergeometric function, and $\lambda = \tilde{Z}_T/k$ with $\tilde{Z}_T$ an effective charge. This charge is chosen according to the energy, ϵ_l, of the molecular orbital as $\tilde{Z}_T = \sqrt{-2n^2\epsilon_l}$, being n the principal quantum number of each atomic compound corresponding to a given molecular orbital into the CNDO description [31] (see Table 1). The continuum function given by (7) corresponds to approximate the potential $V_T(\mathbf{x})$ by an effective Coulomb potential $\left(-\frac{\tilde{Z}_T}{x}\right)$ in the exit channel.

For dressed projectiles, $V_P(s)$ is a non-Coulomb potential that is approximated in the work of Monti et al. [29] by a two-parameter analytic independent-particle Green–Sellin–Zachor (GSZ) model potential [36]. This potential can be written as a Coulomb long-range term plus a short-range one, given by

$$V_P(s) = -\frac{q}{s} - \frac{1}{s}(Z_P - q)\Omega(s) \tag{8}$$

with $\Omega(s) = [H(\exp(s/d) - 1) + 1]^{-1}$. The H and d parameters depend on the projectile nuclear charge Z_P and its degree of ionization q. These values of H and d are tabulated in [36]. Following the recent work by Esponda et al. [37], the initial and final channel distortions are chosen as

$$\mathcal{L}_i^+(\mathbf{s}) = \exp[-i\nu \ln(vs + \mathbf{v}.\mathbf{s})] \tag{9}$$

$$\mathcal{L}_f^-(\mathbf{s}) = N^*(\zeta)_1F_1\left[-i\zeta;\, 1;\, -i(ps + \mathbf{p}.\mathbf{s})\right] \tag{10}$$

where $\mathbf{v}$ is the projectile velocity, $\nu = q/v$, $\zeta = q(\mathbf{K})/p$, and $\mathbf{p} = \mathbf{k} - \mathbf{v}$ is the electron momentum in the projectile reference frame. It can be seen that in the final channel distortion, a dynamic effective projectile charge is introduced, which depends on the momentum transfer $\mathbf{K}$. The effective charge is written as

$$q(\mathbf{K}) = Z_P - F(\mathbf{K}) \tag{11}$$

with $F(\mathbf{K})$ being the projectile form factor [38]. For small momentum transfer $q(\mathbf{K})$ tends to $Z_P - N$, whereas for large momentum transfer $q(\mathbf{K})$ goes to Z_P [37], therefore even for neutral projectiles (with $q = 0$) the final channel distortion never vanishes.

The final scattering amplitude is expressed as the sum of two terms: the first one corresponding to the Coulomb term, representing the ionization by a bare projectile of charge q, and a second term corresponding to the contribution of the short-range potential [16, 29]:

$$\mathcal{A}_{if}^{-}(\boldsymbol{\rho}) = \mathcal{A}_{if}^{-(a)}(\boldsymbol{\rho}) + \mathcal{A}_{if}^{-(s)}(\boldsymbol{\rho}) \tag{12}$$

By employing the Fourier transform, the $\mathcal{A}_{if}^{-}(\boldsymbol{\rho})$ can be expressed as a function of the transverse momentum transfer $\boldsymbol{\eta}$

$$\mathcal{R}_{if}^{-}(\boldsymbol{\eta}) = \int d\boldsymbol{\rho}\, \exp(i\boldsymbol{\rho}.\boldsymbol{\eta})\mathcal{A}_{if}^{-}(\boldsymbol{\rho}) \tag{13}$$

This allows us to calculate the cross sections using the $\mathcal{R}_{if}^{-}(\boldsymbol{\eta})$ for which analytical expression can be obtained [39]. For ionization by bare projectiles, $\mathcal{A}_{if}^{-(s)}(\boldsymbol{\rho}) = 0$, since only the Coulomb contribution affects the transition amplitude. In the case of dressed projectiles, both terms must be considered.

Finally, the doubly differential cross sections are given by

$$\frac{d^2\sigma_{if}}{dE_e d\Omega_e} = k \int d\boldsymbol{\eta} |\mathcal{R}_{if}^{-}(\boldsymbol{\eta})|^2 \tag{14}$$

Then, by numerical integration of the DDCS Eq. 14 over the solid angle Ω_e, we calculate the SDCS as a function of the emission energy. Finally, the TCS is computed by further integrating the SDCS over the electron energy E_e.

The electron loss process can be addressed by reversing the order of collision aggregates with the help of Lorentz transformation of target and projectile reference frames [40]. For example, in a collision system where the water molecule is impacted by neutral hydrogen, the water is then assumed to act as a neutral projectile, with $Z_P = 10$, colliding with the hydrogen atom. Furthermore, it is worthy to mention that the corresponding $V_P(s)$ potential is calculated under a MO-LCAO-CNDO scheme as detailed in Appendix. In this case, the projectile electron loss description model is characterized by a high Sommerfeld factor, Z_P/v, a region where the perturbative theories lose in accuracy. Thus, we have introduced a correction charge into the CDW-EIS approximation depending on the impact energy: the equilibrium projectile mean charge. In the literature, many works exist regarding such equilibrium charge, some of them have proposed a formula based on experimental data [41]. In this work, we proposed a formula based on the Bezt model where the charge of a bare projectile is affected by a factor depending on the projectile velocity, [42]:

$$\begin{aligned} \bar{Z}_P &= Z_P\,\alpha(v) \\ &= Z_P(1 - \exp(-125\,v/c)) \end{aligned} \tag{15}$$

The equilibrium charge given by Eq. 15 is a simple model which depends only on the velocity and charge of the projectile, being c the speed of light ($c = 137$ a.u.). However, more complex formulas have been reported, some of them take into account the target state (solid or gas) [43], and other models include improvements, such as shell, polarization, and screening corrections [44]. In Eq. 15, the mean projectile charge $\bar{Z}_P$ tends to Z_P as the projectile velocity increases. In Fig. 1, we depict the Sommerfeld factor, Z_P/v and $\bar{Z}_P/v$ (red lines), as a function of incident energy, computed for a charge $Z_P = 10$. When the mean projectile charge, Eq. 15, is considered, the Sommerfeld factor moves closer to the region where the perturbative approximation will produce reliable results. We will adopt this correction only in the case of collision systems involving neutral projectiles, since, in these cases, the CDW-EIS approximation reduces to the First Born Approximation (FBA). This happens because the initial and final channel distortions, Eqs. 9 and 10, do not contribute to the initial and final distorted wave functions (see Eqs. 3 and 4). It is well known that the cross sections, computed within the FBA, exhibit a ${Z_P}^2$ dependency. Thus, when considering this term, $\alpha(v)$, the DDCS and SDCS profiles and each point of the TCS are affected by a factor $\alpha^2(v)$.

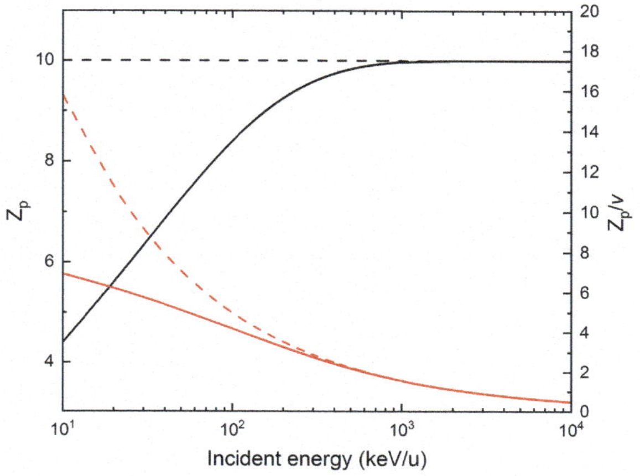

Fig. 1 Fixed projectile charge (dashed black line) and mean charge as function of incident energy (solid black line). Sommerfeld parameter as function of incident energy considering a fixed charge (dashed red line) and mean charge (solid red line) for a projectile charge $Z_P = 10$

3 Results

The theoretical CDW-EIS cross sections are compared with the experimental and existing data. The target ionization and the electron loss (projectile ionization) processes are investigated for the H^0, He^+, and He^0 projectiles impact on the water molecule. For more results regarding H^+ and He^{2+} impinging on water molecules, we refer the reader to our previous works [16, 21, 24, 45, 46]. The theoretical results obtained with the CDW-EIS model including both corrections, mean equilibrium charge and dynamic effective projectile charge, will be compared with the one without corrections. Also, the projectile charge correction, $q(\mathbf{K})$, will be considered only in the final channel as in [37]. For the electron loss process, the water molecule acts as a projectile with a central potential, which is defined in Appendix. For the sake of clarity, we rename the corrections with the indices as $\mathrm{i}_{\tilde{Z}_P}$ for the mean equilibrium charge and $\mathrm{i}_{q(\mathbf{K})}$ for the dynamic effective charge. Thus, when the index value is equal to 1, it means that the correction is included in the CDW-EIS calculations, otherwise its value will be 0.

3.1 H^+ and H^0 Projectiles

In Fig. 2, the theoretical predictions for the target ionization and electron loss processes of water molecules impacted by neutral-hydrogen atom are presented. The results are given in terms of DDCS as a function of ejected electron energy and for the following electron emission angle: $\theta_e = 10°$ and 50°. For $\theta_e = 10°$, the computed DDCS including the corrections shows the presence of the ECC (electron capture to continuum) peak in target ionization induced by a neutral projectile, whereas no evidence of the ECC cusp can be observed when no corrections are considered. Without the $q(\mathbf{K})$ correction, the CDW-EIS model, for neutral projectiles, falls down into a FBA, which does not reproduce the two-center effects in the ion–atom collision, particularly the ECC peak [37, 47]. In general, we observe a better agreement with the experimental data when the charge corrections are included. For $\theta_e = 50°$, theoretical DDCS without corrections largely overestimates the experimental results, while when corrections are considered a much better agreement is observed. Also, the small hump at around 75 eV is due to the electron loss process.

In Fig. 3, we report the theoretical TCSs obtained within the present CDW-EIS approximation. For the proton impact case (see Fig. 3a), we observe a good agreement between the theory and the experimental data for impact energies greater than 50 keV. The sets of experimental data reported are consistent among them. The data of Rudd [48] and Bolarizadeh and Rudd [49] refers to electron target emission cross section (σ_-). Whereas those of Luna et al. [11] and those of Gobet et al. [10] correspond to the sum of the fragment ions, namely, H_2O^+, OH^+, H^+, and O^+. For the hydrogen impact case (see Fig. 3b), three sets of experimental data are reported. The measurements reported by Gobet et al. [10] and Luna et al. [11] are consistent between them. These

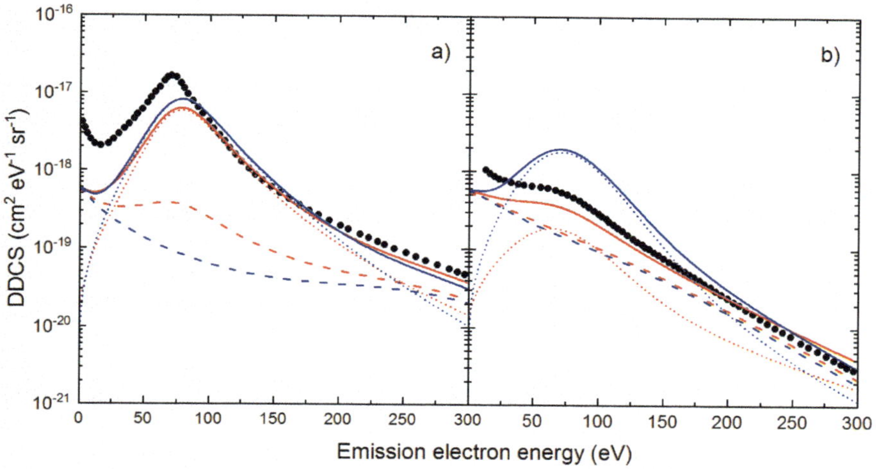

Fig. 2 Double differential cross sections as a function of ionized electron energy of water vapor impacted by H^0 at 150 keV/u. Results for different ejected electron angles, θ_e, equal to 10° (**a**) and 50° (**b**). CDW-EIS ($i_{\tilde{Z}_P} = 0, i_{q(\mathbf{K})} = 0$) (blue line), CDW-EIS ($i_{\tilde{Z}_P} = 1, i_{q(\mathbf{K})} = 1$) (red line). Target ionization (dashed line), electron loss (dotted line), and sum of both (solid line). Experimental data (circles) are given by [9]

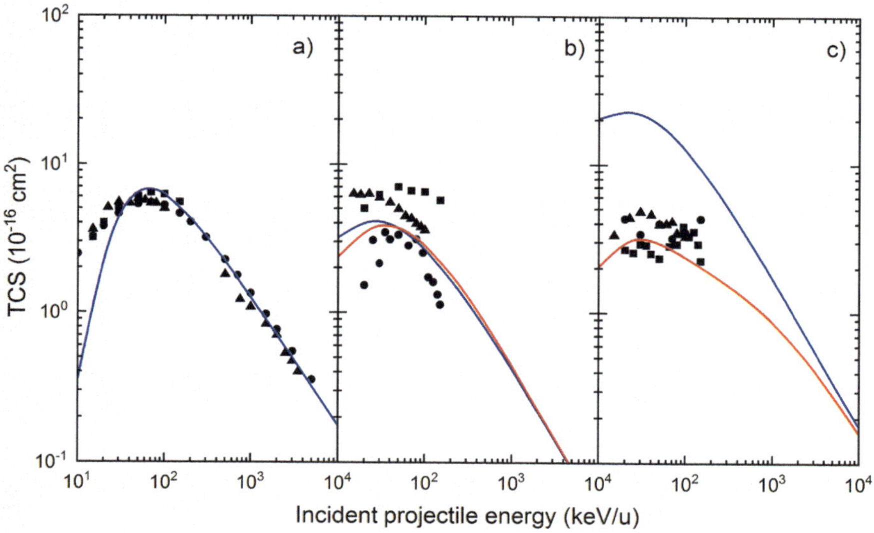

Fig. 3 Total target ionization cross sections of water vapor impacted by H^+ (**a**) and H^0 (**b**) projectiles. Total electron loss of H^0 projectile (**c**) impact on water vapor. Theoretical models: CDW-EIS ($i_{\tilde{Z}_P} = 0, i_{q(\mathbf{K})} = 0$) (blue), CDW-EIS ($i_{\tilde{Z}_P} = 1, i_{q(\mathbf{K})} = 1$) (red). Experimental data in figure (**a**) are given by circles [48], squares [49], and triangles [11], whereas in figures (**b**) and (**c**) circles [9], squares [10], and triangles [11]

experimental data have been measured with the time-of-flight-based mass analysis of charged water fragment production (H_2O^+, OH^+, H^+, and O^+). The data of Gobet et al. was taken in coincidence in such a way that in the final channel both target and projectile charge states were detected. Whereas in the experiment of Luna et al. only the charge state of the target was detected in the final channel. Thus, the final projectile charge state can be 0 or 1 (i.e., projectile ionization). Consequently, the measurements of Luna et al. are also affected by the simultaneous ionization of both target and projectile. In the present work, the reported experimental data has been obtained by subtracting the simultaneous cross sections from the ionization measurements [11]. Another set of measurements is taken from Bolorizadeh and Rudd [9]. We report the experimental target ionization cross section as the subtraction of σ_- - σ_{ELC} (see Fig. 3b). Here σ_- refers to the total electron production both from target and projectile ionization, while σ_{ELC} denotes the Electron Loss to the Continuum (ELC) given by the detachment of the projectile electrons. These sets of total cross sections have been obtained by numerical integration of the double differential cross sections, which have been measured for a few ejected emission electron angles [9]. As shown in Fig. 3b, for the case of water ionization induced by neutral hydrogen, the measurements reported by Bolorizadeh and Rudd are approximately twice greater than those of Gobet et al. and Luna et al.. The present theoretical predictions, namely, CDW-EIS with the charge corrections and without them (red and blue lines, respectively), are very similar between them for energy greater than 40 keV and they are in better agreement with the experimental set of [10, 11].

In Fig. 3c, the theoretical prediction cross sections for the ionization of the neutral-hydrogen projectile (electron loss) have been reported and compared with three series of measurements. The latter taken from [10, 11], while the σ_{ELC} is given by [9]. These experimental data are in accord with them. We observed that the theoretical prediction obtained without corrections largely overestimates the measurements in the presented energy range. However, when the charge corrections are included in the CDW-EIS model, the obtained total cross sections are in reasonable quantitative accord with the experimental data. In general, minor discrepancies still exist between experimental and theoretical results.

3.2 He^{2+}, He^+, and He^0 Projectiles

In Fig. 4, we present the total doubly differential cross sections of electron removal, namely, the sum of target ionization plus projectile ionization, as a function of the electron emission energy. The DDCSs are given for an impact energy of 500 keV/u for the three helium projectiles, namely, He^{2+}, He^+, and He^0. The theoretical predictions have been compared with the measurements given by Toburen et al. [25]. For the He^+ and He^0 cases, and at emission energy of 272 eV, it can be observed, both in experiments and theoretical calculations, the structure corresponds to the ELC peak, which is produced by the electrons emitted from the projectiles. The present DDCSs reproduce well the ELC peak. However, the CDW-EIS results for He^+ tend to

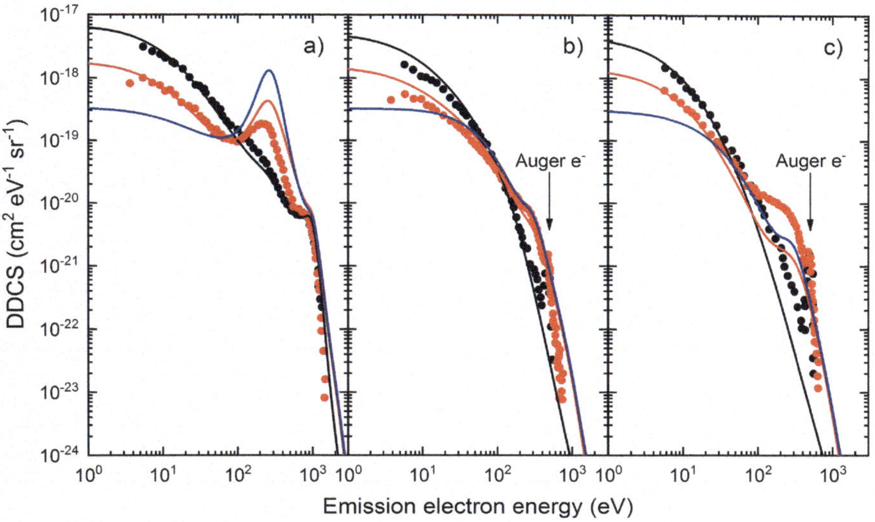

Fig. 4 Double differential cross sections as a function of emission energy for water vapor impacted by Helium ions projectile at 500 keV/u, for a given emission angles θ_e equal to 15° (**a**), 90° (**b**), and 125° (**c**). Solid lines theoretical models: CDW-EIS ($i_{\tilde{Z}_P} = 0, i_{q(\mathbf{K})} = 0$) for He^{2+} (black) and CDW-EIS ($i_{\tilde{Z}_P} = 1, i_{q(\mathbf{K})} = 1$) for He^+ (red) and He^0 (blue). Experimental data are given by circles [25]

overestimate the ELC peak with respect to the measurements. While we can observe a good agreement with the measurements outside the ELC peak region. Also, from the experiments, it is possible to observe a narrow peak located at almost 500 eV electron energy (see Fig. 4b and c). This peak arises from the Auger emission electrons from the innermost orbital of a water molecule. The present theoretical model does not take into account the Auger emission electron process. In addition, the DDCSs prediction reported for the He^{2+} projectile is in accordance with the experimental data in all the presented energy range, except for back scattering at large emission electron energies (Fig. 4c). For comparison, the DDCSs for the He^0 projectile are also reported (see Fig. 4). In this case, the theoretical predictions show a higher ELC peak, compared to the He^+ one, and the lowest yield of low-energy electrons.

We report the theoretical predictions in terms of total cross section as a function of the projectile impact energy for target ionization (Fig. 5) and projectile electron loss (Fig. 6). The results were compared with experimental data sets available in the literature. The measurements reported by Rudd et al. [13, 18], for He^{2+} and He^+ projectiles, respectively, refer to electron target emission cross section (σ_-). Whereas the total cross-sectional data reported by Toburen et al. [25] are obtained by numerical integration of the DDCSs measurements. The latter, for the case of He^+ projectile, are affected by the contribution of the electrons arising from the projectile ionization [25]. The measurements reported by Garcia et al. [26], for the

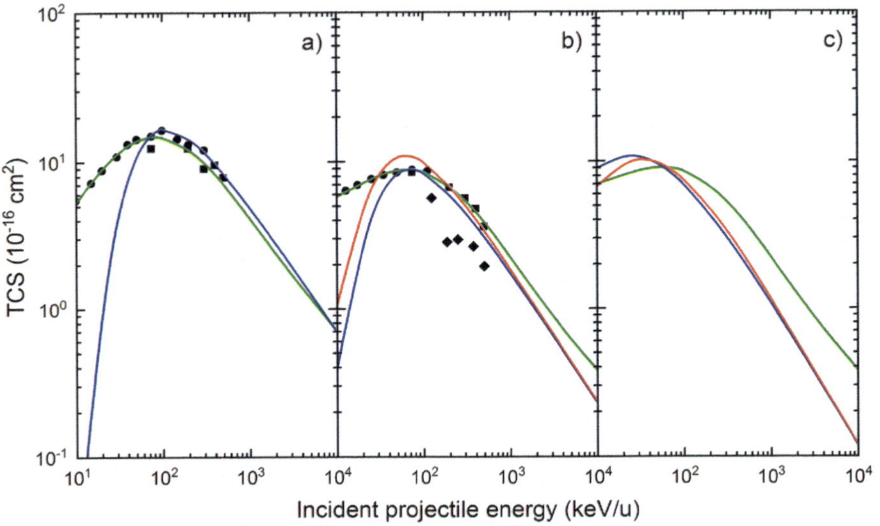

Fig. 5 Total target ionization cross sections of water vapor impacted by He^{2+} (**a**), He^{+} (**b**), and He^{0} (**c**) projectiles. Theoretical models: CDW-EIS ($i_{\tilde{Z}_P} = 0, i_{q(\mathbf{K})} = 0$) (blue), CDW-EIS ($i_{\tilde{Z}_P} = 1$, $i_{q(\mathbf{K})} = 1$) (red) and Uehara's fitting (green) [17]. Experimental data are given by circles [18], squares [25], diamonds [26] (**a**) and circles [13] and squares [25] (**b**)

target ionization and projectile electrons loss resulting from He^{+} collision with water, were obtained from the sum of the fragment ions, namely, H_2O^{+}, OH^{+}, H^{+}, and O^{+}.

For the case of He^{2+} projectile, the target ionization total cross sections are computed without any corrections in the CDW-EIS model, since the interaction potential is pure a Coulomb one. Whereas for the case of He^{+} and He^{0} projectiles, the inclusion of charge corrections in the CDW-EIS approximation introduces minor modifications to the TCS profiles. In the case of He^{+}, the TCS, computed without taking into account the corrections terms, exhibits a slightly better agreement with the experiments. In general, for He^{2+} and He^{+} impact at collision energies larger than 50 keV/u, the present TCSs are in good agreement with the experimental data sets and the semiempirical model of Uehara et al. [17]. As it is shown in Fig. 5c, for the case of the target ionization induced by neutral helium, we can observe some discrepancies between the CDW-EIS predictions and the semiempirical one. These differences are due to the fact that Uehara et al. adopted the same cross sections obtained in He^{+} collision for the case of He^{0} one [17].

The electron loss cross section for He^{+} projectile is depicted in Fig. 6a. The present predictions did not describe well the reported measurements. We observe an overestimation of almost a factor 2 among the CDW-EIS models and the set of reported data [17, 26]. Also, the theoretical results did not reproduce correctly the peak position of the cross-sectional profile. For the case of He^{0}, (see Fig. 6b), the theoretical electron loss TCS computed without any charge corrections overestimates

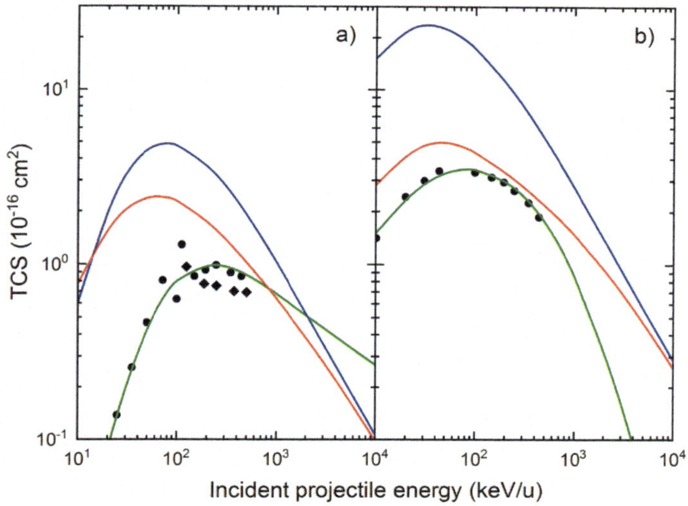

Fig. 6 Total electron loss cross sections of He^+ (**a**) and He^0 (**b**) projectiles impact on water vapor. Theoretical models: CDW-EIS ($i_{\tilde{Z}_P} = 0$, $i_{q(\mathbf{K})} = 0$) (blue), CDW-EIS ($i_{\tilde{Z}_P} = 1$, $i_{q(\mathbf{K})} = 1$) (red) and Uehara's fitting (green) [17]. Experimental data are taken from circles [17] and diamonds [26] (**a**) and circles [17] (**b**)

the experimental data. Whereas when the charge corrections are included in the theoretical model, the resulting cross sections are qualitative well in agreement with the literature data. However, some discrepancies are still observed with the data reported by [17].

3.3 *Influence of Corrections on the TCSs*

It is noteworthy to point out how the charge corrections introduced into the CDW-EIS approximation have an effect on the results. In Fig. 7, we report the target ionization (Fig. 7a) and electron loss (Fig. 7b) total cross section for the collision system H^0 + H_2O.

For target ionization, in the low energy range, we observe an increase of TCS when ($i_{\tilde{Z}_P} = 0, i_{q(\mathbf{K})} = 1$) and a decreasing of TCS when ($i_{\tilde{Z}_P} = 1, i_{q(\mathbf{K})} = 0$), this is mainly due to the $\alpha^2(v)$ factor, with respect to the results computed without any corrections ($i_{\tilde{Z}_P} = 0$, $i_{q(\mathbf{K})} = 0$). When both corrections are included ($i_{\tilde{Z}_P} = 1$, $i_{q(\mathbf{K})} = 1$), their effects are neutralized. As a result, the computed TCSs are very close to the ones computed without corrections. The TCSs computed for these four cases converge at high-impact energy.

For electron loss, we observe that, in both cases, when each correction is considered individually in the CDW-EIS model, i.e., ($i_{\tilde{Z}_P} = 0$, $i_{q(\mathbf{K})} = 1$) or ($i_{\tilde{Z}_P} = 1$,

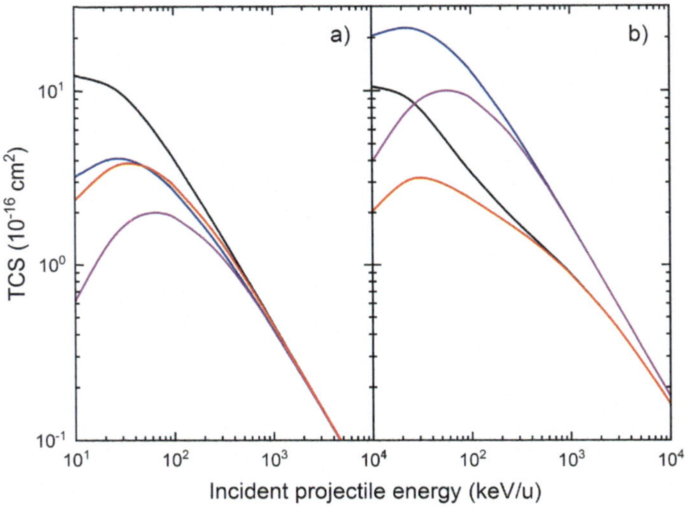

Fig. 7 Total target ionization cross section (**a**) and total electron loss cross section (**b**) given by H^0 projectile impact on water vapor. Theoretical models CDW-EIS: $i_{\tilde{Z}_P} = 0$, $i_{q(\mathbf{K})} = 0$ (blue); $i_{\tilde{Z}_P} = 0$, $i_{q(\mathbf{K})} = 1$ (dark); $i_{\tilde{Z}_P} = 1$, $i_{q(\mathbf{K})} = 0$ (magenta); $i_{\tilde{Z}_P} = 1$, $i_{q(\mathbf{K})} = 1$ (red)

$i_{q(\mathbf{K})} = 0$) the TCS decreases with respect to one computed without any corrections. Consequentially, when both corrections are included a remarkable reduction of a factor of about 10 is observed. Finally, it appears that the corrections introduce significant improvement for the case of electron loss process.

4 Conclusions

Target ionization and electron loss processes have been investigated within the CDW-EIS approximation for the collisions of stripped and neutral-hydrogen and neutral-helium projectiles' impact on water vapor. The results presented in terms of double differential and total cross sections have been compared with existing measurements and theoretical approaches present in the literature. Two charge corrections were introduced into the CDW-EIS approximation: the equilibrium projectile mean charge and the dynamic projectile charge corrections. The first one depends on the projectile energy, while the second one depends on the momentum transfer. We have observed that the present theoretical predictions in terms of the double differential and the total cross sections show a better accordance with the existing data set. However, for the case of the electron loss process in the system He^+ with water, there are differences between the present theoretical total cross sections and the experimental data. Further investigations are needed in order to understand such discrepancies.

Acknowledgements The authors M A Q, N J E, R D R and J M M like to acknowledge the Consejo Nacional de Investigaciones Científicas y Técnicas de la República Argentina through the Project PIP 2021-3245. The results presented in this work have been obtained by using the facilities of the CCT-Rosario Computational Center, a member of the High Performance Computing National System (SNCAD, MincyT-Argentina).

Appendix: The $V_P(s)$ Calculation for the Water Molecule

As stated in Table 1, each molecular orbital is considered as a linear combination of atomic orbitals weighted by the electron population. Therefore, the water molecule potential $V_P(s)$ is calculated by the linear combination of atomic potentials, using its corresponding wave functions. Precisely, it is the electron repulsion $V_{ap}(s)$ between the target-active one and those of the atomic-orbital compound what needs to be calculated since the coulomb part of the H_2O potential is already taken to be Z_P/s. Subsequently:

$$V_P(s) = Z_P/s + V_{ap}(s) \tag{16}$$

with

$$V_{ap}(s) = \sum_{j}^{5} V_{ap,j}(s) \tag{17}$$

where j refers to each molecular orbital. Being $c_{l,j}$ the electron population of the lth atomic orbital $|\varphi_{l,j}\rangle$, each one belonging to the jth molecular orbital description of Table 1, and considering an independent electron framework, we have

$$V_{ap,j}(s) = \sum_{l}^{N_l} c_{l,j} \langle \varphi_{l,j} | \frac{1}{|\mathbf{s} - \mathbf{s}'|} | \varphi_{l,j} \rangle \tag{18}$$

where $\mathbf{s}$ is the active-electron coordinate and $\mathbf{s}'$ the passive ones, over which we make the integration. Each matrix element gives the mean repulsive interaction between the active electron and that passive one bounded to the respective atomic orbital. Being n, λ, and α the principal, angular, and magnetic quantum numbers, let us denote this repulsive potential as $V_{ap}^{n\lambda\alpha}(s)$.

To perform the calculation is useful to express the interaction potential as [40]

$$\frac{1}{|\mathbf{s} - \mathbf{s}'|} = 4\pi \sum_{lm} \frac{1}{2l+1} \frac{s_<^l}{s_>^{l+1}} Y_l^{m}(\Omega') Y_l^{m}(\Omega) \tag{19}$$

where $s_< = \min[s, s']$, $s_> = \max[s, s']$, and Y_l^m are the normalized complex spherical harmonics depending on the passive electron angular coordinates Ω' or the active ones Ω. Using the Clementi-Roetti [34] atomic orbitals, we have that each $n\lambda\alpha$ state is described as a basis of $n_\nu\lambda\alpha_\nu$ Slater orbitals. After some little algebra, that each active-passive repulsive interaction among each atomic orbital is given by

$$
\begin{aligned}
V_{ap}^{n\lambda\alpha}(s) = & (-1)^{\alpha+1}\sqrt{4\pi}(2\lambda+1)\sum_{\nu\mu} C_\nu C_\mu \sum_{l=0}^{2\lambda}\left\{\frac{1}{\sqrt{2l+1}}\begin{pmatrix}\lambda & \lambda & l\\ -\alpha & \alpha & 0\end{pmatrix}\begin{pmatrix}\lambda & \lambda & l\\ 0 & 0 & 0\end{pmatrix}\right.\\
& \times Y_l^0(\Omega)\left[\frac{a!}{(Z_\nu+Z_\mu)^{a+1}}\frac{1}{s^{l+1}}\left(1+\sum_{g=0}^{a-1}\frac{(Z_\nu+Z_\mu)^{a-g}}{(a-g)!}s^{a-g}\right)\right.\\
& \left. + (b-1)!s^l\left(\sum_{g=0}^{b-1}\frac{(Z_\nu+Z_\mu)^{-g-1}}{(b-1-g)!}s^{b-1-g}\right)\right]\\
& \left.\times e^{-(Z_\nu+Z_\mu)s}\frac{(2Z_\nu)^{n_\nu+\frac{1}{2}}(2Z_\mu)^{n_\mu+\frac{1}{2}}}{\sqrt{(2n_\nu)!(2n_\mu)!}}\right\}
\end{aligned} \tag{20}
$$

with $a = (n_\nu + n_\mu + l)$ and $b = (n_\nu + n_\mu - l)$, where the ν and μ indexes run over the elements of the mentioned basis. Then C_ν and C_μ are the coefficients and Z_ν and Z_μ are the νth and μth coefficients and charges of the Slater orbitals of the expansion. Finally, the matrices are the 3j Wigner symbols.

References

1. V.I. Shematovich, D.V. Bisikalo, C. Diéval, S. Barabash, G. Stenberg, H. Nilsson, Y. Futaana, M. Holmstrom, J.-C. Gérard, Proton and hydrogen atom transport in the Martian upper atmosphere with an induced magnetic field. J. Geophys. Res. **116**(A11320), A11320 (2011)
2. J.-P. Pouget, J. Constanzo, Revisiting the radiobiology of targeted alpha therapy. Front. Med. **8** (2021)
3. L.J. Puckett, G.O. Taylor, D.W. Martin, Cross sections for ion and electron production in gases by 0.15–1.00 mev hydrogen and helium ions and atoms. Phys. Rev. **178**, 271–287 (1969)
4. R.J. McNeal, D.C. Clark, R.A. Klingberg, Cross sections for positive-ion and electron production in collisions of 1–25 keV hydrogen atoms with atomic and molecular gases. Phys. Rev. A **2**(1), 131–138 (1970)
5. M.M. Duncan, M.G. Menendez, Ionization of 0.5 MeV H by Ar and a comparison with the H^-/Ar system. Phys. Rev. A **19**, 49–54 (1979)
6. L.H. Andersen, L.B. Nielsen, J. Sorensen, Single and double Jonisation of He and Ar by far H^0 and H-projectiles. J. Phys. B: At. Mol. Opt. Phys. **21**(9), 1587–1602 (1988)
7. L. Sarkadi, P. Herczku, S.T.S. Kovács, A. Kövér, Multiple ionization of rare gases by hydrogen-atom impact. Phys. Rev. A **87**, 062705 (2013)
8. L.H. Toburen, M.Y. Nakai, R.A. Langley, Measurement of high-energy charge-transfer cross sections for incident protons and atomic hydrogen in various gases. Phys. Rev. **171**(1), 114–122 (1968)
9. M.A. Bolorizadeh, M.E. Rudd, Angular and energy dependence of cross sections for ejection of electron from water vapor. III. 20–150 keV neutral-hydrogen impact. Phys. Rev. A **33**, 893–896 (1986)

10. F. Gobet, S. Eden, B. Coupier, J. Tabet, B. Farizon, M. Farizon, M.J. Gaillard, Ouaskit S, M. Carré, T.D. Märk,-loss and target ionization cross sections for water vapor by 20–150 keV neutral atomic hydrogen impact. Chem. Phys. Lett. **421**, 68–71 (2006)
11. H. Luna, A.L.F. de Barros, J.A. Wyer, S.W.J. Scully, J. Lecointre, P.M.Y. Garcia, G.M. Sigaud, A.C.F. Santos, V. Senthil, M.B. Shah, C.J. Latimer, E.C. Montenegro, Water-molecule dissociation by proton and hydrogen impact. Phys. Rev. A **75**, 042711 (2007)
12. M. Dingfelder, M. Inokuti, H. Paretzke, Inelastic-collision cross sections of liquid water for interactions of energetic protons. Radiat. Phys. Chem. **59**, 255 (2000)
13. M.E. Rudd, A. Itoh, T.V. Goffe, Cross sections for ionization, capture, and loss for 5–450 keV He^+ on water vapor. Phys. Rev. A **32**(4), 2499–2500 (1985)
14. L.H. Toburen, W.E. Wilson, Energy and angular distributions of electrons ejected from water vapor by 0.3–1.5 MeV protons. J. Chem. Phys. **66**, 5202–5213 (1977)
15. S. Uehara, L.H. Toburen, W.E. Wilson, D.T. Goodhead, H. Nikjoo, Calculations of electronic stopping cross sections for low- energy protons in water. Radiat. Phys. Chem **59**, 1–11 (2000)
16. M.A. Quinto, J.M. Monti, C. Champion, R.D. Rivarola, Neutral hydrogen versus proton-induced ionization in water vapor. Phys. Rev. A **100**(4), 042704 (2019)
17. S. Uehara, H. Nikjoo, Monte Carlo track structure code for low-energy alpha-particles in water. J. Phys. Chem. B **106**(42), 11051–11063 (2002)
18. M.E. Rudd, T.V. Goffe, A. Itoh, Ionization cross sections for 10–300 keV/u and electron-capture cross sections for 5–150 keV/u $^3He^{2+}$ ions in gases. Phys. Rev. A **32**(4), 2128–2133 (1985)
19. D. Ohsawa, Y. Sato, Y. Okada, V.P. Shevelko, F. Soga, 6.0 − 10.0-MeV/u He^{2+}-ion-induced electron emission from water vapor. Phys. Rev. A **72**, 062710 (2005)
20. C. Champion, O. Boudrioua, C. Dal Cappello, Y. Sato, D. Ohsawa, Theoretical and experimental investigations of electron emission in He^{2+}+H_2O collisions. Phys. Rev. A **75**, 032724 (2007)
21. C.A. Tachino, J.M. Monti, O.A. Fojón, C. Champion, R.D. Rivarola, Ionization of water molecules by ion beams: on the relevance of dynamic screening and the influence of the description of the initial state. J. Phys. B: At. Mol. Opt. Phys. **47**, 035203 (2014)
22. S. Bhattacharjee, S. Biswas, C. Bagdia, M. Roychowdhury, S. Nandi, D. Misra, J.M. Monti, C.A. Tachino, R.D. Rivarola, C. Champion, L.C. Tribedi, Double differential distribution of electron emission in the ionization of water molecules by fast bare oxygen ions. J. Phys. B: At.Ic, Mol. Opt. Phys. **49**(6), 065202 (2016)
23. S. Bhattacharjee, S. Biswas, J.M. Monti, R.D. Rivarola, L.C. Tribedi, Double-differential cross section for ionization of h_2O molecules by 4 mev/u c^{6+} and si^{13+} ions. Phys. Rev. A **96**, 052707 (2017)
24. A. Bhogale, S. Bhattacharjee, M.R. Chowdhury, C. Bagdia, M.F. Rojas, J.M. Monti, A. Jorge, M. Horbatsch, T. Kirchner, R.D. Rivarola, L.C. Tribedi, Electron emission from water vapor under the impact of 250 kev protons. Phys. Rev. A **105**, 062822 (2022)
25. L.H. Toburen, W.E. Wilson, R.J. Popowich, Secondary electron emission from ionization of water vapor by 0.3–2.0 MeV He^+ and He^{2+} Ions. Radiat. Res. **82**(1), 27 (1980)
26. P.M.Y. Garcia, G.M. Sigaud, H. Luna, A.C.F. Santos, E.C. Montenegro, M.B. Shah, Water-molecule dissociation by impact of He^+ ions. Phys. Rev. A **77**, 052708 (2008)
27. S. Kovács, P. Herczku, Z. Juhász, L. Sarkadi, L. Gulyás, B. Sulik, Ionization of small molecules induced by H^+, He^+ and N^+ projectiles: comparison of experiment with quantum and classical calculations. Phys. Rev. A **94**, 07 (2016)
28. P.D. Fainstein, V.H. Ponce, R.D Rivarola, A theoretical model for ionisation in ion-atom collisions: application for the impact of multicharged projectiles on helium. J. Phys. B: At. Mol. Opt. Phys. **21**, 287–299 (1988)
29. J.M. Monti, R.D. Rivarola, P.D. Fainstein, Distorted wave theories for dressed-ion-atom collisions with GSZ projectile potentials. J. Phys. B: At. Mol. Opt. Phys. **44**, 195206 (2011)
30. P.D. Fainstein, V.H. Ponce, R.D. Rivarola, Two-centre effects in ionization by ion impact. J. Phys. B: At. Mol. Opt. Phys. **24**, 3091–3119 (1991)
31. D. Belkic, R. Gayet, A. Salin, Electron capture in high-energy ion-atom collisions. Phys. Rep. **56**, 279 (1979)

32. A. Salin, Helium ionisation by high-energy ions at a function of impact parameter and projectile scattering angle. J. Phys. B: At. Mol. Opt. Phys. **223**, 3901 (1989)
33. K. Siegbahn et al., *ESCA Applied to Free Molecules* (North-Holland, Amsterdam, 1965)
34. E. Clementi, C. Roetti, Roothaan-Hartree-Fock atomic wavefunctions: basis functions and their coefficients for ground and certain excited states of neutral and ionized atoms, $Z <= 54$. At. Data Nucl. Data Tables **14**, 177 (1974)
35. L. Gulyás, S. Egri, H. Ghavaminia, A. Igarashi, Single and multiple electron removal and fragmentation in collisions of protons with water molecules. Phys. Rev. A **93**(3), 032704 (2016)
36. P.P. Szydlik, A.E.S. Green, Independent-particle-model potentials for ions and neutral atoms with $Z<18$. Phys. Rev. A **99**, 1885–1894 (1974)
37. N.J. Esponda, M.A. Quinto, R.D. Rivarola, J.M. Monti, Dynamic screening and two-center effects in neutral and partially dressed ion-atom collisions. Phys. Rev. A **105**(3), 032817 (2022)
38. J.H. McGuire, N. Stolterfoht, P.R. Simony, Screening and antiscreening by projectile electrons in high-velocity atomic collisions. Phys, Rev. A **24**(1), 97–102 (1981)
39. M.S. Gravielle, J.E. Miraglia, Some Nordsieck integral of interest in radiation and atomic collision theories. Comput. Phys. Commun. **69**(1), 53–58 (1992)
40. N. Stolterfoht, R.D. DuBois, R.D. Rivarola, *Electron Emission in Heavy Ion-Atom Collisions* (Springer, Berlin Heidelberg, 1997)
41. C. Schmitt, J. Laverne, D. Robertson, M. Bowers, W. Lu, P. Collon, Equilibrium mean charge states for low-Z ions at $\leq$1MeV/u in carbon. Phys. Rev. A **80**(5), 052711 (2009)
42. H.D. Betz, G. Hortig, E. Leischner, Ch. Schmelzer, B. Stadler, J. Weihrauch, The average charge of stripped heavy ions. Phys. Lett. **22**(5), 643–644 (1966)
43. G. Schiwietz, P.L. Grande, Improved charge-state formulas. Nucl. Instrum. Methods Phys. Res. B **175**, 125–131 (2001)
44. A. Schinner, P. Sigmund, Polarization effect in stopping of swift partially screened heavy ions: Perturbative theory. Nucl. Instrum. Methods Phys. Res. B **164**, 220–229 (2000)
45. M.E. Galassi, C. Champion, P.F. Weck, R.D. Rivarola, O. Fojón, J. Hanssen, Quantum-mechanical predictions of DNA and RNA ionization by energetic proton beams. Phys. Med. Biol. **57**(7), 2081–2099 (2012)
46. S. Nandi, S. Biswas, A. Khan, J.M. Monti, C.A. Tachino, R.D. Rivarola, D. Misra, L.C. Tribedi, Double-differential cross sections for ionization of H_2O by fast bare O ions: comparison with continuum-distorted-wave eikonal-initial-state calculations in prior and post forms. Phys. Rev. A **87**(5):052710 (2013)
47. N. Stolterfoht, H. Platten, G. Schiwietz, D. Schneider, L. Gulyás, P.D. Fainstein, A. Salin, Two-center electron emission in collisions of fast, highly charged ions with He: experiment and theory. Phys. Rev. A **52**(5), 3796–3802 (1995)
48. M.E. Rudd, T.V. Goffe, R.D. DuBois, L.H. Toburen, Cross sections for ionization of water vapor by 7–4000 keV protons. Phys. Rev. A **31**, 492–494 (1985)
49. M.A. Bolorizadeh, M.E. Rudd, Angular and energy dependence of cross sections for ejection of electrons from water vapor. II. 15–150 keV proton impact. Phys. Rev. A **33**, 888–892 (1986)

Classical-Trajectory Time-Dependent Mean-Field Theory for Ion–Molecule Collision Problems

Alba Jorge, Marko Horbatsch, and Tom Kirchner

Abstract A mean-field model to describe electron transfer processes in ion–molecule collisions at the $\hbar = 0$ level is presented and applied to collisions involving water and ammonia molecules. Multicenter model potentials account for the molecular structure and geometry. They include charge screening parameters which in the most advanced version of the model depend on the instantaneous degree of ionization so that dynamical screening effects are taken into account. The work is implemented using the classical-trajectory Monte Carlo method, i.e., Hamilton's equations are solved for classical statistical ensembles that represent the initially populated orbitals. The time-evolved trajectories are sorted into ionizing and electron capture events, and a multinomial analysis of the ensuing single-particle probabilities is employed to calculate differential and total cross sections for processes that involve single- and multiple-electron transitions. Comparison is made with experimental data and some previously reported calculations to shed light on the capabilities and limitations of the approach.

1 Introduction

Atoms and molecules are objects of the microscopic world. Hence, it seems natural to use quantum mechanics to describe their structure and interactions. What is natural is however not necessarily easy and perhaps not strictly required, at least as long as typical actions are much larger than Planck's constant. Put in more quantitative terms, in the $\hbar \to 0$ limit the quantum-mechanical Schrödinger equation can be replaced

---

A. Jorge
Departamento de Química, Universidad Autónoma de Madrid, 28049 Cantoblanco, MD, Spain

M. Horbatsch · T. Kirchner (✉)
Department of Physics and Astronomy, York University, Toronto, ON M3J 1P3, Canada
e-mail: tomk@yorku.ca

M. Horbatsch
e-mail: marko@yorku.ca

L. C. Tribedi (ed.), *Advances in Atomic Molecular Collisions*,
https://doi.org/10.1007/978-981-97-7063-2_10

by classical equations of motion, which are easier to solve, and the time evolution of the quantum system be understood at the level of classical statistical mechanics [1].

In the context of atomic collisions, this insight forms the basis of the classical-trajectory Monte Carlo (CTMC) method. It was introduced a long time ago [2] and has since been used in many different variants for a multitude of scattering problems, starting from the prototypical one-electron proton-hydrogen system, carrying over to multicharged-ion many-electron-atom systems, and more recently also finding application in problems involving molecular targets. A very large number of studies have been carried out and in many cases provided the first (and sometimes the only) theoretical results capable of describing complex experimental data [3].

It is beyond the scope of this chapter to provide a comprehensive overview of all these works. Interested readers are referred to [4, 5] for reviews and lists of references to the original literature. Our objective is more modest and specific: We describe a recently proposed CTMC method that extends the standard approach by incorporating dynamical screening effects in the spirit of a time-dependent mean-field model, and show by way of a few illustrative examples what can be learned from it when it is applied to the problem of single and multiple electron removal in ion–molecule collisions. We study total and differential cross sections. Both are needed for an accurate understanding of, e.g., the problem of radiation damage of biological tissue, which has been a major driver of contemporary ion–molecule collision studies. Differential data for electron emission provide more detailed information than total ionization cross sections but are harder to calculate on the basis of quantum-mechanical methods, especially those which do not make use of perturbative arguments. This makes CTMC calculations all the more important and motivated us to pursue this work.

Ion–molecule collision systems are many-particle problems. We address them in the frameworks of the semiclassical impact parameter approximation and the independent electron model (IEM), which also form the basis of most quantum-mechanical approaches to the problem at hand. For us this means that the $\hbar \to 0$ limit is applied to a one-electron time-dependent Schrödinger equation whose Hamiltonian includes a mean-field potential to account for the electron–electron interactions in an effective manner and which has to be solved for the molecular orbitals (MOs) which form the molecular ground state, i.e., the initial state of the target before it is approached by the projectile ion.

The mean-field potential model used has a multicenter structure reflecting the molecular geometry and takes time-dependent screening effects into account. Our published work focuses on ion collisions with water (H_2O) molecules [6–9], which are of paramount interest for the radiation damage problem mentioned above, but more recently we have also looked at the proton-ammonia (NH_3) system [10], and we discuss it alongside the ion-water problem in this chapter.

The layout is as follows. We begin in Sect. 2 with a short summary of the general theoretical assumptions and ideas on which our model description is built. This is followed by somewhat more detailed discussions of the (time-dependent) mean-field potential models (in Sect. 2.1) and the procedures used to extract the desired information, i.e., collisional cross sections for various electron removal processes, from the solutions of the equations of motion (in Sect. 2.2). Results are presented and compared with experimental data and selected previous calculations in Sect. 3, first for H_2O target molecules (in Sects. 3.1 and 3.2) and then for NH_3 (in Sect. 3.3). The chapter ends with a few concluding remarks in Sect. 4. Atomic units, characterized by $\hbar = m_e = e = 4\pi\epsilon_0 = 1$, are used unless otherwise stated.

2 Theory

We are concerned with relatively high-impact-energy ion–molecule collisions so that we can assume the projectile ion to travel on a straightline trajectory and the target molecule to remain rigid during the (short) interaction time. This is to say that we are using the semiclassical impact parameter approximation and keep the rotational and vibrational degrees of freedom of the target frozen. To deal with the many-electron nature of the problem we apply the IEM, i.e., we assume the Hamiltonian to be of single-particle form ($H \approx \sum_j h^{(j)}$) so that the equations of motion have one-electron character. Solving them for the MOs that form the molecular ground state provides all accessible information on the electronic transitions taking place during the collision. The single-electron Hamiltonian has the standard form

$$h = \frac{\mathbf{p}^2}{2} + v_{\mathrm{mod}} + v_{\mathrm{p}}, \tag{1}$$

where v_{mod} denotes the multicenter target model potential and v_{p} the projectile potential. Both are specified in Sect. 2.1 and together define the (time-dependent) mean-field model used.

Applying the $\hbar \to 0$ limit amounts in practice to solving Hamilton's equations for large sets of initial conditions which represent discretized versions of microcanonical ensembles of trajectories corresponding to the ionization energies of the MOs. For H_2O, a more detailed discussion of how the initial distributions are generated (along with plots to show them) is provided in [11]. An analogous procedure is used for NH_3.

The trajectories are time evolved using a small step size of $\Delta t = 0.05$ a.u. until the projectile has reached a distance of about 500 a.u. to the target at the final time $t = t_f$ after passing it at a specific impact parameter vector. For each trajectory, we look at the energy of the electron with respect to the projectile and the target in order to determine its fate: If one of them is negative the electron is considered bound to that center, i.e., it is either captured by the projectile or it remains bound to the target. If both energies are positive, the trajectory contributes to a transition to the continuum (ionization).

Calculations are carried out for many orientations of the target molecule characterized by randomly chosen sets of Euler angles, and orientation averages are calculated at the end in order to allow comparisons with orientation-insensitive experimental data. The models used to extract the differential and total cross sections for the processes of interest are described in Sect. 2.2.

2.1 Mean-Field Potential Models

The ideas that go into setting up the (time-independent and time-dependent) mean-field potentials are the same for water and ammonia. Hence, we describe them for both target molecules in tandem.

2.1.1 Static Screening Model

The multicenter target (T) potentials are approximated as sums of (time-independent) central potentials for each atom of the molecule:

$$v_{\text{mod}} = \sum_i v_{\text{X}}(r_{\text{X}_i}) = \begin{cases} v_{\text{O}}(r_{\text{O}}) + \sum_{i=1}^{2} v_{\text{H}}(r_{\text{H}_i}), & (T = \text{H}_2\text{O}), \\ v_{\text{N}}(r_{\text{N}}) + \sum_{i=1}^{3} v_{\text{H}}(r_{\text{H}_i}), & (T = \text{NH}_3). \end{cases} \tag{2}$$

In (2), X labels the atoms (X = O, N, H) and r_{X_i} are the distances from the active electron to the $i = 1, \ldots, M$ nuclei of the molecule. Note that H_2O is planar while NH_3 is umbrella shaped. We use the following nuclear geometry parameters: The O-H bond lengths in H_2O are fixed at 1.8 a.u. and the angle between the position vectors for the two protons is 105°. The N-H bond lengths in NH_3 are 1.928 a.u.; the azimuthal angles of the proton position vectors are 90°, 210°, and 330°; and their polar angles are 108.9° [12]. A planar geometry would correspond to right polar angles. The central potentials in (2) are assumed to be of the form:

$$v_{\text{X}}(r_{\text{X}}) = -\frac{Z_{\text{X}} - N_{\text{X}}}{r_{\text{X}}} - \frac{N_{\text{X}}}{r_{\text{X}}}(1 + \alpha_{\text{X}} r_{\text{X}}) \exp(-2\alpha_{\text{X}} r_{\text{X}}), \tag{3}$$

where $Z_{\text{X}} = 8, 7, 1$ are the nuclear charge numbers of X=O, N, H, while the other parameters have the values $N_{\text{O}} = 7.185$, $\alpha_{\text{O}} = 1.602$, $N_{\text{N}} = 6.2775$, $\alpha_{\text{N}} = 1.525$, $N_{\text{H}} = 0.9075$, and $\alpha_{\text{H}} = 0.6170$. These values were found as follows: At very large distances the total potential experienced by any one electron of a neutral atom or molecule behaves like $-1/r$. This condition results in the relations $N_{\text{O}} + 2N_{\text{H}} = 9 = N_{\text{N}} + 3N_{\text{H}}$. Using the first equality as a constraint, the parameters for H_2O were determined by minimizing the differences of the valence MOs' energy eigenvalues found for the (quantum-mechanical) Hamiltonian $\frac{\mathbf{p}^2}{2} + v_{\text{mod}}$ with respect to self-consistent field (SCF) results [11]. For NH_3 we use the same parameters N_H and α_H, find $N_N = 9 - 3N_H = 6.2775$, and determine the remaining parameter α_N such that the energy eigenvalues of $\frac{\mathbf{p}^2}{2} + v_{mod}$ are in reasonable agreement with the SCF results from [12].

If dynamical screening is excluded, the projectile potential of a fully stripped ion is purely Coulombic: $V_p = -Z_p/r_p$ with Z_p and r_p being the charge number of the projectile and the distance from it to the active electron. We treat the partially stripped Si^{13+} ion for which we present results in Sect. 3.1 in the same way, i.e., by a bare Coulomb potential corresponding to the asymptotic charge.

2.1.2 Dynamical Screening Model

Consider a small impact parameter collision of a highly charged projectile ion. Around the distance of closest approach the projectile penetrates the molecule and exerts a strong force on the electrons. The likely outcome is *multiple* electron removal which increases the pull of the now partially ionized target on the electrons.

This dynamical effect can be modeled in a simple fashion by replacing the static screening charge parameters N_X in (3) by time-dependent ones in a way that ensures that the screening decreases with increasing degree of ionization q. The instantaneous degree of ionization corresponds to the (fractional) average number of removed electrons, the net electron removal P_{net}^{rem}. Hence, we can feed information on the degree of ionization into the potential by making the charge screening parameters dependent on P_{net}^{rem}. We use the simple ansatz

$$N_X(P_{net}^{rem}) = \begin{cases} N_X^c & P_{net}^{rem} \leq 1, \\ \frac{N_X^c N_T}{N_T - 1}\left(1 - \frac{P_{net}^{rem}}{N_T}\right) & 1 \leq P_{net}^{rem} \leq N_T, \end{cases} \tag{4}$$

where N_T is the total number of electrons in T (i.e., $N_T = 10$ for both H_2O and NH_3) and we have renamed the constant screening parameter of (3) by adding a superscript c. The idea of this model is that time-dependent screening[1] should not affect single ionization but kick in only after the $q = 1$ threshold has been overcome. It then changes linearly with the degree of ionization until all N_T target electrons are removed, at which point $N_X(P_{net}^{rem} = N_T) = 0$ for all X and the total potential reduces to a pure multicenter Coulomb potential [cf. (3) and (2)]. The model is similar in spirit to the *atomic* dynamical screening model proposed in [13]. We explain in Sect. 2.2.2 how P_{net}^{rem}, the only time-dependent ingredient, is calculated in the present CTMC-IEM approach.

One can take the dynamical screening model to the next level by also including time-dependent charge screening effects on the projectile ion. Here the rationale is that the screening of the projectile nucleus increases with an increasing number of

[1] We use the terms time-dependent screening and dynamical screening interchangeably in this chapter.

captured electrons and hence the strength of the projectile potential should be down-regulated accordingly. This can be accomplished by similar modeling as described above with the (fractional) net electron capture $P_{\rm net}^{\rm cap}$ taking the role of $P_{\rm net}^{\rm rem}$. We have reported on the implementation of such a model in [7]. For the collision systems studied, the effects of dynamical projectile screening were found to be relatively small compared to those of the dynamical target screening, except in situations in which capture into multiply charged ions is the dominating electron removal process (see Sect. 3.2). We skip the technical details here and refer the interested reader to [7].

2.2 Analysis of Electron Capture and Ionization Processes

The ingredients of an IEM-based final-state analysis are single-particle probabilities for the processes of interest and multinomial statistics to combine them. We are interested in electron capture (cap) and ionization (ion) and in the latter case in total and differential probabilities and cross sections. The total probabilities for the jth MO ($j = 1, \ldots, m$) are calculated as the ratios of the number of trajectories contributing to process i = cap or i = ion to the total number of trajectories

$$p_j^i = \frac{n_j^i}{n_{j,\rm tot}}. \tag{5}$$

The differential probabilities are calculated by binning the ionized electrons in small intervals with respect to emission energy $\Delta E_{\rm el}$ and emission angle $\Delta\Omega_{\rm el} = 2\pi[\cos(\theta_{{\rm el}_{i+1}}) - \cos(\theta_{{\rm el}_i})]$ [6]. To arrive at this equation, we have exploited the cylindrical symmetry and integrated over the azimuthal angle. The polar angles $\theta_{{\rm el}_i}$ and $\theta_{{\rm el}_{i+1}}$ define the interval for a given $\theta_{\rm el}$: $\theta_{{\rm el}_i} \le \theta_{\rm el} < \theta_{{\rm el}_{i+1}}$. In extension of (5), the single-particle ionization probability associated with $\Delta E_{\rm el}$ and $\Delta\Omega_{\rm el}$ is

$$\frac{{\rm d}^2 p_j^{\rm ion}}{{\rm d}E_{\rm el}{\rm d}\Omega_{\rm el}} = \frac{n_j^{\rm ion}}{n_{j,\rm tot}\Delta E_{\rm el} 2\pi[\cos(\theta_{{\rm el}_{i+1}}) - \cos(\theta_{{\rm el}_i})]}. \tag{6}$$

2.2.1 Differential q-fold Ionization

A situation of interest is the detection of one electron at $\{\Delta E_{\rm el}, \Delta\Omega_{\rm el}\}$, while a total of q electrons are ionized in the same event. In the simplest case of $q = 1$, the multinomial combination of single-particle probabilities yields

$$\frac{{\rm d}^2 P_{q=1}^{\rm ion}}{{\rm d}E_{\rm el}{\rm d}\Omega_{\rm el}} = 2\sum_{j=1}^{m} \frac{{\rm d}^2 p_j^{\rm ion}}{{\rm d}E_{\rm el}{\rm d}\Omega_{\rm el}}(1 - p_j^{\rm ion})\prod_{k\neq j}^{m}(1 - p_k^{\rm ion})^2. \tag{7}$$

One can read this expression as follows: The differential single-electron probability for ionization from the doubly occupied jth MO (6) is multiplied by a non-ionization probability from the same MO and by a factor of two to account for the spin degeneracy and ensure that exactly one out of two electrons of that MO end up in the continuum. The resulting binomial probability is further multiplied by non-ionization probabilities from all other MOs and summed over j since the detected continuum electron can originate from any of the initially occupied MOs. For higher q, the expressions follow the same logic but become more complicated. The cases $q = 2$ and $q = 3$ are shown explicitly in [6]. It is useful to define a *net* probability as the weighted sum of the q-fold probabilities

$$\frac{\mathrm{d}^2 P_{\mathrm{net}}^{\mathrm{ion}}}{\mathrm{d}E_{\mathrm{el}}\mathrm{d}\Omega_{\mathrm{el}}} = \sum_{q=1}^{N_T} q \frac{\mathrm{d}^2 P_q^{\mathrm{ion}}}{\mathrm{d}E_{\mathrm{el}}\mathrm{d}\Omega_{\mathrm{el}}}. \tag{8}$$

The net probability corresponds to a measurement in which one detects one electron at $\{\Delta E_{\mathrm{el}}, \Delta\Omega_{\mathrm{el}}\}$ but does not determine if and how many additional electrons are ionized in the same event. It can also be calculated directly from the differential single-particle probabilities

$$\frac{\mathrm{d}^2 P_{\mathrm{net}}^{\mathrm{ion}}}{\mathrm{d}E_{\mathrm{el}}\mathrm{d}\Omega_{\mathrm{el}}} = 2 \sum_{j=1}^{m} \frac{\mathrm{d}^2 p_j^{\mathrm{ion}}}{\mathrm{d}E_{\mathrm{el}}\mathrm{d}\Omega_{\mathrm{el}}}, \tag{9}$$

which provides a useful consistency check.

2.2.2 Charge-State Correlated and Inclusive Probabilities

The total probabilities (5) are calculated at each time step, concurrently with the ensembles of trajectories, using the energy criterion described above. At the final time $t = t_f$ they form the basis of the analysis of single and multiple ionization and capture events within the IEM. The most straightforward approach is standard multinomial statistics for all charge-state correlated channels in which a number of k electrons are captured and l electrons are ionized to the continuum. The corresponding transition probabilities P_{kl} (which are functions of the impact parameter vector) can be written as

$$\begin{aligned} P_{kl} = & \sum_{k_1,\ldots,k_m=0}^{M_1,\ldots,M_m} \sum_{l_1,\ldots,l_m=0}^{M_1,\ldots,M_m} \delta_{k,\sum_i k_i} \delta_{l,\sum_i l_i} \prod_{i=1}^{m} \binom{M_i}{k_i+l_i} \binom{k_i+l_i}{k_i} \\ & (p_i^{\mathrm{cap}})^{k_i} (p_i^{\mathrm{ion}})^{l_i} (1 - p_i^{\mathrm{cap}} - p_i^{\mathrm{ion}})^{M_i-k_i-l_i}, \end{aligned} \tag{10}$$

where $\delta_{k,\alpha}$ is the Kronecker delta symbol and $M_1 = M_2 = \cdots = M_m = 2$ refer to the number of electrons in each MO. One problem of this analysis is that the N_T target electrons are distributed over the regions of space purely statistically, regardless of whether the projectile can accommodate a given number of electrons or not. This problem is most prominent for proton projectiles ($Z_p = 1$), which can capture one or two electrons, but not more than that. In fact, double capture leading to the formation of a negatively charged hydrogen ion is already an extremely rare process which is known to involve electron correlations and which cannot be described at the level of the IEM, i.e., the multinomial evaluation of P_{kl} fails for $k \geq 2$ in the case of proton projectiles. One can try to correct for this by combining the single-particle probabilities in alternative ways such that the unphysical capture channels are closed [7, 13, 14], but such approaches can be criticized for their ad hoc character and they may show deficiencies in some of the open channels as a consequence of re-distributing the total capture flux among the physically allowed k, l combinations. In other words, there is no easy fix to the problem of unphysical multiple capture in the IEM and we will not consider it further in this contribution.

The more inclusive k-fold capture probabilities P_k^{cap} are given as

$$P_k^{\text{cap}} = \sum_{i=0}^{N_T-k} P_{ki}, \tag{11}$$

and can be used to calculate the net capture $P_{\text{net}}^{\text{cap}}$

$$P_{\text{net}}^{\text{cap}} = \sum_{k=1}^{N_T} k P_k^{\text{cap}}. \tag{12}$$

This is the analog of (8) for differential ionization. The analog of (9) is

$$P_{\text{net}}^{\text{cap}} = 2 \sum_{j=1}^{m} p_j^{\text{cap}}, \tag{13}$$

and similar equations for the total l-fold and net ionization probabilities can be established as well. We use (13) and its ionization counterpart to calculate the time-dependent net capture $P_{\text{net}}^{\text{cap}}$, net ionization $P_{\text{net}}^{\text{ion}}$, and net removal $P_{\text{net}}^{\text{rem}} = P_{\text{net}}^{\text{cap}} + P_{\text{net}}^{\text{ion}}$, which feed into the dynamical screening models described in Sect. 2.1.2.

As a final comment, we note that within the semiclassical impact parameter approximation differential and total cross sections are obtained by integrating the probabilities for the processes of interest over the impact parameter vector.

3 Results and Discussion

In this section, we present a selection of the results obtained from the CTMC mean-field approach described above. Our goal is to provide the reader with an overview and a general idea of what has been and can be accomplished in this framework, and where gaps in our understanding remain. The results are taken from our original research papers [6–8, 10] to which we refer the reader for more detailed discussions.

3.1 Collisions with Water Molecules: Differential Cross Sections

We begin with differential electron emission from water molecules impacted by highly charged high-velocity projectile ions. A useful parameter for this discussion is the ratio of the projectile charge to the projectile speed $\eta = Z_p/v_p$ measured in atomic units. Low values of η correspond to weak perturbations associated with little ionization of the target molecule. Electron capture is even less likely and can be ignored in the situations studied in this subsection. With increasing η one finds an increasing net ionization cross section, a growing fraction of which is due to multiple ionization events. Accordingly, we can expect time-dependent screening effects to increase in importance with η. The four collision systems studied here comprise the range $0.47 \leq \eta \leq 1.03$. They are 4 MeV/amu C^{6+} ($\eta = 0.47$), 3.75 and 3.0 MeV/amu O^{8+} ($\eta = 0.65$ and 0.73), and 4 MeV/amu Si^{13+} ($\eta = 1.03$) projectiles and were studied experimentally in [15–17]. Along with the data, perturbative quantum-mechanical calculations were reported, and we compare our results with those as well. The latter comparison is interesting since the two approaches complement each other: Ours is based on the $\hbar \to 0$ limit of quantum mechanics and uses a three-center target potential; the approach of [15–17] uses the quantum-mechanical continuum distorted wave with eikonal initial-state (CDW-EIS) approximation coupled with a model in which molecular cross sections are given as linear combinations of atomic contributions. We note that a recently reported multicenter CDW-EIS model overcomes some of the limitations of the earlier distorted-wave calculations, but the first results are somewhat inconclusive [18]. Just like our approach, all these CDW-EIS models are based on single-electron Hamiltonians, i.e., are of IEM type.

In Fig. 1, doubly differential cross sections (DDCSs) for net ionization are shown for the four collision systems. The electron emission energy is fixed at $E_{\mathrm{el}} = 200$ eV and the DDCSs are shown as functions of the polar emission angle θ_{el} [cf. (6)]. As expected, there is an overall increase of the DDCS with η (note that the y-axis scales are different in the four panels). The time-dependent screening results are lower than those obtained with the static model potential, which is a reflection of the fact that the target potential becomes more attractive with decreasing charge screening parameters. At backward angles $\theta_{\mathrm{el}} > 90°$, the gap between both sets of results appears to widen with increasing η.

Truly quantum-mechanical effects are known to be the strongest at low and less important at higher electron emission energies, and indeed we find good overall agreement between the CTMC calculations and the experimental data at $E_{\mathrm{el}} = 200$ eV. One may deem this energy value as medium given that it corresponds to a velocity

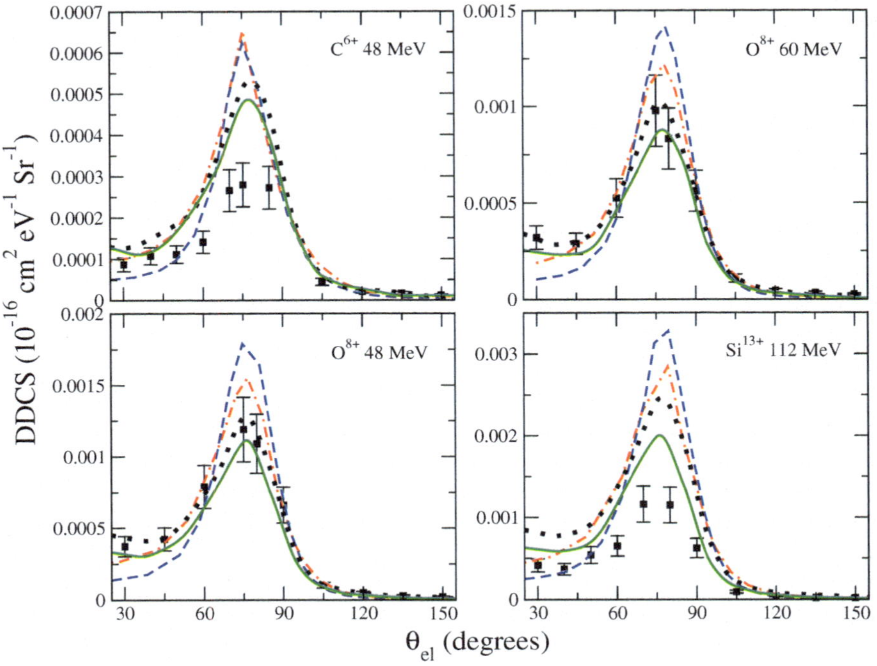

Fig. 1 DDCS for net ionization of water molecules as a function of electron emission angle in degrees, for the electron emission energy $E_{\rm el} = 200$ eV. The CTMC results (obtained with angular resolution $\Delta\theta_{\rm el} = 10°$) are shown as dotted black (static screening) and solid green (dynamical screening) lines. The experimental data are shown as solid squares with error bars, and the prior and post CDW-EIS results are given by dashed-dotted red lines and dashed blue lines, respectively [15–17]. Adapted from [6]

of 3.8 a.u., to be contrasted with projectile speeds between 11.0 and 12.6 a.u. However, the level of agreement in the sequence of the four systems is not as expected: It is almost perfect for the O^{8+} projectiles, i.e., for the two intermediate η values, but the CTMC data appear to overestimate the measurements for the other two cases, C^{6+} and Si^{13+} impact at 4 MeV/amu, regardless of whether the static or time-dependent screening models are used. The overestimation appears to be independent of the emission angle, which might indicate that it is due to a problem with the overall normalization of the experimental data of [16].

Regarding the comparison with the CDW-EIS calculations, we note that two versions of the latter are shown, post and prior, which differ in the way the interaction between the ionized electron and the residual target ion is taken into account. In the post version, this interaction is purely Coulombic (using an effective charge), while in the prior form the influence of the non-ionized electrons on the dynamics of the continuum electron is included more explicitly. Hence, from a formal perspective the prior form is deemed superior [16].

As a general trend across all four systems, we find the CTMC results at small (forward) angles to be slightly higher and in the peak region (around $\theta_{el} = 60 - 90°$) lower than both CDW-EIS data sets, while there are no significant differences at backward angles. This gives the CTMC results a slight edge, since they match the angular shape of the experimental data very well in all cases. The CDW-EIS model for ion–atom collisions is known to be very good at accounting for two-center effects, i.e., the combined role of both the projectile and the residual target ion in shaping the angular distribution of medium-energy electrons [19, 20]. The present CTMC approach considers projectile and target interactions on the same footing as well, but in addition takes the three-center geometry of the target molecule fully into account. We are thus led to conclude that the angular distributions mirror a *many-center* electron emission mechanism. Of the available calculations at $E_{el} = 200$ eV only the present CTMC approach is able to describe this properly.

The same conclusion can be drawn from Fig. 2, which displays singly differential cross sections (SDCSs) as functions of the emission angle for the four collision systems. A closer analysis of the angular distributions in terms of a comparison of ratios of cross sections at forward versus intermediate scattering angles reveals that the CTMC results give a better representation of the experimental data than CDW-EIS, the prior version of which fares better than the post form, as expected [6].

On a quantitative level, there appears to be an inconsistency in the SDCSs of Fig. 2 and the DDCSs of Fig. 1 regarding the comparison between the CTMC and the experimental results: For the O^{8+} projectiles, we found very good agreement for the DDCSs at $E_{el} = 200$ eV, but the SDCSs are underestimated. Likewise, for the other two projectiles, the CTMC data appear to overestimate the measured DDCSs, but they are relatively close to the experimental SDCSs. These observations are, however, quite consistent, since the CTMC approach is known to fall short for low-energy electron emission. This is discussed in more detail in [6] in which doubly differential results at $E_{el} = 5$ eV are shown in addition to the $E_{el} = 200$ eV data. One may conclude that an underestimation of the experimental SDCS is to be expected, and the fact that we do not observe it for C^{6+} and Si^{13+} projectiles points again to a possible overall normalization problem of the experimental data.

As for the static versus time-dependent screening model comparison, we once again observe that the latter results in less ionization than the former and that this effect increases with η. In order to shed more light on the role of time-dependent screening we show in Fig. 3 for both CTMC models the contributions from the first five $q \frac{d\sigma_q}{d\Omega_{el}}$ terms to the net SDCS [cf. (8)] as well as their sum and the full net cross section for the Si^{13+} projectile for which η assumes the largest value among the four collision systems studied ($\eta = 1.03$).

For the σ_1 and $2\sigma_2$ terms, the static and time-dependent screening results are similar, but higher order contributions are suppressed in the time-dependent screening model and increasingly so with increasing electron multiplicity q. As a result, the sum truncated at $q = 5$ is very close to the full net cross section, i.e., contributions from $q > 5$ account for at most 3% in the time-dependent screening model. For the

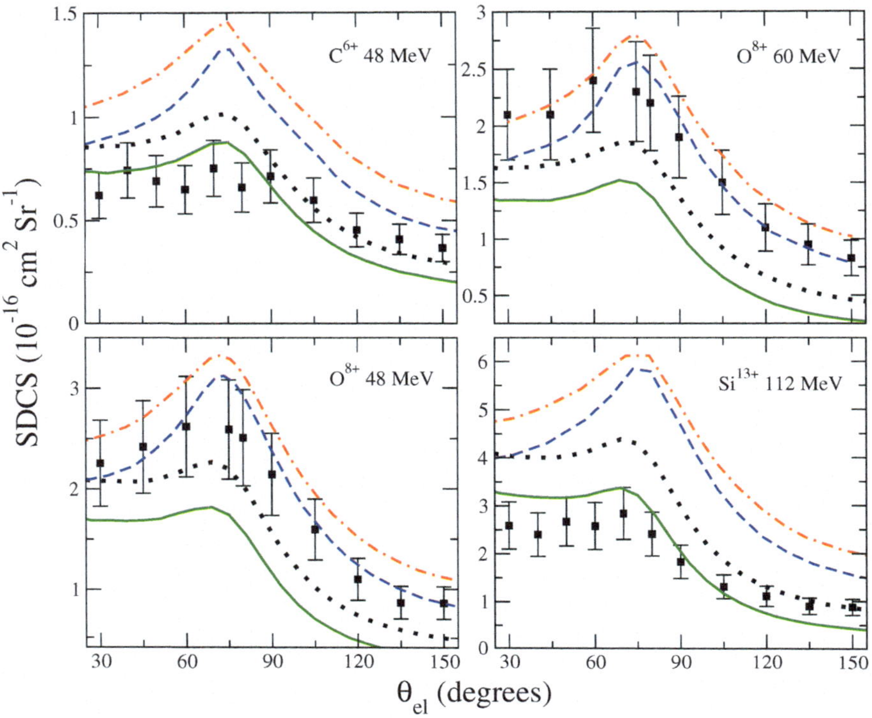

Fig. 2 SDCS for net ionization of water molecules as a function of electron emission angle in degrees, for the electron emission energy $E_{\mathrm{el}} = 200$ eV. The CTMC results (obtained with angular resolution $\Delta\theta_{\mathrm{el}} = 6°$) are shown as dotted black (static screening) and solid green (dynamical screening) lines. The experimental data are shown as solid squares with error bars, and the prior and post CDW-EIS results are given by dash-dotted red lines and dashed blue lines, respectively [15–17]. Adapted from [6]

static case, those contributions range between 13% at small emission angles and 18% at backward angles.

A complementary view is provided by a similar comparison of cross sections differential in electron emission energy instead of emission angle. Such a comparison is shown in Fig. 4. At low electron energies, the contribution from the excluded $q > 5$ terms is very small in both screening models, but the situation is different at higher E_{el}. At the highest energies shown, the contribution from those high-q terms is almost as large as that from $q \leq 5$ in the case of the static model, whereas it stays below 10% if time-dependent screening is turned on. It would be of great interest if differential experimental data for fixed q would become available to test these predictions.

For proton-H_2O collisions a comparison of the present CTMC approach was carried out with recent measurements and also with CDW-EIS theory (including a new model) [9]. This comparison demonstrates that the multicenter CTMC approach

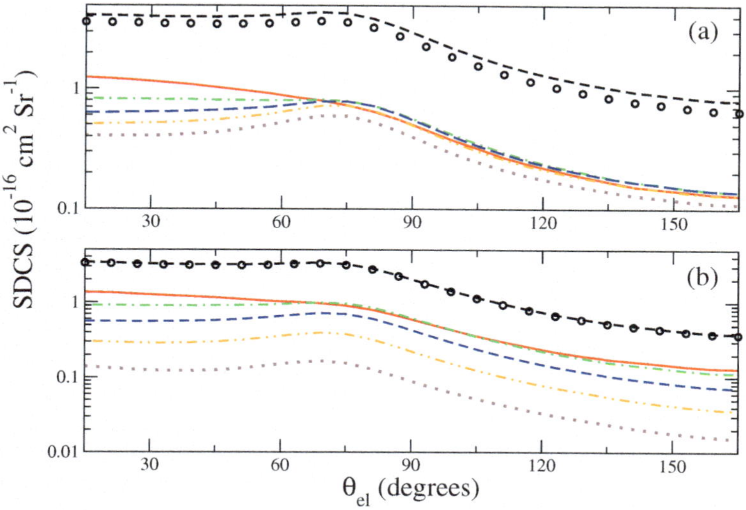

Fig. 3 SDCS for net ionization of water molecules as a function of electron emission angle in degrees for 112 MeV Si^{13+} projectiles with (a) static and (b) dynamical screening. (– – –) Net, (—) $\frac{d\sigma_1}{d\Omega_{el}}$, (· – ·) $2\frac{d\sigma_2}{d\Omega_{el}}$, (– – –) $3\frac{d\sigma_3}{d\Omega_{el}}$, (· · – · ·) $4\frac{d\sigma_4}{d\Omega_{el}}$, (· · ·) $5\frac{d\sigma_5}{d\Omega_{el}}$, (○) $\sum_{q=1}^{5} q\frac{d\sigma_q}{d\Omega_{el}}$. Adapted from [6]

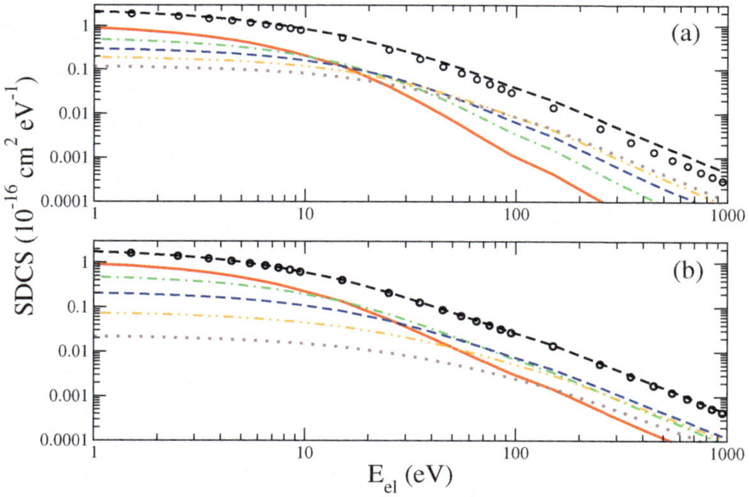

Fig. 4 SDCS for net ionization of water molecules as a function of electron emission energy for 112 MeV Si^{13+} projectiles with (a) static and (b) dynamical screening. (– – –) Net, (—) $\frac{d\sigma_1}{dE_{el}}$, (· – ·) $2\frac{d\sigma_2}{dE_{el}}$, (– – –) $3\frac{d\sigma_3}{dE_{el}}$, (· · – · ·) $4\frac{d\sigma_4}{dE_{el}}$, (· · ·) $5\frac{d\sigma_5}{dE_{el}}$, (○) $\sum_{q=1}^{5} q\frac{d\sigma_q}{dE_{el}}$. Adapted from [6]

works very well when dealing with mostly single ionization, and complements the multiply charged projectile results discussed in this section.

3.2 Collisions with Water Molecules: Total Cross Sections

We now turn our attention to total cross sections for processes that involve the removal of one or several electrons. Figure 5 displays the pure ionization cross sections σ_{01} and σ_{02} calculated with the trinomial formula (10) for the Li^{3+}-H_2O collision system. Three CTMC data sets are shown: static screening results, results obtained from using the time-dependent target charge screening parameters (4), and results from calculations in which time-dependent projectile screening effects are also included. Obviously, the three models give very similar results over the range of impact energies shown and they do not agree very well with the experimental data. The underestimation of σ_{01} can, once again, be explained by a known weakness of the CTMC approach, namely, a shortfall at large impact parameters which dominate singly ionizing collisions. Since ionization at large impact parameters is mostly associated with low-energy electron emission, this is the same weakness as the one identified in Sect. 3.1.

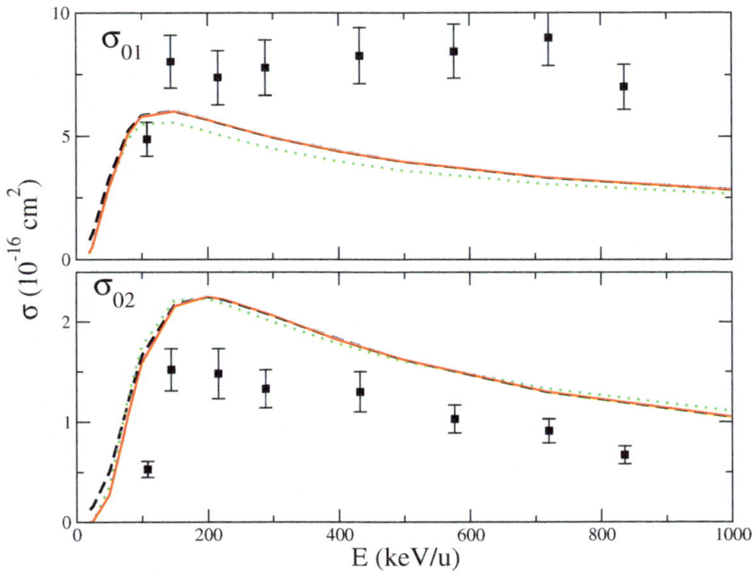

Fig. 5 Pure single and double ionization cross sections for Li^{3+}-H_2O collisions as functions of impact energy. The CTMC results are shown as dashed black (dynamical target and projectile screening), solid red (dynamical target screening), and dotted green (static screening) lines. The experimental data are shown as solid squares with error bars [21]. Adapted from [7]

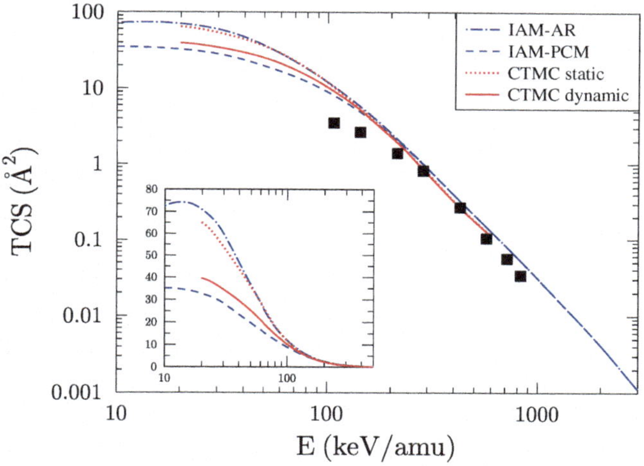

Fig. 6 Net electron capture cross section for Li^{3+}-H_2O collisions as a function of impact energy. The highest curve (dash-dotted blue) shows the additivity rule result (IAM-AR), the lowest (dashed blue) curve the IAM-PCM result. In between are the CTMC results, namely, the (dotted red) static screening, and below it (solid red) the dynamical screening result. The experimental data are shown as black solid squares [21]. Adapted from [8]

Figure 6 shows the net capture cross section for the same collision system. Only static screening ("CTMC static") and time-dependent target *and* projectile screening ("CTMC dynamic") results are included, but we note that the target-only time-dependent screening cross section is very close to the "CTMC dynamic" data shown.

At high impact energies, the static and dynamic results are very similar—there simply is not enough time for the mean-field potential to adapt and thereby change the character of the trajectories. Below $E \approx 100$ keV/amu, however, time-dependent screening effects are strong, as can be seen most clearly in the inset which displays the cross section on a linear scale.

In addition to the CTMC results, the figure includes experimental data from [21] and results from two sets of quantum-mechanical independent-atom-model (IAM) calculations: "IAM-AR", in which the simple additivity rule is used to sum up atomic cross sections, and "IAM-PCM", which is a much more sophisticated model which takes the geometric overlap of the atomic cross sections into account [22]. Interestingly, the static-potential CTMC results agree well with the IAM-AR data, while the time-dependent screening calculations are closer to the IAM-PCM cross sections. A similar pattern was observed for He^{2+} impact [8]. At $E \geq 200$ keV/amu the results from all four models merge and show good agreement with the experimental data. As discussed in [8], it is difficult to understand from a modeling perspective why all calculated cross-sectional curves for Li^{3+} impact are significantly above the two experimental data points at $E < 200$ keV/amu. This observation is also in striking

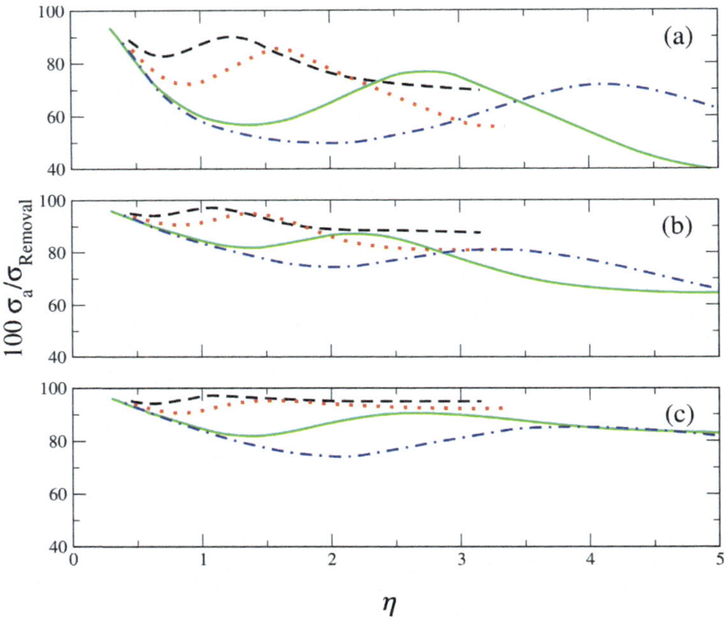

Fig. 7 Percentage of $\sum_{q=1}^{3} q\sigma_q^{\text{rem}} = \sum_{q=1}^{3} q(\sigma_q^{\text{cap}} + \sigma_q^{\text{ion}})$ with respect to the total net removal cross section $\sigma_{\text{net}}^{\text{rem}}$ as a function of $\eta = Z_p/v_p$ in the case of **a** static screening, **b** dynamical target screening, **c** dynamical target and projectile screening potentials. The systems shown are He^{2+} (— — —), Li^{3+} (· · ·), C^{6+} (—), Ne^{10+} (· — ·). Adapted from [7]

contrast to what was found for He^{2+} impact and warrants further investigation, ideally aided by additional experimental work.

We end this subsection with a comparison between static and dynamic screening model results which uses similar ideas as in Figs. 3 and 4, but this time on the basis of *total* cross sections: Fig. 7 displays for four different projectiles (He^{2+}, Li^{3+}, C^{6+}, Ne^{10+}) the percentage contribution of $\sigma_a = \sum_{q=1}^{3} q\sigma_q^{\text{rem}}$ to the total net removal cross section $\sigma_{\text{net}}^{\text{rem}}$ as a function of the parameter $\eta = Z_p/v_p$. A value of close to 100% signals that the $q \geq 4$ contributions to $\sigma_{\text{net}}^{\text{rem}}$ are negligibly small. This is the situation observed in the perturbative small-η regime. As η increases, all the curves in Fig. 7 show a decreasing trend, eventually reaching a first minimum located at an η value which is larger, the larger the projectile charge is. The minimum is followed by a local maximum and then another decrease. A closer analysis of the data shows that for each projectile the minimum occurs in a region in which ionization is the dominant electron removal process. Capture becomes competitive around the maximum and takes over as the dominant process at even higher η values.

For the static screening results displayed in panel (a), the minima are deep, in particular, for the highly charged projectiles, indicating that high-multiplicity (i.e., $q > 3$-fold) ionization is predicted to be very strong. Once time-dependent target screening is included [panel (b)], high-q processes are suppressed and the minima become shallower. Additional inclusion of time-dependent projectile screening [panel (c)] affects the data mostly at large η values, i.e., in the region in which capture becomes dominant.

There is no apparent scaling with respect to η: When moving from one projectile to the next, the positions of the minima and maxima change in nontrivial ways. This is perhaps not surprising in the nonperturbative $\eta \geq 1$ region, but based on the present $\hbar = 0$ mean-field calculations alone we cannot draw definitive conclusions. Ultimately, experimental data would be needed for a better understanding of the strength of the high-multiplicity electron removal processes as a function of η.

3.3 Collisions with Ammonia Molecules

For the case of ammonia target molecules, we focus on collisions with protons for which differential and total cross-sectional data are available for comparison. As can be expected for a singly charged projectile, we found dynamical screening effects to be rather weak, and thus we discuss results obtained from the static screening model only. Net differential electron emission was measured some time ago [23], while total cross-sectional data for single and multiple ionization events were reported more recently [24]. Comparing the latter with calculations requires some caution since the measurements are based on counts for charged fragments and do not directly correspond to the charge-state correlated cross sections for capture and ionization discussed in Sect. 2.2.2.

We show in Fig. 8 DDCS results for $E = 250$ keV proton impact as functions of the polar emission angle for a set of electron energies between $E_{\text{el}} = 10$ eV and $E_{\text{el}} = 500$ eV. We have calculated both net DDCSs and DDCSs for $q = 1$ according to (9) and (7) and display them as solid and dashed lines, respectively. The agreement with the experimental net DDCS results is very good over the entire range of electron angles and energies. Comparison with previously obtained quantum calculations is made in [10], which demonstrates that the present CTMC method which is based on a multicenter effective potential provides an improved description at backward electron emission angles.

The comparison between the solid and dashed lines provides an indication of the significance of multiple ionization. The net and $q = 1$ results are of similar shape and typically differ by a factor of two, i.e., the calculations predict that double and higher multiple ionization contribute about equally to net ionization as the $q = 1$ channel. In the absence of q-specific DDCS data, we can only speculate whether this is indeed the case in this system or if the result signals a deficiency of the IEM. We do note that NH_3 is an extended object with a number of relatively loosely bound electrons, i.e., multiple ionization might indeed be quite strong.

We end this section with a look at the total cross sections for net and $q = 1, 2, 3$-fold ionization in Fig. 9. The present net results are calculated according to $\sigma_{\text{net}}^{\text{ion}} = \sigma_1^{\text{ion}} + 2\sigma_2^{\text{ion}} + 3\sigma_3^{\text{ion}}$, i.e., contributions from $q \geq 4$ events are neglected. IAM-PCM

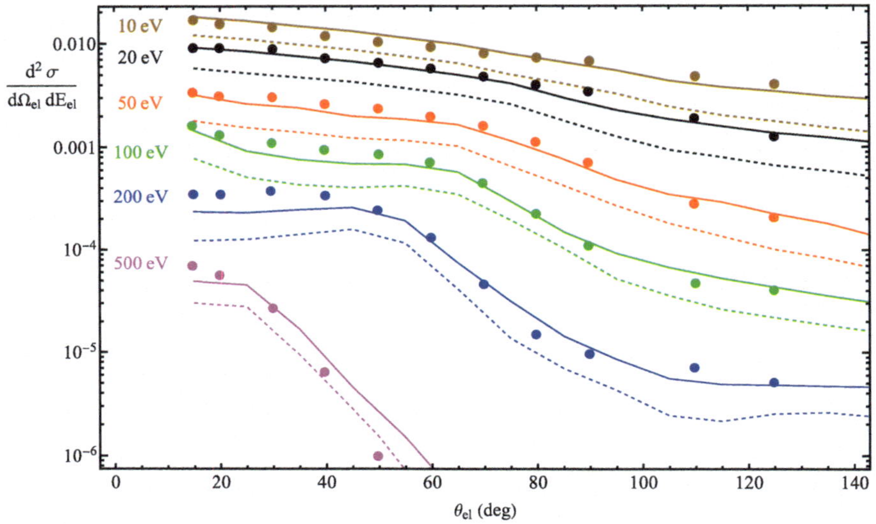

Fig. 8 DDCS in units of Å^2/(eV srad) for p-NH_3 collisions at $E = 250$ keV for ionized electron energies of $E_{el} = 10, 20, 50, 100, 200, 500$ eV. Solid lines: CTMC net ionization results obtained with (9); dashed lines are for single ionization obtained with (7). The data points are the experimental results of [23], as reported in [25, 26]. Adapted from [10]

results from [27] are also included in the figure, and so are net measurements from [23] and σ_q^{ion} data from [24] obtained by summing up fragmentation yields.

The present net cross section is quite close to the data point at $E = 250$ keV, but falls below the measurements at higher energies. The calculation for $q = 1$ produces a cross-sectional curve which matches the shape of the experimental data of [24] quite well, but underestimates them by about 30%. This is different from the IAM-PCM prediction which agrees well with the $q = 1$ data at intermediate and high energies and is marginally below the two data points at $E = 125$ keV and $E = 250$ keV. Also for $q = 2$ and $q = 3$ do the two theories differ—both in shape and in magnitude.

From a comparison of the experimental net data of [23] and the $q = 1$ data of [24], one would conclude that the net cross section is almost entirely due to single ionization, and indeed the experimental $q = 2$ and $q = 3$ data included in Fig. 9 appear to confirm this. However, the fragment measurements of [24] may miss some contributing channels, since certain coincidences cannot be detected with the set-up used. This is to say that the reasons for the apparent very strong overestimation of double and triple ionization by the present CTMC as well as the previous IAM-PCM calculations remain somewhat unclear. We refer the reader to [10] for a more detailed discussion of this issue.

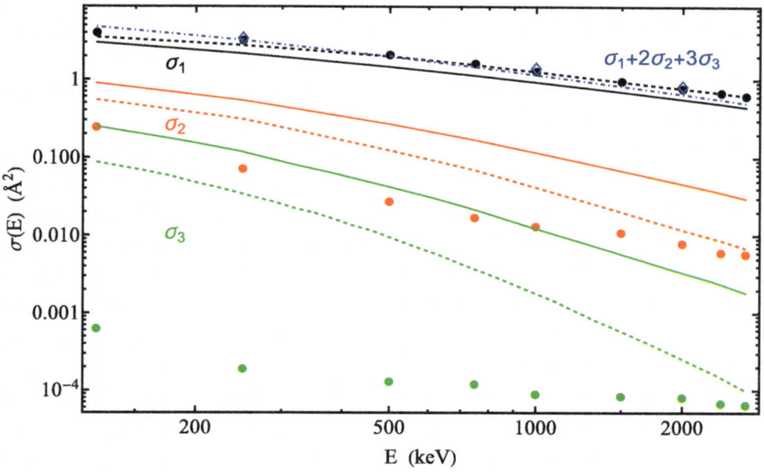

Fig. 9 Total cross sections for p-NH_3 collisions as functions of impact energy. Dash-dotted blue line: CTMC approximate net ionization result, $\sum_{q=1}^{3} q\sigma_q^{\rm ion}$. Blue open diamonds: net ionization cross sections from [23]. Solid lines (black, red, green) from top to bottom: CTMC results for $\sigma_q^{\rm ion}$ with $q = 1, 2, 3$, respectively. Dashed lines (black, red, green): IAM-PCM results shown in [24, 27] for $\sigma_q^{\rm ion}$. Dots (black, red, green): experimental values for $\sigma_q^{\rm ion}$ from fragmentation yields [24]. The results for $q = 3$ are associated solely with H^+ + N^{2+} coincidences. Adapted from [10]

4 Concluding Remarks

We have presented a mean-field model applied at the $\hbar = 0$ level to deal with the problem of capture and ionization in ion–molecule collisions. Both differential and total cross sections were calculated, and a selected set of results for various ions impinging on water and ammonia molecules have been discussed in comparison with experimental data and some of the existing previous calculations. These comparisons shed light on the strengths and weaknesses of the model and on open questions.

To summarize our findings, we start with noting that the use of multicenter potentials to describe the molecular targets appears to be a plus compared to previous calculations in which the molecular geometry was modeled in less sophisticated fashions. This is most obvious in angular electron distributions. The comparisons provided suggest that the notion of two-center electron emission, extensively discussed for ion–atom collision problems, has to be replaced by many-center electron emission for molecular targets.

The IEM framework used includes the straightforward calculation of multiple capture and ionization contributions via multinomial statistics. We have found that contributions from multiplicities higher than $q = 1$ can be quite substantial, even for singly charged ion impact. To which extent this reflects the physics at play or a shortcoming of the IEM remains to be seen, given that the existing coincident experiments for molecular targets are based on the counting of charged fragments and may suffer from incomplete detection, i.e., certain contributions might be missed and others misidentified. Further experimental investigation and also theoretical efforts which do not rely on the IEM are needed to shed more light on this problem.

One limitation of the present CTMC model is its inability to describe correctly the forward emission of low-energy electrons. This problem cannot be fixed easily, certainly not without going beyond the $\hbar = 0$ level of the theory. This, however, appears to be challenging to say the least, unless one is willing to compromise on other aspects of the problem at hand, such as its multicenter nature. It is thus safe to conclude that CTMC models, despite their imperfections, will continue to play an important role in aiding the understanding of ion–molecule collision problems.

Acknowledgements Financial support from the Natural Sciences and Engineering Research Council of Canada (NSERC) (RGPIN-2017-05655 and RGPIN-2019-06305) is gratefully acknowledged.

References

1. A. Messiah, *Quantum Mechanics I* (North Holland, Amsterdam, 1961) Chap. 6
2. R. Abrines, I.C. Percival, Proc. Phys. Soc. **88**, 861 (1966). https://doi.org/10.1088/0370-1328/88/4/306
3. J. Ullrich, R. Moshammer, R. Dörner, O. Jagutzki, V. Mergel, H. Schmidt-Böcking, L. Spielberger, J. Phys. B: Atomic, Mol. Opt. Phys. **30**, 2917 (1997). https://doi.org/10.1088/0953-4075/30/13/006
4. C.L. Cocke, R.E. Olson, Phys. Rep. **205**, 153 (1991). https://doi.org/10.1016/0370-1573(91)90072-T
5. F. Frémont, *Classical Treatment of Collisions Between Ions and Atoms or Molecules* (Springer Nature, Cham, 2021). https://doi.org/10.1007/978-3-030-89428-3
6. A. Jorge, M. Horbatsch, C. Illescas, T. Kirchner, Phys. Rev. A **99**, 062701 (2019). https://doi.org/10.1103/PhysRevA.99.062701
7. A. Jorge, M. Horbatsch, T. Kirchner, Phys. Rev. A **102**, 012808 (2020). https://doi.org/10.1103/PhysRevA.102.012808
8. H.J. Lüdde, A. Jorge, M. Horbatsch, T. Kirchner, Atoms **8**, 59 (2020). https://www.mdpi.com/2218-2004/8/3/59
9. A. Bhogale, S. Bhattacharjee, M.R. Chowdhury, C. Bagdia, M.F. Rojas, J.M. Monti, A. Jorge, M. Horbatsch, T. Kirchner, R.D. Rivarola, L.C. Tribedi, Phys. Rev. A **105**, 062822 (2022). https://doi.org/10.1103/PhysRevA.105.062822
10. A. Jorge, M. Horbatsch, T. Kirchner, J. Phys. B: Atomic, Mol. Opt. Phys. **57**, 045201 (2024). https://doi.org/10.1088/1361-6455/ad285c
11. C. Illescas, L.F. Errea, L. Méndez, B. Pons, I. Rabadán, A. Riera, Phys. Rev. A **83**, 052704 (2011). https://doi.org/10.1103/PhysRevA.83.052704
12. R. Moccia, J. Chem. Phys. **40**, 2176 (1964). https://doi.org/10.1063/1.1725490
13. T. Kirchner, M. Horbatsch, H.J. Lüdde, R.M. Dreizler, Phys. Rev. A **62**, 042704 (2000). https://doi.org/10.1103/PhysRevA.62.042704
14. M. Murakami, T. Kirchner, M. Horbatsch, H.J. Lüdde, Phys. Rev. A **85**, 052704 (2012). https://doi.org/10.1103/PhysRevA.85.052704
15. S. Bhattacharjee, S. Biswas, C. Bagdia, M.R. Chowdhury, S. Nandi, D. Misra, J.M. Monti, C.A. Tachino, R.D. Rivarola, C. Champion, L.C. Tribedi, J. Phys. B: Atomic, Mol. Opt. Phys. **49**, 065202 (2016). http://stacks.iop.org/0953-4075/49/i=6/a=065202

16. S. Bhattacharjee, S. Biswas, J.M. Monti, R.D. Rivarola, L.C. Tribedi, Phys. Rev. A **96**, 052707 (2017). https://doi.org/10.1103/PhysRevA.96.052707
17. S. Bhattacharjee, C. Bagdia, M.R. Chowdhury, J.M. Monti, R.D. Rivarola, L.C. Tribedi, Eur. Phys. J. D **72**, 15 (2018). https://doi.org/10.1140/epjd/e2017-80265-8
18. L. Gulyás, Phys. Rev. A **108**, 032815 (2023). https://doi.org/10.1103/PhysRevA.108.032815
19. P.D. Fainstein, V.H. Ponce, R.D. Rivarola, J. Phys. B: Atomic, Mol. Opt. Phys. **24**, 3091 (1991). https://doi.org/10.1088/0953-4075/24/14/005
20. N. Stolterfoht, R.D. DuBois, R.D. Rivarola, *Electron Emission in Heavy Ion–Atom Collisions* (Springer, Berlin, 1997)
21. H. Luna, W. Wolff, E.C. Montenegro, A.C. Tavares, H.J. Lüdde, G. Schenk, M. Horbatsch, T. Kirchner, Phys. Rev. A **93**, 052705 (2016). https://doi.org/10.1103/PhysRevA.93.052705
22. H.J. Lüdde, A. Achenbach, T. Kalkbrenner, H.-C. Jankowiak, T. Kirchner, Eur. Phys. J. D **70**, 82 (2016). https://doi.org/10.1140/epjd/e2016-70097-5
23. D. Lynch, L. Toburen, W. Wilson, J. Chem. Phys. **64**, 2616 (1976). https://doi.org/10.1063/1.432515
24. W. Wolff, H. Luna, E.C. Montenegro, L.C. Rodrigues Jr., Phys. Rev. A **102**, 052821 (2020). https://doi.org/10.1103/PhysRevA.102.052821
25. B. Senger, Zeitschrift für Phys. D Atoms, Mol. Clust. **9**, 79 (1988). https://doi.org/10.1007/BF01384450
26. A. Mondal, S. Halder, S. Mukherjee, C.R. Mandal, M. Purkait, Phys. Rev. A **96**, 032710 (2017). https://doi.org/10.1103/PhysRevA.96.032710
27. H.J. Lüdde, M. Horbatsch, T. Kirchner, Phys. Rev. A **106**, 022813 (2022). https://doi.org/10.1103/PhysRevA.106.022813

Perturbed Relativistic Coupled-Cluster Calculations of the Properties of Ar^{13+}

B. K. Mani, Palki, Suraj Pandey, and D. Angom

Abstract We have developed a perturbed relativistic coupled-cluster theory to study the parity nonconservation in atoms and ions. The method we have developed is quite robust and could be easily modified to compute other atomic properties just by changing the perturbation Hamiltonian. As benchmark test calculations, we have computed the electric dipole polarizability of ground and first excited state of Ar^{13+}. Further, as the clock transition property in Ar^{13+}, we have computed the M1 transition rate for $2p_{1/2} \rightarrow 2p_{3/2}$ transition. Our calculated results for both the properties are in very good agreement with the previous theoretical and experimental results.

1 Introduction

Highly charged ions (HCIs) are important atomic systems which provide a roadmap to probe a plethora of fundamental and technological implications. Some key examples where HCIs could provide a better probe include parity and time reversal violations [1], atomic clocks [2], variation in the fundamental constants [3], search for new physics beyond the Standard model (BSM) of particle physics [2], etc.

Parity nonconservation (PNC) in atomic systems arises due to the weak interaction between nucleus and electrons, mediated through the exchange of the Z_0 bosons. The atomic PNC experiments when combined with accurate PNC calculations provide an alternate and a simpler way to probe the BSM physics [4]. It is to be mentioned that the conventional way to examine BSM physics is through combining the expensive high energy collider experiments with phenomenological calculations [5]. Experimentally, PNC has been observed in neutral systems like Cs [6, 7], Tl [8–10], Bi [11, 12], Pb [11, 13], Yb [14], and several others. Highly charged ions are reported to be promising candidates for exploring PNC and electroweak sector of Standard model

B. K. Mani (✉) · Palki · S. Pandey
Department of Physics, Indian Institute of Technology, Hauz Khas, New Delhi 110016, India
e-mail: bkmani@physics.iitd.ac.in

D. Angom
Department of Physics, Manipur University, Canchipur 795003, Manipur, India

L. C. Tribedi (ed.), *Advances in Atomic Molecular Collisions*,
https://doi.org/10.1007/978-981-97-7063-2_11

due to their simpler structures [1]. PNC in HCIs varies as Z^5 [1] (Z is the atomic number), and therefore, HCIs have several merits over the neutral systems. The reasons for the enhanced PNC effects in HCIs could be attributed to a large overlap of electron density with nucleus and nearly degenerate opposite parity states [15]. The nearly degenerate opposite parity states in HCIs lead to a strong PNC mixing caused by the weak interaction. In addition, the other key advantage with HCIs is that, due to their simple structure and smaller size, it is easier to perform accurate theoretical calculations of PNC observable, which is a crucial parameter for experiments. Till date, PNC in HCIs has been reported in H, He, Li, Be, and B-like HCIs [15–17]. In terms of theoretical calculations, not many studies have been done, and this is one of the main long-term goals of the present work.

Optical atomic clocks are the most precise time measurement device till date [18] and, due to unprecedented accuracy, they are a powerful tool for probing several fundamental research and technological applications, such as frequency metrology [19], navigation systems [20–22], telecommunications [23], and exploration of geopotential differences [18]. The highly charged ions are reported to have a fractional frequency shift of the order of 10^{-20} or smaller [2], and therefore, are the promising candidates for optical atomic clocks. The reasons for such a small fractional frequency error could be attributed to the stronger nuclear potential, highly forbidden transitions, and very low linewidths [2]. The other key advantage with the HCIs-based clocks is the sensitivity of these ions for detecting the variations in the fine structure constant [24–27]. In literature, Ar^{13+} [28], Pr^{9+} [29], and Ni^{12+} [30] highly charged ions are reported to be promising candidates for optical atomic clocks.

In this work, we report the development of a perturbed relativistic coupled-cluster (PRCC) theory for the study of nuclear spin-dependent (NSD) and spin-independent (NSI) PNC in atomic systems. It is to be emphasized that relativistic coupled-cluster (RCC) theory is one of the most accurate many-body theories for accurate atomic structure and properties calculations. In our previous works, we have developed a series of RCC and PRCC methods which are used to study a plethora of properties in several atoms and ions [31–35]. Further, in the present work, as benchmark test calculations, we have computed the electric dipole polarizability, α, of the ground state, $^2P_{1/2}$, and the fine-structured excited state $^2P_{3/2}$ of Ar^{13+}. To compute α, we employ PRCC theory for PNC discussed in the paper with a small change to replace PNC operator with a dipole operator. It is to be emphasized that the basic formulation of the theory and the Goldstone diagrams remains the same. In addition, as a probe to clock properties, we have computed the magnetic dipole (M1) transition amplitudes and transition rate for $^2P_{1/2} \rightarrow {}^2P_{3/2}$ transition in Ar^{13+}. It is to be mentioned that $^2P_{1/2} \rightarrow {}^2P_{3/2}$ M1 clock transition in Ar^{13+} is the first ever developed HCIs-based atomic clock [28].

The paper is organized as follows. In Section II, we have provided a brief description of the RCC and PRCC theory. Coupled-cluster working equations are discussed and some dominant contributing PRCC diagrams are also provided for the reference. In the same section, we have also discussed the calculation of PNC observable using PRCC theory and have demonstrated the coupling of electronic and nuclear sectors of the PNC matrix elements. Results from our calculations are presented and analyzed in Section III of the paper. The summary of our findings is presented in the last section. The results and equations presented in this work are in atomic units ($\hbar = m_e = e = 1/4\pi\epsilon_0 = 1$).

2 Theoretical Methods

2.1 RCC Theory

One of the main prerequisites for obtaining accurate properties results of a many-electron atom or ion is the use of accurate many-body wavefunction in the properties calculation. For a one-valence atomic system, the many-body wavefunction is obtained by solving the eigenvalue equation

$$H^{\mathrm{DC}}|\Psi_v\rangle = E_v|\Psi_v\rangle, \tag{1}$$

where $|\Psi_v\rangle$ is the exact atomic state of the one-valence system and E_v is the corresponding energy. The Hamiltonian H^{DC} refers to the Dirac-Coulomb Hamiltonian and is expressed as [36],

$$\begin{aligned} H^{\mathrm{DC}} &= \Lambda_+ \sum_{i=1}^{N} \left[c\boldsymbol{\alpha}_i \cdot \mathbf{p}_i + (\beta_i - 1)c^2 - V_N(r_i) \right] \\ &+ \sum_{i<j} \frac{1}{r_{ij}} \Lambda_+. \end{aligned} \tag{2}$$

Here, $\boldsymbol{\alpha}_i$ and β are the Dirac matrices, $\mathbf{p}$ is the linear momentum, $V_N(r)$ is the nuclear Coulomb potential, and the last term is the electron-electron Coulomb interaction. The operator Λ_+ projects on the positive energy eigenstates to avoid the negative energy continuum solutions.

In RCC, $|\Psi_v\rangle$ is expressed as

$$|\Psi_v\rangle = e^{T^{(0)}} \left[1 + S^{(0)}\right] |\Phi_v\rangle, \tag{3}$$

where $T^{(0)}$ and $S^{(0)}$ represent the unperturbed RCC operators for closed-shell and one-valence sectors, respectively. The wavefunction $|\Phi_v\rangle$ is the Dirac-Fock reference state for one-valence systems and obtained by adding an electron to the closed-shell reference state, as $|\Phi_v\rangle = a_v^\dagger|\Phi_0\rangle$. In CCSD approximation, the $T^{(0)}$ and $S^{(0)}$ cluster operators are expressed as $T^{(0)} = T_1^{(0)} + T_2^{(0)}$ and $S^{(0)} = S_1^{(0)} + S_2^{(0)}$, respectively. In the second quantized representation, they can be expressed as

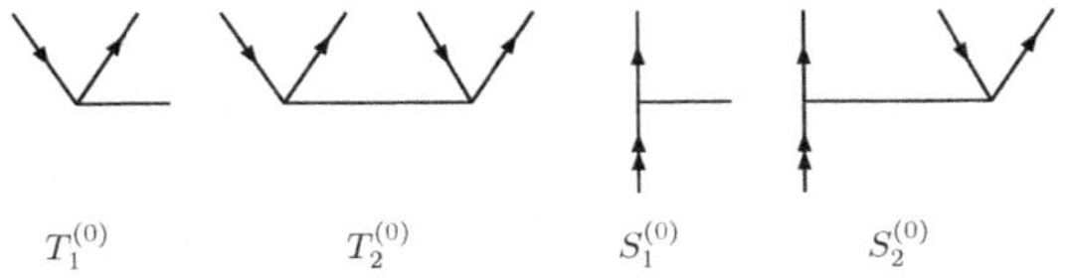

Fig. 1 Diagrammatic representations of the unperturbed single and double cluster operators for closed-shell and one-valence sectors

$$T_1^{(0)} = \sum_{a,p} t_a^p a_p^\dagger a_a, \text{ and } T_2^{(0)} = \frac{1}{2!} \sum_{a,b,p,q} t_{ab}^{pq} a_p^\dagger a_q^\dagger a_b a_a, \tag{4a}$$

$$S_1^{(0)} = \sum_{p} s_v^p a_p^\dagger a_v, \text{ and } S_2^{(0)} = \sum_{a,p,q} s_{va}^{pq} a_p^\dagger a_q^\dagger a_a a_v. \tag{4b}$$

Here, $t_{\cdots}^{\cdots}$ and $s_{\cdots}^{\cdots}$ are the cluster amplitudes and indexes $a, b, c \ldots (p, q, r \ldots)$ represent the core (virtual) states and $v, w, x \ldots$ represents the valence states. The operators T_1 (S_1) and T_2 (S_2) generate the single and double electron replacements after operating on the closed(open)-shell reference states. The diagrammatic representation of these operators are shown in Fig. 1. The open-shell cluster operators $S_1^{(0)}$ and $S_2^{(0)}$ are obtained by solving nonlinear coupled-cluster equations [31, 37]

$$\langle \Phi_v^p | \bar{H}_N + \{\overline{\bar{H}_N S^{(0)}}\} | \Phi_v \rangle = E_v^{\text{att}} \langle \Phi_v^p | S_1^{(0)} | \Phi_v \rangle, \tag{5a}$$

$$\langle \Phi_{va}^{pq} | \bar{H}_N + \{\overline{\bar{H}_N S^{(0)}}\} | \Phi_v \rangle = E_v^{\text{att}} \langle \Phi_{va}^{pq} | S_2^{(0)} | \Phi_v \rangle, \tag{5b}$$

where $H_{\text{N}} = H^{\text{DC}} - \langle \Phi_0 | H^{\text{DC}} | \Phi_0 \rangle$ is the normal ordered Hamiltonian and $\bar{H}_{\text{N}} = e^{-T^{(0)}} H_{\text{N}} e^{T^{(0)}}$ represents the similarity transformed Hamiltonian. E_v^{att} is the attachment energy of the valence electron, expressed as $E_v^{\text{att}} = \epsilon_v + \Delta E_v$, where ϵ_v is the single-particle energy and ΔE_v is the electron correlation energy. The operators $T_1^{(0)}$ and $T_2^{(0)}$ are the solutions of the coupled nonlinear equations [31, 38]

$$\langle \Phi_a^p | \bar{H}_{\text{N}} | \Phi_0 \rangle = 0, \tag{6a}$$

$$\langle \Phi_{ab}^{pq} | \bar{H}_{\text{N}} | \Phi_0 \rangle = 0, \tag{6b}$$

where $|\Phi_a^p\rangle$ and $|\Phi_{ab}^{pq}\rangle$ are the single and double excitation Slater determinants.

2.2 PRCC Theory for PNC

In the presence of PNC interaction, atomic states get modified, as they are mixed with opposite parity states, and are no longer the eigenstates of parity operator. We refer to these modified states as the perturbed states and they are obtained by solving the eigenvalue equation

$$H_t|\widetilde{\Psi}_v\rangle = \widetilde{E}_v|\widetilde{\Psi}_v\rangle, \tag{7}$$

$\widetilde{E}_v$ is the corresponding perturbed energy. To the first-order in λ, $|\widetilde{\Psi}_v\rangle = |\Psi_v\rangle + \lambda|\bar{\Psi}^1_v\rangle$ and $\widetilde{E}_v = E_v + \lambda E^1_v$, where the bar in $|\bar{\Psi}^1_v\rangle$ denotes an opposite parity to $|\Psi_v\rangle$. The H_t is the total Hamiltonian which is expressed as

$$H_t = H^{DC} + \lambda H_{PNC}. \tag{8}$$

Here, H_{PNC} is the PNC interaction Hamiltonian and it has two components, NSI and NSD-PNC, which can be expressed as

$$H^{NSI}_{PNC} = \frac{G_F Q_W}{2\sqrt{2}} \sum_i \gamma_5 \rho_N(r_i), \tag{9a}$$

$$H^{NSD}_{PNC} = \frac{G_F \mu'_W}{\sqrt{2}I} \sum_i \boldsymbol{\alpha}_i \cdot \mathbf{I} \rho_N(r), \tag{9b}$$

where $G_F = 2.22 \times 10^{-14} a.u.$ is the Fermi coupling constant, Q_W and μ'_W are, respectively, the weak nuclear charge and weak nuclear moment of the nucleus. $\boldsymbol{\alpha}$ and γ_5 are the Dirac matrices, $\rho_N(r)$ is the normalized nuclear density, and $\mathbf{I}$ is the nuclear spin. To derive the PRCC equations to incorporate PNC we use the NSD-PNC interaction Hamiltonian. Compared to NSI-PNC, NSD-PNC requires two important considerations due to nuclear spin operator $\mathbf{I}$. First, the cluster operators in electrons sector are rank-one operators, and second, atomic states are the eigenstates of the total angular momentum $\mathbf{F} = \mathbf{I} + \mathbf{J}$. Using Eq. (9b) in Eq. (7), we write

$$\left(H^{DC} + \lambda \mathbf{H}^{NSD}_{elec} \cdot \mathbf{I}\right)|\widetilde{\Psi}_v\rangle = E_v|\widetilde{\Psi}_v\rangle, \tag{10}$$

where $H^{NSD}_{elec} = (G_F\mu'_W/\sqrt{2})\sum_i \boldsymbol{\alpha}_i \rho_N(r)$ represents the electronic part of the H^{NSD}_{PNC}. Here, we have used $E^1_v = \langle\Psi_v|H^{NSD}_{PNC}|\Psi_v\rangle = 0$, since H^{NSD}_{PNC} is an odd parity operator it connects states with opposite parities.

In PRCC theory, the perturbed wave function $|\widetilde{\Psi}_v\rangle$ is expressed as

$$|\widetilde{\Psi}_v\rangle = e^{T^{(0)}}\left[1 + \lambda \mathbf{T}^{(1)} \cdot \mathbf{I}\right]\left[1 + S^{(0)} + \lambda \mathbf{S}^{(1)} \cdot \mathbf{I}\right]|\Phi_v\rangle, \tag{11}$$

where $\mathbf{T}^{(1)}$ and $\mathbf{S}^{(1)}$ are the closed-shell and one-valence perturbed operators, respectively. The superscript (1) is used here to indicate the perturbation. Like the case of unperturbed operators $T^{(0)}$ and $S^{(0)}$, in the CCSD approximation, we take $\mathbf{T}^{(1)} = T^{(1)}_1 + T^{(1)}_2$ and $\mathbf{S}^{(1)} = S^{(1)}_1 + S^{(1)}_2$. And, for example, in second quantized representation, $\mathbf{S}^{(1)}$ can be expressed as

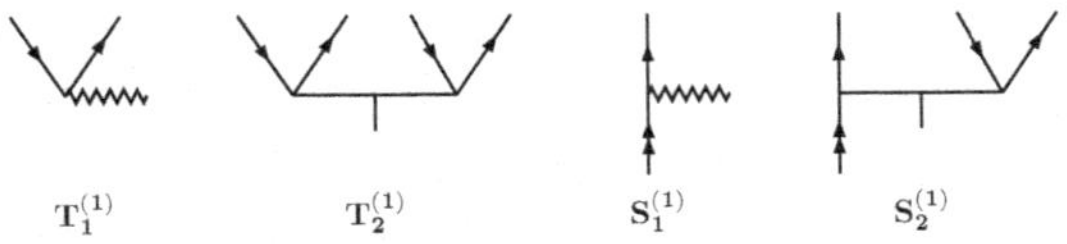

Fig. 2 Diagrammatic representations of NSD-perturbed cluster operators in closed-shell and one-valence sectors. The extra line in $T_2^{(1)}$ and $S_2^{(1)}$ is to indicate the multipole structure of the operators

$$\mathbf{S}_1^{(1)} = \sum_p \xi_v^p \mathbf{C}_1(\hat{r}) a_p^\dagger a_v, \tag{12a}$$

$$\mathbf{S}_2^{(1)} = \sum_{apq} \sum_{lk} \xi_{va}^{pq}(l,k) \mathbf{C}_l(\hat{r}_1) \mathbf{C}_k(\hat{r}_2) a_p^\dagger a_q^\dagger a_a a_v. \tag{12b}$$

Here, $\xi_{\cdots}^{\cdots}$ represents the cluster amplitude for the operator $\mathbf{S}^{(1)}$. The one-body operator $\mathbf{S}_1^{(1)}$ is an odd parity operator and expressed in terms of a rank-one **C**-tensor. It satisfies the orbital-parity and orbital-triangular selection rules, $(-1)^{l_v+l_p} = -1$ and $|j_v - j_p| \leqslant 1 \leqslant (j_v + j_p)$, respectively. The tensor structure of the two-body operator $\mathbf{S}_2^{(1)}$ involves two **C**-tensors with ranks l and k associated with its two vertices. These two **C**-tensors are coupled to give a rank-one operator, $\mathbf{S}_2^{(1)}$. The allowed orbital-parity and and orbital-triangular selections rules for $\mathbf{S}_2^{(1)}$ are $(-1)^{l_v+l_p} = -(-1)^{l_a+l_q}$ and $|j_v - j_p| \leqslant l \leqslant (j_v + j_p)$, $|j_a - j_q| \leqslant k \leqslant (j_a + j_q)$, respectively. The diagrammatic representations of $\mathbf{T}^{(1)}$ and $\mathbf{S}^{(1)}$ are given in Fig. 2.

Using Eq. (11) in Eq. (10), we can rewrite the eigenvalue equation as

$$\begin{aligned}&\left(H^{\rm DC} + \lambda \mathbf{H}_{\rm elec}^{\rm NSD} \cdot \mathbf{I}\right) e^{T^{(0)}} \left[1 + \lambda \mathbf{T}^{(1)} \cdot \mathbf{I}\right] \left[1 + S^{(0)}\right. \\ &\left.+ \lambda \mathbf{S}^{(1)} \cdot \mathbf{I}\right] |\Phi_v\rangle = E_v e^{T^{(0)}} \left[1 + \lambda \mathbf{T}^{(1)} \cdot \mathbf{I}\right] \left[1 + S^{(0)}\right. \\ &\left.+ \lambda \mathbf{S}^{(1)} \cdot \mathbf{I}\right] |\Phi_v\rangle.\end{aligned} \tag{13}$$

Projecting this equation on $e^{-T^{(0)}}$ from left and retaining the terms linear in λ, we get

$$\begin{aligned}&\left[\bar{H}_{\rm N} \mathbf{S}^{(1)} + \bar{H}_{\rm N} \mathbf{T}^{(1)} (1 + S^{(0)}) + \bar{\mathbf{H}}_{\rm elec}^{\rm NSD} (1 + S^{(0)})\right] |\Phi_v\rangle \\ &\quad = \left[\Delta E_v \mathbf{S}^{(1)} + \Delta E_v \mathbf{T}^{(1)} (1 + S^{(0)})\right] |\Phi_v\rangle,\end{aligned} \tag{14}$$

where $\Delta E_v = E_v - \langle \Phi_v | H^{\rm DC} | \Phi_v \rangle$ is the correlation energy of the one-valence system. We have used the normal ordered Hamiltonian $H_{\rm N} = H^{\rm DC} - \langle \Phi_v | H^{\rm DC} | \Phi_v \rangle$. And, like $\bar{H}_{\rm N}$ introduced earlier, $\bar{\mathbf{H}}_{\rm elec}^{\rm NSD} = e^{-T^{(0)}} H_{\rm elec}^{\rm NSD} e^{T^{(0)}}$ is the similarity transformed NSD-PNC Hamiltonian in the electronic space. The PRCC equations of $\mathbf{S}_1^{(1)}$ and $\mathbf{S}_2^{(1)}$ can be obtained by projecting Eq. (14) with single and double excited determinants $\langle \Phi_v^p |$ and $\langle \Phi_{va}^{pq} |$, respectively, as

$$\langle \Phi_v^p | \{\overline{\bar{H}_{\rm N} \mathbf{S}^{(1)}}\} + \{\overline{\bar{H}_{\rm N} \mathbf{T}^{(1)}}\} + \{\overline{\bar{H}_{\rm N} \mathbf{T}^{(1)} S^{(0)}}\} + \bar{\mathbf{H}}_{\rm elec}^{\rm NSD} + \{\overline{\bar{\mathbf{H}}_{\rm elec}^{\rm NSD} S^{(0)}}\} | \Phi_v \rangle = E_v^{\rm att} \langle \Phi_v^p | \mathbf{S}_1^{(1)} | \Phi_v \rangle, \tag{15a}$$

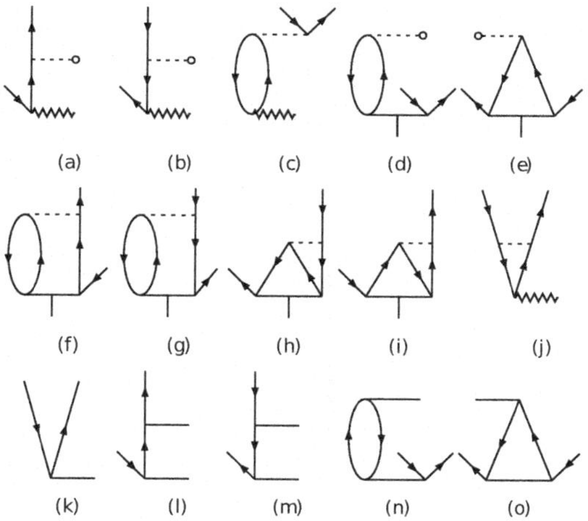

Fig. 3 The PRCC diagrams which contribute to the linearized perturbed CC equations of $\mathbf{T}_1^{(1)}$. The diagrams for $\mathbf{S}_1^{(1)}$ can be obtained by converting the core line to the valence line

$$\langle\Phi_{vb}^{pq}|\{\overline{\bar{H}_{\rm N}\mathbf{S}^{(1)}}\} + \{\overline{\bar{H}_{\rm N}\mathbf{T}^{(1)}}\} + \{\overline{\bar{H}_{\rm N}\mathbf{T}^{(1)}S^{(0)}}\} + \bar{\mathbf{H}}^{\rm NSD}_{\rm elec} + \{\overline{\bar{\mathbf{H}}^{\rm NSD}_{\rm elec}S^{(0)}}\}|\Phi_v\rangle = E_v^{\rm att}\langle\Phi_{vb}^{pq}|\mathbf{S}_2^{(1)}|\Phi_v\rangle. \quad (15b)$$

In deriving these equations we have used the relations: $\langle\Phi_v^p|\mathbf{T}^{(1)}|\Phi_v\rangle = 0$ and $\langle\Phi_v^p|\mathbf{T}^{(1)}S|\Phi_v\rangle = 0$. These follows, since $\mathbf{T}^{(1)}$ is an operator of closed-shell sector it does not contribute to the PRCC equation of $\mathbf{S}_1^{(1)}$ and $\mathbf{S}_2^{(1)}$. The closed-shell perturbed operators $\mathbf{T}^{(1)}$ are the solutions of the similar set of coupled equations [32, 38]

$$\langle\Phi_a^p|\{\overline{\bar{\mathbf{H}}_{\rm N}\mathbf{T}^{(1)}}\}|\Phi_0\rangle = -\langle\Phi_a^p|\bar{\mathbf{H}}^{\rm NSD}_{\rm elec} - \Delta E_0\mathbf{T}^{(1)}|\Phi_0\rangle, \quad (16a)$$

$$\langle\Phi_{ab}^{pq}|\{\overline{\bar{\mathbf{H}}_{\rm N}\mathbf{T}^{(1)}}\}|\Phi_0\rangle = -\langle\Phi_{ab}^{pq}|\bar{\mathbf{H}}^{\rm NSD}_{\rm elec} - \Delta E_0\mathbf{T}^{(1)}|\Phi_0\rangle. \quad (16b)$$

These equations are solved using Goldstone diagrams. The diagrams contributing to the linearized perturbed CC equations of $T_1^{(1)}$ and $T_2^{(1)}$ are shown in Figs. 3 and 4, respectively. We can also derive a similar set of PRCC equations for the NSI-PNC interaction Hamiltonian. One major difference for NSI-PNC is, however, the cluster operators are rank zero operators.

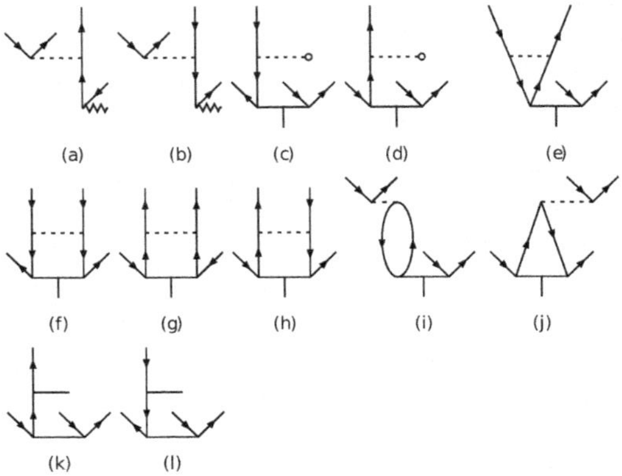

Fig. 4 The PRCC diagrams which contribute to the linearized perturbed CC equations of $\mathbf{T}_2^{(1)}$. The diagrams for $\mathbf{S}_2^{(1)}$ can be obtained by converting one of core-lines with valence line

2.3 Calculation of E1PNC Using PRCC Wavefunction

The perturbed wavefunction obtained by solving eigenvalue Eq. (7) is used to calculate the PNC-induced electric dipole transition amplitude (E1PNC). In PRCC theory, E1PNC is expressed as

$$\text{E1PNC} = \langle \widetilde{\Psi}_w \| \mathbf{D} \| \widetilde{\Psi}_v \rangle, \tag{17}$$

where **D** is the dipole operator. This expression, unlike the conventional sum-over-states approach [39], implicitly accounts for all the possible intermediate states. Using Eq. (11), we can write for NSD-PNC

$$\begin{aligned} E1^{\text{NSD}}_{\text{PNC}} &= \langle \Phi_w \| e^{T^{(0)\dagger}} \left[1 + \lambda \mathbf{T}^{(1)} \cdot \mathbf{I}\right]^\dagger \left[1 + S^{(0)} + \lambda \mathbf{S}^{(1)} \cdot \mathbf{I}\right]^\dagger \\ &\quad \mathbf{D} e^{T^{(0)}} \left[1 + \lambda \mathbf{T}^{(1)} \cdot \mathbf{I}\right] \left[1 + S^{(0)} + \lambda \mathbf{S}^{(1)} \cdot \mathbf{I}\right] \| \Phi_v \rangle. \end{aligned} \tag{18}$$

Retaining the terms linear in λ and CC operators up to second order, we can write the electronic component of E1PNC as

$$\begin{aligned} E1^{\text{NSD}}_{\text{elec}} \approx \langle \Phi_w \| &\mathbf{D}\mathbf{T}^{(1)} + T^{(0)\dagger}\mathbf{D}\mathbf{T}^{(1)} + \mathbf{T}^{(1)\dagger}\mathbf{D}T^{(0)} \\ &+ \mathbf{T}^{(1)\dagger}\mathbf{D} + \mathbf{D}\mathbf{T}^{(1)}S^{(0)} + \mathbf{T}^{(1)\dagger}S^{(0)\dagger}\mathbf{D} \\ &+ S^{(0)\dagger}\mathbf{D}\mathbf{T}^{(1)} + \mathbf{T}^{(1)\dagger}\mathbf{D}S^{(0)} + \mathbf{D}\mathbf{S}^{(1)} + \mathbf{S}^{(1)\dagger}\mathbf{D} \\ &+ S^{(0)\dagger}\mathbf{D}\mathbf{S}^{(1)} + \mathbf{S}^{(1)\dagger}\mathbf{D}S^{(0)} \| \Phi_v \rangle. \end{aligned} \tag{19}$$

We use diagrammatic techniques to identify the Goldstone diagrams contributing to these terms. The diagrams arising from higher order terms are expected to have negligible contributions. This we have shown in our previous calculation [37]. Some dominant PNC diagrams are given in Fig. 5.

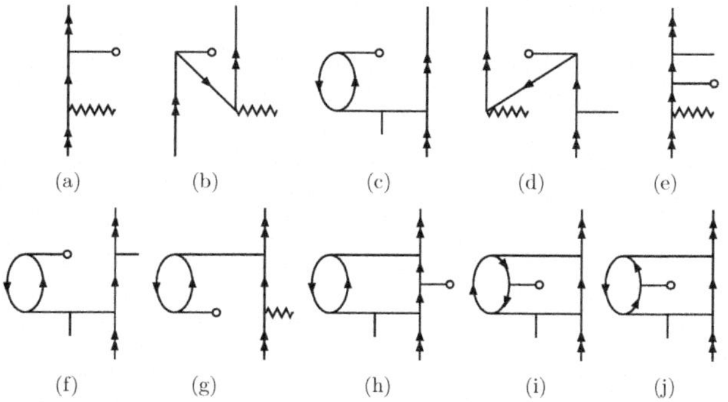

Fig. 5 Some dominant contributing diagrams to $E1^{\rm PNC}_{\rm elec}$ for one-valence systems

2.4 Coupling with Nuclear Spin

The PRCC diagrams of $E1^{\rm NSD}_{\rm elec}$ given in Fig. 5 are defined in the electronic subspace. To calculate $E1^{\rm NSD}_{\rm PNC}$, however, the nuclear spin part of the operator has to be coupled with $E1^{\rm NSD}_{\rm elec}$, and therefore, the $E1^{\rm NSD}_{\rm PNC}$ diagrams must include a vertex which operates on the nuclear spin states. This coupling with nuclear part of the operator is done through the angular momentum algebra using the conventions used by Lindgren and Morrison [40]. The general idea is to reduce the $E1^{\rm NSD}_{\rm elec}$ diagrams to a common form and then couple with the nuclear part.

To elaborate on the coupling schemes, we discuss a specific case where we consider a simple diagram involving the single PRCC operator. For this, we choose diagram in Fig. 5(a). This arises from the CC operator $\mathbf{S}_1^{(1)}$ and, as contains the Dirac-Fock contribution, it is the most dominant $E1^{\rm NSD}_{\rm elec}$ diagram. The angular momentum diagram showing coupling with nuclear part is given in Fig. 6. In the diagram, j_v and j_w are the total angular momentum of the single-particle states for initial and final atomic states, respectively. The line k_1 is a rank-one multipole line which denotes the angular part of the dipole operator. The line k_2 is also a rank-one multipole line in electronic as well as the nuclear sectors. In the electronic part, it represents the angular part of the operator which connects to a transition from j_v to j_p, whereas in the nuclear sector it represents the operator $\mathbf{I}$. Since $\mathbf{I}$ is a diagonal operator the k_2 line does not change the nuclear spin. The electronic and nuclear angular momenta, j_v, j_w, and $\mathbf{I}$, respectively, are then coupled to give hyperfine states, $|F_v\rangle$ and $|F_w\rangle$ (shown in Fig. 6). Using the Wigner-Eckart theorem, we can write the algebraic expression of the diagram in Fig. 6 as

$$
\begin{aligned}
\langle F_w m_w | D\mathbf{S}_1^{(1)} \cdot \mathbf{I} | F_v m_v \rangle = {} & (-1)^{F_w - m_w} \\
& \times \begin{pmatrix} F_w & 1 & F_v \\ -m_w & q & m_v \end{pmatrix} \\
& \times \langle F_w || D_{\rm eff} || F_v \rangle,
\end{aligned}
\tag{20}
$$

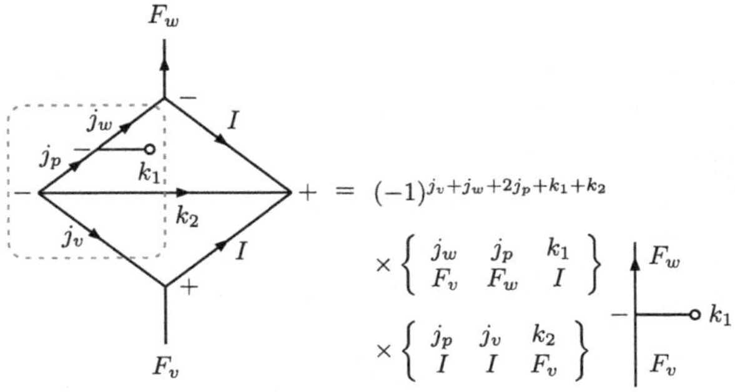

Fig. 6 Angular factor calculation of the $E1^{\text{NSD}}_{\text{PNC}}$ diagram shown in Fig. 5(a). The portion within the rectangled dash line indicates the electronic part. Remaining portion is from the nuclear spin and electronic angular momentum coupling

where $D_{\text{eff}} = D\mathbf{S}_1^{(1)} \cdot \mathbf{I}$ is a rank one operator, m_i is the hyperfine magnetic quantum numbers, and q is the component of D_{eff}. The phase factor and 3j-symbol in the above equation correspond to the free lines on the right-hand side of the diagram in Fig. 6. The remaining expression on the right-hand side, phase factor and 6j-symbols, is the m-independent angular component and corresponds to the reduced matrix element of D_{eff}.

3 Results and Discussions

3.1 Single-Particle Basis

The single-particle basis which provides a good description of the single-electron wavefunctions and energies is essential to obtain the accurate and reliable values of PNC and other atomic properties using PRCC theory. In the present work, for all calculations, we have used Gaussian-type orbitals (GTOs). The single-particle orbitals used in the present calculations are of the form

$$\psi_{n\kappa m}(\boldsymbol{r}) = \frac{1}{r} \begin{pmatrix} P_{n\kappa}(r)\, \chi_{\kappa m}(\hat{\boldsymbol{r}}) \\ i\, Q_{n\kappa}(r)\, \chi_{-\kappa m}(\hat{\boldsymbol{r}}) \end{pmatrix}, \tag{21}$$

where $P_{n\kappa}(r)$ and $Q_{n\kappa}(r)$ are the large and small component radial wave functions, κ is the relativistic total quantum number, and $\chi_{\kappa m}(\hat{\boldsymbol{r}})$ are the spinor-spherical harmonics. The radial components are defined as the linear combination of GTOs [41, 42] as

$$\begin{aligned} P_{n\kappa}(r) &= \sum_p C^L_{\kappa p} g^L_{\kappa p}(r), \\ Q_{n\kappa}(r) &= \sum_p C^S_{\kappa p} g^S_{\kappa p}(r). \end{aligned} \tag{22}$$

Table 1 The orbital energies (in Hartree) from GTO using the Dirac-Coulomb Hamiltonian compared with the GRASP2K [43] and B-Spline [44] results. The optimized α_0 and β parameters of the even tempered GTO basis used in our calculations are also listed

Orbitals	GTO	GRASP2K	B-Spline
$1s_{1/2}$	−144.93621	−144.93621	−144.93622
$2s_{1/2}$	−31.25439	−31.25439	−31.25439
Parameter	s	p	d
α_0	0.00549	0.00577	0.00504
β	1.905	1.892	1.955

Here, the index $p = 1, 2, \ldots, m$, where m is the number of basis functions and $C^{\cdots}_{\kappa p}$ are the coefficients of linear combination. For large components, we choose

$$g^L_{\kappa p}(r) = C^L_{m_{\kappa i}} r^{n_\kappa} e^{-\alpha_p r^2}, \tag{23}$$

where n_κ is an integer and $C^L_{m_{\kappa i}}$ is the normalization constant. The small components are derived from the large components using kinetic balance condition. The α_p in the above equation satisfies the relation

$$\alpha_p = \alpha_0 \beta^{p-1}, \tag{24}$$

where the parameters α_0 and β are optimized to reproduce the single-particle energies and wavefunctions of the core and valence orbitals from the numerical results obtained by GRASP2K [43]. In Table 1, we have compared the single-particle energy from GTO with the GRASP2K and B-Spline [44] data. As evident from the table, our computed energy of GTOs is in excellent agreement with numerical results. The relative difference is of the order of 10^{-5} Hartree or smaller. The optimized α_0 and β parameters are also listed in the table for reference and use in future calculations.

For nuclear potential, we have used the finite size Fermi density distribution of nucleus, expressed as

$$\rho_{\rm nuc}(r) = \frac{\rho_0}{1 + e^{(r-c)/a}}, \tag{25}$$

where $a = t4 \ln 3$. The parameter c is the half-charge radius, that is $\rho_{\rm nuc}(c) = \rho_0/2$ and t is the skin thickness.

3.2 *Electric Dipole Polarizability*

Here we present and analyze our results on dipole polarizability of the ground state and first excited state of Ar^{13+}. To compute α we have used the PRCC theory for NSD-PNC discussed above. For α, in PRCC theory, the contributing diagrams will be topologically the same as those for NSD-PNC. The only difference is the PNC operator will be replaced with a dipole operator. The results on α from our calculation are tabulated in the Table 2. To quantify the electron correlations from various stages of theory, we have provided the term-wise contributions. And for comparison, we have listed the data from other theoretical calculations. To the best of our knowledge, in the literature, there are no experimental results on the polarizability of the ground state and first excited state of Ar^{13+}.

Examining the nature of electron correlations, as can be seen from the table, one common observation is that the terms with close-shell CC operators, $T^{(0)}$ and $\mathbf{T}^{(1)}$, has more contributions than those with open-shell operators, $S^{(0)}$ and $\mathbf{S}^{(1)}$. This trend of contribution to α is very different than in neutral atoms observed in our previous calculation [35], where, for one-valence systems, the terms with operator $\mathbf{S}^{(1)}$ are found to contribute more dominantly. Among all the terms, for both the states, the term $DT^{(1)}$ + H.c. has the dominant contribution. It contributes $\approx 85\%$ and 81% of the total α value for ${}^2P_{1/2}$ and ${}^2P_{3/2}$ states, respectively. The next leading order term is $T^{(0)\dagger}DS^{(1)}$ + H. c. for both states. It contributes $\approx 19\%$ and 21%, respectively, of the total value for ${}^2P_{1/2}$ and ${}^2P_{3/2}$ states. The next dominant contributions come from the terms $T^{(0)\dagger}DT^{(1)}$ + H. c. and $S^{(0)\dagger}DT^{(1)}$ + H. c. They have the similar but opposite contributions of ≈ -10 and 9%, respectively, for ${}^2P_{1/2}$ state. For ${}^2P_{3/2}$ state, they contribute ≈ -9 and 9%, respectively. The terms with only open-shell CC operators contribute smaller value to α.

Comparing our results with the previous literature data, to the best of our knowledge, there is only one result, Ref. [45], from the previous theoretical calculations. The method used in Ref. [45] is similar to ours. As evident from the table, our computed value of α is in good agreement with the value in Ref. [45] for both the states. The reason for small difference could be attributed to the use of larger basis in the present calculation and the inclusion of Breit and QED corrections in Ref. [45]. In the present work, we have used a large basis of 157 ($23s19p17d11f11g9h$) orbitals comprising of orbitals up to $h-$symmetry.

3.3 M1 Transition Amplitudes

Next we present our results on M1 transition amplitudes and rates for $2p_{1/2} \rightarrow 2p_{3/2}$ transition in Ar^{13+}. As mentioned earlier, this transition is important from the atomic clock viewpoint. In Table 3, we have given the M1 reduced matrix element for $2p_{1/2} \rightarrow 2p_{3/2}$ transition. To understand the nature of correlation effects from different terms in RCC, these are listed separately. We have also given an exhaustive list of previous calculations for comparison.

As we observed from the table, the most dominant contribution is from the DF. It contributes $\approx 96\%$ of the total value. It is to be noticed that this trend of contribution is very different than in the case of α. The reason for this is the topologically different DF diagram, as here, unlike the case in α, there is no external perturbation. For

Table 2 Electric dipole polarizability (in a. u.) of the ground state, $^2P_{1/2}$, and first excited state, $^2P_{3/2}$, of Ar^{13+} using RCC theory. For assessing electron correlations, we have listed the dominant contributions from different PRCC terms separately

Term + H. c.	$^2P_{1/2}$	$^2P_{3/2}$
Dirac-Fock	−0.01204	−0.01226
$DS^{(1)}$	0.00753	0.00801
$DT^{(1)}$	0.10196	0.09101
$S^{(0)\dagger}DS^{(1)}$	0.00013	0.00147
$T^{(0)\dagger}DT^{(1)}$	−0.01232	−0.01046
$T^{(0)\dagger}DS^{(1)}$	0.02191	0.02353
$S^{(0)\dagger}DT^{(1)}$	0.01125	0.01094
Normalization	0.00014	0.00006
Total	0.11856	0.11229
Others work	0.1205[a]	0.1200[a]

[a]Ref. [45]

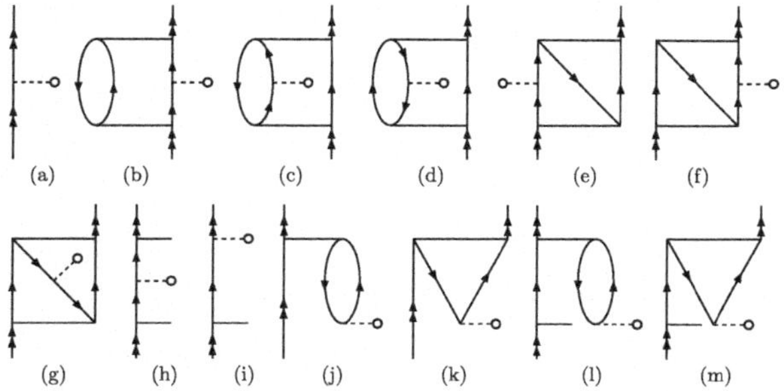

Fig. 7 Some dominant contributing RCC diagrams to $M1$ transition amplitudes. Diagram (**a**) is the leading order Dirac-Fock contribution

M1, the DF contribution is just the bare M1 matrix element and is represented as diagram (a) in Fig. 7. For α, however, the DF diagram is topologically equivalent to Fig. 5(a), with a difference that the operator $\mathbf{S}_1^{(1)}$ is computed up to only first-order in perturbation. The next leading order contribution is from the $T^{(0)\dagger}M1T^{(0)}$ term. It contributes about 4% of the total value. The term with only open-shell CC operators, $S^{(0)\dagger}M1S^{(0)}$, gives the next dominant contribution. The terms with only $S^{(0)}$ or $T^{(0)}$ operator have negligible contributions.

Comparing our results with previous theoretical calculations, except Ref. [45], all other results, Refs. [46–50], are using the multiconfiguration Dirac-Fock (MCDF) and its variations. Based on its mathematical formulation, in terms of accounting for the electron correlation effects in many-body calculations, MCDF is a less accurate method than RCC. And therefore, as to be expected, the results reported from these calculations are very close to each other. Compared with RCC calculation [45], our

Table 3 Magnetic dipole (M1) transition reduced matrix elements (in a.u.) for $2p\,^2P_{1/2} \rightarrow 2p\,^2P_{3/2}$ transition using RCC theory. To assess the nature of electron correlation, some dominant contributions in RCC are also listed

Terms + H. c.	Contribution
Dirac-Fock	1.15340
$M1S^{(0)}$	−0.00005
$T^{(0)\dagger}M1T^{(0)}$	0.04297
$S^{(0)\dagger}M1S^{(0)}$	0.00289
$S^{(0)\dagger}M1T^{(0)}$	−0.00015
Total	1.19997
Others work	1.15424[a], 1.15634[b], 1.15637[c]
	1.15625[d], 1.15628[e], 1.15628[f]

[a]Ref. [45]—RCC, [b]Ref. [46]—MCDHF, [c]Ref. [47]—MCDHF, [d]Ref. [48]—MCDFS, [e]Ref. [49]—MCDF, [f]Ref. [50]—MCDF

Table 4 Transition rate (in s^{-1}) for $2p_{1/2} \rightarrow 2p_{3/2}$ transition using RCC theory

Transition	Our Cal.	Other works	Exp.
$2p_{1/2} \rightarrow 2p_{3/2}$	104.3 (DF)	101[a], 103[b], 104[c], 104.1[d]	110[j], 114.9[k]
	112.9 (RCC)	106.8[e], 105.2[f], 104.5[g],	104[l]
		104.4[i], 104.9[h]	

[a]Ref. [51], [b]Ref. [52]—MCDF+CI [c]Ref. [53],
[d]Ref. [21]—MCDF [e]Ref. [54], [f]Ref. [55]—MCHF
[g]Ref. [49]—RCI, [h]Ref. [46]—MCDHF
[i]Ref. [56]—MRRCI (Multireference Relativistic Configuration Interaction),
[j]Ref. [57]—Exp., [k]Ref. [58]—Exp., [l]Ref. [59]—NIST

result is slightly on the higher side. The reason for this could be attributed to the use of relatively large basis size in our calculations and missing contributions from Breit and QED corrections. Our DF value is, however, in good agreement with all the reported values.

In Table 4, we have provided the transition rate for $2p_{1/2} \rightarrow 2p_{3/2}$ transition from our calculation and have compared that with the previous theoretical and experimental results. We have used the experimental energy in the calculation of transition rate. Though most of the theoretical results are from the MCDF and based calculations, there is a variation in the reported values. The smallest value, 101 [51], is about 6% larger than the highest value, 106.8 [54], listed in the table. The same trend is also observed in the experimental data. There is a variation of about 6% between the smallest, Ref. [60], and largest value, Ref. [57]. Our RCC value is the largest among all the previous theoretical calculations. The reason for this could be attributed to the use of a large basis in our calculation to account for electron correlations from higher energy electrons more accurately. Our DF transition rate, 104.32, is in excellent agreement with the previous theoretical results. Our RCC value, 112.9, is in good agreement to experiments [57, 58].

4 Conclusions

We have developed a perturbed relativistic coupled-cluster theory to compute the observables for parity nonconservation in atomic systems, which we have demonstrated in the context of one-valence systems. The theory we have developed is very general and could be easily modified to calculate other atomic properties just by changing the perturbation interaction Hamiltonian. For benchmark test calculations, by replacing PNC operator with a dipole operator, we have computed the electric dipole polarizability of ground state $2p_{1/2}$ and first excited state $2p_{3/2}$ of Ar^{13+}. Our computed α is in very good agreement with the previous calculation. Further, as a probe to atomic clock properties, we have computed M1 transition amplitude and corresponding transition rate for $2p_{1/2} \rightarrow 2p_{3/2}$ clock transition. Our calculated value of transition rate is in excellent agreement with the experimental results [57, 58]. Based on the sources of uncertainty discussed in detail in our previous calculations [32–35], the maximum uncertainty in the results presented in the present work is 1%.

Acknowledgements The authors wish to thank Ravi Kumar for the useful discussions. B. K. M acknowledges the funding support from the SERB (ECR/2016/001454). The results presented in the paper are based on computations using the High Performance Computing cluster, Padum, at the Indian Institute of Technology Delhi, New Delhi.

References

1. L. Labzowsky, A. Prozorov, Parity nonconservation effects in the highly charged ions. J. Phys.: Conf. Ser. **72**, 012010 (2007) (IOP Publishing)
2. M.G. Kozlov, M.S. Safronova, J.R. Crespo López-Urrutia, P.O. Schmidt, Highly charged ions: optical clocks and applications in fundamental physics. Rev. Mod. Phys. **90**, 045005 (2018)
3. A. Ong, Highly charged ions and the search for the variation of fundamental constants. J. Proc. R. Soc. N. S. W. **146**, 136–141 (2013)
4. I.B. Khriplovich, *Parity Nonconservation in Atomic Phenomena* (Gordon and breach, New York, 1991)
5. J. Beacham, C. Burrage, D. Curtin, A. De Roeck, J. Evans, J.L. Feng, C. Gatto, S. Gninenko, A. Hartin, I. Irastorza, et al., Physics beyond colliders at CERN: beyond the standard model working group report. J. Phys. G: Nucl. Part. Phys. **47**, 010501 (2019)
6. S.G. Porsev, K. Beloy, A. Derevianko, Precision determination of weak charge of ^{133}Cs from atomic parity violation. Phys. Rev. D **82**, 036008 (2010)
7. S.L. Gilbert, C.E. Wieman, Atomic-beam measurement of parity nonconservation in cesium. Phys. Rev. A **34**, 792 (1986)
8. M.G. Kozlov, S.G. Porsev, W.R. Johnson, Parity nonconservation in thallium. Phys. Rev. A **64**, 052107 (2001)

9. V.A. Dzuba, V.V. Flambaum, P.G. Silvestrov, O.P. Sushkov, Calculation of parity nonconservation in thallium. J. Phys. B: At. Mol. Phys. **20**, 3297 (1987)
10. P.A. Vetter, D.M. Meekhof, P.K. Majumder, S.K. Lamoreaux, E.N. Fortson, Precise test of electroweak theory from a new measurement of parity nonconservation in atomic thallium. Phys. Rev. Lett. **74**, 2658 (1995)
11. V.A. Dzuba, V.V. Flambaum, P.G. Silvestrov, O.P. Sushkov, Relativistic many-body calculations of parity nonconservation in lead and bismuth atoms. Europhys. Lett. **7**, 413 (1988)
12. J.H. Hollister, G.R. Apperson, L.L. Lewis, T.P. Emmons, T.G. Vold, E.N. Fortson, Measurement of parity nonconservation in atomic bismuth. Phys. Rev. Lett. **46**, 643 (1981)
13. D.M. Meekhof, P. Vetter, P.K. Majumder, S.K. Lamoreaux, E.N. Fortson, High-precision measurement of parity nonconserving optical rotation in atomic lead. Phys. Rev. Lett. **71**, 3442 (1993)
14. D. Antypas, A. Fabricant, J.E. Stalnaker, K. Tsigutkin, V.V. Flambaum, D. Budker, Isotopic variation of parity violation in atomic ytterbium. Nat. Phys. **15**, 120–123 (2019)
15. M. Maul, A. Schäfer, W. Greiner, P. Indelicato, Prospects for parity-nonconservation experiments with highly charged heavy ions. Phys. Rev. A **53**, 3915 (1996)
16. M. Zolotorev, D. Budker, Parity nonconservation in relativistic hydrogenic ions. Phys. Rev. Lett. **78**, 4717 (1997)
17. A. Surzhykov, A.V. Maiorova, V.M. Shabaev, T. Stöhlker, S. Fritzsche, Parity violation in beryllium-like heavy ions. Phys. Scr. **2013**, 014027 (2013)
18. A.D. Ludlow, M.M. Boyd, J. Ye, E. Peik, P.O. Schmidt, Optical atomic clocks. Rev. Mod. Phys. **87**, 637 (2015)
19. F. Riehle, P. Gill, F. Arias, L. Robertsson, The CIPM list of recommended frequency standard values: guidelines and procedures. Metrologia **55**, 188 (2018)
20. M.S. Grewal, A.P. Andrews, C.G. Bartone, *Global Navigation Satellite Systems, Inertial Navigation, and Integration* (Wiley, 2020)
21. C.Z. Dong, S. Fritzsche, B. Fricke, W.D. Sepp, Ab-initio calculations for forbidden M1 transitions in Ar^{13+} and Ar^{14+} ions. Phys. Scr. **2001**, 294 (2001)
22. J. White, R. Beard, Space clocks-why they're different, in *Proceedings of the 33th Annual Precise Time and Time Interval Systems and Applications Meeting* (2001), pp. 7–17
23. N.F. Ramsey, Application of atomic clocks, in *Laser Pysics at the Limits* (2002), pp. 3–8
24. V.A. Dzuba, V.V. Flambaum, Highly charged ions for atomic clocks and search for variation of the fine structure constant, in *TCP 2014: Proceedings of the 6th International Conference on Trapped Charged Particles and Fundamental Physics, held in Takamatsu, Japan, 1–5, Dec 2014* (Springer, Berlin, 2015) pp. 79–86
25. D.K. Nandy, B.K. Sahoo, Highly charged W^{13+}, Ir^{16+}, and Pt^{17+} ions as promising optical clock candidates for probing variations of the fine-structure constant. Phys. Rev. A **94**, 032504 (2016)
26. A. Ong, J.C. Berengut, V.V. Flambaum, Optical transitions in highly charged ions for detection of variations in the fine-structure constant. Fundam. Phys. Part. Traps 293–314 (2014)
27. M.S. Safronova, The search for variation of fundamental constants with clocks. Annalen der Physik **531**, 1800364 (2019)
28. S.A. King, L.J. Spieß, P. Micke, A. Wilzewski, T. Leopold, E. Benkler, R. Lange, N. Huntemann, A. Surzhykov, V.A. Yerokhin, et al., An optical atomic clock based on a highly charged ion. Nature **611**, 43–47 (2022)
29. H. Bekker, A. Borschevsky, Z. Harman, C.H. Keitel, T. Pfeifer, P.O. Schmidt, J.R. Crespo López-Urrutia, J.C. Berengut, Detection of the 5 p–4 f orbital crossing and its optical clock transition in Pr^{9+}. Nat. Commun. **10**, 5651 (2019)
30. S.-Y. Liang, T.-X. Zhang, H. Guan, Q.-F. Lu, J. Xiao, S.-L. Chen, Y. Huang, Y.-H. Zhang, C.-B. Li, Y.-M. Zou, et al., Probing multiple electric-dipole-forbidden optical transitions in highly charged nickel ions. Phys. Rev. A **103**, 022804 (2021)
31. B.K. Mani, S. Chattopadhyay, D. Angom, RCCPAC: a parallel relativistic coupled-cluster program for closed-shell and one-valence atoms and ions in FORTRAN. Comput. Phys. Commun. **213**, 136–154 (2017)

32. R. Kumar, S. Chattopadhyay, B.K. Mani, D. Angom, Electric dipole polarizability of group-13 ions using perturbed relativistic coupled-cluster theory: importance of nonlinear terms. Phys. Rev. A **101**, 012503 (2020)
33. R. Kumar, S. Chattopadhyay, D. Angom, B.K. Mani, Fock-space relativistic coupled-cluster calculation of a hyperfine-induced $^1S_0 \rightarrow ^3P_0$ clock transition in Al^+. Phys. Rev. A **103**, 022801 (2021)
34. R. Kumar, S. Chattopadhyay, D. Angom, B.K. Mani, Relativistic coupled-cluster calculation of the electric dipole polarizability and correlation energy of CN, Nh^+, and OG: correlation effects from lighter to superheavy elements. Phys. Revi. A **103**, 062803 (2021)
35. R. Kumar, D. Angom, B.K. Mani, Fock-space perturbed relativistic coupled-cluster theory for electric dipole polarizability of one-valence atomic systems: application to al and in. Phys. Rev. A **106**, 032801 (2022)
36. J. Sucher, Foundations of the relativistic theory of many-electron atoms. Phys. Rev. A **22**, 348–362 (1980)
37. B.K. Mani, D. Angom, Atomic properties calculated by relativistic coupled-cluster theory without truncation: hyperfine constants of Mg^+, Ca^+, Sr^+, and Ba^+. Phys. Rev. A **81**, 042514 (2010)
38. B.K. Mani, K.V.P. Latha, D. Angom, Relativistic coupled-cluster calculations of ^{20}Ne, ^{40}Ar, ^{84}Kr, and ^{129}Xe: Correlation energies and dipole polarizabilities. Phys. Rev. A **80**, 062505 (2009)
39. G. Gopakumar, B.P. Das, R.K. Chaudhuri, D. Mukherjee, K. Hirao, Relativistic coupled-cluster calculations of parity nonconservation in Ba^+ by the sum-over-states approach. J. Chem. Phys. **126**, 014301 (2007)
40. I. Lindgren, J. Morrison, *Atomic Many-Body Theory*, 2nd edn. (Springer, Berlin, 1986)
41. A.K. Mohanty, E. Clementi, Kinetically balanced geometric gaussian basis set calculations for relativistic many-electron atoms with finite nuclear size. Chem. Phys. Lett. **157**, 348–352 (1989)
42. R.K. Chaudhuri, P.K. Panda, B.P. Das, Hybrid approach to relativistic gaussian basis functions: theory and applications. Phys. Rev. A **59**, 1187–1196 (1999)
43. P. Jönsson, G. Gaigalas, J. Bieroń, C. Froese Fischer, I.P. Grant, New version: grasp2k relativistic atomic structure package. Comp. Phys. Commun. **184**, 2197–2203 (2013)
44. O. Zatsarinny, C.F. Fischer, DBSR_HF: a B-spline Dirac-Hartree-Fock program. Comput. Phys. Commun. **202**, 287–303 (2016)
45. Y.-M. Yu, B.K. Sahoo, Investigating ground-state fine-structure properties to explore suitability of boronlike $S^{11+} - K^{14+}$ and galliumlike $Nb^{10+} - Ru^{13+}$ ions as possible atomic clocks. Phys. Rev. A **99**, 022513 (2019)
46. M. Bilal, A.V. Volotka, R. Beerwerth, S. Fritzsche, Line strengths of GED-sensitive forbidden transitions in B-, Al-, F- and Cl-like ions. Phys. Rev. A **97**, 052506 (2018)
47. C.F. Fischer, I.P. Grant, G. Gaigalas, P. Rynkun, Lifetimes of some $2s^2 2p^2 P_{3/2}$ states from variational theory. Phys. Rev. A **93**, 022505 (2016)
48. I.I. Tupitsyn, A.V. Volotka, D.A. Glazov, V.M. Shabaev, G. Plunien, J.R. Crespo López-Urrutia, A. Lapierre, J. Ullrich, Magnetic-dipole transition probabilities in B-like and Be-like ions. Phys. Rev. A **72**, 062503 (2005)
49. P. Rynkun, P. Jönsson, G. Gaigalas, C.F. Fischer, Energies and E1, M1, E2, M2 transition rates for states of the $2s^2 2p$, $2s2p^2$, and $2p^3$ configurations in boron-like ions between N iii and Zn xxvi. Atomic Data Nucl. Data Tables **98**, 481–556 (2012)
50. J.P. Marques, P. Indelicato, F. Parente, Relativistic multiconfiguration calculations of the $2s^2 2p$ $^2P_{3/2}$ level lifetime along the boron isoelectronic sequence. Eur. Phys. J. D, Atomic, Mol., Opt. Plasma Phys. **66** (2012)
51. S.W. Rachedi, M.K. Inal, M. Benmouna, Density dependence of intensity ratios of the 3d→2p x-ray anisotropic emission from electron-beam excited Ar^{13+} ions. J. Phys. B: At. Mol. Opt. Phys. **53**, 185701 (2020)
52. K.M. Aggarwal, F.P. Keenan, S. Nakazaki, Energy levels and radiative rates for transitions in Ar xiii, Ar xiv and Ar xv. Astron. Astrophys. **436**, 1141–1150 (2005)

53. G. Del Zanna, K.P. Dere, P.R. Young, E. Landi, H.E. Mason, Chianti-an atomic database for emission lines, version 8. Astron. Astrophys. **582**, A56 (2015)
54. A.K. Bhatia, U. Feldman, J.F. Seely, Atomic data and spectral line intensities for the beryllium isoelectronic sequence (Ar xv through Kr xxxiii). At. Data Nucl. Data Tables **35**, 449–472 (1986)
55. C.F. Fischer, Multiconfiguration hartree-fock breit-pauli results for $^2P_{1/2}$-$^2P_{3/2}$ transitions in the boron sequence. J. Phys. B: Atomic Mol. Phys. **16**, 157 (1983)
56. K. Koc, Relativistic multireference configuration interaction calculations of lifetime of $2s^22p$ $^2P_{3/2}$ level along boron isoelectronic sequence. Eur. Phys. J. D **53**, 9–14 (2009)
57. D.P. Moehs, D.A. Church, Magnetic dipole transition rates from measured lifetimes of levels of Be-like and B-like argon ions. Phys. Rev. A **58**, 1111–1114 (1998)
58. F.G. Serpa, J.D. Gillaspy, E. Träbert, Lifetime measurements in the ground configuration of and using an electron beam ion trap. J. Phys. B: At. Mol. Opt. Phys. **31**, 3345 (1998)
59. A. Kramida, Y. Ralchenko, J. Reader, NIST ASD Team, NIST atomic spectra database (vertion 5.10) [Online]. https://physics.nist.gov/asd [2023, March 8]. National Institute of Standards and Technology, Gaithersburg, MD (2022)
60. K. Koc, Multireference relativistic configuration interaction calculations for $2s^22p$ $^2P_{3/2}$ - $^2P_{1/2}$ M1 and E2 transitions in boron isoelectronic sequence. J. Phys. B: At. Mol. Opt. Phys. **36**, L93 (2003)

Intermolecular Coulombic Decay: Geometric and Electronic Structures of Ionized Water

Philippe F. Weck and Eunja Kim

Intermolecular Coulombic decay (ICD) in liquid water is a relatively novel type of nonlocal electronic decay mechanism, competing with the traditional mechanism of proton transfer between neighboring water molecules. Key features of ICD are its ultrafast non-radiative decay process and ultralong-range for excess energy transfer from the excited atom/molecule to its neighbors. Since detecting unambiguous ICD signatures in bulk liquid water is technically challenging, small water clusters have often been utilized to gain insights into ICD and other ionization processes in aqueous environment. Here, we present results from quantum mechanical calculations of the electronic structures of neutral to multiply-ionized water monomer, dimer, trimer, and tetramer. Core-level electrons of water are also considered here since recent studies demonstrated that emission site and energy of the electrons released during resonant-Auger-ICD cascade can be controlled by coupling ICD to resonant core excitation. Previous studies of ICD and electronic structures of neutral and ionized small water clusters and liquid water are briefly discussed.

1 Introduction

Energy and charge transfer in liquid water has been the subject of widespread research in contemporary science and technology owing to the ubiquity of water on Earth. Water in the liquid state is a well-known transporter of its autoionization products, i.e., protons (H^+) and hydroxide ions (OH^-), and as such is often a central component

P. F. Weck (✉)
Sandia National Laboratories, Albuquerque, NM 87185, USA
e-mail: pfweck@sandia.gov

E. Kim
Department of Physics, The University of Texas at El Paso, El Paso, TX 79968, USA
e-mail: ekim4@utep.edu

L. C. Tribedi (ed.), *Advances in Atomic Molecular Collisions*,
https://doi.org/10.1007/978-981-97-7063-2_12

in numerous applications ranging from modern electrochemical energy production and storage technologies to biomembranes and biochemical processes [1–3]. Since the seminal work of de Grotthuss on charge transfer in liquid water via proton hopping [4], investigations of water structure and dynamics have revealed the complexity of charge transfer in liquid water, often proceeding through the formation of highly excited OH radicals and hydronium cations H_3O^+ or Eigen-type ($H_9O_4{}^+$) or Zundel-type ($H_5O_2{}^+$) cations [5, 6]. The intricate weak and directional hydrogen-bonded network of liquid water and its rearrangement with charge transfer is responsible for numerous anomalous properties in this medium [7–10].

Competing with this traditional mechanism of proton transfer between neighboring water molecules, intermolecular Coulombic decay (ICD) in liquid water is a relatively novel type of nonlocal electronic decay mechanism [11], originally proposed in 1997 by Cederbaum et al. in weakly bonded matter such as van der Waals clusters and hydrogen-bonded systems [12]. Key features of ICD are its ultrafast non-radiative decay process and ultralong-range for excess energy transfer from the excited atom/molecule to its neighbors [13]. In the case of ICD in a water dimer $(H_2O)_2$, for example, after (photo)ionization of an inner-valence-shell electron of one of the water molecules, ICD is initiated upon filling this inner-valence vacancy with an outer-valence electron of the same molecule, with the excess energy gained in this electronic relaxation transferred through a virtual photon mechanism to the neighboring water molecule, which is subsequently ionized. As a result of ICD, a low-energy electron is ejected from this water molecule neighbor and the water dimer eventually dissociates into a pair of water cations by Coulomb explosion. For a water dimer, this sequence of processes from initial photoionization to final fragmentation can be summarized prototypically as follows:

$$H_2O \cdots H_2O \xrightarrow{h\nu} H_2O \cdots \left(H_2O^+ + \overline{e}\right) \rightarrow \left(\overline{e} + H_2O^+\right) \cdots H_2O^+ \rightarrow H_2O^+ + H_2O^+,$$

and is represented schematically in Fig. 1. Let us note that, as a consequence of initial inner-valence ionization, emission of a secondary ICD electron is only possible if the inner-valence ionized state is energetically above the double ionization threshold of the system.

Although ICD can be described in much the same way as Auger decay, a major difference with this local autoionization decay process is the nonlocal nature of ICD. In this respect, ICD belongs to the same class of nonlocal electronic decay mechanisms within loosely bound systems as Electron Transfer Mediated Decay (ETMD), Intermolecular Coulombic Electron Capture (ICEC), and Collective AutoIonization (CAI) with electron emission; or Förster Resonant Energy Transfer (FRET), and Dexter (electron exchange) energy transfer with intersystem excitation; or (Radiative) Charge Transfer ((R)CT) [14]. Closely related to IDC, Penning ionization is another nonlocal ionization mechanism, which occurs by collisional interaction [14].

The first successful observations of ICD in water clusters were achieved in 2010 using coincidence detection techniques [15, 16], due to the difficulty in obtaining unambiguous results with simple electron spectroscopy experiments [17]. A COLd

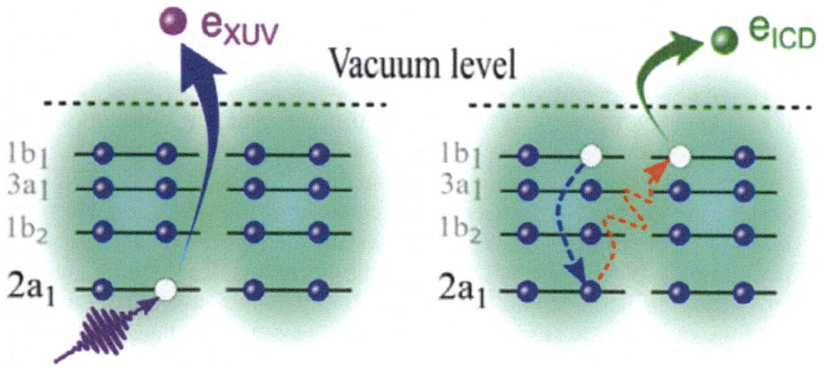

Fig. 1 Schematic representation of intermolecular Coulombic decay (ICD) process initiated by inner-valence ionization of $2a_1$ orbital in water dimer $(H_2O)_2$. *Left*: The inner-valence vacancy is created by the absorption of an extreme-ultraviolet (XUV) photon and ejection of a $2a_1$ inner-valence electron; *Right*: an outermost $1b_1$ valence electron fills up the inner-valence vacancy via releasing the extra energy to the neighboring water molecule causing a further ionization with ejection of an ICD electron from its outer-valence $1b_1$ orbital [adapted from Ref. [20]; Creative Commons Attribution 4.0 International license (CC BY 4.0), American Physical Society]

Target Recoil Ion Momentum Spectroscopy (COLTRIMS) experiment on ICD in water dimers was successfully conducted by Janhke et al. to avoid issues with background of scattered secondary electrons [15]. The ultrafast nature of the ICD mechanism in $(H_2O)_2$ across the hydrogen bridge was demonstrated by the creation of a cationic H_2O^+ pair upon dimer dissociation, compared to typical protonated fragments $(H_2O)_nH^+$ and OH groups detected in previous water cluster ionization experiments [18, 19]. Concurrently, Mucke et al. reported electron-pairs ICD signatures in electron–electron-coincidence spectra of free water cluster jets, featuring water clusters with mean sizes of 45 and 200 molecules [16]. Recently, in 2022, the first observation of ICD in bulk liquid water following inner-valence ionization was reported [20]. This latest experimental observation provided confirmation of the importance of ICD as a source of low-energy electrons in bulk liquid water and yielded quantitative information on the velocity distribution of slow electrons in water, which are thought to be responsible for radiation damage in biological matter (e.g., DNA constituents) in aqueous environments [21, 22]. In particular, recent experiments suggested that up to half of the single-strand breaks in DNA may be caused by ICD electrons [23]. ICD-type electrons with kinetic energies larger than ~5 eV can produce irreversible DNA double-strand breaks [21].

Since detecting unambiguous ICD signatures in bulk liquid water is technically challenging and simulating realistic and accurate bulk liquid water using quantum mechanical methods is computationally expensive, small water clusters have often been utilized in both experiments and calculations to gain insights into ICD and other ionization processes in aqueous environment [11, 12, 15, 18, 19, 24–29]. Such studies demonstrated that small cationic water clusters can serve as efficient models to gain insights into the nature of ionized bulk liquid water and help analyze complex water spectra resulting from ICD and from multiple ionization collisions, as well as the underlying electronic structure of ionized water.

In this chapter, we present results from ab initio quantum mechanical calculations of the electronic structures of neutral to multiply-ionized water monomer, dimer, trimer, and tetramer, using a similar methodology as in previous investigations to describe DNA/RNA targets subjected to ionization by energetic protons [30, 31]. Core-level electrons of water are also considered here since recent studies demonstrated that emission site and energy of the electrons released during resonant-Auger-ICD cascade can be controlled by coupling ICD to resonant core excitation [32]. Previous studies of ICD and electronic structures of neutral and ionized small water clusters and liquid water are briefly discussed.

2 Geometric Structures

The structures of small water cluster isotopomers and their terahertz laser vibration–rotation-tunneling spectra and mid-IR laser spectra were systematically compiled and analyzed experimentally, in conjunction with calculations, to untangle intricacies in cooperative hydrogen bonding and provide insights into the molecular-level description of liquid water (see Ref. [33] and references therein). Early ab initio studies of water clusters with two to several monomers can also be found, for example, in Refs. [34–36].

Figure 2 shows the equilibrium structures of an isolated H_2O monomer (C_{2v} symmetry), a mirror-symmetry water dimer $(H_2O)_2$ (C_s symmetry), a chiral cyclic trimer $(H_2O)_3$ (quasi-D_3 symmetry), and highly symmetric homodromic tetramer $(H_2O)_4$ (S_4 symmetry) relaxed using a similar computational approach as described in details in Refs. [30, 31]. Total-energy calculations for all targets were performed at the unrestricted Hartree–Fock UHF/3-21G level of theory [37]. The equilibrium geometries were obtained without symmetry constraints applied. For H_2O, the calculated O–H bonds of 0.967 Å and H–O–H bond angle of 107.7° are in line with the values of 0.960 Å and 106.5° calculated at the same level in the multiple ionization study of H_2O by Bachi et al. [26]. Computed O–H bonds range from 0.966 to 0.974 Å in the dimer, 0.965–0.985 Å in the trimer, and 0.964–1.000 Å in the tetramer, while the O $\cdots$ H hydrogen bonds contract from 1.822 Å in the dimer to 1.721–1.738 Å in the trimer and 1.592 Å in the tetramer. As an additional effect of clustering, the average H–O–H bond angle opens up to 108.3° in the dimer, 100.0° in the trimer, and 110.2° in the tetramer. Let us note that using higher level methods with more diffuse basis sets, such as second-order Moller–Plesset MP2/aug-cc-pVDZ or MP2/(aug)-cc-pVDZ utilized by Müller and Cederbaum [27], results in slight differences in the dimer optimization, such deviations tend to be reduced in the trimer and tetramer and larger clusters, which tend to exhibit cyclic structures as their global minima. In addition, the impact of such small geometric deviations on their respective ICD and

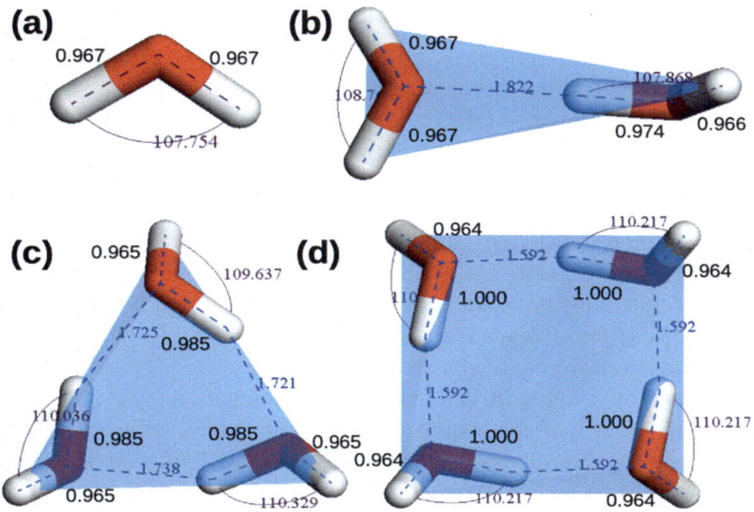

Fig. 2 Geometries of **a** an isolated H_2O monomer (C_{2v} symmetry) and water clusters **b** $(H_2O)_2$ (C_s symmetry), **c** $(H_2O)_3$ (quasi-D_3 symmetry), and **d** $(H_2O)_4$ (S_4 symmetry) optimized at the UHF/3-21G level of theory. The arrangements of adjacent H_2O molecules forming the clusters are represented by blue polyhedra. Calculated bond lengths (in Å) and angles (in degrees) are displayed

multiple ionization spectra is small, as mentioned in previous ICD computational studies [27].

3 Electronic Structures

Using the equilibrium geometries displayed in Fig. 2, the electronic structures of an isolated H_2O molecule and water clusters $(H_2O)_2$, $(H_2O)_3$, and $(H_2O)_4$, as well as their singly and multiply-ionized counterparts, were analyzed at the UHF/3-21G level of theory. Let us note that, consistent with previous computational studies [26, 27], all ionization potentials (IP) for ionized species reported below assume that only vertical transitions occur during the ionization process. Since the collisional/photoionization time scale of multiple ionization dynamics is typically much shorter than the time scale needed for molecular fragmentation or dissociation, ionic states discussed hereafter correspond only to the target charge states during sequential vertical transitions the target is experiencing, without rearrangement of interatomic distances within the ionized species.

3.1 Water Molecule H_2O

The ionization potentials and Mulliken population analysis for closed-shell $(H_2O)^{q+}$ (q = 0, 2, 4, 6) and open-shell $(H_2O)^{q+}$ (q = 1, 3, 5, 7) monomers calculated with UHF/3-21G are listed in Tables 1 and 2, respectively. The present values essentially match the predictions by Bachi et al. (α-spin only) at the same level of theory for the multiple ionization of the H_2O molecule [26] and are in line with the values predicted by Müller and Cederbaum [27] (α-spin only) using MP2/aug-cc-pVDZ to describe neutral H_2O. The calculated IP are also in close agreement with the experimental single-ionization potentials relevant to IDC of liquid water reported by Winter et al. [38]. IP and characteristics for core electrons of $(H_2O)^{q+}$ monomers are also summarized here since electrons released during resonant-Auger-ICD cascade can be controlled by coupling ICD to resonant core excitation [32].

As shown in Table 1, in the neutral H_2O molecule, the outermost valence electrons composing the $1b_1$ molecular orbital (MO) are the O 2p electron lone pair perpendicular to the molecular plane, with a calculated IP of 12.980 eV (experiment in liquid water: 12.60 eV). The second outermost $3a_1$ MO is made of O 2p outer-valence electrons included in the molecular plane, with some slight degree of O 2p—H 1s hybridization and a relatively small O 2s contribution, and a computed IP of 14.422 eV (experiment in liquid water: 14.84 eV). The third outermost $1b_2$ MO is dominated by O 2p—H 1s hybridization and is responsible for the creation of O–H bonds, with a computed IP of 18.640 eV (experiment in liquid water: 18.78 eV). The $2a_1$ MO consists mostly of inner-valence O 2s electrons involved in the initial IDC vacancy states [16], with minor contributions from H 1s and O 2p electrons, and exhibits a computed IP of 35.946 eV (experiment in liquid water: 32.62 eV). The inner-most $1a_1$ MO, corresponding to an IP of 555.847 eV, is made of O 1s core electrons.

Since previous computational studies only focused on ionization potential for α-spin [26, 27] and energy differences between α- and β-spin orbitals are relatively small in general, the discussion in the following sections will be limited to IPs corresponding to α-spin only.

3.2 Water Dimer $(H_2O)_2$

Molecular orbitals and ionization potentials of $[(H_2O)_2]^{q+}$ (q = 0–6) dimers (C_s symmetry) calculated at the UHF/3-21G level of theory are summarized in Table 3 and depicted in Fig. 3, along with previous calculations for the $(H_2O)_2$ dimer [27] and results for $(H_2O)^{q+}$ (q = 0–4) [26, 27] discussed above. The calculated IPs of $(H_2O)_2$ in this study are in overall good agreement with previous predictions by Müller and Cederbaum using MP2/aug-cc-pVDZ [27], especially for the description of outermost O 2p valence electrons composing the six lowest-energy MOs. The largest differences of ~6–8% between both sets of calculated IPs are observed for

Table 1 Molecular orbitals (MO), ionization potentials (IP), and Mulliken population analysis (Pop.) for closed-shell $(H_2O)^{q+}$ (q = 0, 2, 4, 6) monomers calculated at the UHF/3-21G level of theory. Computational results from Refs. [26, 27] and experimental data from Ref. [38] are also shown

	This study					Ref. [26]	Ref. [27]	Expt. [38]
				α or β spin		α or β spin	α or β spin	
Species	MO(char.)	Atom	n, l	Pop	IP (eV)	IP (eV)	IP (eV)	IP(eV)
H_2O	$1(1b_1)$	O1	2p	1	12.980	12.98	12.85	12.60
	$2(3a_1)$	O1	2p	0.78	14.422	14.41	15.16	14.84
		O1	2 s	0.11				
		H2	1 s	0.06				
		H3	1 s	0.06				
	$3(1b_2)$	O1	2p	0.61	18.640	18.65	19.11	18.78
		H3	1 s	0.2				
		H2	1 s	0.2				
	$4(2a_1)$	O1	2 s	0.84	35.946	35.94	33.37	32.62
		H2	1 s	0.07				
		H3	1 s	0.07				
		O1	2p	0.03				
	$5(1a_1)$	O1	1 s	1	555.847			
$(H_2O)^{2+}$	$1(3a_1)$	O1	2p	0.87	44.518	44.56		
		O1	2 s	0.08				
		H2	1 s	0.02				
		H3	1 s	0.02				
	$2(1b_2)$	O1	2p	0.81	48.218	48.47		
		H3	1 s	0.1				
		H2	1 s	0.1				
	$3(2a_1)$	O1	2 s	0.91	65.198	65.3		
		O1	2p	0.05				
		H2	1 s	0.02				
		H3	1 s	0.02				
	$4(1a_1)$	O1	1 s	1	591.276			
$(H_2O)^{4+}$	$1(1b_2)$	O1	2p	0.93	84.246	84.56		
		H3	1 s	0.04				
		H2	1 s	0.04				
	$2(2a_1)$	O1	2 s	0.98	98.777	98.96		
		O1	2p	0.04				
	$3(1a_1)$	O1	1 s	1	636.692			
$(H_2O)^{6+}$	$1(2a_1)$	O1	2 s	1	140.547	140.72		
		O1	2p	0.04				
		H2	1 s	0.02				
		H3	1 s	0.02				
	$2(1a_1)$	O1	1 s	1	697.155			

Table 2 Molecular orbitals (MO), ionization potentials (IP), and Mulliken population analysis (Pop.) for open-shell $(H_2O)^{q+}$ (q = 1, 3, 5, 7) monomers calculated at the UHF/3-21G level of theory. Results from Ref. [26] are also shown

	This study							Ref. [26]
				α spin		β spin		α spin
Species	MO(char.)	Atom	n, l	Pop	IP (eV)	Pop	IP (eV)	IP (eV)
$(H_2O)^+$	1(3a_1)	O1	2p	0.85	29.470	0.82	27.864	29.44
		O1	2 s	0.08		0.11		
		H2	1 s	0.04		0.03		
		H3	1 s	0.04		0.03		
	2(1b_1)	O1	2p	1	30.667			30.66
	3(1b_2)	O1	2p	0.74	33.198	0.69	32.082	33.46
		H3	1 s	0.13		0.15		
		H2	1 s	0.13		0.15		
	4(2a_1)	O1	2 s	0.91	52.082	0.84	47.783	52.24
		O1	2p	0.04		0.05		
		H2	1 s	0.03		0.06		
		H3	1 s	0.03		0.06		
	5(1a_1)	O1	1 s	1	572.772	1	571.629	
$(H_2O)^{3+}$	1(1b_2)	O1	2p	0.89	65.770	0.87	64.110	65.49
		H3	1 s	0.05		0.06		
		H2	1 s	0.05		0.06		
	2(3a_1)	O1	2p	0.94	65.879			66.12
		O1	2 s	0.03				
		H2	1 s	0.02				
		H3	1 s	0.02				
	3(2a_1)	O1	2 s	0.98	82.913	0.92	78.967	83.06
		O1	2p	0.02		0.06		
	4(1a_1)	O1	1 s	1	612.528	1	611.113	
$(H_2O)^{5+}$	1(1b_2)	O1	2p	0.96	109.362			109.67
		H3	1 s	0.02				
		H2	1 s	0.02				
	2(2a_1)	O1	2 s	1.01	120.818	0.98	115.866	121.68
		O1	2p	0.03		0.06		
		H2	1 s	0.02		0.02		
		H3	1 s	0.02		0.02		
	3(1a_1)	O1	1 s	1	664.883	1	663.332	
$(H_2O)^{7+}$	1(2a_1)	O1	2 s	0.92	167.350			167.34
		O1	2p	0.12				

(continued)

Table 2 (continued)

	This study							Ref. [26]
				α spin		β spin		α spin
Species	MO(char.)	Atom	n, l	Pop	IP (eV)	Pop	IP (eV)	IP (eV)
		H2	1 s	0.02				
		H3	1 s	0.02				
	$2(1a_1)$	O1	1 s	1	731.115	1	728.258	

Table 3 Molecular orbitals (MO) and ionization potentials (IP) of $[(H_2O)_2]^{q+}$ (q = 0–6) dimers (C_s symmetry) calculated at the UHF/3-21G level of theory

MO	$(H_2O)_2$	$[(H_2O)_2]^+$	$[(H_2O)_2]^{2+}$	$[(H_2O)_2]^{3+}$	$[(H_2O)_2]^{4+}$	$[(H_2O)_2]^{5+}$	$[(H_2O)_2]^{6+}$
1	11.756	19.565	25.389	41.335	53.580	61.743	71.947
2	12.953	21.253	27.375	42.314	56.002	64.355	79.213
3	14.232	25.089	30.858	44.437	57.825	74.723	82.832
4	15.565	28.028	42.369	48.192	59.049	74.832	95.186
5	17.797	29.117	45.879	52.029	74.669	81.118	98.642
6	19.838	31.675	48.246	63.158	75.893	91.948	626.107
7	34.641	42.369	62.832	68.954	600.801	607.631	631.740
8	37.117	50.478	568.909	584.256	602.134	621.617	
9	554.410	562.650	588.610	594.923			
10	557.150	570.977					

MOs made mostly of inner-valence O 2s electrons involved in the initial IDC vacancy states, as a possible result of diffuse basis functions added in the calculations of Ref. [27].

3.3 Water Trimer $(H_2O)_3$

Table 4 lists the ionization potentials corresponding to the MOs of the $[(H_2O)_3]^{q+}$ (q = 0–6) trimers (quasi-D_3 symmetry) calculated at the UHF/3-21G level of theory, with their graphical representation in Fig. 4, with computational results for the $(H_2O)_3$ trimer [27] and predictions for $(H_2O)^{q+}$ (q = 0–4) [26, 27].

3.4 Water Tetramer $(H_2O)_4$

Ionization potentials calculated at the UHF/3-21G level of theory for the MOs of $[(H_2O)_4]^{q+}$ (q = 0–8) tetramers (S_4 symmetry) are given in Table 5 and shown in

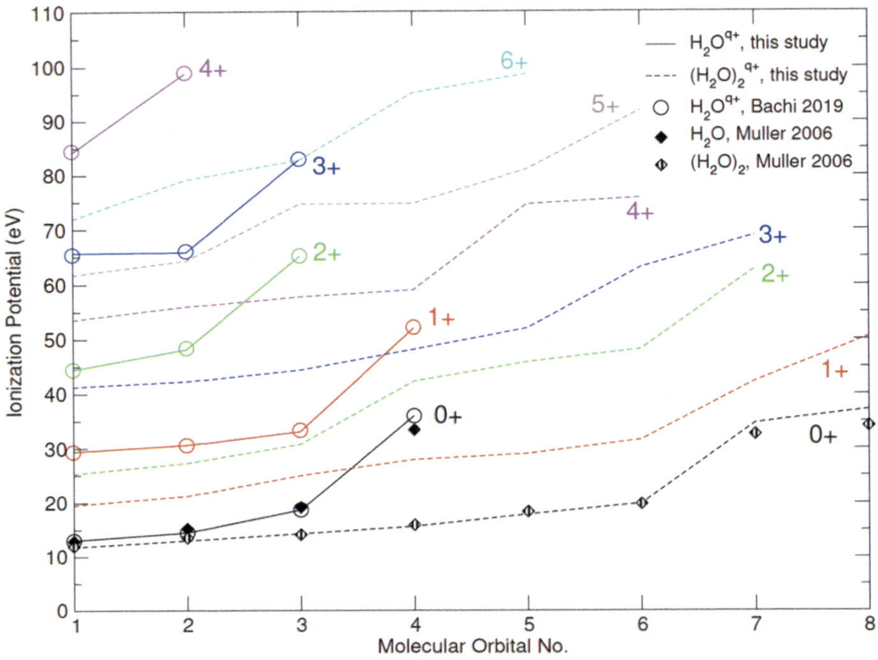

Fig. 3 Ionization potentials for the molecular orbitals of $[(H_2O)_2]^{q+}$ (q = 0–6) dimers (C_s symmetry) calculated at the UHF/3-21G level of theory. Calculations for the $(H_2O)_2$ dimer [27] and for $(H_2O)^{q+}$ (q = 0–4) [26, 27] are also shown

Table 4 Molecular orbitals (MO) and ionization potentials (IP) of $[(H_2O)_3]^{q+}$ (q = 0–6) trimers (quasi-D_3 symmetry) calculated at the UHF/3-21G level of theory

MO	$(H_2O)_3$	$[(H_2O)_3]^+$	$[(H_2O)_3]^{2+}$	$[(H_2O)_3]^{3+}$	$[(H_2O)_3]^{4+}$	$[(H_2O)_3]^{5+}$	$[(H_2O)_3]^{6+}$
1	12.708	18.640	26.151	38.260	40.273	50.314	65.471
2	12.790	18.940	26.913	39.784	47.729	55.893	66.451
3	12.953	19.946	30.314	40.246	48.845	56.981	66.532
4	13.960	20.953	31.511	41.280	49.008	59.131	69.444
5	14.994	24.110	31.919	41.443	52.056	60.328	69.471
6	15.157	24.926	34.940	42.967	52.219	60.464	70.396
7	18.531	28.709	35.158	43.620	53.035	63.566	86.233
8	18.586	29.824	36.219	44.546	69.009	77.281	86.288
9	19.702	32.219	51.947	58.260	69.145	77.907	87.077
10	35.593	41.416	52.246	62.233	69.961	80.437	612.801
11	35.647	42.151	53.199	62.777	593.426	601.236	612.855
12	36.464	50.886	574.188	582.188	593.562	601.916	612.882
13	555.793	561.725	574.242	583.277	593.698	606.325	
14	555.820	562.052	575.168	583.358			
15	555.875	571.358					

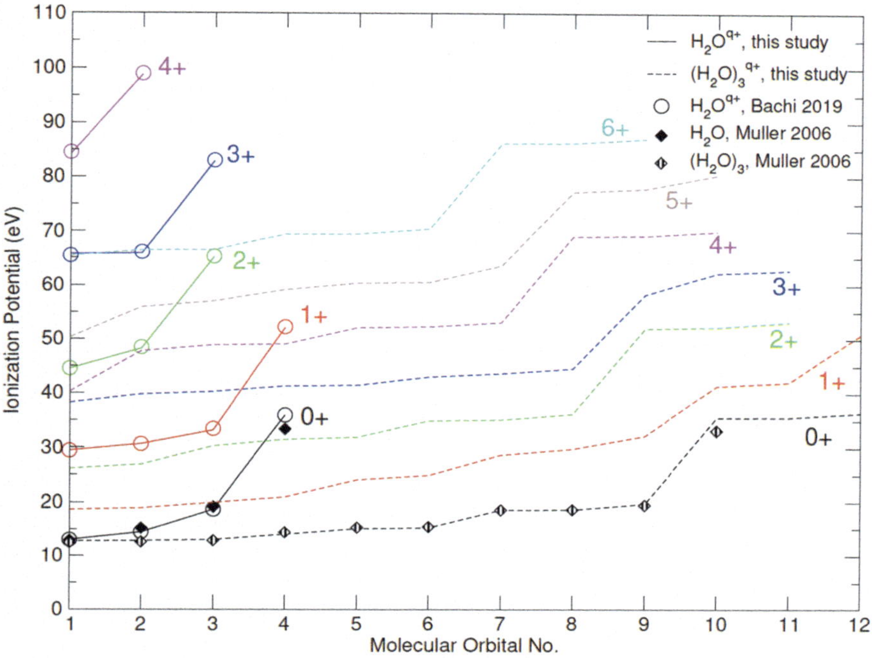

Fig. 4 Ionization potentials for the molecular orbitals of $[(H_2O)_3]^{q+}$ (q = 0–6) trimers (quasi-D_3 symmetry) calculated at the UHF/3-21G level of theory. Calculations for the $(H_2O)_3$ trimer [27] and for $(H_2O)^{q+}$ (q = 0–4) [26, 27] are also shown

Fig. 5, along with previous calculations for the $(H_2O)_4$ tetramer [27] and results for $(H_2O)^{q+}$ (q = 0–4) [26, 27].

4 ICD Ionization Spectra

While the ionization spectra of outer-valence shells ($1b_1$, $3a_1$ and $1b_2$) of molecular and liquid water are characterized by typical main lines/peaks originating from contributions of one-hole configurations (1 h), inner-valence ionization spectra of water exhibit a breakdown of the traditional molecular orbital picture [39], which becomes more pronounced as system size increases from an isolated water molecule to bulk liquid water [27]. As shown in previous experimental investigations of the photoelectron spectrum of water [20, 38, 40, 41], ionization of the inner-valence $2a_1$ band of liquid water typically produces a broad photoelectron spectrum signal in the kinetic energy range ~40–58 eV (see Fig. 6), which reflects the fact that the intensity attributed to ionization of this inner-valence state is distributed over several ionic states, without featuring the local peak-like signals of one-hole contributions.

Table 5 Molecular orbitals (MO) and ionization potentials (IP) of $[(H_2O)_4]^{q+}$ (q = 0–8) tetramers (S_4 symmetry) calculated at the UHF/3-21G level of theory

MO	$(H_2O)_4$	$[(H_2O)_4]^+$	$[(H_2O)_4]^{2+}$	$[(H_2O)_4]^{3+}$	$[(H_2O)_4]^{4+}$	$[(H_2O)_4]^{5+}$	$[(H_2O)_4]^{6+}$	$[(H_2O)_4]^{7+}$	$[(H_2O)_4]^{8+}$
1	12.436	20.083	24.219	25.307	37.743	48.627	48.927	57.335	73.199
2	12.681	20.708	24.219	33.824	37.852	48.791	57.661	65.498	74.288
3	12.708	20.708	25.498	34.205	42.505	51.974	58.913	66.750	74.288
4	12.708	20.817	27.484	36.899	43.267	51.974	58.913	66.750	75.485
5	13.497	21.225	28.953	37.117	44.192	55.022	60.165	67.920	77.362
6	14.559	21.960	28.953	39.294	45.117	55.104	62.315	70.124	77.716
7	14.559	21.960	30.287	39.294	47.131	55.621	62.614	70.260	77.716
8	15.865	23.212	32.899	39.920	47.239	55.811	62.614	70.260	78.043
9	18.586	25.770	33.062	40.137	48.301	56.981	62.940	70.478	93.635
10	18.858	26.123	33.062	41.933	48.437	57.008	78.696	86.642	94.288
11	18.858	26.123	33.443	41.988	63.784	68.138	79.322	87.267	94.288
12	19.321	26.640	49.280	54.369	64.111	68.192	79.322	87.267	94.914
13	35.022	42.722	49.988	54.478	65.172	76.002	80.002	87.920	620.774
14	35.675	43.375	49.988	60.491	65.498	76.029	604.148	612.474	620.801
15	35.675	43.375	50.641	60.546	587.168	592.420	604.175	612.474	620.801
16	36.355	44.001	571.521	576.746	587.168	592.420	604.175	612.474	620.801
17	555.630	563.63	571.548	576.746	588.583	597.753	604.175	612.502	
18	555.660	563.657	571.548	581.208	588.583	597.753			
19	555.660	563.657	571.548	581.236					
20	555.660	563.657							

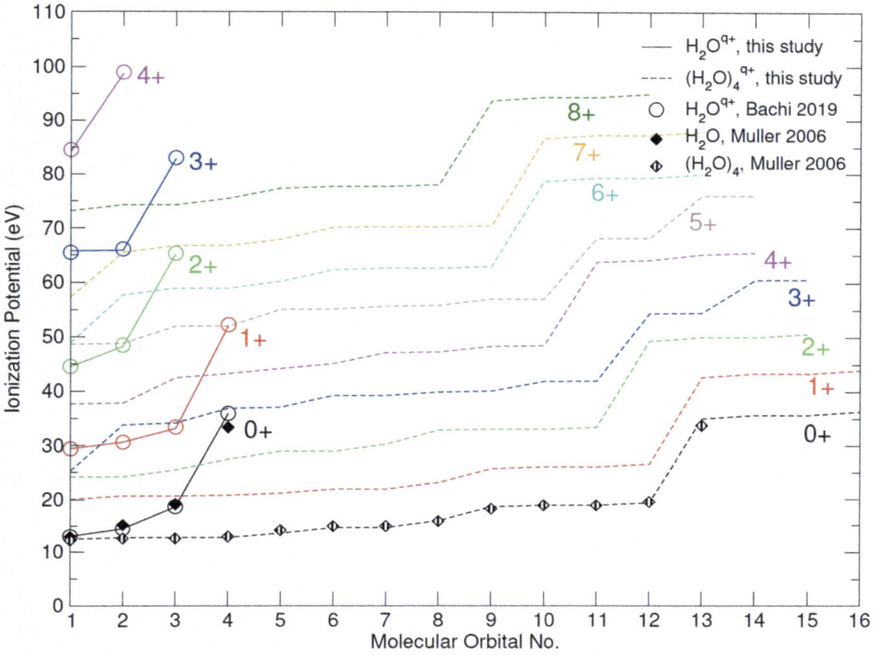

Fig. 5 Ionization potentials for the molecular orbitals of $[(H_2O)_4]^{q+}$ (q = 0–8) tetramers (S_4 symmetry) calculated at the UHF/3-21G level of theory. Calculations for the $(H_2O)_4$ tetramer [27] and for $(H_2O)^{q+}$ (q = 0–4) [26, 27] are also shown

As shown in the inset of Fig. 6, the noncoincident photoelectron spectrum of liquid water using XUV pulses centered at 79.0 eV enables to resolve clear signals above ~58.0 eV from the ionization of the outer-valence $1b_1$, $3a_1$, and $1b_2$ bands, with some overlapping signals from water vapor, and the broad signal from inner-valence $2a_1$ band centered around 48 eV. However, for the slowest electrons with kinetic energies below ~10 eV, electron–electron-coincidence spectroscopy is needed to identify ICD electron signatures, which overlap with signals originating from secondary electrons produced by electron-impact ionization of water molecules by primary photoelectrons, as well as with photoelectrons that have lost energy in electronically or vibrationally inelastic collisions.

Figure 7 depicts representative measured energy spectra of slow ICD electrons from 12- and 241-water clusters [11] and liquid–water microjet recorded with XUV pulses centered at 63.5 and 79.0 eV and integrated over the kinetic energy of fast electrons from electron–electron-coincidence spectra of detected electron pairs [20]. Signal for ICD electrons from the 12-water cluster features a maximum of its normalized intensity in the vicinity of zero kinetic energy, followed by an exponential-like decay, as predicted from theoretical studies of small water clusters [27] and measured in experiments [11, 15, 16]. Although this trend in signal decrease is preserved above

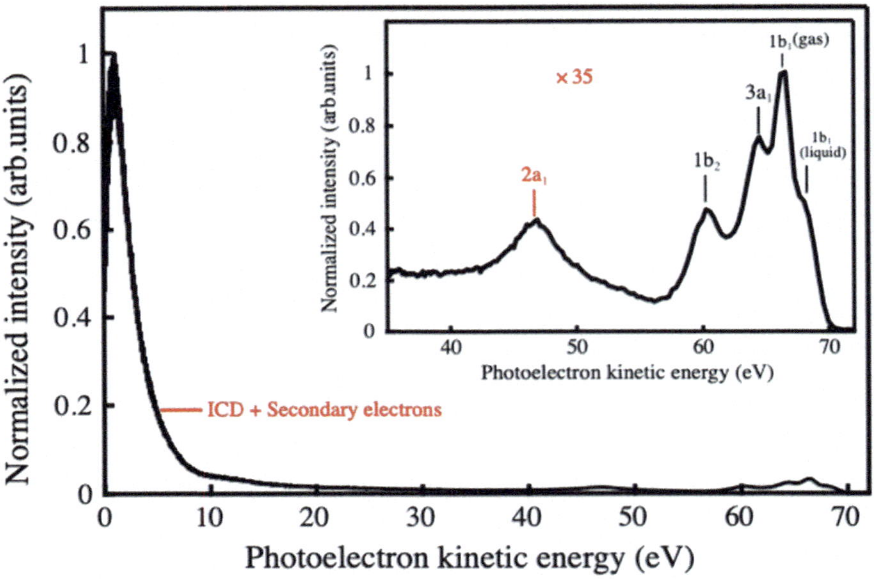

Fig. 6 Photoelectron spectrum (noncoincident) of a liquid–water microjet recorded with XUV pulses centered at 79.0 eV. *Inset*: magnified signals originating from the ionization of the inner-valence $2a_1$ band (below kinetic energy of ~58 eV) and outer-valence $1b_1$, $3a_1$, and $1b_2$ bands (above kinetic energy of ~58 eV) of liquid water; a peak attributed to $1b_1$ electrons of gaseous water is also observed [adapted from Ref. [20]; Creative Commons Attribution 4.0 International license (CC BY 4.0), American Physical Society]

~2.5 eV for larger size water clusters and liquid water, the ICD intensity peaks at ~0.4 and ~0.8 eV for the larger 241-water cluster and bulk liquid water, respectively, as a consequence of low-energy inelastic electron scattering and the existence of an escape barrier on the surface [41]. For kinetic energies of the ICD electron above ~2.5 eV, the high degree of similarity between the energy tails of the 241-water cluster and bulk liquid water indicates that the distribution of energy between the initially created $2a_1$ vacancy and the final 1h1h pair of outer-valence (i.e., $1b_1$, $3a_1$, or $1b_2$) vacancies are nearly identical.

5 Conclusion and Perspectives

Since intermolecular Coulombic decay (ICD) was theoretically proposed in 1997 by Cederbaum et al., ICD in liquid water has been a topic of very active research. This unconventional nonlocal electronic decay mechanism, featuring ultrafast non-radiative decay process and ultralong-range for excess energy transfer from the

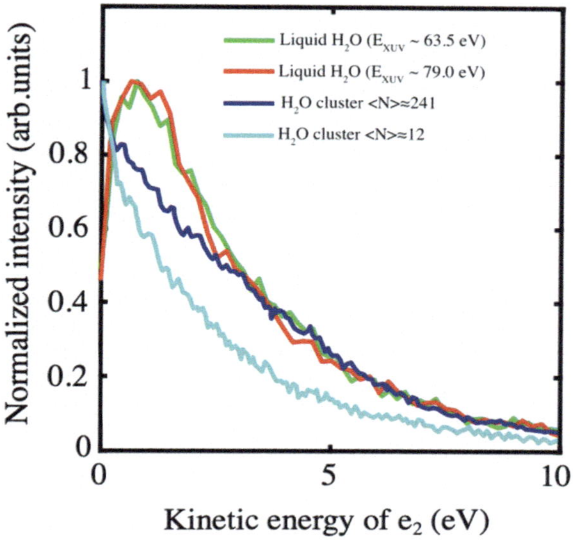

Fig. 7 Comparison of the energy spectra of ICD slow electrons (e_2) from 12- and 241-water clusters (Ref. [11]) and liquid–water microjet recorded with XUV pulses centered at 63.5 and 79.0 eV. The green and red solid curves are the energy spectra integrated over the kinetic energy of fast electrons (e_1) in the ranges 25.5–39.5 eV and 41.0–55.0 eV, respectively, from the electron–electron-coincidence spectra of detected electron pairs (e_1, e_2). All spectra are normalized to their maximum for comparison [adapted from Ref. [20]; Creative Commons Attribution 4.0 International license (CC BY 4.0), American Physical Society]

excited atom/molecule to its neighbors, has triggered a remarkable combined experimental and theoretical research effort. This stems in part from the importance of ICD as a source of low-energy electrons in bulk liquid water and its role in radiation damage in biological matter in aqueous environments. Rapid progress in experimental techniques in electron–electron-coincidence spectroscopy led recently to the first unambiguous observation of ICD in bulk liquid water following inner-valence ionization [20]. In recent studies, control of the emission site and energy of the electrons released during resonant-Auger-ICD cascade was also demonstrated by coupling ICD to resonant core excitation [32].

Both computational and experimental investigations have revealed novel intriguing characteristics of ICD, outlining the need to go beyond the standard picture of traditional molecular orbital picture [39], with typical one-hole configurations, and only vertical transitions occurring during complex ionization processes. With the transformational role of GPU parallel computing and advances in computational methods, several studies provided new insights into the femtosecond dynamics of ICD and energy transfer in water in recent years. These include, for example, the investigation of the reactions of multi-valence ionized states of water dimer using ab initio molecular dynamics [42], the assessment of Real-Time Time-Dependent

Density Functional Theory (RT-TDDFT) for the ionized water dimer [43], the proton-transfer dynamics in ionized water chains using RT-TDDFT [44], the reactivity and energy level of a localized hole in liquid water [45], or electron-induced fragmentation of water droplets [46]. The combination of such computational approaches and experimental techniques opens up new pathways for the accurate and direct determination of the ICD lifetime in liquid water and small water clusters, which is of importance for modeling radiation damage in living tissues and fine-tuning strategies and dosimetry for radiation-based cancer therapies.

Acknowledgements Sandia National Laboratories is a multi-mission laboratory managed and operated by National Technology and Engineering Solutions of Sandia, LLC., a wholly owned subsidiary of Honeywell International, Inc., for the U.S. Department of Energy's (DOE) National Nuclear Security Administration under contract DE-NA0003525. The views expressed in this manuscript do not necessarily represent the views of the U.S. DOE or the United States Government.

References

1. O. Savadogo, J. New Mater. Electrochem. Syst. **1**, 47 (1998)
2. E. Kim, P.F. Weck, N. Balakrishnan, C. Bae, J. Phys. Chem. B (Lett.) **112**, 3283 (2008)
3. J.M.J. Swanson, C.M. Maupin, H. Chen, M.K. Petersen, J. Xu, Y. Wu, G.A. Voth, J. Phys. Chem. B **111**, 4300 (2007)
4. C.J.T. de Grotthuss, Ann. Chim. B **LVIII**, 54 (1806)
5. M. Eigen, Angew. Chem., Int. Ed. Engl. **3**, 1 (1964)
6. G. Zundel, H. Metzger, Z. Phys, Chem. **58**, 225 (1968)
7. D. Marx, M.E. Tuckerman, J. Hutter, M. Parrinello, Nature **397**, 601 (1999)
8. A.A. Kornyshev, A.M. Kuznetsov, E. Spohr, J. Ulstrup, J. Phys. Chem. B **107**, 15 (2003)
9. E.V. Silletta, M.E. Tuckerman, A. Jerschow, Phys. Rev. Lett. **121**, 076001 (2018)
10. P.B. Calio, C. Li, G.A. Voth, J. Am. Chem. Soc. **143**, 18672 (2021)
11. C. Richter, D. Hollas, C.-M. Saak, M. Förstel, T. Miteva, M. Mucke, O. Björneholm, N. Sisourat, P. Slavíček, U. Hergenhahn, Nat. Comm. **9**, 4988 (2018)
12. L.S. Cederbaum, J. Zobeley, F. Tarantelli, Phys. Rev. Lett. **79**, 4778 (1997)
13. N. Sisourat, N.V. Kryzhevoi, P. Kolorenč, S. Scheit, T. Jahnke, L.S. Cederbaum, Nat. Phys. **6**, 508 (2010)
14. T. Jahnke, U. Hergenhahn, B. Winter, R. Dörner, U. Frühling, P.V. Demekhin, K. Gokhberg, L.S. Cederbaum, A. Ehresmann, A. Knie, A. Dreuw, Chem. Rev. **120**, 11295 (2020)
15. T. Jahnke, H. Sann, T. Havermeier, K. Kreidi, C. Stuck, M. Meckel, M. Schöffler, N. Neumann, R. Wallauer, S. Voss, A. Czasch, O. Jagutzki, A. Malakzadeh, F. Afaneh, T. Weber, H. Schmidt-Böcking, R. Dörner, Nat. Phys. **6**, 139 (2010)
16. M. Mucke, M. Braune, S. Barth, M. Förstel, T. Lischke, V. Ulrich, T. Arion, U. Becker, A. Bradshaw, U. Hergenhahn, Nat. Phys. **6**, 143 (2010)
17. U. Hergenhahn, J. Electron Spectrosc. Relat. Phenom. **184**, 78 (2011)
18. V. Hermann, B.D. Kay, A.W. Castleman, Chem. Phys. **72**, 185 (1982)
19. H. Shiromaru, H. Shinohara, N. Washida, H.-S. Yoo, K. Kimura, Chem. Phys. Lett. **141**, 7 (1987)
20. P. Zhang, C. Perry, T.T. Luu, D. Matselyukh, H.J. Wörner, Phys. Rev. Lett. **128**, 133001 (2022). https://doi.org/10.1103/PhysRevLett.128.133001
21. E. Alizadeh, T.M. Orlando, L. Sanche, Annu. Rev. Phys. Chem. **66**, 379 (2015)
22. B. Boudaiffa, P. Cloutier, D. Hunting, M.A. Huels, L. Sanche, Science **287**, 1658 (2000)

23. G.A. Grieves, T.M. Orlando, Phys. Rev. Lett. **107**, 016104 (2011)
24. S. Bhattacharjee, C. Bagdia, M.R. Chowdhury, J.M. Monti, R.D. Rivarola, L.C. Tribedi, Eur. Phys. J. D **72**, 15 (2018)
25. S. Bhattacharjee, S. Biswas, J.M. Monti, R.D. Rivarola, L.C. Tribedi, Phys. Rev. A **96**, 052707 (2017)
26. N. Bachi, S. Otranto, G.S. Otero, R.E. Olson, Phys. Med. Biol. **64**, 205020 (2019)
27. I.B. Müller, L.S. Cederbaum, J. Chem. Phys. **125**, 204305 (2006)
28. S.D. Stoychev, A.I. Kuleff, L.S. Cederbaum, J. Chem. Phys. **133**, 154307 (2010)
29. S. Barth, M. Ončák, V. Ulrich, M. Mucke, T. Lischke, P. Slavíček, U. Hergenhahn, J. Phys. Chem. A **113**(48), 13519 (2009)
30. C. Champion, M.A. Quinto, J.M. Monti, M.E. Galassi, P.F. Weck, O.A. Fojón, J. Hanssen, R.D. Rivarola, Phys. Med. Biol. **60**, 7805 (2015)
31. M.E. Galassi, C. Champion, P.F. Weck, R.D. Rivarola, O. Fojón, J. Hanssen, Phys. Med. Biol. **57**, 2081 (2012)
32. K. Gokhberg, P. Kolorenč, A.I. Kuleff, L.S. Cederbaum, Nature **505**, 661 (2014)
33. F.N. Keutsch, R.J. Saykally, Proc. Natl. Acad. Sci. U.S.A. **98**, 10533 (2001)
34. K.S. Kim, B.J. Mhin, U.-S. Choi, K. Lee, J. Chem. Phys. **97**, 6649 (1992)
35. S.S. Xantheas, T.H. Dunning Jr., J. Chem. Phys. **99**, 8774 (1993)
36. M. Yang, P. Senet, C. van Alsenoy, Int. J. Quantum Chem. **101**, 535 (2005)
37. Gaussian 09, Revision A.02, M.J. Frisch, G.W. Trucks, H.B. Schlegel, G.E. Scuseria, M.A. Robb, J.R. Cheeseman, G. Scalmani, V. Barone, G.A. Petersson, H. Nakatsuji, X. Li, M. Caricato, A. Marenich, J. Bloino, B.G. Janesko, R. Gomperts, B. Mennucci, H.P. Hratchian, J.V. Ortiz, A.F. Izmaylov, J.L. Sonnenberg, D. Williams-Young, F. Ding, F. Lipparini, F. Egidi, J. Goings, B. Peng, A. Petrone, T. Henderson, D. Ranasinghe, V.G. Zakrzewski, J. Gao, N. Rega, G. Zheng, W. Liang, M. Hada, M. Ehara, K. Toyota, R. Fukuda, J. Hasegawa, M. Ishida, T. Nakajima, Y. Honda, O. Kitao, H. Nakai, T. Vreven, K. Throssell, J.A. Montgomery, Jr., J.E. Peralta, F. Ogliaro, M. Bearpark, J.J. Heyd, E. Brothers, K.N. Kudin, V.N. Staroverov, T. Keith, R. Kobayashi, J. Normand, K. Raghavachari, A. Rendell, J.C. Burant, S.S. Iyengar, J. Tomasi, M. Cossi, J.M. Millam, M. Klene, C. Adamo, R. Cammi, J.W. Ochterski, R.L. Martin, K. Morokuma, O. Farkas, J.B. Foresman, D.J. Fox, Gaussian, Inc., (Wallingford CT 2016)
38. B. Winter, R. Weber, W. Widdra, M. Dittmar, M. Faubel, I.V. Hertel, J. Phys. Chem. A **108**, 2625 (2004)
39. L.S. Cederbaum, W. Domcke, J. Schirmer, W. von Niessen, Adv. Chem. Phys. **65**, 115 (1986)
40. K. Nishizawa, N. Kurahashi, K. Sekiguchi, T. Mizuno, Y. Ogi, T. Horio, M. Oura, N. Kosugi, T. Suzuki, Phys. Chem. Chem. Phys. **13**, 413 (2011)
41. T. Gadeyne, P. Zhang, A. Schild, H.J. Wörner, Chem. Sci. **13**, 1675 (2022)
42. T. Takada, H. Tachikawa, J. Phys. B: Atomic, Mol. Opt. Phys. **54**, 145103 (2021)
43. J.C. Chalabala, F. Uhlig, P. Slavíček, J. Phys. Chem. A **122**, 3227 (2018)
44. V. Sharma, M. Fernández-Serra, Phys. Rev. Res. **2**, 043082 (2020)
45. F. Ambrosio, A. Pasquarello, Phys. Chem. Chem. Phys. **20**, 30281 (2018)
46. J. Suchan, J. Kolafa, P. Slavíček, J. Chem. Phys. **156**, 144303 (2022)

Young-Type Electron Interference in Molecular Double Slit: A Brief Overview

Lokesh C. Tribedi

Abstract We have reviewed the recent investigations on the Young-type electron interference in ionization of diatomic molecules which are considered as molecular double slits based on the original proposition by Cohen and Fano. The examples of interference oscillations in electron spectrum, in ion-induced ionization, photoionization, and electron-impact ionization, are presented. The techniques of deriving such oscillations from electron spectrum, forward–backward asymmetry in electron spectroscopic investigation are emphasized besides mentioning about the ColTRIMS and vibrational spectroscopy and electron time-of-flight-based techniques. A series of data obtained from different research groups are discussed which deal with a number of issues, such as the first- and second-order interference, forward–backward difference in the frequency parameter, and the frequency doubling. In addition, the applications of the interference studies toward fundamental physics are also touched upon. The major experimental techniques to study the interference are also reviewed. However, this is not a comprehensive elaborate review, rather a brief overview, which can be used as a reference article.

1 Introduction

The ionization of atoms is one of the most fundamental processes in atomic collision studies. The study of electron emissions in ionization is quite a convenient way to investigate the underlying mechanisms in such collisions involving atoms, molecules or clusters under the impact of energetic photons, electrons, ions. The two-center effect, post-collision interactions, binary collision, elastic and super-elastic processes, ELC (electron loss to continuum) or ECC (electron capture to continuum) are well-studied mechanisms of ion-atom ionization. The small atoms or molecules are generally the best entities for probing such basic processes. Besides the study of fundamental processes involved in ionization there has been renewed interest in these studies due to its potential applications in various related fields. For example,

L. C. Tribedi (✉)
Tata Institute of Fundamental Research, Colaba, Mumbai 400005, India
e-mail: Ltribedi@gmail.com

L. C. Tribedi (ed.), *Advances in Atomic Molecular Collisions*,
https://doi.org/10.1007/978-981-97-7063-2_13

the recent studies on the ionization of water by fast ions have provided valuable data which are important for understanding the charged particle-induced radiation damage in living matter which is composed mostly of water. The electron spectrum is shown to provide rich information regarding the collision mechanisms involving large bio-molecules, particularly DNA/RNA bases, sugar-phosphate backbone, halo-uracils, etc. The other class of large molecules of recent interest are the varieties of the PAH (polycyclic aromatic hydrocarbon) molecules. The investigations regarding bio-molecules are aimed to provide inputs for modeling the radiobiology of radiation damage during hadron therapy of tumors or cancers in human body. The studies on the PAH molecules are of interest due to their role in astrochemistry of interstellar medium. In addition, the low-energy spectrum of electrons provides information on the giant dipole plasmon resonance in the PAH molecules. Similarly in case of collisions involving C_{60}-fullerene the characteristic plasmon electrons are detected arising from the giant dipole resonance. Such GDPR has been addressed earlier via the study of recoil ions of C_{60} in photo-ionization and particle-impact ionization. A number of theoretical mechanisms are developed which are centered around many of the experimental investigations. Besides these processes another interesting and fundamentally important mechanism is the Young-type electron interference in case of ionization spectrum arising from homonuclear diatomic molecules which is the topic of the present review.

2 Historical Perspective

The famous double-slit experiment with light to provide the interference fringe was initially introduced by Thomas Young more than 200 years ago, which unambiguously proved the wave nature of light. In such experiments, light wave was made to pass through two slits and the separation between the slits is comparable to the wavelength of the light. The double slits, being the coherent sources of light, produces the interference pattern. On the other hand, the Compton effect provided the evidence of corpuscular nature of light. The wave-particle duality was proposed in the quantum theory not only for the light but for any quantum particle. Davisson and Germer [1] as well as Thomson and Reid [2] independently used electron-beam scattering from solid to provide experimental evidence of the wave nature of electron. One may cite the example of the famous experiment of Jonsson (1961) to observe the electron-wave interference involving electrons scattering from solids [3]. The interference pattern for the electron beam, similar to those for light waves, clearly indicated its wave nature. The wave-particle duality of electrons was also demonstrated by Tonomura et al. in a two slit-type interference experiment using an electron microscope [4]. Since then, several experiments have been performed on the interference using neutrons, atoms, and even C_{60} fullerenes [5]. It may be noted that the electron interference in a double slit was considered as the most beautiful experiment in physics [6] by the readers of the Physics World (2002).

3 Diatomic Molecule as a Double Slit

Cohen and Fano [7] analyzed the oscillatory nature of the photo-ionization cross-sectional data of [8] N_2 and O_2 in terms of a model assuming electron scattering from a molecular double slit. Such an interference mechanism was also predicted by Kaplan and Markin in 1969 [9]. A homonuclear diatomic molecule, such as H_2, O_2, or N_2, can be used as a double slit or double source of electrons in photo-ionization. In such cases, the electrons are emitted coherently from two spatially separated sources (i.e., the two atoms) in such a fashion that the electron waves are in phase or out of phase, leading to interference maximum or minimum in the electron intensity spectrum. The homonuclear diatomic molecule acts as the source of spatial coherence which reveals through the interference oscillation in the electron spectrum. The internuclear distance (about an Å) makes these molecules as suitable candidates to study the interferences of the ejected electrons.

4 Interference in Fast Ion Collisions with Molecules

It took about four decades to reveal clearly such an interference effect in fast-ion collisions with H_2, by two different groups, i.e., Stolterfoht and co-workers and Tribedi and co-workers [10, 11]. The two-center interference oscillation was revealed very clearly in electron emission spectrum arising from the ionization of H_2 under fast heavy-ion collisions. It may be indicated that the periodic variation in the electron emission cross-sectional ratio (H_2-to-2H) was reported even earlier [12] which, however, was not explained in terms of interference. Thereafter, many experiments have been carried out to investigate the interference effect in case of diatomic molecules [10, 11, 13–25]. Some of the experiments involve the study of cross sections for electron emission, differentials in both ejection energy and angle (commonly named as the double differential cross sections (DDCS)), for the H_2 molecule under the impact of very fast, heavy ions. A prominent oscillatory structure was observed in the DDCS spectrum (rather DDCS ratio) due to the interference. The DDCS spectrum varies by a few orders of magnitude over the energy range of 0 to 300 eV over which one may expect the oscillatory behavior. The variation of the oscillation amplitude is much smaller since it is characterized by a damped sine function. For example, according to Cohen and Fano (1966), i.e., DDCS $\sim A + B[sin(kcd)/(kcd)]$ (with A = additive constant and B = amplitude) and where k = electron velocity in a.u, c = frequency parameter (function of electron emission angle), d = internuclear distance in the diatomic molecule. In the initial studies of the Cohen–Fano-type interference, the frequency parameter "c" was not considered due to an inherent assumption that the frequency is independent of the emission angle. But experimentally "c" was observed to be angle-dependent function in heavy-ion and electron-induced interference oscillation for the H_2 molecule. Initially, it was taken to be $c = \cos\theta$ [14] which implies the magnitude of the parameter is the same for emission angle θ and $\pi - \theta$.

However, it was observed that the frequency parameter at extreme backward angles was almost double to that for extreme forward angles for fast-ion collisions. The angle-dependent frequency parameter derived from heavy-ion collisions as well as electron-H_2 collisions implied that $c = cos\theta$ does not represent the correct angular dependence and the Cohen–Fano (CF) equation was remodeled as *sin(kcd)/(kcd)* and the c(θ) is to be derived from experiments, in absence of an well-accepted analytical expression. Galassi et al. (see below) [13, 15], however, derived a more comprehensive expression for the frequency implying that the simple-minded expressions used in CF-type function may be questionable.

5 Interference Oscillations in DDCS Ratio in Electron Spectroscopy

The question is how to derive the oscillation from the measured DDCS spectrum. It is crucial to divide the molecular cross sections by the corresponding atomic cross section to derive the oscillatory function. For example, to derive the oscillations in case of ion-H_2 collisions one divides the DDCS of H_2 by the DDCS of atomic H (rather 2H), often derived from theory. In such cases, some theoretical parameters are used (such as effective atomic number, Z_{eff}) to derive the oscillation. The shape of the oscillation was found to depend strongly on the choice of the values of the parameter Z_{eff} of the target molecule. Reference [11] demonstrated this aspect using two values of $Z_{eff} = 1.05$ and 1.19 (as used in the models) for H_2 molecule. Therefore, in order to avoid the use of such parameters, it is desirable to have experimentally measured cross section (DDCS) for the atomic H, which is normally a non-trivial task from experimental point of view. There exists only one study in which the experimentally measured data for the atomic H was used to derive the fully measured interference oscillations [11]. The ion– or electron–atom scattering experiments with an atomic target (H or N or O) are non-trivial and hence are very rare in the literature [12, 26]. Particularly the derivation of the absolute cross sections from an atomic jet target needs careful measurements of the dissociation fraction of the molecule, e.g., H_2, since there will be some fraction of un-dissociated/or recombined molecules which interact with the ion beam in a crossed beam experiment. In such experiment, the atomic H is produced in a cavity in which an RF field [27] is applied, such as Slevin type or "surfatron"-type source. The procedure of *in situ* measurements of the dissociation fraction of H_2 and the derivation of the absolute cross section (DDCS) in case of an atomic H as a target have been discussed in a few earlier papers [12, 28]. The dissociation fraction was typically 80% in such experiments and it was deduced by using 9-eV proton detection technique. In this way [11, 20], the oscillation in the electron DDCS spectrum could be derived without using the theory. Another measurement in which the oscillations were derived without using any model [29] is discussed below.

There has been a rapid progress in the study of the interference effect in a molecular double slit: both experimentally [10, 11, 17, 20–23, 25] and theoretically [13, 14]. In the case of ionization of H_2, the existence of a double-frequency component in the interference oscillation has also been observed. Such a double-frequency component was expected based on the two successive intra-molecular scattering from two atoms which was experimentally observed by two independent experiments [30, 31] but did not get support from an *ab initio* theoretical model of Sisourat et al. [32] (see below).

6 Interference for Multi-electron Di-atomic Molecules

Initially the heavy-ion-induced interference studies were mostly limited to the ionization of H_2. For the multi-electron systems, the studies on interference effect were relatively scarce. However, by now there have been several experimental investigations on this aspect using different multi-electron molecules (such as N_2 and O_2, C_2H_2 and CO) [19, 29, 33–41]. Rivarola and co-workers have also studied this aspect theoretically [42, 43] on the ionization of N_2/O_2 by fast-proton or highly charged O-ion impact. Ionization of N_2 under 1–5 MeV proton impact has been studied by Baran et al. [19]. The DDCS ratios of $N_2/2N$ cross sections exhibit nearly constant frequency of oscillations independent of electron emission angle or collision velocity. The results suggest suppression of primary Young-type interference and dominance of second-order oscillations. It's needless to mention that the DDCS for N-atom was taken from the theoretical models.

Similar to the case of N_2, Winkoworth et al. [33–35] have also observed the second-order interference oscillation in collisions of 30 MeV O^{5+} and O^{8+} ions with O_2. Experimentally the oscillations in the DDCS ratio plots [e.g., in Fig. 2 of [33]] are not so obvious, the oscillation could be seen after dividing the e-DDCS ratio with a fitted straight line. This structure was fitted with a function $f(k) = \sin(kcd + w)$, where *k, d, c* are the electron velocity (in a.u.), the internuclear distance (2.28 a.u.), and an adjustable frequency parameter, respectively. The parameter w is an adjustable phase shift. Here, oscillations were reported to have the same frequency parameter for each electron emission angle without damping and hence it was assumed to be dominated by second-order interference. The damping is characteristic of the first order or primary interference as represented by Cohen–Fano equation. The reason for the absence of primary oscillation and dominance of secondary or multi-scattering mechanism is not clearly understood. It may imply that the double or inter-atomic scattering is dominant mechanism. The authors [19] qualitatively explained this by the orbitals' delocalization of the L-electrons over the molecule which might be the reason for the suppression of the primary Young-type interference for both the molecules. But it does not explain the presence of the second-order interference structure. Tachino et al. [43] have studied the interference effect theoretically for the similar collision systems as in Ref. [19], i.e., in collisions of protons with N_2 and showed the presence of the fingerprints of coherent electron emission for the individual molecular orbitals. However, the oscillation structures of electron emission from these molecules were reported to be diminished due to the cancellation of different phases of different MOs. The double differential cross sections (DDCS) and the DDCS ratios obtained by averaging over all possible molecular orientations are also calculated using the CDW-EIS approximation. The ratios for the complete molecule nor the ratios for each MO show the regular oscillating structures rather irregular oscillatory structures at higher energy. However, the results for the two most internal orbitals of the nitrogen molecule, namely, σ_g (1s) and σ_u (1s) orbitals, show interference patterns with opposite phase.

In order to explore this aspect in detail, Nandi et al. [37, 44] studied the interference effect in collisions of 51 and 42 MeV bare carbon ions with O_2. A prominent oscillation structure was not evident in the DDCS ratios. These authors did not see the second-order oscillations either. These are in agreement with the theoretical prediction of Rivarola and co-workers [43] and the reason for the no oscillation was explained in terms of the sum of all interference contributions arising from the ionization of different molecular orbitals in O_2. The resulting oscillations being phase shifted from each other tend to cancel partially or fully when added up. In synchrotron-based photo-ionization experiments by Ilchen et al. [45] the clear existence of interference oscillation was demonstrated for the $1\sigma_g$ (garade) and $1\sigma_u$ un-garade) orbitals of the N_2 molecule, with relatively opposite phase between them which is in good agreement with the theoretical prediction [42]. The non-existence of the primary oscillations in heavy-ion-induced ionization of N_2/O_2 molecules and observation of oscillations in photo-ionization widens the scope of further experiments.

7 Electron-Impact Ionization of N_2 and O_2

To explore more about the so-called ambiguity between the results with photons and heavy ions, experimental studies were performed with fast (keV energy) electrons which impart negligible perturbation to the target molecule as compared to heavy ions and hence it is expected that the probability for multiple ionization will be very less. From various studies on inner shell ionization, it is well known that heavy ions impart huge momentum to the target atom which causes the simultaneous multiple ionization of outer shells. Thus, in such collisions, the simultaneous ionization from different orbitals complicates the process, such that it becomes difficult to extract the interference oscillations. A comparative study of the multiple L-ionization using keV energy electron and MeV energy heavy ions via the study of various satellite K-X-rays was demonstrated in [46, 47]. Fast electrons having keV energies were chosen to probe both the N_2 and O_2 molecules. The experimentally obtained molecular e-DDCS was divided by twice the theoretical e-DDCS of the atomic target. This experimental-to-theoretical e-DDCS shows clear signature of interference oscillations both for N_2 and O_2 [48, 49]. The ratios were further fitted by the first-order Cohen–Fano-type model to obtain the frequency and amplitude of oscillations. The theoretical work on similar e-impact ionization can also be found in the work of Madison and co-workers [50].

Theoretically the DDCS ratios have been obtained using the effective charge Z_T corresponding to the residual target seen by the active electron. This choice is to evaluate the influence of the passive electrons in the final channel. One option is to use the asymptotic charge $Z_T = 1$, corresponding to a total screening of the nucleus. Another option frequently used by the theorists is $Z_T = Z_{bk} = (-2n^2\epsilon_j)^{1/2}$ [51]. The quantity ϵ_j is the Roothaan–Hartree–Fock energy [52] in case of atomic DDCS calculations and for estimating the MO energies for the molecular (N_2) DDCS computations. In both the cases, n_j represents the principal quantum number of the atomic orbital. Further experimental and theoretical details can be found in Ref. [48, 49, 53]. Figures 3, 4, 5 in Ref. [49] clearly demonstrated the influence of two different values of Z_T on the observed oscillation patterns. For example, a half sinusoidal oscillatory structure is observed about a horizontal line. The CF model explained the shape quite well except for the large backward angle, e.g., 145°. By using $Z_T = Z_{bk}$ the DDCS ratios show an oscillatory structure not about an horizontal line but overriding on a straight line of negative slope. Half sinusoidal oscillatory structures are deduced around a horizontal line only after dividing by the fitted straight line. The Cohen–Fano function has a good agreement with the derived DDCS ratios. The choice of the charge Z_T for calculating the atomic cross sections plays a crucial role in determining the shape of the oscillations in the ratios.

It should be mentioned that in another experimental work [50] with 250 eV electron impact on N_2 molecule the triple differential cross section (TDCS) ratio shows a clear structure, although the agreement with the Cohen–Fano factor is a bit poor at such lower energy (250 eV) collisions, compared to that at the 7 keV e-impact. It is remarkable to observe the interference oscillation in such finer measurement of highly (triple) differential cross sections. The oscillation was derived for a particular molecular orbital, i.e., $3\sigma_g$ from a study of triple differential distribution compared to the high-energy (7 keV) e-scattering experiment in which the double differential distribution in Ref. [48, 49] was studied. In spite of the possibility of shadowing the interference effect due to the cancellation of oscillations for multiple orbitals, the observation of residual oscillations in high-energy e-impact is interesting which is due to the fact that simultaneous multiple ionization of different orbitals is much less compared to that in heavy-ion collisions. We may also refer to another two sets of experiments in which fast electron-impact ionization of H_2 and N_2 targets was investigated in order to study the interference oscillation.

8 Angle-Dependent Frequency Parameter

It was further observed that the frequency parameter $c(\theta)$ was higher for backward angles compared to the forward angles, as observed in collisions with heavy ions and electrons. The actual functional form of $c(\theta)$ for these two types of projectiles are, however, different. In case of the e-impact on N_2, the frequency parameter is seen to be almost constant (about 1.2) for all the forward angles. For the backward angles, the parameter then increases steadily and changes from a value of 1.2–1.45 giving an enhancement of a factor (backward to forward) of about 1.2. It can be mentioned here that the shape of the angular dependence of $c(\theta)$ is quite different than that for H_2 [22, 54, 55]. In case of H_2 target, the frequency showed a distribution having a dip at 90°and the large backward angle has the frequency parameter which is almost a factor of two higher than forward angles for heavy-ion impact. These details can be found in Fig. 2 and Table 1 of Ref. [54] which summarizes the frequency data for different projectiles (heavy ions and electrons) taken from Refs. [22, 31, 55, 55].

9 Forward–Backward Asymmetry

This behavior of c(θ), i.e., difference in its values for extreme forward and backward angles was exploited elegantly to avoid the use of the DDCS for atomic target for which one often has to use the theory. The forward–backward asymmetry ($\alpha(k)$) in the DDCS values was studied for this purpose. The asymmetry parameter is defined as the ratio of the difference and sum of the DDCS measured at emission angles θ and $\pi - \theta$, i.e., the forward–backward difference as a fraction of the total DDCS for these two supplementary angles. According to Fainstein et al. [17], the non-Coulombic potential, the target wave function for a two or multi-electronic target contributes to the forward–backward angular asymmetry apart from the two-center (target and projectile center) effect and post-collision interaction. The asymmetry parameter also depends on the ejected electron velocity (k). However, in case of electrons as projectile, the two-center effect is negligible. The angular asymmetry parameter ($\alpha(k)$) was derived from the e-DDCS taken for two supplementary angles and a clear sinusoidal oscillation was observed when plotted as a function of ejected electron velocity which was first shown by Tribedi and co-workers [23, 55] for fast (80 MeV) bare C^{6+}-ions and 95 MeV F^{9+} colliding on H_2. The experiment was also conducted for a He target in the same experiment and no such structure in the asymmetry parameter was observed. The higher frequency of oscillations in the backward angles resulted to the oscillations in the asymmetry parameter, $\alpha(k)$.

The similar experimental observation has also been reported by Chatterjee et al. [55] in which the experiment was conducted using 8 keV electron beam colliding on H_2/He target. The signature of Young-type interference is clearly observed in the $\alpha(k)$ derived from the "only" molecular DDCS measured for a forward and the supplementary backward angle. The phase and frequency of the oscillation are well supported by the theoretical prediction based on the CDW-EIS approximation. A comparison is made between the shape of the $\alpha(k)$ in case of heavy-ion and electron collisions. For the first case where the post-collision and the two-center effects lead to a stronger slope in the $\alpha(k)$ on increasing ejection velocity. The oscillation rides on a line with steeper slope. In contrary, for the electron collision on H_2, the oscillation of $\alpha(k)$ about the horizontal axis ($\alpha(k) = 0$) was observed with positive and negative asymmetry. This again experimentally signifies the negligible role of the two-center effect in fast e-molecule collisions. The e-He collision does not show any oscillation as expected for the one-center atom.

These studies using heavy ions and electron beams imply that the interference plays a major role in the asymmetry parameter. The $\alpha(k)$ itself can be a sensitive tool for the measurement of interference oscillation. This reduces the experimental constraints significantly to derive the interference oscillation. As it is well known that multi-electronic atomic targets, e.g., O, N, etc., the experimental (as well as theoretical) investigations are a challenging task and thereby the present technique may be unique for interference studies for such molecules of H_2, N_2, O_2, etc. In addition it is now obvious that the asymmetry parameter also depends on the interference process besides the other well-known factors mentioned above, i.e., wave function, two-center effect, and non-Coulomb potential for a multi-electron target.

The results for the N_2 obtained for the 7 keV e were further confirmed by varying the incident electron energy. It is noteworthy to mention that similar procedures to obtain the interference oscillations were followed for the heavy-ion impact on N_2 and O_2 (in references [37, 38]), but no oscillations were observed. It may thus be inferred that the fast electrons are a much cleaner probe to observe interference oscillations.

10 Photo-Ionization and Interference

As discussed above that the molecular double-slit effect and the observed interference pattern in e-emission is due to the in-built spatial coherence in the electron emission from homonuclear diatomic molecules and this must be revealed in photo-ionization experiments. The slit distance and the wavelength of the photons correspond to the bond length in the molecule and the de Broglie wavelength of the electron wave, respectively. Several such studies can be found in [8, 16, 45, 56–60]. It is revealed that most of the existing related studies focus on the inner shells of inversion-symmetric homonuclear diatomic molecules in which the core electrons are chemically localized but quantum mechanically nonlocalized. It has been pointed out that it is the absence of "which-way information" that allows the interference to be observed. The non-localization allows coherent electron emission from the two atoms which are the fixed or spatially well-defined scattering centers (the "slits"). However, the fundamental formalism describing the molecular double slit was discussed for the chemically highly delocalized valence states of N_2 and O_2 [7, 8]. It was expected that the basic physics of the work of Samson et al. [8] and the numerous 1s-photo-ionization studies are similar for the prediction of the interference oscillations in the photo-ionization of partial cross sections. However, the oscillations in the partial cross sections were never measured. Another study of the molecular double-slit phenomenon in the vibrational branching ratios of molecular photo-ionization provided the experimental evidence that the basic model of Cohen and Fano is also valid for the valence states [29, 61, 62]. However, most of the earlier studies either in photo-ionization or charged particle-impact ionization, the oscillation in the total electron emission cross sections (integrated over different partial waves) was observed but not in the partial cross sections. This was explored first time by Becker and co-workers [45]. In fact, the difference between almost unexplored valence shells and the well-studied K-shell double-slit experiments and the resulting impact on basic quantum physics was discussed in this work.

The e-interference in the molecular double slit has been a topic to study the fundamental quantum mechanical aspects of the non-localization and coherence. Becker and co-workers have worked on this topic which are published in a series of publications. For example, Roles et al. [57] investigated the effect of different N-isotopes in the N_2 molecules, on the interference oscillations. They used the naturally most abundant $^{14,14}N_2$ nitrogen molecule and two different isotopomers: singly substituted $^{14,15}N_2$ and doubly substituted $^{15,15}N_2$. This isotope effect on the electronic structure of a diatomic molecule was probed by photoelectron spectroscopy, for the first time. This work has revealed the inversion symmetry of a system indeed causes non-local coherence of the otherwise localized *1s* core holes in a homonuclear diatomic molecule. It was argued that the distortion in the symmetry induced by the isotope substitution neither conserved nor destroyed the non-local behavior and coherence completely but instead caused a change in a continuous way into partially localized behavior owing to the gradual breakdown of inversion symmetry. This was inferred from the observed loss of interference and the parity mixing of the outgoing photoelectron waves.

In another experiment using high-brilliance third-generation light sources (such as the Advanced Light Sources), the vibrationally resolved photoelectron spectra were detected deep in the continuum. The authors, Canton et al. [29], reported the vibrationally resolved valence-shell photo-ionization spectra of H_2, N_2, and CO. They revealed that in the all cases, the Cohen–Fano interference is readily observable in the corresponding vibrational ratios (so-called ν-ratios). In this work (as well as in the work by Misra et al. [23]), the derivation of oscillation did not need to use any normalization by a theoretical cross section. Interestingly, the interference was also observed in the case of the heteronuclear CO molecule in which oscillations are visible to a lower extent compared to that for N_2 and H_2, due to the fact that CO has two different, i.e., non-identical centers. As mentioned above, the interference has been predicted in the theory, even for the very asymmetric molecule HeH^+ (Rivarola and co-workers) [63], under ionizing collisions which has not yet been observed in an experiment.

11 Typical Experimental Techniques

The experiments are carried out by different groups using different techniques, and different ion beams using different accelerator facilities. For example, a few groups use standard continuum electron (DDCS) spectroscopy for which the measurement techniques are broadly similar. Tribedi and co-workers have used a 14 MV Pelletron accelerator facility at TIFR, Mumbai for experiments with tens of MeVs of bare ions, as well as keV energy electron beams from an electron-beam gun. The details of the experimental setup can be found in Refs. [11, 23, 36, 44, 49, 55]. The major parts of the system are a high vacuum scattering chamber, a turntable, and a hemispherical electrostatic energy analyzer. In the scattering chamber, the base vacuum was about 1×10^{-7} to 5×10^{-8} mbar or better while the working pressure of the target gas in the flooded chamber was typically about 0.1 mTorr. The turntable allows us to detect the electrons for about 12 different emission angles in the range of (20° to 160° in steps of about 10–15°). The hemispherical electrostatic energy analyzer along with a channel electron multiplier (CEM) is used to analyze and detect the electrons in the selected energy range of 1–600 eV, with an energy resolution of about 5–6%. The front of the CEM is biased to +100 V in order to increase the efficiency of the CEM for low-energy electrons. For each angle, the number of ejected electrons at different energies are detected for a specified amount of collection of incident projectile charge on a Faraday cup. The data acquisition system was based on the LabVIEW software and the related hardware. The stray electric field had to be minimized in order to avoid any deflection of lowest energy electrons (about 1 eV). A pair of thin μ-metal shielding was used in the chamber to reduce the earth's magnetic field to a level of 5–10 mG.

The experiments carried out by Stolterfoht et al. used a setup in which electrons ejected from the target region were analyzed by an electrostatic 90° parallel-plate spectrometer whose solid angle was 1.8×10^{-3} sr, angular acceptance was 3-degree, and energy resolution FWHM was 5–7.5%. A similar electron analyzer was also used by Kamalou et al. [64] for the electron-impact interference study for a D_2 molecule. The experiment was carried out at the GANIL accelerator facility in Caen, France. A high-energy beam of 60 MeV/u Kr^{34+} ions was used for the collisions. At this high-energy heavy-ion collisions, the low-energy electron spectrum was mostly dominated by dipole transitions. The group of John Tanis at Western Michigan University using the 6-MV tandem Van de Graaff accelerator for these experiments. They have also used a parallel-plate electron analyzer. The fast (2.4 and 7 keV) e-impact ionization experiments were carried out using e-gun setups [49, 64].

The photo-ionization experiments [45] carried out at the synchrotron radiation facility BESSY II, Berlin as well as at DESY in Hamburg, were performed with angle-resolving photoelectron spectrometer setup employing 16 or 22 electron time-of-flight spectrometers aligned in the plane of the linear photon beam polarization. The spectrometer setup covered about 4% of the full solid angle. The energy resolution was sufficient to resolve all valence states of interest even up to 600 eV. The other photo-ionization and interference-related experiments by Uwe Becker and co-workers can be found in [57, 58, 65].

The experiment of Canton et al. [29] used the Advanced Light Source (ALS) in which the effusive molecular beams were ionized by linearly polarized radiation from ALS. A hemispherical electrostatic analyzer (Scienta SES-200) [66] mounted at the so-called magic angle with respect to the light polarization axis was used to detect the photoelectrons. The high resolution of the spectrometer helped to resolve the vibrational lines throughout the photon energy range covered (20–300 eV).

The much celebrated cold target recoil-ion-momentum spectroscopy (COLTRIMS) technique has also been used for finer measurements on the double-slit interference effect. Such as experiment involving the study of the double photo-ionization of H_2, Dorner and co-workers [56, 67] used 130–240 eV photon beams delivered by the "Advanced Light Source (ALS)" at the Lawrence Berkeley National Laboratory to cross a well-collimated cooled supersonic H_2 jet target. The electrons and ions after being guided by an electric field and a superimposed magnetic field were collected by two micro-channel plate detectors with delay line anodes.

In another set of photo-ionization experiment [40], the interference was studied using three hydrocarbon molecules C_2H_2, C_2H_2 and C_2H_6. These experiments have been carried out at the soft X-ray beamline at the Synchrotron SOLEIL in which the photons of energies 70–700 eV were used. The measurements were carried out using a wide-angle lens VG-Scienta R4000 electron spectrometer installed at a fixed position, with the electron detection axis perpendicular to the storage ring plane. X-ray light polarization was set at the magic angle of 54.7° with respect to the electron detection axis. The instrumental broadening was 200 meV. A platinum-coated diffraction grating with 1,600 L/mm along with a fine exit slit was used to maintain the photon energy resolution at about 200 meV for the energy range used in this study.

Several experiments [68–70] by the Stockholm group were performed at the ion storage ring CRYRING which was situated at Stockholm University at that time, which has been relocated at GSI, Darmstadt recently. In one of the experiments, the He^{2+} ions were injected, accelerated, stored, and electron cooled at ~MeV energies which interacted with a supersonically cooled H_2 jet. Resulting H_2^{2+} which was produced in a double capture goes through a Coulomb explosion to produce two 9 eV protons. These 9 eV protons were then extracted perpendicular to the projectile beam and the gas jet in a recoil-ion-momentum spectrometer [71–73] equipped with a resistive anode-type position-sensitive detector which detected one of the protons in coincidence with the neutral He projectiles after double e-capture.

In pioneering experiment by Schmidt et al. [69] of Stockholm group, the direct observation of the interference effect in Young's double-slit experiment where the interfering waves due to the two spatially separated components of the de Broglie wave of single hydrogen atoms was reported. The hydrogen atoms are formed due to e-capture from H_2. An RIMS spectrometer was used to perform this experiment using the CRYRING. They measured the momentum vectors for slow H^+ ions from the dissociation of the excited H_2^+ from the time of flight and the position of hitting the detector. In their standard RIMS setup, they also introduced a suppressor foil on the grid to cut down the strong random background of the low-energy H_2^+ ions from single ionization as well as H^+ ions from the dissociation. The experimental parameters were chosen to make sure the detection of two neutral H atoms was negligible which ensured that they indeed probed the interference of individual single neutral H atoms with themselves. From the proton momentum vector, one can identify the instantaneous orientation of the H_2 molecule during collision as dissociation is very fast compared to the rotation.

12 Second-Order Effect on Interference and Frequency Doubling

If we consider a second scattering mechanism in which the ejected electron produced in first scattering is re-scattered from the second atom in the molecule, there will be an additional path difference compared to the first-order scattering. If such a second-order process is considered then one may expect a higher frequency component in the oscillation. Such a formalism has been initially introduced by Messiah [74] in the context of electron scattering which was then used by a few others in heavy-ion collisions.

Double scattering mechanism is known to give rise to many interesting phenomena in atomic collision physics. One famous example is the Thomas scattering predicted as early as in [75] and detected much later by different groups. This double collision mechanism for the electron capture was experimentally explored much later in 1980s [76] and later using the ColTRIMS technique by Mergel et al. [77].

A close inspection of the DDCS ratios showed the presence of a double-frequency component riding over the first-order oscillatory structure. The periodic deviation from the Cohen–Fano model was attributed, by Stolterfoht et al. [31], to the second-order process in the electron DDCS spectrum measured for a fast highly charged ion (Kr^{34+}) collision with H_2. The second-order component arises due to the interference between the ionized electron wave from one atomic center and the re-scattered of the electron from the second identical atomic center. Tanis and co-workers have also demonstrated similar structures [18, 19] for MeV energy protons colliding with H_2 and N_2. However, the theoretical work by Sisourat et al. [32] differed from such an interpretation in terms of second-order Young-type interference.

In order to explore the issue further another set of experiments was conducted by Tribedi and co-workers [30] with an improved statistics, such that the observed structures can be seen beyond statistical fluctuations. A completely different collision system was used at much lower energy than those used previously [18, 19, 31]. The observed oscillatory structure was fitted by an analytical formula derived for the total DDCS ratio which includes the contribution from both the first- and second-order scattering effects. This formula on the ion-impact ionization is derived by extending the model used for the electron-impact ionization of H_2 [74]. The doubling of the frequency of oscillation is clearly observed in [30] due to the second scattering of the emitted electron, in agreement with the interpretation of Stolterfoht et al. [31]. The derived equation (Eq. (4) in Ref. [30]) is essentially the same as that in Ref. [31], where the quantities D, F, and S represent the amplitudes of the direct, first- and second-order contributions, respectively. The first-order term, i.e., *sin (kcd)/(kcd)* is termed as the Cohen–Fano (CF) term. We have used the frequency parameter c in the first-order term following Ref. [23]. In a similar way, the c is introduced in the second-order term, i.e., *sin (kcd/kcd)* instead of *sin(kd/kd)* in the second term (see equation (Eq. (4) in Ref. [30]). The second-order term has oscillation frequency that is twice of the first-order contribution and it is also $\pi/2$ out of phase with respect to the first-order term.

This frequency doubling is not only related to the heavy-ion collisions with H_2 but this has also been observed in fast electron collisions with the N_2 molecule [49]. In this case, the forward–backward asymmetry (as discussed above) parameter was plotted which also revealed the interference oscillation. The first-order CF model fits very well with the experimental data. A closer look at Fig. 7 of Ref. [49] shows that though the first-order model fitting matches well with the data, a periodic deviation is also obvious. In order to study the deviations, the data divided by the CF model clearly reveals an oscillatory structure which is further fitted by a model (solid line), similar to the Cohen–Fano-type formalism with oscillation frequency nearly twice the frequency of the primary oscillation. This is another example indicating the presence of a second-order scattering mechanism.

Sisourat et al. [32] have also investigated the interference effects theoretically using a non-perturbative semiclassical approach, without any approximation. Angle- and energy-differential results were analyzed as a function of the projectile charge, velocity, as well as the geometric characteristics of the molecules, such as for parallel and perpendicular alignment. These calculations agreed qualitatively with the available experimental and theoretical interpretations regarding the presence of the primary interference oscillations and validate the observed primary interference patterns. However, contrary to the experimental observation [30, 31] these new calculations did not show such secondary oscillations, in spite of the inclusion of such multi-scattering mechanism in the model.

13 Applications of Interference Studies Toward Fundamental Physics

It is well known that Young's double-slit experiment played vital role in the development of quantum mechanics. It was one of the most crucial experiments to demonstrate the wave–particle duality by studying the interference using the light, electron, atoms, or even molecules. The probabilistic nature of the particles and the dependence of the outcome on the act of measurement were other interesting fundamental aspects which were highlighted. The electron scattering experiments of Davisson and Germer [1], Thomson and Reid [2], and Jonsson [3], two slit-type interference experiments using electron microscope by Tonomura et al. [4] have put the wave–particle duality on firm ground for electrons. Such experimental evidences are also demonstrated for even a large molecule like fullerenes (C_{60}) [5] besides atoms. The interference effect continues to attract attention of researchers to address a fundamental quantum mechanical issue. Accordingly the recent studies on the electron interference in a single molecular double slit have attracted many researchers to relook into several fundamental quantum mechanical issues, such as localization and loss of coherence, entanglement, influence of particle–environment coupling on the wave phase, etc.

Schmidt et al. [69] demonstrated the direct experimental evidence of the two-center quantum interference in a parameter region where the classical limit is expected to be approached and thereby one expected to have a dominating effect of classical physics. They have provided the experimental evidence of the wave–particle duality for fast neutral H-atom using the two-center double-silt scattering technique. This conclusion was arrived from the fact that the position distribution of the neutral atoms clearly depended on the orientation of the target molecule during collision which only can be accounted for if the wave properties of the propagating fast H atoms are taken into account. The outgoing hydrogen atom wave is a superposition of contributions from either proton in the molecule. Further details can be found in [69]. Using almost similar RIMS technique Misra et al. [68] studied the interference effect in the double e-capture from H_2 in collisions with He^{2+} ions. In this work, the authors observed a strong oscillatory behavior of the double-capture cross section as a function of the molecular orientation which was attributed to the Young-type interference effect. They also deliberated on the frequently asked question of the sequential versus correlated process of double capture and concluded in favor of double capture in a correlated fashion for ion trajectories close to either target nucleus.

The interference in molecular double slit (H_2) has been studied in photo-ionization experiments by Dorner and co-workers [56, 67] to deal with fundamental questions, such as entanglement issues. With the de Broglie wavelengths close to the size of their surroundings, the coupling of the particles to its environment causes a shift in the wave phase. The resulting decoherence can be a main cause of the transition from quantum to classical behavior. The question they address is the amount of interaction needed to induce this transition. They show that a photoelectron and two protons form a minimum slit system while a single additional electron constitutes a minimum environment. These authors used the ColTRIMS technique elegantly to reveal the fact that the interference fringes as observed in the angular distribution of the electron are lost through its interaction with a second electron, while the correlated momenta of the entangled electron pair continue to exhibit the quantum interference. More details can be found in Ref. [56]. The interplay of the electron correlation and double-slit interference in double photo-ionization of H_2 are dealt in Ref. [67]. Their data show a clear interference pattern in Jacobi coordinate K^+ where $K^+ = k_1 + k_2$ which corresponds to center-of-mass motion of the electron pair. By momentum conservation this corresponds to the recoil momentum imparted onto the two protons. Their analysis reveals that the interference arises due to the fact that the electron pair corresponds to the same proton, which constitutes only a small fraction to the ground-state wave function of the H_2. Further experimental and analytical details can be found in Ref. [67].

Uwe Becker and co-workers [58] have addressed the issue of localization and loss of coherence in molecular double-slit experiments by looking into the photo-ionization of two diatomic molecules. In this work, they used homonuclear as well as heteronuclear molecules, such as N_2 and CO. They explored this behavior for photo-electrons emitted from these molecules in interaction with soft X-rays, and showed that intra-molecular scattering can also lead to electron interference between the scattered and unscattered waves in addition to the undisturbed emission process. It implies the interference arises in two ways, i.e., two-center interference between two spatially separated coherent emitters and one-center self-interference. They argued that the one-center interference is the signature of a loss of spatial coherence and the transition from two-center to one-center coherence occurs over a scale of the de Broglie wavelength of the scattered photoelectron. Zimmermann et al.[58] have presented an evidence for a scattering-induced transition of quantum coherence in molecular 1 s photo-ionization from the observation of the oscillation with frequency in momentum space that is caused by the self-interference of photoelectrons emitted from only one atomic site which is the fingerprint of localization. Further details can be found in [58]. In another experiment, Rolles et al. [57] investigated the isotope-induced partial localization of core electrons in the homonuclear diatomic molecule N_2. They explored the gradual transition to a symmetry-broken system of localized electrons by comparing different isotope substituted species. This phenomenon is analogous of having partial "which-way" information in macroscopic double-slit experiments. The experimental techniques, further results, and analysis can be found in [57].

In another photo-ionization experiment [40], the interference technique was successfully applied for the poly-atomic hydrocarbon molecules. By considering inner-valence shell ionization of C_2H_2, C_2H_4, C_2H_6, they showed that double-slit interference is able to provide quantitative information on geometry, orbital composition, and accurate bond length.

14 Conclusions

In this article, we have reviewed the recent studies on the Young-type electron interference due to scattering from a molecular double slit in which two atoms are the centers (slits) of e-scattering. This is not meant to be an exclusive and elaborate review of the topic, rather a brief overview, which can be used as a reference article. The results on the interference oscillations as observed in fast-ion-induced ionization, photo-ionization, and electron-impact ionization are presented. The techniques of deriving the oscillations from the ejected electron spectroscopy, ColTRIMS, vibrational spectroscopy, and the electron time of flight are indicated. The second-order interference, frequency doubling, angular dependence of frequency for different projectiles, and explaining the series of data in terms of the Cohen–Fano model are included in brief. The applications of interference studies to unfold several issues in fundamental physics are also touched upon. The article has also discussed the major experimental techniques involved in these studies.

References

1. C. Davisson, L.H. Germer, Diffraction of electrons by a crystal of nickel, Phys. Rev. **30**, 705–740 (1927). https://doi.org/10.1103/PhysRev.30.705, https://doi.org/10.1103/PhysRev.30.705
2. G.P. Thomson, A. Reid, Diffraction of cathode rays by a thin film. Nature **119**(3007), 890 (1927). https://doi.org/10.1038/119890a0
3. C. Jonsson, Zeitschrift ftir Physik **161**, 454–474 (1961)
4. A. Tonomura, J. Endo, T. Matsuda, T. Kawasaki, H. Ezawa, Am. J. Phys. **57**, 117–120 (1989)
5. M. Arndt, L. Hackermüller, K. Hornberger, A. Zeilinger, *Organic molecules and decoherence: experiments in a molecule interferometer* (Birkhäuser, Boston, 2004)
6. The double-slit experiment. Phys. World **15**(9), 15 (2002). https://doi.org/10.1088/2058-7058/15/9/20
7. H.D. Cohen, U. Fano, Phys. Rev. **150**, 30 (1966)
8. J.A.R. Samson, R.B. Cairns J. Opt. Soc. Am. **55**, 1035 (1965)
9. I.G. Kaplan, A.P. Markin, Sov. Phys. Dokl. **14**, 36 (1969)
10. N. Stolterfoht, B. Sulik, V. Hoffmann, B. Skogvall, J.Y. Chesnel, J. Rangama, F. Frémont, D. Hennecart, A. Cassimi, X. Husson, A.L. Landers, J.A. Tanis, M.E. Galassi, R.D. Rivarola, Evidence for interference effects in electron emission from H_2 colliding with 60MeV$/u$ kr^{34+} ions. Phys. Rev. Lett. **87**, 023201 (2001). https://link.aps.org/doi/10.1103/PhysRevLett.87.023201, https://doi.org/10.1103/PhysRevLett.87.023201
11. D. Misra, U. Kadhane, Y.P. Singh, L.C. Tribedi, P.D. Fainstein, P. Richard, Phys. Rev. Lett. **92**, 153201 (2004)
12. L.C. Tribedi, P. Richard, W. DeHaven, L. Gulyás, M.W. Gealy, M. Rudd, J. Phys. B: At. Mol. Opt. Phys. **31**, L369 (1998)
13. M.E. Galassi, R.D. Rivarola, P. Fainstein, N. Stolterfoht, Phys. Rev. A **66**, 052705 (2002)
14. L. Nagy, L. Kocbach, K. Pora, J.P. Hansen, J. Phys. B: At. Mol. Opt. Phys. **35** (2002)
15. M.E. Galassi, R.D. Rivarola, P.D. Fainstein, Phys. Rev. A **70**, 032721 (2004)
16. O.A. Fojon, J. Fernandez, A. Palacios, R.D. Rivarola, F. Martín, J. Phys. B: At. Mol. Opt. Phys. **37**, 3035 (2004)
17. J.A. Tanis, S. Hossain, B. Sulik, N. Stolterfoht, Phys. Rev. Lett. **95**, 079301 (2005)
18. S. Hossain, A.L. Landers, N. Stolterfoht, J.A. Tanis, Interference phenomena associated with electron-emission from h_2 by $(1--5)$-MeV h^+ impact, Phys. Rev. A **72**, 010701 (2005). https://link.aps.org/doi/10.1103/PhysRevA.72.010701, https://doi.org/10.1103/PhysRevA.72.010701
19. J.L. Baran, S. Das, F. Jarai-Szabo, K. Pora, L. Nagy, J.A. Tanis, Phys. Rev. A **78**, 012710 (2008)
20. D. Misra, U. Kadhane, Y.P. Singh, L.C. Tribedi, P.D. Fainstein, P. Richard, Phys. Rev. Lett. **95**, 079302 (2005)
21. F. Fremont, A. Hajaji, A. Naja, C. Leclercq, J. Soret, J.A. Tanis, B. Sulik, J.Y. Chesnel, Phys. Rev. A **72**, 050704 (2005)
22. J.A. Tanis, J.-Y. Chesnel, B. Sulik, B. Skogvall, P. Sobocinski, A. Cassimi, J.-P. Grandin, L. Adoui, D. Hennecart, N. Stolterfoht, Phys. Rev. A **74**, 022707 (2006)
23. D. Misra, A. Kelkar, U. Kadhane, A. Kumar, L.C. Tribedi, P.D. Fainstein, Phys. Rev. A **74**, 060701 (2006)
24. J. Fernandez, O. Fojon, A. Palacios, F. Martín, Phys. Rev. Lett. **98**, 043005 (2007)
25. L.P.H. Schmidt, S. Schossler, F. Afaneh, M. Schöffler, K.E. Stiebing, H. Schmidt-Bocking, R. Dörner, Phys. Rev. Lett. **101**, 173202 (2008)
26. T.W. Shyn, Doubly differential cross sections of secondary electrons ejected from atomic hydrogen by electron impact. Phys. Rev. A **45**, 2951–2956 (1992). https://link.aps.org/doi/10.1103/PhysRevA.45.2951, https://doi.org/10.1103/PhysRevA.45.2951
27. J. Slevin, W. Stirling, Radio frequency atomic hydrogen beam source. Rev. Sci. Instrum. **52**(11), 1780–1782 (1981). arXiv:https://pubs.aip.org/aip/rsi/article-pdf/52/11/1780/8370705/1780_1_online.pdf, https://pubs.aip.org/aip/rsi/article-pdf/52/11/1780/8370705/1780_1_online.pdf, https://doi.org/10.1063/1.1136497

28. M. Rudd, T. Goffe, R.D. DuBois, L. Toburen, Phys. Rev. A **31**, 492 (1985)
29. S.E. Canton, E. Plesiat, J. Bozek, B. Rude, P. Decleva, F. Martín, Proc. Natl. Acad. Sci. U. S. A. **108**, 7302–7306 (2011)
30. D. Misra, A.H. Kelkar, S. Chatterjee, L.C. Tribedi, Phys. Rev. A **80**, 062701 (2009)
31. N. Stolterfoht, B. Sulik, L. Gulyás, B. Skogvall, J.Y. Chesnel, F. Frémont, D. Hennecart, A. Cassimi, L. Adoui, S. Hossain, J.A. Tanis, Interference effects in electron emission from h_2 by 68-mev/u kr^{33+} impact: dependence on the emission angle. Phys. Rev. A **67**, 030702 (2003). https://link.aps.org/doi/10.1103/PhysRevA.67.030702, https://doi.org/10.1103/PhysRevA.67.030702
32. N. Sisourat, J. Caillat, A. Dubois, P.D. Fainstein, Phys. Rev. A **76**, 012718 (2007)
33. M. Winkworth, P.D. Fainstein, M.E. Galassi, J.L. Baran, B.S. Dassanayake, S. Das, A. Kayani, J.A. Tanis, Nucl. Instrm. Meth. Phys. Res. B **267**, 373 (2009)
34. M. Winkworth, P.D. Fainstein, M.E. Galassi, J. Baran, S. Das, B.S. Dassanayake, A. Kayani, J.A. Tanis, Interference effects in electron emission spectra for 3 mev/u h+ + o2 collisions. J. Phys.: Conf. Ser. **163**(1), 012044 (2009). https://doi.org/10.1088/1742-6596/163/1/012044
35. M. Winkworth, P.D. Fainstein, M.E. Galassi, J. Baran, S. Das, B.S. Dassanayake, A. Kayani, J.A. Tanis, Interferences associated with electron emission from O2 by fast ions compared with H2 and N2. J. Phys.: Conf. Ser. **194**(10), 102044 (2009). https://doi.org/10.1088/1742-6596/194/10/102044
36. S. Nandi, A.N. Agnihotri, C.A. Tachino, R.D. Rivarola, F. Martín, L.C. Tribedi, J. Phys. B: At. Mol. Opt. Phys. **45**, 215207 (2012)
37. S. Nandi, A.N. Agnihotri, S. Kasthurirangan, A. Kumar, C.A. Tachino, R.D. Rivarola, F. Martín, L.C. Tribedi, Phys. Rev. A **85**, 062705 (2012)
38. S. Nandi, S. Biswas, C. Tachino, R. Rivarola, L.C. Tribedi, Eur. Phys. J. D **69**, 192 (2015)
39. N. Watanabe, X. Chen, M. Takahashi, Phys. Rev. Lett. **108**, 173201 (2012)
40. R. Kushawaha, M. Patanen, R. Guillemin, L. Journel, C. Miron, M. Simon, M.N. Piancastelli, C. Skates, P. Decleva, Proc. Natl. Acad. Sci. USA **110** (2013)
41. B. Zimmermann, D. Rolles, B. Langer, R. Hentges, M. Braune, S. Cvejanovic, O. Geßner, F. Heiser, S. Korica, T. Lischke, A. Reinköster, J. Viefhaus, R. Dörner, V. McKoy, U. Becker, Localization and loss of coherence in molecular double-slit experiments. Nat. Phys. **4**(8), 649–655 (2008). https://doi.org/10.1038/nphys993
42. C.A. Tachino, M. Galassi, F. Martín, R.D. Rivarola, J. Phys. B: At. Mol. Opt. Phys. **45**, 025201 (2011)
43. C.A. Tachino, F. Martín, R.D. Rivarola, Theoretical study of interference effects in single electron ionization of N2 molecules by proton impact. J. Phys. B: Atomic, Mol. Opt. Phys. **45**(2), 025201 (2011). https://doi.org/10.1088/0953-4075/45/2/025201
44. S. Nandi, A.N. Agnihotri, C.A. Tachino, R.D. Rivarola, F. Martín, L.C. Tribedi, Phys. Scr. **T156**, 014038 (2013)
45. M. Ilchen, L. Glaser, F. Scholz, P. Walter, S. Deinert, A. Rothkirch, J. Seltmann, J. Viefhaus, P. Decleva, B. Langer, A. Knie, A. Ehresmann, O.M. Al-Dossary, M. Braune, G. Hartmann, A. Meissner, L.C. Tribedi, M. AlKhaldi, U. Becker, Angular momentum sensitive two-center interference. Phys. Rev. Lett. **112**, 023001 (2014). https://link.aps.org/doi/10.1103/PhysRevLett.112.023001, https://doi.org/10.1103/PhysRevLett.112.023001
46. A. Kumar, D. Misra, K. Thulasiram, L. Tribedi, A. Pradhan, High resolution studies of al satellite lines and SI-projectile X-rays in electron-atom and ion-atom collisions. Nucl. Instrum. Methods Phys. Res. Sect. B: Beam Interact. Mater. Atoms **248**(2), 247–252 (2006). https://www.sciencedirect.com/science/article/pii/S0168583X06006422, https://doi.org/10.1016/j.nimb.2006.04.154
47. A. Kumar, D. Misra, A.H. Kelkar, U.R. Kadhane, K.V. Thulasiram, L.C. Tribedi, Pramana **68**, 983–994 (2007)
48. M.R. Chowdhury, L.C. Tribedi, J. Phys. B: At. Mol. Opt. Phys. **50**, 155201 (2017)
49. M.R. Chowdhury, C.R. Stia, C.A. Tachino, O.A. Fojon, R.D. Rivarola, L.C. Tribedi, Phys. Rev. A **94**, 052703 (2016)

50. H. Chaluvadi, Z.N. Ozer, M. Dogan, C. Ning, J. Colgan, D. Madison, Observation of two-center interference effects for electron impact ionization of N2. J. Phys. B: Atomic, Mol. Opt. Phys. **48**(15), 155203 (2015). https://doi.org/10.1088/0953-4075/48/15/155203
51. D. Belkić, R. Gayet, A. Salin, Electron capture in high-energy ion-atom collisions. Phys. Rep. **56**(6), 279–369 (1979). https://www.sciencedirect.com/science/article/pii/0370157379900358, https://doi.org/10.1016/0370-1573(79)90035-8
52. E. Clementi, C. Roetti, Roothaan-hartree-fock atomic wavefunctions: basis functions and their coefficients for ground and certain excited states of neutral and ionized atoms, Z54. Atomic Data Nucl. Data Tables **14**(3), 177–478 (1974). https://www.sciencedirect.com/science/article/pii/S0092640X74800161, https://doi.org/10.1016/S0092-640X(74)80016-1
53. M.F. Ciappina, O.A. Fojón, R.D. Rivarola, Coherent electron emission from simple molecules by impact of energetic charged particles*. J. Phys. B: Atomic, Mol. Opt. Phys. **47**(4), 042001 (2014). https://doi.org/10.1088/0953-4075/47/4/042001
54. S. Chatterjee, D. Misra, P.D. Fainstein, L.C. Tribedi, Electron interference in fast ion collisions with H2 and the frequency parameter. Phys. Scr. **2011**(T144), 014040 (2011). https://doi.org/10.1088/0031-8949/2011/T144/014040
55. S. Chatterjee, D. Misra, A.H. Kelkar, L.C. Tribedi, C.R. Stia, O.A. Fojon, R.D. Rivarola, Phys. Rev. A **78**, 052701 (2008)
56. D. Akoury, K. Kreidi, T. Jahnke, T. Weber, A. Staudte, M. Schöffler, N. Neumann, J. Titze, L.P. Schmidt, A. Czasch, O. Jagutzki, R.A. Costa Fraga, R.E. Grisenti, R. Díez Muiño, N.A. Cherepkov, S.K. Semenov, P. Ranitovic, C.L. Cocke, T. Osipov, H. Adaniya, J. Thompson, M. Prior, A. Belkacem, A. Landers, R. Schmidt-Böcking HDörner, Science **318**, 949 (2007)
57. D. Rolles, M. Braune, S. Cvejanović, O. Geßner, R. Hentges, S. Korica, B. Langer, T. Lischke, G. Prümper, A. Reinköster, J. Viefhaus, B. Zimmermann, V. McKoy, U. Becker, Nature (London) **437**, 711 (2005)
58. B. Zimmermann, D. Rolles, B. Langer, R. Hentges, M. Braune, S. Cvejanovic, O. Geßner, F. Heiser, S. Korica, T. Lischke, A. Reinköster, J. Viefhaus, R. Dörner, V. McKoy, U. Becker, Nat. Phys. **4**, 649 (2008)
59. E. Shigemasa, J. Adachi, M. Oura, A. Yagishita, Phys. Rev. Lett. **74**, 359 (1995)
60. X.-J. Liu, N.A. Cherepkov, S.K. Semenov, V. Kimberg, F. Gel'mukhanov, G. Prümper, T. Lischke, T. Tanaka, M. Hoshino, H. Tanaka, K. Ueda, Phys. B **39**, 4801 (2006)
61. E. Plésiat, P. Decleva, F. Martín, Phys. Chem. Chem. Phys. **14**, 10853 (2012)
62. U. Becker, Nature (London) **474**, 586 (2011)
63. C.A. Tachino, M.E. Galassi, F. Martín, R.D. Rivarola, Coherence and partial localization in electron emission from asymmetric diatomic molecules. J. Phys.: Conf. Ser. **288**, 012026 (2011). https://doi.org/10.1088/1742-6596/288/1/012026
64. O. Kamalou, J.-Y. Chesnel, D. Martina, J. Hanssen, C.R. Stia, O.A. Fojón, R.D. Rivarola, F. Frémont, Evidence for interference effects in both slow and fast electron emission from D_2 by energetic electron impact. Phys. Rev. A **71**, 010702 (2005). https://link.aps.org/doi/10.1103/PhysRevA.71.010702, https://doi.org/10.1103/PhysRevA.71.010702
65. A.S. Baltenkov, U. Becker, S.T. Manson, A.Z. Msezane, Interference in the molecular photoionization and young's double-slit experiment. J. Phys. B: Atomic, Mol. Opt. Phys. **45**(3), 035202 (2012). https://doi.org/10.1088/0953-4075/45/3/035202
66. N. Berrah, B. Langer, A. Wills, E. Kukk, J. Bozek, A. Farhat, T. Gorczyca, High-resolution angle-resolved measurements in atoms and molecules using advanced photoelectron spectroscopy at the ALS. J. Electron Spectrosc. Relat. Phenom. **101–103**, 1–11 (1999). https://www.sciencedirect.com/science/article/pii/S0368204898004642, https://doi.org/10.1016/S0368-2048(98)00464-2
67. K. Kreidi, D. Akoury, T. Jahnke, T. Weber, A. Staudte, M. Schöffler, N. Neumann, J. Titze, L.P.H. Schmidt, A. Czasch, O. Jagutzki, R.A. Costa Fraga, R.E. Grisenti, M. Smolarski, P. Ranitovic, C.L. Cocke, T. Osipov, H. Adaniya, J.C. Thompson, M.H. Prior, A. Belkacem, A.L. Landers, H. Schmidt-Böcking, R. Dörner, Interference in the collective electron momentum in double photoionization of h_2. Phys. Rev. Lett. **100**, 133005 (2008). https://link.aps.org/doi/10.1103/PhysRevLett.100.133005, https://doi.org/10.1103/PhysRevLett.100.133005

68. D. Misra, H.T. Schmidt, M. Gudmundsson, D. Fischer, N. Haag, H.A.B. Johansson, A. Källberg, B. Najjari, P. Reinhed, R. Schuch, M. Schöffler, A. Simonsson, A.B. Voitkiv, H. Cederquist, Two-center double-capture interference in fast $\mathbf{He}^{2+} + \mathbf{h}_2$ collisions. Phys. Rev. Lett. **102** (2009) 153201. https://doi.org/10.1103/PhysRevLett.102.153201
69. H. T. Schmidt, D. Fischer, Z. Berenyi, C. L. Cocke, M. Gudmundsson, N. Haag, H.A.B. Johansson, A. Källberg, S.B. Levin, P. Reinhed, U. Sassenberg, R. Schuch, A. Simonsson, K. Støchkel, H. Cederquist, Evidence of wave-particle duality for single fast hydrogen atoms. Phys. Rev. Lett. **101**, 083201 (2008). https://link.aps.org/doi/10.1103/PhysRevLett.101.083201, https://doi.org/10.1103/PhysRevLett.101.083201
70. K. Støchkel, O. Eidem, H. Cederquist, H. Zettergren, P. Reinhed, R. Schuch, C.L. Cocke, S.B. Levin, V.N. Ostrovsky, A. Kälberg, A. Simonsson, J. Jensen, H.T. Schmidt, Two-center interference in fast proton–h_2-electron transfer and excitation processes. Phys. Rev. A **72**, 050703 (2005). https://link.aps.org/doi/10.1103/PhysRevA.72.050703, https://doi.org/10.1103/PhysRevA.72.050703
71. H.T. Schmidt, A. Fardi, R. Schuch, S.H. Schwartz, H. Zettergren, H. Cederquist, L. Bagge, H. Danared, A. Källberg, J. Jensen, K.-G. Rensfelt, V. Mergel, L. Schmidt, H. Schmidt-Böcking, C.L. Cocke, Double-to-single target ionization ratio for electron capture in fast p-he collisions. Phys. Rev. Lett. **89**, 163201 (2002). https://link.aps.org/doi/10.1103/PhysRevLett.89.163201, https://doi.org/10.1103/PhysRevLett.89.163201
72. H. Schmidt, A. Fardi, J. Jensen, P. Reinhed, R. Schuch, K. Støchkel, H. Zettergren, H. Cederquist, C. Cocke, Transfer ionization in p+he collisions. Nucl. Instrum. Methods Phys. Res. Sect. B: Beam Interact. Mater. Atoms **233**(1), 43–47 (2005). https://www.sciencedirect.com/science/article/pii/S0168583X05003083, https://doi.org/10.1016/j.nimb.2005.03.084. Fast Ion-Atom Collisions
73. H.T. Schmidt, J. Jensen, P. Reinhed, R. Schuch, K. Støchkel, H. Zettergren, H. Cederquist, L. Bagge, H. Danared, A. Källberg, H. Schmidt-Böcking, C.L. Cocke, Recoil-ion momentum distributions for transfer ionization in fast proton-he collisions. Phys. Rev. A **72**, 012713 (2005). https://link.aps.org/doi/10.1103/PhysRevA.72.012713, https://doi.org/10.1103/PhysRevA.72.012713
74. A. Messiah, *Quantum Mechanics*, vol. 2, Quantum Mechanics. (Elsevier Science, 1961)
75. H.L. Thomas, R. Soc. Publ. **114** (1927) 561–576
76. E. Horsdal-Pedersen, C.L. Cocke, M. Stockli, Experimental observation of the Thomas peak in high-velocity electron capture by protons from he. *Phys. Rev. Lett.* **50**, 1910–1913 (1983). https://link.aps.org/doi/10.1103/PhysRevLett.50.1910, https://doi.org/10.1103/PhysRevLett.50.1910
77. V. Mergel, R. Dörner, M. Achler, K. Khayyat, S. Lencinas, J. Euler, O. Jagutzki, S. Nüttgens, M. Unverzagt, L. Spielberger, W. Wu, R. Ali, J. Ullrich, H. Cederquist, A. Salin, C.J. Wood, R.E. Olson, D.C.V. Belkić, C.L. Cocke, H. Schmidt-Böcking, Intra-atomic electron-electron scattering in p-he collisions (Thomas process) investigated by cold target recoil ion momentum spectroscopy. Phys. Rev. Lett. **79** (1997) 387–390. https://link.aps.org/doi/10.1103/PhysRevLett.79.387, https://doi.org/10.1103/PhysRevLett.79.387

The manufacturer's authorised representative in the EU is Springer Nature Customer Service Centre GmbH, Europaplatz 3, 69115 Heidelberg, Germany. If you have any concerns regarding our products, please contact ProductSafety@springernature.com

Printed and bound by CPI Group (UK) Ltd, Croydon, CR0 4YY
01/10/2026
02225962-0008

SÄCHSISCHE AKADEMIE DER WISSENSCHAFTEN
ZU LEIPZIG

JAHRBUCH
1973–1974

MIT 25 ABBILDUNGEN AUF 16 TAFELN

AKADEMIE-VERLAG · BERLIN
1976

Nach dem Stand vom 31. Dezember 1974
Manuskript eingeliefert am 13. Oktober 1975
Druckfertig erklärt am 13. November 1976

Im Auftrage der Akademie
bearbeitet und herausgegeben
von Gerald Wiemers

Erschienen im Akademie-Verlag, 108 Berlin, Leipziger Straße 3—4

Lizenznummer: 202 · 100/93/76 · P 127/76
Gesamtherstellung: VEB Druckhaus „Maxim Gorki", 74 Altenburg
Bestellnummer: 752 940 3 (1011/73–74) · LSV 0725
Printed in GDR
DDR 68,— M

INHALT

PRÄSIDIUM

Präsident
Schwabe, Kurt, em. Prof. Dr.-Ing., Dr. rer. nat. h. c. mult.

Vizepräsident
Lehmann, Edgar, em. Prof. Dr. phil., Dr.-Ing. e. h.

Sekretar der mathematisch-naturwissenschaftlichen Klasse
Neef, Ernst, em. Prof. Dr. phil.

Stellvertr. Sekretar der mathematisch-naturwissenschaftlichen Klasse
Emmrich, Rolf, Prof. Dr. med. †
Drischel, Hans, Prof. Dr. med.

Sekretar der philologisch-historischen Klasse
Große, Rudolf, Prof. Dr. phil.

Stellvertr. Sekretar der philologisch-historischen Klasse
Jahn, Johannes, em. Prof. Dr. phil., Dr. phil. h. c. †

Lösche, Artur, Prof. Dr. phil.

PRÄSIDIUM

nach der Wahl vom 21. Mai 1976

Präsident

Schwabe, Kurt, em. Prof. Dr.-Ing., Dr. rer. nat. h. c. mult.

Vizepräsident

Lehmann, Edgar, em. Prof. Dr. phil., Dr.-Ing. e. h.

Sekretar der mathematisch-naturwissenschaftlichen Klasse

Neef, Ernst, em. Prof. Dr. phil.

Stellvertr. Sekretar der mathematisch-naturwissenschaftlichen Klasse

Drischel, Hans, Prof. Dr. med.

Sekretar der philologisch-historischen Klasse

Große, Rudolf, Prof. Dr. phil.

Stellvertr. Sekretar der philologisch-historischen Klasse

Brüning, Eberhard, Prof. Dr. phil.

Růžička, Rudolf, Prof. Dr. phil.

DIE PRÄSIDENTEN, VIZEPRÄSIDENTEN, SEKRETARE UND STELLVERTRETENDEN SEKRETARE

seit der Wiedereröffnung am 8. Dezember 1948

Präsidenten

1948—1965 Theodor Frings
seit 1965 Kurt Schwabe

Vizepräsidenten

1948—1958 Ludwig Binder
1958—1959 Anton Arland in Stellvertretung
1959 Robert Schröder
1959—1962 Arthur Simon
1960—1965 Anton Arland
1962—1965 Ernst Neef in Stellvertretung
1965—1971 Ernst Neef
1966—1970 Siegfried Morenz
seit 1971 Edgar Lehmann

Die Sekretare der Klassen

Mathematisch-naturwissenschaftliche Klasse

1949—1960 Anton Arland
1960—1965 Ernst Neef
1965—1971 Rolf Emmrich
seit 1971 Ernst Neef

Philologisch-historische Klasse

1948—1965 Erwin Jacobi
1965—1971 Johannes Jahn
seit 1971 Rudolf Große

Die stellvertretenden Sekretare der Klassen

Mathematisch-naturwissenschaftliche Klasse

1949—1956 wurde das Amt nicht besetzt
1957—1958 Heinrich Bredt
1959—1960 Ernst Neef
1960—1965 Erich Strack
1965—1971 Peter Friedrich Matzen
1971—1974 Rolf Emmrich
seit 1974 Hans Drischel

Philologisch-historische Klasse

1949—1956	wurde das Amt nicht besetzt
1957—1958	Heinz Ladendorf
1958—1961	Siegfried Morenz
1961—1965	Edgar Lehmann
1965—1971	Gertrud Schubart-Fikentscher
1971—1976	Johannes Jahn
seit 1976	Eberhard Brüning

SATZUNG DER AKADEMIE

siehe Jahrbuch 1969—1970 (Berlin 1972) S. 9—13

EHRUNG

Die Moritz-Wilhelm-Drobisch-Medaille der Akademie erhielten das ordentliche Mitglied, Herr Walter Baetke aus Anlaß seines 90. Geburtstages und Herr Prof. Dr. Hans Salié in Anerkennung seiner Verdienste um die Herausgabe von J. C. Poggendorff, Biographisch-literarisches Handwörterbuch der exakten Naturwissenschaften.

ARCHIV

701 Leipzig, Goethestraße 3—5, Tel. 292886
Archivare: Dipl. phil. Elisabeth Lea und Diplom-Archivar/Diplom-Historiker Gerald Wiemers.

BIBLIOTHEK

Bibliothekar im Nebenamt: Dr. Dietmar Debes, wissenschaftlicher Bibliothekar an der Universitätsbibliothek Leipzig, 701 Leipzig, Beethovenstraße 6, Tel. 34391
Bibliothekarin im Nebenamt: Jutta Sporleder
Magazinverwalter: Paul Tottermusch

VERWALTUNG UND SEKRETARIAT

701 Leipzig, Goethestraße 3—5, Tel. 296090
Kaderleiterin: Ortrud Sperling
Verwaltungsleiterin: Gerda Rüprich
Hauptsachbearbeiterin: Ruth Keitel
Angestellte: Hilda Hilse, Dieter Frede

VERLAG DER AKADEMIE

Akademie-Verlag GmbH, 108 Berlin, Leipziger Straße 3—4, Tel. 2 20 04 41

VERTRAGSHEIM DER AKADEMIE

Kurhaus Masserberg, Tel. Masserberg/Thür. 2 21

MITGLIEDER

(Die eingeklammerten Zahlen bedeuten das Datum der Zuwahl)

Ordentliche Mitglieder der mathematisch-naturwissenschaftlichen Klasse

Arland, Anton, Dr. phil., Dr. agr. h. c. (Leipzig), em. Professor mit Lehrstuhl für Acker- und Pflanzenbau an der Karl-Marx-Universität Leipzig, Träger des Vaterländischen Verdienstordens, der Ehrennadel der Karl-Marx-Universität Leipzig und der Goldenen Medaille der Fürstlich-Jablonowskischen Gesellschaft der Wissenschaften, Mitglied der Kommission für spezielle Umweltprobleme und der Poggendorff-Kommission der Akademie, em. o. Mitglied der Akademie der Landwirtschaftswissenschaften der DDR, Mitglied des Präsidiums der Deutsch-Arabischen Gesellschaft in Berlin, des Wissenschaftlichen Rates der Internationalen Gesellschaft in Berlin, des Wissenschaftlichen Rates der Internationalen Gesellschaft für Nahrungs- und Vitalstoff-Forschung in Hannover, aktives Mitglied der Société Française des Physiologie Végétale in Paris, geb. 20. VII. 1895 in Eisenbrod, 705 Leipzig, Holsteinstraße 27, Tel. im Institut 68500 (1. VII. 1948). † 23. II. 1975.

Berg, Hermann, Dr. rer. nat., Prof. für physikalische Chemie an der Akademie der Wissenschaften der DDR, Zentralinstitut für Mikrobiologie und experimentelle Therapie in Jena, Leiter der Abt. Biophysikochemie und URANIA Gruppe der Akademieinstitute in Jena, Inhaber der Heyrovsky-Medaille des Heyrovsky-Instituts für Polarographie der Tschechoslowakischen Akademie der Wissenschaften in Prag, Mitglied der Gesellschaft für mathematische und physikalische Biologie der DDR, der International Society of Electrochemistry (ISE) und des Comité International de Photobiologie (CIP), geb. 16. VII. 1924 in Greifswald, 69 Jena, Greifbergstraße 15, Tel. 24817 (28. VI. 1971).

Brednow, Walter, Dr. med., Dr. med. h. c. (Tübingen), em. Professor mit Lehrstuhl für innere Medizin an der Friedrich-Schiller-Universität Jena, Obermedizinalrat, Mitglied der Deutschen Akademie der Naturforscher Leopoldina in Halle, der Gesellschaft Deutscher Naturforscher und Ärzte in Bonn und der Internationalen Gesellschaft für Innere Medizin, Ehrenmitglied der Gesellschaft für Innere Medizin an der Friedrich-Schiller-Universität Jena, der Deutschen Gesellschaft für Innere Medizin in Wiesbaden, der Nordwestdeutschen Gesellschaft für Innere Medizin in Hamburg, der Deutschen Gesellschaft für Klinische Medizin, Sektion Innere Medizin, in Leipzig und der Medizinisch-wissenschaftlichen Gesellschaft Eichsfelder Ärzte, geb. 13. II. 1896 in Berlin, 69 Jena, Dornbluthweg 1, Tel. 25148 (5. XII. 1960). † 11. IX. 1976

Buchheim, Wolfgang, Dr. phil., Professor mit Lehrstuhl für theoretische Physik und

Geophysik der Bergakademie Freiberg, Mitglied der Kommission für spezielle Umweltprobleme der Akademie und der Deutschen Akademie der Naturforscher Leopoldina in Halle, Mitglied im Wissenschaftlichen Beirat des Zentralinstituts für Physik der Erde der Akademie der Wissenschaften der DDR und ständiger Gast der Klasse „Umweltschutz und Umweltgestaltung“ dieser Akademie, Mitglied der Permanenten Kommission für die Festerdegezeiten der Internationalen Association für Geodäsie in Strasbourg und des Nationalkomitees für Geodäsie und Geophysik der DDR, geb. 18. X. 1909 in Leipzig, 92 Freiberg (Sachsen) Richard-Wagner-Straße 25, Tel. 21 32 (7. XII. 1959).

Dörter, Klaus, Dr. agr. (Wissenschaftsgebiet Grünland und Meliorationswesen sowie Landeskultur), o. Professor für landwirtschaftliche Meliorationen an der Martin-Luther-Universität Halle-Wittenberg, Inhaber der Verdienstmedaille der DDR und der Erinnerungsmedaille zum 20. Jahrestag der Demokratischen Bodenreform, Mitglied der Agrarwissenschaftlichen Gesellschaft der DDR, der Internationalen Bodenkundlichen Gesellschaft in Amsterdam/Holland und der Sektion Landeskultur und Naturschutz der Akademie der Landwirtschaftswissenschaften der DDR, Mitarbeiter der Arbeitsgruppe „Ökologie“ der Klasse „Umweltschutz und Umweltgestaltung“ bei der Akademie der Wissenschaften der DDR, geb. 10. I. 1922 in Breslau, 402 Halle/S., Elsa-Brandström-Straße 190, Tel. 4 52 11 (18. VI. 1974).

Drischel, Hans, Dr. med., o. Professor für Physiologie an der Karl-Marx-Universität Leipzig, Nationalpreisträger, o. Mitglied der Akademie der Wissenschaften der DDR, der Deutschen Akademie der Naturforscher Leopoldina in Halle, der New York Academy of Sciences, Member of the Council IUPS (International Union of Physiological Sciences), Ehrenmitglied der Allunionsgesellschaft der Physiologen der UdSSR „I. P. Pawlow“, geb. 8. VI. 1915 in Köln, 7027 Leipzig, Güldengossaer Straße 33, Tel. 8 09 23 (7. XII. 1964).

Eßbach, Hasso, Dr. med., o. Professor für pathologische Anatomie an der Medizinischen Akademie Magdeburg und an der Deutschen Akademie für Ärztliche Fortbildung Berlin, Träger des Vaterländischen Verdienstordens, Inhaber des Rudolf-Virchow-Preises des Ministeriums für Gesundheitswesen in der DDR und der Otto-v.-Guericke-Plakette der Stadt Magdeburg, Mitglied der Deutschen Gesellschaft für Pathologie in Münster, der Gesellschaft für Experimentelle Medizin in der DDR in Berlin, der Gesellschaft für Pathologie der DDR in Berlin, der Gesellschaft für Perinatale Medizin der DDR in Halle, der Gesellschaft für Neuropathologie der DDR in Leipzig, der Deutschen Gesellschaft für Elektronenmikroskopie in Düsseldorf, des Nationalkomitees der DDR zur Erforschung und Bekämpfung der Poliomyelitis und verwandter Erkrankungen in Berlin, geb. 9. III. 1909 in Chemnitz, 3271 Möser über Burg bei Magdeburg, Kastanienallee 3, Tel. 2 74 (24. V. 1965).

Freimuth, Ulrich, Dr. rer. nat., o. Professor für Lebensmittelchemie an der TU Dresden, Inhaber der Verdienstmedaille der DDR, der Kekulé-Medaille der Chemischen Gesellschaft und der Pettenkofer-Medaille dieser Gesellschaft, Vorsitzender des Fachverbandes Lebensmittelchemie der Chemischen Gesellschaft in der DDR, Mitglied der Biochemischen Gesellschaft, der Gesellschaft für Ernährung und der Gesellschaft für Allgemeine und kommunale Hygiene in der DDR, Ehrenmitglied des Ungarischen Wissenschaftlichen Vereins für Lebensmittelindustrie, geb. 16. X. 1914 in Wismar, 8020 Dresden, Berner Straße 1, Tel. 40644 (28. VI. 1971).

Gersch, Manfred, Dr. phil., em. o. Professor für Zoologie an der Friedrich-Schiller-Universität Jena, Nationalpreisträger, Inhaber des Heinrich-Greif-Preises der DDR, Mitglied der Kommission für spezielle Umweltprobleme der Akademie, o. Mitglied der Akademie der Wissenschaften der DDR, Mitglied der Deutschen Akademie der Naturforscher Leopoldina in Halle, Ehrenmitglied der Entomologischen Gesellschaft der UdSSR, geb. 12. VIII. 1909 in Dresden, 69 Jena, Beethovenstraße 8, Tel. 23518 (11. XII. 1961).

Görlich, Paul, Dr.-Ing., Dr. rer. nat. h. c. (Leipzig), Professor, Nationalpreisträger, Verdienter Techniker des Volkes, Inhaber des Vaterländischen Verdienstordens in Silber, der Silbernen Ehrennadel der Kammer der Technik und der Lenin-Erinnerungsmedaille des Obersten Sowjets der UdSSR, Ehrensenator der TH Karl-Marx-Stadt, Mitglied der Kommission für spezielle Umweltprobleme der Akademie, o. Mitglied der Akademie der Wissenschaften der DDR, em. Direktor am Institut für Optik und Spektroskopie dieser Akademie, Mitglied der Deutschen Akademie der Naturforscher Leopoldina in Halle, Vorstandsmitglied der Physikalischen Gesellschaft der DDR und der Deutschen Gesellschaft für Meßtechnik und Automatisierung in der DDR, Mitglied der Royal Microscopical Society in London-Oxford, geb. 7. X. 1905 in Dresden, 69 Jena, Schillbachstraße 24, Tel. 24496 (15. III. 1965).

Hanson, Horst, Dr. med., o. Professor für Biochemie, Direktor des physiologisch-chemischen Instituts der Martin-Luther-Universität Halle—Wittenberg, Inhaber der Carus-Medaille der Deutschen Akademie der Naturforscher Leopoldina und des Carus-Preises der Stadt Schweinfurt (1970), Mitglied der Deutschen Akademie der Naturforscher Leopoldina in Halle, Generalsekretär und Präsidiumsmitglied dieser Akademie, geb. 5. VI. 1911 in Lauchhammer, 4014 Halle (Saale), Straße der Deutsch-Sowjetischen Freundschaft Nr. 93, Tel. 29324 (7. XII. 1959).

Hein, Franz, Dr. phil., Dr. rer. nat. h. c. (Freiburg i. Br. und Rostock), em. Professor mit Lehrstuhl für anorganische Chemie an der Friedrich-Schiller-Universität Jena, Nationalpreisträger, Hervorragender Wissenschaftler des Volkes, Inhaber der Clemens-Winkler-Medaille der Chemischen Gesellschaft in der DDR und der Alexander-von-Humboldt-Gedenkmedaille der DDR, o. Mitglied der Akademie der Wissenschaften der DDR, em. Direktor der Forschungsstelle für Komplexchemie dieser Akademie, Mitglied der Deutschen Akademie der Naturforscher Leopoldina in Halle, korresp. Mitglied der Heidelberger Akademie der Wissenschaften, Ehrenmitglied der Chemischen Gesellschaft in der DDR und Ehrenmitglied auf Lebenszeit der New York Academy of Sciences, geb. 30. VI. 1892 in Grötzingen b. Karlsruhe (Baden), 69 Jena, Gutenbergstraße 5, Tel. im Institut 23472 u. 22578, App. 262 (7. V. 1951). † 26. III. 1976.

Hertz, Gustav, Dr. phil., Dr. rer. nat. h. c. (Berlin und Leipzig), Dr.-Ing. E. h. (Dresden), em. Professor mit Lehrstuhl für Physik an der Karl-Marx-Universität Leipzig, Nobel-, Staats-(UdSSR) und Nationalpreisträger, Hervorragender Wissenschaftler des Volkes, Träger des Vaterländischen Verdienstordens und des Ordens „Banner der Arbeit", der Max-Planck-Medaille des Verbandes Deutscher Physikalischer Gesellschaften, der Helmholtz-Medaille der Akademie der Wissenschaften der DDR, der Alexander-von-Humboldt-Gedenkmedaille der DDR und der Ernst-Moritz-Arndt-Medaille des Nationalrates der Nationalen Front des Demokratischen Deutschland, Ehrenmitglied des Forschungsrates der DDR, o. Mitglied der Akademie der Wissenschaften der DDR und der Tschechoslowakischen Akademie der Wissenschaften in Prag, korresp. Mitglied der Akademie der Wissenschaften in Göttingen, auswärt. Mitglied der Akademie der Wissenschaften der UdSSR in Moskau, Mitglied der Deutschen Akademie der Naturforscher Leopoldina in Halle, Ehrenmitglied der Ungarischen Akademie der Wissenschaften in Budapest und der Rumänischen Akademie der Wissenschaften in Bukarest, geb. 22. VII. 1887 in Hamburg, 117 Berlin, Lienhardweg 47, Tel. 6 714 483 (24. I. 1955). †30. X. 1975.

Jordan, Herbert, Dr. med., Honorarprofessor für Physiotherapie an der Friedrich-Schiller-Universität Jena, Direktor des Forschungsinstituts für Balneologie und Kurortwissenschaft in Bad Elster, Obermedizinalrat, Träger der Hufeland-Medaille in Gold und der Medaille für treue Dienste im Gesundheits- und Sozialwesen in Silber, Mitglied der Gesellschaft für Innere Medizin, Kardiologie, Physiotherapie und Rheumatologie in der DDR, Mitglied der Internationalen Gesellschaft für Biologische Rhythmusforschung, Ehrenmitglied der Medizinisch-Wissenschaftlichen Allunionsgesellschaft für Physiotherapie und Kurortologie der UdSSR, der Bulgarischen Gesellschaft für Physikalische Therapie, Kurortologie und Rehabilitation und der Polnischen Gesellschaft für Balneologie, Bioklimatologie und Physikalische Therapie, geb. 8. XII. 1919 in Markneukirchen (Vogtland), 9933 Bad Elster, Karl-Marx-Straße 30, Tel. 4 67 (18. VI. 1974).

Keller, Ott-Heinrich, Dr. phil. nat., em. o. Professor für Mathematik an der Martin-Luther-Universität Halle—Wittenberg, Nationalpreisträger, Mitglied der Deutschen Akademie der Naturforscher Leopoldina in Halle, geb. 22. VI. 1906 in Frankfurt (Main), 402 Halle (Saale), Dölauer Straße 82, Tel. 31480 (15. IX. 1958).

Lösche, Artur, Dr. phil. o. Professor für experimentelle Physik an der Karl-Marx-Universität Leipzig, Nationalpreisträger, Inhaber des Vaterländischen Verdienstordens in Bronze, o. Mitglied der Akademie der Wissenschaften der DDR, Mitglied der Deutschen Akademie der Naturforscher Leopoldina in Halle, Mitglied des Vorstandes und bis 1974 Präsident der Groupement AMPERE (Internationale

Gesellschaft für HF-Spektroskopie), geb. 20. X. 1921 in Leipzig, 7027 Leipzig, Egon-Erwin-Kisch-Weg 35, Tel. 81932 (24. V. 1965).

Lueken, Bernd Lüder, Dr. med., o. Professor für Physiologie an der Martin-Luther-Universität Halle—Wittenberg, Mitglied der Deutschen Akademie der Naturforscher Leopoldina in Halle, der Gesellschaft für theoretische Medizin und Grenzgebiete Sachsen-Anhalt und der Gesellschaft der Physiologen in der DDR, geb. 8. VIII. 1908 in Rüstringen (Oldenburg), 402 Halle (Saale), Rathenauplatz 14, Tel. 26501 (7. XII. 1964).

Matzen, Peter Friedrich, Dr. med., em. o. Professor für Orthopädie an der Karl-Marx-Universität Leipzig, Verdienter Arzt des Volkes, Inhaber der Priorev-Medaille und des Erich-Lexer-Preises der Deutschen Gesellschaft für Chirurgie bzw. der Deutschen Gesellschaft für Orthopädie und Traumatologie 1974, Mitglied der Deutschen Akademie der Naturforscher Leopoldina in Halle und Senator dieser Akademie, Mitglied der Société Internationale de Chirurgie orthopédique et de Traumatologie in Brüssel, Ehrenmitglied der Bulgarischen Orthopädischen Gesellschaft und der Vereinigung der Orthopäden Österreichs in Wien, korresp. Mitglied der Società Italiana di Ortopedia e Traumatologia in Rom und der Polnischen Medizinischen Gesellschaft, Fellow of the International College of Surgeons in Chicago, geb. 11. X. 1909 in München, 701 Leipzig, Semmelweisstraße 10, Tel. 85848 (7. XII. 1964).

Mothes, Kurt, Dr. phil., Dr. med. h. c. (Halle—Wittenberg), Dr. agr. h. c. (Kiel), Dr. phil. h. c. (Wien), Dr. rer. nat. h. c. (Halle—Wittenberg, Greifswald), Dr. pharm. h. c. (Szeged/Ungarn), em. Professor mit Lehrstuhl für Botanik, Pharmakognosie und Biochemie der Pflanzen an der Martin-Luther-Universität Halle—Wittenberg, Nationalpreisträger, Hervorragender Wissenschaftler des Volkes, Träger des Vaterländischen Verdienstordens und des Österreichischen Ehrenzeichens für Wissenschaft und Kunst, Inhaber des Ordens „Banner der Arbeit“, der Cothenius-Medaille der Deutschen Akademie der Naturforscher Leopoldina, der Döbereiner-Medaille der Pharmazeutischen Gesellschaft der DDR, der Otto-Warburg-Medaille der Gesellschaft für Physiologische Chemie in Berlin-Dahlem, der Carl-Mannich-Medaille der Deutschen Pharmazeutischen Gesellschaft Berlin, der Hoest-Madsen-Medaille der Fédération Internationale Pharmaceutique, der goldenen Gregor-Mendel-Medaille der Tschechoslowakischen Akademie der Wissenschaften, der Pazmany-Medaille der Universität Budapest, der Paul-Karrer-Medaille der Universität Zürich, der August-Kekulé-Medaille der Chemischen Gesellschaft in der DDR, der Charles Reid-Barnes Honorary Life Membership of the American Society of Plant Physiologists, Mitglied der Kommission für spezielle Umweltprobleme und der Poggendorff-Kommission der Akademie, Alt-Präsident der Deutschen Akademie der Naturforscher Leopoldina in Halle, o. Mitglied der Akademie der Wissenschaften der DDR und em. Direktor des Instituts für Biochemie der Pflanzen dieser Akademie, korresp. Mitglied der Heidelberger Akademie der Wissenschaften, der Akademie der Wissenschaften und der Literatur in Mainz, der Bayerischen Akademie der Wissenschaften in München, der Rumänischen Akademie der Wissenschaften in Bukarest, der Rajasthan Academy of Scienses in Pilani (Indien) und der kungl. Vetenskaps-Societen in Uppsala, ausländisches Mitglied der Akademie der Wissenschaften der UdSSR und der Royal Society in London, Ehrenmitglied der math.-nat. Klasse der Österreichischen Akademie der Wissenschaften in Wien, der Japanischen Akademie der Wissenschaften in Tokio, der Ungarischen Akademie der Wissenschaften in Budapest, des Gesamtverbandes der Ungarischen Medizinisch-Wissenschaftlichen Gesellschaften in Budapest und der American Society of Plant Physiologists in Ithaca, Mitglied der American Society of Pharmacognosy in Storrs/Conn., der American Chemical Society in Washington, der Fédération Internationale Pharmaceutique in Amsterdam, der British Society for Experimental Biology in London, der Japanese Society of Plant Physiologists in Tokio, der Societas Physiologiae Plantarum Scandinavica in Lund, der Deutschen Arzneibuch-Kommission in Berlin und des Kuratoriums der Flückiger-Stiftung in Bern, Ehrenmitglied der Deutschen Vereinigung für angewandte Botanik in Berlin-Dahlem, der Indian Society of Plant Physiologists in New Delhi, der Indian Botanical Society in Bombay, der Sowjetischen Botanischen Allunionsgesellschaft in Leningrad, der Biologischen Gesellschaft der DDR, der Gesellschaft für Arzneipflanzenforschung (Camberg, BRD), der Gesellschaft Deutscher Chemiker der Bundesrepublik Deutschland, der Biochemischen Gesellschaft der DDR und der Pharmazeutischen Gesellschaft der DDR, Ehrensenator der Martin-Luther-Universität Halle—Wittenberg, 3. XI. 1900 in Plauen (Vogtland), 402 Halle (Saale), Hoher Weg 23, Tel. 21717 (10. V. 1954).

Neef, Ernst, Dr. phil., em. o. Professor für Geographie an der Technischen Universität Dresden, Inhaber der Verdienstmedaille der DDR, der Hermann-Haack-Medaille der Geographischen Gesellschaft in der DDR und der Verdienstmedaille der Polnischen Geographischen Gesellschaft, Vorsitzender der Kommission für spezielle Umweltprobleme und der Poggendorff-Kommission der Akademie, Mitglied der Deutschen Akademie der Naturforscher Leopoldina in Halle, Ehrenmitglied der Geographischen Gesellschaft in Wien und der Ungarischen Geographischen Gesellschaft in Budapest, korresp. Mitglied der Geographischen Gesellschaft in München, geb. 16. IV. 1908 in Dresden, 8019 Dresden, Florian-Geyer-Straße 44/032 (17. XI. 1956).

Obenaus, Fritz, Dr.-Ing., em. Professor mit Lehrstuhl für Hochspannungstechnik an der Technischen Universität Dresden, Nationalpreisträger, Verdienter Techniker des Volkes, Träger des Vaterländischen Verdienstordens, o. Mitglied der Akademie der Wissenschaften der DDR, geb. 9. XII. 1904 in Niedersedlitz, 8020 Dresden, Donndorfstraße 37, Tel. 41331 (7. XII. 1959).

Rammler, Erich, Dr.-Ing., Dr.-Ing. E. h. (Aachen), em. Professor mit Lehrstuhl für Brennstofftechnik, Wärmewirtschaft und Brikettierung an der Bergakademie Freiberg, Nationalpreisträger, Hervorragender Wissenschaftler des Volkes, Verdienter Bergmann der DDR, Träger des Vaterländischen Verdienstordens in Gold, Inhaber der Carl-Exter-Medaille, der Karl-Kegel-Medaille der ehemaligen Brennstofftechnischen Gesellschaft in der DDR und der Medaille für Verdienste in der Kohlenindustrie der DDR in Gold, o. Mitglied der Akademie der Wissenschaften der DDR, Ehrenmitglied der ehemaligen Gesellschaft Deutscher Berg- und Hüttenleute an der ehemaligen Brennstofftechnischen Gesellschaft in der DDR, geb. 9. VII. 1901 in Tirpersdorf (Vogtland), 92 Freiberg, Richard-Wagner-Straße 12, Tel. 3388 (15. IX. 1958).

Schmalfuß, Karl, Dr. rer. nat., Dr. agr. h. c. (Leipzig) em. Professor mit Lehrstuhl für Pflanzenernährung und Bodenkunde an der Martin-Luther-Universität Halle—Wittenberg, Nationalpreisträger, Träger des Vaterländischen Verdienstordens, Inhaber der Erwin-Baur-Medaille der Akademie der Landwirtschaftswissenschaften der DDR, o. Mitglied der Akademie der Landwirtschaftswissenschaften der DDR, Mitglied der Deutschen Akademie der Naturforscher Leopoldina in Halle, geb. 5. XI. 1904 in Wedruschitz (Kr. Saaz), 4021 Halle (Saale), Weidenweg 7, Tel. 612452 (10. V. 1954).

Schwabe, Kurt, Dr.-Ing., Dr. rer. nat. h. c. (Dresden, Karl-Marx-Stadt, Merseburg), em. Professor mit Lehrstuhl für physikalische Chemie und Elektrochemie an der TU Dresden, Nationalpreisträger, Hervorragender Wissenschaftler des Volkes, Verdienter Hochschullehrer der DDR 1976, Held der Arbeit, Träger des Vaterländischen Verdienstordens, des Ordens „Banner der Arbeit", der Goldenen Ehrennadel und der Ernst-Abbe-Medaille der Kammer der Technik, der Hans-Clemm-Gedenkmünze des Vereins der Zellstoff- und Papierchemiker und -ingenieure, der Clemens-Winkler-Medaille der Chemischen Gesellschaft in der DDR und der Arthur-Becker-Medaille der Freien Deutschen Jugend in Gold, Mitglied des Forschungsrates der DDR in Berlin, Vorsitzender des wissenschaftlich-technischen Rates des Koordinationszentrums der RGW-Länder für Korrosionsschutz, Direktor des Forschungsinstituts Meinsberg (Entwicklungslabor der Technischen Universität Dresden), Mitglied der Kommission für spezielle Umweltprobleme der Akademie, o. Mitglied und Vizepräsident der Akademie der Wissenschaften der DDR, Mitglied der Deutschen Akademie der Naturforscher Leopoldina in Halle, korresp. Mitglied der Jugoslavischen Akademie der Wissenschaften und Künste in Zagreb und der mathematisch-naturwissenschaftlichen Klasse der Österreichischen Akademie der Wissenschaften in Wien, auswärtiges Mitglied der Bulgarischen Akademie der Wissenschaften in Sofia, Mitglied der Electrochemical Society in New York, der Faraday Society in London, der International Society of Elektrochemistry (ISE) in Genf und nationaler Sekretär dieser Society, Mitglied des International Corrosion Council und der Chemischen Gesellschaft in der DDR in Berlin, Ehrenmitglied der Kammer der Technik in Berlin und der Gesellschaft für Strahlenschutz der DDR in Berlin, geb. 29. V. 1905 in Reichenbach (Vogtland), Meinsberg/Sa., Post 7305 Waldheim, Tel. Waldheim 665 (11. XII. 1950).

Strack, Erich, Dr. med., Dr. phil., Dr. med. h. c. (Leipzig und Rostock), Dr. rer. nat.

h. c. (Leipzig), Diplom-Chemiker, em. Professor mit Lehrstuhl für physiologische Chemie an der Karl-Marx-Universität Leipzig, Nationalpreisträger, Mitglied der Kommission für spezielle Umweltprobleme und der Poggendorff-Kommission der Akademie, Mitglied der Deutschen Akademie der Naturforscher Leopoldina in Halle und der National Representation für „The International Conferences on the Biochemistry of Lipids", Ehrenmitglied der Biochemischen Gesellschaft in der DDR, geb. 20. XI. 1897 in Wollin (Pommern), 701 Leipzig, Philipp-Rosenthal-Straße 19, Tel. 35081 (14. II. 1949).

Stubbe, Hans, Dr. agr., Dr. agr. h. c. (Berlin, Jena und Kraków), Dr. Sc. h. c. (Brno), em. Professor mit Lehrstuhl für allgemeine und spezielle Genetik an der Martin-Luther-Universität Halle—Wittenberg, Nationalpreisträger, Hervorragender Wissenschaftler des Volkes, Träger des Vaterländischen Verdienstordens, des Ordens „Banner der Arbeit", der Helmholtz-Medaille der Akademie der Wissenschaften der DDR, der Darwin-Plakette der Deutschen Akademie der Naturforscher Leopoldina, der Alexander-Humboldt-Gedenkmedaille der DDR, der Gregor-Mendel-Medaille der Tschechoslowakischen Akademie der Wissenschaften, der Erwin-Baur-Medaille der Akademie der Landwirtschaftswissenschaften der DDR und der Purkyně-Medaille der Universität Brno, Ehrenpräsident der Akademie der Landwirtschaftswissenschaften der DDR, o. Mitglied der Akademie der Wissenschaften der DDR, em. Direktor des Zentralinstituts für Genetik und Kulturpflanzenforschung Gatersleben dieser Akademie, Mitglied der Deutschen Akademie der Naturforscher Leopoldina in Halle, ausländ. Mitglied der W. I. Lenin-All-Unions-Akademie für Landwirtschaftswissenschaften in Moskau, der Rumänischen Akademie der Wissenschaften in Bukarest, der Bulgarischen Akademie der Wissenschaften und der Bulgarischen Akademie der Landwirtschaftswissenschaften in Sofia, der Polnischen Akademie der Wissenschaften in Warschau, Ehrenmitglied der Ungarischen Akademie der Wissenschaften in Budapest, der All-Unions-Botanischen Gesellschaft der UdSSR, der Schwedischen Saatzuchtgesellschaft Svalöf, der Mendel-Gesellschaft in Lund, der Biologischen Gesellschaft und der Agrarwissenschaftlichen Gesellschaft in der DDR, geb. 7. III. 1902 in Berlin, 4325 Gatersleben, Schmiedestraße 1, Tel. 212 (13. II. 1950).

Uhlmann, Dietrich, Dr. rer. nat., Prof. für Hydrobiologie an der TU Dresden, Mitglied der Kommission für spezielle Umweltprobleme der Akademie und des International Council der International Association of Theoretical and Applied Limnology, Leitungsmitglied der Sektion Ökologie und Vorsitzender der Arbeitsgruppe Hydrobiologie der Biologischen Gesellschaft der DDR in Berlin, Mitglied des Nationalkomitees der DDR für das Internationale Hydrologische Programm und des Herausgeberstabes der „Internationalen Revue der gesamten Hydrobiologie" in Berlin, geb. 4. VIII. 1930 in Chemnitz, 8021 Dresden, Schlüterstraße 19, Tel. 32856 (28. VI. 1971).

Watznauer, Adolf, Dr. rer. nat., em. o. Professor für Geologie an der Bergakademie Freiberg, Hervorragender Wissenschaftler des Volkes, Inhaber des Fourmarier-

Preises (1972) in Gold der Académie Royale des Sciences des Lettres et des Beaux-Arts de Belgique in Brüssel, Mitglied der Kommission für spezielle Umweltprobleme der Akademie, o. Mitglied der Akademie der Wissenschaften der DDR und Mitglied der problemgebundenen Klassen „Umweltschutz und Umweltgestaltung" und Physik in Naturwissenschaft und Technik dieser Akademie, Mitglied der Deutschen Akademie der Naturforscher Leopoldina in Halle, korresp. Mitglied der Geologischen Gesellschaft in Wien und Ehrenmitglied der Geologischen Gesellschaft der ČSSR und der Gesellschaft für Geologische Wissenschaften der DDR, geb. 29. IV. 1907 in Reichenberg in Böhmen, 9030 Karl-Marx-Stadt, Dieselstraße 8, Postschließfach 30 (18. XII. 1972).

Wolf, Friedrich, Dr. rer. nat., Dr. rer. nat. h. c. (Poznań), o. Professor für Technische Chemie an der Martin-Luther-Universität Halle—Wittenberg, Nationalpreisträger 1959 und 1966, Träger des Ordens „Banner der Arbeit" und der Ehrenmedaillen der Universitäten Bratislava und Bagdad, korresp. Mitglied der Akademie der Wissenschaften der DDR in Berlin, Mitglied der Chemischen Gesellschaft der DDR in Berlin und der Faraday Society in London, geb. 9. II. 1920 in Leipzig, 7031 Leipzig, Karl-Heine-Straße 4b, Tel. 484415 (18. VI. 1974).

Auswärtige Mitglieder der mathematisch-naturwissenschaftlichen Klasse

Bredt, Heinrich, Dr. med., o. Professor für allgemeine Pathologie und pathologische Anatomie an der Johannes-Gutenberg-Universität Mainz, Nationalpreisträger, Mitglied und Präsident der Akademie der Wissenschaften und der Literatur in Mainz, auswärt. Mitglied der Akademie der Wissenschaften der DDR, Mitglied der Deutschen Akademie der Naturforscher Leopoldina in Halle, geb. 29. I. 1906 in Oberneudorf (Siebenbürgen), 65 Mainz, Fintherlandstraße 58, Tel. 91933 (12. II. 1951).

Buchner, Paul, Dr. phil., Dr. med. h. c. (München), Dr. Sc. biol. h. c. (Pavia), Dr. rer. nat. h. c. (Greifswald), em. o. Professor für Zoologie an der Ludwig-Maximilians-Universität München, Träger des Großen Verdienstkreuzes der Bundesrepublik Deutschland, der Carus-Medaille der Deutschen Akademie der Naturforscher Leopoldina in Halle und der Erzherzog-Rainer-Medaille der Österreichischen Zoologisch-Botanischen Gesellschaft, Ehrenbürger von Ischia, Ehrenpräsident des Centro di Studio sull'Isola d'Ischia, korresp. Mitglied der Bayerischen Akademie der Wissenschaften in München und der Dansk Naturhistorisk Forening, Mitglied der Deutschen Akademie der Naturforscher Leopoldina in Halle und der Accademia delle Scienze in Neapel, Ehrenmitglied des Naturwissenschaftlichen Vereins für Vorpommern und Rügen und der Deutschen Zoologischen Gesellschaft in Tübingen, geb. 12. IV. 1886 in Nürnberg, Porto d'Ischia (Napoli), Via S. Alessandro (5. VI. 1935).

Budevski, Evgeni Bogdanov, Dr. rer. nat., Professor für Chemie und Elektrochemie bei der Bulgarischen Akademie der Wissenschaften in Sofia, Zentrallaboratorium für elektrochemische Stromquellen, Inhaber des Staatspreises „Georgi Dimitrov" für Wissenschaft und der Auszeichnung „100 Jahre Bulgarische Akademie der Wissenschaften", Träger des Ordens „Kirill und Metodi" 1. und 2. Stufe, Mitglied der Bulgarischen Akademie der Wissenschaften in Sofia, Mitglied und Vizepräsident 1975—1977 der International Society of Elektrochemistry (ISE), geb. 29. VII. 1922 in Sofia, Sofia 1113, Bul. Samokov 82, tel. Institut 72 34 54 (9. XI. 1974).

Dabelow, Adolf, Dr. med. et phil. et. nat., Dr. med. h. c. (Salamanca), em. o. Professor für Anatomie an der Johannes-Gutenberg-Universität Mainz, o. Mitglied der Akademie der Wissenschaften und der Literatur in Mainz, korresp. Mitglied der Bayerischen Akademie der Wissenschaften in München, geb. 28. VI. 1899 in Magdeburg, 78 Freiburg/Breisgau, Sautierstraße 75, Postfach 352 (24. VI. 1944).

Fejes Tóth, László, Dr. phil., Professor für Mathematik an der Ungarischen Akademie der Wissenschaften in Budapest, Kossuth-Preisträger, Mitglied der Ungarischen Akademie der Wissenschaften in Budapest, geb. 12. III. 1915 in Szeged, Budapest II, Pusztaszeri út. 32, Tel. 688636 (13. VI. 1966).

Frumkin, Alexander Naumowitsch, Dr. chem., Dr. rer. nat. h. c. (Dresden, Brno), Professor für Elektrochemie an der Lomonossow-Universität Moskau, Held der sozialistischen Arbeit, Staats- und Leninpreisträger, Inhaber der Medaille „Sichel und Hammer", des Leninordens und der Palladium Medal der Electrochemical Society, o. Mitglied der Akademie der Wissenschaften der UdSSR in Moskau, der Deutschen Akademie der Naturforscher Leopoldina in Halle, ausländ. Mitglied der Polnischen Akademie der Wissenschaften in Warschau, der Bulgarischen Akademie der Wissenschaften in Sofia, der Koninklijke Nederlandse Akademie van Wetenschappen in Amsterdam und der Jugoslawischen Akademie der Wissenschaften und Künste in Zagreb, Ehrenmitglied der Ungarischen Akademie der Wissenschaften in Budapest, Foreign Honorary Member of the American Academy of Arts and Sciences, Boston, Foreign associate National Academy of Sciences, Washington, Ehrenmitglied der All-Union Mendelejew Chemischen Gesellschaft in Moskau, des Council of the National Institute of Sciences of India in New Delhi, der Société Chimique de France in Paris und der Société Chemique de Belgique in Brüssel, geb. 24. X. 1895 in Kischinev, Moskau W-97, Leninsky Prospekt 13, Tel. 232-14-26 (24. X. 1966). †25. V. 1976.

Gerasimov, Innokentii Petrovich, Dr. der geographischen Wissenschaften, em. Professor für Geographie an der Lomonossow-Universität in Moskau, Vizepräsident der International Geographical Union, Inhaber des Lenin-Ordens, des Ordens Roter Stern und des Dimitroff-Preises der VR Bulgarien, Träger der Goldenen Dokučaev-Medaille, der Goldenen Przenalski-Medaille und der Moritz-Wilhelm-Drobisch-Medaille der Sächsischen Akademie, Mitglied der Deutschen Akademie der Naturforscher Leopoldina in Halle, der Akademie der Landwirtschaftswissenschaften der DDR, der Bulgarischen Akademie der Wissenschaften in Sofia, auswärtiges Mitglied der Akademie der Wissenschaften der DDR und

Ehrenmitglied der Geographischen Gesellschaften in Columbien, England, Finnland, Frankreich, Italien, Österreich, Polen, Schottland, Serbien und Ungarn, geb. 9. XII. 1905 in St. Kostroma, Moskau-W 3, Leninsky Prospekt 13, Tel. 231-61-44 (12. XI. 1971).

Heisenberg, Werner, Dr. phil., Dr.-Ing. E. h. (Karlsruhe), D. Sc. h. c. (Brüssel), Dr. rer. nat. h. c. (Budapest), Dr. phil. h. c. (Kopenhagen), Dr. h. c. (Athen), em. Direktor des Max-Planck-Instituts für Physik und Astrophysik in München, Hon.-Professor für theoretische Physik an der Ludwig-Maximilians-Universität München, Präsident der Alexander-von-Humboldt-Stiftung in Bad Godesberg, Mitglied des Vorstandsrates des Deutschen Museums in München, Nobelpreisträger, Träger des Ordens Pour le mérite für Verdienste um die Wissenschaft und um die Künste, des Großen Verdienstkreuzes mit Stern und Schulterband der Bundesrepublik Deutschland, Inhaber des Ordens der Aufgehenden Sonne II. Klasse der japanischen Regierung, des Komturkreuzes des Ordens Georg I., Griechenland, des Großkreuzes des Ordens „Alfonso X el Sabio“, Spanien, des Kultur-Ehrenpreises und der goldenen Ehrenmünze der Stadt München, des Sigmund-Freud-Preises für wissenschaftliche Prosa der Deutschen Akademie für Sprache und Dichtung in Darmstadt, des Romano-Guardini-Preises 1973 der Katholischen Akademie in Bayern, Inhaber der Goldenen Niels-Bohr-Medaille, der Hugo-Grotius-Medaille der International Grotius Foundation for the Propagation of the Law of Nations in Genf, der Max-Planck-Medaille des Verbandes Deutscher Physikalischer Gesellschaften, der Barnard-Medaille der Columbia-University New York, der Matteucci-Medaille der Accademia Nazionale dei Quaranta in Rom, der Goldenen Medaille der Medizinisch-naturwissenschaftlichen Gesellschaft in Jena und der Bronzemedaille der National Academy of Sciences in Washington, o. Mitglied der Akademie der Wissenschaften der DDR, der Bayerischen Akademie der Wissenschaften in München und der Akademie der Wissenschaften in Göttingen, Mitglied der Deutschen Akademie Naturforscher Leopoldina in Halle, der Kongelike Danske Videnskabernes Selskab in Kopenhagen, der Suomaleinen Tiedeakatemia in Helsinki, der Koninklijke Nederlandse Akademie van Wetenschappen in Amsterdam, der Norske-Videnskaps Akademi in Oslo, der Royal Society in London, der Rumänischen Akademie der Wissenschaften in Bukarest, der Jugoslawischen Akademie der Wissenschaften und Künste in Zagreb, der Svenska Akademien in Stockholm, der Real Academie de Ciencias Exactas, Físicas y Naturales in Valverde, der Accademia Nazionale dei Lincei in Rom, der Pontificia Academia Scientiarum, Vatikanstadt, der American Academy of Arts und Sciences in Boston, der National Academy of Sciences in Washington, der American Philosophical Society in Philadelphia, der Société Philomatique in Paris, der Wissenschaftlichen Kommission des Institut International de Physique Solvay in Brüssel, der Kongelige Norske Videnskabers Selskab in Drontheim, der Kungl. Vetenskaps-Societeten in Uppsala, der Medizinisch-naturwissenschaftlichen Gesellschaft in Jena, Ehrenmitglied der New York Academy of Science, der Royal Irish Academy in Dublin und der Societas Scientiarum Fennica-Finska Vetenskaps Societeten-Suomen

Tiedeseura in Helsinki, Ehrenrat des Consejo Superior de Investigaciones Científicas in Madrid, geb. 5. XII. 1901 in Würzburg, 8 München 40, Föhringer Ring 6, Tel. 327001 (1. VII. 1930). †1. II. 1976.

Helferich, Burckhardt, Dr. phil., Dr. rer. nat. h. c. (Stuttgart), em. o. Professor für Chemie an der Rheinischen Friedrich-Wilhelms-Universität Bonn, Träger des Großen Verdienstkreuzes des Verdienstordens der Bundesrepublik Deutschland, Inhaber der Emil-Fischer-Medaille der Gesellschaft Deutscher Chemiker, Mitglied der Deutschen Akademie der Naturforscher Leopoldina in Halle und der Rheinisch-Westfälischen Akademie der Wissenschaften in Düsseldorf, Ehrenmitglied der Chemischen Gesellschaft der DDR, geb. 10. VI. 1887 in Greifswald, 53 Bonn, Bonner Talweg 66, Tel. 223285 (23. II. 1931).

Hertwig, Paula, Dr. phil., Dr. med. h. c. (Halle—Wittenberg und Heidelberg), em. Professor mit Lehrstuhl für allgemeine Biologie an der Martin-Luther-Universität Halle-Wittenberg, Nationalpreisträgerin, Hervorragender Wissenschaftler des Volkes, Trägerin des Vaterländischen Verdienstordens, Inhaberin des Theobald-Christ-Preises der Senkenbergischen Naturforschenden Gesellschaft in Frankfurt (Main), Mitglied der Deutschen Akademie der Naturforscher Leopoldina in Halle, geb. 11. X. 1889 in Berlin, 773 Villingen, Brandenkopfweg 7 (14. III. 1955).

Hill, Archibald Vivian, Sc. D., LL. D. h. c. (Edinburgh, Belfast), D. Sc. h. c. (Algier, Baltimore Maryland, Bristol, Columbia, Exeter, John Hopkins, Liège, Manchester, Oxford, Pennsylvania, Rio de Janeiro, Rochester- und Rockefeller Universities), M. D. h. c. (Louvain, Bruxelles, Toulouse), em. Professor für Physiologie, Hon. Research Associate und Hon. Fellow am University College London, Hon. Fellow am Trinity College und King's College Cambridge, Nobelpreisträger, Inhaber der Copley Medal der Royal Society und der Goldenen Cothenius-Medaille der Deutschen Akademie der Naturforscher Leopoldina in Halle, Mitglied der Royal Society in London, der National Academy of Sciences in Washington, der Kongelige Danske Videnskabernes Selskap in Kopenhagen und der Deutschen Akademie der Naturforscher Leopoldina in Halle, ausländ. Ehrenmitglied der Rumänischen Akademie der Wissenschaften in Bukarest, geb. 26. IX. 1886 in Bristol, 11a Chaucer Road, Cambridge, Tel 54551 (1. VII. 1929).

Hölder, Ernst, Dr. phil., o. Professor der Mathematik an der Johannes-Gutenberg-Universität Mainz, Inhaber der Silbernen Medaille der Universität Helsinki, auswärt. Mitglied der Akademie der Wissenschaften der DDR, Mitglied der Deutschen Akademie der Naturforscher Leopoldina in Halle, geb. 2. IV. 1901 in Leipzig, 65 Mainz-Gonsenheim, Friedrichsstraße 36 (1. VII. 1948).

Hopf, Eberhard, Dr. phil., Forschungs-Professor für Mathematik an der Indiana University, Bloomington, Indiana (USA), korresp. Mitglied der Bayerischen Akademie der Wissenschaften in München, geb. 17. IV. 1902 in Salzburg, Bloomington/Indiana, USA, 512 S. Swain Avenue (26. XI. 1938).

Hund, Friedrich, Dr. phil., Dr. phil. nat. h. c. (Frankfurt [Main]), Dr. phil. h. c. (Uppsala), o. Professor für theoretische Physik an der Georg-August-Universität Göttingen, Nationalpreisträger, Träger der Max-Planck-Medaille der Deutschen Physikalischen Gesellschaft, der Cothenius-Medaille der Deutschen Akademie der Naturforscher Leopoldina, des Otto-Hahn-Preises für Chemie und Physik und des Großen Verdienstkreuzes des Verdienstordens der Bundesrepublik Deutschland, auswärt. Mitglied der Akademie der Wissenschaften der DDR und der Akademie der Wissenschaften in Göttingen, Mitglied der Deutschen Akademie der Naturforscher Leopoldina in Halle, Ehrenmitglied der Deutschen Bunsengesellschaft für Physikalische Chemie, geb. 4. II. 1896 in Karlsruhe, 34 Göttingen, Charlottenburgerstraße 19, Tel. 58589 (18. XI. 1933).

Kähler, Erich, Dr. phil., o. Professor für Mathematik an der Universität Hamburg, Hon.-Professor an der Technischen Universität Berlin, auswärt. Mitglied der Akademie der Wissenschaften der DDR, Mitglied der Deutschen Akademie der Naturforscher Leopoldina in Halle, ausländ. Mitglied der Accademia Nazionale dei Lincei in Rom, geb. 16. I. 1906 in Leipzig, 2 Wedel (Holstein), Mozartstraße 42, Tel. (04103) 86535 (14. II. 1949).

Kaischev, Rostislav, Dr.-Ing., o. Professor für physikalische Chemie an der Universität Sofia, Vizepräsident der Bulgarischen Akademie der Wissenschaften in Sofia, Mitglied der Deutschen Akademie der Naturforscher Leopoldina in Halle, auswärt. Mitglied der Akademie der Wissenschaften der DDR, geb. 16. II. 1908 in Petersburg, Universität Sofia, ul. „7. noemvri“ 1 (19. II. 1968).

Karšulin, Miroslav, Dr. techn., o. Professor für physikalische Chemie an der Universität Zagreb, Träger des Staatspreises für Chemie, des Republikanischen Preises, des Staatsordens II. Klasse und des Ordens für Arbeit mit der roten Fahne, o. Mitglied der Jugoslavischen Akademie der Wissenschaften und Künste in Zagreb, der Makedonischen Akademie der Wissenschaften, der Faraday Society in London, der Electrochemical Society in New York, der Kroatischen Chemischen Gesellschaft, des Vereins der Ingenieure und Techniker Jugoslaviens, der Deutschen Keramischen Gesellschaft und des Internationalen Komitees zur Erforschung der Bauxite, Aluminium-Hydroxide und Oxyde, geb. 4. IV. 1904 in Przemysl (Polen), 41000 Zagreb, Palmotićeva 4/I, Tel. 36761 (9. XI. 1974).

Keldysch, Mstislav V., Dr. phil., Dr. h. c. (Lagos/Nigeria), Professor für Mathematik, Präsident der Akademie der Wissenschaften der UdSSR in Moskau (bis 1975), Held der sozialistischen Arbeit, Staats- und Leninpreisträger, Inhaber von Leninorden und Orden des Roten Arbeitsbanners, auswärt. Mitglied der Akademie der Wissenschaften der DDR, Mitglied der Tschechoslowakischen Akademie der Wissenschaften in Prag, ausländ. Ehrenmitglied der Rumänischen Akademie der Wissenschaften in Bukarest und der National Academy of Sciences in Washington, geb. 10. XI. 1911 in Riga, Moskau, Leninsky Prospekt 14, Tel. V-1-02-06 (24. X. 1966).

Kersten, Martin, Dr.-Ing., Hon.-Professor für Physik an der Technischen Hochschule Carolo-Wilhelmina Braunschweig, Vizepräsident der Deutschen Physikalischen Gesellschaft, Mitglied der Internationalen Kommission für Maß und Gewicht, geb. 28. IV. 1906 in Zittau, 33 Braunschweig, Bundesallee 100, Tel. 592-2201 (14. II. 1949).

Klimaszewski, Mieczysław, Dr. phil., Dr. rer. nat. h. c. (Jena, Kiew, Bratislava, Alliance College, Cambridge Springs [USA], St. Andrews), o. Professor für physische Geographie an der Jagiellonischen Universität Kraków, Stellvertreter des Vorsitzenden des Staatsrates der VR Polen 1965—1972, Träger des Offizierskreuzes, des Kommandeur-Kreuzes mit Stern Polonia Restituta und der Standarte der Arbeit I. Klasse, Grand officier du l'ordre de Mérite, Inhaber der Alexander-von Humboldt-Gedenkmedaille der DDR und der Karl-Sapper-Medaille der Geographischen Gesellschaft in Würzburg, o. Mitglied der Polnischen Akademie der Wissenschaften in Warschau, Mitglied der Deutschen Akademie der Naturforscher Leopoldina in Halle, der Societas Academia Scientiarum Fennica, der Jugoslawischen Akademie der Wissenschaften und Künste in Zagreb, der Wissenschaftlichen Gesellschaft in Wrocław, der Geographischen Gesellschaft der UdSSR in Leningrad, der Koninklijk Nederlandsch Aardrijkskundig Genootschap in Amsterdam, der Société Belge de Geologie, de Paléontologie et d'Hydrologie in Brüssel, der Suomen Maantieteellinen Seura-Geografiska Sällskapet i Finland in Helsinki und der Magyar Fóldrajzi Társaság in Budapest, Ehrenmitglied des Wissenschaftlichen Rates der Universitäten Kiew und Jena, geb. 26. VII. 1908 in Stanisławów, Kraków, Wyspiańskiego 3, Tel. 37573 (13. VI. 1966).

König, Robert, Dr. phil., em. o. Professor für Mathematik an der Ludwig-Maximilians-Universität München, o. Mitglied der Bayerischen Akademie der Wissenschaften in München, geb. 11. IV. 1885 in Linz (Österreich), 8 München 40, Adelheidstraße 21, Tel. 376005 (31. VII. 1934).

Kolotyrkin, Jakov Mikhailovitsch, Dr. der chemischen Wissenschaften, Professor für Elektrochemie und Korrosionsschutz, Direktor des Chemischen Instituts „L. Ja. Karpov" in Moskau, Inhaber des Lenin-Ordens und des Ordens des Roten Arbeitsbanners, Träger der Goldmedaille der Tschechoslowakischen Gesellschaft für Internationale Zusammenarbeit und der Gedenkmedaille zu Ehren des 50. Jahrestages des Bestehens der Kommunistischen Partei der ČSSR, o. Mitglied der Akademie der Wissenschaften der UdSSR, der Internationalen Elektrochemischen Gesellschaft, des Staatlichen Komitees für Wissenschaft und Technik beim Ministerrat der UdSSR und Vertreter der UdSSR im wissenschaftlich-technischen Rat des Koordinationszentrums der RGW-Länder für Korrosionsschutz, geb. 14. XI. 1910 im Dorf Zanino des Jarzevsker Bezirks im Smolensker Gebiet, 107120 Moskva B-120, ul. Obucha 10, Karpov-Institut, Tel. 297-32-57 (12. XI. 1971).

Kramer, Kurt, Dr. med., o. Professor für Physiologie an der Ludwig-Maximilians-Universität München, o. Mitglied der Akademie der Wissenschaften in Göttingen, Mitglied der Deutschen Akademie der Naturforscher Leopoldina in Halle und der Gesellschaft zur Förderung der Naturwissenschaften in Marburg (Lahn), geb.

16. VI. 1906 in Münster, 8 München 15, Schubertstraße 4, Tel. 531199 (24. VI. 1944).

Lange, Erich, Dr.-Ing. em., o. Professor für physikalische Chemie an der Friedrich-Alexander-Universität Erlangen—Nürnberg, Mitglied der Deutschen Bunsengesellschaft für physikalische Chemie in Frankfurt (Main), der Gesellschaft Deutscher Chemiker in Frankfurt (Main), der Gesellschaft Deutscher Naturforscher und Ärzte in Bonn, ehemals Mitglied des Comité International de Thermodynamique et de Cinétique Électrochimiques (Nomenklaturkommission) in Brüssel und der American Electrochemical Society in New York, geb. 26. XI. 1896 in Hartha b. Waldheim, 852 Erlangen, Hindenburgstraße 61, Postfach 142, Tel. 852558 (10. VI. 1961).

Maier, Wilhelm, Dr. phil., em. Professor mit Lehrstuhl für Mathematik an der Friedrich-Schiller-Universität Jena, korresp. Mitglied der Heidelberger Akademie der Wissenschaften, geb. 4. I. 1896 in Neuenbürg (Schwarzwald), 65 Mainz 31, Rubensallee 48 (5. XII. 1960).

Müller, Karl Otto, Dr. phil., Professor für Phytopathologie, Oberregierungsrat a. D., Mitglied der American Phytopathological Society, der Deutschen Botanischen Gesellschaft in Berlin-Zehlendorf und der Australian Academy of Science in Canberra, geb. 16. XII. 1897 in Berlin, 75 Karlsruhe-Waldstadt, Schneidemühler Straße 22a (13. VI. 1966).

Myrberg, Pekka Juhana, Dr. phil., Dr. techn. h. c. (Helsinki), Dr. phil. h. c. (Turku und Stockholm), em. o. Professor für Mathematik an der Universität Helsinki, ehem. Kanzler der Universität Helsinki, Mitglied der Suomen Tiedeseura in Helsinki, ausländ. Mitglied der Kongelige Danske Videnskabernes Selskap in Kopenhagen und der Norske Videnskaps-Akademi in Oslo, Ehrenmitglied der Suomalainen Tiedeakatemia in Helsinki, geb. 30. XII. 1892 in Wiipuri, Helsinki, Mannerheimintie 75 A (5. XII. 1942).

Nissen, Rudolf, Dr. med., Dr. med. h. c. (Humboldt-Universität Berlin, Technische Universität München, Hacettepe Universität Ankara), em. oö. Professor für Chirurgie an der Universität Basel, 1956 Master of Surgery des International College of Surgeons, Inhaber der Rudolf-Virchow-Medaille in Gold der Virchow-Society in New York, der Vesalius-Medaille der Akademie für ärztliche Fortbildung in Stuttgart, des Ordens Civil de Sanidad Medico, Categoria Europienda, der Cron of Academic Honor of the Academic Council of the American International Academy, des Gimbernat-Preises der Asociacion de Cirurgia de Barcelona (1971), des Bayerischen Verdienstordens und der Paracelsus-Medaille der Deutschen Ärzteschaft, Mitglied der Schweizerischen Akademie der Medizinischen Wissenschaften in Basel (1961—1967 Senatsmitglied), der Deutschen Akademie der Naturforscher Leopoldina in Halle, korresp. Mitglied der Van Swieten-Gesellschaft in Wien, Mitglied der Rudolf-Virchow-Society of the City of New York, früher deren Präsident, Mitglied der Mittelrheinischen Chirurgen-Vereinigung der Bayerischen Chirurgen-Vereinigung, der Deutschen Gesellschaft für Chirurgie in Berlin, der Schweizerischen Gesellschaft für Angiologie und der Schweizerischen Gesellschaft für Chirurgie, ehemaliger Vorsitzender des Beirates der Fakultät für Medizin der Technischen Hochschule München, Ehrenmitglied und früherer Präsident des International College of Surgeons, Ehrenmitglied der Accademia Lancisiana in Rom, der International Mark Twain Society, St. Louis, der American International Academy, New York, des American College of Gastroenterology, der Physikalisch-medizinischen Sozietät Erlangen, der Schweizerischen Gesellschaft für Anaesthesiologie in Zürich, der Société Hellénique de Chirurgie, Athen, der Società Piemontese di Chirurgia, der Nordwestdeutschen Chirurgen-Vereinigung in Hamburg, der Society of Thoracic Surgeons of Great Britain and Ireland, in London, der Deutschen Gesellschaft für Anaestesiologie in Freiburg (Br.), der Deutschen Gesellschaft für Verdauungs- und Stoffwechselkrankheiten in Gießen, der Deutschen Gesellschaft für klinische Medizin, Sektion Chirurgie in Berlin, der Österreichischen Gesellschaft für Chirurgie und Traumatologie in Wien, der Berliner Chirurgen-Gesellschaft, der Schweizerischen Gesellschaft für Gastroenterologie, der Società Italiana di Chirurgia in Rom, der Deutschen Röntgengesellschaft in Mainz, des American College of Chest Physicians in Chicago, der International Society of University Colon and Rectal Surgeons, der Société Européenne de Chirurgie Cardio-Vasculaire, der Deutschen Gesellschaft für Angiologie, der Société Internationale de Chirurgie in Bruxelles und der Asociacion de Cirurgia de Barcelona, geb. 9. IX. 1896 in Neisse (Schlesien), CH 4125 Riehen b. Basel, Höhenstraße 45, Tel. 492323 (20. IV. 1963).

Răduleţ, B. Remus, Dr. phil., Professor für Grundlagen der Elektrotechnik an der Technischen Hochschule „Gh. Gheorgiu-Dej“ in Bukarest, Staatspreisträger, o. Mitglied der Rumänischen Akademie der Wissenschaften in Bukarest, ehemaliger Direktor des Forschungsinstituts ICPET in Bukarest, Mitglied der Académie du Monde Latin in Paris, Präsident des Comités d'Etudes Nr. 1 der Internationalen Elektrotechnischen Kommission in Genf, geb. 3. V. 1904 in Hendorf-Brădeni, Bucureşti, Strada Dimitrie Orbescu 11, Tel. 12-56-29 (24. X. 1966).

Schmid, Erich, Dr. phil., Dr. mont. h. c. (Montanistische Hochschule Leoben) Dr. rer. nat. h. c. (TU Clausthal), em. o. Professor für Physik an der Universität Wien, Träger des Österr. Ehrenzeichens für Wissenschaft und Kunst und der Moritz-Wilhelm-Drobisch-Medaille der Sächsischen Akademie, wirkliches Mitglied der Österreichischen Akademie der Wissenschaften in Wien, Präsident, seit 1. Okt. 1973 Vizepräsident, dieser Akademie und Vorsitzender der mathematisch-naturwissenschaftlichen Klasse, Mitglied der Deutschen Akademie der Naturforscher Leopoldina in Halle und der Academy of Sciences in New York, auswärtiges Mitglied der Akademie der Wissenschaften der DDR, korresp. Mitglied der Accademia di Studi Economici e Sociali in Triest, auswärtiges wiss. Mitglied des Max-Planck-Institutes für Metallforschung in Stuttgart, geb. 4. V. 1896 in Bruch Mur (Österreich), A-1090 Wien, Universitätsstraße 10, Tel. 4240062 (12. XI. 1971).

Śmiałowski, Michał, Dr.-Ing., Prof. für physikalische Chemie, Mitglied der polnischen Pugwash-Kommission, Staatspreisträger, Träger des Ordens Sztandar Pracy I. und II. Klasse und Inhaber des Kommandeurkreuzes des Ordens Polonia Restituta, o. Mitglied der Polnischen Akademie der Wissenschaften, Mitglied der Polnischen Chemischen Gesellschaft in Warschau, der Electrochemical Society in New York, der Faraday Society in London und der Société de Chimie Physique in Paris, geb. 14. XII. 1906 in Hordynia (Ost-Galizien), Warszawa, ul. Marszałkowska 73 m. 18, Tel. 283772 (12. XI. 1971).

Spitzin, Viktor Iwanowitsch, Dr. chem., Dr. h. c. (Leipzig und Wrocław), Professor für anorganische Chemie an der Lomonossow-Universität in Moskau, Held der Sozialistischen Arbeit der UdSSR, Inhaber von drei Leninorden, des Ordens der Oktoberrevolution und des Ordens des Roten Arbeitsbanners, o. Mitglied der Akademie der Wissenschaften der UdSSR in Moskau, Direktor des Instituts für Physikalische Chemie dieser Akademie, geb. 25. IV. 1902 in Moskau, Moskau, Leninsky Prospekt 31 (24. X. 1966).

Staude, Herbert, Dr. phil., em. o. Professor für physikalische Chemie an der Johann-Wolfgang-Goethe-Universität Frankfurt (Main), Mitglied der Deutschen Bunsengesellschaft für Physikalische Chemie in Frankfurt (Main), der Gesellschaft Deutscher Chemiker in Frankfurt (Main), der Faraday Society in London und der Chemischen Gesellschaft in der DDR, geb. 20. III. 1901 in Schmölln, 6 Bergen-Enkheim, Ostpreußenstraße 13, Tel. (06194) 33636 (17. XI. 1956).

Ţiţeica, G. Şerban, Dr. phil., Professor für Physik an der Universität Bukarest, o. Mitglied der Rumänischen Akademie der Wissenschaften in Bukarest, Vizepräsident dieser Akademie, korresp. Mitglied der Akademie der Wissenschaften der UdSSR in Moskau, geb. 27. III. 1908 in Bukarest, Bucureşti, B-dul-Magheru 35, Tel. 15-81-99 (24. X. 1966).

Treibs, Wilhelm, Dr. phil., Dr. rer. nat. h. c. (Humboldt-Universität Berlin), Gastprofessor für organische Chemie an der Ruprecht-Karl-Universität Heidelberg, Nationalpreisträger, Hervorragender Wissenschaftler des Volkes, Träger des Vaterländischen Verdienstordens, Inhaber der August-Kekulé-Medaille der Chemischen Gesellschaft in der DDR, auswärt. Mitglied der Akademie der Wissenschaften der DDR, Mitglied der Deutschen Akademie der Naturforscher Leopoldina in Halle, der Gesellschaft Deutscher Chemiker in Frankfurt (Main), der Schweizerischen Chemischen Gesellschaft in Basel, der Chemical Society in London, der International Society for Fat Research in Hannover, Mitglied des Editorial Advisory Board des „Tetrahedron" (International Journal of Organic Chemistry) und der „Tetrahedron Letters", des Herausgeberstabes der „Chemischen Berichte" und des Expertenbeirates von „Parfümerie und Kosmetik" in Heidelberg, Ehrenmitglied der Chemischen Gesellschaft in der DDR, geb. 6. XI. 1890 in Oberstein (Nahe), 69 Heidelberg-Neuenheim, Kastellweg 28 (14. III. 1955).

van der Waerden, Bartel L., Dr. phil., Dr. h. c. (Athen), o. Professor für Mathematik an der Universität Zürich, Inhaber der Goldenen Cothenius-Medaille der Deutschen Akademie der Naturforscher Leopoldina, Mitglied der Deutschen Akademie der Naturforscher Leopoldina in Halle, korresp. Mitglied der Bayerischen Akademie der Wissenschaften in München, der Akademie der Wissenschaften in Göttingen und der Heidelberger Akademie der Wissenschaften, ausländ. Mitglied der Koninklijke Nederlandse Akademie van Wetenschappen in Amsterdam, geb. 2. II. 1903 in Amsterdam, CH 8053 Zürich, Carl-Spitteler-Straße 186, Tel. 552179 (31. VIII. 1934).

Wagner, Carl, Dr. phil., Dr. rer. nat. h. c. (Darmstadt), Dr.-Ing. E. h. (Clausthal-Zellerfeld), Hon.-Professor für physikalische Chemie an der Georg-August-Universität Göttingen, Inhaber der Goldmedaille der American Soc. of Metals, der Cavallaro-Medaille der Europ. Förderrat. für Korrosion, der Palladium Medal der Electrochem. Soc., der Wilhelm-Exner-Medaille des Österreichischen Gewerbevereins, der Heyn-Denkmünze der Deutschen Ges. für Metallkunde und der Carl-Friedrich-Gauß-Medaille der Braunschweigischen Wissenschaftlichen Gesellschaft, o. Mitglied der Akademie der Wissenschaften in Göttingen, Mitglied der Deutschen Akademie der Naturforscher Leopoldina in Halle, des Max-Planck-Institutes für Physikalische Chemie in Göttingen und American Academy of Arts and Sciences in Boston, auswärt. Mitglied der National Academy of Sciences in Washington, Mitglied der Deutschen Bunsengesellschaft für physikalische Chemie in Frankfurt (Main), der Deutschen Gesellschaft für Metallkunde in Stuttgart, der Gesellschaft Deutscher Chemiker in Frankfurt (Main), der Faraday Society in Washington, der American Physical Society in New York, der National Association of Corrosion Engineers in Houston (Texas) und des American Institute of Mining, Metallurgical and Petroleum Engineers in New York, Ehrenmitglied der Österreichischen Akademie der Wissenschaften in Wien und des American Institute of Mining, Metallurgical and Petroleum Engineers, geb. 25. V. 1901 in Leipzig, 34 Göttingen-Geismar, Charlottenburger Straße 19, Wohnstift C-309 (9. III. 1959).

Zhdanov, German Stepanowitsch, Dr. phys. math., Professor für Metallphysik an der Lomonossow-Universität Moskau, Verdienter Wissenschaftler und Techniker der UdSSR, Inhaber des A. N. Bach-Preises und des D. I. Mendelejew-Preises, Vorsitzender der Kommission für Röntgenographie der Akademie der Wissenschaften der UdSSR in Moskau, Mitglied des Nationalkomitees für Kristallographie in Moskau, geb. 11. VIII. 1906 in Petersburg, Moskau B-234, Moskauer Staatliche Universität, Block I, W. 52, Tel. 139-16-34 (13. VI. 1966).

Ordentliche Mitglieder der philologisch-historischen Klasse

Baetke, Walter, Dr. phil., Dr. theol. h. c. (Leipzig), em. Professor mit Lehrstuhl für nordische Philologie und Religionsgeschichte an der Karl-Marx-Universität Leipzig, Hervorragender Wissenschaftler des Volkes, Träger des Vaterländischen Verdienstordens und der Moritz-Wilhelm-Drobisch-Medaille der Sächsischen Akademie, geb. 28. III. 1884 in Sternberg (Neum.), 703 Leipzig, Turmweg 18, Tel. 383881 (19. VI. 1943).

Bahner, Werner, Dr. phil., Professor, Bereichsdirektor am Zentralinstitut für Literaturgeschichte bei der Akademie der Wissenschaften der DDR, Träger des Vaterländischen Verdienstordens, Inhaber des Friedrich-Engels-Preises der Akademie der Wissenschaften der DDR, Mitglied der Sprachwissenschaftlichen Kommission der Sächsischen Akademie, o. Mitglied der Akademie der Wissenschaften der DDR, Mitglied des Büros des Comité Internationale Permanent des Linguistes, der Société de Linguistique Romane in Paris und Strasbourg, des Comité International der Internationalen Vereinigung für Südosteuropäische Studien (AJESEE) in Bukarest und der Asociación Internacional de Hispanistas, geb. 18. VI. 1927 in Neuwiese/Erzgeb., 1138 Berlin-Kaulsdorf, Scharnauer Straße 23, Tel. 6561782 (24. V. 1965).

Blumenthal, Lieselotte, Dr. phil., em. Professor an der Akademie der Wissenschaften der DDR, Inhaberin der Goethe- und Schiller-Plakette der Nationalen Forschungs- und Gedenkstätten der klassischen deutschen Literatur in Weimar, geb. 30. IX. 1906 in Bielefeld, 7027 Leipzig, Kommandant-Prendel-Allee 107, Tel. 81614 (5. XII. 1960).

Brüning, Eberhard, Dr. phil., o. Professor für Amerikanistik an der Karl-Marx-Universität Leipzig, Verdienter Lehrer des Volkes, Inhaber der Johann-Heinrich-Pestalozzi-Medaille in Bronze und Silber und der Ehrennadel der Karl-Marx-Universität, Mitglied der Deutschen Shakespeare-Gesellschaft in Weimar und der Association Internationale de Littérature Comparée, Vorsitzender der Zentralen Fachkommission Englisch der DDR, geb. 4. VI. 1925 in Chemnitz, 703 Leipzig, Scheffelstraße 53, Tel. 384593 (28. VI. 1971).

Buchda, Gerhard, Dr. jur., em. Professor mit Lehrstuhl für deutsche Rechtsgeschichte und Handelsrecht an der Friedrich-Schiller-Universität Jena, Träger des Vaterländischen Verdienstordens, Inhaber der Deutschen Friedensmedaille des Friedensrates der DDR in Berlin, der Gedächtnisplakette der Humboldt-Universität Berlin und der Friedrich-Schiller-Medaille der Universität Jena, Leiter der Historischen Kommission der Akademie, Mitglied der Vereinigung der Deutschen Rechtshistoriker und der Société Jean Bodin pour l'histoire comparative des institutions in Brüssel, geb. 22. X. 1901 in Stadtroda (Thür.), 654 Stadtroda (Thür.), Straße der Jugend 5, Tel. Stadtroda 798 (15. IX. 1958).

Coblenz, Werner, Dr. phil., Direktor des Landesmuseums für Vorgeschichte in Dresden, Inhaber der Verdienstmedaille der DDR, der Leibniz-Medaille der Akademie der Wissenschaften der DDR und der Johannes-R.-Becher-Medaille des Kulturbundes der DDR, Mitglied und stellvertretender Vorsitzender der Historischen Kommission der Akademie, o. Mitglied des Deutschen Archäologischen Instituts in Berlin-Dahlem, Mitglied im Conseil Permanent der Union Internationale des Sciences Protohistoriques et Préhistoriques, des Nationalkomitees für Ur-

und Frühgeschichte der DDR, des wissenschaftlichen Beirats für Heimatforschung, des geographischen Instituts der Akademie der Wissenschaften der DDR und des Problemrates für Alte Geschichte und Archäologie beim Wissenschaftlichen Rat für Geschichtswissenschaften der DDR, geb. 24. V. 1917 in Pirna/Elbe, 8020 Dresden, Geystraße 6, Tel. 40916 (18. XII. 1972).

Große, Rudolf, Dr. phil., o. Professor für deutsche Philologie an der Karl-Marx-Universität Leipzig, Leiter der Sprachwissenschaftlichen Kommission und Mitglied der Historischen Kommission der Akademie, o. Mitglied der Akademie der Wissenschaften der DDR, Vorsitzender der Germanisten-Kommission der DDR und der VR Polen, Ausschußmitglied der Internationalen Vereinigung für Germanische Sprach- und Literaturwissenschaft, Mitglied des Comité International de Dialectologie in Louvain, geb. 28. VII. 1924 in Heldrungen, 703 Leipzig, Aschenbrödelweg 36, Tel. 383802 (24. V. 1965).

Hiebsch, Hans, Dr. phil., o. Professor für Psychologie an der Friedrich-Schiller-Universität Jena, Inhaber der Verdienstmedaille der DDR und der Verdienstmedaille der Nationalen Volksarmee der DDR in Silber, Mitglied der Sprachwissenschaftlichen Kommission der Akademie und der Gesellschaft für Psychologie in der DDR, stellvertretender Vorsitzender dieser Gesellschaft, und Mitglied der European Association of Experimental Social Psychology in Hague, geb. 10. VIII. 1922 in Bodenbach/ČSR, 69 Jena, Professor-Ibrahim-Straße 35, Tel. 24907 (28. VI. 1971).

Jahn, Johannes, Dr. phil., Dr. phil. h. c. (Leipzig), em. Professor mit Lehrstuhl für Kunstgeschichte an der Karl-Marx-Universität Leipzig, Ehrenpräsident des Nationalen Museumsrates, Träger des Vaterländischen Verdienstordens, des Ordens „Banner der Arbeit", der Moritz-Wilhelm-Drobisch-Medaille der Sächsischen Akademie und der Winckelmann-Medaille der Stadt Stendal, Ehrenmitglied des Verbandes Bildender Künstler in Berlin, geb. 22. XI. 1892 in Orlandshof (Krs. Wissek), 701 Leipzig, Auenstraße 13, Tel. 208634 (11. XII. 1961). † 17. II. 1976.

Krauss, Werner, Dr. phil., Dr. phil. h. c. (Aix-Marseille und Leipzig), em. Professor mit Lehrstuhl für romanische Philologie an der Karl-Marx-Universität Leipzig, Nationalpreisträger, Hervorragender Wissenschaftler des Volkes, Träger des Vaterländischen Verdienstordens, Inhaber des Friedrich-Engels-Preises der Akademie der Wissenschaften der DDR und der Medaille für Kämpfer gegen den Faschismus 1933—1945, o. Mitglied der Akademie der Wissenschaften der DDR und em. Direktor des Instituts für romanische Sprachen und Kultur dieser Akademie, korresp. Mitglied der Academia Nacional de Derecho y Ciencias Sociales in Buenos Aires, ausländ. korresp. Mitglied der Rumänischen Akademie der Wissenschaften in Bukarest, Vizepräsident der Asociacion Internacional de Hispanistas und Mitglied des Büros der Association Internationale de Littérature Comparée, geb. 7. VI. 1900 in Stuttgart, 1167 Berlin-Hessenwinkel, Kanalstraße 35, Tel. 6589636 (14. II. 1949). † 28. VIII. 1976

Kühnert, Friedmar, Dr. phil., o. Professor für klassische Philologie, mit besonderer Berücksichtigung der Lateinischen Sprache und Literatur an der Friedrich-Schiller-Universität Jena, Inhaber der Theodor-Neubauer-Medaille in Silber, Mitglied der Winckelmann-Gesellschaft der DDR, geb. 16. XI. 1924 in Dreißigacker (Kr. Meiningen), 6905 Jena-Göschwitz, August-Bebel-Straße 6, Tel. 32321 (18. IV. 1974).

Lehmann, Edgar, Dr. phil., Dr.-Ing. e. h. (Eidgenössische Technische Hochschule Zürich), em. Professor, em. Direktor des Geographischen Instituts der Akademie der Wissenschaften der DDR, Träger des Vaterländischen Verdienstordens in Silber, Inhaber der Alexander-von-Humboldt-Gedenkmedaille der DDR, der Hermann-Haack-Medaille der Geographischen Gesellschaft der DDR, der Lajós-Loczy-Gedenkmedaille der Ungarischen Geographischen Gesellschaft und der Medaille für wissenschaftliche Verdienste der Polnischen Geographischen Gesellschaft, Mitglied der Historischen Kommission und der Kommission für spezielle Umweltprobleme der Akademie, Präsident der „Kommission National- und Regionalatlanten“ der Internationalen Geographischen Union (IGU), o. Mitglied der Akademie der Wissenschaften der DDR, Vorsitzender der Klasse „Umweltschutz und Umweltgestaltung“, Vorsitzender der „Kommission zur Herausgabe des Atlas der DDR“, Vorsitzender des Wissenschaftlichen Beirates für Heimatforschung, Mitglied des Wissenschaftlichen Beirates des Forschungsbereiches „Kosmische Physik“ und des Wissenschaftlichen Rates für Umweltforschung bei dieser Akademie, Vorstandsmitglied der Geographischen Gesellschaft der DDR, Mitglied des Nationalkomitees für Geographie und Kartographie der DDR, Ehrenmitglied der Geographischen Gesellschaft der UdSSR, geb. 25. III. 1905 in Berlin, 703 Leipzig, Prinz-Eugen-Straße 44, Tel. 31219 (7. XII. 1959).

Lieberwirth, Rolf, Dr. jur., o. Professor für deutsche Rechtsgeschichte an der Martin-Luther-Universität Halle—Wittenberg, Inhaber der Verdienstmedaille der DDR und der Wilhelm-Külz-Ehrennadel der LDPD, Mitglied der Gesellschaft für Völkerrecht der DDR und der International Law Association in London, geb. 1. XII. 1920 in Halle (Saale), 402 Halle (Saale), Rainstraße 3b, Tel. 30465 (18. XII. 1972).

Markov, Walter, Dr. phil., Dr. phil. h. c. (Leipzig), em. o. Professor für mittelalterliche und neuere Geschichte an der Karl-Marx-Universität Leipzig, Nationalpreisträger, Träger des Vaterländischen Verdienstordens, Inhaber der Medaille für Kämpfer gegen den Faschismus 1933—1945, der Bronzemedaille der Universität Helsinki und der Ehrennadel für Verdienste um die Freundschaft der Völker in Gold, Mitglied der Historischen Kommission der Akademie, o. Mitglied der Akademie der Wissenschaften der DDR, Vorstandsmitglied der Société des Études Robespierristes in Paris und der Association Europénne d’Histoire Contemporaine in Strasbourg, Mitglied der Société d’Histoire Moderne et Contemporaine in Paris und der Société Française d’Étude du XVIIIe Siècle in Paris, der Historical Society of Nigeria in Ibadan, Vorstandsmitglied des Nationalkomitees der Historiker der DDR und der Historikergesellschaft der DDR, geb. 5. X. 1909 in Graz, 7124 Holzhausen (Sachsen), Weißdornweg 4, Tel. 61384 (7. XII. 1964).

Meyer, Rudolf, Dr. theol., Dr. theol. h. c. (Humboldt-Universität Berlin), em. o. Professor für Altes Testament und rabbinische Literatur an der Friedrich-Schiller Universität Jena, Sektion Theologie, Leiter des Fachbeirates Altes Testament, Inhaber der Schiller-Medaille der Friedrich-Schiller-Universität, Kurator der Hilprecht-Sammlung Vorderasiatischer Altertümer, geb. 8. IX. 1909 in Leipzig, 69 Jena, Hornstraße 1, Tel. 24382 (15. IX. 1958).

Müller, Joachim, Dr. phil., em. o. Professor für Deutsche Literatur an der Friedrich-Schiller-Universität Jena, Inhaber der Schiller-Plakette der Deutschen Schiller-Stiftung, der Schiller-Medaille in Silber und der Ehrennadel in Gold der Friedrich-Schiller-Universität sowie der Johann-Heinrich-Pestalozzi-Medaille in Bronze, Mitglied der Sprachwissenschaftlichen Kommission der Akademie, korresp. Mitglied des Adalbert-Stifter-Institutes des Landes Oberösterreich in Linz, Mitglied des PEN-Clubs, geb. 12. VII. 1906 in Oberwiesenthal (Sachsen), 69 Jena, Schillbachstraße 39, Tel. 23808 (7. XII. 1959).

Neubert, Albrecht, Dr. phil., o. Professor an der Sektion Theoretische und angewandte Sprachwissenschaft an der Karl-Marx-Universität Leipzig, Inhaber der Ehrennadel der Karl-Marx-Universität Leipzig und der Johann-Heinrich-Pestalozzi-Medaille in Silber, Mitglied der Sprachwissenschaftlichen Kommission der Akademie, o. Mitglied der Akademie der Wissenschaften der DDR, Vorsitzender des Wissenschaftlichen Rates für Sprachwissenschaft bei dieser Akademie, Mitglied der Deutschen Shakespeare-Gesellschaft in Weimar, Mitglied und Vizepräsident der International Association of Applied Linguistics, geb. 3. III. 1930 in Hartenstein (Sachsen), 7022 Leipzig, Poetenweg 5, Tel. 50131 (18. XII. 1972).

Neumann, Werner, Dr. phil., Professor, Direktor des Bach-Archivs Leipzig, Nationalpreisträger, Träger des Vaterländischen Verdienstordens und der Verdienstmedaille der DDR, Mitglied der Neuen Bachgesellschaft in Leipzig und Mitglied des Verwaltungsrates dieser Gesellschaft, geb. 21. I. 1905 in Königstein (Sachsen), 7113 Markkleeberg, Robert-Schuhmann-Straße 3, Tel. 385525 (18. VI. 1974).

Peek, Werner, Dr. phil., em. o. Professor für klassische Philologie an der Martin-Luther-Universität Halle—Wittenberg, Nationalpreisträger, Inhaber der Verdienstmedaille der Martin-Luther-Universität, o. Mitglied der Akademie der Wissenschaften der DDR, korresp. Mitglied der Heidelberger Akademie der Wissenschaften und des Deutschen Archäologischen Instituts in Berlin-Dahlem, auswärt. Mitglied der Kommission für die archäologische Erforschung Kleinasiens bei der Österreichischen Akademie der Wissenschaften in Wien, geb. 6. VI. 1904 in Bielefeld, 403 Halle (Saale), Am Krähenberg 21, Tel. 32822 (7. XII. 1959).

Peukert, Herbert, Dr. phil., em. o. Professor für slawische Sprachen an der Friedrich-

Schiller-Universität Jena, Inhaber der Johann-Heinrich-Pestalozzi-Medaille in Bronze und Silber und der Johann-Gottfried-Herder-Medaille in Gold, Mitglied und stellvertretender Leiter der Sprachwissenschaftlichen Kommission der Akademie, Mitglied des Nationalkomitees der Slawisten der DDR bei der Akademie der Wissenschaften der DDR, ausländ. Mitglied der Makedonischen Akademie der Wissenschaften und Künste zu Skopje, geb. 29. IV. 1907 in Reichenau, 69 Jena, Okenstraße 33, Tel. 24186 (24. V. 1965).

Rudolph, Kurt, Dr. theol., Dr. phil., o. Professor für Religionsgeschichte an der Karl-Marx-Universität Leipzig, auswärtiges Mitglied der Königlich-Dänischen Akademie der Wissenschaften in Kopenhagen (Det Kongelige Danske Videnskabernes Selskab), korresp. Mitglied des Institute for Antiquity and Christianity in Claremont/Californien (USA) und der Internationalen Association for the History of Religions (IAHR), geb. 3. IV. 1929 in Dresden, 701 Leipzig, Dufourstraße 38, Tel. 30057 (18. VI. 1974).

Růžička, Rudolf, Dr. phil., o. Professor für Slawistik und Allgemeine Sprachwissenschaft an der Karl-Marx-Universität Leipzig, o. Mitglied der Akademie der Wissenschaften der DDR, Mitglied des Wissenschaftlichen Rates für Sprachwissenschaft bei dieser Akademie, Mitglied der Internationalen Terminologischen Kommission und der Internationalen Grammatikkommission des Internationalen Slawistenkomitees, geb. 20. XII. 1920 in Löbau, 701 Leipzig, Gottschedstraße 12, Tel. 293852 (18. XII. 1972).

Schubart-Fikentscher, Gertrud, Dr. jur., em. Professor mit Lehrstuhl für Rechtsgeschichte und Zivilrecht an der Martin-Luther-Universität Halle—Wittenberg, Trägerin des Vaterländischen Verdienstordens, Mitglied der Historischen Kommission der Akademie, korresp. Mitglied der Monumenta Germaniae Historica in München, geb. 23. XII. 1896 in Zwickau (Sachsen), 403 Halle (Saale), Marsstraße 10, Tel. 32710 (7. XII. 1959).

Straub, Werner, Dr. phil., em. Professor mit Lehrstuhl für Psychologie an der Technischen Universität Dresden, Träger des Vaterländischen Verdienstordens in Silber und Bronze, des Ordens „Banner der Arbeit“ und der Verdienstmedaille der DDR, Ehrenmitglied der Gesellschaft für Psychologie in der DDR, geb. 14. VII. 1902 in Stuttgart, 8051 Dresden-Bad Weißer Hirsch, Collenbuschstraße 10, Tel. 36036 (16. V. 1949).

Weller, Friedrich, Dr. phil., em. Professor mit Lehrstuhl für Indologie an der Karl-Marx-Universität Leipzig, Nationalpreisträger, Hervorragender Wissenschaftler des Volkes, Mitglied der Deutschen Gesellschaft für Natur- und Völkerkunde Ostasiens in Tokio, korresp. Mitglied der School of Oriental and African Studies in London, Ehrenmitglied der Deutschen Morgenländischen Gesellschaft in Mainz und der Kőrösi Csoma-Gesellschaft in Budapest, geb. 22. VII. 1889 in Markneukirchen, 701 Leipzig, Stephanstraße 12, Tel. 295801 (19. VI. 1943).

Werner, Ernst, Dr. phil. o. Professor für Geschichte des Mittelalters an der Karl-Marx-Universität Leipzig, Nationalpreisträger, Träger des Vaterländischen Verdienstordens in Bronze, Inhaber der Verdienstmedaille der DDR und der Erinnerungsmedaille zum 50. Jahrestag der Großen Sozialistischen Oktoberrevolution der Universität Kiew, Mitglied der Sprachwissenschaftlichen Kommission der Akademie, o. Mitglied der Akademie der Wissenschaften der DDR, geb. 20. XI. 1920 in Tyssa/ČSR, 7022 Leipzig, Braunschweiger Straße 19, Tel. 54390 (18. VI. 1971).

Auswärtige Mitglieder
der philologisch-historischen Klasse

Aebischer, Paul, Dr. phil., em. Professor für romanische Philologie an der Universität Lausanne, auswärt. Mitglied der Akademie der Wissenschaften der DDR und der Accademia delle Scienze in Turin, korresp. Mitglied der Academia espagñola in Madrid, der Academia de Buenas Letras in Barcelona, des Institut d'Estudis Catalans in Barcelona und der Accademia della Crusca in Florenz, geb. 8. XII. 1897 in Hauterive (Kanton Fribourg), CH 1680 Romont (Fribourg), route d'Arruffens, 573 (29. XI. 1943).

Aland, Kurt, D., D. theol. h. c. (Göttingen), D. D. h. c. (St. Andrews), D. Litt. h. c. (Wartburg College, Iowa), o. Professor für Geschichte der Alten Kirche und Neutestamentliche Textforschung im Fachbereich Evangelische Theologie der Westfälischen Wilhelms-Universität Münster, Direktor des Instituts für neutestamentliche Textforschung, Vizepräsident der Association Internationale d'Etudes Patristiques und Geschäftsführer der Stiftung zur Förderung der neutestamentlichen Textforschung, Leiter der Arbeitsstelle Münster der Patristischen Kommission, Träger des Athos-Kreuzes in Gold, verliehen vom Ökumenischen Patriarchen von Konstantinopel und der Burkitt-Medal for Biblical Studies der British Academy, korresp. Mitglied der British Academy und der Göttinger Akademie der Wissenschaften, Mitglied der Studiorum Novi Testamenti Societas (England), des Kuratoriums der Vetus Latina in Beuron (Hohenzollern), Ehrenmitglied auf Lebenszeit der American Bible Society und Ehrenmitglied der Society of Biblical Literature and Exegesis (USA), geb. 28. III. 1915 in Berlin, 44 Münster (Westf.), Roxeler Straße 12, Tel. 22409 (14. III. 1955).

Besevliev, Vešelin, Dr. phil., em. ao. Professor für Altgriechische Philologie an der Universität Sofia, Professor für Altgriechische Philologie am Institut für Balkanistik der Bulgarischen Akademie der Wissenschaften in Sofia, Inhaber des Gottfried-von-Herder-Preises der Universität Wien, Träger des Ordens für Wissenschaft „Kirill i Metodi", I. Stufe, korresp. Mitglied der Bulgarischen Akademie der Wissenschaften in Sofia, Mitglied des Bulgarischen Archäologischen Instituts in Sofia und der Ungarischen Orientalischen Gesellschaft in Budapest, geb. 25. III. 1900 in Sofia, Sofia, L. Karavelov 2^{v}, Tel. 87-96-08 (9. XI. 1974).

Bischoff, Karl, Dr. phil., em. o. Professor für deutsche Philologie und Volkskunde an der Johannes-Gutenberg-Universität Mainz, Mitglied der Historischen Kommission der Akademie, o. Mitglied der Akademie der Wissenschaften und der Literatur in Mainz, korresp. Mitglied der Akademie der Wissenschaften in Göttingen und buitenlands erelid der Koninklijke Vlaamse Academie voor Taal- en Letterkunde in Gent, geb. 19. V. 1905 in Aken (Elbe), 65 Mainz, In der Meielache 15, Tel. 32174 (14. III. 1955).

Bruford, Walter Horace, M. A., LL. D. h. c. (Aberdeen), D. Litt. h. c. (Newcastle, Edinburg), em. Professor für Germanistik an der Universität Cambridge (England), Inhaber der Goethe-Medaille in Gold des Goethe-Instituts in München, Mitglied der British Academy in London, korresp. Mitglied der Deutschen Akademie für Sprache und Dichtung in Darmstadt, Mitglied der Modern Humanities Research Association in Cambridge, 1965 deren Präsident, Mitglied der Internationalen Vereinigung für germanische Sprach- und Literaturwissenschaft, 1955—1960 deren Vizepräsident, geb. 14. VII. 1894 in Manchester (England), Moorhouse, Abbey St. Bathans, Duns, Berwickshire/Great Britain, Tel. Abbey St. Bathans 235 (21. VI. 1965).

Droz, Jacques, Dr. phil., o. Professor für neuere Geschichte an der Sorbonne (Panthéon) in Paris, Direktor des Instituts für Geschichte der Gewerkschaftsbewegung, Träger des Bandes der Ehrenlegion, Mitglied der Gesellschaft Frankreich—DDR, geb. 12. III. 1909 in Paris, Paris, 55 rue de Varenne 7 (9. XI. 1974).

Erben, Johannes, Dr. phil., o. Professor für deutsche Sprache und ältere deutsche Literatur an der Leopold-Franzens-Universität Innsbruck, Nationalpreisträger, Inhaber des Konrad-Duden-Preises 1969 der Stadt Mannheim, auswärt. Mitglied der Akademie der Wissenschaften der DDR, Kuratoriumsmitglied des Instituts für deutsche Sprache in Mannheim, geb. 12. I. 1925 in Leipzig, A-6020 Innsbruck, Reithmannstraße 20 I, Tel. 522623 (11. XII. 1961).

Gadamer, Hans-Georg, Dr. phil., o. Professor für Philologie an der Ruprecht-Karl-Universität Heidelberg, Mitglied des Ordens Pour le mérite für Wissenschaft und Kunst, Träger des Großen Verdienstkreuzes mit Stern der Bundesrepublik Deutschland und Inhaber des Reuchlinpreises 1971 der Stadt Pforzheim, o. Mitglied der Heidelberger Akademie der Wissenschaften, Vorsitzender der Cusanus-Kommission dieser Akademie, korresp. Mitglied der Deutschen Akademie für Sprache und Dichtung in Darmstadt, der Akademie der Wissenschaften in Athen und der Accademia Nazionale dei Lincei in Rom, geb. 11. II. 1900 in Marburg, 69 Heidelberg, Am Büchsenackerhang 53, Tel. 59536 (15. VI. 1940).

Georgiev, Vladimir I., Dr. phil., Dr. phil. h. c. (Berlin und Prag), o. Professor für allgemeine und indoeuropäische Sprachwissenschaft, Slawistik und Balkanistik an der Universität Sofia, Inhaber des Nationalpreises für Wissenschaft, Mitglied und Vizepräsident der Bulgarischen Akademie der Wissenschaften in Sofia, Mitglied des Institut de France/Académie des Inscriptions et Belles-Lettres in Paris, korresp. Mitglied der Societas Scientiarum Fennica-Finska Vetenskaps-Societeten-

Souomen Tiedeseura in Helsinki und der Académie Royale des Sciences, des Lettres et des Beaux-Arts de Belgique in Brüssel, geb. 3. II. 1908 in Gabare (Bulg.), Sofia, ul. Čapaev, 52, Tel. 72-20-58, Privatanschrift: Sofia, Oboriste 11, Tel. 44-16-40 (19. II. 1968).

Havránek, Bohuslav, Dr. phil., Dr. phil. h. c. (Leipzig, Brno, Kraków und Olmomouc), o. Professor für Slawistik an der Karls-Universität Prag, Träger des Ordens der Tschechoslowakischen Sozialistischen Republik, Inhaber der J. A. Comenius-Medaille, der Josef-Dobroský-Medaille und der Medaille für Verdienste um Wissenschaft und Menschheit, o. Mitglied der Tschechoslowakischen Akademie der Wissenschaften in Prag, auswärt. Mitglied der Akademie der Wissenschaften der DDR, korresp. Mitglied der philosophisch-historischen Klasse der Österreichischen Akademie der Wissenschaften in Wien, ausländ. Mitglied der Bulgarischen Akademie der Wissenschaften in Sofia, der Serbischen Akademie der Wissenschaften in Belgrad, der Jugoslawischen Akademie der Wissenschaften in Zagreb und des Comité International Permanent des Linguistes in Louvain, stellv. Vorsitzender des Internationalen Komitees der Slawisten, geb. 30. I. 1893 in Prag, 16000 Praha 6 — Hradčany, Na Valech 8, Tel. 325-103 (10. XII. 1962).

Heimpel, Hermann, Dr. phil., Dr. jur. h. c. (Freiburg i. Br.), em. o. Professor der mittelalterlichen und neueren Geschichte an der Georg-August-Universität Göttingen, ehem. Direktor des Max-Planck-Instituts für Geschichte in Göttingen, Inhaber des großen Verdienstkreuzes des Verdienstordens der Bundesrepublik Deutschland und des Kulturpreises der Stadt Goslar für das Jahr 1965, o. Mitglied der Akademie der Wissenschaften in Göttingen, korresp. Mitglied der Deutschen Akademie für Sprache und Dichtung in Darmstadt, der Bayerischen Akademie der Wissenschaften in München, der Heidelberger Akademie der Wissenschaften, der Akademie der Wissenschaften und der Literatur in Mainz, der Österreichischen Akademie der Wissenschaften in Wien und der Royal Historical Society in London, o. Mitglied der Zentraldirektion der Monumenta Germaniae Historica in München und der Historischen Kommission bei der Bayerischen Akademie der Wissenschaften in München, Mitglied des Wissenschaftlichen Beirates des Deutschen Historischen Instituts in Rom, Ehrenmitglied des Instituts für Österreichische Geschichtsforschung in Wien, geb. 19. IX. 1901 in München, 34 Göttingen, Dahlmannstraße 14, Tel. 59382 (5. VI. 1935).

Iordan, Iorgu, Dr. phil., Dr. phil. h. c. (Humboldt-Universität Berlin, Gent, Jassy, Montpellier), em o. Professor für romanische Philologie an der Universität Bukarest, Verdienter Wissenschaftler, Staatspreisträger I. und II. Klasse, Inhaber des „Ordinul Muncii" I. Klasse, des Ordens „Steaua Republicii Populare Române" I. Klasse, des Ordens „Meritul Ştiintific" I. Klasse, des Ordens „23. August" II. Klasse, des Ordens „Tudor Vladimirescu" II. Klasse, der Medaille „40 de ani de la întemeierea Partidului Comunist din România", der Medaille „A XX-a aniversare a Eliberării Patriei", der Medaille „A cincizecea aniversare a P. C. R.", der Medaille „25 de ani de la proclamarea Republicii" und der Medaille „30 de ani de la Eliberarea de sub dominatio fascistă", Inhaber des Titels „Eron al Muncii Socialiste", o. Mitglied der Rumänischen Akademie der Wissenschaften in Bukarest, früher Vizepräsident dieser Akademie und Präsident der Klasse für Sprache, Literatur und Kunst, auswärt. Mitglied der Akademie der Wissenschaften der DDR, der Bayerischen Akademie der Wissenschaften in München, der Österreichischen Akademie der Wissenschaften in Wien, Präsident der Rumänischen Gesellschaft für romanische Sprachforschung und des Rumänischen Nationalkomitees für Volksmusik, Vizepräsident der Rumänischen Gesellschaft für Beziehungen mit der UdSSR, Ehrenpräsident der Société de Linguistique Romane in Paris und Strasbourg, Mitglied der Société de Linguistique in Paris, der Asociación Internacional de Hispanistas, der Junta directiva dieser Asociación, seit 1971 deren Vizepräsident, der Asociación de Lingüística y Filología Hispanoamericana, des Comité International des Sciences Onomastiques in Louvain und des Comité International Permanent des Linguistes (CIPL), korresp. Mitglied des Instituts Mexicano de Cultura in Mexico City, geb. 29. IX. 1888 in Tecuči (Rumänien), București, str. Sofia 12, Tel. 331886 (10. III. 1958).

Kumaniecki, Kazimierz, Dr. phil., Dr. phil. h. c. (Debrecen und Warszawa), o. Professor für klassische Philologie an der Universität Warschau, Staatspreisträger, Inhaber des Kommandeur- und Kavalierkreuzes des Ordens Polonia Restituta und des Tapferkeitskreuzes, o. Mitglied der Polnischen Akademie der Wissenschaften in Warschau, auswärt. Mitglied der Koninklijke Vlaamse Academie voor Taal-en Letterkunde in Gent, o. Mitglied des Centro di Studi Ciceroniani in Rom, ehem. Präsident der Union Académique Internationale in Brüssel und Vizepräsident des Conseil International de la Philosophie et des Sciences Humaines in Paris, Präsident und Ehrenmitglied der Polnischen Philologischen Gesellschaft, geb. 18. V. 1905 in Krakau, Warszawa, Sewerynów 6/10, Tel. 269305 (21. VI. 1965).

Ladendorf, Heinz, Dr. phil., o. Professor für Kunstgeschichte an der Universität Köln, Träger des Ordens Officier des Palmes Académiques, o. Mitglied des Deutschen Archäologischen Instituts in Berlin-Dahlem, geb. 29. VI. 1909 in Leipzig, 5 Köln 41, Kunsthistorisches Institut, Albertus-Magnus-Platz, Tel. 4702349 und 5 Köln-Lindenthal, Kerpener Straße 4, Tel. 416058 (4. III. 1955).

Ligeti, Lajos, Dr. phil., Dr. lit., em. Professor mit Lehrstuhl für innerasiatische Studien an der Eötvös-Loránd-Universität Budapest, Träger des Arbeitsordens, Inhaber des Kossuth-Preises, des Fahnenordens I. und II. Klasse der Ungarischen Volksrepublik, des mongolischen Nairamdalordens, der Goldmedaille der Ungarischen Akademie der Wissenschaften in Budapest, der Goldmedaille der Permanent International Altaistic Conference und des Ehrendiploms der Türkischen Republik, o. Mitglied der Ungarischen Akademie der Wissenschaften in Budapest, Präsident des Bibliotheksausschusses und des Ausschusses für Orientalistik dieser Akademie, o. Mitglied der Akademie der Wissenschaften der Mongolischen Volksrepublik in Ulan-Bator, korresp. Mitglied des Instituts de France/Académie des Inscriptions et Belles-Lettres in Paris und der Suomalaisurgilainen Seura in

Helsinki, Ehrenmitglied der Société Asiatique in Paris, der Royal Asiatic Society in London und der Türk Dil Kurumu in Ankara, Vizepräsident des Permanent Comittee of the International Congress of Mongolists, geb. 26. X. 1902 in Balassagyarmat, Budapest V., Belgrád-rkp. 26, Tel. 187—529 (18. I. 1965).

Lissa, Zofia, Dr. phil., Dr. d. Musikwissenschaft, o. Professor für Musikwissenschaft an der Universität Warschau, Staatspreisträgerin, Träger des Ordens Sztandar Pracy II. Klasse, des Goldenen Verdienstkreuzes, des Ordens Polonia Restituta IV. und V. Klasse, des Kavalier- und Offiziersordens, Inhaber von Verdienstmedaillen, korresp. Mitglied der Akademie der Wissenschaften und der Literatur in Mainz und der Akademie der Künste der DDR, Mitglied des Direktoriums der Internationalen Gesellschaft für Musikwissenschaft, der Gesellschaft für Musikforschung in Kiel, des Polnischen Komponisten-Verbandes und des Komitees für Theorie und Geschichte der Kunst der Polnischen Akademie der Wissenschaften in Warschau, früher Vizepräsident des Polnischen Nationalkomitees des Musikrates des Conseil International de Musique/UNESCO und des Wissenschaftlichen Rates der Chopin-Gesellschaft, geb. 19. X. 1908 in Lwów, 02637 Warszawa 87, ul. Miączyńska 37, Tel. 44-09-85 (21. VI. 1965).

Mayrhofer, Manfred, Dr. phil., o. Prof. für Allgemeine und Indogermanische Sprachwissenschaft an der Universität Wien, wirkliches Mitglied der Österreichischen Akademie der Wissenschaften in Wien und Sekretär der philosophisch-historischen Klasse dieser Akademie, Kommandeur des kaiserl. iranischen Humāyūn-Ordens, Träger des Ritterkreuzes I. Klasse des finnländischen Ordens der Weißen Rose, auswärtiges Mitglied der Accademia di Scienze e Lettere in Mailand, der Königl. Schwedischen Akademie für Literatur, Geschichte und Altertümer in Stockholm und der Suomalais-Ugrilainen Seura [Finnisch-ugrische Gesellschaft] in Helsinki, korresp. Mitglied des Instituts de France/Académie des Inscriptions et Belles-Lettres in Paris, gewähltes Mitglied des Council of „Corpus Inscriptionum Iranicarum“ in London, Corresponding Member of the Staff of the School of Oriental and African Studies, Universität London, Ehrenmitglied der Ungarischen Akademie der Wissenschaften in Budapest und der Linguistic Society of India in Calcutta, geb. 26. IX. 1926 in Linz a. d. Donau (Österreich), A-1190 Wien, Bauernfeldgasse 9/II/6, Tel. 3639354, Post: Institut für Allgemeine und Indogermanische Sprachwissenschaft der Universität Wien, A-1010 Wien, Dr.-Karl-Lueger-Ring 1, Tel. 427611/440 (12. XI. 1971).

Michałowski, Kazimierz, Dr. phil., Dr. phil. h. c. (Strasbourg und Cambridge), o. Professor für Mittelmeer-Archäologie an der Universität Warschau, stellv. Direktor des Nationalmuseums Warschau, Leiter des Polnischen Zentrums für Mittelmeer-Archäologie in Kairo und des Zentrums für Mittelmeer-Archäologie der Polnischen Akademie der Wissenschaften, Träger des Silbernen Kreuzes des Ordens Virtuti Militari, zweifacher Träger des Golenen Verdienstkreuzes, der Fahne der Arbeit I. Klasse, des Kommandeurkreuzes Polonia Restituta, des Kommandeurkreuzes mit Stern Polonia Restituta, des Kommandeurkreuzes

Corona d'Italia und des Offizierskreuzes der Ehrenlegion, Inhaber des Polnischen Nationalpreises I. und II. Klasse, der Medaille XXX-Jahre VR Polen und der Medaille George Perrot des Institut de France/Académie des Inscriptions et Belles-Lettres in Paris, Herder-Preisträger 1971 in Wien, o. Mitglied und Präsidiumsmitglied der Polnischen Akademie der Wissenschaften in Warschau, Mitglied der Accademia Nazionale dei Lincei in Rom und der British Academy in London, auswärt. Mitglied der Akademie der Wissenschaften der DDR und der Heidelberger Akademie der Wissenschaften, o. Mitglied des Deutschen Archäologischen Instituts in Berlin-Dahlem, der Association Internationale des Égyptologues und des Tschechoslowakischen Orientalischen Instituts, Ehrenmitglied des Institut d'Égypte in Kairo, des Archaeological Institute of America in Washington und des Archaeologiki Hetairia in Athen, stellv. Präsident der Association Internationale d'Epigraphie Latine, Präsident des Internationalen Komitees des Experten für Abu Simbel bei der UNESCO und der Society for Nubian Studies, geb. 14. XII. 1901 in Tarnopol, Warszawa, Sewerynów 6, Tel. 263979 (5. XII. 1960).

Németh, Julius, Dr. phil., em. o. Professor für Turkologie an der Eötvös-Loránd-Universität Budapest, o. Mitglied der Ungarischen Akademie der Wissenschaften in Budapest und em. Direktor des Sprachwissenschaftlichen Instituts dieser Akademie, Träger des Arbeiterordens, des Ordens „Kirill i Metodi" (Sofia), Inhaber des Kossuth-Preises und des Indiana University Prize for Altaic Studies, auswärt. Mitglied der Akademie der Wissenschaften der DDR und der Bulgarischen Akademie der Wissenschaften in Sofia, Ehrenmitglied der Royal Asiatic Society in London, der Suomalaisugrilainen Seura in Helsinki, der Societas Uralto-altaica in Göttingen, der Türk Dil Kurumu und der Türk Tarih Kurumu in Ankara und der Polnischen Orientalischen Gesellschaft in Warschau, geb. 2. XI. 1890 in Karcag (Ungarn), Budapest XI., Karinthy út. 24. III. 1., Tel. 257-382 (13. X. 1958).

Orlandos, Anastasios K., Dr. philol., em. o. Professor für Baugeschichte an der Technischen Hochschule Athen, em. o. Professor für Byzantinische Archäologie an der Universität Athen, Inhaber des Preises „Zographos" de l'Association pour l'Encouragement des Etudes Grecques en France und des Gottfried-von-Herder-Preises der Universität Wien, Träger der Verdienstmedaille der Stadt Athen, der Medaille d'Argent de la Ville de Paris, der Orden Commendatore de la Corona d'Italia, Chevalier de la Légion d'Honneur, Etayrós, Anotéron Taxiarchov toy Tágmatos toy Phoinikos, Cavaliere Ufficale dell'Ordine all merito della Republica Italiana, Anótatos Taxiárches Geogion und Jugoslovenske Zastave sa zlatnim vengem, o. Mitglied der Akademie der Wissenschaften in Athen, korresp. Mitglied de l'Académie des Inscriptions et Belles-Lettres in Paris und corresponding Fellow of the British Academy in London, auswärtiges Mitglied der königlich-belgischen Akademie der Wissenschaften in Brüssel und der Akademie der Wissenschaften in Palermo, Mitglied und Generalsekretär der Archäologischen Gesellschaft in Athen, o. Mitglied des Deutschen Archäologischen Instituts in Berlin-Dahlem und der Historisch-Ethnologischen Gesellschaft in Athen, korresp.

glied des Österreichischen Archäologischen Instituts in Wien, Mitglied und Präsident der Christlichen Archäologischen Gesellschaft in Athen und der Gesellschaft für Byzantinische Studien in Athen, Präsident und Generalsekretär des Aufsichtskomitees des Istituto Hellenico di Studi Bizantini e Post-Bizantini in Venedig, Präsident des griechischen Komitees der Association des Etudes Byzantines, Ehrenmitglied der Society for the Promotion of Hellenic Studies, des Archaeological Institute of America, der Griechischen Architektonischen Gesellschaft und der griechischen Wissenschaftlichen Gesellschaft „Athena“ in Athen, geb. 23. XII. 1887 in Athen, Athen 138, Koumbaristraße 2, Tel. 61 62 96 (9. XI. 1974).

Pée, Willem, Dr. phil., em. Professor für Niederländische Linguistik, Dialektologie und Experimentelle Phonetik an der Universität Gent (Belgien), Inhaber der belgischen Leopolds- und Kronorden, des Niederländischen Kommandeurordens für Oranien-Nassau, des Sterns der Völkerfreundschaft in Gold und der Ehrennadel der Liga für Völkerfreundschaft der DDR in Gold für Verdienste um die Völkerfreundschaft, Mitglied der Maatschappij der Nederlandse Letterkunde in Leiden, der Fryske Akademy in Leeuwarden, des Provinciaal Utrechts Genootschap van Kunsten en Wetenschappen, der Commissie belast met de voorbereiding van de Nederlandse tekst van de Grondwet, de wetboeken en van de voornaamste wetten en besluiten, der Centrale Commissie voor de Nederlandse rechtstaal en bestuurstaal in België, Vizepräsident dieser Kommission seit 1969, Mitglied der Niederländisch-Belgischen Stiftung für niederländische Lexikologie, der Belgische Spellingcommissie und der Belgische Woordenlijstcommissie, Präsident dieser Kommission, Mitglied der Belgische Vaste Commissie van Advies voor de Spelling van de Nederlandse taal, Präsident dieser Kommission seit 1970, Präsident der Terminologische Commissie bei het Ministerie van Binnenlandse Zaken, der Belgischen Abteilung der Belgisch-Nederlandse Commissie van Advies voor de Spelling van de Bastaardwoorden, Vizepräsident des Belgisch Centrum voor Neerlandistiek, Generalsekretär dieser Vereinigung, Präsident der Gesellschaft Belgien-DDR in Brüssel, Ehrenmitglied der Koninklijke Commissie voor Toponymie & Dialectologie in Brüssel, geb. 9. IV. 1903 in Brügge (West-Flandern), 1170 Brüssel, Bosvoorde, Vuurkeienweg 3, Tel. 02/672 2649 (12. XI. 1971).

Petschow, Herbert, Dr. phil. et jur., o. Professor für antike Rechtsgeschichte an der Ludwig-Maximilians-Universität München, Gastprofessor für altorientalische Rechtsgeschichte an der Karl-Marx-Universität Leipzig, o. Mitglied der Bayerischen Akademie der Wissenschaften in München, Mitglied der Deutschen Morgenländischen Gesellschaft in Mainz, geb. 26. XII. 1909 in Dresden, 8033 Planegg, Rudolfstr. 36, Tel. 859 87 96 (10. XII. 1962).

Poulík, Josef, Dr. phil., o. Professor an der Universität Brno, Direktor des Archäologischen Instituts der Tschechoslowakischen Akademie der Wissenschaften, Staatspreisträger, Ehrenbürger von Mikulčice, o. Mitglied und Vizepräsident der Tschechoslowakischen Akademie der Wissenschaften in Prag, auswärt. Mitglied der Akademie der Wissenschaften der DDR, Präsident der Union International d'Acheologie Slave, Mitglied des Conseil Permanet des Sciences Pré- et Protohistoriques, geb. 6. VIII. 1910 in Jiřikovice, Archäologisches Institut der Tschechoslowakischen Akademie der Wissenschaften (CSAV), Brno, Sady osvobozeni 19—23 (9. XI. 1974).

Préaux, Claire, Dr. phil., Dr. phil.h. c. (Strasbourg), Dr. jur. h. c. (Freiburg i. Br.), o. Professor für Klassische Altertumswissenschaft an der Universität Brüssel, Abteilungsleiter in der Fondation Égyptologique Reine Elisabeth, Inhaberin des Prix de la Société pour l'Encouragement des Études grecques en France und des Prix Francqui, Mitglied der Académie Royale des Sciences, des Lettres et des Beaux-Arts de Belgique in Brüssel, korresp. Mitglied der British Academy in London, des Institut de France/Académie des Inscriptions et Belles-Lettres in Paris, der Koninklijke Nederlandse Akademie van Wetenschappen in Amsterdam, geb. 21. XII. 1904 in Liège, 1050 Bruxelles, 109 rue de l'Abbaye, Tel. 3442753 (14. III. 1955).

Průšek, Jaroslav, Dr. phil., Dr. Sc., Dr. phil. h. c. (Berlin, Olmouc, Stockholm), Professor für Sinologie und Japanologie an der Karls-Universität Prag, zweifacher Träger des Staatspreises für Wissenschaft und Kunst, Inhaber der Medaille für Verdienste um Wissenschaft und Menschheit der Tschechoslowakischen Akademie der Wissenschaften, o. Mitglied der Tschechoslowakischen Akademie der Wissenschaften in Prag, auswärt. Mitglied der Akademie der Wissenschaften der DDR und der Bayerischen Akademie der Wissenschaften in München, Mitglied der Société Asiatique in Paris und der American Oriental Society in New Haven, geb. 14. IX. 1906 in Prag, Praha 1, Gogolova 2 (21. VI. 1965).

Schmidt, Eberhard, Dr. jur., Dr. med. h. c. (Göttingen), em. o. Professor der Rechte (Straf- und Strafprozeßrecht, Rechtsgeschichte) an der Ruprecht-Karl-Universität Heidelberg, Träger des Großen Verdienstkreuzes des Verdienstordens mit Stern der Bundesrepublik Deutschland, Inhaber des Ehrenzeichens der deutschen Ärzteschaft, o. Mitglied der Heidelberger Akademie der Wissenschaften, korresp. Mitglied der Österreichischen Akademie der Wissenschaften in Wien und der Deutschen Gesellschaft für Psychiatrie und Nervenheilkunde in Marburg (Lahn), geb. 16. III. 1891 in Jüterbog, 69 Heidelberg, Werderplatz 9, Tel. 40796 (14. VI. 1941).

Tackenberg, Kurt, Dr. phil., em. o. Professor für Vor- und Frühgeschichte an der Westfälischen Wilhelms-Universität Münster, Träger des Großen Verdienstkreuzes des Niedersächsischen Verdienstordens 1974, Mitglied der Historischen Kommission für Niedersachsen in Hannover, der Archäologischen Kommission für Niedersachsen und des Deutschen Archäologischen Instituts in Berlin-Dahlem, Ehrenmitglied des Niedersächsischen Landesvereins für Urgeschichte in Hannover, geb. 30. VI. 1899 in Tschammendorf (Schlesien), 44 Münster/Westf., St. Mauritz, Parkallee 2, Tel. Münster 315158 (19. X. 1959).

Thieme, Paul, Dr. phil., em. o. Professor für Indologie und vergleichende Religionswissenschaft an der Eberhard-Karls-Universität Tübingen, geb. 18. III. 1905 in Berlin, 74 Tübingen, Spemannstraße 14 (12. XI. 1949).

Vossler, Otto, Dr. phil., em. o. Professor für mittlere und neuere Geschichte an der Johann-Wolfgang-Goethe-Universität Frankfurt am Main, Inhaber der Jusserand Medal der American Historical Association in Washington, geb. 14. II. 1902 in Heidelberg, 60 Frankfurt (Main), Grüneburgweg 143, Tel. 726269 (20. VI. 1942).

von Weizsäcker, Carl Friedrich, Freiherr, Dr. phil., Honorar-Professor für Philosophie an der Universität München, Direktor des Max-Planck-Instituts zur Erforschung der Lebensbedingungen der wissenschaftlich-technischen Welt in Starnberg (Obb.), Träger des Ordens Pour le mérite für Verdienste um die Wissenschaft und um die Künste, des Großen Verdienstkreuzes mit Stern der Bundesrepublik Deutschland und des Österreichischen Ehrenzeichens für Wissenschaft und Kunst, Inhaber der Goldenen Wilhelm-Bölsche-Medaille der Kosmos-Gesellschaft der Naturfreunde in Stuttgart, der Max-Planck-Medaille des Verbandes Deutscher Physikalischer Gesellschaften in Braunschweig, der Hans-Thoma-Medaille der Hans-Thoma-Gesellschaft in Reutlingen, des Goethe-Preises der Stadt Frankfurt (Main), des Prix Arnold Reymond der Universität Lausanne, des Friedenspreises des Deutschen Buchhandels und des Erasmus-Preises der Stiftung „Prämium Erasmianum" der Niederlande, o. Mitglied der Akademie der Wissenschaft in Göttingen, Mitglied der Deutschen Akademie der Naturforscher Leopoldina in Halle, der Deutschen Akademie für Sprache und Dichtung in Darmstadt und der Joachim-Jungius-Gesellschaft der Wissenschaften in Hamburg, korresp. Mitglied der Bayerischen Akademie der Wissenschaften in München und der Österreichischen Akademie der Wissenschaften in Wien, Senator der Max-Planck-Gesellschaft zur Förderung der Wissenschaft, Vorsitzender des Verwaltungsrates des Deutschen Entwicklungsdienstes (DED), Mitglied des Verwaltungsausschusses des Deutschen Museums in München und des PEN-Zentrums in Darmstadt, geb. 28. VI. 1912 in Kiel, 813 Starnberg/Obb., Riemerschmidstraße 7, Tel. 7169 und 8135 Söcking, Alpenstraße 15.

Wieacker, Franz, Dr. jur., Dr. phil. h. c. (Freiburg i. Br.), Dr. jur. h. c. (Uppsala), LL. D. h. c. (Glasgow), o. Professor für römisches und bürgerliches Recht an der Georg-August-Universität Göttingen, o. Mitglied der Akademie der Wissenschaften in Göttingen, korresp. Mitglied der Bayerischen Akademie der Wissenschaften in München, der Heidelberger Akademie der Wissenschaften, der Accademia Nazionale dei Lincei in Rom und des Instituto Lombardo in Mailand, geb. 5. VIII. 1908 in Stargard (Pommern), 34 Göttingen, Prinzenstraße 21 (14. VI. 1941).

TODESFÄLLE

Ordentliche Mitglieder
der mathematisch-naturwissenschaftlichen Klasse

Emmrich, Rolf, Dr. med., o. Professor für innere Medizin an der Karl-Marx-Universität Leipzig, Mitglied der Kommission für spezielle Umweltprobleme der Akademie, der Deutschen Akademie der Naturforscher Leopoldina in Halle, der Internationalen Gesellschaft für Innere Medizin in Basel, der Gesellschaft für Experimentelle Medizin in der DDR in Berlin, des International College of Angiology in New York, Ehrenmitglied der Gesellschaft für Innere Medizin in der DDR, geb. 26. VIII. 1910 in Freiberg (Sachsen), 7027 Leipzig, Naunhofer Straße 12 (11. XII. 1961). †5. V. 1974.

Runge, Franz, Dr. phil., em. Professor mit Lehrstuhl für technische Chemie an der Martin-Luther-Universität Halle—Wittenberg, Nationalpreisträger, Mitglied der Kommission für spezielle Umweltprobleme und der Poggendorff-Kommission der Akademie, Mitglied der Deutschen Akademie der Naturforscher Leopoldina in Halle, der Deutschen Chemischen Gesellschaft in der DDR, der Deutschen Bunsen-Gesellschaft für physikalische Chemie in Frankfurt (Main), der Gesellschaft Deutscher Naturforscher und Ärzte in Bonn, des Deutschen Museums in München, der Faraday Society in London und des Plastics Institut in London, geb. 5. XII. 1893 in Göttingen, 402 Halle (Saale), Spechtweg 8 (14. III. 1965). †29. VIII. 1973.

Schulze, Gustav, E. R., Dr. phil., o. Professor für Experimentalphysik an der Technischen Universität Dresden, Nationalpreisträger, Träger der Verdienstmedaille der DDR, Vorstandsmitglied der Deutschen Vereinigung für Kristallographie in Berlin, Mitglied des Nationalkomitees für Kristallographie, der Physikalischen Gesellschaft in der DDR und des Institute of Metals in London, geb. 24. II. 1911 in Berlin, 8036 Dresden, Kauschaer Straße 6 (5. XII. 1960). †6. X. 1974.

Stather, Fritz, Dr. phil., em. Professor für Gerbereichemie und -technologie an der Technischen Universität Dresden, em. Direktor des Deutschen Lederinstituts, Hervorragender Wissenschaftler des Volkes, Preisträger 1958 des Vereins für Gerbereichemie und -technik, Träger des Goldenen Ehrenzeichens der Kammer der Technik und der Edmund-Stiasny-Medaille des Vereins für Gerbereichemie und -technik, o. Mitglied der Akademie der Wissenschaften der DDR, Ehrenmitglied der Kammer der Technik und des Vereins österreichischer Gerbereitechniker, geb. 5. XI. 1901 in Offenburg (Baden), 92 Freiberg, Goethestraße 11 (15. IX. 1958). †30. III. 1974.

Wolf, Leopold, Dr. phil., em. Professor mit Lehrstuhl für anorganische Chemie an der Karl-Marx-Universität Leipzig, Nationalpreisträger, Träger des Vaterländischen Verdienstordens, korresp. Mitglied der Akademie der Wissenschaften der DDR, Ehrenmitglied der Chemischen Gesellschaft in der DDR, geb. 23. XI. 1896 in Niederlungwitz (Sachsen), 7022 Leipzig, Quedlinburger Straße 27 (5. XII. 1960) †6. XI. 1974.

Auswärtige Mitglieder der mathematisch-naturwissenschaftlichen Klasse

Chadwick, James, Sir, C. H., Ph. D., D. Sc. h. c. (Birmingham, Exeter Dublin, Leeds, McGill, Oxford, Reading), LL. D. h. c. (Edinburgh, Liverpool), Fellow of Gonville and Caius College in Cambridge, Nobelpreisträger, Inhaber der Hughes Medal und der Copley Medal der Royal Society, der Franklin Medal des Franklin Institute of the State of Pennsylvania, Mitglied der Royal Society in London und der Pontificia Academia Scientiarum, Vatikanstadt, ausländ. Mitglied der Kongelige Danske Videnskabernes Selskab in Kopenhagen, der Koninklijke Nederlandse Akademie van Wetenschappen in Amsterdam, der Académie Royale de Belgique in Brüssel, der American Philosophical Society in Philadelphia, Ehrenmitglied des American Institute of Physics in New York, ausländ. Mitglied des Ordens pour le mérite für Wissenschaften und Künste, geb. 20. X. 1891 in Cheshire, 16 Grange Court, Pinehurst, Cambridge, England (1. VII. 1932). †24. VII. 1974.

Eggert, John, Dr. phil., Dr.-Ing. E. h. (Karlsruhe), Dr. rer. nat. h. c. (Freie Universität Berlin), em. o. Professor für Photochemie und Photographie an der Eidgenössischen Technischen Hochschule Zürich, Inhaber der Goldenen Medaille und der Goldenen Voigtländer-Medaille (1972) der Photographischen Gesellschaft in Wien, der Goldenen Medaille der Universität Padua, der Traill Taylor Medal 1935 und der Progress Medal 1949 der Royal Photographic Society of Great Britain, der Médaille Péligot der Société Française de Photographie et Cinématographie, der Plakette der Stadt Paris, der Gauss-Weber-Medaille der Mathematisch-naturwissenschaftlichen Fakultät der Georg-August-Universität Göttingen, der Special Citation of the Photographers' Association of America, der Photokina-Plakette Köln, der Goldenen Adolf-von-Baeyer-Denkmünze der Gesellschaft Deutscher Chemiker, des Kulturpreises der Deutschen Gesellschaft für Photographie in Köln, korresp. Mitglied der Bayerischen Akademie der Wissenschaften in München, der Heidelberger Akademie der Wissenschaften und der Société Royale des Sciences in Liège, Ehrenmitglied der Royal Photographic Society of Great Britain in London, der Photographic Society of America in Philadelphia, der Society of Photographic Scientists and Engineers in Washington, der Japanischen Gesellschaft für wissenschaftliche Photographie in Tokio, der Photographischen Gesellschaft in Wien (1971), der Fédération internationale de l'art photographique in Bern, der Deutschen Bunsengesellschaft für Physikalische Chemie in Frankfurt (Main), des Schweizerischen Photographenverbandes und der Physikalischen Gesellschaft Zürich, Mitglied der Gesellschaft Deutscher Chemiker in Frankfurt (Main) und der Deutschen Physikalischen Gesellschaft in Frankfurt (Main), geb. 1. VIII. 1891 in Berlin, CH 8702 Zollikon, Höhestraße 44 (12. V. 1958). †29. IX. 1973.

Leutwein, Friedrich, Dr. phil. nat., Professor, Directeur de Recherches (C. N. R. S.) am Centre de Recherches Pétrographiques et Géochimiques in Nancy-Vandoeuvre (Frankreich), Hon. Professor (Hamburg), Nationalpreisträger, korresp. Mitglied der Akademie der Wissenschaften der DDR, Mitglied der Geologischen Vereinigung in Bonn, der Deutschen Mineralogischen Gesellschaft in Bonn, der Gesellschaft Deutscher Metallhütten- und Bergleute in Clausthal, der Geochemical Society in Washington, der Société Française de Minéralogie et de Cristallographie in Paris, der Société Géologique de France in Paris und der Société de Géologie Appliquée aux Gites Minéraux, geb. 9. VIII. 1911 in Berlin, 54500 Vandœuvre-Nancy (Frankreich), 2, Rue de la Louvière (17. XI. 1956). †28. XII. 1974.

Ordentliche Mitglieder
der philologisch-historischen Klasse

Eißfeldt, Otto, Dr. phil., D. theol. h. c. (Berlin), DD. (Glasgow), D. theol. h. c. (Budapest), em. Professor mit Lehrstuhl für Altes Testament und semitische Religionsgeschichte an der Martin-Luther-Universität Halle—Wittenberg, Ehrensenator dieser Universität, Träger des Vaterländischen Verdienstordens in Silber und des Ordens „Banner der Arbeit", o. Mitglied der Akademie der Wissenschaften der DDR, auswärt. Mitglied des Institut de France/Académie des Inscriptions et Belles-Lettres in Paris, Mitglied der Kungl. Humanistiska Vetenskapssamfundet in Lund (Regia societas humaniorum litterarum Lundensis), korresp. Mitglied der Akademie der Wissenschaften und der Literatur in Mainz und der British Academy in London, o. Mitglied des Deutschen Archäologischen Instituts in Berlin-Dahlem, Ehrenmitglied der Deutschen Morgenländischen Gesellschaft in Mainz, der Society for Old Testament Study (Great Britain) und der Society of Biblical Literature and Exegesis (USA), geb. 1. IX. 1887 in Northeim (Hann.), 402 Halle (Saale), Steffensstraße 7 (1. VII. 1948). †23. IV. 1973.

Fück, Johann, Dr. phil., em. Professor mit Lehrstuhl für semitische Philologie und Islamkunde an der Martin-Luther-Universität Halle—Wittenberg, Nationalpreisträger, korresp. Mitglied der Akademie der arabischen Sprache der VAR in Kairo, Mitglied des Executive Committee of the Union für Scientific Collaboration in Islamic Studies, der Association pour l'Avancement des Études Islamiques in Paris, Ehrenmitglied des Instituto Peruano de Altos Estudios Islámicos in Lima und der Deutschen Morgenländischen Gesellschaft in Mainz, geb. 8. VII. 1894 in Frankfurt (Main), 402 Halle (Saale), Karl-Liebknecht-Straße 23 (1. VII. 1948). †24. XI. 1974.

Lau, Franz, D., D. theol. h. c. (Rostock), em. o. Professor für Kirchengeschichte an der Karl-Marx-Universität Leipzig, Mitglied und stellv. Vorsitzender der Historischen Kommission der Akademie, Domherr und Domdechant des Hochstiftes Meißen, Präsident des Gustav-Adolf-Werkes in der DDR, Vorsitzender der Arbeitsgemeinschaft für sächsische Kirchengeschichte und der Vereinigten Arbeitsgemeinschaften für Kirchengeschichte in den evangelischen Landeskirchen der DDR, Mitglied der Kommission zur Herausgabe der Werke Martin Luthers, geb. 18. II. 1907 in Leipzig, 7113 Markkleeberg I, Mittelstraße 8 (18. I. 1965). †6. VI. 1973.

Auswärtige Mitglieder
der philologisch-historischen Klasse

van Giffen, Albert Egges, Dr. phil., Dr. phil. h. c. (Köln), em. o. Professor für Prä- und Protohistorie an der Universität Groningen, em. o. Professor an der Universität Amsterdam, Rijksadviseur voor de Bescherming en de Instandhouding van de Hunnebedden en van gerestaureerde archaeologische Monumenten, Direktor a. D. des Biologisch-Archäologischen Instituts der Rijksuniversiteit in Groningen, Direktor a. D. des Instituts für Prä- und Protohistorie (IPP) der Universität Amsterdam, Konservator a. D. des Museums der Provinz Drente, Chevalier de l'Ordre National de la Légion d'Honneur, Ridder in de Orde van de Nederlandsche Leeuw, Inhaber der Gold Medal of the Society of Antiquaries in London, Mitglied der Koninklijke Nederlandse Akademie van Wetenschappen in Amsterdam, der Koninklijke Vlaamse Academie voor Taal- en Letterkunde in Gent und der Akademie der Wissenschaften in Göttingen, korresp. Mitglied der Académie Royale de Belgique in Brüssel, Mitglied der Society of Antiquaries in London und der Society of Antiquaries of Scotland, der Hollandsche Maatschappij der Wetenschappen in Haarlem, o. Mitglied des Deutschen Archäologischen Instituts in Berlin-Dahlem, ausländ. Mitglied der Soumen Muinaismuistoyhdistys-Finska Fornminnesföreningen in Helsinki, korresp. Mitglied des Institute of Prehistory and Archaeology, University Aarhus, Ehrenmitglied des Marschenrates zur Förderung der Forschung im Küstengebiet, geb. 14. III. 1884 in Noordhorn (Holland), Zwolle (Holland), Zalnéflat 62, Weteringpark 7 (5. XII. 1960). †31. IV. 1973.

Jahn, Martin, Dr. phil., em. Professor mit Lehrstuhl für Vor- und Frühgeschichte an der Martin-Luther-Universität Halle—Wittenberg, em. Direktor des Landesmuseums für Vorgeschichte in Halle, o. Mitglied des Deutschen Archäologischen Instituts in Berlin-Dahlem, Mitglied der Deutschen Akademie der Naturforscher Leopoldina in Halle, geb. 20. IX. 1888 in Berlin, 852 Buckenhof b. Erlangen, Hallerstraße 33 (1. VII. 1948). †11. IX. 1974.

Kühn, Johannes, Dr. phil., em. o. Professor für neuere Geschichte an der Ruprecht-Karl-Universität Heidelberg, Mitglied der Historischen Kommission bei der Bayerischen Akademie der Wissenschaften in München, geb. 24. I. 1887 in Bogschütz, 69 Heidelberg, Mühlingstraße 23 (19. VI. 1943). †24. II. 1973.

Schadewaldt, Wolfgang, Dr. phil., em. o. Professor für klassische Philologie und für Geschichte der Fortwirkung des Griechentums an der Eberhard-Karls-Universität Tübingen, Visiting Professor an der University of Michigan Ann Arbor, USA, 1968 Träger des Ordens Pour le mérite für Verdienste um die Wissenschaft und die Künste, des Großen Verdienstkreuzes mit Stern des Verdienstordens der Bundesrepublik Deutschland und des Österreichischen Ehrenzeichens für Wissenschaft und Kunst, Inhaber des Reuchlin-Preises der Stadt Pforzheim 1963 und des Übersetzerpreises der Deutschen Akademie für Sprache und Dichtung in Darmstadt 1965, o. Mitglied der Akademie der Wissenschaften der DDR und der Heidelberger Akademie der Wissenschaften, Ehrenmitglied der Österreichischen Akademie der Wissenschaften in Wien, Mitglied der Deutschen Akademie für Sprache und Dichtung in Darmstadt, der Zentraldirektion des Deutschen Archäologischen Instituts in Berlin-Dahlem und des PEN-Clubs, geb. 15. III. 1900 in Berlin, 74 Tübingen, Stauffenbergstraße 65 (5. VI. 1935). †10. XI. 1974.

Zucker, Friedrich, Dr. phil., em. Professor mit Lehrstuhl für klassische Philologie und Papyruskunde an der Friedrich-Schiller-Universität Jena, Nationalpreisträger, o. Mitglied der Akademie der Wissenschaften der DDR und des Deutschen Archäologischen Instituts in Berlin-Dahlem, geb. 30. VI. 1881 in Fürth (Bayern), 2 Wedel/Holst., Hans-Böckler-Platz 15, Graf-Luckner-Haus, App. 225 (1. VII. 1948). †4. IV. 1973.

ZUWAHLEN

In der Plenarsitzung vom 18. Juni 1974 wurden gewählt:

Ordentliche Mitglieder
der mathematisch-naturwissenschaftlichen Klasse

Dörter, Klaus, Dr. agr. (Wissenschaftsgebiet Grünland und Meliorationswesen sowie Landeskultur), o. Professor für landwirtschaftliche Meliorationen an der Martin-Luther-Universität Halle—Wittenberg, Inhaber der Verdienstmedaille der DDR und der Erinnerungsmedaille zum 20. Jahrestag der Demokratischen Bodenreform, Mitglied der Agrarwissenschaftlichen Gesellschaft der DDR, der Internationalen Bodenkundlichen Gesellschaft in Amsterdam (Holland) und der Sektion Landeskultur und Naturschutz der Akademie der Landwirtschaftswissenschaften der DDR, Mitarbeiter der Arbeitsgruppe „Ökologie" der Klasse „Umweltschutz und Umweltgestaltung" bei der Akademie der Wissenschaften der DDR, geb. 10. I. 1922 in Breslau, 402 Halle (Saale), Elsa-Brandström-Straße 190, Tel. 45211.

Jordan, Herbert, Dr. med., Honorarprofessor für Physiotherapie an der Friedrich-Schiller-Universität Jena, Direktor des Forschungsinstituts für Balneologie und Kurortwissenschaft in Bad Elster, Obermedizinalrat, Träger der Hufeland-Medaille in Gold und der Medaille für treue Dienste im Gesundheits- und Sozialwesen in Silber, Mitglied der Gesellschaften für Innere Medizin, Kardiologie, Physiotherapie und Rheumatologie in der DDR, Mitglied der Internationalen Gesellschaft für Biologische Rhythmusforschung, Ehrenmitglied der Medizinisch-Wissenschaftlichen Allunionsgesellschaft für Physiotherapie und Kurortologie der UdSSR, der Bulgarischen Gesellschaft für Physikalische Therapie, Kurortologie und Rehabilitation und der Polnischen Gesellschaft für Balneologie, Bioklimatologie und Physikalische Therapie, geb. 8. XII. 1919 in Markneukirchen (Vogtland), 9933 Bad Elster, Karl-Marx-Straße 30, Tel. 467.

Wolf, Friedrich, Dr. rer. nat., Dr. rer. nat. h. c. (Poznań), o. Professor für Technische Chemie an der Martin-Luther-Universität Halle—Wittenberg, Nationalpreisträger 1959 und 1966, Träger des Ordens „Banner der Arbeit" und der Ehrenmedaillen der Universitäten Bratislava und Bagdad, korresp. Mitglied der Akademie der Wissenschaften der DDR, Mitglied der Chemischen Gesellschaft der DDR in Berlin und der Faraday Society in London, geb. 9. II. 1920 in Leipzig, 7031 Leipzig, Karl-Heine-Straße 4b, Tel. 484415.

In der Plenarsitzung vom 9. November 1974 wurden gewählt:

Auswärtige Mitglieder
der mathematisch-naturwissenschaftlichen Klasse

Budevski, Evgeni Bogdanov, Dr. rer. nat., Professor für Chemie und Elektrochemie bei der Bulgarischen Akademie der Wissenschaften in Sofia, Zentrallaboratorium für elektrochemische Stromquellen, Inhaber des Staatspreises „Georgi Dimitrov" für Wissenschaft und der Auszeichnung „100 Jahre Bulgarische Akademie der Wissenschaften", Träger des Ordens „Kirill und Metodi" 1. und 2. Stufe, Mitglied der Bulgarischen Akademie der Wissenschaften in Sofia, Mitglied und Vizepräsident 1975—1977 der International Society of Electrochemistry (ISE), geb. 29. VII. 1922 in Sofia, Sofia 1113, Bul. Samokov 82, Tel. Institut 723454.

Karšulin, Miroslav, Dr. techn., o. Professor für physikalische Chemie, an der Universität Zagreb, Träger des Staatspreises für Chemie, des Republikanischen Preises, des Staatsordens II. Klasse und des Ordens für Arbeit mit der roten Fahne, o. Mitglied der Jugoslavischen Akademie der Wissenschaften und Künste in Zagreb, der Makedonischen Akademie der Wissenschaften, der Faraday Society in London, der Electrochemical Society in New York, der Kroatischen Chemischen Gesellschaft, des Vereins der Ingenieure und Techniker Jugoslaviens, der Deutschen Keramischen Gesellschaft und des Internationalen Komitees zur Erforschung der Bauxite, Aluminium-Hydroxide und Oxyde, geb. 4. IV. 1904 in Przemysl (Polen), 41000 Zagreb, Palmotićeva 4/I, Tel. 36761.

Ordentliche Mitglieder
der philologisch-historischen Klasse

Kühnert, Friedmar, Dr. phil., o. Professor für klassische Philologie, mit besonderer Berücksichtigung der Lateinischen Sprache und Literatur an der Friedrich-Schiller-Universität Jena, Inhaber der Theodor-Neubauer-Medaille in Silber, Mitglied der Winckelmann-Gesellschaft der DDR, geb. 16. XI. 1924 in Dreißigacker (Krs. Meiningen), 6905 Jena-Göschwitz, August-Bebel-Straße 6, Tel. 32321.

Neumann, Werner, Dr. phil., Professor, Direktor des Bach-Archivs Leipzig, Nationalpreisträger, Träger des Vaterländischen Verdienstordens und der Verdienstmedaille der DDR, Mitglied der Neuen Bachgesellschaft in Leipzig und Mitglied des Verwaltungsrates dieser Gesellschaft, geb. 21. I. 1905 in Königstein (Sachsen), 7113 Markkleeberg, Robert-Schumann-Straße 3, Tel. 385525.

Rudolph, Kurt, Dr. theol., Dr. phil., o. Professor für Religionsgeschichte an der Karl-Marx-Universität Leipzig, korresp. Mitglied des Institute for Antiquity and Christianity in Claremont/Californien (USA) und der Internationalen Association for the History of Religions (IAHR), geb. 3. IV. 1929 in Dresden, 701 Leipzig, Dufourstraße 38, Tel. 30057.

Auswärtige Mitglieder
der philologisch-historischen Klasse

Besevliev, Vešelin, Dr. phil., em. ao. Professor für Altgriechische Philologie an der Universität Sofia, Professor für Altgriechische Philologie am Institut für Balkanistik der Bulgarischen Akademie der Wissenschaften in Sofia, Inhaber des Gottfried-von-Herder-Preises der Universität Wien, Träger des Ordens für Wissenschaft „Kirill i Metodi", I. Stufe, korresp. Mitglied der Bulgarischen Akademie der Wissenschaften in Sofia, Mitglied des Bulgarischen Archäologischen Instituts in Sofia und der Ungarischen Orientalischen Gesellschaft in Budapest, geb. 25. III. 1900 in Sofia, Sofia, L. Karavelov 2ᵛ, Tel. 87-96-08.

Droz, Jacques, Dr. phil., o. Professor für neuere Geschichte an der Sorbonne (Panthéon) in Paris, Direktor des Instituts für Geschichte der Gewerkschaftsbewegung, Träger des Bandes der Ehrenlegion, Mitglied der Gesellschaft Frankreich—DDR, geb. 12. III. 1909 in Paris, Paris, 55 rue de Varenne 7.

Orlandos, Anastasios K., Dr. philol., em. o. Professor für Baugeschichte an der Technischen Hochschule Athen, em. o. Professor für Byzantinische Archäologie an der Universität Athen, Inhaber des Preises „Zographos" de l'Association pour l'Encouragement des Etudes Grecques en France und des Gottfried-von-Herder-Preises der Universität Wien, Träger der Verdienstmedaille der Stadt Athen, der Medaille d'Argent de la Ville de Paris, der Orden Commendatore de la Corona d'Italia, Chevalier de la Légion d'Honneur, Etayrós Anotéron Taxiarchov toy Tágmatos toy Phoinikos, Cavaliere Ufficale dell'Ordine all merito della Republica Italiana, Anótatos Taxiárches Geogion und Jugoslovenske Zastave sa zlatnim vengem, o. Mitglied der Akademie der Wissenschaften in Athen, korresp. Mitglied de l'Académie des Inscriptions et Belles-Lettres in Paris und corresponding Fellow of the British Academy in London, auswärtiges Mitglied der königlich-belgischen Akademie der Wissenschaften in Brüssel und der Akademie der Wissenschaften in Palermo, Mitglied und Generalsekretär der Archäologischen Gesellschaft in Athen, o. Mitglied des Deutschen Archäologischen Instituts in Berlin-Dahlem und der Historisch-Ethnologischen Gesellschaft in Athen, korresp. Mitglied des Österreichischen Archäologischen Instituts in Wien, Mitglied und Präsident der Christlichen Archäologischen Gesellschaft in Athen und der Gesellschaft für Byzantinische Studien in Athen, Präsident und Generalsekretär des Aufsichtskomitees des Istituto Hellenico di Studi Bizantini e Post-Bizantini in Venedig, Präsident des griechischen Komitees der Association des Etudes Byzantines, Ehrenmitglied der Society for the Promotion of Hellenic Studies, des Archaeological Institute of America, der Griechischen Architektonischen Gesellschaft und der griechischen Wissenschaftlichen Gesellschaft „Athena" in Athen, geb. 23. XII. 1887 in Athen, Athen 138, Koumbaristraße 2, Tel. 61 6296.

Poulík, Josef, Dr. phil., o. Professor an der Universität Brno, Direktor des Archäologischen Instituts der Tschechoslowakischen Akademie der Wissenschaften, Staatspreisträger, Ehrenbürger von Mikulčice, o. Mitglied und Vizepräsident der

Tschechoslowakischen Akademie der Wissenschaften in Prag, auswärt. Mitglied der Akademie der Wissenschaften der DDR, Präsident der Union International d'Archeologie Slave, Mitglied des Conseil Permanent des Sciences Pré- et Protohistoriques, geb. 6. VIII. 1910 in Jiřikovice, Archäologisches Institut der Tschechoslowakischen Akademie der Wissenschaften (ČSAV), Brno, Sady osvobozeni 19—23.

In der Plenarsitzung vom 17. Oktober 1975 wurden gewählt:

Ordentliche Mitglieder
der mathematisch-naturwissenschaftlichen Klasse

Beckert, Herbert, Dr. phil., o. Professor für Mathematik (Analysis) an der Karl-Marx-Universität Leipzig, Nationalpreisträger, Mitglied der Deutschen Akademie der Naturforscher Leopoldinain Halle, der Mathematischen Gesellschaft der DDR in Berlin und der Österreichischen Mathematischen Gesellschaft in Wien, geb. 12. X. 1920 in Colmnitz, Kreis Freital, 7113 Markkleeberg, Raschwitzerstraße 12, Tel. Leipzig 31205.

Bethge, Heinz, Dr. rer. nat., Professor mit vollem Lehrauftrag für Physik an der Martin-Luther-Universität Halle—Wittenberg, Professor für Physik an der Akademie der Wissenschaften der DDR, Institut für Festkörperphysik und Elektronenmikroskopie in Halle, Nationalpreisträger, Inhaber der Verdienstmedaille der DDR und der Carus-Medaille der Deutschen Akademie der Naturforscher Leopoldina in Halle, o. Mitglied der Akademie der Wissenschaften der DDR, Mitglied und Präsident der Deutschen Akademie der Naturforscher Leopoldina in Halle, Mitglied der Physikalischen Gesellschaft der DDR, der International Organization for Crystal Growth (IOCG) und des International Thin Film Committee (ITFC), geb. 15. XI. 1919 in Magdeburg, 402 Halle (Saale), Ginsterweg 15, Tel. 24659.

Oelßner, Wilhelm, Dr. med., o. Professor für Radiologie an der Karl-Marx-Universität Leipzig, Verdienter Arzt des Volkes, Inhaber der Ehrenplakette der Deutschen Gesellschaft für klinische Medizin und der Ehrennadel der Karl-Marx-Universität Leipzig, Mitglied der Gesellschaft für Medizinische Radiologie der DDR und der Gesellschaft für Geschwulstbekämpfung der DDR, korresp. Mitglied der Deutschen Röntgengesellschaft und Ehrenmitglied der Tschechoslovakischen Radiologischen Gesellschaft, geb. 3. III. 1920 in Leipzig, 7124 Holzhausen (Sachsen), Richard-Wagner-Str. 15, Tel. Leipzig 2012363.

Ordentliche Mitglieder der philologisch-historischen Klasse

Diesner, Hans-Joachim, Dr. phil., o. Professor für Alte Geschichte an der Martin-Luther-Universität Halle—Wittenberg, Inhaber der Deutschen Friedensmedaille des Friedensrates der DDR in Berlin, der Johann-Heinrich-Pestalozzi-Medaille in Bronze und Silber und der Verdienstmedaille der DDR, Mitglied der Association

Internationale des Études Patristiques in Paris, geb. 21. I. 1922 in Böhlitz-Ehrenberg bei Leipzig, 402 Halle (Saale), August-Bebel-Straße 69, Tel. 31788.

Dietze, Walter, Dr. phil., o. Professor für Geschichte der deutschen Literatur an der Karl-Marx-Universität Leipzig, Generaldirektor der Nationalen Forschungs- und Gedenkstätten der Klassischen Deutschen Literatur in Weimar, Nationalpreisträger, Kunstpreisträger der Stadt Leipzig, Inhaber der Johannes-R.-Becher-Medaille in Silber und der Johann-Heinrich-Pestalozzi-Medaille in Bronze und Silber, o. Mitglied der Akademie der Wissenschaften in Berlin, Mitglied und Vorstandsmitglied der Goethe-Gesellschaft in Weimar, Mitglied der Modern Language Association of America (MLA), der Internationalen Vereinigung für germanische Sprach- und Literaturwissenschaften (IVG) und der Association Internationale de la Litterature Comparée (AILC), geb. 22. V. 1926 in Leipzig, 53 Weimar, Hermann-Löns-Straße 10.

Forberger, Rudolf, Dr. rer. pol., Professor für Wirtschaftsgeschichte an der Akademie der Wissenschaften der DDR, Institut für Wirtschaftsgeschichte in Berlin, Honorarprofessor für Wirtschaftsgeschichte an der Bergakademie Freiberg, Träger der Ehrennadel für heimatkundliche Leistungen in Gold des Kulturbundes der DDR, geb. 13. IV. 1910 in Carlsfeld (Kreis Aue), 8019 Dresden, Eibenstocker Straße 14, Tel. 31261.

Träger, Claus, Dr. phil., o. Professor für Allgemeine Literaturwissenschaft an der Karl-Marx-Universität Leipzig, Inhaber des Lessingpreises der DDR, der Verdienstmedaille der DDR und der Ehrennadel der Karl-Marx-Universität Leipzig, Mitglied des Hoch- und Fachschulrates beim Minister für Hoch- und Fachschulwesen der DDR, des Nationalkomitees für Literaturwissenschaft bei der Akademie der Wissenschaften der DDR, der Association Internationale de Litterature Comparée (AILC) und Ausschußmitglied der Internationalen Vereinigung für germanische Sprach- und Literaturwissenschaft (IVG), geb. 4. II. 1927 in Leipzig, 7113 Markkleeberg-Ost, Heinrich-v.-Kleist-Straße 19.

EHRUNGEN

Nobelpreisträger

Das ordentliche Mitglied

Hertz, Gustav †	1925	Physik

Auswärtige Mitglieder

Chadwick, James †	1935	Physik
Heisenberg, Werner †	1932	Physik
Hill, Archibald Vivian	1922	Physiologie und Medizin

Nationalpreisträger der DDR für Wissenschaft und Technik

Ordentliche Mitglieder

Drischel, Hans	1967
Fück, Johann †	1956
Gersch, Manfred	1958
Görlich, Paul	1954
Hein, Franz †	1952
Hertz, Gustav †	1955
Keller, Ott-Heinrich	1966
Krauss, Werner †	1949
Lösche, Artur	1958
Markov, Walter	1961
Mothes, Kurt	1953
Neumann, Werner	1973
Obenaus, Fritz	1972
Peek, Werner	1961
Rammler, Erich	1951
Runge, Franz †	1959
Schmalfuß, Karl	1953
Schulze, Gustav E. R. †	1967
Schwabe, Kurt	1954, 1961
Strack, Erich	1960
Stubbe, Hans	1949, 1960
Weller, Friedrich	1955
Werner, Ernst	1966
Wolf, Friedrich	1959, 1966
Wolf, Leopold †	1957

Auswärtige Mitglieder

Bredt, Heinrich	1954
Erben, Johannes	1961
Hertwig, Paula	1956
Hund, Friedrich	1949
Leutwein, Friedrich †	1956
Treibs, Wilhelm	1954
Zucker, Friedrich †	1954

Träger des Vaterländischen Verdienstordens

Ordentliche Mitglieder

Arland, Anton †
Baetke, Walter
Bahner, Werner
Buchda, Gerhard
Eißfeldt, Otto †
Eßbach, Hasso
Görlich, Paul
Hertz, Gustav †
Jahn, Johannes †
Krauss, Werner †
Lehmann, Edgar
Lösche, Artur
Markov, Walter
Mothes, Kurt
Neumann, Werner
Obenaus, Fritz
Rammler, Erich
Schmalfuß, Karl
Schubart-Fikentscher, Gertrud
Schwabe, Kurt
Simon, Arthur †
Straub, Werner
Stubbe, Hans
Werner, Ernst
Wolf, Leopold †

Auswärtige Mitglieder

Hertwig, Paula
Treibs, Wilhelm

Träger des Ordens „Banner der Arbeit“

Ordentliche Mitglieder

Eißfeldt, Otto †
Hertz, Gustav †
Jahn, Johannes †
Mothes, Kurt
Schwabe, Kurt
Straub, Werner
Stubbe, Hans
Wolf, Friedrich

Hervorragende Wissenschaftler des Volkes

Ordentliche Mitglieder

Baetke, Walter
Hein, Franz †
Hertz, Gustav †
Krauss, Werner †
Mothes, Kurt
Rammler, Erich
Schwabe, Kurt
Stather, Fritz †
Stubbe, Hans
Weller, Friedrich
Watznauer, Adolf

Auswärtige Mitglieder

Hertwig, Paula
Treibs, Wilhelm

Verdienter Bergmann der DDR

Das ordentliche Mitglied

Rammler, Erich

Verdienter Arzt des Volkes

Das ordentliche Mitglied

Matzen, Peter Friedrich

Verdiente Techniker des Volkes

Ordentliche Mitglieder

Görlich, Paul
Obenaus, Fritz

Held der Arbeit

Das ordentliche Mitglied

Schwabe, Kurt

Medaille für Kämpfer gegen den Faschismus 1933—1945

Ordentliche Mitglieder

Krauss, Werner
Markov, Walter

Verdienstmedaille der DDR

Ordentliche Mitglieder

Coblenz, Werner
Dörter, Klaus
Freimuth, Ulrich
Hentschel, Walter †
Hiebsch, Hans
Klinkowski, Maximilian †
Lieberwirth, Rolf
Neef, Ernst
Neumann, Werner
Schulze, Gustav E. R. †
Straub, Werner
Werner, Ernst

Friedensmedaille des Friedensrates in der DDR

Das ordentliche Mitglied

Buchda, Gerhard

Johann-Heinrich-Pestalozzi-Medaille

Ordentliche Mitglieder

Brüning, Eberhard
Müller, Joachim
Neubert, Albrecht
Peukert, Herbert

GEDENKTAGE

100. Geburtstage

Mathematisch-naturwissenschaftliche Klasse

1976	Gildemeister, Martin	21. Februar
	Kowalewski, Gerhard	27. März
	Kegel, Karl	19. Mai
	Heidebroek, Enno	15. November
	Reinhard, Rudolf	28. November
1977	Gros, Oskar	18. August
	Pfeiffer, Richard Arwed	21. November
1978	Ruhland, Wilhelm	7. August
	König, Walter	26. September

Philologisch-historische Klasse

1977	Pușcariu, Sextil	4. Januar
	Junker, Hermann	29. November
1978	Schücking, Levin L.	29. Mai
	Pinder, Wilhelm	25. Juni
	Bertoni, Giulio	24. August

150. Geburtstage

Mathematisch-naturwissenschaftliche Klasse

1976	Scheibner, Wilhelm	8. Januar
	Gegenbaur, Carl	14. Juni
	Wiedemann, Gustav Heinrich	2. Oktober
1978	Zeuner, Gustav Anton	30. November

Philologisch-historische Klasse

1976	Peschel, Oscar Ferdinand	17. März
	Overbeck, Johannes Adolf	27. März
	Voigt, Moritz	10. September
1977	Voigt, Georg	5. April

50. Todestage

Mathematisch-naturwissenschaftliche Klasse

1976	Joest, Ernst	7. Juli
	Boehm, Rudolf	19. August
	Behrend, Robert	15. September
	des Coudres, Theodor	8. Oktober
1977	Wiener, Otto	18. Januar
	Ambronn, Hermann	28. März
1978	Wiedemann, Eilhard	7. Januar
	Marchand, Felix	4. Februar

Philologisch-historische Klasse

1976	Kluge, Friedrich	21. Mai
1977	Boehmer, Heinrich	25. März

100. Todestage

Mathematisch-naturwissenschaftliche Klasse

1977	Hofmeister, Wilhelm	12. Januar
	Volkmann, Alfred Wilhelm	21. April
1978	Weber, Ernst Heinrich	26. Januar

Philologisch-historische Klasse

1976	Albrecht, Wilhelm Eduard	22. Mai
	Ritschl, Friedrich Wilhelm	9. November
1977	Brockhaus, Hermann	5. Januar
1978	Hänel, Gustav	18. Oktober

KOMMISSIONEN

Mathematisch-naturwissenschaftliche Klasse

Kommission für spezielle Umweltprobleme

701 Leipzig, Goethestraße 3—5, Tel. 296090

Leiter: Neef

Mitglieder: Arland †, Buchheim, Dörter, Freimuth, Gersch, Görlich, Jordan, Lehmann, Schwabe, Strack, Uhlmann, Watznauer, F. Wolf

Die Kommission für spezielle Umweltprobleme faßt gemäß Statut vom 3. II. 1971 die wissenschaftlichen Einzelunternehmen der mathematisch-naturwissenschaftlichen Klasse zusammen, koordiniert ihre Tätigkeit und berät über deren weitere Entwicklung. Sie unterhält Verbindungen zu anderen wissenschaftlichen Institutionen und einzelnen Wissenschaftlern, die an Umweltproblemen arbeiten. Sie fördert die Programmgestaltung der Klassensitzungen und unterstützt die Öffentlichkeitsarbeit der Sächsischen Akademie der Wissenschaften. Die Leiter der nachstehend aufgeführten Arbeitsgruppen, die der Kommission zugehören, tragen für die Durchführung der Forschungsarbeiten die volle Verantwortung.

Düngung der Pflanzen

(bis 31. XII. 1975)

701 Leipzig, Johannisallee 19, Tel. 68500

Leiter: Arland †

Mitarbeiter: Dr. Ruth Zwicker, Herbert Pührer

Strukturprobleme und Umweltaspekte des Wassers, insbesondere der therapeutisch nutzbaren Wässer im Süden der DDR

92 Freiberg (Sachsen), Gustav-Zeuner-Straße, Tel. 512591

Leiter: Buchheim, Jordan, Watznauer

Wiss. Arbeitsleiter: Dr. Manfred Klose

Wiss. Assistenten: Dipl.-Phys. Henning Lemke (bis 1975), Dipl.-Phys. Gunter Kabisch, Dipl.-Geol. Mathias Ruchholz

Mitarbeiter: Dr. Ruth Zwicker, Herbert Pührer

Laboratorium für Alternsforschung

701 Leipzig, Johannisallee 32, Tel. 30254

Leiter: Emmrich †

Untersuchungen über die Wirkungsmechanismen von Neurohormonen

69 Jena, Erbertstraße 1, Tel. 27122, App. 316

Leiter: Gersch
Wiss. Arbeitsleiter: Dr. sc. Klaus Richter
Wiss. Assistenten: Dr. Eckehard Baumann, Dr. Gustav-Adolf Böhm, Dr. Heiner Birkenbeil, Dr. Hermann Eibisch
WTA: Rita Leser, Rosemarie Meißner, Gisela Radtke

Untersuchungen über Naturhaushalt und Gebietscharakter

8027 Dresden, Zellescher Weg 19, Tel. 4633275 und 493387

Leiter: Neef
Wiss. Arbeitsleiter: Dr. Arnd Bernhardt
Wiss. Oberassistent: Dr. Karl Mannsfeld
Wiss. Assistenten: Dr. Inge Hartsch, Dr. Klaus-Dieter Jäger
Ingenieur-Kartograph: Joachim Bieler
WTA: Karin Kießling
Mitarbeiter: Dipl.-Geograph Ernst Benedict, Dipl.-Forst-Ingenieur Felix Lampadius

Reinigung von Abwässern mit Hilfe von Adsorbentien

402 Halle (Saale), Schloßberg 2, Tel. 21191

Leiter: F. Wolf
Wiss. Oberassistentinnen: Dr. Erika Laqua, Dr. Sabine Lindau
Wiss. Assistentinnen: Dipl.-Chemikerin Christine Frank, Dipl.-Chemikerin Magdalene Weise
WTA: Friederike Frahm, Ingeborg Schwab
Mitarbeiterin: Bibliothekarin Bärbel Runge

Wirkung von Carnitinester auf den physiologischen und pathologischen Stoffwechsel sowie auf den Tumorstoffwechsel

701 Leipzig, Liebigstraße 16, Tel. 7164379

Leiter: Strack
Wiss. Oberassistent: Dr. Detlef Müller, Dr. Heinz Löster
Med.-technische Assistentin: Helga Landmann

Düngung der Pflanzen

In Fortsetzung der unter klimatisierten Bedingungen eingeleiteten photoperiodischen Untersuchungen wurde die Jarowisation von Winterweizensorten geprüft. Bekanntlich erfordert das Erreichen der generativen Phase (Formierung der embryonalen Ähre, 4. Etappe der Entwicklung nach Kuperman) ausreichend lange Einwirkung niedriger Temperatur in der 1. und 2. Etappe der Entwicklung.

Bei Jarowisation im Langtag (14-Stunden-Tag) und Einwirkung von Temperaturen im Bereich von 4—6 °C benötigte die westeuropäische Sorte Poros 60 Tage, die sowjetische Kavkas nur 40, um die Entwicklung bis zur generativen Phase fortzusetzen.

Ein Einfluß der Düngung (variierte N- und K-Gaben) und der Qualität des Lichtes auf die Dauer der Jarowisation und die Evapotranspiration war nicht nachzuweisen.

Bei sortenspezifischer (Poros 60, Kavkas 40 Tage) wie auch bei 60tägiger Jarowisation erreichten beide Sorten die 4. Etappe der Entwicklung nahezu gleichzeitig 18 bis 20 Tage nach Abschluß der Jarowisation, wenn das Wachstum unter Blaulicht erfolgte. Bei Einwirkung von Rotlicht reagierte Poros mit 10 bis 12 Tagen Verzögerung.

Die Evapotranspiration der Sorte Poros war sowohl unter Blau- als auch Rotlicht höher als die von Kavkas. Höhere Evapotranspiration der Sorte Poros korrelierte mit niedrigerem Kornertrag, niedrigere Evapotranspiration der Sorte Kavkas mit höherem. Im Vergleich zu Blaulicht reagierte Kavkas auf Rotlicht wie auch bei 40tägiger Jarowisation mit weiterer Senkung der Evapotranspiration.

Weiteren Untersuchungen bleibt vorbehalten zu ermitteln, wie sich beide Sorten bei Jarowisation im Kurztag (9-Stunden-Tag) verhalten.

Dr. Ruth Zwicker

Strukturprobleme und Umweltaspekte des Wassers, insbesondere der therapeutisch nutzbaren Wässer im Süden der DDR

Die unbezweifelbare therapeutische Wirkung äußerst schwach mineralisierter Thermalwässer ist physikalisch noch immer unverständlich [21]. Nachdem sich bisher weder theoretische noch experimentelle Anhaltspunkte für stabile, spezifisch akratothermale Wasserstrukturen haben finden lassen [4, 6, 22, 25, 27], bleibt kaum eine andere physikalische Erklärung übrig als die, daß eine verschieden schnelle Annäherung an das thermodynamische Gleichgewicht für dissoziierte und undissoziierte Lösungsbestandteile bei Absinken der Temperatur auf das Niveau beim Quellenaustritt stattfindet. Der Aufklärung dieses Komplexes können auch Untersuchungen von mehr oder weniger konzentrierten wäßrigen Lösungen [34] durch Raman-Effekt [39, 40] und inelastische Streuung thermischer Neutronen [14] dienen.

In den Gesteinen des Grundgebirges im Süden der DDR sind radioaktive Elemente weit verbreitet. Sie erscheinen in weiter Streuung als Elementspuren oder in konzentrierter Form als Gangfüllungen oder als kompakte Gesteinsimprägnationen.

Aus dieser geologischen Situation resultiert das Auftreten von Uran-Lagerstätten, radioaktiven Wässern und radioaktiver Bodenluft [35, 36, 37, 41].

Soweit die Radioaktivität der beiden letzteren im Bereich der therapeutischen Anwendung liegt, werden sie medizinisch gewertet bzw. sind sie für eine medizinische Nutzung prädestiniert [15]. Eine Aktivität, die über der oberen Grenze der therapeutischen Breite liegt, ist medizinisch bedenklich; die untere Grenze der therapeutischen Wirksamkeit steht derzeit in Diskussion. Viele Erfahrungen sprechen dafür, daß die intensivste und nachhaltigste Wirkung der radioaktiven Quellwässer nicht durch epidermale Resorption, sondern inhalativ erreicht wird [1, 16]. Die Hauptrolle spielt dabei das Radon über seine aktiven, z. T. im Organismus inkorporierbaren Nachfolgeprodukte wie das langlebige RaD. Mit dieser Erkenntnis gewinnen die Ra- und Rn-haltigen Wässer geringer Aktivität ein größere Bedeutung. Damit tritt auch der epidermatologische Aspekt schwacher, aber langlebiger Kontaminierung von Lokalitäten und Gebieten wieder stärker in den Vordergrund.

Auch die obere Grenze der therapeutischen Anwendung bedarf einer Neufestlegung. Es erscheint darum nötig, den gesamten Komplex „radioaktive Mineralwässer" einer Revision zu unterziehen. Diese Revision erfordert eine interdisziplinäre Zusammenarbeit von Medizin, Geowissenschaften und Physik.

Da für das Territorium Bad Elster—Bad Brambach [12] vorrangig balneologische Aspekte und solche des Strahlenschutzes im Vordergrund stehen, wurden das Forschungsinstitut für Balneologie und Kurortwissenschaft (FBK) sowie das Staatliche Amt für Atomsicherheit und Strahlenschutz der DDR (SAAS) kooperativ an dem Forschungsvorhaben beteiligt.

A. Strukturphysikalischer Teil

Über Problematik und Nachweismöglichkeiten spezifischer „Wasserstrukturen" in Thermalwässern geringster Mineralisation (sog. Akratothermen) wurde schon verschiedentlich in den Jahrbüchern der Sächsischen Akademie [4] und andernorts berichtet. Strukturen 1. Art [6] sind, sofern sie in der vadosen Bildungsphase überhaupt existenzfähig sind, beim Austritt derartiger Thermalwässer an der Erdoberfläche nicht mehr zu erwarten. So ist z. B. in der Akratothermae vom Warmbad bei Wolkenstein kein Kieselsäuregel nachweisbar. Strukturen 2. Art [6], zu denen auch Änderungen der H-Brücken zählen, relaxieren, soweit sie nach der Theorie wäßriger Elektrolyte von Debye, Falkenhagen und Hückel auf elektrostatischer Wechselwirkung gelöster Radikale beruhen, in Zeiten der G. —O. 10^{-8} s. Wesentlich langsamer dürften aber Komplexbildungen ungeladener Gruppen mit fallender Temperatur auf Grund von kräftefreier Diffusion vor sich gehen, wenn die Konzentrationen nur gering sind (z. B. 10^{-6} bis 10^{-7} g/l, „Spurenelemente", [6]). Ob der Austritt von gelösten Gasen oder ein langsame Umbildung des H-Brückengefüges während oder sogar noch nach dem Aufstieg solcher Wässer für die von M. Klose beobachteten Alterungseffekte bei Akroatothermen [23] verantwortlich zu machen sind, kann möglicherweise durch inelastische Neutronenstreuung oder durch Raman-Spektroskopie mit sehr intensiver Laser-Anregung entschieden werden. Dem gelegentlich in der Literatur zu findenden Pessimismus gegenüber der Anwendung solcher „aufwendigen“ Methoden in der Wasserstrukturforschung können wir uns nicht anschließen. Gerade die Streuung thermischer Neutronen dürfte zu den „schonendsten“ Sondierungen der labilen Protonenanordnungen und H-Brückenstrukturen in Wasser gehören. Die bisher nur bescheidenen Fortschritte in der Erforschung der Struktur des Wassers und von wäßrigen Lösungen in den letzten drei Jahrzehnten beruhen nicht darauf, daß die zur Sondierung eingesetzte Strahlung (γ-Strahlung, optische Photonen, thermische Neutronen) die „zarten“ Protonen-Anordnungsstrukturen zerstörte, sondern darauf, daß die Wirkungsquerschnitte der Protonen sehr klein und überdies die Linienverbreiterung durch Diffusion und zwischenmolekulare Wechselwirkung im Schwingungsspektrum des flüssigen H_2O relativ groß ist. Es scheint daher zunächst sinnvoll, sich dem Problem von der Seite höherer Konzentrationen gerade solcher Ionen zu nähern, die deutliche Anomalien der Wasserstrukturparameter verursachen [3, 8, 13, 18, 19, 28, 29, 33]. Von M. Klose wurden am sowjetischen Zentralinstitut für Physikalische Chemie Moskau ramanspektrometrische Messungen mit Ar^+- und Kr^+-Laser-Anregung und am Zentralinstitut für Kernforschung Dubna gemeinsam mit J. Natkaniec inelastische Neutronen-Streumessungen an wäßrigen $(UO_2)(ClO_4)_2$-Lösungen [26], in denen die Wasserstruktur durch Hydratation und Wasserstoffbrückenbildung besonders stark geändert wird, durchgeführt. Beide Arten von Untersuchungen lassen hoffen, daß vielleicht auch an schwach mineralisierten Wässern Änderungen des H-Brückengefüges durch „Fervorisierung“ nachgewiesen werden könnten, falls sie eine hinreichend lange Lebensdauer zeigen. Die Beschaffung eines Ar^+—Kr^+-Lasers hoher Intensität an der Sektion Physik der Bergakademie Freiberg bietet u. a. für diese Untersuchungen günstige Voraussetzungen.

Das Studium der künstlichen Fervorisierung von Wasser (mehrstündige Erhitzung auf 130°—150°C unter mehreren Atü) nach V. Vouk [38] wurde wieder aufgenommen, wobei auf die Pflanze als Indikator zurückgegriffen wurde und an die letzten Ergebnisse von C. Job [20] aus dem Balneologischen Forschungsinstitut Bad Gastein der Österreichischen Akademie der Wissenschaften angeschlossen wurde. Im Labor für Wasserhaushalt der Pflanzen an der Karl-Marx-Universität Leipzig wurden auf Veranlassung von Herrn A. Arland (†) eine Reihe von diesbezüglichen pflanzenphysiologischen Tests durchgeführt, die noch nicht abgeschlossen sind.

B. Geologischer Teil

I. Allgemeine Situation

Aus dem umfangreichen Untersuchungsgebiet des Forschungsvorhabens wurde als wirtschaftlich wichtigster Teilbereich das Territorium von Bad Elster und Bad Brambach [12] als erstes in den Kreis der Betrachtung gezogen. Diese Entscheidung wurde außerdem deshalb getroffen, da für dieses Gebiet bereits eine größere Zahl von Vorarbeiten vorliegt und es geeignet erschien, als Testobjekt für weitere, territorial ausgedehntere Gebiete zu dienen.

1. Die geologische Situation von Bad Elster
   Bad Elster liegt im äußeren, nur schwach entwickelten Kontakthof des Fichtelgebirgsgranits. Die Metamorphose in unmittelbarer Umgebung des Bades geht nicht über die Chloritoidzone hinaus.
2. Die geologische Situation von Bad Brambach.
   Bad Brambach liegt im inneren Kontakthof des Fichtelgebirgsgranits. Die Metamorphose geht bis zur Entwicklung von Schiefergneisen der Amphibolitfacies. Südlich von Bad Brambach greifen diese hochkontaktmetamorphen Gesteine als Dachpartien auf den Fichtelgebirgsgranit direkt über [2, 36].

II. Das Auftreten der Heilwässer

Die Heilwässer des genannten Territoriums sind in ihrem Auftreten zum weitaus größten Teile an Bruchstörungen gebunden. Die Lage der Bruchstörungen [2] ist im Gebiet von Bad Elster von der im Gebiet von Bad Brambach verschieden. Im ersteren liegt ein NW—SO/NO—SW-System vor, während in letzterem als vorherrschend eine markante NNW-Richtung erkennbar ist, neben der eine NNO-Richtung nur schwach in Erscheinung tritt. Der Unterschied der Bruchtektonik kann im unterschiedlichen Stockwerkanschnitt, im unterschiedlichen Materialverhalten oder, jedoch mit einem geringeren Wahrscheinlichkeitsgrad, in einem geänderten Beanspruchungsplan liegen. Die Klärung dieser Frage ist notwendig, um einen Einblick in das Quellregime des gesamten Territoriums zu erhalten, denn es steht außer Zweifel, daß die heute bekannten Quellzentren nicht die einzigen ihrer Art sind, und sich jeder eventuell geplante künstliche Eingriff in die allgemeine Quellensituation auf das gesamte Quellensystem auswirken wird. Nur eine regionale Übersicht wird die Anlage eines Quellenschutzbezirkes möglich machen.

Die Mineralisation der Quellwässer ist als genetisch unterschiedlich zu betrachten.

Es ist mit großer Wahrscheinlichkeit anzunehmen, daß die Radioaktivität an die Existenz des fichtelgebirgischen Granitkörpers gebunden ist. Es wird notwendig sein, zu prüfen, wie sich die im äußeren Kontakthof auftretenden Wässer von Bad Elster von denen im inneren Kontaktbereich liegenden von Bad Brambach in Bezug auf ihren Ra- und Th-Gehalt bzw. deren Zerfallsprodukte unterscheiden. Die CO_2-Führung, die ursprünglich wohl als die Mineralisation überhaupt angesehen wurde, ist juvenilen Ursprungs. Das CO_2 benutzt als Aufstieg in vielen Fällen das gleiche Spaltensystem wie die radioaktiven Wässer, doch ist diese Bindung keine primärgenetische. Die Kenntnis des CO_2-Systems ist jedoch für das gesamte Quellregime von größter Bedeutung, da es im weitesten Sinne und auf große regionale Entfernung das Druckniveau bestimmt. Aus der Tatsache der Notwendigkeit einer genetischen Trennung der radioaktiven von den CO_2-Wässern ergibt sich die Möglichkeit des getrennten Auftretens beider Agenzien, d. h. von CO_2-freien radioaktiven Wässern einschließlich trockenen Radon-Austritten einerseits und trockenen Kohlendioxyd-Austritten andererseits, d. h. von reinen emanationsfreien Säuerlingen, wobei bei letzteren die H_2O-Komponente als vadoses Oberflächenwasser aufzufassen sein wird. In welchem Umfange auch die Wässer der Radon-Quellen als vados anzusehen sind, muß noch geklärt werden. Wahrscheinlich ist der größte Teil dieses Ursprungs, doch wird es notwendig sein, Tiefenwässer von den oberflächennahen Wildwässern zu trennen.

In bezug auf die radioaktiven Wässer sei noch hervorgehoben, daß eine Akkumulation ihrer wirksamen Substanzen an Material mit großer Oberfläche, wie an Moor oder Mn—Fe-Schlämmen, erfolgen kann. Auch radioaktive Smog-Erscheinungen sowie Ansammlungen von Radon unter Frostboden muß in den Kreis der Möglichkeiten gezogen werden.

Die weiteren Forschungsaufgaben werden sich auf eine Analyse der tektonischen, insbesondere der bruchtektonischen Verhältnisse des Territoriums sowie auf die Herausarbeitung der Unterschiede in der Mineralisation der Heilwässer von Bad Elster und Bad Brambach konzentrieren.

C. Geophysikalischer Teil

Für das Quellgebiet Bad Elster—Bad Brambach existiert ein beträchtlicher geowissenschaftlicher, insbesondere auch geophysikalischer Vorlauf in der Literatur [2, 9, 10, 11, 30, 31, 32, 35]. Entsprechend der in AI) dargestellten Situation schlossen sich einige im Jahre 1974 durchgeführte geophysikalische Untersuchungen an diese älteren Arbeiten eng an. Dabei wurde zunächst die Problematik von Bad Brambach aufgegriffen.

I. Geomagnetische Arbeiten

Zur Klärung der in AII) erläuterten Verknüpfung von CO_2-und Rn-Gehalt wurden zwei der kleineren Basaltvorkommen nördlich Bad Brambach, nämlich das südlich Raun und das östlich Frauengrün [12] gelegene, einer engräumigen geomagnetischen Spezialuntersuchung unterzogen.

Dabei sollte entschieden werden, ob es sich um schlotartige, spaltenförmige oder deckenförmige Basaltvorkommen handelt. Das Schwergewicht lag dabei auf der Bestimmung der etwaigen Streichrichtung. In beiden Fällen zeigte sich eindeutig ein NNW—SSO-Streichen der magnetischen Anomalien, was auf gangartige Ausbildung hinweist. Diese Streichrichtung ist aber zugleich diejenige, an die die bis jetzt bekannten Rn-Austritte bei Bad Brambach vorwiegend gebunden sind. Damit wird sehr wahrscheinlich, daß der Gastrieb der Brambacher Rn-Quellen, die, soweit genutzt, alle dem sog. „Vogtschen Quellspaltensystem" angehören, auf einem im gleichen System wirksamen CO_2-Austritt beruht.

II. Radiologische Arbeiten

Das größtenteils auf die radiologischen Untersuchungen des sächsischen Erzgebirges von vor und unmittelbar nach dem 1. Weltkrieg (C. Schiffner u. M. Weidig [35], H. Witt [41], P. Ludewig [30], W. Vogt [37] u. a.) zurückgehende Material bedarf angesichts der steigenden Beanspruchung von Bad Brambach einer Überprüfung mit neuen Meßmethoden und einer Erweiterung des Untersuchungsgebietes. Mit elektronischen transportablen RFT-Geräten (Zähl-Meßplatz VA-W 141 und Szintillations-Curiemeter VA-J-05) wurde daher damit begonnen, das Quellspaltensystem von Brambach und einer größeren Zahl von Hausbrunnen im Ort sowie auch die Kläranlage, in die die vom Badebetrieb genutzten Radonwässer abfließen, einer eingehenderen Durchmusterung auf γ-Strahlen-Intensität zu unterziehen. Dabei konnten die alten, größtenteils noch fontaktoskopisch bzw. bodenluftemanometrisch gewonnenen Resultate bestätigt, die von Vogt [37] 1935 angegebenen Quellspalten im Kurpark von Bad Brambach nach N verfolgt und die γ-dosimetrische Unbedenklichkeit der neuen Kanalisation und Kläranlage erhärtet werden. Als wichtigstes neues Ergebnis zeigte sich, daß es offenbar noch unbekannte aktive Quellen östlich des Ortes außerhalb des Vogtschen Spaltensystems gibt.

Charakteristisch scheint der starke Abfall der γ-Intensität radial zu den gefaßten Quellen (Obere und Untere Grenzquelle, Eisenquelle, Schiller-, Radon-, Wiesenquelle), die selbst Gesamt-Aktivitäten bis zu 30 μr/h aufweisen, während sich eine nur schwache γ-Anomalie von wenigen μr/h über dem Röthenbach zeigt, der den größten Teil des radioaktiven Zuflusses wegführt.

In die Untersuchungen des tektonischen Bruchsystems einbezogen wurden auch die Resultate der früheren geoelektrischen Turam-Kartierungsarbeiten des VEB Geophysik [10, 17] und der Sektion Physik der Bergakademie [11] im gesamten Quellengebiet, die in südlicher Richtung bis in die Gegend westlich von Oberbrambach reichen. Der sich dort findende „Lohsäuerling“ ist zwar nur schwach aktiv, aber unverkennbar im Kreuzungspunkt mehrerer tektonischer, geoelektrisch belegter Störungen gelegen (s. A II). Eine dieser Störungen läuft genau auf die Brambacher Quellen zu. Es bedarf deshalb noch einer eingehenderen geophysikalischen Untersuchung, ob sie dem Vogtschen Quellspaltensystem direkt zuzurechnen ist.

Wolfgang Buchheim   Adolf Watznauer

Literatur

[1] Andrejew, S. W.: Die Bestrahlung des menschlichen Organismus bei der Radonbehandlung. Z. f. Physiotherapie *25* (1973) 161–171.

[2] Blüher, H.-J.: Zur Tektonik des Mineralquellengebietes von Brambach im Vogtland und dessen Beziehungen zur nordwestböhmisch-fichtelgebirgischen Quellprovinz. Z. f. d. Deutsch. Geol. Ges. *88* (1936) 547–557.

[3] Bonner, O. D., Ch. F. Jumper: Effects of Ions on Water Structure. Infrared Physics *13* (1973) 233–242.

[4] Buchheim, W.: Untersuchungen zur Physik der Thermalwässer. Jahrbuch d. Sächs. Akad. d. W. *1969–70* (1972) 113–116; *1971–72* (1974) 58–59.

[5] Buchheim, W.: Ergänzende Diskussionsbemerkungen zum Vortrag von W. Michler. International Symposium on Protection of Mineral Waters against undesirable human Interference. Karlovy Vary (Carlsbad) 15.–19. 5. 1972. Prag 1973.

[6] Buchheim, W.: Probleme der Physik der Thermalwässer. Arbeitstagung „Probleme der flüssigen Phase“ Rostock 1973, S. 141–145.
[7] Buchheim, W.: Übersichtsbericht zur Problematik der physikalischen Parameter von Wasser in kapillaren Hohlräumen. Acta Hydrophysica *18* (1974) 77–88.
[8] Enderby, J. E., W. S. Howells, R. A. Hewe: The Structure of Aqueous Solutions. Chem. Phys. Letters *21* (1973) 109–113.
[9] Fröhlich, K., W. Michler und D. Tauchert: Messungen natürlicher Tritiumkonzentrationen an vogtländischen Heilwässern. Z. f. Physiotherapie *26* (1974) 267–273.
[10] Gast, H.: Geoelektrische Untersuchungen im Gebiet Gürth bei Bad Brambach. Unveröff. Bericht bei VEB Geophysik Leipzig.
[11] Gast, H.: Geoelektrische Erkundung von tektonischen Störungslinien im Heilquellgebiet des Staatsbades Bad Elster. Unveröffentl. Forschungsbericht bei der Sektion Physik d. Bergakademie Freiberg, 1971.
[12] Geolog. Spezialkarte 1:25000 des Sächs. Geol. Landesamtes, Sektion Elster nebst Schönberg, Blatt 154 mit Erläuterungen. 1. Auflage; Bearbeiter R. Beck, Leipzig 1884/1885.
[13] Golikov, V. V., Shukovskaja, I., Shapiro, F. L., Shkatula, A.: Rassejanije cholodnych nejtronov na vode i nekotorych organičeskych večestvach. Report P-1903. Joint Inst. Nucl. Res. Dubna (1964).
[14] Gurevich, I. I. and Tarasov, L. V.: Low-Energy Neutron Physics, North-Holl. Publ. Comp., Amsterdam (1968) p. 164–165.
[15] Gusarov, J. J., Andrejew, S. W.: Vortragsreihe über aktuelle Fragen der Radontherapie. Wiedergabe von D. Tauchert u. G. Plötner. Z. f. Physiotherapie *24* (1972) 473–475.
[16] Gusarov, J. J., Obrosov, A. N.: Über den Mechanismus der biologischen und therapeutischen Wirkung von Radonbehandlungen. Z. f. Physiotherapie *26* (1974) 27–35.
[17] Hedström, H.: Phasenmessungen mit dem „Turam“-Verfahren. ATM V 65.
[18] Janik, J. M., Rachwalska, M., Janik, J. A.: Proton Jumps in Crystalline Hydronium Perchlorate. Report Nr. 846/PS. Inst. Nucl. Phys. Cracow (1973).
[19] Janik, J. M., Pytasz, G., Rachwalska, M., Janik, J. A., Natkaniec, I., Nawrocik, W.: Dynamical Aspects of Phase Transistions in Perchloric Acid Monohydrate. Acta Polon. *A 43* (1973) 419–424.
[20] Job, C.: Beiträge zur Kenntnis der Heilquellenwirkung auf Pflanzen. Sitz.-Ber. d. Österr. Akad. d. W., Math.-nat. Klasse II *175* (1966) 5.–7. Heft.
[21] Jordan, H.: Grundriß der Balneologie und Balneoklimatologie S. 145. VEB Georg Thieme, Leipzig 1964.
[22] Klose, M.: Beiträge zur D- und V-Struktur von Wasser und speziellen wäßrigen Elektrolytlösungen. Abhandl. d. Sächs. Akad. d. Wiss. Math.-nat. Kl. Band 52. H. 4 Akademie-Verlag 1974; als zusammenfassender Bericht dasselbe in Acta Hydrophysica, Berlin, *18* (1974) H. 2, 95–122.
[23] Klose, M.: Beiträge über Langzeitänderungen spezieller Thermalwässer. Z. f. Physiotherapie *26* (1974) 275–277.
[24] Klose, M.: Über „magnetisiertes Wasser“ (zusammenfassender Literaturbericht. Acta Hydrophysica *17* (1972) H. 1, 53–55.
[25] Klose, M.: Investigation of hydration and vibrational structure of concentrated aqueous $UO_2(ClO_4)_2$ by means of inelastic photon (Raman) and neutron scattering. IV. Intern. Ramankonf. Brunswick U. S. A. August 1974 (Paper zum Druck eingereicht).
[26] Klose, M.: Contributions to the Raman spectrometric studies of the quasi-Fermi-

resonance of water and aqueous electrolyte solutions (abstract Nr. 188. IVth European Congress on Molecular Spectroscopy, Tallinn USSR, 1973).
[27] Klose, M.: Schwingungsspektrometrische Untersuchungen an speziellen wäßrigen Elektrolyten (Arbeitstagung „Probleme der flüssigen Phase" Rostock 1973, 147—150).
[28] Larsson, K.-E., Bergstedt, L.: Proton Motions in Complex Hydrogenous Liquids. I. A Cross Section for Quasi-Elastic Scattering of Slow Neutrons. Phys. Rev. *151* (1966) 117—125.
[29] Leung, P. S., Safford, G. J.: A Neutron Inelastic Scattering Investigation of the Diffusion Kinetics of M_2O Molecules and Hydration Complexes in Concentrated Ionic Solutions. J. Solut. Chem. *2* (1973) 525—546.
[30] Ludewig, P., Witte, H.: Radioaktive Messungen im Quellgebiet von Brambach. Z. f. Geophys. *1* (1925) 242—250; *2* (1926) 70—77.
[31] Michler, W.: Beiträge zur Hydrogeologie des oberen Vogtlandes unter Berücksichtigung der Mineralquellen. Unveröff. Dissertation an der Martin-Luther-Universität Halle—Wittenberg 1973.
[32] Nosske, G., Särchinger, H.: Geophysikalische Untersuchungen zwischen den Mineralquellgebieten von Bad Elster und Bad Brambach. Intern. Geol. Congr. Copenhagen 1961. Report of the 21th Session: Norden. Part II: Proceedings of Session 2; Geol. Results of appl. Geochem. a. Geophys.
[33] Peleg, W.: A Raman Spectroscopic Investigation of the Magnesium Nitrate—Water System. J. Phys. Chem. *76* (1972) 1019—1025.
[34] Samoilov, O. Ya.: Structure of Aqueous Elektrolyte Solutions and the Hydration of Ions (Consultants Bureau Enterprises Inc., New York 1965).
[35] Schiffner, C., Weiding, W.: Radioaktive Wässer in Sachsen, Teil IV. Craz und Gerlach Freiberg 1912.
[36] Stettner, G.: Die Beziehungen der kohlensäureführenden Mineralwässer Nordostbayerns und der Nachbargebiete zum rhegmatischen Störungssystem des Grundgebirges. Geologische Bavarica *64* (1971) 305—374.
[37] Vogt, W.: Radiologische Untersuchungen im Radiumbad Brambach. Z. f. Geophys. *11* (1935) 29—35.
[38] Vouk, V.: Ein neuer Versuchsweg zur Klärung der Frage der balneologischen Wirkung des Thermalwassers. Der Balneologe *8* (1941) 71—73.
[39] Walrafen, G. E.: Raman Spectral Studies of the Effects of Perchlorate Ion on Water Structure. J. Chem. Phys. *52* (1970) 4176—4193.
[40] Walrafen, G. E.; Blatz, L. A.: Weak Raman bands from water. J. Chem. Phys. *59* (1973) 2646—2650.
[41] Witte, H.: Bestimmung des Radium-Emanationsgehaltes von Erd- und Quellgasen in Bad Brambach. Z. f. Geophys. *2* (1926) 181—187.

Untersuchungen über die Wirkungsmechanismen von Neurohormonen

Die von der Arbeitsgruppe seit einigen Jahren verfolgte Gesamtthematik, nämlich die Aufklärung der Funktion des neurohormonalen Systems insbesondere bezogen auf die Häutungsdrüse der Insekten und Prüfung einer gezielten Einflußnahme auf die endokrine Regulation der Insektenentwicklung und -häutung wurde im Berichtszeitraum kontinuierlich fortgesetzt. In Anbetracht der bei anderen Arthropodenklassen, vornehmlich bei Krebsen, vorliegenden Art der Steuerung der Entwicklung mittels eines Hemmhormons wurden erste Untersuchungen zur Charakterisierung des Hemmhormons und seiner Aktivität bei Krebsen unternommen. Außerdem wurden die laufenden Arbeiten zur Isolierung und Reindarstellung von Neurohormon D durch Anwendung von Mikro-Auftrennungsverfahren weitergeführt.

1. Untersuchungen zur Funktion des neurohormonalen Systems auf die Insektenentwicklung

In den früheren Untersuchungen konnte gezeigt werden, daß die beiden entwicklungsphysiologisch aktiven Fraktionen des sog. Gehirnhormons (Aktivationsfaktor I und II) unterschiedliche Wirkungsmechanismen auf zellulärer Ebene besitzen. Aktivationsfaktor I stimuliert die RNS-Synthese der Prothoracaldrüsenzellen, Aktivationsfaktor II beeinflußt dagegen die Permeabilität der Zellmembran. Von diesen Befunden ausgehend, wurde ein in-vitro-Testverfahren für die RNS-Syntheserate der Prothoracaldrüse (sog. Prothoracaldrüsentest) entwickelt (Abb. 1). Es bietet sowohl gegenüber den entwicklungsphysiologischen als auch den in-vivo-Testverfahren eine Anzahl entscheidender Vorteile. Das betrifft die wesentlich höhere Konstanz der Testparameter, die höhere Spezifität der Wirkung durch Testung im zellulären bzw. molekularen Bereich und die exaktere quantitative Auswertung auf der Grundlage absoluter Versuchswerte. Als Maß für die zelluläre Aktivität der Drüsen dient der Einbau von ^{3}H-5-Uridin in die RNS der Prothoracaldrüsen. Die Einbaurate ist dabei eine Funktion der zugegebenen Uridinmenge und der Inkubationszeit. Der Prothoracaldrüsentest eignet sich besonders für Untersuchungen zur Wirkung von biologisch aktiven Substanzen auf das endokrine System der Insekten.

In Fortführung der Untersuchungen über die Beeinflussung des Membranpotentials der Prothoracaldrüsenzellen der Wachsmotte Galleria mellonella wurden 1. das Verhalten des Membranpotentials im zeitlichen Ablauf nach Applikation unterschiedlicher Konzentrationen des Aktivationsfaktors II und 2. die Wirkung des Aktivationsfaktors II auf das Membranpotential der Prothoracaldrüsen in den einzelnen 5 Augenklassen der Galleria-Raupen untersucht. Es zeigte sich, daß Aktivationsfaktor II eine konzentrationsabhängige, biphasische Wirkung auf das Membranpotential der Prothoracaldrüsenzellen von Galleria besitzt. Durch niedrigere Konzentration werden Hyperpolarisationen, durch höhere Depolarisationen verursacht. Weiterhin konnte nachgewiesen werden, daß Aktivationsfaktor II auf Prothoracaldrüsen des letzten Raupenstadiums nur im Augenstadium 1 und 2 eine signifikante Steigerung der Membranpotentiale verursacht (Abb. 2). Das steht in guter Übereinstimmung mit der für diese Phase einsetzenden hormonalen Aktivität der Drüse.

Eingehende elektronenmikroskopische Untersuchungen der Prothoracaldrüse von Galleria-mellonella-Raupen des letzten Stadiums sollten klären, inwieweit Zusammenhänge zwischen den einzelnen Aktivitätsphasen der Drüse und ihrer Substruktur bestehen. Dabei ergab sich, daß in der Ausbildung und Größe der Zellmembran sowie in der Substruktur der Mitochondrien Unterschiede in den Drüsenzellen der einzelnen Augenstadien, die jeweils bestimmten Aktivitätsphasen entsprechen, auftreten (Abb. 3, 4). Bei anderen Zellorganellen bzw. -strukturen waren keine signifikanten Unter-

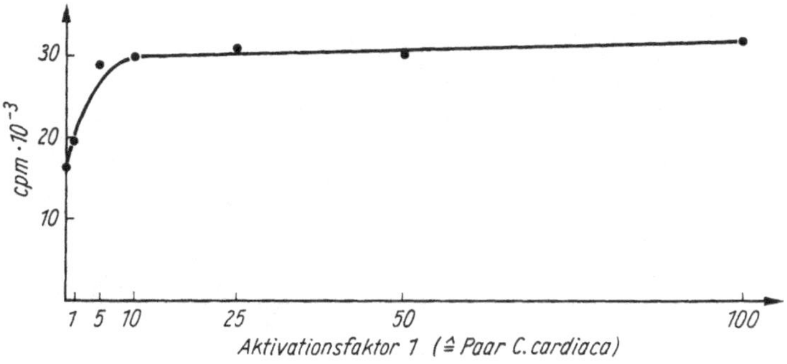

Abb. 1. Dosis-Wirkungs-Beziehung zwischen der Konzentration des Aktivationsfaktors I und der Inkorporation von ^{3}H-5-Uridin in die RNS kultivierter Prothoracaldrüsen 5 Tage alter Larven von *Periplaneta americana*. (Nach Gersch und Bräuer 1974)

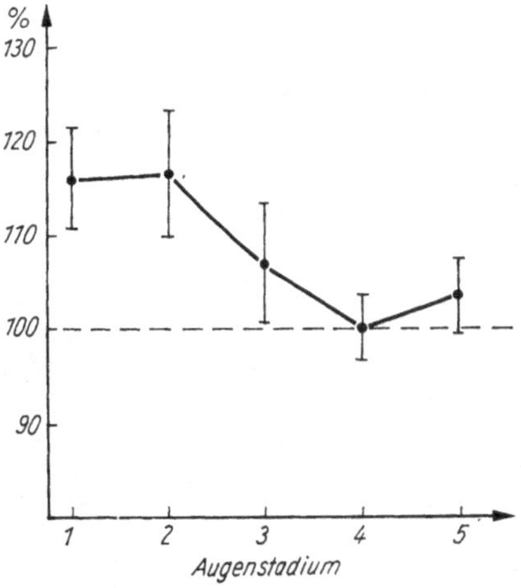

Abb. 2. Prozentuale Veränderung der Membranpotentiale der Prothoracaldrüsenzellen von Wachsmottenraupen des letzten Stadiums der Augenklassen 1 bis 5 nach 1/2stündiger Einwirkung von Aktivationsfaktor II. Konzentration entsprechend 7,5 Paar Corpora cardiaca des Ausgangshomogenates. Von jeder Augenklasse wurden 10 Tiere gemessen. Es ist deutlich erkennbar, daß Aktivationsfaktor II das Membranpotential der Prothoracaldrüsenzellen von Tieren auf den beiden frühen Augenstadien erhöht. Der Faktor spricht bei Augenstadium 4 nicht mehr an, was mit den entwicklungsphysiologischen Daten gut übereinstimmt. (Nach Birkenbeil und Gersch 1974)

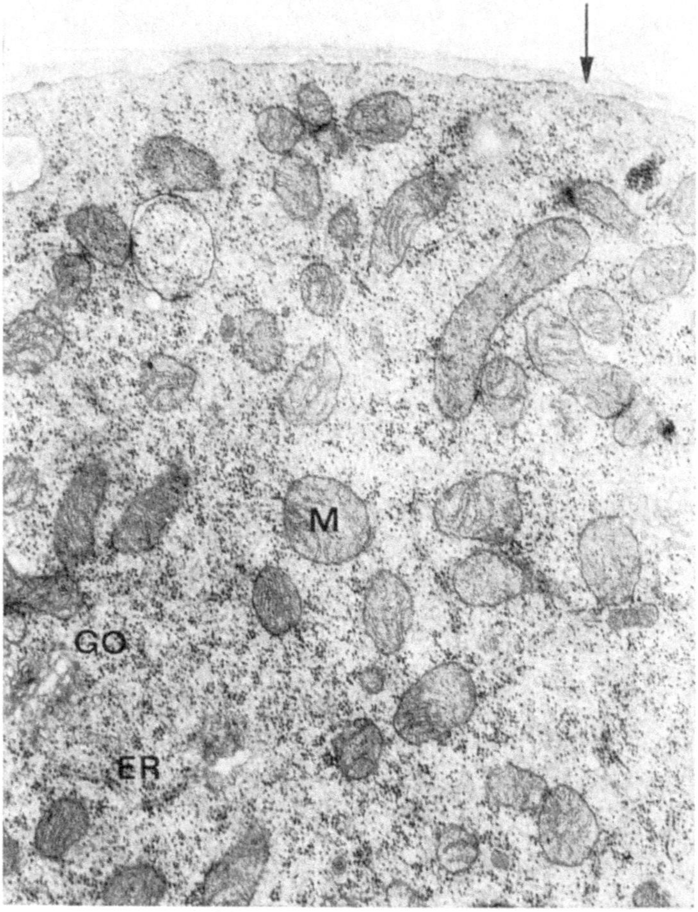

Abb. 3. Elektronenmikroskopisches Bild vom Ausschnitt einer Prothoracaldrüsenzelle der Wachsmotte *Galleria mellonella* einer frisch gehäuteten Raupe des letzten Larvenstadiums. Zu beachten hier die glatte Zellmembran, ohne Einfaltungen (←), weiterhin die Mitochondrien mit klarer Matrix und die große Zahl von Ribosomen. M Mitochondrien, GO Golgi Apparat, Er Endoplasmatisches Retikulum Vergrößerung 1:21000

Abb. 4. Ausschnitt einer Prothoracaldrüsenzelle einer Larve vom Augenstadium 3. Hier zeigt die Membran starke Einfaltungen (←) mit der Ausbildung eines umfangreichen lakunären Systems (LS). N Nukleus; a) Zellquerschnitt; b) Oberflächenschnitt, der besonders die Ausbildung der Einfaltungen mit dem lakunären System veranschaulicht. In beiden Fällen Vergrößerung 1:13000 (Nach Gersch, Birkenbeil und Ude, 1975)

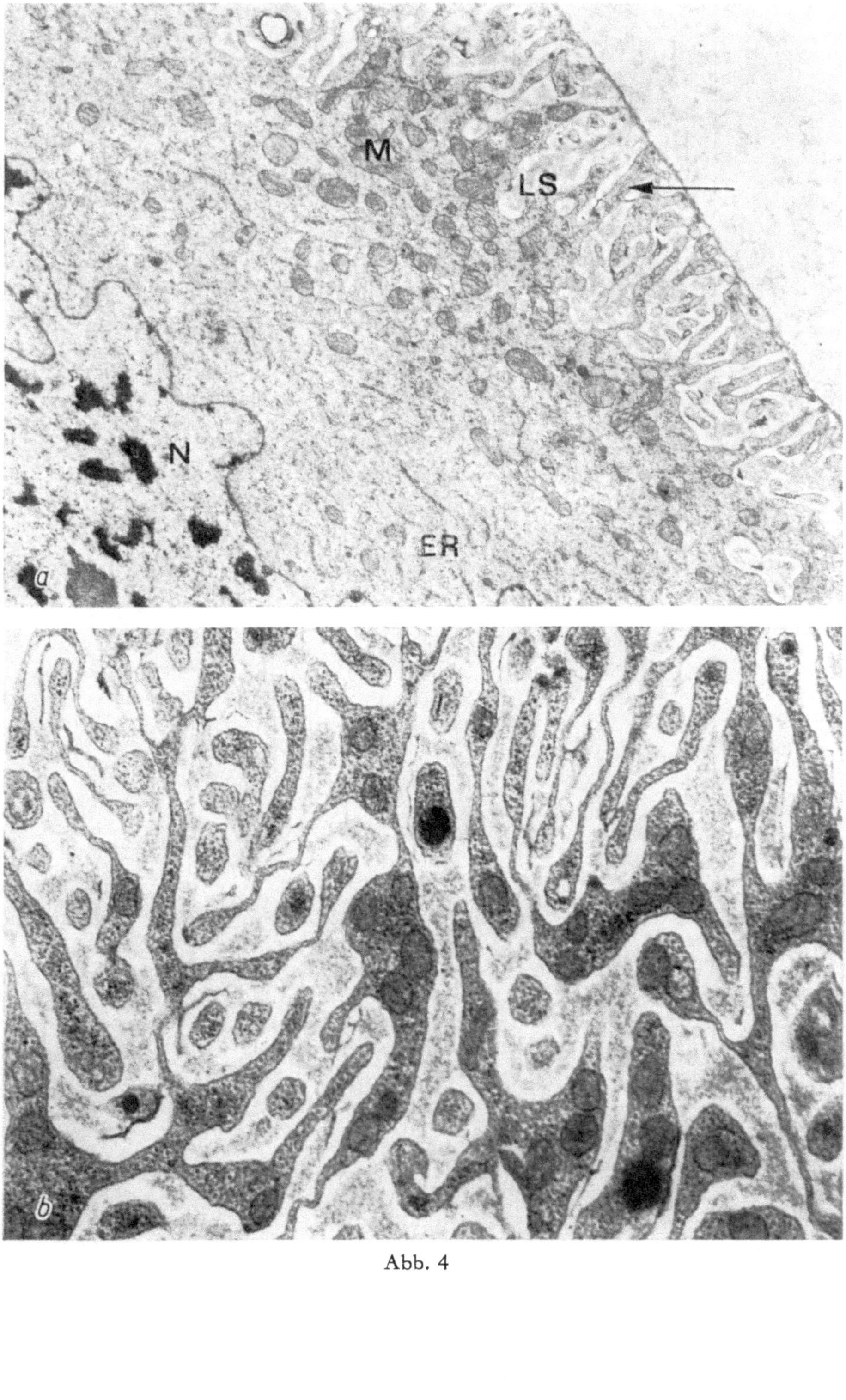

Abb. 4

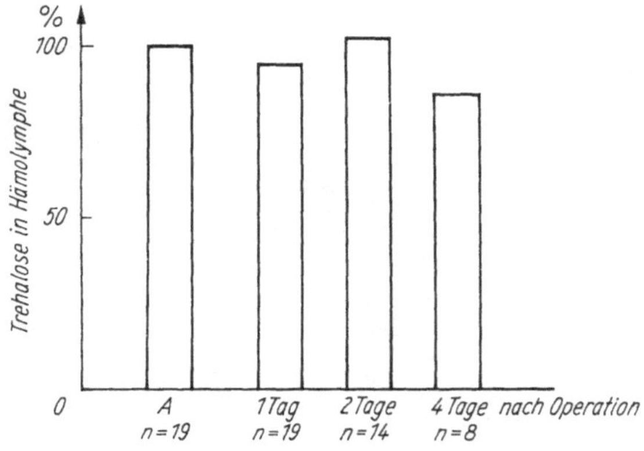

Abb. 5. Übersicht über die Werte des Trehalosegehaltes der Haemolymphe von *Periplaneta americana* nach Durchtrennung von Nervus corporis cardiacus I 1, 2 und 4 Tage nach der Operation. Die Werte an den verschiedenen Tagen nach Operation zeigen keine signifikanten Unterschiede zu den Ausgangswerten. (Nach Gersch 1974)

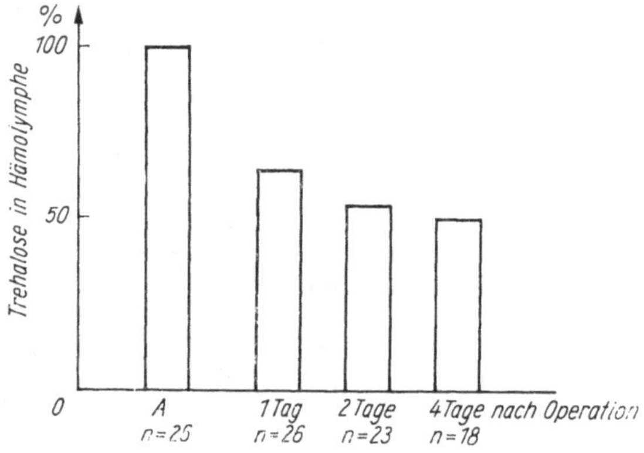

Abb. 6. Übersicht über die Werte des Trehalosegehaltes von *Periplaneta americana* nach Durchtrennung von N. c. c. II

Hier ist zu den verschiedenen Tagen nach der Operation ein deutliches Absinken der Trehalose in der Hämolymphe zu beobachten, woraus geschlossen werden kann, daß die Ausschüttung dieses Neurohormons aus den Corpora cardiaca nervös über diesen Nerven reguliert wird. (Nach Gersch 1974)

schiede festzustellen. Die exzessive Ausbildung der Zellmembran während der Augenstadien 1 bis 3 kann als Anzeichen einer erhöhten Aktivität der Zelle gewertet werden und korreliert mit den früheren Befunden über die Veränderungen im Membranpotential in diesen Phasen.

Weitere physiologische Untersuchungen der Aktivität der Prothoracaldrüse von Galleria mellonella sollten klären, ob bei nervöser Hemmung der Entwicklung über das Unterschlundganglion eine Korrelation zur Aktivität der Häutungsdrüse, gemessen am Membranpotential, erkennbar ist und weiterhin, ob sich die Aufhebung der Hemmung nach Exstirpation des Unterschlundganglions wiederum an der Höhe des Membranpotentials der Drüsenzellen der Häutungsdrüse auswirkt. Es konnte eindeutig nachgewiesen werden, daß auch unter den genannten experimentellen Bedingungen die Höhe des Membranpotentials der Zellen der Häutungsdrüse dem des jeweiligen Augenstadiums entspricht. Bei Aufhebung der Hemmung nach Exstirpation des Unterschlundganglions, was zur Weiterentwicklung des Raupenstadiums und zugleich damit des Augenstadiums führt, ist wiederum die Höhe des Membranpotentials mit dem Augenstadium korreliert. Die Höhe des Membranpotentials ist also auch unter den Bedingungen der nervösen Hemmung durch das Unterschlundganglion und nach dessen Exstirpation ein zuverlässiger Indikator für den Aktivitätszustand der Prothoracaldrüse von Galleria mellonella.

2. *Neurohormonale Steuerung der Krebsentwicklung*

Obwohl das Vorkommen und die Wirkung eines Hemmhormons der Häutung bei Krebsen in allen einschlägigen älteren und neueren Übersichten zur Crustaceen-Endokrinologie erwähnt werden, erwies sich in den eigenen Versuchen, daß die chemische Erfassung und die Testung dieses Faktors einen breiteren Ansatz erfordert, da sich die Literaturangaben als sehr ungenügend herausstellten. Deshalb mußten in wesentlicher Erweiterung der ursprünglich vorgesehenen Aufgabe, nämlich die Einwirkung des Hemmhormons von Krebsen auf die Insektenentwicklung zu untersuchen, die Arbeiten auf Biotests an Krebsen selbst ausgedehnt werden.

Die Wirkung des Hemmfaktors von Krebsen wurde einerseits durch experimentelle Ausschaltung der Produktions- und Speicherorte und andererseits durch Injektion von Gewebshomogenaten wie schließlich auch durch davon aufgetrennte Fraktionen getestet. Die umfangreichen Versuchsserien über die Wirkung dieses möglichen Hemmfaktors aus den Augenstielen von Krebsen, die an Isopoden und Krabben erfolgten, führten nicht zu eindeutigen Ergebnissen. Es traten in nicht wenigen Fällen trotz experimenteller Ausschaltung der Sinusdrüse statt der zu erwartenden Beschleunigung der Entwicklung verlangsamte oder zumindest normale Entwicklungsabläufe ein. Umgekehrt führte die Injektion von Homogenaten der Sinusdrüse wie auch ihrer durch Säulenchromatographie aufgetrennten Fraktionen häufig eher zu einer Beschleunigung statt zu einer Verlängerung der Zwischenhäutungsphase.

Für die Erklärung dieser unsicheren Befunde können verschiedene Gründe diskutiert werden. So erscheint es zur genaueren Aufklärung des Wirkungsmechanismus der jeweiligen Faktoren analog den Verhältnissen bei Insekten angebracht, nicht nur die meist üblichen entwicklungs-physiologischen Teste mit einzelnen Individuen vorzunehmen, sondern neue Testmöglichkeiten auf zellulärer Ebene einzubeziehen. Als Testsystem wurde zunächst die Sternalplattenbildung bei Isopoden benützt. Ausgangsmaterial für die Injektion wie auch für die gelchromatographische Aufarbeitung einzelner Fraktionen, die dann injiziert wurden, bildeten Augenstiele bzw. präparierte Sinusdrüsen von marinen Decapoden und von Isopoden. Hierzu mußte eine erhebliche Materialmenge präparatorisch verarbeitet werden.

Um zu genaueren Aussagen über die Wirkung des Hemmfaktors aus den Augenstielen (bzw. Sinusdrüsen) von Krebsen zu gelangen, wurden folgende weitere Testverfahren entweder neu entwickelt oder zumindest weiter ausgebaut: Röntgenographische Ermittlung der Gastrolithenbildung bei Decapoden, die Einbaurate von ^{3}H-Leucin in die Häutungsdrüse von Normaltieren verglichen mit augenstielexstirpierten Individuen, weiterhin die RNS-Bestimmung im Hepatopankreas von Flußkrebsen der Zwischenhäutungs- im Vergleich zu denen der Vorhäutungsphase und schließlich der Calciumstoffwechsel von Isopoden durch Ca-Bestimmung einzelner Extremitäten in verschiedenen Entwicklungsstadien mittels einer dafür neu eingesetzten Mikrobestimmungsmethode. Obgleich noch keine endgültigen Ergebnisse zu den einzelnen Verfahren vorliegen, ist zu erwarten, daß sich hieraus einige geeignete Teste entwickeln lassen. Die weiteren Versuche sind daher in vollem Gange.

3. Untersuchung einiger weiterer neurohormonaler Regulationen bei Insekten

In Weiterführung einer früheren Arbeit (vgl. Jahresbericht 1969/70 und 1971/72), in der nachgewiesen wurde, daß durch Reizung des Nervus corporis cardiacus II eine selektive Ausschüttung des hyperglykaemischen Neurohormons bei der Schabe Periplaneta americana erfolgt, wurde jetzt die Hormonfreisetzung aus den Corpora cardiaca von Periplaneta americana in vivo geprüft. Hierzu mußten die Trehalosewerte der Hämolymphe von Versuchstieren ermittelt werden, denen entweder N. c. c. I oder N. c. c. II durchtrennt worden waren. Die Versuche erbrachten sehr klare Befunde. Durchtrennung von N. c. c. I beeinflußt die Trehalosewerte nicht (Abb. 6). Dagegen erfolgt eindeutiges Absinken der Werte nach Durchtrennung des N. c. c. II (Abb. 5). Hieraus ist die Schlußfolgerung zu ziehen, daß die Ausschüttung des hyperglykaemischen Faktors aus den Corpora cardiaca in vivo selektiv über den N. c. c. II reguliert wird. Dies steht mit den früheren in-vitro-Ergebnissen in guter Übereinstimmung.

In Verbindung mit der festgestellten selektiven Hormonausschüttung nach elektrischer Reizung von N. c. c. I oder II der Corpora cardiaca wurde die Hormonfreisetzung nach elektrischer Reizung der einzelnen Ganglien der Bauchkette von Periplaneta americana experimentell geprüft. Elektrische Reizung des 6. Abdominalganglions führte zur Ausschüttung eines herzanregenden Faktors. Er konnte auf Grund biochemischer Untersuchungen und physiologischer Teste als Neurohormon D identifiziert werden. Seine Ausschüttung nach elektrischer Ganglionreizung wird durch Atropin gehemmt, durch Eserin erhöht. Das zeigt, daß die Hormonfreisetzung durch cholinerge Komponenten reguliert wird.

Weitere elektrische Reizversuche der anderen Ganglien der Ventralkette von Periplaneta americana ergaben, daß von den 3 Thoracalganglien ein neurohormonaler Faktor ausgeschüttet wird, der im Gegensatz zu Neurohormon D die Herztätigkeit hemmend beeinflußt. Seine höchste Aktivität wurde im 1. Thoracalganglion ermittelt. Die Hemmwirkung der Thoracalganglien-Homogenate zeigte eine Dosis-Wirkungs-Beziehung. Der wirksame Faktor ist temperaturempfindlich und dialysierbar. Nach Behandlung der Homogenate mit Trichloressigsäure ist keine biologische Aktivität mehr nachweisbar. Es handelt sich wahrscheinlich um ein Peptid mit einem MG unter 20000.

4. *Weitere Untersuchungen zur Reindarstellung und Isolierung von Neurohormon D*

Für die in Aussicht stehende Isolierung und Strukturaufklärung von Neurohormon D mußte ein größeres Schabenmaterial gezüchtet werden. Es wurden davon ca. 80000 Paar Corpora cardiaca von Periplaneta americana in minutiöser Präparation für die weitere Aufarbeitung gewonnen.

Im Gegensatz zu den anderen vorangegangenen Trennungs- und Reinigungsverfahren des Neurohormons D wurde jetzt die Technik der Dansylierung mit anschließender Mikrochromatographie der fluoreszierenden Substanzen auf Polyamidfolien angewandt. Die Umsetzung von Neurohormon D, das durch Auftrennung von Extrakten der Corpora cardiaca an Sephadex G 100 gereinigt wurde, mit Dansylchlorid liefert im Mikrochromatogramm einen deutlich fluoreszierenden Fleck. Die Rechromatographie an Sephadex G 100 bestätigt, daß ein wesentlicher Teil der Fluoreszenzintensität auf markiertes Neurohormon D zurückzuführen ist. Da nach Hydrolyse mit 6 N HCl im Mikrochromatogramm mehrere Flecken zu beobachten sind, kann die Verunreinigung durch andere Peptide nicht ausgeschlossen werden. Ein weiterer Fortschritt bedeutet die Markierung der Substanzen mit Fluorescamin, wodurch quantitative Bestimmung von Aminosäuren im Nanomol-Bereich möglich wird. Dies ist zugleich eine Voraussetzung für die geplante Massenspektrometrie. Außerdem wird die in Aussicht genommene Mikrofluorometrie genauere Angaben über die Reinigung des Faktors ermöglichen. Während bisher die biologische Aktivität auf die Menge von Corpora cardiaca des Ausgangsmaterials bezogen werden mußte, kann durch fluorometrische Messung der Gehalt an freien NH_2-Gruppen quantitativ bestimmt werden. Auf diese Weise sind genaue Angaben über den Reinheitsgrad des Faktors möglich. Damit ist zugleich der Fortgang der laufenden Arbeiten zur Reinigung von Neurohormon D bis zur Inangriffnahme der Strukturaufklärung vorgezeichnet.

Literatur

Baumann, E., und M. Gersch: Untersuchungen zur Stabilität des Neurohormons D. Zool. Jb. Physiol. *77* (1973) S. 153–160.

Baumann, E., und M. Gersch: Versuche zur Markierung von Neurohormon D aus Periplaneta americana mit Hilfe von Dansylchlorid. Zool. Jb. Physiol *78* (1974) S. 533–541.

Birkenbeil, H., und M. Gersch: Weitere Untersuchungen zur hormonalen Beeinflussung des Membranpotentials der Prothorakaldrüsenzellen von Galleria mellonella. Zool. Jb. Physiol. *78* (1974) S. 150—156.

Böhm, G.-A.: Neuere Befunde zur hormonalen Steuerung einiger Verhaltensweisen von Insekten. Biol. Rdsch. *12* (1974), S. 382—392.

Gersch, M.: General aspects and problems in comparative endocrinology. Gen. comp. Endocrinol. *20* (1973) S. 425—435.

Gersch, M.: Effects of hormones in invertebrates. International Encyclopedia of Pharmacology and Therapeutics Section 85 Comp. Pharm. 1973, S. 711—798.

Gersch, M.: Experimentelle Untersuchungen zur Ausschüttung von Neurohormonen aus Ganglien des Bauchmarks von Periplaneta americana nach elektrischer Reizung in vitro. Zool. Jb. Physiol. *78* (1974), S. 138—149.

Gersch, M.: Selektive Freisetzung des hyperglykämischen Faktors aus den Corpora cardiaca von Periplaneta americana in vivo. Experientia *30* (1974) S. 767.

Gersch, M., H. Birkenbeil und J. Ude: Ultrastructure of the prothoracic gland cells of the last instar of Galleria mellonella in relation to the state of development Cell. Tiss. Res. *160* (1975) S. 389—397.

Gersch, M., und R. Bräuer: In-vitro-Stimulation der Prothoracaldrüsen von Insekten als Testsystem (Prothoracaldrüsentest). J. Insect Physiol. *20* (1974) S. 735—741.

Gersch, M., E. Hentschel und J. Ude: Aminerge Substanzen in lateralen Herznerven und im stomatogastrischen Nervensystem der Schabe Blaberus craniifer BURM. Zool. Jb. Physiol. *78* (1974) S. 1—15.

Richter, K.: Struktur und Funktion der Herzen wirbelloser Tiere. Zool. Jb. Physiol. *77* (1973) S. 477—668.

Richter, K.: Zum Problem der hormonalen Modulation neuronaler Vorgänge. Biol. Rdsch. *12* (1974), S. 326—334.

Richter, K. , und M. Gersch: Untersuchungen über die Wirkung cholinerger, aminerger und peptiderger Substanzen am lateralen Herznerven und den funktionellen Zusammenhang der Wirkungen bei der Herzregulation von Blaberus craniifer BURM. (Insecta: Blatteria). Zool. Jb. Physiol. *78* (1974) S. 16—32.

Ude, J., und K. Richter: The submicroscopic morphology of the heart ganglion of the spider Tegenaria atrica (C. L. KOCH) and its neuroendocrine relations to the myocard. Comp. Biochem. Physiol. *48* (1974) S. 301—308.

Manfred Gersch

Naturhaushalt und Gebietscharakter

Die früheren Untersuchungen der Dresdner Arbeitsgruppe führten bereits zu den Problemen der gesellschaftlichen Beurteilung und Wichtung von Inhalt und räumlicher Differenzierung der Naturräume unterer Dimensionsstufen. Im Bericht des Jahrbuches 1971/72 heißt es (S. 69):

„Die Aussagen der Naturraumerkundung bleiben .. für die praktische Anwendung unbestimmt, solange die hochintegrierten Naturkomplexe nicht im Rahmen der Auseinandersetzung von Gesellschaft und Natur gesehen werden. Die Gesellschaft setzt jeweils die Maßstäbe für die Nutzung der Naturressourcen und gibt durch ihre Entwicklungsprognosen eine Reihe von Vorgaben für die naturwissenschaftliche Erkundung.“

Diese Vorgaben, von gesellschaftlichen Zielen und Aktivitäten abgeleitet, lassen diejenigen Komponenten des Naturdargebotes erkennen, die für die Beherrschung des Stoffwechsels zwischen Gesellschaft und Natur vorrangige Bedeutung haben. Mit ihrer Hilfe gelingt es, die Relativierung naturräumlicher Aussagen hinsichtlich der gesellschaftlichen Sachverhalte durchzuführen und Form und Auswahlprinzipien der relevanten Elemente in den Naturraumanalysen zu differenzieren. Diese Relativierung naturräumlicher Aussagen erfordert auch klarere terminologische Unterscheidungen. Es werden unterschieden:

Natur- oder Naturraumausstattung = neutraler Begriff der physisch-geographischen Analyse, enthält reine Sachverhaltsfeststellungen ohne Bezug auf die Nutzung;
Natur- oder Naturraumdargebot = zusammenfassender Begriff für die Gesamtheit der von der Natur bereitgestellten materiellen Reichtümer einschließlich der Lagebeziehungen und ist auf die Nutzbarkeit durch die Gesellschaft orientiert;
Natur- oder Naturraumpotentiale = die Teile des Naturdargebots, die für bestimmte gesellschaftliche Anliegen oder Funktionen von Interesse sind und daür ein feststellbares Leistungsvermögen aufweisen;
Naturressourcen = der Teil der Naturpotentiale, der in ökonomische Beziehungen eintritt oder dafür vorgesehen ist.

Die Gesellschaftsrelevanz der Naturraumpotentiale läßt sich nur durch eine genaue Analyse der gesellschaftlichen Anforderungen an die einzelnen Naturraumpotentiale bestimmen. Der Ausarbeitung von „Anforderungsbildern" für die einzelnen Naturraumpotentiale und die Klärung der Anforderungsstrukturen war eine der Hauptaufgaben der Berichtsjahre.

Dabei stellte sich heraus, daß zwar die Elemente der Naturraumausstattung durch die physisch-geographische Forschung weitgehend bekannt oder durch erprobte Methoden jederzeit erfaßt werden können, daß jedoch Formen, Verteilung und Einflußfelder der landeskulturellen Elemente in der Landschaft als die Ergebnisse der historischen Entwicklung zur Kulturlandschaft entweder nur sehr allgemein oder ohne Beziehung zu Struktur und Dynamik der Naturräume dargestellt worden sind. So machten sich spezielle und systematische *Untersuchungen zur landeskulturellen Ausstattung* erforderlich.

Weiterhin erwies sich, daß die verschiedenen Anwendungsbereiche und Aufgabenstellungen in Abhängigkeit vom Maßstab verschiedene *Intensitätsstufen* der Analyse verlangen, die in Methoden, Auswahlprinzipien und Aussageformen der Naturraumerkundung aufeinander abgestimmt sein müssen.

Alle diese Studien sind jedoch nur auf einen Teil der Beziehungen im Stoffwechsel zwischen Gesellschaft und Natur gerichtet, nämlich auf die „Aneignung der Naturreichtümer für die Befriedigung gesellschaftlicher Bedürfnisse", also auf die produktive und konsumtive Nutzung als erstrebtes Hauptziel der gesellschaftlichen Aktivitäten. Die „Nebenwirkungen", die notwendigerweise mit der Wirtschaftstätigkeit verbunden und vielfach umweltstörend, z. T. potentialmindernd sind sowie das Wohlbefinden der Menschen beeinträchtigen können, werden damit nicht erfaßt. Sie nehmen an Bedeutung ständig zu und sind zu einem Hauptproblem des Umweltschutzes geworden. Eine konzeptionelle Studie „zu landeskultureller Funktion und Dynamik chorischer Strukturen" dient in erster Linie der systematischen Erfassung von Folgeprozessen im Naturraum, die durch die Eingriffe der Gesellschaft ausgelöst werden.

Die Ergebnisse der Untersuchungen sind in Forschungsberichten niedergelegt; sie werden im folgenden kurz referiert.

1. Gesellschaftliche Anforderungen an die Naturraumpotentiale

Ein besonderer Aufgabenschwerpunkt für die Forschungen war im Berichtszeitraum die detaillierte und systematische Analyse der gesellschaftlichen Anforderungen an die Ausstattung und Differenzierung von Naturräumen in den unteren Dimensionsstufen. Als deren Voraussetzung, Grundlage und Ausgangspunkt konnten vorangehende Untersuchungen von E. Neef (1966, 1969) über das allgemeine Naturraum-Potential sowie von G. Haase (1973) öber dessen nutzungsbezogene Differenzierung in mehrere sog. partielle Naturraum-Potentiale herangezogen werden.[1] Mit diesen partiellen Naturraum-Potentialen erfaßt Haase (1973, S. 84) gesellschaftliche „Anforderungen an den Naturraum, die sich aus den verschiedenen Aspekten des Reproduktionsprozesses ergeben". Als volkswirtschaftlich relevante partielle Naturraum-Potentiale sind das Rohstoffpotential, das Bebauungspotential, das Ertragspotential, das Wasserdargebotspotential, das Entsorgungspotential sowie das Rekreationspotential anzuführen.

Erstes Ziel ihrer Untersuchung bildete die systematische Erfassung der von den einzelnen partiellen Naturraum-Potentialen beanspruchten bzw. belasteten Eigenschaften und Komponenten der natürlichen Ausstattung sowie der gesellschaftlich bewirkten Überprägung konkreter Landschaftseinheiten. Dieses Ziel ergab sich aus dem Erfordernis der geoökologischen Erkundung, für die Inhaltscharakteristik und Abgrenzung naturräumlicher Einheiten zur vorrangigen Verwendung von Kriterien mit hoher Praxisbezogenheit zu gelangen, um dem Erkundungsergebnis ein Höchstmaß an nutzungsspezifischer Transkriptionsfähigkeit, Vergleichbarkeit und Differenzierbarkeit unter räumlichem wie unter funktionalem Aspekt abzugewinnen.

Eine geeignete Methodik für die Lösung dieser Aufgabe wurde am Beispiel des Bebauungspotentials erarbeitet und erstmalig in einer Veröffentlichung von K.-D Jäger & K. Hrabowski (1973) anläßlich eines RGW-Symposiums in Smolenice (ČSSR) einem weiteren Kreis von Fachvertretern vorgelegt. Von diesen methodischen Erfahrungen ausgehend, erfolgten vergleichbare Untersuchungen seither auch für andere partielle Naturraum-Potentiale, und zwar teils innerhalb der Dresdener Arbeitsgruppe (besonders durch I. Hartsch für das Rekreationspotential), teils in

---

[1]) E. Neef (1966), Zur Frage des gebietswirtschaftlichen Potentials, in: Forschungen und Fortschritte *40*, S. 65—70 (Berlin); ders. (1969), Der Stoffwechsel zwischen Gesellschaft und Natur als geographisches Problem, in: Geographische Rundschau *21*, S. 453—459 (Braunschweig); G. Haase (1973), Zur Zielstellung geoökologischer Forschungsarbeiten unter landeskulturellem Aspekt, in: Informacionnyj bjulleten' o naučno-issledovatel'skoj probleme SÄV-8. VI, *2*, S. 77—89 (Brno).

anderen Forschungseinrichtungen und Betrieben der volkswirtschaftlichen Praxis in enger Zusammenarbeit mit der Arbeitsgruppe. Insgesamt erstreckten sich die bisherigen Bearbeitungen partieller Naturraum-Potentiale auf das Bebauungspotential, das biotische Ertragspotential, das Rekreationspotential, das Wasserdargebotspotential und das mineralische Rohstoffpotential. Für das Entsorgungspotential, mit dessen Analyse nicht nur in praktischer Hinsicht, sondern auch unter theoretischem Aspekt besondere Probleme verbunden sind, wurde deren Durchführung bislang lediglich konzipiert. Ungeachtet der Unterschiede im Bearbeitungsstand belegen jedoch die bisherigen Untersuchungen die Realisierbarkeit eines vergleichbaren methodischen Vorgehens bei der Analyse unterschiedlicher partieller Naturraum-Potentiale.

Für jedes der genannten partiellen Naturraum-Potentiale geht das Verfahren von einer Liste der jeweiligen Anforderungskriterien aus, deren Gruppierung und Anordnung in allen Fällen einem vergleichbaren Schema folgt:

1. Potentialspezifisch leistungsbestimmende Anforderungen an die Flächenausstattung;
2. Schad- und Störanfälligkeit
2.1. der potentialspezifisch leistungsbestimmenden Flächenausstattung,
2.2. der aktuellen oder potentiellen potentialspezifischen Flächennutzung;
3. Potentialspezifische Erschließungsfähigkeit der Flächen
3.1. durch Verbesserungsfähigkeit der flächenwirksamen Nutzungsprozesse (z. B. bei Einsatz technischer Hilfsmittel),
3.2. durch Verbesserungsfähigkeit der nutzungswirksamen Flächenausstattung (z. B. Flächenmelioration).

Die Zusammenhänge der so gruppierten Anforderungskriterien eines partiellen Naturraum-Potentials mit den Eigenschaften der natürlichen Ausstattung sowie der gesellschaftlich bewirkten Überprägung konkreter Landschaftseinheiten werden in einer ersten Bearbeitungsstufe der Analyse mit Hilfe eines zunächst qualitativ erarbeiteten Beziehungsschemas erfaßbar, das als „Anforderungsbild" bezeichnet wird. In diesem Beziehungsschema ist, konkretisiert für das jeweilige partielle Naturraum-Potential, die Liste seiner Anforderungskriterien als Leitspalte enthalten, während in der Kopfzeile, gruppiert nach Partialkomplexen, alle natürlich und gesellschaftlich bedingten Eigenschaften des landschaftlichen Dargebotes enthalten sind, für die Abhängigkeitsbeziehungen zu einem oder mehreren Anforderungskriterien erkennbar sind. Damit wird es möglich, derartige Abhängigkeitsbeziehungen, für jedes der partiellen Naturraum-Potentiale gesondert, jeweils in einem übersichtlichen Schema zu erfassen und darzustellen. Einheitliche Regeln nicht nur für die Anordnung der Anforderungskriterien in der Leitspalte, sondern auch für diejenige der Partialkomplexe der Landschaftsausstattung in der Kopfzeile der Beziehungsschemata ermöglichen deren Vergleichbarkeit auch für unterschiedliche partielle Naturraum-Potentiale. In dieser Bearbeitungsstufe werden die erkennbaren Zusammenhänge in jedem der Schemata zunächst generell durch eine einheitliche Signatur gekennzeichnet, d. h. Relevanzunterschiede zwischen den einzelnen Relationen kommen in dem so gewonnenen qualitativen Beziehungsschema noch nicht zum Ausdruck. Diese Bearbeitungsstufe bildet jedoch die Voraussetzung für die Ableitung differenzierterer Aussagen.

Eine systematische Weiterführung der Analyse für ein einzelnes partielles Naturraum-Potential wurde über die beschriebene Bearbeitungsstufe hinaus bisher nur für das Bebauungspotential erreicht. Dabei waren in einer zweiten Bearbeitungsstufe die im qualitativen Beziehungsschema enthaltenen Relationen zwischen potentialbezogenen Anforderungskriterien und landschaftlicher Ausstattung mehrereren Relevanzstufen zuzuordnen, bei deren Aufstellung sowohl die technologische Bedeutung als auch die mittlere Kostenwirksamkeit der erfaßten Zusammenhänge Berücksichtigung erforderte (Tab. 1).

Tabelle 1

Relevanzstufen für die numerische Bearbeitung von Anforderungsbildern in bezug auf den Beispielfall des Bebauungspotentials

Relevanzstufe	Bedeutung für die standörtliche Anpassung baulicher Maßnahmen: Technologische Bedeutung	Bedeutung für die standörtliche Anpassung baulicher Maßnahmen: Kostenrelevanz	Beispielfall oder mögliche Überlegung im Bauwesen
0	Ohne Beziehungen zur Bautechnologie	Ohne Kostenwirksamkeit	
1	Auswirkungen auf die Bautechnologie (Stufen 1–3)	Gering und im Regelfall zu vernachlässigen	Standortalternative von Sander- u. Geschiebemergelfläche
2		Erheblich, jedoch im Regelfall (volkswirtschaftlich) vertretbar	Inkaufnahme hoher Wasserhaltungskosten im Baugeschehen bei Industriebau, dafür Anschluß an Schiffahrtsweg nach Inbetriebnahme
3		Sehr erheblich, nur in Sonderfällen (volkswirtschaftlich) vertretbar	Fernsehturm und Metro trotz hoher Erdbebengefährdung in Taškent (UdSSR)
(4)	Technologische Realisierung mit vertretbarem Risiko unmöglich	Kosteneinsatz (technologisch und volkswirtschaftlich) nicht vertretbar	Atomkraftwerk auf der San-Andreas-Spalte (USA). Bemerkung: Für Mitteleuropa und damit auch für die DDR sind die Bedingungen dieser Relevanzstufe praktisch nicht gegeben. Für die vorliegende Untersuchung kommt ihr somit lediglich theoretische Bedeutung zu.

Die gesellschaftliche Flächennutzung im Territorium ist stets mit der Inanspruchnahme partieller Naturraum-Potentiale verbunden, jedoch erfolgt diese Inanspruchnahme für unterschiedliche territoriale Teilstrukturen in unterschiedlichen Kombinationen von partiellen Naturraum-Potentialen. Umgekehrt wird ein und dasselbe partielle Naturraum-Potential durch die Nutzungsziele unterschiedlicher territorialer Teilstrukturen beansprucht, so das Bebauungspotential für die Nutzungsziele von Industrie, Landwirtschaft, Siedlungs- und Erholungswesen, Verkehrswesen, technische Kommunikationssysteme (Rohr- und Leitungstrassen) sowie Maßnahmen der Wasserversorgung (wasserwirtschaftlicher Speicherbau). Das beschriebene Anforderungsbild ist jedoch sowohl in der qualitativen als auch in der numerischen Bearbeitungsstufe als Beziehungsschema unabhängig von seiner jeweiligen Kombination mit anderen partiellen Naturraum-Potentialen und deren Anforderungsbildern in den unterschiedlich zusammengesetzten Anforderungskomplexen verschiedener territorialer Teilstrukturen. So bestehen beispielsweise die Anforderungsbeziehungen des Wasserdargebotspotentials an die Naturraumausstattung unabhängig davon, ob das gewinnbare Wasser im Rahmen der territorialen Teilstrukturen für Industrie, Landwirtschaft oder Siedlung genutzt wird bzw. werden soll. Die partiellen Naturraum-Potentiale mit ihren Anforderungsbildern gehen also als formal gleichbleibende Bausteine in die Anforderungskomplexe verschiedener territorialer Teilstrukturen ein. Andererseits erhalten die für ein bestimmtes partielles Naturraum Potential kennzeichnenden Anforderungskriterien in Abhängigkeit von der jeweiligen Zuordnung der Potentialbeanspruchung zu unterschiedlichen territorialen Teilstrukturen auch unterschiedliche Relevanz. So bestehen im Beispielfall des Bebauungspotentials unterschiedliche Ansprüche an die als Anforderungskriterium des Bebauungspotentials bedeutsame Standsicherheit etwa einerseits im Verkehrsbau und andererseits im Industriebau, wobei in diesem Beispiel als Besonderheit des Verkehrsbaues der Umstand wesentlich ist, daß bei Verkehrsbauten der Baugrund zwar überwiegend verhältnismäßig geringen, dessen ungeachtet aber nicht nur dynamischen, sondern zugleich auch bewegten und überdies ständig wechselnden Belastungen ausgesetzt wird. Industrie-, Landwirtschafts-, Siedlungs- und Verkehrsbau sind deshalb Beispiele unterschiedlicher Anforderungsziele für das Bebauungspotential, denen für das biotische Ertragspotential im gleichen Sinne besonders Land- und Forstwirtschaft an die Seite zu stellen sind.

Die unterschiedliche Relevanz der Anforderungskriterien eines partiellen Naturraum-Potentials für dessen unterschiedliche Anforderungsziele konnte nach dem bereits beschriebenen Vorbild der Bearbeitung von Anforderungsbildern abgeleitet werden: Zunächst waren die bestehenden Zusammenhänge in einem Beziehungsschema qualitativ zu erfassen und nachfolgend den bereits charakterisierten Relevanzstufen zuzuordnen.

Eine erste systematische Erfassung von Anforderungszielen für die derzeit bekannt gewordenen partiellen Naturraum-Potentiale versuchten an Hand einer tabellarischen Übersicht in einem zur Veröffentlichung vorbereiteten Vortrag bereits D. Kopp & K.-D. Jäger (1973).

Die multiplikative Verknüpfung beider Wertungsansätze vermittelt schließlich die Möglichkeit, zu jedem der partiellen Naturraum-Potentiale für die einzelnen Dargebotsmerkmale der Landschaftsausstattung deren Relevanz für konkrete Anforderungsziele differenziert zu ermitteln. Somit bietet eine dritte Bearbeitungsstufe für die Anforderungsbilder die Gelegenheit, die in den vorangehenden Entwürfen zunächst noch enthaltene Redundanz unter Beibehaltung der besonders wesentlichen Zusammenhänge weitgehend zu reduzieren und für die Inhaltscharakteristik und Abgrenzung naturräumlicher Einheiten die Kriterien mit der relativ höchsten Relevanz für die gesellschaftliche Praxis abzuleiten. Die bei diesem Verfahren gewonnene Auswahl von Kartierungsmerkmalen stellt einen unmittelbaren Beitrag zur „Rahmenmethodik für die geotopologische und geochorologische Naturraumerkundung" dar. Sie erfordert und ermöglicht einen kritischen Vergleich mit den bisher verfügbaren und angewandten methodischen Möglichkeiten, Verfahren und Erfahrungen der naturräumlichen Erkundung in den unteren Dimensionsstufen sowie hinsichtlich der gesellschaftlich bewirkten Anteile an der Landschaftsausstattung eine Überprüfung der Verfügbarkeit und Quellen benötigter Daten. Umgekehrt ergeben sich Ansätze für eine mehr systematische Ermittlung der Potentialeigenschaften bereits auskartierter naturräumlicher Einheiten, denen in einem Teil des Beispielgebietes nördlich von Dresden K. Mannsfeld (1974) nachging. Schließlich sind im Beispielfall des Bebauungspotentials mit der differenzierten Analyse von Anforderungsbeziehungen zur natürlichen und gesellschaftlichen Flächenausstattung auch quantitativ handhabbare Ansätze für eine fundierte Methodik zur Inwertsetzung naturräumlicher Gegebenheiten gewonnen worden, die eine erste systematische Auswertung in Zusammenarbeit mit der Dresdener Arbeitsgruppe im Rahmen eines Forschungsberichtes von K. Hrabowski (1974) im VEB Baugrund Berlin erfuhren.[2]

Die besonderen Erfahrungen der Dresdener Arbeitsgruppe bei der Analyse partieller Naturraum-Potentiale begründeten im Rahmen der Zusammenarbeit mit dem Geographischen Institut der Akademie der Wissenschaften der DDR die Regelung, wonach im Auftrage dieses Institutes die weitere Bearbeitung und Auswertung von Anforderungsbildern in verschiedenen Forschungseinrichtungen der DDR durch Angehörige dieser Arbeitsgruppe koordiniert und angeleitet werden soll. Ein entsprechender Arbeitskreis, der geeignete Fachvertreter aus mehreren Akademie- und Hochschuleinrichtungen sowie aus volkseigenen Betrieben und staatlichen Dienststellen zusammenführt, konstituierte sich am 11. 6. 1974 in Leipzig.

Im Rahmen dieser Koordinierungsaufgabe wurde 1974 eine „Regelempfehlung für die formale Gestaltung von sogenannten Anforderungsbildern partieller Naturraum-Potentiale" in der Dresdener Arbeitsgruppe entwickelt und den mitarbeitenden Forschungseinrichtungen übermittelt. Ebenfalls im Rahmen dieser Koordinierungsaufgabe erfolgte in der Arbeitsgruppe die Zusammenstellung und kritische Auswertung einer Stichwortkartei zur Erfassung der bisher angewandten Terminologie bei der

---

[2]) K. Hrabowski (1974), Untersuchungen zur aktuellen und prognostischen Praxisrelevanz des Anforderungsbildes für das Bebauungspotential unter besonderer Berücksichtigung technologischer und ökonomischer Gesichtspunkte. Forschungsbericht für das Geographische Institut der AdW der DDR, Maschinenschrift, unveröffentlicht, Oktober 1974.

Untersuchung von partiellen Naturraui-Potentialen und deren Anforderungsbildern. Diese Stichwortkartei veranlaßte — unter Einbeziehung kompetenter Fachvertreter aus verschiedenen kooperierenden Einrichtungen — eine Serie klärender Diskussionen zu bestimmten Unklarheiten im bisherigen Sprachgebrauch und ermöglichte schließlich noch 1974 deren weitgehende Bereinigung.

2. Untersuchungen zur landeskulturellen Ausstattung

Diese Arbeiten (s. Jahrbuch 1971/72, S. 74) waren zunächst auf eine hinreichende Testung des topographischen Kartenwerkes 1:10000 (Ausgabe für die Volkswirtschaft = AV-Karte) bezüglich ihres Quellenwertes orientiert. Die im Teilabschlußbericht vom 30. V. 74 mitgeteilten Ergebnisse legen die Methoden der Materialgewinnung, den Katalog der ableitbaren Parameter sowie Beziehungen der landeskulturellen Ausstattung zu den naturräumlichen Teilkomplexen dar. Da die Karte nicht alle Elemente der landeskulturellen Ausstattung abbilden und über die funktionalen Beziehungen keine Auskunft geben kann, müssen zusätzliche Informationsquellen herangezogen werden. Diese wurden in einer Liste zusammengestellt und ihre Auswertung in Angriff genommen.

Die Ergebnisse erlauben eine Reihe von Schlußfolgerungen, die neben der Einschätzung des Kartenwerkes als Materialgrundlage besonders den Grundansatz des Forschungsweges betreffen und deshalb hier stärker hervorzuheben sind:

1. Das gewonnene landeskundliche Datenmaterial ist nur sinnvoll auswertbar, wenn es mit gesellschaftlichen Nutzungszielen oder Funktionen in Beziehung gesetzt wird. Anderenfalls sind nur formalstatistische Auswertungen möglich, die mehrdeutige oder sehr allgemeine Aussagen ergeben. Es kommt daher darauf an, die Bezugssysteme für die verschiedenen Nutzungsformen oder für definierte gesellschaftliche Funktionen oder Prozesse klarzustellen, in denen die einzelnen landeskulturellen Elemente oder ihre Gruppierungen wirksam werden. Dabei dürften zentrierte, d. h. auf das konkrete Siedlungsnetz bezogenen Untersuchungen notwendig sein, um zu vermeiden, daß die Beziehungen zwischen Naturraum und landeskultureller Ausstattung allein auf die Flächennutzung reduziert werden.

2. Natur und Landeskultur sind zwei verschieden determinierte Systeme, die sich mit Wechselwirkungen verschiedener Art überlagern. Die daraus resultierenden Interferenzen können ohne vorherige Kenntnis der einzelnen Prozesse nicht interpretiert werden.

3. Die Verschiedenartigkeit dieser beiden Systeme verlangt zwei voneinander getrennte Raumgliederungen, nämlich eine Naturraumgliederung und eine Raumgliederung nach landeskulturellen Gebietseinheiten. Letzterer gebührt dabei insofern die Vorrangstellung, als sie das ständige Einwirken der Gesellschaft auf den Naturraum verkörpert. Mit der Dynamik des gesellschaftlichen Geschehens wird sich daher auch die landeskulturelle Ausstattung ändern, dementsprechend sind raschere Grenzverschiebungen zu erwarten. Demgegenüber bezieht sich die Naturraumgliederung allein auf den relativ stabilen naturgesetzlichen Seinsbereich als potentiellen Produktions- und Lebensbereich der Gesellschaft.

Naturräume und landeskulturelle Gebietseinheiten koinzidieren daher nur teilweise miteinander. So können mehrere Mikrochoren eine landeskulturelle Gebietseinheit verkörpern oder aber auch Teile einzelner Mikrochoren verschiedenen landeskulturellen Einheiten angehören (s. Abb. 1). Manche gesellschaftliche Funktionen bevorzugen sogar die Kontraste an naturräumlichen Grenzsäumen, wie z. B. in vielen Fällen die Erholung.

4. Obwohl die landeskulturelle Entwicklung eine ganze Reihe enger Beziehungen zur Ausstattung der Naturräume erkennen läßt, darf nie außer acht gelassen werden, daß die Verschiedenartigkeit der Teilsysteme mit besonderer Sorgfalt beachtet werden muß, soll nicht geodeterministischen Fehldeutungen Vorschub geleistet werden.

5. Die typologische Bearbeitung landeskultureller Gebietseinheiten steht erst in den Anfängen und bereitet wegen der Unterschiedlichkeit der Elemente und der Diversität der Zusammenhänge erhebliche Schwierigkeiten. Es muß geprüft werden, ob eine solche Typisierung zu einem wirklichen Erkenntnisgewinn beiträgt. Bisher ist jedenfalls noch keine praktikable Grundlage für die Typenbildung gefunden worden. Es empfiehlt sich, diese Frage mit der Erörterung der Anforderungsbilder zu verbinden. Das gleiche gilt für die Anordnungsmuster landeskultureller Elemente, die nur dann sinnvoll einsetzbar erscheinen, wenn sie auf funktionale und historische Zusammenhänge zurückgeführt werden können.

6. Ein Nachteil der kartographischen Auswertung ist, daß die Elemente der landeskulturellen Ausstattung zwar nach Zahl und Standort nachweisbar sind, über ihre Leistung aber nichts ausgesagt werden kann. Die qualitativen Kennzeichnungen sind allgemein schwerer zu beschaffen, jedoch unerläßlich für die Kenntnis der funktionalen Beziehungen. Der Aufwand für die Auswertung der Karten, auch wenn sie zunächst nur für die Speisung einer Datenbank vorgesehen ist, dürfte in keinem günstigen Verhältnis zur Nutzbarkeit stehen. Erst die Kombination mit Angaben über die Qualität der Funktionsfähigkeit landeskultureller Elemente gestattet praktisch anwendbare Schlußfolgerungen.

7. Die Trennung naturräumlicher und landeskultureller Ausstattung ergibt sich auch aus der Tatsache, daß landeskulturelle Maßnahmen oftmals für einen oder mehrere Wirtschaftszweige recht förderlich sein können, für andere jedoch nachteilig, ja in extremen Fällen empfindlicher Störfaktor sind. Aus diesem Grunde sollte die Interferenzanalyse stärker herangezogen werden als bisher.

8. Die landeskulturelle Ausstattung eines Gebietes gibt zu jedem beliebigen Zeitpunkt (u. a. auch dem der Kartenaufnahme) ein Augenblicksbild einer Entwicklung, den historisch gewordenen Zustand, wider. Er enthält Elemente, die sich aus früheren gesellschaftlichen Entwicklungsphasen ergeben haben und im Zuge der weiteren Entwicklung, zeitlich recht unterschiedlich, dem Umbau oder Abbau anheimfallen, abhängig von der Intensität, mit der die Entwicklung der Produktivkräfte, insbesondere auch der Technik, in den einzelnen Landschaftsgebieten vorangetrieben wird. Er weist aber auch Elemente aus, die neue Entwicklungen anzeigen, sich weiterhin ausweiten und vervollständigen werden. Die Interpretation der landeskulturellen Ausstattung muß daher immer im Hinblick auf die gesellschaft-

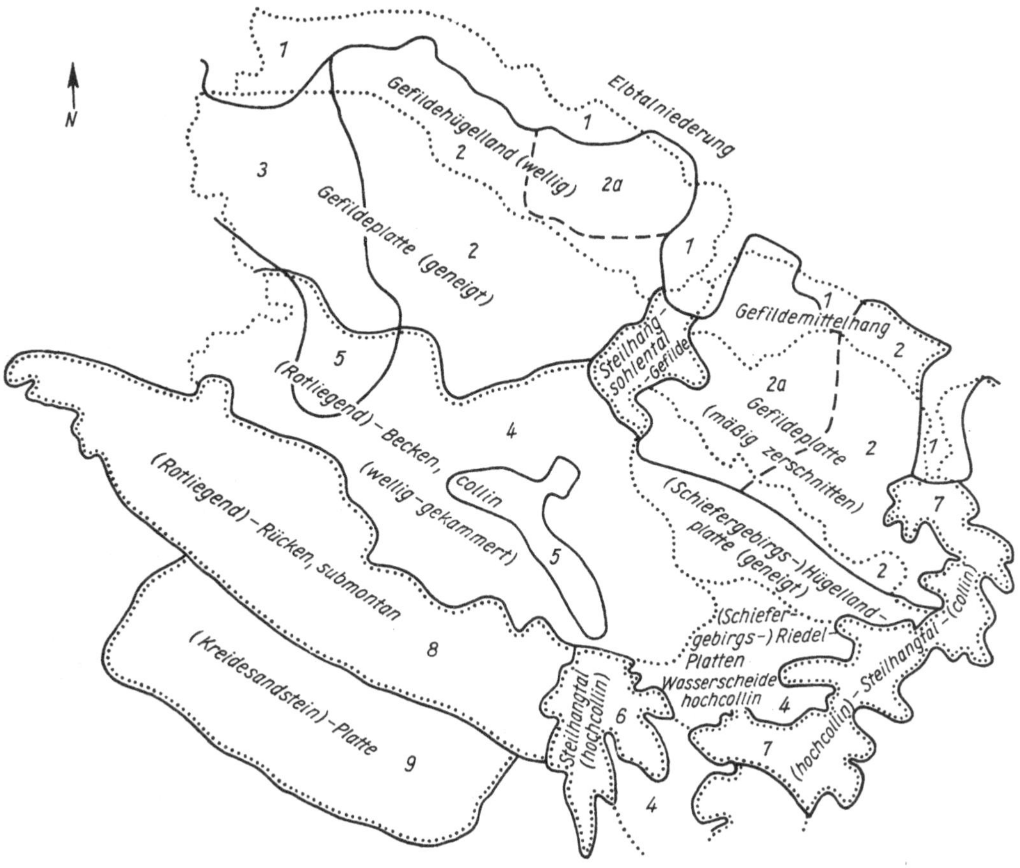

Abb. 1. Beziehungen zwischen landeskulturellen Gebietseinheiten und Naturraumeinheiten (Mikrochoren)

.... Mikrochorengrenzen 1:100000

Schrift = Mikrochoren-Kurzbezeichnung

——— ungefähre Grenzen landeskultureller Gebietseinheiten bzw. —-— Untereinheiten

Informative Bezeichnung dieser Einheiten (noch keine terminologische Konzeption):

1 = Stadtrandgebiet (Wohnbebauung, durchsetzt von Gärten u. etwas Kleinindustrie)
2 = reiche Agrargebiete (geringe Grünlandanteile)
2a = z. Z. bereits überwiegend in Intensivobstbaugebiete, durchsetzt von Hopfen- u. Rosenkulturen, umgewandelt
3 = reiche Agrargebiete mit stärkerer Wohnbebauung und etwas Industrie
4 = mittlere Agrargebiete (mäßige Grünlandanteile)
5 = mittlere Agrargebiete mit Lokalzentren bzw. Sekundärzentren
6 = Grünland-Waldtal (steilhängig)
7 = Grünland-Waldtal (steilhängig), mäßig industriell durchsetzt u. wichtiger Verkehrsträger
8 = mäßiges (Wald)-Grünland-Agrargebiet
9 = Nadelwaldgebiet

liche Dynamik erfolgen. Nur dann wird die Auswertung der landeskulturellen Gebietsanalyse den Forderungen der Gesellschaft entsprechen, wenn sie das Zustandsbild nach der historischen Stellung ihrer Elemente und Elementgruppierungen differenziert und den Tendenzen des Ausbaus hinreichend Aufmerksamkeit widmet.

3. Intensitätsstufen der geochorologischen Erkundung

Die bisherigen Auffassungen zur Theorie der geographischen Regionen gingen davon aus, daß zwei methodische Grundprinzipien anwendbar sind, nämlich

1. die geochorologische Erkundung auf der Grundlage topologischer Aufnahmen, also ein induktives Verfahren, oft als „Weg von unten" bezeichnet;
2. die geochorologische Gliederung auf der Basis der Interpretationen übergreifender geowissenschaftlicher Merkmale, also ein deduktives Verfahren und in der Analogie auch „Weg von oben" genannt.

Die Darstellung geographischer Gebiete unterschiedlicher Dimension muß beide Wege miteinander verknüpfen, d. h. sich sowohl der Ableitung der Gebietschrakteristika aus der Differenzierung größerer Einheiten als auch Kennzeichnung charakteristischer Merkmale aus der induktiven Forschung bedienen. Feste Regeln hierfür schienen im Rahmen physisch-geographischer Deskription nicht erforderlich. Im Blickfeld landeskultureller Zielsetzung für verschiedene gesellschaftliche Anwendungsbereiche und Aufgabenstellungen, die eine einheitliche Methodik der Naturraumerkundung voraussetzen, mußten jedoch differenzierte Wege der chorischen Naturraumerkundung festgelegt werden. Letztlich war die Frage entscheidend, in welchem Maße von der Anforderungsseite her allgemeinere Auswertungsbefunde oder unmittelbare Kartierungsergebnisse gefordert werden und in welchem Umfang sich diese Forderungen praktisch realisieren lassen.

Im Jahre 1973 wurde eine Folge von *vier Intensitätsstufen der chorischen Naturraumerkundung* aufgestellt. Sie ist nach einer gewissen Präzisierung von allen Mitgliedern des Forschungsverbandes übernommen worden und hat damit Eingang in die entsprechenden Forschungsdokumente gefunden.

Für die einzelnen Stufen kommen folgende Maßstabsbereiche als Schwerpunkte ihrer Anwendung in Betracht:

Intensitätsstufe	A	1: 5000—1: 25000
„	B	1: 10000—1: 50000
„	C	1: 50000—1:200000
„	D	1:200000—1:1 Mio.

Gleichzeitig wurde geprüft, welche praktischen Anwendungsmöglichkeiten den einzelnen Intensitätsstufen zukommt. Sie sind in erster Linie von den Forderungen nach Informationen der einzelnen Anwendungsziele und damit stark vom Maßstab abhängig. Tabelle 2 zeigt eine Übersicht über den Informationsgehalt der Intensitätsstufen bezüglich verschiedener Aussagen der Naturraumcharakteristik. Sie läßt erkennen, daß sowohl nach der kartographischen Aufnahme und Darstellung wie nach der Verarbeitung des anfallenden Datenmaterials die Stufen A und B sehr erheblich

Tabelle 2

Informationsgehalt der vier Intensitätsstufen zur geochorologischen Erkundung

			Inhaltliche Merkmale	Stufe A	Stufe B	Stufe C	Stufe D
Inventar	komplexe Grundlagen		Erfassung von Öko/Physiosystem-Typen	●	●	○	—
			Kartierung dieser Typen als Öko/Physiotope	●	●	—	—
	Kennzeichnung der Partialkomplexe	Relief	Oberflächenformen und -eigenschaften	●	●	●	○
			(Formtypen, Neigungsverh., Zersch.grad, Taldichte usw.)				
		Geol. Bau und Boden	Gesteinsverhältnisse und Substrate	●	●	⊗	○
			Bodenformen, Trophieverhältn.				
		Wasser und Wasserhaushalt	Wasserhaushalt d. Bodens	●	○	•	•
			Wasserhaushalt d. Oberflächengewässer	⊗	•	•	—
			Grundwasserverhältnisse	○	○	•	•
		Klima	Makroklima	⊗	⊗	⊗	○
			Mesoklima	○	•	•	—
		Bios	Vegetationsgliederung	⊗	•	•	•
Muster			Frequenz, Deckungsgrad, Flächengröße, Verbreitungsdichte, D/F Quotient usw.	●	●	—	—
untere chorische Dimension			Elementargefüge	●	●	○	—
			Mikrochoren	●	●	⊗	•

● = vollständig und umfassend ⊗ = ausreichend, teilw. umfassend ○ = näherungsweise, teilw. ausreichend • = grobe Einschätzung, teilweise näherungsweise — = Bestimmung fehlend von den Stufen C und D unterschieden sind. Das soll an zwei Beispielen erläutert werden.

Bearbeitungen auf vollständiger topologischer Grundlage (Stufe A) sind vornehmlich zur direkten Anwendung und Unterstützung von Projektierungsvorhaben geeignet. Sie basieren die Forstliche Standorterkundung als Vorstufe der Forsteinrichtung und zahlreiche Nutzungsverfahren in der Forstwirtschaft sowie Naturraumerkundungen zur Vorbereitung von Meliorationsvorhaben, Landschaftspflege-Vorhaben, Generalbebauungsplänen von Großstädten usw. auf Untersuchungen in der Stufe A.

Mit der deutlichen Maßstabzäsur zwischen den Stufen A und B einerseits und der Stufe C ist die starke Betonung des Überblickscharakters der letztgenannten verbunden.

Mit Erkundungen der Stufe C lassen sich in den Maßstäben 1:100000 und 1:200000 (als Herausgabemaßstab) vor allem Übersichten zur naturräumlichen Ausstattung und ihren Potentialeigenschaften für Kreise, Bezirke bzw. diesen Einheiten entsprechende Räume erstellen. Der Hauptnutzungsbereich liegt damit in Planungsbehörden von Räten der Bezirke und Kreise, der Räte für Land-, Forst- und Nahrungsgüterwirtschaft, der Wasserwirtschaftsdirektionen und Oberflußmeistereien, von Meliorationsverbänden und Verwaltungen von LSG bzw. großen NSG u. a. m.

Die Intensitätsstufe C der Naturraumerkundung eignet sich weiterhin für die Klärung wissenschaftlicher Fragen, wie z. B. der Analyse landschaftsgenetischer Zusammenhänge. Sie besitzt Bedeutung als Vorstudie zu komponentenbezogenen Teilanalysen der Naturraumausstattung, z. B. für morphographisch-morphogenetische Karten oder Bodenkartierungen. Sie wird mit großem Gewinn als Vorerkundungsstufe bei großmaßstäbigen Naturraumuntersuchungen eingesetzt.

Grundsätzlich lassen sich in allen Stufen der chorischen Erkundung enge Beziehungen zu gesellschaftlich notwendigen Aufgaben erkennen. Besonders die Kombination mit wichtigen Anwendungsbereichen wie

- Landeskulturelle Entwicklung des Territoriums
- Nutzung von Naturressourcen (in Bereichen des Bergbaus, der Land- und Forstwirtschaft, der Wasserwirtschaft, des Erholungswesens usw.)
- Umweltforschung und Umweltschutz
- Entwicklung gesamtstaatlicher Planungsunterlagen, vor allem für Fragen der Siedlungsstruktur, der Standortverteilung der Produktivkräfte u. ä.

sind hierbei zu nennen.

Aus der Analyse bisheriger Erkundungspraxis in der Naturraumforschung und der Bestimmung von Intensitätsstufen ergaben sich auch Anhaltspunkte, in welchen Stufen zur raschen Befriedigung der erwähnten gesellschaftlichen Anforderungen gearbeitet werden muß.

Zur Lösung der meisten Aufgaben wäre die Stufe B wegen ihres hohen Informationsgehaltes das am besten geeignete Verfahren, aber aus Kapazitätsgründen läßt sich das nur etappenweise realisieren, da das topologische Inventar hinreichend erkundet sein muß. Für gesamtstaatliche Übersichten und mittelmaßstäbige Karten sind Untersuchungen nach Stufe C angemessen. Die Dresdner Arbeitsgruppe hält nach Erfahrungen aus den von ihr bearbeiteten Gebieten eine Kombination der Stufen C (als Vorerkundung) und B (als Intensivierungsstufe) für einen sehr zweckmäßigen und erstrebenswerten Hauptweg.

Mit dieser Differenzierung der methodischen Wege wird nicht nur eine Grundlage für die zu erarbeitende Rahmenmethodik für das Gesamtgebiet der DDR bereitgestellt, sondern zugleich ein höheres Maß von Praxisrelevanz erreicht und das Angebot an geoökologischen Informationen für Probleme der Auseinandersetzung von Gesellschaft und Natur präzisiert.

4. Zur Frage der „Nebenwirkungen"

Jeder Arbeitsprozeß, der von der Gesellschaft im geographischen Raum durchgeführt wird, führt zu Änderungen der Größen des Naturhaushaltes, denn jeder gesellschaftliche Arbeitsprozeß ist mit Stoffzufuhr oder Stoffentnahme, sowie mit dem Einsatz von Energie verbunden. Diese Änderungen von Größen des Stoff- und Energiezustandes sind in jedem Falle zu verzeichnen. Sie bedeuten einen *Impuls*, der auf das Natursystem ausgeübt wird. Der Impuls (oder besser: die Impulse, denn meistens sind mehrere Veränderungen von Haushaltgrößen beteiligt) kann langsam oder rasch, ständig oder intermittierend erfolgen. Auf jeden Fall muß sich für jeden geoökologischen Wirkungsimpuls eine konkrete Aussage nach Art und Wirkungsweise finden lassen, indem die einzelnen Veränderungen stofflicher oder energetischer Art bestimmt werden.

Auf einen Impuls reagiert das Natursystem gemäß seiner Struktur und Dynamik. Man könnte von einer „*Impulsantwort*" sprechen. Eine solche liegt auch vor, wenn ein Natursystem dank seines Puffervermögens den Impuls zu einem gewissen Grade abfangen kann. In vielen Fällen, in denen ein solches Pufferungsvermögen nicht vorhanden ist oder überfordert wird, werden Folgeprozesse ausgelöst, wobei die freigesetzten oder hinzugefügten Kräfte Art und Tendenz der Folgeprozesse bestimmen, während die Struktur der topischen und chorischen Einheiten die Bedingungen schafft, unter denen die einzelnen Folgeprozesse im Territorium ablaufen.

In der Regel sind diese Folgeprozesse, zumindest in ihrem Initialstadium, verdeckt oder unsichtbar. Erst an den Ergebnissen, Zustandsänderungen des Natursystems, erkennt man schließlich den eingetretenen Wandel. Im allgemeinen wird der gesellschaftliche Eingriff als Gesamtvorhaben ohne Aufgliederung in einzelne als Impulse wirkende stoffliche und energetische Eingriffe dargestellt, und ebenso wird das schließlich zutagetretende Ergebnis als Indikator für eingetretene Veränderungen beschrieben, ohne im einzelnen auf die auslösenden Momente und die zugrundeliegenden Folgeprozesse einzugehen. Es handelt sich also um eine recht grobe Form der input-output-Beziehungen. Will man aber die Nebenwirkungen gesellschaftlicher Aktivitäten im Territorium vorausschauend einschätzen und die zu erwartenden Reaktionen des Natursystems als naturgesetzlich notwendig zur Vorhersage der Nebenwirkungen durch Folgeprozesse einsetzen, so muß der Ablauf der Prozesse von ihrem Initialstadium an verfolgt werden. Daraus ergibt sich eine Konzeption bzw. die Leitlinie einer Konzeption für die systematische Prognose von Folgeprozessen und deren Wirkungen.

Durch die Interpretation einer Reihe von Beispielen landschaftlicher Veränderungen, die auf Nutzungsumstellung zurückzuführen sind, ergeben sich Ansätze für den Weg der Analyse, die in mehreren Stufen erfolgen muß.

Für die erste Stufe der Analyse sind drei Informationen erforderlich,

1. über die Art der Impulse, die durch gesellschaftliche Arbeit ausgelöst werden, also Art und Menge der zugeführten oder entnommenen Stoffe bzw. von Energie,
2. über die Kräfte, die freigesetzt werden und die Folgeprozesse tragen, also z. B. Gravitationsprozesse bei Hangunterschneidungen, chemische Reaktionen usw.,
3. über die Bedingungen der Naturraumausstattung, unter denen sie verschiedene Formen oder Intensität annehmen und zur räumlichen Differenzierung beitragen.

Die zweite Stufe untersucht die einzelnen Folgeprozesse selbst, die insgesamt die Reaktion bzw. die Impulsantwort des Natursystems ausmachen, und deren Ergebnisse. Das muß unter drei Gesichtspunkten geschehen, nämlich

1. nach der Art und Intensität des einzelnen Vorgangs,
2. nach seiner räumlichen Einordnung in das Landschaftsgefüge unter Berücksichtigung der Lagebeziehungen und der Wirkungsfelder,
3. nach seinem zeitlichen Ablauf wie Gleichförmigkeit, Beschleunigung, aber auch nach Spätwirkungen, Summationseffekten usw.

Auf der dritten Stufe der Analyse sind komplexe Erscheinungen zu untersuchen, wie

1. die Kombination verschiedener Folgeprozesse unter dem Gesichtspunkt der Hemmung oder Verstärkung der Wirkungen,
2. die Interferenz der Wirkungsfelder im gegebenen Landschaftsraum einschließlich der räumlichen Übertragung der Störungen durch bewegte Medien (Wasser, Luft, Sickerwasserbewegung im Boden),
3. nach landschaftsbezogenen kritischen Schwellen, Belastungsgrenzen und ihrer Lokalisierung an bestimmten Örtlichkeiten oder bestimmten Arealen.

Eine solche Analyse der Folgeprozesse und ihrer Wirkungen bietet eine Möglichkeit, die eintretenden Potentialänderungen genauer zu bestimmen, die zwar in allgemeinen Worten immer wieder genannt, aber kaum einmal mit größerer Genauigkeit ermittelt werden. Das ist wiederum von größter Bedeutung für das Weiterverfolgen der Konsequenzen und den Aufbau weitreichender Konsequenzreihen oder Reaktionsketten, in deren Ablauf immer neue Entscheidungen gefordert werden. Die systematische Erforschung der Nebenwirkungen und Folgeprozesse und die Klarlegung weiterer gesellschaftlicher Konsequenzen in Form von weiteren Maßnahmen ergibt eine nützliche Grundlage für die Erfüllung der Forderung nach gesamtvolkswirtschaftlicher bzw. gesamtgesellschaftlicher Effektivität. Sie ist damit eine wichtige Entscheidungshilfe für langfristige landeskulturelle Maßnahmen.

Mit der gleichrangigen Berücksichtigung der geplanten Hauptwirkungen und ihrer Nebenwirkungen wird schließlich die dialektische Einheit aller Vorgänge, die den Stoffwechsel zwischen Gesellschaft und Natur ausmachen, zur Grundlage der theoretischen und praktischen Arbeiten der Landschaftsplanung. Einseitigkeit der Planvorbereitung, die zu Fehllösungen und später zu ökonomisch oder gesellschaftlich belastenden Regulierungsmaßnahmen führen können, werden vermindert, die Qualität der Entscheidungsgrundlagen für die Planungsorgane erheblich verbessert.

Nach dreijähriger Mitarbeit im Forschungsverband „Geoökologische Grundlagen für die Planung komplexer landeskultureller Maßnahmen" (Hauptauftragnehmer dieses Staatsplanthemas beim Ministerium für Wissenschaft und Technik ist das Geographische Institut der AdW), dürfte es angebracht sein, kurz Rechenschaft über positive und negative Wirkungen dieses Schrittes zur Vertragsforschung abzulegen.

Insgesamt überwiegen eindeutig die Vorteile. Will man die Notwendigkeit, daß die gesamte Arbeitsgruppe oder einzelne Mitarbeiter zusätzlich in Aufgaben eingespannt werden, die zeitweise erhebliche Belastungen darstellen, als negativ bezeichnen, so steht dem eine beträchtliche Erhöhung des Wirkungsgrades der eigenen Forschung gegenüber. Denn die Zusammenarbeit in speziellen zeitweiligen Arbeitsgruppen zur Klärung einzelner Fragen, wie zur Kartenanalytik (J. Bieler), zur Ausarbeitung der „Anforderungsbilder" (Dr. Jäger), zur Ausarbeitung größerer zentraler Forschungsberichte (Dr. Bernhardt, Dr. Mannsfeld, Dr. Jäger), und die monatlichen Diskussionen im Themenleiterkollektiv erbringt nicht nur eine große Zahl von Informationen, sondern zugleich eine sehr frühzeitige Kenntnis der Forschungsvorhaben, methodischen Ansätze und Ergebnisse anderer Institutionen, die am gleichen Forschungsgegenstand arbeiten. In der Regel werden die eigenen Forschungsergebnisse innerhalb weniger Monate zur Diskussion gestellt, und dadurch gewinnt nicht nur die eigene Forschungsarbeit an Sicherheit, sondern sie gelangt auch bald zur Auswertung durch einen relativ breiten Kreis von Fachleuten. Darüber hinaus sind alle organisatorischen Fragen der Verbindung zu anderen Forschungsstellen, zu staatlichen Organen, zu internationalen Organisationen und auch zur raschen Überleitung der Ergebnisse in die Praxis erheblich erleichtert. Dementsprechend haben sich die Kontakte der Arbeitsgruppe der Sächsischen Akademie der Wissenschaften ständig erweitert, sowohl innerhalb der DDR als auch in der internationalen bilateralen Zusammenarbeit und im Rahmen des RGW, Komplexthema 12, „Maßnahmen zum Schutz der Natur". So fand der Bericht über die „Anforderungsbilder" auf dem erweiterten RGW-Symposium in Smolenice (ČSSR) im Dezember 1973 großes Interesse und löste Kontakte zu dem Prager Zentralen Geologischen Institut aus, das Kartierungen von Umweltfaktoren durchführt. Noch in einer weiteren Richtung ist diese Erweiterung der Kontakte von besonderer Bedeutung. Immer mehr spezielle Wissensgebiete, deren Erkenntnisse geoökologisch von Interesse sind, werden angeregt, ihre Materialien für die Umweltforschung, speziell für die Differenzierung der geographischen Bedingungen in dem Stoffwechsel zwischen Gesellschaft und Natur, auszuwerten und teilweise durch Kooperation unmittelbar nutzbar zu machen. Das gilt z. B. für die Baugrundforschung, in besonderem Maße aber auch für die im Rahmen der Internationalen Quartär-Organisation INQUA durchgeführte Holozänforschung.

Die beiden Berichtsjahre haben auch in dieser Hinsicht einen merklichen Fortschritt gebracht, der von der Arbeitsgruppe „Naturhaushalt und Gebietscharakter" der Sächsischen Akademie der Wissenschaften durch die Aktivitäten mehrerer Mitarbeiter mit getragen worden ist. Zugleich hat diese Arbeitsgruppe damit zu ihrem Teil die im Statut der Akademie festgelegte Zusammenarbeit mit der Akademie der Wissenschaften der DDR realisiert.

Literatur

1. Forschungsberichte:

Neef, E., Bernhardt, A., Jäger, K.-D., und Mannsfeld, K.: Bericht über die Auswertemöglichkeiten großmaßstäbiger Naturraumerkundungen (bis 1:50000). (Verfaßt auf der Basis veröffentlichter und allgemein zugänglicher Arbeiten zur Diskussion der Möglichkeiten, topologische Inhalte für chorische Strukturen nutzbar zu machen)
31. 3. 1973, 24 S., 5 Tabellen.

Neef, E., Bernhardt, A., Hartsch, I., Jäger, K.-D., und Mannsfeld, K.: Bericht: Wege und Methoden zur Erfassung chorischer Strukturen im mittleren Maßstab einschließlich des Entwurfs von Rahmenbedingungen zur Kartierung im Berg- und Hügelland.
30. 6. 1973, 69 S., 7 Tabellen.

Neef, E., Bernhardt, A., Bieler, J., und Mannsfeld, K.: Zwischenbericht: Wege und Methoden zur Erfassung chorischer Strukturen im mittleren Maßstab — Vorlage von Beispielen aus dem Bezirk Dresden nach Weg B und Weg C.
30. 11. 1973, 17 S., 4 Anlagen (zusammen 58 S.), 3 Tab. und 2 Karten

Neef, E., Richter, H., Barsch, H., Haase, G. unter Mitwirkung von: Hubrich, H., Mannsfeld, K., Hentschel, P., und Kugler, H.: Beiträge zur Klärung der Terminologie in der Landschaftsforschung (vorbereitet für die Diskussion auf dem Symposium zu Inhalt und Objekt der komplexen Landschaftsforschung im Rahmen des RGW-Problems „Schutz der Ökosysteme (Biogeozönosen) und der Landschaft" vom 28. 11.—1. 12. 1973 in Smolenice/ČSSR).
Geogr. Inst. d. Akad. d. Wiss. d. DDR. Leipzig, November 1973.

Neef, E., Bernhardt, A., Bieler, J., Hartsch, I., Jäger, K.-D., und Mannsfeld, K.: Teilabschlußbericht: Untersuchungen zur Erfassung der landeskulturellen Ausstattung und deren Beziehungen zur Naturraumausstattung (im Rahmen der Aufgabenstellung B 1). Teil A: Analyse großmaßstäbigen topographischen Kartenmaterials (1:10000 — AVW-Karte) an Beispielen aus dem Bezirk Dresden
30. 6. 1974, 48 S., 3 Anlagen (zusammen 21 S.), 2 Abb., 5 Tabellen

Neef, E., Bernhardt, A., Jäger, K.-D., Hartsch, I., und Mannsfeld, K.: Konzeptionelle Studie: Untersuchungen zu landeskultureller Funktion und Dynamik chorischer Strukturen in theoretischer Sicht und an Hand von Beispielen.
30. 11. 1974, 17 S.

Hartsch, I.: Zur Charakteristik der gesellschaftlichen Anforderungen für das Rekreationspotential an das Naturraumdargebot.
Anlage zum Jahresabschlußbericht 1974, 10 S.

Haase, G., Mannsfeld, K., Jäger, K.-D.: Erfassung, Kennzeichnung und Kartierung von Naturraumtypen, Interpretation von Naturraum-Potentialen und Ermittlung von Bewertungsmöglichkeiten für volkswirtschaftlich bedeutsame Formen der Flächennutzung. F/E-Bericht B 1/Dezember 1974 des Forschungsverbandes „Geoökologische Grundlagen der sozialistischen Landeskultur".
Leipzig 1974, 123 S.

Bernhardt, A.: Konzeptionelle Vorarbeiten für den Aufbau von Landeskultur- und Naturraumkatastern.
Anlage 2 des F/E-Berichtes B 1/Dezember 1974
12 S., 1 Beilage (13 S.), 1 Abb.

Mannsfeld, K.: Erfassung von Naturraumtypen unter Einbeziehung von Elementargefügen im Gebiet nördlich von Dresden.
Anlage 5 des F/E-Berichtes B 1/Dezember 1974
9 S., 3 Beilagen (12 S.), 9 Tab., 3 Abb.

Bernhardt, A.: Erfassung von Naturraumtypen und Kennzeichnung von Mikrochoren nach dem Weg C im Beispielsgebiet südlich Dresden.
Anlage 6 des F/E-Berichtes B 1/Dezember 1974
9 S., 3 Tab., 4 Beispiele (10 S.).

Neef, E.: Untersuchungen zu landeskultureller Funktion und Dynamik chorischer Strukturen (Entwurf).
Anlage 8 des F/E-Berichtes B 1/Dezember 1974. — 9 S.

Jäger, K.-D., Hrabowski, K.: Analyse der gesellschaftlichen Anforderungen an den Naturraum.
Beispiel: Anforderungsbild zum Bebauungspotential und Untersuchungen zur Bewertung von Naturraumeigenschaften für die Bebauung.
Anlage 9 des F/E-Berichtes B 1/Dezember 1974.
27 S., 7 Tab.

Mannsfeld, K.: Vorschlag zur Bestimmung der potentialbezogenen Nutzungseignung für chorische Naturraumtypen.
Anlage 14 des F/E-Berichtes B 1/Dezember 1974.
14 S. mit 4 Tab., 3 Abb.

2. Veröffentlichungen:

Bernhardt, Arnd: Das mittlere Erzgebirge — eine geographische Komplexbetrachtung in einem Exkursionsprofil. In: Exkursionen um Aue. Geogr. Ges. d. DDR, Fachverband Schulgeographen, VI. Wiss. Arbeitstagung 7.—10. 5. 73 in Aue, S. 36—44.

Jäger, K.-D.: Holozänstratigraphische Befunde als Zeugnisse für den Landesausbau im slawischen und deutschen Mittelalter. In: Berichte über den II. Internationalen Kongreß für Slawische Archäologie, 3., 1973, S. 75—88 (Berlin).

Jäger, K.-D., und Hrabowski, K.: Zum Verhältnis von geoökologischer Erkundung zu nutzungsspezifischen Anforderungsbildern am Beispiel des Bebauungspotentials. In: IIIrd International Symposium „Content and Object of the Complex Landscape Research in the Protection and Formation of Human Environment" 16 S. mit 3 Beilagen, Bratislava 1973.

Jäger, K.-D., und Lange, E.: Symposium on Changes in the Palaeogeography of Valley Floors in the Holocene. In: Ethnographisch-Archäologische Zeitschrift, 14., 1973, S. 367—369, Berlin.

Lampadius, Felix: Schädigung des Waldes durch Waschmittelstaub. In: Mitt. d. forstl. Bundes-Versuchsanstalt Wien, H. 97, 1972, S. 451–471.

Lampadius, Felix: Beitrag zum Nachweis der Wertminderung des Waldes als Folge von Immissionswirkung. In: Abh. d. Sächs. Akad. d. Wiss. zu Leipzig, Math.-Nat. Klasse, Bd *52*, H. 3, 1974, 14 S., 4 Abb., 8 Tab.

Mannsfeld, Karl: Zu einigen geographischen Aspekten bei der Erforschung des Bodenwasserhaushaltes durch mehrjährige Messungen. In: Wiss. Zeitschrift d. TU Dresden, *22*, H. 1, 1973, S. 147–152.

Mannsfeld, Karl, zusammen mit Barthel, H., und Sandner, E.: Flächen gleicher Abflußbereitschaft bei sommerlichen Starkregen (dargestellt am Beispiel des Einzugsgebietes der Flöha im Erzgebirge). In: Peterm. Geogr. Mitt., *117*, H. 2, 1973, S. 107–116.

Ernst Neef

Reinigung von Abwässern mit Hilfe von Adsorbentien

Im ersten Teil des Berichtszeitraumes fanden die Arbeiten zur Synthese von Adsorbentien auf Organopolymer-Siloxan-Basis, die unter der Leitung von Prof. F. Runge angefertigt worden waren, ihren Abschluß. In einem Patent „Verfahren zur Herstellung von Adsorberharzen" kommt der Erfolg bei der Synthese und den anwendungstechnischen Möglichkeiten dieser Polymeren zum Ausdruck. Danach eignen sich mit Vinyltrimethoxysilan vernetzte organische Polymere, die zur Wasserstoffbrückenbildung befähigte funktionelle Gruppen enthalten, besonders zur Adsorption phenolischer Verbindungen bei einfacher Wasserdampf- oder Lösungsmittelregenerierung.

Die Möglichkeit der Copolymerisation von Vinylalkoxysilanen unter Anwendung der Suspensionsmethodik wurde unter dem gleichen Aspekt der Adsorbersynthese auf das System Styrol/Vinyltrimethoxysilan übertragen. Die Harzsynthese war insofern erfolgreich, als es gelang, gut ausgebildete Perlpolymerisate zu gewinnen. Diese erwiesen sich jedoch bei der Sulfonierung als nicht ausreichend stabil. Als Ursache wird einmal der die Harzmatrix sprengende, von der Zahl der eingeführten Sulfogruppen abhängige Hydratisierungsdruck, zum anderen hydrolytische Aufspaltung der die Vernetzung bewirkenden Si—O—Si-Brücken angenommen. Es konnten folglich nur Ionenaustauschkapazitäten von maximal 1,9 mval/g trockenes Harz erzielt werden. Damit bilden diese Harze keine Konkurrenz zu den handelsüblichen organischen Ionenaustauschern mit Kationenaustauschkapazitäten bis über 5 mval/g trockenes Harz. Auch ihre Adsorptionseigenschaften, die mit dem Antibiotikum Turimycin zu Vergleichsuntersuchungen am hiesigen Institut getestet wurden, befriedigten nicht.

Unter Beibehaltung der Arbeitskonzeption wurde nach Übernahme der Leitung der halleschen Arbeitsgruppe der SAW durch Prof. F. Wolf auf die Austestung bekannter, z. T. kommerzieller nichtsiliciumhaltiger Organopolymerer für die Moleküladsorption orientiert. Als Adsorbate wurden die Verbindungsklassen gewählt, die noch immer als Noxen die Umwelt stark belastende Faktoren darstellen, wie: Chlorkohlenwasserstoffe, Phenole, Huminsäuren. Alle für die Eliminierungsuntersuchungen herangezogenen Adsorber waren Copolymerisate auf Styrol-Divinylbenzol-Basis, die sich hinsichtlich ihrer Struktur, der Art und Anzahl ihrer funktionellen Gruppen voneinander unterschieden. Es handelte sich dabei um Substanzen, die bisher noch nicht für Moleküladsorptionszwecke eingesetzt worden waren. Sie umfaßten Polymere ionischen und nichtionischen Charakters und wurden teils in ursprünglicher (d. h. in der vom Herstellerbetrieb Chemiekombinat Bitterfeld gelieferten), teils in von uns modifizierter Form verwendet.

Ein großer Teil unserer Untersuchungen bezog sich auf die Adsorption von Lindan und Toxaphen als Vertreter der stark toxischen Gruppe der Chlorkohlenwasserstoff-Insektizide an stark basischen Anionenaustauscherharzen verschiedener Gegenionenformen, an Styrol-Divinylbenzol-Chlormethylaten — Zwischenprodukte der Ionenaustauscherharzsynthese —, an reinen Styrol-Divinylbenzol-Copolymerisaten als nichtionische Vorprodukte der Ionenaustauscherharze sowie an ihrer modifizierten Form: kernchlorierte Styrol-Divinylbenzol-Copolymerisate, ebenfalls nichtionische Copolymere. Bei allen genannten Harztypen wurde Wert darauf gelegt, sowohl gelartige als auch makroporöse Strukturen in die Untersuchungen einzubeziehen. Erschwerend für die analytische Bestimmung der beiden genannten Insektizide war ihre außerordentlich geringe Wasserlöslichkeit, was ein Arbeiten in Wasser-Lösungsmittel-Mischungen bedingte. Bei den genannten Adsorptionsversuchen von Lindan und Toxaphen zeigten alle nichtionischen Styrol-Divinylbenzol-Copolymere gute, zum Teil sehr gute Adsorptionseigenschaften, wobei die Überlegenheit der makroporösen Harzstrukturen deutlich zum Ausdruck kam. Dabei ist besonders die ausgezeichnete Insektizid-Aufnahmefähigkeit des makroporösen, unmodifizierten, nichtionischen Styrol-Divinylbenzol-Copolymerisats Y 29/11 hervorzuheben. Die mit diesem Harz erzielten Eliminierungseffekte (= Adsorptionseffektivität in %), gemessen unter statischen Bedingungen durch die Beladungshöhen in mg adsorbierter Verbindung/g trockenes Harz und in Beziehung gesetzt zur Adsorbatausgangskonzentration, sind bei entsprechender Berücksichtigung der Versuchsbedingungen ca. 5 mal höher als die mit den kommerziellen basischen Wofatit-Ionenaustauscherharzen ermittelten Eliminierungseffekte. Eine Mittelstellung nehmen die Styrol-Divinylbenzol-Chlormethylate und die kernchlorierten Copolymerisate ein.

Neben Adsorptionsversuchen mit den genannten vorwiegend nichtionischen Styrol-Divinylbenzol-Copolymeren war ein zweites Hauptarbeitsgebiet die Austestung von Ionenaustauscherharzen für die Moleküladsorption, die ihren ausgesprochen ionischen Charakter der Einführung zweier unterschiedlicher, funktioneller Gruppierungen — Sulfo- und Aminogruppen — in das Harzgerüst verdanken. Diese Harztypen sind in der Literatur unter dem Aspekt eines selektiven Ionenaustauschs bereits erwähnt, haben aber unter dem Aspekt der Moleküladsorption noch keine Anwendung gefunden. Für uns bestand als erste Aufgabe die Synthese dieser bifunktionellen Ionenaustauscherharze. Sie wurde auf zwei Wegen realisiert, die sich durch die Reihenfolge bei der Einführung der funktionellen Gruppen in die Styrol-Divinylbenzol-Harzmatrix unterschieden. Dabei konnte festgestellt werden, daß die Art der zuerst eingeführten funktionellen Gruppe bestimmend ist für den ionischen Charakter (vorwiegend kationisch oder vorwiegend anionisch) des resultierenden Harzes. Offensichtlich sind sterische und induktive Effekte bei der Substitution am Benzolkern die entscheidenden Faktoren. Die gegenwärtigen Arbeiten sind darauf gerichtet, zu entscheiden, welcher Synthesemethodik der Vorzug zu geben ist, um spezielle Adsorptionseffekte zu erzielen.

Nach orientierenden Adsorptionsversuchen von Insektiziden der Chlorkohlenwasserstoffgruppe (Lindan, Toxaphen s. o.) an diesen bifunktionellen Austauscherharzen, die negativ verliefen, wandten wir uns der Adsorption phenolischer Verbindungen zu. Als Modellverbindung wurde p-Chlorphenol gewählt, das bekannt ist für eine relativ gute Adsorbierbarkeit, andererseits im Falle von Phenoleinbrüchen eine akute Gefahr in Trinkwasseraufbereitungsanlagen darstellen kann. Es bot gleichzeitig die Möglichkeit eines Leistungsvergleiches mit dem früher synthetisierten Vinylacetatsilanharz. In diesen Leistungsvergleich wurden auch die oben erwähnten nichtionischen Copolymerisate einbezogen:
kernchloriertes Styrol-Divinylbenzol-Copolymerisat Y 29/11,
chlormethyliertes makroporöses Styrol-Divinylbenzol-Copolymerisat,
chlormethyliertes gelartiges Styrol-Divinylbenzol-Copolymerisat.

Nach den Adsorptionsisothermen (Adsorberaufnahmevermögen in Abhängigkeit von Adsorbatkonzentration) und den ermittelten Eliminierungseffekten lassen sich die genannten Adsorberharze entsprechend ihrer zunehmenden Adsorptionsfähigkeit für p-Chlorphenol in folgender Reihe anordnen: gelartiges Chlormethylat $\ll$ makroporöses Chlormethylat $<$ kernchloriertes Copolymerisat y 29/11 $<$ sulfoniert-aminiertes bifunktionelles Austauscherharz gleich mit Vinylacetatsilan-Suspensionsharz $<$ aminiert-sulfoniertes bifunktionelles Austauscherharz. Danach liegt das neuartige bifunktionelle, aminiert-sulfonierte Austauscherharz mit einem Eliminierungseffekt von mehr als 90% eindeutig an der Spitze aller untersuchten Harztypen, während sich vergleichsweise das gelartige Chlormethylat mit 4% EE (Eliminierungseffekt) als völlig uninteressant für diese Art der Moleküladsorption erwies.

Eine endgültige Einschätzung der Eignung der getesteten Harze für einen Einsatz als Moleküladsorber läßt sich erst geben nach Abschluß der gegenwärtigen Arbeiten zur Prüfung ihrer Regenerierfähigkeit im Säulenverfahren, die auch Aufschluß geben sollen über ihr Verhalten unter Dauerbelastung (Be- und Entladungscyclen). Dies trifft auch zu für die erfolgversprechende Adsorption von Huminsäuren an den bifunktionellen Austauscherharzen. Die Ergebnisse erster Adsorptionsversuche, wiederum ausgedrückt in Eliminierungseffekten (ermittelt im statischen Verfahren), die im Huminsäurekonzentrationsbereich bis 100 mg HS/l über 90% und im Bereich bis 300 mg HS/l zwischen 80 und 90% liegen, lassen bereits auf eine Konkurrenzfähigkeit mit dem hochporösen Adsorberharz Wofatit ES schließen.

Literatur

1. H. Sontheimer: Forschung in Wasserfach-Aufgaben und Ergebnisse. Gas- u. Wasserfach *113* (1972) 513.
2. D. Hempel, R. Liebmann, A. Hellwig: Vortrag, gehalten auf der Jahrestagung des Fachverb. Wasserchem. d. Chem. Gesellsch. in der DDR, Leipzig, April 1969.

3. D. C. Kennedy: Ind. Eng. Chem. Prod. Res. Develop. *12* (1973) 56.
4. Technical bulletin „Amberlite XAD Macroreticulare Adsorbens". Rohm and Haas Co, Philadelphia.
5. R. L. Gustafson, R. L. Albright, J. Heisler, J. A. Lirio, O. T. Reid: Ind. Eng. Chem. Prod. Res. Develop., *7* (1968) 107.
6. R. Corsin: Vortrag, gehalten auf dem „3. Symposium über Ionenaustausch" Balatonfüred, Ungarn, Mai 1974, Rohm and Haas Co., Paris.
7. H. Krill, H. Wirth: Chemie-Ing.-Techn. *46* (1974) 757.
8. J. A. Mikes, L. J. Kovacs: J. Polym. Science *59* (1962) 209.
9. R. Obenaus: WWT *16* (1966) 10.
10. F. Runge, E. Laqua: WP C 08f/174136.

Friedrich Wolf

Carnitinesterwirkung

Im Forschungsauftrag „Wirkung von Carnitinester auf den physiologischen und pathologischen Stoffwechsel sowie auf den Tumorstoffwechsel" (kurz: Carnitinesterwirkung) wurden Probleme über das Verhalten und die Wirkungen der beiden optisch isomeren Carnitine und einige ihrer O-Acylester in Fortsetzung früherer Arbeiten (1, 2, 3) in 4 Schwerpunkten beforscht: 1. Chemische Darstellung der Stoffe, 2. biologische Wirkungen der Substanzen auf Organe und Gesamttierkörper, 3. Entstehung und Ausscheidung von Carnitinmetaboliten im Harn, 4. therapeutischer Einsatz von L(—)-Carnitin bei Erkrankungen.

Chemie: Die Haupttätigkeit unseres Arbeitskollektivs lag in den laufenden Racematspaltungen von D,L-Carnitinchlorid, um das körpereigene L(—)-Carnitin für den klinischen Einsatz zu beschaffen. Die beengten Verhältnisse erschwerten eine Großproduktion. Weitere Versuche sollten ergründen, wie die Darstellungsweise rationeller gestaltet werden kann, was jedoch bisher nicht im gewünschten Umfang erreicht werden konnte. — Die optisch aktiven Camphersulfonsäuren sind ausgezeichnete Trennsäuren für viele medizinisch interessante, basische Substanzen, die bei der Synthese als Racemat anfallen. Die D,L-Camphersulfonsäure läßt sich in ihre optischen Isomeren auftrennen durch eine Umkehrung der Verfahrensweise, wie sie bei der Racematspaltung von D,L-Carnitinnitril von uns geübt wurde. Mit einem „zweistufigen" Verfahren werden die optisch aktiven Carnitinnitrile abwechselnd nacheinander eingesetzt und das jeweils schwer lösliche Diastereomere gewonnen. Durch gezielten Einsatz von Ionenaustauschern werden die beiden Camphersulfonsäuren rationell, quantitativ und sogleich analysenrein erhalten. Die Carnitinnitrile werden quantitativ zurückgewonnen. — Weitere Racemattrennungen, wie die mit den optisch aktiven Carnitinamiden, erwiesen sich als nicht günstig. Eine spezielle Darstellungsweise für Carnitinamide wurde ausgearbeitet.

Für intensive Studien über die Verhaltensweise der Carnitine und ihrer Ester im Tierkörper mit Isotop-markierten Substanzen wurden die notwendigen Vorarbeiten fortgeführt. Einige Versuche mit Tritium-markiertem Carnitin konnten bisher bereits durchgeführt werden. Diese Arbeiten, wie auch solche mit ^{15}N- und ^{14}C-Markierungen sollen vordringlich in Angriff genommen werden.

Die Studien über die Bindungsenergie ΔG der O-Acylgruppe in den O-Acylestern (3) wurden mit einer 3. Publikation abgeschlossen (4). (O. A. Dr. rer. nat. D. M. Müller, der dieses Problem bearbeitete, schied am 30. 11. 1974 aus unserem Kollektiv aus. Seine Stelle wurde am 1. 12. 74 mit Dipl.-Chemiker Heinz Löster besetzt).

Einige weitere Synthesen und Nachweise wurden erarbeitet. Acetonyltrimethylammonium$^{\oplus}$ = ATMA, ein Carnitinmetabolit, wurde aus D(+)-Carnitin synthetisiert, das bei Racematspaltungen von D,L- Carnitin als Nebenprodukt anfällt: Carnitin wird mit Kaliumbichromat in Eisessig erhitzt. Nach Entfernen des Eisessigs und Chromhydroxids wird das Filtrat eingedampft, aus dem Aethanolextrakt hiervon das ATMA fast quantitativ als Perchlorat erhalten. Für ATMA wurde ein photometrischer Nachweis zur quantitativen Bestimmung als 2,4-Dinitro-phenylhydrazon ausgearbeitet. Für die dünnschichtchromatographischen Bestimmungen aller im Harn ausgeschiedenen Carnitinmetaboliten wurden umfangreiche Studien über die günstigsten Beschichtungssubstanzen und Laufmittelzusammensetzungen erprobt. Für Nachweise zur Sicherung der am Rf-Wert erkannten Metaboliten wurden einige spezielle Bestimmungsverfahren auf der Dünnschichtplatte ausgearbeitet.

Biologie: Ein Teil der in der Berichtszeit bearbeiteten Probleme sind Fortsetzungen von Studien, die in den früheren Jahren begonnen wurden (5). — Die *Toxicität* des therapeutisch eingesetzten L(—)-Carnitins wurde an 850 Mäusen bestimmt. Die LD_{50} betrug bei subcutaner Injektion 8,9 mg/g Maus (Vertrauensbereich 8,1 bis 9,8 mg/g), bei intraperitonealer Injektion 6,6 mg/g (Vertrauensbereich 6,3—7,0 mg/g). Für D(+)-Carnitin war die subcutane LD_{50} 10,3 mg/g, für D,L-Carnitin 10,0 mg/g, für γ-Butyrobetain 4,2 mg/g, für Acetonyltrimethylammoniumchlorid (ATMA) 0,4 mg/g. — Bei verteilter Carnitinzufuhr als Zusatz zum Trinkwasser wurden erheblich größere Mengen täglich schadlos vertragen. Da durch Umbauvorgänge im Tierkörper aus den Carnitinen das toxischere γ-Butyrobetain gebildet wurde — im Harn werden vom zugeführten Carnitin nach Tagen 1/3 als Carnitin und 2/3 als γ-Butyrobetain im Harn ausgeschieden — wurde geprüft, ob sich die Toxicitäten beider Stoffe summieren. Das ist der Fall. Erhielten Mäuse 1 Woche lang 3 proz. L(—)-Carnitinlösung als Trinkwasser, nahmen sie täglich 10,4 mg/g auf, vom D(+)-Carnitin 9,5 mg/g. Bis auf eine Diarrhoe waren keinerlei Folgeerscheinungen sichtbar. Aus einer 2 proz. L(—)-Carnitinlösung nahmen Mäuse 4 Wochen lang täglich 5 mg/g auf. Die anfängliche Diarrhoe verschwand in wenigen Tagen, die Tiere blieben ohne irgendwelche Erscheinungen, ihr Lebendgewicht vergrößerte sich um ~ 8%. Bei der Sektion fiel eine kräftige Vermehrung des Eingeweidefettes auf, Ratten verhielten sich gleichartig. — Die größere Toxicität des natürlich vorkommenden L(—)-Carnitins gegenüber der des unphysiologischen D(+)-Carnitins wird auf die Effekte des L(—)-Carnitins im intermediären Stoffwechsel zurückgeführt. — Bei Injektion von D(+)-Carnitin erschien ATMA nach 2—3 Stunden im Harn und erreichte nach 4—6 Stunden seine höchste Konzentration. Da die absolute Menge aber gegenüber dem ausgeschiedenen D(+)-Carnitin gering blieb, kann das im Tierkörper entstandene ATMA die LD_{50} von D(+)-Carnitin nur unwesentlich beeinflußt haben. — *Rattenschwimmversuche* hatten in früheren Untersuchungen ergeben, daß Ratten unter L(—)-Carnitin etwa 10% länger schwimmen konnten als Kontrolltiere. Dieser nicht signifikante Unterschied sollte auf Signifikanz geprüft werden. Es zeigte sich jedoch, daß die Schwankungsgrößen zwischen den einzelnen Tieren sehr groß waren. Auch in mehrfachen Versuchen mit den gleichen Tieren schwankten die Schwimmzeiten stark, anscheinend hatte auch die wechselnde Wetterlage in den wochenlangen Versuchen Einfluß, so daß bei grundsätzlicher Bestätigung der 10% höheren Schwimmleistung unter Carnitin eine Signifikanz nur durch sehr hohe Versuchszahlen zu erreichen gewesen wäre. Der benötigte Zeitaufwand und die Kosten standen mit dem Wert der zu erwartenden Erkenntnis nicht mehr im Einklang, zumal Begünstigungen der muskulären Leistung durch L(—)-Carnitin inzwischen mit anderen Versuchsanordnungen meßbar wurden. Die Versuche wurden deshalb abgebrochen. — Die *spontane Kontraktion* von isolierten Rattenlebermitochondrien auf O-Palmitoyl-L(—)-carnitin = (—)-PC wurden nach Abschluß der vorwiegend von Doz. Dr. sc. med. R. Dargel in Zusammenarbeit mit unserem Kollektiv durchgeführten Studien (1, 3) nochmals aufgegriffen. Beide optischen Isomeren erzeugen einen etwa gleich großen Schwellungseffekt (Tensidwirkung), aber nur (—)-PC bewirkt eine spontane Kontraktion. Mit Streulichtmessungen wurden Schwellung und Spontankontraktion an Rattenlebermitochondrien bei verschieden hohen (+)- und (—)-PC-Konzentrationen gemessen. 0,9 μM erzeugt noch keine Schwellung, jedoch schon eine Kontraktion. Größere Dosen erzeugen verstärkte Schwellungen und Kontraktionen. 9,0 μM läßt nur noch eine starke Schwellung der Mitochondrien zu. Mischungen von (+)- und (—)-PC bei insgesamt 3 μM PC lassen bei gleicher Schwellungsgröße die spontane Kontraktion nur entsprechend dem (—)-PC-Anteil auftreten. Die wirksamen Konzentrationen liegen deutlich unterhalb der kritischen Micellbildungskonzentration = CMC. Serumalbumin bindet Monomere von (+)- und von (—)-PC. Wenn in Mischungen der beiden Isomeren erst (+)-PC und anschließend (—)-PC zur Mitochondriensuspension gegeben wird, tritt jedoch kein meßbarer Effekt einer größeren Kontraktion durch das (—)-PC auf, die als Ausdruck von Eiweißbindungskapazitäten für das (—)-PC zu werten wäre. — Äquimolare Wirkungsvergleiche von Myristoylcarnitin mit PC zeigen, daß das Myristoyl-D(+)-carnitin keine, sein (—)-Isomeres noch eine Kontraktionsphase hervorruft, die aber wesentlich schwächer ist als die vom (—)-PC. Lauroyl-L(—)-carnitin zeigt bei 60 μM noch eine schwache, spontane Kontraktion. — Die Schwellungsphasen wurden gleichfalls eingehend studiert. Sie sind abhängig von der Größe der tensiden Wirkung des jeweiligen O-Acylesters, die mit der Kettenlänge des Acylrestes ansteigt. Aus der Literatur und nach unseren Befunden kommt die spontane Kontraktion dadurch zustande, daß allein die (—)-Isomeren in der Lage sind, die Acylgruppe in die Mitochondrien einzuschleusen, wo aus ihrer durch Abbau gewonnenen Energie die bei der Schwellung eingeströmten Ionen wieder herausbefördert werden. — Die *Oberflächenaktivität* der höheren O-Acylcarnitine bedingen die 3 hydrophilen Gruppen und die hydrophobe Kohlenwasserstoffkette. Mit ihren starken tensiden Eigenschaften gehören sie zu den Ampholytseifen vom Betaintyp. Die Herabsetzung der Oberflächenspannung wurde tensiometrisch an der Grenzfläche Wasser/

Luft nach 30 Minuten gemessen. Mit steigender Zahl der Methylengruppen nimmt die Oberflächenaktivität in der homologen Reihe der O-Acylcarnitine mit geradzahligem Acylrest bis zum Palmitoylcarnitin zu, die kritische Micellbildungskonzentration (CMC) logarithmisch ab. In 0,9%iger NaCl-Lösung liegt die CMC vom Dekanoylcarnitin bei $8{,}6 \times 10^{-3}$, vom Lauroylcarnitin bei $8{,}4 \times 10^{-4}$, vom Myristoylcarnitin bei $7{,}9 \times 10^{-5}$ und vom Palmitoylcarnitin bei $9{,}6 \times 10^{-6}$ Mol/l. Am Stalagmometer wird die Tropfenbildungszeit zum ausschlaggebenden Faktor, so daß die Literaturangabe, Lauroylcarnitin sei der aktivste O-Acylester des Carnitins, nur bei sehr kurzen Tropfenbildungszeiten zutrifft. Bei Messungen mit der quasi-statischen Bügelabreißmethode zeigen sich die C-14- und C-16-Ester oberflächenaktiver. Die Ergebnisse erlauben es, die Wirkung der höheren O-Acylcarnitine auf biologisches Material auf der Grundlage hydrophober Wechselbeziehungen zu diskutieren. Sowohl die Wirkung auf die Kontraktionskraft des isolierten Froschherzens als auch die schwellungsinduzierende Wirkung an isolierten Rattenlebermitochondrien nimmt mit steigender C-Zahl im Acylrest teilweise analog der CMC zu. Die relativ geringe Wirkungssteigerung des Palmitoylcarnitins zum Myristoylcarnitin wird in der kleineren, effektiven Monomerenkonzentration durch Überschreiten der CMC bzw. durch unspezifische Bindung der Acylcarnitine an Proteine gesehen. — Eine deutliche *muskarinartige* Wirkungsweise soll Acetylcarnitin literaturgemäß entfalten können. Die verwandte Konfiguration von L(—)-Carnitin und Cholin und das fast gleiche chemische Verhalten ihrer Acetylverbindungen hat zu jener Annahme geführt. Diese Vorstellung hatten wir bereits früher nicht bestätigen können, aber es tauchen im Schrifttum immer wieder erneut experimentell belegte Angaben für diese Wirkungsart der O-Acylcarnitinester auf. Sie bekommt besonderes Interesse dadurch, daß beide Substanzarten sich in ihren Funktionen stören können. Andernfalls können sie nebeneinander am gleichen Ort wirken, sich sogar in der Acetylversorgung fördern. Im Tierkörper treten größere Konzentrationen an Acetylcarnitin bei exogener Zufuhr von Carnitin auf, ohne sichtliche Störungen hervorzurufen. Um diese Beziehungen zu klären, haben wir ältere Versuche am Froschherzen nachgeprüft und nach den möglichen Fehlern bei den Experimenten der verschiedenen Autoren gesucht. Wir benutzten hierfür das Straub'sche Froschherz und studierten vorwiegend die inotrope Wirkung der O-Acylcarnitinester. Bei den experimentellen Untersuchungen im Schrifttum ist vielfach die Reinheit der verwendeten O-Acylcarnitinester zweifelhaft. Unsere Substanzen wurden nach unseren Verfahren als reinste Verbindungen dargestellt. Wir fanden, daß Carnitin und Acetylcarnitin bis zu 0,3% in der Perfusionslösung keine negative oder positive inotrope Reizeffekte am Froschherzen auslösen. Höhere Konzentrationen von ihnen erzeugten typische positiv-inotrope Wirkungen. Konzentrationen bis zu 4% in der Perfusionslösung steigern mit der Konzentration die positiv-inotrope Wirkung. Konzentrationen über 4% erzeugen eine unregelmäßige Schlagfolge. Als Ursache der im Schrifttum experimentell ausgewiesenen muskarinartigen Wirkung von Carnitin und Acetylcarnitin fanden wir: 1) In frühen Versuchen, in denen Carnitinchlorid und Acetylcarnitinchlorid mit Natronlauge neutralisiert werden mußten, verursachten die entstandenen hohen Konzentrationen an NaCl die gefundenen negativ-inotropen Effekte. 2) Bei der Gewinnung aus tierischem Material blieb das Carnitin mit Cholin verunreinigt, von dem nach der Acetylierung Spuren genügen, um eine negativ-inotrope Wirkung vorzutäuschen. 3) Bei Reinigungen der sauren Carnitin- und Acetylcarnitinsalze in Alkoholen tritt leicht eine Veresterung der Carboxylgruppe ein, die durch die Acylgruppen gefördert wird. Diese Carboxylester haben eine negativ-inotrope Wirkung, die zudem durch Acetylierung kräftig ansteigt. 4) Bei der Darstellung von Carnitin entsteht durch Wasserabspaltung leicht Crotonsäurebetain, von dem schon Spuren von Carboxylestern eine kräftige muskarinartige Wirkung hervorrufen. 5) Die auf Grund der gleichen spezifischen Konfiguration von (—)-Acetylcarnitin mit Acetylcholin geschlossene acetylcholinähnliche Aktivität für das natürliche L(—)-Carnitin und seine O-Acylester ist ein Irrschluß. Beide optisch aktiven Carnitinester wirken gleichartig und gleich stark. — Es ist demzufolge festzuhalten: Solange die freie Carboxylgruppe besteht, ist jegliche negativ-inotrope Wirkung unterdrückt. — Carnitin hat bis 42 ± 12 mg/ml, γ-Butyrobetain bis 15 ± 8 mg/ml, Crotonsäurebetain bis 22 ± 7 mg/ml eine positiv-inotrope Wirkung. Mit höherer Konzentration erzeugen sie Rhythmusänderungen. Wir haben die Versuche auf höhere Acylcarnitine erweitert. Sie wirken gleichartig wie Carnitin und Acetylcarnitin. Bis zu den Butyrylcarnitinen fanden wir eine dem Carnitin gleich starke positiv-inotrope Wirkung. Längerkettige O-Acylcarnitine könnten mit ihrer oberflächenaktiven Wirkung auch noch andere Effekte zeigen. Wir fanden, daß die in ihnen gebundenen Acyle als freie Fettsäuren erheblich stärker positiv-inotrop wirken als die O-Acylcarnitine. Wird als Meßgröße eine anfänglich gleichgroße positiv-inotrope Wirkung gewählt, so ist die Dekansäure 18mal stärker als das O-Dekanoylcarnitin. Letzteres wirkt jedoch langsamer als die Fettsäure. Nimmt man die gleiche Höhe der positiv-inotropen Wirkung als Maßstab, die die Dekansäure in wenigen Sekunden, das Dekanoylcarnitin aber erst in 10 Minuten erreicht, so ist die Dekansäure 6mal stärker als das Dekanoylcarnitin. Die Wirkung der Carnitinester ist lang anhaltend. Die O-Acylcarnitine werden mit steigender C-Zahl in ihrem positiv-inotropen Effekt wirksamer. Palmitoylcarnitin ist in äquipotenter Menge 1240fach wirksamer als Carnitin.— Ein 20 Stunden in feuchter Kammer gehaltenes Froschherz, das keine sichtbaren Kontraktionen mehr ausführte, und auch nicht durch mehrfaches Spülen mit Göthlin'scher Lösung dazu gebracht werden konnte, entwickelte allmählich ansteigend innerhalb von 30 Minuten eine erhebliche und andauernde Kontraktion, ausgelöst durch verschieden hohe Myristoylkonzentrationen. Konzentrationen von Palmitoyl-(—)-carnitin ($7{,}5 \times 10^{-6}$ M), die noch keinen positiv-inotropen Effekt erkennen lassen, schwächten bereits die Wirkung von 20 ng Acetylcholinchlorid um 90%. Bekannt ist, daß in der homologen Reihe der Alkylsulfate an Kaltblüterherzen die Lauroylkette am aktivsten ist. Bei den O-Acylcarnitinen ist es die Palmitoylkette. — Die Wirkung von L(—)-Carnitin, das Arrhythmien bei Kranken mit Stoffwechselstörungen behob, könnte darauf beruhen, daß durch die exogenen Carnitinzufuhren ein störend hoher Fettsäuregehalt durch Bindung an Carnitin ausgeschaltet wird, denn freie, langkettige Fettsäuren erzeugen bereits in niederen Konzentrationen Arrhythmien, die durch Mangel an Carnitin verursacht sein könnten. — Die *Carnitinabbaustudien* (3) wurden erweitert und vertieft. Der L(—)-Carnitinumsatz im Tierkörper wird von Zufuhr, Eigensynthese, Verbrauch und Ausscheidung bestimmt. Der L(—)-Carnitingehalt hängt von der jeweiligen Stoffwechsellage des Organismus und gegebenenfalls von Stoffwechselstörungen ab. Wir studierten an normalen Mäusen und Ratten die Ausscheidung der Carnitinmetaboliten im Harn durch hohe Zufuhren der beiden optischen Carnitine und ihrer O-Acylderivate. 10 Metaboliten, die einen quaternären Stickstoffkern enthalten, können entstehen oder im Harn vorkommen: Carnitin, Acetylcarnitin, Crotonsäurebetain, γ-Butyrobetain, Glycinbetain, Methylcholin, Acetonyl-trimethylammonium = ATMA, Trimethylamin, Trimethylaminoxid und das zersetzliche Dehydrocarnitin, das im Säugerkörper bisher nicht nachgewiesen wurde. Alle sind seit langem bekannt, aber noch unzureichend studiert. Die Metaboliten wurden von uns durch hohe Zufuhren von Carnitin untersucht, weil anzunehmen ist, daß durch die große Zufuhr Umbauwege und Abwehrreaktionen gut erkennbar werden und die maximalen Umwandlungen besser erfaßt werden können. — Wir fanden bei normal gehaltenen Tieren, daß β-Methylcholin nur in äußerst kleiner Menge entstanden sein konnte, die dünnschichtchromatographisch für uns nicht mehr nachweisbar war. 2) ATMA wurde nur bei den D(+)-Verbindungen ausgeschieden. Nach Injektion von 4,8 mg/g Maus D(+)-Carnitin wurden 1,5% ATMA ausgeschieden, nach 9,2 mg/g gingen 6,2% in ATMA über. Zugeführtes β-Methylcholin bildete kein ATMA. — Die O-Acylcarnitine werden wahrscheinlich von einer der im Tierkörper vorkommenden Esterasen gespalten. Nach Zufuhr von L(—)-Carnitin werden außer Acetylcarnitin keine anderen O-Acylcarnitine ausgeschieden. — Bei langdauernder Zufuhr größerer Mengen von L(—)-Carnitin wurde aus ihm steigend γ-Butyrobetain gebildet. Nach dem 10. bis 12. Tag bleibt die Ausscheidung konstant bei 2/3 γ-Butyrobetain und 1/3 L(—)-Carnitin + Acetylcarnitin. Die Enzyme, die bei diesen Umwandlungen tätig sind, sind noch unbekannt. Sie sind stark adaptationsfähig. γ-Butyrobetain kommt normalerweise in Gewebe und Harn nur in Spuren vor. Vermehrt wird es bei einer Reihe von Krankheiten des Stoffwechsels ausgeschieden. Auch bei Zufuhren von D(+)-Carnitin und Crotonsäurebetain entsteht γ-Butyrobetain. Crotonsäurebetain könnte somit als Zwischenprodukt in diesen Umwandlungen angesehen werden. — γ-Butyrobetain kann durch eine Hydroxylase in L—)-Carnitin umgebildet werden. Dadurch bietet es sich für den Carnitinhaushalt als das zentrale Betain an, welches sowohl aus L(—)-Carnitin entsteht, aber auch in L(—)-Carnitin übergeführt werden kann. — Bei den hohen L(—)-Carnitinzufuhren enthält der Kot neben L(—)-Carnitin auch γ-Butyrobetain. Um auszuschließen, daß es die Darmbakterien sind, die diese Umwandlung vollziehen, der Darm die Stoffe resorbiert, und sie über den Kreislauf in den Harn gelangen, wurden zwei Versuchsreihen angestellt: a) Die Darmbakterien wurden durch hohe Gaben von Antibiotika weitgehend ausgeschaltet. Die Ausscheidungsgröße von γ-Butyrobetain verringerte sich nur wenig. b) Lösungen von L(—)-Carnitin, D(+)-Carnitin, Crotonsäurebetain und γ-Butyrobetain wurden mit Bakteriengemisch aus Kot angeimpft. Nur L(—)-Carnitin förderte das Bakterienwachstum stark und wurde zum γ-Butyrobetain umgewandelt. Der Tierkörper vermag jedoch auch D(+)-Carnitin und Crotonsäurebetain in γ -Butyrobetain zu überführen. D(+)-Carnitin wird im Tierkörper signifikant geringer als L(—)-Carnitin in γ-Butyrobetain umgewandelt. Von mäßigen Zufuhren, etwa 2 mg/g L(—)-Carnitin, scheidet die Maus 90% als quaternäre Metaboliten aus, bei 2monatiger Zufuhr von 0,2 proz. Carnitin im Trinkwasser wurden im Sammelharn nur 45% in den Metaboliten wiedergefunden. Bei den langdauernden Fütterungen wurden in den späten Tagen Trimethylamin und Trimethylaminoxid vermehrt ausgeschieden. — Die literaturbekannte Überführung von γ-Butyrobetain durch die γ-Butyrohydroxylase in L(—)-Carnitin findet vermutlich in der Leber statt. Wir fanden bei Fütterung von γ-Butyrobetain mit steigender Menge eine erhöhte, aber begrenzte Carnitinausscheidung. Bei 8,2 mg täglicher γ-Butyrobetainzufuhr erschienen 3,5 mg, bei 140 mg täglichem γ-Butyrobetain 4,2 mg L(—)-Carnitin im Harn. Die Mäuseleber dürfte danach maximal am Tage ~4 mg L(—)-Carnitin aus γ-Butyrobetain bilden können. — Literaturangaben, daß radioaktiv markiertes L(—)-Carnitin kein markiertes γ-Butyrobetain im Harn finden ließ, können durch ungeeignetes Nachweisverfahren (beide Stoffe haben bei dünnschichtchromatographischer Bestimmung vielfach den gleichen Rf-Wert) oder dadurch, daß die umwandelnden Enzyme nicht genügend adaptiert gewesen sind, zustande gekommen sein. Einige unserer früheren Versuche weckten Zweifel an der schnellen Mischbarkeit von exogen zugeführtem L(—)-Carnitin mit dem gesamten Körperbestand, was für Umsatzbestimmungen mit Isotopmarkiertem Carnitin von Bedeutung sein kann.

Zusammenarbeit besteht mit der Hautklinik, der Medizinischen Klinik, der Medizinischen Poliklinik und der Chirurgischen Klinik der KMU Leipzig. Eigenversuche am Altersherzen laufen weiter.

Literatur

[1] Strack, E.: Jahrbuch *1966/68* der Sächsischen Akademie der Wissenschaften zu Leipzig, Akademie-Verlag Berlin, S. 181—184.

[2] Strack, E.: desgl. *1969/70*, S. 136—140.

[3] Strack, E.: desgl. *1971/72*, S. 78—84.

[4] Müller, D. M., und E. Strack (1972): Z. physiol. Chem. *354*, 1091—96.

[5] Seim, H. (1974): Dissertation A, Math.-Naturwiss. Fak. Karl-Marx-Universität Leipzig, 155 S.

Erich Strack

Komission für J. C. Poggendorff, Biographisch-literarisches Handwörterbuch der exakten Naturwissenschaften

701 Leipzig, Goethestraße 3—5, Tel. 296090

Vorsitzender: Neef

Mitglieder: Arland †, Mothes, Strack

Prof. Dr. Hans Salié als außerakademisches Mitglied

Leiter der Redaktion: Prof. Dr. Hans Salié
701 Leipzig, Beethovenstraße 6, Tel. 34391, App. 88 und 89
Wiss. Arbeitsgruppenleiter: Dr. Lebrecht Weichsel
Wiss. Assistentinnen: Dipl.-Chem. Margot Köstler, Dipl. phil. Winnifred Schenk
Bibliothekarin: Gudrun Scheibe
Techn. Assistentin: Lona Meischner
Forschungs-Facharbeiterinnen: Annelies Körner, Gertrud Lindemann, Margot Mohs, Christa Siedentopf
Mitarbeiterinnen: Dr. Hildegard Delling, Dipl. phil. Lili Schneyer, Dipl.-Chem. Ursel Gerhardt

J. C. Poggendorff: Biographisch-Literarisches Handwörterbuch der exakten Naturwissenschaften

Herausgegeben von der Sächsischen Akademie der Wissenschaften zu Leipzig
Leitung der Redaktion: Prof. Dr. Hans Salié

Da nach mehrjähriger Unterbrechung die Sächsische Akademie der Wissenschaften im Jahre 1949 die Bearbeitung des „Poggendorff" mit dem VII. Bande in Angriff genommen hatte, war der Anlaß gegeben, 1974 das 25jährige Bestehen der neuen Redaktion und ihre Leistungen zu würdigen, über die im einzelnen jeweils die Jahrbücher der Akademie (zuletzt ausführlicher in Jahrbuch 1971/72, S. 84—86) berichtet hatten. Auf Beschluß von Präsidium und Plenum vom 18. 3. 1974 verlieh die Akademie dem verdienstvollen Herausgeber und Leiter der Redaktion, Herrn Prof. Dr. Hans Salié, die Drobisch-Medaille, in Würdigung seiner hervorragenden Verdienste um die planmäßige Weiterführung des Werkes, insbesondere um die Fertigstellung des Bandes VIIa und des Supplementbandes, in Anerkennung seines unermüdlichen Einsatzes, der alle sachlichen und personellen Schwierigkeiten überwinden ließ und das regelmäßige Erscheinen sicherstellte, und als Ausdruck des Dankes für die Förderung des internationalen Ansehens der Sächsischen Akademie der Wissenschaften zu Leipzig, zu dem er mit seinen Mitarbeitern durch die hohe Wertschätzung dieses Standardwerkes beigetragen hat.

Auf der öffentlichen Festsitzung anläßlich des 25jährigen Bestehens der Deutschen Demokratischen Republik, die die Sächsische Akademie der Wissenschaften am 9. November 1974 zur Rechenschaftslegung über ihre Tätigkeit verwendete, konnte Prof. Dr. Salié einen Bericht über den „Poggendorff, ein Standardwerk zur Geschichte der Naturwissenschaften", seine Geschichte, seine inhaltliche Gestaltung, den jetzigen Stand und die Zukunft des Poggendorff geben. Da dieser Bericht in den Schriften der Akademie erscheinen wird, kann für die Einzelheiten auf diesen verwiesen werden.

Die Redaktion konnte im Berichtszeitraum mit der 4. Lieferung den 4. Teilband

(Hen-K) abschließen. Weiter sind vom 5. Teilband (L—M) drei Lieferungen erschienen. Die vierte Lieferung von Teil 5 ist im Satz. Die Bearbeitung von Manuskripten des Buchstaben N hat begonnen.

Ernst Neef

Philologisch-historische Klasse

Historische Kommission

701 Leipzig, Goethestraße 3—5, Tel. 29 28 86

I. Personalstand

Nach vierjähriger Amtszeit wurde der bisherige Vorstand im April 1974 erneut vorgeschlagen und entsprechend der Arbeitsordnung vom 22. I. 1973 gewählt und bestätigt.

Leiter: Buchda

Stellvertretender Leiter: Lau (†6. VI. 1973), Coblenz (seit 19. X. 1973)

Beisitzer: Dr. Hans Eberhardt

Mitglieder: Die ordentlichen Mitglieder der Philologisch-historischen Klasse: Große, Lehmann, Lieberwirth (seit 15. V. 1973), Markov, Schubart-Fikentscher, Werner (seit 15. V. 1973); das auswärtige Mitglied: Bischoff; weitere Mitglieder: Dr. Friedrich Beck, Dr. phil. habil. Karlheinz Blaschke, Prof. Dr. Karl Czok, Dr. Dietmar Debes, Prof. Dr. Ernst Eichler, Dr. Hanns Gringmuth-Dallmer, Prof. Dr. Gerhard Heitz, Prof. Dr. Dr.-Ing. Hans-Joachim Mrusek, Prof. Dr. Erich Neuß, Prof. Dr. Paul Nowotny, Dr. Horst Schlechte, Prof. Dr. Max Steinmetz, Prof. Dr. Dr. h. c. Leo Stern, Dr. Manfred Unger, Dr. phil. habil. Hans Walther, Prof. Dr. Dr. Hermann Weidhaas.

Verwaltung: Ruth Keitel (bis 30. IX. 1974).

II. Unternehmungen

Bibliographie zur Sächsischen Geschichte

Leiter: Buchda

Dresden-Bibliographie, Ergänzungen, Biographie (in Verbindung mit der Sächsischen Landesbibliothek)

Leiter: Dr. Dietmar Debes

Bearbeiter: Dr. Christian Alschner (bis 31. X. 1974), Rosemarie Wünsche

Leipzig-Bibliographie (in Verbindung mit dem Stadtarchiv)

Leiter: Dr. Manfred Unger

Bearbeiterin: Hildegard Heilemann (bis 18. IV. 1973)

Akten und Briefe zur Kirchenpolitik Herzog Georgs von Sachsen
Leiter: Buchda und Lau (†)
Bearbeiterin: Dr. Elisabeth Werl

Politische Korrespondenz des Herzogs und Kurfürsten Moritz von Sachsen
Leiter: Buchda und Lau (†)
Bearbeiter: Dr. Johannes Herrmann, Dr. Günther Wartenberg, Dipl. theol. Renate Rochler (bis 28. II. 1974), Dipl. theol. Wolfgang Rochler (seit 1. V. 1974)

Historischer Atlas von Sachsen
Leiter: Lehmann
Bearbeiter: Dr. phil. habil. Karlheinz Blaschke (unter Mitwirkung kartographischer Fachkräfte)

Quellen und Forschungen zur Sächsischen Geschichte
Leiter: Buchda mit Dr. phil. habil. Karlheinz Blaschke und Dr. phil.habil. Hans Walther

Jahrbuch für Regionalgeschichte
Leiter: Buchda
Redaktion: Prof. Dr. phil. habil. Karl Czok (und Mitarbeiter)

Codex diplomaticus Saxoniae regiae
Hauptteil 1, Bd. 3 Urkunden der Markgrafen von Meißen und Landgrafen von Thüringen. 1196–1234.
Leiter: Manfred Kobuch
Bearbeitung des Registers: Dr. Elisabeth Boer und Manfred Kobuch

Arbeiten zur Erschließung der Handschriftenbestände im Archiv der Franckeschen Stiftungen in Halle
Leiter: Lieberwirth
Bearbeiter: Dipl. theol. Bernd Stephan

III. Fortgang der Arbeiten

An allen wissenschaftlichen Unternehmungen wurde planmäßig weitergearbeitet.

Von der Leipzig-Bibliographie ging der zweite Hauptband an den Verlag; er befindet sich im Satz.

Die Dresden-Bibliographie ist inhaltlich so weit abgeschlossen, daß nur noch Schreibarbeiten am Druckmanuskript auszuführen sind.

Die Absicht, die „Quellen und Forschungen zur Sächsischen Geschichte“ als Beihefte zum „Jahrbuch für Regionalgeschichte“ herauszubringen, hat sich nicht verwirklichen lassen. Die Reihe wird in der alten Form fortgesetzt. Zu erwarten ist ein Band über Zwickauer Frühdrucke.

Vom „Jahrbuch für Regionalgeschichte“ konnten wir das Manuskript des fünften Bandes zum Druck geben.

Die Erschließung der Handschriftenbestände im Archiv der Franckeschen Stiftungen rückte bis zum Buchstaben S vor. Über die etwa 900 Träger mit S beginnender Namen sind Ende 1974 gegen 300 Kurzbiographien fertiggestellt.

Gerhard Buchda

Sprachwissenschaftliche Kommission

701 Leipzig, Goethestr. 3—5, Tel. 296090

Leiter: Große

Stellv. Leiter: Peukert

Beisitzer: Prof. Dr. phil. Ernst Eichler

Mitglieder: Prof. Dr. phil. Werner Bahner, Prof. Dr. phil. Wolfgang Fleischer, Prof. Dr. phil. Hans Hiebsch, Dr. phil. Gerhard Ising, Prof. Dr. phil. Gotthard Lerchner, Prof. Dr. phil. Heinz Mettke, Prof. Dr. phil. Joachim Müller, Prof. Dr. phil. Albrecht Neubert, Prof. Dr. phil Heinz Schuster, Dr. phil. Hans Walther, Prof. Dr. phil. Ernst Werner

Wiss. Sekretär: Dr. Gunter Bergmann

Althochdeutsches Wörterbuch

701 Leipzig, Goethestr. 3—5, Tel. 7197, App. 326

Leiter: Große

Wiss. Arbeitsleiter: Dr. Siegfried Blum, Dr. Heinrich Götz (zugleich als Geschäftsführer), Dr. Rolf Heller, Dr. Gertraud Müller, Dr. Ernst Otto, Dr. Gerhard Wolfrum

Wiss. Oberassistentin: Dr. Ingeborg Köppe

Wiss. Assistent(in): Dipl. phil. Sybille Blum, Dipl. phil. Bernhard Langer

Mitarbeiter: Dr. Walter Krug

Brandenburg-Berlinisches Wörterbuch

108 Berlin, Zimmerstr. 94, Tel. 2202116, App. 23 und 24

Bereichsleiter für das Brandenburg-Berlinische, das Mecklenburgische Wörterbuch und das Pommersche Wörterbucharchiv:

Dr. Gerhard Ising

Wiss. Arbeitsleiter: Dr. Joachim Wiese

Wiss. Assistentin: Dipl. phil. Annemarie Wiese

Wiss.-techn. Assistentin: Charlotte Müller

Mecklenburgisches Wörterbuch

253 Rostock-Warnemünde, Seestr. 15, Tel. Warnemünde 341, App. 344—347

Leiter der Arbeitsstelle: Dr. Jürgen Gundlach

Wiss. Arbeitsleiter: Dr. Jürgen Gundlach, Dr. Eva-Sophie Dahl

Wiss. Oberassistent: Dr. Christian Rothe
Wiss. techn. Assistentin: Dipl. phil. Erika Krackow
Sekretärin u. Sachbearbeiterin: Irma Horak
Angestellte: Gertrud Fröhlich

Pommersches Wörterbucharchiv

22 Greifswald, R.-Petershagen-Allee 1
Verwaltung des Archivs: Dr. Renate Herrmann

Wörterbuch der obersächsischen Mundarten

701 Leipzig, Goethestr. 3—5, Tel. 7197, App. 334 u. 338
Leiter der Arbeitsstelle: Dr. Gunter Bergmann
Wiss. Arbeitsleiter: Dr. Gunter Bergmann
Wiss. Oberassistentin: Dr. Dagmar Helm
Wiss. Assistenten: Dipl. phil. Ingrid Eichler, Dipl. phil. Ingrid Pak, Dipl. phil. Lothar Wezel, Dipl. phil. Hartmut Wittkowski
Wiss. Mitarbeiterin: Dr. Isolde Neumann

Thüringisches Wörterbuch

69 Jena, Universitätshochhaus, 8. Etage, Tel. 834633
Leiter der Arbeitsstelle: Dr. Karl Spangenberg
Wiss. Arbeitsleiter: Dr. Karl Spangenberg, Dr. Heinz Rosenkranz
Sachbearbeiterin: Herta Rädlein

Mitarbeiter am Siebenbürgisch-Sächsischen Wörterbuch: Dr. Helmut Protze, wiss. Arbeitsleiter, 7126 Mölkau bei Leipzig, Dr. Wilhelm-Külz-Str. 4, Tel. 67552

Sprachwissenschaftliche Kommission

Regelmäßig im Frühjahr und Herbst trat die Sprachwissenschaftliche Kommission der Sächsischen Akademie der Wissenschaften zu Beratungen zusammen. Die Arbeitspläne aller ihr zugehörenden wissenschaftlichen Institutionen lagen ihr zur Begutachtung und Genehmigung vor. Die Kontrolle über ihre Einhaltung erfolgte jeweils am Jahresende. Sie ergab, daß an allen Arbeitsstellen die Pläne erfüllt oder übererfüllt wurden. Dabei ist besonders hervorzuheben, daß an den Dialektwörterbüchern auch die Arbeiten am „Atlas Linguarum Europae“ bisher termingerecht vorangetrieben wurden, obwohl sie erst nachträglich in die Pläne aufgenommen werden mußten.

Die Dialektwörterbuch-Unternehmen befinden sich seit 1974 im Wettbewerb. Seine abrechenbaren, konkreten Zielstellungen üben auf die Bemühungen um hohe Leistungen eine stimulierende Wirkung aus.

Althochdeutsches Wörterbuch

Die Arbeiten zum Althochdeutschen Wörterbuch wurden gleichzeitig am zweiten Band mit den Buchstaben C, D und am dritten Band mit den Buchstaben E, F weitergeführt. Erschienen sind im Berichtszeitraum die zweite Lieferung des zweiten Bandes (thannân — thanne), die zweite und dritte Lieferung (edouuân — gi-eiscôn, Doppellieferung) und die vierte Lieferung (ir-eiscôn — gi-entôn) des dritten Bandes. Das Manuskript für die sechste Lieferung dieses Bandes ist abgeschlossen, am Manuskript für die siebente Lieferung (erste Lieferung im Buchstaben F) wird gearbeitet, desgl. am Material für weitere Lieferungen im Buchstaben F und für eine weitere Lieferung im Buchstaben D.

Aus dem Material konnten Auskünfte auf Anfragen aus der DDR und dem Ausland erteilt werden. Besucher aus Dänemark, der Schweiz und den USA suchten die Arbeitsstelle auf, um Material einzusehen, sich mit der Arbeitsmethode bekannt zu machen und Erfahrungen in der Arbeit an historischem Wortschatz auszutauschen.

Brandenburg-Berlinisches Wörterbuch

Die Bearbeitung des Brandenburg-Berlinischen Wörterbuchs wurde planmäßig fortgesetzt. Im Berichtszeitraum wurden die Manuskripte für die 8. Lieferung (mit 3 Sprachkarten) und die 9. Lieferung (mit 5 Sprachkarten) fertiggestellt und dem Akademie-Verlag zugeleitet. Publiziert wurden die 7. und 8. Lieferung, mit dem Erscheinen der 9. Lieferung ist 1975 zu rechnen. Damit ist die Publikation des Wörterbuchs bis zum Ende des Buchstaben D gelangt.

Der Archivbestand des Wörterbuchs wurde durch Erschließung einiger neuer Quellen weiter ergänzt.

Die Druckkorrekturen an dem unter der Gesamtredaktion von Gerhard Ising stehenden Sammelband „Aktuelle Probleme der sprachlichen Kommunikation. Soziolinguistische Studien zur sprachlichen Situation in der DDR" wurden 1973 abgeschlossen.

Von der Neuausgabe der Niederdeutschen Bibelfrühdrucke erfolgte die Drucklegung des 5. Bandes. Das Manuskript des 6. Bandes wurde fertiggestellt und dem Akademie-Verlag übergeben.

Mecklenburgisches Wörterbuch

Im Berichtszeitraum erschienen die 52., 53. und 54. Lieferung, die 55. Lieferung befindet sich im Druck. Die Arbeit am Manuskript wurde planmäßig fortgeführt und die 56. Lieferung abgegeben. Das vorhandene Belegmaterial wurde kontinuierlich ergänzt durch die Übernahme weiteren Wortguts aus dem sprachlichen Nachlaß Richard Wossidlos und durch Exzerptionsarbeiten, besonders aus neu publizierter volkskundlicher Mundartliteratur.

Auch die als zusätzliche Aufgabe übernommenen und 1974 durchgeführten Erhebungen für den Atlas Linguarum Europae im mecklenburgischen Sprachraum kamen der Vervollständigung des Wörterbucharchivs zugute.

Pommersches Wörterbucharchiv

Das Zettelarchiv wurde weiterhin durch exzerpiertes Material aus historischen Quellen und durch neues mundartliches Wortgut ergänzt. Für das ehemalige vorpommersche Gebiet wurden wortgeographische Verbreitungskarten erarbeitet und gezeichnet. Auf zahlreiche Anfragen aus dem In- und Ausland wurden Auskünfte erteilt.

Wörterbuch der obersächsischen Mundarten

Die Ziele, die der detailliert aufgegliederte Arbeitsplan vorsieht, wurden termingerecht erreicht. 1973 wurde die Befragsungaktion zu Ende geführt, bei der in Form von Stichwortlisten etwa 35 Gewährspersonen im Land nach besonders undurchsichtigen Wörtern gefragt wurden. Damit ist die gesamte Sammeltätigkeit abgeschlossen, und das Material wird jetzt nur noch durch das Auszetteln der erscheinenden wissenschaftlichen und belletristischen Mundartliteratur ergänzt. Das Jahr 1974 brachte den Abschluß der Fragebogenauswertung. Damit wird der Schwerpunkt der Arbeit auf das Lemmatisieren verlagert. Kurzfristig eingeschoben wurde die Bearbeitung der Fragebücher des „Atlas Linguarum Europae“ für die vorgesehenen 12 Orte.

In regelmäßigen Beratungen wurden die neuesten Forschungen auf dem Gebiet der Lexikologie und Dialektologie vorgestellt und diskutiert.

Thüringisches Wörterbuch

Im Berichtszeitraum erschienen die 9. und 10. Lieferung des Thüringischen Wörterbuchs. Im Druck befindet sich die 11. Lieferung (Abschluß Bd. IV); ferner liegt die 1. Lieferung (Bd. V) beim Verlag vor, so daß das Thüringische Wörterbuch planmäßig ab 1975 mit je 2 Lieferungen pro Jahr erscheinen kann. Weitgehend fertiggestellt wurden 1974 die Manuskriptfassungen zur 3. und 4. Lieferung des Thüringischen Wörterbuchs.

Das Wörterbucharchiv ist um ca. 50000 Wörterbuchzettel bereichert worden; der Kartenbestand wurde um 250 Karten auf insgesamt 2849 Sprachkarten erhöht. In Arbeit ist eine Manuskriptfassung „Laut- und Formenlehre der Mundart und Umgangssprache in Thüringen“. Abfragungen zum Atlas Linguarum Europae fanden in 12 thüringischen Gemeinden statt.

Für das Siebenbürgisch-Sächsische Wörterbuch wurde der 5. Band revidiert (Buchstabe K, Manuskriptseiten 1—1211; Überprüfung der Belege besonders zur alten Sprache anhand der Quellen, Einarbeitung der Nachträge). Mit den Vertretern aus der Sozialist. Republik Rumänien wurde das Manuskript des K-Bandes besprochen, bevor es an den Verlag gegeben wurde.

Schriftenreihen

Dialektologische Studien zur Sprachgeschichte und Sprachsoziologie, hrsg. von Rudolf Große, Akademie-Verlag Berlin (geplant als Fortführung der Mitteldeutschen Studien)

Deutsch-Slawische Forschungen zur Namenkunde und Siedlungsgeschichte, begründet von Theodor Frings und Rudolf Fischer, hrsg. von Rudolf Fischer und Rudolf Große bis Nr. 26, dann von: Ernst Eichler, Wolfgang Fleischer, Rudolf Große, Albrecht Neubert und Hans Walther

Nr. 27: Der Name in Sprache und Gesellschaft. Berlin 1973.

Veröffentlichungen

Althochdeutsches Wörterbuch

Auf Grund der von Elias von Steinmeyer hinterlassenen Sammlungen im Auftrag der Sächsischen Akademie der Wissenschaften zu Leipzig

Bd II C u. D Zweite Lieferung

Bd III E u. F Zweite und dritte Lieferung, vierte Lieferung

Begründet von Elisabeth Karg-Gasterstädt u. Theodor Frings,

herausgegeben von Rudolf Große,

bearbeitet von Siegfried Blum, Sybille Blum, Heinrich Götz, Rolf Heller, Ingeborg Köppe, Bernhard Langer, Gertraud Müller, Ernst Otto, Elfriede Ulbricht, Gerhard Wolfrum

Akademie-Verlag Berlin 1973. 1974

R. Heller: (Aufsatz) Laxdœla saga und Landnáma bók, in: Arkiv för nordisk filologi 89, Lund 1974, S. 84—145.

G. Wolfrum: (Aufsatz) Mengenlehre als Hilfswissenschaft der Lexikographie, in: Linguistische Arbeitsberichte 10, S. 120—124, Leipzig 1974.

Brandenburg-Berlinisches Wörterbuch

Begründet und angelegt von Anneliese Bretschneider, unter Einschluß der Sammlungen von Hermann Teuchert; bearbeitet unter der Leitung von Gerhard Ising

I. Band, 7. Lieferung (Brathering bis Bürgerheide)

bearbeitet von Heinz Gebhardt, Gerhard Ising und Joachim Wiese. Akademie-Verlag Berlin, in Arbeitsgemeinschaft mit Karl Wachholtz Verlag, Neumünster 1973

I. Band, 8. Lieferung (Bürgermeister bis derweil)

bearbeitet von Gerhard Ising und Joachim Wiese. Akademie-Verlag Berlin, in Arbeitsgemeinschaft mit Karl Wachholtz Verlag, Neumünster 1974

G. Ising: Die Niederdeutschen Bibelfrühdrucke, Band V Jeremia-Maleachi, herausgegeben von Gerhard Ising, bearbeitet unter Mitwirkung von Charlotte Müller. Akademie-Verlag Berlin 1974. VI, 629 S., 8° (in: Deutsche Texte des Mittelalters, hrsg. von der Akademie der Wissenschaften der DDR, Zentralinstitut für Sprachwissenschaft, Band LIV/V).

(Gesamtredaktion) Aktuelle Probleme der sprachlichen Kommunikation. Soziolinguistische Studien zur sprachlichen Situation in der DDR. Akademie-Verlag Berlin 1974. 433 S. 8° (Reihe Sprache und Gesellschaft der Akademie der Wissenschaften der DDR, Zentralinstitut für Sprachwissenschaft, Band 2)

(Aufsatz) Struktur und Funktion der Sprache in der gesamtgesellschaftlichen Entwicklung, in: Aktuelle Probleme der sprachlichen Kommunikation, Berlin 1974, S. 9—36.

(Aufsatz) Schritte auf dem Wege zur Reformation. Niederdeutsche Bibeln vor Martin Luther. In: Neue Zeit v. 1. 7. 74, S. 16

J. Wiese: (Aufsatz) Sprachwandel auf dem Teltow in den letzten hundert Jahren. In: Heimatkalender für den Kreis Zossen 1973, S. 47—50.

— u. K. Spangenberg: (Aufsatz) Sprachwirklichkeit und Sprachverhalten sowie deren Auswirkungen auf Leistungen im muttersprachlichen Unterricht der Allgemeinbildenden Polytechnischen Oberschule, in: Aktuelle Probleme der sprachlichen Kommunikation, Berlin 1974, S. 258—337

Wossidlo-Teuchert: Mecklenburgisches Wörterbuch.
Herausgegeben von der Sächsischen Akademie der Wissenschaften zu Leipzig aus den Sammlungen Richard Wossidlos und aus den Ergänzungen und nach der Anlage Hermann Teucherts.
Zweiundfünfzigste Lieferung (VI, 5), Sodmäuhm bis Spir.
Bearbeitet unter der Leitung von Jürgen Gundlach unter Mitarbeit von Eva-Sophie Dahl, Christian Rothe und Erika Krackow. Akademie-Verlag Berlin, in Arbeitsgemeinschaft mit Karl Wachholtz Verlag, Neumünster 1973
Dreiundfünfzigste Lieferung (VI, 6), spirböken bis Stapel[2].
Bearbeitet unter der Leitung von Jürgen Gundlach unter Mitarbeit von Eva-Sophie Dahl, Christian Rothe und Erika Krackow. Akademie-Verlag Berlin, in Arbeitsgemeinschaft mit Karl Wachholtz Verlag, Neumünster 1973
Vierundfünfzigste Lieferung (VI, 7), stapeldörlik bis Stoeker.
Bearbeitet unter der Leitung von Jürgen Gundlach unter Mitarbeit von Eva-Sophie Dahl, Christian Rothe und Erika Krackow. Akademie-Verlag Berlin, in Arbeitsgemeinschaft mit Karl Wachholtz Verlag, Neumünster 1974

E.-S. Dahl: (Aufsatz) Interferenz und Alternanz — zwei Typen der Sprachschichtenmischung im Norden der Deutschen Demokratischen Republik. In: Aktuelle Probleme der sprachlichen Kommunikation. Berlin 1974, S. 339—387.

C. Rothe: (Aufsatz) Ciceros humanitas und römische Provinzverwaltung. In: WC Rostock XXIII 1974, G 3, S. 233—236.

Pommersches Wörterbucharchiv

R. Herrmann-Winter: (Aufsatz) Auswirkungen der sozialistischen Produktionsweise in der Landwirtschaft auf die sprachliche Kommunikation in den Nordbezirken der DDR, in: Aktuelle Probleme der sprachlichen Kommunikation, Berlin 1974, S. 135—190.

Wörterbuch der obersächsischen Mundarten

G. Bergmann: (Aufsatz) Vorgänge an Dialektgrenzen, in: Linguistische Arbeitsberichte 10, S. 13—23, Leipzig 1974.

(Aufsatz) Sprachschichtung und Sprachwandel, dargestellt an den Entwicklungen der Vokalphoneme im Gebiet um Karl-Marx-Stadt, in: Beiträge zur Soziolinguistik, Halle 1974, S. 84—98.

I. Neumann: (Aufsatz) Offizielle und nichtoffizielle Personenbenennungen, in: Namenkundliche Informationen Nr. 23, Leipzig 1973.

(Aufsatz) Das slawische -š-Suffix in Familiennamen des obersächsischen Sprachbereichs, in: Onomastica Slavogermanica VII, Berlin 1973.

(Aufsatz) Zur Herausbildung des anthroponymischen Prinzips in der Doppelnamigkeit, in: Der Namen in Sprache und Gesellschaft (Deutsch-Slawische Forschungen zur Namenkunde und Siedlungsgeschichte Nr. 27), Berlin 1973.

(Aufsatz) Namen im Wörterbuch der obersächsischen Mundarten, in: Namenkundliche Informationen Nr. 24, Leipzig 1974.

I. Pak: (Übersetzung) Die Leute aus den Ostfjorden. Altisländische Erzählungen. Leipzig 1973.

Thüringisches Wörterbuch

Auf Grund der von V. Michels begonnenen und von H. Hucke fortgeführten Sammlungen, bearbeitet unter der Leitung von Karl Spangenberg an der Sektion Sprachwissenschaft der Friedrich-Schiller-Universität Jena, IV. Band, 9. Lieferung (p — Pfingstmaie); bearbeitet von Herbert Schrickel, Heinz Rosenkranz. Akademie-Verlag Berlin 1973. IV. Band, 10. Lieferung (Pfingstmaien — plispeln); bearbeitet von Heinz Rosenkranz. Akademie-Verlag Berlin 1974.

K. Spangenberg: (Aufsatz) Versuch einer sprachlichen und gesellschaftlichen Grundlegung des gegenwärtigen Lautwandels ch zu sch in Thüringen, in: Ideologie und Sprache (Wiss. Beiträge der Friedrich-Schiller-Universität Jena 1974. Hrsg.: Der Rektor, Prof. Dr. Franz Bolck), S. 166—183.

(Rezension) F. Hager/H. Haberland/R. Paris, Soziologie + Linguistik, Stuttgart 1973, in: Deutsche Literaturzeitung (1974), H. 1.

— u. J. Wiese: Sprachwirklichkeit und Sprachverhalten sowie deren Auswirkungen auf Leistungen im muttersprachlichen Unterricht der Allgemeinbildenden Polytechnischen Oberschule, in: Aktuelle Probleme der sprachlichen Kommunikation. Akademie-Verlag Berlin 1974, S. 285—337.

H. Rosenkranz: (Aufsatz) Veränderungen der sprachlichen Kommunikation im Bereich der industriellen Produktion und ihre Folgen für die Sprachentwicklung in der Deutschen Demokratischen Republik, in: Aktuelle Probleme der sprachlichen Kommunikation. Akademie-Verlag Berlin 1974, S. 75—134.

(Aufsatz) Die Kommunikation im sozialistischen Industriebetrieb in ihrer Bedeutung für die Entwicklung der Sprache, in: Ideologie und Sprache (Wiss. Beiträge der Friedrich-Schiller-Universität Jena 1974. Hrsg.: Der Rektor, Prof. Dr. Franz Bolck), S. 184—207.

Siebenbürgisch-Sächsisches Wörterbuch
H. Protze: (Aufsatz) Typische Erscheinungen deutsch-sorbischer Interferenz unter Berücksichtigung ihrer sozialen Bedingtheit, in: Aktuelle Probleme der sprachlichen Kommunikation, Akademie-Verlag Berlin 1974, S. 389—416.

— u. S. Michalk: Studien zur sprachlichen Interferenz, II, Deutsch-sorbische Dialekttexte aus Radibor, Kreis Bautzen, Domowina Verlag Bautzen 1974, 211 S.

(Rezension) Frankfurter Wörterbuch. Auf Grund des von Joh. Jos. Oppel (1815 bis 1894) u. Hans Ludw. Rauh (1892—1945) gesammelten Materials hrsg. v. Inst. f. Volkskunde der Joh.W.Goethe-Universität Frankfurt a. M. in Verb. mit der Frankf. Histor. Kommission, 1. Lieferung, bearb. von R. Alsheimer, 2. u. 3. Lieferung bearb. von R. Schanze (A — Eva, 624 S.), Frankfurt a. M. 1971—73; in: Deutschen Literaturzeitung 95 (1974), Sp. 591—594.

Rudolf Große

ZUSCHÜSSE

Es werden Zuschüsse geleistet für die Veröffentlichung des

Thesaurus Linguae Latinae

Für die Unternehmungen innerhalb der *Union Académique Internationale* ist je ein Mitarbeiter eingesetzt:

für das Corpus Vasorum Antiquorum: Dr. phil. Eberhard Paul
für das Mittellateinische Wörterbuch: Dr. phil. Anneliese Grau

Das Corpus Vasorum Antiquorum Leipzig

Die Bearbeitung der attisch rotfigurigen Keramik des Antikenmuseums Leipzig für den dritten Band des Corpus Vasorum wurde mit einer zeitweiligen krankheitsbedingten Unterbrechung weitergeführt.

Die Arbeit erstreckte sich wie bisher sowohl auf die konservatorische Behandlung der Gefäße und Fragmente als auch auf ihre archäologisch-kunsthistorische Klassifizierung.

Die für die Wiederherstellung der Gefäße oder Teilgefäße notwendigen Restaurierungsarbeiten wurden im Zusammenhang mit der Konservierungswerkstatt des Museums für Ur- und Frühgeschichte Thüringens durchgeführt. Dabei bestätigte sich wiederum die Richtigkeit der an der Leipziger Keramik entwickelten und bisher geübten neutralen Ergänzungsmethode. Dagegen brachten Versuche, in ausgleichender Weise restaurierte Partien zu behandeln, sowohl unter ästhetischem Gesichtspunkt unbefriedigende als auch in wissenschaftlicher Hinsicht unvertretbare Ergebnisse.

Bei der systematischen Anordnung des Materials wirkte erschwerend, daß ein Teil der umfangreichen Bestände an Einzelscherben keine eindeutige Identifizierung des Gefäßtypus erlaubt.

Unter den technisch und wissenschaftlich bearbeiteten Objekten befinden sich wiederum neben Werken von durchschnittlicher Qualität zahlreiche Erzeugnisse von bedeutenden Vasenmalern der klassischen Zeit.

Veröffentlichung:
Union Académique Internationale
Corpus Vasorum Antiquorum
Deutsche Demokratische Republik Leipzig, Antikenmuseum der Karl-Marx-Universität Band 2 (DDR, Band 2 Leipzig, Band 2) Akademie-Verlag-Berlin 1973

Dr. Eberhard Paul

Die Arbeit für das Mittellateinische Wörterbuch

In dem Berichtszeitraum 1973 und 1974 standen — im Gegensatz zu früheren Jahren — die Lektüre und Exzerpierung von umfangreichen Einzelwerken durchaus im Vordergrund. Fast alle der (zehn) durchgearbeiteten Monographien erbrachten nicht nur eine große Anzahl, sondern die meisten von ihnen auch besonders ausführliche Exzerpte. Hervorgehoben seien in diesem Sinne die Werke von: F. Graus, „Volk, Herrscher und Heiliger im Reich der Merowinger", Prag 1965, (46 Exz.); A. Angenendt, „Monachi peregrini", München 1972, (70 Exz.); G. Koch, „Auf dem Wege zum Sacrum Imperium", Berlin 1972, (117 Exz.); W. Eggert, „Das ostfränkisch-deutsche Reich in der Auffassung seiner Zeitgenossen", Berlin 1973, (144 Exz.), oder E. Pitz, „Papstreskript und Kaiserreskript im Mittelalter", Tübingen 1971, (143 Exz.). Insgesamt konnten der Gruppe der Einzelwerke 735 Exzerpte und außerdem sechs Quellenverweise entnommen werden.

Eine Zwischenstellung zwischen Monographien und Zeitschriften nahm der F. Blatt zum 70. Geburtstag dargebrachte Band der Classica et Mediaevalia, (Dissertationes IX), Kopenhagen 1973, ein. Von den hierin enthaltenen 57 Arbeiten wurden 23, die für das Wörterbuch wichtig erschienen, gelesen und aus dreizehn von ihnen 108 Exzerpte gewonnen.

Aus dem Kreise der Leipzig seit langem als hauptsächliches Aufgabengebiet zugewiesenen Zeitschriftenreihen wurden fünf Bände bearbeitet. 53 Aufsätze aus ihnen wurden gelesen und 35 davon exzerpiert. Am ergiebigsten erwies sich hierbei das Archiv für Diplomatik 17 (1971) mit 224 Exzerpten, wovon allein 203 auf die Abhandlung von R. Falkowski, „Studien zur Sprache der Merowingerdiplome", entfielen. An zweiter Stelle stand im Blick auf die Anzahl der Exzerpte (82) der 78. Band (1970) der Mitteilungen des Instituts für österreichische Geschichtsforschung, deren 79. Band (1971) noch 18 Exzerpte bot. Der 88. Band (1971) der Zeitschrift für Rechtsgeschichte, Kanonistische Abteilung, lieferte 39 Exzerpte. Dagegen erbrachte der 89. Jahrgang (1969) des Historischen Jahrbuchs, der wegen des Aufsatzes von G. Köbler, „Zur Frührezeption der consuetudo in Deutschland", herangezogen worden war, aus diesem und einem anderen Aufsatz nur neun Exzerpte. Zusammen ergaben die genannten Zeitschriften also 372 Exzerpte, dazu noch fünfzehn Quellenverweise.

Dem Material der Arbeitsstelle des Mittellateinischen Wörterbuchs bei der Akademie der Wissenschaften der DDR in Berlin wie dem der Münchener Arbeitsstelle bei der Bayerischen Akademie der Wissenschaften konnten demnach in der Berichtszeit — neben den für jede exzerpierte Arbeit nötigen Bibliographiezetteln und rund zwanzig Quellenverweisen — 1215 Exzerpte neu zugeführt werden.

Dr. Anneliese Grau

VERTRETER DER AKADEMIE

Internationale Thesaurus-Kommission
Wolfgang Schadewaldt †

Redaktionsausschuß der Deutschen Literaturzeitung
Gerhard Buchda
Ernst Neef

Kommission für die Herausgabe einer Enzyklopädie
der mathematischen Wissenschaften
Ernst Hölder

Unternehmungen der Akademien der Wissenschaften
innerhalb der Union Académique Internationale

Corpus Vasorum Antiquorum
Werner Peek

Mittellateinisches Wörterbuch
Wolfgang Schadewaldt †

SITZUNGSKALENDER FÜR DIE JAHRE 1973 UND 1974

Die Vorträge werden nacheinander gehalten und vom Plenum entgegengenommen. Im Anschluß an die Vorträge finden getrennte Geschäftssitzungen der Klassen statt, an diese schließen Gesamtsitzungen an.

1973

		Mathematisch-naturwissenschaftliche Klasse	Philologisch-historische Klasse
Montag,	22. Januar	Neef	Hiebsch
Montag,	12. Februar	Mothes	Peek
Montag,	19. März	Buchheim	—
Sonnabend,	14. April	Öffentliche Gesamtsitzung zur Erinnerung an den Geburtstag von Leibniz	Meyer
Montag,	14. Mai	Schulze	Müller
Montag,	18. Juni	Schwabe	Blumenthal
Montag,	15. Oktober	Emmrich	Große
Sonnabend,	17. November	Öffentliche Gesamtsitzung zur Erinnerung an den Todestag von Leibniz	Drischel
Montag,	10. Dezember	Eßbach	Lehmann

1974

Montag,	14. Januar	Maier	Jahn
Montag,	11. Februar	Hanson	Werner
Montag,	18. März	Strack	Lieberwirth
Sonnabend,	20. April	Öffentliche Gesamtsitzung zur Erinnerung an den Geburtstag von Leibniz	Schubart-Fikentscher
Montag,	20. Mai	Matzen	Markov
Montag,	17. Juni	Gersch	Pée
Sonnabend,	9. November	Öffentliche Gesamtsitzung zur Erinnerung an den Todestag von Leibniz	Görlich
Montag,	16. Dezember	Watznauer	Buchda

ANSPRACHE DES PRÄSIDENTEN, HERRN SCHWABE

bei der öffentlichen Gesamtsitzung am 14. April 1973

Unsere Sitzung hat einmal die Aufgabe, über die Tätigkeit der Akademie im vergangenen Jahr zu berichten. Diesen Bericht wird der Vizepräsident, Herr Lehmann, geben. Andererseits soll ein wissenschaftlicher Vortrag eines Mitgliedes unserer Akademie aus seinem Arbeitsbereich die Bedeutung solcher Arbeit für Kultur- und Geistesleben oder für die Volkswirtschaft unserer sozialistischen Gesellschaft sichtbar machen.

Gestatten Sie mir zu diesen beiden Anliegen wenige Gedanken zu äußern. Die Sächsische Akademie hat als gelehrte Gesellschaft ihre Arbeit in vier Kommissionen konzentriert, deren jede von einem ordentlichen Mitglied geleitet wird, in der aber Fachwissenschaftler verschiedener Einrichtungen mitarbeiten. Dadurch wird, so glauben wir, die Kooperation und Information über organisatorische Schranken hinweg gefördert.

Eine Akademie mit großen, modernen Forschungseinrichtungen wie etwa die Sowjetische Akademie der Wissenschaften oder die Akademie der Wissenschaften der DDR sind vor allem dazu aufgerufen, ihre große Forschungskapazität für den wissenschaftlich-technischen und gesellschaftlichen Fortschritt zum Wohle aller Menschen in einer friedlichen Welt einzusetzen.

Eine gelehrte Gesellschaft wie unsere Akademie kann und sollte sich in unserer Zeit und in unserem sozialistischen Staat der Erhaltung und Pflege der kulturellen und materiellen Güter besonders widmen. Die vier Kommissionen unserer Akademie, über deren Tätigkeit der Herr Vizepräsident nachher berichten wird, haben solche Aufgaben.

Die Historische Kommission unter Herrn Buchda widmet sich in erster Linie der Geschichte des Landes Sachsen im Hinblick auf seine Bedeutung für die kulturelle und gesellschaftliche Entwicklung in Europa.

Die Sprachwissenschaftliche Kommission unter Herrn Große, Sekretär der philologisch-historischen Klasse, widmet sich den Problemen der Sprachentwicklung, vor allem von den Dialekten her, und hellt dabei wichtige kulturelle und soziologische Fragen auf.

Die Poggendorff-Kommission unter Herrn Neef, Sekretar der mathematisch-naturwissenschaftlichen Klasse, betreut die Herausgabe von Poggendorffs bibliographischem Handbuch der exakten Naturwissenschaften und trägt dazu bei, daß die Erinnerung an die Bedeutung großer Naturwissenschaftler wachgehalten wird.

Die Kommission für spezielle Umweltprobleme, ebenfalls unter Herrn Neef, bemüht sich in enger Kooperation mit der AdW die Voraussetzungen dafür zu schaffen, daß die materiellen Güter, Wasser, Luft, mineralische, pflanzliche Rohstoffe usw., die für die Existenz der rapide zunehmenden Menschheit Voraussetzung sind, in ausreichendem Umfang und in einer ihr zuträglichen Form erhalten werden. Ich glaube, daß die Naturwissenschaftler, Ingenieure und Gesellschaftswissenschaftler in Zukunft mindestens ebenso viel Aktivität und Kreativität brauchen werden, um die materiellen Güter unserer Erde für die friedliche Verwendung und die Versorgung der Menschen zu erhalten, wie dafür, sie in immer steigendem Maße für die ständige Erhöhung des Lebensstandards einzusetzen.

Dieser Lebensstandard wird heute meistens daran gemessen, wieviel man über den notwendigen Lebensbedarf an Nahrung und Kleidung hinaus ausgeben kann, um möglichst oft neue Bekleidungsmittel, Möbel, Autos usw. anschaffen zu können. Ist das der Fortschritt, der uns glücklicher macht? Der Sorgfalt der Produktion solcher Güter und ihrer Pflege ist eine solche Einstellung auf jeden Fall nicht dienlich. Eine internationale Rohstoffkommission hat kürzlich folgendes festgestellt, wenn man den Lebensstandard der hoch entwickelten kapitalistischen Industrieländer, wie der Schweiz oder der USA auf alle Menschen, die heute auf der Erde leben (etwa 3,5 Milliarden) sofort ausdehnen wollte, wären die heute bekannten Vorräte an Eisenerzen durch den enorm ansteigenden Stadtbedarf in etwa 20 Jahren erschöpft.

Wenn die Wissenschaft ihrer Rolle als Produktivkraft gerecht werden soll, dann muß sie auch um eine sinnvoll gelenkte Produktion im Materiellen und Geistigen bemüht sein. Sinnvoll im sozialistischen Sinne heißt aber doch wohl so zu produzieren, daß möglichst alle Menschen, vor allem in Zukunft, an den Gütern unserer Erde teilhaben und daß die Ergebnisse geistiger Arbeit möglichst vielseitig genutzt werden. Werden aber die Leistungen aus wissenschaftlicher Arbeit, die sich in einer exponentiell ansteigenden Flut von wissenschaftlichen Publikationen ausdrücken, noch ausreichend genutzt? Durch die Verzweigung der Wissenschaften in zahllose Disziplinen und die Spezialisierung in diesen wird der Wirkungsgrad wissenschaftlicher Arbeit stark beeinflußt. Ihn möglichst zu erhalten, sollte eines der Hauptanliegen der Wissenschaftler selbst und gelehrter Gesellschaften sein. Goethe hat einmal prophezeit:

> „Die Wissenschaften zerstören sich auf doppelte Weise selbst:
> durch die Breite in die sie gehen
> und durch die Tiefe in die sie sich versenken“.

Diese Gefahr ist heute, wo sie wirklich akut geworden ist, kaum so klar erkannt worden.

Es liegt nur bei den Wissenschaftlern selbst, diese Zerstörung zu verhüten. Das sicherste und vielleicht einzige Mittel dafür ist wohl: Vorbehaltlose Zusammenarbeit und aufrichtige Verständigung zwischen den Wissenschaftlern aller Disziplinen in unserer Welt. Die ungeheuren Erfolge wissenschaftlicher Arbeit in der Technik, aber auch in Biologie und Medizin erfüllen uns immer wieder mit Staunen, Bewunderung und Stolz. Wir sollten aber nicht vergessen, daß all diese Leistungen auf der wissenschaftlichen Arbeit von Jahrhunderten beruhen. Hanns Eisler hat einmal gesagt:

„Altes ist natürlich alt,
wenn ihr etwas Neues seht.
Vorsicht, man vergißt zu bald,
daß es auf dem Alten steht."

Der Vortrag von Herrn Meyer über Kultur und Sprache im frühantiken Syrien wird sicher deutlich machen, daß die Menschen vor Jahrtausenden bereits ein Denk- und Abstraktionsvermögen bewiesen haben, daß dem der heutigen kaum nachsteht. Ihnen fehlten dabei alle Hilfsmittel, die unsere geistige Arbeit so wirksam unterstützen, dafür nahmen sie sich wohl aber mehr Zeit und Muße.

Heute ist es wichtiger als je zuvor, aus den Ergebnissen geistiger Arbeit die notwendigen Schlußfolgerungen für unser Verhalten als Mitglieder der menschlichen Gesellschaft in einer Welt zu ziehen, die durch Wissenschaft, Technik und Industrialisierung ständig verändert und kleiner wird.

Die für das Verhalten in der Gesellschaft wesentlichen Prinzipien: Freiheit, Gleichheit, Brüderlichkeit, die eine Grundforderung des Marxismus sind, werden lange genug proklamiert, um sie nun endlich weltweit zu befolgen, ehe es zu spät ist.

ANSPRACHE DES PRÄSIDENTEN, HERRN SCHWABE

bei der öffentlichen Gesamtsitzung am 17. November 1973

Als die Sächsische Akademie der Wissenschaften durch die Bombenangriffe auf Leipzig im zweiten Weltkrieg die ihr von der Universität Leipzig zur Verfügung gestellten Räume verloren hatte, war sie bei ihrer Wiedereröffnung vor 25 Jahren nur so notdürftig untergebracht, daß dadurch ihre Arbeitsfähigkeit wesentlich eingeschränkt war. Nun ist es mit Unterstützung der Karl-Marx-Universität Leipzig, die inzwischen große neue Bauten erhielt, gelungen, im ersten Stock des Franz-Mehring-Hauses in der Goethestraße Räume zu erhalten, die eine erhebliche Verbesserung darstellen.

Es wird möglich sein, die normalen, nichtöffentlichen Sitzungen in diesen Räumen durchzuführen, während bisher nur durch das freundliche Entgegenkommen von Herrn Drischel Räume in seinem Institut, getrennt vom Sitz der Akademie, für die monatlichen Sitzungen verwendet werden konnten. Die neuen Zimmer im Franz-Mehring-Haus stehen uns vielleicht schon zu Beginn des nächsten Jahres dauernd zur Verfügung. Sie können daher auch für Kolloquien der Kommissionen unserer Akademie benutzt werden; und so hoffen wir, daß dadurch die wissenschaftliche Arbeit, vor allem auch die Kooperation zwischen den verschiedenen Arbeitsgebieten, neue Impulse erhält. Darüber hinaus können mindestens zwei Forschungsgruppen auf philologischem Gebiet, die Gruppe des Althochdeutschen Wörterbuches und die Gruppe des Wörterbuches der Obersächsischen Mundarten, in den neuen Räumen neben der Verwaltung untergebracht werden. Beide Gruppen stehen unter der Leitung des Sekretars der philologisch-historischen Klasse, Herrn Große. So wird ihm durch die räumliche Konzentration die Betreuung dieser Gruppen erleichtert werden; das wird sicher den Wirkungsgrad der wissenschaftlichen Arbeiten weiter steigern. Auch die Akademie-Leitung selbst hat jetzt bessere Möglichkeiten, ihre Aktivität zu verstärken. Ich möchte daher Magnifizenz Winkler auch von dieser Stelle aus sehr herzlich für die Unterstützung bei der Bereitstellung dieser Räume danken und meine Hoffnung auf eine weitere fruchtbare Zusammenarbeit mit der Karl-Marx-Universität aussprechen.

Da uns durch den Faschismus, den Krieg und seine Nachwirkungen so viele hervorragende Mitglieder verlorengegangen sind und wir so lange Zeit über keine geeigneten Räume für Zusammenkünfte mit dem Ziele eines disziplinären und interdisziplinären Erfahrungsaustausches verfügen konnten, haben wir viel nachzuholen, um durch die Ergebnisse unserer Arbeit unseren Ruf als gelehrte Gesellschaft zu rechtfertigen.

Gestatten Sie mir in diesem Zusammenhang noch einige Bemerkungen zum Problem der Kooperation und Information.

Das exponentielle Wachstum des Wissens und der Zahl der Wissenschaftler im Zuge der wissenschaftlich-technischen Revolution hat dazu geführt, daß die verschiedenartigsten wissenschaftlichen Organisationen mit dem Ziele einer verstärkten Kooperation und Information zwischen ihren Mitgliedern entstanden sind.

In die besondere Obhut aller wissenschaftlichen Akademien ist es gegeben, die gerade heute so bedeutsame Aufgabe der interdisziplinären Kooperation und Integration der Wissenschaft über alle Schranken hinweg zu pflegen. Daneben gibt es nationale und internationale wissenschaftliche Fachverbände, die sich immer weiter in Sektionen oder anders benannte Spezialorganisationen aufteilen. Dementsprechend nimmt auch Zahl und Spezialisierung der wissenschaftlichen Zeitschriften exponentiell zu. Wenn man alle Veranstaltungen auf einem Fachgebiet besuchen wollte, wäre man kaum im eigenen Institut, und wenn man alle Fachzeitschriften des eigenen Arbeitsgebietes sorgfältig studieren wollte, dann bliebe keine Zeit für schöpferische Tätigkeit, und trotzdem würden Verluste an wertvollen Anregungen für das eigene Fachgebiet unvermeidlich, die aus anderen Disziplinen kommen.

Alle Wissenschaftler klagen über die unerhörte Fülle von wissenschaftlichen Tagungen und die beängstigend ansteigende Flut von Veröffentlichungen.

Wissenschaftsorganisatoren und Informationswissenschaftler haben zwar die Zahl der Informationsorgane und Mittel erheblich vergrößert, nämlich auf ihrem Wissenschaftsgebiet. Einen Ausweg aus dieser Überflutung mit Informationen, die weder vollständig konzipiert noch genutzt werden, können sie uns nicht zeigen; denn den Regen an Informationen, der diese unübersehbare Flut verursacht, erzeugen die Wissenschaftler selbst. Sie führt um so mehr Überflüssiges, Bekanntes und Falsches mit, je größer sie wird, und die Abtrennung der wertvollen Informationen wird daher immer schwieriger. Die Kooperation wird durch diese Flut um so stärker belastet, je mehr sich die zahllosen Wissenschaftsdisziplinen ihre eigene Terminologie schaffen ohne Rücksicht darauf, ob die verwendeten Termini bereits mit einer ganz anderen Bedeutung gebraucht werden.

Culpa nostra est — wir können uns nur selbst anklagen. Wenn die wissenschaftlich-technischen Leistungen für die Gesellschaft als Ergebnis schöpferischer Arbeit nicht im gleichen Maße wachsen wie die Zahl der Wissenschaftler, dann liegt das nicht an einem Mangel an Information, sondern höchstens an ihrer Qualität und ihrem Übermaß. Jeder einzelne Wissenschaftler ist aufgerufen, bei sich selbst zu prüfen, ob diese oder jene wissenschaftliche Tagung nötig ist oder unbedingt besucht werden muß, und ob seine Publikationen eine nützliche und wichtige Information sind. Die wissenschaftlichen Organisationen, insbesondere die Akademien, müssen an ihre Publikationen und Tagungen einen sehr strengen Maßstab anlegen. Ihre besondere Aufgabe bei der Eindämmung des die wissenschaftliche Arbeit überschwemmenden Informationsstromes ist aber die Erziehung der Wissenschaftler zur kritischen Einschätzung ihrer eigenen Informationsaktivität in Publikationen und Vorträgen. Das Urteil eines Herausgeberkollegiums kann diese Prüfung kaum ersetzen. Das rapide Wachstum des Wissens und der Zahl der Wissenschaftler hat dazu geführt, daß wir, wie heute in manchen anderen Bereichen menschlicher Tätigkeit, Gefahr laufen, die Kontrolle über den Nutzen dieser Tätigkeit für die Gesellschaft zu verlieren, wenn nicht jeder einzelne sich selbst kontrolliert. Dabei sollte man auch an den Aufwand an Material, Papier und Arbeitskräften denken.

So kann ein Übermaß an Information sich selbst entwerten und die Kooperation zwischen den Disziplinen hemmen, weil man schon die Literatur auf dem eigenen Fachgebiet nicht mehr bewältigen kann. Weltweite Kooperation und Integration ist aber für die wissenschaftlich-technischen Aufgaben der Zukunft in doppelter Hinsicht genau so wichtig wie die Information. Die unablässigen Bemühungen um Frieden auf dem ganzen Erdball, wie die Weltfriedenskonferenz in Moskau und die Abrüstungskonferenz in Wien, lassen hoffen, daß kriegerische Auseinandersetzungen in Zukunft vermieden und politische Spannungen immer mehr abgebaut werden. Dann wird auch die internationale wissenschaftliche Kooperation weiter ausgedehnt werden. Vorbehaltlose Zusammenarbeit muß aber auch zwischen den Wissenschaftsdisziplinen immer stärker gepflegt werden, denn die Aufgaben der Zukunft, so die Gestaltung unserer Umwelt oder die Energieprobleme, können nur durch gemeinsame Anstrengungen der Wissenschaftler und Ingenieure vieler Wissenschaftsgebiete gelöst werden. Diese Gemeinsamkeit zu erhalten und zu stärken ist, so meine ich, Aufgabe der wissenschaftlichen Akademien.

ANSPRACHE DES PRÄSIDENTEN, HERRN SCHWABE

bei der öffentlichen Gesamtsitzung am 20. April 1974

Wir führen heute die öffentliche Frühjahrssitzung der Sächsischen Akademie der Wissenschaften durch, die wie in jedem Jahr dem Andenken von Gottfried Wilhelm Leibniz gewidmet ist, fast 34 Dezennien nach seiner Geburt in Leipzig und im 25. Gründungsjahr der Deutschen Demokratischen Republik.

In Vorbereitung der 125 Jahrfeier unserer Akademie (1971) hatten wir die Herstellung einer neuen Amtskette beschlossen, nachdem die alte in den Wirren der letzten Kriegstage verlorengegangen war. Es scheint uns ein würdiger Anlaß, sie erstmalig im Jubiläumsjahr unserer Republik anzulegen, denn ihr ist auch die Sächsische Akademie der Wissenschaften zu Leipzig für wohlwollende Förderung in diesem Vierteljahrhundert in Dankbarkeit verbunden.

Ihre Form, die wir Volker Knauf verdanken, bringt unsere Verbundenheit mit der Stadt Leipzig zum Ausdruck. Sie soll gleichzeitig durch das einfache und edle Material, das verwendet wurde, die progressiven Aufgaben einer gelehrten Gesellschaft in einem modernen sozialistischen Staat veranschaulichen: Schönes und Zweckmäßiges mit dem dafür gerade nötigen Aufwand zu schaffen, wird für den Wohlstand aller Menschen in einer friedlichen Welt die Voraussetzung sein. Möge allen künftigen Trägern dieser Kette die Bürde ihres damit verbundenen Amtes ebenso leicht sein,wie es ihr Gewicht entsprechend diesem Postulat ist.

Eben war von dem Aufwand die Rede, der zur Schaffung von schönen und zweckmäßigen Gütern optimiert werden sollte. Das zu prüfen, bedarf für die industrielle Warenproduktion sorgfältiger wissenschaftlicher Untersuchungen; dabei sind die ökonomischen Probleme wohl die schwierigsten, vor allem wenn Qualität und Produktivität miteinander korreliert sind. Es wird leicht übersehen, daß hohe Qualität die Nutzungsdauer und damit auch die Produktivität im ganzen erhöht, die Umweltbelastung durch Müll senkt.

Aber die Wissenschaft selbst sieht sich immer dringlicher vor der Frage, ob ihr Wirkungsgrad als Produktivkraft im richtigen Verhältnis zum Aufwand steht. Als die Zahl der Wissenschaftler im Verhältnis zur Gesamtzahl aller Werktätigen verschwindend klein und der Forschungsaufwand pro Kopf gering war, spielte dieser Aufwand im Verhältnis zu allen anderen Ausgaben keine Rolle. Heute kann der Aufwand pro Wissenschaftler je nach dem Arbeitsgebiet zwischen 50000 und mehr als 500000 Mark pro Jahr geschätzt werden. Die Zahl der Wissenschaftler nimmt zur Zeit in der Welt relativ stärker zu als die Weltbevölkerung pro Jahr. Man hat daher ausgerechnet, daß in 200 bis 300 Jahren mehr Wissenschaftler auf der Erde sein würden als Menschen. Rechnungen dieser Art haben natürlich keinen konkreten Aussagewert. Immerhin wird damit auf einen Trend hingewiesen, der bei unverminderter Fortsetzung die Wissenschaft unproduktiv machen müßte. Denn auch in Zukunft werden die Ergebnisse wissenschaftlicher Forschung durch Menschen in die Produktion überführt werden müssen, die dabei zwar weitgehend automatisierte, aber keine eigentlich wissenschaftliche Arbeit leisten. Schon jetzt muß jeden Wissenschaftler und erst recht eine gelehrte Gesellschaft die Feststellung der Bibliotheken beunruhigen, daß noch nicht einmal 10% des publizierten wissenschaftlichen Schrifttums jemals gelesen werden. Dies zeigt: der immer wieder beklagte Mangel an Information ist nicht eine Frage der Informationstechnik, sondern der Selektion, was zu publizieren und zu lesen sich lohnt. Diese Frage kann nur der Wissenschaftler selbst beantworten, nicht der Computer oder der Dokumentalist. Der große sowjetische Physiker Kapiza hat von einer Krise der wissenschaftlichen Information gesprochen.

Diese Krise ist eine Herausforderung an die Vernunft der Wissenschaftler. Sie steht im Zusammenhang mit dem zukünftigen Wachstum der menschlichen Gesellschaft überhaupt. Die Fruchtbarkeit dieses Wachstums ist nur durch die Vernunft der Menschen, insbesondere ihrer Führungskräfte, zu gewährleisten.

Bisher waren auch die Proportionen der Kapazitäten an Wissenschaftlern und wissenschaftlichen Einrichtungen für die verschiedenen Disziplinen nicht relevant. Mit dem rasch wachsenden Anteil der Wissenschaftler an der Gesamtzahl der Beschäftigten gewinnen diese Relationen für den Wirkungsgrad wissenschaftlicher Arbeit ganz wesentliche Bedeutung. Bisher wurde er vorwiegend daran gemessen, wie diese Arbeit zur Steigerung unseres materiellen Wohlstandes und zur Bequemlichkeit beiträgt. Daher hat die naturwissenschaftlich-technische, medizinische und ökonomische Forschung die größte Steigerungsrate, aber selbst hier sind Disproportionen zwischen Grundlagenforschung und der nicht so hoch im Kurs stehenden Technologie deutlich erkennbar. Der englische Nobelpreisträger Norrish hat kürzlich darauf hingewiesen, daß wir in den letzten 10 Jahren „reine" Wissenschaftler in einer übermäßigen Anzahl zu schaffen versuchen.

Um im Wohlstand glücklich zu leben, bedarf es der kulturellen Betätigung auch außerhalb des Berufs. Daher ist es notwendig, den Kulturwissenschaften in Zukunft einen Platz einzuräumen, der das Interesse an solchen Wissenschaftsgebieten verstärkt. Relationen dieser Art zu untersuchen und sich um optimale Proportionen zu bemühen, sind wichtige Aufgaben, die einer gelehrten Gesellschaft nach meiner Meinung in unserer Zeit gestellt sind, weil sie ein hohes Maß an Übersicht, Einsicht und Selbstbeschränkung erfordern.

Unser Mitglied Frau Schubart-Fikentscher wird heute über einen Hochschullehrer sprechen, der vor 300 Jahren einige Irrlehren der Kirche unerschrocken bekämpfte und für ein den Naturgesetzen entsprechendes allgemeingültiges Recht eingetreten ist. Er hat auf seine Weise dazu beigetragen, daß an unseren akademischen Bildungsanstalten die Wissenschaften ohne doktrinäre Beschränkung gelehrt werden können. Ihre Entfaltung vor schädlicher Selbstüberschätzung und Verwahrlosung zu bewahren, sind wir heute alle, besonders die Akademien und Hochschulen, aufgerufen. Mir scheint, daß bereits in den Schulen der Sinn für das Ganze und das Wesentliche stärker geweckt und zum bleibenden Besitz gemacht werden muß. Das wird den

Schüler weniger überfordern als die Unterrichtung in den jeweils aktuellen Wissensgebieten und das Studium an den Hochschulen erfolgreicher machen. Diejenigen Formen des Unterrichts, die vor wenigen Jahren von sowjetischen Pädagogen auf folgende Weise charakterisiert wurden, sollten nun endlich verschwunden sein:

„Der Unterricht gleicht einer Verdauung nach Portionen. Heute hat man sich eine Portion Kenntnisse angeeignet, morgen sie dem Lehrer mitgeteilt und sich gleichsam davon befreit. — Man darf nicht zulassen, daß der Jugendliche von früh bis spät über den Lehrbüchern sitzt und sich in eine Lehrmaschine verwandelt".

Die wissenschaftlichen Lehr- und Forschungsstätten gelangen am besten zu großen Ergebnissen, wenn sie zur Achtung vor den Leistungen der Vergangenheit erziehen, die Integration der Wissenschaft durch vorbehaltlose Kooperation zwischen den Disziplinen international fördern und die wissenschaftliche Leistung als Maßstab der Treue zur Gesellschaftsordnung in unserer Republik gilt.

Organisatorische Maßnahmen und Reformen der Studienpläne zur Verbesserung der Forschungs- und Studienergebnisse sind in den letzten Jahren wohl genügend und vielgestaltig durchgeführt worden. Nun sind kontinuierliche, ungestörte Arbeit und ein Berufsethos nötig, wie es Thomasius zu revolutionierenden Ideen und Bekennermut befähigt hat.

Erlauben Sie mir zum Schluß, die eben ausgeführten Gedanken noch einmal zusammenzufassen unter dem Namen des Mannes, dessen 250. Geburtstages wir in diesen Tagen gedenken: Immanuel Kant. Ein Jahrhundert nach Thomasius nennt er Faulheit und Feigheit die Ursachen geistiger Unmündigkeit und fordert Entschlußkraft und Mut für jede wissenschaftliche Erkenntnis. Den antiken Ausspruch „Sapere aude — wage es, weise zu sein" übersetzt er in seiner Schrift „Was ist Aufklärung" mit den Worten „Habe Mut, dich deines eigenen Verstandes zu bedienen" und bezeichnet sie als den Wahlspruch der Aufklärung.

Eine Akademie sollte als eine Gemeinschaft von Forschern und Lehrern keine Gelegenheit versäumen, zur unermüdlichen und unerschrockenen Hingabe an die Wissenschaft im Dienste unserer sozialistischen Gesellschaft als dem sichersten Mittel für den Erfolg der Arbeit aufzurufen, und bemüht sein, dafür die besten Voraussetzungen zu schaffen. Denn, um mit einer Stelle aus Kants „Logik" zu enden: „Der praktische Philosoph, der Lehrer der Weisheit durch Lehre und Beispiel, ist der eigentliche Philosoph".

BERICHT DES VIZEPRÄSIDENTEN, HERRN LEHMANN, ÜBER DEN STAND DER ARBEITEN DER AKADEMIE

Erstattet in der öffentlichen Gesamtsitzung am 14. April 1973

Der Bericht, den ich über die Arbeiten der Akademie im Zeitraum vom April 1972 bis März 1973 zu erstatten habe, schließt sich einer langen Kette ähnlicher Darlegungen an. Bei diesem traditionell gewordenen Programmpunkt der in jedem Jahre stattfindenden öffentlichen Frühjahrssitzung der Akademie kann es sich weder um eine Einschätzung oder Bewertung noch um eine Rechenschaftslegung handeln. Die Arbeiten der Akademie sind so vielfältig, daß es notwendig ist, den für eine breitere Öffentlichkeit bestimmten Bericht auf die formelhaft kurze Kennzeichnung der Hauptergebnisse zu konzentrieren, zugleich aber auf einen Einblick in das Umfassende und Zusammenführende abzustellen, das Akademien vom Typ der Sächsischen Akademie unter dem Signum einer „Gelehrten Gesellschaft" kennzeichnet. Schon immer, schon lange vor der modernen, bewußt gepflegten notwendigen Kooperation der Wissenschaften war der Arbeitsstil der Akademie auf die Verbindung und Ergänzung klar unterschiedener Fachdisziplinen gerichtet. Schon immer zielte die Akademie auf eine weitgehende Kommunikation, das heißt auf die Verstärkung eines Verbindungsprozesses zwischen den Akademiemitgliedern selbst einschließlich ihrer Beziehungen zu den Akademien anderer Länder, insbesondere der Sowjetunion auf der einen Seite und dem praktischen Teilhaben der die Wissenschaft Nutzenden unserer sozialistischen Gesellschaft auf der anderen Seite. Diese Wechselbeziehung bedarf der bewußten Pflege und Förderung besonders in einer Zeit, in der die wissenschaftliche Erkenntnis und ihre Anwendung von Stufe zu Stufe sehr schnell fortschreitet und überdies überhaupt nur noch in kleinen Teilbereichen überschaubar ist.

Aber noch aus einem anderen Grunde sieht sich die Akademie in relativ kurzen Zeitabständen zu einer Darlegung ihrer Arbeitsergebnisse veranlaßt. Schon von ihrem gesellschaftlichen Auftrag her erscheint es für die Akademie notwendig, mittels einer Übersicht über ihre Aktivitäten sich zu vergegenwärtigen, ob die möglichst flexibel gestaltete Einzelforschung ihrer Mitglieder oder kleiner, von ihnen geleiteter Forschungsgruppen mit dem sich durch die Wirtschaftsentwicklung selbst ständig verändernden Gesamtkomplex von Zielstellungen und Praxisanforderungen in Einklang gehalten werden konnte.

Im neuen Statut der Sächsischen Akademie der Wissenschaften zu Leipzig vom 3. Februar 1971 wurde festgelegt, daß die Bearbeitung der wissenschaftlichen Aufgaben der Akademie in Kommissionen erfolgt. Diese Form der Arbeitsweise ist ebenfalls nicht neu. Sie erwies sich in der Tradition gelehrter Gesellschaften schon immer als zweckmäßig und dürfte auch in Zukunft die Wirkungsaktualität der Akademie fördern. Denn solche Kommissionen vermögen sich aus ihrer jeweiligen Zielsetzung und aus ihrer personalen, einen bestimmten Erkenntnisstand verkörpernden Zusammensetzung heraus von theoretisch nicht fundierten Vorstellungen ebenso fernzuhalten wie von einer Loslösung von den Aufgaben der Praxis. In der Akademie haben sich Vertreter sehr verschiedener Wissenschaften zur Gemeinschaftsarbeit bisher in 4 Kommissionen zusammengeschlossen, von denen sich jede in einer Art interdisziplinärer Wahlverwandtschaft um einen komplexen Fragenkreis gruppiert. So arbeitet zum Beispiel in der Kommission „Spezielle Probleme der Umweltforschung", die unter der Leitung von Herrn Neef steht, neben dem Biologen, dem Mediziner und dem Geographen der Geologe, der Chemiker, der Geophysiker und der Agrarwissenschaftler.

Wenn ich den Bericht mit den Arbeiten der in dieser Kommission tätigen Akademiemitglieder eröffne, so geschieht dies nicht nur im Hinblick auf die gegenwärtig häufige Konfrontation sehr vieler Bereiche der Natur- und Gesellschaftswissenschaften mit differenzierten Aufgaben der Umweltforschung. Es geschieht auch nicht allein, weil Herr Schwabe als Präsident unserer Akademie wie zugleich als Vizepräsident der AdW der DDR die Umweltforschung durch eigene, vielfältige Untersuchungen wie durch grundlegende organisatorische Maßnahmen, zum Beispiel auch durch die Initiative zur Begründung des im folgenden kurz als „Umweltkommission" bezeichneten Gremiums unserer Akademie sehr wirksam förderte. Es geschieht, weil hier in besonders klarer Weise der Charakter von Korrelationen solcher Wissenschaftsdisziplinen im Akademiebereich sinnfällig wird, die vom Standpunkt ihrer speziellen Forschungsrichtungen aus relativ weit voneinander entfernt sind und dennoch die Umweltforschung durch gemeinsames, im optimalen Fall integriertes Vorgehen zu fördern vermögen.

So lassen zum Beispiel die von Herrn Gersch und seinen Mitarbeitern durchgeführten Untersuchungen über den „Wirkungsmechanismus von Neurohormonen" zunächst nicht vordergründig erkennen, welche große Bedeutung auch für die Umweltforschung den von Herrn Gersch erarbeiteten grundlegenden Erkenntnissen auf dem Gebiet der Neurophysiologie und Endocrinologie von Insekten zukommt. Die komplizierten Untersuchungen der Nervendrüsenzellen einer Schabe, der Periplaneta americana, sowie einer Fliege aus der Familie der Drosophilidae erlaubten es, Rückschlüsse auf hormonale, im Gehirn, aber auch in anderen Körperteilen von Insekten wirkende Faktoren, die als Aktivitationsfaktoren bezeichnet wurden, zu ziehen. Außer einer ganzen Reihe generell wichtiger Ergebnisse dieser Untersuchungen wie z. B. der unterschiedlichen zellulären Wirkungsmechanismen des Gehirnhormons der Insekten, des Nachweises einer direkten Einwirkung eines der beiden experimentell isolierten Aktivitationsfaktoren auf das genetische System oder der Erkennung der Beeinflussung der Zellmembran durch das Verhalten eines anderen, zweiten ausgesonderten Aktivitationsfaktors mittels elektrophysiologischer Analysen — außer diesen grundlegenden Erkenntnissen ist die Bedeutung dieser Grundlagenforschungen für die Umweltproblematik darin zu sehen, daß hier neue Prinziplösungen zur biologischen Kontrolle — nicht zuletzt zur Ausschaltung von Schadinsekten durch die Steuerung der biologischen Entwicklung solcher Insekten erkundet werden, die als Schädlingsbekämpfer, als nicht toxisch wirkende Insektizide zu wirken vermögen.

Bei der Sicht auf die verschiedenen Formen von Umweltschäden und der vielen, gegen sie punktuell oder auf weiten Nutzflächen getroffenen Schutzmaßnahmen tritt eine Erscheinung nur wenig in das äußere Blickfeld: das ist der Mensch selbst als sicher wichtigstes Glied der Umwelt. Die schnell wechselnden und neuartigen Faktoren, die auf den Menschen als einen höchst entwickelten Organismus mit großer Anpassungsbreite einwirken, setzen die Aufklärung der physikalisch-chemischen Vorgänge in und an verschiedenen Organen des menschlichen Körpers voraus. Nur auf der Grundlage exakter Empirie in Form umfassender Beobachtungsreihen kann das Analytisch-Besondere mit dem Synthetisch-Allgemeinen verbunden werden, können zum Beispiel die Funktionen und Leistungen des menschlichen Herzens in die größeren Zusammenhänge des zeitlichen Ablaufes eines Menschenlebens, also der in den einzelnen Altersstufen sich vollziehenden physiologischen Wandlungen einbezogen werden. Alternsbedingte Komplikationen können bei chirurgischen Eingriffen schwerer wiegen als das operationspflichtige Leiden selbst —, ganz abgesehen von der bekannten Tatsache, daß sich unter den sich sehr schnell verändernden Umweltbedingungen das Bild ganzer Krankheitsgruppen wandelt und die Bekämpfung der Herz- und Gefäßkrankheiten neben den Krebskrankheiten den ersten Platz bei allen klinischen Behandlungen einnimmt. In den letzten Jahren hat sich unter solchem Aspekt Herr Emmrich mit einer kleinen Forschungsgruppe speziellen Untersuchungen der chemischen Zusammensetzung des menschlichen Herzmuskels sowie der unterschiedlichen biochemischen Struktur und Funktion von rechter und linker Herzkammer zugewendet. Konnte in vorangegangenen öffentlichen Sitzungen der Akademie über wichtige Ergebnisse der Untersuchungen des Herrn Emmrich berichtet werden, die durch Bestimmung der Mineralien Kalium und Natrium und des Lipidgehaltes im rechten und linken Herzmuskel erzielt wurden, so ist heute auf die weitere Bearbeitung der gleichen Fragestellung hinzuweisen, die nunmehr die Bestimmung des Magnesiumgehaltes der Herzmuskeln in seiner Bedeutung als Co-Faktor von Enzymen für die Erregbarkeit von Nerven und Muskeln zum Inhalt hatte. Von den Ergebnissen, die aus einem auf 4 Altersgruppen aufgegliederten Sektionsmaterial erzielt wurden, sei als Beispiel nur die Erkenntnis herausgegriffen, daß der Magnesiumgehalt der weiblichen Herzen von der Jugend bis zum 60. Lebensjahr höher als bei den männlichen Herzen sowohl im linken wie im rechten Muskel ist. Erst bei den 61—80-Jährigen wurden die höheren Magnesiumswerte bei den männlichen Individuen gefunden.

Im Dienst der klinischen Medizin steht die Chemie auch in dem von Herrn Strack und seinen Mitarbeitern weiterbearbeiteten Untersuchungskomplex der Carnitinesterwirkung. Die Wirkung von Carnitinester auf den physiologischen und pathologischen Stoffwechsel sowie auf den Tumorstoffwechsel konnte bis an die Schwelle des praktischen therapeutischen Einsatzes weiter aufgehellt werden. Die Zusammenarbeit mit der Hautklinik der KMU wurde intensiviert. An der Chirurgischen Klinik der KMU wurden mit Carnitin kombinierte Therapien bei bindegewebigen Erkrankungen mit ausgezeichnetem Erfolg erprobt. In Eigenversuchen wurde die positive Wirkung des Carnitins auf das Altersherz geprüft.

Teils mittelbar, teils unmittelbar sind auch die Arbeiten der Herren Buchheim, Runge und Arland auf spezielle Probleme des Umweltschutzes bezogen. Der immer kostbarer werdende Bodenschatz Wasser ist es, den Herr Buchheim unter dem speziellen Gesichtspunkt der Heilwässer, insbesondere ihrer physiologischen und biologischen Regeneration, nicht zuletzt im Hinblick auf Veränderungen des Spurenelemente-Spektrums und auf eine geophysikalisch-hydrologisch und klimatologisch fundierte Abgrenzung von Quellschutzgebieten untersuchte. Ausgangspunkte für die Untersuchung der physikalischen Besonderheit verschiedener Heilwässer waren sogenannte Akratothermen, d. h. einige mäßig temperierte Quellen von nur schwacher Mineralisierung, wie sie z. B. in Warmbad Wolkenstein, in Jánske Lázně und anderen Quellen der ČSSR vorhanden sind. Es konnten mittels verschiedener physikalischer, z. B. spektrometrischer und refraktometrischer Methoden bäderheilkundlich wichtige Alterungseffekte nachgewiesen werden, die sich als Folge von Gasaustritt verbunden mit langsamen Änderungen der Ionenassoziation verstehen lassen.

Nicht dem Grundwasser wie in den Untersuchungen Herrn Buchheims, sondern dem Oberflächenwasser, insbesondere den im Bereich des Leipzig-Hallenser Ballungsgebietes superverschmutzten Flüssen sind die Untersuchungen von Herrn Runge und seinen Mitarbeitern über synthetische Harze zur Adsorption von Phenol aus Abwässern gewidmet. Es wurden in Modifizierung vorangegangener Forschungen Präparate erarbeitet, mittels deren noch Spuren von Phenol, aber auch anderer Verunreinigungen, z. B. organische Säuren, besonders Salicylsäure, beseitigt werden können. Wird bedacht, daß selbst im tiefsten Süßwassersee der Welt, dem Baikalsee, Schutzmaßnahmen gegen die Abwässer der Holz- und Papierindustrie in der Akademie der Wissenschaften der Sowjetunion zur Diskussion stehen, so ist leicht die Notwendigkeit und Dringlichkeit der Arbeiten zu ermessen, mit denen sich Herr Runge in Zusammenarbeit mit den Chemischen Werken Wolfen und dem VEB Leuna-Werke einer ganzen Reihe verwandter, von Instituten der Akademie der Wissenschaften der DDR durchgeführten Forschungen zur Rehabilitierung mit Schadstoffen höchst belasteter Gebiete einfügt.

Bei den Arbeiten des Herrn Arland, in deren Rahmen sich die Pflanze als feines biologisches Umwelt-Meßinstrument auffassen läßt, handelt es sich um Unternehmungen, die das Reagieren von Winterweizensorten auf Licht zum Inhalt haben. Herrn Arland gelang es im Berichtszeitraum in Weiterführung seiner Versuche zur Ertragssteigerung von Kulturpflanzen, mittels Einsatz verschiedener, in ihrer Wirkung miteinander verglichener Leuchtstofflampen neben verschiedenen anderen Ergebnissen festzustellen, daß der N(= Stickstoff)-Ausnutzungsgrad unter Lumoflorlampen geringer ist als unter Lampen eines anderen Typs. Im weiteren Sinne fügen sich diese Forschungen in eine Umweltproblematik, die darauf gerichtet ist, die unmittelbar nach der Keimung von der jungen Pflanze zu durchlaufende temperaturempfindliche Entwicklungsphase genau zu beobachten und aus ihrem Verhalten unter experimentell klimatisierten Bedingungen Schlüsse für Maßnahmen zur Überwindung kritischer Keimsituationen in der offenen physischen, vom Menschen weithin umgestalteten Umwelt zu ziehen.

Manche solche wie andere isoliert vorgehende und notwendig punktuell sachbezogene Forschungen pflegen eines Teiles ihres Erfolges verlustig zu gehen, wenn sie sich nicht auf einer Reflexionsstufe bewegen, die die Einfügung ihrer Ergebnisse in territoriale Zusammenhänge ermöglicht. Hilfe hierzu, das heißt zu einem Gesamthandeln zum mindesten in Teilbereichen der Umweltforschung bietet nicht zuletzt jene Arbeitsgruppe, die unter der Bezeichnung „Naturhaushalt und Gebietscharakter" von Herrn Neef selbst zugleich, wie ich bereits erwähnte, als Vorsitzender der gesamten Umweltkommission der Akademie geleitet wird. Die Arbeitsgruppe hat nach Abschluß einer Reihe von Vorstudien zum physisch-geographischen Charakter der Naturräume und zur funktionellen Raumgliederung des Siedlungsnetzes vorrangig Fragen der Wechselbeziehungen zwischen gesellschaftlichen Prozessen bei der Nutzung des Naturdargebotes aufgegriffen.

Die Ziele, die sich Herr Neef gesetzt hatte, machten es erforderlich, den heterogenen Verbänden natürlicher territorialer Grundeinheiten, den sogenannten chorischen Einheiten, besonderes Augenmerk zu schenken. Dem Naturdargebot der räumlichen Einheiten, das in erster Linie auf die Landschaftsgenese und die Landschaftsdynamik zurückgeht, wird das Anforderungsbild der Gesellschaft gegenübergestellt und dadurch die Möglichkeit gewonnen, die planmäßige und nachhaltige Nutzung der natürlichen Ressourcen, die Impulse der menschlichen Tätigkeit auf die Naturressourcen und damit eine Reihe spezieller Umweltprobleme zu überschauen. Die Ergebnisse werden zur Zeit in einer Rahmenmethode zusammengefaßt. Sie gestatten, die Naturreichtümer und ihre Bewirtschaftung besser als bisher in ihrer territorialen Differenzierung in die ökonomische Planung einzubeziehen und künftige Auswirkungen technischer Maßnahmen durch rechtzeitige Klärung ihrer Konsequenzen für die Wandlung der Landschaftspotentiale zu erkennen.

In diesem Vorgehen wird offenkundig, daß die bisher im wesentlichen naturwissenschaftlich basierten Forschungen der Akademie zur Umweltproblematik sich nicht auf ein analytisches Befragen der Natur beschränken, sondern zielbewußt die Anwendung speziell erarbeiteter Methoden auf hochkomplexe Umweltbeziehungen und -zusammenhänge im physischen wie im technisch-wirtschaftlichen Bereich anstreben. Eine auf solche Weise verfahrende Umweltforschung, die Herr Neef unter dem Aspekt der von ihm vertretenen geographischen Wissenschaft auch in Bereichen außerhalb der Akademie richtungweisend zu fördern vermochte, trägt zugleich zur Erkenntnis der Gesetze bei, nach denen das Leben der menschlichen Gesellschaften als Folge des Zusammenwirkens wirtschaftlicher, sozialer, kultureller und natürlicher Faktoren abläuft. Der notwendige Brückenschlag zwischen den Gesellschafts- und Naturwissenschaften, also die Verklammerung kategorial grundverschiedener Gegenstandsbereiche innerhalb übergreifender Systemzusammenhänge bedarf einer weiteren Vervollkommnung der Erkenntnismittel gleichsam auf beiden Seiten der Brücke.

Im gesellschaftswissenschaftlichen Bereich der Akademiearbeiten, dem ich mich mit diesem Hinweis zuwende, bestehen drei Kommissionen, von denen zwei sich historischen Aufgaben widmen, während eine sprachwissenschaftlich orientiert ist. Ein gemeinsamer Grundzug in der Behandlung der gesellschaftswissenschaftlichen Stoffgebiete kann im Rahmen der Akademie darin gesehen werden, daß versucht wird, alle Merkmale, und zwar einzelne, einfache Tatsachen wie vielseitige, komplexe Erscheinungen eines gegenwärtigen gesellschaftlichen oder eines historischen Sachverhaltes als Quelle für die wissenschaftliche Nutzung zu erschließen oder zum mindesten leicht erschließbar zu machen. Eine solche Erkundungs- und Materialgrundlagenbearbeitung kennzeichnet in charakteristischer Weise einen wesentlichen Teil des Aufgabenbereiches der ältesten Arbeitsgruppe der Akademie, der unter Leitung von Herrn Buchda stehenden „Historischen Kommission“. Die Unternehmungen dieser Kommission haben überdies sämtlich einen regionalen, vorwiegend auf das Einzugsgebiet der Akademie bezogenen Charakter. Das spiegelt sich in der Herausgabe des umfassenden Unternehmens der Sächsischen Bibliographie wider, von der im Berichtsjahr die Leipzig-Bibliographie im wesentlichen als abgeschlossen angesehen werden kann. Das gilt ebenso von dem Historischen Kartenwerk über das ehemalige Sachsen, dessen Publikation jedoch auf erhebliche, im Berichtsjahr noch nicht überwundene Schwierigkeiten der technischen Reproduktion und des Druckes stieß, nachdem von den insgesamt 50 Kartenblättern bereits 28 Blätter im Entwurf — und von diesen wiederum 25 kartographisch-technisch bearbeitet, 16 in Form von Andrucken vorliegen. Das gilt ferner von der Bearbeitung und Herausgabe der Politischen Korrespondenz des Herzogs und Kurfürsten Moritz, deren 3. Band für den Druck vorbereitet werden konnte, während der 4., den Zeitraum vom 15. 5. 1548 bis zum 9. 11. 1551 umfassende Band weitgehend gefördert werden konnte. Das gilt schließlich für eine Reihe neuer Quellen, besonders über Strömungen des Pietismus, die aus den Handschriftenbeständen des Archivs der Franckeschen Stiftungen in Halle erschlossen werden konnten. Die Historische Kommission, die überdies den 4. Band des „Jahrbuches für Regionalgeschichte“ veröffentlichen konnte, suchte auch im Berichtsjahr weiterhin der Geschichtsforschung dadurch zu dienen, daß sie in klarer Sicht auf die möglichen Irrwege eines traditionellen Historismus an die großen editorischen Leistungen einer vergangenen Epoche in Neubesinnung bewußt anknüpfte. Der Rückgriff auf die Quellen der Geschichte im regionalen Rahmen, der neue Erkenntnisse und Einsichten auf der Grundlage bisher unbekannter Zeugnisse erschließt, geschieht im Wissen um die Verfeinerung und Vertiefung der historischen Methoden, die gerade in der Regionalgeschichte große Fortschritte zu verzeichnen hatten. Sie geschieht im Wissen um die Rangordnung der historischen Tatsachen, die erst durch die Analyse ihrer gesellschaftlichen Herkunft und ihres gesellschaftlichen Charakters ihr Eigengewicht erhalten. Die Veröffentlichung der politischen Korrespondenz des Herzogs und Kurfürsten Moritz von Sachsen zum Beispiel wird eine kritische Würdigung nicht nur dieser vielumstrittenen, bedeutenden Persönlichkeit ermöglichen, sondern sie öffnet auch den Blick für die Verflechtung regionaler Politik in die Weite damaligen großräumigen Handelns. Jede Gesellschaftswissenschaft ist methodisch durch ein Denken in Fronten gekennzeichnet. Das bestimmt auch die Position der Historischen Kommission wie jeder fortschrittlichen, weltaufgeschlossenen, humanistischen Geschichtswissenschaft.

Zwischen der Geschichtswissenschaft und der Sprachwissenschaft bestanden in der vergangenen Zeit der Akademieentwicklung noch bis weit in unser Jahrhundert hinein enge Beziehungen. In der Gegenwart aber, in der sich die weitere Entwicklung der natürlichen Sprachen in stärkerem Ausmaß als in der Vergangenheit als Kommunikationsmittel vollzieht, um die Anforderungen einer sich immer mehr komplizierenden, z. T. ganz neuartigen Umwelt zu erfüllen, hat sich in Dialektik zum Historischen, jedoch nicht aus einem ahistorischen Denken, die systematische Bearbeitung sprachlicher Sachverhalte und Entwicklungen zum Teil auf die Klärung von Strukturproblemen der Sprache verlagert. Die moderne Linguistik sucht dort, wo es der Gegenstandsbereich erlaubt, Prinzipien auf die Erkenntnis der Sprache und ihrer Entwicklung anzuwenden, die bisher nur in den als exakt bezeichneten Wissenschaften angewendet wurden. Aber das verengt nicht die Aufgabenstellung der unter Leitung des Sekretars der Philologisch-historischen Klasse, Herrn Große, stehenden „Sprachwissenschaftlichen Kommission“, sondern vermochte im Gegenteil die Weiterführung der Arbeit an den verschiedenen sprachwissenschaftlichen Unternehmungen der Akademie fruchtbar zu beeinflussen, wie die Bearbeitung des Wortschatzes der althochdeutschen Periode, die Erfassung, Ordnung und Untersuchung des Wortschatzes der Mundarten in der DDR, die Untersuchungen zur soziolinguistischen Entwicklung in der sozialistischen Gesellschaftsordnung sowie die Untersuchungen zur Geschichte der Personen- und Ortsnamen und der Namengebung in der DDR unter besonderer Berücksichtigung der soziolinguistischen Zusammenhänge. Über die Fortschritte in der Bearbeitung dieser Aufgaben, die ja neben einer vielfältigen praktischen Hilfe und Information zugleich der allgemeinen theoretischen Grundlagenforschung dienen, sei in einer sehr begrenzten Auswahl nur berichtet, daß die erste Lieferung des dritten Bandes des „Althochdeutschen Wörterbuches“ erscheinen konnte, daß von der Redaktion des Brandenburg-Berlinischen Wörterbuches die 7. Lieferung mit 7 Sprachkarten und das Manuskript eines Sammelbandes „Aktuelle Probleme der sprachlichen Kommunikation. Soziolinguistische Studien zur sprachlichen Situation in der DDR“ abgeschlossen wurde, daß vom „Mecklenburgischen Wörterbuch“ die 51. Lieferung druckfertig wurde, daß von der Arbeitsgruppe „Wörterbuch der obersächsischen Mundarten“ das Planziel von 30000 Stichwortansetzungen in der Berichtszeit erfüllt wurde und zugleich die wissenschaftliche Aufbereitung und Systematisierung besonders in Richtung auf die Erfassung alter Wortschichten, nicht zuletzt durch Auswertung von Karten des Deutschen Wortatlasses wie durch Erkundung neuen Wortgutes mittels Tonbandaufnahmen wesentlich gefördert wurde.

Die dialektologischen Arbeiten der „Sprachwissenschaftlichen Kommission“ gelten im Endziel nicht nur einer quellenkritischen Aufhellung der Sprachgeschichte, auch nicht allein einer statischen Lexikologie, sondern der Erkenntnis der Dynamik des Sprachlebens. Es wird an diesen Arbeiten ein Prinzip deutlich, das die philologisch-historischen Untersuchungen der Akademie im weitergesteckten Rahmen gesellschaftlich-geisteswissenschaftlicher Arbeit überhaupt kennzeichnet, — nämlich die breite Durchdringung wissenschaftlicher Forschung und operativ-publizistischer Tätigkeit. Die Altertumswissenschaft, deren Aktivitäten in der Akademie von Herrn Peek vertreten werden, gab durch die schon im vorigen Jahrhundert begonnene Großtat systematischer Sammlung von Inschriften als älteste und seinerzeit am meisten gepflegte Gesellschaftswissenschaft hierfür das Vorbild. Die Mitarbeit am Corpus Vasorum Antiquorum, einem großangelegten Unternehmen der Union Académique Internationale, wie am „Mittellateinischen Wörterbuch“ erscheint heute als Folge einer bemerkenswerten historischen Tiefenwirkung, und dies in einem Sinne, der auch die bekannte, im Auftrage der Akademie von Herrn Prof. Dr. H. Salié geleitete Redaktion und Herausgabe des von J. C. Poggendorff begründeten „Biographisch-Literarischen Handwörterbuches der exakten Naturwissenschaften“ nicht nur als traditionelle Verpflichtung rechtfertigt, sondern als Ausdruck der Bezogenheit von Persönlichkeiten und Ereignissen auf das in sich höchst differenzierte gesellschaftliche Ganze gebietet. Die Poggendorff-Redaktion konnte die Arbeit am 4. Teil des Bandes VIIb (F—K) vollenden. Etwa die Hälfte des bedeutenden Gesamtwerkes steht damit einer breiten wissenschaftlichen Öffentlichkeit zur Verfügung. Die wissenschaftliche und dokumentarische Fundierung, die durch den „Poggendorff“ erstrebt und, nach dem internationalen Widerhall zu urteilen, auch mit größtem Erfolg erreicht wurde, enthüllt besonders sinnfällig den allgemeinen Wert der philologisch-historischen wie der naturwissenschaftlich-historischen Aktionen der Akademie. Sie zeigt das Gewicht des Begriffes Geschichte an, der mit der bloßen Bezeichnung einer in der Gegenwart in schnelle Rotation geratenen Zeitlichkeit, einer bloßen Chronologie, häufig verwechselt wird. In der Kartographie zum Beispiel als einer sehr alten Wissenschaft und Technik vollziehen sich gegenwärtig durch den Einfluß der Automatisation und der Computertechnik auf die schnelle Herstellung besonders von isarithmischen und choropletischen Karten als Forschungsmittel tiefgreifende Wirkungen. Wie am Beispiel der Kartographie, so könnte man sicher in sehr vielen anderen fachbezogenen Rückblicken nachweisen, daß wir einer Täuschung unterliegen, wenn wir annehmen, daß die geschichtliche Entwicklung sich gegenwärtig schneller vollzöge als in vergangener Zeit. Das Grundwissen aktualisiert sich in unserer Zeit mittels des Großeinsatzes der Technik auf allen Gebieten zwar schneller, — aber es entwickelt sich auch in der Gegenwart nur sehr langsam, nur in ganz kleinen, mühsamen Schritten. Daher auch zielt das organisierende Prinzip, nach dem im Poggendorff'schen Handbuch vorgegangen wird, darauf, auf biographischem Wege Werte zu erschließen, die als wissenswürdig für den Fortschritt der Forschung erscheinen und zugleich geschichtliches Gewicht haben. Es gilt, einen leichten Zugang zu schaffen zur Durchdringung, zur Durcharbeitung des Bedeutsamen in der Entfaltung naturwissenschaftlichen Denkens. Es gilt, der Originalität eines einzelnen Gedankens nachzuspüren wie ebenso jener Originalität, die gerade in der Verknüpfung, im Aufbau und in einer Gesamtsicht liegen kann. Das historische Haupt- und Grundanliegen des Poggendorff'schen Handbuches wird im übrigen in Zukunft in einer anderen, seine ursprüngliche Zielfunktion aber konsequent weiterführenden äußeren Form erfüllt werden: an die Stelle der lexikalischen Anordnung wird die Herausgabe in einzelnen, disziplingebundenen Bänden treten.

Im Berichtszeitraum konnte mit der Vorbereitung des Bandes Mathematik begonnen werden, während gleichzeitig das Register zu den erwähnten Poggendorff-Bänden I—VII bearbeitet wurde. Es erscheint sicher, daß durch diese redaktionell notwendige Veränderung die historische Funktion des Werkes, die Verbindung von strenger Wissenschaftlichkeit mit biographisch-literarischem Verständnis keinesfalls beeinträchtigt wird.

Die Akademie ist nur ein relativ kleines, aber wesentliches wissenschaftliches Wirkungszentrum starker eigener Prägung. Die Leistungen, die ihre Mitglieder im Berichtszeitraum in der verschiedensten Weise, z. B. auch durch die hier nicht zu behandelnde Mitarbeit in den beiden Klassen und im Plenum erbrachten, lassen erkennen, daß es bei den Aktivitäten der Akademie wie bei ihrer Reaktion auf sehr berechtigte, notwendige Anforderungen des Staates und der in ihm und durch ihn organisierten sozialistischen Gesellschaft nicht nur auf den Gegenstand der Forschung, sondern zugleich auf die Art der Betrachtung ankommt. In den Natur- wie in den Gesellschaftswissenschaften ist, soweit sie im Rahmen der Akademie gefördert werden konnten, die gesammelte Kraft darauf gerichtet gewesen, mitzuwirken an den großen Aufgaben, die nicht zuletzt darin bestehen, daß die in unsere Umwelt durch die Arbeit, durch die Wissenschaft und Technik einfließenden Kräfte aus Verantwortung für das Ausmaß von Veränderungen in eine kontrollierbare, Aktionen mitunter begrenzende, ja mit Verzichten für den Einzelnen wie die Gesamtheit verbundene Form gebracht werden. In solchem Sinn mögen die Arbeiten, über die berichtet wurde, insgesamt den Standort kennzeichnen, von dem aus die Akademie in Pflicht und guter Zuversicht ihre Funktion auch in Zukunft mit Erfolg zu erfüllen sucht.

BERICHT DES VIZEPRÄSIDENTEN, HERRN LEHMANN, ÜBER DEN STAND DER ARBEITEN DER AKADEMIE

Erstattet in der öffentlichen Gesamtsitzung am 20. April 1974

Der Bericht, den ich über die Arbeiten der SAW im Zeitraum vom 1. IV. 73 bis zum 31. III. 74 zu erstatten habe, soll der allgemeinen Information, zugleich aber auch der Urteilsbildung über den Beitrag der Akademie zur Entwicklung unserer sozialistischen Gesellschaft dienen. Ein Bericht dieser Art möchte im Idealfall ein wirkliches „Teilhaben" an den Erkenntnissen oder sogar an den Erfahrungen ermöglichen, die etwa bei der Lösung theoretisch bedeutender Aufgaben oder ihrer wissenschaftlich-schöpferischen Anwendung gewonnen wurden. Aber einer solchen Einsicht in die Arbeiten der Akademie stellen sich Hindernisse in den Weg. Zum einen sind die Ergebnisse, über die zu berichten ist, aus einem kontinuierlichen Forschungsprozeß gleichsam herausgeschnitten. Zum anderen bestimmt der Berichtszeitpunkt, nicht die jeweils erreichte Reife der Ergebnisse die Form der Berichterstattung. Und drittens sind durch die knappe Berichtszeit enge Grenzen gezogen. Eine Auswahl aber aus dem umfänglichen Berichtsmaterial käme einer nicht zulässigen Wertung gleich. So bleibt als erlaubtes Mittel nur die Möglichkeit, über die wissenschaftlichen Aktivitäten der Akademie unter einem bestimmten übergreifenden Gesichtspunkt zu referieren. Im vergangenen Jahr war es der Umweltaspekt, der in diesem Sinne leitend war. Heute sei der Bericht in einer Richtung geboten, die der Herr Präsident soeben in einem der Punkte seiner Ansprache anklingen ließ: Es sei der Blick auf die Proportionen zwischen Grundlagenforschung, angewandter Wissenschaft und der kulturell-gesellschaftlichen Entwicklung gelenkt. Es gibt keine genau bestimmbare Grenzlinie zwischen diesen drei Bereichen. Sie stehen in einer ständigen, sich gegenseitig bedingenden Veränderung. Die Beobachtung ihres Verhältnisses zueinander wird es überdies erleichtern, möglichst allgemeinverständlich die Problematik anzudeuten, die hinter dem eigentlichen Gegenstand des Berichtes, den Forschungsergebnissen selbst, wirkt.

Herr Präsident Schwabe berührte in seiner Ansprache auch die Bedeutung, die der Wirkungsgrad wissenschaftlicher Arbeit im Zuge eines rasch wachsenden Anteils der Wissenschaftler an der Gesamtzahl der Beschäftigten hat. Die Notwendigkeit und Dringlichkeit, bestimmte ökonomische und gesellschaftliche Ziele schnell zu erfüllen, erfordert die Zusammenarbeit und den Zusammenschluß einer großen Zahl von Wissenschaftlern in technisch und organisatorisch hervorragend ausgerüsteten Forschungszentren. Darüber besteht nirgendwo in der Welt ein Zweifel. Die verantwortliche Leitung, Koordinierung und zu einem bedeutenden Anteil auch die Durchführung eines Teiles der Grundlagenforschung vollzieht sich, wie jedermann weiß, in unserer Republik in der AdW der DDR zu Berlin und in ihren auf das ganze

Land verteilten Zweigeinrichtungen. Das hat aber die Arbeit kleinerer Forschungsgruppen in Akademien und Universitäten keinesfalls in ihrem Wert gemindert oder unfruchtbar gemacht. Im Gegenteil, eine Interpretation der Leistungen, die die SAW auch im vergangenen Jahr aufzuweisen hat, würde sichtbar machen, daß individuelle, auch in der Themenbildung freie, zugleich aber auf die praktische Anwendung von Anbeginn zielgerichtete Arbeiten kleiner Arbeitsgruppen oder sogar einsam vorangehender Einzelner sich günstig auf den Fortgang der Forschung und auf ihre volkswirtschaftliche Produktivität auswirken können. Eine Kommentierung der Leistungen des Berichtsjahres würde ferner erweisen, daß in den kleinen Forschungskollektiven der SAW, deren Untersuchungen sich nicht selten als Initiativforschungen enthüllen, eine flexible Aufgabengestaltung oft leichter gewährleistet ist als in großen Forschungseinrichtungen, in denen das formalistische Element einen erheblichen Zeit- und Kraftaufwand benötigt. Ein ursprünglich eingeschlagener Forschungsgang kann nicht zuletzt auf Grund eines Effektes, für den die Kybernetik den sinnfälligen Begriff der „Rückkoppelung" geprägt hat, eine schnelle Abänderung durch die Aufnahme neuer, bei Beginn der Untersuchungen mehr geahnter als konkret formulierter Folgethemen erfahren, — zum Beispiel durch Einflüsse, die sich aus der Verbindung von Erkenntnissen der Grundlagenforschung mit kreativen technologischen Arbeiten oder im übergreifenden Sinn mit den niemals aus der Sicht zu verlierenden gesellschaftlichen Gesamtzielen ergeben. Wer sich der Mühe unterzöge, den diesjährigen Bericht über die Arbeiten der Akademie mit vorangegangenen zu vergleichen, würde — zum Teil an leichten Abwandlungen der Thematik, zum Teil an kleinen Schwerpunktverlagerungen der Forschungen — diese Elastizität feststellen. Es geht in diesem Sinn auch im Bereich der SAW um die höchstmögliche Effektivität bei geringstmöglichem Aufwand. Jeder Erfolg bei der Annäherung an dieses Ziel gibt denen, die sich in den Dienst der von der SAW betriebenen Forschungen gestellt haben, die Kraft und das Selbstvertrauen, durch die Bewältigung ihrer wissenschaftlichen Aufgaben zur weiteren Entwicklung unserer sozialistischen Gesellschaft beizutragen.

Als ein erstes Beispiel für die Bedeutung einer induktiven, vom Experiment ausgehenden und zur Anwendung fortschreitenden, fest in den theoretischen Erkenntnisprozeß eingefügten Grundlagenforschung sei auf die von Herrn Gersch und seinen Mitarbeitern fortgeführten Untersuchungen über den Wirkungsmechanismus von Neurohormonen gewiesen. Herr Gersch, der durch zahlreiche Originalarbeiten über Invertebraten-Hormone hervorgetreten ist, also über die Bildungs- und Speicherorte von Hormonen bei wirbellosen Tieren, insbesondere über die Neurosekretion, das heißt über die Bildung von Inkreten, von Stoffen der inneren Sekretion, in Nervenzellen, konnte mittels neuerarbeiteter Mikromethoden die Voraussetzungen zur Reindarstellung kleinster Hormonmengen und damit auch zur Strukturaufklärung erfüllen. Die weitere Bearbeitung des Problemkreises, mit dem Ziel, allgemein gültige, schließlich praktisch anwendbare Funktionsprinzipien zu gewinnen, wird in Zusammenarbeit mit dem Institut für Naturstoffe der Akademie der Wissenschaften der UdSSR in Moskau erfolgen. Im Zusammenhang mit diesen Untersuchungen konnte ferner das funktionelle Zusammenwirken von Neurohormon D mit aminergeen und cholinergeen Faktoren am Insektenherzen erstmalig experimentell nachgewiesen werden. Auch die Beziehungen, die zu den Enzym- und Strukturproteinen und deren Dynamik bestehen, also zu den hochmolekularen chemischen Substanzen von besonderer Struktur, den Ribonukleinsäuren, die als entscheidende Träger des Mechanismus der Vererbung bekannt sind, konnten mittels verschiedener Testsysteme, zum Beispiel durch Ausarbeitung des sogenannten Prothoracaldrüsentestes, zu einem Teil-Abschluß gebracht werden.

Die biologischen Arbeiten des Herrn Gersch und seiner Mitarbeiter sind auf die Erkenntnis bestimmter physikalischer und chemischer Seiten der Lebenserscheinungen gerichtet, die sich in ihrer praktischen Auswirkung bis hin zur Medizin auswirken können. Ergebnisse der allgemeinen Hormonforschung veränderten beispielsweise die Auffassung der Medizin über viele Krankheiten. Die Proportion dieser ausgesprochenen Grundlagenforschung zur Anwendung wird sich in Zukunft mit dem Fortschritt dieser Untersuchungen wahrscheinlich durch eine erhebliche Steigerung ihres praxis- und technologiebezogenen Anteils verändern, so wie es bei den von der Akademie geförderten Arbeiten auf dem Gebiet der chemischen Physiologie und der Medizin, durch den anderen Charakter der Untersuchungen bedingt, zum Teil von Anbeginn festzustellen ist. Die von Herrn Strack durchgeführten Untersuchungen über die Wirkung von Carnitinester auf den physiologischen und pathologischen Stoffwechsel und auf den Tumorstoffwechsel (Kurzbezeichnung „Carnitinesterwirkung“) sowie die von Herrn Emmrich initiierten Studien über die Hyperlipoproteinämie, eine heute weit verbreitete, durch den erhöhten Gehalt des Blutes an Fettbestandteilen (Lipoiden) gekennzeichnete Krankheit von großer epidemiologischer Bedeutung, zeigen, so verschieden die Forschungswege und -inhalte dieser beiden Arbeitsrichtungen sind, in ihrer Eigenart als medizinische Grundlagenforschung wie in ihrer Anwendung, in diesem Fall der klinischen Zielrichtung, verwandte Züge. Beides sind experimentelle Untersuchungen auf der Basis theoretischer Arbeiten mit einem klar definierten praktischen Endzweck. Beide stehen im Dienste der Diagnostik. Die Arbeiten von Herrn Strack sind vom Objekt her an das Carnitin gebunden, das im Energiestoffwechsel des menschlichen und tierischen Körpers eine der Kernsubstanzen darstellt. Carnitinzuführungen, — das ergaben die Untersuchungen von Herrn Strack und seiner Mitarbeiter —, können bei Stoffwechsel- und Bindegewebserkrankungen in verschiedener, zu steuernder Richtung wirksam werden. Carnitin, das verändernd auf den Stoffwechsel einwirkt, also als Metabolit angesehen werden kann, wurde therapeutisch in mehreren Kliniken, auch in der chirurgischen Klinik der KMU eingesetzt.

Herr Emmrich wertete das im Laboratorium für Alternsforschung der von ihm geleiteten Medizinischen Universitätsklinik der KMU von 100 Patienten mit primärer Hyperlipoproteinämie gewonnene Material unter anderem mittels Lipoproteinelektrophorese im Agar-Agarosegel für eine Einteilung in Hyperlipoproteinämietypen aus. Die Bedeutung dieser Typisierung, die insbesondere die prozentuale Altersverteilung der Patienten bezogen auf ihre Gesamtzahl berücksichtigt, wird klar im Hinblick auf den engen Zusammenhang zwischen Hyperproteinämie und Arteriosklerose: Patienten mit Koronarinsuffizienz, mit Schäden der Kranzgefäße des Herzens, weisen bis zu 80% eine Hyperlipoproteinämie auf.

Die von Herrn Emmrich in Verbindung mit der klinischen Kennzeichnung des Verhältnisses zwischen Cholesterol- und Triglyzeridwerten erzielten Ergebnisse einer chemisch orientierten, medizinischen Grundlagenforschung dienen praktisch der Erhöhung der Lebenserwartung, die in den letzten 100 Jahren, besonders aber in der Gegenwart so große Erfolge erzielte und zugleich als allgemeinste Kennziffer des Gesundheitszustandes einer Bevölkerung angesehen werden kann. Es ist allerdings ein merkwürdiges, in solchem Zusammenhang wohl der Erwähnung wertes Phänomen, daß mit der Verlängerung des Zeitraumes, der zwischen den beiden Polen Geburt und Tod für die Reifung und Gestaltung des Lebens in den hochindustrialisierten Ländern liegt, der Gedanke an den Tod vom modernen Menschen so weit wie möglich an die Peripherie des Lebens verdrängt wird. Rainer Maria Rilke läßt dies Malte Laurids Brigge kritisch aussprechen, wenn er sagt, daß man einen Tod sterbe, der zur Krankheit gehöre — nicht zu dem Menschen, für dessen Denken und Handeln doch die innere Einstellung zum Tod wie zur Geburt eine große Bedeutung hat.

Selbstverständlich muß die Krankheit selbst, die Störung des vitalen Gleichgewichts, zentrales wissenschaftliches Untersuchungsobjekt sein. Nicht selten spielen die medizinisch nutzbaren, räumlich und quantitativ ja stets begrenzten Heilwässer in diesem Zusammenhang eine wichtige Rolle. Die Physik der Thermalwässer, ein Kapitel der exakten Grundlagenforschung für die therapeutische Wirkung und technische Behandlung unserer Bäder, wurde von Herrn Buchheim dadurch gefördert, daß die Untersuchung örtlicher natürlicher Kontaminierung mit Nachfolgeprodukten Rn-haltiger Quellen als Teilglied einer geplanten größeren interdisziplinären Forschungsthematik zwischen Medizin, Physik und Geowissenschaften weiter vorangetrieben werden konnte. Die Forschungsergebnisse des Herrn Buchheim und seiner Mitarbeiter fanden ihren Niederschlag unter anderem darin, daß die Struktur, die thermodynamischen und Transporteigenschaften spezieller konzentrierter wäßriger Elektrolyte in Zusammenarbeit mit dem Zentralinstitut für Kernforschung Rossendorf und dem Institut für physikalische Chemie der AdW der UdSSR untersucht wurden. Die Ergebnisse dieser Forschungen wurden (von Herrn Buchheim im Rahmen einer Arbeitstagung über „Probleme der flüssigen Phase" in Rostock, Oktober 1973, vorgetragen und anschließend, wie auch die Arbeit seines Mitarbeiters Dr. M. Klose über „Schwingungsspektronomische Untersuchungen an speziellen wäßrigen Elektrolyten") bereits veröffentlicht. Eine Kooperationsvereinbarung mit den Staatsbädern Bad Elster und Bad Brambach wurde getroffen und eine Abstimmung mit dem Staatlichen Amt für Atomsicherheit und Strahlenschutz in der DDR angebahnt.

Ebenfalls physikalisch in der Methode, jedoch biologisch und wirtschaftlich in der Auswirkung sind die Untersuchungen des Herrn Arland über die Beziehungen einzelner physikalischer Faktoren zum Pflanzenwachstum. Es ist leicht verständlich, daß in einer Zeit, in der ganz neue Licht- und Strahlungsquellen, z. B. Lumineszenzlampen, zur Verfügung stehen, ihre Anwendung in der Wissenschaft und Technik auch für die Beeinflussung der Ertragsfähigkeit von Nutzpflanzen herangezogen wird. Bei seinen vergleichenden Untersuchungen über die Wirkung photoperiodischer Effekte konnte von Herrn Arland festgestellt werden, daß die unter „Blaulicht“ erfaßten Evatranspirationswerte Beziehungen zum Gesamtkornertrag erkennen lassen, wobei niedrige Evatranspirationswerte hohen, hohe Evatranspirationswerte hingegen niedrigen Kornmassen je Gefäß entsprechen.

Einem einzelnen chemischen Faktorenkomplex, der Moleküladsorption auf dem Sektor der Silanharze, waren Forschungen des 1973 der Akademie durch den Tod entrissenen Herrn Runge gewidmet. Herr Prof. Dr. Wolf, Direktor der Sektion Chemie der Martin-Luther-Universität Halle, der die Arbeitsgruppe der Sächsischen Akademie in seine Obhut nahm, legte einen umfassenden Abschlußbericht vor, aus dem hervorgeht, daß die Untersuchungen über die Reinigung von Abwässern mittels Adsorbentien, also mittels der Anlagerung gelöster Stoffe an feste Stoffe durch molekulare Bindung, Erkenntnisse unter anderem über die Eliminierung von Giften durch selektiv wirkende anorganische Adsorbentien sowie über die Vernetzwirkung von Vinyltrimethoxysilan lieferten. Der technologische Aspekt der Untersuchungen war insbesondere auf den Einsatz von hochdispersen Ionenaustauschern und das Leistungsverhalten von Adsorbertrennsäulen gerichtet.

Die naturwissenschaftlichen Grundlagenforschungen der SAW sind, wie überall dort, wo die Wissenschaft am Werk ist, stärker an gesellschaftlichen und wirtschaftlichen Leitbildern orientiert als es gemeinhin den Anschein hat, — schon deshalb, weil Strukturen der Natur und Strukturen der Wirtschaft in Wechselwirkung aufeinander bezogen sind. Viele Fragen der Wissenschaft münden letzten Sinnes in das zentrale Problem der optimalen Gestaltung der Lebensumwelt ein. Man kann, sehr vereinfacht, sagen: der übergeordnete Gesichtspunkt, unter dem die Ergebnisse der Grundlagenforschung erzielt werden, ist die viel berufene Einheit der Wissenschaft, die aber erst gewährleistet ist, wenn Natur-, Gesellschafts- und Wirtschaftswissenschaften eng zusammenarbeiten. Eine wichtige Voraussetzung, sich diesem Ziel mit den der Wissenschaft eigenen kleinen Schritten stetig und zäh zu nähern, ist die theoretische Festlegung von Verfahrensgrundlinien, die geeignet sind, Beziehungen zwischen den Ergebnissen der modernen Naturwissenschaften und den grundlegenden Wertsystemen des Menschen, wie sie von den Wirtschafts- und Sozialwissenschaften erarbeitet werden, quantitativ und qualitativ zu ermitteln. Die großen Schwierigkeiten, die hier zu bewältigen sind, hängen mit der Stellung und Bedeutung der Theoriebildung in den Natur- wie in den Gesellschaftswissenschaften zusammen. Wissenschaftliche Erfahrungen sind wesentlich durch die ihnen zugrunde liegenden theoretischen Voraussetzungen bestimmt. Theorien sind als offene Denksysteme in beiden großen Bereichen der Wissenschaft, den Natur- wie den Gesellschaftswissenschaften, an den ständig neu gewonnenen Ergebnissen zu überprüfen. Die dialektische materialistische Methodik hat hier ihre Wurzeln. Da naturwissenschaftliche Theorien nicht in erster Linie auf das praktische Handeln ausgerichtet sind, sondern darauf, auf Grund bereits erkannter Naturgesetze oder auf Grund von Hypothesen mit eigens ersonnenen Methoden Neues zu erkennen und in eine systematische, Gesetzmäßigkeiten unterworfene Ordnung zu bringen, so müssen die auf verschiedenen gedanklichen Ebenen strukturanalytisch gewonnenen Erkenntnisse der Natur- und Gesellschaftswissenschaften homologisiert werden, das heißt:

sie sind für Vergleiche und für begriffliche Identifizierungen geeignet zu machen.

Wechselnde Bedingungen, sei es in der Natur selbst oder auf Grund menschlicher Eingriffe in die Natur, sei es im wirtschaftlichen oder politischen Sektor gesellschaftlichen Handelns, lösen neue, unterschiedliche Gesetzmäßigkeiten aus. In einer Zeit, die durch umwälzende Änderungen der natürlichen wie der gesellschaftlichen Gleichgewichtsbedingungen gekennzeichnet ist, in der reine Luft, sauberes Wasser oder unverseuchter Boden zur schutzbedürftigen Ressource, zum immer wertvoller werdenden Naturpotential werden, — in einer Zeit, in der die Humboldtsche Vorstellung der Einheit — und zugleich Schönheit — der in sich so komplizierten, leicht verletzbaren Naturphänomene ganz neu begriffen werden muß — in dieser Zeit, in der der Bauplan unserer Kulturlandschaft unter dem Einfluß neuer Sozialkräfte, neuer Technologien und der Dynamik des sozialistischen Wirtschaftssystems eine tiefe Strukturveränderung erfährt, sind fast alle Wissenschaften, besonders aber jene, die sich mit Fragen des Stoffaustausches zwischen Natur und Gesellschaft beschäftigen, aufgerufen, die theoretische Tragfähigkeit ihrer Ergebnisse für die Anwendung in der Praxis zu überprüfen.

In solchem Sinne hat sich der Vorsitzende der mathematisch-naturwissenschaftlichen Klasse, Herr Neef, mit der von ihm geleiteten Arbeitsgruppe „Naturhaushalt und Gebietscharakter" in speziellen Analysen und in Auswertung der Ergebnisse differenzierter territorialer Strukturforschungen, die mit geoökologischen Übersichtscharakterisierungen und Kartierungen im Maßstab 1:200000 bis 1:500000 verbunden waren, der Untersuchung der Relationen zwischen Naturdargebot und gesellschaftlicher Anforderung zugewandt. Schon im Bericht über die Arbeiten Herrn Neefs und seiner Arbeitsgruppe über die Forschungsarbeiten im Jahre 1972 konnte darauf hingewiesen werden, daß Herr Neef angeregt hatte, der Analyse des Naturdargebots sogenannte Anforderungsbilder gegenüberzustellen, die sich auf einzelne Landschaftspotentiale beziehen. Das wurde in den Plan 1973 aufgenommen. Die Ergebnisse der inzwischen durchgeführten Beispieluntersuchungen wurden auf einem RGW-Symposium in Bratislava zur Diskussion gestellt und als erste Schritte in einer bisher nicht entwickelten Arbeitsrichtung als bedeutsam eingeschätzt. In entsprechender Weise sind neue Ansätze entwickelt worden, die darauf abzielen, die Naturhaushaltsproblematik durch das Bezugssystem Siedlungsnetz—Naturdargebot zu erweitern. Herr Neef hat in einer für die SAW charakteristischen, individuell geprägten Arbeitsweise und mit einer für die SAW wiederum kennzeichnenden sehr kleinen Schar von Mitarbeitern einen Beitrag zur Auffindung eines Lösungsweges geliefert, der zweifellos eine der möglichen Durchbruchsstellen bezeichnet für die Bewältigung jener bisher immer erneut programmatisch geforderten, aber durch schöpferische wissenschaftliche Arbeit bisher nur wenig geförderten Gesamtproblematik der Wechselbeziehungen zwischen Natur und Gesellschaft, spezieller: der im gesellschaftlichen Reproduktionsprozeß wirksamen Wechselbeziehungen zwischen Naturraumstruktur und der territorialen Produktions-, Siedlungs- und Infrastruktur sowie der Veränderungsmöglichkeiten ihrer naturbedingten Belastbarkeitsgrenzen. Die Proportion zwischen Grundlagenforschung, Technologie und landeskultureller

Wissenschaftsauswertung ist hier wie in allen Bereichen, in denen Mensch und Erde nicht getrennt betrachtet werden können, in ständiger Veränderung. Schon in der Konzeption, die Leibniz für die Aufgaben der Akademien entwarf, wird die Welt als ein in fortdauernder Entwicklung befindliches, zusammenhängendes Ganzes mit dem zur Abstraktion, zur Mathematik, zur Kunst, zu berechneten Vorstellungen über Raum und Zeit befähigten Menschen als Zentralphänomen betrachtet. Spätestens seit Hegel und Marx steht das Problem des Sinnes und der Richtung der Geschichte im Vordergrund gesellschaftlichen Interesses. Und Immanuel Kant, dessen 250. Geburtstag in diesen Tagen in der ganzen Welt Anlaß zu wissenschaftlichen Gedenkfeiern, u. a. zu großen, von den Akademien der DDR und UdSSR veranstalteten Symposien war, sah die Geschichte als einen Prozeß an, der nur glücklich verlaufen kann, wenn der einzelne so handelt, daß sein Tun Prinzip einer allgemeinen Gesetzgebung sein könnte. Die Verantwortlichkeit des Einzelnen wie ganzer Staaten und Völker, die in diesem „kategorischen Imperativ" zum Ausdruck gelangt, setzt voraus, daß auch im historischen Bereich, weiter und tiefer als nur bis zur Oberfläche der Ereignisse gesehen wird. Angesichts der wachsenden Komplexität, aber auch der Gefahr, den „Überbau" nur als eine Art Balkenverstärkung für ein maximales, rein ökonomisch bedingtes Wachstumsdenken anzusehen, ist es in unserer Gegenwart mindestens ebenso notwendig wie im vergangenen Jahrhundert, als Akademien und Universitäten philologisch-historische Klassen und Fakultäten begründeten, die verborgenen, die Evolution bestimmenden Faktoren des Geistes- und Kulturlebens zu erforschen. Die Noosphäre, ein Begriff, den der bedeutende sowjetische Mineraloge und Geochemiker V. J. Vernadsky als Ausdruck für die gegenwärtig vom menschlichen Geist tief umgeprägte Umwelt der Geo- und Biosphäre als eine neue Schichtung gleichsam aufsetzte, erfordert als logisches Korrelat eine Noogenese, die gesellschaftlich-kulturelle und physische Wachstumsprozesse in den wissenschaftlichen Griff zu bekommen versucht.

Solcher Erfassung evolutiver Tendenzen, die sich in der Originalität eines wissenschaftlichen Gedankens und seiner Umsetzung in die Tat am klarsten spiegeln, dient auch das international in höchstem Maße geschätzte Großunternehmen der Akademie, die von Herrn Prof. Salié geleitete, von seiten des Präsidiums unter der Obhut des Herrn Neef stehende Herausgabe des von dem Physiker Johann Christian Poggendorff 1863 mit dem genialen Blick für die Wirksamkeit echter Probleme begründeten „Biographisch-Literarischen Handwörterbuches zur Geschichte der exacten Naturwissenschaften". In einer Plenarveranstaltung, die die Akademie anläßlich des 25. Jahrestages unserer Republik im September vorgesehen hat, wird dieses Werk, dessen jetzige Redaktion gleichfalls 25 Jahre besteht, eine besondere Würdigung erfahren, und zwar zusammen mit der Behandlung der umfassenden sprachwissenschaftlichen Erkundungs- und Grundlagenbearbeitungen der Akademie. Das Poggendorff'sche Handbuch, das mit jeder neuen Lieferung einen höchst beachtenswerten, starken internationalen Widerhall auslöst, ist weit mehr als ein bloßer Leitfaden naturwissenschaftlicher Chronologie. Dieses Standardwerk der Wissenschaftsgeschichte bringt durch den verantwortungsvollen, stillen Einsatz seiner Bearbeiter das gewiß Zeitbedingte, aber zugleich das Einmalige, nicht selten Irreversible in ein unablösbares Verhältnis zur Gegenwart. Das Hauptziel der Bearbeiter des Poggendorff ist darauf gerichtet, Werte von geschichtlichem Gewicht für den wissenschaftlichen Fortschritt durch Herausstellung des Bedeutsamen in der Entfaltung naturwissenschaftlichen Denkens zu erschließen. Die Poggendorff-Redaktion konnte bis zum März dieses Jahres vom 5. Teilband von VIIb die erste Lieferung veröffentlichen. Seit Beginn des Druckes der Abteilung VII im Jahre 1954 konnten von der Redaktion 10 Bände mit einem Gesamtumfang von über 7000 Seiten herausgegeben werden.

Die Biographien, die historische Fortschritte der Naturwissenschaften den gegenwärtigen, auf Zeitgewinn und rationelles Arbeiten besonders angewiesenen Wissenschaftlern leicht zugänglich machen, vermögen die Rolle eines Brückenschlages zu den geschichtlichen Arbeiten im engeren Sinn zu bilden, das heißt zu Unternehmungen, die die von Herrn Buchda geleitete „Historische Kommission" im Sinne einer Aufbereitung historischer Quellenbefunde oder bedeutender historischer Zeugnisse leistet — wie zum Beispiel der Bearbeitung der „Politischen Korrespondenz des Herzogs und Kurfürsten Moritz von Sachsen", von der der 3. Band zum Druck bereitliegt. Von den Bibliographien sächsischer Städte konnte der 2. Hauptband über Leipzig in Druck gegeben werden. Weitere, im wesentlichen editorische Vorhaben der Kommission, wie die Vorbereitung eines Bandes über „Zwickauer Frühdrucke", die „Erschließung der Handschriftenbände des Archivs der Franckeschen Stiftung" in Halle und die Arbeiten an einem wichtigen Ergänzungsband zum Codex Diplomaticus Saxoniae Regiae konnten zügig gefördert werden. Für den „Historischen Atlas von Sachsen", der insbesondere auf der kreativen historisch-kartographischen Quellenarbeit seines Hauptbearbeiters, des Herrn Dr. Blaschke und einer Reihe anderer Autoren beruht, wird zur Zeit eine der veränderten technischen und publikationsorganisatorischen Situation entsprechende Form der Herausgabe erarbeitet.

Der Geschichtsschreibung, die nach einem häufig mißverstandenen Wort Gorkis die Gegenwart vom Standort der Zukunft zu sehen habe, wird durch die historischen, quellenkritischen Erschließungsbeiträge der „Historischen Kommission" Material in die Hand gegeben, das den umprägenden Durchgang durch das Tor gegenwärtigen historischen Schaffens wissenschaftlich sichert. Es gilt, durch historische Quellenerfassung und -aufbereitung zur Erforschung des Verhältnisses politischer und ökonomischer Interessen zur gesellschaftlichen Entwicklung, mithin zur Bestimmung des Platzes beizutragen, der den aus solchen Interessen entspringenden Ideen und Begriffen des gesellschaftlichen Bewußtseins entspricht. In solchem Sinne sind auch die unter der Obhut von Herrn Peek stehenden Arbeiten, die die Akademie durch Mitbeteiligung an der Bearbeitung des Polybios-Lexikons, des „Mittellateinischen Wörterbuches" sowie insbesondere am „Corpus Vasorum Antiquorum Leipzig" des Antikenmuseums der Karl-Marx-Universität übernahm, nicht nur für den Spezialisten von Bedeutung. Die griechische Vasenmalerei reflektiert einschneidende gesellschaftliche Veränderungen, die sich in der Kunstentwicklung durch Wandlungen in der bildlichen Raumerscheinung bis hin zur Perspektive spiegeln. Im Berichtsjahr konnten Vasen und Scherben von rotfigurigen Bauch- und Halsamphoren in chronologischer Reihenfolge zusammengestellt und somit zugleich ein Beitrag zur Systematik der Gefäßtypen geliefert werden.

Ein Resonanzboden anderer Art, der das verstärkt, was die Entwicklung vorwärtstreibt, ist die leicht in ihrer Auswirkung unterschätzte Macht der Sprache. Ohne den Sinn für Geschichtlichkeit und die gesellschaftlichen Tendenzen, die die Sprache spiegelt, ist ihr Wesen als eine die moderne Mentalität durchaus mitprägende Kraft nicht verstehbar.

Als Grundlagenforschungen im Bereich der Sprachwissenschaft sind die Aufgaben anzusehen, die sich die „Sprachwissenschaftliche Kommission" unter der Leitung des Sekretars der Philologisch-historischen Klasse, Herr Große, gestellt hat. Die Sprache, die der Beobachtung und daher einer Vielzahl linguistischer Fragestellungen und Methoden zugänglich ist, kann wie ein Phänomen der Natur wissenschaftlich interpretiert, in ihrem Ursprung, in ihren Wandlungen und Wirkungen bis hin zu ihrer Funktion und Verflechtung als Kommunikationsmittel begriffen werden. Die Erfassung des unmittelbar gegebenen oder als Relikt vorhandenen Wortgutes, das überdies durch konkrete Sammlung einer dynamischen Strukturdeutung zugänglich gemacht werden kann, erfordert Aktionspläne, Strategien des linguistischen Herangehens, die in der durch die AdW der DDR unterstützte Konzentration der Mundartforschungen unserer Republik in der SAW einen weithin wirkenden Ausdruck finden.

Schon G. W. Leibniz regte die Sammlung mundartlichen Wortschatzes unter dem Gesichtspunkt der „Ausübung und Verbesserung der deutschen Sprache" an. Heute kann berichtet werden, daß die Bearbeitung der stattlichen Reihe mundartlicher Wörterbücher kontinuierlich fortgesetzt werden konnte, nämlich des Brandenburg-Berlinischen Wörterbuches, dessen 7. Lieferung sich im Druck befindet, des „Mecklenburgischen Wörterbuches", dessen 52. Lieferung erscheinen konnte, des „Thüringischen Wörterbuches", für das die Manuskripte der 11. Lieferung abgeschlossen wurden, des „Siebenbürgisch-Sächsischen Wörterbuches", dessen Revision in Zusammenarbeit mit der Akademie der Rumänischen Sozialistischen Republik und der SAW planmäßig bis zum Band mit dem Buchstaben K vorgenommen werden konnte, des „Obersächsischen Wörterbuches", für das die Materialsammlung mittels Fragelisten abgeschlossen wurde, und des „Althochdeutschen Wörterbuches", für das 2 Lieferungen als Manuskript fertiggestellt und eine Doppellieferung im Druck erscheinen konnte. Wörter, zumal Mundartwörter, die für viele Begriffe eine größere Kraft der Prägung und auch eine größere Anzahl sinnverwandter Bezeichnungen entwickelten als die Hochsprache, sind mehr als bloße Instrumente der Mitteilung oder der Kennzeichnung von Gegenständen. Wörter spiegeln die Wirklichkeit nicht als abgehobenes Ideal, sondern in oft bildhafter, lebendiger Klarheit. In den Naturwissenschaften — und auch in manchen Bereichen der Gesellschaftswissenschaften — sind gewisse Sachlagen in der Umgangssprache nicht mehr ausdrückbar. Aber in letzter Instanz ist auch der moderne Theoretiker, der zum Beispiel lediglich bildhaft, nicht auf Grund unmittelbarer Beobachtung von der Wellennatur des Lichtes spricht, bei aller Notwendigkeit zur Abstraktion und Formalisation doch auf die natürliche Sprache angewiesen. Die kognitive, auf sinnenhaftes Verstehen gerichtete Funktion der Sprache und ihre kommunikative, reine Sachmitteilungen in Wörter umsetzende Wirksamkeit sind eng miteinander verbunden. Zur Ausarbeitung des Brandenburgisch-Berlinischen Wörterbuches zum Beispiel, das seit 1959 von Dr. Gerhard Ising geleitet wird, werden niederdeutsche Bibelfrühdrucke als wichtige Quelle für die sprachgeschichtliche Beurteilung von Luthers Bibelübersetzung sowie der Herausbildung des schriftsprachlichen deutschen Wortschatzes herangezogen. Am Beispiel des gleichen Brandenburg-Berlinischen Wörterbuches, also eines Gebietes, in dem die Landbewohner dem Herrn Fontane auf seinen Wanderungen durch die Mark in der Mitte des vorigen Jahrhunderts noch in ihrer Mundart zu antworten pflegten, sei darauf hingewiesen, daß die Bearbeitung einzelner Mundartwörter und ihrer Lautvarianten sich mehr an Verwaltungsbezirke unterschiedlicher Größe hält als an mundartlich-landschaftlich umgrenzte Gebiete. So ragen von Norden her die vom Mecklenburgischen beeinflußten Mundarten der Priegnitz und der Uckermark in das brandenburgische Gebiet, in dessen Mitte und Süden sich das Mittelmärkische auf niederdeutscher, von holländischen Einwanderern aus dem 12. Jahrhundert beeinflußter Grundlage entwickelte. Berlin aber nahm sprachlich ähnlich wie in der Gegenwart eine Sonderstellung ein, da sich hier auf eben der gleichen niederdeutschen Grundschicht bereits im 15. Jahrhundert eine stark vom Sächsischen beeinflußte hochdeutsche Umgangssprache herausbildete, aus der sich, vor allem seit im 19. Jahrhundert viele Menschen in der Industrie der schnell wachsenden Stadt Arbeit suchten, die Mundart des heutigen „Berlinisch“ mit seiner starken Ausstrahlungskraft entwickelte.

Die Förderung solcher auf Beobachtungserfahrungen beruhenden Arbeiten, die die SAW in gleichem Maße zu betreiben sucht wie die der Naturwissenschaften, erscheinen nicht unwesentlich in einer Zeit, in der Formalisierungen und Mathematisierungen, wie sie zur Überprüfung naturwissenschaftlicher und ökonomischer Theorien unabdingbar notwendig sind, im Vordergrund stehen müssen. Die Arbeiten beider Klassen der Akademie, über die zu berichten war, stützen sich auf alles übergreifende Erkenntnismethoden. Beide, Naturwissenschaftler und Gesellschaftswissenschaftler, suchen die lebendige Berührung mit dem Ganzen nicht nur aufrecht zu erhalten, sondern bewußt zu pflegen und zu fördern — ganz so wie es Akademien immer als eine ihrer wesentlichen Aufgaben angesehen haben. Die Ergebnisse der Akademiearbeiten wollen, jedes für sich, von dem Ganzen des Zusammenhangs aus begriffen sein. Sie wollen gesehen sein als Ausdruck für die untrennbare Einheit von Erkenntnis, persönlicher Willensbildung und Weltanschauung, von Geschichte als lebendiger Tradition, von der Wechselwirkung zwischen Natur- und Gesellschaftswissenschaften — und nicht zuletzt als Ausdruck für die Einheit der Haupttriebkräfte, die das Antlitz unseres Landes prägen, — der Wissenschaft und Wirtschaft und der mit ihr verbundenen, sozialistisch bestimmten Veränderung des Charakters unserer Arbeit.

KURZFASSUNGEN DER VORTRÄGE UND AUSZÜGE AUS DEN PROTOKOLLEN ÜBER DIE SITZUNGEN DER JAHRE 1973–1974

Mathematisch-naturwissenschaftliche Klasse

1973

Sitzung am 22. Januar

Vortrag des Herrn Ernst Neef: „Aspekte der Umweltforschung". — Die gewaltige Ausweitung der Literatur zu Umweltfragen, nach Inhalt und Tendenz recht unterschiedlich, gestattet kaum noch einen klaren Überblick über die Problemlage. Daher wird der Versuch gerechtfertigt, eine Systematisierung vorzunehmen, die sich angesichts der Überfülle der Objekte nicht auf die sachlichen Inhalte, sondern auf die verschiedenen Aspekte der Umweltforschung richten muß. Zum Unterschied von den Umweltbeziehungen in der Biologie, also deren ökologischer Arbeitsrichtung, weitet sich bei den Mensch-Umweltbeziehungen das Feld der Probleme ganz erheblich aus. Die Befriedigung der Lebensbedürfnisse des Menschen wird durch ihn als gesellschaftlich organisiertes und bewußt handelndes Wesen durch zahlreiche Kulturmaßnahmen in seiner materiellen Umwelt gesichert, und die Gestaltung der Umwelt wird sowohl unter dem Gesichtspunkt seiner biologischen Existenz, als insbesondere seiner Gesundheit, als auch hinsichtlich der gesellschaftlichen Zielsetzung zu beurteilen sein. Daraus ergeben sich zwei Aspekte:

1. der biologische, der durch die Humanwissenschaften Anthropologie, Medizin und Psychologie behandelt wird und
2. der gesellschaftliche Aspekt, der sich von der Gesellschaftsorganisation und ihrer philosophisch-weltanschaulichen Basis über die breite Palette menschlicher Tätigkeiten bis zu Fragen der Ästhetik erstreckt.

Beiden Aspekten ist gemeinsam, daß sie die Anforderungen der entwickelten sozialistischen Gesellschaft begründen, also Postulate für die biologisch-medizinische Seite einerseits, für die Entwicklung der Produktivkräfte, ihrer Standortsverteilung und für die planmäßige und proportionale Entwicklung der Territorialstruktur anderseits formulieren. Im Rahmen der Ziel-Mittel-Relation bestimmen die auf den Menschen bezogenen Aspekte die gesellschaftspolitischen Ziele.

Die Umwelt erscheint als das materielle System, in dem diese gesellschaftlichen Postulate realisiert werden müssen, als Territorium, und alle Maßnahmen und Gestaltungen im Territorium, einschließlich der ökonomischen, sind als die Mittel anzusprechen, mit deren Hilfe das vorgegebene Ziel immer besser erreicht wird. Die Effektivität der Maßnahmen hängt natürlich in jedem konkreten Einzelfalle von der Kenntnis der Strukturgesetze und dynamischen Entwicklung ab, die im geographischen Raum wirksam sind.

Trotz der umfangreichen territorial bezogenen Literatur besteht ein erheblicher Rückstand in der Theorie, und es ist kennzeichnend, daß erst Ende 1972 zum ersten Male im Rahmen einer philosophischen Dissertation erkenntnistheoretisch-weltanschauliche Grundfragen der geographischen Wissenschaften behandelt worden sind. Wenn diese Dissertation zu dem Schluß kommt, daß die materiellen Systeme an der Erdoberfläche als eine besondere „geographische Bewegungsform der Materie" aufgefaßt werden müssen, so bedeutet das, daß sie auch eine spezifische Theorienbildung und Gesetzesformulierung haben müssen. Das aber sind die entscheidenden Ansätze für die Fragestellungen im Rahmen des territorialen Aspektes der Umwelt, zugleich aber auch wichtige Grundlagen für die Koordinierung der vielfältigen Einzelforschungen verschiedenster Fachgebiete im Territorium.

Da sich im Territorium Erscheinungen und Prozesse aus verschiedenen Kausalbereichen vorfinden, miteinander in Wechselwirkung stehen und damit die Charakteristika eines Territoriums prägen, ergeben sich zwei für die Umweltforschung wesentliche Orientierungen:

1. auf fachbezogene Beiträge, die für verschiedene Umweltfragen unmittelbar oder auch mittelbar als Grundlage benötigt werden;
2. auf die Integration verschiedenartiger Phänomene in der Territorialstruktur.

Während in der ersten Gruppe Forschungen erscheinen, die einzelne Elemente und Verfahren zum Gegenstand haben, aber nur indirekt Integrationsprobleme ansprechen, stehen bei der zweiten Gruppe gerade die Integrationsphänomene im Vordergrund. Für die Lösung von Umweltproblemen sind beide unerläßlich, insbesondere die planmäßige Verknüpfung beider Forschungsansätze. Nach dem Grad der Integration ergeben sich drei große Forschungsbereiche, nämlich die Erforschung der Stoffwandlungsprobleme in der Physiosphäre, die Ökosystemforschung in der Biosphäre und die territoriale Strukturforschung in der Geosphäre.

Die in der Umwelt ausgelösten Prozesse lassen sich in einer Reihe von analytischen und teilsynthetischen Arbeitsstufen genauer abbilden, dabei ist von besonderer Bedeutung, daß die einzelnen Impulse in allen ihren Folgewirkungen in Raum und Zeit erfaßt werden, wobei chemisch-physikalische Veränderungen einzelner Umweltfaktoren sowohl für Pflanzen- und Tierwelt als auch für die Gesundheit des für Menschen die Wirtschaft und die äußeren Lebensbedingungen des Menschen in Siedlung und Verkehr Bedeutung erlangen.

Sowohl die verschiedenen praktischen Aufgaben als auch die unterschiedlichen wissenschaftlichen Aspekte sollten dazu dienen, innerhalb der Umweltforschung durch bessere terminologische Unterscheidungen die Zielsetzung der einzelnen Beiträge der Umweltforschung zu bestimmen. — Erscheint in den Sitzungsberichten.

Vorgelegt werden Veröffentlichungen von: Wolfgang Buchheim, Paul Görlich, Horst Hanson, Ernst Hölder, Erich Lange, Artur Lösche, Ernst Neef, Rudolf Nissen, Erich Rammler, Erich Strack, Dietrich Uhlmann.

Sitzung am 12. Februar

Vortrag des Herrn Kurt Mothes: „Über einen neuen rauschgiftfreien Arzneimohn". — Der Schlafmohn (Papaver somniferum) ist eine sehr alte Kulturpflanze unbekannter Herkunft. Wahrscheinlich wurde er aus einem Wildmohn selektioniert in der Richtung starker narkotischer und berauschender Wirkung. Diese Wirkung beruht auf dem Gehalt an Morphin. Es ist nur eines von einer größeren Zahl von Alkaloiden, die alle dem Benzylisochinolintyp angehören. Die Natur hat offenbar nur in zwei verschiedenen Mohnarten die Synthese des Morphins verwirklicht. Heute sind wir therapeutisch am Morphin nur wenig interessiert. Bedeutend wichtiger ist das Hustenmittel Codein. Dieses Alkaloid wird heute chemisch aus Morphin gewonnen. Das bedeutet aber, daß die Produktion von Morphin weiterlaufen muß. Dies hat sehr nachteilige Folgen, denn ein wesentlicher Teil des Morphins fließt auf Schleichwegen in den Rauschgifthandel ab. Es wäre gut, einen Mohn zu haben, der Codein liefert unter Umgehung von Morphin. Die Biosyntheseforschung hat gezeigt, daß alle Mohnalkaloide aus der Aminosäure Tyrosin gebildet werden. Der sehr komplizierte Syntheseweg läuft über Thebain, Codeinon, Codein zum Morphin. Wir suchten nach einem Mohn, bei dem diese Synthesekette bei Codein abbricht. Merkwürdigerweise fanden wir einen solchen Mohn nicht. Aber wir selektionierten einen neuen Mohn, der tiefrote große Blüten hat (Papaver bracteatum Halle III), der praktisch nur ein Alkaloid besitzt: Thebain. Thebain kann leicht auf chemischem Wege in Codein verwandelt werden.

Dieser Mohn hat das Interesse der ganzen Welt gefunden. Die UNO schuf eine spezielle Arbeitsgruppe zur Einbürgerung und Weiterzüchtung dieses „rauschgiftfreien Arzneimohnes".

Vorgelegt werden Veröffentlichungen von: Hermann Berg, John Eggert, Burckhardt Helferich, Erich Schmid, Gustav E. R. Schulze, Michał Śmiałowski, Erich Strack, Carl Wagner.

Sitzung am 19. März

Vortrag anläßlich der 500. Wiederkehr des Geburtstages von Nicolaus Copernicus von Wolfgang Buchheim: „Die kopernikanische Wende und das Problem der Gravitation". — Die 500. Wiederkehr des Geburtstages von Nicolaus Copernicus (19. 2. 1473) ließ den Begründer des astronomischen Weltbildes der Neuzeit, der, an Wirkung seinen Zeitgenossen Columbus und Luther durchaus vergleichbar, das Abendland aus mittelalterlich-dogmatischer Beschränkung und aristotelisch fixierter Überlieferung in eine neue Phase des Naturverständnisses führte, unter den verschiedensten Aspekten erscheinen. Hier sei nur der astronomie- und physikgeschichtliche aufgegriffen, der mit der Person der am engsten verbundene, wenn auch vielleicht nicht der umfassendste ist.

Die „kopernikanische Wende", der Übergang von der geozentrischen zur heliozentrischen Sicht unseres Planetensystems, war, streng genommen, die Wiedererweckung eines Gedankens der Antike, der sich bei Aristarch von Samos im 3. Jahrhundert v. Chr. findet, aber in der streng geozentrisch orientierten Überlieferung des Claudius Ptolomäus, die über den Almagest schließlich in die aristotelisch-scholastische Tradition des europäischen Mittelalters integriert wurde, völlig verdrängt worden war. Aber auch Copernicus hatte noch eine rein kinematisch-beschreibende, keine erklärende Astronomie im Sinne, die am Kreis als vollkommenste Bewegungsform (Apollonische Doktrin) festhielt und das ptolomäische System der „Deferenten" und „Epizyklen" durch die Verlegung des Bezugssystems in der Sonne in erster Linie nur zu vereinfachen trachtete. In Wahrheit aber führte Copernicus mit dem Begriff der „Weltmitte" (im „Commentariolus" etwa 1510) intuitiv und noch unklar und verdeckt ein physikalisch-materialistisches Element in die Astronomie ein, das sich als äußerst weittragend erwies und in der Folge über das Phänomen der Gravitation eine immer enger werdende Verbindung dieser uralten Wissenschaft mit der neuen galileischen Physik herstellte.

Diese Verbindung entwickelte sich langsam und schubweise in einem Prozeß von über vier Jahrhunderten Dauer, der im wesentlichen mit der Herausbildung der Mechanik als Teildisziplin der Physik Hand in Hand geht und zweifellos von Copernicus eingeleitet wurde. Dieser Vorgang läßt sich in sechs deutlich unterscheidbare Phasen gliedern, nämlich in

1. die Phase des Kompromisses mit Ptolemäus, eingeleitet durch das Erscheinen von Copernicus' Hauptwerk mit dem Titel „De revolutionibus" 1543 und gekennzeichnet durch einen nur bedingt heliozentrischen Standpunkt (Tycho de Brahe) mit dem Festhalten an der Apollonischen Doktrin und damit an den Epizyklen, die auch der strenge Heliozentrismus von Copernicus nicht ganz zu entbehren vermochte (Tychonische Phase).

2. die Phase der noch immer rein kinematischen Beschreibung der Planetenbewegung, aber unter Verzicht auf den Kreis, eingeleitet durch Kepler mit der Entdeckung seiner drei Gesetze (1610—1619) und durch die Erfindung des Fernrohres (Galilei und Kepler 1609). Gleichzeitig weist Galilei um 1600 durch seine berühmten Fallversuche nach, daß alle Körper durch die Schwere gleichstark beschleunigt werden, wenn Reibung wegfällt. Bei Kepler verdichtet sich die Ahnung einer heliozentrisch wirkenden Kraft, die das Planetensystem zusammenhält (Keplersche Phase).

3. die Phase des beginnenden dynamisch-kausalen Verständnisses des Planeten- und Mondumlaufs mit Hilfe des von Newton 1666 entdeckten Gravitationsgesetzes und der Begründung einer „Mechanik des Himmels" auf dieser Grundlage. Sie umfaßt auch die mechanisch-deterministische Theorie der Gezeiten und der Figur der Erde, die durch die ersten Gradmessungen 1735—1741 bestätigt wird. Newton selbst erfindet um 1670 das Spiegelteleskop, das den Blick über die Grenzen des Milch-

straßensystems hinaus ermöglicht. Kant versucht 1755 auf der Basis der Newtonschen Mechanik und des Gravitationsgesetzes, allerdings nur qualitativ, eine erste mechanistische Kosmogonie („Naturgeschichte und Theorie des Himmels") (Newtonsche Phase).

4. die Phase der mathematischen Durcharbeitung und Anwendung der Newtonschen Gravitationstheorie, eingeleitet 1773 durch Lagrange mit der Konzeption des Potentialbegriffs und dessen Einführung in die Himmelsmechanik, gekennzeichnet durch die großen astronomischen Arbeiten von Laplace (seit 1784) und die terrestrische Bestimmung der Gravitationskonstanten durch Cavendish 1798. Die mechanische Stabilität des Planetensystems wird von Laplace mathematisch nachgewiesen, sein 1825 abgeschlossener „Traité de mécanique céleste" ist das astronomische Standardwerk der Epoche (Laplacesche Phase).

5. die Phase des Eindringens des Prinzips der stationären Wirkung, 1834/35 von Hamilton entdeckt, in die theoretische Astronomie und Kosmologie, zunächst in Form einer neuen, die Laplaceschen Gedanken systematisierenden Behandlung der astronomischen Störungstheorie, bestehend in der Anwendung der kanonischen Bewegungsgleichungen und Transformationen auf die Himmelsmechanik. Bei Einteilung der Variablen in Winkel- und Wirkungsvariable erweist sich das Planetensystem auf Grund der nur schwachen gravitativen Wechselwirkung der Planeten miteinander in hoher Näherung als vielfachperiodisch. Bessel entdeckt 1838 die erste Fixstern-Parallaxe, Galle 1846 den auf Grund der Störungstheorie von Leverrier 1845 vorausgesagten Planeten Neptun an angegebener Stelle — der bis dahin größte Triumph der rechnenden Astronomie auf den Spuren Newtons. Die Gleichheit von träger und schwerer Masse wird von Eötvös mit höchster Präzision bestätigt (Hamiltonsche Phase).

6. die letzte Phase, von Einstein 1915 mit seiner Gravitationstheorie (ursprünglich als „allgemeine Relativitätstheorie" bezeichnet) eingeleitet, mit der die Gegenwart eingeholt wird. Sie kann gekennzeichnet werden durch die Verschmelzung von Gravitation und Trägheit auf Grund der Äquivalenz von träger und schwerer Masse mittels des geometrisch umgedeuteten Wirkungsprinzips von Hamilton. Die Kontingenz mit der aus der Elektrodynamik abgeleiteten speziellen Relativitätstheorie (Einstein 1905) zwingt zur Identifizierung aller Gravitations- bzw. Trägheitsbeschleunigungen mit nichteuklidischer Metrik des Raum-Zeit-Kontinuums im Großen. Bewegungen in Schwerefeldern verlaufen demgemäß nun auch nach dem Galileischen Trägheitsprinzip. Viele kosmologischen Probleme bekommen damit über Newton, Kant und Laplace hinaus ein neues Gesicht: Ein unbegrenzter, aber endlicher, expandierender Kosmos scheint möglich, die Erhaltungssätze der klassischen Physik indessen sind im Großen in Frage gestellt, Materie kann somit entstehen und vergehen. Masse-Energie und Substanz sind nicht mehr identisch (Einsteinsche Phase).

Wissenschaftsgeschichtlich bemerkenswert ist, daß das Gravitationsproblem der Astronomie und Kosmologie in der letzten Phase auf höherer Ebene gleichsam über der copernicanischen Ausgangsposition anlangt, nämlich bei einer kräftefreien

Kinematik der Massenbewegung im Weltraum, wie sie ein Astronaut tatsächlich erlebt. Die apollonische Doktrin ist darin durch das Prinzip der kürzesten Verbindung („kleinste Wirkung“) ersetzt, Newtons Fernkraft Gravitation wieder eliminiert, seine Dynamik, was die Bewegung aller trägen Massen, also auch der Gestirne anlangt, in nichteuklidische Geometrie konvertiert. Dieses gilt freilich nur in einem Kosmos, der, im Gegensatz zu dem, den Copernicus vor Augen hatte, „ohne Mitte“ ist. — Erschienen in den Sitzungsberichten. Bd 111 (1975) H. 5.

Vorgelegt werden Veröffentlichungen von: Jakov Mikhailovitsch Kolotyrkin, Wilhelm Treibs.

Sitzung am 14. Mai

Vortrag des Herrn Gustav E. R. Schulze: „Zur Rolle des Einfachheitsprinzips im physikalischen Weltbild“. — Nach Gegenüberstellung verschiedener Formulierungen des Einfachheitsprinzips wird seine Bedeutung für das Auffinden neuer physikalischer Gesetze am Beispiel der Planetenbewegung und der Hohlraumstrahlung dargelegt. Dabei wird besonders auf die Notwendigkeit der Idealisierung der wirklichen Verhältnisse und auf ihre Folgen hingewiesen. Durch Vergleich der elastischen und plastischen Verformung wird dann gezeigt, daß der Idealisierung und damit auch der Erreichbarkeit einfacher Gesetze enge Grenzen im Hinblick auf die Verwertbarkeit der Ergebnisse gesetzt sein können. Besonders deutlich wird dies bei Randgebieten der Physik, z. B. der musikalischen Akustik, wo sich physikalisch einfache Gegebenheiten wie „natürliche“ Stimmung, obertonarme Klänge, zeitlich wohldefinierte Klangabläufe als durchaus unzureichend zur physikalischen Deutung der musikalischen Erfahrungen erwiesen haben. Die Rolle des Einfachheitsprinzips ist also im physikalischen Weltbild größer als in der realen Welt. — Erschienen in den Sitzungsberichten Bd 110 (1974) H. 6.

Vorgelegt werden Veröffentlichungen von: Walter Brednow, Hans Drischel, Ulrich Freimuth, Werner Heisenberg.

Sitzung am 18. Juni

Vortrag des Herrn Kurt Schwabe: „Analytische Probleme des Umweltschutzes“. — Die Erkennung von Schadwirkungen durch Noxen in der Atmosphäre, im Wasser und in Nahrungsmitteln setzt ihre analytische Bestimmung voraus. Es ist durchaus denkbar, daß Schadstoffe bei der industriellen Produktion emittiert werden oder in Erzeugnissen vorhanden sind, die wegen der Unmöglichkeit einer analytischen Bestimmung in niedrigen Konzentrationsbereichen nicht bekannt sind.

Daher ist die analytische Erfassungsgrenze für die Ermittlung von Noxen von großer Bedeutung. Wenn sie höher liegt als die maximal zulässigen Konzentrationen (MIK) der Immission, kann eine Anreicherung die Ermittlung möglich machen. Zwischen Emissionskonzentration und Immissionskonzentration besteht wegen der zeitlich und örtlich variablen Verdünnungsverhältnisse keine konstante Korrelation.

Es werden die Anforderungen an kontinuierlich arbeitende Analysenverfahren erörtert und die entwickelten elektrochemischen Methoden zur registrierenden Bestimmung von gasförmigen Noxen sowie Schadstoffen im Wasser beschrieben.

Die pH-Wert-Messung und -Regelung erfolgt nach dem bekannten potentiometrischen Prinzip auch in Oberflächengewässern.

Ionensensitive Elektroden zur Bestimmung von Cyanid-, Sulfid-, Nitratgehalten sind wegen zu geringer Empfindlichkeit und Selektivität für die Wasserkontrolle nicht einsetzbar.

Für die Emissionskontrolle galvanischer Betriebe in bezug auf Cyanid und Chromat dient ein amperostatisches Verfahren mit potentiostatischer Regelung. Für die Bestimmung des Sauerstoffgehaltes in größeren Meerestiefen wurde eine Sauerstoffelektrode mit eingebautem Verstärker entwickelt.

Für die Immission von SO_2, H_2S, Cl_2 wurde ein coulometrisches Gerät entwickelt, das nach einem Differenzprinzip arbeitet und den Momentanwert sowie Mittelwerte über vorgegebene Zeiten zu registrieren gestattet.

Zur Bestimmung des biologischen Sauerstoffbedarfs wurde ein Gerät entwickelt, das nach einem Leitfähigkeitsprinzip arbeitet und die Sauerstoffzehrung mit der Zeit zu verfolgen gestattet. Die Vor- und Nachteile verschiedener analytischer Meßprinzipien wurden erörtert. — Erschienen in den Sitzungsberichten Bd 111 (1975) H. 4.

Vorgelegt wird eine Veröffentlichung von Franz Hein.

Sitzung am 15. Oktober

Vortrag des Herrn Rolf Emmrich: „Zwischen Leben und Tod — ärztliche Probleme der Thanatologie". — Obwohl der Arzt mit seinem Denken und Handeln zutiefst dem Leben zugewandt ist, um Gesundheit zu erhalten und gestörte Gesundheit wiederherzustellen, findet er sich konfrontiert mit Fragen, die Tod und Sterben angehen. Wenn der Tod bereits eingetreten ist, befassen sich der Pathologe und der Arzt retrospektiv mit dem, was geschehen ist, wie es geschehen ist und was wann erfolgte. Jeder Gedanke, der dem Tode gewidmet ist, stellt gleichzeitig auch ein Bekenntnis zum Leben dar.

Als Pathologe scheute sich Heinrich Bredt, der Biologie als der Lehre vom Leben eine ebenso gegliederte Lehre vom Tode, also die Thanatologie, an die Seite oder gar gegenüberzustellen. Der Arzt von heute kann sich dem jedoch nicht anschließen, für ihn enthalten Tod und Sterben, insbesondere die „Schwebezeit" zwischen Leben und Tod eine Vielzahl von Problemen. In den letzten Jahren wurde die Erörterung dieser Probleme angeregt durch die Fortentwicklung der klinischen Medizin, die Reanimation im Rahmen der Intensivmedizin und durch die rechtlichen und medizinischen Fragen, die mit der Organtransplantation zusammenhängen. Nicht nur die Feststellung und die Definition des Todes gewinnen eine neue Bedeutung, sondern auch die Übergangszeit vom Leben zum Tod. Wenn der Tod biologisch als Negativum des Lebens definiert wird, vom lebenden Organismus aus als irreversibles Aufhören der Lebenserscheinungen, so gibt es, abgesehen von der Einwirkung stark formzerstörender Kräfte auf den ganzen Menschen, *reversible Vorphasen*, die in die Irreversibilität führen oder aber überwunden werden können.

Als Wissensgebiet hat die Thanatologie verschiedene Aspekte, einen wissenschaftlich und praktisch medizinischen Aspekt, den der Rechtsprechung, Ethik und Philosophie, aber auch einen religiösen und historischen Aspekt, wenn man sich mit dem befaßt, was unsere Vorfahren über den Tod und über das Sterben dachten.

Noch bis vor wenigen Jahren galten Atemstillstand und Herzstillstand allein als untrügliche Zeichen des Todes, die nicht nur von Juristen und Theologen, sondern auch in der praktischen Medizin von den Ärzten anerkannt wurden. Seitdem es jedoch gelingt, Atmung und Kreislauf notfalls wenigstens über einen gewissen Zeitraum aufrechtzuerhalten, ist eine Definition des Todeszeitpunktes zunächst nur für einzelne Organe oder Zellgruppen möglich. Für den Gesamtorganismus kann eine Definition erst dann gegeben werden, wenn eine irreversible Schädigung des Gehirns das Leben beendet. Früher schloß man aus Atem- und Kreislaufstillstand auf den eingetretenen oder kurz darauf eintretenden Hirntod. Wenn man derzeit den Hirntod als beweisend anerkennt, so kann es auch hierbei Schwierigkeiten geben, denn das Dezerebrationssyndrom mit tiefer Bewußtlosigkeit ohne Reaktion auf starke Reize, mit fehlender Spontanatmung und mit weiten lichtstarren Pupillen beweist nur eine Funktionshemmung des Gehirns, aber nicht seinen irreversiblen Ausfall.

Zu den akuten krankhaften Zuständen, die eine unmittelbare Bedrohung vitaler Funktionen darstellen, gehören Störungen der Zirkulation (Herz-Kreislauf-System), des alveolären Gasaustausches (Atmung), des Elektrolyt- und Säure-Basenhaushaltes, des Stoffwechsels und der Nierenleistung. Zirkulation und Atmung sind funktionell aufs engste miteinander verbunden, das Versagen einer dieser beiden Funktionen führt zwangsläufig zum Zusammenbruch der anderen. Der Bewußtlose ist nicht erweckbar, die völlige Bewußtlosigkeit, das Koma, wird zumeist über den an Tiefe zunehmenden Schlaf erreicht. Im Schockzustand kann das Gehirn auch bei niedrigen Blutdruckwerten und/oder bei erheblichen Volumenverlusten als Folge der Zentralisation noch ausreichend durchblutet sein, so daß das Bewußtsein erhalten bleibt. Eine Trübung des Bewußtseins, kombiniert mit motorischer Unruhe, spricht für das beginnende Versagen der Hirndurchblutung. Zwischen dem plötzlichen Bewußtseinsverlust und völliger Klarheit gibt es Übergänge, je nach Herzminutenvolumen, Blutdruck und zerebraler Durchblutung. — Erschienen in den Sitzungsberichten Bd 11 (1974) H. 7.

Vorgelegt werden Veröffentlichungen von: Horst Hanson, Kurt Mothes, Kurt Schwabe.

Sitzung am 10. Dezember

Vortrag des Herrn Hasso Eßbach: „Über angeborene Herz- und Gefäßformfehler, ihre Prognose und Korrekturmöglichkeiten“. — In der Darstellung wird zunächst darauf hingewiesen, daß unter den angeborenen Anormogenesen die Herzformfehler 40,5% ausmachen. Im eigenen Obduktionsgut von 52400 Fällen sind sie zu 2,07% vertreten.

Im Kinderobduktionsgut kommen sie in 7,6% vor, bei Erwachsenen über 14 Jahren nur noch in 2,4%.

Der Unterschied erklärt sich daraus, daß 79% bereits im 1. Lebensjahr sterben und weitere 12,8% das 14. Jahr nicht überleben. So entfallen nur noch 8% der Herz- und Gefäßformfehler auf das Erwachsenenalter.

Die Analyse zeigt, daß die Sterblichkeit von der Schwere der hämodynamischen Funktionsstörung des Herzfehlertypes abhängt, und daß hochgradig lebenswidrige Fehler zur Frühsterblichkeit führen. Da schwerwiegende Herzfehler überwiegen, ergibt sich eine Sterblichkeit im Säuglingsalter am Herzfehler in fast 80%.

Die Relation zu den wichtigsten Herzfehlertypen wird in Tabellen dargestellt.

Zum Verständnis der Kompliziertheit und Vielfalt der Herzfehlertypen wird kurz auf die Embryonalentwicklung und auf einige Erscheinungsbilder eingegangen. Es wird auf die Häufigkeit intracardialer variabler Zusatzanomalien verwiesen, welche die Fehlertypen zu Komplexanomalien morphologischer und hämodynamischer Art erweitern und die bei vielen Fehlern in großer Regelmäßigkeit vorliegen.

Vermerkt werden auch extracardiale Anormogenesen in anderen Organsystemen, am ZNS, Urogenitaltrakt, Verdauungsapparat und Skelettsystem, die nach dem Schrifttum bis zu 37% beobachtet werden.

Bezüglich der möglichen Operabilität ergibt sich die Einsicht, daß eine optimale praeoperative Diagnostik mit allen zur Zeit verfügbaren Mitteln unter gedanklicher Berücksichtigung aller möglichen Formfehlerkomplexe eine conditio sine qua non für eine operative Korrektur sein muß.

Nicht immer ist, was sich leicht diagnostizieren und operieren läßt, dem Patienten dienlich. Die beliebten Verschlüsse persistierender Foraminae oder des Ductus Botalli können schwere Kunstfehler sein, wenn diese Gebilde hämodynamisch lebenswichtige Teilkomponenten eines Fehlerkomplexes sind, der nicht erkannt oder nicht gleichzeitig korrigiert werden konnte. Dafür werden Beispiele gebracht.

Zur Operabilität wurde vermerkt, daß extracardiale Korrekturen relativ günstig durchzuführen sind, und daß intracardiale Eingriffe nach Ruhigstellung des Herzens unter Herz-Lungen-Maschine erheblich erweitert werden konnten.

Die Hauptproblematik stellt dabei zur Zeit noch das Herz des Neugeborenen und des Säuglings dar. Hier müßte an einem Organ operiert werden, daß sehr klein ist und hämodynamisch schwierig ausgelotet werden kann und einem raschen Wachstumsprozeß unterliegt, so daß vielfach in mehreren zeitlichen Etappen korrigiert werden müßte. Vorher sind vorerst nur Paliativ-Eingriffe zur Ermöglichung einer Vita minima zumutbar.

Das Operationsrisiko wird vom Fehlertyp, vom Lebensalter und vom Eingriff her bestimmt.

Viele Prognosen haben sich entscheidend gebessert. Man hat gelernt, Patienten mit Herzfehlern durch klinisch-therapeutische Vorbehandlung bis ins günstigste Operationsalter zu bringen.

Auch zahlreiche Operationsmethoden sind verbessert worden. Nicht erkannte extracardiale Zusatzmißbildungen können den Erfolg einer gelungenen Herzfehler-korrektur zunichte machen. Auf postoperative Komplikationen, die den Enderfolg eines Eingriffes mitbestimmen können, wurde nicht eingegangen. — Erscheint in den Sitzungsberichten.

Vorgelegt werden Veröffentlichungen von: Manfred Gersch, Paul Görlich.

1974

Sitzung am 14. Januar

Vortrag des Herrn Wilhelm Maier: „Vom Erbe Bernhard Riemanns". — Der 1826 geborene Mathematiker B. Riemann war als schaffender Geist richtungsweisend für die Entwicklung der mathematischen Wissenschaft des 19. und 20. Jahrhunderts. Als Göttinger Doktorand schuf er die begriffliche Grundlegung für Funktionen einer komplexen Veränderlichen. Die Bedingung $w_x + iw_y = 0$ für differenzierbare Funktionen $w(x + iy)$ kann in potentialtheoretischer Deutung den Aufbau gewisser Geschwindigkeitsfehler aus Quell-, Senk- und Wirbelpunkten vermitteln. In abbildungsgeometrischer Sicht gibt die Existenz der Ableitung $\frac{dw}{dz} \neq 0, \infty$ lokal eine gestaltstreue Abbildung der z-Ebene. Die Nicht-Existenz der Winkeltreue für $z : \{w'(z) = 0, \infty\}$ führt zur Funktionsverzweigung auf mehrblättrigen Flächen als Träger der unabhängigen Argumente, zu dem Begriff der mehrwertigen analytischen Funktion.

Zur vieldimensionalen Verallgemeinerung der konform abgebildeten ebenen Gebiete, nämlich zur inneren Geometrie gekrümmter n-dimensionaler Mannigfaltigkeiten kommt Riemann durch Benützung einer maßbestimmenden quadratischen Differentialform:

$$\sum_{\mu,\upsilon \leq n} dx_\mu \, dx_\upsilon \, g_{\mu\upsilon} = (ds)^2.$$

Die Abweichung des Maßtensors vom Einheitstensor bemißt die Abweichung zwischen gekrümmter und ebener Berührungsmannigfaltigkeit. Mit der Konstruktion des Krümmungstensors gelangt Riemann zur einheitlichen Behandlung aller Mannigfaltigkeiten mit freier Beweglichkeit in sich. Die später entwickelte Theorie der sogenannten Lie'schen Transformationsgruppen hat gezeigt, daß unter Zugrundelegung einer hyperkomplexen Algebra die freie Beweglichkeit auch durch Spinoren, d. h. Elemente der Quantenphysik zu erreichen ist.

Im Bereich der Zahlentheorie hat Riemann die Anzahl $\pi(x)$ der Primzahlen $p < x$ mit Hilfe der meromorphen Funktion $\eta(s) = \sum n^{-s}$, $\mathrm{Re}(s) < 1$ gefunden. Es besteht der Grenzwert

$$\lim_{x\to\infty} \frac{\pi(x) \log x}{x} \overset{0<n}{=} 1;$$

zur genaueren Bestimmung der Anzahlen $\pi(x)$ benötigt man die nicht reellen Nullstellen von $\eta(s)$. Die von Riemann ausgesprochene Vermutung, daß für diese stets $Re(s) = \frac{1}{2}$ bleibe, ist entscheidend für das Wachstum des Betrages

$$\left| \pi(x) - \frac{x}{\log x} \right|.$$

Die genauere Untersuchung der komplexen Nullstellen von $\eta(s)$ gilt als besonders schwierige Aufgabe der zeitgenössischen Mathematik. — Erschienen in den Sitzungsberichten Bd 111 (1975) H. 1.

Vorgelegt werden Veröffentlichungen von Rolf Emmrich.

Sitzung am 11. Februar

Vortrag des Herrn Horst Hanson: „Immobilisierte Enzyme und ihre Anwendung in Biotechnologie und Medizin". — Einleitend werden die allgemeine Stellung und Rolle der Enzyme in Biosystemen sowie ihr Charakter als Protein erörtert. Der Weg, der über die Reinigung und Isolierung einzelner Enzyme aus Körperflüssigkeiten, Organen und Geweben zu ihrer Verknüpfung mit festen Trägern führt, wird dargestellt. Dabei wird die Zielsetzung derartiger Untersuchungen besonders berücksichtigt: Schaffung von Modellen für das Verständnis von In-vivo-Vorgängen in zellulären Strukturen, Einsatz in Biotechnologie und Medizin. Die verschiedenen Prinzipien, nach denen Enzyme an unlösliche Träger fixiert werden, werden dargelegt (physikalisch-chemisch absorbiert, kovalent gebunden, im Netzwerk des Trägers eingeschlossen, mikroinkapsuliert). Anhand des Berichtes über eigene Untersuchungen mit trägerfixierten, eiweißspaltenden Enzymen, speziell der Leucinaminopeptidase aus Augenlinsen werden Anwendungsbeispiele und einige Eigenschaften sowie Vor- und Nachteile der verschiedenen Bindungstypen für Aktivität und Stabilität der Enzyme wie auch ihre Anwendung in biochemisch-technologischen Verfahren erörtert (Kinetik, pH-Profile, Temperaturstabilität, Trennung racemischer Aminosäuren, Sequenzanalyse von Proteinen, Proteolysereaktor, Affinitätschromatographie). — Erscheint in den Sitzungsberichten.

Sitzung am 18. März

Vortrag des Herrn Erich Strack: „Das Carnitin im Tierkörper". — L(—)-Carnitin ist im Intermediärstoffwechsel aller Zellen wirksam. Es ist Carrier für Fettsäuren und ist mit der Tätigkeit des Coenzym A eng verbunden. Seine O-Acylester hemmen und fördern Enzyme, die im Energiehaushalt und im Umbau der Nährstoffe tätig sind. Dies bereits erklärt die überraschende Vielfalt des Einflusses von therapeutischen Zufuhren von L(—)-Carnitin auf Stoffwechselstörungen beim Menschen. Bei Hyperthyreosen, bei bindegewebigen Störungen, beim Altersherzen, bei Dysfunktionen der Leber, bei Dystrophien und weiteren Störungen wurde das Krankheitsbild durch L(—)-Carnitin gebessert. Günstig wirken kombinierte Therapien. Zufuhr von L(—)-Carnitin steigert den physiologischen Stoffwechsel, erhöht die Abwehr-kraft des Organismus, drängt den fehlgeleiteten Stoffwechsel zurück und begünstigt so die Heilungstendenz. Langzeitige Zufuhren von täglich 1 bis 3 g L(—)-Carnitin erhöhen beim Menschen mit seinem L(—)-Carnitinbestand von annähernd 15 g die aktiven Gewebskonzentrationen schon beträchtlich. Über die zweckmäßigste Zufuhr und über das Verhalten von L(—)-Carnitin bei längerdauernder Zufuhr ist bisher wenig bekannt.

Literaturangaben der letzten Jahre berichten über Einflüsse von Nahrung, Umwelt, körperlicher Tätigkeit, krankhaftem Stoffwechsel auf den Carnitinbestand und seinen Verschleiß im Tierkörper. Ein herabgesetzter Carnitinbestand kann entstehen: 1. Durch verminderte Synthese, 2. Durch verstärkten Verlust (Ausscheidung, Um- und Abbau), 3. Durch gestörte Verteilung auf die Organe und die Zellgruppen. Die Eigensynthese läuft über Lysin [1], das permethyliert und zu γ- Butyrobetain abgebaut wird, welches dann durch die γ-Butyrobetainhydroxylase [2] in (L(—)-Carnitin übergeführt wird. Über den Carnitinverlust sind jedoch die experimentellen Grundlagen wenig gesichert, obwohl alle quaternären Umwandlungsprodukte des Carnitins bekannt waren. (Unsere umfangreichen Studien über die Carnitinmetaboliten werden ausführlich in der Dissertationsschrift von H. Seim beschrieben. Teile dieser Arbeit wurden auf der Tagung der Biochemischen Gesellschaft in der DDR im November 1972 vorgetragen).

Einige von den typischen Ergebnissen an Ratte und Maus wurden vorgetragen und eingehender besprochen. Die optisch aktiven Carnitine sind bei einer LD_{50} von etwa 9 mg/g Körpergewicht/Maus wenig toxisch. Große Mengen können deshalb ohne sichtlichen Schaden dem Tierkörper zugeführt werden. Hiermit sollten auch jene Umbaureaktionen deutlich nachweisbar werden, die unter üblichen Bedingungen bei geringer Carnitinbelastung wenig hervortreten. Zugeführt wurden die beiden optisch aktiven Carnitine und einige ihrer O-Acylester. Die Stoffe wurden teils subcutan injiziert, teils bei Trockennahrung im Trinkwasser gelöst. Letzteres Vorgehen stellt eine Art Dauerinfusion per os dar und kann beliebig lange Zeit durchgeführt werden. Wir haben es bis zu 4 Wochen mit Mengen bis zu 6 mg/g/die getan.

10 quaternäre Abkömmlinge vom Carnitin sind bekannt. Sie werden im Harn quantitativ mittels Dünnschichtchromatographie und Jodanfärbung nachgewiesen. Bei subcutaner Injektion von 2 mg/g L(—)-Carnitin wird in den ersten 3 Stunden dieses und Acetyl-L(—)-Carnitin ausgeschieden. Ab 4. Stunde nach der Injektion stellt sich ein Verhältnis der beiden Substanzen von etwa 1:1 ein. Zugeführtes D(+)-Carnitin erscheint als solches und dazu eine mäßige Menge Acetonyltrimethylammonium (= ATMA) im Harn. Zugeführtes Acetyl-L(—)-Carnitin wird wie L(—)-Carnitin ausgeschieden. Das Ausscheidungsverhältnis zwischen Acetyl-L(—)-Carnitin und L(—)-Carnitin gleicht 4 Stunden nach der Zufuhr fast dem nach der Zufuhr von L(—)-Carnitin. Die Acylester (C_4 bis C_{10}) von L(—)-Carnitin lassen wenig Acylester, aber viel L(—)-Carnitin und Acetyl-L(—)-Carnitin im Harn erscheinen. Nur in den ersten Stunden nach der Injektion ist erwartungsgemäß das jeweils zugeführte Carnitinderivat vermehrt im Harn nachweisbar.

Bei langdauernder Fütterung von L(—)-Carnitin erscheinen im Rattenharn am ersten Tag nur L(—)-Carnitin und Acetyl-L(—)-Carnitin. Vom zweiten Tag an vermindert sich dann deren Ausscheidung und entsprechend vermehrt erscheint γ-Butyrobetain. In den späteren Tagen der Zufuhr werden auch beträchtliche Anteile an Trimethylamin und Trimethylaminoxid gefunden. Ab 10. bis 12. Tag bleibt bis zu 4 Wochen das Verhältnis der L(—)-Carnitine und des γ-Butyrobetains gleich groß. 1 ml Harn enthält dann 30 mg γ-Butyrobetain und 11 mg von den L(—)-Carnitinen. Die normale Ratte scheidet im gesamten Tagesharn weniger als 0,1 mg γ-Butyrobetain aus.

Zufuhr von D(+)-Carnitin ergibt schwankende und verminderte Mengen an γ-Butyrobetain, ebenfalls Trimethylamin und Trimethylaminoxid und dazu noch ATMA. Überraschenderweise waren kleine Mengen an Acetylcarnitin bei der D(+)-Carnitinzufuhr nachweisbar. Da Acetylcarnitin jedoch spezifisch nur aus L(—)-Carnitin gebildet wird, muß die Quelle für das Acetylcarnitin das γ-Butyrobetain sein, das aus dem zugeführten D(+)-Carnitin entstanden ist. Ein Teil von diesem ist durch die γ-Butyrobetainhydroxylase in L(—)-Carnitin umgewandelt worden. Die Enzymsysteme für die Umwandlung sind adaptationsabhängig. Sie sind noch unbekannt. Eine Zwischenstufe könnte auch Crotonsäurebetain sein, das aus dem D(+)-Carnitin durch Wasserabspaltung gebildet wurde. Jedoch ist diese Umwandlung als Quelle für das gefundene Acetylcarnitin nicht wahrscheinlich. Crotonsäurebetain kann ebenfalls von Ratten in γ-Butyrobetain umgewandelt werden.

ATMA (Acetonyltrimethylammonium) wird in sehr geringer Menge im menschlichen Harn gefunden. Bei Pseudomonaden entsteht es beim Abbau von Carnitin durch die L(—)-Carnitindehydrogenase aus dem entstehenden Ketocarnitin, das leicht spontan decarboxyliert. Im Tierkörper ist aber kein Enzym dieser Art nachzuweisen gewesen [3]. ATMA entstand unter unseren Bedingungen bei der Ratte nur bei D(+)-Carnitinzufuhr. β-Methylcholin ließ sich bei unseren Versuchen nicht mit Sicherheit im Rattenharn nachweisen. Dieses und ATMA könnten durch Hydrierung und Dehydrierung ineinander umgewandelt werden. β-Methylcholin kann im Herzmuskel aus L(—)-Carnitin durch die Carnitindecarboxylase entstehen. Als Acetylester wirkt es stark muscarinartig. In diesem Zusammenhang wurde diskutiert, daß Rhythmusstörungen bei Hyperthyreose und Sklerodermia progressiva durch L(—)-Carnitinzufuhren eingeregelt werden können. Über die Bedeutung von L(—)-Carnitinzufuhren für den Herzinfarkt und über Eigenversuche am Altersherzen mit positiven Wirkungen wurde gesprochen.

Trimethylamin und Trimethylaminoxid treten nur bei längerdauernder Carnitinzufuhr auf. Vermehrt ausgeschieden und im Verhältnis zueinander verschoben sind sie bei Leberstörungen.

γ-Butyrobetain ist anscheinend die zentrale Substanz im L(—)-Carnitinhaushalt. L(—)-Carnitin entsteht aus ihr, und es kann in γ-Butyrobetain zurückverwandelt werden. Warmblüter und viele Bakterien vermögen Carnitin und Crotonsäurebetain in γ-Butyrobetain umzuwandeln. Bei Kaltblütern stehen die 3 quaternären Stoffe — γ-Butyrobetain, L(—)-Carnitin, Glycinbetain — in enger Beziehung zueinander. Unsere neuen Befunde an Ratten stützen die Hypothese, daß die Stoffwechselgröße von Kaltblütern durch Temperaturveränderungen über diese quaternären Stoffe gesteuert werden kann. In gleicher Richtung, wie die Temperaturänderung, wirkt zudem γ-Butyrobetain durch seine hemmende Wirkung auf die Enzyme der L(—)-Carnitinfunktionen.

Unsere experimentellen Befunde an Maus und Ratte werfen eine Anzahl von Problemen auf, z. B. für den Menschen: Ist die therapeutische Zufuhr wirkungsvoller als bisher zu gestalten; gibt es Einwirkungen von γ-Butyrobetain auf den gestörten Stoffwechsel; ist eine vermehrte Ausscheidung von γ-Butyrobetain ein Krankheitssymptom beim Menschen; ist die veränderte Trimethylamin- und Trimethylaminoxidausscheidung beweisend für Leberstörungen, die Carnitin beeinflussen kann; theoretisches Interesse beanspruchen auch die raschen Acetylierungen, Umacetylierungen und ihre Orte in der Zelle; kann zugeführtes D(+)-Carnitin auch als Quelle für L(—)-Carnitin nutzbar gemacht werden.

Literatur

[1] V. Tanphaichitr. D. W. Horne and Harry P. Broquist: Lysine, a Precursor of Carnitine in the Rat. In: J. biol. Chem. *246* (1971) 6364—6366.
R. A. Cox and Ch. Hoppel: Biosynthesis of Carnitine and 4-N-Trimethylaminobutyrate from Lysine. In: Biochem. J. *136* (1973) 1075—1082.

[2] G. Lindstedt: Hydroxylation of γ-Butyrobetaine to Carnitine in Rat Liver. In: Biochemistry *6* (1967) 1271—1282.

[3] H.-P. Kleber und H. Aurich: Der katabole Carnitinstoffwechsel und seine Regulationsmechanismen. In: Wissenschaftliche Z. der Karl-Marx-Universität, Math.-Naturwiss. R. *17* (1968) 739—751.

Vorgelegt werden Veröffentlichungen von: Ulrich Freimuth, Erich Rammler.

Sitzung am 20. Mai

Vortrag des Herrn Peter Friedrich Matzen: „Probleme der Querschnittsgelähmten". — Es wird in diesem Vortrag ausschließlich auf die Querschnittslähmung als Unfallfolge eingegangen. Es werden die Wandlungen in der Auffassung über die therapeutischen Möglichkeiten bei traumatischer Querschnittslähmung geschildert und dabei das Verdienst Guttmanns in Stoke Mandeville bei der Abkehr vom therapeutischen Nihilismus gewürdigt. Das klinische Bild der Querschnittslähmung in Abhängigkeit von Ort und Schwere der Verletzung des Rückenmarks wird gezeichnet und die das Leben des Gelähmten bedrohenden Komplikationen (Dekubitus, Harnwegsinfektion, Atemstörung, Krampflähmung) und die Möglichkeiten ihrer Vermeidung durch entsprechenden Pflegeaufwand geschildert. Ausführlich wird der Mechanismus der Peitschenhiebverletzung, der Kompressionsfraktur, der Scherverletzung, der Kippverletzung, der Luxation mit und ohne gleichzeitigen Bogenbruch dargestellt.

Es wird auf die verschiedenen Standpunkte zur Frage der Einrenkung und Einrichtung bzw. auf deren Verzicht sowie auf die konservativen und operativen Einrichtungs- und Einrenkungsmöglichkeiten bei Wirbelsäulenverletzungen eingegangen und begründet, weshalb grundsätzlich vom Vortragenden die exakte anatomische Einrichtung und Einrenkung gefordert wird und die Behandlung der Luxation und Luxationsfraktur der Halswirbelsäule im Dauerzug abgelehnt wird. Sie hat nach Auffassung der Leipziger Orthopädischen Universitätsklinik nur ihre Berechtigung bei Trümmerbrüchen der Halswirbelsäule, die sich auf andere Weise nicht stabilisieren lassen. Es wird das in der Leipziger Orthopädischen Klinik geübte Verfahren der Einrenkung von Halswirbelluxationen gezeigt: Rechts- und Linksdrehung des Kopfes unter Längszug bei Kyphosierung der Halswirbelsäule. Die Ergebnisse werden durch zahlreiche Bilderserien dokumentiert. Ausführlich wird auf die operativen Zugangswege zur Wirbelsäule eingegangen und auseinandergesetzt, weshalb die Leipziger Orthopädische Universitätsklinik bis auf Ausnahmen im Gesamtbereich der Wirbelsäule den Zugang von dorsal bevorzugt und den Weg von ventral denjenigen Fällen vorbehält, bei denen ein Kompressionstrümmerbruch im Wirbelkörperbereich besteht, der sich durch Lordosierung der Wirbelsäule nicht exakt einrichten läßt. Auf die Bedeutung der Röntgen-Kontrastdarstellung des Wirbelkanals bei Beurteilung des Repositionsresultats wird eingegangen. Es wird zur Prognose der Querschnittslähmung in Abhängigkeit vom Ausmaß des Erstbefundes Stellung genommen. In Auswertung histologischer Befunde bei frühzeitig gestorbenen Patienten mit Halsmarkverletzung kann gezeigt werden, daß bei einem Teil dieser Verletzten trotz des zunächst kompletten Lähmungsbildes keine völlige Zerstörung des Rückenmarks bestanden hat, so daß es möglich erscheint, daß nach längerer Zeit Remissionen eingetreten wären, wenn es gelungen wäre, die das Leben dieser Frühfälle vor allem bedrohenden Lungenkomplikationen zu beherrschen. Rechtzeitige Tracheotomie und künstliche Beatmung bei gleichzeitiger Dekubitusprophylaxe und Vermeidung der Harnwegsinfektion bei entsprechendem Pflegeaufwand dürfte in manchem zunächst desolat erscheinenden Fall doch noch eine Teilremission der Lähmung auch bei zunächst aussichtslos erscheinendem Befund erwarten lassen. Bei Freilegung des Rückenmarks sollte man völlig zerfetzte und schlecht ernährte Teile der nervalen Substanz resezieren, bei völliger Querdurchtrennung ebenfalls die zerfetzten Gewebsränder sorgfältig glätten.

Auf die Behandlung von Dekubitalulzera, Kontrakturen, Harnwegsinfektionen und Spasmen wird eingegangen. Schließlich werden die Behinderungen bei der beruflichen Wiedereingliederung Querschnittsgelähmter geschildert und die vielfältigen Hilfen an Geräten und Fahrstühlen gezeigt, die zu diesem Zweck entwickelt worden sind. Es wird abschließend auf die Schwierigkeiten organisatorischer und personeller Art, die sich einer optimalen Versorgung von Querschnittsgelähmten noch entgegenstellen, hingewiesen.

Durch zahlreiche Dias wird das Gesagte veranschaulicht. — Erscheint in den Sitzungsberichten.

Vorgelegt werden Veröffentlichungen von: Wolfgang Buchheim, Rolf Emmrich, Paul Görlich, Kurt Schwabe.

Sitzung am 17. Juni

Vortrag des Herrn Manfred Gersch: „Probleme der Insektizide aus heutiger Sicht". — Die Insekten machen gemeinsam mit den meist marinen Krebsen den größten Anteil der heutigen tierischen Biomasse aus. Obgleich noch nicht ein Prozent der Insektenarten in bezug auf die Nahrungsquellen oder als Krankheitsüberträger in direkte Beziehung zum Menschen treten, stellen sie aber infolge der Unermeßlichkeit ihrer Individuenzahl und ihrer zerstörenden Tätigkeit den größten Konkurrenten der Menschheit dar. Die Eindämmung und Bekämpfung schädlicher Insekten sind daher nach wie vor ein vordringliches Problem. Unter den verschiedenen Bekämpfungsmethoden, die in heutiger Zeit zur Anwendung kommen, ist die durch Insektizide am verbreitetsten und am wirksamsten.

Die chemische Schädlingsbekämpfung mittels Insektiziden läßt sich in 3 Phasen einteilen. Die erste, bereits im Altertum beginnend, ist durch den Einsatz von anorganischen Giftstoffen gekennzeichnet. Sie wird durch die Phase der synthetischen Kontaktinsektizide vom Typus des Dichlordiphenyltrichloräthan (DDT) oder der organischen Phosphorverbindungen u. a. abgelöst. Alle chemischen Insektizide sind aber mehr oder weniger starke Gifte. Obgleich eine Vielzahl von Verbindungen meistens auf dem Wege des Screening-Verfahrens entwickelt wurde, haben sich bisher kaum Zusammenhänge zwischen der Konstitution der jeweiligen Verbindung und ihrer Wirkung finden lassen. So ist auch trotz der breiten Nutzanwendung der verschiedenen Insektizide nichts über die zellulären oder molekularbiologischen Wirkungsmechanismen dieser Verbindungen bekannt.

Diese Tatsachen machen das Bemühen um andere Bekämpfungsmittel verständlich. Der Einsatz von Insektenhormonen bzw. deren Mimiksubstanzen erscheint hierbei besonders naheliegend. Während die bisherigen Schädlingsbekämpfungsmittel sämtlich auf dem Prinzip beruhen, die Schadinsekten, in Wirklichkeit meistens weniger selektiv alle Insekten, durch Tötung mittels Giftstoffen auszuschalten, bezwecken die auf der Grundlage der Insektenhormone angestrebten Mittel, die Ausmerzung der Population durch Verhinderung der Weiterentwicklung der Individuen, damit Unterbindung der Metamorphose und in Verbindung damit die Ausschaltung der Fortpflanzung zu erreichen. Das entspricht der dritten Phase der Entwicklung von Insektiziden.

Seit langem ist bekannt, daß an der Steuerung der Insektenentwicklung drei Hormone beteiligt sind: Das aus dem Gehirn-Corporacardiaca-Komplex stammende sogenannte „brain"-Hormone (von uns als Aktivationshormon bezeichnet), das von der im ersten Brustsegment liegenden Prothoraxdrüse produzierte Häutungshormon Ecdyson bzw. Ecdysteron und schließlich das die Art der Häutung bestimmende Juvenilhormon.

Mit den drei für die Insektenentwicklung verantwortlichen Hormonen verbinden sich jeweils bestimmte und im einzelnen unterschiedliche Lösungswege des Einsatzes als Insektizide. Hinsichtlich der Einsatzmöglichkeiten sind bisher zweifellos die Arbeiten mit dem Juvenilhormon und den Juvenoiden am weitesten fortgeschritten. Dabei hat sich jedoch gezeigt, daß nicht das Molekül des Juvenilhormons selbst, sondern andere synthetische Verbindungen die stärksten biologischen Wirkungen aufweisen. Der Einsatz von Steroiden von Häutungshormon-Charakter ist bisher noch unsicher und unklar. Die dritte Möglichkeit betrifft die Steuerung der Entwicklung über das übergeordnete neuroendokrine System. Als günstige Voraussetzung für einen möglich werdenden Eingriff hierdurch sind einerseits die Kenntnis des Ziel-Organs sowie andererseits der Wirkungsmechanismen anzusehen. Außerdem bietet sich mit dem Einsatz der bei Chilopoden und Krebsen vorhandenen Hemmhormone ein neuartiges biologisches Prinzip an. Die damit zusammenhängenden Fragen bedürfen allerdings noch umfassender Untersuchungen.

Keiner der drei Lösungswege erlaubt bisher eine endgültige Beurteilung. Vielmehr bestehen in jedem Falle noch zahlreiche offene und ungelöste zentrale Probleme. In dieser Situation kann nicht nur eine Möglichkeit verfolgt werden. Vielmehr sind weitere Untersuchungen für alle der hier aufgezeigten Richtungen zweckmäßig und sinnvoll. Damit stellt sich dieses Gebiet zugleich als ein besonders anschauliches Beispiel der engen Verbindung zwischen echter Erkundungsforschung und gezielter Grundlagenforschung dar. — Erscheint in den Sitzungsberichten.

Vorgelegt werden Veröffentlichungen von: Hermann Berg, Paul Görlich, Kurt Schwabe.

Sitzung am 16. Dezember

Vortrag des Herrn Adolf Watznauer: „Die geologische Situation in ihrer Wirksamkeit als Umweltfaktor". — Der Begriff Umwelt wird in wörtlichem Sinne, d. h. als die Gesamtheit dessen aufgefaßt, was einen Gegenstand umgibt. Umwelt setzt also eine Bezogenheit auf einen Gegenstand voraus, wobei dieser Gegenstand selbst einen Bestandteil des Systems „Umwelt" bildet. In diesem Sinne gibt es nicht eine Umwelt, sondern so viele, wie Bezugspunkte postuliert werden.

Umweltfaktoren sind jene Parameter der Umwelt, die deren Zustand bestimmen. Da diese Parameter keine Konstanten, sondern mit der Zeit in unterschiedlichem Maße veränderliche Größen sind, kann das System „Umwelt" nie einen zeitinvarianten Gleichgewichtszustand besitzen.

Die Umweltfaktoren sind an sich wertfrei. Ihre Wertung erfolgt erst vom Bezugspunkt aus.

Änderungen des Systems „Umwelt" werden von einem wertenden Bezugspunkt, z. B. vom Menschen, der menschlichen Gesellschaft usw. ausgesehen, als Störungen empfunden. Das System wird versuchen, durch Beeinflussung der Störfaktoren, d. h. der Umweltfaktoren, die Störung im Falle einer positiven, d. h. gewünschten Wertung, zu verstärken, im Falle einer negativen, d. h. ungewünschten Wertung, sie abzuschwächen bzw. auszuschalten. Das Prädikat gewünscht bzw. unerwünscht ist dabei mit dem Begriffspaar notwendig/ nicht notwendig eng verknüpft und zeitabhängig. Als Beispiel wird die bergbauliche Tätigkeit angeführt. Sie ist stets mit einer radikalen Veränderung der Umwelt verknüpft, aber diese Veränderung ist notwendig, denn ohne sie kann eine Wirtschaft, deren Funktion als positiver Faktor der menschlichen Gesellschaft anerkannt werden muß, nicht aufgebaut werden.

Diese Umweltveränderung ist also als Ganzes positiv zu bewerten. Ein negatives Vorzeichen bekäme sie erst dann, wenn die ungewünschte Veränderung der Umwelt durch Nebenfaktoren bedingt sind, die technisch, ohne Änderung des gewünschten Ergebnisses, ausgeschaltet werden können. Man darf dabei nicht übersehen, daß jede Änderung in einem System zwangsläufig andere in diesem selbst nach sich zieht. Es wird also kaum eine heile, sondern im besten Falle eine optimale Umwelt geben.

Anhand einer Reihe von Beispielen wird gezeigt, daß es zwei Gruppen von Umweltfaktoren gibt. Die erste Gruppe umfaßt Faktoren, die in ihrer Wirkung rasch in Erscheinung treten, adaptiv nicht aufgefangen werden können, aber grundsätzlich korrigierbar, d. h. in ihrer Tendenz beeinflußbar sind. Man könnte sie als „akute" Fälle bezeichnen.

Die zweite Gruppe umfaßt langsam verlaufende, in ihrer Tendenz nicht beeinflußbare Prozesse, an die, auf eine menschliche bzw. biologische Ebene bezogen, eine zeitweilige Adaption vielleicht möglich ist. Ein Charakteristikum dieser Gruppe besteht darin, daß die Wirkung ihrer Faktoren kaum bzw. nicht mehr korrigierbar ist, wenn sie erkennbar wird. Man könnte sie als „chronische" Fälle bezeichnen. Diese Faktoren müssen erkannt, ihre Wirkung respektiert und ihre Auswirkungen in perspektivische Planungen als sichere unabänderliche Parameter einbezogen werden. In diese Gruppe fällt die geologische Situation.

Anhand einer Reihe von Beispielen wird die Wirksamkeit der geologischen Situation im einzelnen erläutert.

Wenn man Schwerpunkte setzen will, so sind beide Gruppen als gleichwertig anzusehen. Die akuten Fälle augenfälliger, d. h. in ihrer Tendenz beeinflußbar, die chronischen aber noch weitgehend unbekannt bzw. unerkannt und deshalb gefährlicher. Der Vortrag hatte den Zweck, die geologische Situation als einen Faktor der zweiten Gruppe in ihrer Wirksamkeit zu verdeutlichen. — Erscheint in den Sitzungsberichten.

Vorgelegt werden Veröffentlichungen von: Walter Brednow, Adolf Dabelow, Manfred Gersch, Herbert Jordan, Ott-Heinrich Keller, Kurt Mothes, Franz Runge, Gustav E. R. Schulze.

Philologisch-historische Klasse

1973

Sitzung am 22. Januar

Vortrag des Herrn Hans Hiebsch: „Zum Problem der Sprachhaftigkeit des menschlichen Bewußtseins". — In der Selbstbeobachtung scheint es evident zu sein, daß psychische Erscheinungen wie Gedanken, Wahrnehmungen, Entscheidungen etc. auf irgendeine Weise mit der Sprache verbunden sind. Gegen Belege dieser Art können aber methodische Bedenken geltend gemacht werden, besonders wegen der

Subjektivität der Selbstbeobachtung. Zu fordern ist also eine objektive Untersuchung der Beteiligung der Sprache an Bewußtseinsprozessen.

Psychophysiologische Untersuchungen, die besonders in der Nachfolge Pawlows zum Thema des sogenannten zweiten Signalsystems gemacht wurden (Nebilizin u. a.) legen die Annahme nahe, daß nahezu jedes innere funktionelle System eines psychischen Prozesses beim Menschen das innere Sprachsystem einschließt. Demnach müßten die strukturellen Elemente der inneren Sprache die Ausbildung, die Eigenart und den Verlauf eines psychischen Prozesses mitbestimmen.

Das müßte dann nicht nur für die Resultate psychischer Verarbeitungsprozesse, die psychischen Gebilde, zutreffen, sondern auch für die inneren psychischen Mechanismen oder Funktionen. Läßt sich erweisen, ob und in welchem Maße die innere Sprache an den psychischen Funktionen konstitutiv beteiligt ist, dann kann das Problem der Sprachhaftigkeit des menschlichen Bewußtseins als gelöst gelten. Im Vortrag können nur Ansätze zur Lösung dieses Problems vorgestellt werden.

In Anlehnung an kybernetische Vorstellungen zum Prozeß der organismischen Informationsverarbeitung werden drei Hauptfunktionen unterschieden: 1. die der Strukturbildung, 2. der Bewertung und 3. die Funktion der Voraussage. Diese Funktionen, wirkend im gesamten, komplexen Prozeß der Informationsverarbeitung beim Aufbau des „internen Modells der Außenwelt" (i. e. „Bewußtsein"), ermöglichen realitätsbezogenes und realitätsgerechtes menschliches Handeln.

Psycholinguistische Befunde und Experimente, besonders solche aus der Zeit des Spracherwerbs im Kindesalter, lassen es nun als wahrscheinlich erscheinen, daß das innere funktionelle System der Sprache an allen drei Hauptfunktionen mitbeteiligt ist. So dürfte die lexikalische und die grammatische Struktur der (subjektiven) Sprache für eine Reihe von Phänomenen der Strukturbildung während der Informationsverarbeitung und der Speicherung verantwortlich sein. Ein innerer Zusammenhang zwischen dem konnotativen Aspekt der Sprache und der Bewertungsfunktion scheint sehr plausibel zu sein.

Diese und andere Zusammenhänge können allerdings auch dazu verführen, n u r in der Sprache (d. h. natürlich in dem von der Muttersprache abgeleiteten inneren Sprachsystem) das universelle bewußtseinsbildende Prinzip zu sehen. Eine Diskussion der linguistischen Relativitätstheorie (Whorf und Sapir) erweist die Gefahren einer solchen Hypertrophie. Demgegenüber ist mit G. A. Willer einzuwenden, daß es eine große Anzahl von Invarianzen über alle natürlichen Sprachen hinweg gibt. Auch die psychologische Theorie des semantischen Raumes (Osgood et al.) verweist auf grundlegende Universalien innerhalb der uns bekannten Sprachen. Die Gründe für diese Gemeinsamkeiten liegen 1) in der gemeinsamen psychophysiologischen Struktur der Gattung. In ihrer konkreten funktionellen Wirksamkeit aber dürften sie 2) in der gemeinsamen Struktur des spezifisch menschlichen Verhaltens, nämlich in der Struktur der gegenständlichen Tätigkeit zu suchen sein.

Die eigentliche Wurzel für die Bildung des menschlichen Bewußtseins ist also das konkrete menschliche Handeln, die Arbeit. Die Sprache entwickelte sich als Kommunikationsmittel innerhalb der Bedingungen dieser Tätigkeit und bildet in einem gewissen Ausmaß deren Wesenszüge und deren kognitive Resultate ab. In der Genese

(Phylogenese und Ontogenese) findet also eine unablässige Wechselwirkung zwischen dem in die Tätigkeit einbezogenen Kommunikationsmittel und den zentralen Mechanismen des Bewußtseins statt, die das Verhalten regulieren. Auf diese Weise bildet sich das innere System der Sprache im Zentralnervensystem als integrierender Bestandteil aller Regulationseigenschaften des Handelns heraus. — Erscheint in den Sitzungsberichten.

Vorgelegt werden Veröffentlichungen von: Eberhard Brüning, Karl Bischoff, Johannes Jahn, Manfred Mayrhofer, Claire Préaux, Otto Vossler.

Sitzung am 12. Februar

Vortrag des Herrn Werner Peek: „Bericht über eine Forschungsreise nach Griechenland". — Der Vortragende berichtete über eine von März bis Juli 1972 durchgeführte Forschungsreise nach Griechenland. Im Mittelpunkt des von Lichtbildern begleiteten Referats standen schon im Winter des Jahres 1969 begonnene Forschungen in den Ruinen der antiken Stadt Epidauros („Alt-Epidauros"). Methodische Untersuchung der in eine mittelalterliche Mauer auf der einstigen Akropolis verbauten Inschriften, deren Zahl nach völliger Freilegung der Außenwand sowie durch Schürfungen im Gelände auf beiden Seiten des Akropolis-Hügels von bisher 18 auf über 60 vermehrt werden konnte, führte auf die Spuren eines bis dahin unbekannt gewesenen antiken Theaters. Auch die Stelle, an der es gelegen hatte, konnte durch die Entdeckung mehrerer noch in situ vorgefundener Theater-Sitze auf der Westseite des Hügels ermittelt werden. Der Vortragende wies nach, daß sämtliche Quadern jener Mauer einst Sitzbänke dieses Theaters gewesen waren und die Namen ihrer jeweiligen Stifter tragen. Mehrere Sitze, auf denen zusätzlich noch die Widmung „Für Dionysos" eingraviert ist, zeigen, daß es, wie das berühmte Dionysos-Theater in Athen, nach diesem Gott benannt war. Auf Grund dieser Ergebnisse beschloß die griechische Altertumsverwaltung die Freilegung des Theaters. Aus dem Beginn der Ausgrabung konnten ebenfalls noch einige Bilder vorgeführt werden. Sie ist inzwischen so gut wie abgeschlossen. Die Aufdeckung des Theaters hat mehr als 100 weitere Namen von wohlhabenden Bürgern der Stadt bekannt gemacht und die erstaunliche Tatsache erhärtet, daß der Bau mit dem des bekannten Theaters im Asklepiosheiligtum von Epidauros gleichzeitig erfolgt ist: ein weiterer Beweis für die Leistungsfähigkeit der Stadt im 4. Jh. v. u. Z., die schon aus ihrer dortigen Bautätigkeit abzulesen gewesen war.

Im weiteren berichtete der Vortragende über seine epigraphischen Studien in Sparta, Thessalien und Makedonien, die bei den Ruinen der antiken Stadt Atrax am Mittellauf des Peneios zur Entdeckung eines Heiligtums der Nymphen führten. Von den Einzelfunden zeigte er einen spartanischen Staatsvertrag aus der Zeit um 470 v. u. Z. sowie ein längeres Dialog-Epigramm auf die Nymphen aus dem 4. Jh.; das letztere wurde auch in Übersetzung dargeboten. —

Die Ergebnisse der Forschungen sollen in mehreren Einzelberichten in den Abhandlungen veröffentlicht werden.

Vorgelegt werden Veröffentlichungen von: Werner Coblenz, Johannes Erben, Johannes Jahn, Rolf Lieberwirth, Walter Markov, Manfred Mayrhofer, Joachim Müller, Willem Pée.

Sitzung am 14. Mai 1973

Vortrag des Herrn Joachim Müller: „Zeitgeschehen und Umweltbezug in Goethes Campagne in Frankreich". — Goethes Bericht über seine Teilnahme an der „Campagne in Frankreich" 1792 und an der „Belagerung von Mainz" 1793 wurde erst zwischen 1820 und 1822 niedergeschrieben und als Teil seiner Autobiographie in Fortsetzung von „Dichtung und Wahrheit" und der „Italienischen Reise" veröffentlicht. Der Bericht stellt sich wie diese beiden Titel als episches Kunstwerk dar. Es basiert in seinen Fakten auf Goethes eigenen Erinnerungen, wenn auch nicht auf unmittelbaren Notizen, die er auf der Rückreise verbrannt haben will, mit Ausnahme eines auf Karten eingetragenen Itinerars, auf Briefen, die er aus dem Feldlager an Freunde schrieb, auf dem Tagebuch des Weimarer Kammerdieners Wagner und auf ad hoc erbetenen Mitteilungen seines damaligen Dieners und Reisebegleiters Götze, ferner auf umfangreicher Lektüre von Memoirenwerken. Das Gesamtopus ist durch das fortlaufende Kontrastprinzip strukturiert, das sowohl die epochalen Spannungen wie die von Goethe erlebten Widersprüche seiner Umwelt und seines eigenen Inneren sichtbar macht. In aphoristisch-sentenziöser Zuspitzung manifestiert sich das Epochenbewußtsein im Ausspruch am Abend des 19. September 1792 nach der Kanonade von Valmy, die den strategischen Abbruch der Alliiertenaktion zur Folge hatte: „Von hier und heute geht eine neue Epoche der Weltgeschichte aus, und ihr könnt sagen, ihr seid dabei gewesen". Gegenwendig dazu ist Goethes Berufung des ästhetischen Urphänomens in dem Satz, das Schöne sei „das gesetzmäßig Lebendige". Dieses Wort wird in Münster gesprochen, wo sich Goethe auf dem durch die anhaltenden militärischen Wirren bedingten Umweg der Heimreise aufhält: Der Freundeskreis um die Fürstin Gallitzin ist das friedliche Gegenstück zur Lagergesellschaft. Zu den von Goethe verbuchten ästetischen Eindrücken auf der Campagne wie auf dem Rückzug und vor Mainz gesellen sich die auf allen Stationen der Reise eifrig betriebenen naturwissenschaftlichen Studien, vor allem zur Farbenlehre.

Goethe erkannte die Epochenwende, ohne seine Skepsis der Französischen Revolution gegenüber aufzugeben, deren Ursache er in der Korruptheit des Feudalregimes erblickte. Im Gefolge seines Weimarer Herzoges, der ein preußisches Regiment befehligte, beobachtete er zwar kritisch die Ereignisse, zahlte aber auch seiner antirevolutionären Umwelt manchen Tribut, so, wenn auch er zu Beginn des Feldzuges mit dem Einzug der Alliierten in Paris rechnete. Einerseits sympathisierte er mit der französischen Zivilbevölkerung und tadelte streng das Verhalten der emigrierten Adligen, andererseits hatte er kein Verständnis für die Mainzer Klubisten. In Kontrastspannung zum Kriegsgeschehen stand der monumentale antike Obelisk in Igel bei Trier, und dieser wieder bildete einen Gegensatz zum Filigranwerk der Gemmensammlung, die er in Münster kennenlernte.

Die Kontraststruktur ist der Ausdruck von Goethes Polaritätsanschauung, die ihn in Epochenkritik und Umweltanalyse gegen die Zerstörung den Aufbau, gegen die Verwilderung die Menschlichkeit, gegen die Verzweiflung die Hoffnung, gegen das Beharren die Aktivität, gegen den Tod das Leben aufrufen ließ. — Erschienen in den Sitzungsberichten Bd 117 (1974) H. 3.

Vorgelegt werden Veröffentlichungen von: Gerhard Buchda, Hans-Georg Gadamer, Walter Markov, Gertrud Schubart-Fikentscher, Otto Vossler, Friedrich Weller (Sitzung vom 19. März 1973), Kurt Aland, Hermann Heimpel, Willem Pée, Carl Friedrich von Weizsäcker.

Sitzung am 18. Juni

Vortrag der Frau Lieselotte Blumenthal: „Die verbrannte und die gestohlene Handschrift von Schillers Wilhelm Tell". — In der Nacht vom 21. zum 22. März 1825 brannte das Weimarer Hoftheater ab. Alles wurde vernichtet, darunter auch die Bibliothek, in der sich außer den Büchern die Originalmanuskripte der zur Aufführung bestimmten Werke befanden. Trotzdem konnte aus den Trümmern und der Asche einiges gerettet werden, aber es fehlen genaue Nachrichten. 1960 wurde bekannt, daß sich in der Saltykow-Schtschedrin-Bibliothek in Leningrad drei Blätter von Schillers Weimarer Bühnenbearbeitung seines „Wilhelm Tell" befinden, und 1970 zeigte mir ein Sammler in der Schweiz 64 Blätter desselben Manuskriptes und erlaubte ihre Bekanntmachung und spätere Verwertung in der Schiller-Nationalausgabe. Das umfangreiche Fragment beginnt mit dem von Schiller geschriebenen Personenverzeichnis, in das Goethe die Namen der Schauspieler für eine spätere Aufführung eingetragen hat, und endet in der zweiten Szene des vierten Aktes; der fehlende Rest der Handschrift dürfte verbrannt sein. Eine Analyse von Schillers Korrekturen ergibt die überraschende Tatsache, daß der Autor seine Arbeit mit der Abgabe des Manuskripts durchaus nicht als abgeschlossen ansah, sondern sich bis zur Uraufführung um die Verbesserung des Stückes bemühte. Dann aber stellte sich heraus, daß die Aufführung viel zu lang geraten war und umfangreiche Eingriffe gemacht werden mußten. Sie beschränkten sich aber nicht nur auf die Kürzung langer Reden und entbehrlicher kleiner Szenen, sondern griffen die Struktur des Schauspiels an. Es läßt sich nachweisen, daß Schiller solche Korrekturen wegen bestimmter Theaterverhältnisse oder aus politischen Rücksichten vornehmen mußte.

Die gestohlene Handschrift befindet sich im Großherzoglich Sächsischen Hausarchiv des Staatsarchivs Weimar. Sie galt bisher als Inspektionsbuch des Weimarer Hoftheaters, das irgendwann entwendet worden war und 1881 in einem Münchner Antiquariat von dem Sohn des Weimarer Großherzogs gekauft werden konnte. Seitdem wird es im Hausarchiv verwahrt. Eine Untersuchung der Handschrift ergab, daß sie nicht von Anfang an ein Inspektionsbuch gewesen war, sondern erst später dafür benutzt wurde und überhaupt nicht nach Weimar gehörte. Es ist ein Originalmanuskript Schillers mit seinen eigenhändigen Korrekturen, das er im März 1804 dem Hamburger Theaterdirektor geschickt hatte. Diese Bühnenbearbeitung ist zwar weitgehend eine Abschrift des Weimarer Manuskripts, zeigt aber stellenweise in der ersten Niederschrift eine ältere, bisher unbekannte Fassung. Schillers Korrekturen beweisen, daß er nicht nur die Weimarer Bearbeitung übernahm, sondern mitunter der fremden Bühne andere Anweisungen für ihre Regie gab. Beide Handschriften geben wichtige Aufschlüsse über das Verhältnis von Dichtung und Bühnenbearbeitung bei Schiller und dürften bisherige Meinungen der Forscher korrigieren. — Erscheint in den Sitzungsberichten.

Vorgelegt werden Veröffentlichungen von: Rudolf Große, Joachim Müller, Werner Peek, Friedrich Weller.

Sitzung am 15. Oktober

Vortrag des Herrn Rudolf Große: „Über die Geschichtlichkeit der Sprache". — Nachdem die sprachtheoretische Diskussion der letzten Jahrzehnte die Probleme der Systemhaftigkeit in den Vordergrund hatte treten lassen und dabei gegenwärtige Sprachzustände für die Exemplifizierung bevorzugt hatte, mehren sich in jüngster Zeit die Bemühungen um eine historische Linguistik, die die Erkenntnisse der Strukturforschung auf dem Gebiet der Grammatik und Semantik wie auch die neueren sozio- und psycholinguistischen Einsichten berücksichtigt. Folgende Überlegungen sind dabei hervorzuheben:

1. Der individuelle Sprachgebrauch wie auch das individuelle Sprachgefühl (die Kompetenz des Sprechers) sind für die Entwicklung einer Sprache nicht ausschlaggebend. Entscheidend ist die Sanktionierung einer Veränderung durch die Sprachgemeinschaft.
2. Jeder synchrone Sprachzustand enthält diachronische Spannungen, wie umgekehrt diachronische Prozesse auch systembedingt sind.
3. Im Wesen der Sprache sind Antinomien angelegt (Sprecher — Hörer, Code — Text, Sprachgebrauch — Möglichkeiten des Sprachsystems, Asymmetrie des sprachlichen Zeichens, informatorische und expressive Funktion der Sprache). Sie begründen die Dialektik von Stabilität und Dynamik im Sprachsystem.
4. Die kommunikative Funktion der Sprache erfordert Redundanzen und Variationen in den sprachlichen Äußerungen und erlaubt Heterogenität im System.
5. Damit erhält die soziale und territoriale Differenzierung der Sprache für die historische Linguistik besonderes Gewicht.
6. Für die offene Frage nach dem Übergang von individueller Variabilität und soziologisch fundierter Variationsmöglichkeit in den Status des gesamtsprachlichen Systems sind sozialpsychologische, soziologische und sozialhistorische Erkenntnisse beizuziehen. An dieser Stelle der sprachhistorischen Entwicklungsprozesse werden auch für die sprachintern veranlaßten Veränderungen im System die gesellschaftlichen Momente greifbar. — Erscheint in den Sitzungsberichten.

Vorgelegt werden Veröffentlichungen von: Walter Baetke, Eberhard Brüning, Rudolf Große, Hans Hiebsch, Rudolf Meyer, Claire Préaux, Carl Friedrich von Weizsäcker, Friedrich Weller.

Sitzung am 4. Dezember

Vortrag des Herrn Edgar Lehmann: „Die Kartographie als Hilfsmittel der Umweltforschung". — Der Vortragende behandelt „Umweltkarten", d. h. Karten, die territoriale Auswirkungen von Umweltstörungen, des Umweltschutzes oder der optimalen Gestaltung der Umwelt zum Inhalt haben, als eine spezielle Gruppe von Karten innerhalb der thematischen Kartographie. Es wird auf die methodische Konzeption dieser Karten anhand von praktischen Beispielen eingegangen und gezeigt, wie sich auf ihnen der Vorzug der Anschaulichkeit mit dem Charakter eines graphischen Modells verbindet, dem konkrete, überdies maßstäblich richtige, auf anderen Wegen nicht erzielbare Aussagen entnommen werden können. Voraussetzung sind Messungen und Datensammlungen aus den verschiedensten Bereichen der Natur Wirtschaft und Gesellschaft, die durch ihre Umsetzung in die Karte den Charakter von räumlich basierten Analogmodellen annehmen und somit Informationen über Umwelteinflüsse negativer oder positiver Art zu vermitteln vermögen. Umweltkarten übertreffen daher bloße Datenspeicher, denen räumliche Korrelationen nur schwierig zu entnehmen sind.

Fragen der Eingliederung der „Umweltkarte" in das System der thematischen Kartographie werden nur einleitend berührt. Der Vortragende wendet sich angesichts der bisher sehr geringen Zahl von speziellen Umweltkarten vielmehr der Analyse der ihnen zugrunde liegenden wissenschaftlichen Problematik zu, um Möglichkeiten und Grenzen der jeweils angewendeten kartographischen Entwurfsprinzipien im Rahmen der Umweltforschung herauszuarbeiten. Dabei wird verdeutlicht, daß „Umweltkarten" wie thematische Karten überhaupt nicht mehr lediglich Darstellungsmittel sind, sondern zu Forschungs- und Erkenntnismitteln entwickelt wurden. Auf Ähnlichkeiten dieser Wandlungen mit Entwicklungen im wissenschaftlichen Gerätebau wird eingegangen.

Es werden folgende drei Grundformen von Umweltkarten — ohne Berücksichtigung des Maßstabes — behandelt: einfache Distributionskarten, komplexe Karten, Integrationskarten. Als „Distributionskarten" sind in der Umweltforschung z. B. Luft- oder Wasserbelastungskarten anzusehen. Als komplexe Karten werden solche bezeichnet, auf denen einzelne Elemente aus kategorial verschiedenen Bereichen auf ein und derselben Zeichenfläche zusammengeführt werden, um „Ansatzpunkte" für die nähere Untersuchung ihres möglichen Zusammenwirkens zu gewinnen. Die kritische Überprüfung der den komplexen Karten zugrundeliegenden Parameter führt zu einer dritten Gruppe, die der Vortragende als „Integrationskarten" bezeichnet. Das sind Umweltkarten, die Aussagen über den Grad von Zusammenhängen bestimmter Umweltelemente miteinander ermöglichen. Sie vermitteln Vorstellungen über die funktionelle Verbundenheit kategorial heterogener Elemente, die mittels quantifizierender Synthesen gewonnen werden. Die Zweckmäßigkeit solcher Integrationskarten besonders für Typisierungen, die sich auf die Struktur und Dynamik von Geosystemen gründen, wird näher erörtert und mit einem Blick auf ihren Wert für territoriale Prognosen verbunden. Schließlich wird die Rolle von Umweltkarten als Katalysatoren eines neuen Wirklichkeitsverständnisses, ihre Funktion als Mittel der sozialistischen territorialen Flächenorganisation hervorgehoben. — Erscheint in den Sitzungsberichten.

Vorgelegt werden Veröffentlichungen von: Rolf Lieberwirth, Joachim Müller, Friedrich Weller.

1974

Sitzung am 14. Januar

Vortrag des Herrn Johannes Jahn: „Antike Tradition in der Landschaftsdarstellung bis zum 15. Jahrhundert". — Die antike Landschaftsmalerei hat Naturräume in idealer Sicht von großer Mannigfaltigkeit dargestellt, meist als Schauplatz mythologischen Geschehens. In der von christlichen Vorstellungen bestimmten mittelalterlichen Malerei dagegen erscheinen außermenschliche Bereiche nur in formelhafter Abkürzung und nur soweit, wie sie zum Verständnis des jeweiligen Bildinhaltes notwendig sind. Doch haben sich aus dem Motivvorrat der antiken Landschaftsdarstellung einige Elemente erhalten, die namentlich durch die byzantinische Variante der christlichen Kunst weitergegeben wurden und bei der Erneuerung der abendländischen Malerei durch Duccio und Giotto um 1300 als wichtige Kompositionsmittel Verwendung fanden. Die antike Felsenlandschaft ist zur Landschaft an sich geworden, die mit steil ansteigenden Felsen den vordergründigen Schauplatz der Handlung nach hinten abschließt, womit zugleich der der mittelalterlichen Malerei eigenen Tendenz zur spannungerfüllten Fläche und zur Vermeidung des Tiefraumes Rechnung getragen wird. Von Italien aus treten jene Formelemente im späteren 14. Jahrhundert auch in der Malerei diesseits der Alpen auf, so besonders in der böhmischen und niederländisch-burgundischen Tafel- und Miniaturmalerei. Auch mit zunehmender Beobachtung der Naturwirklichkeit im 15. Jahrhundert verschwinden sie nicht, obwohl sie sich zuweilen unter naturalistischer Hülle verbergen. Sie gehen mit der wirklich beobachteten Natur die verschiedensten Verbindungen ein, am vielseitigsten in der italienischen Malerei des 15. Jahrhunderts bis zu Leonardo.

Man hat die Entwicklung der Landschaftsdarstellung im späteren Mittelalter im allgemeinen an ihren progressiven Elementen verfolgt, d. h. an denen, die auf eine Zunahme der Naturbeobachtung zurückgehen. Hier wurde der umgekehrte Weg eingeschlagen: Das Augenmerk galt einem ganz bestimmten überlieferten Formenmaterial, das mit Erfassung der Naturwirklichkeit nichts zu tun hat, im Gegenteil ein Beharren auf alten Vorstellungen bedeutet, die aber mit den neuen Naturbeobachtungen mehr oder weniger innig verschmolzen sind oder auch schöpferisch zu eigenartigen Wirkungen gesteigert wurden. — Erschienen in den Sitzungsberichten Bd 117 (1975) H. 1.

Vorgelegt werden Veröffentlichungen von: Lieselotte Blumenthal, Walter Markov, Albrecht Neubert.

Sitzung am 11. Februar

Vortrag des Herrn Ernst Werner: „Häresie und Gesellschaft im 11. Jahrhundert". — Nach einer kritischen Wertung moderner bürgerlicher Häreseologie werden mit Hilfe religionshistorischer Methoden folgende Nachrichten über Ketzer durchleuchtet:

1000 Leuthart im Gebiet von Châlons-sur-Marne
1018 „Manichäer" in Aquitanien
1022 „Manichäer" in Toulouse
1022 Orléans
1025 Arras
1028 Monteforte (Mailand)
1030 Westburgund
1042—1048 Châlons-sur-Marne
1049 Synode von Reims
1085 „Liber contra Wolfelmum" des Manegold von Lautenbach

Als Ergebnis zeigt sich eine gnostisch-spiritualistische Grundstruktur, die von einer Vergeistigung religiöser Erfahrungen, einer Entmaterialisierung sichtbarer und greifbarer Religiosität, einer Individualisierung des Glaubens sowie einer Intellektualisierung des Heils- und Vergottungsweges geprägt wurde. Als Mittler des Ideengutes spielte das bulgarisch-byzantinische Bogomilentum eine vordergründige Rolle, das sowohl über Süditalien wie über Venetien in die lateinische Welt einströmte. Die mündliche Mission trugen nicht nur Ketzer, sondern ebenso Kaufleute, Mönche und Pilger.

Als sozialer Hintergrund der lateinischen Häresie bietet sich die neue Stadt an, in welcher im Kampf gegen die feudale Bevormundung ein neuer Geist geboren wurde, der nach einer neuen Ideologie verlangte. Die mobile Lebensweise, die Erweiterung des geistigen Horizontes, das rationale Denken, der Wert des Geldes, die Verwandlung von Produkten in Waren steigert das Selbstbewußtsein, erheischte eine stärker auf das erfolgreiche Individuum zugeschnittene Religion, deren Vertreter persönliche Leistungen über tradierte Weihen stellten, sich gewissermaßen als Geistesträger im täglichen Leben auswiesen. So entsprach der antisakramentale Spiritualismus mehr kaufmännischem Denken als der primitive Sakramentsglauben der Kirche mit den daraus entspringenden materiellen Verpflichtungen. Die Ketzer boten zusammen mit den Dialektikern vom Schlage eines Berengar von Tours und Roscellin von Compiègne dem jungen Bürgertum in der Morgenröte des kommunalen Zeitalters eine adäquate Ideologie, die ihr erwachtes Selbstbewußtsein besser zum Ausdruck brachte als der orthodoxe Spiritualismus der Kirchenreformer.

Darüber hinaus zogen die lateinischen Häretiker auch Bauern in ihren Bann. Die seit dem 10. Jahrhundert in Italien, Frankreich und Westdeutschland zu beobachtende bäuerliche Mobilität weichte das starre ländliche Sozialgefüge auf, begünstigte die Herausbildung einer Kleinfamilie und gab auch auf dem Lande Raum für einen gewissen Individualismus. Galten die kirchlichen Sakramente nichts mehr und waren die

Priester heillos und unrein, dann entfiel für die Plebanen die lästige Unterhaltspflicht für die Pfarrer. Die gnostischen Perfecti versprachen ihren Anhängern schon im Diesseits Erlösung.

Dem gnostischen Dualismus im 11. Jahrhundert kam eine Doppelfunktion zu: Einmal ideologisierte er den sozialen Aufstiegs- und Umschichtungsprozeß im ländlichen und städtischen Sektor und zum anderen zwang er die ecclesia Romana zu einer weit gründlicheren Reform an Haupt und Gliedern, als das ursprünglich ihre Vertreter beabsichtigt hatten. So folgte der Aktion von unten eine Reaktion von oben. Kirche und Klöster, wie vor allem Cluny, versuchten durch einen liturgischen Gegenzug den ketzerischen Antisakramentalismus zu neutralisieren. Mit Kreuzes- und Mariendevotion sowie einer gezielten eucharistischen Frömmigkeit brach die Orthodoxie in die heterodoxe Front ein. So wich der gnostische Spiritualismus der ersten Jahrhunderthälfte einem sakramentellen Materialismus der zweiten. Dennoch gelang der Kirche kein entscheidender Sieg. Bereits im Bauernkreuzzug von 1096 revoltierten die Pauperes gegen die ecclesia Romana und nährten eine latente antipäpstliche Stimmung unter dem Volke. Sie leiteten über zu dem Katharertum des 12. Jahrhunderts, das sich zur Gegenkirche formierte und im 13. Jahrhundert nur mit nackter, brutaler Gewalt niedergeworfen werden konnte. — Erschienen in den Sitzungsberichten Bd 117 (1975) H. 5.

Vorgelegt werden Veröffentlichungen von: Gerhard Buchda, Werner Peek.

Sitzung am 18. März

Vortrag des Herrn Rolf Lieberwirth: „Die rechtshistorischen Grundlagen der Lehre vom Gesellschaftsvertrag". — Unter den vormarxistischen Staatstheorien spielt die Vertragstheorie eine bedeutende Rolle. Im Verlaufe von acht Jahrhunderten war sie die staatstheoretische Grundlage für die politischen Auseinandersetzungen zwischen Kaiser- und Papsttum um die Vorrangstellung der Universalmächte und seit dem 16. Jahrhundert in den Kämpfen der National- und Territorialstaaten um die äußere und innere Unabhängigkeit. Mit ihr wurde der extreme Fürstenabsolutismus ebenso begründet wie seine Einschränkung und schließlich auch die völlige Ablehnung der Feudalgewalt überhaupt. In ihrem Gefolge entwickelten sich staatstheoretische und staatsrechtliche Begriffe, wie die Lehre von der Volkssouveränität, der moderne Souveränitätsbegriff, von dem auch die UNO ausgeht, und die Lehre von den Menschen- und Bürgerrechten, die sich sowohl in den verfassungsmäßigen Grundrechten als auch in der Konvention der Menschenrechte vom 10. Dezember 1948 widerspiegelt. Hinsichtlich der Lehre von der Volkssouveränität ist festzuhalten, daß ihre Anfänge in der bisherigen Literatur zu früh gesehen werden. Bis ins 16. Jahrhundert ging es nicht um eine echte Souveränität des Volkes, sondern nur um Einschränkung der Herrschergewalt. Erst danach trat die Volkssouveränitätslehre im eigentlichen Sinne in den Vordergrund.

Die Vertragstheorie geht grundsätzlich von zwei Verträgen aus. Durch den Gesellschafts- oder Staatsvertrag schließen sich die Individuen, die bisher in einem Natur-zustand lebten, zu einer Gemeinschaft, zum Staat zusammen und bestimmen durch einen weiteren Vertrag (Herrschaftsvertrag), wer die Staatsgewalt für die Gemeinschaft ausüben solle. In der mittelalterlichen Geschichte Europas entwickelte sich zunächst nur die Lehre vom Herrschaftsvertrag. Erst unter dem Einfluß des neuzeitlichen Naturrechts wurde der Gesellschaftsvertrag zum theoretischen Ausgangspunkt genommen, um von einem rational gewonnenen Ursprung des Staates die Herrschaftsgewalt im Sinne des Absolutismus zu begründen, einzuschränken oder völlig abzulehnen. Daraus geht hervor, daß die Vertragstheorie keine echte Lehre vom geschichtlichen Ursprung des Staates ist, sondern nur die Aufgabe hatte, die bestehenden Herrschaftsverhältnisse eines Staates zu rechtfertigen oder zu bekämpfen. Die Vertragslehre gehört also eindeutig zu den Rechtfertigungstheorien. Eine Theorie vom Ursprung des Staates setzt wissenschaftlich begründete historische Untersuchungen voraus. Diese beginnen in umfassender Weise mit Marx und Engels, die aus ihrer wissenschaftlichen Geschichtskonzeption heraus nicht nur den Fragen nach dem Ursprung des Staates nachgehen, sondern auch eine völlig neue Staats- und Rechtstheorie entwickelten, die zwischen Staat und Gesellschaft keinen Widerspruch sieht und die in den sozialistischen Staaten heute in die Praxis umgesetzt wird. — Erscheint in den Sitzungsberichten.

Vorgelegt werden Veröffentlichungen von: Werner Bahner, Manfred Mayrhofer, Joachim Müller, Ernst Werner.

Sitzung am 20. Mai

Vortrag des Herrn Walter Markov: „1789: Die politische Linke und die Theologie der Revolution". — Führte Unbehagen am Verlauf und den Ergebnissen der Französischen Revolution zur Herausbildung einer „äußersten Linken", die entweder für von ihr nicht permanent einbeziehbare soziale Gruppen stand oder auch nur den Einspruch ungeduldiger Revoluzzer wiedergab, so verdient eine Erscheinung Aufmerksamkeit, die wir aliter als den „linken Priester" umschrieben haben. Lange Zeit betrachteten ihn Historiker als ein Kuriosum, ohne sich Rechenschaft abzulegen, daß er in allen vergleichbaren Revolutionen und keineswegs nur als Zierat aufzutreten pflegte.

Die konventionelle katholische Kirchengeschichtsschreibung hat, ohne allerdings dafür den Beweis anzutreten, bis zum Pontifikat Johannes XXIII. unter dem nihil obstat bestritten, daß auch religiös inspiriertes „Ernst-nehmen" zur äußersten Linken vorstoßen kann. Sie überließ es der Profangeschichte, sich zu fragen: Gehörten die curés rouges oder gehörten sie nicht zum soeben noch hilflosen, jetzt von allen Seiten umschmeichelten und bald wieder abgemusterten „guten Volk", dem bon peuple?

Darauf zu erwidern, reichte die staatsbürgerliche Option des Priesters allein nicht aus. Er mußte sich gleichfalls den theologischen Forderungen stellen.

Die Ecclesia Gallicana hat sich dem revolutionären Anruf nicht entziehen können, und sie hat mehrstimmig auf ihn reagiert. Auf beiden Seiten stellte es sich als unmöglich heraus, Temporalia und Spiritualia reinlich zu scheiden. Das führte zu mannigfachen Verschränkungen zwischen politischer Ideologienbildung und Glaubenshaltung, zu Frontziehungen und Gleichschaltungen, zu einer Theologie der Konterrevolution und ebenfalls zu einer Theologie der Revolution. Deren aufrichtiges Bemühen anzuzweifeln, besteht kein Anlaß. Sie hat sich in einer ihrer Auszweigungen sogar angestrengt, mit der Entwicklung Schritt zu halten, die spezifische historische Mission eines plebejischen Radikalismus zu begreifen und geistlich zu rechtfertigen.

Die Frage, ob sie damit zum konstitutiven Bestandteil desselben geworden ist, wäre indessen zu verneinen. Sie verdient Aufmerksamkeit als eine Nebenhandlung sui generis, die einen bisher ziemlich dunklen Winkel der Revolutionsgeschichte ausleuchtet, erweitert jedoch den Katalog der äußersten Linken um keine neue Variante. Das liegt nicht daran, daß — ganz unvermeidlich — alle Seiten Fehler begingen, noch daran, daß sie von Ersatzreligionen wie dem revolutionären Heiligenkult, der „Göttin Vernunft", dem „Höchsten Wesen" oder der Theophilanthropie überschattet wurde. Es liegt vielmehr daran, daß am Ausgang des 18. Jahrhunderts die Emanzipation der bürgerlichen Gesellschaft so weit fortgeschritten war, daß ihr System aus sich selbst heraus rational ausreichend begründbar erschien und sie ihre Denkweise ebenfalls der sozialen Peripherie einprägen konnte. Der von der militanten Aufklärung eingeleitete Säkularisierungsprozeß erwies sich in der Revolution von 1789 als bereits irreversible Schwelle. Damit entfiel das objektive wie das subjektive Bedürfnis nach jener sakralen Legitimation, die zu Zeiten der Fra Dolcino, John Ball, Prokop, Müntzer und Winstanley dem Volk in Aktion noch unentbehrlich erschienen war. — Erscheint in den Sitzungsberichten.

Vorgelegt werden Veröffentlichungen von: Walter Baetke, Joachim Müller, Herbert Peukert, Gertrud Schubart-Fikentscher.

Sitzung am 17. Juni

Vortrag des auswärtigen Mitgliedes Herrn Willem Pée: „Rechtschreibungsprobleme im Niederländischen". — Die Geschichte der Rechtschreibung im Niederländischen ist, wie übrigens die der meisten anderen westeuropäischen Sprachen, lang und kompliziert. Sie vom Mittelalter ab zu behandeln, ist meines Erachtens überflüssig, um einen richtigen Blick auf den heutigen Zustand zu bekommen. Wir können anfangen mit einem kurzgefaßten Überblick über den Zustand in den Niederlanden und in Flämisch-Belgien im Augenblick der Gestaltung der Rechtschreibung De Vries und Te Winkel im Jahre 1863, wobei die Einheit der Rechtschreibung für Nord und Süd eingeleitet wurde.

Diese Rechtschreibung stützt sich auf die folgenden Regeln:

1. Die Regel der gebildeten Aussprache;
2. Die Regel der Einheitlichkeit;
3. Die Regel der Ableitung (oder historische Etymologie);
4. Die Regel der Analogie

Wo die drei obengenannten Regeln schweigen — schreiben De Vries und Te Winkel — handelt man der Analogie gemäß, d. h. die Wörter, deren Rechtschreibung weder durch die Aussprache noch durch die Einheitlichkeit oder durch die Ableitung bestimmt wird, werden auf dieselbe Weise geschrieben wie andere, deren Rechtschreibung mit Bestimmtheit bekannt ist und die anscheinend auf ähnliche Weise gebildet sind.

Die Grammatiken führen noch eine fünfte Regel an, nämlich die Regel des Gebrauchs, die eine nicht zu unterschätzende Rolle spielt.

Auch die Rechtschreibung der ‚Bastardwörter' (Fremdwörter, die sich einigermaßen den Regeln der Sprache, in die sie aufgenommen wurden, angepaßt haben) wurde geregelt.

Am 21. November 1864 wurde in Belgien die Rechtschreibung der niederländischen Sprache geregelt, aber in den Niederlanden wartete man bis Dezember 1888, ehe die Rechtschreibung De Vries und Te Winkel offiziell eingeführt wurde.

Keine zehn Jahre später erschien der in den Niederlanden sehr bekannt gewordene Beitrag von Dr. R. A. Kollewijn über ‚Onze lastige spelling — Een voorstel tot vereenvoudiging' (Unsere schwierige Rechtschreibung — Ein Vorschlag zur Vereinfachung).

Kollewijn wies auf eine Menge von Gegensätzen, die das System, das er unpraktisch und zeitraubend nannte, mit sich brachte.

Eine Kommission wurde 1892 ernannt, und im Januar 1893 machte sie ihre Vorschläge zur Vereinfachung der niederländischen Rechtschreibung und Deklination. Die Grundsätze, von denen die Kommission ausging, waren:

1. das System De Vries und Te Winkel zu entlasten von dem, was unwissenschaftlich und unhaltbar war;
2. soviel wie möglich denselben Sprachlaut durch das gleiche Schriftzeichen darzustellen;
3. den Gebrauch von Schriftzeichen, die keinen Laut darstellen, abzuschaffen;
4. sich in der Deklination nach der gebildeten Aussprache zu richten.

Ob Bastardwörter, die allgemein gebraucht werden, niederländisch geschrieben werden sollen oder nicht, muß man der Einsicht und dem Geschmack des Schreibers überlassen. Wer ein Wort als einen Fremden betrachtet, wird von selbst geneigt sein, es völlig oder größtenteils zu schreiben wie in der Sprache, aus welcher es entlehnt ist. Es ist jedoch vorzuziehen, diese Art von Wörtern, wenn sie in der Volkssprache allgemein in Gebrauch sind, mit den Buchstaben der gewöhnlichen niederländischen Wörter zu schreiben.

Zwischen 1895 und 1934 herrschte sowohl in Belgien als in Holland eine große Anarchie: die einen schrieben De Vries und Te Winkel, die anderen Kollewijn, und dieses Chaos wurde noch größer, als im Jahre 1934 die Rechtschreibungsordnungen von Minister Marchant erschienen und diese nur in Holland zur Regel wurden. Dadurch kam es zu einem Bruch zwischen Nord und Süd, der zehn Jahre dauern sollte.

Nach der Befreiung wurden im Jahre 1945 sowohl in Belgien als auch in Holland eine Rechtschreibungskommission ernannt, um die Einheit wiederherzustellen. Das geschah schon 1946 für die Schreibung des *e*, des *o* und des *sch*. Die Regelung war dieselbe wie die von R. A. Kollewijn und Marchant.

Die Schreibweise der geographischen Namen sollte später genehmigt werden. Die Rechtschreibung De Vries und Te Winkel galt aber weiter für den pronominalen Gebrauch, die zweiten Kasusformen, die Bastardwörter und die Inlaute in Zusammensetzungen, bis eine neue, noch zusammenzustellende Wörterliste in Kraft treten würde.

1947 wurde eine gemischte niederländisch-belgische Kommission eingesetzt. In dieser Kommission gab es verschiedene Meinungen hinsichtlich der Rechtschreibung der Bastardwörter, und das Resultat der vielen und langen Beratungen konnte nur zu einem Kompromiß führen. Trotz der Plädoyers der Fortschrittlichen für eine Vereinfachung der Schreibweise der Bastardwörter wurde daraus ein konservativer Kompromiß mit vielen Doppelformen. Dadurch entstand die größte Verwirrung.

Der allgemeine Wunsch später war: eine einzige Form, wie dem auch sei.

Darum wurden neue Kommissionen gebildet, die letzte 1963. Am 28. Mai 1969 wurden auf einer gemeinsamen niederländisch-belgischen Pressekonferenz in Den Haag die ‚Endvorschläge' zur Publikation freigegeben.

Über diese ‚Endvorschläge' wurde anfänglich viel geredet und geschrieben, aber andere genaue Entwürfe für die Rechtschreibung der Bastardwörter lassen noch immer auf sich warten, und die beiden Regierungen haben bis jetzt noch stets diese ‚Endvorschläge' eingefroren.

Vorgelegt werden Veröffentlichungen von: Walter Baetke, Willem Pée.

Sitzung am 16. Dezember

Vortrag des Herrn Gerhard Buchda: „Das Grundgesetz über die Landständische Verfassung des Großherzogtums Sachsen—Weimar—Eisenach vom 5. Mai 1816 und Okens ‚Isis'". — In diesem Thema verbinden sich miteinander ein Stück Verfassungs- und Verwaltungsgeschichte des Herzog-, seit 1815 Großherzogtums Sachsen—Weimar—Eisenach und ein Stück Lebensgeschichte des aus der Ortenau gebürtigen Mediziners, Naturforschers und Naturphilosophen Lorenz Oken. Mit einbezogen ist die Universität Jena, der Oken von 1807 bis 1819 als Professor angehört hat. Das von Weimar aus regierte ernestinische Herzogtum bestand bis 1815 aus den beiden Fürstentümern Weimar und Eisenach und aus der sogenannten Jenaischen Landesportion als Nachfolgeterritorium des 1690 erloschenen Herzogtums Sachsen—Jena. Jeder dieser drei Landesteile hatte seine eigenen Landstände. Durch eine ausführliche Konstitution wurden 1809 die drei Landständischen Gruppen zu einer gemeinschaftlichen Landschaft vereinigt und für die künftige Tätigkeit der aus zwölf Abgeordneten bestehenden landschaftlichen Deputation unter dem Vorsitz des Generallandschaftsdirektors die erforderlichen gesetzlichen Bestimmungen getroffen. Nachdem Sachsen—Weimar—Eisenach im Verlaufe des Wiener Kongresses beträchtlichen Gebietszuwachs erfahren hatte, gab Großherzog Carl August in dem Besitzergreifungspatent vom 15. November 1815 den Einwohnern der darin verzeichneten Landesteile und Ortschaften folgende Zusicherung: „Wir sichern den, Einwohnern der hierdurch von Uns in Besitz genommenen Länder allen den Schutz die unpartheyische Rechtspflege und Religionsfreiheit zu, deren Unsere Unterthanen in Unsern übrigen Landen zu genießen haben. Wir versprechen ihnen gleichmäßig und gemeinschaftlich mit den letztern eine landständische Verfassung zu geben, welche ihnen das Recht gewähren soll, durch aus allen Classen der Staatsbürger von Unsern Unterthanen selbst erwählten Repräsentanten, bei der Gesetzgebung mitzuwirken, Steuern und Finanzmaßregeln, die das Vermögen des Landes oder der Unterthanen betreffen, nur nach freier Prüfung zu bewilligen, über Mängel und Mißbräuche in der Landesverwaltung oder Gesetzgebung mit gutachtlichen Vorschlägen zu Abstellung derselben Uns Vortrag zu thun, sowie auch bei willkührlichen Eingriffen der Staatsbeamten in die gesetzliche Freiheit, die Ehre und das Eigenthum der Staatsbürger bei uns Klage zu führen und auf rechtliche Untersuchung ihren Antrag zu richten". In Erfüllung dieses Versprechens setzte der Großherzog am 5. Mai 1816 die aus den Verhandlungen einer Beratungsversammlung hervorgegangene neue Landständische Verfassung in Kraft, wobei er am Schluß des Publikationspatentes hinzufügte, daß er „das Recht auf Freiheit der Presse hierdurch ausdrücklich anerkenne und gesetzlich begründe". Diese Freiheit im höchsten Maße für sich in Anspruch nehmend, eröffnete Oken mit dem 1. August 1816 seine „Isis oder Encyclopädische Zeitung", die in Jena gedruckt und von Brockhaus in Altenburg verlegt wurde. Durch ihren Inhalt, soweit er die Politik berührte oder abseits fachwissenschaftlicher Fragen Kritik übte, verursachte Oken mancherlei Verwicklungen, die ihm 1819 schließlich zum Verhängnis geworden sind. Unser Bericht hat es allein mit dem Beginn dieser sich steigernden Konflikte zu tun. In den Vordergrund tritt dabei der Aufsatz, den Oken im ersten Jahrgang seiner Zeitung in den Stücken 9 bis 11 über die neue weimarische Verfassung veröffentlichte. Letzten Endes durch diesen von Oken selbst verfaßten Aufsatz wurde das berühmte Votum Goethes vom 5. Oktober 1816 ausgelöst. In seiner Abhandlung hatte Oken die neue Landständische Verfassung keineswegs in jedem Punkte verworfen, sie vielmehr hier und da wie auch ihre Schöpfer sogar gelobt. Seine beiden Haupteinwände betrafen 1) die Gliederung der Landstände in die drei allein anerkannten Stände der Rittergutsbesitzer, der Bürger und der Bauern, die er für „völlig verfehlt" erklärte und die er durch die Dreigliederung in Nähr-, Lehr- und Wehrstand ersetzt wissen wollte, sowie 2) das Fehlen klar fixierter Grundrechte der Staatsbürger. Oken gebrauchte derbe Worte. Leidenschaftlich rief er die Universität, die er nur, weil sie Rittergüter besaß, nicht aber als gelehrtes Institut, als gelehrten Stand unter den Landständen fand, dazu auf, ihre alte Prälatur zu verteidigen. Den Generalsuperintendenten, die Superintendenten und die Pastoren ermunterte er, sich gegen den Ausschluß der Geistlichkeit aus den Landständen zur Wehr zu setzen. Schließlich sollte auch der General mit einigen Obersten „dem Wehrstand sein Recht, vertreten zu werden, nicht nehmen lassen". Wegen des Mangels an Grundrechten würdigte Oken die neue Verfassung nur als Landstandsordnung („nicht als eine ... gut ausgedachte Wahlordnung der Volksvertreter") und sprach ihr den Charakter eines Staatsgrundgesetzes und einer wirklichen Staatsverfassung entschieden ab („daß wir in diesem

Grundgesetz nicht eine Spur von Staatsgrundgesetz erkennen"). Wer zu Okens Kritik Stellung nehmen will, muß weiter ausholen und außer auf die Ideen Okens auch auf den Werdegang der weimarischen Verfassung, auf die staatsmännischen Absichten des Großherzogs Carl August, die fortschrittlichen Pläne seines Ministers v. Gersdorf, die bisherige Stellung der Universität unter den Landständen, auf das Problem einer politischen Vertretung der Intelligenz, wie wir heute sagen würden, und noch auf anderes eingehen. — Erscheint in den Sitzungsberichten.

Öffentliche Gesamtsitzungen

14. April 1973

Ansprache des Präsidenten, Herrn Schwabe, Bericht des Vizepräsidenten, Herrn Lehmann, über die Arbeiten beider Klassen, — Vortrag des Herrn Meyer: „Kultur und Sprache im frühantiken Syrien".

Die Textfunde von Ugarit/*rās eš-šamra*, Mari/*tell ḥarīri*, daneben etwa die Entdeckungen in Byblos/*ǧebēl* und Alalach/*tell el- ʿaṭšāna* haben die altsyrische Stadtstaatenwelt des 2. Jahrtausends v. Chr. in das helle Licht der Geschichte treten lassen. Die Sprachwissenschaft ist durch diesen Sachverhalt insofern grundsätzlich betroffen, als hierdurch die Frage beantwortet werden kann, wie sich das Semitische als Volkssprache in einem Raume verhält, der durch eine lange Vorgeschichte und damit durch eine Vielfalt kultureller und ethnischer Erscheinungen gekennzeichnet ist, die zugleich den Einfluß fremder Sprachen einschließt.

Das frühantike Syrien umfaßt als Sprachraum den westlichen Schenkel des „Fruchtbaren Halbmondes" samt Scheitel, also die heutigen politischen Einheiten Syrien bis zum Euphrat, der nur in Nordmesopotamien überschritten wird, Libanon, Israel, Königreich Jordanien und einen schmalen türkischen Streifen im Norden. Die Vorgeschichte dieses Raumes als eines Gebietes seßhafter Kultur reicht nach Ausweis der Ortslagen von Ugarit im Norden und Jericho im Süden bis in das 7. Jahrtausend v. Chr. und damit bis in die vorkeramische Jungsteinzeit zurück. In diesem langen Zeitraum hat es ohne Zweifel zahlreiche Bevölkerungsverschiebungen gegeben, ohne daß wir die einzelnen Phasen nachzuzeichnen vermöchten. Eines aber scheint nach den neuesten Erkenntnissen sicher zu sein, nämlich daß die Primärträger des Semitischen, die der orientaliden Rasse angehören, nicht als späte Zuwanderer in Altsyrien anzusehen sind; denn ihre kleinwüchsigen Vorfahren sind bereits im Natufium nachweisbar. Als zweiter Sprachträger kommt eine armenoïde Menschengruppe in Frage, die seit grauer Vorzeit vor allem in Nordsyrien lebte und die schon frühzeitig semitisiert worden sein muß.

Wieweit das Semitische im syrischen Sprachraum zurückreicht, ist unbekannt; wohl aber beweisen die Quellen des 2. Jahrtausends, daß es um diese Zeit bereits auf eine lange Entwicklung zurückblicken kann. Sicher stand es im 3. Jahrtausend in der Form des West- oder Nordwestsemitischen beziehungsweise des Altkanaanäischen in voller Blüte. Da man sich den „Fruchtbaren Halbmond" nicht als sprachlich einheitlich vorstellen darf, ergeben sich für die Entwicklung des Semitischen drei verschiedene Möglichkeiten: a) die Symbiose mit einer völlig andersartigen Sprache, b) das Aufgehen in einem anderen Idiom und c) Weiterleben unter Bewahrung seiner Eigenart. Der erste Fall liegt im ostsemitischen Akkadisch vor; durch Verbund mit dem wesensfremden Sumerisch wird es teilweise geradezu „verfremdet", wie dies etwa am Satzbau besonders deutlich wird. Der zweite Fall begegnet beim östlichen Zweige des Amurritischen oder Ostkanaanäischen. Dieses westsemitische Idiom breitet sich seit Ende des 3. Jahrtausends aus der Steppengegend nördlich Palmyra sowohl nach Syrien als auch nach dem Zweistromland aus. Hier geht es im 2. Jahrtausend infolge Sprachwechsels seiner Träger im Akkadischen auf und ist um 1500 auch aus der Namengebung verschwunden.

Den dritten Fall bietet das Kanaanäische. Wie die ugaritischen Texte vermuten lassen, hat sich das bereits vorhandene Altkanaanäische mit dem zuwandernden Amurritischen zu einem neuen Sprachtypus wesensgleicher Art verbunden, der sich als jungkanaanäisch herauskristallisierte und in der Folge die Vorstufe für die Dialekte und Dialektgruppen darstellte, die um die Jahrtausendwende begegnen und von denen das klassische Hebräisch des Alten Testaments am bekanntesten ist.

Unbeschadet des inneren Wandels etwa in der verbalen Syntax hat das Kanaanäische als Ganzes seine Struktur nicht verändert. Das gilt zum Beispiel vom Verbal- und Nominalsatz ebenso wie von der Kasussyntax. So können auch fremde Einsprengsel im Wortschaft den Sprachcharakter nicht verändern, im Gegenteil, Träger anderer und nicht nur verwandter Idiome geben am Ende ihre Sprache zu Gunsten der übermächtigen Volkssprache Syriens auf.

Dieser Eigenart entspricht die Selbstdarstellung in der Schrift. Die Akkader übernehmen im 3. Jahrtausend von den Sumerern die ihrer Sprache wesensfremde Keilschrift und tragen damit auch zur „Verfremdung" des Sprachbildes bei. Hingegen erfinden die Altsyrer, durch die kulturellen und kommerziellen Umweltbedingungen des Ostmittelmeerraumes angeregt, vielleicht in Byblos, dem antiken Gubla, im frühen 2. Jahrtausend die reine Konsonantenschrift. Sie stellt die vollständige Abstraktion gegenüber den ägyptischen Einkonsonantenzeichen und den Silbenzeichen der Kreter und Akkader dar; diese Erfindung ist somit als Durchbruch zum modernen Schriftwesen zu bezeichnen.

Daß es sich bei dieser Erfindung, die zunächst als sogenanntes „altsinaitisches" Alphabet von 27 Zeichen vorlag, um das Ergebnis gelehrter Reflexion handelt, wird indirekt durch das ugaritische Keilschrift-Alphabet bewiesen. Etwa um die Mitte des 2. Jahrtausends entworfen, knüpft letzteres unter Verwendung von Keilschriftzeichen an das altsinaitische System an, durchbricht jedoch das strenge Konsonantenprinzip zu Gunsten des syllabischen, indem es den Kehlverschlußlaut Aleph einmal als *'a*, zum anderen als *'i* und *'u* wiedergibt. Offenbar sollte dieses Alphabet in dem mehrsprachigen Ugarit allgemein verwendet werden; westsemitisch konnte ein solches Aleph jeweils mit konsonantischem Einsatz gelesen, hingegen akkadisch und hurritisch auch vokalisch als *a*, *i* und *u* gebraucht werden. Beachtenswert ist auch bei der Vokaldarstellung die Abstraktion: nur für die Eckvokale *a*, *i* und *u*, wie heute noch im Arabischen, nicht dagegen für die auch damals vorhandenen Zwischenlaute waren Symbole geschaffen.

Dieses Mehrzweckalphabet stellte aber nur einen Teil umfassender philologischer Arbeit dar. So versuchte man des weiteren, die Probleme der Mehrsprachigkeit Ugarits als internationaler Handels- und Kulturmetropole zu lösen. Fußend auf sumero-akkadischen Wortlistentraditionen, stieß man bis zu einer regelrechten Lexikographie vor, indem man „Wörterbücher" auf Tontafeln schuf, die — mit dem Akkadischen als Basis — in ihrer ausgeprägten Form vier Kolumnen enthielt: sumerisch und akkadisch gleichsam als „klassische" Sprachen sowie hurritisch und kanaanäisch als Umgangsidiome.

Die Sprache von Ugarit wurde in der ihr eigenen Schriftgestalt nicht nur als politisch-ökonomisches und kultisches Kommunikationsmittel verwendet, sondern sie diente auch zur Fixierung der mythischen und epischen Traditionen, die ihrerseits auf die Eigenständigkeit syrischen beziehungsweise kanaanäischen Denkens hinweisen, wobei diese Eigenständigkeit nicht als Abgrenzung verstanden werden darf, sondern im Rahmen fruchtbarer Auseinandersetzungen im Bereiche der gesamten Frühantike zu verstehen ist. Die Verwendung des Kanaanäischen als literarische Hochsprache läßt sich für das 14. Jahrhundert belegen.

Ugarit ist mit seiner Sprach- und Schriftkultur um 1200 zugrunde gegangen; aber es darf als Beispiel angesehen werden, wie sich die Altsyrer im Bereiche der Sprache und ihrer Wiedergabe behauptet und damit zugleich ihr eigenes geistiges Gut bewahrt haben. Die Entwicklung der Schriftkultur als solcher lag freilich in der Linie des altsinaitischen Alphabets; aus ihm ging im Raume von Byblos das sogenannte „phönikische" Alphabet hervor, das mit seinen 22 Konsonantenzeichen über die Griechen seinen Siegeszug durch die Welt antreten sollte und als eine der Großtaten der Menschheit aus jener frühen Zeit in die Gegenwart hereinragt, ohne die die weltweite Schriftkultur undenkbar wäre.

17. November 1973

Ansprache des Präsidenten, Herrn Schwabe. — Vortrag des Herrn Drischel: „Organismus und geophysikalische Umwelt".

Es wird versucht, die verwickelten Beziehungen zwischen dem komplexen hierarchischen System „lebender Organismus" einerseits, den ebenso komplexen „geophysikalische Umwelt" andererseits in einer Übersicht darzustellen. Die biologisch wirksamen geophysikalischen Faktoren werden im einzelnen besprochen: Lufttemperatur, Luftfeuchte, Luftbewegung, atmosphärischer Luftdruck, Sauerstoffpartialdruck, Strahlung, Schwerkraft, mechanische Erderschütterungen, Luftionen, elektrische und elektromagnetische Felder. Desgleichen wird das reagierende System des höheren Organismus, der weitgehend autonom arbeitende nervale (vegetativ-nervöse) und hormonale Regulations- und Anpassungsapparat dargestellt. Auf das zentrale Problem der „physiologischen Adaptation" (insbesondere in der Evolution und der Individualanpassung) wird eingegangen. Danach werden prinzipielle wissenschaftliche Zugangswege zur Untersuchung der biologisch-geophysikalischen Beziehungen erörtert und an charakteristischen Beispielen erläutert: 1. die „geographische" Methode (Klimazonen und Biotope), 2. die „experimentell-empirische"

Methode (Untersuchungen über die individuelle Anpassung des rezenten Menschen an veränderte geophysikalische Bedingungen, z. B. Höhenstreß, 3. die „rhythmologische“ Methode (Vergleich biologischer und geophysikalischer Rhythmen) sowie 4. die „biometeorologische“ Methode (Einfluß des Wetters auf den menschlichen Organismus).

20. April 1974

Ansprache des Präsidenten, Herrn Schwabe, Bericht des Vizepräsidenten, Herrn Lehmann, über die Arbeiten beider Klassen, — Vortrag der Frau Schubart-Fikentscher: „Christian Thomasius — seine Bedeutung als Hochschullehrer am Beginn der deutschen Aufklärung“.

Im Gedächtnis des Volkes verknüpfen sich mit dem Namen Thomasius zweierlei verkürzte Vorstellungen: Kampf für die deutsche Sprache und Kampf gegen Vorurteile, insbesondere bezogen auf Hexenverfolgung. In einem Brief an Goethe vor 175 Jahren spielte Schiller darauf an und wollte Schriften von ihm kennenlernen (29. 5. 1799). Er traf mit wenigen Worten das Wesentliche.

Thomasius wurde am 1. 1. 1655 in Leipzig als Sohn des berühmten Professors für Beredsamkeit Jakob Thomasius geboren. Herkömmlich verlief das juristische Studium in Leipzig und Frankfurt/Oder, wo er 1679 mit der Doktorpromotion abschloß. In Leipzig arbeitete er zunächst als Anwalt und lehrte dann, seiner Neigung und Begabung folgend, an der Universität. Hier geriet er bald durch Angriffe auf Mängel und Vorurteile in Spannung zu der Umwelt: So 1687 durch die Ankündigung (am Schwarzen Brett) mit dem ersten deutschen Vorlesungsprogramm überhaupt, die „Nachahmung der Franzosen ...“, einer Erziehungslehre, verbunden mit der Forderung, auch in der gelehrten Welt die Muttersprache zu pflegen. Gleiches gilt von der ersten deutschen Zeitschrift „Monatsgespräche“ 1688/89. Mit Dissertationen und anderen Schriften (Rechtliches Gutachten für A. H. Francke) erregte er besonderen Anstoß in theologischen Kreisen. Andere Arbeiten gingen über den Universitätsrahmen hinaus und betrafen auch den Landesherrn: z. B. der Streit mit dem dänischen Hofprediger über Religionsabsolutismus und der Streit für eine fürstliche Eheschließung verschiedener Bekenntnisse. Schließlich wurde ihm durch ein Lehr- und Schreibverbot 1690 die Lebensgrundlage entzogen. Er ging nach Brandenburg, wurde wohl aufgenommen und vom Kurfürsten nach Halle gewiesen. Es war keine „Flucht“, sondern ein vorbereiteter Übergang. Eine Universität bestand dort noch nicht, war aber lange geplant und wurde nun allmählich aufgebaut. Thomasius ist später mit Recht als ihr geistiger Begründer bezeichnet worden. Trotz ehrenvoller Rückberufungen (1705/09) blieb er ihr bis zu seinem Tod verbunden (23. 9. 1728).

Sein Werk liegt äußerlich sichtbar in der Bibliographie von Rolf Lieberwirth zusammengefaßt vor (1955). Sie enthält 305 Schriften verschiedenster Größe und verschiedensten Inhalts, viele wurden wiederholt aufgelegt, einige gesammelt herausgegeben. Der Zugang ist nicht leicht, weil grundsätzlich Äußerungen überall zu finden sind, verändert oder ergänzt, und deshalb zu einem Thema stets vielerlei Arbeiten herangezogen werden müssen. Einen sicheren Ausgangspunkt findet man in seiner immer wieder betonten Stellung als Hochschullehrer. Das läßt sich an der Widmung vieler Arbeiten „zum Gebrauch des Thomasianischen Auditorii“ und weiter am Gehalt aus und für Vorlesungen erkennen. Getragen wurde dieses Werk von der großen europäischen Bewegung der Aufklärung, deren berühmter Vertreter er in Deutschland mit seinem Kampf gegen überkommene Autoritäten war. Im Zusammenhang damit steht die rechtsphilosophische Frage nach dem Naturrecht. Es war der Maßstab, nach dem, gegründet auf die menschliche Vernunft, alles zu messen sei, auch das Christentum. Von hieraus sah Thomasius die Aufgabe der Rechtswissenschaft, den Weg zu Änderungen zu zeigen. Das berühmteste Beispiel, stellvertretend für andere, ist sein erfolgreicher Kampf gegen die Hexenverfolgungen und die Folter. Er berichtet selbst, wie er durch seinen „1. Hexenfall“ (1694), befangen in Vorurteilen und übereilt, Falsches vorgeschlagen hatte, aber vom Spruchkollegium der Fakultät genötigt wurde, alles umzuändern. Es ist bezeichnend für ihn, daß er sich von da an bis zum Lebensende immer wieder mit dem gesamten Fragenkreis energisch wirkend befaßt hat.

Seine Arbeit stand unter dem Ziel der Universitäten, „Pflanzgärten für Kirche und Staat“, „mehr (noch) ... des Friedens“ zu sein und damit ein neues Menschenbild in allgemeiner wie fachlicher Hinsicht zu erreichen. Für ihn stand zunächst in der Philosophie die Sittenlehre im Mittelpunkt, begriffen unter dem Decorum der Bewährung in der Gemeinschaft. Gerade damit erregte er aber Anstoß in theologischen Kreisen, weil er sich als Christ und Professor für berechtigt hielt, insbesondere die Bibel selbständig und kritisch mit zu behandeln. Lange Streitigkeiten führten schließlich zum königlichen Verbot solcher Vorlesungen und dem Befehl, sich „auf das jus“ zu richten (1702).

Unter den Rechtsfragen ist besonders sein Kampf gegen das fremde römische und kanonische Recht bemerkenswert. Seit langem war es eingedrungen und auch in deutschen Gerichten angewendet worden. Der Vorrang deutschen Rechts blieb unbestritten, fremdes Recht galt nur hilfsweise, aber es gab noch kein eigenes deutsches System und für Vorlesungen wurde das römische als Grundlage mit jeweiligem Bezug auf deutsche Rechte genommen. Thomasius versuchte mit genauer geschichtlicher Untersuchungsmethode beide Quellenbereiche zu trennen, wies die breitere Wirkung deutscher Rechte nach und legte damit allmählich den Grund zu einem System des deutschen Privatrechts.

Über das Studium und seine Reform in sachlicher und persönlicher Hinsicht spricht er oft. Er versuchte eine genaue Ordnung in den Studienplan zu bringen. Zeiteinteilung und Stoffbewältigung, die Notwendigkeit neuer Lehrstühle (Ökonomie), wie Fragen nach einführenden Vorlesungen und Büchern, nach der Gestaltung von Vorlesungen im einzelnen (Diktat, Beispielen aus der Praxis, Exercitia, Selbststudium), alles, auch seine eigene Erfahrung, wird behandelt. Gleiches gilt für die Forderung nach deutscher Sprache, deren grundsätzliche übergeordnete Bedeutung er für die Hochschule erkämpft hat. Machten die Studenten vernünftige Vorschläge etwas zu ändern, ging er darauf ein. Das Verhältnis zu ihnen baute er auf gegenseitiger Achtung auf und lehnte, wie allgemein auch sich gegenüber, den geringsten Autoritätsglauben ab, denn jeder solle sich seine eigene Meinung schaffen. „Die Wahrheit“, „Weisheit ist nicht an mich gebunden“, wiederholte er oft. — Das alles galt aber nur für junge Männer, von Frauen ist keine Rede. Aber auch hier räumte er mit Vorurteilen auf und fand, daß die „Wahrheit“ von allen erkannt werden könne, wenn man die Menschen vernünftig erziehe, ebenso fand er, daß auch Mädchen Latein zu lernen fähig seien.

Es muß betont werden, daß Thomasius mit seinen Reformforderungen nicht allein stand, wenn er auch andere überragte und in vielem erfolgreicher war. Trotz seiner angreifenden Art war er immer um „Toleranz“ — das Schlagwort des Jahrhunderts — bemüht, und zwar im Sinne der Anerkennung auch des Gegners.

Sein Werk hat heute noch Erkenntniswert, sowohl für jene Zeit wie für die Grundlagen des heutigen Rechts und der Kultur, und es kann für das, was allgemein gültig geblieben ist, von neuen Gesichtspunkten aus noch weiter erschlossen werden.

9. November 1974

Ansprache des Präsidenten, Herrn Schwabe. — Vortrag des Herrn Görlich: „Meßtechnik, Gerätebau und Kybernetik als Grundlagen zur Umweltforschung“.

Wenn durch einen Steuerungsprozeß die Einheit „Leben — Umwelt“ positiv beeinflußt werden soll, muß über eine „Umwelt-Technik“ verfügt werden können. Als wichtige Grundpfeiler einer Umwelt-Technik hat man die Meßtechnik, den Gerätebau und die Kybernetik anzusehen. Diese drei genannten selbständigen Wissenschaftsdisziplinen zeigen eine zunehmende Verschmelzungstendenz, woraus sich als Folgerung die Forderung ergibt, komplette Gerätesysteme für vollständige Problemlösungen unter Sicherung der Kompatibilität in mechanischer, signaltechnischer und programmtechnischer Hinsicht zu schaffen sowie die dazu nötige Programmtechnik bereitzustellen. Das Ziel ist, zu versuchen, die Umweltprobleme mit kybernetischen Methoden zu lösen.

Es zeigt sich beispielsweise am Problem des Wasserhaushaltes, daß es sich um eine sehr komplizierte Problematik handelt, wobei das zweidimensionale Grundwasserproblem vom eindimensionalen Flußwasserproblem zu unterscheiden ist. Hinzu kommt, daß aus Sicherheitsgründen eine Wasser-Qualitäts-Überwachung konzipiert werden muß, die eine Fülle von verschiedenen Meßverfahren, die zu entwickeln sind, nach sich zieht.

Im Schwierigkeitsgrad steht die Luftkontrolle der Wasserhaushaltskontrolle keineswegs nach, im Gegenteil, die kybernetische Behandlung der Luftsteuerungssysteme fordert, daß sowohl die Meß- als auch die Gerätetechnik vorerst voll entwickelt werden, ehe die Kybernetik ihrerseits voll zur Wirkung kommen kann. Luftschadstoffmessungen, Staubmessungen, Stoffwechselkontrolle u. a. Messungen sind Gegenstand intensiver Forschungen.

Es ist aus den angeführten Gründen verständlich, daß nach wirkungsvollen neuen Meßmethoden laufend gesucht wird. In neuerer Zeit tritt die photogrammetrische Meßtechnik für die Umweltforschung sehr stark in den Vordergrund, wobei die Photogrammetrie als ein Verfahren der Meßtechnik definiert ist, womit Form, Größe und Raumlage von Objekten aus photographischen Bildern bestimmt werden. Der Hauptvorteil photogrammetrischer Messungen liegt in der Möglichkeit eines berührungsfreien Messens. Sowohl für eine Aerophotogrammetrie als auch für eine terrestrische Photogrammetrie existieren ganze Gerätesysteme, die in der Umweltforschung wachsend Bedeutung gewinnen.

Aus der Photogrammetrie heraus hat sich in den letzten Jahren die Fernmessung bzw. -erkundung (remote sensing) entwickelt. Die Fernerkundung kann mit den Begriffen Farbenphotographie, multispektrale Photographie, Falschfarbenphotographie, Thermalmeßmethode, Infrarot- und multispektrales Scanner- sowie Radar-Verfahren charakterisiert werden. Es werden alle elektromagnetischen Spektralbereiche vom äußersten UV bis in den Mikrowellenbereich benutzt, d. h. breite Spektralbereiche, die dem „Standard"-menschlichen Beobachter ohne entsprechenden Gerätepark nicht zugänglich sind.

Alle Gebiete der Fernerkundung haben bereits breite Anwendung in der Umweltforschung gefunden.

Mit den gemachten Ausführungen sollte umrissen werden, wie weit bisher die Grunddisziplinen Meßtechnik, Gerätebau und Kybernetik für die Umwelt-Technik zur Nutzung gekommen sind und welche stimulierende Wirkungen weiterhin für die Umweltforschung zu erwarten sein werden.

VERÖFFENTLICHUNGEN DER AKADEMIE IN DEN BERICHTSJAHREN

Jahrbuch 1971—1972. 1974. 483 S., 11 Taf., 4 Abb.

Festschrift. Zur Feier des 125jährigen Bestehens der Sächsischen Akademie der Wissenschaften zu Leipzig. Herausgegeben von Kurt Schwabe, Präsident der Akademie. 1974. 195 S., 34 Abb. im Text, 45 Abb. auf 35 Taf.

Mathematisch-naturwissenschaftliche Klasse

Abhandlungen

Band 52

Heft 2: Dietrich Uhlmann: Störungen des biologischen Gleichgewichts in Gewässern. 1973. 17 S., 16 Abb.

Heft 3: Felix Lampadius: Beitrag zum Nachweis der Wertminderung des Waldes als Folge von Immissionseinwirkung. 1974. 27 S., 4 Abb., 8 Tab.

Heft 4: Manfred Klose: Beiträge zur D- und V-Struktur von Wasser und speziellen wäßrigen Elektrolytlösungen. 1974. 22 Abb., 21 Tab. 63 S.

Heft 5: Horst Hanson: Eiweißumsatz und Differenzierungsvorgänge tierischer Zellen. 1974. 26 S., 13 Abb.

Sitzungsberichte

Band 110

Heft 3: Walter Brednow: Zur Anthropologie des Schwindels. 1973. 17 S., 2 Abb. auf 2 Kunstdrucktaf.

Heft 4: Paul Görlich: Betrachtungen über den wissenschaftlichen Gerätebau. 1973. 39 S.

Heft 5: Erich Rammler: Einige Betrachtungen über Erdgas. 1974. 43 S., 8 Abb., 3 Tab.

Heft 6: Gustav E. R. Schulze: Zur Rolle des Einfachheitsprinzips im physikalischen Weltbild. 1974. 23 S., 4 Abb.

Heft 7: Rolf Emmrich: Zwischen Leben und Tod. Ärztliche Probleme der Thanatologie. 1974. 22 S., 2 Abb., 4 Tab.

Abhandlungen

Band 60

Heft 3: Hans Joachim Mrusek: Gestalt und Entwicklung der feudalen Eigenbefestigungen im Mittelalter. 1973. 176 S., 4 Textfig., 119 Pläne und 79 Abb. auf 96 Taf.

Band 64

Heft 1: Walter Hentschel, Walter May: Johann Christoph Knöffel. Der Architekt des sächsischen Rokokos. 1973. 168 S., 111 Abb. auf 72 Taf.

Heft 2: Onomastica Slavogermanica VII. Rudolf Fischer in memoriam. Herausgegeben von Ernst Eichler und Hans Walther. 1973. 204 S.

Heft 3: Gotthard Neumann: Die Fibeln vom Kleinen Gleichberge bei Römhild. 1973. 74 S., 10 Taf., 1 Tab.

Heft 4: Herbert Petschow: Mittelbabylonische Rechts- und Wirtschaftsurkunden der Hilprecht-Sammlung Jena. Mit Beiträgen zum mittelbabylonischen Recht. 1974. 116 S., 5 Abb. auf 2 Kunstdrucktaf.

Band 65

Heft 3: Werner Peek: Ein neuer spartanischer Staatsvertrag. 1974. 15 S., 8 Faksimilia im Text, 8 Abb. auf 5 Taf.

Band 66

Heft 1: Onomastica Slavogermanica IX. Herausgegeben von Ernst Eichler und Hans Walther. 1974. 128 S., 5 Tab.

Sitzungsberichte

Band 116

Heft 4: Gertrud Schubart-Fikentscher: Studienreform. Fragen von Leibniz bis Goethe. 1973. 109 S.

Heft 5: Ernst Werner: Ketzer und Weltverbesserer. Zwei Beiträge zur Geschichte Südosteuropas im 13. und 15. Jahrhundert. 1974. 57 S.

Heft 6: Eberhard Brüning: Die Entwicklung progressiver und sozialistischer Literatur in den USA. Fortschrittliche Schriftstellerorganisationen und ihr Einfluß auf den nationalen Literaturprozeß von 1917 bis 1945. 1974. 40 S.

Band 117

Heft 2: Joachim Müller: „Verwirrung des Gefühls“. Der Begriff des „Pathologischen“ im Drama Goethes und Kleists. Mit einem Exkurs: Zur dichterischen Pathographie im Drama des 19. Jahrhunderts. 1974. 40 S.

Heft 3: Joachim Müller: Goethes „Campagne in Frankreich“. Epochenkritik, Umweltanalyse und Kontraststruktur. 1974. 42 S.

Heft 4: Günter Schmidt: Die literarische Rezeption des Darwinismus. Das Problem der Vererbung bei Emile Zola und im Drama des deutschen Naturalismus. 1974. 209 S., 1 Ausschlagtaf.

Nachauflage

Band 106

Heft 3: Joseph Aistleitner: Wörterbuch der ugaritischen Sprache. Herausgegeben von Otto Eißfeld. 4. Aufl., Nachdruck der 3., durchgesehenen und ergänzten Aufl. 1974. 364 S.

Mit Unterstützung der Akademie erschienen im Berichtszeitraum:

Thesaurus Linguae Latinae

Vol. VII Pars 2	Fasc. VII	*laqueo-lēgo*
	Leipzig 1973. Sp. 961—1120	
	Fasc. VIII	*lēgo-līber* (adi.)
	Leipzig 1974. Sp. 1121—1280	
Vol. IX Pars 2	Fasc. III	*obvius-oenanthe*
	Leipzig 1973. Sp. 321—480	
	Fasc. IV	*oenanthe-onocrotalus*
	Leipzig 1974. Sp. 481—640	

VERÖFFENTLICHUNGEN VON AKADEMIEMITGLIEDERN IN DEN JAHREN 1973 UND 1974

Mathematisch-naturwissenschaftliche Klasse

Berg, Hermann: Bioelektrochemie — Fortschritt und Risiko. In: Spektrum *8* (1974) S. 25.
— E. Bauer, H. Berg, W. Förster, A. Walter: Inhibition of thermal DNA denaturation kinetics by inter actions. In: studia biophysica *40* (1973) S. 249.
— E. Bauer, H. Berg, H. Schütz: Ionenstärkeeinfluß auf das Komplexbildungsvermögen der DNA. In: Faserforschung u. Textiltechnik *24* (1973) S. 58.
— G. Löber, V. Kleinwächter, H. Berg: Effect of organic solvents on the properties of the complexes of a polyphosphate with acridines. In: studia biophysica *35* (1973) S. 29.
— E. Bauer, H. Berg, G. Löber, K. Weller, M. Hartmann, Ch. Zimmer: Studies on the Structure of Chemically Methylated DNA. In: Biophysical Chemistry *1* (1974) S. 338 bis 348.
— J. Flemming, H. Berg: On the Association of DNA at the Mercury/Electrolyte Interface. In: Bioelectrochem. Bioenerg. *1* (1974) S. 459—465.

Brednow, Walter: Zum Begriff des „Pathologischen" bei Goethe. In: Medizinhistorisches Journal *8* (1973) H. 4, S. 257—289.
— Das Humanum im Spannungsfeld von Natur und Technik. In: Festschrift zur Feier des 125jährigen Bestehens der Sächsischen Akademie der Wissenschaften zu Leipzig. Berlin 1974. S. 187—195.

Buchheim, Wolfgang: Untersuchung zur Physik der Thermalwässer. In: Jahrbuch Sächs. Akad. Wiss. zu Leipzig *1971—1972* (Berlin 1974) S. 58—59.
— Ergänzende Diskussionsbemerkungen zum Vortrag von W. Michler. In: International Symposium on Protection of Mineral Waters against undesirable human Interference. Karlovy Vary (Carlsbad) 15.—19. 5. 1972. Prag 1973.
— Probleme der Physik der Thermalwässer. In: Arbeitstagung „Probleme der flüssigen Phase". Rostock 1973. S. 141—145, 3 Abb.
— Übersichtsbericht zur Problematik der physikalischen Parameter von Wasser in kapillaren Hohlräumen. In: Acta Hydrophysica *18* (1974) S. 77—88.

Budevski, Evgeni Bogdanov. – W. Bostanov, R. Roussinova, E. Budevski: Ausbreitungsgeschwindigkeit von monoatomaren Schichten und der Mechanismus der elektrolytischen Abscheidung des Silbers. In: Chem.-Ing. Technik *1973*, H. 4, S. 179.
— I. Iliev, S. Gamburzev, A. Kaischeva, E. Wakanova, J. Muchovski, E. Budevski: Sauerstoffelektroden auf hydrophobierter Kohlenstoffbasis. I. Strukturelle, physikalische und elektrochemische Eigenschaften von Elektroden, die mit PTFC hydrophobiert sind.

In: Comm. Dept. Chem. Bulg. Acad. Sc. *7* (1974) S. 223; II. Der Einfluß des Hydrophobisators auf die strukturellen, physikalischen und elektrochemischen Eigenschaften der Elektroden. Ebda S. 233.

— I. Iliev, S. Gamburzev, A. Kaischeva, E. Budevski: Sauerstoff-Gasdiffusions-Elektroden auf hydrophobierter Kohlenstoffbasis für Brennstoffzellen, die mit flüssigen oder in Elektrolyten dispergierten festen Brennstoffen arbeiten. Ebda S. 755.

— T. Vitanov, A. Popov, E. Budevski: Mechanism of Elektrocrystallization. In: J. Electrochem. Soc. *121* (1974) S. 207.

Dörter, Klaus: Süßgräser, Riedgras- und Binsengewächse. Leipzig, Berlin, Jena 1974 und Neudamm (BRD) 1974. 222 S.

— Wege von der extensiven Grünlandnutzung zur intensiven Futterproduktion im Harz unter besonderer Berücksichtigung von Forschungs- und Entwicklungsarbeiten im VEG Hasselfelde, Krs. Wernigerode. In: Hercynia NF *12* (1975) H. 1.

— E. Beuschold, K. Dörter, U. Wegener: Zur Problematik des Phosphoreintrages in Trinkwassertalsperren unter besonderer Berücksichtigung von Untersuchungen im Gebiet der Ostharztalsperren. Ebda H. 2/3, S. 185—200.

— E. Cronewitz, K. Dörter, I. Lieberoth, M. Pretzschel: Standortkundliche Beurteilung der wichtigsten Auenböden der DDR als Grundlage für acker- und pflanzenbauliche sowie meliorative Maßnahmen. In: Archiv Acker- und Pflanzenbau und Bodenkunde *18* (1974) S. 121—133.

— E. Beuschold, K. Dörter, U. Wegener: Der Einfluß der landwirtschaftlichen Nutzung von Talsperreneinzugsgebieten auf den Nährstoffeintrag in Trinkwassertalsperren. Vortrag, gehalten auf der Fortbildungstagung „Naturwissenschaftliche Grundlagen der Gütebewirtschaftung von Talsperren und Speichern" in Karl-Marx-Stadt 1974. S. 131 bis 143 [Kurzfassung].

Drischel, Hans: Zur Dynamik von Regulationssystemen des Körpers. In: Z. ges. inn. Med. *28* (1973) S. 7—12.

— Über simultane intermodale Frequenzvergleiche im Bereich von 5 bis 15 Hz. In: Nova Acta Leopoldina NF 211, *38* (1973) S. 177—191.

— Biologische Rhythmen und Anpassung von Lebewesen an die geophysikalische Umwelt. In: Geol. Wiss. Z. *2* (Berlin 1974) S. 905—925.

— Informationsauffindung und -wiederfindung in Nervensystemen von Organismen — Fakten und Hypothesen. In: Kybernetik und Bionik. Berichtswerk über den 5. Kongreß der Deutschen Gesellschaft für Kybernetik in Nürnberg 1973. München, Wien 1974. S. 81—102.

Freimuth, Ulrich: Veränderungen von Proteinen durch thermische Einwirkungen und *Maillard*-Reaktion. In: Periodica Polytechnica, Chemical Engineering *17* (Budapest 1973) Nr. 1, S. 19—28.

— u. E. Gahner, W. Krause: Versuche zur Fraktionierung der Weizenproteine. In: Nahrung *17* (1973) H. 4, S. 451—460.

— I. Tinius, B. Schlegel, H. Nötzold, U. Freimuth: Versuche zur Bestimmung des Molkenproteinanteils in Copräzipitaten. Ebda H. 3, S. 341—348.

— K. D. Schwenke, B. Raab, J. Uhlig, H. Tkocz, J. Behlke, M. Böttger, U. Freimuth: Über Samenproteine. — 3. Mitt. Isolierung und Charakterisierung der Albumine aus Sonnenblumen- und Rapssamen. Ebda H. 8, S. 791—809.

— R. Kretschmar, D. Hübner, R. Schmidt, U. Freimuth: On the interaction between β-Lactoglobulin and reducing sugars studia biophysica *40* (Berlin 1973) S. 213—220.

Freimuth, Ulrich u. B. Schlegel, E. Gahner, H. Nötzold: Die Bildung von Lysinoalanin in alkalibehandelten Milchproteinen. In: Nahrung *18* (1974) H. 1, S. K5–K8.
– u. W. Krause: Auswahl und Prüfung einiger Methoden zur Qualitätsbeurteilung von Rohmilchzucker und gereinigter Laktose. In: Lebensmittel-Industrie *21* (1974) H. 7, S. 319–322.
– u. W. Krause, D.Hübner, A. Beier, R. Schwarzer, D. Grimm: Untersuchungen zur Proteinschädigung durch *Maillard*-Reaktion mittels enzymatischer Methoden. In: Nahrung *19* (1974) H. 1, S. 83–91.
– R. Schröder, W. Rödel, V. Erhardt, U. Freimuth: Beitrag zur Korrelation sensorischer und gas-chromatogr. Daten. Ebda *18* (1974) H. 2, S. 171–172.

Gersch, Manfred: General aspects and problems in comparative endocrinology. In: Gen. comp. Endocrinol. *20* (1973) S. 425–435.
– Experimentelle Untersuchungen zum Freisetzungsmechanismus von Neurohormonen nach elektrischer Reizung der Corpora cardiaca von Periplaneta americana in vitro. In: J. Insect Physiol. *18* (1973) S. 2425–2439.
– Effects of hormones in invertebrates. International Encyclopedia of Pharmacology and Therapeutics Section 85. In: Com. Pharm. *1973*, S. 711–798.
– u. H. Birkenbeil: Das Membranpotential der Prothoracaldrüsenzellen von Galleria mellonella in Beziehung zum Entwicklungsstadium und nach Einwirkung von Neurohormonen. In: Zool. Jb. Physiol. *77* (1973) S. 1–24.
– u. R. Bräuer, H. Birkenbeil: Experimentelle Untersuchungen zum Wirkungsmechanismus der beiden entwicklungsphysiologisch aktiven Fraktionen des „Gehirnhormons" der Insekten (Aktivationsfaktor I und II) auf die Prothoracaldrüse. In: Experientia *29* (1973) S. 425.
– Experimentelle Untersuchungen zur Ausschüttung von Neurohormonen aus Ganglien des Bauchmarks von Periplaneta americana nach elektrischer Reizung in vitro. In: Zool. Jb. Physiol *78* (1974) S. 138–149.
– Selektive Freisetzung des hyperglykämischen Faktors aus den Corpora cardiaca von Periplaneta americana in vivo. In: Experientia *30* (1974) S. 767.
– Untersuchungen über die Wirkungsmechanismen von Neurohormonen. In: Jahrbuch Sächs. Akad. Wiss. zu Leipzig *1971–1972* (Berlin 1974) S. 61–68.
– Das Phyletische Museum. In: Reichtümer und Raritäten. Jena 1974. S. 102–111 (= Jenaer Reden und Schriften der Friedrich-Schiller-Universität).
– E. Baumann u. M. Gersch: Untersuchungen zur Stabilität des Neurohormons D. In: Zool. Jb. Physiol. *77* (1973) S. 153–160.
– u. R. Bräuer: In-vitro-Stimulation der Prothoracaldrüsen von Insekten als Testsystem (Prothoracaldrüsentest). In: J. Insect Physiol. *20* (1974) S. 735–741.
– u. E. Hentschel, J. Ude: Aminerge Substanzen in lateralen Herznerven und im stomatogastrischen Nervensystem der Schabe Blaberus craniifer BURM. In: Zool. Jb. Physiol. *78* (1974) S. 1–15.
– E. Baumann u. M. Gersch: Versuche zur Markierung von Neurohormon D aus Periplaneta americana mit Hilfe von Dansylchlorid. Ebda S. 533–541.
– H. Birkenbeil u. M. Gersch: Weitere Untersuchungen zur hormonalen Beeinflussung des Membranpotentials der Prothorakaldrüsenzellen von Galleria mellonella. Ebda S. 150–156.
– K. Richter u. M. Gersch: Untersuchungen über die Wirkung cholinerger, aminerger und peptiderger Substanzen am lateralen Herznerven und den funktionellen Zusammen-

hang dieser Wirkung bei der Herzregulation von Blaberus craniifer BURM. (Insecta: Blatteria). Ebda S. 16–32.

Görlich, Paul: Kepler und seine Beziehungen zur Optik. In: Sitzungsber. Plenum und Klassen der AdW 1972, Nr. 12, S. 21–31.
— Introductory Remarks to the Round Table Discussion „Photon-Detectors". In: Proc. of the Photon-Detectors Juni 1973. IMEKO-Secretariat Budapest, S. 1–6.
— u. W. Skarus: Zu Problemen der Meßtechnik und Meßdatenverarbeitung für die Automatisierung. In: m–s–r (Messen–Steuern–Regeln) *16* (1973) H. 6, S. 194–201.
— u. Hj. Bernhardt: Investigations of the Coloration Behaviour of Cerium-Oxide-Doped Alkaline-Earth Fluoride Crystals. In: phys. stat. sol. (a) *20* (1973) S. 487–499.
— u. G. Schwotzer, G. Kötitz: Donorno-akceptornye Pary v Ionnych Kristallach. In: Izv. Akad. Nauk SSSR, ser. fizičeskaja *37* (1973) H. 4, S. 812–817.
— u. P. Ullmann: Reversible Photochemical Reactions in Additively Coloured CaF_2 Crystals. In: phys. stat. sol. (b) *58* (1973) S. 371–377.
— Trends in Photon Detectors. In: Trends in Control Components. North-Holland Publishing Comp. 1974. S. 227–242.
— Beziehungen zwischen Feingerätetechnik und Elektronik. In: Feingerätetechnik *23* (1974) H. 11, S. 482–484.
— u. Hj. Bernhardt: Investigations of the Dynamical Coloration Behaviour of Cerium-Oxide-Doped Alkaline Earth Fluoride Crystals. In: phys. stat. sol. (a) *22* (1974) S. 559 bis 568.

Hanson, Horst: Eiweißumsatz und Differenzierungsvorgänge tierischer Zellen. Berlin 1974. 26 S. (= Abhandlungen Sächs. Akad. Wiss. zu Leipzig Bd 52, H. 5).
— Intracellular Proteolysis and its Demonstration with Synthetic and Natural Peptides and Proteins as Substrates. In: The Chemistry of Polypeptides *1973*, S. 237–258.
— S. Ansorge, P. Bohley, H. Kirschke, J. Langner, I. Marquardt, B. Wiederanders, H. Hanson, B. Kaiser: The Identy of the Insulin Degrading Thiol-Protein Disulfide Oxidoreductase (Glutathione-Insulin Transhydrogenase) with the Sulfhydryl-Disulfide Interchange Enzyme. In: FEBS Letters *37* (1973) No. 2, S. 138–140.
— M. Frohne, J. Lasch, H. Hanson: Bovine Lens Leucine Aminopeptidase pH dependence of the kinetic parameters. In: Acta biol. med. germ. *31* (1973) S. 879–884.
— S. Ansorge, P. Bohley, H. Kirschke, J. Langner, B. Wiederanders, H. Hanson: Metabolism of Insulin and Glucagon Glutathione-Insulin Transhydrogenase from Microsomes of Rat Liver. In: Eur. J. Biochem. *32* (1973) S. 27–35.
— G. Damaschun, H. Damaschun, H. Hanson, J. J. Müller, H.-V. Pürschel: Brief an die Redaktion: Zur Quartärstruktur der Leucinaminopeptidase. In: studia biophysica *35* (Berlin 1973) S. 59–60.
— J. Lasch, W. Kudernatsch, H. Hanson: Bovine-Lens Leucine Aminopeptidase — Kinetic Studies with Activators and Competitive Inhibitors —. In: Eur. J. Biochem. *34* (1973) S. 53–57.
— Chr.-A. Gründig, H. Hanson: Teilreinigung und Eigenschaften von Glycyl-L-leucin-Hydolase (EC 3.4.3.2) aus der Frauenmilch. In: Hoppe-Seyler's Z. f. Physiol. Chem. *354* (1973) S. 487–500.
— B. Wiederanders, J. Lasch, H. Kirschke, P. Bohley, S. Ansorge, H. Hanson: The Suitability of Bovine-Lens Leucine Aminopeptidase for Sequence Analysis and Limited Hydrolysis of Polypeptides. In: Eur. J. Biochem. *36* (1973) S. 504–508.

Hanson, Horst. — C. A. Gründig, H. Hanson: Über die Enzymaktivität der Frauenmilch gegenüber Aminosäurearylamiden. In: Acta biol. med. germ. *30* (1973) S. 599—606.
— M. Frohne, H. Hanson: Untersuchungen zum Einfluß von Präinkubationsbedingungen auf die Enzymaktivität der Leucinaminopeptidase. Ebda *31* (1973) S. 453—456.
— H. Damaschun, G. Damaschun, J. J. Müller, H.-V. Pürschel, H. Hanson: Quartärstruktur und Hydration der Leucinaminopeptidase. In: studia biophysica *43* (1974) H. 2, S. 126—127.
— M. Ludewig, H. Hanson, N. Kiselev, V. Stermstuk, F. Lerner: Elektronenmikroskopische Untersuchung der Leucinaminopeptidase. Ebda S. 127—128.
— G. Lassmann, B. Ebert, A. N. Kuznetsov, W. Dammerau, H. Hanson: Untersuchung der Polarität in der Umgebung eines am Protein gebundenen Spinmarkers am Beispiel der Leucinaminopeptidase (LAP). Ebda S. 128—129.
— P. Danew, E. Friedrich, M. Iwig, H. Hanson: Untersuchungen zur Peptidaseaktivität hautpathogener Pilze. Aminopeptidaseaktivität im Pilzmyzel und Kulturmedium von Microsporum gypseum in Abhängigkeit von den Wachstumsphasen. In: mykosen *17* (1974) S. 179—178.

Heisenberg, Werner: Physik und Philosophie. 2. Aufl. Stuttgart 1972. 201 S.
— Schritte über Grenzen. Gesammelte Reden und Aufsätze. 2. Aufl. München 1973. 351 S.
— Der Teil und das Ganze. Gespräche im Umkreis der Atomphysik. München 1973. 287 S. (= dtv 903).
— Wandlungen in den Grundlagen der Naturwissenschaft. 10 Vorträge. 10. Aufl. Stuttgart 1973, 183 S.
— Bemerkungen zur neuen Tamm-Dancoff-Methode. In: Problemy teoreticheskoi Fiziki, Akad. der Wissenschaften der UdSSR *1972*, S. 34—36.
— Naturwissenschaftliche und religiöse Wahrheit. In: Mitteilungen der Max-Planck-Gesellschaft *1973*, S. 73—90; Physikal. Blätter *29* (1973) S. 339—349.
— Construction of Large Accelerators: Scientific and Political Aspects. In: Science *179* (1973) S. 643—647.
— Development of Concepts in the History of Quantum Theory. In: Physicist's Conception of Nature. Hrsg. von Jagdish Mehra. Dordrecht/Holland 1973. S. 264—275 [S. 274—275: Diskussion zwischen W. Heisenberg, P. A. M. Dirac und E. P. Wigner].
— The Philosophical Background of Modern Physics. In: Encyclopaedia Moderna *28* (1974) S. 133—141.
— Tradition in Science. In: Science and Public Affairs *29* (1973) Nr. 10, S. 4—10; Phys. Bull. *25* (1974) S. 231—235.
— The Unified Field Theory of Elementary Particles: Some Recent Advances. In: Naturwissenschaften *61* (1974) S. 1—5.
— [Interview] Die Evolution ist kein Betriebsunfall. 2 Gespräche, aufgezeichnet von Adelbert Reif [neben: Alexander Mitscherlich: Utopie und Wirklichkeit?]. Zürich 1972. 48 S. (= Arche nova).

Hund, Friedrich: Wie haben die Physiker die Chemie verstanden? In: Physikal. Blätter *29* (1973) S. 310—314.
— The History of Quantum Theory. London 1974 [Übersetzung von: Geschichte der Quantentheorie, Mannheim 1967].
— Hätte die Geschichte der Quantentheorie auch anders ablaufen können? In: Nova Acta Leopoldina NF *38* (1974) Nr. 212, S. 15—23.

Jordan, Herbert: Diätkatalog für Kurorte und Heilbäder in der DDR. In: Balneol. pol. *8* (1973) S. 189–192 [polnisch].

— Das rhythmologische Verhalten (Streubreite) der Systolen- und Diastolendauer in Ruhe und unter körperlicher Belastung. In: Z. ges. inn. Med. *28* (1973) S. 394–399.

— Leistungsbezogenes Funktionstraining im Zuge der Kurorttherapie bei Koronarkrankheiten. In: Verh. 25. Int. Ärztl. Fortbildungskurs, 1.–5. Okt. 1973, Karlovy Vary, ČSSR.

— u. D. Reinhold, L. Zelaitis: Fortlaufende Messungen der Diastolen- und Systolendauer des menschlichen Herzens. In: Digest *II* (1973) S. 362 vom X. Internat. Conf. Med. Biol. Engineering, Dresden, 13.–17. Aug. 1973.

— R. Callies, H. Jordan, J. Danz: Physiotherapeutische Richtlinien beim Weichteilrheumatismus. In: Z. Physiother. *25* (1973) S. 381–387.

— R. Callies, H. Aurich, H. Jordan: Bewegungstherapie bei Erkrankungen des rheumatischen Formenkreises. In: Z. ärztl. Fortbild. *67* (1973) S. 977–981.

— u. H. Wagner: Die Kurorttherapie unter geriatrischem Aspekt. In: Z. Alternsforsch. *28* (1974) S. 113–124.

— Stellung und Aufgaben des Kur- und Bäderwesens in der medizinischen Betreuung. In: Z. ärztl. Fortbild. *68* (1974) S. 393–396.

— Professor Albert Kukowa zum 80. Geburtstag. In: Z. Physiother. *26* (1974) S. 165–166.

— Leitlinien der Arbeit des Kur- und Bäderwesens der DDR. Ebda S. 261–265.

— Physiotherapie als adaptive Leistungstherapie. In: Wiss. Z. Univ. Jena, math.-nat. R. *23* (1974) S. 555–564.

— Rehabilitation im Kurort. In: Das ambulante und stationäre Gesundheitswesen *23* (1974).

Keller, Ott-Heinrich: Das Unendliche in der Geometrie. In: Mitteilungen der Deutschen Akademie der Naturforscher Leopoldina *17* (1973) S. 117–133, Reihe 3.

— Vorlesungen über algebraische Geometrie. Leipzig 1974.

— Normalformen für Flächen 3. Ordnung. In: Beiträge zur Algebra und Geometrie *3* (1974) S. 11–21.

Klimaszewski, Mieczysław: Conditions essential to the pleistocene glaciation on the Carpathians. In: Folia Geographica, Series Geographica-Physica *VII* (Kraków 1973) S. 5–26.

— The state and archievements of geographical sciences at the Jagellonian University 1945–1970. In: Prace Instytutu Geograficznego Uniwersytetu Jagiellońskiego *59* (Kraków 1974) S. 95–119.

— Nauki geograficzne w Uniwersytecie Jagiellońskim. Zeszyty Naukowe UJ. In: Prace historyczne *47* (Kraków 1974) S. 379–399.

— Geografia w działalności i pracach Polskiej Akademii Umiejętności. In: Polska Akademia Umiejętności 1872–1952. *I* (Kraków 1974) S. 321–328.

Lösche, Artur: Rezonans magnetyczny w ciekłych kryształach [Magnetische Resonanz in Flüssigkristallen]. In: Postępy Fizyki *29* (1973) H. 4, S. 407–421.

— Some NMR investigations of liquid crystals. In: Proc. XVII. Congress Ampère. Turku *1972* (1973) S. 101–117.

— NMR-Studies of liquid crystals. In: Comments on Solid State Physics *5* (1973) S. 119 bis 131.

— Analyse von kristallin-flüssigen Systemen mit NMR. In: 1. Arbeitstagung „Probleme der flüssigen Phase". Rostock 1973. S. 319–337.

Lösche, Artur u. D. Dietze, S. Grande: Einfaches Verfahren zur Erhöhung der Meßgenauigkeit von Kernspinresonanz-Spektrometern bei shift-Messungen an Festkörpern. In: Exp. Technik der Physik *19* (1971) S. 75—76.

— u. S. Grande, K. Eidner: The smectic phase, an example of high resolution-NMR in solids. In: Proc. 1st Spec. Coll. Ampère. Kraków 1973. S. 103—105.

— Orientational behaviour of liquid crystals, studied by NMR. — Int. Summer-School on phase-transitions, Varenna 1973. In: Il Nuovo Cimento 1974.

— Die Struktur von NMR-Signalen in nematischen Phasen. In: Ann. Physik (7) *31* (1974) S. 182—196.

— Nicht-konventionelle Methoden der paramagnetischen Resonanz. In: Neuere Entwicklungen der Physik (= Festschrift der Exp. Techn. der Physik). Berlin 1974. S. 228—239.

— Möglichkeiten zur Untersuchung der Anisotropie und des Orientierungsverhaltens nematischer Phasen mit Hilfe der NMR. In: Z. physik. Chemie (Leipzig 1974).

— u. S. Grande: Proton chemical shifts in systems of oriented rotating molecules. In: Proc. XVIII. Congress Ampere. Nottingham 1974.

Matzen, Peter Friedrich: Orthopädische Operationen an der Wirbelsäule. In: Bier, Braun, Kümmel: Orthopädische Operationslehre. 8. Aufl. Leipzig 1973. Bd *2/1*, S. 345—384.

— Die extramedulläre Metallosteosynthese in der Orthopädie. In: Beiträge zur Orthopädie und Traumatologie *20* (1973) H. 1, S. 1—19 [Vortrag und Leitung der Podiumsdiskussion auf der Orthopädentagung 1972 in Leipzig].

— u. Peter Felix Matzen: Hüftarthrodese. Ebda H. 2, S. 74—81.

— Behandlung der posttraumatischen Querschnittslähmung. Vortrag gehalten auf dem 9. Kongreß der Gesellschaft für Chirurgie der DDR in Berlin 1972. In: Zentralblatt für Chirurgie *98* (Leipzig 1973) Supplement 1973.

— Symptomatische Coxa vara und varisierende Osteotomie. Herrn Professor Janecek zum 60. Geburtstag. In: Beiträge zur Orthopädie und Traumatologie *20* (1973) H. 10, S. 515—519.

— u. Peter Felix Matzen, H. Nenning: Die operative Rekonstruktion der Dysplasiehüfte. Ebda *21* (1974) H. 11, S. 645—670.

— [Würdig.] Herrn Mr Prof. Dr. med. habil. H. Unger zum 60. Geburtstag. Ebda S. 653.

— [Nachr.] Georg Merrem 21. 9. 1908—8. 7. 1971. In: Jahrbuch Sächs. Akad. Wiss. zu Leipzig *1971—1972* (Berlin 1974) S. 280—282.

Mothes, Kurt: Pflanze und Tier. Ein Vergleich auf der Ebene des Sekundärstoffwechsels. In: Vorträge im Rahmen der 125-Jahr-Feier der Akademie, gehalten am Donnerstag, den 18. Mai 1972. Österr. Akademie der Wissenschaften, math.-nat. Klasse. S. 1—37, 19 Abb.

Myrberg, Pekka Juhana: Über ganze analytische Funktionen zweier Variablen, welche eine schlichte und volumentreue gelöcherte Abbildung vermitteln. Helsinki 1973. 12 S. (= Annales Academiae Scientiarum Fennicae. Series A. I. Mathematica, 540).

Neef, Ernst: Teoretičeskie osnovi landšaftovedenija. (Übersetzung von „Die theoretischen Grundlagen der Landschaftslehre“, Gotha 1967, von A. W. Drozdow, redigiert von D. L. Armand). Moskau 1973. S. 219.

— Funktionsüberlagerung im Territorium als Problem und als Aufgabe der Territorialforschung in der sozialistischen Gesellschaft. In: Wiss. Z. TU Dresden, Bd *22* (1973) H. 4, S. 580—584.

Neef, Ernst: Dry periods in mid-latitude humid regions (Vortrag auf UNESCO-Seminar on Natural Hazards, Budapest 1971). In: Studies in Geography in Hungary, Bd 11: Man and Environment. Budapest 1974. S. 169—171.
— Zur Kartierung von Umweltstörungen. In: Geographische Berichte *70* (1974) S. 1—11.
— Mehrfachnutzung des Bodens. In: Wissenschaft und Fortschritt *1974*, H. 5, S. 208—213.
— Geographie und geologische Entwicklungsprobleme. Die Tätigkeit des Menschen und ihre Bedeutung für die geologische Evolution. In: Z. Geol. Wiss. *2* (1974) S. 919—926.
— [Hrsg.] Schlußbericht der Wissenschaftlichen Tagung der TU Dresden vom 21.—24. Nov. 1972: Sozialistische Umweltgestaltung — Probleme der planmäßigen proportionalen Entwicklung und der Funktionsüberlagerung im Territorium. — Wiss. Z. TU Dresden Bd *22* (1973) H. 4, S. 547—711, darin: Thesen zum Tagungsgegenstand, S. 706 bis 708.
— [Bespr.] K. K. Markow, Einführung in die allgemeine Physische Geographie. Übers. aus dem Russischen, Gotha 1971. In: DLZ *94* (1973) Sp. 428—430.
— [Bespr.] Westermann Lexikon der Geographie, Bd IV und V. Ebda Sp. 275—277.
— [Hrsg.] Petermanns Geographische Mitteilungen.
— [Mithrsg.] Archiv f. Naturschutz und Landschaftsforschung.

Răduleţ, Remus: Le statut des relations et des concepts abstraits spécifiques aux sciences techniques. In: Noesis *I* (Bucureşti 1973) S. 65—75.
— Les débuts de l'energétique roumaine. Ebda S. 225—238.
— Culture des masses et éducations des adultes en Roumanie. In: Bulletin de la Commission Nationale de la R. S. de Roumanie pour l'Unesco *1973*. Nr. 3/4, S. 14—26.
— Zu einigen Erfahrungen der Überführung von Ergebnissen der Grundlagenforschung in die gesellschaftliche Praxis. In: Überführung, Wissenschaft, Produktion. Berlin 1973. S. 60—65.
— Sporirea eficienţei cercetărilor în domeniul ştiinţelor tehnice în lumina sarcinilor trasate de Partidul Communist Român [Erhöhung der Wirksamkeit der Forschung auf dem Gebiet der Technik unter dem Aspekt der von der Rumänischen Kommunistischen Partei beschlossenen Aufgaben]. In: Ahalele Academiei R. S. România pe anul 1969. Bucureşti 1973. S. 158—180.
— u. E. Costin: Methode folosite în prognoza de energie în ţara noastră [Für die Energieprognose des Landes angewandte Methoden]. In: Prognoza tehnologică şi marketing. Bucureşti 1973. S. 89—115.
— u. O. Groza, N. Gheorghiu, V. Nitu: Energy development and its social impact. Detroit 1974. 22 S. (= World Energy Conference, Detroit).
— On the Evolution from Unidiscipline towards. Interdisciplinary sciences. In: The Revolution in Science and Technology and Contemporary Social Development. Bucureşti 1974. 16 S.
— Energetics and its Social Incidence. In: Viitorul social — The Social Future.. Special issue for the 8th World Congress of Sociology. Bucharest 1974. 10 S.
— Direcţii de viitor în des voltarea energeticii. In: Perspective de dezvoltare a energeticii. Bucureşti 1974. 31 S.
— Necesitatea unei politici a ştiinţei şi tehnicii în condiţiile revoluţiei tehnice-ştiinţifice contemporane [Notwendigkeit einer Politik der Wissenschaft und Technik unter den Bedingungen der derzeitigen technisch-wissenschaftlichen Revolution]. In: Revoluţia ştiinţifică — tehnică şi progresul. Bucureşti 1974. 17 S.
— Relationships between Industry and higher technical Education Institutions. In: Bulletin

of the National Commission of the Socialist Republic of Romania for UNESCO. *1974*, Nr. 7, 13 S.
— Classification of Scientific Researches. Ebda Nr. 4, 21 S.

Rammler, Erich: Einige Betrachtungen über Erdgas, Berlin 1974. 43 S. (=Sitzungsber. Sächs. Akad. Wiss. zu Leipzig, math.-nat. Kl. Bd 110, H. 5).
— H. Krug, W. Naundorf, E. Rammler: Beitrag zur Modellierung und Optimierung der Trockenkohlenabsiebung durch Siebkettenförderer. — Teil 2: Durchführung von Optimierungsrechnungen mit linearen Nebenbedingungen. In: Neue Bergbautechn. *3* (1973) S. 36—40.
— u. P. Modde: Zur Zerkleinerungscharakteristik von Schlagkreuzmühlen. In: Freib. Forschungshefte *A 512* (1973) S. 5—25.
— Charakteristiken einer Schlagplattenmühle mit Austragsieb. Ebda S. 27—46.
— u. A. Bahr: Zum Strukturvergleich und zur Berechnung der Korngrößenverteilungsdichte verschiedener Verteilungsarten. Ebda S. 47—69.
— u. A. Bahr: Über die Nebenparameter wichtiger Korngrößenverteilungen und ihre Zusammenhänge mit den Verteilungshauptparametern. Ebda S. 71—112.
— u. A. Bahr: Über die Anwendung von Regressionsrechnungen zur Auswertung von Körnungsanalysen. Ebda *A 515* (1973) S. 7—17.
— u. H. Flachsbart, H. Skadow: Über die Abhängigkeit der hygroskopischen Eigenschaften der Entgasungsrückstände von Braunkohlen von der Entgasungstemperatur. Ebda S. 19—43.
— u. W. Michel: Über die Art der Abhängigkeit der Brikettfestigkeit vom Wassergehalt des Preßgutes und ihre Beziehungen zur Brikettiertheorie. Ebda S. 45—73.
— u. M. Weber: Über die mathematische Abhängigkeit der Brikettfestigkeit vom Preßdruck bei der bindemittellosen Brikettierung. Ebda S. 75—89.
— u. A. Bahr: Zur Ermittlung der spezifischen Oberfläche von Korngrößenverteilungen. Teil 3: Graphische Verfahren der Oberflächenbestimmung für spezielle Korngrößenverteilungen. In: Verfahrenstechnik *7* (1973) S. 273—279.
— Zur Entwicklung der Darstellung von Korngrößenverteilungen, speziell nach der Formel von Rosin, Rammler, Sperling und Benett. Harold Heywood-Memorial Symposium. University of Technology, Loughborough (England), 17.—18. Sept. 1973. — Preprint — 23 S. Text und 12 S. Quellenangaben.
— M. Weber u. E. Rammler: Über die Einflußfaktoren für den ausfallseitigen Unterdruck bei Röhrentrocknern und ihre gegenseitige Beeinflussung. In: Neue Bergbautechn. *3* (1973) S. 698—704.
— Zur Anwendung der logistischen Funktion in der mechanischen und thermischen Verfahrenstechnik. Freiberger Forschungshefte *A 524* (1973) S. 5—38.
— u. A. Bahr: Untersuchung einer Kreislaufschaltung mit Vor- und Nachsiebmaschine für die Rohkohlenaufbereitung. Ebda S. 39—60.
— u. M. Weber: Über die mathematische Abhängigkeit der Brikettfestigkeit vom Preßdruck bei der bindemittellosen Brikettierung, Teil 2. Ebda S. 61—88.
— u. K.-H. Höfer: Zum 75. Geburtstag Georg Bilkenroths. In: Neue Bergbautechn. *3* (1973) S. 247—248; Energietechn. *23* (1973) S. 233—234.
— [als Vorsitzender einer aus 12 Sachverständigen bestehenden Arbeitsgruppe] Begriffe für Braunkohlenbrikettfabriken — Vereinheitlichung und Überarbeitung der Standards 100—6028/01 bis 100—6028/07. In: Mitteilungsblatt der Zentralstelle für Standardisierung Kohle *14* (1973) 5, Teil E, 143 S.

Rammler, Erich u. W. Naundorf, H. Krug: Über die Abhängigkeit des Abscheidegrades von Fliehkraftabscheidern von der Art der Korngrößenverteilung des Staubes. In: Fortschritte der Entstaubungstechnik 1974. S. 19–42 (= Broschürenreihe Technik und Umweltschutz H. 6).

— u. W. Naundorf, H. Krug: Über die Veränderung der Feinheit von Stäuben verschiedener Korngrößenverteilungsarten beim Abscheidevorgang in Fliehkraftabscheidern. Ebda S. 53–64.

— H. Brandt, E. Rammler: Über die Zweckmäßigkeit des Einsatzes von Wirbelschichteinblasevorrichtungen an Röhrentrocknern von Braunkohlenbrikettfabriken. In: Neue Bergbautechnik *4* (1974) S. 859–863.

Schwabe, Kurt: Physikalische Chemie. Bd *1*: Physikalische Chemie. Berlin 1973. XXII, 694 S., 229 Abb., 90 Tab. — Bd *2*: Elektrochemie. Berlin 1974. XVI, 376 S., 103 Abb., 50 Tab. — Bd *3*: Aufgabensammlung. Berlin 1974. VI, 189 S., 22 Abb.

— Passivität und Korrosionsschutz. Berlin 1974. 20 S., 19 Abb. (= Sitzungsber. der AdW der DDR Jg 1973, Nr. 17).

— [Hrsg. unter Mitarbeit eines Autorenkollektivs] Physikalisch-chemische Meßverfahren. Teil 1: Physikochemische und physikalische Meßverfahren. Leipzig 1973. 508 S., 289 Abb., 41 Tab. (= Handbuch der Meßtechnik in der Betriebskontrolle. Bd *IV*: Physikalisch-chemische Meßverfahren).

— u. M. Schwarzbach: Neuere Erkenntnisse der Korrosionsvorgänge. In: Mitt. Chem. Ges. *20* (1973) S. 89–96.

— u. R. Kretschmar, W. Krippner, M. Michalke: Aufbau einer Apparatur zur Bestimmung der Dichte von Flüssigkeiten. In: Z. phys. Chem. *252* (Leipzig 1973) S. 365–373.

— Physikalisch-chemische Probleme der Hydrochemie. In: Acta hydrochim. hydrobiol. *1* (1973) S. 27–41.

— u. Le Dang Anh: Die Inhibition des Lochfraßes an Chrom-Nickel-Stählen. In: Zaščita metallov *9* (1973) S. 541–550.

— u. H.-D. Suschke, M. Thieme: Der Stand der Forschung auf dem Gebiet der Theorie der Metallpassivierung in der DDR. Ebda *10* (1974) S. 491–507.

— Analytische Chemie und Analysenmeßtechnik im Betrieb. In: Chem. Technik *26* (1974) S. 101.

— u. R. Lebet, H. Worch: Über das elektrochemische Verhalten der Eisen-Chrom-Legierungen GX40Cr30 und GX120Cr30 in sulfat- und chloridhaltigen Elektrolyten. In: Werkst. u. Korr. *25* (1974) S. 32–39.

— u. J. Hahn, Ch. Dörfel, J. Eckert: Eine rotierende Ring-Scheiben-Elektrodenanlage zur Untersuchung von Elektrodenprozessen. In: Chem. Technik *26* (1974) S. 170–173.

— pH measurements and their applications. [7. Kapitel] in: Advances in Analytical Chemistry and Instrumentation. Sussex 1974. S. 495–586.

— u. K. Spiegel: Elektrodeneinsatz zur Messung bestimmter Badbestandteile. In: Tagungsband der KdT Suhl, XII. Galvanotechnische Fachtagung in Suhl 1974, 24.–26. 4. 1974, AA „Galvanotechnik" S. 76–85.

— R. Schnoor u. K. Schwabe: Die Wechselwirkung zwischen Schwefelwasserstoff und aufgedampften Gold- und Nickelfilmen. I. Gold. In: Z. phys. Chem. *255* (Leipzig 1974) S. 261–272.

— H. Kelm u. K. Schwabe: Über eine Möglichkeit zur Interpretation von Aktivitätskoeffizienten konzentrierter Elektrolytlösungen. Ebda S. 475–485.

— H.-D. Suschke, K. Schwabe, J. Broers: Der Einfluß der Oberflächenvorbehandlung auf das Adsorptionsverhalten von Stahl. In: Oberfläche — surface *15* (1974) S. 295–300.

Schwabe, Kurt.— H. Worch, K. Schwabe: Eine Korrosionsmeßzelle für Serienuntersuchungen. In: Korrosion *5* (1974) S. 39—41.

Spitzin, Viktor Iwanowitsch und V. V. Gromov: Physikalisch-chemische Eigenschaften radioaktiver fester Körper. Moskau 1973. 192 S. [russ.].
— V. V. Gromov u. V. I. Spitzin: Künstliche radiaoktive Nukleine im Mittelmeer. Moskau 1975. 224 S. [russ.]
— Progress in the Chemistry of Actinide Elements. In: XXIVth International Congress of pure and applied Chemistry *6* (London 1974) S. 105—140.

Strack, Erich: Carnitinesterwirkung. In: Jahrbuch Sächs. Akad. Wiss. zu Leipzig *1971* bis *1972* (Berlin 1974) S. 78—84.
— D. M. Müller u. E. Strack: Zur Bindungsenergie der Estergruppe in *O*-Acylcarnitinen und einigen ihrer Carboxylderivate, III. Hydrolyseenthalpie von *O*-Acylcarnitinen und Betainestern. In: Hoppe-Seyler's Z. Physiol. Chem. *354* (1973) S. 1091—1096.

Stubbe, Hans: History of Genetics. From Prehistoric Times to the Rediscovery of Mendel's Laws. Translated by T. R. W. Waters (= The MIT Press). Cambridge/Mass. and London 1. Aufl. 1973. 356 S. u. 45 Abb.
— Evolution unter dem Einfluß des Menschen. In: Biol. Rundschau *12* (1974) H. 1, S. 1—11.
— Das Problem Mensch und Umwelt aus genetischer Sicht. In: Festschrift zur Feier des 125jährigen Bestehens der Sächsischen Akademie der Wissenschaften zu Leipzig. Berlin 1974. S. 145—156.

Uhlmann, Dietrich: Störungen des biologischen Gleichgewichts in Gewässern. Berlin 1973. 16 S. (= Abhandlung Sächs. Akad. Wiss. zu Leipzig, math.-nat. Kl. Bd 52, H. 2).
— u. H. Klapper: Kurz- und langfristige Auswirkungen der Abwasserbelastung der Gewässer auf ihren Erholungswert. In: Wiss. Z. TU Dresden *22* (1973) S. 682—685.
— u. F. Cramer, E.-A. Arndt: Step-forcing of laboratory-scale aquatic ecosystems as a means to test their response and stability. Paper prepared for 1. International Congress of Ecology Den Haag, 8.—14. 9. 1974. S. 1—7 [als Manuskript gedruckt].
— u. G. Weise, A. Gnauck: Gleichgewicht und Stabilität in Ökosystemen. In: Biol. Rundschau *12* (1974) S. 365—381.
— W. v. Tümpling u. D. Uhlmann [Hrsg.]: Eutrophierung und Gewässerschutz. Kurzfassungen der Vorträge eines Symposiums mit internationaler Beteiligung vom 16. bis 20. 10. 1973 auf Schloß Reinhardsbrunn, veranstaltet von der Biologischen Gesellschaft der DDR und der TU Dresden. 174 S. [als Manuskript gedruckt].
— W. v. Tümpling u. D. Uhlmann [Hrsg.]: Naturwissenschaftliche Grundlagen der Gütebewirtschaftung von Talsperren und Speichern. Kurzfassungen der Vorträge der Fortbildungstagung, veranstaltet von der Biologischen Gesellschaft der DDR in Verbindung mit der TU Dresden, der Chemischen Gesellschaft der DDR und der Gesellschaft für Allgemeine und Kommunale Hygiene der DDR, Karl-Marx-Stadt 1.—3. 10. 1974. 162 S. [als Manuskript gedruckt].

Van der Waerden, Bartel L.: Einführung in die algebraische Geometrie. 2. Aufl. Heidelberg 1973. XII, 280 S., 17 Abb. (= Grundlehren der mathematischen Wissenschaften, Bd 51).
— Sience Awakening II. Leiden 1974. XVI, 347 S., 26 u. 31 Abb.
— Group Theory and Quantum Mechanics. Heidelberg 1974. VIII, 211 S., 10 Fig. (= Grundlehren der mathematischen Wissenschaften, Bd 214).

Van der Waerden, Bartel L.: Die Galois-Theorie von Heinrich Weber bis Emil Artin. In: Archive for History of Exact Sciences *9* (1972) S. 240–248.
– Ein Briefwechsel zwischen Finsler und van der Waerden über die Grundlegung der algebraischen Geometrie. Ebda S. 249–256.
– [Art.] Klaudios Ptolemaios. In: Der kleine Pauly IV (1972) S. 1224–1229.
– Die Erkenntnistheorie der Pythagoreer. In: Rete *1* (1973) S. 209–224.
– Über die Wechselwirkung von Mathematik und Physik. In: Elemente Math. *28* (1973) S. 33–41.
– From Matrix Mechanics and Wave Mechanics to Unified Quantum Mechanics. In: The Physicist's Conception of Nature, ed. J. Mehra. Dordrecht (Holland) 1973. S. 276–293.
– Hamiltons Entdeckung der Quanternionen. Hamburg 1973 (= Veröffentlichung der Jungius-Gesellschaft der Wissenschaften).
– Die Vorgänger des Copernicus im Altertum. In: Philosophia naturalis *14* (1973) S. 407 bis 415.
– The Game of Rectangles. In: Annali di Matematica *98* (1974) 4, S. 63–75.
– The Earliest Form of Epicycle Theory. In: J. for Hist. of Astronomy *5* (1974) S. 175 bis 185.
– [Rez.] Kudlek u. Mickler: Solar and Lunar Elipses 3000 BC to O. In: BO *30* (1973).
– [Rez.] R. A. Parker: Demotic Math. Papyri. In: Isis *65* (1974) S. 110–111.

Wagner, Carl: Methoden der naturwissenschaftlichen und technischen Forschung. Mannheim, Wien, Zürich 1974. 219 S.
– Phenomenal and Thermodynamic Equations of Adsorption. In: Nachrichten der Akademie der Wissenschaften in Göttingen. II. Mathematisch-physikalische Klasse *1973* (Göttingen 1973) Nr. 3, S. 37–63.
– The Formation of Thin Oxide Films on Metals. In: Corrosion Science *13* (1973) S. 23.
– General Aspects of High-Temperature Corrosion of Structural Materials. In: AGARD Conference Proc. 120, High Temperature Corrosion of Aerospace Alloys, 1973. S. 3–7.
– The activity coefficient of oxygen and other nonmetallic elements in binary liquid alloys as a function of alloy composition. In: Acta Metallurgia *21* (1973) S. 1297–1303.
– The enthalpy change due to uptake of the nonmetallic component X by an ionic compound AX as a Function of the activity of X. In: Ber. Bunsenges. physikalische Chemie *77*(1973) Nr. 7, S. 518–521.
– Models for lattice defects in oxide layers on passivated iron and nickel. Ebda Nr. 12, S. 1090–1097.
– Deviations from internal equilibrium in scales. Growing on metals in corroding gases. Ebda *78* (1974) Nr. 6, S. 611–614.
– Corrosion in aqueous solutions and corrosion in gases at elevated temperatures-analogies and disparities. In: Werkstoffe und Korrosion *25* (1974) H. 3, S. 161–165.

Watznauer, Adolf: Beitrag zur Frage des zeitlichen Ablaufes der Granulitgenese (Sächsische Granulitgebirge). In: Krystalinikum *10* (1974) S. 181–192.
– Granulit und Granulitfacies. In: Centenaire de la Société Géologique de Belgique. Géologie des Domaines Cristallins. Liège *1974*. S. 257–265.

Wolf, Friedrich u. U. Koch, S. Keil: Polarographische Untersuchungen an metallkomplexbildenden Azofarbstoffen. In: Melliand Textilberichte *54* (1973) S. 61–66.
– u. S. Kreissl, O. Sonntag: Zur katalytischen Oxydation von HCl. In: Chem. Technik *25* (1973) S. 37–40.

Wolf, Friedrich u. F. Danes: Zum Ionenaustausch von Na-Ionen gegen zweiwertige Kationen am zeolithischen Molekularsieb Typ A. I. Zur Untersuchung der Kinetik des Ionenaustausches. In: Z. phys. Chem. *251* (1972) S. 329—338; II. Gleichgewichte und Thermodynamik des Austausches Mg/Na am Zeolith Typ A. Ebda S. 339—352.

— u. S. Kreissl, O. Sonntag: Zur katalytischen Gasphasenchlorierung von Äthylen. In: Chem. Technik *25* (1973) S. 89—92.

— u. U. Koch, G. Knorr: Zur Herstellung von Dispersionsfarbstoffen mit längeren Glykolätherketten. In: Z. Chem. *13* (1973) S. 15—16.

— u. P. Renger: Zur Adsorption von Chlorcyan an modifizierten kristallinen Zeolithen des Typs 13 X. Ebda S. 31—32.

— u. P. König, M. Gross: Gastrennung mit Hilfe des Zirkulations-Diffusions-Verfahrens. Teil II: Trennung von stabilen Isotopen und technischen Gasmischungen durch Massen- bzw. Sweep-Diffusion. In: Isotopenpraxis *8* (1972) S. 416—422.

— u. S. Kreissl, O. Sonntag: Zur Oxychlorierung von Äthylen. In: Chem. Technik *25* (1973) S. 156—158.

— u. S. Horn: Über den Einfluß organischer Verbindungen auf die hydrothermale Synthese zeolithischer Molekularsiebe der Typen A, X und Y. In: Z. Chem. *13* (1973) S. 109—110.

— u. F. Danes: Zum Ionenaustausch von Na-Ionen gegen zweiwertige Kationen am zeolithischen Molekularsieb Typ A. III. Zur Kinetik des Ionenaustausches — Berechnung der Koeffizienten der inneren Diffusion. In: Z. phys. Chem. *252* (1973) S. 15—32.

— u. F. Danes, K. Pilchowski: — IV. Mitteilung: Zum Mechanismus der Kationendiffusion im Zeolith Typ A. Ebda S. 33—49.

— u. D. Ceacareanu, K. Pilchowski: Zum Ionenaustausch von Na-Ionen gegen zweiwertige Kationen am zeolithischen Molekularsieb Typ X. I. Gleichgewichte und Thermodynamik des Austausches Mg/Na, Ca/Na und Cu/Na am Zeolith Typ X. In: Z. phys. Chem. *252* (1973) S. 50—64.

— u. W. Andreas: Zur Styrolpolymerisation in Abhängigkeit vom Reaktortyp. Über die kontinuierliche Suspensionspolymerisation. In: Plaste und Kautschuk *20* (1973) S. 327 bis 329.

— u. G. Fechtel: Zur Synthese, Charakterisierung und Anwendung von Intermesh-Kationenaustauschermembranen mit Natriumchlorvinylsulfonat als ein Monomeres. Ebda S. 329—332.

— u. B. Maier: Zur Entölung von Wässern: In Acta hydrochim. *1* (1973) S. 93—100.

— u. R. Grohmann: Der Einfluß von Polyelektrolyten auf die Filtration und die Morphologie der Hydroxyniederschläge des Al, Fe und Ca. Ebda S. 103—106.

— u. J. Schmieder: Permeationsmessungen an weichgemachten, pigmenthaltigen PC-Schichten auf Acetylzellulose- und Polyesterfolien. In: Plaste und Kautschuk *20* (1973) S. 685—688.

— u. D. Seidig: Untersuchungen zur Synthese von zeolithischen Molekularsieben aus natürlichen Rohstoffen. In: Tonindustrie-Ztg u. keramische Rundschau *97* (1973) S. 281—285.

— u. R. Gaedeke, L. Hanl: Untersuchungen zur Freifließbarkeit von Harnstoff. In: Wiss. Z. Univ. Halle *22* (1973) S. 105—110.

— u. R. Gaedeke: Über die Beeinflussung der Lagereigenschaften von Harnstoff durch Druck. Ebda S. 111—118.

— u. J. Renning: Untersuchungen zur Hydrothermalen Synthese von Mordenit aus vulkanischen Gesteinen. In: Tonindustrie-Ztg u. keramische Rundschau *97* (1973) S. 293—298.

— u. G. Fechtel, E. Taeger: Zur Synthese, Charakterisierung und Metallionenselektivität von Intermesh-Kationenaustauscherharzen unter Verwendung von Natriumchlorvinylsulfonat als ein Monomeres. In: Plaste und Kautschuk *20* (1973) S. 816–818.

Wolf, Friedrich u. H. John: Herstellung und Eigenschaften dealuminierter synthetischer Mordenite Teil I. In: Chem. Technik *25* (1973) S. 736–739.

— u. D. Ceacareanu, K. Pilchowski: Zum Ionenaustausch von Na-Ionen gegen zweiwertige Kationen am zeolithischen Molekularsieb Typ X. II. Mitteilung: Zur Kinetik des Ionenaustausches — Berechnung der Koeffizienten der inneren Diffusion. In: Z. phys. Chem. *254* (1973) S. 83–95; III. Mitteilung: Zum Mechanismus der Kationendiffusion im Zeolith Typ X. Ebda S. 96–112.

— u. Z. Borhan, K.-H. Mohr: Zum Ionenaustausch in dünnen Austauscherschichten (Powdex-Verfahren). I. Einflußgrößen auf die Lage des lokalen Gleichgewichtes in der Austauscherschicht. In: Vom Wasser *41* (1973) S. 457–463; II. Zur Kinetik des Ionenaustausches in der Austauscherschicht. Ebda S. 464–481.

— u. D. Ceacareanu, K. Pilchowski: Zum Ionenaustausch von Na^{+}-Ionen gegen zweiwertige Kationen am zeol. MS Typ X. II. Mitteilung: Zur Kinetik des Ionenaustausches — Berechnung des Koeffizienten der inneren Diffusion. In: Z.. phys. Chemie *254* (1973) S. 83–95; III. Mitteilung: Zum Mechanismus der Kationendiffusionen im Zeolith Typ X. Ebda S. 96–112.

— u. H. John: Herstellung und Eigenschaften dealuminierter synthetischer Mordenite, Teil I. In: Chem. Technik *25* (1973) S. 736–740.

— u. H. Kunert: Zur Entkieselung alkalischer Lösungen mit stark basischen Anionenaustauscherharzen. In: Acta hydrochim. *1* (1973) S. 493–496.

— u. L. Eckert, S. Vollmann: Fortschritte der Wasserchemie über die Oxydation von Sulfidionen durch Sauerstoff in Gegenwart von Schwefelfarbstoffen. In: Wasserchemie *1973*, H. 15, S. 65–68 u. 69–77.

— u. J. Renning: Untersuchungen zur hydrothermalen Synthese von Mordenit aus vulkanischen Gesteinen. In: Tonindustrie-Ztg u. keramische Rundschau *97* (1973) S. 293 bis 298.

— u. E. Kreter: Zur Diffusion von monomerem Vinylchlorid in Polyvinylchlorid. In: Plaste und Kautschuk *21* (1974) S. 27–28.

— u. H. Schmidtke: Zur Trennung des Heliums und Neons von anderen Gasen mit Hilfe von Zeolithen des Typs A. In: Z. Chem. *14* (1974) S. 71–72.

— u. H. John: Herstellung und Eigenschaften dealuminierter synthetischer Mordenite, Teil II. In: Chem. Technik *26* (1974) S. 159–162.

— u. K.-H. Mohr: Zum Druckeinfluß auf das Quellungsgleichgewicht und die Ionenselektivität von Ionenaustauscherharzen. In: Vom Wasser *42* (1974).

— u. K.-H. Bergk: Zur Adsorption von Alkyl- und Arylmercaptanen an Molekularsieben des Typs A und X. In: Chem. Technik *26* (1974) H. 6.

— u. P. Renger: Zur Synthese von Cyanurchlorid durch diverse Umsetzung von Chlor und Cyanurwasserstoff. Ebda S. 242–243.

— Zur Korngrößenverteilung von Suspensionspolymerisaten. In: Dechema-Monographie Bd *74* (1974).

— u. R. Haase: Untersuchungen zum Verhalten von Vanadinkatalysatoren während der SO_2-Oxidation. In: Chem. Technik *26* (1974) S. 400–402.

— u. H. Fürtig, U. Neef, W. Krüger: Zur Synthese von Molekularsieben des Typs X aus Kaolinen, Tonen und Kraftwerksfilteraschen. Ebda.

— u. K.-H. Bergk: Heterogen-katalysierte Verseifung von Acrylnitril. In: Z. Chem. *14* (1974) S. 272.

Wolf, Friedrich u. H. Schneider: Untersuchungen über Membranen für den Einsatz in der Ultrafiltrationstechnik. In: Acta hydrochim. *2* (1974) S. 363—365.
— u. P. König: Untersuchungen zur Trennung von Stickstoff und Sauerstoff an synthetischen Zeolithen. In: Z. Chem. *14* (1974) S. 320—322.
— u. P. König: Gaschromatographische Trennung von Argon/Sauerstoff/Stickstoff-Gemischen an zeol. Molekularsieben des Typs L. Ebda S. 367—368.
— u. W. Heyer: Zum Ionenaustausch in porösen Gläsern. In: Z. anorg. allg. Chemie *408* (1974) S. 237—246.
— u. K.-H. Bergk: Zur Sorption schwefelhaltiger Verbindungen an zeol. Molekularsieben. In: Z. Chem. *14* (1974) S. 344—348.
— u. U. Koch: Zur Stabilisierung von Farbstoffdispersionen mit Polyvinylpyrrolidon. Ebda H. 10.
— Ionenaustauscherharze auf der Basis ternärer Kopolymerisate des Epichlorhydrins. In: Plaste und Kautschuk *21* (1974) S. 650.

Philologisch-historische Klasse

Aebischer, Paul: Oliveriana et Rolandiana. Sur le résumé du „Girard de Viane“ conservé par la première branche de la Karlamagnús saga. Une ultime mise au point. In: Revue belge de philologie et d’histoire *51* (1973) S. 517—533.
— La mesnie Doon de Mayence et son plus illustre représentant, Ogier le Danois. In: Mélanges de langue et de littérature médiévales offerts à Pierre le Gentil, s. l. n. d. [Paris 1973], S. 13—32.
— Aspects négatifs de la Toponymie valaisanne. In: Revue suisse d’historie, 23e année (1973) S. 479—491.
— Girard d’Anciens. Le Roman du Cheval de Fust, ou de Meliain. Extraits publiés d’aprés le texte du ms. de la Biblioteca Riccardiana de Florence, avec une introduction et un glossaire réduit par Paul Aebischer. In: Textes littéraires francais, 212, Genève, Librairie Droz, 1974, 197 S.
— Pagus „région territoire“ dans le latin médiéval des Pyrénées catalanes. In: Estudis Romànico 15 (1970) [1975] S. 123—127.
— Eléments traditionnels et investions personelles dans les textes annalistiques et historiographiques rélatifs à l’expédition franque de 773 en Espagne. In: Revue suisse d’histoire *24* (1974) S. 373—400.
— [Hrsg.] La „Vita Sancti Marini“. Texte du manuscrit F. III. 16 de la Bibliothèque nationale de Turin précédé d’une introduction. Saint-Martin 1974. 77 S.

Aland, Kurt: Martin Luther in der modernen Literatur. Ein kritischer Dokumentarbericht. Witten und Berlin 1973.
— Martin Luther. In: Die Großen der Weltgeschichte *IV* (Zürich 1973) S. 814—847.
— Huldrych Zwingli. Ebda S. 848—861.
— Der Pietismus und die soziale Frage. In: Arbeiten zur Geschichte des Pietismus *12* (Witten 1974) S. 99—137.
— Der Schluß des Markusevangeliums. L’Évangile selon Marc. In: Bibliotheca Ephemeridum Theologicarum Lovaniensium *24* (Gembloux 1974) S. 435—470.
— Neue neutestamentliche Papyri III. In: New Testament Studies *20* (1974) S. 357—381.
— Der deutsche Pietismus als Wegbereiter für die Arbeit der Bibelgesellschaften. In: On Language, Culture and Religion: in honor of Eugene A. Nida. Den Haag/Paris 1974. S. 3—21.

Aland, Kurt — Jean Calvin. In: Die Großen der Weltgeschichte *V* (Zürich 1974) S. 34—51.
— [Vorw.] zum Katalog: Ignaz Heinrich von Wessenberg 1774—1860, Gedächtnisausstellung 1974. Konstanz 1974.
— [Hrsg.] Synopsis Quattuor Evangeliorum. 8. Aufl. Stuttgart 1973.
— [Hrsg.] E. Nestle, K. Aland: Novum Testamentum Graece et Latine. 24. Aufl. Stuttgart 1974.
— [Hrsg.] E. Nestle, K. Aland: Novum Testamentum Graece et Germanice. 19. Aufl. Stuttgart 1974.
— [Hrsg.] Luther Deutsch, Ergänzungsband: Lutherlexikon. 3. Aufl. Göttingen 1974.

Baetke, Walter: Geschichte, Recht und Religion im germanischen Schrifttum. Hrsg. von Kurt Rudolph und Ernst Walter. Weimar 1973. 387 S.
— Die Isländersaga. Darmstadt 1974. 390 S. (= Wege der Forschung, Bd 51).

Bahner, Werner: „Aufklärung" als Periodenbegriff der Ideologiegeschichte. Einige methodologische Überlegungen und Grundsätze. Berlin 1973. 26 S. (= Sitzungsberichte des Plenums und der Klassen der Akademie der Wissenschaften der DDR 1972, Nr. 9).
— Zur Einordnung der „Aufklärung" in die literarhistorische Periodisierung. In: Genese und Gültigkeit von Epochenbegriffen. Berlin 1974. S. 25—39. (= Sitzungsberichte des Plenums und der Klassen der Akademie der Wissenschaften der DDR 1973, Nr. 1).
— Ein bedeutender Gelehrter an der Schwelle zur Frühaufklärung: Dimitrie Cantemir (1673 bis 1723). In: Sitzungsberichte der Akademie der Wissenschaften der DDR *1973* (Berlin 1974) Nr. 13, S. 7—31.
— Zum wissenschaftlichen Stellenwert von Şăineanus rumänischer Semasiologie. In: Studii şi cercetări lingvistice *24* (1973) S. 485—490.
— Quelques observations sur la notion de Zone Littéraire. In: Neohelicon *1/2* (Budapest 1973) S. 157—161.
— Cantemir si Academia din Berlin. In: Secolul *20* (Bukarest 1973) Nr. 11—12, S. 93—97.
— Aufgaben der vergleichenden Literaturwissenschaft bei der Erschließung des humanistischen Kulturerbes. In: Festschrift zur Feier des 125jährigen Bestehens der Sächsischen Akademie der Wissenschaften zu Leipzig. Berlin 1974. S. 115—128.
— Beispiele rumänisch-deutscher Wechselbeziehungen im 18. Jahrhundert und ihr europäischer Kontext. In: Cahiers roumains d'études littéraires *3* (1974) S. 14—22.
— [Nachr.] Walther von Wartburg 18. 5. 1888—15. 8. 1971. In: Jahrb. Sächs. Akad. Wiss. zu Leipzig *1971—1972* (Berlin 1974) S. 371—374.
— [Bespr.] M. Bucur u. a.: Istoria şi teoria comparatismului în România. In: Referatedienst zur Literaturwissenschaft *5* (1973) H. 4, S. 385—387.
— [Bespr.] Werner Krauss: Die Aufklärung in Spanien, Portugal und Lateinamerika. Ebda H. 5, S. 551—553.
— [Bespr.] Wolfgang Raible: Moderne Lyrik in Frankreich. Darstellung und Interpretationen. Ebda S. 610—611.
— [Bespr.] Hartig/Kurz: Sprache als soziale Kontrolle. 3. Aufl. Frankfurt/Main 1971. In: DLZ *94* (1973) Sp. 196—197.
— [Bespr.] Peter Brockmeier: Lust und Herrschaft. Studien über gesellschaftl. Aspekte der Novellistik: Boccaccio, Sacchetti, Margarete von Navarra, Cervantes. Ebda Sp. 900—902.
— [Bespr.] Peter Ronge: Studienbibliographie Französisch. Beiträge zur Erschließung der französischen Philologie, 2 Bde.
— [Bespr.] Peter Ronge: Studienbibliographie Französisch. Beiträge zur Erschließung der französischen Philologie. 2 Bde, Frankfurt/Main 1971. In BrP *XI* (1972) S. 346.

Beševliev, Veselin: Extrema minoris Scythiae. In: Izvestija okr. ist. muzej i balg. ist. d-vo Tolbychin *I* (1973) S. 139–141.
– Über das Corpus der byzantinischen Inschriften. In: Byzantina *5* (Thessalonike 1973) S. 309–311.
– [Nachruf] Gyula Moravcsik. In: Etudes balkaniques. Sofia 1973. S. 147–148.
– D-r Nikolo Sava Pikolo i grackoto osvoboditelno dviženie. In: Sb. Sto i petdeset godini ot grackoto vastanie 1821–1828 g. Sofia 1973. S. 79–87.
– Zwei neue altchristliche Grabinschriften. In: Izvestija nar. muzej Varna *IX* [XXIV] (1973) S. 301–304.
– Protobulgarische Amuletten. Ebda S. 53–63.
– Die Geschichte von Plisca und Umgebung. In: Izvestija nar. muzej Šumen *VI* (1973) S. 3–14.
– Nov vazspomenatelen parvobalgarski nadpis. In: Sb. v pamet na prof. Al. Burmov. Sofia 1973. S. 412–418.
– Za antičnoto ime na Šumenskata krepost. In: Muzej i pametnici na kulturata *XIV* (1974) H. 1.
– Kam vaprosa za Devtašlarite. In: Izvestija nar. muzej Varna *X* [XXV] (1974).
– Novi starochristijanski nadpisi ot Varna. Ebda.
– Die Botschaften der byzantinischen Kaiser aus dem Schlachtfeld. In: Byzantina *6* (Thessalonike 1974) S. 73–83.
– Zur Bestimmung der protobulgarischen Gebäude. In: Bulletin du musée national de Varna *X* [XXV] (1974) S. 103–107.

Bischoff, Karl: Das Pfälzische Wörterbuch. In: Akademie der Wissenschaften und der Literatur Mainz *1949–1974*. Mainz, Wiesbaden 1974. S. 130–133.
– [Nachr.] Kurt Wagner. In: Jahrbuch Akademie Wiss. u. Literatur zu Mainz *1973* (Mainz, Wiesbaden 1974) S. 107–109.
– [Bespr.] Hermann Teuchert: Die Sprachreste der niederländischen Siedlungen des 12. Jahrhunderts. 2. Aufl. besorgt von R. Olesch u. L. E. Schmidt. Köln, Wien 1972. In: Niederdeutsches Jahrbuch *96* (1973) S. 190–191.
– [Bespr.] Pierre Heßmann: Die Flurnamen des nördlichen und östlichen Kreises Rotenburg (Wümme). Rotenburg 1972 (= Rotenburger Schriften, Sonderband 16 – Name und Wort Bd 4). In: Beiträge zur Namenforschung NF *9* (1974) S. 194–195.

Blumenthal, Lieselotte: Zwei Überbleibsel von Schillers Vorarbeiten für Wilhelm Tell. Aus der Arbeit der Schiller-Nationalausgabe. In: Untersuchungen zur Literatur als Geschichte. Festschrift für Benno von Wiese. Vincent J. Günther, Helmut Koopmann u. a. Berlin 1973. S. 92–111.
– Die verbrannte und die gestohlene Handschrift von Schillers „Wilhelm Tell“. Mit einem Anhang: Zwei Zeilen von Schillers Vorarbeiten für „Wilhelm Tell“. In: Jahrbuch der deutschen Schiller-Gesellschaft *17* (Stuttgart 1973) S. 21–62.
– [Hrsg.] Schillers Werke. Nationalausgabe. Band 7. Erster Teil. Don Karlos. Unter Mitwirkung von L. B. hrsg. von Paul Böckmann und Gerhard Kluge. Weimar 1974. 645 S.

Bruford, Walter Horace: Literary criticism and Sociology. In: Yearbook of Comparative Criticism *V* (Pennsylvania State University 1973) S. 3–20.
– Some early Cambridge links with German scholarship and Literatur. In: Erfahrung und Überlieferung. Festschrift für C. P. Magill. Cardiff 1974. S. 75–87.
– [Bespr.] W. Dobbek: J. G. Herders Weltbild. Köln, Wien 1969. In: German Life and Letters, New Series, *26* (1973) Nr. 3, S. 239–240.

Bruford, Walter [Bespr.] Carmen Kahn-Wallerstein: Die Frau im Schatten, Karoline von Wolzogen. Bern, München 1970. Ebda Nr. 4, S. 321–322.
— [Bespr.] Goethe und die Tradition. Hrsg. von Hans Reiss. Frankfurt/M. 1972. In: Modern Language Review *1974*, S. 705–706.

Brüning, Eberhard: Die Entwicklung progressiver und sozialistischer Literatur in den USA. Fortschrittliche Schriftstellerorganisationen und ihr Einfluß auf den nationalen Literaturprozeß von 1917 bis 1945. Berlin 1974. 40 S. (= Sitzungsber. Sächs. Akad. Wiss. zu Leipzig, phil.-hist. Kl. Bd 116, H. 6).
— Die Konzeption vom „neuen amerikanischen Menschen" bei Charles A. Reich und Erich Segal. In: Wiss. Z. Karl-Marx-Univ. Leipzig, ges.- u. sprachwiss. R., *1973*, H. 3/4, S. 389–399.
— Sämtliche Artikel zur Nordamerikanischen Literatur für Meyers Neues Lexikon, 2. völlig neu erarbeitete Auflage in 18 Bdn, Bd *4* (Leipzig 1972), Bde *5*, *6*, *7* (Leipzig 1973), Bde *8*, *9*, *10*, (Leipzig 1974).
— [Nachw.] Oliver Goldsmith: Der Landpfarrer von Wakefield. Leipzig 1974. S. 219 bis 242.
— [Bespr.] Fay M. Blake: The Strike in the American Novel. Metuchen 1972. In: ZAA *21* (1973) S. 323–325.
— [Bespr.] Das amerikanische Drama von den Anfängen bis zur Gegenwart. Herausgegeben von Hans Itschert. (Ars interpretandi Bd 5). Darmstadt 1972. Ebda S. 437–440.
— [Bespr.] Die amerikanische Short Story. Theorie und Entwicklung. Herausgegeben von Hans Bungert. (Wege der Forschung Bd CCLVI). Darmstadt 1972. Ebda S. 434–437.

Buchda, Gerhard: Enthält die Glosse zum sächsischen Weichbild echte Schöffensprüche? In: Festschr. Heinrich Demelius zum 80. Geburtstag. Erlebtes Recht in Geschichte und Gegenwart. Hrsg. von G. Frotz u. W. Ogris. Wien 1973. S. 25–50.
— Art. Jus evocandi. In: Handwörterbuch zur deutschen Rechtsgeschichte, hrsg. von A. Erler u. E. Kaufmann unter philolog. Mitarbeit von R. Schmidt-Wiegand, mitbegr. von W. Stammler. *10.* Lieferung. Berlin 1973, Sp. 496–498.
— Art. Klage. Ebda *12.* Lieferung. Berlin 1974, Sp. 837–845.
— 75 Jahre Historische Kommission (1896–1971). In: Jahrb. Sächs. Akad. Wiss. *1971* bis *1972* (Berlin 1974) S. 459–461.
— [Ber.] Historische Kommission. Ebda S. 86–90.

Coblenz, Werner: Vier Befestigungen der Lausitzer Kultur am sächsisch-böhmischen Grenzgebirge. In: Časopis Moravského Musea *LVII* (1972) S. 51–60, 8 Abb.
— Ur- und Frühgeschichte (in Überschau) und Einzelbeschreibungen. In: Brockhaus-Reisehandbuch Sächsische Schweiz-Osterzgebirge. 2. neubearb. Aufl. Leipzig 1972, S. 48–54.
— Paul Reinecke, ein Wegbereiter der modernen Ur- und Frühgeschichtsforschung. In: Z. f. Archäologie *6* (1972) S. 240–249.
— Archäologische Slawenforschung in Sachsen. In: Zur archäologischen Frühgeschichtsforschung unserer Heimat. Erläuterungen zur gleichnamigen Ausstellung im Schloßberg-Museum Karl-Marx-Stadt vom 6. Okt. 1973 bis 27. Jan. 1974. Hrsg. Landesmuseum f. Vorgeschichte Dresden in Verbindung mit dem Bezirkssekretariat Natur und Heimat Karl-Marx-Stadt i. Kulturbund der DDR 1973. S. 3–14, Abb. 15.
— Eine Aunjetitzer Vorratsgrube mit Getreide aus Döbeln-Masten (nach einem Fundbericht von Siegfried Ludwig, Leisnig). In: Ausgrabungen und Funde *18* (1973) S. 70–80, 6 Abb., 1/2 Taf.

Coblenz, Werner: Eine mittelalterliche Knochenflöte aus Großstorkwitz, Kr. Borna. In: Ausgrabungen und Funde *18* (1973) S. 98–102, 1 Abb., 1 Taf.

— u. C. Fritzsche: Neolithische Siedlungsbestattung mit drei Skeletten und Resten weiterer Schädel aus Zauschwitz, Kr. Borna. Ebda *18* (1973) S. 276–281, 2 Abb., 2 Taf. u. Umschlagbild.

— „Sammlung über die Geschichte der Arbeit". Zur Entwicklung des Landesmuseums für Vorgeschichte Dresden. Ebda *19* (1974) S. 66–69, 1 Taf.

— Aunjetitzer Grabbeigabe als Urne in einer jüngstbronzezeitlichen Bestattung. Ebda *19* (1974) S. 89–94, 2 Abb., 1 Taf.

— Zu einem Neufund von der Südwestgrenze der Sächsischen Schweiz. Bemerkungen zur Gebirgsarchäologie. Ebda *19* (1974) S. 94–100, 3 Abb., 1/2 Taf.

— 100 Jahre Prähistorische Staatssammlung in Dresden. In: Blick ins Museum (= Mitteilungen aus den Staatlichen Wissenschaftlichen Museen Dresden) *10/11* (1973) S. 38 bis 47, 8 Abb.

— Internationales Kolloquium über den Beginn der Bronzezeit in Zentral- und Osteuropa, Igołomio bei Kraków 1973. In: Ethnographisch-Archäologische Zeitschrift *15* (1974) S. 332–335.

— Ur- und Frühgeschichte (in Überschau) und Einzelbeschreibungen. In: Lößnitz und Moritzburger Teichlandschaft (Werte unserer Heimat 22) Berlin 1973. S. 16–17.

— Ur- und Frühgeschichte (in Überschau) und Einzelbeschreibungen. In: Zwischen Tharandter Wald, Freital und dem Lockwitztal (Werte unserer Heimat 21) Berlin 1973, S. 9–10.

— Ur- und Frühgeschichte (in Überschau) und Einzelbeschreibungen. In: Zwischen Strohmberg, Czorneboh und Kottmar (Werte unserer Heimat 24) Berlin 1974. S. 7–9.

— J. Herrmann, W. Coblenz: III. Dorfsiedlungen, Burgen und frühe Städte. 3. Burgen und Befestigungen. In: J. Herrmann [Hrsg.]: Die Slawen in Deutschland. 3. unv. Auflage. Berlin 1974. S. 147–181, 18 Abb.

— [Tätigkeitsber.] des Landesmuseums für Vorgeschichte Dresden für das Jahr 1972. In: Ausgrabungen und Funde *18* (1973) S. 55–59.

— Gotthard Neumann. 8. 6. 1902–29. 4. 1972. In: Jahrbuch der Sächs. Akademie der Wissenschaften zu Leipzig *1971–1972* (Berlin 1974) S. 316–321.

— [Bespr.] M. Menke: Die jüngere Bronzezeit in Holstein. Neumünster 1972 (Offa-Bücher Bd 25). In: DLZ *95* (1974) Sp. 114–116.

— [Bespr.] A. Mozsolics: Bronze- und Goldfunde des Karpatenbeckens. Budapest 1973. Ebda Sp. 444–446.

— Numismatik und Archäologie. In: Münzen und Medaillen als Dokumente der Verkehrsgeschichte. Münzausstellung im Verkehrsmuseum Dresden v. 14. 9.–13. 10. 1974. Dresden 1974. S. 13–15, 2 Abb.

— [Hrsg.] Veröffentlichungen des Landesmuseums für Vorgeschichte, Bd 12 (H. Hanitzsch, Groitzsch bei Eilenburg. Schlag- und Siedlungsplätze der späten Altsteinzeit) Berlin 1972, 123 S., 21 Abb., 77 Taf., 5 Tab.

— [Hrsg.] in Verbindung mit dem Bezirkssekretariat Natur und Heimat Karl-Marx-Stadt im Kulturbund der DDR: Zur archäologischen Frühgeschichtsforschung unserer Heimat. Erläuterungen zur gleichnamigen Ausstellung im Schloßberg-Museum Karl-Marx-Stadt v. 6. Okt. 1973 bis 27. Jan. 1974. 1973, 31 S., 25 Abb., 1 Kt.

— [Hrsg.] Selimchanow, I. R.: Enträtselte Geheimnisse der alten Bronzen. Berlin 1974, 93 S., 28 Abb., 8 Taf.

Erben, Johannes: Sprechhandlung der Nicht-Verständigung. Bemerkungen zu Gestalt und Leistung dialogischer Texte. In: Studien zur Texttheorie und zur deutschen Grammatik. Festgabe für Hans Glinz zum 60. Geburtstag (= Sprache der Gegenwart. Schriften des Instituts für deutsche Sprache, Bd 30). Düsseldorf 1973. S. 122–129.

— [Neudruck von:] Aufgaben der deutschen Philologie heute. In: Sprachkunst als Weltgestaltung. Festschrift für Herbert Seidler. Salzburg 1966. S. 75–87, in: Methodenfragen der deutschen Literaturwissenschaft, hrsg. von R. Grimm u. J. Hermand. Darmstadt 1973. S. 187–203.

— [Einführung zu:] Deutsche Wortbildung. Typen und Tendenzen in der Gegenwartssprache. 1. Hauptteil: Das Verb. In: Sprache der Gegenwart. Schriften des Instituts für deutsche Sprache *29* (Düsseldorf 1973) S. 7–14.

— Ritter, Bürger und Bauer, oder: Der Bauer im Zweikampf. In: Studien zur deutschen Literatur und Sprache des Mittelalters. Festschrift für Hugo Moser zum 65. Geburtstag. Berlin 1974. S. 238–244.

— Luther und die neuhochdeutsche Schriftsprache. In: Deutsche Wortgeschichte. Hrsg. von Maurer-Rupp. 3. Aufl. Bd *1* (Berlin 1974) S. 509–581.

— [Rez.] H. Glinz: Deutsche Grammatik II (= Studienbücher zur Linguistik und Literaturwissenschaft, Bd 3), Frankfurt/M. 1971; K.-D. Bünting: Einführung in die Linguistik, Frankfurt/M. 1971; J. Lyons: Noam Chomsky, München 1971. In: Sprachkunst, Beiträge zur Literaturwissenschaft *4* (1973) 1/2, S. 170–173.

— [Rez.] H. Wolf: Die Sprache des Johannes Mathesius. Philologische Untersuchung frühprotestantischer Predigten. Einführung und Lexikologie (= Mitteldeutsche Forschungen, Bd 58). Köln 1969. In: Z. f. deutsche Philologie *92* (1973) S. 474–475.

— [Rez.] H. Haller: Übersetzungen im ‚gemeinen Deutsch' (1464). Hrsg. von E. Bauer. Göppingen 1972. In: Germanistik *14* (1973) S. 627–628.

— [Rez.] M. Rössing-Hager: Syntax und Textkomposition in Luthers Briefprosa, 2 Bde, Köln/Wien 1972. In: Beitr. *96* (Tübingen 1974) S. 185–187.

— [Rez.] Veit Dietrich: Etliche Schrifften für den gemeinē man/von vunterricht Christlicher lehr vnd leben/vnnd trost der engstigen gewissen. Hrsg. von O. Reichmann. Assen 1972 (= Quellen und Forschungen zur Erbauungsliteratur des späten Mittelalters und der frühen Neuzeit. Bd 5). Ebda S. 178–181.

Große, Rudolf: Gesellschaftsstruktur und Sprachstruktur. In: Festschrift zur Feier des 125jährigen Bestehens der Sächsischen Akademie der Wissenschaften zu Leipzig. Berlin 1974. S. 103–113.

— Zum Begriff der Nationalität in der Geschichte der deutschen Sprache. In: Linguistische Studien, Reihe A, Akademie der Wissenschaften der DDR, Zentralinstitut für Sprachwissenschaft. Berlin 1973. S. 54–66.

— u. A. Neubert, Thesen zur marxistisch-leninistischen Soziolinguistik. In: Beiträge zur Soziolinguistik, hrsg. von R. Große und A. Neubert. Halle (Saale) 1973. S. 7–24.

— W. Fleischer, R. Große: Ein Vierteljahrhundert germanistische Linguistik als gesellschaftswissenschaftliche Teildisziplin. In: Wiss. Zeitschrift der Karl-Marx-Universität Leipzig, Ges.- und Sprachwiss. R., *23* (1974) S. 431–440.

— [Nachruf] u. E. Eichler: Viktor Maximowitsch Schirmunski. In: Jahrbuch der Sächsischen Akademie der Wissenschaften zu Leipzig *1971–1972* (Berlin 1974) S. 219–222.

— [Hrsg.] Althochdeutsches Wörterbuch. Auf Grund der von E. v. Steinmeyer hinterlassenen Sammlungen im Auftrag der Sächsischen Akademie der Wissenschaften zu Leipzig begründet von E. Karg-Gasterstädt und Th. Frings, hrsg. von R. Große, Bd III, 2. u. 3. Lfg, Berlin 1973; 4. Lfg 1974.

Große, Rudolf: [Hrsg.] Beiträge zur Geschichte der deutschen Sprache und Literatur, hrsg. von G. Schieb, W. Fleischer, R. Große, G. Lerchner, Halle (Saale) *94.* und *95.* Bd (1974).
— [Hrsg.] Linguistische Arbeitsberichte, Mitteilungsblatt der Sektion Theoretische und angewandte Sprachwissenschaft der Karl-Marx-Universität Leipzig, Heft 7—10, 1973 bis 1974.
— [Hrsg.] Der Name in Sprache und Gesellschaft. Deutsch-Slawische Forschungen zur Namenkunde und Siedlungsgeschichte, begründet von Th. Frings u. R. Fischer, hrsg. von E. Eichler, W. Fleischer, R. Große, A. Neubert, H. Walther. Berlin 1973. Nr. 27.

Havránek, Bohuslav: Staročeský slovník [Alttschechisches Wörterbuch]. Prag 1973—1974, H. 5—6, S. 503—912.
— Staroslověnský slovník [Lexikon linguae palaeolovenicae]. Prag 1973—1974, H. 22 –26, S. 1—192.
— Větná struktura v nejstarších fázích slovanských jazyků [Satzstruktur in den ältesten Phasen der slavischen Sprachen]. In: Československé přednašky pro 7. Mezinárodní sjezd slavistů ve Varšavě. Praha 1973. S. 5—13 [deutsches Resumé].
— Maristické řešení problematiky spsiovných jazyků [Marxistische Lösung der Problematik der Schriftsprachen]. In: Naše řeč *56* (Prag 1973) S. 225—228.
— Květen, máj. Ebda *57* (1974) S. 160—174.
— Vztah klaštera na Slovanech k jazyku a literatuře české a hlaholsko-charvátské [Beziehungen des Klosters na Slovanech zur tschechischen und glagolitisch-kroatischen Sprache und Literatur]. In: Z tradic slovanské kultury. Prag 1974.
— Jungmannův význam pro nový rozvoj slovní zásoby spisovné češtiny [Jungmanns Bedeutung für die neue Entwicklung der neuen tschechischen Schriftsprache]. In: Spisovné jazyky slovanské. Prag 1974.

Heimpel, Hermann: Konrad von Soest und Job Vener, Verfasser und Bearbeiter der Heidelberger Postillen zu der Berufung des Konzils von Pisa. In: Z. „Westfalen" *50* (1972) S. 115—124; *51* (1973) S. 115—124.
— Für Bayern. Schicksale der „Bayernhymne". In: Z. f. Bayerische Landesgeschichte *36* (1973) S. 680—718.
— Der verketzerte Matthäus von Krakau. In: Festschrift für Walter Schlesinger *2* (1974) S. 443—455.
— Studien zur Kirchen- und Reichsreform des 15. Jahrhunderts. II. Zu zwei Kirchenreform-Traktaten des beginnenden 15. Jahrhunderts. Heidelberg 1974. 54 S. (= Sitzungsbericht der Heidelberger Akademie der Wissenschaften, Philologisch-historische Klasse, Jahrgang 1974, 1. Abhandlung).
— [Nachr.] Friedrich Baethgen 30. 7. 1890—18. 6. 1972. In: Jahrbuch Bayerische Akad. Wiss. *1973* (München 1973) S. 198—202.

Hiebsch, Hans: Perspektiven und Aufgaben der Psychologie in der entwickelten sozialistischen Gesellschaft. In: H. Hiebsch und L. Sprung (Hrsg.): Aufgaben, Perspektiven und methodologische Grundlagen der marxistischen Psychologie in der DDR. Berlin 1973. S. 11—22.
— Einige Anmerkungen zum Problem des Verhältnisses zwischen dem Biologischen und dem Sozialen. Ebda S. 35—39.

Iordan, Iorgu: Note de lexicologie românească. In: Studii și cercetări lingvistice *24* (1973) S. 41—52.
— Bilingsvism în domeniul romanic. Ebda S. 129—137.

Iordan, Iorgu: Al XI-lea Congres al lingviştilor (Bologna 1972). Ebda S. 338—340.
— Pe marginea volumului „Pagini istorico-filologice“ de Vasile Bogrea. In: Cercetări de lingvistică *XVIII* (1973) S. 5—6.
— Formations romanes à affixes internationaux. In: The Canadian Journal of Romance Linguistics I (1973) S. 26—32.
— Prefaţă la Maria Manoliu Manea. In: Structuralismul lingvistic. Bucureşti 1973. S. 5—6.
— La Société Roumaine de Linguistique Romane à son dixième anniversaire. In: Bulletin de la S. R. L. R. *IX* (1973) S. 5—9.
— Note sur la double valeur du participe passé roman. In: Mélanges de linguistique française et de philologie et littérature médiévales, dédiés à Monsieur Paul Imbs. Strasbourg 1973. S. 401—405.
— Limba lui Cantemir. In: Cronica, 26 octombrie 1973. S. 8, 11.
— Congresul de studii romanice din Perú. In: Secolul *20* (1973) Nr. 4, S. 181—183.
— „Miercuri seară“ sau „Miercuri seara“? In: Limbă şi literatură *II* (1973) S. 37—38.
— Note lingvistice. Ebda S. 741—742.
— Los estudios hispánicos en Rumania. In: Actas de la primera reunión latino-americana de lingüística y filología. Bogotá 1973. 11 S.
— Acad. Viktor Vladimirovici Vinogradov (12 ian. 1895 — 4 oct. 1969). In: Analele Academiei R. S. R., Seria IV, *XIX* (Bucureşti 1973) S. 929.
— Intervenţie la Sesiunea „Statul entităţilor abstracte“ 1972. In: Noesis. Travaux du Comité Roumain d'Histoire et de Philosophie des Sciences. Bucarest 1973. S. 85—86.
— Introdução à linguistica românica. Lisboa 1973. 607 S. [= Traducere portugheză a cărţii Lingvistica romanică. Bucureşti 1962].
— und John Orr: Introduzione alla linguistica romanza. Torino 1973. XVI, 501 S. Traducere italiană a cărţii Iorgu Iordan: An Introduction to Romance Linguistics. Its Schools and Scholars. Revised and translated by John Orr. London 1937.
— Istoricitatea comediilor lui Caragiale. In: Cronica, 13 iulie 1973. S. 8, 13.
— A fi romantic ... [Răspuns la o anchetă]. In: Cinema, februarie 1973. S. 24.
— Schiţă de autobiografie spirituală. In: România literară, 4 octombrie 1973. S. 16—17.
— Ce ştiam la 18 ani, interviu. In: Scînteia tineretului, 27 septembrie 1973.
— Ce înseamană a fi bucureştean? Interlocutor acad. Iorgu Iordan. In: Informaţia Bucureştiului, 29 septembrie 1973.
— u. Maria Manoliu Manea: Linguistica romanza, a cura di A. Limentani, traduzione di Marinella Lörinczi Angioni. Padua 1974. XX, 486 S. [ital. Übers. von: Introducere în lingvistica romanică].
— Elogiul consecvenţei, interviu luat de Florenţa Albu. In: Viaţa românească, matie 1974. S. 68—74 [Colaborare: conţinutul, în general, al meu, redactarea, în general, a ei].
— In chestiunea Congresului de Lingvistică si filologie romanică de la Napoli (polemică cu D. Macrea): 1. Ochiul magic (O scrisoare a mea, fără titlu). In: România literară, 16/V/1974 şi 2. O scrisoare a acad Iorgu Iordan . Ebda 30/V/1974.
— Note de lexicologie românească. In: Studii şi cercetări lingvistice *25* (1974) S. 439—449.
— Limba noastră în epoca actuală. In: Limba Română *23* (1974) S. 263—266.
— Corespondenţa mea cu N. Iorga. In: Manuscriptum *2* (1974) S. 81—87.
— O scrisoare. In: România literară, 8 august 1974. S. 16. [Rectificarea unor de istorie literară universală ale lui Liviu Doineanu].
— Titluri si lucrari. Bucureşti 1974. 60 S.
— Neculce, precursor al lui Creangă. In: Cronica, 6 septembrie 1974. S. 1, 6.
— Calcuri sintactice. In: Limbă şi literatură *III* (1974) S. 47—49.

Iordan, Iorgu: Cuvînt înainte. In: Liceul Internat Costache Negruzzi Iaşi. Reuniunea festivă consacrată evocării unor figuri ale culturii româneşti care şi-au făcut studiile la Liceul Internat. Bucureşti 1974. S. 2.
— u. Maria Manoliu Manea: Linguistica romanza, a cura di Alberto Limentani. Padova 1974. XX, 486 S.
— Cuvînt înainte, la Dan Bădărău şi Ioan Caproşu, Iaşie vechilor zidiri. Iaşi 1974. S. 5—7.
— In ... apărarea lui Neoulce. In: România literară, 26 decembrie 1974.
— Pierre Gardette. In: Studii şi cercetări lingvistice *25* (1974) S. 567—568.
— [Rez.] N. Mihăescu: Carte depre limba românească. In: Limba Română *22* (1973) S. 82—84.
— [Rez.] A. de Vincenz: Traité d'Anthroponymie houtzoule. In: Studii şi cercetări lingvistice *24* (1973) S. 338—340.
— [Rez.] M. Davan, M. Cohen, M. l'Allemand: Dictionaire du français vivant. Ebda *25* (1974) S. 543—545.
— [Rez.] F. de B. Moll: El habla de Mallorca. Ebda S. 551—552.
— [Interv.] Interviu, in: Magazinul, 7 iulie 1973. S. 5 [Titlu dat de reporter: Cuvîntul ştiinţei asupra ştiintei cuvîntului].
— [Interv.] Interviu, in: Flacăra, 7 aprilie 1973, S. 10 [Titlu, dat de revistă: Dialog cu viitorul].
— [Interv.] Interviul nostru cu academicianul Iorgu Iordan. In: Cadran universitar *19* (Craiova 1973).
— [Interv.] Interviu, in: Magazin istoric, noiembrie 1973, S. 25—28 [Titlu dat de redacţie: Intre linvistică şi istorie].
— [Interv.] Interviu, in: Pentru Socialism, Baia Mare, 20 iulie 1973 [Titlu dat de reporter: Plec cu imaginea acestor locuri în care omenia nu are limite].
— [Interv.] Interviu, in: O Seculo, Lisabona, 23 septembrie 1973. [Cu titlul dat de reporter: Na Roménia não ha escritores malditos, nem desconhecidos].
— [Interv.] Interviu, in: Cutezătorii, 29 noiembrie 1973, S. 4—5 [Titlu dat de redacţie: Cutezătorii in vizită la academicianul Iorgu Iordan].
— [Interv.] O vizită în laboratorul intelectual al savantului Iorgu Iordan, de Mariana Filimon. In: Forum *7* (1974) S. 34—35. Pe baza unui interviu dat de mine.
— [Interv.] Limba-creatoare de cultură spirituală, interviu în: Tribuna, 15 august 1974, S. 4 [Titlu dat de Victor Ivanovici, care a stat de vorbă cu mine].
— [Interv.] Geneza unui dicţionar patronimic, interviu (Catinca Muscan), in: Magazinul, 19 ianuarie 1974.
— [Interv.] Un chimist al limbii: acad. Iorgu Iordan, interviue luat de Ana Blandiana, in: Anfiteatru, decembrie 1974, S. 3.

Jahn, Johannes: Zerbster Prunkbibel „Cranachbibel". Die Apokalypse. Begleittext S. I bis XVII. Leipzig 1973.
— Michelangelo. Die Sixtinische Decke. Leipzig 1974. 19 S., 12 Taf.
— Bemerkungen zu Lucas Cranach als Zeichner. In: Lucas Cranach. Künstler und Gesellschaft. Referate des Colloquiums zum 500. Geburtstag Lucas Cranachs des Älteren, Staatliche Lutherhalle Wittenberg 1.—3. Okt. 1972. Wittenberg 1973. S. 87—91.
— Unsere architektonische Umwelt. In: Festschrift zur Feier des 125jährigen Bestehens der Sächsischen Akademie der Wissenschaften zu Leipzig. Berlin 1974. S. 167—177.
— [Rez.] Hildegard Schnabel [Hrsg.]: Lucas Cranach d. Ä. Passional Christi und Antichristi. In: DLZ *94* (1973) Sp. 851—852.
— [Rez.] N. Huse: Studien zu Giovanni Bellini. Ebda *95* (1974) Sp. 140—142.

Jahn, Johannes: [Rez.] Anja Broschek: Michel Erhardt. Ebda Sp. 217–219.
– [Rez.] G. Kraus: Das Bildgedicht in Europa. Ebda Sp. 408–409.
– [Rez.] J. Poeschke: Die Sieneser Domkanzel des Nicola Pisano. Ebda Sp. 649–650.

Krauss, Werner: Die Aufklärung in Spanien, Portugal und Lateinamerika. München 1973.
– Zur strukturalistischen Literaturbetrachtung. 4. erweiterte Aufl. von „Grundprobleme der Literaturwissenschaft" (= Rowohlts deutsche Enzyklopädie von März 1968). Reinbeck 1973.
– Gottsched als Übersetzer französischer Werke. Beitrag zur Reimarus-Tagung der Jungius-Gesellschaft, Hamburg. In: Hermann Samuel Reimarus (1674–1768), ein „bekannter Unbekannter" der Aufklärung. Göttingen 1973. S. 66–74.
– Goethe und die französische Revolution. In: neohelicon. Acta comparationis literarum universarium. *1* (Budapest/The Hague–Paris 1973) H. 3–4, S. 77–90.
– [Rez] R. Stephan: Goldenes Zeitalter und Arkadien. Studien zur französischen Lyrik des ausgehenden 18. und des 19. Jahrhunderts. Heidelberg 1971. In: DLZ *94* (1973) Sp. 35–36.
– [Rez.] F. Venturi: Europe des lumières. Recherches sur le 18ᵉ siècle. Paris 1971. Ebda Sp. 210–211.
– [Rez.] F. Nies: Gattungspoetik und Publikumsstruktur. Zur Geschichte der Sévignébriefe. München 1972. Ebda Sp. 333–335.
– [Rez.] H.-G. Funke: Crébillon fils als Moralist und Gesellschaftskritiker. Heidelberg 1972. Ebda Sp. 568–570.
– [Rez.] H. R. Picard: Die Illusion der Wirklichkeit im Briefroman des 18. Jahrhunderts. Heidelberg 1971. Ebda *95* (1974) Sp. 22–24.
– [Rez.] G. Hoffmeister: Die spanische Diana in Deutschland. Vergleichende Untersuchungen zu Stilwandel und Weltbild des Schäferromans im 17. Jahrhundert. Westberlin 1972. Ebda Sp. 99–100.
– [Rez.] K.-H. Mulagk: Phänomene des politischen Menschen im 17. Jahrhundert. Propädeutische Studien zum Werk Lohensteins unter besonderer Berücksichtigung Diego Saavedra Fajardos und Baltasar Graciáns. Westberlin 1973. Ebda Sp. 331–332.
– [Rez.] S. S. Prawer: Comperative Literary Studies. An Introduction. London 1973. Ebda Sp. 407.
– [Rez.] Spanische Literatur im Goldenen Zeitalter. Fritz Schalk zum 70. Geburtstag. Frankfurt/Main 1973. Ebda Sp. 706–707.
– [Rez.] H.-J. Lope: Die „Cartas marruecas" von José Cadalso. Frankfurt/Main 1973. In: Romanische Forschungen *85* (1973) H. 1/2, S. 235–237.

Kühnert, Friedmar: Der Orientbericht Tacitus Ann. VI 31–37. In: Wiss. Z. Friedrich-Schiller-Universität Jena, ges.- und sprachwiss. R. 22 (1973) H. 3, S. 491–496.

Lehmann, Edgar: Wandlungen im Gegenstand und in der Methode der Geographie in Beziehungen zur Entwicklung des Geographischen Instituts der Akademie der Wissenschaften. In: Akademie der Wissenschaften der DDR. Veröffentlichungen des Forschungsbereichs Kosmische Physik *1973*, H. 2, S. 5–21.
– Hermann Haacks Wirkungsaktualität. In: PGM *116* (1972) H. 4, S. 241–246.
– Zur theoretischen Grundlegung des Begriffes Region. In: Geographische Berichte *18* (1973) H. 1 (66. Heft) , S. 41–48.
– Quantitativ nicht erfaßbare Werte in der Umweltforschung. Ebda H. 3 (68. Heft), S. 168–173.

Lehmann, Edgar, Die Klasse „Umweltschutz und Umweltgestaltung“ der Akademie der Wissenschaften der DDR. Ebda S. 229–234.
— Spezielle Regionalatlanten — erläutert an der kartographischen Konzeption eines Atlasses über die „Optimale Gestaltung der Umweltbedingungen“. In: Geoforum *15* (1973) S. 59–62.
— Immaterielle Werte in der Umweltgestaltung. In: Festschrift zur Feier des 125jährigen Bestehens der Sächsischen Akademie der Wissenschaften zu Leipzig. Berlin 1974. S. 157–166.
— Bericht des Vizepräsidenten, Herrn Lehmann, über den Stand der Arbeiten der Akademie. Mathematisch-naturwissenschaftliche und philologisch-historische Klasse. Erstattet in öffentlicher Gesamtsitzung am 17. April 1971. In: Jahrbuch Sächs. Akad. Wiss. zu Leipzig *1971–1972* (Berlin 1974) S. 112–117.
— Bericht des Vizepräsidenten, Herrn Lehmann, über den Stand der Arbeiten der Akademie. Mathematisch-naturwissenschaftliche und philologisch-historische Klasse. Erstattet in öffentlicher Gesamtsitzung am 15. April 1972. Ebda S. 118–125.
— [Mithrsg.] Internationales Jahrbuch für Kartographie.
— [Mithrsg.] Z. Geoform (Journal of Physical, Human and Regional Geo-Sciences).
— [Mithrsg.] Schriftenreihe „Beiträge zur Alexander-von-Humboldt-Forschung“ der Akademie der Wissenschaften der DDR.

Lissa, Zofia: Zene és csend [Musik und Stille]. Budapest 1973. 271 S.
— Wstęp do muzykologii [Einführung in die Musikwissenschaft]. 2. Aufl. Warszawa 1974. 277 S.
— Klavierpolonaise und -mazurka im 19. Jahrhundert. In: Gattungen der Musik in Einzeldarstellungen. Gedenkschrift für Leo Schrade. Bern und München 1973. S. 813–840.
— Witold Lutosławskis Konzert für Orchester. Analytische Studie. In: Zur musikalischen Analyse. Hrsg. von G. Schumacher. Darmstadt 1974. S. 282–323.
— Musik und Revolution. In: International Review of the Aethetics and Sociology. IMS Symposium. Zagreb 1974. S. 113–125.
— Zur Theorie der musikalischen Rezeption. In: Archiv für Musikwissenschaft *31* (1974) H. 3, S. 157–170.
— Ebenen des musikalischen Verstehens. In: Musik und Verstehen. Aufsätze zur semiotischen Theorie, Ästhetik und Soziologie der musikalischen Rezeption. Hrsg. von P. Faltin und H. P. Reinecke. Köln 1973. S. 217–247; dasselbe [O rozumieniu muzyki] auch in: Studia Estetyczne *11* (Warszawa 1974) S. 281–315.

Markov, Walter und A. Soboul: 1789 — Die Große Revolution der Franzosen. Berlin 1973. 480 S. mit 4 Farbtaf., 16 Abb. u. 2 Ktn.
— Sommario della storia coloniale. Editori Riuniti. Rom 1973.
— Es brennt in der Franc-Zone. In: Die Weltbühne *68* (1973) S. 275–277.
— Wahlkurven. In: Die Weltbühne *68* (1973) S. 520–523.
— Das erste Jahrzehnt der OAU. In: Die Weltbühne *68* (1973) S. 673–676.
— Der letzte Tango in Paris. In: Die Weltbühne *68* (1973) S. 817–819.
— Nachruf auf Nurmi. In: Die Weltbühne *68* (1973) S. 1373.
— Wiederbegegnung mit Jean Jaurès. In: Die Weltbühne *68* (1973) S. 1417–1418.
— Die türkische Brücke. In: Die Weltbühne *68* (1973) S. 1509–1511.
— Mein Film, mein Stück des Jahres. In: Die Weltbühne *68* (1973) S. 1658.
— Wege und Formen der Staatsbildung in Asien und Afrika seit dem zweiten Weltkrieg.

In: Meždunarodnyj komitet istoričeskich nauk, XIII. meždunarodny kongress istoričeskich nauk. Doklady kongress, Bd *I*, T. 2 (1973) S. 139—154.

Markov, Walter: Die illyrische Paradoxie. In: Mitteilungen des Österreichischen Staatsarchivs *25* (1972) S. 587—594.

— Les provinces illyriennes. In: Actes du Colloque Patriotisme et nationalisme en Europe à l'époque de la Révolution française et de Napoléon. XIIIe Congrès international des Sciences historiques, (1973) S. 161—165.

— Auszug aus: Die Freiheiten des Priesters Roux. Berlin 1967. In: Suratteau, J.-R., La Révolution française. Certitudes et controverses. Dossiers Clio. Texte 21. S. 69—70.

— Auszug aus: Jacques Roux. Scripta et Acta. Berlin 1969. S. 141—142. In: Suratteau, J.-R., La Révolution française. Certitudes et controverses. Dossiers Clio. Document 4. S. 38—39.

— H. Piazza und W. Markov: Leninskaja vnešnjaja politika i antikolonial'noe osvoboditel'noe dviženie. In: V. I. Lenin i vnešnjaja politika socialističeskogo gosudarstva (1972) S. 101—132.

— Zu einigen Ergebnissen und Problemen der Erforschung der Geschichte der nationalen Befreiungsbewegung. In: Wissenschaftliche Mitteilungen der Historiker-Gesellschaft der DDR *II* (1974) S. 49—53.

— Toga Party in Rom. In: Die Weltbühne *69* (1974) S. 527—529.

— Pompidou und andere. In: Die Weltbühne *69* (1974) S. 572—573.

— Der letzte Kaiser. In: Die Weltbühne *69* (1974) S. 1125—1128.

— Taifun 18. In: Die Weltbühne *69* (1974) S. 1428—1430.

— Armagnacs und Bourguignons. In: Die Weltbühne *69* (1974) S. 1668—1670.

— La Révolution Française vue par les historiens de la République de Weimar. In: Revue d'Allemagne et des pays de langue allemande. T. *VI*, 2 (1974) S. 58—66.

— Erinnerung an den Beginn. In: Asien. Afrika. Lateinamerika *3* (1974) S. 789—793.

— Il Caso della Dalmazia ⟨1797—1813⟩. In: Settanta, Rivista bimestrale *3* (1974) S. 121 bis 124.

— Dimitrie Cantemir und Rußland. In: Dimitrie Cantemir. Sitzungsberichte der Akademie der Wissenschaften der DDR *13* (1973) S. 53—58.

— W. Loch und W. Markov: Die französischen Revolutionen zwischen 1789 und 1871 im Lichte von Lenins Auffassung über den Revolutionszyklus. In: Studien zur vergleichenden Revolutionsgeschichte 1500—1917 (= Studien zur Revolutionsgeschichte) 1974. S. 74—91.

— M. Kossok und W. Markov: Zur Methodologie der vergleichenden Revolutionsgeschichte der Neuzeit. In: Studien zur vergleichenden Revolutionsgeschichte 1500 bis 1917 (= Studien zur Revolutionsgeschichte) 1974. S. 1—28.

— [Bespr.] I. I. Zil'berfarb: Idei i tradicii Velikoj francuzskoj revoljucii v bor'be sil demokratii i fašizma. Leipzig 1973. In: Wissenschaftliche Zs. d. KMU, Gesellsch.- u. sprachwiss. R. *2* (1973) S. 248.

— [Bespr.] Francuzskij ežegodnik 1970. Stat'i i materialy po istorii Francii (Französisches Jahrbuch 1970. Artikel und Materialien zur Geschichte Frankreichs) Moskau 1972. In: ZfG *21* (1973) S. 1515—1518.

— [Bespr.] A. Z. Manfred: Napoléon Bonaparte. Moscou 1971 [russ.]. In: Annales Historiques de la Révolution Française *4* (1973) S. 623—625.

— [Bespr.] Pierangelo Catalano: Tribunato e resistenza. Torino, Milano, Genova (u. a.) 1971. In: DLZ *94* (1973) Sp. 193—194.

Markov, Walter: [Bespr.] Armando Saitta: Filippo Buonarroti. Contributo alla storia della sua vita e del suo pensiero. 2 Bde. 2. durchges. u. erw. Ausg. Rom 1972. In: ZfG *21* (1973) S. 1390–1392.

— [Anz.] Histoire des Juifs en France (Geschichte der Juden in Frankreich). Toulouse 1972. In: ZfG *21* (1973) S. 98.

— [Anz.] Jürgen Rosenbaum: Frankreich in Tunesien. Die Anfänge des Protektorats 1831–1886. Zürich/Freiburg i. Br. 1971. In: ZfG *21* (1973) S. 109.

— [Anz.] Jean-Jacques Rousseau: Du contrat social (Der Gesellschaftsvertrag). London 1973. In: ZfG *21* (1973) S. 484.

— [Anz.] Artikel aus der von Diderot und d'Alembert herausgegebenen Enzyklopädie. Leipzig 1972. In: ZfG *21* (1973) S. 493.

— [Anz.] Béla Köpeczi: La France et la Hongrie au début du XVIIIe siècle. Etude d'histoire des relations diplomatiques et d'histoire des idées (Frankreich und Ungarn am Anfang des 18. Jh. Studien zur Geschichte der diplomatischen Beziehungen und der Geschichte der Ideen). Budapest 1971. In: ZfG *21* (1973) S. 617.

— [Anz.] Lew S. Gordon: Studien zur plebejisch-demokratischen Tradition in der französischen Aufklärung. Neue Beiträge zur Literaturwissenschaft. Bd 32. Berlin 1972. In: ZfG *21* (1973) S. 735.

— [Anz.] Alfred Körner: Die Wiener Jakobiner. Deutsche revolutionäre Demokraten. Bd 3. Stuttgart 1972. In: ZfG *21* (1973) S. 738.

— [Anz.] Jean-René Suratteau: La Révolution française. Certitudes et controverses. Dossiers Clio. Nr 58. Paris 1973. In: ZfG *21* (1973) S. 878.

— [Anz.] Jean Pinasseau: L'Emigration militaire. Campagne de 1792. Armée des Princes. Compagnies de Saintonge, Angoumois et Aunis. Paris 1971. In: ZfG *21* (1973) S. 878–879.

— [Anz.] B. S. Chlepner: Cent ans d'histoire sociale en Belgique. Brüssel 1973. In: ZfG *21* (1973) S. 1010.

— [anz.] Charles-Robert Ageron: Politiques coloniales am Maghreb. Paris 1972. In: ZfG *21* (1973) S. 1262–1263.

— [Hrsg.] Világtörténelmi kisenciklopédia (Kleine Enzyklopädie Weltgeschichte [ungar.] Hrsg. W. Markov, A. Anderle, E. Werner, H. Wurche). Budapest 1973. 811 S.

— [Anz.] Paule-Marie Duhet: Les femmes et la Révolution 1789–1794 (Die Frauen und die Revolution von 1789–1794). Paris 1971. In: ZfG *22* (1974) S. 126.

— [Anz.] Rudolf von Thadden: Restauration und napoleonisches Erbe. Der Verwaltungszentralismus als politisches Problem in Frankreich (1814–1830). Veröffentlichungen des Instituts für europäische Geschichte Mainz, Bd 63. Wiesbaden 1972. In: ZfG *22* (1974) S. 126.

— [Anz.] Die Französische Revolution. Eine Dokumentation. München 1973. In: ZfG *22* (1974) S. 252–253.

— [Anz.] Benjamin Constant et Goyet de la Sarthe. Correspondance 1818–1822. Genève 1973. In: ZfG *22* (1974) S. 650–651.

— [Anz.] La quantification en histoire (Die Quantifizierung in der Geschichte). Bruxelles 1973. In: ZfG *22* (1974) S. 891–892.

— [Anz.] Dirk Hoeges: François Guizot und die Französische Revolution. Romanistische Versuche und Vorarbeiten. Nr 44. Bonn 1973. In: ZfG *22* (1974) S. 1014.

— [Anz.] Friedrich Sieburg: Das Geld des Königs. Eine Studie über Colbert. Stuttgart 1974. In: ZfG *22* (1974) S. 1142.

— [Anz.] Pierre Goubert: Ludwig XIV. und zwanzig Millionen Franzosen. (West-)Berlin 1973. In: ZfG *22* (1974) S. 1142.

Markov, Walter: [Anz.] Die Französische Revolution. Anlässe und langfristige Ursachen. Darmstadt 1973. In: ZfG *22* (1974) S. 1142—1143.

— [Anz.] Paysannerie française. Paysannerie hongroise. XVIe—XXe siècle (Französische und ungarische Bauernschaft im 16.—20. Jh.). Budapest 1973. In: ZfG *22* (1974) S. 1269.

— [Anz.] Reinhart Koselleck: Kritik und Krise. Eine Studie zur Pathogenese der bürgerlichen Welt. Frankfurt/M. 1973. In: ZfG *22* (1974) S. 1275—1276.

— [Anz.] Vincent Cronin: Napoleon. Eine Biographie. Hamburg/Düsseldorf 1973. In: ZfG *22* (1974) S. 1277.

— [Mithrsg.] Studien zur Revolutionsgeschichte. In Verb. mit W. Markov u. G. Schilfert hrsg. v. M. Kossok. Berlin 1974.

— [Vortrag] Napoleonisches Illyrien und slowenische Nationwerdung. Berlin 1974. In: Jahrbuch der Sächsischen Akademie der Wissenschaften zu Leipzig *1971—1972* (Berlin 1974) S. 149—150.

Mayrhofer, Manfred: Onomastica Persepolitana. Das altiranische Namengut der Persepolis-Täfelchen. Unter Mitarbeit von J. Harmatta, W. Hinz, R. Schmitt und J. Seifert. Sitzungsberichte der österr. Ak. d. Wiss., phil.-hist. Klasse, 286 [= Veröffentlichungen der iranischen Kommission, herausgegeben von M. Mayrhofer, Bd 1]. Wien 1973. 358 S.

— Alltagsleben und Verwaltung in Persepolis. Linguistisch-onomastische Aufgaben aus neuerschlossenen Profantexten. In: Anzeiger der phil.-hist. Klasse der österr. Akademie d. Wiss. *109* (1972[73]) S. 192—202.

— Aus dem perserzeitlichen Ägypten. Eine namenkundliche Notiz. Ebda S. 317—320.

— Der Reiner-Test. Festschrift f. Heinrich Otten. Wiesbaden 1973. S. 191—198.

— Die Arier im Vorderen Orient — ein Mythos? Mit einem bibliographischen Supplement (= Sitzungsberichte der österr. Ak. d. Wiss., phil.-hist. Klasse, 294/3). Wien 1974. 93 S.

— Lieferung 25 [= Band III] des Altindischen etymologischen Wörterbuches, Heidelberg 1974. S. 561—640.

— Das Wiener Projekt eines Iranischen Namenbuches. In: Onoma *17* (1972/73 [1974]) S. 234—236.

— Die arischen Sprachreste in Vorderasien. Eine Abwehr der Hyperkritik. In: Acta Antiqua Academiae Scientiarum Hungaricae *20* (1972[74]) S. 271—282.

— Richard Hauschild. In: Jahrbuch der Sächs. Ak. d. Wiss. *1971—1972* (Berlin 1974) S. 307—311.

— Xerxès, Roi des Rois. In: Commémoration Cyrus *I* (Leiden-Teheran-Liège 1974) S. 108 bis 116.

— Irans Kultur- und Sprachenwelt in der Arbeit der Österreichischen Akademie der Wissenschaften. Ebda S. 328—334.

— Ein neuer Beleg zu der indogermanischen Sippe für 'Halsschmuck'. In: Antiquitates Indogermanicae. Gedenkschrift H. Güntert. Innsbruck 1974. S. 289—291, 1 Tafel.

— Plansprachen und Sprachplanung. Geschichte, Kritik, Perspektiven. In: Festschrift zur Feier des 125jährigen Bestehens der Sächs. Akad. d. Wissenschaften zu Leipzig. Berlin 1974. S. 77—86.

— Zu den Parther-Namen der griechischen Awrōmān-Texte. In: Mémorial Jean de Menasce. Louvain 1974. S. 205—213.

Meyer, Rudolf: Bemerkungen zur syntaktischen Funktion der sogenannten Nota Accusativi. In: Wort und Geschichte, Festschrift für Karl Elliger zum 70. Geburtstag. Neukirchen-Vluyn 1973. S. 137—142.

Meyer, Rudolf: Bemerkungen zum literargeschichtlichen Hintergrund der Kanontheorie des Josephus. In: Josephus-Studien. Festschrift für Otto Michel zum 70. Geburtstag. Göttingen 1974. S. 285—299.
— [Bespr.] Aron Freimann: Katalog der Judaica und Hebraica. Graz 1968. In: OLZ *68* (1973) Sp. 49—50.
— [Bespr.] Alexander A. Di Lella, O. F. M.: The Hebrew Text of Sirach. The Hague 1966. Ebda Sp.148—149.
— [Bespr.] Alexander Sperber: A Historical Grammar of Biblical Hebrew. Leiden 1966. Ebda Sp. 260—263.
— [Bespr.] Matthew Black: Paul Ernst Kahle 1875—1965. Proceedings of the British Academy LI. London 1967. S. 485—495. Ebda Sp. 367—368.
— [Bespr.] Peshitta: The Old Testament in Syriac. Song of Songs-Tobit-4 Ezra. Leiden 1966. In: ThLZ *98* (1973) Sp. 110—112.
— [Bespr.] Joseph Reider: An Index to Aquila. Leiden 1966. Ebda Sp. 506—507.
— [Bespr.] W. Baars: New Syro-Hexaplaric Texts. Leiden 1963. Ebda *99* (1974) Sp. 334—336.
— [Bespr.] Werner Fuß: Die sogenannte Paradieserzählung. Gütersloh 1968. In: OLZ *69* (1974) Sp. 45—49.
— [Bespr.] Rainer Degen: Altaramäische Grammatik. Wiesbaden 1969. Ebda Sp. 471—475.

Michałowski, Kazimierz: Pyramídy. Bratislava 1973. 30 S., 84 Abb.
— Piramisok és Masztabák. Budapest 1973. 30 S., 84 Abb.
— Pyramiden und Mastabas. Wien, München 1973. 31 S., 84 Abb.
— Pyramiden und Mastabas. Leipzig 1973. 31 S., 84 Abb.
— Louxour. Paris 1973. 23 S., 78 Abb.
— Faras. Malowidła ścienne w zbiorach Muzeum Narodowego w Warszawie. Warszawa 1974. 334 S., 102 Abb., 2 Landktn, 16 Zeichnungen, 5 Pläne.
— Od Edfu do Faras. Polskie odkrycia archeologii śródziemnomorskiej. Warszawa 1974. 269 S., 101 Abb., 1 Landkte.
— Teby. Warszawa 1974. 26 S., 48 Abb.
— Galeria Sztuki Faras. Warszawa 1974. 21 S., 24 Abb.
— Tell Atrib. In: Enciclopedia dell'Arte Antica *VIII* (Roma 1973) S. 799—800.
— Archeologia śródziemnomorska-jej rola i znaczenie w nauce polskiej. In: Miesięcznik Literacki *2* (Warszawa 1973) S. 32—40.
— La publication des textes égyptiens des Musées de Pologne. In: Mélanges Champollion. Bulletin de l'Institut d'Archéologie Orientale, Kair 1973.
— La naissance d'une nouvelle discipline historique: La Nubiologie. In: Mélanges Arif Mansel. Ankara 1974. S. 7—12.
— Ausgrabungen in Faras. In: Festschrift zur Feier des 125jährigen Bestehens der Sächsischen Akademie der Wissenschaften zu Leipzig. Berlin 1974. S. 87—99, 16 Taf.
— Ancient Egypt. The Visual Arts. In: Encyclopaedia Britannica. Chicago, London 1974.
— [Vorw.] zur polnischen Ausgabe: F. Daumas: La civilisation de l'Egypte pharaonique. Warszawa 1973. S. 5—7.

Müller, Joachim: Thomas Manns „Doktor Faustus" — Grundthematik und Motivgefüge. In: Ansichten zu Faust. Hrsg. v. Günther Mahal. Stuttgart 1973. S. 199—218.
— Die Relation von Traum und Leben bei Calderon und bei Grillparzer. In: Grillparzer-Forum 1972. Eisenstadt 1973. S. 46—65.
— Der Dichter Thomas Bernhard. In: Universitas *28* (1973) S. 533—541.

Müller, Joachim: Epik Dramatik Lyrik. Analysen und Essays zur deutschsprachigen Literatur. Halle/S. 1974. 452 S.

— „Verwirrung des Gefühls". Der Begriff des „Pathologischen" im Drama Goethes und Kleists. Mit einem Exkurs: Zur dichterischen Pathographie im Drama des 19. Jhdts. Berlin 1974. 40 S. (= Sitzungsbericht Sächs. Akademie der Wissenschaften zu Leipzig. Phil.-hist. Klasse Bd 117, H. 2.)

— Goethes „Campagne in Frankreich". Epochenkritik, Umweltanalyse und Kontraststruktur. Berlin 1974. 42 S. (= Sitzungsbericht Sächs. Akademie der Wissenschaften zu Leipzig. Phil.-hist. Klasse Bd 117, H. 3.)

— Goethes Romantheorie. In: Deutsche Romantheorien Bd 1. Hrsg. von Reinhold Grimm. Frankfurt a. M. 1974. S. 61—104.

— Über Herbert Zand. In: Literatur und Kritik. Mai 1974, Heft 84, S. 229—241.

— Die kulturpolitische Position des Essayisten Heinrich Mann. In: Wiss. Z. Univ. Jena, ges.- u. sprachwiss. R., *23* (1974) H. 1, S. 163—177.

— Schiller, Goethe und Novalis in Jena. Ebda S. 179—200.

— Lessing, Gotthold Ephraim. In: Encyclopaedia Britannica 1974, S. 838—841.

— Sturm und Drang. Ein Lesebuch für unsere Zeit. Zus. m. Klaus Herrmann. 10. Aufl. Berlin u. Weimar 1973. 414 S. Einl. 32 S.

— Stifters Werke in vier Bänden. 3. Aufl. Berlin und Weimar 1973. 512, 327, 421, 378 S. Einl. 23 S. (= Bibliothek deutscher Klassiker).

— Schillers Werke in fünf Bänden. 13. Aufl. Berlin u. Weimar 1974. 371, 435, 307, 309, 580 S. Einl. 36 S. (Bibliothek deutscher Klassiker).

— [Bespr.] Walther Killy: Elemente der Lyrik. München 1972. In: DLZ *94* (1973) Sp. 17 bis 20.

— [Bespr.] Schillers Werke. Nationalausgabe Bd 11: Demetrius. Hrsg. v. Herbert Kraft. Weimar 1971; Bd 36,1 Briefe an Schiller. Hrsg. v. Norbert Oellers. Weimar 1972. Ebda Sp. 127—129.

— [Bespr.] Hans Dietrich Irmscher: Adalbert Stifter. München 1971. Ebda Sp. 220—223.

— [Bespr.] Gert Ueding: Schillers Rhetorik. Tübingen 1971. Ebda Sp. 346—349.

— [Bespr.] Herbert Zeman: Die deutsche anakreontische Dichtung. Ebda Sp. 342—345.

— [Bespr.] Hölderlin, Sämtliche Werke. Bd 7, 2. Stuttgart 1972. Ebda Sp. 475—476.

— [Bespr.] Wilhelm Raabe. Briefwechsel mit Wilhelm Jensen. Göttingen 1970. Ebda Sp. 593—596.

— [Bespr.] Werner Keller: Goethes dichterische Bildlichkeit. München 1972. Ebda Sp. 705 bis 708.

— [Bespr.] Das Grillparzerbild des 20. Jhdts. Wien 1972. Ebda Sp. 788—791.

— [Bespr.] Robert Mühlher: Österreichische Dichter seit Grillparzer. Wien 1973. Ebda Sp. 905—908.

— [Bespr.] Manfred Dierks: Studien zu Mythos und Psychologie bei Thomas Mann. Bern 1972. Ebda *95* (1974) Sp. 33—35.

— [Bespr.] Günther Mahal: Mephistos Metamorphosen. Göppingen 1972. Ebda Sp. 103 bis 107.

— [Bespr.] Reinhard Schuler: Das Exemplarische bei Goethe. München 1973. Ebda Sp. 337—340.

— [Bespr.] Untersuchungen zur Literatur als Geschichte. Festschrift Benno v. Wiese. Berlin 1973. Ebda Sp. 513—516.

— [Bespr.] Friedrich Nemec: Die Ökonomie der „Wahlverwandtschaften". München 1973. Ebda Sp. 600—601.

Müller, Joachim: [Bespr.] Thomas Mann: Briefwechsel mit seinem Verleger G. B. Fischer. Frankf. M. 1973. Ebda Sp. 793–795.
– [Bespr.] Walter Dietze: Erbe und Gegenwart. Berlin 1972. – Ders.: Reden, Vorträge, Essays. Leipzig 1972. Ebda Sp. 868–874.
Ferner: 30 Kurzreferate in: Germanistik Jgg. *14–15* (1973/1974).

Neubert, Albrecht: Zur Interpretation englischer Komposita wie lookout, pick-up, pullback. In: ZAA *21* (1973) S. 64–69.
– Das CPD, ein neues Ausspracheworterbuch des Englischen. Ebda S. 299–306.
– Zur Determination des Sprachsystems. In: Wiss. Z. Karl-Marx-Univ. Leipzig, ges.- u. sprachwiss. R. *22* (1973) S. 193–204; Z. für Phonetik, Sprachwiss. u. Kommunikationsforschung *26* (1973) S. 617–629.
– Name und Übersetzung. In: Der Name in Sprache und Gesellschaft. Beiträge zur Theorie der Onomastik. Hrsg. v. E. Eichler, W. Fleischer, R. Große, A. Neubert u. H. Walther. Berlin 1973. S. 74–79.
– Invarianz und Pragmatik. In: Neue Beiträge zu Grundfragen der Übersetzungswissenschaft. Hrsg. v. A. Neubert u. O. Kade. Leipzig 1973. S. 13–26.
– R. Große, A. Neubert: Thesen zur marxistisch-leninistischen Soziolinguistik. In: Beiträge zur Soziolinguistik. Hrsg. v. R. Große u. A. Neubert. Halle (Saale) 1973. S. 7–24.
– Zu Gegenstand und Grundbegriffen einer marxistisch-leninistischen Soziolinguistik. Ebda S. 25–46.
– Theorie und Praxis für die Übersetzungswissenschaft. In: Proceedings of the Third International Congress of Aplied Linguistics *3* (Heidelberg 1973) S. 38–60.
– Invariance and Textual Types in Translation. Ebda S. 482–490.
– Englische Aussprache. 6. Aufl. Leipzig 1973.
– Überlegungen zum Thema Sprache und Geschichte. In: Linguistische Arbeitsberichte *10* (Leipzig 1974) S. 78–87.
– Sprache, Erkenntnis, Handlungsanweisung. In: Erkenntnis und Handeln. Wissenschaftliche Beiträge der Karl-Marx-Universität Leipzig. Leipzig 1974. S. 119–125.
– The English Language. In: English Grammar. A University Handbook. Potsdam 1974. S. 3–11.
– Englische Intonation. 3. Aufl. Leipzig 1974. 32 S. u. eine Schallplatte.
– Englische Aussprache. 7. Aufl. Leipzig 1974.
– [Bespr.] M. Lehnert: Der englische Grundwortschatz. Leipzig 1971. 263 S. In: ZAA *21* (1973) S. 70–71.
– [Bespr.] L. M. Davis (Hrsg.): Studies in Linguistics in honor of Raven I. McDavid, Jr. The University of Alabama Press 1972. XII. 461 S. In: Z. für Phonetic, Sprachwiss. u. Kommunikationsforschung *27* (1974) S. 360–361.

Orlandos, Anastasios K.: Les graffiti du Parthénon. Inscriptions gravées sur les colonnes du Parthénon a l'époque paléochrétienne et byzantine. Athen 1973. 199 S. (unter Mitarbeit von L. Vranoussis).

Pée, Willem: Een Vlaamse vriendengroet aan Klaas Heeroma. In: Taal en Tongval *XXV* (1973) H. 1–2, S. 9–13.
– Cyriel Buysse en het Westvlaams. Ebda S. 24–34.
– Vlaams, Hollands of Nederlands. In: Nu Nog *XXI* (1973) S. 76–83.
– De riem afleggen – De riemen afgooien. In: Taal en Tongval *XXV* (1973) H. 3–4, S. 122–126.

Pée, Willem: Eugène de Bock vijfentachtig jaar. In: Nu Nog *XXII* (1974) S. 26–29.
– Een week in de D.D.R. In: Informatiebulletin Vereniging België – Duitse Democratische Republiek *10* (1974) Nr. 2, S. 12–16.
– Hoeden-wachten – Koewachter-koe(h)oeder: Een Taalpuzzel! In: Taalkunde – 'N Lewe, studies opgedragen aan prof. dr. W. Kempen by geleentheid van sy 65ste verjaardag, 3 julie 1974, Kaapstad en Johannesburg, Tafelberg-Uitgewers, S. 77–84.
– [Hrsg.] E. Blancquaert† u. W. Pée: Reeks Nederlandse Dialektatlassen, Bd 12: H. Entjes, A. R. Hol†: Dialektatlas van Gelderland en Zuid-Overijsel. Antwerpen 1973. CXXXVIII, 350 S., 150 Ktn.
– u. A. van Elslander, J. van Schoor: Gesamtausgabe der Werke von Herman Teirlinck. Textausgabe und Notizen zu Band VIII. Brüssel 1969. 974 S.; Textausgabe und Notizen zu Band IX. Brüssel 1973. 1278 S.
– [Alfons Vranckx] De Ondervoorzitter van de Commissie voor het opstellen van de Nederlandse tekst van Grondwet en Wetboeken. In: Alfons Vranckx, Antwerpen, Rockox-Uitgaven 1973. S. 33–40.
– [Nachr.] Clara Hammerich-Flensborg (Roskilde 12-6-1894 – Hillerd 14-5-1972). In: Taal en Tongval *XXV* (1973) H. 3–4, S. 101–105.
– [Nachr.] Hermann Teirlinck (Sint-Jans-Molenbeek 24-2-1879 – Beersel 4-2-1967). In: De autotoerist *27* (1974) Nr. 9, S. 409–411.
– [Nachr.] Inna Dimitrievna Sjkoenaieva-Gimpelson (Moskou 12-1-1923 – 12-8-1971). In: Taal en Tongval *XXVI* (1974) H. 1–2, S. 1–4.

Peek, Werner: Inschrift einer Sonnenuhr aus Traianopolis. In: Z. f. Papyrologie u. Epigraphik *10* (1973) S. 98–100.
– Eros poimen laon. Ebda S. 91–93.
– Athenopolis von Priene. Ebda S. 95–97.
– Griechische Vers-Inschriften aus Ägypten. Ebda S. 230–248.
– Didyma II 340. Ebda *11* (1973) S. 25–28.
– Zu einem archaischen Grabepigramm aus Athen. Ebda S. 29–30.
– Leipsanotheke. Ebda S. 30–31.
– Zu der Dreifußbasis IG I^2 673. Ebda *13* (1974) S. 199–200.
– Zwei Gedichte auf den Neuplatoniker Plutarch. Ebda S. 201–204.
– Zwei Weihungen vom Ptoion. Ebda S. 264–265. [S. 19–28.
– Ein Weihgedicht für die Nymphen und drei andere Inschriften aus Atrax. Ebda *14* (1974)
– Anthologia Planudea 322. Ebda *15* (1974) S. 142.
– Hagemon von Rhodos. Ebda S. 223–225.
– Zu einem athenischen „Hymn to Athena". Ebda S. 225–227.
– Zu den Inscriptiones Thessalonicae. In: Maia *25* (1973) S. 198–203.
– Die Inschriften des Dionysos-Theaters von Epidauros. In: Vestigia *17* (1973) S. 562–564 [Bericht über die Entdeckung des Theaters].
– Zu einem Epheben-Epigramm aus Athen. In: Archaiologika Analekta ex Athenon *6* (1973) S. 125–127.
– Ein Grabgedicht aus Apollonia am Pontos. In: Klio *55* (1973) S. 147–150.
– Zu drei Epigrammen aus Kleinasien. In: Philologus *116* (1972) S. 254–260.
– Ein Weihgedicht des Theodoridas. Ebda *117* (1973) S. 66–69.
– Epigramme von der Agora. In: PHOROS, Tribute to B. D. Meritt. J. J. Augustin, Locust Valley, New York 1974. S. 121–131, 13 Abbildungen auf 3 Tafeln.
– Ein neuer spartanischer Staatsvertrag. Berlin 1974. 15 Seiten, 8 Abbildungen auf 5 Tafeln. (= Abhandlung Sächs. Akad. Wiss. zu Leipzig, phil.-hist. Kl. 65, H. 3).

Peek, Werner: Griechische Vers-Inschriften aus Thessalien. Heidelberg 1974. (= Sitzungsber. Heidelberger Akad. d. Wiss. 1974, 3, 39 Seiten. 17 Abbildungen auf 10 Tafeln).
- [Bespr.] Inscript. Graecae X 2, 1. Ed. Ch. Edson 1972. In: DLZ *94* (1973) Sp. 317–322.
- [Hrsg.] Lexikon zu den Dionysiaka des Nonnos. Herausgegeben von den Mitarbeitern der Abt. Klass. Philologie der Sekt. Orient- u. Altertumswiss. an d. Martin-Luther-Univ.unter Leitung von W. Peek. Berlin 1974. Lfg 3: *Λ — Π*. Sp. 897—1432.

Petschow, Herbert: Mittelbabylonische Rechts- und Wirtschaftsurkunden der Hilprecht-Sammlung Jena mit Beiträgen zum mittelbabylonischen Recht. Berlin 1974. 116 S., 2 Taf. (= Abhandlung Sächs. Akad. Wiss. zu Leipzig, phil.-hist. Kl. Bd 64, H. 4).
- Altorientalische Parallelen zur spätrömischen calumnia. In: ZSS, Rom. Abt. *90* (1973) S. 14—35.
- Zur mittelbabylonischen „Buchhaltungstechnik" und zur Tempelwirtschaft der NIN. DINGIR-Priesterinnen. In: Symbolae bibl. et mesopotamicae Francisco Mario Theodoro de Liagre Böhl dedicatae. Leiden 1973. S. 299—307.
- Havarie. In: Reallexikon der Assyriologie *IV* (Berlin 1973) Lfg 2—3, S. 233—237.
- [Hrsg.] M. San Nicolò: Die Schlußklauseln der altbabylonischen Kauf- und Tauschverträge. 2. Aufl. München 1974. 246 S. (= Münchener Beiträge zur Papyrusforschung und antiken Rechtsgeschichte, H. 4).
- [Rez.] Sh. M. Paul: Studies in the book of the Covenant in the light of cuneiform and biblical law. In: BO *XXX* (1973) H. 5—6.

Peukert, Herbert: O značaju Jakoba Grima za Vukovu koncepciju književnog jezika. In: Naučni sastanak slavista u Vukove dane. Referati i saopštenja. *2* (Beograd 1972) S. 147—153.
- O prihvatanju Vuka Stefanovića Karadžića u drugoj polovini XIX veka u Nemačkoj. Ebda *3* (1973) S. 309—321.
- O ritmu teksta i napeva u lirskim narodnim pesmama Kosova i Metohije. In: Rad XIV kongresa Saveza udruženja folklorista Jugoslavije u Prizrenu 1967. Beograd 1974. S. 83—85.
- Gibt es eine ‚slawische' Volksdichtung? In: Jahrb. Sächs. Akad. Wiss. zu Leipzig *1969—1970* (Berlin 1972) S. 102f.
- Rudolf Fischer 18. 12. 1910—2. 8. 1971. Ebda *1971—1972* (1974) S. 288—293.
- Probleme der Beziehungen Jakob Grimms zu den Slawen. Ebda S. 158f.
- [Bespr.] B. Koneski: Jazikot na makedonskata narodna poezija. Skopje 1971. In: ZfSl. *18* (Berlin 1973) S. 297—300.
- [Bespr.] W. Eschker: Untersuchungen zur Improvisation und Tradierung der Sevdalinka an Hand der sprachlichen Figuren. München 1971. Ebda S. 926—928.
- [Bespr.] Đ. Trifunović: Portret u srpskoj srednjovekovnoj književnosti. Kruševac 1971. Ebda S. 928f.
- [Bespr.] Serta slavica in memoriam Aloisii Schmaus. Gedenkschrift für Alois Schmaus. München 1971. Ebda *19* (1974) S. 427f.
- [Bespr.] Russkij fol'klor XIII. Russkaja narodnaja proza. Leningrad 1972. Ebda S. 687 bis 689.
- [Bespr.] M. Radovanović: Vuk Karadžić etnograf i folklorist. Beograd 1973. Ebda S. 689—691.
- [Bespr.] N. Milošević-Đorđević: Zajednička tematsko-sižejna osnova srpskohrvatskih neistorijskih epskih pesama i prozne tradicije. Beograd 1971. In: DLZ *94* (1973) Sp. 800—803.

Peukert, Herbert: [Bespr.] J. Hrabák, Polyglotta. Brno 1971. In: Referatedienst zur Literaturwissenschaft. Berlin 1973. S. 139f.
— [Bespr.] B. P. Gončarov, Zvukovaja organizacija sticha i problemy rifmy. Moskau 1973. Ebda 1974. S. 171f.
— [Bespr.] Z. Kopczyńskia, L. Pszołowska (Hrsg.): Metryka słowiańska. Warszawa, Wrocław, Kraków, Gdańsk 1971. Ebda S. 441—443.

Préaux, Claire: La lune dans la pensée grecque (= Mémoires de la Classe des Lettres et des Sciences Morales et Politiques de l'Académie Royale de Belgique). Brüssel 1973. 312 S.
— La place des papyrus dans les sources de l'histoire hellénistique. In: Akten des XIII. Internationalen Papyruskongresses Marburg/Lahn 2. bis 6. August 1971. Hrsg. von E. Kiessling und H. A. Rupprecht (= Münchener Beiträge zur Papyrusforschung und antiken Rechtsgeschichte, Heft 66). München 1974. S. 1—26.
— Rapport sur la 48e. Session de l'Union Académique Internationale tenue á Bruxelles, du 16 au 22 juin 1974. In: Bulletin de la Classes des Lettres et des Sciences Morales et Politiques de l'Académie Royale de Belgique. 5e série, *LX* (1974) fasc. 6—9, S. 181—195.
— Notice sur Jacques Pirenne. In: L'Annuaire de l'Académie Royale de Belgique *1974*, pp. 1—39 du tirage á part.

Průšek, Jaroslav: [Hrsg.] Dictionary of Oriental Literatures. 3 Bde. London 1974. 630 S.

Rudolph, Kurt: Mandean Sources. In: W. Foerster, Gnosis. A Selection of Gnostic Texts. Engl. Translation es by R. McL. Wilson. II. Coptic and Mandaic Sources. Oxford 1974. S. 121—319.
— Der Koran. Aus dem Arabischen von M. Henning. Einleitung S. 16—32. Textdurchsicht, Anmerkungen und Register. 3. Aufl. Leipzig 1974.
— [Hrsg.] und E. Walter: Walter Baetke: Kleine Schriften. Geschichte, Recht und Religion im altnordischen Schrifttum. Weimar 1973. 387 S.
— Gnosis und Gnostizismus, ein Forschungsbericht (Schluß). In: Theologische Rundschau NF *38* (1973) H. 1, S. 1—25.
— „Kritik einer ungerechtfertigten Kritik". Aus Anlaß einer Rezension: G. Krotkoff über R. Macuch. In: WZKM *63*/4 (1972/73) S. 306—311.
— Das Problem der Autonomie und Integrität der Religionswissenschaft. In: Ned. Theol. Tijdschr. *27* (1973) S. 105—131.
— „Historia Religionum". Bemerkungen zu einigen neueren Handbüchern der Religionsgeschichte. In: ThLZ *98* (1973) Sp. 401—418.
— Zum gegenwärtigen Stand der mandäischen Religionsgeschichte. In: Gnosis und Neues Testament. Hrsg. von K.-W. Tröger. Berlin 1973. S. 121—148.
— Die Bedeutung des Kölner Mani-Codex für die Manichäismusforschung. Vorläufige Anmerkungen. In: Mélanges d'histoire des religions offerts à Henri-Charles Puech. Vendôme 1974. S. 471—486.
— [Rez.] H. Räisänen: Das koranische Jesusbild. Helsinki 1971. In: ThLZ *98* (1973) Sp. 264—266.
— [Rez.] History of Religions, Vol. 11. In: OLZ *68* (1973) Sp. 535—537.
— [Rez.] Werblowsky-Bleeker (Ed.): Types of Redemption. Contributions to the Theme of the Study-Conference held at Jerusalem 14th to 19th July 1968. Leiden 1970. In: ThLZ *98* (1973) Sp. 658—662.
— [Rez.] P. Antes: Zur Theologie der Schi'a. Freiburg i. Br. 1971. In: Bibl. Orient. *XXX* (1973) S. 291—292.

Rudolph, Kurt: [Rez.] D. Krawulsky: Briefe und Reden des Abū Hāmid Muḥammad al-Gazzālī übersetzt und erläutert. Freiburg i. Br. 1971. Ebda S. 297–298.
– [Rez.] E. A. S. Butterworth: The Tree at the Navel of the Earth. Berlin 1970. In: DLZ *94* (1973) Sp. 808–810.
– [Rez.] Ex Orbe Religionum. Studia Geo Widengren Oblata. I und II. Leiden 1972. In: ThLZ *99* (1974) Sp. 93–95.
– [Rez.] A. D. Nock: Essays on Religion and the Ancient World. Oxford 1972. 2 Bde. Ebda Sp. 175–178.
– [Rez.] Th. O'Shaugnessy: Muhammad's Thoughts on Death. Leiden 1969. In: OLZ *69* (1974) Sp. 480–482.

Růžička, Rudolf: Integration slawischer und nichtslawischer Sprachen an der syntaktischen Peripherie. In: ZfS *18* (1973) S. 453–470.
– Reflexive oder nichtreflexive Pronominalisierung im modernen Russischen und in anderen slawischen Sprachen der Gegenwart? Ebda S. 737–779.
– Kto-to und Kto-nibud'. Ebda S. 705–736.
– Über die Einheitlichkeit der Modalität. In: Otázky slovanské syntaxe III, Opera Universitatis Purkynianae Brunensis Facultas Philosophica *180* (Brno 1973) S. 47–59 (Sborník symposia „Modální výstavba výpovědi v slovanských jazycích", Brno 27. bis 30. září 1971).
– Integration of Peripherical Syntactic Phenomena in Modern Standard Slavic and Non-Slavic Languages. In: The Prague Bulletin of Mathematical Linguistics *19* (1973) S. 5–6. Universita Karlova Praha.
– Zur Theorie der vergleichenden Syntax moderner slavischer Literatursprachen (am Material der Gerundien und der Nominalphrase). In: VII. Miedzynarodowy Kongres Slawistów w Warszawie 1973, Streszczenia referatów, S. 270–272.
– Reflexive versus Nonreflexive Pronominalzation in Modern Russian and other Slavic Languages. A Conflict between Domains of Rule Application. In: Generative Grammar in Europe, ed. by F. Kiefer and N. Ruwet. Foundations of Language Supplementary Series *13* (Dordrecht-Holland 1973) S. 445–481.
– Indefinite Pronouns and Quantification. In: Liczba, Ilość, Miara: Materiały Konferencji Naukowej w Jadwisinie, 11–13 maja 1972r. Wrocław, Warszawa, Kraków, Gdańsk 1973. S. 49–75.
– Generative Semantik und Translationsmodell. In: Neue Beiträge zu Grundfragen der Übersetzungswissenschaft. Leipzig 1973. S. 147–152.
– Zur Syntax und Semantik der Verben des modernen Russisch. In: E. Daum, W. Schenk: Die russischen Verben . 9. Aufl. Leipzig 1974. S. 691–742.
– Soobraženija o teoretičeskom u metodologičeskom obosnovanii sopostavite l'noj grammatiki. In: Československá rusistika *19* (1974) 4, S. 145–149.
– K voprosu o suščestvovanii osobych principov i metodov opisanija grammatičeskoj struktury slavjanskich jazykov. In: Grammatičeskoe opisanie slavjanskich jazykov. Moskau 1974. S. 8–12.
– Ob otnošenijach meždu semantičeskimi i grammatičeskimi svojstvami slov. In: Academie Serbe des Sciences et des Arts. Monographies *CDLXXVII* (Belgrade 1974) S. 49–53 (= Classe de Langue et de Litterature No 23).
– O sposobach postroenija opisatel'nych grammatik slavjanskich jazykov. Ebda S. 127 bis 130.
– u. G. Walther: Beziehungen zwischen Bedeutung und Syntax. In: ZfS *19* (1974) S. 460 bis 474.

R ů ž i č k a, Rudolf: [Hrsg.] Taschenwörterbuch Russisch-Deutsch. 11. Aufl. Leipzig 1973; 12. Aufl. Leipzig 1974. 323 S.

S c h m i d t, Eberhard: Die Mark Brandenburg unter den Askaniern (1134—1320). Mitteldeutsche Forschungen *71* (Köln, Wien 1973) XIV, 200 S.
— Zur Entwicklung des Inquisitionsprozesses in Baiern. In: Z. für die gesamte Strafrechtswissenschaft *85* (1973) H. 4, S. 851—866.
— Markgraf Otto I. von Brandenburg. In: ZSS, germ. Abt. *90* (1973) S. 1—9.
— Theodor Fontane und die Askanier. In: Festschrift für Wilhelm Gallas. Berlin 1973.
— Strafrechtliche Vorbeugungsmittel im Preußischen Allgemeinen Landrecht von 1794. In: Z. für die gesamte Strafrechtswissenschaft *86* (1974) H. 3, S. 621—625.
— Vom ältesten Recht Magdeburgs. In: Festschr. Karl Peters. Tübingen 1974. S. 603—609.
— [Rez.] Z. für die gesamte Strafrechtswissenschaft *86* (1974) H. 1, S. 150—159; H. 3, S. 748—757.

S c h u b a r t - F i k e n t s c h e r, Gertrud: Studienreform. Fragen von Leibniz bis Goethe. Berlin. 1973. 109 S. (= Sitzungsber. Sächs. Akad. Wiss. Leipzig, phil.-hist. Klasse Bd 116, H. 4).
— [Rez.] Otto Bardong: Die Breslauer an der Universität Frankfurt/Oder. Ein Beitrag zur schlesischen Bildungsgeschichte 1648—1811 (= Quellen und Darstellungen zur schlesischen Geschichte 14) Würzburg 1970. XXXIII, 368 S. In: ZSS, germ. Abt. *90* (1973) S. 364—365.
— [Rez.] Die Matrikel der Universität Jena II (1652—1723), Lfg 7: Ortsregister Be(rgneustadt)-He(imersdorf). Bearbeitet von Reinhold Jauernig (†), weitergeführt von Marga Steiger (= Veröff. des Historischen Instituts [seit 1968 Fachbereich Geschichte der Sektion Philosophie/Geschichte] der Universität Jena). Weimar 1970. S. 961—1112; III (1723—1764), 1. Teil: Text (chronologischer Abdruck) 1.—3. Lieferung: 1723—1746. Bearb. von Otto Köhler (= Veröff. der Universitätsbibliothek Jena). Halle/Saale 1969. 480 S. Ebda S. 365—366.
— [Rez.] Guido Kisch: Studien zur humanistischen Jurisprudenz. Berlin, New York 1972. 308 S. Ebda S. 366—368.
— [Rez.] Goethes Amtliche Schriften. Goethes Tätigkeit im Geheimen Consilium II: Die Schriften der Jahre 1788—1819, 2. Halbbd: 1798—1819. Bearb. von Helma Dahl. Weimar 1970. 474 S.; III: Erläuterungen zu den Schriften der Jahre 1788—1819. Bearb. von Helma Dahl. Ebda 1972. 508 S. Ebda S. 368—370.

T a c k e n b e r g, Kurt: [Art.] Beusterburg. In: Reallexikon der Germanischen Altertumskunde *2* (1973) Lfg 2/3, S. 322—323.
— u. J. Frechen: Ein Gefäß der danubisch-sudetischen Hügelgräberbronzezeit von der Elbemündung. In: Die Kunde NF *24* (1973) S. 63—67.
— [Bespr.] J. Bergmann: Die ältere Bronzezeit Nordwestdeutschlands. 1970. In: Bonner Jahrbücher *172* (1973) S. 601—605.
— [Bespr.] F. Laux: Die Bronzezeit in der Lüneburger Heide. 1971. In: Prähist. Z. *47* (1973) S. 229—234.
— [Bespr.] W. Korzus: Die Fundmünzen der römischen Kaiserzeit in Deutschland. Abt. VI. — Nordrhein Westfalen 4 — Münster 1971. Ebda S. 252—253.
— [Bespr.] I. i K. Dabrowscy: Cmentarzysko z okresów póznolateúskiego i wplywów rzymskich w Wesólkach. 1967. In: Erasmus *25* (1973) Sp. 811—812.
— [Bespr.] W. John; P. Quinctilius Varus und die Schlacht im Teuteburger Walde, in: Pauly-Wissowa, Realencyclopädie der classischen Altertumswissenschaft, Bd 24, 1963,

S. 907–984. In: Lippische Mitteilungen aus Geschichte und Landeskunde *43* (1974 S. 321–323.

Tackenberg, Kurt: [Hrsg.] u. K. J. Narr: Münstersche Beiträge zur Vor- und Frühgeschichte. Bd *7*: H. G. Stephan: Archäologische Beiträge zur Frühgeschichte der Stadt Hötxer. 35 S., 30 Taf.; Bd *8*: V. Zedelius: Spätkaiserzeitlich — völkerwanderungszeitliche Keramik und römische Münzen von Laatzen, Landkreis Hannover. 56 S., 9 Taf., 1 Karte.
— [Hrsg.] u. A. Alföldi, J. Straub: Antiquitas, Reihe 3: Abhandlungen zur Vor- und Frühgeschichte, zur klassischen und provinzial-römischen Archäologie und zur Geschichte des Altertums. Bd *13*: M. Korfmann: Schleuder und Bogen in Südwestasien. 282 S., 12 Taf. Chronologietabelle, 19 vierfarb. Faltktn.; Bd *14*: A. Alföldi: Die zwei Lorbeerbäume des Augustus. X, 65 S., 32 Taf.; Bd *15*: H. Roth: Die Ornamentik der Langobarden in Italien. Eine Untersuchung zur Stilentwicklung an Hand der Grabfunde. XV, 306 S., 1 Tabelle, 4 Ktn.; Bd *17*: A. Alföldi: Caesar in 44 v. Chr. Bd 2: Das Zeugnis der Münzen. Mit einer Revision der Stempel und Stempelverbindungen von Wendelin Kellner. XIV, 83 S., 156 Taf.

Vossler, Otto: Alexis de Tocqueville. Frankfurt/M. 1973. 279 S. (= Das Abendland — NF 2).

von Weizsäcker, Carl Friedrich: Was wird aus dem Menschen? Zürich 1972. 46 S. (= Arche nova).
— Bedingungen des Friedens. Mit der Laudatio von Georg Picht anläßlich der Verleihung des Friedenspreises des deutschen Buchhandels 1963 an Carl Friedrich von Weizsäcker. 6. Aufl. Göttingen 1973. 37 S.
— Die Einheit der Natur. Studien. 5. Aufl. Stuttgart, Hamburg, München 1973. 491 S.
— Die Tragweite der Wissenschaft. 4., unveränderte Aufl. Stuttgart 1973. XI, 243 S.
— Der Gedanke der Einheit der Natur. Festvortrag in der Feierlichen Sitzung der Österreichischen Akademie der Wissenschaften am 17. Mai 1972. In: Almanach der Österreichischen Akademie der Wissenschaften *1972*. Wien 1973. S. 40–59.
— Information und Imagination. In: Information und Imagination. Vorträge von C. F. von Weizsäcker, Golo Mann, H. Weinrich, Th. Sieverts, L. Kolakowski, gehalten vor der Bayerischen Akademie der Schönen Künste im Mai 1973. München 1973. S. 11–32. (= Serie Piper, 75).
— Die Aktualität der Tradition: Platons Logik. Vortrag, gehalten am 19. März 1973 in München vor der Schelling-Kommission der Bayerischen Akademie der Wissenschaften. In: Philosophisches Jahrbuch *80* (1973) 2. Halbband, S. 221–242.
— Classical and Quantum Descriptions. In: The Physicist's Conception of Nature. Hrsg. von Jagdish Mehra. Dordrecht/Holland 1973. S. 635–667.
— Comment on Dirac's Paper. Ebda S. 55–59.
— Grenzen des Wachstums. In: Naturwissenschaften *60* (1973) H. 6, S. 267–273.
— Die Herstellung des Friedens und die Weltpolitik [Diese Ausführungen wurden für einen Vortrag in Oxford geschrieben.]. In: Universitas *28* (1973) H. 12, S. 1271–1276.
— Physics and Philosophy. In: The Physicist's Conception of Nature. Hrsg. von Jagdish Mehra. Dordrecht/Holand 1973. S. 736–746.
— Probability and Quantum Mechanics. In: British Journal for Philosophy of Science *24* (1973) S. 321–337.
— Reinhold Schneider in unserer Zeit. Vortrag, gehalten am 13. Mai 1973 in Freiburg/Br.

vor der Reinhold-Schneider-Gesellschaft. In: Merkur *27* (Stuttgart 1973) H. 9, S. 813 bis 826.

von Weizsäcker, Carl Friedrich: Über die Kunst der Prognose. In: Kursbuch ins dritte Jahrtausend. Hrsg. von Alwin Toffler. München 1973. S. 187—205.
— [Vorwort] zu J. Jauch: Die Wirklichkeit der Quanten. Ein zeitgenössischer galileischer Dialog. München 1973. S. 7—9.
— Geometrie und Physik. In: Physical Reality and Mathematical Description. Hrsg. von Charles Enz und Jagdish Mehra. Dordrecht/Holland 1974. S. 48—90.
— Der Zusammenhang der Quantentheorie elementarer Felder mit der Kosmologie. In: Grundfragen der Quanten- und Relativitätstheorie. Nova Acta Leopoldina NF *39* (Halle/S. 1974) Nr. 212, S. 61—80.
— Notizen zur Ideologiekritik. In: Merkur *28* (Stuttgart 1974) H. 9, S. 807—809.
— Rede am 20. Juli 1974. Ebda H. 10, S. 953—960.
— Wissenschaftsgeschichte als Wissenschaftstheorie. Zur Frage nach der Rolle der Gesellschaft in der Wissenschaft (= Festvortrag anläßlich der Mitgliederversammlung 1974 in Berlin des Stifterverbandes für die Deutsche Wissenschaft). In: Wirtschaft und Wissenschaft, Sept. 1974. S. 5—11; Merkur *29* (Stuttgart 1975) H. 2.
— Reinhold Schneider in unserer Zeit. In: Reinhold Schneider, Carl Friedrich von Weizsäcker, Immanuel Kant: Der Friede der Welt — Schicksal der Menschheit. Hrsg. von der Reinhold-Schneider-Stiftung. Freiburg/Br. 1974. S. 95—112.
— Die heutige Menschheit von außen betrachtet. Ein Versuch. In: Ev. Komm. *7* (1974) S. 328—336, 392—396; Merkur *28* (Stuttgart 1974) S. 505—518, 607—616.
— Einführung zu dem vom MPIB hrsg. Buch „Elementarmathematik: Lernen für die Praxis?", Stuttgart 1974. S. 11—12.

Weller, Friedrich: Bemerkungen zum 7. Kapitel des Dharmasamuccaya. In: Wissenschaftliche Zeitschrift der Universität Halle *XXII* (1973), Gesellschafts- und Sprachwissenschaftliche Reihe, Heft 3, S. 77—95.
— Der Arme Heinrich in Indien. In: OLZ *68* (1973) Spalte 437—448.
— [Bespr.] Mélanges d' Indianisme à la Mémoire de Louis Renou. Paris: Boccard 1968. XXXIX, 800 S., 1 Taf. 4° = Anniversaire de la fondation de l'Institut de Civilisation Indienne de l' Université de Paris 1967 = Publications de l' Institut de Civilisation Indienne, 28. In: OLZ *68* (1973) Spalte 67—73.
— [Bespr.] Emmerick, R. E.: The Sūtra of Golden Light. Being a Translation of the Suvarṇabhāsottamasūtra. London: Luzac 1970. XIII, 108 S. 8° = Sacred Books of the Buddhists, XXVII. In: OLZ *69* (1974) Spalte 387—393.
— [Bespr.] Groslier, Bernard P., Indochina. Deutsche Bearbeitung: G. Gabbert. München—Genf—Paris: Nagel Verlag [1966]. 286 S. m. 145 Abb. Taf. davon 35 farb., 1 Tab. gr. 8° = Archaeologia Mundi, hrsg. v. J. Marcadé. In: OLZ *69* (1974) Spalte 501—503.
— [Ber.] Harvard Journal of Asiatic Studies 30 1970. In: OLZ *68* (1973) Spalte 212—213.
— [Ber.] Harvard Journal of Asiatic Studies 31 1971. Ebda. Spalte 213—215.
— [Ber.] Oriens Extremus 16 1969, Heft 2. Ebda. Spalte 220—221.
— [Ber.] Indo-Iranian Journal 13 1971. Ebda. 314—316.
— [Ber.] East and West, N. S. 18 1968. In: OLZ *69* (1974) Spalte 303—307.
— [Ber.] Wiener Zeitschrift für die Kunde Südasiens und Archiv für Indische Philosophie 14 1970. Ebda. Spalte 425—426.
— [Ber.] Wiener Zeitschrift für die Kunde Südasiens und Archiv für Indische Philosophie 15 1971. Ebda. Spalte 426—427.

Werner, Ernst: Zwischen Canossa und Worms. Staat und Kirche 1077—1122. Berlin 1973. 203 S., 15 Abb., 2 Ktn.

— Ketzer und Weltverbesserer. Zwei Beiträge zur Geschichte Südosteuropas im 13. und 15. Jahrhundert. Berlin 1974. 57 S. (= Sitzungsber. Sächs. Akad. Wiss. zu Leipzig, phil.-hist. Klasse Bd 116, H. 5).

— Die Armeniergreuel 1915—1916 — ein armenisches Greuelmärchen? In: ZfG *21* (1973) S. 218—225 und in: Vestnik Archivov Armenii *1*, 38, (1974) S. 137—143 [russ.].

— Popular Ideologies in Late Medieval Europe: Taborite Chiliasm and its Antecedents. In: Western Civilization. Mainstream readings and radical critiques, ed. J. Kaplow, vol. *I* (New York 1973) S. 177—191.

— Methoden und Möglichkeiten komparativer Mediävistik. In: ZfG *21* (1973) S. 542 bis 547.

— Grenzscheide 11. Jahrhundert. In: Antičnaja drevnost'i Srednie Veka, Sbornik 10. Sverdlovsk 1973. S. 33—38.

— Tanzimat-Staatsreform und Aufklärung (münevver) 1839—1860. In: Aspekte der Aufklärungsbewegung in Lateinamerika, Deutschland, Rußland und der Türkei. In: Sitzungsber. des Plenums und der Klassen der AdW der DDR *1972* (Berlin 1974) Nr. 10, S. 49 bis 59.

— Spätbogomilisch-adamitische Spekulationen und Praktiken in religionshistorischer Sicht. In: Byzantine Studies *I* (Pittsburgh 1974) S. 40—53.

— Visione marxista del monachesimo bizantino. In: Dizionario degli Istituti di perfezione. Roma 1974. Sp. 1474—1476.

— Ökonomische und militärische Aspekte der Türkeipolitik Österreich—Ungarns 1915 bis 1918. In: Jahrbuch für Geschichte *10* (Berlin 1974) S. 373—401.

Wieacker, Franz: Industriegesellschaft und Privatrechtsordnung. Frankfurt/Main 1974 93 S. (= Fischer Athenäum Taschenbücher 6014) [enthält vier Aufsätze von 1953, 1960 1966 und 1970].

— Zur Topikdiskussion in der zeitgenössischen deutschen Rechtswissenschaft. In: Xenion. Festschrift für Pan. J. Zepos anläßlich seines 65. Geburtstages am 1. Dezember 1973. *I* (Athen 1973) S. 391—415.

— I „libri opinium" (di Ulpiano?). In: Labeo Rassegna di diritto romano *19* (Napoli 1973) 2, S. 196—207.

— Contractus und Obligatio im Naturrecht zwischen Spätscholastik und Aufklärung. In: La seconda scolastica nella formazione del diritto privato moderno. Milano 1973. S. 223—239.

— Bemerkungen über Ihering und den Darwinismus. In: Biologismus im 19. Jahrhundert. Hrsg. von Gunter Mann. Stuttgart 1973. S. 158—163.

— [Bespr.] Klaus Adomeit: Rechtsquellenfragen im Arbeitsrecht. München 1969. XXIX, 176 S. In: Rechtstheorie. Z. für Logik, Methodenlehre, Kybernetik und Soziologie des Rechts *2* (Berlin 1971) S. 118—123.

NACHRUFE
UND
SCHRIFTENVERZEICHNISSE

Rolf Emmrich

26. 8. 1910—5. 5. 1974

Im Alter von 63 Jahren verstarb am 5. Mai 1974 in Leipzig das ordentliche Mitglied und der langjährige Sekretar der mathematisch-naturwissenschaftlichen Klasse der Sächsischen Akademie der Wissenschaften, der ordentliche Professor Dr. sc. med. Rolf Emmrich, Direktor der Medizinischen Klinik der Karl-Marx-Universität Leipzig.

Er wurde am 26. August 1910 in Freiberg/Sa. als Sohn eines Baumeisters geboren. Nach Ablegung der Reifeprüfung im Jahre 1930 am humanistischen Gymnasium seiner Heimatstadt nahm er das Medizinstudium in Freiburg/Breisgau und in Tübingen auf, wo er es auch 1935 mit dem medizinischen Staatsexamen beendete. Während seiner Medizinalpraktikantenzeit war er am Kreiskrankenhaus Freiberg, an der Universitäts-Kinderklinik Tübingen und am Pathologischen Institut der Universität München tätig. Sein Promotionsverfahren schloß er mit der Note sehr gut ab. Auf Grund seines besonderen Interesses an der Chemie ging er 1937 an das Chemische Institut und Physiologisch-Chemische Institut der Universität Leipzig, um fundierte theoretische wissenschaftliche Kenntnisse zu erwerben, die ihm während seiner späteren klinischen Tätigkeit stets zugute kamen. Seine Ausbildung zum Facharzt für Innere Medizin erfolgte an der Medizinischen Universitätsklinik Frankfurt/Main, während seines Kriegsdienstes an verschiedenen Lazaretten und schließlich an der Medizinischen Poliklinik der Martin-Luther-Universität Halle—Wittenberg. Seiner Lehrer, der Professoren Thomas, Hein, Nonnenbruch und Ratschow, gedachte er immer in großer Verehrung. Sie waren ihm Vorbild und Ansporn, ihre wissenschaftlichen Arbeiten durch eigene neue Gedanken und Forschungen fortzuführen.

Mit der Arbeit „Makromolekulare Struktureigenschaften der Serumproteine und ihre Bedeutung für die Klinik“ habilitierte er sich im Jahre 1948 und erhielt am 1. Januar 1949 die Dozentur für das Fachgebiet Innere Medizin. Seine Ernennung zum Professor mit Lehrauftrag erfolgte am 1. April 1953, nachdem ihm Mitte 1952 die kommissarische Leitung der II. Medizinischen Klinik und Poliklinik der Martin-Luther-Universität Halle—Wittenberg übertragen worden war. 1954 wurde er an die neugegründete Medizinische Akademie Magdeburg als Professor mit Lehrstuhl für Innere Medizin berufen. Die damit verbundene Aufgabe, eine städtische Medizinische Klinik in eine moderne, den hohen Anforderungen des Medizinstudiums entsprechende Hochschulklinik umzuprofilieren, bewältigte er in relativ kurzer Zeit mit großer Zielstrebigkeit.

1959 erreichte Rolf Emmrich der ehrenvolle Ruf, das Direktorat der Medizinischen Klinik der Karl-Marx-Universität Leipzig als Nachfolger von Max Bürger zu übernehmen, einer Klinik, deren Leitung so bekannte und hervorragende Internisten wie Wunderlich, Curschmann, Strümpell und Morawitz einmal innehatten. Dieser neuen Tätigkeit widmete er sich mit der ganzen Kraft seiner Persönlichkeit, seinem Können und seinem Fleiße. Die zahllosen Aufgaben, Verpflichtungen und Probleme, die mit der Führung einer so großen und traditionsreichen Klinik verbunden sind, löste er mit der ihm eigenen Umsicht, Gewissenhaftigkeit, Beharrlichkeit und außergewöhnlichen Arbeitsamkeit.

Seine ruhige, ausgeglichene und ausgleichende Art trotz der vielen hohen Anforderungen, die an ihn gestellt wurden und die er auch an sich selbst bis zuletzt stellte, verlangte allen seinen Mitarbeitern und Kollegen Bewunderung und Respekt ab. Sein Rat wurde stets gesucht und geachtet, sei es als Konsilarius bei schwierigen diagnostischen und therapeutischen Krankheitsfällen, sei es in Fragen der Reform des Medizinstudiums, der Facharztausbildung oder der postgraduellen Weiterbildung, sei es in Fragen der Gründung und Entwicklung der medizinischen Fachgesellschaften oder sei es in Fragen der Wissenschaftentwicklung. Seine Tätigkeit war so vielseitig, daß sie nur von denen richtig gewürdigt werden kann, die ihn in seiner täglichen Arbeit ständig erlebten.

In seiner Tätigkeit als Hochschullehrer verstand er es, durch sein ärztliches Vorbild und durch seine wohlvorbereiteten, didaktisch ausgezeichneten Vorlesungen, in denen er neueste wissenschaftliche Erkenntnisse in Verbindung zur klinischen Praxis vermittelte, viele Studenten für die Innere Medizin so zu begeistern, daß viele für ihre ärztliche Tätigkeit dieses Fachgebiet wählten.

In seiner eigenen wissenschaftlichen Arbeit war Rolf Emmrich äußerst vielseitig und bestrebt, die gesamte Innere Medizin zu umfassen. Er verkannte nicht die Notwendigkeit der Spezialisierung und Subspezialisierung der Inneren Medizin für den Fortschritt in Forschung und Betreuung, was sich auch in der Struktur der von ihm geleiteten Klinik ablesen läßt. Er betonte aber auch die Notwendigkeit der Integration und Synthesis in bezug auf die Wissenschaftsorganisation, die Gliederung von Fachbereichen, die Lehre überschaubarer Unterrichtsfächer, die Tätigkeit des Facharztes und des Forschers, die stets den Menschen in seiner Gesamtheit und seiner Umwelt und nicht nur ein Organ oder Organsystem als Ziel und Aufgabe ihres ärztlichen Tuns und Handelns vor sich haben.

Das Ergebnis der umfassenden und fruchtbaren wissenschaftlichen Tätigkeit von Rolf Emmrich kann man annähernd ermessen an seinen über 250 veröffentlichten Originalarbeiten, Übersichten und Referaten sowie seinen zahlreichen Beiträgen zu Hand- und Lehrbüchern, die sich vor allem mit Herz-Kreislaufkrankheiten, Erkrankungen der Leber, Gerinnungsstörungen, dem Eiweißstoffwechsel, der Gerontologie und dem Schmerz und seiner Bekämpfung befaßten. Großes Interesse fanden seine in mehreren Auflagen erschienenen Monographien „Das Bluteiweißbild“, „Die chronischen Krankheiten des Bindegewebes“, „Schock und Schockbehandlung“ sowie die von ihm in mehreren Bänden herausgegebenen „Arbeitsmethoden der Inneren Medizin und ihr verwandter Gebiete“, ein Handbuch, das dem Internisten das notwendige technische Wissen für sein spezielles Fachgebiet in die Hand gibt. Sein ganz besonderes Anliegen war die „Zeitschrift für die gesamte Innere

ROLF EMMRICH

26. 8. 1910–5. 5. 1974

Medizin und ihre Grenzgebiete", für die er seit 1963 als Mitherausgeber tätig war und der er ein Profil gab, das ihr internationale Anerkennung verschaffte und dazu beitrug, die Erfolge unseres sozialistischen Gesundheitswesens publizistisch zu verbreiten.

Entscheidende Anstöße und Beiträge leistete Rolf Emmrich für die Neubegründung der medizinischen Fachgesellschaften. So war er Mitbegründer der Gesellschaft für Innere Medizin der DDR, mehrere Jahre lang ihr erster Vorsitzender und als solcher Initiator der international beachteten Internistenkongresse in Leipzig, was auch durch die Verleihung der Ehrenmitgliedschaft dieser Ge sellschaft gebührend gewürdigt wurde. Internationales Interesse fanden die auf seinen Ideen beruhenden und von ihm viermal veranstalteten Symposien „Gefäßwand und Blutplasma", ein Forschungsgebiet, dem er sich in den letzten Jahren besonders gewidmet hatte.

Für seine Schüler und Mitarbeiter war das Wirken von Rolf Emmrich stets Vorbild, Verpflichtung und Aufgabe, ihm mit eigenen Leistungen nachzueifern. Er erwartete von ihnen Fleiß und Gewissenhaftigkeit in ihrer Arbeit, kritische Überprüfung der erzielten Ergebnisse und exakte klare Gestaltung und Formulierung, die gleichen Maßstäbe, die er auch seiner eigenen wissenschaftlichen und ärztlichen Tätigkeit auferlegte. Mit großer Freude konnte er verzeichnen, daß zehn seiner Schüler zu Professoren ernannt wurden und zahlreiche andere leitende Stellungen im staatlichen Gesundheitswesen bekleiden.

In Würdigung seiner Leistungen und Verdienste wurde Rolf Emmrich 1963 ordentliches Mitglied der Deutschen Akademie der Naturforscher Leopoldina zu Halle (Saale) und im gleichen Jahr auch Fellow of the International College of Angiology.

Die Sächsische Akademie der Wissenschaften zu Leipzig wählte ihn am 11. Dezember 1961 zum ordentlichen Mitglied ihrer mathematisch-naturwissenschaftlichen Klasse, deren Sekretar er von 1965 bis 1972 war. Er leitete das Laboratorium für Alternsforschung, das sich mit den natürlichen Prozessen des Alterns, aber auch mit den pathologischen Vorgängen im Alter, insbesondere den Ursachen, der Erkennung, den Erscheinungen und der Behandlung der Arteriosklerose befaßte. Die wichtigsten Erkenntnisse hat er in den Sitzungsberichten der Akademie veröffentlicht. Als eine der letzten Publikationen, die erst nach seinem Dahinscheiden gedruckt vorlag, verfaßte er die Arbeit „Zwischen Leben und Tod — Ärztliche Probleme der Thanatologie". Darin setzte er sich mit Fragen der modernen Intensivmedizin auseinander. Vielleicht in Vorahnung seines nahen Ablebens schrieb er: „Es ist allem Leben gegeben, einmal zu enden, und kein Mensch kennt zuvor seinen Ausgang aus dem irdischen Dasein. Der plötzliche und schmerzlose Tod wird als Gnade gepriesen, ein lange währendes Siechtum gefürchtet." Und diese Gnade ist ihm zuteil geworden.

Sein Ableben hat eine schwer zu schließende Lücke für die medizinischen Wissenschaften hinterlassen. Mit Rolf Emmrich verlor sie einen wirklich begnadeten Humanisten, Mediziner, Hochschullehrer und Wissenschaftler, dessen überragende Arztpersönlichkeit in jeder Phase seines Wirkens der traditionsreichen und international anerkannten Leipziger Medizinischen Universitätsklinik würdig war.

Prof. Dr. sc. med. Heinz Trenckmann

Bibliographie Rolf Emmrich

I. Monographien und Buchbeiträge

1. Das Bluteiweißbild. Diagnose und Therapie der Bluteiweißstörungen. Stuttgart 1952. 100 S.
2. Das Bluteiweißbild. Diagnose und Therapie der Bluteiweißstörungen. 2., völlig neu bearbeitete Auflage. Stuttgart 1957. 227 S.
3. Gefäßerkrankungen und Durchblutungsstörungen, in: Differentialdiagnose innerer Erkrankungen. Hrsg. von F. Steigerwaldt und H. Braun. Stuttgart—Wien—Zürich 1958. S. 180—200.
4. Chronische Krankheiten des Bindegewebes. Beiträge zur Pathologie und Klinik der Organ- und Systemsklerosen. Leipzig 1959. 280 S.
5. Blutzell- und Bluteiweißveränderungen bei peripheren Durchblutungsstörungen, in: Angiologie. Hrsg. von M. Ratschow, Stuttgart 1959. S. 255—266.
6. Lehrbuch der Inneren Medizin, hrsg. von A. Sundermann. Kapitel: Erkrankungen der Nieren, der ableitenden Harnwege und der männlichen Geschlechtsorgane, Bd I; Kapitel: Periphere Durchblutungsstörungen und Krankheiten der peripheren Gefäße. Bd II. *I* (Jena 1961) S. 637—779; *II* (1962) S. 343—386.
7. Pathophysiologie der inneren Sekretion, in: Lehrbuch der Pathologischen Physiologie. Hrsg. von E. Götze. Jena 1962. S. 607—676.
8. Chronische Krankheiten des Bindegewebes. Beiträge zur Pathologie und Klinik der Organ- und Systemsklerosen. 2. Aufl. Leipzig 1961. 406 S.
9. Taschenbuch klinischer Funktionsprüfungen, hrsg. von A. Gitter und L. Heilmeyer. Kapitel: Funktionsprüfungen der peripheren Gefäße, S. 98—125, Kapitel: Nierenfunktionsprüfungen, S. 345—381. 8. Aufl. Jena 1963.
10. Arbeitsmethoden der Inneren Medizin. Bd I, 1. Lieferung, Kapitel: Die allgemeine klinische Untersuchung. Jena 1964. S. 17—52.
11. Pathophysiologie der inneren Sekretion, in: Lehrbuch der Pathologischen Physiologie. Hrsg. von E. Götze. 2. Aufl. Jena 1964. S. 653—724.
12. Lehrbuch der Inneren Medizin, hrsg. von A. Sundermann. Kapitel: Erkrankungen der Nieren, der ableitenden Harnwege und der männlichen Geschlechtsorgane, Bd I; Kapitel: Periphere Durchblutungsstörungen und Krankheiten der peripheren Gefäße, Bd II. 2. Aufl. *I* (Jena 1964) S. 653—797; *II* (1965) S. 209—254.
13. Handlexikon der Medizinischen Praxis. Bd IV: Internistische Differentialdiagnose, hrsg. von H. Braun. Kapitel: Gefäßerkrankungen und Durchblutungsstörungen. Stuttgart—Wien 1965. S. 189—213.
14. u. H. Petzold: Laparoskopie und Leberpunktion, in: Arbeitsmethoden der Inneren Medizin. Bd *III* (Jena 1966) S. 249—304.
15. Bluteiweiß und Bluteiweißstörungen. Leipzig 1966. 226 S.
16. Lehrbuch der inneren Medizin, hrsg. von A. Sundermann. Bd I, Kapitel:Erkrankungen der Nieren, der ableitenden Harnwege und der männlichen Geschlechtsorgane. 3. Aufl. Jena 1969. S. 661—810, Bd II, Kapitel: Periphere Durchblutungsstörungen und Krankheiten der peripheren Gefäße. Jena 1970. S. 421—470.
17. u. W. Lembcke: Schock und Schockbehandlung. Leipzig 1970. 148 S.
18. Arbeitsmethoden der Inneren Medizin. Bd VIII, Therapeutische Verfahren, 1. Lieferung Allgemeine therapeutische Arbeitsmethoden. Jena 1971. 99 S.

19. Diagnostisch-therapeutisches Vademecum. 41. Aufl. Leipzig 1972. Kapitel: Krankheiten der Nieren und der ableitenden Harnwege, S. 176—195; Kapitel: Rheumatische Erkrankungen, S. 304—312; Kapitel: Stoffwechselkrankheiten, S. 313—328; Kapitel: Endokrines System, S. 329—344.
20. u. W. Lembcke: Schock und Schockbehandlung. 3. überarb. Aufl. Leipzig 1974.

II. Veröffentlichungen in Zeitschriften und Kongreßberichten

1. Spina bifida und Familien mit Spina bifida (= Inaugural-Dissertation) Tübingen 1937. 66 S.
2. Spina bifida und Bettnässen. Kinderhk. *68* (1937) 87—93.
3. Stoffwechselversuche mit einigen methylierten niedermolekularen Dicarbonsäuren. Hoppe-Seyler's Z. physiol. Chem. *261* (1939) 61—70.
4. u. E. Höhne: Bestimmung ätherlöslicher Säuren im Harn. Ber. d. math.-phys. Klasse d. Sächs. Akad. d. Wiss. *XCII* (1940) 1—18.
5. u. E. Nebe: Stoffwechselversuche mit natürlichem Kokosfett und einem aus Kokosfett bereiteten Fett mit ungeradzahligen Fettsäuren Hoppe-Seyler's Z physiol. Chem. *266* (1940) 174—182.
6. u. Ilse Emmrich-Glaser: Stoffwechselversuche mit höheren geradzahligen gesättigten Dicarbonsäuren. Hoppe-Seyler's Z. physiol. Chem. *266* (1940) 183—192.
7. u. Ilse Emmrich-Glaser: Ein Beitrag zur w-Oxydation. Ber. d. math.-phys. Klasse d. Sächs. Akad. d. Wiss. *XCIII* (1941) 1—8.
8. u. P. Neumann, Ilse Emmrich-Glaser: Über das Verhalten einiger alkylierter Malonsäuren, Bernsteinsäuren und Glutarsäuren im Tierkörper. Hoppe-Seyler's Z. physiol. Chem. *267* (1941) 228—241.
9. Über das Verhalten der Sebacinsäure im Stoffwechsel des kranken Menschen. Dtsch. Arch. klin. Med. *187* (1941) 504—518.
10. Leberfunktionsprobe mit p-Oxyphenylbrenztraubensäure. I. Mitt. Z. exper. Med. *109* (1941) 604—617.
11. Leberfunktionsprobe mit p-Oxyphenylbrenztraubensäure. II. Mitt. Z. exper. Med. *110* (1942) 667—677.
12. Anginabehandlung mit Sulfapyridin. Münch. med. Wschr. *89* (1942) 1096—1097.
13. Die Flockungsreaktion nach GROS im Verlaufe der Hepatitis infectiosa. Dtsch. Militärarzt *9* (1944) 8—12.
14. Zur Kenntnis des Erythema nodosum. Dtsch. Ges.wesen *3* (1948) 111—115.
15. Über die makromolekulare Chemie der Serumproteine. I. Mitt. Z. inn. Med. *3* (1948) 162—165.
16. Über die makromolekulare Chemie der Serumproteine. II. Mitt. Z. inn. Med. *3* (1948) 165—170.
17. Über die makromolekulare Chemie der Serumproteine. III. Mitt. Z. inn. Med. *3* (1948) 242—249.
18. Über die makromolekulare Chemie der Serumproteine. IV. Mitt. Z. inn. Med. *3* (1948) 312—315.
19. Über das Vorkommen von Bernsteinsäure im Harn Leberkranker. Klin. Wschr. *26* (1948) 659—660.
20. Das hepatorenale Syndrom und die Hyposthenurie „H" bei Kreislaufkranken. Z. inn. Med. *3* (1948) 718—728.

21. u. M. Ratschow: Differentialtherapie mit Eiweißpräparaten. Dtsch. Ges.wesen *4* (1949) 475—478.
22. Die Bedeutung der Fette für die Ernährung. Urania *12* (1949) 147—149.
23. Humusstoffe zur internen Therapie. Ärztl. Wschr. *4* (1949) 378—379.
24. Zur Pathogenese der diabetischen Glomerulosklerose. Dtsch. Arch. klin. Med. *196* (1949) 116—122.
25. Über die makromolekulare Chemie der Serumproteine. V. Mitt. Z. inn. Med. *4* (1949) 178—181.
26. Strukturveränderungen der Serumproteine. Verh. dtsch. Ges. inn. Med. Wiesbaden *55* (1949) 299—303.
27. Über die makromolekulare Chemie der Serumproteine. VI. Mitt. Z. inn. Med. *4* (1949) 604—610.
28. Über den Cystingehalt normaler und pathologischer Serumproteine. Z. inn. Med. *5* (1950) 402—407.
29. Beeinflussung des Albumin- und Globulincystingehaltes der Serumproteine durch perorale Cystingaben. Klin. Wschr. *28* (1950) 563—565.
30. Hepatitis epidemica und Icterus catarrhalis. Neue Med. Welt *1950.* 1265—1268.
31. u. K. Wandel: Veränderungen der Serumproteine nach Benzol- und Röntgenschädigung der Ratte. Naunyn-Schmiedebergs Arch. exper. Path. *210* (1950) 236—246.
32. u. H. zur Horst-Meyer: Zur Pathogenese und Therapie der Sklerodermie und ihr verwandter Krankheitsbilder. Z. klin. Med. *146* (1950) 578—591.
33. Die Konstellationstypen im Protogramm. Dtsch. Ges.wesen *5* (1950) 871—876.
34. Beeinflussung des Cystingehaltes der Serumalbumine und -globuline durch Methionin. Klin. Wschr. *29* (1951) 82—85.
35. u. Elfriede Pötzsch: Diabetes mellitus, nephrotisches Syndrom und Hypertonie. Dtsch. Z. Verdauungskrkh. *11* (1951) 123—130.
36. Kritisches zur Therapie mit Aminosäuren bei Herzinsuffizienz. Therapiewoche *2* (1951/52) 449—451.
37. u. H.-J. Keutel, A. Morgenstern: Über den Ponndorf-Impfstoff und seine Wirkungsweise. Naunyn-Schmiedebergs Arch. exper. Path. *213* (1951) 94—101.
38. Myokardose und hämodynamisch-dystrophische Insuffizienz. Z. klin. Med. *148* (1951) 279—287.
39. Die Papierelektrophorese zur Trennung der Serumeiweißkörper. Dtsch. Ges.wesen *6* (1961) 585—588.
40. u. H. zur Horst-Meyer: Über die Depotwirkung von Hypophysenhinterlappen-Transplantaten bei Diabetes insipidus. Klin. Wschr. *29* (1951) 553—554.
41. u. H. Fliege: Die Aszites-Reinfusion als Therapie der dekompensierten Leberzirrhose. Med. Welt *1951*, 931—932.
42. H. Bödecker u. R. Emmrich: Über die Beeinflussung der Kapillarfragilität durch Homoseran. Z. inn. Med. *6* (1951) 747—749.
43. Das Protogramm bei peripheren Durchblutungsstörungen. Verh. dtsch. Ges. Kreislaufforsch. *17* (1951) 265—268.
44. Über die makromolekulare Chemie der Serumproteine. VII. Mitt. Z. inn. Med. *7* (1952) 316—318.
45. u. H. Petzold: Über die Wirkung lipotroper Stoffe auf die Regeneration der Rattenleber. Naunyn-Schmiedebergs Arch. exper. Path. *214* (1952) 333—340.
46. Therapie der chronischen Polyarthritis mit Eiweiß und Eiweißabbaustoffen unter Berücksichtigung der vegetativen Ausgangslage. Therapiewoche *2* (1951/52) 325—358.

47. u. H. Scheffler: Die „alkalische“ und die „saure“ Serum-Phosphatase in verschiedenen Lebensaltern. Z. Altersforsch. *VI* (1952) 156—159.
48. u. H. Petzold: Über den Lokalisatoreffekt in der Pathogenese der experimentellen Endangiitis. Klin. Wschr. *30* (1952) 488—492.
49. u. A. Morgenstern: Über die Änderung der anaphylaktischen Potenz artfremden Eiweißes. Klin. Wschr. *30* (1952) 596—598.
50. u. H. Petzold: Experimentelle Untersuchungen zur Frage der Leberschädigung durch TB I. Z. inn. Med. *7* (1952) 625—628.
51. Die Skleropathie. Z. inn. Med. *7* (1952) 987—996.
52. u. G. Heidelmann: Die Behandlung chronischer Leberparenchymschäden mit Corhormon- und Cholininfusionen. Dtsch. Z. Verdauungskrkh. *12* (1952) 261—268.
53. u. H. Petzold: Über die Wirkung des Magnesiums auf die Regeneration der Rattenleber. Naunyn-Schmiedebergs Arch. exper. Path. *217* (1953) 10—12.
54. u. H. Petzold: Tierexperimentelle Untersuchungen zur Frage einer Leberschädigung durch Rimifon (Isonikotinylhydrazin). Klin. Wschr. *30* (1952) 1081—1083.
55. u. Eva Götze, G. Penitzka: Über wachstumbeeinflussende Stoffe des Blutes. Z. Krebsforsch. *59* (1953) 191—208.
56. Bemerkungen zur Arbeit M. Noverraz: Détermination des protéines sériques par électrophorese sur papier filtré. Schweiz. med. Wschr. *83* (1953) 77—78.
57. Über eine neue Form der Lebertherapie mit wachstumanregenden Stoffen. Therapiewoche *3* (1952/53) 524—525.
58. u. Ruth Becker: Tierexperimentelle Untersuchungen über Regulationen des Bluteiweißbildes. Plasma *1* (1953) 49—59.
59. u. Eva Götze, Brigitte Morgenstern: Pflanzen-cytologische Untersuchungen mit Tumorfiltraten. Z. inn. Med. *8* (1953) 181—184.
60. u. E. Perlick: Cytomorphologische Veränderungen des Knochenmarkes und Veränderung des Bluteiweißbildes bei Retikulosen. Z. inn. Med. *8* (1953) 401—410.
61. Das Bluteiweißbild in der ärztlichen Praxis. Dtsch. Ges.wesen *8* (1953) 957 bis 965.
62. u. H. Petzold: Das Bluteiweißbild nach partieller Resektion der Rattenleber, seine Beeinflussung durch Corhormon. Z. exper. Med. *122* (1953) 264—269.
63. Leber und Bluteiweißbild. Dtsch. Med. J. *4* (1953) 202—205.
64. Blutzellbild und Bluteiweißbild. Münch. med. Wschr. *95* (1953) 1289—1293.
65. Das Bluteiweißbild bei Leberparenchymschäden. Plasma *1* (1953) 495—512.
66. u. H. Petzold: Der medikamentöse Heilschlaf in der Behandlung der Sklerodermie und ihr verwandter Krankheitsbilder. Dtsch. med. Wschr. *79* (1954) 1003—1007.
67. u. H. Petzold: Tierexperimentelle Untersuchungen über die Wirksamkeit des Leberparenchymfaktors. Z. inn. Med. *9* (1954) 95—98.
68. W. Blech u. R. Emmrich: Über den Einfluß der partiellen Hepatektonie auf die Formalinarthritis der Ratte. Z. inn. Med. *9* (1954) 146—149.
69. u. G. Folk: Über die LCPC (least coagulable protein concentration). Z. inn. Med. *9* (1954) 393—396.
70. u. J. Huth: Der Histamin-Bindehaut-Test während und nach der Schlaftherapie. Klin. Wschr. *32* (1954) 514—516.
71. Die sklerolytische Wirkung des medikamentösen Heilschlafs. Arbeitstagung über kortiko-viszerale Regulationen. Berlin 1955. 242—248.
72. Die Hypo-γ-Globulinämie im Bluteiweißbild. Dtsch. Arch. klin. Med. *201* (1954) 425—433.

73. Die moderne Schlaftherapie in der inneren Medizin. I. Über Methoden, Rhythmik und Indikationen der Schlaftherapie. Z. inn. Med. *9* (1954) 737—741.
74. Schmerz und Heilanästhesie. Neuralmedizin *2* (1954) 204—216.
75. u. H. Neumann: Erkennung und Behandlung der Gallenblasenerkrankungen. Z. ärztl. Fortbild. *48* (1954) 749—755.
76. u. G. Preuß: Nekrose und Nekrosebehandlung bei peripheren arteriellen Durchblutungsstörungen. Ärztl. Wschr. *9* (1954) 1109—1113.
77. Der sklerolytische Effekt der Schlaftherapie. Proceedings of the third Congress of Internal Medicine, Stockholm 1954. 394—398.
78. u. Christa Pferner: Bluteiweißveränderungen bei experimentell erzeugter toxischer Nephrose. Z. inn. Med. *10* (1955) 355—358.
79. u. H. Petzold: Das Bluteiweißbild bei chronischen interstitiellen Entzündungen der Leber. Dtsch. Arch. klin. Med. *202* (1955) 303—311.
80. Das Bluteiweißbild. Dtsch. Ärztekalender *1956*. 156—160.
81. u. R. Sauerbrey: Über den derzeitigen Stand der Pneumoniebehandlung. Dtsch. Ges.-wesen *11* (1956) 325—328.
82. u. Eva Götze: Papierchromatographische Analyse kleiner Mengen mit Ninhydrin reagierender Substanzen aus alkoholischen Extrakten der Rattenorgane. Z. inn. Med. *10* (1955) 1024—1028.
83. Endoangiitis obliterans und Arteriosklerose. Dtsch. Ges.wesen *11* (1956) 1444—1448.
84. u. Eva Götze: Über das Vorkommen freier Aminosäuren in malignen Tumoren und den dazugehörigen Organen. Z. inn. Med. *10* (1955) 1046—1049.
85. u. H. Petzold: Klinische Erfahrungen mit der Infusionsbehandlung bei chronischen Leberschäden. Z. klin. Med. *154* (1956) 237—251.
86. Die Bedeutung der Blutübertragung bei Hypoproteinämien. Dtsch. Internistentagung Leipzig *1955*. Berlin 1956. 78—80.
87. Die Konstellationstypen im Bluteiweißbild. Münch. med. Wschr. *98* (1956) 481 bis 486.
88. Die Plasmaproteine und Serum-Labilitätsreaktionen in der Leberdiagnostik. 4. Freiburger Symposium über Pathophysiologie. Diagnose und Therapie der Leberkrankheiten *1956*. Berlin—Göttingen—Heidelberg 1957. 77—85.
89. Ergebnisse der Eiweißforschung. Monatskurse ärztl. Fortbildg. München 1956. Nr. 12, 1—6.
90. u. H. Linke, H. Trenckmann, H. E. Schultze: Kryoglobulinämien. Acta haemat. *18* (1957) 191—204.
91. Zerebralsklerose vom internistisch-neurologischen Standpunkt. Dtsch. Med. J. *8* (1957) 297—299.
92. Über die Differenzierung des Herzschmerzes. Dtsch. Med. J. *9* (1958) 301—306.
93. Die Virusmeningitisepidemie 1956/57. Therapiewoche *59* (1958) 99—102.
94. Therapie des Schlaganfalls. Jahreskongreß für ärztl. Fortbildg. Dresden *1958*. Berlin 1959. 169—177.
95. Leberfunktionsprüfungen in der Allgemein- und Fachpraxis. Monatskurse ärztl. Fortbildg. *9* (1959) 23—28.
96. Das Bluteiweißbild. Beiträge aus dem Arbeitsbereich der Orthopädie und der chirurg.-medizin. Technik. *6* (Berlin 1959) 211—215.
97. u. H. Petzold: Histochemische Untersuchungen an Leberpunktaten. Acta hepatosplen. *6* (1959) 328—357.
98. u. S. Zimmermann: Über das Verhalten der Glutaminsäure-Oxalessigsäure-Transami-

nase bei der elektrophoretischen Auftrennung der Serumproteine. Klin. Wschr. *37* (1959) 935—936.

99. u. Eva Götze, H. Petzold, K. Hoffmeister: Über die Beeinflussung des Granulationsgewebes durch Cholin, Acetylcholin und Cortison. Ärztl. Forsch. *13/I* (1959) 553—556.
100. Die Therapie der sogenannten Kollagenkrankheiten. Therap. Gegenw. *98* (1959) 376—380.
101. Biochemie der Gefäßwand. Symposion „Gefäßwand und Blutplasma". Magdeburg 1959. Jena 1961. 15—35.
102. Probleme der Therapie. Symposion „Gefäßwand und Blutplasma" Magdeburg 1959. Jena 1961, 229—240.
103. Klinisch-chemische Problematik der Bindegewebspathologie. Schriftenreihe der ärztl. Fortbild. Berlin 1961. 13—22.
104. u. N. Bannert: Über proteingebundene Kohlehydrate im Serum und in der Interzellularflüssigkeit bei chronischer Polyarthritis rheumatica und bei Arteriosclerosis obliterans. Ärztl. Wschr. *15* (1960) 115—118.
105. Über die Proteinfraktionen von Thromben und Gefäßfüllseln bei arteriellen peripheren Durchblutungsstörungen. Münch. med. Wschr. *102* (1960) 823—826.
106. Antifebrile Therapie und neurale Therapie der Infektionskrankheiten. Therapietagung Weimar 1960, in: Beiträge zur modernen Therapie. *III* (Jena 1961) 120—124.
107. u. H. Petzold: Glykogenspeicherkrankheiten im Erwachsenenalter. Dtsch. Z. Verdauungskrkh. *20* (1960) 162—168.
108. Über einige Probleme der chronischen Krankheiten des Bindegewebes. Münch. med. Wschr. *102* (1960) 1462—1465.
109. Viruspneumonien. Jahreskongreß für ärztl. Fortbildg. Leipzig 1960. Berlin 1961. 126—136.
110. Der Begriff der klinischen Entität in der inneren Medizin. Wiss. Z. Karl-Marx-Univ., Math.-Nat. Reihe *9* (1959/60) 517—522 — Leipziger Universitätsreden, NF, Leipzig 1960, H. 17.
111. u. Eva Götze, H. Petzold, K. Hoffmeister: Über die Beeinflussung des Granulationsgewebes durch Cholinderivate und durch Glykokoll-Betain. Ärztl. Forsch. *15/I* (1961) 45—47.
112. u. S. König: Über die Differenzierung von Exsudat und Transsudat. Med. Klin. *56* (1961) 661—664.
113. Kollagenkrankheiten. Ärztl. Praxis *13* (1961) 1663—1671.
114. Gefahren einer Langzeitbehandlung mit hochwirksamen Kortikosteroiden. Dtsch. Ges.wesen *16* (1961) 952—955.
115. Die Beurteilung der Serumlabilitätsproben in der ärztlichen Praxis. Tägliche Praxis *5* (1964) 29—35; internistische praxis *4* (1964) 51—57; pädiatrische praxis *4* (1965) 189—195.
116. Leberdiagnostik. 1. Kongreß klin. Path. u. klin. Chem. Dresden 1961. Verh. exper. Med. *1* (1962) 182—188.
117. u. H. Petzold: Histochemische Reaktionen am Leberparenchym chronischer Hepatopathien. Acta hepato-splen. *9* (1962) 253—263.
118. Das Postcholezystektomie-Syndrom. Dtsch. Z. Verdauungskrkh. *21* (1962) 245 bis 253.
119. Dysproteinämien und Paraproteinämien. FuF *36* (1962) 193—196.
120. Die klinische Bedeutung der serologischen Befunde beim Rheumatismus. Verh. dtsch. Ges. inn. Med. *68* (1962) 458—478.

121. u. W. Urbaszek: Eine Methode zur elektrophoretischen Trennung der Serum-Mukoide. Z. inn. Med. *17* (1962) 801—803.
122. Zur Klinik der Hepatitis. Therapietagung Weimar 1962. Beiträge zur modernen Therapie V., Jena 1963. 87—98.
123. Diagnose und Therapie arterieller peripherer Durchblutungsstörungen an den unteren Extremitäten. Beiträge zur Orthopädie und Traumatol. *10* (1963) 249—257.
124. Beurteilung chronischer Lebererkrankungen. Z. ärztl. Fortbild. *57* (1963) 377—381.
125. Über die Beeinflussung des Granulationsgewebes durch Carnitin. Dtsch. Z. Verdauungskrkh. *22* (1962) 102—105.
126. Pathophysiologie und medikamentöse Beeinflussung peripherer arterieller Durchblutungsstörungen. Kongreßbericht Pharmakol. Ges. d. DDR 1962. Verh. dtsch. Ges. exper. Med. *5* (1964) 3—16.
127. Formen und Verlaufsformen der Hepatitis. Internat. Lebersymposium Berlin 1962. Z. ärztl. Fortbild. *58* (1964) 8—14.
128. u. Ursula Schade: Über die Beeinflussung des im Tierexperiment erzeugten Granulationsgewebes. Z. inn. Med. *19* (1964) 429—430.
129. u. J. Schwarz: Die Gelenkbeweglichkeit in Abhängigkeit vom Altern und ihr Verhalten bei verschiedenen Krankheiten. Z. Alternsforsch. *16* (1963) 297—303.
130. u. W. Urbaszek: Untersuchungen über die Elektrophorese der Serummukoide. Z. inn. Med. *18* (1963) 501—502.
131. u. H. Nollau: Diagnose und Diagnosen. Dtsch. Ges.wesen *18* (1963) 1901—1906.
132. W. Urbaszek u. R. Emmrich: Die elektrophoretische Trennung der Serummukoide bei rheumatischen Erkrankungen. Z. inn. Med. *18* (1963) 721—724.
133. Hypertonie. Einteilungsprinzipien, essentielle Formen. Jahreskongreß für ärztl. Fortbildung 1963. Berlin 1964. 233—245.
134. Beziehungen zwischen Gefäßwand und Blutplasma — kleinere Arterien und Arteriolen. Symposion „Gefäßwand und Blutplasma", Leipzig 1963. *II* (Jena 1965) 17—28.
135. Die Leberzirrhose als klinisches Problem. Internistentagung Leipzig 1963. Z. inn. Med. *19* (1964) Beilage, 133—143.
136. Zur Problematik der Leberzirrhose. Sitzungsberichte der Sächs. Akad. d. Wiss., Math.-Nat. Reihe *106* (1964) 1—16.
137. Untersuchungen über die Differenzierung von Exsudaten und Transsudaten. Z. inn. Med. *19* (1964) 8—12.
138. Gewebsenzyme bei chronischen Leberkrankheiten. FuF *38* (1964) 134—136.
139. u. H. Gäbert, H. Petzold: Organveränderungen bei experimentell ausgelöstem generalisiertem Schwartzmann-Phänomen. Z. inn. Med. *19* (1964) 193—197.
140. Wissen und Gewissen in der Medizin. Wissenschaft aus nationaler Verantwortung, Karl-Marx-Universität Leipzig 1963. 119—126.
141. Erythematodes visoeralis und Tuberkulose. Beitr. Klin. Tub. *128* (1964) 266 bis 276.
142. u. H. Köhler: Ergebnisse einer Umfrage über bronchologische Arbeitsmethoden in der DDR. Dtsch. Ges.wesen *19* (1964) 1763—1767.
143. Therapie der chronischen Hepatitis und Cholangiohepatitis. 10. Therapietagung Weimar 1964. Bibliograph. Berichte der Deutschen Staatsbibliothek, Berlin 1964. 222—230.
144. Quellen und Bedeutung des Vertrauensverhältnisses zwischen Arzt und Patient. Medizin und Philosophie, Arzt und Gesellschaft, Sonderband der Wissenschaftl. Z. d. Karl-Marx-Universität *V* (1965) 232—242.

145. Prophylaxe und innere Medizin. Internistentagung Leipzig 1964. Z. inn. Med. *20* (1965) Beilage 61 (= H. 13) — 65 (= H. 14).
146. Struktur und Aufgaben einer modernen inneren Klinik. Internistentagung Leipzig 1964. Z. inn. Med. *20* (1965) 193—203.
147. Medizinische Klinik und Ambulanz. Sonderband der Wissenschaftl. Z. d. Karl-Marx-Universität Leipzig „550 Jahre Medizinische Fakultät", 1965. 120—131.
148. Entzündungsstoffe in Exsudaten. Z. inn. Med. *20* (1965) 393—396.
149. Zur Chronizität der Entzündung. Berichte der Dtsch. Akad. d. Naturforscher Leopoldina *11* (1965) 103—115.
150. Der Stand der Arteriosklerosetherapie. 11. Therapietagung Weimar 1965. Bibliograf. Berichte d. Dtsch. Staatsbibliothek, Berlin 1966. 177—184.
151. Vegetative Dystenie und Gallenwege. Dtsch. Z. Verdauungskrkh. *25* (1965) 155—157.
152. Akuter Herzstillstand bei internen Erkrankungen. Wissenschaftl. Z. d. Karl-Marx-Universität Leipzig, Math.-Nat. Reihe *14* (1965) H. 3, 589—593.
153. u. Helga Mertens: Intermittierende Behandlung der primär chronischen Polyarthritis. Z. inn. Med. *20* (1965) 553—558.
154. Die klinische Diagnose der Arteriosklerose. Z. inn. med. *20* (1965) 641—653.
155. Zur Differentialdiagnose peripherer arterieller Verschlußkrankheiten. Münch. med. Wschr. *107* (1965) 2277—2281.
156. u. Jutta Fehres: Tierexperimentelle Untersuchungen über die Trypsinempfindlichkeit isolierter normaler und geschädigter Leberzellen. Acta hepato-splen. *14* (1967) 1—8.
157. u. F.-B. Spancker: Zum Verhalten des sogenannten leichtgebundenen Hydroxyprolins in der menschlichen Haut. Z. inn. Med. *21* (1966) 257—260.
158. u. Barbara Herberger: Vegetative Stigmata bei Kranken mit peripheren arteriellen Durchblutungsstörungen. Dtsch. Ges. wesen *21* (1966) 1353—1356.
159. Periphere arterielle Durchblutungsstörungen. Einzelheft „Kleine Gesundheitsbücherei" des Deutschen Hygiene-Museums Dresden. Berlin 1967. H. 88, 20 S.
160. Realität und Theorien des Alterns. Berichte d. Sächs. Akad. d. Wiss., Math.-Nat. Klasse *107* (1966) 5, 1—20.
161. Alternsforschung und Innere Medizin. Z. inn. Med. *21* (1966) 513—517.
162. Walter Brednow zum 70. Geburtstag. FuF *40* (1966) 60—61.
163. Angeborene Anomalien der Plasmaproteine. (Z. ges. inn. Med.) Ber. Sekt. Inn. Med. *4* (1966) 62—72.
164. Zur Pathophysiologie peripherer arterieller Verschlußkrankheiten. Zbl. Chir., Sonderband 1967, Teil II, 1489—1493.
165. Arteriosklerose und Hochdruck. Dtsch. Ges.wesen. *22* (1967) 1249—1253.
166. Permeationssteigernder Effekt von Exsudatproteinen. Z. inn. Med. *22* (1967) 129—132.
167. u. H. Paasch: Chylomikronenzahlen im Nüchternserum bei Arteriosclerosis obliterans und bei Diabetes mellitus mit und ohne Hypertonie. Z. inn. Med. *22* (1967) 231—233.
168. u. H.-J. Häntzschel, H. Häntzschel: Die klinische Bedeutung der Hydroxyprolinausscheidung im Harn. Z. inn. Med. *22* (1967) 193—199.
169. u. F.-J. Edelmann, H. Gäbert: Zur Diagnostik des Magenkarzinoms mit Hilfe des Tetrazyklin-Fluoreszenz-Testes. Med. Klin. *62* (1967) 493—495.
170. u. M. G. Wähmer, E. Liebaug: Endogenes Pyrogen im „Arthus-Phänomen" und dessen Verhalten nach Prednisolon. Z. inn. Med. *22* (1967) 358—362.
171. Das Bluteiweißbild — sein Wert für Diagnose und Therapie. Paracelsus Beihefte (Vorträge des 19. Ärztetreffens in Kärnten—Velden 4.—7. 5. 67). Wien 1967. 1—10.

172. La biochimie des changements inflammatoires de la paroi vasculaire. Angéiologie *19* (1967) 29–33.
173. Akuter Herzstillstand. FuF *41* (1967) 353–357.
174. Neue Methoden zur Diagnostik und Therapie der Hyperthyreosen. Münch. med. Wschr. *109* (1967) 2217–2221.
175. Molekularbiologie und Innere Medizin. Z. inn. Med. *22* (1967) 441–447.
176. Die Colitis ulcerosa und ihre Behandlung. Wiss. Z. d. KMU, Math.-Nat. Reihe *16* (1967) 49–53.
177. u. H. Köhler: Erste Ergebnisse einer Intensivbehandlungsstation in der Medizinischen Klinik. Z. inn. Med. *22* (1967) 616–620.
178. Therapie der Arteriosklerose. (Z. inn. Med.) Ber. Ges. inn. Med. *5* (1967) 28–32.
179. Kollagenosen der Lunge. Z. Tuberk. *128* (1968) 280–285.
180. Molekularstrukturen der γ-Globulinkomponenten und ihre Bedeutung für die Klinik. Z. inn. Med. *23* (1968) 129–136.
181. Pathophysiologie des Alterns. Z. inn. Med. *23* (1968) 481–485.
182. u. J. Schaaf: Über die Altersabhängigkeit der 5-Hydroxyindolessigsäure-Ausscheidung im Harn. Z. inn. Med. *23* (1968) 551–553.
183. Therapie der Arteriosklerose mit Ansatzpunkt an den sog. Risikofaktoren. Münch. med. Wschr. *110* (1968) 1918–1923.
184. Zur Thematik Gefäßwand und Blutplasma. Symposion-Ber. Gefäßwand und Blutplasma *III* (Jena 1970) 25–30.
185. H. Marek, H. Koch u. R. Emmrich: Autoradiographische Untersuchungen über die Aufnahme von tritiummarkiertem Hydroxyprolin und Prolin in der Rattenaorta. Symposion-Ber. Gefäßwand und Blutplasma *III* (Jena 1970) 157–164.
186. u. P. Drechsler: Hypalbuminämien und ihre prognostische Bedeutung. Z. inn. Med. *24* (1969) 161–167.
187. Die Bedeutung der Risikofaktoren Hypertonie und Hyperlipidämie für die Pathogenese der Arteriosklerose. Sympos. Bericht „Arterielle Hypertonie und Koronarinsuffizienz im Altern und im Alter" [russisch]. Kiew 1969. 250–255.
188. u. V. Kleint: Kritische Betrachtungen zur Therapie und Diagnostik in der inneren Medizin. Z. inn. Med. *24* (1969) 359–363.
189. u. J. Wagner, E.-H. Axthelm: Die Bindungskapazitäten der Serumproteine für Herzglykoside, speziell für Pengitoxin (zum 75. Geburtstag von Herrn Prof. Dr. Bennhold). Münch. med. Wschr. *111* (1969) 401–404.
190. Kardiosklerose. „Herz und Altern" (Tagungsbericht). Dresden 1971. 39–50.
191. Therapie der Arteriosklerose (La terapia dell'arteriosclerosi). Panorama medico (Sandoz) Maggio Giugno 3, Milano 1969. p. 8–12.
192. Gefäßstoffwechsel. Deine Gesundheit *7* (1969) 202–204.
193. Diät zur Prophylaxe und Therapie der Arteriosklerose. 15. Weimarer Therapie-Tagung [unveröffentlicht].
194. Die Chronizität der Entzündung, Prinzipien und Mechanismen. Dtsch. Ges.wesen *25* (1970) 56–61.
195. Pathokinese und Kybernetik. Z. inn. Med. *24* (1969) 593–599.
196. Die Hydroxyprolinausscheidung im Harn bei Endoangiitis und Arteriosclerosis obliterans. Z. inn. Med. *24* (1969) 747–748.
197. u. R. Hafermalz, D. Jäger, R. Klaus, Petra Schneider: Vergleichende Bausteinanalysen zwischen großen Gefäßen und dem Skelettsystem in Abhängigkeit vom Altern. Z. Alternsforsch. *23* (1970) 29–33.

198. Arzt und Arzneimittel. Z. ärztl. Fortbild. *64* (1970) 1092—1096.
199. Zur Methodik der Alternsforschung. Z. Alternsforsch. *23* (1970) 97—103.
200. Das hepatorenale Syndrom. Z. inn. Med. *25* (1970) 377—380.
201. Infarktgefährdung und Präinfarkt. Dtsch. Ges.wesen *25* (1970) 1072—1077.
202. Hochdruck und Hyperlipidämie (Hypercholesterinämie) als Risikofaktoren für die Entstehung der Arteriosklerose. Berichte Sächs. Akad. d. Wiss. Leipzig, Math.-Nat. Klasse *109* (1971) H. 4, 23 S.
203. Die Therapie der Angina pectoris und des Herzmuskelinfarktes bei älteren Patienten. „Scriptum Geriatricum". Wien 1970, 163—171.
204. u. Helga Marek, H. Koch: Experimentelle autoradiographische Untersuchungen verschiedener Gewebe mit tritiummarkiertem Hydroxyprolin und Prolin. Z. inn. Med. *25* (1970) 889—895.
205. Die Hydroxyprolinausscheidung im Harn bei chronischen Krankheiten des Bindegewebes. Z. inn. Med. *25* (1970) 1062—1065.
206. H. Koch, R. Emmrich u. Helga Marek: Tierexperimentelle Untersuchungen über die Kinetik von tritiummarkiertem Prolin. Z. inn. Med. *26* (1971) 166—170.
207. u. A.-J. Hegedüsch: Hydroxyprolinausscheidung im Harn nach operativer Behandlung und nach Radiojodtherapie von Hyperthyreosen. Z. inn. Med. *26* (1971) 241—242.
208. Schock in der inneren Medizin. Tagungsbericht 7. Internistentagung der DDR 1970 in Leipzig, *7* (1970) 154—159.
209. Helga Marek, H. Koch u. R. Emmrich: Tierexperimentelle Untersuchungen über die Verteilung und Ausscheidung von tritiummarkiertem Hydroxyprolin und Prolin. Z. inn. Med. *25* (1970) 830—834.
210. Zur Problematik der akuten Entzündung. Dtsch. Ges.wesen *26* (1971) 627—631.
211. u. Helga Marek: Das Verhalten von markiertem Prolin und Hydroxyprolin im Bindegewebsstoffwechsel. Symposion-Bericht „Biomorphose des Bindegewebes." Wiss. Z. d. Humboldt-Univ. Berlin, Math.-Nat. R. *XXII* (1973) H. 5, 530—531.
212. Die klinische Bedeutung des Kollateralkreislaufes bei peripheren arteriellen Verschlußkrankheiten. Symposion-Ber. „Gefäßwand und Blutplasma III". Jena 1970. 197—200.
213. u. G. Schmidt: Symptome und Verlauf der Pneumonien im Alter. Z. Alternsforsch. *25* (1971) 151—161.
214. Intensivmedizin. Z. inn. Med. *26* (1971) 473—475.
215. Angeborene und erworbene Bluteiweißstörungen. Bericht des 23. Ärztetreffens in Kärnten. Wien 1971. 22—34.
216. Klinische Forschung. Dtsch. Ges.wesen *26* (1971) 2249—2252.
217. Aktuelle klinische Probleme der akuten Hepatits. Dtsch. Z. Verdauungskrkh. *32* (1972) 29—35.
218. Angeborene und erworbene Bluteiweißstörungen. Wien. med. Wschr. *121* (1971) 761—765.
219. Technisierung in der inneren Medizin. Unveröffentlicht.
220. Der Facharzt für Innere Medizin — Ausbildung und Weiterbildung. Tagungsbericht der Ges. inn. Med. d. DDR 1970. Ber. Ges. inn. Med. *7* (1971) 314—315.
221. M. G. Wähmer, R. v. Baehr, E. Liebaug, I. Wagner u. R. Emmrich: Relation von endogenem Pyrogen zu permeabilitätsverändernden Substanzen in den frühen Stadien des Arthus-Phänomens. Z. inn. Med. *26* (1971) 799—802.
222. Leber und Eiweißstoffwechsel. Z. inn. Med. *27* (1972) 369—374.
223. Herzinfarkt und Wiederaufnahme der Arbeit. Z. inn. Med. *27* (1972) 548—554.

224. Nachruf auf Prof. Eberhard Perlick. Z. inn. Med. *26* (1971) 757–758.
225. u. H. Koch, M. Rödenbeck, Helga Marek, A. Ziegler: Mathematische Modellierung des Kollagenstoffwechsels. Tracerkinetische Untersuchungen mit ^{3}H-Prolin am Organbindegewebe gesunder Ratten. Z. inn. Med. *27* (1972) 685–693.
226. Vorwort zum Symposiumband „Gefäßwand und Blutplasma IV“. Jena 1974. S. 5–6.
227. Zur Problematik der Beziehungen zwischen Gefäßwand und Blutplasma. Ebda S. 17–23.
228. The activity of arteriosclerosis in old age. 9th Internat. Congress of Gerontology Kiev 1972. Plenary and Section Sessions, Reports — Introductory Lectures — Vol. *1* (1972) 240–243.
229. Der heutige Stand der Arteriosklerose-Therapie. Z. Alternsforsch. *28/I* (1974) 31–38.
230. P.-F. Mahnke, R. Emmrich, Helga Marek, A. Ziegler, A. Neugebauer: Klinisch-morphologische Befunde bei Muskelbiopsien unter besonderer Berücksichtigung der Hypertonie. Ber. Ges. Inn. Med. *8* (1973) 226–229.
231. W. Storch, R. Emmrich, H. Petzold: Zur Bedeutung histologischer Befunde für die Indikation zur immunsuppressiven Therapie der chronisch aggressiven Hepatitis. Ber. Ges. Inn. Med. *8* (1973) 174–176.
232. Zur endokrinen Regulation des Kollagenstoffwechsels. Wiss. Z. d. KMU, Math.-Nat. Reihe *21* (1972) 500–506.
233. Franz Volhard und seine Bedeutung für die Entwicklung der Nephrologie. Z. inn. Med. *27* (1972) 962–964.
234. Klinische Pharmakologie. Z. inn. Med. *28* (1973) 213–216.
235. Zur Differentialdiagnose Arteriosclerosis obliterans — Thromboangiitis obliterans. Z. inn. Med. *28* (1973) 389–394.
236. Cholostase als klinisches Syndrom. Z. Verdauungskrkh. *34* (1974) 85–92.
237. W. Storch, R. Emmrich: Anwendungsmöglichkeiten der Immunfluoreszenz bei chronisch-aggressiver Hepatitis. Z. Verdauungskrkh. *33* (1973) 119–123.
238. Zwischen Leben und Tod. Ärztliche Probleme der Thanatologie. Sitzungsber. Sächs. Akad. Wiss. Leipzig. *110* (Berlin 1974) H. 7, 22 S.
239. Immunologie. Vnitřni lékařstoi *19* (1973) 1060–1065.
240. Prof. Dr. med. habil. Hans Petzold 60 Jahre alt. Laudatio. Z. inn. Med. *28* (1973) 656.
241. Entwicklungstendenzen der inneren Medizin. Z. inn. Med. *29* (1974) 177–179.
242. Zur Differentialdiagnose der Arteriitis. Z. inn. Med. *29* (1974) 265–272.
243. Das Erkennen von Störungen der vitalen Funktionen in der Notfallsituation aus klinischer Sicht (vor dem Einsatz von Hilfsuntersuchungen). Bulletin de la Societé des Sciences Médicales du Grand-Duché de Luxembourg, Numéro special *2* (1973) 217 bis 224.
244. W. Herrmann, R. Emmrich u. I. Wagner: Die permeabilitätssteigernde Wirkung von Gefäßwandphosphatiden atherosklerotisch veränderter Humanaorten. Atherosclerosis *20* (1974) 117–130.
245. Aspekte der Herz-Kreislauf-Forschung am Bereich Medizin der Karl-Marx-Universität Leipzig. „Diagnostik und Therapie der ischämischen Herzkrankheit“ —Sonderheft im Rahmen der Wissenschaftlichen Beiträge der KMU Leipzig, Reihe Biowissenschaften — Medizin, 1974. 7–10.
246. u. M. Seige: Zur Wertigkeit von Krankheiten in der Inneren Medizin. Z. inn. Med. *29* (1974) 605–611.
247. Prävention der Arteriosklerose. Z. inn. Med. *29* (1974) 645–650.
248. u. M. Gülzow, H. Trenckmann: 25 Jahre DDR. Z. inn. Med. *29* (1974) 733.
249. Diagnose und diagnostische Prozesse. Z. inn. Med. *29* (1974) 734–739.

III. Herausgeber-Tätigkeit

Arbeitsmethoden der inneren Medizin und ihr verwandter Gebiete.

Bd I: Spezielle physikalische Arbeitsmethoden. 1. Lieferung, mit einem Beitrag über die allgemeine klinische Untersuchung. Jena 1964. S. 1—228. 2. Lieferung. Herzkatheterismus. Elektrokardiographische Untersuchungsmethoden. Jena 1967. S. 1—308. 3. Lieferung. Lungenfunktionsprüfungen. Jena 1971. S. 1—276.

Bd II: Röntgendiagnostische Untersuchungsmethoden. 1. Lieferung. Jena 1965. S. 1 bis 386.

Bd III: Endoskopische Arbeitsmethoden. Jena 1966. S. 1—436.

Bd IV: Arbeitsmethoden der Neurologie. Jena 1965. S. 1—208.

Bd V: Klinisch-chemische Labormethoden. 1. Lieferung. Allgemein-chemische Laboratoriumsverfahren. Das klinisch-chemische Routinelaboratorium. Spezielle klinisch-chemische Laboratoriumsmethoden. Jena 1969. S. 1—244. 2. Lieferung. Klinisch-chemische Labormethoden. Das Elektrophoreselabor. Das Enzymlabor. Das klinisch-serologische Labor. Das Hormonlabor. Jena 1969. S. 1—235.

Bd VI: Mikroskopische Untersuchungsmethoden und Arbeitsmethoden auf verschiedenen Spezialgebieten. 3. Lieferung. Klinisch-parasitologische Untersuchungsmethoden. Jena 1970. S. 1—160. 4. Lieferung. Spezielle Arbeitsmethoden der klinischen Endokrinologie. Jena 1973. S. 1—200.

Bd VIII: 1. Lieferung. Therapeutische Verfahren. Allgemeine therapeutische Arbeitsmethoden der Inneren Medizin. 2. Lieferung Physiotherapie. Jena 1970. S. 1—176. 3. Lieferung. Das Nierenzentrum. Jena 1971. S. 1—95.

u. E. Perlick: Gefäßwand und Blutplasma. I. Symposion 2./3. Oktober 1959 in Magdeburg. Jena 1961. 250 S.

u. E. Perlick: Gefäßwand und Blutplasma. II. Symposion 8./9. Oktober 1963 in Leipzig. Jena 1965. 368 S.

u. E. Perlick: Gefäßwand und Blutplasma. III. Symposion 17./18. September 1968 in Leipzig. Jena 1970. 408 S.

Gefäßwand und Blutplasma. IV. Symposion 27./29. April 1972 in Leipzig. Jena 1974.

3. Internistentagung der med.-wiss. Ges. Halle—Jena—Leipzig 14./15. Oktober 1960 in Leipzig. Leipzig 1963. S. 1—299.

u. M. Gülzow: Zeitschrift für die gesamte innere Medizin. Leipzig 1964—1974.

u. E. Perlick: Tagungsbericht der Sektion Innere Medizin 1963. Redaktion: H. Trenckmann. Leipzig 1964. 362 S.

Tagungsbericht der Sektion Innere Medizin 1964. Leipzig 1964. 200 S.

Tagungsbericht der Sektion Innere Medizin 1965. Leipzig 1965. 292 S.

u. H. Trenckmann: Tagungsbericht der Gesellschaft für Innere Medizin 1966. Leipzig 1967. 320 S.

Tagungsbericht der Gesellschaft für Innere Medizin der DDR 1967. Leipzig 1968. 316 S.

Tagungsbericht der Gesellschaft für Innere Medizin der DDR 1968. Leipzig 1969. 220 S.

Tagungsbericht der Gesellschaft für Innere Medizin der DDR 1970. Leipzig 1971. 347 S.

Tagungsbericht der Gesellschaft für Innere Medizin der DDR 1972. Leipzig 1973. 325 S.

IV. Redaktionelle Mitarbeit

1. Archiv für klinische Medizin. München, Berlin, Heidelberg. 1. 1. 1966 — 31. 12. 1969. (Erscheinen eingestellt).
2. Acta hepato-splenologica. Stuttgart 1961—1974.
3. Zeitschrift für Physiotherapie. Leipzig 1971—1974.
4. Dtsch. Z. Verdauungs- u. Stoffwechselkrankheiten. Leipzig 1960—1974.

Zur Biobibliographie

Schriftenverzeichnisse. Siehe vorliegendes Jahrbuch S. 230—242. — Jahrbuch Sächsische Akademie der Wissenschaften zu Leipzig *1960—1962* (Berlin 1964) S. 275—276; *1963—1965* (Berlin 1967) S. 200—202; *1966—1968* (Berlin 1970) S. 205—206; *1969—1970* (Berlin 1972) S. 153—154; *1971—1972* (Berlin 1974) S. 165—167. — Kürschners Deutscher Gelehrten-Kalender *7* (Berlin 1950) Sp. 411; *8* (Berlin 1954) Sp. 455; *9* (Berlin 1961) S. 385—386; *10* (Berlin 1966) S. 473; *11* (Berlin 1970) S. 588—589. — Poggendorffs Handwörterbuch VII_1^a (Berlin 1956) S. 503—504.

Autobiographisches. Medizinische Klinik und Ambulanz. In: 550 Jahre Medizinische Fakultät. Wissenschaftliche Z. Karl-Marx-Universität Leipzig, math.-nat. Reihe *14* (1965) H. 1, S. 126—127.

Würdigungen. Zum 60. Geburtstag: o. V., in: UZ Leipzig *14* (1970) Nr. 34/36 vom 17. Sept. — Allgemein: K. Seidel: Zentrum der inneren Medizin. In: Leipziger Volks-Ztg *1963*, 25. Apr. — H. W. St.: „Noch näher an die praktische Arbeit binden ..." Ebda *1961*, 7. Dez. — Ebda *1963*, 24. Aug. — UZ Leipzig *17* (1973) Nr. 1 vom 4. Jan. — Der neue Weg *27* (Halle/S. 1972) Nr. 117 vom 17. Mai.

Nachrufe. M. Gülzow, H. Trenckmann, in: Z. inn. Med. *29* (1974) S. 557—558. — Z. Kojecký, in: Vnitřni lékařdkví 1974, S. 924—925. — W. Ries, 2 S. [Ms. in den Akten der SAW]. — H. Trenckmann, siehe vorliegendes Jahrbuch S. 227—229. — Allgemein: UZ Leipzig *18* (1974) Nr. 41 vom 14. Nov.

G. W.

Für ihre Unterstützung bei der Zusammenstellung der Bibliographie dankt der Bearbeiter Frau Ilse Schreiter, der langjährigen Chefsekretärin von Herrn Prof. Dr. Emmrich.

Franz Runge

5. 12. 1893—29. 8. 1973

80 Jahre wäre am 5. Dezember 1973 Prof. Dr. phil. Franz Runge geworden, hätte nicht am 29. August dieses Jahres ein jäher Tod sein Leben beendet. Ein Leben, gekennzeichnet durch schöpferische Zielstrebigkeit und charakterliche Geradheit, widerspiegelt den Werdegang eines vorbildlichen Forschers und Lehrers.

Sein Vater, der Geh. Med.-Rat Prof. Dr. Max Runge, leitete die Universitäts-

Frauenklinik in Göttingen. So wuchs der junge Runge in einer Atmosphäre auf, die vom Geist der Universität geprägt war. Er besuchte das humanistische Gymnasium zu Göttingen. Nach dem Abitur 1914 begann er das Studium der Chemie in Kiel, u. a. bei Prof. Otto Mumm. Der Krieg unterbrach sein Studium, 1919 nahm er es in Göttingen wieder auf. Hier waren u. a. seine akademischen Lehrer A. Windaus, G. Tamann, W. Borsche und R. Pohl. Er schloß sein Studium 1922 mit einer Dissertation über „o-Dioxybenzol-Derivate" bei Walter Borsche ab.

Nach erfolgreicher Assistentenzeit begann er seine berufliche Laufbahn in verschiedenen chemischen Großbetrieben (z. B. in Außig, Oppau-Ludwigshafen, Bitterfeld, Wolfen). Diese Zeit bildete für Franz Runge die Grundlage für sein späteres fruchtbares Wirken als Hochschullehrer für Technische Chemie an der Martin-Luther-Universität Halle—Wittenberg (von 1947 bis 1963).

In dieser Zeit war er z. B. an der Ausarbeitung eines Verfahrens zur technischen Darstellung von Butadien beteiligt. In Bitterfeld führte er die technische Chlorierung von PVC zu PeCe ein. Ferner war er mit der Herstellung von Farbstoffen und Zwischenprodukten befaßt sowie mit dem Aufbau einer Destillationsanlage für Methanol.

Während dieser langen Industrietätigkeit sind verständlicherweise nur wenig Publikationen von ihm erschienen. 1931 veröffentlichte er eine vielbeachtete Monographie über „Organometallverbindungen", die 1937 auch ins Russische übersetzt wurde. Später erweiterte er die Monographie zu dem bekannten zweibändigen Werk über „Organo-Metallverbindungen", das 1944 erschien.

1947 erhielt er einen Ruf an die Martin-Luther-Universität Halle—Wittenberg als ord. Prof. mit Lehrstuhl und Direktor des Institutes für Technische Chemie. Seit dem Tode von Moritz Dolch 1931 war dieser Lehrstuhl praktisch verwaist. So galt es, den Lehrbereich „Chemische Technologie" völlig neu aufzubauen. Hier war ihm ein Aufgabenbereich übertragen, in dem Beharrlichkeit und Ideenreichtum beim Forschen, Erkennen von Entwicklungstendenzen der modernen chemischen Technik, Geschick bei der Überführung der Ergebnisse der Grundlagenforschung in die Produktion, gepaart mit pädagogischem Einfühlungsvermögen, voll zur Geltung kamen. Durch diese Eigenschaften wurde das Profil des Institutes für Technische Chemie ebenso geprägt, wie sie sich positiv auf seine Tätigkeit als Dekan der Naturwissenschaftlichen Fakultät und später der Fachrichtung Chemie auswirkten. Sie führten zeitig zu Ausbildungsformen (Betriebspraktika, technologisches Praktikum), die als nachahmenswert erkannt wurden. Damals gehörte Halle zu den wenigen Universitäten, an denen Chemische Technologie gelehrt wurde. Es galt hier in Halle eine verpflichtende Tradition fortzusetzen, denn dieser Lehrstuhl für Chemische Technologie besteht seit dem Jahre 1900 und war der älteste seiner Art in Deutschland.

Charakteristisch für Franz Runges wissenschaftliche Arbeiten war seine von Anfang an enge Verbindung mit der chemischen Industrie. So bearbeitete er zusammen mit seinen zahlreichen Schülern Probleme, die für eine Nutzung in der chemischen Industrie wichtig waren. Dabei wurden aber auch alte Interessengebiete wieder neu aufgegriffen.

So arbeitete er mit einer Reihe von Mitarbeitern über metallorganische Verbindungen, z. B. über indiumorganische Verbindungen.

Ein Arbeitsgebiet, das ihn bis zu seinem Tode beschäftigte, war das Gebiet der siliziumorganischen Verbindungen. Als Ziel standen temperaturbeständige Ionenaustauscher im Blickpunkt, die vor allem als Katalysatoren eingesetzt werden sollten. Andere Untersuchungen galten den Inhaltsstoffen der hydrierten Braunkohle.

In zahlreichen Arbeiten beschäftigte sich Franz Runge mit Problemen der Hochpolymeren-Chemie im weitesten Sinne.

Zusammen mit P. Fijolka wurden Probleme der Polyester-Synthese bearbeitet. Hierbei handelt es sich vorwiegend um analytische Untersuchungsmethoden, z. B. die Bestimmung der endständigen Carboxylgruppen und die papierchromatographische Analyse der Dicarbonsäuren und Glykole, die bei der Hydrolyse von Polyesterharzen anfallen. Des weiteren wurde ein technisches Verfahren zur Herstellung von Terephthalsäure erarbeitet.

Zahlreiche Untersuchungen befaßten sich ferner mit Disulfimiden. Praktisch angewandt wurden diese Disulfimide zur Isolierung und Identifizierung organ. Basen in Braunkohlehydrierprodukten und zur Bestimmung basisch organischer Wirkstoffe in Arzneimitteln. Weiterhin eignen sich Disulfimide als Textilhilfsmittel; sie erhöhen z. B. die Bauschelastizität von Dederonfasern. — Zahlreiche schwefelhaltige Verbindungen, vor allem Disulfide und Abkömmlinge der Trithiokohlensäureester wurden synthetisiert und auf ihre insektizide Wirkung hin geprüft.

Während seiner Lehrtätigkeit an der Martin-Luther-Universität erschien von ihm eine „Einführung in die Chemie und Technologie der Kunststoffe". Sie wurde auch in das Polnische übersetzt. Zusammen mit Prof. W. K. Schwarze gab er die 27. Auflage des Lehrbuches „Chemische Technologie" von Ost-Rassow in neuer Bearbeitung heraus.

Seine Tätigkeit als Hochschullehrer und Wissenschaftler sowie seine hohe menschliche Persönlichkeit fanden vielfache Anerkennung. In Würdigung seiner Verdienste wurde ihm 1959 der Nationalpreis der DDR verliehen. 1953 wurde er zum Mitglied der Deutschen Akademie der Naturforscher „Leopoldina" und 1955 zum Mitglied der Sächsischen Akademie der Wissenschaften gewählt. Von 1962—65 gehörte er dem ständigen Ausschuß der Deutschen Bunsengesellschaft für Physikalische Chemie an.

Mit seiner Emeritierung im Jahre 1963 bot sich die Möglichkeit zur weiteren wissenschaftlichen Betätigung, die er bis kurz vor seinem Tode als Leiter einer Forschungsgruppe der Sächsischen Akademie der Wissenschaften zu Leipzig mit der aktuellen Thematik des Umweltschutzes ausübte.

Groß ist die Zahl seiner Schüler, von denen viele verantwortungsvolle, leitende Positionen in Industrie, Forschung und Lehre innehaben. Beeindruckt von seiner durch Ausgeglichenheit, Gerechtigkeitssinn und menschliche Wärme geprägten Arbeitsatmosphäre erinnern sie sich seiner in Dankbarkeit und Verehrung.

Friedrich Wolf

Bibliographie Franz Runge

I. Selbständige Veröffentlichungen

1. Organometallverbindungen, Chemie in Einzeldarstellungen. I. Teil: Organomagnesiumverbindungen. Wissenschaftliche Verlagsgesellschaft, Stuttgart 1932. 328 Seiten. Hrsg. I. Schmidt.
   Titel der russischen Ausgabe: MAGNIORGANITSCHESKIE COEDINENIJA, Leningrad 1937.
2. Organo-Metallverbindungen, Die organische Synthese mit Hilfe von Organometallverbindungen. I. Teil: Organoverbindungen der Alkalimetalle, II. Teil: Organoverbindungen von Magnesium und Zink. Wissenschaftliche Verlagsgesellschaft, Stuttgart 1944. 738 Seiten.
3. Was die Chemie vermag. Aufbau-Verlag, Berlin 1954. 98 Seiten.
4. Einführung in die Chemie und Technologie der Kunststoffe. Akademie-Verlag, Berlin, 1. Auflage 1952, 156 Seiten, 2. Auflage 1959, 246 Seiten, 3. Auflage 1963, 300 Seiten, 4. Auflage 1974, zusammen mit E. Taeger.
   Polnische Übersetzung der 1. Auflage: Wstep do chemii i technologii tworzyw sztucznych. Panstwowe wydawnictwo naukowe, Warszawa, 1956, 184 Seiten.

II. Beiträge in Zeitschriften, Sàmmelwerken usw.

1. F. Runge: Über o-Dioxybenzol-Derivate einiger heterocyclischer Ringsysteme, Dissertation, Göttingen 28. 6. 1922, 41 Seiten, auszugsweise im „Jahrbuch der Mathem.-Naturw. Fakultät der Universität Göttingen“ Juli/Dez. 1922, S. 139 (1923).
2. G. Tammann und F. Runge: Abhängigkeit des Potentials der Sauerstoffelektroden vom Druck. Z. anorg. Chemie *156* (1926) 85.
3. F. Runge: Über die kontinuierliche Darstellung von Enoläther. Angew. Chemie *62* (1950) 88.
4. F. Runge, W. Zimmermann und Irmtraud Pfeiffer: Indiumorganische Verbindungen. Z. anorg. u. allg. Chemie *267* (1951) H. 1—3, S. 39—48.
5. F. Runge und H. Hummel: Die katalytische Umsetzung von Azetylen mit Ammoniak. Chem. Tech. *3* (1951) Nr. 6, S. 163—168.
6. F. Runge: Stoffliche Grundlagen der organischen chemischen Industrie. Chem. Tech. *4* (1952) H. 1, S. 4—6.
7. F. Runge. Entwicklungstendenzen in der amerikanischen Kunststoffindustrie. Chem. Tech. *4* (1952) H. 9, S. 429—431.
8. F. Runge: Die Reaktion von Oxalylchlorid mit aliphatischen Kohlenwasserstoffen im ultravioletten Licht. Z. für Elektrochem. *56* (1952) H. 8, S. 779—781.
9. F. Runge und P. Maass: Untersuchungen über die BECKMANN-Umlagerung des Cyclohexanonoxims. Chem. Tech. *5* (1953) H. 8, S. 421—24; Wissenschaftl. Zeitschr. d. Martin-Luther-Universität Halle—Wittenberg *II* (1952/53), Naturwissenschaftl. Reihe H. 1, S. 1—9.
10. F. Runge: Darstellung von Tetracarbonsäuren aus Dioxynaphthalinen. Wissenschaftl. Zeitschr. d. Martin-Luther-Universität Halle—Wittenberg *II* (1952/53), Naturwissenschaftl. Reihe H. 1, S. 80—81.
11. F. Runge, O. Bankowski und G. Hoffmann: Die Hydratation von Propylen an festen Katalysatoren. Brennstoff-Chemie *34* (1953) H. 21/22, S. 330—333.

12. F. Runge und A. Meckelburg: Auffindung und Synthese von 4,9-Dimethyl-pyren. Chem. Ber. *86* (1953) H. 3, S. 373—376.
13. F. Runge und K. Goldann: Darstellung von Polycarbonsäuren. I. Mitteilg.: Darstellung der Äthan-1.1.2.2-tetraessigsäure. Chem. Ber. *86* (1953) H. 3, S. 377—380.
14. F. Runge, H.-J. Engelbrecht und G. Preusser: Die Kondensation von Sulfimidsalzen mit basisch substituierten Alkylhalogeniden. Chem. Ber. *86* (1953) H. 12, S. 1571—1576.
15. F. Runge und H. J. Philipp: Silikonisolierstoffe und ihre Anwendung in der Elektrotechnik. Die Technik, Messesonderheft *1953*, S. 158—160.
16. F. Runge: Die Entwicklung der Technischen Chemie in der Universität Halle—Wittenberg. Chem. Tech. *5* (1953) H. 1, S. 3—5; 450 Jahre Martin-Luther-Universität Halle—Wittenberg *III* (1945—1952) S. 155—160.
17. F. Runge und G. Adolphi: Vorschlag zur Destillation korrodierender Flüssigkeiten. Chemiker-Ztg *77* (1953) Nr. 14, S. 468—470.
18. F. Runge und A. Meckelburg: Isolierung und Identifizierung chemischer Individuen aus den hochsiedenden Rückständen der Hochdruckhydrierung von mitteldeutscher Braunkohle. Freiberger Forschungshefte, *1953*, H. A 17. S. 54—58.
19. F. Runge: Vollsynthetische Faserstoffe. Wissenschaftl. Zeitschrift der Martin-Luther-Univ. Halle—Wittenberg *III* (1953—1954) math.-nat. Reihe, H. 2, S. 507—514.
20. F. Runge und W. Zimmermann: Über Chlormethyl-chlorsilane. Chem. Ber. *87* (1954) H. 2, S. 282—287.
21. F. Runge und H. Laske: Untersuchungen zur Isolierung von Einzelkomponenten aus der Vorlauffettsäure der Paraffinoxydation. Fette und Seifen *56* (1954) H. 3, S. 145—149.
22. F. Runge, G. Geiseler und R. Kaufhold: Über die Polymerisation des Äthylens zu Schmierölen. 1. Untersuchungen über die Zusammensetzung der niedermolekularen Anteile des Polymerisationsproduktes mittels Ramananalyse. Erdöl und Kohle *7* (1954) S. 281—286.
23. F. Runge und K.-H. Zepf: Über die Wirksamkeit der verschiedenen Alkalimetalle im Isobutylölkontakt, Brennstoffchemie *35* (1954) S. 167—170.
24. G. Geiseler, H. Herold und F. Runge: Über die Polymerisation des Äthylens zu Schmierölen. 2. Untersuchungen über die Zusammensetzung der höhermolekularen Anteile des Polymerisationsproduktes durch Adsorptionsanalyse. Erdöl und Kohle *7* (1954) Nr. 6, S. 357—362.
25. F. Runge, R. Hueter und H.-D. Wulf: Darstellung von Polycarbonsäuren. II. Mitteilg.: Die Aufspaltung des α-substituierten Tetrahydrofuran-Ringes in der Alkalischmelze. Chem. Ber. *87* (1954) H. 10, S. 1430—1439.
26. F. Runge, J. Freytag und J. Kolbe: Inhaltsstoffe der hydrierten Braunkohle. I. Mitteilg. Chem. Ber. *87* (1954), H. 6, S. 873—881.
27. E. Profft, F. Runge und A. Jumar: Zur Kenntnis der Falicaine. IV. Über 4-Alkoxyphenylvinylketone. J. prakt. Chem. Reihe 4, Bd *1* (1954) H. 1—2, S. 57—86.
28. E. Profft, F. Runge und H. J. Blanke: Über alkoxyphenylierte heterocyclische Verbindungen. J. prakt. Chem. Reihe 4, Bd *1* (1954) H. 1—2, S. 110—128.
29. F. Runge, H.-J. Engelbrecht und H. Franke: Disulfimide als Reagenzien für die Isolierung bzw. Charakterisierung basischer organischer Stickstoffverbindungen. Chem. Ber. *88* (1955) H. 4, S. 533—541.
30. F. Runge und H. Laske: Chemische Technologie als Unterrichtsfach an Hochschulen. Chem. Tech. *7* (1955) H. 7, S. 381—382.
31. F. Runge und H. John: Chlorkautschuk aus Buna. Chem. Tech. *7* (1955) H. 9, S. 526 bis 529.

32. F. Runge und W. Zimmermann: Organisch substituierte Kieselsäuregele. J. prakt. Chem. Reihe 4, Bd *1* (1955) H. 5–6, S. 283–287.
33. F. Runge, E. Profft und R. Drux: Über Tetrahydrothiophen und seine Derivate. J. prakt. Chem. Reihe 4, Bd *2* (1955) H. 5–6, S. 279–292.
34. F. Runge und H. Herbst: Schutz von Hydroxyl-Gruppen. Angew. Chem. *68* (1956) Nr. 19, S. 618.
35. F. Runge: Die Reaktionsketten bei dem photochemischen Umsatz von Oxalylchlorid und von Phosgen mit Kohlenwasserstoffen. Z. Elektrochem. *60* (1956) H. 9/10, S. 956 bis 958.
36. F. Runge, H. Reinhardt und G. Kühnhanss: Über ein Verfahren zur Herstellung von Terephthalsäure aus Toluol und Harnstoffchlorid. Chem. Tech. *8* (1956) H. 11, S. 644–649.
37. F. Runge, H.-J. Engelbrecht und H. Franke: Isolierung, Charakterisierung und quantitative Bestimmung basischer organischer Wirkstoffe aus verschiedenen Arzneimitteln mit Hilfe von Disulfimiden. Die Pharmazie *12* (1957) S. 8–13.
38. F. Runge und F. Pfeiffer: Inhaltsstoffe der hydrierten Braunkohle. II: Isolierung und Identifizierung primärer und sekundärer Basen mit Hilfe von 4.4′-Dichlordiphenyldisulfimid. Chem. Ber. *90* (1957) Nr. 9, S. 1757–1760.
39. F. Runge, J. Behrens und A. Ernst: Diaryl-disulfimide als Urtitersubstanzen. Z. anal. Chem. *158* (1957) H. 4, S. 267–270.
40. F. Runge und R. Aust: Untersuchungen zur kationischen Polymerisation konjugierter Diolefine. Chem. Tech. *9* (1957) H. 12, S. 701–704.
41. P. Fijolka, J. Lenz und F. Runge: Analytische Untersuchungsmethoden für Polyester. Über Umwandlungen der Doppelbindung bei der Verseifung ungesättigter Polyester. Die Makromolekulare Chemie *26* (1957) H. 1/2, S. 61–66.
42. P. Fijolka, I. Lenz und F. Runge: Analytische Untersuchungsmethoden für Polyester. Carbonylgruppenbestimmung. Die Makromolekulare Chemie *23* (1957) H. 1, S. 60–70.
43. F. Runge und U. Koch. Darstellung von Polycarbonsäuren. III: Die Reaktion von Oxalylchlorid mit Acetylchlorid und einigen phenylsubstituierten Carbonsäurechloriden. Chem. Ber. *91* (1958) Nr. 6, S. 1217–1224.
44. F. Runge, Zaki El-Hewehi und D. Hempel: Zur Kenntnis des s-Trithians. J. prakt. Chem., Reihe 4, Bd 8 (1969), H. 1–2, S. 1–16.
45. F. Runge und Zaki El-Hewehi: Zur Kenntnis der Trithiokohlensäure, I. Über aliphatische und araliphatische Trithiokohlensäure-diester, J. prakt. Chem. Reihe 4, Bd *7* (1959) H. 5–6, S. 268–278.
46. F. Runge, Zaki El-Hewehi und E. Taeger: Zur Kenntnis der Trithiokohlensäure. II. Darstellung von Trithiokohlensäure-di-ester durch Umsetzung von Thiophosgen mit Thiolen und Thiophenolen, J. prakt. Chem. Reihe 4, Bd *7* (1959) H. 5–6, S. 279–285.
47. F. Runge, A. Jumar und P. Held: Über einige unsymmetrische Disulfide. J. prakt. Chem. Reihe 4, Bd 8 (1959) H. 1–2, S. 44–54.
48. Zaki El-Hewehi und F. Runge: Zur Kenntnis der Trithiokohlensäure. III. Über die Verwendbarkeit von Trithiokohlensäure-di-ester als „Additives" und Sparbeizmittel. J. prakt. Chem. Reihe 4, Bd *9* (1959) H. 1–2, S. 33–42.
49. Zaki El-Hewehi und F. Runge: Über die Herstellung und Verwendbarkeit organischer Schwefel-Verbindungen als Schädlingsbekämpfungsmittel. Wissenschaftl. Zeitschr. der Martin-Luther-Univ. Halle–Wittenberg *VIII*/6 (1959) Okt., S. 1013–1024.
50. F. Runge: Reinigung von Abwässern mit Hilfe von Adsorbentien. Jahrbuch der Sächsischen Akademie der Wissenschaften zu Leipzig *1957–1959* (Berlin 1961) S. 189–194.

51. F. Runge: Methylolverbindungen des Melamins. Plaste und Kautschuk *6* (1959) H. 12, S. 580.
52. F. Wolf, F. Runge und R. Korn: Untersuchungen über die katalytische Oxydation von Chlorwasserstoff mit Sauerstoff zu Chlor (DEACON-Prozeß). Z. anorg. u. allg. Chem. *304* (1960) H. 1–2, S. 48–57.
53. F. Runge, Zaki El-Hewehi und H.-J. Renner: Zur Kenntnis der Trithiokohlensäure. IV. Über die Herstellung und Eigenschaften von Trithiokohlensäure-di-estern und ihre Verwendbarkeit als Schädlingsbekämpfungsmittel. J. prakt. Chem. Reihe 4, Bd *11* (1960) H. 5–6, S. 284–308.
54. Zaki El-Hewehi und F. Runge: Sulfonsäure-Derivate; Herstellung und Verwendbarkeit als Mottenschutzmittel. J. prakt. Chem. Reihe 4, Bd *16* (1962) H. 5–6, S. 297 bis 336.
55. F. Runge und E. Taeger: Methylen durch α-Eliminierung aus grignardiertem Chlordimethyläther. Z. Chem. *2* (1962) H. 3, S. 84–85.
56. F. Runge und U. Koch: Über die Darstellung von Dihydroxydiphenyl-disulfimiden, Z. Chem. *2* (1962) H. 4, S. 114–115.
57. F. Runge, F. Wolf und R. Bachmann: Beziehungen zwischen Struktur und elektrochemischen Eigenschaftskennwerten von Ionenaustauschermembranen (Anionenaustauschermembranen). Dechema-Monographien Bd *47* (1962) S. 465–476.
58. F. Runge, Zaki El-Hewehi und E. Taeger: Zur Kenntnis der Trithiokohlensäure. V. Über die Synthese und biologischen Eigenschaften von heterocyclischen Trithiokohlensäure-di-estern. J. prakt. Chem. Reihe 4, Bd *18* (1962) H. 5–6, S. 262–268.
59. E. Taeger, Zaki El-Hewehi und F. Runge: Zur Kenntnis der Trithiokohlensäure. VI. Über die Reaktion von aromatischen und heterocyclischen Trithiokohlensäure-di-estern mit Diazomethan und Phenyldiazomethan. J. prakt. Chem. Reihe 4, Bd *18* (1962) H. 5–6, S. 269–274.
60. Zaki El-Hewehi, E. Taeger und F. Runge: Zur Kenntnis der Trithiokohlensäure. VII. Über die Reaktion von aromatischen und heterocyclischen Trithiokohlensäure-di-estern mit primären und sekundären Aminen, Hydrazin und Phenylhydrazin. J. prakt. Chem. Reihe 4, Bd *18* (1962) H. 5–6, S. 275–282.
61. Zaki El-Hewehi, E. Taeger und F. Runge. Zur Kenntnis der Trithiokohlensäure. VIII. Über die Einwirkung von Halogen auf Trithiokohlensäure-di-ester. J. prakt. Chem. Reihe 4, Bd *18* (1962) H. 5–6, S. 283–286.
62. F. Runge: Reinigung von Abwässern mit Hilfe von Adsorbentien. Jahrbuch der Sächsischen Akademie der Wissenschaften zu Leipzig *1960–1962* (Berlin 1964) S. 236–242.
63. F. Runge, A. Jumar und F. Koehler: Über einige unsymmetrische heterocyclische Disulfide, II. J. prakt. Chem. Reihe 4, Bd *21* (1963) H. 1–2, S. 39–49.
64. F. Runge, E. Taeger, Ch. Fiedler und E. Kahlert: Über die Einwirkung von Metallen auf α-Halogenäther, I. J. prakt. Chem. Reihe 4, Bd *19* (1963) H. 1–2, S. 37–44.
65. P. Fijolka, Y. Shahab und F. Runge: Beiträge zur kalorimetrischen Untersuchung der Kopolymerisation von ungesättigten Polyestern mit Styrol. Plaste und Kautschuk *10* (1963) H. 8, S. 460–461.
66. P. Fijolka, W. Radowitz und F. Runge: Beiträge zur papierchromatografischen Analyse der Dikarbonsäuren und Glykole in Polyesterharzen. Plaste und Kautschuk *10* (1963) H. 9, S. 521–525.
67. F. Runge und H. Kraus: Das Verhalten von Alkylaminen mit gerad- und ungeradzahliger Kohlenstoffkette bei der Flotation von Kalisalzen. Fette, Seifen, Anstrichmittel *66* (1964) H. 7, S. 505–507.

68. K.-H. Mohr, F. Runge und F. Wolf: Die Kinetik der Isopropanolhydrierung in einer Katalysatorwirbelschicht. Chem. Tech. *16* (1964) H. 8, S. 457–463.
69. F. Runge, H. Bendix und H. Breuer: Über die Technologie und Chemie des Hexachlorcyclopentadiens. Chem. Tech. *16* (1964) H. 4, S. 203–208.
70. F. Runge: Das Institut für Technische Chemie der Universität Halle–Wittenberg. ACHEMA-Jahrbuch Bd *I* (1962/64) S. 298–299.
71. F. Runge: Reinigung von Abwässern mit Hilfe von Adsorbentien. Jahrbuch der Sächsischen Akademie der Wissenschaften zu Leipzig *1963–1965* (Berlin 1967) S. 155–162.
72. F. Runge, H. Ehrhardt und G. Penndorf: Über vinylsubstituierte Kieselsäuregele. Makromolekulare Chemie *81* (1965) S. 68–77.
73. F. Runge: Membranen aus vinylsubstituierten Kieselgelen. Desalination *3* (1967) H. 3, S. 304–307.
74. F. Runge: Reinigung von Abwässern mit Hilfe von Adsorbentien. Jahrbuch der Sächsischen Akademie der Wissenschaften zu Leipzig *1966–1968* (Berlin 1970) S. 161 bis 177.
75. F. Runge und W. Abel: Zur Polymerisation von Alkenyloxysilanen mit organischen Vinylverbindungen. Die Makromolekulare Chemie *120* (1968) S. 148–153.
76. F. Runge, G. Wilde, P. H. Berthold und H. Seumel: Vorschlag eines komplexen Analysenverfahrens zur Charakterisierung technischer Kohlenwasserstoffgemische im Siedebereich 160–250° C. J. prakt. Chem. *312* (1970) H. 1, S. 55–68.
77. F. Runge und E. Laqua: Silanvernetztes Polystyrol als katalytisch wirksames Kationenaustauscherharz. Die Makromolekulare Chemie *136* (1970) S. 315–317.
78. F. Runge, E. Laqua und H. Thielemann: Siloxanvernetztes Polyvinylacetat als Adsorbens für Phenole I. J. prakt. Chem. Bd *312* (1970) H. 4, S. 734–736.
79. F. Runge: Reinigung von Abwässern mit Hilfe von Adsorbentien. Jahrbuch der Sächsischen Akademie der Wissenschaften zu Leipzig *1969–1970* (Berlin 1972) S. 128 bis 135.
80. E. Laqua, F. Runge und H. Thielemann: Experimentelle Untersuchungen zur adsorptiven Eliminierung und Identifizierung organischer Wasserinhaltsstoffe unter Verwendung unterschiedlicher Filtermedien, Acta hydrochim. hydrobiol. *1* (1973) H. 4, S. 345 bis 351.
81. F. Runge und J. Ackermann: Disiloxanvernetzte Polymere als Phenoladsorber. Acta hydrochim. hydrobiol. *1* (1973) H. 6, S. 573–576.
82. F. Runge und E. Laqua: Siloxanvernetztes Polyvinylacetat-Suspensionsharz ein Adsorbens für Phenole, Acta hydrochim. hydrobiol. *2* (1974) H. 3, S. 241–248.

III. Gedruckte Vorträge

1. F. Runge: Metallorganische Verbindungen des Galliums und Indiums. Tagungsbericht der Chemischen Gesellschaft in der DDR, Hauptjahrestagung. Berlin 1955, S. 74–76.
2. F. Runge, G. Kühnhanss und H. Seumel: Über die Kohlenwasserstoffe des Espenhainer Braunkohlenleichtöles. Freiberger Forschungshefte *A 131* (Jan. 1960) S. 84–97.
3. F. Runge: Kristallinität der Hochpolymeren in den Kunststoffen. Forschungen und Fortschritte Bd *35* (1961) H. 10, S. 294–296.
4. F. Runge: Über die Polymerisation ungesättigter organischer Verbindungen. Sitzungsberichte der Sächsischen Akademie der Wissenschaften zu Leipzig, Math.-nat. Klasse Bd *102* (1957) H. 3, 13 Seiten.
5. F. Runge: Grignard und die nach ihm benannte Synthese. Sitzungsberichte der Sächsi-

schen Akademie der Wissenschaften zu Leipzig, Math.-nat. Klasse Bd *107* (1966) H. 2, 17 Seiten.

6. F. Runge: Organische Disulfimide in Wissenschaft und Technik. Sitzungsberichte der Sächsischen Akademie der Wissenschaften zu Leipzig, Math.-nat. Klasse, Bd *108* (1970) H. 7, 24 Seiten.

IV. Ungedruckte Vorträge und Schriften

F. Runge: Chlor aus Chlorwasserstoff — der DEACON-Prozeß. Vortrag gehalten auf Sitzung der Deutschen Akademie der Naturforscher „Leopoldina", Halle, am 17. 5. 1960.

F. Runge: Die Silikone, ein zusammenfassender Bericht, 126 Seiten, Anfang d. 50er Jahre.

V. Nachrufe

F. Runge: Costin D. Nenitzescu 15. 7. 1902 — 28. 7. 1970. Jahrbuch der Sächsischen Akademie der Wissenschaften zu Leipzig *1969—1970* (Berlin 1972) S. 281—284.

VI. Mitherausgebertätigkeit

Ost-Rassow: Lehrbuch der Chemischen Technologie Bd I und II. 27., neubearbeitete Auflage, Johann-Ambrosius-Barth-Verlag, Leipzig 1965, 1727 Seiten (Zusammen mit W. K. Schwarze).

Desalination. The International Journal on the Science and Technology of Water Desalting. Elsevier Publishing Company, Amsterdam, 1967—1968.

VII. Patente

1. Darstellung von Butanelen, DRP 536362 (1930).
2. Darstellung von Diolefinen, DRP 578038 (1930).
3. Buten-2 aus Buten-1, DRP 583790 (1935).

Nichtveröffentlichte Patentanmeldungen während des 2. Weltkrieges:

4. Darstellung von Enoläthern.
5. Darstellung von gesättigten Äthern.
6. Verfahren zum Betreiben gasbeheizter Öfen.
7. Ausführung von Grignard-Reaktionen.
8. Verfahren zur Trennung von Gasgemischen
   F. Runge, F. Wolf, R. Grießbach, DWP 7758, 12. 12. 52/11. 2. 54.
9. Organisch substituierte Kieselgele
   F. Runge, W. Zimmermann, WP 12 e/22227 (1952).
10. Verfahren zur Durchführung katalytischer Reaktionen mittels Wasserstoffaustauschern
    F. Runge, W. Zimmermann, G. Naumann, (WP 12 g/2778) DWP 8560, 29. 12. 53/18. 2. 55.
11. Verfahren zur Herstellung von substituierten Malonsäurederivaten
    F. Runge, J. Teubel, G. Kühnhanns, DWP 8600, 12. 3. 53/15. 11. 54.
12. Verfahren zur Herstellung von Terephthalsäure
    F. Runge, G. Kühnhanns, H. Reinhardt, DWP 10866, 12. 3. 53/3. 12. 55.

13. Verfahren zur Umwandlung von 1,2-Dichlorbuten-(3) in 1,4-Dichlorbuten-(2)
H. Stemmler, F. Runge, R. Bemmann, DWP 14502, 27. 8. 54.
14. Verfahren zur Durchführung von Stoff- und/oder Wärmeaustauschprozessen unter Verwendung von Füllkörpern
F. Runge, F. Wolf, W. Günther, (WP 12 e/39487) DWP 22425, 22. 10. 55/22. 12. 61.
15. Verfahren zur Herstellung von Vinylverbindungen stickstoffhaltiger Heterocyclen
F. Runge, G. Naumann, M. Morgner, DWP 13099, 26. 9. 55/25. 4. 57.
16. Verfahren zur Herstellung von Kationenaustauschern
F. Runge, J. Behrends, DWP 14852, 13. 12. 55.
17. Verfahren zur Herstellung reiner, monomerer Methylolverbindungen des Melamins
F. Runge, WP 12p/49131 (1957).
18. Verfahren zur Herstellung von phosphorsäurehaltigen Siliconharzen
F. Runge, K. Helke, (WP 39c/49644) DWP 16966, 25. 6. 57/2. 2. 59.
19. Verfahren zur Verbesserung der Bauschelastizität von Polyamidfasern
W. Roth, F. Runge, DWP 21776, 6. 7. 59/16. 8. 61.
20. Verfahren zur Herstellung von siliciumhaltigen Hochpolymeren
F. Runge, W. Abel, (WP 39c/129577) DWP 65457, 15. 1. 68/5. 2. 69.
21. Verfahren zur Herstellung von Adsorberharzen
F. Runge, E. Laqua, WP C 08f/174136, 18. 10. 73.

Zur Biobibliographie

Schriftenverzeichnisse. E. Laqua, siehe vorliegendes Jahrbuch S. 245—251. — Archiv der Univ. Halle, Rep. 31, Nr. 439: Vorschläge für staatliche Auszeichnungen (1958). — Jahrbuch der Sächsischen Akademie der Wissenschaften zu Leipzig *1954—1956* (Berlin 1958) S. 188—189; *1960—1962* (Berlin 1964) S. 289; *1963—1965* (Berlin 1967) S. 223—224; *1966—1968* (Berlin 1970) S. 227; *1969—1970* (Berlin 1972) S. 166; *1971—1972* (Berlin 1974) S. 177. — Kürschners Deutscher Gelehrten-Kalender *10*$_2$ (Berlin 1966) S. 2060; *11* (Berlin 1970) S. 2506. — Poggendorffs Handwörterbuch *VII*$_3^a$ (Berlin 1959) S. 853—854.

Festschrift. Franz Runge zum 65. Geburtstag. Journal für praktische Chemie *7* (1958) H. 3—4, S. 117—240.

Würdigungen. Zum 60. Geburtstag: R. Zaunick, in: Hochschulwesen *1* (1953/54) H. 4, S. 38—39. — Derselbe, in: Neuer Weg, Halle, vom 5. Dez. 1953. — Zum 65. Geburtstag: E. Leibnitz, in: J. prakt. Chem. *7* (1958) S. 117—118. R. Zaunick, in: LDZ *13* (Halle 1958) Nr. 283 vom 5. Dez. — Neuer Weg, Halle, vom 5. Dez. 1958. — Zur Verleihung des Nationalpreises: H. Laske, in: Mitteilungsbl. Chem. Ges. DDR *6* (1959) S. 180—181. — Forum *13* (1959) Nr. 40. — Zum 70. Geburtstag: H. Laske, in: Mitteilungsbl. Chem. Ges. DDR *10* (1963) S. 236—237. — Neues Deutschland *18* (1963) Nr. 334 vom 5. Dez. — Zum 75. Geburtstag: F. Wolf, in: Ber. Bunsenges. physik. Chem. *72* (1968) H. 9—10, S. 1081 bis 1082. — Neuer Weg, Halle, vom 6. Dez. 1968.

Nachrufe. F. Wolf, siehe vorliegendes Jahrbuch S. 242—244 (mit Bildnis). — Freiheit *28* (Halle 1973) 14. Sept. — Ber. Bunsenges. physik. Chem. *77* (1973) H. 12 (mit Bildnis).

Dr. rer. nat. Erika Laqua

Fritz Stather

5. 11. 1901—31. 3. 1974

Am 31. März 1974 verstarb im 73. Lebensjahr der langjährige Direktor des Deutschen Lederinstitutes in Freiberg, Prof. Dr. phil. habil. Fritz Stather. Sein Name ist auf das engste mit dem von ihm begründeten und geleiteten Institut und der wissenschaftlichen Entwicklung der Gerbereichemie in Theorie und Praxis verbunden. Auf einem wirtschaftlich bedeutungsvollen Gebiet der angewandten Chemie hat er es verstanden, mit klarem Blick die zur Lösung technologischer Probleme oder zur Entwicklung technischer Verfahren erforderlichen wissenschaftlichen Grundlagen zu schaffen.

Fritz Stather wurde am 5. 11. 1901 in Offenburg/Baden geboren und besuchte dort die Volksschule und das humanistische Gymnasium. Nach der Ablegung der Reifeprüfung in Lörrach studierte er von 1920 bis 1924 Chemie an der Universität Freiburg/Br. und der Technischen Hochschule Dresden. Hier geriet er in den Wirkungskreis des zu seiner Zeit weltbekannten Eiweißforschers Max Bergmann am damaligen Kaiser-Wilhelm-Institut für Lederforschung, unter dessen Leitung er seine Dissertation mit dem Thema: „Eine neue Art der Disaccharidbildung" ausführte. Die Promotion erfolgte an der Universität Berlin mit dem Prädikat „magna cum laude".

Von 1924 bis 1930 blieb Fritz Stather bei Max Bergmann als Assistent am Institut für Lederforschung und vertiefte seine Kenntnisse auf dem Gebiet der Eiweißchemie durch experimentelle Arbeiten in chemischer und histologischer Richtung. Im Jahre 1929 habilitierte er sich an der Technischen Hochschule Dresden für organisch-technische Chemie, insbesondere Gerbereichemie, und wurde zum Privatdozenten ernannt.

Mit der Übernahme der Leitung der Deutschen Versuchsanstalt für Lederindustrie in Freiberg/Sa. im Jahre 1930 begann seine selbständige Schaffensperiode, die sich für die wissenschaftlich-technische Entwicklung der Lederherstellung und auch der bis dahin rein empirisch betriebenen Pelzfellerzeugung als überaus fruchtbar erweisen sollte. Durch die Vereinigung der Versuchsanstalt mit der Deutschen Gerberschule entstand 1938 die „Deutsche Versuchsanstalt und Fachschule für Lederindustrie, Freiberg", das spätere „Deutsche Lederinstitut", das Fritz Stather in wenigen Jahren zur Weltgeltung brachte. Die Funktion als Direktor dieses Institutes behielt er bis zu seiner Emeritierung im Jahre 1966.

Der von ihm vorgenommenen systematischen Ausgestaltung der Studienpläne ist es ohne Zweifel zuzuschreiben, daß in einem zunächst 2-, dann 4- und schließlich 6-semestrigen Studium den Schülern der „Gerberschule" in ausgewogenem Gleichgewicht von theoretischer und praktischer Ausbildung alle notwendigen Kenntnisse vermittelt wurden, wie sie den Erwartungen der Lederindustrie und den mit ihr verbundenen Industriezweigen entsprachen. Der gute Ruf des Freiberger Institutes zog viele Schüler aus dem In- und Ausland herbei, um von dieser zugleich modernen und praxisgerechten Ausbildung zu profitieren.

Diese Lehrtätigkeit innerhalb des „Deutschen Lederinstituts" erfuhr bald eine Ergänzung durch einen Lehrauftrag an der Chemischen Abteilung der Technischen Hochschule Dresden, wo er 1935 zum a. o. Professor ernannt wurde. Hierdurch konnte er eine Reihe von begabten Hochschulabsolventen für die Gerbereichemie interessieren und gewinnen. Für seine Wirksamkeit als akademischer Lehrer kam ihm sein außergewöhnliches Rednertalent zustatten, das es ihm ermöglichte, auch komplizierte Zusammenhänge in zugleich fesselnder, überzeugender und klarer Weise darzustellen. Diese besondere Begabung machte auch sein Auftreten in Vorträgen oder in Diskussionsbeiträgen auf Kongressen besonders eindrucksvoll und trug wesentlich zum Ansehen des von ihm vertretenen Gebietes und seines Institutes bei. 1941 ernannte ihn die Technische Hochschule Dresden in Anerkennung seiner wissenschaftlichen Leistungen zum apl. Professor für Gerbereichemie.

Da Stather dem Faschismus ablehnend gegenübergestanden hatte, wurde er 1945 sowohl als Direktor der Deutschen Versuchsanstalt und Fachschule für Lederindustrie in Freiberg wie als apl. Professor an der Technischen Hochschule Dresden bestätigt.

Seine Publikationen wissenschaftlicher und allgemein verständlicher Art umfassen das gesamte Gebiet der Gerbereichemie, der Gerbereitechnologie und der Werkstoffe Leder und Rauchwaren. Schon seine Arbeiten über die Rohhautkonservierung und Häuteschäden haben ein bis dahin wissenschaftlich vernachlässigtes Gebiet durch systematische Grundlagenforschung erschlossen und hierdurch einen außerordentlichen volkswirtschaftlichen Nutzen erzielt. Seine wissenschaftlichen Untersuchungen wie seine etwa 350 Publikationen, zahlreichen Patente und mehrere Lehrbücher sichern ihm ein bleibendes Andenken. Ganz besonders gilt dies von dem 1948 erschienenen und mehrfach wiederaufgelegten Lehrbuch „Gerbereichemie und Gerbereitechnologie“, das international als Standardwerk und als umfassendstes Kompendium für dieses Fachgebiet angesehen wird. Ebenso werden auch die Monographien über „Haut- und Lederfehler“ und „Leder und Kunstleder“ als grundlegende Werke geschätzt und sowohl zur Information des Fachmannes in der Praxis wie zur Ausbildung des Nachwuchses in vielen Ländern verwendet.

Durch seine Beherrschung der chemischen Grundlagen und die vollkommene Kenntnis aller technologischen Einzelheiten konnte Fritz Stather wissenschaftlich exakte Methoden mit Klarblick und Zielstrebigkeit auf die praktischen Probleme der Gerbereitechnologie orientieren, womit er zum Pionier und Bahnbrecher für sein Wissenschaftsgebiet wurde. Von größter Bedeutung war dieses Arbeitsprinzip für die Entwicklung und Erzeugung der synthetischen Gerbstoffe. Durch seine systematischen Untersuchungen über die Herstellung, chemische Konstitution und die Wirkungsweise der pflanzlichen Gerbstoffe hat er die Grundlagen für die großtechnische Produktion synthetischer Austauschstoffe und damit die Voraussetzungen geschaffen, daß ein bedrohlicher wirtschaftlicher Engpaß überwunden werden konnte.

Als besonders verdienstvoll sind auch die von ihm bald nach Beginn seiner Tätigkeit als Direktor des Deutschen Lederinstituts eingeleiteten Untersuchungen zur Ermittlung objektiver Qualitätsmerkmale des Leders anzusehen, die vor allem die physikalischen Eigenschaften betrafen. Auf Grund verbesserter Methoden und tausender von Untersuchungen führten diese Bemühungen zur Aufstellung exakter Güterichtlinien, die in die internationale n Festlegungen von Qualitätsnormen eingegangen sind.

Diese vielfältigen wissenschaftlichen Leistungen wären undenkbar, wenn nicht Fritz Stather durch rastlose Tätigkeit, durch eine völlige Identifizierung mit seiner Aufgabe und eine außerordentliche Arbeitsleistung sowohl die organisatorischen Pflichten als Institutsdirektor wie die wissenschaftlichen Aufgaben voll ausgefüllt hätte. Höchsten Einsatz forderte er nicht nur von seinen Mitarbeitern, sondern vor allem und zuerst von sich selbst. Dabei war die organisatorische Belastung sehr stark, da das Institut nach dem von ihm vollzogenen Ausbau bei einer Fläche von 3750 m^2 etwa 180 Wissenschaftler, Techniker und Facharbeiter aufwies — die Schüler nicht gerechnet.

Seine wissenschaftlich und wirtschaftlich gleichermaßen bedeutenden Arbeiten haben Fritz Stather naturgemäß eine Reihe hoher Auszeichnungen eingetragen. In Anerkennung seiner wissenschaftlichen Leistungen berief ihhn die Sächsische Akademie der Wissenschaften zu Leipzig 1958 zu ihrem ordentlichen Mitglied, womit — nach seinen eigenen Worten — „erstmalig die Gerbereiwissenschaft zum akademiewürdigen Zweig am Baume der Wissenschaft“ erklärt wurde.

Auch sonst sind ihm zahlreiche Ehrungen zuteil geworden: 1955 wurde er mit dem Ehrentitel „Hervorragender Wissenschaftler des Volkes“ ausgezeichnet. Er war ordentliches Mitglied der Akademie der Wissenschaften der DDR, Ehrenmitglied der KdT und Träger ihres Goldenen Ehrenzeichens sowie Ehrenmitglied des Vereins Österreichischer Gerbereitechniker. Auch sein Buch „Gerbereichemie und Gerbereitechnologie“ erfuhr nach dem Erscheinen der 3. Auflage durch die Verleihung eines Jahrespreies schon 1958 eine besondere Anerkennung. 1967 erhielt er die BERZELIUS-Medaille der Königlich Schwedischen Akademie der Wissenschaften zu Stockholm sowie die Edmund-STIASNY-Medaille des Vereins für Gerberei-Chemie und -Technik.

Seine zahlreichen Schüler, Mitarbeiter und alle, die ihn näher kennenlernen durften, werden nicht nur den Verlust eines erfolgreichen und bahnbrechenden Wissenschaftlers beklagen, sondern auch den eines aufrechten und warmherzigen Menschen. Möge die ihm eigene Dynamik und Ausstrahlungskraft noch lange in seinem Sinne weiterwirken!

Ulrich Freimuth

Bibliographie Fritz Stather

Bücher

1. Wilson-Stather-Gierth: Die Chemie der Lederfabrikation. 2 Bände, 1040 Seiten, Julius Springer, Wien, 1930 und 1931.
2. F. Stather: Haut- und Lederfehler. 178 Seiten, Julius Springer, Wien, 1. Auflage 1934, 2. Auflage 1952.
3. F. Stather: Das Leder. 36 Seiten, Freiberg, 1. Auflage 1936, 2. Auflage 1939.

4. F. Stather: Gerbereichemie und Gerbereitechnologie. 919 Seiten, Akademie-Verlag Berlin, 1. Auflage 1948, 4. Auflage 1967. Die 2. Auflage (1951) ist in das Chinesische übersetzt. 925 Seiten, Erscheinungsjahr 1958.
5. F. Stather: Leder und Kunstleder. 139 Seiten, Akademie-Verlag Berlin, 1. Auflage 1949, 3. Auflage 1959.

Beitrag für ein Handbuch

F. Stather: Die Rohhautschäden. Im Handbuch der Gerbereichemie und Lederfabrikation, herausgegeben von M. Bergmann und W. Graßmann, Springer-Verlag Wien, 1. Band, 1. Teil, 1944, S. 896—950.

Wissenschaftliche und technische Experimentalarbeiten

1. M. Bergmann, A. Miekeley und F. Stather: Über das Anhydrid eines Disaccharids aus 4-Oxy-4-aceto-butylalkohol. Ber. Dtsch. Chem. Ges. *56* (1923) S. 2255—2257.
2. F. Stather: Über das Tri-[m-nitro-benzoyl]-glycerin. Ber. Dtsch. Chem. Ges. *57* (1924) S. 1392—1393.
3. M. Bergmann, A. Miekeley und F. Stather: Über das Anhydrid eines Disaccharids aus 4-Oxy-4-aceto-butylalkohol (II). Ber. Dtsch. Chem. Ges. *58* (1925) S. 82—85.
4. M. Bergmann und F. Stather: Über die Veränderung von Keratin durch Alkalien. Collegium *1925*, S. 109—110.
5. M. Bergmann und F. Stather: Umwandlung eines cystinhaltigen Diketopiperazins. Hoppe-Seyler's Zschr. f. physiol. Chem. *152* (1926) S. 189—201.
6. M. Bergmann und F. Stather: Studien zum Äschervorgang: Über die Veränderung von Keratinen durch Alkalien (2. Mitteilung). Collegium *1926*, S. 249—256.
7. M. Bergmann und F. Stather: Über isomere Dioxopiperazine: Iso-3-methylen-6-iso-butyl-2,5-dioxopiperazin. Liebigs Ann. d. Chem. *448* (1926) S. 32—37.
8. F. Stather: Untersuchungen über Salzflecken (1. Mitteilung über Häuteschäden). Collegium *1928*, S. 567—599.
9. M. Bergmann und F. Stather: Rote und blaue Stockflecken auf feuchtem Chromleder (2. Mitteilung über Häute- und Lederschäden). Collegium *1929*, S. 326—327.
10. M. Bergmann, F. Stather und L. Seligsberger: Über die Schwellungsänderungen der Haut während der Chromgerbung. Collegium *1929*, S. 397—404.
11. F. Stather und E. Liebscher: Über das Rotwerden gesalzener Rohhäute (3. Mitteilung über Häuteschäden) Collegium *1929*, S. 427—437.
12. F. Stather und E. Liebscher: Zur Bakteriologie des Rotwerdens gesalzener Rohhäute (4. Mitteilung über Häute- und Lederschäden). Collegium *1929*, S. 437—450.
13. F. Stather: „Rote Verfärbung" und „Rote Erhitzung" auf gesalzehen Häuten (5. Mitteilung über Häute- und Lederschäden). Collegium *1930*, S. 151—153.
14. F. Stather, G. Schuck und E. Liebscher: Violette Flecken auf gesalzenen Rohhaut (6. Mitteilung über Häute- und Lederschäden). Collegium *1930*, S. 153—161.
15. F. Stather und G. Schuck: Untersuchungen über Salzflecken II. Narbenschäden auf geäscherter Haut und chromgarem Leder (7. Mitteilung über Häute- und Lederschäden). Collegium *1930*, S. 161—169.
16. F. Stather und E. Liebscher: Zur Bakteriologie violettverfleckter gesalzener Rohhaut (8. Mitteilung über Häute- und Lederschäden). Collegium *1930*, S. 170—175.
17. F. Stather: Vorschläge zur Neufestsetzung der Vorschriften für das Filterverfahren der quantitativen Gerbstoffbestimmung. Collegium *1930*, S. 480—493.

18. F. Stather: Zur Kenntnis weniger bekannter tropischer Gerbmaterialien. Collegium *1931*, S. 256–262.
19. F. Stather: Über den mittleren Wassergehalt pflanzlich gegerbter Unterleder. Collegium *1931*, S. 254–256.
20. F. Stather und R. Lauffmann: Zum qualitativen Nachweis einer Sulfitierung von Gerbextrakten. Collegium *1931*, S. 314–315.
21. F. Stather und H. Machon: Über den Einfluß der Temperatur auf den Abbau von Kollagen durch Pankreatin. Biochem. Zschr. *239* (1931) S. 430–433.
22. M. Bergmann, F. Stather, W. Hausam und E. Liebscher: Über die sogenannten Salzstippen (11. Mitteilung über Häute- und Lederschäden) Collegium *1931*, S. 538–544.
23. F. Stather: Hautpulver. Kommiss. Ber. d. Gemeinschaftskommiss. der internat. Gerbereichem. Verbände. Collegium *1931*, S. 833–835.
24. F. Stather und H. Machon: Beizuntersuchungen mit einem Enzymgemisch aus Fischeingeweiden. Collegium *1931*, S. 721–723.
25. F. Stather: Sind gefroren gewesene flüssige Gerbextrakte zur Musterziehung noch geeignet? Collegium *1931*, S. 874–876.
26. F. Stather und R. Lauffmann: Über die Veränderung von Harzen im Hinblick auf das Ausharzen von Leder. Collegium *1932*, S. 74–83.
27. F. Stather und H. Sluyter: Über das Hart- und Brüchigwerden von Schuhoberleder beim Tragen am Schuh. Ledertechn. Rundschau *1932*, Nr. 5.
28. F. Stather und R. Lauffmann: Über das Verhalten sulfonierten Rizinusöls beim Lickern von Chromleder I. Collegium *1932*, S. 391–404.
29. F. Stather: Über die Zerkleinerung des Leders für die chemische Analyse. Ledertechn. Rundschau *1932*, Nr. 5.
30. F. Stather und R. Lauffmann: Zum qualitativen Nachweis der Gegenwart von Sulfit bzw. einer Sulfitierung von Gerbextrakten II. Collegium *1932*, S. 511–513.
31. F. Stather und R. Lauffmann: Über das Verhalten sulfonierten Rizinusöls beim Lickern von Chromleder II. (2. Mitteilung über Untersuchungen zum Lickerprozeß). Collegium *1932*, S. 672–679.
32. F. Stather: Gerbmittel- und Gerbextrakt-Statistik der Deutschen Versuchsanstalt für Lederindustrie Freiberg, 1927–1931. Ledertechn. Rundschau *1932*, Nr. 10 u. 12, *1933*, Nr. 2.
33. F. Stather und R. Lauffmann: Über das Lickern von Chromleder mit Fettgemischen aus sulfoniertem Rizinusöl und unsulfonierten Ölen (3. Mitteilung über Untersuchungen zum Lickerprozeß). Collegium *1932*, S. 940–946.
34. F. Stather: Über die Diffusion von Kastaniengerbstoff und Quebrachogerbstoff in tierische Haut (Untersuchungen zur Charakteristik pflanzlicher Gerbstoffe I). Collegium *1933*, S. 9–22.
35. F. Stather und R. Lauffmann: Über das Verhalten von Seife und Seife-Fettgemischen beim Lickern von Chromleder (4. Mitteilung über Untersuchungen zum Lickerprozeß). Collegium *1933*, S. 129–137.
36. F. Stather und R. Lauffmann: Über das Verhalten von Eigelb und Eigelb-Fettgemischen beim Lickern von Chromleder (5. Mitteilung über Untersuchungen zum Lickerprozeß). Collegium *1933*, S. 138–141.
37. F. Stather: Vergleichende Untersuchungen über das Diffusionsvermögen verschiedener pflanzlicher Gerbstoffe in tierische Haut (Untersuchungen zur Charakteristik pflanzlicher Gerbstoffe II). Collegium *1933*, S. 316–327.
38. F. Stather und R. Lauffmann: Vergleichende Untersuchungen über das Verhalten

FRANZ RUNGE

5. 12. 1895 – 29. 8. 1973

FRITZ STATHER

5. 11. 1901 – 31. 3. 1974

GUSTAV E. R. SCHULZE
24. 2. 1911 – 5. 10. 1974

LEOPOLD WOLF
23. 11. 1896 – 6. 11. 1974

verschiedener Fettlicker beim Lickern von chromgegerbtem Hautpulver und Chromleder aus intakter Haut. Collegium *1933*, S. 394—401.

39. F. Stather: Zur Frage der Sumachanalyse. Ledertechn. Rundschau *1933*, Nr. 7.
40. F. Stather: Hautpulver II. Kommiss. Ber. d. Gemeinschaftskommiss. der internat. Gerbereichem. Verbände. Collegium *1933*, S. 451—453.
41. F. Stather und H. Sluyter: Über den Einfluß der Konservierungsart der Rohhaut auf ihr Weich-, Blößen- und Lederrendement. Ledertechn. Rundschau *1933*, Nr. 8.
42. F. Stather: Über den Einfluß des Alterns von pflanzlich gegerbtem Leder auf seinen Gehalt an gebundenem Gerbstoff (Untersuchungen zur Charakteristik pflanzlicher Gerbstoffe III). Collegium *1933*, S. 612—617.
43. F. Stather, H. Sluyter und R. Lauffmann: Tranausharzungen auf pflanzlich gegerbtem Leder (7. Mitteilung über Untersuchungen zum Fettungsprozeß). Collegium *1933*, S. 617—628.
44. F. Stather und R. Lauffmann: Über die Veränderung von Lickerfetten in Chromleder (8. Mitteilung über Untersuchungen zum Fettungsprozeß). Collegium *1933*, S. 723 bis 726.
45. F. Stather und H. Herfeld: Beiträge zur Kenntnis der Salzkonservierung tierischer Haut (Untersuchungen zum Konservierungsprozeß I). Collegium *1934*, S. 166—180.
46. F. Stather und H. Herfeld: Über den Gehalt tierischer Haut an löslichen Eiweißstoffen und dessen Beeinflussung durch den Konservierungs-, Weich- und Äscherprozeß (Untersuchungen zum Konservierungsprozeß II). Collegium *1934*, S. 317—324.
47. F. Stather und K. Löchner: Untersuchungen über die gerbereitechnologischen Eigenschaften einiger synthetischer Gerbstoffe (Untersuchungen zur Charakteristik von Gerbstoffen IV). Collegium *1934*, S. 374—399.
48. F. Stather und R. Lauffmann: Über den Einfluß der Nichtgerbstoffe bei der Gerbung mit pflanzlichen Gerbmaterialien (Untersuchungen zur Charakteristik pflanzlicher Gerbstoffe V). Collegium *1934*, S. 495—512.
49. F. Stather und H. Herfeld: Über den Einfluß von Neutralsalzen auf die von wasser- und neutralsalzlöslichen Eiweißstoffen befreite tierische Haut (Untersuchungen zum Konservierungsprozeß III). Collegium *1934*, S. 512—518.
50. F. Stather und R. Schubert: Über den spezifischen Einfluß pflanzlicher Gerbextrakte auf die Eigenschaften des damit gegerbten Leders (Untersuchungen zur Charakteristik pflanzlicher Gerbstoffe VI). Collegium *1934*, S. 609—621.
51. F. Stather und H. Herfeld: Einfache Methode zur vergleichenden Prüfung der Wasserdichtigkeit von Leder. Collegium *1935*, S. 13—15.
52. F. Stather und H. Sluyter: Über den Einfluß der chemischen und physikalischen Beschaffenheit von Tranen auf ihre Eignung zum Ausharzen auf pflanzlich gegerbtem Leder (9. Mitteilung über die Untersuchungen zum Fettungsprozeß). Collegium *1935*, S. 51—62.
53. F. Stather und H. Herfeld: Verbrennungserscheinungen an pflanzlich gegerbtem Leder. Collegium *1935*, S. 118—126.
54. F. Stather und H. Herfeld: Über die Aufnahme, schichtweise Verteilung und entwässernde Wirkung des Salzes bei der Konservierung tierischer Haut (Untersuchungen zum Konservierungsprozeß IV). Collegium *1935*, S. 333—344.
55. F. Stather: Bericht der internationalen Kommission des IVLIC für quantitative Gerbstoffanalyse 1934—35. Collegium *1935*, S. 353—356.
56. G. Parker und F. Stather: Gemeinschaftsbericht der Generalberichterstatter der paneuropäischen Kommission für quantitative Gerbstoffanalyse. Collegium *1935*, S. 359—360.

57. F. Stather und R. Lauffmann: Beiträge zur Kenntnis der Bindungsgröße, Bindungsgeschwindigkeit und Bindungsfestigkeit pflanzlicher Gerbmaterialien an tierische Hautsubstanz (Untersuchungen zur Charakteristik pflanzlicher Gerbstoffe VII). Collegium *1935*, S. 420—433.
58. F. Stather und H. Herfeld: Über den Einfluß der Nichtgerbstoffe bei der Gerbung auf die Eigenschaften des ausgegerbten Leders (Untersuchungen zur Charakteristik pflanzlicher Gerbstoffe VIII). Collegium *1935*, S. 463—470.
59. F. Stather und R. Lauffmann: Zur Kenntnis der Bindungsgeschwindigkeit pflanzlicher Gerbmaterialien an tierische Hautsubstanz. Collegium *1935*, S. 470—471.
60. F. Stather und R. Lauffmann: Metallschädigungen durch freie Fettsäuren und Lederschädigungen durch fettsaure Metallverbindungen. Collegium *1935*, 541—545.
61. F. Stather, R. Lauffmann und T. Bau Miau: Über die Löslichkeit pflanzlicher Extrakte in organischen Lösungsmitteln und Gemischen dieser mit Wasser und die gerbende Wirkung solcher Gerbextraktlösungen (Untersuchungen zur Charakteristik pflanzlicher Gerbmaterialien IX). Collegium *1936*, S. 66—74.
62. F. Stather und H. Herfeld: Über den Einfluß der Lagerdauer der Haut vor der Konservierung auf ihr Verhalten beim Konservierungsprozeß und der Lederfabrikation (Untersuchungen zum Konservierungsprozeß V). Collegium *1936*, S. 215—229.
63. F. Stather und R. Schubert: Über den Dispersitätsgrad pflanzlicher Gerbextraktlösungen und seinen Einfluß auf deren Gerbvermögen (Untersuchungen zur Charakteristik pflanzlicher Gerbstoffe X). Collegium *1936*, S. 380—412.
64. F. Stather und R. Lauffmann: Löslichkeitszahlen als Hilfsmittel zur Unterscheidung pflanzlicher Gerbextrakte (Untersuchungen zur Charakteristik pflanzlicher Gerbstoffe XI). Collegium *1936*, S. 437—439.
65. F. Stather und H. Herfeld: Über die Korrosionsbeständigkeit von Leichtmetallen gegenüber pflanzlichen Gerbstoffen, synthetischen Gerbstoffen, Celluloseextrakten und verschiedenartig gefettetem lohgarem Leder. Collegium *1937*, S. 9—23.
66. F. Stather und H. Herfeld: Über den Einfluß der Konservierungsart der Rohhaut auf ihr Verhalten bei der Ledererzeugung und die Eigenschaften des Leders (Untersuchungen zum Konservierungsprozeß VI). Collegium *1937*, S. 129—136.
67. F. Stather und O. Endisch: Beiträge zur Kenntnis des gerberischen Verhaltens von Sulfitcellulosegerbextrakten (Untersuchungen zur Charakteristik pflanzlicher Gerbmaterialien XII). Collegium *1937*, S. 193—216.
68. F. Stather: Bericht der Kommission des IVLIC für quantitative Gerbstoffanalyse 1936—1937. Collegium *1937*, S. 340—343.
69. F. Stather: Gerbmittel- und Gerbextrakt-Statistik der Deutschen Versuchsanstalt für Lederindustrie Freiberg 1932—1936. Ledertechn. Rundschau *1937*, Nr. 6 u. 7.
70. F. Stather und R. Schubert: Über die Veränderung und Bindung der Fettstoffe beim Fetten pflanzlich gegerbten Leders (10. Mitteilung über Untersuchungen zum Fettungsprozeß). Collegium *1937*, S. 456—466.
71. F. Stather und H. Herfeld: Untersuchungen über einige neuere synthetische Gerbstoffe (Zur Charakteristik pflanzlicher und synthetischer Gerbstoffe XIII). Collegium *1937*, S. 570—590.
72. F. Stather und R. Lauffmann: Über den Einfluß der Extraktionsart pflanzlicher Gerbmaterialien auf die Eigenschaften der erhaltenen Extraktlösungen (Untersuchungen zur Charakteristik pflanzlicher und synthetischer Gerbmaterialien XIV). Collegium *1938*, S. 164—169.
73. F. Stather und H. Herfeld: Über den Einfluß der Gerbart auf die chemischen und physi-

kalischen Eigenschaften pflanzlich gegerbten Unterleders (Untersuchungen zur Qualitätsbeurteilung von Leder I). Collegium *1938*, S. 321—335.

74. F. Stather und H. Herfeld: Austauschwerkstoffe für Kupfer und Messing in der Lederindustrie. Collegium *1938*, S. 578—590.
75. F. Stather und H. Herfeld: Unterschiede zwischen altgrubengegerbten und beschleunigt extraktgegerbten Unterledern hinsichtlich physikalischer Eigenschaften, insbesondere Porositätseigenschaften. (Untersuchungen zur Qualitätsbeurteilung von Leder II). Collegium *1938*, S. 656—662.
76. F. Stather und H. Herfeld: Abnutzungsprüfung von Leder. Einfacher Apparat zur vergleichenden Bestimmung der Widerstandskraft von Leder und Lederaustauschwerkstoffen gegen Abnutzung. (Untersuchungen zur Qualitätsbeurteilung von Leder III). Collegium *1939*, S. 1—6.
77. F. Stather und H. Herfeld: Über die Korrosionsbeständigkeit von Leichtmetallen und einigen nichtmetallischen Werkstoffen gegenüber pflanzlichen und synthetischen Gerbstoffen und Zelluloseextrakten II. Collegium *1939*, S. 11—30.
78. F. Stather und H. Herfeld: Über den Einfluß von Alter und Geschlecht des Tieres und der Hautstelle auf die chemische Zusammensetzung der tierischen Haut, ihren Gehalt an löslichen Proteinen, ihr Verhalten bei der Salzkonservierung und die Arbeiten der Wasserwerkstatt. (Untersuchungen zum Konservierungsprozeß VII). Collegium *1939*, S. 307—318.
79. F. Stather: Quantitative Gerbstoffanalyse. Bericht der gemeinschaftlichen Kommission des IVLIC und der ISLTC (Kommission I) für den Zeitraum 1938—39. Collegium *1939*, S. 397—413.
80. F. Stather und H. Herfeld: Bestimmung des Gerbwerts und Bindungswerts pflanzlicher Gerbmaterialien. (Untersuchungen zur Charakteristik pflanzlicher und synthetischer Gerbmaterialien XV). Collegium *1939*, S. 453—456.
81. F. Stather und H. Herfeld: Einfluß der Gerbart auf Zusammensetzung, Festigkeitseigenschaften und Abnutzungswiderstand des Leders in verschiedenen Schichten. (Untersuchungen zur Qualitätsbeurteilung von Leder IV). Collegium 1939, S. 609 bis 630.
82. F. Stather, H. Herfeld und H. Schöpel: Über die gerbereitechnologischen Eigenschaften von Fichtenrindenauszügen. (Untersuchungen zur Charakteristik pflanzlicher und synthetischer Gerbmaterialien XVI). Collegium *1939*, S. 562—581.
83. F. Stather und H. Herfeld: Untersuchungen über einige neuere synthetische Gerbstoffe III. (Untersuchungen zur Charakteristik pflanzlicher und synthetischer Gerbmaterialien XVII). Collegium *1940*, S. 49—69.
84. F. Stather und H. Herfeld: Über die Korrosionsbeständigkeit von metallischen und nichtmetallischen Werkstoffen gegenüber pflanzlichen und synthetischen Gerbstoffen und Zelluloseextrakten III. Collegium *1940*, S. 69—89.
85. F. Stather, H. Herfeld und R. Schubert: Austauschmöglichkeiten für Glyzerin bei der Herstellung von Transparentleder. Collegium *1940*, S. 341—350.
86. F. Stather, H. Herfeld, R. Schubert und O. Gottlieb: Über das gerbereitechnologische Verhalten von Gemischen von Fichtenrindenauszügen und synthetischen Austauschgerbstoffen. (Untersuchungen zur Charakteristik pflanzlicher und synthetischer Gerbmaterialien XVIII). Collegium *1941*, S. 1—23.
87. F. Stather und H. Herfeld: Untersuchungen über neuere synthetische Gerbstoffe IV. (Untersuchungen zur Charakteristik pflanzlicher und synthetischer Gerbmaterialien XIX). Collegium *1941*, S. 211—230.

88. F. Stather und H. Herfeld: Untersuchung zur Frage der Imprägnierung von Unterleder. Collegium *1941*, S. 247–257.
89. F. Stather, H. Herfeld und O. Gottlieb: Über das gerbereitechnologische Verhalten von Gemischen von Eichenrindenauszügen und synthetischen Austauschgerbstoffen. (Untersuchungen zur Charakteristik pflanzlicher und synthetischer Gerbmaterialien XX). Collegium *1941*, S. 297–318.
90. F. Stather und H. Herfeld: Über flächenmäßige Unterschiede im Aufbau tierischer Haut, gemessen an Zugfestigkeit, Dehnung und Raumgewicht daraus hergestellter Leder. Collegium *1942*, S. 1–13.
91. F. Stather und H. Herfeld: Über neuere Produkte für die Lederfettung I. (11. Mitteilung über Untersuchungen zum Fettungsprozeß). Collegium *1942*, S. 81–113.
92. F. Stather und H. Herfeld: Über neuere Produkte für die Lederfettung II. (12. Mitteilung über Untersuchungen zum Fettungsprozeß). Collegium *1942*, S. 121–145.
93. F. Stather und H. Schöpel: Über die Messung der Wärmeleitfähigkeit von Leder und Lederaustauschwerkstoffen. (Untersuchungen zur Qualitätsbeurteilung von Leder VI). Collegium *1942*, S. 259–269.
94. F. Stather und H. Herfeld: Über neuere Produkte für die Lederfettung III. (13. Mitteilung über Untersuchungen zum Fettungsprozeß). Collegium *1942*, S. 313–334.
95. F. Stather, H. Herfeld, R. Schubert und R. Bellmann: Über den Einfluß von Bindemittelart und Bindemittelmenge auf die Eigenschaften von Lederfaserwerkstoffen. (Untersuchungen zum Herstellungsprozeß von Lederfaserwerkstoffen I). Sonderdruck der Fachabteilung Lederfaserwerkstoffe *1943*, Heft 1, 34 Seiten.
96. F. Stather und H. Herfeld: Über die Lagerbeständigkeit von mit Austauschgerbstoffen gegerbten Ledern. (Untersuchungen zur Charakteristik pflanzlicher und synthetischer Gerbmaterialien XXII). Collegium *1943*, S. 210–222.
97. F. Stather, H. Herfeld und R. Schubert: Über die Veränderungen der Eigenschaften von Lederfaserwerkstoffen bei längerer Lagerung in Abhängigkeit von Bindemittelart, Bindemittelmenge und Faserart. (Untersuchungen zum Herstellungsprozeß von Lederfaserwerkstoffen II). Sonderdruck der Fachabteilung Lederfaserwerkstoffe *1944*, Heft 2, 25 Seiten.
98. F. Stather und H. Herfeld: Chemisch-technische Probleme zur Gerbstoffversorgung der deutschen Lederwirtschaft. Beiheft 53 zur Ztschr. des Vereins Deutscher Chemiker *1945*, S. 1–24.
99. F. Stather und H. Herfeld: Über Diffusion und Bindung von synthetischen Austauschgerbstoffen. (Untersuchungen zur Charakteristik pflanzlicher und synthetischer Gerbmaterialien XXIII). Ges. Abh. DLI Heft 1 (1949) S. 5–18.
100. F. Stather und H. Herfeld: Zur Methodik der Bestimmung der Gerb- und Bindungswerte pflanzlicher und synthetischer Gerbstofflösungen. Ges. Abh. DLI Heft 1 (1949) S. 32–46.
101. F. Stather und H. Herfeld: Über den Einfluß des Alterns von pflanzlich und synthetisch gegerbtem Leder auf seinen Gehalt an gebundenem Gerbstoff II. (Untersuchungen zur Charakteristik pflanzlicher und synthetischer Gerbstoffe XXIV). Ges. Abh. DLI Heft 1 (1949) S. 47–49.
102. F. Stather und H. Herfeld: Güterichtlinien für pflanzlich gegerbte Lederarten. (Untersuchungen zur Qualitätsbeurteilung von Leder VII). Ges. Abh. DLI Heft 1 (1949) S. 50–71.
103. F. Stather und R. Schubert: Technische Analyse von Lederfettungsmitteln. Ges. Abh. DLI Heft 1 (1949) S. 84–90.

104. F. Stather und H. Herfeld: Güterichtlinien für mineralgare und kombiniert gegerbte Lederarten. (Untersuchungen zur Qualitätsbeurteilung von Leder VIII). Ges. Abh. DLI Heft 2 (1949) S. 3—20.
105. F. Stather und H. Herfeld: Über den Einfluß langsamerer oder schnellerer Gerbart auf chemische Zusammensetzung, physikalische Eigenschaften und Lagerbeständigkeit pflanzlich gegerbter Unterleder. (Untersuchungen zur Qualitätsbeurteilung von Leder IX). Ges. Abh. DLI Heft 2 (1949) S. 31—51.
106. F. Stather und H. Herfeld: Über den Einfluß der Temperatur bei der Gerbung mit pflanzlichen und synthetischen Gerbstoffen. (Untersuchungen zur Charakteristik pflanzlicher und synthetischer Gerbmaterialien XXV). Ges. Abh. DLI Heft 3 (1950) S. 3—30.
107. F. Stather und R. Schubert: Über den Einfluß von Gerbart und Gerbintensität auf die Eigenschaften pflanzlich gegerbten Leders. (Untersuchungen zur Qualitätsbeurteilung von Leder X). Ges. Abh. DLI Heft 3 (1950) S. 31—43.
108. F. Stather und H. Herfeld: Über die Lagerbeständigkeit von mit Austauschgerbstoffen und mit Ligninextrakten gegerbten Ledern. (Untersuchungen zur Charakteristik pflanzlicher und synthetischer Gerbmaterialien XXVI). Ges. Abh. DLI Heft 3 (1950) S. 44—56.
109. F. Stather und H. Herfeld: Über den Einfluß der Fettung auf die Eigenschaften von Schuhoberleder. (17. Mitteilung über Untersuchungen zum Fettungsprozeß). Ges. Abh. DLI Heft 3 (1950), S. 57—78.
110. F. Stather: Zur mikroskopischen Anatomie einiger einheimischer Pelzfelle. (Untersuchungen zur Pelzzurichtung und Pelzveredlung I). Ges. Abh. DLI Heft 4 (1950) S. 3—16.
111. F. Stather, H. Herfeld und K. Sohre: Über den Einfluß der wichtigsten Pelzzurichteverfahren auf die Eigenschaften des Pelzleders. (Untersuchungen zur Pelzzurichtung und Pelzveredlung II). Ges. Abh. DLI Heft 4 (1950) S. 17—38.
112. F. Stather und H. Herfeld: Vergleichende Auswertung statistischen Analysenmaterials von pflanzlich gegerbten Unterledern langsamer Grubengerbung und beschleunigter Gruben-Faßgerbung. (Untersuchungen zur Qualitätsbeurteilung von Leder XI). Ges. Abh. DLI Heft 4 (1950) S. 39—44.
113. F. Stather: Langsame oder schnelle Gerbverfahren bei der Herstellung pflanzlich gegerbten Unterleders. (Untersuchungen zur Qualitätsbeurteilung von Leder XII). Ges. Abh. DLI Heft 4 (1950) S. 45—68.
114. F. Stather und R. Schubert: Über den Einfluß von Netz- und Dispergiermitteln auf die Beschaffenheit pflanzlicher Gerbstofflösungen und deren gerberisches Verhalten. (Untersuchungen zur Charakteristik pflanzlicher und synthetischer Gerbmaterialien XXVII). Ges. Abh. DLI Heft 4 (1950) S. 69—83.
115. F. Stather und H. Herfeld: Über den Einfluß der Trocknung und mechanischen Zurichtung auf die Eigenschaften von Schuhoberleder. Ges. Abh. DLI Heft 4 (1950) S. 84—95.
116. F. Stather, H. Herfeld, R. Schubert und R. Bellmann: Herstellung, Konstitution und Gerbwirkung synthetischer Gerbstoffe. Ges. Abh. DLI Heft 5 (1950) S. 3—77.
117. F. Stather, H. Herfeld und K. Sohre: Vergleichende Untersuchungen über den Einfluß der Wasserwerkstattarbeiten und der Gerbung auf die Flächenausbeute und Eigenschaften von Chromoberleder. Ges. Abh. DLI Heft 6 (1951) S. 3—23.
118. F. Stather und H. Herfeld: Über den Einfluß der Zurichtung mit Deckfarben und

Appreturen auf die Eigenschaften von Schuhoberleder. Ges. Abh. DLI Heft 6 (1951) S. 24—39.
119. F. Stather und K. Sohre: Unterschiedlicher Gerbverlauf bei der Ledererzeugung und der Pelzzurichtung. Ges. Abh. DLI Heft 6 (1951) S. 40—49.
120. F. Stather und R. Schubert: Über die Farblackbildung beim Färben von Haaren mit Oxydationsfarbstoffen. (Untersuchungen zur Pelzzurichtung und Pelzveredlung V). Ges. Abh. DLI Heft 6 (1951) S. 50—64.
121. F. Stather und G. Königfeld: Über den Einfluß des Bleichens von naturfarbenen Pelzfellen auf die Eigenschaften der Haare und des Pelzleders. (Untersuchungen zur Pelzzurichtung und Pelzveredlung VI). Ges. Abh. DLI Heft 7 (1951) S. 5—20.
122. F. Stather und H. Nebe: Über den Einfluß verschiedener kernverknüpfender Mittel auf die Gerbwirkung kondensierter phenolischer synthetischer Gerbstoffe. (Untersuchungen über Konstitution und Gerbwirkung synthetischer Gerbstoffe II). Ges. Abh. DLI Heft 7 (1951) S. 34—47.
123. F. Stather und K. Schmidt: Über den Einfluß der Gerbart, Gerbintensität und Fettung des Leders auf dessen Zugfestigkeit und Dehnbarkeit. Ges. Abh. DLI Heft 7 (1951) S. 66—79.
124. F. Stather: Mikroskopische Untersuchungen zum Bleich- und Färbeprozeß in der Rauchwarenveredlung. Ges. Abh. DLI Heft 8 (1952) S. 3—11.
125. F. Stather: Zur Frage des Brühens von Schweinshäuten. Ges. Abh. DLI Heft 8 (1952) S. 12—20 und Dtsche Schuh- und Lederzeitschr. *1952*, Heft 5, S. 124, 125, 127—129.
126. F. Stather und H. Herfeld: Zur Frage der gerberischen Bewertung synthetischer Austauschgerbstoffe. Ges. Abh. DLI Heft 8 (1952) S. 21—44.
127. F. Stather: Langsame oder schnelle Gerbmethoden bei der Herstellung pflanzlich gegerbter Unterleder? Ges. Abh. DLI Heft 8 (1952) S. 45—50 und Das Leder *1951*, S. 245—247.
128. F. Stather: Trane zur Sämischgerbung. Ges. Abh. DLI Heft 8 (1952) S. 51—52.
129. F. Stather, H. Herfeld und W. Hartung: Vergleichende Untersuchungen über den Einfluß von Neutralisation, Fettung und mechanischer Zurichtung auf die Flächenausbeute und Eigenschaften von Chromoberleder. Ges. Abh. DLI Heft 9 (1953) S. 3—21.
130. F. Stather und H. Herfeld: Vergleichende Untersuchungen über die Geeignetheit verschiedener antiseptischer Mittel für die Rohhautkonservierung, Ledererzeugung und Lederlagerung. Ges. Abh. DLI Heft 9 (1953) S. 22—45.
131. F. Stather, H. Herfeld und G. Reich: Über den Einfluß der Vorgeschichte von Sulfitzelluloseablauge auf ihr gerberisches Verhalten und die Eigenschaften daraus hergestellter synthetischer Austauschgerbstoffe. Ges. Abh. DLI Heft 10 (1954) S. 5—15.
132. F. Stather, H. Herfeld und W. Hartung: Über den Einfluß von Rasse, Geschlecht, Gewicht und Hautstelle der Schweinshäute auf die Eigenschaften von Schweinsvelourleder. (Untersuchungen zur Herstellung von Schweinsleder I). Ges. Abh. DLI Heft 10 (1954) S. 16—33.
133. H. Herfeld und F. Stather: Über den Einfluß chrom-pflanzlicher Kombinationsgerbungen auf die Ledereigenschaften. Ges. Abh. DLI Heft 10 (1954) S. 34—51.
134. F. Stather, H. Herfeld und S. Beyer: Über Wirkung und zweckmäßige Durchführung des Säurepickels bei der Zurichtung von Pelzfellen. (Untersuchungen zur Pelzzurichtung und Pelzveredlung VIII) Ges. Abh. DLI Heft 10 (1954) S. 52—66.
135. F. Stather, H. Herfeld und H. Moser: Über den Einfluß der Arbeiten der Wasserwerk-

statt auf die Eigenschaften von Schweinsoberleder. (Untersuchungen zur Herstellung von Schweinsleder II) Ges. Abh. DLI Heft 11 (1955) S. 3—34.

136. F. Stather, H. Herfeld und S. Beyer: Vergleichende Untersuchungen über verschiedene Verfahren der Zurichtung von Pelzfellen. (Untersuchungen zur Pelzzurichtung und Pelzveredlung IX) Ges. Abh. DLI Heft 11 (1955) S. 60—78.

137. F. Stather, H. Herfeld, H. Moser und K. Härtewig: Über den Einfluß der Nachgerbung, Neutralisation, Fettung und Zurichtung auf die Eigenschaften von Schweinsoberleder mit korrigiertem Narben (Konaleder). (Untersuchungen zur Herstellung von Schweinsleder III). Ges. Abh. DLI Heft 12 (1955) S. 3—30.

138. H. Herfeld, F. Stather, K. Härtewig und H. Moser: Über die zweckmäßigste Gerbung und Nachgerbung von Schweinsvelourleder. (Untersuchungen zur Herstellung von Schweinsleder IV). Ges. Abh. DLI Heft 12 (1955) S. 31—55.

139. F. Stather und G. Reich: Über den Einfluß einer Alkalibehandlung von Sulfitzellulose-ablauge auf deren gerberische Eigenschaften. Ges. Abh. DLI Heft 12 (1955) S. 56—68.

140. F. Stather und G. Reich: Über die Gewinnung von Gerbstoffen aus Lignophenolen. Ges. Abs. DLI Heft 12 (1955) S. 69—81.

141. F. Stather, S. Walther und U. Stather: Über den Einfluß unterschiedlichen Beizens auf die Färbbarkeit und Farbintensität von Kaninhaaren. (Untersuchungen zur Pelzzurichtung und Pelzveredlung X). Ges. Abh. DLI Heft 12 (1955) S. 82—86.

142. F. Stather und G. Weinhold: Chemische Zusammensetzung und physikalische Eigenschaften einiger einheimischer Pelztierhäute. (Untersuchungen zur Pelzzurichtung und Pelzveredlung XI). Ges. Abh. DLI Heft 12 (1955) S. 87—90.

143. F. Stather, S. Walther und U. Stather: Über die Widerstandsfähigkeit von Leder gegen enzymatischen Abbau durch Trypsin. Das Leder *1956*, S. 13—17.

144. F. Stather: Zur Frage des enzymatischen Abbaus von Kollagen und Leder. Das Leder *1956*, S. 187—188.

145. F. Stather und G. Königfeld: Faktis-Bildung als Ursache von Fettausscheidungen auf Leder. (18. Mitteilung über Untersuchungen zum Fettungsprozeß). Das Leder *1956*, S. 35—36.

146. H. Herfeld, F. Stather, H. Moser und K. Härtewig: Über die zweckmäßigste Fettung und mechanische Zurichtung von Schweinvelourleder. (Untersuchungen zur Herstellung von Schweinsleder V). Ges. Abh. DLI Heft 13 (1956) S. 3—36.

147. F. Stather, H. Herfeld, K. Härtewig und H. Moser: Über die Deckfarbenzurichtung von Schweinsoberleder mit korrigierten Narben (Konaleder) unter Mitverwendung synthetischer Binder. (Untersuchungen zur Herstellung von Schweinsleder VI). Ges. Abh. DLI Heft 13 (1956) S. 37—68.

148. F. Stather und G. Reich: Neue Methoden und Ergebnisse der Wertbeurteilung vegetabilischer und synthetischer Gerbstoffe. Das Leder *1958*, 178—182.

149. G. Reich und F. Stather: Über den Wasserhaushalt vegetabilisch-synthetisch gegerbter Leder. Das Leder *1958*, S. 249—253.

150. F. Stather und G. Reich: Über die Fällung von synthetischen Gerbstoffen mit aromatischen Aminen. Das Leder *1959*, S. 1—4. Italienische Übersetzung: Sulla precipitazione di concianti sintetici con amine aromatiche. Cuoio, Pelli, Materie Concianti *1959*, S. 113—121.

151. F. Stather, H. Herfeld, G. Reich und H. Moser: Über die Verbesserung der Schweißbeständigkeit von Rinds- und Schweins-Brandsohlleder. Ges. Abh. DLI Heft 14 (1959) S. 5—26.

152. F. Stather, G. Reich, J. Überla und H. Moser: Über die Verkürzung der Gerbdauer

von Unterleder durch Chromvorgerbung und pflanzlich-synthetische Flottengerbung bei erhöhter Temperatur. (Untersuchungen zur Unterledergerbung I). Ges. Abh. DLI Heft 14 (1959) S. 53—59.

153. F. Stather, H. Herfeld, H. Moser und K. Härtewig: Über das Färben von Schweinsvelourleder. (Untersuchungen zur Herstellung von Schweinsleder VII). Ges. Abh. DLI Heft 14 (1959) S. 61—76.

154. F. Stather, H. Herfeld, H. Moser und K. Härtewig: Über die Herstellung von Schweinshandschuhleder. (Untersuchungen zur Herstellung von Schweinsleder IX). Ges. Abh. DLI Heft 14 (1959) S. 87—102.

155. F. Stather und G. Wiegand: Über die Untersuchung und Beurteilung von Fettlickerprodukten und Lederhilfsmitteln. Ges. Abh. DLI Heft 14 (1959) S. 103—113.

156. F. Stather, H. Herfeld, S. Walther und U. Stather: Über die Schwarzfärbung von Persianerfellen. (Untersuchungen zur Pelzzurichtung und Pelzveredlung XIII). Ges. Abh. DLI Heft 14 (1959) S. 115—129.

157. F. Stather und S. Walther: Backsteinblattern als Schweinshautschaden. Das Leder *1959*, S. 73—74.

158. F. Stather, H. Herfeld, H. Moser und K. Härtewig: Über das Lüstern von Schweinsvelourleder. (Untersuchungen zur Herstellung von Schweinsleder VIII). Ges. Abh. DLI Heft 14 (1959) S. 77—86.

159. F. Stather, G. Reich und J. Überla: Über die Herstellung von Unterleder unter Verwendung hoher Anteile synthetischer Gerbstoffe. (Untersuchungen zur Unterledergerbung II). Ges. Abh. DLI Heft 15 (1959) S. 5—15.

160. F. Stather, G. Reich, J. Überla und W. Barthel: Über das unterschiedliche Verhalten von Syntanen und Pflanzengerbstoffen im kontinuierlichen Einsatz bei der Grubengerbung. (Untersuchungen zur Unterledergerbung III). Ges. Abh. DLI Heft 15 (1959) S. 16—33.

161. F. Stather, G. Reich und G. Zimmermann: Über die Anwendung aromatischer Syntane bei der Gerbung bei erhöhter Temperatur. (Untersuchungen zur Unterledergerbung IV). Ges. Abh. DLI Heft 15 (1959) S. 34—39.

162. F. Stather, G. Reich und S. Walther: Über beschleunigte Unterledergerbung durch Vorgerbung mit Metallkomplexgerbstoffen. (Untersuchungen zur Unterledergerbung V). Ges. Abh. DLI Heft 15 (1959) S. 40—49.

163. G. Reich, S. Walther und F. Stather: Über eine einfache Methode zur Messung der Komplexaffinität aromatischer Gerb- und Farbstoffe gegenüber Chrom(III)-salzen. Ges. Abh. DLI Heft 15 (1959) S. 50—53.

164. F. Stather und G. Wiegand: Zur Bindungsfähigkeit aromatischer Gerbstoffe. (Untersuchungen zur Charakteristik pflanzlicher und synthetischer Gerbmaterialien XXIX). Ges. Abh. DLI Heft 15 (1959) S. 86—92.

165. G. Reich, M. Wassiljew und F. Stather: Über die Abhängigkeit der Diffusionsgeschwindigkeit aromatischer Gerbstoffe in Blöße von Gerbstoffkonzentration und Art der Vorgerbung. (Untersuchungen zur Charakteristik pflanzlicher und synthetischer Gerbmaterialien XXX). Ges. Abh. DLI Heft 15 (1959) S. 93—104.

166. F. Stather, G. Reich und S. Wünsch: Über die Adsorption aromatischer Syntane und deren Vor- und Zwischenprodukte an Hautsubstanz und Polyamid. (Untersuchungen über Konstitution und Gerbwirkung synthetischer Gerbstoffe VI). Ges. Abh. DLI Heft 15 (1959) S. 112—128.

167. F. Stather und G. Reich: Chrom-Syntan-Einbadgerbung. Ges. Abh. DLI Heft 15 (1959) S. 137—144.

168. F. Stather und H. Herfeld: Untersuchung und Gütebewertung von wäßrigen Eiweiß-Lederdeckfarben. Ges. Abh. DLI Heft 15 (1959) S. 145—156.
169. F. Stather, G. Wiegand und H. Nebe: Untersuchungen zur Leistungsprüfung und Beurteilung von Leder-, Textil- und Gummi-Treibriemen. Ges. Abh. DLI Heft 15 (1959) S. 157—173.
170. F. Stather, G. Reich und R. Steinhardt: Über die Wechselwirkung von tierischer Haut mit Brenzkatechin und Formaldehyd. Ein Beitrag zum Wesen der Kondensationsgerbung. Das Leder *1959*, S. 5—11.
171. F. Stather, G. Reich und J. Überla: Über die Herstellung von Unterleder aus Rindshäuten in ausschließlicher Faßgerbung (3-Phasengerbung). (Untersuchungen zur Unterledergerbung VI). Ges. Abh. DLI Heft 16 (1961) S. 5—16.
172. F. Stather, G. Reich und J. Überla: Über die Herstellung von Brandsohlleder aus Schweinshäuten durch ausschließliche Faßgerbung (3-Phasengerbung). (Untersuchungen zur Unterledergerbung VII). Ges. Abh. DLI Heft 16 (1961) S. 17—22.
173. F. Stather, G. Reich und H. Moser: Beschleunigte Unterledergerbung durch Vorgerbung mit Metallkomplexgerbstoffen II. (Untersuchungen zur Unterledergerbung VIII). Ges. Abh. DLI Heft 16 (1961) S. 23—37.
174. F. Stather, G. Reich und J. Überla: Über die Herstellung von Rindsjuchten- und Schweinsjuchtenleder in 1-Phasengerbung. Ges. Abh. DLI Heft 16 (1961) S. 38—43.
175. F. Stather, G. Reich und G. Weigelt: Über die Anwendung aromatischer Syntane sowie ihrer Vor- und Zwischenprodukte bei der Pelzzurichtung. (Untersuchungen zur Pelzzurichtung und Pelzveredlung XV). Ges. Abh. DLI Heft 16 (1961) S. 75—78.
176. F. Stather und G. Wiegand: Über einige Beziehungen zwischen Gerbstoffbindung und Gerbstoffverlusten bei der pflanzlichen Gerbung. (Untersuchungen zur Charakteristik pflanzlicher und synthetischer Gerbmaterialien XXXII). Ges. Abh. DLI Heft 16 (1961) S. 79—83.
177. F. Stather und G. Wiegand: Zur sulfitierenden Extraktion von Fichtenrinde. (Untersuchungen zur Charakteristik pflanzlicher und synthetischer Gerbmaterialien XXXIII). Ges. Abh. DLI Heft 16 (1960) S. 98—103.
178. F. Stather, M. Gierth und A. Meißner: Neue Güterichtlinien für Leder. Ges. Abh. DLI Heft 16 (1960) S. 104—136.
179. F. Stather, K. Knoll und G. Zimmermann: Untersuchung und Gütebewertung von wäßrigen Eiweiß-Lederdeckfarben. Ges. Abh. DLI Heft 16 (1960) S. 137—143.
180. F. Stather und M. Wassiljew: Über die Eignung der Häute verschiedener Schweinsrassen zur Herstellung von Schweinsvelour- und Schweinskonaleder. (Untersuchungen zur Herstellung von Schweinsleder X). Ges. Abh. DLI Heft 16 (1960) S. 152—156.
181. F. Stather und G. Wiegand: Über Wechselwirkungen zwischen Erdalkali-Ionen und pflanzlichen Gerbstoffen. (Untersuchungen zur Charakteristik pflanzlicher und synthetischer Gerbmaterialien XXXIII) Ges. Abh. DLI Heft 17 (1961) S. 25—32.
182. F. Stather, G. Reich und G. Wiegand: Zur Verwendung von Gerbextrakten als kesselsteinverhütende Zusätze im Kesselspeisewasser. (Untersuchungen zur Charakteristik pflanzlicher und synthetischer Gerbstoffe XXXIV). Ges. Abh. DLI Heft 17 (1961) S. 33—36.
183. F. Stather und K. Pauligk: Welche Mindestgerbstoffmengen sind zur Umwandlung von Haut in Leder notwendig? Ges. Abh. DLI Heft 17 (1961) S 37—55.
184. F. Stather, G. Reich und L. Müller: Konstitution und Gerbwirkung kondensierter Sulfonsäuren des 2-Naphthols. (Untersuchungen über Konstitution und Gerbwirkung synthetischer Gerbstoffe VII). Ges. Abh. DLI Heft 17 (1961) S. 69—87.

185. F. Stather, G. Reich und J. Überla: Untersuchungen zur stationären Gerbung von Unterleder. (Untersuchungen zur Unterledergerbung IX). Ges. Abh. DLI Heft 17 (1961) S. 88—98.
186. F. Stather, G. Reich und M. Wassiljew: Über den Einfluß des Spaltens bzw. Falzens auf die Eigenschaften von Rind- und Mastboxleder aus Häuten höherer Gewichtsklassen. Ges. Abh. DLI Heft 17 (1961) S. 99—108.
187. F. Stather und G. Zimmermann: Untersuchungen über den Einfluß von Weichmachern, Lösungs- bzw. Verschnittmitteln und Pigmenten auf die Beschaffenheit von Nitrozellulosefilmen. Ges. Abh. DLI Heft 17 (1961) S. 109—126.
188. G. Reich, S. Walther und F. Stather: Untersuchungen zur Gelatinegewinnung aus Hautkollagen. Ges. Abs. DLI Heft 17 (1961) S. 127—133.
189. G. Reich, S. Walter und F. Stather: Über einige Gesetzmäßigkeiten bei der Herstellung von Gelatine aus Schweinshaut nach dem sauren Aufschlußverfahren. (Untersuchungen über Kollagen und Gelatine III). Ges. Abh. DLI Heft 18 (1962) S. 15—23.
190. G. Reich, S. Walther und F. Stather: Über den Einfluß des Alters von Rinds- und Schweinshäuten auf Ausbeute und Eigenschaften daraus nach dem sauren Aufschlußverfahren gewonnener Gelatine. (Untersuchungen über Kollagen und Gelatine IV). Ges. Abh. DLI Heft 18 (1962) S. 24—30.
191. F. Stather und G. Wiegand: Möglichkeiten zur Rationalisierung der Chromgerbung: Veränderung der Aziditätsverhältnisse und Verbesserung der Chromausnutzung. (Untersuchungen zur Technologie der Chromgerbung I) Ges. Abh. DLI Heft 18 (1962) S. 70—97.
192. F. Stather und G. Wiegand: Veränderung der Aziditätsverhältnisse und Verbesserung der Chromausnutzung bei der Herstellung von chromgarem Kalb-, Rind- und Schweinsleder. (Untersuchungen zur Technologie der Chromgerbung II). Ges. Abh. DLI Heft 18 (1962) S. 98—107.
193. F. Stather, G. Reich und R. Steinhardt: Konstitution und Gerbvermögen eines Kondensationsprodukts aus 2-Naphtholsulfonsäure, p-Kresol und Formaldehyd. (Untersuchungen über Konstitution und Gerbwirkung synthetischer Gerbstoffe VIII) Ges. Abh. DLI Heft 18 (1962) S. 108—125.
194. F. Stather und M. Schlöffel: Über die Glanzmessungen an Oberflächen und den Einfluß verschiedener Faktoren auf die Glanzwerte unpigmentierter Kaseinoberflächen. (Untersuchungen über Deckfarben und Deckfarbenzurichtung V). Ges. Abh. DLI Heft 18 (1962) S. 134—152.
195. F. Stather und A. Meißner: Über die Eignung verschiedener Deckfarben für die Zurichtung von Schuhoberleder und Futterleder. (Untersuchungen über Deckfarben und Deckfarbenzurichtung VI). Ges. Abh. DLI Heft 18 (1962) S. 153—162.
196. G. Wiegand und F. Stather: Dauerbiegeprüfungen zur Ermittlung der mechanischen Haltbarkeit von Leder. Ges. Abh. DLI Heft 18 (1962) S. 163—165.
197. F. Stather und G. Wiegand: Möglichkeiten zur Rationalisierung der Chromgerbung: Steuerung der Gerb- und Nachgerbintensitäten von Chromleder. (Untersuchungen zur Technologie der Chromgerbung III). Ges. Abh. DLI Heft 19 (1964) S. 107—122.
198. F. Stather, G. Reich und F. Pallas. Über die Reaktion aromatischer Sulfochloride mit Aminosäuren, Kollagen und Gelatine. (Untersuchungen zur Charakteristik pflanzlicher und synthetischer Gerbmaterialien XXXV). Ges. Abh. DLI Heft 19 (1964) S. 138—155.
199. F. Stather und A. Meißner: Über Prägetiefe und Haltbarkeit künstlicher Narbenprägungen auf Leder. Ges. Abh. DLI Heft 19 (1964) S. 156—167.

200. L. Müller, G. Reich und F. Stather: Infrarotspektroskopische Untersuchungen von Polymerisatbindemitteln für die Deckfarbenzurichtung von Leder. Das Leder *1966*, S. 221–226.
201. G. Hebestreit, G. Reich und F. Stather: Elektronenmikroskopische Untersuchungen an Polymerisatbindemitteln für die Deckfarbenzurichtung von Leder. (Untersuchungen über Deckfarben und Deckfarbenzurichtung VIII) Das Leder *1966*, S. 227 bis 230.
202. G. Reich, W. Gierth, G. Hebestreit und F. Stather: Über die Umsetzung von Tropokollagen mit gerbenden Substanzen. Hoppe-Seyler's Z. physiol. Chem. Bd *349* (1968) S. 271–274.

Allgemeine Veröffentlichungen in Zeitschriften

1. F. Stather: Die Bedeutung wissenschaftlicher Forschung für die Lederindustrie. Lederarbeiter-Zeitung vom 29. Juni 1928, 4 Seiten.
2. F. Stather: Rückblicke und Rundblicke in der Ledererzeugung. Vortrag auf dem Verbandstag des Deutschen Sattler-, Tapezierer- und Portefeuiller-Verbandes am 5. Juni 1929 in Dresden, S. 1–8.
3. F. Stather: Die Aufgaben der Deutschen Versuchsanstalt für Lederindustrie. Vortrag auf der Hauptversammlung des Centralvereins der Deutschen Lederindustrie am 31. Mai 1930 in Wiesbaden. 5 Seiten.
4. F. Stather: Analytische Tätigkeit in der Gerberei. Jahrbuch der Deutschen Lederindustrie, Berlin *1930*, 12 Seiten.
5. F. Stather: Forschung, Untersuchung, Unterricht. Die wissenschaftlichen Institute der deutschen Lederindustrie auf der Internationalen Lederschau. Der Ledermarkt, Frankfurt a. M., Nr. 240 vom 14. Okt. 1930.
6. F. Stather: Gerbstoffflecken an getragenem Schuhwerk. Der Ledermarkt Frankfurt a. M., Nr. 2 vom 3. Januar 1931, 3 Seiten.
7. F. Stather: Defects in raw hides and their resulting effects on the finished leather. Journal of the International Society of Leather Trades' Chemists, Redbourn, Engl. January 1931, 10 Seiten.
8. F. Stather: Rohhautschäden und ihre Auswirkung auf Leder. Collegium *1931*, S. 3–12 [mit der vorangehenden Veröffentlichung 7 identisch].
9. F. Stather: Wirtschaftsnot und wissenschaftliche Forschung. Der Ledermarkt, Frankfurt a. M., Nr. 130 vom 6. Juni 1931, 1 Seite.
10. F. Stather: Enzyme, ein wichtiger Faktor bei der Ledererzeugung. Vortrag auf der 52. Hauptversammlung des Verbandes Sächsischer Gerber in Chemnitz am 27. Juni 1931; Ledertechn. Rundschau *1931*, Nr. 9 und 10.
11. F. Stather: Neue Hilfsmittel für die Ledererzeugung I. Ledertechn. Rundschau *1932*, Nr. 2, 4 Seiten.
12. F. Stather: Rohhautkonservierung und Rohhautschäden. Tropenpflanzer *35* (1932) S. 91–108.
13. F. Stather: Käfer als Lederschädlinge. Ledertechn. Rundschau *1932*, Nr. 3, 2 Seiten.
14. F. Stather: Die Beurteilung pflanzlich gegerbten Leders auf Grund seiner chemischen und physikalischen Analyse. Ledertechn. Rundschau *1932*, Nr. 8 und 9, Sonderdruck 24 Seiten.
15. F. Stather: Beiträge zur Charakteristik pflanzlicher Gerbstoffe. Vortrag am 21. Sept. in Amsterdam gehalten. Ledertechn. Rundschau *1933*, Nr. 10, S. 1–5.

16. F. Stather: Haut, Gerbstoffe, Leder. Vortrag am 29. August 1933 auf der Leipziger Herbstmesse gehalten. Die Schuhreparatur *4.* u. *5.* Jahrgang, Sonderdruck 11 Seiten.
17. F. Stather: Lederfehler und ihre Ursachen. Chemiker-Zeitung *58* (1934) Nr. 25, S. 253—254.
18. F. Stather: Sachgemäße Betriebskontrolle in der Gerberei. Die Lederindustrie Nr. 74 vom 27. März 1934, Sonderdruck 3 Seiten.
19. F. Stather: Können Leder und Lederabfälle bei Lagerung durch Selbstentzündung zu Bränden Veranlassung geben? Feuerpolizei Zschr. für Feuerschutz- und Rettungswesen, München, *38* (1936) Nr. 5, 2 Seiten.
20. F. Stather: Eine moderne Versuchsanstalt. Bildbericht aus der Deutschen Versuchsanstalt für Leder-Industrie. Süddeutsche Schuh- und Lederzeitung *1936*, Nr. 226, 2 Seiten.
21. F. Stather: Julius von Schroeder, ein deutscher Gerbereichemiker. Collegium *1936*, S. 229—231.
22. F. Stather und H. Herfeld: Chemie und Technologie der pflanzlichen Gerbung. Fortschrittsbericht 1926—1936. Collegium *1936*, S. 497—532.
23. F. Stather: Beiträge zur Kolloidchemie der vegetabilischen Gerbung. Kolloid-Zeitschrift *77* (1936) S. 201—204.
24. F. Stather: Fett und Säure im Leder. Die Lederindustrie Nr. 292 vom 5 Dezember 1936, 2 Seiten.
25. F. Stather: Möglichkeiten vermehrter einheimischer Gerbstoffgewinnung. Häute und Leder, Berlin 1937, Nr. 7, 3 Seiten.
26. F. Stather: Die Prüfung und Beurteilung von Kunstleder und Lederersatzmaterialien. Chemiker-Zeitung *61* (1937) S. 445—447.
27. F. Stather und H. Herfeld: Probleme der Rohhautkonservierung. Stiasny-Festschrift *1937*, S. 381—389.
28. F. Stather: Probleme der Fettung chromgaren und pflanzlich gegerbten Leders. Der Gerber *1937*, Heft 1503—1504.
29. F. Stather und H. Herfeld: Prüfung und Beurteilung von Kunstleder. Der Schuhmarkt *46* (1937) Nr. 142, Sonderdruck 8 Seiten.
30. F. Stather: Die Verwendung von Walöl in der Lederindustrie. Fette und Seifen *45* (1938) S. 86—87.
31. F. Stather: Neue Werkstoffe als Lederersatz. Collegium *1938*, S. 497—509.
32. F. Stather: 50 Jahre Freiberger Gerberei-Institute 1889—1939. Festschrift. 89 Seiten.
33. F. Stather und H. Herfeld: Austauschwerkstoffe für Leder. Lederkalender *1939*, 10 Seiten.
34. F. Stather und H. Herfeld: Kunststoffe an Stelle von Leder. Kunststoffe *30* (1940) S. 253—260.
35. F. Stather und H. Herfeld: Aktuelle technische Probleme der Lederwirtschaft im Lichte neuerer Untersuchungen der Deutschen Versuchsanstalt und Fachschule für Lederindustrie, Freiberg/Sa. Die Lederindustrie, Berlin, *84* (1941) Nr. 89/91, 92 u. 93, Sonderdruck 20 Seiten.
36. F. Stather und H. Herfeld: Über die Qualitätsbewertung pflanzlich gegerbter Leder auf Grund chemischer und physikalischer Prüfungen und die Ursache der häufigst festzustellenden Fehler. Die Lederindustrie, Berlin, *84* (1941) Nr. 179/180 u. 181/182, Sonderdruck 15 Seiten.
37. F. Stather und H. Herfeld: Welche Eigenschaften können von einem einwandfreien

Schweinsleder verlangt werden? Die Lederindustrie, Berlin, *84* (1941) Nr. 195, Sonderdruck 11 Seiten.
38. F. Stather: Technische Probleme der Kriegslederwirtschaft. Ledertechn. Rundschau *34* (1942) S. 3–8.
39. F. Stather und R. Schubert: Kunststoffe als Austauschmaterialien für Leder. Österreichische Chemiker-Zeitung *45* (1942) S. 1–18.
40. F. Stather und J. A. Sagoschen: Sulfitierende Extraktion bei der Herstellung von Fichtenrindenauszügen. Ledertechn. Rundschau *34* (1942) Nr. 2, Sonderdruck 8 Seiten.
41. F. Stather und H. Herfeld: Über Chromlederqualität und Möglichkeiten der Chromeinsparung bei der Chromledererzeugung. Ledertechn. Rundschau *34* (1942) Nr. 2, Sonderdruck 12 Seiten.
42. F. Stather und H. Herfeld: Analytische Mindestanforderungen für pflanzlich gegerbte Leder. Ledertechn. Rundschau *34* (1942) Nr. 4.
43. F. Stather: Lederwirtschaft und Lederforschung im Zweijahresplan. Forum *2* (1948) Heft 9, S. 5–6.
44. F. Stather: Gütesicherung in der Lederwirtschaft. Chemische Technik *1* (August 1949) Nr. 2, S. 64–66.
45. F. Stather: Synthetische Gerbstoffe in der modernen Lederfabrikation. Vortrag auf der Landestagung der Kammer der Technik Dresden Juni *1949*, S. 22–26.
46. F. Stather: Dr. Hermann Sluyter 1869–1951 [Nachruf]. Ges. Abh. DLI Heft 7 (1951) S. 3–4.
47. F. Stather: Deutsches Lederinstitut — Deutsche Lederwirtschaft. Deutsche Schuh- und Lederzeitschr. *1* (1952) S. 149–152.
48. F. Stather: Kampf der Dasselfliege. Die Bedeutung der Wiederaufnahme der Bekämpfung der Dasselfliege in der Deutschen Demokratischen Republik für die Volkswirtschaft. Beilage zur Dtsch. Schuh- und Lederzeitschr. *1952*, Heft 7, 3 Seiten.
49. F. Stather: Lederdeckfarben in neuer Entwicklung und ihre Bedeutung für die Lederindustrie der Deutschen Demokratischen Republik. Deutsche Schuh- u. Lederzeitschr. *1952*, Heft 7, S. 181–182.
50. F. Stather: Hautparasiten, Läuse, Haarlinge, Milben, Zecken sind Schädlinge der Volkswirtschaft! Deutsche Schuh- und Lederzeitschr. *1952*, Heft 8, S. 223–226.
51. F. Stather: Boemotan SK, ein neuer synthetischer Austauschgerbstoff. Deutsche Schuh- und Lederzeitschr. *1952*, Heft 9, S. 241–242.
52. F. Stather: Lederfehler durch starkes Hervortreten der Blutadern. Deutsche Schuh- und Lederzeitschr. *1952*, Heft 10, S. 281–283.
53. F. Stather: Fleckenbildungen und Verfärbungen auf salzkonservierten Rohhäuten und Rohfellen. Deutsche Schuh- und Lederzeitschr. *1952* Heft 11, S. 307–309.
54. F. Stather: Stark wirkende freie Säuren in Leder, ein ausgesprochener Lederfehler. Dtsche Schuh- u. Lederzeitschr. *1952*, Heft 12, S. 337–338.
55. F. Stather: Gewichtsausbeute und Flächenausbeute bei der Ledererzeugung. Deutsche Schuh- u. Lederzeitschr. *1953*, Heft 1, S. 2–3.
56. F. Stather: Der Qualitätsbegriff in der ledererzeugenden Industrie. Deutsche Schuh- u. Lederzeitschr. *1953*, Heft 2, S. 32–34.
57. F. Stather: Neuere Forschungsergebnisse auf dem Gebiete der Pelzzurichtung und Pelzveredlung. Vorgetragen auf der Hauptversammlung des Vereins für Gerbereichemie und -technik am 13. 9. 1952 in Konstanz. Das Leder *1953*, Heft 1, S. 9–14.
58. F. Stather: Gesundheitsgefahren in der Leder- und Pelzindustrie. Deutsche Schuh- und Lederzeitschr. *1953*, Heft 3, S. 64–65.

59. F. Stather: Lederfärberei im Lichte neuerer elektrochemischer Anschauungen. Deutsche Schuh- und Lederzeitschr. *1953*, Heft 4, S. 91—94.
60. F. Stather: Zur Frage der Haltbarkeit der Laufleder auf Kammstühlen der Textilindustrie. Deutsche Schuh- und Lederzeitschr. *1953*, Heft 5, S. 122—124.
61. F. Stather: Ledereigenschaften und Schuhqualität. Deutsche Schuh- und Lederzeitschr. *1953*, Heft 6, S. 152—155.
62. F. Stather: Oberflächenaktive Substanzen als Hilfsmittel der ledererzeugenden Industrie. Deutsche Schuh- und Lederzeitschr. *1953*, Heft 7, S. 180—183.
63. F. Stather: Sachgemäße Lederbehandlung und Lederpflege. Deutsche Schuh- und Lederzeitschr. *1953*, Heft 8, S. 211—214.
64. F. Stather: Zur Frage des Brühens von Schweinshäuten II. Deutsche Milch—Fett—Fleisch—Fisch-Zeitschr. *1953*, Heft 10, S. 276—278 und Deutsche Schuh- und Lederzeitschr. *1953*, Heft 9, S. 262.
65. F. Stather: Polymerisationsgerbung von Haut und Polymerisatimprägnierung von Leder. Deutsche Schuh- und Lederzeitschr. *1953*, Hefte 10, S. 288 und 11, S. 323. 6 Seiten.
66. F. Stather: Über die Schädigung der Schweinshaut beim Brühen nach dem Dresdner Brühverfahren. Das Leder *1953*, Heft 11, S. 274—279.
67. F. Stather: Über die Schädigung der Schweinshaut beim Brühen nach dem Dresdner Brühverfahren. Deutsche Schuh- und Lederzeitschr. *1953*, Heft 12, S. 349—353.
68. F. Stather: 65 Jahre Deutsches Lederinstitut Freiberg/Sa. 1889—1954. Deutsche Schuh- und Lederzeitschr. *1954*, Heft 5, S. 146—148; Heft 6, S. 186—188; Heft 7, S. 220—222.
69. F. Stather: Schweinsleder, richtig beurteilt. Deutsche Schuh- und Lederzeitschr. *1954*, Heft 5, S. 146—151.
70. F. Stather: Ing. Chem. Reinhold Lauffmann 1876—1952 [Nachruf]. Ges. Abh. DLI Heft 10 (1954) S. 3—4.
71. F. Stather: Zur Frage der sachgemäßen Erzeugung und des richtigen Einsatzes von Kammstuhl-Laufledern in der Textilindustrie. Deutsche Schuh- und Lederzeitschr. *1955*, Heft 8, S. 251, 4 Seiten.
72. F. Stather: „Bügeln" des Oberleders mit Infrarotstrahlen. Deutsche Schuh- und Lederzeitschr. *1956*, Heft 5, S. 137.
73. F. Stather: Der Narben, das natürliche Gesicht des Leders. Deutsche Schuh- und Lederzeitschr. *1956*, Heft 10, S. 291—294.
74. F. Stather: 2. Internationales Colloquium für Gerbereichemie in Otrokowice (CSR). Deutsche Schuh- und Lederzeitschr. *1957*, Heft 4, S. 109—110.
75. F. Stather: Maschineller Abzug von Schweinshaut-Croupons. Das Leder *1957*, S. 63 bis 64.
76. F. Stather: Der Ledernarben künstlich verändert. Deutsche Schuh- und Lederzeitschr. *1957*, Heft 5, S. 147—150.
77. F. Stather und E. Seidel: [Beiheft zur Lichtbildreihe BR 73] Rohhaut und Rohhautschäden I, Struktur, Einteilung und Erkennungsmerkmale. Deutsches Zentralinstitut für Lehrmittel *1957*, 24 Seiten.
78. F. Stather und E. Seidel: [Beiheft zur Lichtbildreihe BR 74] Rohhaut und Rohhautschäden II, Rohhautschäden. Deutsches Zentralinstitut für Lehrmittel *1957*, 24 Seiten.
79. F. Stather: Schneckenfässer als Mittel der Mechanisierung in der Lederindustrie. Deutsche Schuh- und Lederzeitschr. *1957*, Heft 6, S. 169—171 und Das Leder *1957*, S. 145—147.

80. F. Stather: Schweinshaut und Schweinsleder. Das Leder *1958*, S. 35—39; in französischer Übersetzung: Peau de porc et cuir de porc. Revue Technique des Industries du Cuir *1957*, S. 263—266.
81. F. Stather: Das Deutsche Lederinstitut Freiberg in neuer Form vor neuen Aufgaben. Deutsche Schuh- und Lederzeitschr. *1958*, Heft 1, S. 1—4.
82. F. Stather: Zur Charakteristik pflanzlicher und synthetischer Gerbstoffe und deren kombinierte Verwendung bei der Gerbung. Berichte über die Verhandlungen der Sächsischen Akademie der Wissenschaften zu Leipzig, Math.-nat. Kl., Bd *104* (1960) Heft 3, 19 Seiten.
83. F. Stather: Neue Lederstandards in der DDR. Deutsche Schuh- und Lederzeitschr. *1960*, Heft 12, S. 349—357.
84. F. Stather: Nové Standardy usní Německé Demokratické Republice. Kožařstvi *XI* (1961) S. 66—67 [Auszug aus der vorherigen Veröffentlichung].
85. F. Stather: Zur Problematik der Gerbzeitverkürzung bei der pflanzlich-synthetischen Gerbung. Das Leder *1961*, Heft 8, S. 189—194.
86. F. Stather: Deutsches Lederinstitut 1950—1960 (62.—72. Jahresbericht). Ges. Abh. DLI Heft 17 (1961) S. 7—24.
87. F. Stather: Narbenplatzen an Schuhoberleder. Deutsche Schuh- und Lederzeitschr. *1963*, Heft 3, S. 58—60.
88. F. Stather: Imprägnieren von Leder. Deutsche Schuh- und Lederzeitschr. *1963*, Heft 4, S. 99—101.
89. F. Stather, G. Reich und W. Barthel: Über die Anwendung chromatographischer Methoden zur Charakterisierung aromatischer Syntane. Das Leder *1963*, Heft 3, S. 53—59.
90. F. Stather: Sachgemäße Lederbehandlung in der Lederwarenindustrie. Innendekoration und Lederwaren *1963*, Heft 5, S. 112—113.
91. F. Stather: Sachgemäße Rohhautkonservierung durch Stapel- oder Lakesalzung? Der Fleischermeister *1963*, Heft 6, S. 147—149.
92. F. Stather: Probleme der modernen Unterledergerbung. Vortrag, gehalten auf dem 2. Internationalen Kongreß des Ung. Wissenschaftlichen Vereins für die Leder-, Schuh- und lederverarbeitende Industrie am 5. 10. 62 in Budapest. Ges. Abh. DLI Heft 18 (1962) S. 126—133; Übersetzung ins Ungarische: A korszerü keménybörgyártás problémái. Bör- és Cipötechnika *XIII* (1962) S. 135—139.
93. F. Stather: Wissenschaftliche Grundlagen der Rauchwarenzurichtung. Das Pelzgewerbe *1963*, Nr. 2 u. 3, S. 53, 7 Seiten.
94. F. Stather: 75 Jahre Deutsches Lederinstitut Freiberg 1889—1964. Berlin 1964. 106 Seiten.
95. F. Stather: Chromatographie und Elektrophorese bei der Untersuchung synthetischer Gerbstoffe. FuF *40* (1966) Heft 3, S. 81—82.
96. F. Stather: Einige Untersuchungsergebnisse zur Deckfarbenzurichtung von Leder. Das Leder *1966*, Heft 9, S. 217—220.
97. F. Stather: Prof. Dr. Wolfgang Graßmann — 70 Jahre. Leder- und Häutemarkt *1968*, Heft 3, S. 26—27.
98. F. Stather: Techn. Rat Prof. Ing. J. A. Sagoschen beendet berufliche Tätigkeit. Das Leder *1968*, S. 184.
99. F. Stather: Professor Dr. Adolf Küntzel zum 70. Geburtstag am 4. Dezember 1968. Leder- und Häutemarkt *1968*, Heft *46* v. 17. 11. 68.

Patentschriften

1. DRP. 475301 vom 3. 7. 1924	Verfahren zum Enthaaren und Äschern von Häuten und Fellen Inhaber: M. Bergmann, E. Immendörfer u. H. Loewe Erfinder: F. Stather
2. DRP. 482418 vom 17. 6. 1925	Verfahren zum Enthaaren von Häuten und Fellen Inhaber: M. Bergmann Erfinder: F. Stather
3. DRP. 496922 vom 13. 2. 1925	Verfahren zur Behandlung von Häuten und Fellen Inhaber: M. Bergmann, E. Immendörfer u. H. Loewe Erfinder: F. Stather
4. DRP. 582919 vom 28. 8. 1931	Verfahren zum Entgerben von ungegerbtem Chromleder Inhaber: Deutsche Gold- und Silber-Scheideanstalt vormals Roessler in Frankfurt a. M. Erfinder: F. Stather
5. DDR-P. 98 vom 23. 11. 1951	Verfahren zur Herstellung von synthetischen Austauschgerbstoffen aus unsulfonierten, einwertigen, einkernigen Phenolen, Sulfitzellstoffablauge und Aldehyden in einstufiger Arbeitsweise Erfinder: F. Stather, H. Herfeld, R. Schubert Inhaber: F. Stather, H. Herfeld
6. DDR-P. 99 vom 12. 11. 1951	Verfahren zur Herstellung von synthetischen Austauschgerbstoffen aus sulfoniertem β-Naphtol, unsulfonierten ein- oder mehrkernigen einwertigen Phenolen und Aldehyden in einstufiger Arbeitsweise Erfinder: F. Stather, H. Herfeld, R. Schubert Inhaber: F. Stather, H. Herfeld
7. DDR-P. 4418 vom 29. 7. 1952	Verfahren zur Herstellung von synthetischen Austauschgerbstoffen Erfinder zugleich Inhaber: F. Stather u. H. Herfeld
8. DDR-P. 9989 vom 19. 9. 1954	Verfahren zur Herstellung von synthetischen Austauschgerbstoffen Erfinder zugleich Inhaber: F. Stather u. G. Reich
9. DDR-P. 10016 vom 19. 9. 1954	Verfahren zur Herstellung von synthetischen Austauschgerbstoffen Erfinder zugleich Inhaber: F. Stather u. G. Reich
10. DDR-P 12290 vom 27. 5. 55	Verfahren zur Herstellung von synthetischen Vollgerbstoffen Erfinder zugleich Inhaber: F. Stather und G. Reich
11. DDR-P. 28739 Ausgabetag: 25. 5. 1964	Verfahren zur Herstellung von Leder Erfinder zugleich Inhaber: F. Stather u. G. Reich

Dr. Martin Gierth

Zur Biobibliographie

Schriftenverzeichnisse. 50 Jahre Freiberger Gerberei-Institute *1889–1939* (Freiberg 1939) S. 73–89. – M. Gierth, siehe vorliegendes Jahrbuch S. 254–272. – Jahrbuch Sächs. Akad. Wiss. *1957–1959* (Berlin 1961) S. 258–259; *1960–1962* (Berlin 1964) S. 293–294; *1963–1965* (Berlin 1967) S. 227; *1966–1968* (Berlin 1970) S. 230. – Kürschners Dtsch. Gelehrten-Kalender *4* (Beriln, Leipzig 1931) Sp. 2874; *5* (Berlin, Leipzig 1935) Sp. 1343; 6_2 (Berlin 1940/41) Sp. 835–836; 10_2 (Berlin 1966) S. 2389 [ohne Literaturangaben]; *11* (Berlin 1970) S. 2914. – Poggendorffs Handwörterbuch VI_4^a (Berlin 1938) S. 2525; VII_4^a (Berlin 1961) S. 491–494.

Würdigungen. Zur Habilitation: Dresdner Anzeiger, 30. Jan. 1930. — Dresdner Neueste Nachrichten 31. Jan. 1930. — Zur Leitung der Deutschen Versuchsanstalt für Lederindustrie in Freiberg: Der Ledermarkt, Frankfurt/M., 17. März 1930. — Freiberger Anzeiger, 26. März 1930. — Die Lederindustrie *73* (Berlin 1930) S. 1 v. 24. März. — Der Gerber, Teplitz-Schönau *1930*, H. 1325. — The Footwear Organiser, London, Juni 1930. — Zur Leitung der Deutschen Versuchsanstalt und Fachschule für Lederindustrie: Die Lederindustrie, Berlin, 6. März 1938. — Häute und Leder, 6. März 1938. — Leder-Ztg, 6. März 1938. — 25 Jahre Leiter der Freiberger Gerbereiinstitute: H. Herfeld, in: Deutsche Schuh- und Lederzeitschrift *9* (Berlin 1955) Nr. 3, S. 68—70. — A. K[üntzel], in: Das Leder *6* (1955) S. 71. — J. A. Sagoschen, in: Österreichische Lederztg, Wien 1955, Nr. 3. — E. Simoncini, in: Cuoio-Pelli-Materie, Concinati, Neapel 1955, Nr. 3. — Das ABC der Schuhfabrikation, Frankfurt/M. 1955, S. 111. — Bör- és Cipötechnika, Budapest 1955, S. 63. — Angewandte Chemie, Weinheim 1955, Nr. 6. — Felle und Seifen, Münster/Westf. 1955, Nr. 1, S. 86. Deutsche Schuh- und Lederindustrie, Berlin 1955, Nr. 5. — Przegląd Skorzany, Warschau 1955, S. 69—70. — Revue Technique des Industries du Cuir, Paris 1955, S. 30. — Volksstimme, Karl-Marx-Stadt, 22. März 1955. — Zur Verleihung desTitels „Hervorragender Wissenschaftler des Volkes“: ND, 15. Dez. 1955. — Volksstimme, Karl-Marx-Stadt, 16. Dez. 1955; Freiberg, 22. Dez. 1955. — Deutsche Schuh- und Lederzeitschrift, Berlin, Jan. 1956. — Mitteilungsblatt TH Dresden, 1. Jan. 1956. — Das Leder, Darmstadt, Jan. 1956. — Berliner Ztg, 17. Febr. 1956. — Zur Wahl in die Sächsische Akademie der Wissenschaften: Neue Zeit, Berlin, 23. Apr. 1959. — Volksstimme, Freiberg, 2. Mai 1959 . — Leder-Kurier, Wien 1959, H. 5, S. 86. — Deutsche Schuh- und Lederzeitschrift *7* (Berlin 1959) S. 201. — Revue Technique des Industries du Cuir, Paris 1959, S. 142. — Przegląd Skorzany, Warschau 1959, Nr. 5. — Zur Wahl in die Deutsche Akademie der Wissenschaften zu Berlin: Volksstimme, Freiberg, 17. Juli 1961. — Technische Gemeinschaft (KdT), Berlin 1961, H. 4. — Leder- und Häutemarkt, Frankfurt/M. Juli 1961. — Kolloidzeitschrift *178* (Darmstadt 1961) H. 1, S. 56. — Zum 60. Geburtstag: P. Gourlay, in: Revue Technique des Industries du Cuir, Paris Okt. 1961. — A. Küntzel, in: Das Leder, Darmstadt Nov. 1961, S. 284. — H. Moser, in: Technische Gemeinschaft (KdT) *9* (1961) H. 12, S. 449. — J. A. Sagoschen, in: Leder-Kurier, Wien Okt. 1961. — P. Schöps, in: Pelzgewerbe, Leipzig 1961, Nr. 5. — R. Schubert, in: Leder- und Häutemarkt, Frankfurt/M. Okt. 1961. — Deutsche Schuh- und Lederzeitschrift *1961*, H. 11. — Die Union, Karl-Marx-Stadt, 5. Nov. 1961. — Volksstimme, Freiberg, 4. Nov. 1961. — Innendekoration und Lederwaren, Leipzig 1961, H. 11, S. 202. — Schuh-Technik, Heidelberg 1961, Dez.-Nr. — Bild-Ztg Freiberg, 8. Nov. 1961. — Chem. Technik *13* (1961) H. 11, S. 699. — Deutsche Schuh- und Lederzeitschrift *1962*, H. 1. — Die Wirtschaft *17* (Berlin 1961) Nr. 3. — Przegląd Skorzany 1961. — Bör- és Cipötechnika, Budapest 1962, S. 26. — Gesammelte Abhandlungen des Deutschen Lederinstituts, Freiberg 1961, H. 17, S. 5. — [Interview] Wochenpost, Berlin, 15. Apr. 1961. — Zum 65. Geburtstag: B. Anton, in: Kožařství, Prag 1966, H. 11, S. 354. — Baran, in: Das Pelzgewerbe, Leipzig 1966, Nr. 5, S. 232. — [E. Heidemann], in: Das Leder, Darmstadt Okt. 1966, S. 259. — H. Moser, in: Leder-Schuhe-Lederwaren, Leipzig 1967, H. 2, S. 74. — G. Reich, in: Mitteilungsblatt der Chem. Gesellschaft der DDR, Nov. 1966. — [R. Schubert], in: Leder- und Häutemarkt, Frankfurt/M. 1966, H. 43. — Freie Presse, Karl-Marx-Stadt/Freiberg, 5. Nov. 1966. — Leder-Schuhe-Lederwaren, Leipzig 1966, H. 11, S. 424. — Bör- és Cipötechnika, Budapest 1966, H. 16. — Revue Technique des Industries du Cuir, Paris Nov. 1966. — Brühl, Leipzig Jan./Febr. 1967, S. 31. — Die Schuh-Technik, Heidelberg Jan. 1967. — Kozi i Obuvki, Sofia Dez. 1966. — Przegląd Skorzany, Warschau 1966, Nr. 11, S. 244. — Zur Verleihung der Edmund-Stiasny-Medaille: Aachener Ztg, 6. Mai 1967. — Leder- und Häutemarkt, Frankfurt/M. 1967, S. 141—142. — Leder-Schuhe-Lederwaren, Berlin Juli 1967. — Z. f. angewandte Chemie, Weinheim Juni 1967. — Zum 70. Geburtstag: W. Graßmann, in: Das Leder, Darmstadt 1971, S. 243—244. — H. Herfeld, in: Leder- und Häutemarkt, H. 42, 17. Okt. 1971. — J. A. Sagoschen, in: Österreichische Leder- und Häutewirtschaft *7* (Wien 1971) Okt.-Nr. — Freie Presse, Freiberg/Karl-Marx-Stadt, 5. Okt. [irrtümlich] 1971. — Z. f. angewandte Chemie, Weinheim Okt. 1971. — Bör- és Cipötechnika, Budapest Sept. 1971, S. 140. — Kožařství, Prag 1971, S. 281. — Leder-Schuhe-Lederwaren, Leipzig Nov. 1971, H. 11. — Mitteilungsblatt der Chem. Gesellschaft der DDR, Okt. 1971. — Przegląd Skorzany, Warschau 1971, Nr. 26, S. 12. — Allgemein: Wissenschaftliche Forschungen in der Lederindustrie, in: Süddeutsche Schuh- und Lederztg, Freiburg/Br., 13. Juli 1928. — Der Gerber, 1928, Nr. 148. — Freiberger Anzeiger, 10. Dez. 1932. — Der Gerber, 1932, Nr. 124. — Norsk Skolvi, Kopenhagen Mai 1935. — Freiberger Anzeiger und Tageblatt, 15. Mai 1937, Nr. 111. — Häute und Leder, 23. Juli 1937, Nr. 30. — Freiberger Anzeiger und Tageblatt, Mai 1939, Nr. 122. — Allgemeine Lederindustrie-Ztg, Wien, 27. Mai 1939, Nr. 21. — Häute und Leder, 26. Mai 1939. — Freiberger Anzeiger und Tageblatt, 3. Juni 1939, Nr. 127. — Die Lederindustrie *82* (Berlin 1939) Nr. 130 v. 1. Juni. — Ebda Nr. 133 v. 5. Juni. — Häute und Leder, 9. Juni 1939, Nr. 23. — Volksstimme, 4. Febr. 1946. — Tägliche Rundschau 17. Dez. 1946. — Die Schuhwirtschaft, Berlin, 1. Febr. 1948. — Der Schuh, Berlin, 14. Apr. 1948. — Leder-Journal 1949, H. 3. — Die Union, Dresden 30. Mai 1952. — Die Neue Zeit, Berlin, 19. Juni 1952. — Sächsische Neueste Nachrichten, Chemnitz, 11. Dez. 1952. — ND, 13. April 1954, Nr. 87. — Sächsisches Tageblatt, Leipzig, 14. April 1954. — H. Hälbig, in: Sächsische Neueste Nachrichten, Dresden, 21. Sept. 1956. — Mitteilungsblatt der Chemischen Gesellschaft in der DDR, Sept. 1958. — Das Leder, Darmstadt Juni 1958. — Leder-Kurier, Wien 1958, Nr. 6. — Das Leder, Darmstadt 1964, Nr. 6, S. 154. — Freie Presse, Karl-Marx-Stadt/Freiberg, 31. Dez. 1964.

Nachrufe. U. Freimuth, siehe vorliegendes Jahrbuch S. 252—254. — H. Herfeld, in: Das Leder *25* (1974) S. 108. — H. János, in: Bör- és Cipötechnika *24* (Budapest 1974) S. 80. — G. Reich, in: Leder-Schuhe-Lederwaren, Leipzig 1974, H. 6, S. 210. — J. A. Sagoschen, in: Österreichische Leder- und Häutewirtschaft *10* (Wien 1974) Nr. 9.

G. W.

Gustav E. R. Schulze

24. 2. 1911—5. 10. 1974

Am 5. Oktober 1974 starb im 64. Lebensjahr das ordentliche Mitglied der mathematisch-naturwissenschaftlichen Klasse Gustav E. R. Schulze, emeritierter ordentlicher Professor für Experimentalphysik an der Technischen Universität Dresden.

Mit ihm verliert die Sächsische Akademie der Wissenschaften einen Gelehrten, der durch seine Arbeiten auf den Gebieten der Röntgen- und der Metallphysik auch international ein hohes Ansehen erworben hatte. Am 24. 2. 1911 in Berlin als Sohn eines Verlagsbuchhändlers geboren, nahm er nach seiner Reifeprüfung 1929 das Physikstudium zunächst in Berlin auf und setzte es dann in Göttingen fort. Während des Wintersemesters 1931/32 begann er im Mineralogisch-petrographischen Institut der Georg-August-Universität kristallographische Arbeitsverfahren — insbesondere röntgenographische — zu erlernen. Am 29. 11. 1933, also im 23. Lebensjahr, promovierte er bei V. M. Goldschmidt zum Dr. phil. mit einer Arbeit über „Die Kristallstruktur von BPO_4 und $BAsO_4$", die als beste Arbeit eines Jahres preisgekrönt wurde. — In dieser Untersuchung zeigt sich bereits seine besondere Begabung, Strukturverwandtschaften zu erkennen, die ihm auch später zahlreiche wissenschaftliche Erfolge bescheren sollte. Hier erkannte er die Verwandtschaft der beiden Strukturen mit der des Hoch-Cristobalits. Goldschmidt schreibt danach u. a. „... insbesondere möchte ich seine ungewöhnliche Veranlagung für Kristallstruktur-Untersuchungen hervorheben ..."

Nach kurzer Assistentenzeit bei Goldschmidt empfiehlt ihm dieser, die Möglichkeit zu nutzen, als Stipendiat der Deutschen Forschungsgemeinschaft im damaligen von Otto Hahn geleiteten Kaiser-Wilhelm-Institut für Chemie in Berlin-Dahlem zu arbeiten. G. Schulze folgt diesem Rat und arbeitet im Dahlemer Institut vom 1. 9. 1934 bis 31. 8. 1935. Während dieser Zeit führte er dort die damals noch junge Debye-Scherrer-Methode ein, nahm Röntgenspektren einer Anzahl von Metalloxiden auf und befaßte sich mit der Aufklärung einiger anomaler Erscheinungen, die beim Einbau kleinster Mengen radioaktiver Elemente in auskristallisierenden Salzen beobachtet werden. Hahn ist von seinen gründlichen kristallographischen und röntgenographischen Kenntnissen sehr angetan und stellt ihm das Zeugnis eines „sehr selbständigen, gewissenhaften und kenntnisreichen Mitarbeiters" aus.

Anschließend bietet sich ihm eine Stelle als planmäßiger wissenschaftlicher Assistent am Physikalischen Institut der Universität Greifswald bei F. Krüger, die er bis zum 28. 2. 1938 einnimmt. Hier findet er Muße, sich mit magnetischen Eigenschaften von Kristallen zu beschäftigen, einer Problematik, auf die er auch nach dem 2. Weltkrieg wieder zurückkommt.

Für seinen weiteren wissenschaftlichen Werdegang wohl von nachhaltigstem Einfluß war die Begegnung mit Ulrich Dehlinger in Stuttgart. Vom 1. 3. 1938 bis zum 31. 10. 1939 kam er dort als Forschungsassistent in der Abteilung Dehlinger des Institutes für Metallphysik (Direktor R. Glocker) am K. W. I. für Metallforschung mit metallphysikalischen Fragestellungen in Berührung und begründete mit der Habilitationsschrift „Zur Kristallchemie der intermetallischen AB_2-Verbindungen (Laves-Phasen)" sein späteres Hauptarbeitsgebiet. Darin untersucht er die Existenzbedingungen dieser homöotekten Gruppe intermetallischer Verbindungen, die auch heute noch die zahlenmäßig bedeutendste ist, und kommt zu dem bemerkenswerten Resultat, daß die stöchiometrische Zusammensetzung bei dieser Gruppe durch geometrische Strukturargumente beherrscht wird, d. h. die üblichen Valenzregeln der Chemie nicht wirksam sind. Ein Gedanke, der sich später als noch von größerer Tragweite erweisen sollte, spielt bei seiner Argumentation eine bedeutende Rolle, daß es nämlich zweckmäßig ist, die zwölf nächsten B-Nachbarn und die vier nur wenig weiter entfernten A-Nachbarn zu einer Koordinationssphäre mit der Koordinationszahl 16 zusammenzufassen. Da Kasper später auf die viel allgemeinere Bedeutung solcher Koordinationspolyeder überhöhter Koordinationszahl hingewiesen hat, werden sie heute Kasper-Polyeder genannt.

Am 1. 11. 1939 wechselt G. E. R. Schulze an die TH Dresden über, wo er infolge des Assistentenmangels nahezu vollständig mit Lehraufgaben ausgefüllt ist. Er bewirbt sich um eine Dozentur für Experimentalphysik, die ihm am 6. 4. 1940 verliehen wird. Am 1. 10. 1941 erfolgt die Ernennung zum Oberingenieur im Physikalischen Institut bei H. A. Stuart. Er leitet das physikalische Anfängerpraktikum und baut u. a. einen dreisemestrigen Vorlesungszyklus über höhere Experimentalphysik auf. Dazu entfaltet er eine vielseitige Vortragstätigkeit. Wiederholt werden seine pädagogischen Fähigkeiten hervorgehoben. Willers schreibt z. B. als Leiter der Abteilung für Mathematik und Physik über ihn: „... Er erfreute sich seines klaren Vortrages wegen bei den Studenten großer Beliebtheit und war bei seinem bescheidenen Wesen von den anderen Herren des Lehrkörpers gern gesehen ...“

1943 schließt er die bis zuletzt ungetrübte Ehe mit Dora Witte, die ihm nach seinen eigenen Worten immer wieder Quell neuer Kraft ist.

Sein Interesse für Musik ist in jener Zeit schon besonders ausgeprägt und auch auf den Zusammenhang mit physikalischen Grundlagen gerichtet. Davon zeugt u. a. auch ein Brief Max Plancks an ihn aus dem Jahre 1944, in dem dieser seine Ansicht über die natürliche Stimmung auseinandersetzt. Erst dreißig Jahre später erscheint eine Publikation, in der G. E. R. Schulze seine Überlegungen dazu abschließt.

Einige Monate nach Kriegsende findet er eine Stellung als Leiter der Thermodynamischen Abteilung der Fa. Junkers-Motorenbau Dessau und siedelt mit diesem Betrieb 1946 in die Sowjetunion über, wo er in gleicher Stellung bis 1954 arbeitet.

Nach seiner Rückkehr erhält er einen Ruf an die TH Dresden, wird am 15. 8. 1954 zum Professor mit Lehrauftrag für Sondergebiete der Physik und schon wenig später, am 1. 12. 1954, zum Professor mit vollem Lehrauftrag ernannt. Er übernahm als Nachfolger von Wiedmann die Leitung des Institutes für Röntgenographie, dessen Arbeitsrichtung er in metallphysikalischer Hinsicht erweiterte und das deshalb 1957 die Bezeichnung „Institut für Röntgenkunde und Metallphysik“ erhielt. 1958 wird ihm der neugeschaffene Lehrstuhl für Röntgenkunde und Metallphysik übertragen.

Im gleichen Jahre wird er für eine Periode von 3 Jahren zum Dekan der Fakultät für Mathematik und Naturwissenschaften gewählt, eine Verpflichtung, die ihm 1967 ein zweites Mal auferlegt wird, dann aber nur bis zur Auflösung der alten Fakultäten 1968 im Zuge der 3. Hochschulreform.

Auch über den Hochschulrahmen hinaus setzt er sich bald als Mitglied der Sektion Physik der DAW Berlin für die Förderung der Metallphysik ein. 1957 wird er Vorsitzender der Unterkommission für Metallphysik.

Im Dezember 1960 wird G. E. R. Schulze zum ordentlichen Mitglied der Sächsischen Akademie der Wissenschaften gewählt. In seiner Mitarbeit zeigte er, daß sein Wirken und seine Gedanken weit über das unmittelbare Fachgebiet hinausgingen. Hier sei nur an den letzten Vortrag vor der Akademie im Jahre 1973 „Zur Rolle des Einfachheitsprinzips im physikalischen Weltbild“ erinnert, in dem er Resultate jahrzehntelangen Forschens zusammenfaßt.

Kehren wir zurück zu seiner Tätigkeit als Hochschullehrer. Er baut in jahrelangem Bemühen je einen Vorlesungszyklus über Metallphysik (später kommt noch Festkörperphysik hinzu) und Röntgenstrahlinterferenzen auf, hält daneben noch Vorlesungen über ausgewählte Kapitel der Metallphysik, in denen sich ein Wesenszug besonders widerspiegelt, nämlich die aktuellsten wissenschaftlichen Entwicklungen der heranwachsenden Jugend in Permanenz nahezubringen. Eine Folge davon war, daß er praktisch niemals die gleiche Vorlesung zweimal gehalten hat. Seine Lehrerfahrungen schlagen sich u. a. in seinem Lehrbuch „Metallphysik" nieder. Nach der 3. Hochschulreform übernimmt er zusätzlich die Einführungsvorlesung Physik (Teil Mechanik) für das 1. Semester. Die Beschäftigung damit bereitet ihm trotz der enormen zusätzlichen Belastung großen Spaß und regt ihn zu einer Reihe historischer Untersuchungen an.

Seine besonderen Verdienste um die Darstellung der Historie des Fachgebietes reichen bis Anfang der 60er Jahre zurück (siehe Bibliographie). Zu Unrecht in Vergessenheit geratene Entwicklungsetappen und historische Zusammenhänge werden von ihm aufgezeigt, teilweise erstmals erkannt.

Daneben werden unter seiner Leitung umfangreiche Forschungsarbeiten zur Metallphysik intermetallischer Verbindungen durchgeführt. Auf seinem früheren engeren Arbeitsgebiet, der Kristallchemie intermetallischer Verbindungen, gelingt ihm die Entdeckung des Prinzips maximaler Dichte. Sein Hauptaugenmerk gilt nun der Aufklärung der physikalischen Grundlagen von Eigenschaften intermetallischer Verbindungen (mechanischen, magnetischen und elektronischen) mit dem Blick auf ihre mögliche technische Anwendung. Dazu wird ein breites Methodenspektrum — Röntgenstrahl- und Neutronendiffraktometrie, Röntgenspektrometrie, Mössbauerspektrometrie, Elektronenstrahlmikroanalyse, Kosseltechnik, Röntgenprimäranalyse. Röntgenfluoreszenzanalyse, Metallographie und andere — eingesetzt. Ein wesentliches Resultat dieser Untersuchungen besteht im Nachweis und der Deutung von Eigenschaftsänderungen im Homogenitätsbereich intermetallischer Verbindungen.

Das von G. E. R. Schulze geleitete Institut für Röntgenkunde und Metallphysik wurde am 1. 8. 1965 im Zusammenhang mit wachsenden Verpflichtungen für die allgemeine Physikausbildung in II. Institut für Experimentalphysik umbenannt. Im Zuge der 3. Hochschulreform erhielt es 1968 den Status eines Bereiches (Experimentalphysik II) innerhalb der Sektion Physik. Am 1. 9. 1971 wurde der Bereich in „Arbeitsgruppe Experimentalphysik 2" umbenannt, um eine intersektionelle Strukturänderung zum Ausdruck kommen zu lassen.

Im Herbstsemester des Jahres 1973 verschlechtert sich der Gesundheitszustand G. E. R. Schulzes so bedeutend, daß er seine Lehrtätigkeit einstellen muß. Zum 15. 5. 1974 wird er deshalb vorzeitig emeritiert.

In Anerkennung seiner Verdienste um die praxisorientierte Lehre und Forschung werden ihm 1959 die Verdienstmedaille und 1967 der Nationalpreis (gemeinsam mit K. Kleinstück, G. Försterling, R. Gnauck) verliehen.

Übei den engeren Arbeitskreis hinaus war er aktives Mitglied zahlreicher wissenschaftlicher Gesellschaften und Gremien (z. B. auch der Vereinigung für Kristallographie der DDR, deren Vorstand er angehörte; des Nationalkomitees für Kristallographie; des Kooperationsrates Metallforschung Dresden—Freiberg, des Institute of Metals, London) sowie von Redaktionskollegien der Zeitschriften „Kristall und Technik" und „Experimentelle Technik der Physik".

Die Hoffnung, daß Gustav E. R. Schulze trotz früher Erkrankung noch lange Jahre an der Arbeit der Akademie teilnehmen kann, hat sich leider nicht erfüllt. Die Sächsische Akademie der Wissenschaften wird ihm stets ein ehrendes Andenken bewahren.

Dr. sc. nat. Peter Paufler, Artur Lösche

Frau Dora Schulze gebührt herzlicher Dank für ihre freundliche Unterstützung des Quellenstudiums.

Bibliographie Gustav E. R. Schulze

Die Kristallstruktur von BPO_4 und $BAsO_4$. In: Naturwiss. *21* (1933) 562.
Die Kristallstruktur von BPO_4 und $BAsO_4$. In: Z. Phys. Chem. *B 24* (1934) 215–240.
Kristallform und Raumgitter von ZrF_4 und HfF_4. In: Z. Kristallographie *A 89* (1934) 477–480.
H. Hauptmann, G. E. R. Schulze: Über ein neues Mikrovolumenometer. In: Z. Phys. Chem. *A 171* (1934) 36–40.
Kristallstrukturelle Untersuchungen an Ergosterin- und Cholesterinderivaten. In: Z. Phys. Chem. *A 171* (1935) 436–444.
Boron Arsenate. In: Journ. Am. Chem. Soc. *57* (1935) 883.
Die Kristallstruktur von Radiumfluorid. In: Z. Phys. Chem. *B 32* (1936) 430–432.
Über das magnetische Verhalten diamagnetischer Salze bei Gitterumwandlungen. In: Z. Phys. Chem. *B 40* (1938) 308–317.
Zur Kristallchemie der intermetallischen AB_2-Verbindungen (Laves-Phasen). In: Z. Elektrochem. *45* (1939) 849–865.
Bemerkung zum Zustandsbild Magnesium-Kadmium. In: Z. Metallkunde *32* (1940) 252.
U. Dehlinger, G. E. R. Schulze: Kristallchemie der Verbindungen vom Typ $MgCu_2$ und $MgZn_2$. In: Z. Krist. *102* (1940) 377–390.
U. Dehlinger, G. E. R. Schulze: Das einfachste Prinzip in der Kristallchemie der Legierungen. In: Z. Metallkunde *33* (1941) 157–160.
Kristallstruktur von $LaAl_2$. In: Z. Krist. *104* (1942) 257–260.
U. Dehlinger, G. E. R. Schulze: Aufbau der Kristalle 3. In: Physik in regelm. Ber. *10* (1942) 35–49.
Das Kristallgitter von Nitropentaminkobaltiperchlorat. In: Naturwiss. *31* (1943) 511.
Auswertung von Röntgengoniometer-Aufnahmen höherer Schichtlinien auf ebenem Film bei schiefer Inzidenz. In: Z. Krist. *105* (1944) 324–329.
Lehrbriefe für das Fernstudium. „Physik der Röntgenstrahlen" Nr. 1–4. 283 S. Selbstverlag der TH Dresden 1956/57.
u. J. Henke: Weiterentwicklung der Rahmenkammer zur Aufnahme von Röntgen-Kleinwinkelstreuung. In: Exp. Technik Phys. *5* (1957) 180–183.
Die technischen Eigenschaften und der atomistische Aufbau unserer Werkstoffe. In: Berichtsband „I. Polytechn. Tagung der TH Dresden 1956" 33–40.
J. Henke, G. E. R. Schulze: Eine verbesserte Rahmenkammer für Röntgen-Kleinwinkelstreuung. In: Z. Naturforschg. *12a* (1957) 346.
Erforschung der Orientierung magnetischer Atommomente in Kristallen mittels Neutronen-Interferenzen. In: Fortschr. Min. *35* (1958) 135–150.

u. J. Henke: Eine einfache Röntgen-Kleinwinkelkammer hoher Präzision. In: Wiss. Zeitschr. d. TH Dresden *7* (1957/58) Heft 6, 1—6.

Fortschritte auf dem Gebiet der Röntgenstrahl-Kleinwinkelstreuung. In: Sbornik Referátů Mezinárodní Konference Českéno vysokéno Učení Technického v Praze, II, Praha 1958, 239—248.

u. J. Henke: Eine lichtstarke Planglas-Endmaß-Kammer für Kleinwinkelstreuung. In: Z. Kristallogr. *111* (1959) 172—184.

Über die Atomabstände in Laves-Phasen. In: Z. Kristallogr. *111* (1959) 249—258.

Über geometrische Einflußgrößen bei der Bildung intermetallischer Verbindungen. In: Freiberger Forschungshefte *B 37* (1959) 78—83.

Nachprüfung der Atomabstands-Gesetzmäßigkeiten in Laves-Phasen. In: Monatsberichte der DAW *2* (1960) 11.

Lehrbriefe für das Fernstudium „Physik der Röntgenstrahlen" Nr. 1—4, veränderter Nachdruck 1961; Ergänzung „Die dynamische Theorie der Röntgeninterferenzen" TU Dresden 1962.

Über Stapelfehler in Laves-Phasen. In: Festkörperphysik *1961*, 182—184.

Dichte und Raumerfüllung bei intermetallischen Verbindungen, insbesondere Laves-Phasen. In: Z. Kristallogr. *115* (1961) 261—268.

Über intermetallische Verbindungen. In: FuF *35* (1961) 262—264.

u. J. Wieting: Über Bauprinzipien des $CaZn_2$-Gitters. In: Z. Metallkunde *52* (1961) 743—746.

Die physikalischen Grundlagen der Ordnungserscheinungen und deren Nachweis. In: Ordnungsvorgänge in Legierungen. Berlin 1962. 10—36.

Ehrenpromotion von Prof. Dr. U. Dehlinger. In: Wiss. Zeitschr. TU Dresden *11* (1962) 427—428.

D. Sippel, K. Kleinstück, G. E. R. Schulze: Nachweis der anomalen Absorption thermischer Neutronen bei Interferenz am Idealkristall. In: physica status solidi *2* (1962) K 104 bis 105.

u. U. Werncke: Dampfdruckkurve und Verdampfungswärme von $CaZn_2$. In: Monatsberichte der DAW *5* (1963) 68.

u. H.-J. Ullrich, P. Paufler: Über den gegenseitigen Ausschluß intermetallischer Phasen in binären Systemen. In: Naturwiss. *50* (1963) 515—516.

Geometrische Prinzipien beim Gitteraufbau intermetallischer Verbindungen. Berlin 1963. 18 S. (= Abhandlungen Sächs. Akademie der Wiss. zu Leipzig, Math.-nat. Klasse, Bd 48, H. 4).

Die Entdeckung der Raumgitterinterferenzen und ihre Auswirkungen in der Metallphysik. In: Spektrum *9* (1963) 392—400.

Ergebnisse magnetischer Neutronenbeugung. In: Wiss. Zeitschr. der TU Dresden *12* (1963) 1159—66.

Neuere Erkenntnisse über den Magnetismus. In: FuF *37* (1963) 321—325.

Chr. Kirsten, P. Paufler, G. E. R. Schulze: Zur plastischen Verformung intermetallischer Verbindungen. In: Monatsberichte der DAW *6* (1964) 140—147.

D. Sippel, K. Kleinstück, G. E. R. Schulze: Neutron diffraction of ideal crystals using a double crystal spectrometer. In: physics letters *8* (1964) 241—242.

Metallphysik für die Praxis. In: Spektrum *10* (1964) 107—109.

Einführung in die Konferenz „Ausscheidungsvorgänge in Legierungen". In: Ausscheidungsvorgänge in Legierungen. Berlin 1964. 9—10.

Artikel „Intermetallische Verbindungen" und „Einlagerungsverbindungen". In: Brockhaus-ABC-Chemie. Leipzig 1965.

Einführung in die Konferenz „Grundlagen des Festigkeitsverhaltens von Metallen". In: Grundlagen des Festigkeitsverhaltens von Metallen. Berlin 1965. 7–10.

H. Giegengack, H. Schott, G. E. R. Schulze, H.-J. Ullrich: Die linearen thermischen Ausdehnungskoeffizienten einiger intermetallischer Verbindungen von Laves-Phasen-Typ. In: phys. stat. sol. *14* (1966) K 189–193.

Einführung zur Konferenz „Rekristallisation metallischer Werkstoffe". In: Rekristallisation metallischer Werkstoffe. Leipzig 1966. 7–9.

W. Brückner, K. Kleinstück, G. E. R. Schulze: Untersuchungen von Gittereigenschaften im System LiH-LiD. In: phys. stat. sol. *14* (1966) 297–302.

Zur Kristallchemie intermetallischer Verbindungen. In: Kristall und Technik *1* (1966) 375–386.

P. Paufler, G. E. R. Schulze: Zwillingsbildung in $MgZn_2$. In: Naturwiss. *54* (1967) 69.

Untersuchung der Realstruktur von Kristallen mittels Neutroneninterferenzen. In: 2. Internat. Symposium „Reinststoffe in Wissenschaft und Technik" Dresden 1965. Berlin 1967. Teil 3, 185–200.

P. Paufler, G. E. R. Schulze: Zur Zwillingsbildung in $MgZn_2$. In: Kristall u. Technik *2* (1967) 231–244.

W. Brückner, K. Kleinstück, G. E. R. Schulze: Atomic Arrangement in the Homogeneity Range of the Laves Phases $ZrFe_2$ and $TiFe_2$. In: phys. stat. sol. *23* (1967) 475 bis 480.

Metallphysik, 1. Aufl. Akademie-Verlag Berlin 1967. 458 S., 30 Tab., 227 Abb.

W. Blau, H. Schreiber, G. E. R. Schulze, H.-J. Ullrich: Zur experimentellen Bestimmung der Streuphasen in der Kristallstrukturanalyse. In: Naturwiss. *54* (1967) 535–536.

P. Paufler, G. E. R. Schulze: Plastic Deformation of the Intermetallic compound $MgZn_2$. In: phys. stat. sol. *24* (1967) 77–87.

Vergleichende Strukturanalyse intermetallischer Phasen. In: Jahrb. Sächs. Akad. Wiss. zu Leipzig *1963–1965* (Berlin 1967) 165.

W. Hase, K. Kleinstück, G. E. R. Schulze: Kristallstrukturuntersuchungen an Sesquioxiden mit α-Mn_2O_3-Struktur. In: Z. Kristallogr. *124* (1967) 428–451.

Einführung zur Konferenz „Magnetismus". In: Magnetismus. Leipzig 1967. 7–10.

P. Paufler, G. E. R. Schulze: Gleitsysteme intermetallischer Verbindungen. In: Kristall u. Technik *2* (1967) K 11–14.

P. Paufler, G. E. R. Schulze: Zum Ätzen von Versetzungen in $MgZn_2$. In: Kristall u. Technik *3* (1968) 113–124.

Die physikalischen Grundlagen der Röntgenfluoreszenzanalyse. In: Mitteilungen der Kammer der Technik *3* (1968) 3–7.

Die Entwicklung der Vorstellung vom plastischen Gleitvorgang. In: Wiss. Z. TU Dresden *17* (1968) 781–786.

Prof. Dr. Ing. Günther v. Hornuff. In: Wiss. Z. TU Dresden *17* (1968) 1167.

U. Krämer, G. E. R. Schulze: Gittergeometrische Betrachtung der plastischen Verformung von Laves-Phasen. In: Kristall u. Technik *3* (1968) 417–420.

H.-J. Ullrich, G. E. R. Schulze: Die Kossel-Technik als mikroanalytische Methode. In: Microchimica Acta (Wien) Suppl. *III* (1968) 188–200.

W. Brückner, R. Perthel, K. Kleinstück, G. E. R. Schulze: Magnetic Properties of $ZrFe_2$ und $TiFe_2$ within their Homogeneity Range. In: phys. stat. sol. *29* (1968) 211–216.

Einführung zur Konferenz „Kristallisation metallischer Werkstoffe". In: Kristallisation metallischer Werkstoffe, Leipzig 1969, 7–11.

F. Eichhorn, M. Kosmowski, H.-G. Schöpf, G. E. R. Schulze: Dynamical neutron diffraction from Germanium Crystals containing randomly distributed dislocations of low density. In: Acta Cryst. *A 25* (1969) S. XVI–2.

P. Paufler, G. E. R. Schulze: Wechselbeziehung zwischen Kristallstruktur und plastischen Eigenschaften. In: Monatsberichte DAW *11* (1969) 426–440.

P. Paufler, G. E. R. Schulze: Zum Einfluß der Biegung auf den Schubversuch. In: Exp. Technik d. Physik *17* (1969) 543–552.

D. Hinz, P. Paufler, G. E. R. Schulze: Temperature Change Experiments during Secondary Creep of the Intermetallic Compound $MgZn_2$. In: phys. stat. sol. *36* (1969) 609 bis 615.

P. Paufler, E. Wieser, K. Kleinstück, G. E. R. Schulze: Anwendung des Mössbauer-Effekts in der Metallphysik. In: Exp. Technik d. Physik *17* (1969) 459–475.

W. Brückner, K. Kleinstück, G. E. R. Schulze: Mössbauer Study of the Laves-Phase $Ti_{1-x}Fe_{2+x}$. In: phys. stat. sol. (a) *1* (1970) K 1–4.

P. Paufler, G. E. R. Schulze: Advances in the field of mechanical properties of intermetallic compounds at the Department of Physics of the Technical University, Dresden. In: Monatsber. DAW *12* (1970) 67–82.

P. Paufler, J. Marschner, G. E. R. Schulze: The Mobility of Grown-in Dislocations in the Intermetallic Compound $MgZn_2$. I. Stress Dependence for Edge Dislocations in Prism Slip at 390 °C. In: phys. stat. sol. *40* (1970) 573–579.

G. E. R. Schulze: Einführung zur Konferenz „Diffusion in metallischen Werkstoffen". In: Diffusion in metallischen Werkstoffen. Leipzig 1970. 11–15.

R. Günzel, P. Paufler, G. E. R. Schulze: Zn^{65}-Diffusion in der intermetallischen Verbindung $MgZn_2$. In: Diffusion in metallischen Werkstoffen. Leipzig 1970. S. 183–189.

P. Paufler, K. Eichler, G. E. R. Schulze: Einfluß der Abweichungen von der Stöchiometrie intermetallischer Verbindungen auf deren plastische Verformbarkeit. In: Monatsber. DAW *12* (1970) 950–958.

P. Paufler, J. Marschner, G. E. R. Schulze: The Mobility of Grown-in Dislocations in the Intermetallic Compound $MgZn_2$. II. Stress Dependence of Basal Slip at 390 °C. In: phys. stat. sol. (b) *43* (1971) 279–282.

F. Eichhorn, M. Kosmowski, H.-G. Schöpf, G. E. R. Schulze: Experimental and Theoretical Investigations of Dynamical Neutron Diffraction by Using Crystals with a Low Dislocation Density. In: phys. stat. sol. (a) *4* (1971) 445–457.

P. Paufler, G. E. R. Schulze: Über einige physikalische Eigenschaften von Laves-Phasen, insbesondere $MgZn_2$ (Mit einem Anhang: Zur Bezeichnung „Laves-Phasen"). In: Z. Kristallogr. *133* (1971) 7–17.

G. Leitner, G. E. R. Schulze: Crystal Chemical Stability Conditions for Intermetallic Compounds(I), Frequency Distributions of Laves phases. In: Kristall u. Technik *6* (1971) 449–463.

Metallofizika [Übersetzung aus dem Deutschen]. Moskau 1971. 502 S.

u. P. Paufler: Die plastische Verformung „spröder" intermetallischer Verbindungen und ihre Elementarprozesse. Berlin 1972. 24 S. (= Abh. Sächs. Akad. Wiss. Leipzig, mat.-nat. Klasse, Bd 51, H. 5).

Plastic Deformation of Intermetallic Compounds. In: Proceedings of the 11th Science Week. Syrian Arabic Republic Supreme Council of Sciences *2* (Damaskus 1971) 11–22.

Neutron Diffraction of Nearly Perfect Crystals. Ebda S. 1–10.

Über Homogenitätsbereiche in intermetallischen Verbindungen und ihre Bedeutung für die physikalischen Eigenschaften. In: Reinststoffe in Wissenschaft u. Technik. Berlin 1972. 641–653.

Einführung zur Konferenz „Elektronenstruktur“. In: Elektronenstruktur. Leipzig 1972. 11–15.

H.-J. Ullrich, G. E. R. Schulze: Röntgenographische Mikrobeugungs-Untersuchungen an kristallinen Festkörpern mittels Gitterquelleninterferenzen (Kossel-Linien) und Weitwinkelinterferenzen (Pseudo-Kossel-Linien). In: Kristall u. Technik *7* (1972) 207–220.

Helmuth Kuhlenkampff. In: Jahrb. Akad. Wiss. Leipzig *1971–1972* (Berlin 1974) 284 bis 286.

P. Paufler, K. Eichler, S. Siegel, G. E. R. Schulze: Messungen zur Elektronenstruktur der intermetallischen Verbindung $MgZn_2$. In: Elektronenstruktur. Leipzig 1972. 73–80.

W. Meyer, G. E. R. Schulze: Messung und Auswertung von langwelligen Röntgenemissionsbanden an FeAl und CoGa. Ebda 139–143.

W. Blau, G. E. R. Schulze: Einfluß von Aluminium auf die d-Bandstruktur in Fe–Al-Legierungen. Ebda 249–253.

Th. Müller, H. Krahl, P. Paufler, A. Lamsa, G. E. R. Schulze: Gleitbanduntersuchungen während und nach Verformung der intermetallischen Verbindung $MgZn_2$. In: Kristall u. Technik *7* (1972) 1249–1264.

H. Kubsch, P. Paufler, G. E. R. Schulze: The Mobility of Grown-in Dislocations in the Intermetallic Compound $MgZn_2$. III. Dependence of Basal Slip on Chemical Composition within the Homogeneity Range and on Temperature. In: phys. stat. sol. (b) *56* (1973) 231–234.

u. G. Leitner, P. Paufler: On Laves Phase Families Containing Lanthanides and Other Metals. In: Redkozemelnije metally, splavy i soedineniya. Moskau 1973. 137–142.

M. v. Laue und die Geschichte der Röntgenfeinstrukturuntersuchung. In: Kristall u. Technik *8* (1973) 527–543.

u. D. Stephan: Zur anomalen Absorption bei der Röntgen-Strahl-Beugung an Realkristallen. In: Z. Naturforschung *28a* (1973) 647–653.

Umstellung auf Supraleiter. In: Spektrum *4* (1973) 8.

Einführung zur Konferenz „Phasenumwandlungen im festen Zustand“. In: Phasenumwandlungen im festen Zustand. Leipzig 1973. 9–14.

Metallphysik. 2. Aufl. Berlin 1974 und Wien 1974.

Zur Rolle des Einfachheitsprinzips im physikalischen Weltbild. Berlin 1974. 23 S. (= Sitzungsber. Sächs. Akad. Wiss. Leipzig, math.-nat. Klasse, Bd 110, H. 6).

H. Kubsch, P. Paufler, G. E. R. Schulze: On the Comparison between Grown-in and Fresh Dislocations in the Intermetallic Compound $MgZn_2$. In: phys. stat. sol. (a) *24* (1974) K 53–54.

H. Kubsch, P. Paufler, G. E. R. Schulze: The Mobility of Grown-in Dislocations in the Intermetallic Compound $MgZn_2$ during Prismatic Slip. In: phys. stat. sol. (a) *25* (1974) 269–275.

D. Stephan, W. Blau, H.-J. Ullrich, G. E. R. Schulze: Helldunkel-Struktur der Kossel-Interferenzlinien und Kristallstrukturanalyse. (I) Theoretisches Linienprofil unter Berücksichtigung der Absorption. In: Kristall u. Technik *9* (1974) 707–727.

S. Siegel, P. Paufler, G. E. R. Schulze: On grain-boundary sliding during creep of the intermetallic compound $MgZn_2$. In: Kristall u. Technik *9* (1974) K 101–K 102.

Einführung zur Konferenz „Magnetismus“. In: Magnetische Eigenschaften von Festkörpern. Leipzig 1975. 11–18.

H. J. Ullrich, D. Stephan, G. E. R. Schulze: Nachweis der Verletzung der Friedelschen Regel beim Kossel-Effekt (Bragg-Fall). In: 3. Arbeitstagung Mikrosonde 1975 in Berlin. Hrsg. von der Physikalischen Gesellschaft in der DDR.

D. Stephan, H. J. Ullrich, G. E. R. Schulze: Helldunkel-Struktur der Kossel-Inferenzlinien und Kristallstruktur-Analyse (II). In: Kristall und Technik *11* (1976) 475—500.
Beiträge zur Entwicklung der Metallphysik. Hrsg. von P. Paufler. Berlin 1977.
P. Paufler, G. E. R. Schulze: Physikalische Grundlagen mechanischer Festkörpereigenschaften. Berlin 1978.

Zur Biobibliographie Gustav E. R. Schulze

Schriftenverzeichnisse. P. Paufler, siehe vorliegendes Jahrbuch S. 278—283. — Jahrbuch Sächsische Akademie der Wissenschaften zu Leipzig *1957—1959* (Berlin 1961) Beilage, S. 9; *1960—1962* (Berlin 1964) S. 290; *1963—1965* (Berlin 1967) S. 224—225; *1966—1968* (Berlin 1970) S. 227—228; *1969—1970* (Berlin 1972) S. 166; *1971—1972* (Berlin 1974) S. 178. — Kürschners Deutscher Gelehrten-Kalender 6_2 (Berlin 1940/41) Sp. 1271; *9* (Berlin 1961) S. 1893; 10_2 (Berlin 1966) S. 2271; *11* (Berlin 1970) S. 2766. — Poggendorffs Handwörterbuch VI_4^a (Berlin 1961) S. 320.

Widmung. Zum 60. Geburtstag: Wissenschaftliche Zeitschrift TU Dresden *20* (1971) H.2 S. 381—469.

Würdigungen. Zum 50. Geburtstag: J. Henke, Chr. Kirsten, in: Hochschul-Ztg TH Dresden *1961*, Nr. 5, S. 3. — Zum 60. Geburtstag: P. Paufler, S. Braebny, K. Kleinstück, H.-J. Ullrich: 16 Jahre Metallphysik in Lehre und Forschung unter Leitung von Professor Dr. phil. habil. Schulze. In: Wiss. Z. TU Dresden *20* (1971) S. 381—385. — Mitteilungen der Deutschen Vereinigung für Kristallographie *6* (1971) S. 28—30. — Allgemein: Sächsisches Tageblatt *17* (1962) Nr. 287 vom 11. Dez.

Nachrufe. K. Kleinstück, P. Paufler, in: Exp. Technik der Physik *23* (1975) S. 209 bis 210. — H. Neels, P. Paufler, K. Kleinstück, in: Kristall und Technik *10* (1975) S. 95—96. — P. Paufler, in: Mitteilungen der Vereinigung für Kristallographie *9* (1974) H. 2. — P. Paufler, K. Kleinstück: Gustav E. R. Schulze — 35 Jahre Forschung an intermetallischen Verbindungen. In: Kristall und Technik *10* (1975) S. 801—803. — P. Paufler, A. Lösche, siehe vorliegendes Jahrbuch S. 274—278.

Dr. sc. nat. Peter Paufler

Leopold Wolf

23. 11. 1896—6. 11. 1974

Kurz vor Vollendung seines 78. Lebensjahres verstarb am 6. 11. 1974 das ordentliche Mitglied der Sächsischen Akademie der Wissenschaften zu Leipzig Prof. emer. Dr. phil. habil. Leopold Wolf, der langjährige Direktor des Institutes für Anorganische Chemie an der Karl-Marx-Universität Leipzig.

Leopold Wolf wurde am 23. November 1896 im sächsischen Niederlungwitz als Sohn eines Färbermeisters geboren. Die Eindrücke, die er im väterlichen Betrieb erhielt, bewogen ihn, nach dem Abitur an ein Chemiestudium zu denken. Durch den 1. Weltkrieg war jedoch eine Ausführung des Planes nicht möglich. Von 1915—18 nahm er als Unteroffizier am Kriege teil und kehrte erst 1920 aus der französischen Gefangenschaft zurück. Er begann im gleichen Jahr mit dem Chemiestudium an der

Universität Leipzig, ging später nach München, kehrte aber bald nach Leipzig zurück, wo er 1925 unter Hantzsch über „Die Natur der Salpetersäure“ und „Beiträge zur Esterverseifung“ promovierte. Er blieb als Assistent bei Hantzsch und beteiligte sich maßgebend an der Forschung über die Säure-Basen-Theorie.

Im Jahre 1930 wechselte er innerhalb des Leipziger Chemischen Laboratoriums zur Abteilung für angewandte technologische Chemie über, deren Leiter damals B. Rassow war. Hier beschäftigte er sich mit praktischen elektrochemischen Problemen, u. a. mit der visuellen konduktometrischen Analyse. 1936 habilitierte er mit „Beiträgen zur elektrolytischen Abscheidung von Chrom aus Chromsäurelösungen“ und „Über die Passivität des Eisens, Kobaltes und Nickels“. Aus diesem und den nachfolgenden Jahren stammen zahlreiche Patente zum Korrosionsschutz. Sein besonderes Interesse galt der Theorie der ultrareinen Metalle.

1938 wurde er zum Dozenten für das Fach Chemie und 1945 zum a. o. Professor für Chemie an der Universität Leipzig ernannt.

Nach den Wirren des Krieges fiel ihm, als einem der wenigen zurückgebliebenen Professoren die Aufgabe zu, in den schwer zerstörten naturwissenschaftlichen Instituten der Leipziger Universität die Lehre und Forschung neu zu beleben. Bei der Überwindung der enormen Schwierigkeiten erhielt er die tatkräftige Unterstützung von der SMAD und der damaligen Landesregierung. In Würdigung seiner Verdienste wurde er 1948 zum Professor mit vollem Lehrauftrag für Chemische Technologie ernannt, und im gleichen Jahr mit dem Amt des Fachrichtungsleiters für Chemie, das er bis zu seiner Emeritierung 1961 innehatte, betraut.

Ein weites neues Betätigungsfeld eröffnete sich für L. Wolf und seine Mitarbeiter, als er die Feit'sche Sammlung der Seltenen Erden übernahm und einen Forschungsauftrag über die Trennung und Gewinnung der Seltenen Erden erhielt. Dieses Gebiet war besonders im Hinblick auf die Isolierung der Spaltprodukte bei Kernreaktionen von großer wissenschaftlicher und volkswirtschaftlicher Bedeutung. Die Untersuchungen verfolgten das Ziel, die zeitraubenden klassischen Fraktioniermethoden durch moderne automatisierbare Verfahren zu ersetzen. Als Methoden wurden der Ionenaustausch und die fraktionierte Zweiphasenverteilung gewählt. Zusammen mit R. Hering stellte er neuartige Ionenaustauscher mit komplexbildenden Ankergruppen auf Basis von Polyiminoessigsäuren her. Er erzielte bedeutende Ergebnisse und erarbeitete die Grundlage für die industrielle Gewinnung von Seltenen Erden aus Rohphosphaten. Noch heute wird im VEB Düngemittelkombinat Piesteritz nach diesem Verfahren der Bedarf unserer Volkswirtschaft an reinsten Seltenen Erden gedeckt. Internationale Anerkennung erfuhren diese Arbeiten durch die Vergabe des Internationalen Symposiums zur Chemie seltener Elemente an die Karl-Marx-Universität Leipzig im Jahre 1956. Ein Jahr später wurde ihm für diese Leistungen der Nationalpreis (II. Klasse) der DDR verliehen.

Angeregt von den Arbeiten über Seltene Erden fand er noch Zeit, sich intensiv mit der Komplexchemie in Lösung zu befassen, besonders mit den Komplexen von β-Diketonen und verwandten Verbindungen.

Als Hochschullehrer hielt er die Grundvorlesung über Allgemeine und Anorganische Chemie. Besonders beliebt waren seine Spezialvorlesungen, z. B. die Anorganische Chemie für Fortgeschrittene, die Koordinationschemie u. a. Seine unkonventionelle Betrachtungsweise vermittelte seinen Zuhörern viele Denkanstöße. Er war ein scharfsinniger, den Dingen auf den Grund gehender Diskussionsredner. Auch nach seiner Emeritierung 1961 behielt er den wissenschaftlichen und persönlichen Kontakt mit seinen ehemaligen Mitarbeitern.

Seine wissenschaftlichen Leistungen, die in über 50 Publikationen niedergelegt sind, und seine Persönlichkeit fanden vielseitige Anerkennung. 1955 wurde er Mitglied des wissenschaftlichen Rates für die Anwendung der Kernenergie für friedliche Zwecke. 1959 ernannte ihn die Deutsche Akademie der Wissenschaften zu Berlin zum korrespondierenden Mitglied. Ein Jahr später ernannte ihn die Sächsische Akademie der Wissenschaften zu Leipzig zum ordentlichen Mitglied der mathematisch-naturwissenschaftlichen Klasse. Im gleichen Jahr übernahm er den Vorsitz der Arbeitsgruppe „Seltene Erden".

Von seiner Ausstrahlungskraft als Forscher- und Lehrerpersönlichkeit zeugen seine zahlreichen Schüler, die heute verantwortungsvolle Positionen an Hochschulen und in der Industrie bekleiden.

Friedrich Wolf

Bibliographie Leopold Wolf

1. Berechnung der Salpetersäure-Gleichgewichte. In: Chem. Berichte *58* (1925) S. 953 bis 955 [Teil der Dissertation im Auszug].
2. B. Rassow u. L. Wolf: Über die Zusammensetzung der Alkalizellulose aus Baumwolle. Ebda *62* (1929) S. 2949–2953.
3. B. Rassow u. L. Wolf: Zur Kenntnis der Herstellung von Viskose aus Holzzellstoffen. In: Cellulose-Chemie, Fest- und Auslandsheft 1930. S. 77–102.
4. B. Rassow u. L. Wolf: Untersuchungen über das Chromprotekt. In: Chem. Ztg *55* (1931) S. 73–76; Metallwaren-Ind. Galvano-Techn. *29* (1931) S. 73–75, 117–119.
5. B. Rassow u. L. Wolf: Apparatur zur Absorption von Wasserstoff mittels Silberpermanganat-Lösungen nach Fr. Hein und W. Daniel. In: Chem. Fabrik *4* (1931) S. 407.
6. B. Rassow u. L. Wolf: Chromprotekt in der Praxis. In: Metallwaren-Ind. Galvano-Techn. *29* (1931) S. 166–167.
7. B. Rassow u. L. Wolf: Das Hochsäure/Chrom-Verfahren. In: Metallwaren-Ind. Galvano-Techn. *30* (1932) S. 225–227.
8. B. Rassow u. L. Wolf: Über eine neue Form von elektrolytisch abgeschiedenenem Chrom. In: Angew. Chemie *46* (1933) S. 141–142.
9. Gitterenergie. — Born'scher Kreisprozeß für Salzbildung. In: Physikalisch-chemisches Taschenbuch, Bd *II*, 12 (Leipzig 1933).
10. Wärmetönung. — Bildungswärme. Ebda Bd *II*, 13 (Leipzig 1933).
11. Zur Methodik der visuellen Konduktometrie I u. II. In: Chem. Fabr. *9* (1936) S. 46 bis 49, 139–140.
12. Passivierbare Metalle in bimetallischen Elektrodenpaaren. In: J. prakt. Chem. NF *147* (1936) S. 133–144.
13. Neues Elektrolysiergefäß für die quantitative Elektroanalyse. Ebda *148* (1937) S. 205 bis 209.

14. [1] Über die Stromausbeuten bei der Abscheidung des Chroms aus Chromsäurelösungen und ihre Bedeutung für Theorie und Praxis der galvanischen Verchromung. [2] Über den Nachweis des passiven Zustands der Metalle durch ihre Detektorwirkung. Beitrag zur chemischen Passivität der Metalle Chrom, Eisen und Nickel. Gräfenhainichen 1937. 125 S. [= Habil.-Schrift].
15. Zur Potentiometrie mittels mehrerer in Serie geschalteter Elektrodenpaare. In: Z. analyt. Chem. *107* (1937) S. 338—342.
16. Zur Verwendung der Hempelschen Vierkugel-Gaspipetten. In: J. prakt. Chem. NF *153* (1939) S. 263—264; *154* (1940) S. 284.
17. M. Schenck u. L. Wolf: Über eine Reaktion der Hydroxamsäuren. In: Chem. Ber. *73* [B] (1940) S. 25—28.
18. Spezifische Korrosionsmöglichkeiten bei der Anwendung von Anstrich-Emulsionen. In: Berichteband der Emulsionstagung. Leipzig 1943.
19. Zur Theorie des Periodischen Systems (Über ein Verfahren der Korn-Quantelung). In: WZ Univ. Leipzig *1951/52*, H. 4, S. 1—29.
20. u. W. Forberg: Zur Kenntnis der Chrom (III)-Triäthanolamin-Komplexe. In: Z. anorg. allgem. Chem. *268* (1952) S. 327—350.
21. u. P. Christofzik: Löslichkeits-, Leitfähigkeits- und Viskositätsmessungen im System wäßriger Lösungen von CrO_3 und $HClO_4$. In: J. prakt. Chem. *4* (1954) S. 237—247.
22. u. H.-J. Jahn: Über Komplexverbindungen des Thoriums I. Ebda S. 257—276.
23. u. Dorothée Stather: Über Komplexverbindungen des Thoriums II. Ebda S. 329—336.
24. u. F. Seidel, H. Krause: Über die Bildungsweise der 2-Oxynaphtoesäure-(3). Ebda *2* (1955) S. 53—83.
25. u. J. Massone: Beiträge zur Trennung der Seltenen Erden I. Ebda *3* (1956) S. 178 bis 182.
26. u. H. Bärnighausen, J. Massone: Beiträge zur Trennung der Seltenen Erden II. Ebda S. 211—227.
27. u. H. Bärnighausen: Beiträge zur Trennung der Seltenen Erden III. Ebda S. 228—231.
28. u. J. Massone: Beiträge zur Trennung der Seltenen Erden IV. Ebda *5* (1958) S. 14 bis 2.0.
29. u. J Massone: Beiträge zur Trennung der Seltenen Erden V. Ebda S. 21—33.
30. Beiträge zur Trennung der Seltenen Erden VI. Ebda S. 288—290.
31. u. G. Wolff, H. Bärnighausen: Beiträge zur Trennung der Seltenen Erden VII. Ebda *6* (1958) S. 259—265.
32. u. H. Schwab: Beiträge zur Trennung der Seltenen Erden VIII. Ebda *13* (1961) S. 337 bis 344.
33. u. H. Schwab: Beiträge zur Trennung der Seltenen Erden IX. Ebda *24* (1964) S. 293 bis 304.
34. u. K. Wetzel: Untersuchungen über Komplexverbindungen der β-Ketosäureamid-Reihe. In: Chem. Ber. *90* (1957) S. 1007—1023.
35. u. H. Bärninghausen: Kristalldaten von Thorium-tetrakis-(dibenzoylmethan) und den isomeren Verbindungen des vierwertigen Cers und Urans. In: Acta cryst. *10* (1957) S. 605—606.
36. u. M. Beer: Ferrocenoylaceton. In: Naturwiss. *44* (1957) S. 442.
37. u. M. Beer: Furoyl-(3)-aceton. Ebda *45* (1958) S. 164.
38. u. J. Massonne: Über eine schnelle Ionenaustauschtrennung größerer Mengen Ceriterden unter Verwendung von Ammonium-Nitrilotriacetat als Elutionsmittel. In: Chem. Techn. *10* (1958) S. 290—296.

39. u. R. Hering: Über Ionenaustauscherharze mit komplexbildenden Ankergruppen. Ebda S. 661—662.
40. Komplexbildung und Komplexelution der Seltenen Erden. In: Österr. Chem.-Ztg. *60* (1959) S. 99—106.
41. u. E. Butter, H. Weinelt: Über die Darstellung höherer β-Diketonkomplexe des Eisens, Rutheniums und Rhodiums. In: Z. anorg. allg. Chem. *306* (1960) S. 87—93.
42. Über den räumlichen Bau von β-Diketonkomplexen vierwertiger Kationen mit der Koordinationszahl 8. Acta cryst. *13* (1960) S. 778.
43. u. H. Hennig: Ferrocenometrie. In: Z. Chem. *1* (1961) S. 220—221.
44. u. E. Uhlig: Zur Salzbildung von Derivaten von Tris(β-hydroxyäthyl)-amins und des Tetrakis(β-hydroxyäthyl)-ammoniumhydroxydes. In: J. prakt. Chem. *14* (1961) S. 99 bis 107.
45. u. E. Butter, H. Weinelt: Zwei Formen von Nickel(II)-β-Diketoverbindungen. In: Naturwiss. *48* (1961) S. 378.
46. [Nachr.] Arthur Simon. In: Z. physik. Chem. *221* (Leipzig 1962) S. 285—287.
47. u. E. Uhlig: Die Zersetzung von Tetrakis[β-(p-phenylcarboxy)äthyl-ammonium-Salzen in alkalischer Lösung. In: J. prakt. Chem. *17* (1962) S. 241—249.
48. u. Chr. Tröltzsch: Über Chelate von Zirkonium und Hafnium mit 1.3-Dicarbonylverbindungen. Ebda S. 78—91.
49. u. E. Uhlemann: Isomere Kupferchelate von β-Diketonanilen. In: Z. Chem. *2* (1962) S. 219—221.
50. Koordinationschemie von 2-Cyclohexyliminodiessigsäure (I) und o-Hydroxyanilindiessigsäure (II), Theory Struct. In: Complex Compounds, Papers Symposium Wrocław, Poland *1962*. S. 349ff.
51. u. G. Kozlowski, E. Hoyer: Kationische Kupferkomplexe Schiffscher Basen des Äthylendiamins. In: Z. Chem. *3* (1963) S. 68.
52. u. H. Hennig: Ferrocensubstituierte 1.3-Diketone. Ebda S. 469—470.
53. u. Chr. Tröltzsch: Über polymere Kupferchelate von Bis-(1.3.-diketonen). In: J. prakt. Chem. *24* (1964) S. 305—311.
54. u. H. Henning: Über das Komplexbildungsvermögen der isomeren Pyridoyl-benzoylmethane. In: Z. anorg. allg. Chem. *329* (1964) S. 301—308.
55. u. W. Matzel, H. Schreiber: Die Extraction von Seltenen Erden durch langkettige tertiäre Amine. Ih: WZ Univ. Leipzig, math.-nat. R. *14* (1965) S. 551—553.
56. u. E. Jäger: Metallchelate mit formaler Koordinationszahl 3 und deren reaktives Verhalten. In: Z. Chem. *5* (1965) S. 317—319.
57. u. E. Butter: Zur Struktur der Nickel (II)-β-diketonate. In: Z. anorg. allg. Chem. *339* (1965) S. 191—207.
58. u. H. Hennig: Über das Komplexbildungsvermögen der isomeren Pyridoyl-ferrocenoyl-methane. Ebda *341* (1965) S. 1—10.
59. u. H. Schwab, J. Sieler: Röntgenographische Untersuchungen am System Neodymoxid-Ytterbiumoxid. In: J. prakt. Chem. *32* (1966) S. 113—117.
60. u. E.-G. Jäger: α (Aminomethylen)-β-dicarbonyl-Verbindungen als Komplexliganden. I: Kupfer- und Nickelchelate von Diaminderivaten. In: Z. anorg. allg. Chem. *346* (1966) S. 76—91.
61. u. H. Henning, I. P. Sereda: Über die Säurebasenfunktion der isomeren Pyridoyl-benzoyl-methane und der analogen Pyridoyl-ferrocenoyl-methane. In: J. prakt. Chem. *32* (1966) S. 105—112.
62. V. V. Suchan, E. Uhlemann u. L. Wolf: Bestimmung der Stabilitätskonstanten von

Silberkomplexen mit einigen β-Diketonen in einem wäßrigen Dioxan-Medium. In: Sh. njeorg. Chim. *12* (Moskau 1967) S. 455—458.
63. u. W. Matzel, H. Schreiber: Die Extraktion der Seltenen Erden durch langkettige tertiäre Amine. In: Vestn. Kiewsk. Univ. Spets. Vyp. Nauch. Statei Uch. Kiev 1967, S. 89.
64. u. H. Hennig, I. P. Sereda: Potentiometrische Bestimmung der Komplexbildungskonstanten von Kupfer(II), Nickel(II) und Zink mit den isomeren Pyridoylferrocenoylmethanen und Pyridoylbenzoylmethanen. In: Sh. njeorg. Chim. *12* (Moskau 1967) S. 450—454.

Patente und Patentanmeldungen

1. DRP Nr. 751074
Verfahren zum Schutz von Behältern aus Schwarzblech gegen Korrosion.
2. DRP Nr. 762517
Verfahren zum Schutz von Eisen- und Stahlgegenständen gegen Korrosion; 1. Zusatz zum DRP Nr. 751074, Verallgemeinerung und Erweiterung des DRP 751074.
3. Patentanmeldung D 90098 VIa/48d; 2. Zusatz zu DRP 751074
Verfahren zum Schutze von Gegenständen der verschiedensten Art aus Metallen und Legierungen gegen Korrosion. Anwendung von WO-Emulsionen auf *nicht* entwässerte Zwischenschichten: *Gemeinsames* Trocknen von Zwischenschicht und Film.
4. Patent-Anmeldung D 91086 VIa/48d, 3. Zusatz zu DRP 751074
Verfahren zur Erzeugung einer Oberflächenrostung auf Eisen bzw. Stahl. Phosphorsäure-Beize bestimmten pH-Wertes.
5. Patent-Anmeldung S 160075 VIa/48d, 4. Zusatz zu DRP 751074
Verfahren zur Erzeugung einer Oberflächenrostung.
Oberchlorsäure-Essigsäure-Beize bei Gegenwart von NH_4-Ion und H_2O_2.
6. Patent-Anmeldung S 161536 VIa/48d 5. Zusatz zu DRP 751074
Verfahren zum Oberflächenschutz von Eisen- und Stahlgegenständen verschiedenster Art gegen Korrosion. Pigmentierte Filme mit Fe(III)-Pigmenten und Erdalkali-ferriten vom Spinelltyp.
7. DRP 727964
Verfahren zur Erzeugung von Phosphatschichten auf Metallen. Ausnützung der im Ultraschallgebiet liegenden magnetostriktiven Druckwellen doppelter Frequenz von Wechselfeldern zur gestörten Kristallisation von 200—500 kHz in Verbindung mit Temperaturgefälle.
8. Patent-Anmeldung D 90066 VIa/48d
Oberflächenbehandlung durch Entphosphatieren; Matrizenwirkung; Prägungseffekt.
9. Patent-Anmeldung D 91050 VIa/48d
Verfahren zur Vorbehandlung von Schwarzblech. Oberflächenbehandlung durch übersäuerte Phosphatierungslösung.
10. Patent-Anmeldung D 90115 Ib/7c
Verfahren zur Erleichterung der Verformbarkeit von Eisen bzw. Stahl.
11. Patent-Anmeldung D 84140 IVd/15c; *Patent erteilt, DRP Nr. nicht mehr bekanntgegeben*
Verfahren zur induktiven Lacktrocknung.
12. Patent-Anmeldung D 8786 VIa/48d
Verfahren zur Entwässerung von unselbständigen Zwischenschichten unter Erhaltung der genuinen Struktur durch Infrarot-Bestrahlung (10—16.103 A) [Infrarottrocknung auf anderem Gebiet wird im Ausland verwendet].
13. DRP Nr. 751074 in Kombination mit DRP 762517.

14. und R. Hering: Komplexbildende Austauschharze auf Iminodiessigsäure-Basis.
15. DDR-Patent 38340 vom 25. Juni 1965
Anreicherung und Trennung von komplexbildenden Metallionen.

Zur Biobibliographie

Schriftenverzeichnisse. E. Hoyer, H. Hennig, siehe vorliegendes Jahrbuch S. 285 bis 289. — E. Uhlemann: Die Arbeiten von L. Wolf auf dem Gebiete der Seltenen Erden. In: Z. Chem. *1* (1961) S. 322—328. — Jahrbuch Sächs. Akad. Wiss. zu Leipzig *1957—1959* (Berlin 1961) Beilage S. 9—10; *1960—1962* (Berlin 1964) S. 300; *1963—1965* (Berlin 1967) S. 233—234; *1966—1968* (Berlin 1970) S. 235. — Poggendorffs Handwörterbuch VII_4^a (Berlin 1962) S. 1067—1068.

Festschriften: Journal für praktische Chemie *34* (1966) Heft 1—4, Widmungsband zum 70. Geburtstag. — WZ Univ. Leipzig, math.-nat. Reihe *21* (1972) Heft 1, Widmungsheft zum 75. Geburtstag: Zur Analytik und Komplexchemie der 3d und 4f Elemente.

Würdigungen. Zum Nationalpreis: E. Hoyer, in: Mitteilungsbl. Chem. Ges. in der DDR *4* (1957) S. 153. — D. Wg., in: Kulturspiegel der Messestadt Leipzig *1957*, S. 1059—1060 (mit Bildnis). — Zum 65. Geburtstag: E. Hoyer, in: Z. Chem. *1* (1961) S. 321. — G. Schott, in: Mitteilungsbl. Chem. Ges. in der DDR *8* (1961) S. 173—174 (mit Bildnis). — E. Uhlemann, in: Leipziger Volks-Ztg *16* (1961) Nr. 330 vom 30. Nov. (mit Bildnis). — UZ Leipzig *5* (1961) Nr. 48, S. 3. — Zum 70. Geburtstag: H. Holzapfel, in: Leipziger Volks-Ztg *21* (1966) vom 23. Nov. — Zum 75. Geburtstag: H. Holzapfel, E. Hoyer, in: WZ Leipzig, math.-nat. Reihe *21* (1972) S. 1. — Allgemein: H. Holzapfel: Institut für Anorganische Chemie — Zehn Jahre Forschung über Seltenerdmetallverbindungen. In: Lehre — Forschung — Praxis. Sonderband „Die Karl-Marx-Universität Leipzig zum zehnten Jahrestag ihrer Namensgebung" am 5. Mai 1963. S. 44. — Leipziger Volks-Ztg *16* (1961) vom 21. Dez.

Nachrufe. E. Hoyer, in: WZ Univ. Leipzig *24* (1975) H. 4, S. 449—450. — F. Wolf, siehe vorliegendes Jahrbuch S. 283—285 (mit Bildnis). — UZ Leipzig *19* (1975) Nr. 2 vom 23. Jan., S. 2.

Prof. Dr. sc. Eberhard Hoyer
Doz. Dr. sc. Horst Hennig

John Eggert

1. 1. 1891—29. 9. 1973

John Eggert kam am 1. Aug. 1891 als einziges Kind des Bücherrevisors Emil Eggert und seiner Frau Martha, geb. Zwingmann, in Berlin zur Welt. Er hat aber immer betont, daß dies auf einem Zufall beruhte; beide Eltern stammten aus Ostpreußen, und auch er fühlte sich dort zugehörig und konnte den dortigen Dialekt treffend nachahmen, obwohl normalerweise der Berliner bei ihm nicht zu verkennen war. Seine ganze Schulzeit verbrachte er in Berlin und sein Studium der Naturwissenschaften an der dortigen Universität.

Er war sich nach dem Abitur über seine Bestimmung nicht klar und belegte im ersten Semester neben seiner Chemie auch Architektur. Seine künstlerische Ader ist nie versiegt; um jede Jahreswende erfreute er seine Freunde mit minuziös ausgeführten Glückwunschkärtchen. Sein Studium wurde ihm durch eine Schwester seines Vaters, der Frau des bekannten und wohlhabenden Kunstmalers Emil Neide, ermöglicht, die kinderlos war.

Schmunzelnd hat er gelegentlich berichtet, daß er bei seinem ersten Versuch, das Staatsexamen für Oberlehrer zu bestehen, durchfiel, weil er, neben anderen Schwächen, sich an den Namen des Lehrers des Apostel Paulus nicht erinnern konnte: charakteristisch für unseren jungen Mann, der in einer freisinnig-liberalen Familie aufgewachsen war. Kennzeichnend für diese aufgeschlossene Haltung der Aufruhr, den er bei den Eltern, und die Begeisterung, die er bei ihren Töchtern hervorrief, als er, junger Lehrer, auf seinen Wunsch nicht nur für Naturwissenschaften, sondern auch für Deutsch eingesetzt, solch „aufrührische" Dramen wie Emilia Galotti besprach. Diese Ereignisse resultierten in einer Verwarnung von seiten der gestrengen Frau Direktorin und in der baldigen Aufkündigung dieser seiner ersten Verdienstquelle und stellten wohl ein Nebengleis während seines Studiums dar, denn bald finden wir den jungen Forscher, schon vor der Promotion im Jahre 1914, als Vorlesungsassistenten bei seinem hochverehrten Lehrer und Doktorvater Nernst. Seiner Doktorarbeit „Über die Absorptionsgeschwindigkeit von Wasserstoff und Sauerstoff durch Metallsalzlösungen" folgte eine steile akademische Karriere. Er habilitierte sich 1921 mit einer Arbeit „Über die Sensibilität hochempfindlicher Sprengstoffe" und schon 1924 begegnen wir ihm als a. o. Professor und bald darauf als Begründer und Leiter des wissenschaftlichen Zentrallabors der AGFA in Wolfen.

1921 verheiratete er sich mit der Chemikerin Dr. Margarethe Ettisch. Sie hatten zwei Töchter. Er verlor seine Frau im Herbst 1973, und damit schien das Licht und der Sinn aus seinem Leben verschwunden zu sein.

Eggert hatte während der Nazi-Zeit der Abstammung seiner Frau wegen die größten Schwierigkeiten. Die Größe seiner Persönlichkeit zeigte sich daran, daß er allen Versuchungen, eine Veränderung seiner Verhältnisse zu schaffen, auf das deutlichste widerstand. Beinahe wie durch ein Wunder hat die Familie Eggert auch die Kriegszeit überstanden. 1946 wurde Eggert zum ordentlichen Professor ad personam und Leiter des Photographischen Instituts der Eidgenössischen Technischen Hochschule in Zürich erwählt, und noch einmal waren ihm und seinen Schülern eine Zeit der wissenschaftlichen, schöpferischen Blüte gewährt, die aber selbst mit seiner Pensionierung 1961 noch nicht ihren Abschluß fand. Noch viele Jahre danach wurde sein Rat überall gesucht; wurden seine Dienste als begnadeter Vortragender gewünscht, der es verstand, auch trockene Themata durch Anekdoten zu würzen; nahm er regen Anteil am Schicksal seines alten Instituts, dessen wissenschaftlichen Ruf er sowohl innerhalb der Hochschule wie auch weltweit etabliert hatte; war er regelmäßiger Teilnehmer an wissenschaftlichen Tagungen und den periodischen Kolloquien des Instituts, die er seinerzeit eingeführt hatte.

Es wurden ihm viele Ehrungen zuteil (s. Liste). Unter diesen waren ihm die Ehrendoktorengrade, die Ehrenmitgliedschaften wissenschaftlicher Gesellschaften und insbesondere Mitgliedschaften wissenschaftlicher Akademien von großer Wichtigkeit; seiner Freude über diese Art der Anerkennung von seiten der Sächsischen Akademie der Wissenschaften hat er oft deutlichen Ausdruck verliehen.

In einem Versuch, Eggerts wissenschaftliche Tätigkeit zu würdigen, sollte man über 5 Lehrbücher sowie über einen Schatz von ca. 122 Originalarbeiten und 179 Berichten referieren; von den ersteren war er bei nicht weniger als 105 Veröffentlichungen Ko-Autor, was deutlich aufzeigt, wie er es verstand, Mitarbeiter zu schöpferischer Tätigkeit anzuregen.

Am bekanntesten ist Eggert wohl durch sein „Lehrbuch der physikalischen Chemie" geworden, das er zuerst allein und in späteren Auflagen mit Hock, und mit Hock und Schwab, herausgebracht hat und das u. a. auf englisch und russisch erschienen ist. Die letzte, IX. Auflage (1968) hat ihn mehrere Jahre seines Lebensabends in stimulierenden Anspruch genommen.

Seine photographischen Arbeiten umfassen ein weites Spektrum der wissenschaftlichen Photographie: der photographische Elementarprozeß; der Mechanismus der Entwicklung; die Eigenschaften des entwickelten Bildes; nicht-Silber-Verfahren; verschiedene Gebiete der angewandten Photographie. Wenn im folgenden seine Forschung über den Elementarprozeß herausgestellt wird, so deswegen, weil sie einmal von grundlegender Bedeutung ist, dann aber, weil sich hier am klarsten seine Fähigkeit zeigte, sofort zum Kernpunkt einer Situation vorzustoßen.

Seine erste photographische Arbeit (mit Noddack) war seine neunte und befaßte sich mit der Quantenausbeute der Photolyse, also der direkten Zersetzung der Silberhalogenide durch Licht.

Um die grundlegende Bedeutung dieser Arbeit zu würdigen, muß man versuchen, sich in das Denken, den Zustand der Erkenntnis der damaligen Physiker zurückzuversetzen. Die Quantentheorie stak noch in den Kinderschuhen; zwar hatte Planck schon um die Jahrhundertwende die Quantennatur des Lichts erkannt; zwar hatte Einstein die quantenhafte Absorption und Emission des Lichts als grundlegend beschrieben, aber diese Konzepte waren doch noch nicht völlig in die Denkweise der Wissenschaftler eingedrungen. Noch gab es keine Wellenmechanik; noch hatte kein Heisenberg oder Schrödinger es verstanden, eine Brücke zwischen der Beschreibung der Ausbreitung des Lichts als Welle einerseits und als Energiequant in Emission oder Absorption andererseits zu schlagen.

Die Grundfrage war ja die folgende. Licht ist Energie. Die Energie eines Lichtquants ist $h \cdot \nu = h \cdot c/\lambda$, ist also höher, je kürzer die Wellenlänge λ. Die gleiche Energiemenge kann einem System in der Form von vergleichsweise wenigen Quanten höherer Energie, d. h. von kurzwelligem Licht, oder vielen kleinerer Energie als langwelliges Licht zugeführt werden. Wenn nun aber nach der klassischen Vorstellung das Licht einen kontinuierlichen, also nicht quantisierten, Energiestrom darstellt, käme es in der Photochemie nur auf die absorbierte Gesamtenergie, nicht aber auf ihre Qualität, d. h. die Wellenlänge des Lichtes an. Nach der Quantenvorstellung jedoch ist die Anzahl der absorbierten Quanten entscheidend und muß mit dem Produkt, in unserem Falle mit der Anzahl der gebildeten Silberatome, verglichen werden.

Eggert hat sich nie mit der Diskrepanz der Beschreibung der elektromagnetischen Strahlung versöhnen können. Um so bemerkenswerter ist es, daß sich seine erste photographische Arbeit gerade mit der Frage der Quantenausbeute in der Photographie beschäftigt hat. Dieses Problem war fundamental sowohl vom Standpunkt der Photographie wie der Quantenvorstellung, da man Quantenausbeuten zwar in Gasreaktionen und flüssigen Systemen, jedoch bislang nie an festen Körpern bestimmt hatte.

Man kann sich den Reiz der Aufgabe für den stürmischen Geist des jungen Forschers gut vorstellen. Das Resultat würde die Allgemeinheit der quantisierten Wechselwirkung von Strahlung und Materie betreffen und uns dem Verständnis des rätselhaften latenten Bildes der Photographie näher führen. Könnte man sich unter diesen Umständen ein schöneres Thema denken?

Die Untersuchung erforderte eine tiefgehende experimentelle Akribie. Die Bestimmung der Quantenausbeute erfordert ja die Kenntnis der vom Silberhalogenid in der Emulsion absorbierten Zahl der Quanten. Die Situation ist bei einer Emulsion durch die Streuung des Lichts kompliziert. Das vielleicht gerichtet einfallende Licht läßt sich auf einfache Weise messen. Nun müssen wir, um die Zahl der im Silberhalogenid absorbierten Quanten (L_A) zu bestimmen,

einmal das reflektierte (L_R)

dann das hindurchgelassenen (L_T)

Licht bestimmen, wobei beide gestreut sind, und drittens

den Anteil des in der Gelatine absorbierten Lichtes (L_G) erfassen.

Die Differenz

$$L_{ein} - L_R - L_T - L_G = L_A.$$

Hierbei ist die genaue Erfassung von L_G das schwierigste Problem, das zu seiner genauen Bearbeitung Computermethoden erfordert, und das bis zum heutigen Tage nur in Annäherung gelöst ist. Unsere jungen Forscher mußten sich mit einer Abschätzung begnügen, die auf Messungen bei längeren Lichtwellenlängen beruhte, bei der das AgBr mit Sicherheit nicht absorbierte und nur streute. Die Unsicherheit hierbei beruht auf der Wechselwirkung zwischen Absorption und Streuung in einem Medium wie der photographischen Emulsion: je geringer die Absorption, um so länger der Lichtweg im streuenden Medium und damit die Wahrscheinlichkeit, daß ein schwach absorbierender Anteil noch wirksam wird, und umgekehrt.

Wie gut die Abschätzung war, zeigte das Resultat: die Photolyse hat eine Quantenausbeute von der Größenordnung 1, was später völlig bestätigt worden ist. Die Quantenausbeute ist jedenfalls nicht größer, und darauf kam es jetzt an: es konnte sich nicht um eine vom Lichtquant ausgelöste Kettenreaktion handeln. Und noch viel grundlegender: die Quantenvorstellung war zum erstenmal an der Photochemie eines Festkörpers aufs schönste bestätigt.

Die Tatsache, daß die Quantenausbeute für stärkere Belichtungen abfiel, führten die Forscher auf eine Regression zurück: photolytisches Silber und Brom reagieren miteinander zu Silberbromid. Sie zeigten, daß diese Rückreaktionen durch Zugabe eines geeigneten Halogenakzeptors vermieden werden konnte, unter Verbesserung der Quantenausbeute.

So führte diese Erstarbeit in das Zentralproblem der Photographie: die Natur des latenten Bildes. Wenn man von der Absorption eines Lichtquants nur *ein* Ag-Atom erwarten kann, dann kann man aus dem Verhältnis der Belichtungen, die für das latente Bild und für die Photolyse notwendig sind, die Anzahl der Ag-Atome abschätzen, die zur Erzeugung des latenten Bildes notwendig sind. Auch diesen Schritt taten die jungen Forscher sogleich und fanden eine Zahl von der Größenordnung 100 — verschwindend klein, verglichen mit den 10^8 bis 10^9 Ag-Ionen in einem einzigen Körnchen einer photographischen Emulsion.

Die weiteren Schritte der Erkenntnis und des Experiments erfolgten Schlag auf Schlag. 1926 berichtet Eggert in der Z. f. Elektrochemie (*32*, 491) über seine Ansichten betr. des latenten Bildes, und wenn man diese Arbeit liest, dann steht man in fassungslosem Staunen vor der Breite und Tiefe der im Laufe so weniger Jahre entwickelten Vorstellungen. Was gab es danach für die spätere Generation noch zu tun, außer dem Ausfüllen von Einzelheiten? Selbst wenn sich nicht alle seine Vorstellungen bewährt haben, so war dies nicht das Wesentliche, das in der Anregung bestand, den klaren Vorstellungen nachzugehen. Eine gute Theorie hat ihren größten Wert als Arbeitshypothese, nicht darin, daß sie in allen Aspekten recht hat.

Da jedenfalls mehrere Ag-Atome am Aufbau des latenten Bildzustandes beteiligt sind, folgerte Eggert, daß das latente Bild ein Aggregat von Silberatomen darstellt, im Anklang an frühere Vorstellungen, und daß eine kritische Minimalgröße der Keime zur Einleitung der Entwicklung notwendig sei. Ein Keim, kleiner als diese kritische Größe, könnte noch immer stabil sein und bei der Nachreifung entstehen, die sozusagen die Anfangsstufe des Aufbaus des Keimes darstellen würde: eine einfache Erklärung der chemischen Sensibilisierung, in Anlehnung an die Ideen von Shephard, der dem Silbersulfid die gleiche Rolle von Konzentrationskeimen zuschrieb.

Eggert fordert seine Leser auf, sich das auskopierte Ag in Emulsionskörnern genau anzusehen. Quanten fallen überall ein; Silber scheidet sich jedoch nur an wenigen, offensichtlich prädestinierten Stellen ab. Es ist ihm klar, daß dies eine Wanderung von Ag durch den Kristall hindurch bedeutet, und er erklärt dies, ähnlich wie Fajans und Frankenburger, durch den „nineren photoelektrischen Effekt". Absorption von Licht löst Elektronen aus; deren Wanderung durch den Kristall ist de facto eine Wanderung von Silberatomen. Er realisiert aber auch, daß nach Tubandt Ag^+-Ionen im Kristall beweglich sind und führt den Aufbau der Silberteilchen auf die Beweglichkeit sowohl der Photoelektronen wie der Ionen zurück. So greift er den modernen Vorstellungen um 12 Jahre vor; jedoch hat damals niemand diese Vorstellungen ernst genommen, am wenigsten er selbst; merkwürdigerweise hat er sich später lebhaft gegen die Theorie von Gurney und Mott gesträubt. Er hat mir selbst erklärt, daß ihm die damaligen Festkörper-Physiker die Vorstellung einer Ionenwanderung ausgeredet haben: alle Vorgänge in Kristallen mußten elektronischer Natur sein.

Wir finden in dieser Arbeit auch Gedanken über die Verteilung des latenten Bildes zwischen der Oberfläche und dem Inneren der Emulsionskörner und über die Abhängigkeit dieser Verteilung und des Disperitätsgrades der Bildkeime von den Belichtungsbedingungen. So bilden sich bei Blitzbelichtungen viele kleine Keime; dies erklärt die niedrige Entwicklungsgeschwindigkeit von mit Blitzbelichtungen entstandenen Bildern. Der Herscheleffekt: die Zerstörung eines latenten Bildes durch eine Nachbelichtung mit langwelliger Strahlung wird durch eine Umverteilung der Bildkeime erklärt. All dies sind Konzepte, die noch in unseren Vorstellungen eine ausschlaggebende Rolle spielen.

Die Arbeiten über die Quantenausbeute wurden systematisch fortgesetzt. Es zeigte sich, daß energiereiche Quanten wie die der Röntgenstrahlen von der Größenordnung 1000 Ag-Atome pro Emulsionskorn erzeugten. Er schloß, daß Emulsionen für Röntgenstrahlen keiner Nachreifung bedürfen sollten.

Eine weitere, äußerst wichtige Erkenntnis bezog sich auf die spektrale Sensibilisierung durch Farbstoffe. Es wurde festgestellt, daß in mit geeigneten Farbstoffen sensibilisierten Emulsionen die Anzahl der photolytisch erzeugten Ag-Atome um den Faktor 100 größer sein konnte als die Gesamtzahl der vorhandenen Farbstoffmoleküle. Das heißt, daß ein Farbstoffmolekül ca. 100 mal ein Lichtquant absorbieren und ein Silberatom erzeugen kann, ohne dabei zerstört zu werden. Diese Tatsache bereitet uns noch heute Schwierigkeiten im Verstehen der spektralen Sensibilisierung.

Vieles wäre noch von späteren Arbeiten zu berichten. Von ganz besonderer theoretischer Tragweite waren die Untersuchungen mit Biltz und Kleinschrod über die spektrale Empfindlichkeit von Emulsionen (ohne Farbstoffe). Sie zeigten, daß die Empfindlichkeit einen langwelligen Ausläufer besitzt, dessen Stärke von dem Grad der Nachreifung abhängt. Diese Untersuchungen lieferten die Grundlage, mit Hilfe derer die späteren Festkörperphysiker die Elektronenniveaus festlegen konnten, die die sensibilisierende Substanz charakterisieren.

Wir überschlagen die zahlreichen späteren Abenteuer und kommen zu den letzten Jahrzehnten. Vor nunmehr $13^1/_2$ Jahren trat ich seine Nachfolge am Photographischen Institut an; ohne akademische Lehrerfahrung, jedenfalls ohne Erfahrungen in Schweizer Belangen und ohne Kenntnis der ETH. So war vielleicht seine letzte Originalarbeit die Einführung seines Nachfolgers in ein Amt, das für ihn die Krönung seiner Lebensaufgabe bedeutet hatte. War er es doch gewesen, der unser Institut zu einer weltbekannten Stellung geführt hatte.

Ohne Zweifel hatte er für seinen Nachfolger ein latentes Bild im Sinne, das er nun vorsichtig, Schritt für Schritt entwickelte. Wie leicht hätte eine solche Situation zu persönlichen Schwierigkeiten führen können. Aber davon konnte nicht die Rede sein. Wenn mir die Anlaufzeit verkürzt worden und mir manche Schwierigkeit durch rechtzeitigen Rat erspart geblieben ist, wenn das Institut wachsen und gedeihen konnte, dann dank seiner stetigen gütigen Fürsorge und seines Einsatzes und Interesses.

So bin ich ihm einen Dank schuldig, den ich nie werde abtragen können. Nicht, daß mich dies bedrückt; ich sehe darin eine freudige Verpflichtung der zukünftigen Generation gegenüber und hoffe nur, daß mir sein Beispiel ermöglichen möge, dies in seinem Sinne zu tun.

Prof. Dr. phil. Wolfgang Friedrich Berg

Zürich, im Februar 1975

Ehrungen

Datum	Auszeichnung	Verliehen durch
16. 2. 31	Goldene Medaille der Photographischen Gesellschaft Wien	Photographische Gesellschaft in Wien
15. 10. 35	Traill Taylor Medal	Royal Photographic Society of Great Britain
31. 3. 50	Progress Medal für 1949	Royal Photographic Society of Great Britain
27. 10. 50	Doktor Ing. e. h.	Fakultät für Natur- und Geisteswissenschaften der T. H. Karlsruhe
21. 9. 51	Médaille Péligot	Société française de Photographie
März 52	Korrespondierendes Mitglied	Bayerische Akademie der Wissenschaften
9. 3. 53	Honorary Fellow	Royal Photographic Society of Great Britain
23. 9. 54	Plakette der Stadt Paris	Stadt Paris
19. 10. 55	Honorary Fellow	Photographic Society of America
23. 2. 55	Gauss-Weber-Medaille	Mathematisch-Naturwissenschaftliche Fakultät der Georg-August-Universität Göttingen
17. 8. 55	Special Citation	Photographers' Association of America
1. 10. 56	Photokina-Plakette	Direktorium der Organisation Photokina, Köln
20. 2. 57	Ehrenmitglied der FIAP	Fédération internationale de l'art photographique
21. 2. 57	Korrespondierendes Mitglied der mathematisch-naturwissenschaftlichen Klasse	Heidelberger Akademie der Wissenschaften
30. 5. 57	Ehrenmitglied der Deutschen Bunsengesellschaft	Deutsche Bunsen-Gesellschaft für physikalische Chemie
23. 6. 57	Ehrenmitglied des Schweiz. Photographenverbandes	Schweizerischer Photographenverband
12. 5. 58	Korrespondierendes Mitglied	Sächsische Akademie der Wissenschaften zu Leipzig
19. 6. 58	Korrespondierendes Mitglied	Société Royale des Sciences de Liège
30. 5. 59	Honorary Member	Society of Scientific Photography of Japan
Sept. 57	Honorary Member	Society of photographic Scientists and Engineers, USA
13. 6. 61	Kulturpreis	Deutschen Gesellschaft für Photographie in Köln
14. 6. 61	Adolf-von-Baeyer-Denkmünze	Gesellschaft Deutscher Chemiker in Frankfurt/M.

23. 9. 63	Goldene Medaille	Universität Padua
7. 5. 64	Dr. rer. nat. h. c.	Freie Universität Berlin
28. 6. 64	Ehrenmitglied	Physik. Ges. Zürich
1. 3. 71	Ehrenmitglied	Photogr. Ges. Wien
1. 8. 71	Goldene Voigtländer-Medaille	Photogr. Ges. Wien

Bibliographie John Eggert

I. Selbständige Veröffentlichungen

1. Lehrbuch der Physikalischen Chemie. 9. Aufl. Stuttgart 1968. 934 S., 248 Abb., gemeinsam bearbeitet mit L. Hock, Giessen und G. M. Schwab, München. 1. Aufl. in englischer Übersetzung 1932. 3. Aufl. in russischer Übersetzung 1939. 4. Aufl. in italienischer Übersetzung 1941. 2. Aufl. in spanischer Übersetzung 1943. 1. Aufl. in französischer Übersetzung 1944. 1. Aufl. in kroatischer Übersetzung 1945.
2. Einführung in die medizinische Röntgenphotographie. 7. Aufl. S. Hirzel, Zürich 1951.
3. Einführung in die Technische Röntgenphotographie (mit H. Gajewski) 2. Aufl. S. Hirzel, Leipzig 1945.
4. Einführung in die Tonphotographie (mit R. Schmidt) S. Hirzel, Leipzig 1932.
5. Ergebnisse der Technischen Röntgenkunde, Band 1—6, gemeinsam hrsg. mit E. Schiebold, Akad. Verl. Ges. Leipzig, 1930, 1931, 1933, 1934, 1935, 1938.
6. Herausgeber der Veröffentlichungen der Wissenschaftlichen Photo-Laboratorien AGFA, Band 1—6, S. Hirzel, Leipzig 1930, 1931, 1933, 1935, 1937, 1939.

II. Wissenschaftliche Aufsätze

1914—1924

Dissertation: Über die Absorptionsgeschwindigkeit von Wasserstoff und Sauerstoff durch Metallsalzlösungen. März 1914. Auszug in: Z. f. Elektrochem. *20* (1914) 370—381.

Zur Aktivierung von H_2 und O_2 durch Platin. Z. f. Elektrochem. *21* (1915) 349—352.

Über die Landoltreaktion I. Z. f. Elektrochem. *23* (1917) 8—19.

u. H. Schimank: Einige Vorlesungsversuche zur Theorie der Explosivstoffe. Z. f. Elektrochem. *23* (1917) 189—192; Z. f. d. ges. Schieß- u. Sprengstoffwesen *12* (1917) 1—3; Phys. Z. *18* (1917) 7—8; Chemiker Ztg *41* (1917) 11.

u. H. Schimank: Über einige Vorlesungsversuche mit Acetylensilber. Ber. d. dt. Chem. Ges. *51* (1917) 454—456.

Über Acetylensilber. Z. f. Elektrochem. *24* (1918) 150—154 = Vortrag zur 24. Bunsenges. in Berlin 8.—10. 4. 1918 sowie am 22. 10. 1917 in der Dt. Chem. Ges; Chemiker Ztg 1918, H. 49; Z. f. d. ges. Schieß- u. Sprengstoffwesen *13* (1918) 153—154.

u. L. Zipfel: Über ein cyanometrisches Verfahren zur Bestimmung von Silber und Halogen in amm. Lösung. Ber. d. dt. Chem. Ges. *52* (1919) 1177—1185.

Über den Dissoziationszustand der Fixsterngase. Phys. Z. *20* (1919) 570—574.

u. W. Noddack: Über die Prüfung des photochem. Äquivalentgesetzes an der photogr. Trockenplatte. Ber. Preuß. Akad. Wiss. *39* (1921) 631—635; Phys. Z. *22* (1921) 673—674.

u. B. Scharnow: Über die Landoltreaktion II. Einige Analogiefälle zur Landoltreaktion. Z. f. Elektrochem. *27* (1921) 455–470.

u. B. Scharnow: Einige Vorlesungsversuche zur Reaktionskinetik in Lösungen (angestellt an d. Landoltreaktion). Ber. d. Dt. Chem. Ges. *54* (1921) 2521–2525.

Über die Sensibilität hochempfindlicher Sprengstoffe = Habilitationsschrift; nur im Manuskript vorhanden, 1921. Auszug: Z. f. Elektrochem. *27* (1921) 547–558.

u. W. Noddack: Zur Prüfung des photochem. Äquivalentgesetzes an Trockenplatten II. Ber. Preuß. Akad. Wiss. *25* (1923) 116–122.

u. G. Book: Über die Photochlorierung des Toluols. Z. f. Elektrochem. *29* (1923) 521 bis 527.

Ein cyanometrisches Verfahren zur quantitativen Silberbestimmung in photogr. Präparaten. Z. f. wiss. Photogr. *22* (1923) 209–213.

u. H. Scheffers: Studien über die Solarisation. Z. f. Phys. *20* (1923) 109–128.

u. W. Noddack: Zur Prüfung des photochem. Äquivalentgesetzes an Halogensilberemulsionen. Phys. Z. *24* (1923) 488–490.

u. W. Noddack: Zur Prüfung des photochem. Äquivalentgesetzes an Trockenplatten. Z. f. Phys. *20* (1923) 299–314; *21* (1924) 264.

u. W. Meidinger: Ein sicheres und einfaches Verfahren zur Silberbestimmung in photographischen Präparaten. Photogr. Industrie 1924, S. 647–649.

u. W. Borinski: Über die photochem. Sensibilisation der Maleinesterumlagerung durch Brom. Phys. Z. *25* (1924) 19–21.

u. G. Archenhold: Das optische Streuvermögen photogr. entwickelter Silberschichten. Z. f. phys. Chem. *110* (1924) 497–513.

u. L. Pfeffermann: Über die Landoltreaktion III. Z. f. anorgan. u. allgem. Chem. *139* (1924) 310–316.

Über die Landoltreaktion IV. Zur Kinetik des Natriumthiosulfats (nach Versuchen von E. Heisenberg). Z. f. Elektrochem. *30* (1924) 501–503; Chemiker Ztg *48* (1924) 492.

1925–1927

u. W. Meidinger: Untersuchungen über die photographische Schwärzungskurve. Z. f. phys. Chem. *114* (1925) 89–113.

Über einige Reaktionen des durch Licht angeregten Broms (nach Versuchen von F. Wachholtz und R. Schmidt). Phys. Z. *26* (1925) 865–868.

O. Becker u. J. Eggert: Ein neueres Verfahren zur Demonstration der Mischung von Lichtarten. Z. f. d. phys. u. chem. Unterricht *38* (1925) 142–143.

u. J. Reitstötter: Die photographische Wirksamkeit von Methylenblau als Adsorptionseffekt. Zsigmondy-Festschrift. Erg.-Bd d. Koll. Z. *XXXVI* (1925) 298–305.

u. W. Noddack: Zur Prüfung des Quanten-Äquivalentgesetzes an einigen Halogensilberemulsionen. Z. f. Phys. *31* (1925) 922–940.

u. W. Noddack: Zur Photochemie der Silberverbindungen. Bemerkungen zu einigen Arbeiten gleichen Titels von F. Weigert, W. Schöller und E. Schubert. Z. f. Phys. *31* (1925) 942–948.

u. W. Noddack: Zur Photochemie photographischer Schichten. Bemerkungen zu der vorangehenden Arbeit von F. Weigert. Z. f. Phys. *34* (1925) 917–920.

u. W. Meidinger: Silberbestimmung in photographischen Präparaten. Z. f. wiss. Phot. *23* (1925) 282–285.

u. W. Meidinger: Erwiderung (Steigmann). Z. f. wiss. Phot. *24* (1925) 16–17.

G. Book u. J. Eggert: Zur Photochlorierung des Toluols (Erwiderung auf die den gleichen Gegenstand behandelnde Veröffentlichung von F. Bergel). Ber. d. dt. Chem. Ges. *59* (1926) 1192—1193.

Margarete Eggert u. J. Eggert: Untersuchungen an Bauschen, Kammzügen und im Stapel. Einzeldarstellungen aus dem Kaiser-Wilhelm-Institut für Faserstoffchemie. 1926, 69—87.

Zusammenfassender Bericht über die Vorgänge bei der Belichtung der Silberhalogenide. Z. f. Elektrochem. *32* (1926) 491—501.

H. Arens u. J. Eggert: Über Dunkelkammerbeleuchtung unter besonderer Berücksichtigung der psychologischen Empfindlichkeit des menschlichen Auges. Z. f. wiss. Phot. *24* (1926) 229—248.

u. J. Reitstötter: Über das Molekulargewicht und den Gelzustand der Gelatine. Z. f. phys. Ch. *123* (1926) 363—382.

u. W. Leszynski: Studien über die Sensibilisation photographischer Platten. Z. f. wiss. Phot. *24* (1926) 261—275.

u. W. Leszynski: Studien über den Herscheleffekt. Z. f. wiss. Phot. *24* (1926) 275—291.

u. W. Noddack: Quantentheorie und Photographie. Naturwiss. *15* (1927) 57—69.

u. F. Wachholtz: Die photochemische Einwirkung des Broms auf Malein- und Fumarsäurediäthylester. Z. f. phys. Chem. *125* (1927) 1—48.

u. J. Reitstötter: Beiträge zur Kenntnis des latenten Bildes. Z. f. wiss. Phot. *24* (1927) 350—361.

H. Beck u. J. Eggert: Eine Methode zur zeitlichen photometrischen Verfolgung des Verbrennungsvorganges von Blitzlicht. Z. f. wiss. Phot. *24* (1927) 367—379.

u. W. Noddack: Über die Quantenausbeute bei der Wirkung von Röntgenstrahlen auf Silberbromid. Z. f. Phys. *43* (1927) 222—229.

Vorsicht beim Versilbern. Chem. Z. *51* (1927) 935.

Die photochemische Einwirkung des Broms auf Malein- und Furmarsäurereste. Z. f. Elektrochem. *33* (1927) 542—545.

u. F. Wachholtz: Die Umwandlung von Malein- in Furmarsäure durch Bromatome. Z. f. Elektrochem. *33* (1927) 545—547.

H. Arens u. J. Eggert: Über die Schwärzungsfläche photographischer Schichten. Z. f. phys. Chem. *131* (1927) 297—309.

1928—1929

H. Beck u. J. Eggert: Erwiderung. Z. f. wiss. Phot. *25* (1928) 7 u. 8, 262—263.

u. L. Hock, W. Barth: Strukturuntersuchung am gedehnten synthetischen Kautschuk. Z. f. Chem. *134* (1928) 3/4, 271—278.

u. F. Wachholtz: Die Umwandlung von Malein- in Fumarsäure durch Bromatome. Z. f. phys. Chem. *135* (1928) 3/4, 147—181.

u. R. Schmidt: Zur Lichtabsorption von Silberbromid- und Silberchloridschichten. Z. f. Phys. *48* (1928) 7/8, 541—544.

u. W. Roman: Über die Landoltreaktion V. Zur katalytischen Einwirkung des Natriumthiosulfats auf die Dushman-Reaktion. Z. f. Elektrochem. *34* (1928) 333—372.

u. W. Barth: Die Röntgenstereoskopie vom Standpunkte der physiologischen Optik. Fortschr. a. d. Geb. d. Röntgenstr. *XXXVIII* (1928) 299—311.

Zur Quantenausbeute bei der photochemischen Zersetzung lichtempfindlicher Diazoverbindungen. Z. f. Elektrochem. *9* (1928) 602—605.

u. H. Kieser: Zur Photolyse des bindemittelfreien Silberbromids (Silberbestimmung) Z. f. wiss. Phot., Photophys. u. Photochem. *26* (1928) 1—45.

H. Arens u. J. Eggert: Zur Farbtemperatur von Magnesium und Blitzlicht. Z. f. wiss. Phot. *26* (1928) 95—96.
u. R. Schmidt: Die photochemische Einwirkung des Broms auf Malein- und Fumarsäuredimethylester. Z. f. phys. Chem. Abt. B, *1* (1928) 3/4, 205—238.
H. Beck u. J. Eggert: Schlußwort. Z. f. wiss. Phot. *25* (1928) H. 12.
u. H. Meyer, R. Walter: Quecksilbercyanid als Desensibilisator. Z. f. wiss. Phot. *26* (1928) 4, 165—170.
H. Arens u. J. Eggert: Zur Sensitometrie desensibilisierter Schichten. Z. f. wiss. Phot. *26* (1928) 4—6, 111—126.
Diskussion zu Weigert u. Lühr „Über einige Eigenschaften photographischer Schichten". Z. f. Elektrochem. *34* (1928) S. 605.
u. W. Noddack: Über die Quantenausbeute bei der Wirkung von Röntgenstrahlen auf Silberbromid, II. Z. f. Phys. *51* (1928) 796—804.
u. E. Mutter: Zur Photolyse des bindemittelfreien Silberbromids (Brombestimmung). Z. f. wiss. Phot. *26* (1928) 193—234.
W. Barth u. J. Eggert: Photographische Studien an Kalziumwolframatverstärkungsfolien. Fortschr. a. d. Geb. d. Röntgenstr. *XXXIX* (1929) 1, 88—100.
u. W. Barth, A. Hälssig: Über die Betrachtung stereoskopischer Röntgenbilder und über die Herstellung von Anaglyphen. Forschr. a. d. Geb. d. Röntgenstr. *XXXIX* (1929) 2, 319—326.
u. W. Barth: Erwiderung auf die Arbeit „Über Stereoskopie" von L. Drüner, Quierschied. Fortschr. a. d. Geb. d. Röntgenstr. *XXXIX* (1929) 5, 902—905.
u. R. Schmidt: Photographische Probleme der Dichteschrift. Filmtechnik Nr. 9 (1929) 1—4.
u. H. Mediger: Fortschritte in der Photographie in den letzten 10 Jahren. Z. f. angew. Chem. *42* (1929) 653—659, 684—687, 700—703.
u. W. Noddack: On Secondary Processes in the Action of Light on Silver Halides. The Proceedings of the Seventh International Congress of Photography, July 1928. Cambridge 1929. p. 34—41.
H. Arens u. J. Eggert: Die Schwärzungsflächen zweier gering empfindlicher Emulsionen. Z. f. wiss. Phot. *27* (1929) 3/4, 79—84.
H. Arens u. J. Eggert: Das Wachstum des kolliden Silbers in Gelatineschichten. Z. f. Elektrochem. *9* (1929) 728—733.
u. F. Luft: Ein neuartiges Röntgen-Filmdosimeter, I. Röntgenpraxis *1* (1929) 4, 188 bis 189.
u. F. Luft: Ein neuartiges Röntgen-Filmdosimeter, II. Röntgenpraxis *1* (1929) 14, 655 bis 659.
u. W. Urban: Seltene Fehlerscheinungen auf Röntgenfilm. Röntgenpraxis *1* (1929) 4, 190 bis 191.
u. H. Tollert: Quantitative Untersuchungen einiger photographischer Effekte. Z. f. phys. Chem. A, *140* (192) 355—378.
Der Laue-Film. Naturw. *17* (1929) 41, 805.
u. R. Schmidt: Sensitometrische Untersuchung der Tonaufzeichnung in Dichteschrift. Filmtechnik *16* (1929) 1—3.
u. W. Noddack: Über die Gültigkeit des Einsteinschen Äquivalentgesetzes bei sensibilisierten und nicht sensibilisierten Platten. Z. f. Phys. *58* (1929) 11/12, 861—862.
Bericht über den 7. Internationalen Kongreß in London vom 7.—14. Juli 1928. Naturwiss. *17* (1928) 51, 1009—1010.

1930—1931

H. Arens u. J. Eggert: Die Aktinität und Farbtemperatur des Magnesiumband- und Blitzlichtes. Z. f. wiss. Phot. *28* (1930) 6/7, 169—177.

H. Arens u. J. Eggert: Zur Frage des Ursilbergehaltes und der Keimbildung in photographischen Emulsionen. Z. f. wiss. Phot. *28* (1930) 6/7, 178—190.

H. Arens u. J. Eggert: Sensitometrie mit Spektrum und Farbentafel. Z. f. wiss. Phot. *29* (1930) 239—261.

u. F. Luft: Zur Struktur der B-Cellulose. Z. f. phys. Chem. Abt. B, *7* (1930) 6, 468—470.

Die Photographie in der technischen Röntgenkunde. Ergebn. d. techn. Röntgenkd. Bd *I* (1930) 49—60.

Die Entwicklung des kinematographischen Rohfilms. Kinotechnik *XII* (1930) 7, 185—187.

Photographische Erfordernisse des Tonfilms. Kinotechnik *XII* (1930) 20, 549—556.

Die Empfindlichkeit photographischer Emulsionen für Röntgenstrahlen in Abhängigkeit von der Korngröße. Z. f. Elektrochem. *36* (1930) 9, 750—753.

H. Arens, J. Eggert u. E. Heisenberg: Zusammenhang zwischen Schwärzung, Silbermenge, Deckkraft, Kornzahl und Korndimension enwickelter photographischer Schichten. Z. f. wiss. Phot. *28* (1931) 12, 356—366.

H. Arens u. J. Eggert: Die Schwärzungsfläche des Villard-Effektes. Z. f. wiss. Phot. *30* (1931) 3—6, 121—126.

A comparative study of the photographic process in different experimental conditions. Transact. of the Faraday Soc., 123, *XXVII* (1931) 8, p. 528—533.

u. F. Luft: Untersuchungen zur photographischen Wirkung der α-Strahlen. Z. f. phys. Chem. (= Bodenstein-Festband) 1931, 745—754.

u. A. Küster, R. Schmidt: Gegenwärtiger Stand der photographischen Technik bei der Lichttonaufzeichnung. Z. f. techn. Phys. *XII* (1931) 12, 639—644.

u. H. Arens: Zur Photochemie endothermer Silberverbindungen. Photogr. Korr. *8* (1931) 17—21.

$1\ \mathrm{cm}^{-1} = 1$ Balmer. Naturwiss. *19* (1931) 860.

1932—1933

u. H. Bincer: Über die Abhängigkeit des osmotischen Druckes und des Micellargewichtes von Gelatinelösungen von der Temperatur bzw. von der Vorgeschichte der Lösungen. Biochem.-Z. *247* (1932) 85.

u. H. Bincer: Erwiderung auf Frankel. Biochem.-Z. *256* (1932) 239—240.

(Kollektiv) Terminus technicus, I. Kinotechnik *14* (1932) 300—301.

Infrarotphotographie. FuF *9* (1933) Nr. 8.

Die Leistungsfähigkeit photographischer Schichten in der Spektrographie aller Wellenlängen. Z. techn. Physik *14* (1933) 177—186.

(Kollektiv) Terminus technicus, II. Kinotechnik *15* (1933) H. 10.

1934—1937

u. A. Küster: Über die Körnigkeit photographischer Schichten. Kinotechnik *16* (1934) 127—130.

Fortschritte auf dem Gebiete der Herstellung photographischer Filme. Chemiker-Ztg *58* (1934) 397—404.

u. A. Küster: Bemerkungen zu der Arbeit von A. Narath „Über die Körnigkeit photographischer Schichten. Kinotechnik *XVI* (1934) 291—293.

Neuere Anwendungen der Photographie. FuF *10* (1934) Nr. 32, S. 395—396.

Die technisch wichtigsten lichtempfindlichen Systeme. Z. f. techn. Phys. *15* (1934) 436—443.

M. Biltz u. J. Eggert: Über die Beziehung von DIN-Graden zu Scheiner-Graden. Photogr. Industrie, 1934, H. 49, 1296—1298; 1935, H. 3, 51—53.

Neuere Anwendungen der Photographie. Naturwiss. *23* (1935) H. 3, 49—50.

Über Farbensensitometrie. Z. f. wiss. Phot. *34* (1935) 54—58.

Einige neue Anwendungen der Infrarotphotographie. Naturwiss. *23* (1935) H. 18, 281 bis 286.

u. H. Franke, F. Luft: Vereinheitlichung der Nomenklatur in der röntgenphotographischen Aufnahmetechnik. Fortschr. a. d. Geb. d. Röntgenstr. *52* (1935) 82—87.

Frage und Antwort betr. Schnellentwicklung von Röntgenfilmen. Röntgenpraxis *7* (1935) H. 4.

Die Bedeutung der physikalischen Chemie für die Photoindustrie. Z. f. Elektrochem. *41* (1935) 461—466.

Some Important Light-Sensitive Systems. The Photographic Journal *LXXVI* (1936) 17—26.

H. Arens u. J. Eggert: Etude d'une Emulsion pour obtention directe de contretypes. — Bericht über den IXe Congrès International de Photographic Scientifique & appliquée, Paris, Juillet 1935, S. 135—139.

M. Biltz u. J. Eggert: Projet de détermination de la Sensibilité chromatique des emulsions négatives pour photographic usuelle. Bericht über den IXe Congrès Internationale de Photographic Scientifique & appliquée, Paris, Juillet 1935, S. 478—483.

La photographie au service de la paléontologie. IXe Congrès Internationale de Photographic Scientifique et appliquée, Paris, Juillet 1935, S. 737—741.

u. A. Küster: Über die sogenannte photometrische Konstante. Z. f. wiss. Phot. Bd *35* (1936) H. 9—10, 233—234.

H. Arens u. J. Eggert: Rückgewinnung von Silber aus gebrauchten Fixierbädern. Röntgenpraxis *8* (1936) 753—763.

Über das neue Agfacolor-Verfahren. FuF *13* (1937) 72—74.

Über den gegenwärtigen Stand der Theorie des latenten photographischen Bildes. Phot. Korr. Bd *73* (1937) Nr. 3, 42—46.

Die photographischen Erfordernisse für verschiedene Aufnahmeverfahren. Fortschr. auf d. Geb. d. Röntgenstr. *56* (1937) Sonderheft = Kongreßheft; Verhandlg. d. Dt. Röntgen-Ges. *31* (1937) 72—73.

Die Photogtaphie in der Medizin. Sonderdruck aus den Verhandlungen des 3. Internationalen Kongresses für Lichtforschung, Wiesbaden 1. — 7. 9. 1936, 206—218.

u. G. Heymer: Der Stand der Farbenphotographie. Naturwiss. *25* (1937) H. 43, 689—699.

1938—1940

u. A. Küster: Grain size determination and other applications of the Callier effect. J. of the soc. of Motion Picture Engineers *30* (1938) No. 2, 181—190.

Die photographischen Grundlagen der Materialprüfung mit Röntgenstrahlen. Berg- u. Hüttenmännische Monatshefte *86* (1938) 174—181.

u. M. Biltz: The Spectral Sensitivity of Photographic Layers. Transactions of the Faraday Society *34*, Part 8 (1938) Nr. 208, p. 892—901.

u. E. Schopper: Körnigkeit photographischer Schichten bei Bestrahlung mit energiereichen Quanten. Z. wiss. Phot. *37* (1938) 221—232.
u. G. Heymer: Das Agfa-Panchrom-Verfahren, ein Aufnahme- und Kopierverfahren für subtraktiven Farbenfilm. FuF *15* (1939) 49—52.
Lochkameraaufnahmen mit Röntgenstrahlung. Z. f. angew. Phot. *1* (1939) 18—20.
K. F. Bonhoeffer u. J. Eggert: Photographische Untersuchungen an der Bunsenflamme. Z. f. angew. Phot *1* (1939) 43—47.
Die Ultrarotphotographie in der Medizin. Strahlentherapie *66* (1939) 663—671.
u. A. Küster: Zur Frage der Körnigkeit entwickelter photographischer Schichten mit verschiedener Kornverteilung. Z. f. wiss. Phot. *39* (1940) 61—66.
u. A. Küster: Die Empfindlichkeit photographischer Schichten bei Glühlampenbeleuchtung unter Berücksichtigung der Aktivität der Sperrschicht-Photozellen. Photogr. Industrie *38* (1940) 516—518.
u. M. Biltz: Zur spektralen Empfindlichkeit photographischer Schichten. Z. f. wiss. Phot. *39* (1940) 140—155.
u. F. G. Kleinschrod: Zur spektralen Empfindlichkeit photographischer Schichten II; Einfluß verschiedener Emulsions- und Gelatinearten. Z. f. wiss. Phot. *39* (1940) 155—165.
u. F. G. Kleinschrod: Zur spektralen Empfindlichkeit photographischer Schichten III; Einfluß verschiedener Bindemittel. Z. f. wiss. Phot. *39* (1940) 165—174.

1941—1946

H. Arens u. J. Eggert: Über einige Eigenschaften der Direkt-Duplikatschicht. Z. f. wiss. Phot. *40* (1941) 9—18.
Die photographischen Grundlagen des Röntgenbildes. Albers-Schönberg „Die Röntgenkunde" *1* (1941) 664—690.
u. F. G. Kleinschrod: Zur Temperaturabhängigkeit des photographischen Prozesses II. Z. f. phys. Chem. (A) *189* (1941) 1—9.
H. Arens, J. Eggert u. F. G. Kleinschrod: Zur spektralen Empfindlichkeit unsensibilisierter photographischer Schichten IV. Z. f. wiss. u. angew. Phot. *42* (1943) 33—42.
Erinnerungen an Walther Nernst. Z. f. d. phys. u. chem. Unterricht *56* (1943) 43—50.
Walther Nernst's Lehrbuch. Naturwiss. *31* (1943) 412—415.
Robert Luther. Z. f. Naturforschung *1* (1946) 357—359.

1947—1948

Grenzleistungen der Photographie (Einführungsvorlesung ETH, 30. 11. 46). Schweiz. Photo-Rundschau *11* (1946) 615—618; *12* (1947) 4—9.
Erdaufnahmen aus der V 2-Rakete. Schweiz. Photo-Rundschau *12* (1947) 181—183.
Kleine historische Betrachtung über die wichtigsten phototechnischen Prozesse. Camera *26* (1947) 3—5.
Schnellphotographie. Camera *26* (1947) 129—133.
Photographische Curiosa. Schweiz. Photo-Rundschau *12* (1947) 253—257 u. 273—276.
Dr. Momme Andresen — 90 Jahre alt. Schweiz. Photo-Rundschau *12* (1947) 321.
Dr. Momme Andresen zum 90. Geburtstag. Camera *26* (1947) 242—243.
Zur Katalytischen Abscheidung von Silber. Helv. Chim. Acta *30* (1947) 2114—2119.
Prof. Dr. E. Rüst 70 Jahre alt. Schweiz. Photo-Rundschau *13* (1948) 22—23.

Quelques expériences statistiques sur la sensation de contraste des images avec ou sans éclairage du champ périphérique. Sc. et Ind. Phot. [2] *19* (1948) 41–45.

Zur Bestimmung der Lichtempfindlichkeit von Negativmaterialien für bildmäßige Aufnahmen (Kollektivversuche AfS der D. G. P. F. 1939–1941). Schweiz. Photo-Rundschau *13* (1948) 193–200.

u. W. Meidinger, H. Arens: Zum Mechanismus der photographischen Sensibilisation. Helv. Chim. Acta *31* (1948) 1163–1168.

Erich Stenger zur Vollendung seines siebzigsten Lebensjahres 5. 8. 1948. Camera *27* (1948) 250.

u. E. Schopper: Zur Energieausbeute der Röntgenstrahlung bei der Fluoreszenz von Calciumwolframat. Annalen der Physik 6, *3* (1948) 270–279.

Samuel Edward Sheppard †. Camera *27* (1948) 380–381.

Die menschlichen Sinnesorgane und die Photographie. Schweiz. Photo-Rundschau *12* (1948) 421–426; *13* (1948) 442–445.

1949–1951

Die photographische Wirkung der Röntgen- und Korpuskularstrahlen. Bulletin de la Soc. Fribourgeoise *39* (1949) 1–3.

Les développateurs photogr. en tant que réducteurs dans la réaction de Landolt. Sc. et Ind. Phot. *20* (1949) 161–167.

La brillance du champ périphérique et son influence sur la représent. subject. correcte. – Sc. et Ind. Phot. *20* (1949) 204–205.

Über die Landoltreaktion VI. Die photographischen Entwicklersubstanzen als Reduktionsmittel bei der Landoltreaktion. Helv. Chim. Acta *32* (1949) 692–702.

Probleme aus dem Gebiet der Photographie. Neue Zürcher Zeitung Nr. 1123 (29) Blatt 5 vom 1. Juni 1949.

Die Natur des latenten photographischen Bildes und seine Entwicklung. Röntgen-Blätter *2* (1949) 249–260.

Naturaleza y revelado de la imagen fotografica latente. Anales de la Real Soc. Espanola de Fisica y Quimica. Ser. A – Fisica, Tomo *XLV* (1949) 203–214.

Los procedimentos mas importantes de la fotografia en color. Anales de la Real Soc. Espanola de Fisica y Quimica, Ser. A – Fisica, Tomo *XLV* (1949) 291–300.

Photographische Wirkung der Röntgen- und Korpuskularstrahlen. Röntgenphotographie *2* (1949) 163–165.

H. Dehio, J. Eggert, M. Honerjäger-Sohm, H. Hörmann u. H. Kaiser: Begriffliche Überlegung zum Problem des Plattenfehlers. Spectrochimica Acta *3* (1949) 488–497.

Aus Forschung und Technik der neuzeitlichen Photographie. Phys. Blätter *6* (1950) 56 bis 61.

A. Theorie des Röntgenbildes, B. Schutzmaßnahmen. Schinz/Baensch/Friedl/Ühlinger – Lehrbuch der Röntgendiagnostik *5* (1950) 3–71.

Das Verhalten von Selenphotozellen bei kurz dauernden Bestrahlungen. Z. f. phys. Chem. *196* (1950) 35–39.

Die wichtigsten Verfahren der Farbenphotographie. Experientia *6* (1950) 401–410.

L. P. Clerc zum 75. Geburtstage. Camera *29* (1950) 258.

Die Landoltreaktion mit Photochemikalien und zur Photochemie des Stickstoffs. Schweiz. Photo-Rundschau *15* (1950) 357–360.

Aus Forschung und Technik der neuzeitlichen Photographie. Viertelj. Z. d. Naturf. Ges. in Zürich *XCV* (1950) 265–272.

The behaviour of selenium photocells during iradiation of short duration. PSA-Journal, Sect. b, *17* B (1951) 98–101.
Nachruf auf Dr. Momme Andresen. Camera *30* (1951) 71.
Contribution to the photochemistry of endothermic compounds. Fundamental Mechanisms of Photographic Sensitivity, London 1951, p. 94–98.
Schweizer Pioniere der Photographie. Echo *31* (1951) H. 11 u. 12.
Deutsche Pioniere der Photographie. Camera *30* (1951) 404–406.

1952–1953

Photographie in der Wissenschaft. Naturwissenschaftl. Rundschau *5* (1952) 133–134.
Verheißungsvolle Perspektiven in der Phototechnik. Photo-Technik u. Wirtschaft *3* (1952) 172–174.
u. W. Großmann: Das Auflösungsvermögen von Dreischichten-Farbfilmen. Naturwiss. *39* (1952) 132–133.
Das latente photographische Bild und seine Entwicklung. Umschau *52* (1952) 275–278.
Martin Biltz †. Z. wiss. Phot. *47* (1952) 184–187.
Stand und Probleme der Farbenphotographie. Z. f. Elektrochem. *56* (1952) 712–716.
Bemerkungen zum Schwarzschild-Effekt. Z. f. angew. Physik *4* (1952) 445–447.
J. Berchtold u. J. Eggert: Über die Zündfähigkeit von Sprengstoffen durch Strahlung hoher Intensität. Naturwiss. *40* (1953) 55–56.
Paul Günter 60 Jahre. Z. f. wissenschaftl. Phot. *47* (1953) 255–256.
u. Th. Rohr: Über die Landoltreaktion VII; weitere Befunde an Typ I, II und III der Landoltreaktion. Helv. Chim. Acta *36* (1953) 855–868.
Über die Landoltreaktion VIII; theoretische Betrachtungen zu den Befunden der vorangehenden Arbeit. Helv. Chim. Acta *36* (1953) 868–875.
u. R. Zemp: Über Silberazid-Gelatine-Emulsionen. Z. f. Naturf. *86* (1953) 389–395.

1954–1955

u. H. Amsler: Silver Bromide Photo Cells. Photographic Science and Technique, PSA Quarterly May *1954*, S. 52–56.
u. B. Koch: Bestimmung der Korngröße unentwickelter photographischer Schichten mit Hilfe ihres Streuvermögens. Z. wiss. Phot. *49* (1954) 104–114.
Photochemische und photothermische Reaktionen bei hoher Lichtintensität. Physikalische Blätter *10* (1954) 549–556.
Albert Einstein, Nekrolog. Camera *34* (1955) 348–349.
u. R. v. Wartburg: Ein Sensitometer für Serienbestimmungen des Schwarzschildeffektes (Reziprozitätsfehlers) in weiten Bereichen. Z. f. Elektrochemie *59* (1955) 353–361.
L. P. Clerc. Camera *34* (1955) 393–396.
The behaviour of endothermic compounds on illumination with high intensities of light. III. – Science and Applications of Photography, Proceedings of the R. P. S. Centenary Conference, London *1953* (1955) 248–249.
u. E. Zünd: The influence of pressure on the photographic properties of emulsions in the wet and dry states. – Science and Applications of Photography, Proceedings of the R. P. S. Centenary Conference, London *1953* (1955) 141–144.
Photographie. ETH-Festschrift, *1955*, 342–343.
Photographisches Institut. ETH-Festschrift, *1955*, 645–647.

JOHN EGGERT

1. 8. 1891 – 29. 9. 1973

SIR JAMES CHADWICK
20. 10. 1891—24. 7. 1974

FRIEDRICH LEUTWEIN
9. 8. 1911—28. 12. 1974

Robert Luthers „Sechs Vorträge, Leipzig 1898" und seine „Antrittsvorlesung Dresden 1908". Z. f. wiss. Phot. *50* (1956) 4–14.

Dr. phil., Dr. rer. nat. h. c. Fritz Gajewski zur Vollendung seines siebzigsten Lebensjahres, 13. Oktober 1955. Z. f. Elektrochm. *59* (1955) 901.

Lothar Hock zur Vollendung seines fünfundsechzigsten Lebensjahres am 12. November 1955. Z. f. Elektrochem. *60* (1956) 1–2.

Neues auf phot. Gebiet in Wissenschaft und Technik. Vortrag gehalten i. d. Chem. Ges. Karlsruhe am 28. 6. 56. Angewandte Chemie *68* (1956) 582–583.

u. Y. Okamoto: Sensitometrische Studien in weiten Belichtungsbereichen II. Z. f. Elektrochem. *60* (1956) 71–85.

u. R. v. Wartburg: Sensitometrische Studien in weiten Belichtungsbereichen III. Z. f. Elektrochem. *60* (1956) 479.

u. R. v. Wartburg: Sensitometrische Studien in weiten Belichtungsbereichen IV. – Z. f. Elektrochem. *60* (1956) 487.

u. Y. Okamoto: Sensitometrische Studien in weiten Belichtungsbereichen V. Z. f. Elektrochem. *60* (1956) 500–507.

Über die Haltbarkeit von Dokumenten und deren Reproduktionen, insbesondere auf Film. Nachrichten SVB und SVD 1956, S. 82–86; Nachrichten f. Dokumentation *7* (1956) H. 3, 149–150.

Teaching Experiments on the Correlations between the Density Curves of photographic Layers of Negatives and Positives. Prof. E. Goldberg zum 75. Geburtstag. Bulletin of the Research Council of Israel, Section C Technology *56* (1957) 299–305.

u. R. v. Wartburg: The Behaviour of 16 mm Film Emulsions at very short Exposure Times. Studies on Sensitometry over a wide Range of Exposure Times VI. High Speed Photography, Third International Congress, London 10.–15. Sept. 1956, London 1957, 211–213.

u. H. Fischer: Sur le mécanisme de la sensibilisation par l'or. Remarques sur un travail de même titre du Dr. P. Faelens. Sc. et Ind. Phot. *28* (1957) 190–191.

Prof. Dr. Emanuel Goldberg zur Vollendung des 75. Lebensjahres. Camera *35* (1956) 608.

Die internationalen photographischen Veranstaltungen im Herbst 1956, Camera *36* (1957) 34.

u. R. v. Wartburg: Sensitometrische Studien in weiten Belichtungsbereichen IX. Über das Verhalten einer Silberbromidemulsion bei Variation der Fällungs- und Reifzeiten sowie des Silberjodidgehaltes. Z. f. Elektrochem. *61* (1957) 693–715.

Ist Photographie als Kunst oder Technik zu werten? Photo-Amateur *24* (1957) Juliheft, 135–136.

Nachrufe auf Ernst Rüst und Rudolf Fischer. Camera *36* (1957) 362–363.

Die Photographie als Wissenschaft und als Technik. Physikalische Blätter *13* (1957) 486–492.

u. K. Pfister: Entwicklung mit „Phenidon". Schweiz. Photo-Rundschau *22* (1957) 370 bis 372, 393–397, 413–417.

u. K. Pfister: Photography in Switzerland. „The Focal Encyclopedia of Photography", Focal Press, London 1956, S. 1147–1149.

1958–1959

Radioaktive Konferenz (14. u. 15. Nov. 1957 im Photogr. Institut ETH Zürich). Physikalische Blätter *14* (1958) 130.

Ein Unterrichtsversuch über die gegenseitigen Beziehungen der Schwärzungskurven von photographischen Negativ- und Positivschichten. Z. wiss. Phot. *52* (1957) 57–63.

The present status of photographic development in theory and practice. Phot. Sci. and Eng. 1 (1958) 93—103.

Nuovi aspetti della fotografia nel campo della scienza e della tecnica. Rivista Ferrania Mai 1958, Nr. 5, 2—5.

Die Stabilität des latenten Bildes und des photolytischen Silbers. Z. f. Elektrochem. *62* (1958) 48—49.

u. H. Fischer: Über die Konvertierung photographischer Schichten, insbesondere bei Anwesenheit eines latenten Bildes. Z. f. Elektrochem. *62* (1958) 230—235, 393 bis 400.

Die Photographie als Wissenschaft und als Technik. Heidelberger Akademie der Wissenschaften, Jahresheft 1956/57. Heidelberg 1958. 49—56.

Prof. Dr. Walter Noddack zur Vollendung seines 65. Lebensjahres. Camera *37* (1958) 435.

u. Y. Ohyama: Weitere Versuche über den Weigert-Effekt III. Z. f. Elektrochem. *62* (1958) 520—529.

The initiation of explosion by light. Proc. Royal Soc. A *246* (1958) 240—247.

Stand der photographischen Entwicklung in Theorie und Praxis. Wiss. Photogr. Köln 24.—27. Sept. 1956. Darmstadt 1958. 357—367.

Karl Friedrich Bonhoeffer 13. 1. 1899—15. 5. 1957 (Nekrolog). Jahrbuch der Bayerischen Akademie der Wissenschaften *1958* (München 1958) 189—196.

2. Radioaktive Konferenz (Zürich 13. u. 14. Nov. 1958 im Photogr. Inst. der ETH). Physikalische Blätter *15* (1959) 40—41.

u. Y. Yücelen: Über das Verhalten photographischer Schichten bei Einwirkung von Wasserstoffperoxyd-Dampf. Z. f. Elektrochem. *63* (1959) 18—29.

The Ignition of Explosives by Radiation. J. of Physical Chemistry *63* (1959) 11—15.

u. A. Büchler: Weitere Versuche über den Weigert-Effekt II. Ergebnisse der Intern. Konf. f. wiss. Photogr. Köln 24.—27. Sept. *1956* (1958) 145—154.

u. H. Hopff: Über den Angriff von Polystyrol durch Insektenlarven. Kunststoffe *49* (1959) 323.

Zur spektralen Empfindlichkeit photographischer Schichten. Annalen der Physik 7. Folge, *4* (1959) 140—144. Prof. Dr. W. Gerlach zum 70. Geburtstage gewidmet.

Louis Philippe Clerc †. Schweiz. Photogr. Rundschau *24* (1959) 436—439.

1960

Korngrößen und Empfindlichkeit photographischer Schichten. Freiberger Forschungshefte B *38* (1960) 226—230 (= Schiebold-Festschrift).

Photochemische Reziprozität. Phys. Bl. *16* (1960) 178—181.

Die Zündung von Sprengstoffen durch Strahlung. Bericht über den IV. Internat. Kongreß f. Kurzzeitphotographie u. Hochfrequenzkinematographie, Köln Sept. 1958. Darmstadt 1959, 50—58.

Die Anschauung über das latente Bild im Wandel der Zeiten. Eine intersäkulare internat. Konferenz. Festband 100 Jahre Schleußner. Hrsg. G. Haase u. H. Schleußner. Frankfurt/M. 1960, 13—31.

u. K. Pfister: Bericht über einen Vortrag auf dem 1. Internat. Kongreß für med. Photographie: „Methoden der Bildaufzeichnung mit Ausnahme der klassischen photogr. Verfahren". Ärztl. Panorama (Sandoz) Dez. 1960, S. 13.

An Automatic Sensitometer for Large Exposure Ranges and some of its Applications. Vortrag gehalten in Tokyo, 26. 9. 1960 [jap.]. Tokyo, Soc. Sci. Phot. Japan, 1960 (Eggert, J., Summary of Lectures, p. 1—28).

Silver Bromide Photo Cells. Vortrag gehalten in Kyoto, 4. 10. 1960 [jap.]. Tokyo, Soc. Sci. Phot. Japan, 1960 (Eggert, J., Summary of Lectures, p. 29—50).

1961

u. H. Heimann: Einige Versuche über den Herschel-Effekt. Photo. Korr. *96* (1960) 167 bis 174.

Neue Methoden der Bildaufzeichnung. Phys. Bl. *17* (1961) 4—13.

Zum Tode von W. Noddack. Phot. Korr. *97* (1961) 27—28.

3. Radioaktive Konferenz in Zürich am 17./18. 11. 60. Phys. Bl. *17* (1961) 41.

Die Photographie als Teilgebiet der Photochemie. Chimia *15* (1961) 283—286.

u. H. Pestalozzi: Zur spektralen Empfindlichkeit photogr. Schichten II. Z. f. Elektrochem. *65* (1961) 50—61.

Einladung nach Japan. Phys. Bl. *17* (1961) 128—131.

u. K. Pfister: Methoden der Bildaufzeichnung mit Ausnahme der klassischen photogr. Verfahren. Chimia *15* (1961) 296—300.

u. K. Pfister: Die Photographie in der Schweiz. Schweiz. Photo-Rundschau *26* (1961) 254—256.

Über Silberbromid-Photozellen. Angew. Chemie *73* (1961) 417—423.

Neues in der Photographie seit 15 Jahren. Schweiz. Photo-Rundschau *26* (1961) 391—396.

ORR Dr. W. Meidinger 60 Jahre alt. Z. f. wiss. Phot. *55* (1961) 77—79.

Grundlagenforschung. Vortrag zur Verleihung des Kulturpreises. Photo. Korr. *97* (1961) 105—106.

1962

Zürich 1961 — Intern. Kolloquium über wissenschaftliche Photographie. Phys. Bl. *18* (1962) 136—137.

Photochemisches Symposium vom 18.—22. Juni 1962 in Brüssel. Phys. Bl. *18* (1962) 427.

Photographische Schichten mit ungewöhnlichen Reziprozitätsfehlern. Z. angew. Phys. *14* (1962) 713—715.

1963

Mysteriöse Kriställchen. Z. Naturforsch. *18a* (1963) 251—252.

Klaus Clusius 60 Jahre alt. Chimia *17* (1963) 92—93.

Walter Noddack in memoriam. Chem. Berichte *96* (1963) XXVIII—LI.

Vorgeschichte und Organisation. Photographic Science, Symposium Zürich 1961. Ed. W. F. Berg. London 1963, 15—17.

Die Photographie mit unsichtbaren Strahlen. Photographic Theory. Liège Summer School 1962. Ed. A. Hautot. London 1963. 245—266.

1964

Reprographische Verfahren unter Verwendung von Silberverbindungen. Reprographie, Hrsg. v. O. Helwich, Darmstadt 1964, 14—18; Reprographie *3* (1963) 171—173.

Walther Nernst. Zur 100. Wiederkehr seines Geburtstages. Angew. Chem. *76* (1964) 445—455.

Die Bedeutung Berlins für die Photochemie in Wissenschaft und Technik. Phot. Korr. 6. Sonderheft 1964, 21–29.

125 Jahre Photographie. Bild der Wissenschaft 1964, H. 3, S. 19–29.

W. F. Berg, J. Eggert, E. Klein, A. Langner u. E. Moisar: The Colloquio Internazionale sulla Scienza della Fotografia – Turin 23.–28. Sept. 1963. 1963. J. Phot. Sci. *12* (1964) 241–268.

1965

Über den Weigert-Effekt. Phys. Bl. *20* (1964) 559–562.

Max Volmer †. Phys. Bl. *21* (1965) 324–325.

1967

A. Büchler u. J. Eggert: Über Silberbromid-Photozellen. Z. wiss. Photogr. *60* (Leipzig 1967) S. 21–32, 17 Abb.

Prof. Dr. Hans Ammann zur Vollendung seines 60. Lebensjahres. Chimia *21* (1967) 262, m. Abb.

u. W. Jost, H. Witte: Prof. Paul Günther zum 75. Geburtstag. In: Ber. Bunsenges. *71* (1967) 933–934, m. Abb.

1968

Die Stabilität des latenten photographischen Bildes und der Auskopierschwärzung. Z. f. wiss. Phot. *62* (1968) 25–31, 7. Abb.

[Interview] Abermals nach Japan eingeladen. Interview mit Prof. J. Eggert. In: Phys. Bl. *24* (1968) 173–178.

Wolfgang Friedrich Berg zur Vollendung seines 60. Lebensjahres am 30. 6. 1968. Phot. Korr. *104* (1968) H. 4, 57–58; Chimia *22* (1968) 147.

1969

u. B. Högg: Weitere Versuche über den Weigert-Effekt IV. Z. phys. Chem. *64* (1969) 115–121, 5 Abb.

1970

The stability of the latent image and of the print-out silver „The photographic image", Symposium Tokyo Sept. 1967. London 1970, 130–136 [dt. Übersetzung in: Z. wiss. Phot. *62* (1968) 25–31, 7 Abb.].

u. H. J. Elster, W. Gerlach, W. Haupt, M. Himmelheber, A. Huning, W. A. P. Luck, H. Schäfer, H. Schimank, V. Weidemann: Die „Baiersbronner Erklärung" [Zur „Lindauer Rede" von Kardinal König]. Internationale Dialog-Zeitschrift *3* (1970) H. 1, 3–5.

[Vortr.] Fifty years Weigert-Effekt. Internationaler Kongreß für fotografische Wissenschaft. Moskau, 29. Juli–5. August 1970. Plenarvortrag Section A, 1–9.

Herrn Dr. Hans Arens zur Vollendung seines achtzigsten Lebensjahres (zum 28. Februar 1970). Phot. Korr. *106* (1970) 46.

1971

Der Werdegang der photographischen Schicht bis zum Beginn des 20. Jahrhunderts. In: Geschichte der Naturwiss. und der Technik im 19. Jahrhundert. 7. Gespräch der Georg-

Agricola-Gesellschaft vom 26.—28. Februar 1970. 55—79, 20 Abb. (= Technikgeschichte in Einzeldarstellungen, H. 19).
50 Jahre Weigert-Effekt. Ber. Bunsenges. *75* (1971) 725—728.
[Nachr.] Othmar Helwich. Phot. Korr. *107* (1971) Nr. 10, 87—88.

1973

„Ceterum censeo" — ein zeitgeschichtlicher Rückblick auf dreihundert „Letzte Seiten". Phys. Bl. *29* (1973) 51—58.

Barbara Ramuz
Bibliotheksassistentin

Zur Biobibliographie

Schriftenverzeichnisse. B. Ramuz, siehe vorliegendes Jahrbuch S. 296—309. — Jahrbuch Sächsische Akademie der Wissenschaften zu Leipzig *1957—1959* (Berlin 1961) S. 243—244; *1966—1968* (Berlin 1970) S. 205; *1969—1970* (Berlin 1972) S. 153; *1971—1972* (Berlin 1974) S. 165. — Kürschners Deutscher Gelehrten-Kalender *3* (Berlin—Leipzig 1928/29) Sp. 441; *4* (Berlin—Leipzig 1931) Sp. 557; *5* (Berlin—Leipzig 1935) Sp. 267; *6_1* (Berlin 1940/41) Sp. 352; *7* (Berlin 1950) Sp. 390—391; *9* (Berlin 1961) S. 365; *10* (Berlin 1966) S. 445; *11* (Berlin 1970) S. 553. — Poggendorffs Handwörterbuch *VI_1* (Leipzig—Berlin 1936) S. 641; *$VII_1{}^a$* (Berlin 1956) S. 472—473.

Festschrift. Zum 75. Geburtstag. Sonderheft der Photogr. Korrespondenz *102* (1966) Nr. 8, Nr. 1225 der ganzen Folge (mit Bildnis).

Würdigungen. Zum 60. Geburtstag: H. Amman, Kl. Clusius, in: Z. Naturforschung *6a* (1951) S. 337. — H. Arens, in: Z. wiss. Photogr. *46* (1951) S. 114—118. — E. Brüche, in: Physikal. Bl. *7* (1951) S. 411—412. — P. Günther, in: Z. Elektrochem. *55* (1951) S. 341—342 (mit Bildnis). — H. J. Lummerzheim, in: Photo-Technik *2* (1951) S. 7 und 296. — o. V., in: Z. Naturforschung *6b* (1951) S. 104. — o. V., in: Schweizer Photo-Rundschau *16* (1951) S. 57. — Zum 65. Geburtstag: o. V., in: Z. wiss. Photogr. *51* (1956) S. 130. — Zum 70. Geburtstag: H. Arens, in: Z. wiss. Photogr. *55* (1961) S. 241. — R. Brill, in: Physikal. Bl. *17* (1961) S. 375—376. — K. V. Čibisov, in: Z. wiss. Photogr. u. Kinomatographie 1961, Nr. 4 [russ.]. — H. Frieser, E. Klein, R. Matejec: J. Eggerts bedeutende Beiträge zu photographischen Problemen. In Photogr. Korrespondenz *97* (1961) S. 115—121. — A. Küster, in: Chemiker-Ztg *85* (1961) S. 516. — W. Meidinger, in: Photogr. Korrespondenz *97* (1961) S. 113—114. — G.-M. Schwab, in: Z. Elektrochem. *65* (1961) S. 491—492. — K. Schwabe, in: FuF *35* (1961) S. 253—254. — o. V.: Laudatio für Herrn Prof. Dr. Dr. h. c. J. Eggert. In: Photogr. Korrespondenz *97* (1961) S. 108—109. — Zum 75. Geburtstag: H. Arens, in: Z. wiss. Photogr. *59* (1965) H. 9—12, S. 190—191. — W. F. Berg, in: Photogr. Korrespondenz *102* (1966) Nr 8, S. 117. — u. r., in: Zolliker Bote *56* (Zürich 1966) Nr. 42 vom 21. Okt., S. 5 (mit Bildnis). — Zum 80. Geburtstag: H. Raupach, W. Kunkel, Glückwunschschreiben. In: Jahrbuch Bayer. Akademie der Wissenschaften *1972* (München 1972) S. 207—208.

Nachrufe. W. F. Berg, siehe vorliegendes Jahrbuch S. 289—294; Photographic Science and Engineering *18* (1974) Nr. 1, S. 3; ETH-Bulletin Zürich *7* (1973) Nr. 77 v. 10. Dez., S. 41; Manuskript für die Royal Photographic Society London. — E. Brüche, in: Physikal. Bl. *30* (1973) S. 551. — A. C. H. van Peski, in: Chemisch Weekblad *69* (Den Haag 1973)

Nr. 51/52 vom 21. Dez., S. 7. — G.-M. Schwab, in: Jahrbuch Bayer. Akademie der Wissenschaften *1974* (München 1974) S. 212—215. — [H.] Zw[icky], in: Chimia *27* (1973) Nr. 11, S. 612—613. — o. V., in: Neue Zürcher Zeitung Nr. 463 vom 6. Okt. 1973, S. 14.

G. W.

James Chadwick

20. 10. 1891—24. 7. 1974

Am 24. Juli 1974 starb im 83. Lebensjahr das auswärtige Mitglied der mathematisch-naturwissenschaftlichen Klasse Sir James Chadwick, C. H., Ph. D., D. Sc. h. c. (Birmingham, Exeter, Dublin, Leeds, McGill, Oxford, Reading), LL. D. h. c. (Edingburgh, Liverpool), Fellow of Gonville and Caius College in Cambridge, Nobelpreisträger und Inhaber zahlreicher weiterer Auszeichnungen, Mitglied der Royal Society, London, der Pontificia Academia Scientiarum, Vatikanstadt, ausländisches Mitglied der Kongelige Danske Videnskabernes Selskab, Kopenhagen, der Koninklijke Nederlandse Akademie van Wetenschappen, Amsterdam, der Académie Royale de Belgique, Brüssel, der American Philosophical Society, Philadelphia, Ehrenmitglied des American Institute of Physics, New York, und ausländisches Mitglied des Ordens pour le mérite für Wissenschaften und Künste.

James Chadwick wurde am 20. Oktober 1891 in Cheshire, England, als Sohn von John Joseph Chadwick und Anne Mary Knowles geboren. Er begann das Studium 1908 an der Universität Manchester. Entscheidend für sein späteres Wirken war 1911 sein Eintritt in das Physikalische Laboratorium, wo er unter der Leitung von Professor (dem späteren Lord) Rutherford zwei Jahre lang an verschiedenen Problemen des damals noch sehr jungen Gebietes der natürlichen Radioaktivität arbeitete. Dabei erwarb er 1913 den Grad eines Master of Science. Ein Stipendium ermöglichte ihm, anschließend seine Arbeiten an der Physikalisch-Technischen Reichsanstalt in Berlin-Charlottenburg unter Prof. H. Geiger fortzusetzen, bis ihn der erste Weltkrieg überraschte und ihm einen längeren Aufenthalt in einem Internierungslager einbrachte.

Erst 1919 konnte er die Zusammenarbeit mit Rutherford weiterführen, der inzwischen die Leitung des Cavendish Laboratoriums in Cambridge übernommen hatte; es war das Jahr, in dem Rutherford die erste künstlich angeregte Atomspaltung gelang, indem er Stickstoffatome mit Alpha-Teilchen beschoß. Die Entwicklung der experimentellen Technik der künstlichen Atomumwandlungen, insbesondere bei leichteren Elementen, und das Studium der dabei auftretenden Teilchen blieben viele Jahre das Hauptarbeitsgebiet von J. Chadwick. Hierbei wurden die Grundlagen für unsere heutigen Kenntnisse über die Struktur von Atomkernen mit gelegt. Er wurde zum Fellow des Gonville und Caius College (1921—1935) gewählt und war von 1923—1935 stellvertretender Direktor für Forschung im Cavendish Laboratorium; bereits 1922 wurde er Mitglied der Royal Society.

In jener Zeit kannte man nur Protonen und Elektronen als Bausteine der Atome. Um die experimentellen Werte für Kernladung und Kernmasse erklären zu können, mußte man auch im Innern der Atomkerne noch Elektronen annehmen, die einen Teil der positiven Ladungen der Protonen kompensieren sollten. Dieses Modell konnte nicht alle experimentellen Ergebnisse erklären. So fanden Bothe und Becker

1930 beim Beschuß mit Helium-Kernen eine neue, sehr durchdringende Strahlung. Die erste Vermutung, es handelte sich um eine elektromagnetische Wellenstrahlung, also um sehr harte γ-Strahlung, ließ sich kaum aufrechterhalten, da diese dann eine unwahrscheinlich hohe Energie haben müßte. J. Chadwick studierte derartige Strahlungen auch bei anderen Prozessen und kam bald zu der Überzeugung, daß es sich um eine neue Art Teilchenstrahlung handeln mußte, und zwar aus ungeladenen Teilchen der Masse des Protons. Rutherford hatte die Existenz solcher Neutronen bereits 1920 postuliert, sie konnten aber noch nicht gefunden werden. Das Verdienst Chadwicks bestand darin, die Masse dieser Teilchen bestimmt zu haben. Er beobachtete dazu in einer Nebelkammer Zusammenstöße mit anderen Atomen; die Bahnänderungen ergaben nicht nur die Neutronenmasse, sondern gleichzeitig auch genauere Werte für die Massen der anderen beteiligten Atomkerne. Damit war die Existenz der Neutronen 1932 endgültig bewiesen, und ein intensives Studium der Quellen dieser neuen Strahlung konnte beginnen.

Für diese epochemachende Entdeckung erhielt J. Chadwick 1935 den Nobelpreis für Physik; die Sächsische Akademie der Wissenschaften hatte ihn bereits am 1. 7. 1932 zum auswärtigen Mitglied gewählt.

Die Entdeckung des Neutrons hat nicht nur die Vorstellungen über die Struktur der Atomkerne verändert und viele bis dahin bestehende Diskrepanzen beseitigt; sie gab dem Physiker auch neue Teilchen in die Hand, die, da sie ungeladen sind, ungehindert elektrische Potentialwälle durchdringen und damit auch in die Kerne schwerster Atome gelangen können. Man besaß ein neues Werkzeug zur Untersuchung der Kernspaltung, der Weg zur Uran-Spaltung, zur Gewinnung von Energie und auch zum Bau der Atombombe, aber auch zu neuen Methoden zur Untersuchung der Struktur kondensierter Phasen war frei.

1935 ging J. Chadwick nach Liverpool an den Lyon Jones Lehrstuhl für Physik der Universität. Von 1943 bis 1946 leitete er die britische Gruppe, die in USA am Manhattan-Projekt zur Entwicklung der Atombombe mitarbeitete.

1945 wurde er in den Adelsstand erhoben. Er arbeitete zwar von 1957 bis 1962 noch in der Atomenergie-Kommission des Vereinigten Königreiches mit, aber bereits 1948 zog er sich von der aktiven Arbeit als Physiker und von seinem Amt in Liverpool zurück. Bei der Bewirtschaftung seines Gartens, beim Angeln erholte er sich von seiner umfangreichen Arbeit.

Mit Sir James Chadwick verliert die Menschheit einen Wissenschaftler, der Wesentliches zu unseren heutigen Kenntnissen über die Mikrostruktur der Materie beigetragen hat. Die Sächsische Akademie der Wissenschaften zu Leipzig rechnet es sich als hohe Ehre an, daß sie ihn zu den ihren rechnen durfte.

Artur Lösche

Bibliographie Sir James Chadwick

1. Selbständig erschienene Schriften

u. E. Rutherford, C. D. Ellis: Radiations from radioactive substances. London 1930. 12 und 588 S.; New York 1952. 588 S.

Radioactivity and radioactive substances; an introduction to the study of radioactive sub-

stances and their radioactions; the nature of radioactivity and the bearing of radioactive transformations on the structure of the atom. 1. Aufl. London 1921; 2. Aufl. London 1931; 3. Aufl. New York 1947; 4. Aufl. New York 1953, 1962. 12 und 116 S.
Prospects for atomic power. Toronto 1954. 16 S.

2. Aufsätze

u. E. S. Bieler: Collis of a particles w. hydrogen nuclei. In: Philos. Magazine *42* (1921) 18 S.

u. C. D. Ellis: Investigation of the intensity contribution in the β-ray spectra of radium B & C. In: Proc. Cambridge philos. Soc. *21* (1922) 7 S.

u. P. H. Mercier: Scattering of beta-rays. In: Philos. Magazine *50* (1925) 17 S.

u. K. G. Emeléus: δ-rays produced by α-particles in different gases. Ebda *1* (1926) 12 S.

Observat. concerning the artificial disintegrat. of elem. Ebda *2* (1926) 20 S.

Scattering of α-particles in helium. In: Proc. Roy. Soc. London (A) *128* (1930) 9 S.

u. J. E. R. Constable, E. C. Pollard: Artificial disintegration by α-particles. Ebda *130* (1931) 27 S.

u. G. Gamow: Artificial disintegration by α-particles. In: Nature *126* (London 1930) 2 S.

Possible existence of a neutron. Ebda *129* (1932) S. 312—313.

Neutronen. In: Usp. fizič Nauk *12* (1932) S. 557 [russ.].

Über das Neutron. In: Z. Elektrochem. und angew. physik. Chem. *38* (1932) S. 546.

u. J. E. R. Constable: Artificial disintegration by α-particles, 2: Fluorine and aluminium. In: Proc. Roy. Soc. London (A) *135* (1932) S. 48—68.

The existence of a neutron. Ebda *136* (1932) S. 692—708.

The neutron. Ebda *142* (1933) S. 1—25.

The neutron and its properties. In: British J. Radiol *6* (1933) S. 24—32.

u. P. M. S. Blackett, G. P. S. Occhialini: New evidence for the positive electron. In: Nature *131* (1933) S. 473.

Anomale Diffusion von α-Teilchen, Umwandlung von Elementen durch α-Teilchen, das Neutron. In: Institut international de physique Solvay. 7e Conseil de physique: Structures et propriétés des noyaux atomiques, Bruxelles 1933. Rapports et discussions. Paris 1934. S. 81—120.

u. M. Goldhaber: A „nuclear photo-effect": Disintegration of the deuton by γ-rays. In: Nature *134* (1934) S. 237—238.

u. D. E. Lea: An attempt to detect a neutral particle of small mass. In: Proc. Cambridge philos. Soc. *30* (1934) S. 59—61.

u. N. Feather, W. T. Davies: Evidence for a new type of disintegration produced by neutrons. Ebda S. 357—364.

u. P. M. S. Blackett, G. P. S. Occhialini: Some experiments on the production of positive electrons. In: Proc. Roy. Soc. London (A) *144* (1934) S. 235—249.

Das Neutron. In: Usp. fizič. Nauk *14* (1934) S. 183. [russ.].

u. M. Goldhaber: Der Kern-Photoeffekt (Zerfall des Deutons durch γ-Strahlung). Ebda S. 953 [russ.].

u. M. Goldhaber: Disintegration by slow neutrons. In: Nature *135* (1935) S. 65 und in: Proc. Cambridge philos. Soc. *31* (1935) S. 612—616.

u. M. Goldhaber: The nuclear photoelectric effect. In: Proc. Roy. Soc. London (A) *151* (1935) S. 479—493.

The neutron and its properties. In: Les Prix Nobel en *1935* (Stockholm 1937) S. 1—10.

The elementary particles of matter. In: J. Instn. electr. Engr. *81* (1937) S. 697–700.
u. N. Feather, E. Bretscher: Measurements of range and angle of projection for the protons produced in the photo-disintegration of deuterium. In: Proc. Roy. Soc. London (A) *163* (1937) S. 366–375.
The cyclotron and its applications. In: Nature *142* (1938) S. 630 bis 634.
u. C. F. Powell, A. N. May, T. G. Pickavance: Excited states of stable nuclei. Ebda *145* (1940) S. 893–894.
Excited states of stable nuclei. Ebda *146* (1940) S. 401.
New applications of physics to medicine. In: Memoirs and Proc. Manchester lit. and philos. Soc. *84* (1942) S. 9–22.
u. A. N. May, T. G. Pickavance, C. F. Powell: An investigation of the scattering of high-energy particles from the cyclotron by the photographic method, 1: The experimental method. In: Proc. Roy. Soc. London (A) *183* (1944) S. 1–25.
Atomic energy and its applications. In: J. Inst. Fuel *20* (1946) S. 34–41 und in: Nature *159* (1947) S. 421–425.
The beginnings of the atomic bomb. In: J. Franklin Inst. *252* (1951) S. 453–455.
u. W. J. Arrol, J. Eakins: Preparation from irradiated uranium of iodine-131 and certain other fission products. In: Progress in nuclear energy, Ser. 3: Process chemistry. New York 1956. S. 356–362.
u. E. Rutherford, N. Feather: Natural radioactivity. In: Encyclopaedia Britannica. A new survey of universal knowledge *18* (Chicago–London–Toronto ... 1965) S. 896F/P-902.

3. Ausgaben

Rutherford of Nelson — The collected papers of Lord Rutherford of Nelson ... *1* (London 1962); *2* (London 1963).

4. Würdigungen, Nachrufe

Lord Rutherford †. In: Nature *140* (1937) S. 749–751.
J. F. Cameron †. Ebda *169* (1952) S. 777.
The Rutherford Memorial Lecture, 1953. In: Proc. Roy. Soc. London (A) *224* (1954) S. 435–447.
Irène Joliot-Curie †. In: Nature *177* (1956) S. 964–965.
Frederick John Marrian Stratton, 1881–1960. In: Biographical Mem. Fellows Roy. Soc. *7* (1961) S. 281–293.

Zur Biobibliographie

Schriftenverzeichnisse. Siehe vorliegendes Jahrbuch S. 311–313. — Poggendorffs Handwörterbuch *VI_1* (Berlin 1936) S. 421; *VII_2*^{b} (Berlin 1968) S. 762–763. — A. Hermann: Lexikon Gesc hichte der Physik A–Z. Köln 1972. S. 54. — The Int. Who's who? *18* (London 1954) S. 149; *20* (London 1956) S. 144; *26* (London 1962/63) S. 166; *35* (London 1971/72) S. 281.

Würdigungen. N. de V. Heathcote, in: Nobel Prize winners in physics *1901–1950* (New York 1953) S. 331–338 (m. Bildnis). — A. Hermann, Die Nobelpreise und Nobelpreisträger für Physik und Chemie. In: Physikal. Blätter *20* (1964) S. 476. — Derselbe, in: Lexikon Geschichte der Physik A–Z. Köln 1972. S. 53–54. — F. G. Hopkins, in:

Proc. Roy. Soc. London (A) *139* (1933) S. 25. — R. A. Millikan, in: Scientific Monthly *42* (1936) S. 87—88 (m. Bildnis). — R. Robinson, in: Proc. Roy. Soc. London (A) *205* (1951) S. 1. — N. K. Saha, in: Science and Culture *1* (1935) S. 402—403. — G. J. Sizoo, in: Nederlandsch Tijdschr. Natuurkunde *3* (1936) S. 1—8. — E. Schneider, Von Röntgen zu Einstein, von Planck zu Heisenberg. Berlin—München 1953. S. 200—205. — Les Prix Nobel en *1935* (Stockholm 1937) S. 53 (m. Bildnis, Faksimile). — The Nobel Prize winners and the Nobel Foundation *1901—1937* (Zürich 1938) S. 100 (m. Bildnis S. 432). — Nature *130* (1932) S. 898—890; *135* (1935) S. 463; *136* (1935) S. 824; *161* (1948) S. 964; *165* (1950) S. 99; *166* (1950) S. 924. — Who's who? *4*. Ed. (London 1965) S. 527. — Who's who? An annual biogr. Dictionnary with which is incorporated „Men and women of the time". *91* (London 1939) S. 551; *96* (1944) S. 476; *105* (1953) S. 506; *111* (1959) S. 523; *116* (1964) S. 522; *117* (1965) S. 527; *118* (1966) S. 528; *119* (1967) S. 529; *125* (1973) S. 558—559.

Nachruf. A. Lösche, siehe vorliegendes Jahrbuch S. 310—311 (m. Bildnis).

G. W.

Friedrich Leutwein

9. 8. 1911—28. 12. 1974

Am 28. Dezember 1974 verstarb an den Folgen einer Gehirnblutung in Nancy — Frankreich — das auswärtige Mitglied der Akademie Friedrich Leutwein. Friedrich Leutwein wurde am 9. August 1911 in Berlin geboren. Er entstammte einer Offiziersfamilie. Sein Vater wurde schon 1914 ein Opfer des 1. Weltkrieges und die Mutter übersiedelte nach Crossen a. O. Hier legte Friedrich Leutwein sein Abitur ab. Sein Hochschulstudium begann er 1929 in Freiburg i. B., setzte es 1931 in Gießen fort und schloß es, wieder in Freiburg, 1936 mit der Promotion zum Dr. phil. ab. Hatten während seiner Ausbildungszeit Physik, Chemie und Mineralogie gleichmäßig im Mittelpunkt seines Interesses gestanden, so bildeten sich, wohl unter dem Einfluß seines Freiburger Lehrers, H. Schneiderhöhn, bald Geochemie und Lagerstättenkunde als eindeutiger Schwerpunkt heraus. Von 1936 bis 1939 arbeitete Friedrich Leutwein am Mineralogischen Institut der Freiburger Universität als wissenschaftlicher Mitarbeiter an der erwähnten Thematik weiter. Diese Tätigkeit führte ihn 1939 nach Freiberg, wo ihm, erst als wissenschaftlicher Mitarbeiter des Oberbergamtes, dann als Leiter des Forschungslaboratoriums des Bergreviers, die Aufgabe oblag, die Spurenelementverteilung bzw. deren Weg in den Hüttenprodukten und in der Hüttentechnologie zu klären. Nach 1945 war Friedrich Leutwein zwei Jahre beratender Ingenieur des technischen Stabes des Kommissariates für Buntmetalle der UdSSR. Nach der Wiedereröffnung der Bergakademie Freiberg erfolgte 1947 seine Berufung zum o. Prof. für Mineralogie und zum Direktor des Mineralogisch-Lagerstättenkundlichen Institutes. Er richtete das Institut zielbewußt auf geochemische Belange aus. Im Vordergrund seiner Arbeiten standen zuerst Untersuchungen über Elementspuren und Spurenelemente in Gesteinen und Hüttenprodukten sowie in Sapropeliten und Kohlen; später standen geochronologische Themen im Vordergrund seines Interesses. Obwohl die Geochemie dabei stets im Vordergrund stand, vergaß Friedrich Leutwein nie, daß seine Untersuchungen letzten Endes stets der Lösung geologischer Aufgaben dienen sollten. Er sah die Geochemie, ähnlich der Stratigraphie, Tektonik u. a., stets als einen methodischen Weg zur Lösung geologischer Probleme an. Am schärfsten kommt diese ihm eigene Tendenz in seinem Bemühen um die Schaffung einer Abteilung für Geochronologie zum Ausdruck. Hier sollte auch später, in Frankreich, der Schwerpunkt seiner Forschungen liegen. 1949 wurde Friedrich Leutwein zum Rektor gewählt. Das Amt hatte er vier Jahre inne. In Hamburg, wo er bis 1960 als Honorar-Professor tätig war, setzte er seine bereits in Freiberg begonnenen Arbeiten über die Geochemie der Kohlen bzw. der Sapropelite fort. 1960 übersiedelte er nach Nancy, wo er zuerst als Prof.-Associé, später als Directeur de Recherche du Centre National de la Recherche Scientifique an der Abteilung Geochronologie und Geochemie tätig war. Hier nahm er seine bereits in Freiberg begonnenen Arbeiten über radiometrische Altersbestimmungen von Gesteinen wieder auf. Sein Hauptinteresse galt dabei der Altersdatierung des Pentevrien, als der anscheinend ältesten geologischen Einheit Mitteleuropas. Diese Arbeiten brachten ihn in enge Beziehungen zu den wissenschaftlichen Fachkreisen in England. Die Problematik über die Stellung des Pentevriens ist zur Zeit noch offen. Wie auch die endgültige Lösung erfolgen wird: Die von Friedrich Leutwein erarbeiteten geochronologischen Daten werden ein festes Fundament jeder Synthese darstellen.

Die erfolgreiche wissenschaftliche Tätigkeit Friedrich Leutweins fand breite internationale Anerkennung. 1956 wählte ihn die Sächsische Akademie der Wissenschaften zu Leipzig, 1957 die Deutsche Akademie der Wissenschaften zu Berlin zu ihrem ordentlichen Mitglied. 1958 verlieh ihm die Deutsche Demokratische Republik den Nationalpreis III. Klasse.

Friedrich Leutwein hat frühzeitig die große Bedeutung geochemischer Forschung erkannt, sie maßgeblich gefördert und methodisch weiterentwickelt. Sein Name wird mit der Entwicklung dieser Forschungsrichtung stets verbunden bleiben.

Adolf Watznauer

Bibliographie Friedrich Leutwein

Diese Bibliographie ist nach Jahren geordnet. Es werden zuerst die selbständigen Veröffentlichungen angeführt; es folgen Beiträge in Festschriften, Aufsätze in Zeitschriften und Nachrufe. Am Schluß des jeweiligen Jahres sind redaktionelle Mitarbeit und Herausgebertätigkeit verzeichnet.

1937

Untersuchungen über die Gesteine des Untermünstertales und der Culmstreifen von Badenweiler im südlichen Schwarzwald. In: Neues Jahrbuch für Mineral., Geol. Paläontol., Beil.-Band *72* (1937) S. 232–264 [= Dissertation].

1939

Über das Vorkommen einiger seltener Elemente in metamorphen Gesteinen des Schwarzwaldes. In: Zentralblatt Mineralogie *1939*. S. 123–128.

1940

Untersuchungen über die Titerkonstanz hochverdünnter Vergleichslösungen für die Spektralanalyse. In: Zentralblatt Mineralogie *1940*. S. 129–133.

Ein neues Verfahren zur Bestimmung von Zinn in Erzen und Aufbereitungsprodukten. In: Z. analyt. Chem. *120* (1940) S. 233–242.

1941

Die Geochemie und Vorkommen des Vanadiums. In: Berichte Freiberger geol. Gesellschaft *18* (1941) S. 73–83.

[Mitarbeit] H. Schneiderhöhn: Lehrbuch der Erzlagerstättenkunde. Band *1*: Die Lagerstätten der magmatischen Abfolge. Jena 1941. 858 S.

1942

[Mitarbeit] Die Analyse der Metalle. Band *1*: Schiedsverfahren. Berlin 1942. VIII, 506 S., herausgegeben vom Chemiker-Fachausschuß der Gesellschaft Deutscher Metallhütten- und Bergleute. 2. Auflage Berlin 1949.

1944

Photometrische Bestimmung von Wismut in Erzen und Aufbereitungsprodukten. In: Z. Verein Deutscher Chemiker *48* (1944) Beiheft 4, S. 110–112.

[Mitherausgeber] Bergakademie [Freiberg], mit Beilage Freiberger Forschungshefte *1* (1944) – *9* (1953).

1948

Spektrographische Analysenmethode im Bereich des Erzbergbaues und Metallhüttenwesens. In: Archiv Metallkunde *2* (1948) S. 75–82.

[Mitherausgeber] Sowjetwissenschaft. Naturwissenschaftliche Abteilung *1* (1948) – *3* (1950).

1949

Die Bedeutung russischer Forscher für die Geochemie. In: Urania *12* (1949) Heft 12, S. 454–458.

H. Schneiderhöhn und G. Claus, F. Leutwein, G. Prell, A. Schinzinger, W. Spitz: Das Vorkommen von Titan, Vanadium, Chrom, Molybdän, Nickel und einigen anderen Spurenelementen in deutschen Sedimentgesteinen. In: Neues Jahrbuch für Mineral., Geol. Paläontol. *1949*. S. 50–72.

1950

Über die Verteilung einiger Metalle zwischen Stein und Speise (Sulfid- und Arsenidphase) im Schmelzfluß. In: Z. Erzbergbau und Metallhüttenwesen *3* (1950) S. 11–32.

Über das Vorkommen von Vanadin und Molybdän in den Kieselschiefern Thüringens. In: Freiberger Forschungshefte (Sonderheft) *1* (1950) S. 125–128.

Über den Selbstbau spektrographischer Geräte. In: Technik *5* (1950) S. 238–242.

1951

Geochemische Untersuchungen an den Alaun- und Kieselschiefern Thüringens. (= Habil.-Schrift). Berlin 1951. 45 S. (= Archiv für Lagerstättenforschung, 82).

u. Charles Sommer-Kulaszewski: Allgemeine Mineralogie. Lehrbriefe 1—10. Berlin und Leipzig 1951—1955. 640 S. (= Lehrbriefe für das Fernstudium an der Bergakademie Freiberg).

Über die Herkunft der Amethyste in den cenomanen Schottern des Erzgebirges. In: Urania *14* (1951) S. 373—375.

Ein neues Polarisations-Mikroskop für Gesteins- und Erzuntersuchungen. In: Bergbautechnik *1* (1951) S. 226—228.

1952

Über das Vorkommen von Gold in thüringischen Alaun- und Kieselschiefern. In: Hallesches Jahrbuch für Mitteldeutsche Erdgeschichte *I* (1952) Heft 2, S. 83—85; 2è édition: Sur la présence d'or dans des schistes siliceux et alunifères de Thüringe.

Über die Zinn-Wolfram-Provinz des Erzgebirges. In: Freiberger Forschungshefte *3* (1952) S. 7.

Die Wolframitgruppe. Mineralogisch-lagerstättenkundliche Untersuchungen. In: Freiberger Forschungshefte *3* (1952) S. 8—19.

Zur Entwicklung eines mineralogischen Kursmikroskopes. In: Bergakademie [Freiberg] *4* (1952) S. 59—62.

Über die Entwicklung eines neuartigen Funkenerzeugers mit extrem kurzer Funkendauer und sehr hoher Impulsleistung. In: Chem. Technik *4* (1952) S. 512—513.

Die chemische Zusammensetzung der Wolframite und ihre lagerstättenkundliche Bedeutung. In: Acta geologica Hungarica *1* (1952) S. 133—141.

Das Vorkommen von Spurenmetallen in organogenen Sedimenten. In: Acta geologica Hungarica *1* (1952) S. 143—157.

Die chemische Zusammensetzung und die Entstehungsbedingungen der Wolframite. In: Chem. Technik *4* (1952) S. 115.

Vorkommen von Bakterien unter Tage. In: Zentralblatt Chirurgie *77* (1952) S. 546.

Bericht über die Vorträge der geologischen und mineralogischen Sektion der Jahrestagung der Ungarischen Akademie der Wissenschaften in Budapest, 9.—14. Dezember 1951. In: Geologie *1* (1952) S. 215—221.

[Herausgeber] Geologie *1* (1952) — *5* (1956).

1953

Über die Anwendung der Spektrochemie in der Metallurgie und Montanindustrie. Berlin 1953. 32 S. (= Wiss. Berichte 1, 62).

Ein elektronischer Funkenerzeuger hoher Anregungsenergie und -konstanz. In: Experiment. Technik d. Physik *1* (1953) S. 173—180.

u. Lore Doerffel-Sachse: Vergleichende Untersuchungen zur Spektralanalyse mit dem Impulsfunkenerzeuger. In: Experiment. Technik d. Physik *1* (1953) S. 181—192.

Karl-Marx-Jahr 1953. In: Bergakademie *5* (1953) S. 1—2.

Berufsbild des Mineralogen. In: Bergakademie *5* (1953) S. 375.

Die Entwicklung der Bergakademie 1952 und 1953. In: Bergakademie *5* (1953) S. 437—440.

Wladimir Afanasjewitsch Obrutschew (zu seinem 90. Geburtstag). In: Geologie *2* (1953) S. 399—400.

[Mitarbeit] Die Analyse der Metalle. Band *2*: Betriebsanalysen. Berlin 1953. XI, 716 S., herausgegeben vom Chemiker-Fachausschuß der Gesellschaft Deutscher Metallhütten- und Bergleute.

[Wissenschaftliche Redaktion] A. Saukov: Geochemie. Berlin 1953. 311 S.

1954

u. Fr. Edelmann, E. Krüger: Allgemeine Lötrohrprobierkunde. Lehrbrief 1. Berlin 1954. 63 S. (= Lehrbriefe für das Fernstudium an der Bergakademie Freiberg).

u. M. Sommer, Charles Sommer-Kulaszewski: Spezielle Mineralogie. Lehrbriefe 1—2. Berlin 1954 und 1956. 60 S. und 66 S. (= Lehrbriefe für das Fernstudium an der Bergakademie Freiberg).

Spektralanalytische Geräte und Spektrallaboratorien in der Sowjetunion. In: Feingerätetechnik *3* (1954) S. 211—215.

u. G. Herrmann: Kristallchemische und Geochemische Untersuchungen über Vorkommen und Verteilung des Wismuts im Bleiglanz der kiesig-blendigen Formation des Freiberger Gangreviers. In: Geologie *3* (1954) S. 1039—1056.

u. L. Pfeiffer: Ergebnisse und Anwendungsmöglichkeiten geochemischer Prospektionsmethoden auf hydrosilikatische Nickelerze. In: Geologie *3* (1954) S. 950—1008.

u. H. J. Rösler: Spurenelemente in Steinkohlenaschen (Autorreferat). In: Freiberger Forschungshefte *10* (1954) S. 68—70.

Walther Gothan zum 75. Geburtstag. In: Geologie *3* (1954) S. 491.

Im Dienste der geologischen Wissenschaft. Zum 70. Geburtstag von Professor Dr. Kurt Pietzsch. In: Geologie *3* (1954) S. 679.

[Wissenschaftliche Redaktion] A. N. Zavaricki: Einführung in die Petrochemie der Eruptivgesteine. Berlin 1954. XIII, 445 S., 107 Abbildungen, 32 Tabellen.

[Wissenschaftliche Redaktion] A. G. Betechtin: Lehrbuch der Mineralogie. 1. Auflage Berlin 1954, 2. Auflage Berlin 1957. 681 S.

[Wissenschaftliche Redaktion] S. S. Smirnov: Die Oxydationszone sulfidischer Lagerstätten. Berlin 1954. 312 S. mit Abbildungen.

1955

Geochemie und Lagerstättenkunde. In: Freiberger Forschungshefte *17* (1955) S. 5—25.

Der Universaldrehtisch nach Fedorow. In: Bergakademie [Freiberg] *7* (1955) S. 236—241.

u. K. Doerffel: Zur Methodik der Lösungsspektralanalyse. In: Bergakademie [Freiberg] *7* (1955) S. 426—433.

A. Schüller: Die Eigenschaften der Minerale, Bd I und II (Referat). In: Geologie *4* (1955) S. 204—207.

Der ROW-Kursspektrograph. Ein Instrument zur Ausbildung von Spektrochemikern. In: Feingeräte-Technik *9* (1955) S. 417—418.

[Wissenschaftliche Redaktion] A. Zaidel', Prokof'ew, Rajskij: Spektraltabellen (deutsch), Tables des raies spectrales (französisch), Tables of spectrum lines (englisch). Berlin 1955. LI, 550 S. 2. Auflage Berlin 1961.

1956

u. Fr. Edelmann: Spezielle Lötrohrprobierkunde. Lehrbrief 1. Berlin 1956. 64 S. (= Lehrbriefe für das Fernstudium an der Bergakademie Freiberg).

u. K. Krüger. Bodenkunde. Lehrbriefe 1—3. Berlin 1956 (= Lehrbriefe für das Fernstudium an der Bergakademie Freiberg).

u. H. J. Rösler: Geochemische Untersuchungen an paläozoischen und mesozoischen Kohlen Mittel- und Ostdeutschlands. In: Freiberger Forschungshefte *19* (1956) 196 S.
u. K. Doerffel: Über einige Verfahren zur theoretischen Klärung geochemischer Prozesse, unter besonderer Berücksichtigung der Gitterenergie. In: Geologie *5* (1956) S. 65–100.
Untersuchungen über das Vorkommen von Spurenmetallen in Torfen und Braunkohlen. In: Freiberger Forschungshefte *30* (1956) S. 28–48.
[Vorwort] Georgius Agricola: Ausgewählte Werke in 10 Bänden. Band *1*: Berlin 1956. 335 S., 70 Tafeln mit Geleitwort S. V–VII.

1957

Über die Anwendbarkeit geochemischer Untersuchungen zur stratigraphischen Einordnung von Kieselschiefern. In: Neues Jahrbuch Mineralogie, Abhandlung *91* (Stuttgart 1957) S. 257–270 (= Festband Hans Schneiderhöhn).
Geochemische Prospektion. In: Z. für angewandte Geologie *3* (1957) S. 178–182.
Zur Ausbildung von Fachkräften für spektrochemische Arbeiten — ein Diskussionsbeitrag. In: Bergakademie [Freiberg] *9* (1957) S. 161–163.
u. R. Starke: Über die Möglichkeit der geochemischen Prospektion auf Selen-Mineralien, untersucht am Beispiel des Kupferschiefers und des Tikeröder Erzbezirkes. In: Geologie *6* (1957) S. 349–378.
Alter und paragenetische Stellung der Pechblende erzgebirgischer Lagerstätten. In: Geologie *6* (1957) S. 797–805.

1958

Das Alter der erzgebirgischen Uranvorkommen. Ein Beitrag zur Methodik geochronologischer Arbeiten. In: Jahrbuch Sächsische Akademie der Wissenschaften zu Leipzig *1957* bis *1959* (Berlin 1961) S. 137–138; Deutsche Literatur-Zeitung *79* (1958) Spalte 563–564 [= Kurzfassung eines Vortrages vor der Sächsischen Akademie am 10. Februar 1958].
Anregungsbedingungen und Lichtquellen für die spektrochemische Emissionsanalyse. In: Fünf Vorträge über Spektroskopie. Gehalten auf der Arbeitstagung der Physikalischen Gesellschaft in der DDR vom 14.–16. November 1957 in Jena. Berlin 1958. S. 53–65.

1960

Geochemische Prospektion. In: Metall und Erz *XIII* (1960) S. 493–498.

1962

u. R. Waskowiak: Geochemische Untersuchungen an rezenten marinen Molluskenschalen. In: Neues Jahrbuch Mineralogie, Abhandlung *99* (1962) S. 45–78.
u. L. Weise: Hydrogeochemische Untersuchungen an erzgebirgischen Gruben- und Oberflächenwassern. In: Geoch. Cos. Acta *26* (1962) S. 1333–1348.

1963

G. Kaplan, F. Leutwein: Contribution à l'étude géochronologique du massif granitique de Vire (Normandie). In: C. R. Acad. Sc. *256* (1963) S. 2006–2008.

u. G. Kaplan: Quelques recherches sur l'aptitude de certain cristaux de néoformation à capturer de l'argon radiogénique. In: C. R. Acad. Sc. *257* (1963) S. 1315—1317.
M. Roubault, F. Leutwein, J. Sonet: Mise en évidence de quatre „âges apparents" dans les massifs granitiques et les séries cristallophylliennes des Pyrénées. In: C. R. Acad. Sc. *257* (1963) S. 3786—3789.
Spurenelemente in rezenten Cardien verschiedener Fundorte. In: Fortschritte Geologie Rheinland Westfalen *10* (1963) S. 283—292.
Contribution à la connaissance de la distribution des oligoeléments dans les coquilles actuelles. In: Sciences de la Terre *IX* (Nancy 1962—1963) S. 63—70.

1964

La géochronologie des roches. In: Bull. Soc. Franc. de Céramique *1964*, Nr. 63, A 6D/683, S. 49—64.
Les caractéristiques géochimiques de l'évolution des gisements d'étain et de wolfram de l'Erzgebirge (Allemagne). In: Sciences de la Terre *X* (Nancy 1964—1965) S. 37—78.
u. J. Sonet: Contribution à la connaissance de l'évolution geochronologique de la partie nord-est du Massif Armoricain français. In: Sciences de la Terre *X* (Nancy 1964—1965) S. 345—366.

1965

M. Roubault, R. Delafosse, F. Leutwein, J. Sonet: Premières données géochronologiques sur les format ions granitiques et cristallophylliennes de la République Centre-Africaine. In: C. R. Acad. Sc. *260* (1965) S. 4787—4792.
Etude géochimique et géochronologique d'un cristal de phlogopite de Myari près de Bekily (Madagascar). In: C. R. Serv. Géol. Madagascar *1965*, S. 95—97.
R. Coppens, F. Leutwein: Radioactivité des gisements stratiformes de cuivre et d'uranium des grés bigarrés de Wrexen (Hesse) Allemagne. Déséquilibre radioactif dans les charbons et les grés. In: 89ème Congrès des Sociétés Savantes, 1965, S. 219—230.
u. C. Levy, H. G. Oehlschlegel: Etude géochimique du gisement d'étain et de tungstène de Montbelleux. In: Bull. du BRGM *1965*, Nr. 6, S. 97—110.
Geochemische Charakteristica mariner Einflüsse in Torfen und anderen quarternären Sedimenten. In: Geologische Rundschau *55* (1966) S. 97—112.

1966

M. Bonhomme, J. Cogné, F. Leutwein, J. Sonet: Données nouvelles sur l'âge des séries rouges du golfe normanno-breton. In: C. R. Acad. Sc. *262* (1966) S. 606—609.
Tableau comparatif des échelles géochronologiques récemment publiées pour les temps phanérozoïques. Notice explicative. In: Sciences de la Terre *XI* (Nancy 1966) S. 367 bis 373.

1967

[Nachruf] Karl Hermann Scheumann 25. 2. 1881—28. 4. 1964. In: Jahrbuch Sächsische Akademie der Wissenschaften zu Leipzig *1963—1965* (Berlin 1967) S. 310—313.

1968

u. J. Sonet, J.-L. Zimmermann: Géochronoligie et évolution orogénique précambrienne et hercynienne de la partie Nord-Est du Massif Armoricain. Nancy 1968. 84 S., 8 Abb., 6 Taf., 1 Karte (= Sciences de la Terre, Mémoire n° 11).

Contribution á la connaissance du Précambrien récent en Europe Occidentale et développement géochronologique du Briovérien en Bretagne (France). In: Canadian Journal of Earth Sciences *5* (1968) S. 673–682.

Geochemie. In: Vom Erdkern bis zur Magnetosphäre. Frankfurt/M. 1968. S. 173–190.

u. J. P. von Eller, J. Sonet, J.-L. Zimmermann: Identification géochronologique des schistes de Villé (Vosges) comme Briovérien (Protérozoïque Supérieur). In: Bull. Serv. Carte géol. Als. Lorr. *21* (1968) S. 47–52.

Analyse spectrale des objets microscopiques dans le domaine de la géologie. In: Colloques nationaux du CNRS, n° 293, Dosage des éléments à l'état de traces dans les roches et les autres substances minérales naturelles. Nancy 1968. S. 453–464.

J.-L. Zimmermann, F. Leutwein: Contribution à l'étude de la libération de l'argon contenu dans les minéraux: application à la méthode de datation K-A. In: C. R. Acad. Sc. *267* (1968) S. 22–24.

1969

u. L. Chauris, J. Sonet, J.-L. Zimmermann: Principaux résultats de mesures géochronologiques dans le Nord-Ouest de la Bretagne. In: C. R. Acad. Sc. *268* (1969) S. 2552–2555.

u. L. Chauris, J. Sonet, J.-L. Zimmermann: Etudes géochronologiques et géotectoniques dans le Nord-Finistère (Massif Armoricain). In: Sciences de la Terre *XIV* (Nancy 1969) S. 329–358.

1970

Contribution à l'étude chronostratigraphique du Précambrien Supérieur et du Permien Français — Rapport à la Délégation Générale à la Recherche Scientifique et Technique. Vandoeuvre-les-Nancy Avril 1970. 13 S., 6 Tab.

Preliminary remarks on some geochronological analyses of irish granites and gneisses. In: Irish naturalists journal *16* (1970) S. 306–308.

u. B. Poty, J. Sonet, J.-L. Zimmermann: Age des cavités à cristaux du granite du Mont Blanc. In: C. R. Acad. Sc. *271* (1970) S. 156–158.

Y. Fuchs, F. Leutwein, J.-L. Zimmermann: Etude géochronologique et géochimique des roches volcaniques du Stéphanien et Autunien du Détroit de Rodez. In: C. R. Acad. Sc. *270* (1970) S. 2415–2417.

J. Blaise, P. Bordet, J. Lang, A. F. de Lapparent, F. Leutwein, J. Sonet: Mesures géochronologiques des quelques roches cristallines d'Afghanistan central. In: C. R. Acad. Sc. *270* (1970) S. 2772–2775.

u. F. Saupé, J. Sonet, E. Bouyx: Première mesure géochronologique en Sierra Morena. La granodiorite de Fontanosas (Province de Ciudad Real, Espagne). In: Geol. en Mijnbouw *49* (1970) S. 297–304.

M. Roubault, P. Bordet, F. Leutwein, J. Sonet; J.-L. Zimmermann: Ages absolus formations volcaniques du Tanneron-Esterel. In: C. R. Acad. Sc. *271* (1970) S. 1157–1160.

M. Roubault, P. Bordet, F. Leutwein, J. Sonet, J.-L. Zimmermann: Ages absolus des formations cristallophylliennes des Massifs des Maures et du Tanneron. In: C. R. Acad. Sc. *271* (1970) S. 1067–1070.

[Nachruf] Fritz Deubel 9. 9. 1898–16. 12. 1966. In: Jahrbuch Sächsische Akademie der Wissenschaften zu Leipzig *1966–1968* (Berlin 1970) S. 340–343.

1971

J. Boissonnas, F. Leutwein, J. Sonet: Age du granite hyperalcalin de la Gara Adjèmamaye (Ahaggar du Sud-Est, Sahara algérien). In: C. R. somm. des Séances de la Soc. Géol. de France *7* (1971) S. 251–252.

1972

u. J. Sonet, J.-L. Zimmermann: Dykes basiques du Massif Armoricain septentrional. Contribution à leur étude géochronologique. In: C. R. Acad. Sc. *275* (1972) S. 1327–1330.

u. J. Sonet, M. D. Max: The age of the Carnsore granodiorite. In: Geol. Surv. Ireland Bull. *1* (1972) S. 303–309.

Etude géochronologique de l'évolution du Permien continental en France. Colloque Orsay en Octobre 1970. In: Mém. BRGM, Fr. *77* (1972) S. 973–977.

J. Blaise, F. Leutwein, J. Sonet: Sur la datation á 140–150 MA de la granodiorite d'Alishah et sur ses conséquences pour la géologie des montagnes à l'Ouest de Kaboul (Afghanistant central). In: C. R. Acad. Sc. *274* (1972) S. 1457–1460.

C. Berthomier, A. Lacour, F. Leutwein, J. Maillot, J. Sonet: Sur quelques trhondhjemites de Norvège: étude géchronologique et géochimique. In: Sciences de la Terre *XVII* (1972) S. 341–351.

H. & G. Termier, F. Leutwein, J. Sonet: Sur l'âge varisque des plutonites du massif du Tichka (Maroc). In: Bull. de la Soc. franc. de Minér. et de Cristallogr. *95* (1972) S. 124 bis 127.

Tellurium. In: Handbook of Geochemistry *II/3* (1972) Nr. 52, S. 52B1–52 O1.

Selenium. In: Handbook of Geochemistry *II/3* (1972) Nr. 34.

1973

u. G. Power, R. Roach, J. Sonet: Quelques résultats géochronologiques obtenus sur des roches d'âge précambrien du Cotentin. In: C. R. Acad. Sc. *276* (1973) S. 2121–2124.

D. Billaud, A. Herold, F. Leutwein: Sur l'effet isotopique observé au cours de l'insertion du rubidium dans le graphite. In: C. R. Acad. Sc. *277* (1973) S. 419–420.

1974

u. J. Sonet: Geochronologische Untersuchungen im Südschwarzwald. In: N. Jb. Miner. Abh. *121* (1974) S. 252–271.

Geochemische und geochronologische Untersuchungen an sterilen und lagerstättenführenden Graniten insbesondere der Vendée (Frankreich). — In: Metallogenetische und Geochemische Provinzen, Symposiumsbericht Leoben, Schriftenreihe Erdwissenschaftliche Kommission der Österreichischen Akademie der Wissenschaften *1* (Wien 1974) S. 128 bis 139.

Zur Biobibliographie

Schriftenverzeichnisse. Jahrbuch Sächsische Akademie der Wissenschaften zu Leipzig *1954* bis *1956* (Berlin 1958) S. 186–187; *1957–1959* (Berlin 1961) S. 246; *1963–1965* (Berlin 1967) S. 213–214; *1966–1968* (Berlin 1970) S. 216; *1969–1970* (Berlin 1972) S. 161; *1971* bis

1972 (Berlin 1974) S. 173. — Kürschners Deutscher Gelehrten-Kalender *7* (Berlin 1950) Spalte 1208; *10* (Berlin 1966) S. 1415; *11* (Berlin 1970) S. 1744. — Poggendorffs Handwörterbuch *VII_3*[a] (Berlin 1959) S. 86—87. — A. G. Herrmann, in: Fortschritte der Mineralogie *53* (1975) H. 1, S. 5—10.

Würdigungen. Zum 55. Geburtstag: Oelsner, in: Bergakademie [Freiberg] *8* (1956) S. 535 bis 536 (mit Bildnis). — Zur Verleihung des Nationalpreises der DDR: F. Stock, in: Z. für angewandte Geologie *2* (1956) S. 535. — Allgemein: E. Rammler, in: Bergakademie [Freiberg] *4* (1952) S. 90. — ohne Verfasser, in: Geologie *6* (1957) S. 99 (m. Bildnis). — Wer ist wer? Herausgeber W. Habel, *14.* Ausgabe (Berlin 1962) S. 901.

Nachrufe. A. Watznauer, in: Jahrbuch Sächsische Akademie der Wissenschaften zu Leipzig *1973—1974* (Berlin 1976) S. 314—315. A. G. Herrmann, in: Fortschritte der Mineralogie *53* (1975) H. 1, S. 1—5.

G. W.

Otto Eißfeldt

1. 9. 1887—23. 4. 1973

Am 23. April 1973 verstarb nach längerem Leiden im 86. Lebensjahr das ordentliche Mitglied der philologisch-historischen Klasse, Dr. theol. et phil. Otto Eißfeldt, emeritierter ordentlicher Professor für Altes Testament und semitische Religionsgeschichte an der Martin-Luther-Universität Halle—Wittenberg, Ehrensenator dieser Universität, Ehrendoktor der Theologie der Humboldt-Universität Berlin (1922), der University of Glasgow (1951) und der Reformierten Theologischen Akademie Budapest (1955), Träger des Ordens „Banner der Arbeit" (1964) und des Vaterländischen Verdienstordens in Silber (1967).

Otto Hermann Leonhard Wilhelm Eißfeldt wurde am 1. 9. 1887 im damals hannoverschen Northeim als ältester Sohn des Juristen Otto Eißfeldt geboren. Nach dem Besuch des Progymnasiums in seiner Geburtsstadt und des Gymnasiums in Dortmund nahm er 1905 in Göttingen das Studium der Theologie und Orientalistik auf; 1906 ging er nach Berlin, kehrte jedoch bereits 1907 wieder nach Göttingen zurück. Als ihm Ende des gleichen Jahres die Stelle eines Konviktsinspektors am „Johanneum" in Berlin angeboten wurde, war damit zugleich der Grundstein für seine wissenschaftliche Laufbahn gelegt. 1911 wurde er in Berlin zum Licentiaten der Theologie (Dr. theol.) promoviert und 1913 erfolgte daselbst seine Habilitation auf Grund der Monographie „Der Maschal im Alten Testament". In seiner Privatdozentenzeit entfaltete er von 1913 bis 1922 eine reiche wissenschaftliche Tätigkeit; so promovierte er 1916 in Göttingen zum Dr. phil. mit einem Teil seiner größeren Abhandlung „Erstlinge und Zehnten im Alten Testament, ein Beitrag zur Geschichte des israelitisch-jüdischen Kultus", die 1917 vollständig erschien. 1918 wurde er zum Titularprofessor ernannt. Während seiner Universitätstätigkeit wirkte er zugleich als Frühprediger an der Jerusalems- und Neuen Kirche in Berlin. Aus diesem praktischen Dienst schied er mit dem Weggang von Berlin aus; doch als im Jahre 1936 nach dem Zusammenbruch der deutschchristlichen Herrschaft die kirchlichen Verhältnisse in der Provinz Sachsen neu geordnet werden mußten, stellte er seine Dienste der Magdeburger Kirchenbehörde zur Verfügung und wirkte für über zwei Jahrzehnte als nebenamtlicher Konsistorialrat.

Im Jahre 1922 ging Otto Eißfeldt als Nachfolger von Albrecht Alt, der nach kurzer Lehrtätigkeit einem Rufe nach Leipzig gefolgt war, als Ordinarius für Altes Testament und semitische Religionsgeschichte nach Halle. Hier entfaltete er eine Tätigkeit als Hochschullehrer und Forscher, die ihn sehr bald weit über die Grenzen seiner Universität hinaus bekannt machen sollte. Nach seinem ersten Dekanat im Jahre 1926/27 übernahm er in krisenreicher Zeit, als der Nationalsozialismus schon seine

Schatten vorauswarf, 1929/30 aus hoher ethischer Verantwortung das Rektorat, nachdem er im Zusammenhang hiermit einen Ruf nach Gießen als Nachfolger W. Baumgartners abgelehnt hatte. In seiner von 1930 bis 1932 währenden Amtsperiode als Prorektor hatte er die rechtsradikalen Studentenunruhen in Verbindung mit der Berufung des ehemaligen religiösen Sozialisten Günther Dehn auf den Lehrstuhl für Praktische Theologie durchzustehen. Er, der aus seiner antinationalsozialistischen Haltung kein Hehl machte, bekleidete von da an bis zum Ende des Zweiten Weltkrieges kein akademisches Amt mehr; auch lehnte er 1942 einen Ruf nach Göttingen ab. Dafür aber ist die Wiedereröffnung der Martin-Luther-Universität mit seinem Namen für immer verbunden; vom 26. 6. 1945 bis zum 31. 7. 1948 hatte er das Amt des Rektors inne, nachdem er nur kurze Zeit Dekan der Theologischen Fakultät gewesen war. In seiner Rektoratszeit mühte er sich mit aller Kraft um die Schaffung eines neuen akademischen Lebens in vertrauensvoller Zusammenarbeit mit der damaligen Sowjetischen Militäradministration und den antifaschistischen Kräften. Seine Universität hat es ihm, einem weltweit anerkannten Vertreter jenes anderen Deutschlands, das der Barbarei des Nationalsozialismus nicht erlegen war, mit der Ernennung zum Ehrensenator im Jahre 1956 gedankt.

Die wissenschaftliche Bedeutung Otto Eißfeldts hat nicht zuletzt darin ihren Grund, daß sein Werdegang alles andere als einseitig verlief. So war er zunächst Schüler von Julius Wellhausen (1844—1918) und damit der letzte große und international anerkannte Vertreter der literarkritischen Methode; er vertrat sie in der von seinem anderen Lehrer Rudolf Smend (1851—1913) modifizierten Form, wonach die älteste der vier Pentateuch- beziehungsweise Hexateuchquellen aus zwei Erzählungsfäden bestehen soll; für die ältere Schicht prägte er den Namen *Laienquelle*, während er für die jüngere die traditionelle Bezeichnung *Jahwist* beibehielt. Diese Grundkonzeption hat in der „Hexateuch-Synopse“ mit dem bezeichnenden Untertitel „Die Erzählung der fünf Bücher Mose und des Buches Josua mit dem Anfange des Richterbuches in ihre vier Quellen zerlegt ...“ (Leipzig 1922; Nachdruck Darmstadt 1962) ihren bleibenden Niederschlag gefunden; ihr ist Otto Eißfeldt zeit seines Lebens unbeschadet unterschiedlicher Gegenströmungen treu geblieben, und in seiner „Einleitung in das Alte Testament“ (Tübingen 1934; ²1956; ³1964) nimmt sie einen entscheidenden Platz ein.

Aber Otto Eißfeldt war nicht nur ein Vertreter der literarkritischen oder analytischen Methode, sondern er ging auch bei Hermann Gunkel (1862—1932) in die Schule. Er eignete sich die form- und gattungsgeschichtliche Methode als kritisches Gegenüber und als notwendige Ergänzung zur Literarkritik an und verstand es, beides zu einem methodisch ausgewogenen Ganzen zu vereinen. So beginnt die bereits erwähnte „Einleitung“ mit einem ersten Teil: „Die vorliterarische Stufe: Die kleinsten Redeformen und ihr Sitz im Leben“, der die Voraussetzung darstellt für die folgenden Teile: „Die literarische Vorgeschichte der Bücher des Alten Testaments“ und — als Hauptteil — „Die Analyse der Bücher des Alten Testaments“.

Den dritten entscheidenden Impuls empfing Otto Eißfeldt durch Wolf Wilhelm Graf Baudissin (1847—1926), der ihm den Weg zur Religionsgeschichte wies mit dem speziellen Ziel, die Religion des Alten Testaments mit den Glaubensvorstellungen der Umwelt zu vergleichen, um umgekehrt von hier aus wieder die Besonderheit israelitischen Glaubenslebens zu verdeutlichen.

So stellt sich Otto Eißfeldt als ein Gelehrter dar, der nicht auf ein spezielles Fachgebiet eingeengt gewesen ist, sondern als ein Wissenschaftler, der die ganze Weite der Disziplin Altes Testament samt ihren Hilfswissenschaften übersah und beherrschte. War er dabei — wie es nicht anders sein konnte — mit seinen literaranalytischen und seinen form- und gattungsgeschichtlichen Arbeiten auf Grund der bereits vorliegenden Ergebnisse in bestimmte Bahnen gewiesen, auf denen es galt, möglichst zu allgemein gültigen oder doch zumindest weithin anerkannten Resultaten zu gelangen, so eröffnete sich ihm seit Ende der zwanziger Jahre auf dem Gebiet der westsemitischen Religionsgeschichte bis dahin ungeahntes Neuland. Es ist gekennzeichnet zunächst durch die epochemachende Entdeckung der keilschriftalphabetischen Texte von Ugarit (*rās eš-šamra*) in Nordsyrien, die 1928 durch einen Zufallsfund ausgelöst wurde, sodann durch die Handschriftenfunde aus der Wüste Juda, die seit 1947 bekannt geworden sind und ebenfalls durch einen Zufallsfund eingeleitet wurden. Werden wir durch erstere in die Zeit kurz nach der Mitte des 2. Jahrtausends v. Chr. geführt, so gehören die letzteren in ihrer Hauptmasse den beiden Jahrzehnten um die Zeitenwende herum an. Es ist kennzeichnend für die wissenschaftliche Spannweite Otto Eißfeldts, daß er das jeweils anfallende Material — sei es aus dem vorisraelitischen Kanaan, oder sei es aus dem palästinischen Judentum in römischer Zeit — sofort schöpferisch zu verarbeiten begann. Als Symbol und zugleich als Zeichen internationaler Anerkennung einer solchen hervorragenden Leistung ist die Festschrift anzusehen, die ihm 1957 anläßlich seines 70. Geburtstages mit dem Titel „Von Ugarit nach Qumran" (Berlin 1958; ²1961 [BZAW 77]) überreicht wurde. Dabei durfte es für ihn eine große Genugtuung sein, daß er viele seiner Thesen, die er hinsichtlich des Verhältnisses zwischen Altisrael und seiner syrischen Umwelt vor Bekanntwerden der neuen Texte vertreten hatte, weithin bestätigt fand.

Religionsgeschichtliche Forschung war für ihn in letzter Konsequenz ebensowenig Selbstzweck wie für seinen Lehrer auf diesem Gebiet, Wolf W. Graf Baudissin; sie stand immer zugleich in Bezug zur alttestamentlichen Theologie. Das markanteste Beispiel hierfür stellen seine Untersuchungen zum Begriffe *El* dar. In einem Akademievortrag „El im ugaritischen Pantheon" vom 21. 3. 1949 (BAL 98,4) geht er in einer sorgfältigen philologischen Analyse den bis dahin bekannten Belegen nach. Dabei kommt er zu dem Ergebnis, daß El — **'ilu* ist mit Ausnahme des Äthiopischen gemeinsemitisch und sowohl Appellativum »Gott« als auch Eigenname — in spezifischer Bedeutung der Eigenname des obersten Gottes im Pantheon ist, daß nur ihm der Titel »König« zukommt und daß, unbeschadet der in der Diskussion oft herausgestellten Spannungen zwischen ihm und den anderen großen Göttern Baʿal, Mot und Yamm, in der ugaritischen Mythologie ein ausgesprochen monarchisches Prinzip herrschte. Diesem Problem ist Otto Eißfeldt 1956 in dem Aufsatz „El and Yahwe" (JSS 1,25—37; deutsch: Kleine Schriften III, 386—397) weiter nachgegangen mit dem Ergebnis, daß Jahwe bei der israelitischen Landnahme mit dem durch El verkörperten Schöpfungs- und Erhaltungsprinzip konfrontiert wurde. Hierdurch wurde eine Entwicklung eingeleitet, in der sich jene für das Alte Testament charakteristische Gottesvorstellung herausgebildet hat, in der sich in polarer und fruchtbarer Spannung der Glaube an Jahwe als den drängenden Gott der Geschichte und der Verheißung auf der einen und der Glaube an ihn als den Schöpfer und Erhalter des Kosmos auf der anderen Seite gegenüberstehen. Hier — neben anderen Forschern — den positiven Beitrag Kanaans zur Religionsgeschichte Altisraels herausgestellt zu haben, ist eines seiner besonderen Verdienste.

Auch im Bereiche der Grundlagenforschung, also der kritischen Arbeit an den Texten, hat Otto Eißfeldt Entscheidendes geleistet. Als Rudolf Kittel, der Begründer der BIBLIA HEBRAICA (Stuttgart 1906; ²1909) im Jahre 1929 starb, wurde ihm zusammen mit Albrecht Alt die Herausgabe dieses Standardwerkes in der dritten Auflage, die nunmehr auf dem von Paul Kahle besorgten Codex Leningradensis B 19 A fußte, übertragen. Im Jahre 1937 war diese Arbeit abgeschlossen. Als 1947 in den Rollen von Qumran Varianten zu Jesaja und Habakuk zum Vorschein kamen, die den masoretischen Text wesentlich berührten, war es Otto Eißfeldt, der sie alsbald in einem Ergänzungsheft zur BIBLIA HEBRAICA erfaßte (1951) und sie in einem Zusatzapparat in die anschließenden Nachdrucke einfügte. Mittlerweile war es notwendig geworden, die BIBLIA HEBRAICA philologisch neu zu fundieren, da ohnehin die technischen Voraussetzungen für weitere Nachdrucke erschöpft waren und vor allem die Editionsmethode, die noch aus dem Beginn unseres Jahrhunderts stammte, modernen wissenschaftlichen Ansprüchen nicht mehr genügte. Auch jetzt gehörte Otto Eißfeldt zu den Mitarbeitern an der neuen Ausgabe mit dem Titel BIBLIA HEBRAICA STUTTGARTENSIA; er edierte noch im hohen Alter die Genesis und erlebte deren Erscheinen im Jahre 1969.

Ebenso hat er sich an der Erschließung und Ordnung der Alphabettexte von Ugarit beteiligt; so hat er ein eigenes Benennungssystem entwickelt (Kleine Schriften II, 330—415 und: „Neue keilalphabetische Texte aus Ras Schamra-Ugarit [SAB 1965, Heft 6]), das sich zwar nicht durchgesetzt hat, aber seinen historischen Platz in: M. Dietrich-O. Loretz, Konkordanz der ugaritischen Texterzählungen“ (Neukirchen-Vluyn 1972 [AOAT 19]) gefunden hat. Sein besonderes Verdienst um die ugaritischen Texte hat er sich dadurch erworben, daß er Joseph Aistleitners „Untersuchungen zur Grammatik des Ugaritischen“ 1954 in den Sitzungsberichten unserer Akademie hat erscheinen lassen (BAL 100,6), ebenso wie das „Wörterbuch der ugaritischen Sprache“, das nach dem Tode des Verfassers am 9. 9. 1960 im Jahre 1963 erschien und das 1965 eine zweite verbesserte Auflage unter seiner Herausgeberschaft erlebte (BAL 106,3 — ³1967, ⁴1974). Damit verdankt die Wissenschaft Otto Eißfeldt ein lexikographisches Hilfsmittel, das dem Glossar von: Cyrus H. Gordon, Ugaritic Textbook (Rom 1965 [Analecta Orientalia 38]) als wertvolle Ergänzung an die Seite zu stellen ist.

Damit ist zugleich die wissenschaftsorganisatorische Seite von O. Eißfeldt berührt. Bereits zu Ende der zwanziger Jahre war er als kritischer Herausgeber des von seinem Lehrer Wolf W. Graf Baudissin hinterlassenen vierbändigen Monumentalwerkes „Kyrios als Gottesname im Judentum und seine Stelle in der Religionsgeschichte“ (Gießen 1929) hervorgetreten. 1932 erschien das erste Heft seiner „Beiträge zur Religionsgeschichte des Altertums“ (Halle) mit dem Titel „Baal Zaphon,

Zeus Kasios und der Durchzug der Israeliten durchs Meer"; 1934 übernahm er, zusammen mit dem Hallenser Systematiker G. Heinzelmann, die Herausgabe der „Theologischen Studien und Kritiken", die er bis zu deren Einstellung im Jahre 1942 fortführte; von 1936 bis 1945 gab er für die Vorderasiatisch-Ägyptische Gesellschaft die Reihe „Der Alte Orient. Gemeinverständliche Darstellungen", Band 34 bis 43 heraus. Seine bedeutendste editorische Leistung ist jedoch das „Handbuch zum Alten Testament" (Tübingen), das er ins Leben rief und dessen erste Bände 1936 erschienen; längst nimmt dieses Standardwerk im Rahmen der exegetischen Sammelwerke einen hervorragenden Platz ein. Daneben bedarf seine Tätigkeit in der Redaktion der „Orientalistischen Literaturzeitung" besonderer Hervorhebung. Im Jahre 1922 trat er in den Kreis der Mitarbeiter ein, und 1953 übernahm er als Mitherausgeber das Referat „Altes Testament, Neues Testament, Spätjudentum, Urchristentum", das er bis zum Jahre 1972 leitete und im Rahmen dessen er eine Leistung vollbrachte, die schon fast die Grenzen des Menschenmöglichen erreichte (vgl. die Würdigungen in: OLZ 67 [1972], 437f.; 68 [1973], 325f.).

Das Bild von der wissenschaftlichen Persönlichkeit Otto Eißfeldts wird dadurch abgerundet, daß er ein ausgesprochenes Traditionsbewußtsein im guten Sinne des Wortes besaß. Dies kommt einmal in seinen wertvollen Nachrufen etwa auf Julius Wellhausen (Kleine Schriften I, 56—71) oder Wolf W. Graf Baudissin (ebenda, 115—142) zum Ausdruck, zum anderen darin, daß er in seiner „Einleitung" die behandelten Themen und Probleme von ihren Anfängen bis zum jeweils gegenwärtigen Stand der Forschung heranführt und somit disziplingeschichtlich unschätzbare Dienste geleistet hat, indem er für die Jüngeren gleichsam zum Brückenbauer in eine große Vergangenheit geworden ist. In diesen Zusammenhang gehört auch, daß er selbst fernere Zeiten, die auch ihm entrückt waren, gern lebendig werden ließ. So hat er 1966 eine hübsche dokumentarische Abhandlung mit dem bezeichnenden Titel „Franz Delitzsch und Wolf Graf Baudissin. Dokumente über die Anfänge ihrer Freundschaft aus dem Erlanger Wintersemester 1866/67" erscheinen lassen (SAL 112,2), die uns ein beeindruckendes Augenblicksbild aus dem ersten Semester seines nachmaligen Lehrers vermittelt. Der gesamte Briefwechsel zwischen beiden wurde von ihm und K. H. Rengstorf unter dem Titel „Briefwechsel zwischen Franz Delitzsch und Wolf Wilhelm Graf Baudissin 1866—1890" in den „Abhandlungen der Rheinisch-Westfälischen Akademie der Wissenschaften", Band 43 (Köln-Opladen 1973) herausgegeben; das Erscheinen hat er jedoch nicht mehr erlebt.

Nach dem Gesagten nimmt es nicht wunder, wenn die Leistungen Otto Eißfeldts durch zahlreiche Ehrungen anerkannt wurden. Von 1927 gehörte er der Deutschen Akademie für allgemeinnützliche Wissenschaften in Erfurt bis zu deren Auflösung im Jahre 1945 an. Unter dem 1. Juli 1948 wurde er in die philologisch-historische Klasse unserer Akademie als ordentliches Mitglied gewählt; 1949 kam er als ordentliches Mitglied in die Deutsche Akademie der Wissenschaften (jetzt Akademie der Wissenschaften der DDR), nachdem er ihr vorher kurze Zeit als korrespondierendes Mitglied angehört hatte. Seit 1955 war er korrespondierendes Mitglied der Akademie der Wissenschaften und der Literatur in Mainz, seit 1957 Associé étranger d' Institut de France: Académie des Inscriptions et Belles-Lettres, Paris, seit 1958 korrespondierendes Mitglied sowie Träger der Burkitt-Medal for Biblical Studies 1965 der British Academy, London; in gleicher Eigenschaft gehörte er seit 1969 der Kungliga Humanistika Vetenskapssamfundet (Regia societas humaniorum litterarum) Lund an. Außerdem war er ordentliches Mitglied des Deutschen Archäologischen Instituts in Berlin-Dahlem seit 1963, Ehrenmitglied der Society of Old Testament Study (Großbritannien) seit 1933, der Society of Biblical Literature and Exegesis (USA) seit 1937 und der Deutschen Morgenländischen Gesellschaft (Mainz) seit 1918 (Vorstandsmitglied seit 1948, Ehrenmitglied seit 1957).

Als Otto Eißfeldt im 61. Lebensjahr in die Sächsische Akademie gewählt wurde, brachte er ein umfangreiches wissenschaftliches Werk mit ein. In den folgenden fast 25 Jahren hat er dieses Werk zu einer Vollendung führen können, wie sie nur wenigen vergönnt ist, und in seinen zahlreichen Vorträgen und Diskussionsbeiträgen hat er die Mitglieder daran teilnehmen lassen. So gedenkt denn die Akademie eines der Großen in ihren Reihen, der mit seiner wissenschaftlichen Persönlichkeit, seinem hohen ethischen Einsatz sowie seiner ausgleichenden Menschlichkeit unter allen, die ihn kannten und mit ihm zusammenarbeiten durften, unvergessen bleiben wird.

Für wertvolle biographische Hinweise danke ich Prof. Dr. Dr. G. Wallis, Martin-Luther-Universität Halle/S.

Rudolf Meyer

Bibliographie Otto Eißfeldt

zusammengestellt von Karl-Martin Beyse

1910

Die Rätsel in Jdc 14, in: ZAW 30, 1910, S. 132–135.

1911

Engert, Th.: Das Alte Testament im Lichte modernistisch-katholischer Wissenschaft, 1910, in: KZ 21, 1910/11, Sp. 87.

Meinhold, H.: Die Weisheit Israels in Spruch, Sage und Dichtung, 1908, in: KZ 21, 1910/11, Sp. 114.

Strack, H. L.: Jesus, die Häretiker und die Christen nach den ältesten jüdischen Angaben, 1910, in: KZ 21, 1910/11, Sp. 16.

1912

Schneider, Th.: Vom Ursprung des Christentums. Drei Vorträge, 1910, in: KZ 22, 1911/12, Sp. 151.

1913

Der Maschal im Alten Testament. Gießen 1913, 71 S. (BZAW 24 = Diss. theol. Berlin 1913)

Mose und seine Zeit (zu H. Gressmann: Mose und seine Zeit, 1913), in: Prbl 46, 1913, S. 557–559. 583–585. 614–616. 647–649. 685–687.

Die Verwertbarkeit der Vätergeschichten in der Genesis für die Rekonstruktion der vormosaischen hebräischen Profan- und Religionsgeschichte, in: PrM 17, 1913, S. 329–345.

BEER, G.: Mose und sein Werk. Ein Vortrag, 1912, in: KZ 23, 1912/13, Sp. 180–181.

KITTEL, R.: Geschichte des Volkes Israel. I. Band, ²1912, in: Prbl 46, 1913, Sp. 888–889.

POTT, A.: Der griechisch-syrische Text des Matthäus ε 351 im Verhältnis zu Tatian s^sc Ferrar, 1912, in: ThLZ 38, 1913, Sp. 233–234.

SMEND, R.: Die Erzählung des Hexateuch auf ihre Quellen untersucht, 1912, in: KZ 23, 1912/13, Sp. 180.

1914

Israels Geschichte, Tübingen 1914, 52 S. (Religionsgeschichtliche Volksbücher, Reihe 6' Heft 4).

Julius Wellhausen: Zu seinem 70. Geburtstag, in: KZ 24, 1913/14, Sp. 185–187 und Prbl 47, 1914, Sp. 522–524.

Eine künstliche Religionsbildung vor 3000 Jahren, in: Prbl 47, 1914, Sp. 300–303. 373 bis 375. 402–404.

Jahve und Baal, in: PrJb 155, 1914, S. 257–270 = Kl. Schr. I, S. 1–12.

GUNKEL, H.: Reden und Aufsätze, 1913, in: Prbl 47, 1914, S. 432.

1915

Krieg und Bibel. Tübingen 1915, 84 S. (Religionsgeschichtliche Volksbücher, Reihe 5, Heft 15/16).

Krieg und Neues Testament, in: Prbl 48, 1915, Sp. 214–218. 227–230.

Vom alten Gesenius, in: Prbl 48, 1915, Sp. 779–782.

Die Chetiter (zu E. MEYER: Reich und Kultur der Chetiter, 1914), in: Prbl 48, 1915, Sp. 994–999.

DELITZSCH, F.: Die Welt des Islam, 1915, in: Prbl 48, 1915, Sp. 802.

KITTEL, R.: Die Psalmen, übersetzt und erklärt, ¹⁻²1914, in: Prbl 48, 1915, Sp. 397.

STEINDORFF, G.: Ägypten, 1915, in: Prbl 48, 1915, Sp. 802.

1916

Erstlinge und Zehnten im Alten Testament, ein Beitrag zur Geschichte des israelitisch-jüdischen Kultus, 1. Teil. Göttingen 1916, 32 S. (Diss. phil. Göttingen).

Altorientalische Symbolik und antike Symbolik überhaupt (zu H. PRINZ: Altorientalische Symbolik, 1915), in: ZMR 31, 1916, S. 117–124.

HELL, J.: Die Religion des Islam. I. Von Mohammed bis Ghazali, 1915, in: Prbl 49, 1916, Sp. 302.

1917

Die Einführung der Reformation und der Union in der Jerusalemskirche in Berlin. Berlin 1917, 34 S.

Erstlinge und Zehnten im Alten Testament. Ein Beitrag zur Geschichte des israelitisch-jüdischen Kultus. Leipzig 1917, VIII, 172 S. (BWAT 22; Teildruck als Diss. phil. Göttingen 1916).

DUHM, B.: Israels Propheten, 1916, in: DLZ 38, 1917, Sp. 1032–1034 und Prbl 50, 1917 Sp. 385–387. 401. 403. 424.

Jeremias, A.: Das Alte Testament im Lichte des Alten Orients, [3]1916, in: PrJb 169, 1917, S. 270–275.

Prinz, H.: Altorientalische Symbolik, 1915, in: PrM 21, 1917, S. 28–29.

Schmidt, Hans: Der Prophet Amos, 1917, in: DLZ 38, 1917, Sp. 1308–1309.

1918

Zum Zehnten bei den Babyloniern, in: BZAW 33, 1918, S. 163–174 (Abhandlungen zur Semitischen Religionskunde und Sprachwissenschaft Wolf Wilhelm Grafen von Baudissin zum 26. September 1917) = Kl. Schr. I, S. 13–22.

Verzeichnis der Schriften Wolf Wilhelm Grafen von Baudissin. Nachtrag zum Verzeichnis der Schriften von Wolf Wilhelm Grafen von Baudissin, in: BZAW 33, 1918, S. 1–16, 419–420.

Die Bedeutung der Märchenforschung für die Religionswissenschaft, besonders für die Wissenschaft vom Alten Testament (zu H. Gunkel: Das Märchen im Alten Testament, 1917), in: ZMR 33, 1918, S. 65–71. 81–85 = Kl. Schr. I, S. 23–32.

Zum Tode Julius Wellhausens, in: Tägliche Rundschau, 1918, Nr. 8 vom 10. Januar.

Gunkel, H.: Die Propheten, 1917, in: DLZ 39, 1918, Sp. 230–231.

Kellermann, B.: Der ethische Monotheismus der Propheten und seine soziologische Würdigung, 1917, in: DLZ 39, 1918, Sp. 386–390.

König, E.: Das Deuteronomium, 1917 (KAT 3), in: KZ 28, 1917/18, S. 43.

Schmidt, Hans: Psalmen, 1917, in: DLZ 39, 1918, Sp. 845–846.

Stuhlfauth, G.: Die „ältesten Porträts“ Christi und der Apostel, 1918, in: KZ 28, 1917/18, S. 81.

Theis, J.: Die Weissagung des Abdias, 1917, in: DLZ 39, 1918, Sp. 992.

Lohmann, P.: Die selbständigen lyrischen Abschnitte in Jes. 24–27. Hrsg. von O. Eissfeldt, in: ZAW 37, 1918, S. 1–58.

1919

Luther und die deutsche Kultur, in: KZ 29, 1918/19, S. 2–6.

Das Alte und das Neue Testament in ihrer Stellung zur Kultur, in: PrJb 178, 1919, S. 373 bis 384 = Kl. Schr. I, S. 44–55.

Die Schichten des Hexateuch als vornehmste Quelle für den Aufriß einer israelitisch-jüdischen Kulturgeschichte, in: PrM 23, 1919, S. 173–185 = Kl. Schr. I, S. 33–43.

Zum gegenwärtigen Stand der Pentateuch-Kritik, in: ZMR 34, 1919, S. 113–123.

Gressmann, H.: Vom reichen Mann und armen Lazarus. Eine literargeschichtliche Studie, 1918 (AAB 1917, Nr. 17), in: DLZ 40, 1919, Sp. 311–313.

König, E.: Die Genesis, 1919, in: KZ 29, 1918/19, S. 60.

Volz, P.: Der Prophet Jeremia, 1918, in: DLZ 40, 1919, Sp. 850–851.

1920

Julius Wellhausen (geb. am 17. Mai 1844 in Hameln, gest. 7. Januar 1918 in Göttingen), in: Internationale Monatsschrift für Wissenschaft, Kunst und Technik 14, 1920, Sp. 193–208. 325–338 = Kl. Schr. I, S. 56–71.

Beer, G.: Die soziale und religiöse Stellung der Frau im israelitischen Altertum, 1919 (SGV 88), in: KZ 30, 1919/20, S. 27.

BUDDE, K.: Das Lied Moses Deut. 32 erläutert und übersetzt, 1920, in: KZ 30, 1919/20, S. 110.

HEILER, F.: Das Gebet, ²1920, in: KZ 30, 1919/20, S. 109.

JEREMIAS, A.: Allgemeine Religionsgeschichte, 1918, in: ZMR 35, 1920, S. 31–32.

KIESLING, H. v.: Damaskus. Altes und Neues aus Syrien, 1920, in: KZ 30, 1919/20, S. 43.

KÖNIG, E.: Hermeneutik des Alten Testaments, 1916, in: KZ 30, 1919/20, S. 27.

KRETZSCHMER, P. (Hrsg.): Neugriechische Märchen, 1917, in: KZ 30, 1919/20, S. 43.

SACHSSE, E.: Die Propheten des Alten Testaments und ihre Gegner, 1920, in: KZ 30, 1919/20, S. 53.

UNGNAD, A.: Briefe König Hammurapis, 1919, in: KZ 30, 1919/20, S. 70.

1921

Christentum und Altes Testament. Eine Bemerkung zu Harnacks Marcion, in: KZ 31, 1920/21, S. 82–83 = Adolf von Harnack zum 70. Geburtstag, 1921, S. 29–33 = Kl. Schr. I, S. 72–75.

Zu Friedrich Delitzsch's „Die große Täuschung“, in: KZ 31, 1920/21, S. 5–6.

DELITZSCH, F.: Die Lese- und Schreibfehler im Alten Testament nebst den dem Schrifttexte einverleibten Randnoten klassifiziert, 1920, in: KZ 31, 1920/21, S. 32–33.

DERS.: Die große Täuschung II, 1921, in: KZ 31, 1920/21, S. 96–97.

DERS.: Babel und Bibel. Vortrag, gehalten am 13. Januar 1902. Neu bearbeitete Ausgabe, 1921, in: KZ 31, 1920/21, S. 164.

DÖLLER, J.: Die Reinheits- und Speisegesetze des Alten Testaments in religionsgeschichtlicher Beleuchtung, 1917 (ATA 7, 2/3), in: DLZ 42, 1921, Sp. 511–513.

KITTEL, R.: Die Religion des Volkes Israel, 1921, in: KZ 31, 1920/21, S. 129.

KÖNIG, E.: Friedrich Delitzsch's „Die große Täuschung“ kritisch beleuchtet, 1920, in: KZ 31, 1920/21, S. 8–9.

MEINHOLD, J.: Einführung in das Alte Testament. Geschichte, Literatur und Religion Israels, 1919, in: DLZ 42, 1921, Sp. 86–87.

VOLZ, P.: Hiob und Weisheit, ²1921 (SAT 3, 2), in: KZ 31, 1920/21, S. 115–116.

1922

Hexateuch-Synopse. Die Erzählung der fünf Bücher Mose und des Buches Josua mit dem Anfange des Richterbuches in ihre vier Quellen zerlegt und in deutscher Übersetzung dargeboten samt einer in Einleitung und Anmerkungen gegebenen Begründung. Leipzig 1922, XV, 108, 285* S. [Fotomechanischer Nachdruck: Darmstadt 1962].

Das erste Buch der Könige. Das zweite Buch der Könige. Das Buch Jesaja Kap. 36–39 ins Deutsche übersetzt, in: HSAT I, 1922, S. 492–585. 649–653.

Luthers „Ein feste Burg“ und der 46. Psalm, in: ThBl 1, 1922, Sp. 54–59 = Kl. Schr. I, S. 67–83.

BENZINGER, I.: Jahvist und Elohist in den Königsbüchern, 1921 (BWAT 27), in: ThBl 1 1922, Sp. 232–233.

DELITZSCH, F.: Die Lese- und Schreibfehler im Alten Testament nebst den dem Schrifttext einverleibten Randnoten klassifiziert, 1920, in: DLZ 43, 1922, Sp. 247–248.

HEINISCH, P.: Personifikationen und Hypostasen im Alten Testament und im Alten Orient, 1921, in: ThLZ 47, 1922, Sp. 31.

KITTEL, R.: Die Zukunft der alttestamentlichen Wissenschaft (Vortrag), 1921, in: OLZ 25, 1922, Sp. 411–414.

SZERUDA, J.: Das Wort Jahwes. Eine Untersuchung zur israelitisch-jüdischen Religionsgeschichte, 1921 (Diss. theol. Warschau), in: ThLZ 47, 1922, Sp. 162–163.

1923

Stammessage und Novelle in den Geschichten von Jakob und seinen Söhnen, in: Eucharisterion. Studien zur Religion und Literatur des Alten und Neuen Testaments. H. Gunkel zum 60. Geburtstage, 1923, S. 56–77 (FRLANT 36, I) = Kl. Schr. I, S. 84–104.

Ein neuer Jeremia-Kommentar (zu P. VOLZ: Der Prophet Jeremia, 1922, [KAT 10]), in: ThBl 2, 1923, Sp. 93–95.

Babylonische und israelitische Religion, in: ZMR 38, 1923, S. 289–303.

GRESSMANN, H.: Die Anfänge Israels (Von 2. Mose bis Richter und Ruth), 1922 (SAT 1, 2), in ThLZ 48, 1923, Sp. 154–157.

HÄNEL, J.: Das Erkennen Gottes bei den Schriftpropheten, 1923 (BWAT 29), in: ThBl 2, 1923, Sp. 148–149.

HÖLSCHER, G.: Geschichte der israelitischen und jüdischen Religion, 1922, in: ThBl 2, 1923, Sp. 122–123.

OESTREICHER, TH.: Das Deuteronomische Grundgesetz, 1923 (BFChTh 27, 4), in: ThBl 2, 1923, Sp. 172–173.

SACHSSE, E.: Die Bedeutung des Namens Israel, 1922, in: ThBl 2, 1923, Sp. 69.

SCHAEFFER, H.: Hebrew Tribal Economy and the Jubilee as illustrated in Semitic and Indo-European Village Communities, 1922, in: ThLZ 48, 1923, S. 196–197.

SCHLÖGL, N.: Die heiligen Schriften des Alten Bundes ... übersetzt und kurz erläutert. I.: Das Buch Mosche, Jehoschua u. Schophetim, 1922, in: ThBl 2, 1923, Sp. 148.

Die Heilige Schrift des Alten Testaments in Verbindung mit ... übersetzt von E. KAUTZSCH. Vierte, umgearbeitete Auflage in Verbindung mit den früheren Mitarbeitern und Professor EISSFELDT hrsg. von A. BERTHOLET. Tübingen I, 1922; II, 1923.

1924

BEWER, J. A.: Der Text des Buches Esra. Beiträge zu seiner Wiederherstellung, 1922, in: DLZ 45, 1924, Sp. 573.

BUDDE, K.: Der Segen Moses Deut. 33. Erläutert und übersetzt, 1922, in: DLZ 45, 1924, Sp. 334–335.

HORST, F.: Die Anfänge des Propheten Jeremia (ZAW 41, 1923, S. 94–153). – Die Kultusreform des Königs Josia (II. Reg. 22–23) (ZDMG N. F. 2, 1923, S. 220–238), in: ThLZ 49, 1924, Sp. 223–225.

JAHNOW, HEDWIG: Das hebräische Leichenlied im Rahmen der Völkerdichtung, 1923 (BZAW 36), in: ThLZ 49, 1924, Sp. 81–82.

LÖHR, M.: Untersuchungen zum Hexateuchproblem. I.: Der Priesterkodex in der Genesis, 1924 (BZAW 38), in: ThLZ 49, 1924, Sp. 148–151.

OESTERLY, W. O. E.: The Sacred Dance, 1923, in: ThLZ 49, 1924, Sp. 104–105.

PÉRENNES, H.: Les Psaumes. Traduits et commentés, 1922, in: ZS 2, 1924, S. 282–284.

1925

Die Quellen des Richterbuches. In synoptischer Anordnung ins Deutsche übersetzt samt einer in Einleitung und Noten gegebenen Begründung. Leipzig 1925, X, 116, 66* S.

Bruppacher, H.: Die Beurteilung der Armut im Alten Testament, 1924, in: ThLZ 50, 1925, Sp. 320.

Causse, A.: Israël et la Vision de l'Humanité, 1924, in: ThLZ 50, 1925, Sp. 269–270.

Herrmann, J.: Ezechiel, 1924 (KAT 11), in: DLZ 46, 1925, Sp. 101–104.

Hertzberg, H. W.: Prophet und Gott. Eine Studie zur Religiosität des vorexilischen Prophetentums, 1923 (BFChTh 28, 3), in: OLZ 28, 1925, Sp. 671–672.

Nikel, J.: Grundriß der Einleitung in das Alte Testament, 1924, in: ThLZ 50, 1925, Sp. 149–150.

Procksch, O.: Die Genesis, $^{2-3}$1924 (KAT 1), in: OLZ 28, 1925, Sp. 665–670.

Ders.: König und Prophet in Israel, 1924 (Greifswalder Universitätsreden 10), in: ThLZ 50, 1925, Sp. 128–129.

Schulz, A.: Das Buch Josue, 1924 (HS 2, 3), in: OLZ 28, 1925, Sp. 880–882.

Wiener, H. M.: Das Hauptproblem des Deuteronomiums, 1924, in: ThLZ 50, 1925, Sp. 127–128.

1926

Vom Lebenswerk eines Religionshistorikers (Wolf Wilhelm Graf von Baudissin, † am 6. Februar 1926). Leipzig 1926, 42 S. (Sonderabdruck).

Graf Baudissin's Kyrios-Werk, in: OLZ 29, 1926, Sp. 783–788.

Israelitisch-jüdische Religionsgeschichte und alttestamentliche Theologie, in: ZAW 44, 1926, S. 1–12 = Kl. Schr. I, S. 105–114.

Vom Lebenswerk eines Religionshistorikers (Wolf Wilhelm Graf Baudissin, † am 6. Februar 1926). 2. Nachtrag zum Verzeichnis der Schriften von Wolf Wilhelm Grafen von Baudissin; in: ZDMG 80, 1926, S. 89–130 = Kl. Schr. I, S. 115–142 [ohne Nachtrag].

Bauer, Th.: Die Ostkanaanäer, 1926, in: ZMR 41, 1926, S. 312.

Comba, E.: I libri dei Profeti d'Israele, 1924; I libri poetici dell'Antico Testamento, 1924, in: ThLZ 51, 1926, Sp. 8–10.

Jirku, A.: Das Alte Testament im Rahmen der altorientalischen Kulturen, 1926, in: ThLZ 51, 1926, Sp. 556.

Gressmann, H.: Altorientalische Texte zum Alten Testament, ^{2}1926, in: ZMR 41, 1926, S. 311–312.

Karge, P.: Rephaim, die vorgeschichtliche Kultur Palästinas und Phöniziens, ^{2}1925, in: ZMR 41, 1926, S. 310–311.

Landauer, G. (Hrsg.): Palästina. Mit ausführlich beschreibendem Text, 1925, in: ZMR 41, 1926, S. 61–62.

Lindblom, J.: Die literarische Gattung der prophetischen Literatur, 1924, in: OLZ 29, 1926, Sp. 475–477.

Löhr, M.: Untersuchungen zum Hexateuchproblem. II.: Das Deuteronomium, 1925 (SKG 1, 6), in: ThLZ 51, 1926, Sp. 391–393.

Loewe, H.: Catalogue of the manuscripts in the Hebrew character, collected and bequeathed to Trinity College Library by the late William Aldis Wright, 1926, in: DLZ 47, 1926, Sp. 2560–2561.

Lofthouse, W. F.: Jeremiah and the new covenant, 1925, in: ThLZ 51, 1926, Sp. 179 bis 181.

LONGO, T.: I libri storici dell'Antico Testamento, 1924, in: ThLZ 51, 1926, Sp. 8–10.

NESTLE, E. † (Hrsg.): Das Buch Jeremia. Griechisch und hebräisch, 1924, in: DLZ 47, 1926, Sp. 601–603.

POPPER, W.: Parallelism in Isaiah. Chapters 1–35 and 37, 22–35. The Reconstructed Text (Hebrew), 1923, in: DLZ 47, 1926, Sp. 937–938.

SELLIN, E.: Wolf Wilhelm Graf v. Baudissin (Gedächtnisrede), 1926, in: DLZ 47, 1926, Sp. 1828–1830.

WUTZ, F.: Die Psalmen textkritisch untersucht, 1925, in: OLZ 29, 1926, Sp. 992–998.

1927

Überblick über die israelitisch-jüdische Geschichte. Zur Landeskunde Palästinas, in: A. FALCKE und TH. FÖRSTER. Religionsbuch für evangelische Schulen. Neubearbeitung. Halle 1927, S. 92–99.

The Smallest Literary Unit in the Narrative Books of the Old Testament, in: Old Testament Essays, 1927, S. 85–93.

Die alttestamentliche Wissenschaft in den letzten 5 Jahren, in: Evangelischer Religions-Unterricht 38, 1927, S. 4–11. 73–81.

Text-, Stil- und Literarkritik in den Samuelisbüchern (zu W. CASPARI: Die Samuelisbücher, 1926 [KAT 7]), in: OLZ 30, 1927, Sp. 657–664.

Theologische Wissenschaft und internationale Freundschaftsarbeit der Kirche (Ansprache am 18. Mai 1927), in: ThBl 6, 1927, Sp. 169–171.

Internationale Alttestamentlertagung in Oxford (27.–30. September 1927), in: ThBl 6, 1927, Sp. 307–310.

Die kleinste literarische Einheit in den Erzählungsbüchern des Alten Testaments (Vortrag), in: ThBl 6, 1927, Sp. 333–337 = Kl. Schr. I, S. 143–149.

Zur Religionsgeschichte des Spätjudentums. Ein Bericht, in: ZMR 42, 1927, S. 89–94.

Jahwe-Name und Zauberwesen. Ein Beitrag zur Frage „Religion und Magie", in: ZMR 42, 1927, S. 161–186 = Kl. Schr. I, S. 150–171.

Die Religion in Geschichte und Gegenwart, hrsg. von H. GUNKEL und L. ZSCHARNACK. Bd. I, Tübingen [2]1927: Basan Sp. 783; Beerseba Sp. 849–850; Bethanien Sp. 952; Bethel Sp. 952–953; Bethsaida Sp. 954–955; Caesarea Philippi Sp. 1418; Chorazin Sp. 1522; Dekapolis Sp. 1819–1820.

BENTZEN, A.: Die Josianische Reform und ihre Voraussetzungen, 1926, in: ThLZ 52, 1927, Sp. 314–315.

FISCHER, J.: Zur Septuaginta-Vorlage im Pentateuch, 1926 (BZAW 42), in: DLZ 48, 1927, Sp. 197–198.

GRESSMANN, H.: Altorientalische Bilder zum Alten Testament, 1927, in: ZMR 42, 1927, S. 156.

GUTHE, H.: Bibelatlas, [2]1926, in: DLZ 48, 1927, Sp. 526–527.

HONOR, L. L.: Sennacherib's Invasion of Palestine, 1926, in: ThLZ 52, 1927, Sp. 412.

JEPSEN, A.: Untersuchungen zum Bundesbuch, 1927 (BWANT 41), in: ThLZ 52, 1927, Sp. 553–555.

JIRKU, A.: Der Kampf um Syrien-Palästina im orientalischen Altertum, 1926 (AO 24, 4), in: ThLZ 52, 1927, Sp. 364.

KÖNIG, E.: Der doppelte Wellhausenianismus im Lichte meiner Quellenforschung, 1927, in: ThLZ 52, 1927, Sp. 486–488.

MEINHOLD, J.: Der Dekalog (Rektoratsrede), 1927, in: DLZ 48, 1927, Sp. 1043–1045.

DERS.: Einführung in das Alte Testament. Geschichte, Literatur und Religion Israels, ²1926, in: ThLZ 52, 1927, Sp. 313–314.
PEDERSEN, J.: Israel. Its Life and Culture. Vol. I, II, 1926, in: DLZ 48, 1927, Sp. 946 bis 948.
RAHLFS, A. (Hrsg.): Septuaginta Societatis Scientiarum Gottingensis auctoritate edidit A. RAHLFS. Bd. I: Genesis, 1926, in: DLZ 48, 1927, Sp. 452–453.
SCHMIDT, HANS: Gott und das Leid im Alten Testament, 1926, in: AfO 4, 1927, S. 164.
SOSKIN, A. (Hrsg.): Album Tel-Aviver Ansichten von A. SOSKIN, Photograph in Tel-Aviv, o. J., in: ZMR 42, 1927, S. 157.
STEINER, H.: Die gēʾîm in den Psalmen, 1925 (Diss. phil. Bern), in: OLZ 30, 1927, Sp. 172 bis 173.

1928

Feasts and Festivals in Israel, in: Bible and Modern Religious Thought 2, 1928, S. 12–21.
Jahwe als König, in: Deutsche Theologie. Bericht über den ersten deutschen Theologentag zu Eisenach [Herbst 1927]. Hrsg. von A. TITIUS, 1928, S. 87–91.
Noch einmal: Text-, Stil- und Literarkritik in den Samuelisbüchern (zu L. ROST: Die Überlieferung von der Thronnachfolge Davids, 1926 [BWANT 42]), in: OLZ 31, 1928, Sp. 801–802.
Jahwe als König, in: ZAW 46, 1928, S. 81–105 = Kl. Schr. I, S. 172–193.
Die Erwählungstraditionen Israels (Zu K. GALLING: Die Erwählungstraditionen Israels, 1928 [BZAW 48]), in: ThBl 7, 1928, Sp. 123–125.
Der fünfte Deutsche Orientalistentag in Bonn (21.–25. August 1928) und der Siebzehnte Internationale Orientalistenkongreß in Oxford (27. August bis 1. September 1928), in: ThBl 7, Sp. 303–309.
Die Religion in Geschichte und Gegenwart. Bd. II, Tübingen ²1928: Eiferopfer Sp. 55; El Sp. 97–98; Eli Sp. 104; Elohim Sp. 117–118; Emmaus Sp. 130; Erstlinge, 2 Sp. 294; Ewiges Leben I. Im AT und im Judentum. 1. Im AT Sp. 456–457; Feste und Feiern II. In Israel Sp. 550–558; Galiläa Sp. 848; Garizim Sp. 859–860; Gaza Sp. 866–867.
[GRESSMANN, H. †:] A. TITIUS, TH. H. ROBINSON, E. SELLIN, J. HEMPEL: Hugo Greßmann, gestorben 7. April 1927. Gedächtnisworte (ZAW 45, 1927, S. I–XXIV), in: OLZ 31, 1928, Sp. 378–379.
GUTHE, H.: Palästina, ²1927, in: OLZ 31, 1928, Sp. 571–572.
KAHLE, P.: Masoreten des Westens, 1927 (BWANT 33), in: ThLZ 53, 1928, Sp. 149–151.
NIELSEN, D. (Hrsg.): Handbuch der Altarabischen Altertumskunde. 1. Bd.: Die Altarabische Kultur, 1927, in: ThBl 7, 1928, Sp. 229–230.
OESTERLEY, W. O. E.: The Wisdom of Egypt and the OT in the Light of the newly discovered „Teaching of Amen-em-ope", 1927, in: ThBl 7, 1928, Sp. 182.
Old Testament Essays. Papers read before the Society for OT Study at its eighteenth meeting held at Keble College, Oxford, September 27th to 30th 1927, 1927, in: ThBl 7, 1928, Sp. 182.
RICHTER, G.: Textstudien zum Buche Hiob, 1927 (BWANT 43), in: ThLZ 53, 1928, Sp. 410–411.
WEISER, A.: Religion und Sittlichkeit der Genesis in ihrem Verhältnis zur alttestamentlichen Religionsgeschichte, 1928, in: ThBl 7, 1928, Sp. 203–204.
WELCH, A. C.: The Code of Deuteronomy, 1924, in: ThLZ 53, 1928, Sp. 1–2.
WELLHAUSEN, J.: Reste arabischen Heidentums, 2. Ausg., Neudruck, 1927, in: OLZ 31, 1928, Sp. 36.

1929

Vom Werden der biblischen Gottesanschauung und ihrem Ringen mit dem Gottesgedanken der griechischen Philosophie. Halle 1929, 18 S. (Hallische Universitätsreden 42).

Zum Problem des Spätjudentums (zu G. F. MOORE: Judaism in the first Centuries of the Christian Era. The Age of the Tannaim. Vol. I, II, 1927), in: OLZ 32, 1929, Sp. 731 bis 741.

Dievu vardai ir Dievo ivaizdai semituose = De nominibus deorum et dei informatione apud populos semiticos [litauisch], in: Soter 6, 1929, S. 118–127.

Götternamen und Gottesvorstellung bei den Semiten (Vortrag), in: ZDMG 83, 1929, S. 21–36 = Kl. Schr. I, S. 194–205.

Die Religion in Geschichte und Gegenwart. Bd. III, Tübingen [2]1929: Jericho Sp. 81; Jesreel Sp. 103; Joppe Sp. 371–372; Jordan Sp. 372; Ituräa Sp. 456; Judäa Sp. 459–460; Judentum I. Vom Exil bis Hadrian Sp. 469–476; Kadesch Sp. 578; Kana Sp. 597; Kanaan I. Geographisch Sp. 597–600; Kapernaum Sp. 617–618; Karmel Sp. 635; Kison Sp. 1048; Libanon Sp. 1607; Machpela Sp. 1811; Mahlzeiten 2. Im AT und NT Sp. 1856–1857; Mamre Sp. 1950; Megiddo Sp. 2066; Menschenopfer 2. Sp. 2114 bis 2115.

BEER, G.: Welches war die älteste Religion Israels?, 1927, in: OLZ 32, 1929, Sp. 469.

JUNKER, H.: Prophet und Seher in Israel, 1927, in: OLZ 32, 1929, Sp. 470–471.

KAHLE, P.: Die hebräischen Bibelhandschriften aus Babylonien mit Faksimiles von 70 Handschriften (ZAW 46, 1928, S. 113–137. Taf. 1–70), in: ThBl 8, 1929, Sp. 42–43.

MENES, A.: Die vorexilischen Gesetze Israels im Zusammenhang seiner kulturgeschichtlichen Entwicklung, 1928 (BZAW 50), in: OLZ 32, 1929, Sp. 657–659.

PETERS, N.: Das Buch Job, 1928 (EH 21), in: DLZ 50, 1929, Sp. 1273–1276.

RAHLFS, A.: Paul de Lagardes wissenschaftliches Lebenswerk im Rahmen einer Geschichte seines Lebens dargestellt, 1928, in: DLZ 50, 1929, Sp. 409–411.

Reallexikon der Assyriologie, hrsg. von E. EBELING und B. MEISSNER, Bd. 1, Liefg. 1–2, 1929, in: ThBl 8, 1929, Sp. 162–163.

ROTHSTEIN, J. W. †: Kommentar zum ersten Buch der Chronik. Nach des Verfassers Tod bearb., abgeschlossen und eingel. von J. HÄNEL, 1927 (KAT 18, 2), in: ThLZ 54, 1929, Sp. 581–583.

SANDERS, H. A. und SCHMIDT, C.: The Minor Prophets in the Freer Collection and the Berlin Fragment of Genesis, 1927, in: DLZ 50, Sp. 1468–1469.

BAUDISSIN, W. W. GRAF: Kyrios als Gottesname im Judentum und seine Stelle in der Religionsgeschichte, hrsg. von OTTO EISSFELDT, Bd. I–IV. Gießen 1929. XVI, 602; VII, 316; XII, 710; IV, 228.

1930

Der Gott Bethel, in: ARW 28, 1930, S. 1–30 = Kl. Schr. I, S. 206–233.

Evangelische Charakterköpfe: Franz Delitzsch und Wolf Graf Baudissin, in: Das Evangelische Deutschland 7, 1930, S. 444–446 = Kl. Schr. I, S. 234–238.

Die Religion in Geschichte und Gegenwart. Bd. IV, Tübingen [2]1930: Nazareth Sp. 475; Nebo Sp. 480; Ölberg Sp. 641; Opfer II A. Im AT Sp. 711–717; Ophra Sp. 730; Pella Sp. 1062; Pentapolis Sp. 1064; Penuel Sp. 1064; Peräa Sp. 1065; Petra Sp. 1101; Rama Sp. 1691–1692; Religionsgeschichtliche Schule Sp. 1898–1905; Rotes Meer Sp. 2115 bis 2116.

DÜRR, L.: Die Heilige Schrift des Alten Testaments, 1929, in: ThBl 9, 1930, Sp. 356.

LIPMAN, A.: Authenticité du Pentateuque ou la Critique devant la Tradition, 1929, in: ThLZ 55, 1930, Sp. 345—346.

VOLZ, P.: Der Prophet Jeremia, ^{2}1928 (KAT 10), in: ThBl 9, 1930, Sp. 119.

WEISER, A.: Die Profetie des Amos, 1929 (BZAW 53), in: DLZ 51, 1930, Sp. 1394—1398.

YAHUDA, A. S.: Die Sprache des Pentateuch. 1. Buch, 1929, in: ThLZ 55, 1920, Sp. 342 bis 345.

1931

Vom Werden der biblischen Gottesanschauung und ihrem Ringen mit dem Gottesgedanken der griechischen Philosophie. Halle, 2. Abdr. 1931, 18 S. (Hallische Universitätsreden 42).

Die Komposition der Samuelisbücher. Leipzig 1931, 63 S.

Der Text von I. Sam. 3, 21 im Licht der literarischen Analyse von c. 1—7, in: AfO 6, 1930/31, S. 17—19.

Die Entzifferung der alphabetischen Keilschriften von Ras Schamra, in: Hallische Nachrichten Nr. 220 vom 19. September 1931 = RuS, S. 1—4.

Ezechiel als Zeuge für Sanheribs Eingriff in Palästina, in: PJB 27, 1931, S. 58—66 = Kl. Schr. I, S. 239—246.

Das Alte Testament auf dem 18. Internationalen Orientalistenkongreß zu Leiden, in: ThBl 10, 1931, S. 273—283.

Gott und Götzen im Alten Testament, in: ThStKr 103, 1931, S. 151—160 = Kl. Schr. I, S. 266—273.

Zwei Leidener Darstellungen der israelitischen Religionsgeschichte (A. KUENEN und B. D. EERDMANS), in ZDMG 85, 1931, S. 172—195.

יְהֻד Jos. 19, 45 und ἡ 'Ιουδαία 1. Makk. 4, 15 = *el-jehūdīje*, in: ZDPV 54, 1931, S. 271 bis 278 = Kl. Schr. I, S. 274—279.

Werden, Wesen und Wert geschichtlicher Betrachtung der israelitisch-jüdisch-christlichen Religion, in: ZMR 46, 1931, S. 1—24 = Kl. Schr. I, S. 247—265.

Die Religion in Geschichte und Gegenwart. Bd. V, Tübingen ^{2}1931: Samaria Sp. 97; Samuelisbücher Sp. 106—108; Sarepta Sp. 117; Saron Sp. 117; Schephela Sp. 151; Seir Sp. 397; Sichem Sp. 471—472; Silo Sp. 493; Tabor Sp. 973—974; Tekoa Sp. 1034; Tiberias Sp. 1172; Tirza Sp. 1185; Totes Meer Sp. 1244—1245; Zehnten, 2. Im AT Sp. 2088—2089.

BÖHL, F. M. TH.: Nieuwjaarsfeest en Koningsdag in Babylon en in Israël (Rede), 1927, in: AfO 6, 1930/31, S. 28—29.

BORÉE, W.: Die alten Ortsnamen Palästinas, 1930, in: ThLZ 56, 1931, Sp. 121—122.

CRAMER, K.: Amos, 1930 (BWANT 51), in: ThLZ 56, 1931, Sp. 122—125.

MEYER, E.: Geschichte des Altertums. 2. Bd., 2. Abt.: Der Orient vom zwölften bis zur Mitte des achten Jahrhunderts. Aus dem Nachlaß hrsg. von H. E. Stier, ^{2}1931, in: ThBl 10, 1931, Sp. 226—227.

RAD, G. V.: Das Geschichtsbild des chronistischen Werkes, 1930 (BWANT 54), in: OLZ 34, 1931, Sp. 951—952.

Reallexikon der Assyriologie, hrsg. von E. EBELING und B. MEISSNER, Bd. 1, Liefg. 3—5, 1929—1931, in: ThBl 10, 1931 Sp. 208.

SCHAEDER H. H.: Esra der Schreiber, 1930 (BHTh 5); Iranische Beiträge I, 1930 (SKG 6, 5), in: ThLZ 56, 1931, Sp. 241—246.

SELLIN, E.: Das Zwölfprophetenbuch, $^{2-3}$1929 (KAT 12), in: OLZ 34, 1931, Sp. 45—46.

SMITH, J. M. P.: The Psalms. Translated, 1926, in: AfO 6, 1930/31, S. 28.

TORREY, CH. C.: The Second Isaiah. A new interpretation, 1928, in: AfO 6, 1930/31, S. 119–120.

1932

Baal Zaphon, Zeus Kasios und der Durchzug der Israeliten durchs Meer. Halle 1932, 71 S. (Beiträge zur Religionsgeschichte des Altertums, Heft 1).

Die religionsgeschichtliche Bedeutung der Funde von Ras Schamra, in: FF 8,1932, S. 314 bis 315.

Ein religionsgeschichtlicher Beitrag zur Frage der Gottesoffenbarung in vorchristlicher Zeit, in: ZMR 47, 1932, S. 193–207.

EERDMANS, B. D.: De Godsdienst van Israël. Deel I, 1930, in: OLZ 35, 1932, Sp. 567 bis 568.

MOORE, G. F. †: Judaism in the First Centuries of the Christian Era. The age of the Tannaim, Vol. 3, 1930, in: OLZ 35, 1932, Sp. 777–778.

PROCKSCH, O.: Jesaja I, 1930 (KAT 9,1), in: OLZ 35, 1932, Sp. 475–476.

RABIN, I.: Studien zur vormosaischen Gottesvorstellung. Erster Abschn.: Untersuchungen der entwicklungsgeschichtlichen Grundlagen, 1929, in: ZS 8, 1932, S. 268–270.

RAHLFS, A. (Hrsg.): Septuaginta Societatis Scientiarum Gottingensis auctoritate editit A. RAHLFS. Bd. 10: Psalmi cum Odis, 1931, in: ThLZ 57, 1932, Sp. 153–155.

Reallexikon der Assyriologie, hrsg. von E. EBELING und B. MEISSNER, Bd. 1, Liefg. 6, 1932, in: ThBl 11, 1932, Sp. 215.

WUTHNOW, H.: Die semitischen Menschennamen in griechischen Inschriften und Papyri des vorderen Orients, 1930, in: AfO 7, 1931/32, S. 289.

1933

Der Gottesknecht bei Deuterojesaja (Jes. 40–55) im Lichte der israelitischen Anschauung von Gemeinschaft und Individuum. Halle 1933, 27 S. (Beiträge zur Religionsgeschichte des Altertums, Heft 2).

The Ebed-Jahwe in Is. 40–55 in the Light of the Israelite Conceptions of the Community and the Individual, the Ideal and the Real, in: ET 44, 1932/33, S. 261–268.

Das Datum der Belagerung von Tyrus durch Nebukadnezar, in: FF 9, 1933, S. 421–422 = RuS, S. 4–8 = Kl. Schr. II, S. 1–3.

Eine neue Darstellung des Judentums zur Zeit Jesu Christi (zu P. M.-J. LAGRANGE: Le Judaïsme avant Jésus-Christ, 1931), in: OLZ 36, 1933, S. 478–482.

Die Bedeutung der Funde von Ras Schamra für die Religionsgeschichte, in: Türkische Post. Tageszeitung für den Nahen Osten. Nr. 207 vom 3. 9. 1932, S. 2.

AUERBACH, E.: Wüste und gelobtes Land. Geschichte Israels von den Anfängen bis zum Tode Salomons, 1932, in: ZDMG 87, 1933, S. 82–86.

BAUER, H.: Das Alphabet von Ras Schamra. Seine Entzifferung und seine Gestalt, 1932, in: DLZ 54, 1933, Sp. 2167–2168.

DHORME, P.: Langues et Écritures sémitiques, 1930, in: AfO 8, 1932/33, S. 146–147.

EERDMANS, B. D.: De Godsdienst van Israël. Deel II, 1930, in: OLZ 36, 1933, Sp. 100 bis 101.

MEINHOLD, J.: Einführung in das Alte Testament. Geschichte, Literatur und Religion Israels, ^{3}1932, in: ThLZ 58, 1933, Sp. 100–101.

OLMSTEAD, A. T. E.: History of Palestine and Syria to the Macedonian Conquest, 1931, in: AfO 8, 1932/33, S. 243–245.

POPPER, W.: The Prophetic Poetry of Isaiah, Chapters 1–37, 1931, in: AfO 8, 1932/33, S. 324–325.

SCHULZ, A.: Kritisches zum Psalter, 1932 (ATA 12, 1), in: DLZ 54, 1933, Sp. 341–342.

TORREY, CH. C.: Pseudo-Ezekiel and the Original Prophecy, 1930, in: AfO 8, 1932/33, S. 69–71.

1934

Einleitung in das Alte Testament unter Einschluß der Apokryphen und Pseudepigraphen. Entstehungsgeschichte des Alten Testaments. Tübingen 1934, XVI, 752 S. (Neue Theologische Grundrisse, hrsg. von R. BULTMANN).

Der Gott des Tabor und seine Verbreitung, in: ARW 31, 1934, S. 14–41 = Kl. Schr. II, S. 29–54.

Eine antike literarische Bezeugung des Ras-Schamra-Alphabets, in: FF 10, 1934, S. 164 bis 165 = RuS, S. 8–12.

Die Wanderung palästinisch-syrischer Götter nach Ost und West im zweiten vorchristlichen Jahrtausend, in: JPOS 14, 1934, S. 294–300 = RuS, S. 12–19 = Kl. Schr. II, S. 55–60.

Der Siebente Deutsche Orientalistentag vom 29. August bis 1. September 1934 in Bonn, in: ThBl 13, 1934, Sp. 273–285.

Die religionsgeschichtliche Bedeutung der Funde von Ras Schamra, in: ZDMG 88, 1934, S. 173–184 = RuS, S. 19–31.

Ernst von Dobschütz † 20. Mai 1934. Zum Reformationstag. in: Saale-Zeitung. Halle/S., 31. Oktober 1934 = Kl. Schr. II, S. 61–63.

CANAAN, T.: The Palestine Arab House. Its Architecture and Folklore, 1933, in: ThLZ 59, 1934, Sp. 298–299.

COOK, ST. A.: The Place of the OT in the Modern Research, 1932, in: OLZ 37, 1934, Sp. 166–167.

DALMAN, G.: Arbeit und Sitte in Palästina III. Von der Ernte zum Mehl, 1933, in: ZMR 49, 1934, S. 250–251.

EULER, K. F.: Die Verkündigung vom leidenden Gottesknecht aus Jes. 53 in der griechischen Bibel, 1934 (BWANT 66), in: DLZ 55, 1934, Sp. 2449–2452.

FISCHER, J.: Die hebräischen Bibelzitate des Scholastikers Odo, 1934, in: DLZ 55, 1934, Sp. 1641–1642.

FRIEDRICH, J.: Ras Schamra. Ein Überblick über die Funde und Forschungen, 1933 (AO 33, 1/2), in: DLZ 55, 1934, Sp. 419–421.

The Mount Sinai Manuscript of the Bible. London, for the Trustees of the British Museum (Codex Sinaiticus), 1934, in: DLZ 55, 1934, Sp. 1057–1058.

ROBINSON, TH. H. and OESTERLEY, W. O. E.: A History of Israel. Vol. I, II, 1932, in: AfO 9, 1933/34, S. 211–216.

RUDOLPH, W.: Jesaja 24–27, 1933 (BWANT 62), in: DLZ 55, 1934, Sp. 675–677.

VOLZ, P.: Jesaja II, 1932 (KAT 9, 2), in: OLZ 37, 1934, Sp. 100–104.

VOLZ, P. und RUDOLPH, W.: Der Elohist als Erzähler. Ein Irrweg der Pentateuchkritik? An der Genesis erläutert, 1933 (BZAW 63), in: DLZ 55, 1934, Sp. 1297–1305.

Handbuch zum Alten Testament, hrsg. von OTTO EISSFELDT. Erste Reihe, Band 15: HANS SCHMIDT, Die Psalmen. Tübingen 1934.

1935

Molk als Opferbegriff im Punischen und Hebräischen und das Ende des Gottes Moloch. Halle 1935, 71 S. 4 Tafeln (Beiträge zur Religionsgeschichte des Altertums, Heft 3).

Der geschichtliche Hintergrund der Erzählung von Gibeas Schandtat (Richter 19—21), in: Festschrift für G. Beer, 1935, S. 19—40 = Kl. Schr. II, S. 64—80.

Karl Budde † am 20. Januar 1935 in Marburg, in: FF 11, 1935, S. 91—92 = Kl. Schr. II, S. 98—100.

Molochs Glück und Ende, in: FF 11, 1935, S. 285—286.

Der Zweite Internationale Alttestamentlertag vom 4. bis 10. September 1935 in Göttingen, in: ThBl 14, 1935, Sp. 233—247.

Vom 19. Internationalen Orientalistenkongreß vom 23.—29. September 1935 in Rom, in: ThBl 14, 1935, Sp. 247—249.

Das Berufungsbewußtsein der Propheten als theologisches Gegenwartsproblem, in: ThStKr 106, 1934/35, S. 124—156 = Kl. Schr. II, S. 4—28.

Neue Zeugnisse für die Aussprache des Tetragramms als Jahwe; in: ZAW 53, 1935, S. 59 bis 76 = Kl. Schr. II, S. 81—96.

Ergänzungen zu: Karl Buddes Schrifttum (ZAW Beiheft 54, 1930). Auf Grund von Aufzeichnungen des Verstorbenen zusammengestellt, in: ZAW 53, 1935, S. 286—289.

Dobschütz, E. v.: Die Bibel im Leben der Völker, 1934, in: ThStKr 106, 1934/35, S. 74—76.

Jonat, F. K.: Die Heiligen Bücher des Alten Bundes. In deutscher Übersetzung und mit Anmerkungen, 2 Bde. 1934, in: ThStKr 106, 1934/35, S. 411.

Reallexikon der Assyriologie, hrsg. von E. Ebeling und B. Meissner, Bd. 2, Liefg. 1, 1933; Liefg. 2, 1934, in: ThStKr 106, 1934/35, S. 172.

Vanutelli, P.: Libri Synoptici Veteris Testamenti seu Librorum Regum et Chronicorum loci paralleli, quos hebraice graece et latine critice edidit. Tomus prior, 1931, in: ThStKr 106, 1934/35, S. 73—74.

Theologische Studien und Kritiken. Eine Zeitschrift für das gesamte Gebiet der Theologie. In Verbindung mit ..... hrsg. von O. Eissfeldt und G. Heinzelmann, Bd. 106—109, 1934/35—1947.

1936

Philister und Phönizier. Leipzig 1936, 42 S. 1 Textkarte (Der Alte Orient Bd. 34, Heft 3).

Altertumskunde und Altes Testament, in: BZAW 66, 1936, S. 155—161.

Das Alte Testament als literarische Urkunde und als Gottesoffenbarung, in: Deutsches Pfarrerblatt 40, 1936, S. 877—879.

Die ältesten Bezeugungen von Baalbek als Kultstätte, in: FF 12, 1936, S. 51—53 = RuS, S. 31—36.

Baal Hammon und Tanit, in: FF 12, 1936, S. 378—379 = RuS, S. 36—39.

Zeus Amon, in: FF 12, 1936, S. 407—408 = RuS, S. 39—42.

Hesekiel Kap. 16 als Geschichtsquelle, in: JPOS 16, 1936, S. 286—292 = Kl. Schr. II, S. 101—106.

Die Zuverlässigkeit der Bibel. Ein Gutachten über die neuesten Angriffe gegen sie (O. Eissfeldt, E. Klostermann, F. K. Schumann), in: ThStKr 107, 1936, S. 354—370.

Bruno, A.: Das hebräische Epos. Eine rhythmische und textkritische Untersuchung der Bücher Samuelis und Könige, 1935, in: OLZ 39, 1936, Sp. 231—232.

Dalman, G.: Arbeit und Sitte in Palästina. IV. Brot, Öl und Wein, 1935, in: ZMR 51, 1936, S. 62.

GINSBERG, H. L.: The Ugaritic Texts, 1936, in: OLZ 39, 1936, S. 735–736.

GOODENOUGH, E. R.: By Light, Light. The Mystic Gospel of Hellenistic Judaism, 1935, in: OLZ 39, 1936, Sp. 311–314.

HEMPEL, J.: Die althebräische Literatur und ihr hellenistisch-jüdisches Nachleben, 1930, in: DLZ 57, 1936, Sp. 47–50.

HESCHEL, A.: Die Prophetie, 1936, in: ThStKr 107, 1936, S. 214.

JEAN, CH.-F.: La bible et les récits babyloniens, 1933, in: OLZ 39, 1936, Sp. 159–160.

LODS, A.: Les prophètes d'Israël et les débuts du Judaïsme, 1935, in: OLZ 39, 1936, Sp. 22 bis 24.

MONTGOMERY, J. A. and HARRIS, Z. S.: The Ras Shamra Mythological Texts, 1935, in: OLZ 39, 1936, Sp. 735–736.

Reallexikon der Assyriologie hrsg. von E. EBELING und B. MEISSNER. Bd. 2, Liefg. 3, 1935, in: ThStKr 107, 1936, S. 214; Liefg. 4, 1935, in: ThStKr 107, 1936, S. 466.

SCHAEFFER, CL. F.-A., THUREAU-DANGIN, F., VIROLLEAUD, CH. et al., La Cinquième Campagne de Fouilles à Ras Shamra (Printemps 1933), 1934; La Sixième Campagne... (Printemps 1934), 1935, in: OLZ 39, 1936, Sp. 610–611.

Nachwort zu A. L. WILLIAMS, The Tetragrammaton — Jahwe, Name or Surrogate? (ZAW 54, 1936, S. 262–269), in: ZAW 54, 1936, S. 269 = Kl. Schr. II S. 97.

Der Alte Orient. Gemeinverständliche Darstellungen. Herausgegeben für die Vorderasiatisch-Ägyptische Gesellschaft von OTTO EISSFELDT, Band 34–43. Leipzig 1936–1945.

Handbuch zum Alten Testament, hrsg. von OTTO EISSFELDT. Erste Reihe, Band 13: A. BERTHOLET, Hesekiel. Mit einem Beitrag von K. GALLING. Tübingen 1936; Band 14: TH. H. ROBINSON und F. HORST, Die Zwölf Kleinen Propheten [TH. H. ROBINSON: Hosea bis Micha (Aus dem Englischen übersetzt vom Herausgeber); F. HORST: Nahum bis Maleachi]. Tübingen 1936; darin S. VII–IX: Das Zwölfprophetenbuch: Einleitung.

1937

O. EISSFELDT, E. KLOSTERMANN, F. K. SCHUMANN, Die Entstehung der Bibel. Ein Gutachten über die neuesten Angriffe auf ihre Zuverlässigkeit. Gotha 1937, 19 S. (= ThStKr 107, 1936, S. 354–370).

Hans Bauer †, in: AfO 11, 1936/37, S. 405–406 = RuS, S. 46–48 = Kl. Schr. II, S. 110 bis 111.

Eine Einschmelzstelle am Tempel zu Jerusalem, in: FF 13, 1937, S. 163–164 = RuS, S. 42–46 = Kl. Schr. II, S. 107–109.

'Anat oder Tanit?, in: OLZ 40, 1937, Sp. 598–600 = RuS, S. 49–51.

Ras Schamra und das Alte Testament (zu R. DUSSAUD: Les découvertes de Ras Shamra (Ugarit) et l'Ancien Testament, 1937), in: OLZ 40, 1937, Sp. 721–725.

A foundry at the Temple of Jerusalem, in: Research and Progress 3, 1937, S. 59–62.

PAULYS Real-Encyclopädie der classischen Altertumswissenschaft. Neue Bearbeitung begonnen von G. WISSOWA. Bd. 19, 1, 1937: Pentateuch Sp. 513–524.

BAUER, H. (Hrsg.): Die alphabetischen Keilschrifttexte von Ras Schamra, 1936 (Kl T 168), in: DLZ 58, 1937, Sp. 912 und ZDMG 91, 1937, S. 532–533.

BIELER, L.: *ΘΕΙΟΣ ΑΝΗΡ*. Das Bild des „göttlichen Menschen" in Spätantike und Frühchristentum. Bd. I, 1935, in: OLZ 40, 1937, Sp. 18–19.

DIETRICH, A.: Phönizische Ortsnamen in Spanien, 1936 (AKM 21, 2), in: ZDMG 91, 1937, S. 237–238.

HAEFELI, L.: Sprichwörter und Redensarten aus der Zeit Christi, 1935, in: ZDMG 91, 1937, S. 236.

KAPPLER, W. (Ed.): Maccabaeorum libri I—IV, fasc. 1: Maccabaeorum liber I, 1936 (Septuaginta. Vetus Testamentum Graecum auctoritate Societatis Litterarum Gottingensis editum, vol. IX, 1), in: ThLZ 62, 1937, Sp. 46—47.

LEVY, K. †: Zur masoretischen Grammatik. Texte und Untersuchungen, 1936, in: ZDMG 91, 1937, S. 236—237.

LODS, A.: Les prophètes d'Israël et les débuts du Judaïsme, 1935, in: ZDMG 91, 1937, S. 236.

MAYANI, Z.: L'arbre sacré et le rite de l'alliance les anciens Sémites, 1935, in: OLZ 40, 1937, Sp. 526—527.

MITTELMANN, J.: Der altisraelitische Levirat. Eine rechtshistorische Studie, 1934, in: ZDMG 91, 1937, S. 236.

NIELSEN, D.: Ras Schamra-Mythologie und Biblische Theologie, 1936 (AKM 21, 4), in: ZDMG 91, 1937, S. 238—239.

SCHMIDT, J.: Studien zur Stilistik der alttestamentlichen Spruchliteratur, 1936 (ATA 13, 1), in: ZDMG 91, 1937, S. 237.

VOLZ, P.: Die Eschatologie der jüdischen Gemeinde im neutestamentlichen Zeitalter nach den Quellen der rabbinischen, apokalyptischen und apokryphen Literatur, 1934, in: ZDMG 91, 1937, S. 235.

Biblia Hebraica adjuvantibus... edidit R. KITTEL. Textum Masoreticum curavit P. KAHLE. Editionem tertiam denuo elaboratam ad finem perduxerunt A. ALT et O. EISSFELDT. Stuttgart 1937, XL, 1434 S.

Editio quarta, quinta, sexta typis editionis tertia expressa (1949—1950).

Editio septima aucta et emendata typis editionis tertia expressa, Stuttgart 1951, LIII, 1434 S.

Editio octava, nona, decima, undecima, duodecima, tertia decima, quarta decima typis editionis septimae expressa (1952—1966).

Handbuch zum Alten Testament, hrsg. von OTTO EISSFELDT. Erste Reihe, Band 1: K. GALLING, Biblisches Reallexikon. Tübingen 1937; Band 16: B. GEMSER, Sprüche Salomos. Tübingen 1937; Band 17: G. HÖLSCHER, Das Buch Hiob. Tübingen 1937; Band 19: A. BENTZEN, Daniel. Tübingen 1937.

1938

Neue Götter im Alten Testament, in: Atti del XIX. Congresso Internazionale degli Orientalisti ... Roma (23—29 Settembre 1935), 1938, S. 478—479 = Kl. Schr. II, S. 145—146.

Modern Criticism, in: Record and Revelation, ed. by H. W. ROBINSON, 1938, S. 74—109.

Zur Frage nach dem Ursprung unseres Alphabets, in: FF 14, 1938, S. 4—5 = RuS, S. 51 bis 57 = Das Alphabet. Entstehung und Entwicklung der griechischen Schrift. Hrsg. von G. PFOHL, 1968, S. 214—220.

Die Herkunft der 3 Zeichen für Aleph im Alphabet von Ras Schamra, in: FF 14, 1938, S. 125—126 = RuS, S. 58—62.

Schamemrumim, „Hoher Himmel", ein Stadtteil von Groß-Sidon, in: FF 14, 1938, S. 171 bis 173 = RuS, S. 62—67 = Kl. Schr. II, S. 123—126.

Zur Frage nach dem Alter der Phönizischen Geschichte des Sanchunjaton, in: FF 14, 1938, S. 251—252 = RuS, S. 67—71 = Kl. Schr. S. 127—129.

Alfred Bertholet zum 70. Geburtstag (9. November 1938), in: FF 14, 1938, S. 364 = Kl. Schr. II, S. 147—149.

Neue Belege für אדת „Herrin", in: OLZ 41, 1938, Sp. 489—491 = RuS, S. 71—74.

Hegel-Kritik und Pentateuch-Kritik. Historisch-kritische Analyse und systematisch-praktische Synthese in der Bibelforschung, in: ThBl 17, 1938, Sp. 33–41 = Kl. Schr. II, S. 112–122.

Religionsdokument und Religionspoesie, Religionstheorie und Religionshistorie. Ras Schamra und Sanchunjaton, Philo Byblius und Eusebius von Caesarea, in: ThBl 17, 1938, Sp. 185–197 = RuS, S. 75–95 = Kl. Schr. II, S. 130–144.

Das Alte Testament auf dem 20. Internationalen Orientalistenkongreß in Brüssel (5.–10 September 1938), in: ThBl 17, 1938, Sp. 299–306.

Die literarkritische Arbeit am Alten Testament in den letzten 12 Jahren, in: ThR N. F. 10, 1938, S. 255–291.

Bissing, Fr. W. v.: Die Kirche von Abd el Gadir bei Wadi Halfa und ihre Wandmalereien, 1937, in: ThStKr 108, 1937/38, S. 333.

Bratsiotis, P. I.: *Εἰσαγωγὴ εἰς τὴν Παλαιὰν Διαθήκην*, 1937, in: ThStKr 108, 1937/38, S. 179–180.

[Cumont, F.:] Mélanges Franz Cumont. Fasc. I. II, 1936 (= AIPhHOS IV, 1936), in: OLZ 41, 1938, Sp. 14–17.

Dhorme, É.: L'Évolution Religieuse d'Israël. Tome I: La Religion des Hébreux Nomades, 1937, in: ThStKr 108, 1937/38, S. 334–335.

Dürr, L.: Die Wertung des göttlichen Wortes im Alten Testament und im Antiken Orient, 1938 (MVÄG 42, 1), in: ThLZ 63, 1938, Sp. 414–415.

Grimme, H.: Altsinaitische Forschungen. Epigraphisches und Historisches, 1937, in: ThLZ 63, 1938, Sp. 466–468.

Hertz, Amelja: Besprechung von H. Bauer, Der Ursprung des Alphabets [AO 36, 1/2, 1937] (ZDMG 92, 1938, S. 670–674), in: ZDMG 92, 1938, S. 674–675.

Jeremias, J.: Jerusalem zur Zeit Jesu. Kulturgeschichtliche Untersuchungen zur neutestamentlichen Zeitgeschichte. II. Teil, 1. u. 2. Liefg., 1929. 1937, in: ThBl 17, 1938, Sp. 133 bis 134.

Reallexikon der Assyriologie, hrsg. von E. Ebeling und B. Meissner, Bd. 2, Liefg. 5, 1938 in: ThStKr 108, 1937/38, S. 333–334.

Schaeffer, Cl. F.-A., Virolleaud, C., Dussaud, R.: La Septième Campagne de Fouilles à Ras Shamra (Printemps 1935), 1935/6, in: OLZ 41, 1938, Sp. 617–618.

Tallqvist, K.: Akkadische Götterepitheta, 1938 (StOr 7), in: ThLZ 63, 1938, Sp. 265 bis 266.

Thiersch, H.: Ependytes und Ephod. Gottesbild und Priesterkleid im alten Vorderasien, 1936, in: DLZ 59, 1938, Sp. 650–656.

Unger, E.: Altindogermanisches Kulturgut in Nordmesopotamien, 1938, in: ThLZ 63, 1938, Sp. 241.

Ders.: Das antike Hakenkreuz als Wirbelsturm, 1937; Wahrsage-Symbolik, 1937; Orientierungssymbolik, 1937, in: ThLZ 63, 1938, Sp. 297–298.

Vincent, A.: La Religion des Judéo-Araméens d'Éléphantine, 1937, in: OLZ 41, 1938, Sp. 624–626.

Handbuch zum Alten Testament, hrsg. von Otto Eissfeldt. Erste Reihe, Band 7: M. Noth, Das Buch Josua. Tübingen 1938; Zweite Reihe, Band 6: J. Fichtner, Weisheit Salomos. Tübingen 1938.

1939

Ras Schamra und Sanchunjaton. Halle 1937, 157 S. (Beiträge zur Religionsgeschichte des Altertums, Heft 4).

Die geistesgeschichtliche Bedeutung der Funde von Ras Schamra, in: FF 15, 1939, S. 217 bis 219.

Linos und Alijan, in: Mélanges Syriens offerts à Monsieur René Dussaud. Tome II, 1939, S. 161—170 = RuS, S. 96—108 = Kl. Schr. II, S. 150—159.

Himmel und Erde als Bezeichnung phönizischer Landschaften, in: RuS 1939, S. 109 bis 127 = Kl. Schr. II, S. 227—240.

Phönizische Überlieferungen als Quelle für die Bücher 40—43 der Dionysiaca des Nonnos von Panopolis, in: Ras Schamra und Sanchunjaton, S. 128—151 = Kl. Schr. II, S. 241—257.

Die Komposition von Exodus 1—12. Eine Rettung des Elohisten, in: ThBl 18, 1939, Sp. 224—233 = Kl. Schr. II, S. 160—170.

Ras Schamra, in: ThLZ 64, 1939, Sp. 401—406.

Ba'alšamēm und Jahwe, in: ZAW 57, 1939, S. 1—31 = Kl. Schr. II, S. 171—198.

Die Komposition der Bileam-Erzählung. Eine Nachprüfung von Rudolphs Beitrag zur Hexateuchkritik, in: ZAW 57, 1939, S. 212—241 = Kl. Schr. II, S. 199—226.

PAULYS Real-Encyclopädie der classischen Altertumswissenschaft. Neue Bearbeitung begonnen von G. WISSOWA. Bd. 18, 1: Onias Sp. 474—477; Ophel Sp. 629.

COOK, ST. A.: The „Truth" of the Bible, 1938, in: DLZ 60, 1939, Sp. 1010—1013.

SCHAEFFER, CL. F.-A., THUREAU-DANGIN, F., VIROLLEAUD, C.: La Huitième Campagne de Fouilles à Ras Shamra-Ugarit (Printemps 1936), 1937, in: OLZ 42, 1939, Sp. 95—96.

TORCZYNER, H., HARDING, L., LEWIS, A., STARKEY, J. L.: The Lachish Letters (Lachish I), 1938, in: DLZ 60, 1939, Sp. 403—407.

VIROLLEAUD, C.: La Déesse 'Anat, 1938 (BAH 28), in: ThLZ 64, 1939, Sp. 407—409.

WILHELM II.: Das Königtum im alten Mesopotamien, 1938, in: ThLZ 64, 1939, Sp. 36—37.

Handbuch zum Alten Testament, hrsg. von OTTO EISSFELDT. Erste Reihe, Band 3: G. BEER, Exodus. Mit einem Beitrag von K. GALLING. Tübingen 1939.

1940

Das Chaos in der biblischen und in der phönizischen Kosmogonie, in: FF 16, 1940, S. 1—3 = Kl. Schr. II, S. 258—262.

Neues und Altes zur Paradieseserzählung (zu: P. Humbert, Études sur le Récit du Paradis et la Chute dans la Genèse, 1940), in: OLZ 43, 1940, Sp. 401—409.

Zum geographischen Horizont der Ras-Schamra-Texte, in ZDMG 94, 1940, S. 59—85 = Kl. Schr. II, S. 263—281.

BIDEZ, J. et CUMONT, F.: Les Mages Hellénisés. Zoroastre Ostanès et Hystaspe. Tome I. II, 1938, in: OLZ 43, 1940, Sp. 293—295.

Handbuch der Archäologie im Rahmen des Handbuches der Altertumswissenschaft, hrsg. von W. OTTO. Erster Text- und Tafelband, 1939, in: ThBl 19, 1940, Sp. 231—234.

JANSEN, H. L.: Die Henochgestalt, 1939 (SNVAO 1939, No. 1), in: DLZ 61, 1940, Sp. 457 bis 460.

LANGHE, R. DE: Les Textes de Ras Shamra-Ugarit et leurs apports à l'Histoire des Origines israélites, 1939, in: ZDMG 94, 1940, S. 110—115.

OBERMANN, J.: The Archaic Inscription from Lachish, 1938, in: OLZ 43, 1940, Sp. 308 bis 309.

SCHAEFFER, CL. F.-A.: La Neuvième Campagne de Fouilles à Ras Shamra-Ugarit (Printemps 1937), 1938, in: OLZ 43, 1940, Sp. 184—187.

DERS.: The Cuneiform Texts of Ras Shamra Ugarit, 1939, in: ZDMG 94, 1940, S. 110—115.

Schneider, N.: Die Götternamen von Ur III, 1939, in: ThLZ 65, 1940, Sp. 184–185.

Stein, B.: Der Begriff Kebod Jahweh und seine Bedeutung für die alttestamentliche Gotteserkenntnis, 1939, in: DLZ 61, 1940, Sp. 1177–1180 und BL 1947, S. 33 = BLs 1957, S. 95.

Weill, R.: La Phénicie et l'Asie Occidentale, 1939, in: ZDMG 94, 1940, S. 115–117.

Handbuch zum Alten Testament, hrsg. von Otto Eissfeldt. Erste Reihe, Band 18: M. Haller und K. Galling, Die fünf Megilloth. Tübingen 1940.

Theologische Gegenwartsfragen, behandelt von K. Galling, G. Heinzelmann, O. Michel und E. Wolf. Hrsg. von Otto Eissfeldt. Halle 1940 IV, 46 S.; darin: S. III–IV Vorwort von Otto Eissfeldt.

1941

Tempel und Kulte syrischer Städte in hellenistisch-römischer Zeit. Leipzig 1941, 160 S. 38 Abb. 16 Taf. (AO 40).

Tempel und Kulte syrischer Städte in hellenistisch-römischer Zeit, in: FF 17, 1941, S. 5–7 = Kl. Schr. II, S. 306–308. Taf. I–II.

Die letzten Dinge in religionsgeschichtlicher Bedeutung (zu T. Andrae: Die letzten Dinge. Deutsch von H. H. Schaeder, 1940), in: OLZ 44, 1941, Sp. 97–101.

Zu syrischen Tempeln und Kulten in hellenistisch-römischer Zeit, in: OLZ 44, 1941, Sp. 433–441 = Kl. Schr. II, S. 309–316.

Ruinen und Heiligtümer in Irak und Iran, in: ThBl 20, 1941, Sp. 209–216 = Kl. Schr. II, S. 321–329.

Lade und Stierbild, in: ZAW 58, 1940/41, S. 190–215 = Kl. Schr. II, S. 282–305.

Abdalonymus und *'lnm*, in: ZAW 58, 1940/41, S. 248–251 = Kl. Schr. II, S. 317–320.

Paulys Real-Encyclopädie der classischen Altertumswissenschaft. Neue Bearbeitung begonnen von G. Wissowa. Bd. 20, 1: Phoiniker und Phoinikia Sp. 350–380.

Gordon, C. H.: Ugaritic Grammar, 1940, in: OLZ 44, 1941, Sp. 289–291.

Parrot, A.: Malédictions et Violations de Tombes, 1939, in: OLZ 44, 1941, Sp. 11–13.

Schmidt, J.: Der Ewigkeitsbegriff im Alten Testament 1940 (ATA 13, 5), in: DLZ 62, 1941, Sp. 869–870 und BL 1947, S. 32 = BLs 1957, S. 94.

Stamm, J. J.: Erlösen und Vergeben im Alten Testament, 1940 (Diss. theol. Basel), in: DLZ 62, 1941, Sp. 434–436.

1942

Die Flügelsonne als künstlerisches Motiv und als religiöses Symbol, in: FF 18, 1942, S. 145 bis 147 = Kl. Schr. II, S. 416–419. Taf. III–V.

W. Gesenius als Archäologe. Ein Wort zur 100. Wiederkehr seines Todestages, in: FF 18, 1942, S. 297–299 = Kl. Schr. II, S. 430–434.

Otto Kern † 31. Januar 1942, in: HZ 166, 1942, S. 219–220 = Kl. Schr. II, S. 428–429.

Mémorial Lagrange, in: OLZ 45, 1942, Sp. 425–433 = Kl. Schr. II, S. 420–427.

Mythus und Sage in den Ras-Schamra-Texten, in: ZDMG 96, 1942, S. 35*–38*.

Bestand und Benennung der Ras-Schamra-Texte, in: ZDMG 96, 1942, S. 505–539 = Kl. Schr. II, S. 330–355.

W. Gesenius und die Palästinawissenschaft. Zur 100. Wiederkehr seines Todestages († 23. Oktober 1842), in: ZDPV 65, 1942, S. 105–112 = Kl. Schr. II, S. 435–440.

Andrae, W.: Alte Feststraßen im nahen Orient, 1941, in: DLZ 63, 1942, Sp. 439–441 und BL 1947, S. 6 = BLs 1957, S. 68.

BIRKELAND, H.: Akzent und Vokalismus im Althebräischen, 1940 (SNVAO 1940, No. 3), in: ZDMG 96, 1942, Sp. 141–144.

GROBBA, F.: Irak, 1941, in: ZDMG 96, 1942, S. 148–149.

NILSSON, M. P.: Geschichte der Griechischen Religion. I. Bd.: Bis zur griechischen Weltherrschaft, 1941 (Handbuch der Altertumswissenschaft 5, 2, 1, hrsg. von W. OTTO), in: HZ 166, 1942, S. 329–336.

RATSCHOW, C. H.: Werden und Wirken. Eine Untersuchung des Wortes *hajah* als Beitrag zur Wirklichkeitserfassung des Alten Testaments, 1941 (BZAW 70), in: DLZ 63, 1942, Sp. 101–103 und BL 1947, S. 41–42 = BLs 1957, S. 103.

1943

Die Bedeutung der Funde von Ras Schamra für die Geschichte des Altertums, in: HZ 168, 1943, S. 457–486 = Kl. Schr. II, S. 464–488.

Neue Forschungen zum 'Ebed Jahwe-Problem, in: ThLZ 68, 1943, Sp. 273–280 = Kl. Schr. II, S. 443–452.

Zu Abdalonymus und *'lnm* (vgl. ZAW 58, 1941, S. 248–251), in: ZAW 59, 1942/43, S. 219 = Kl. Schr. II, S. 320.

Israelitisch-philistäische Grenzverschiebungen von David bis auf die Assyrerzeit, in: ZDPV 66, 1943, S. 115–128 = Kl. Schr.II, S. 453–463.

DUSSAUD, R.: Les Découvertes de Ras Shamra (Ugarit) et l'Ancien Testament, [2]1941, in: OLZ 46, 1943, Sp. 115–116.

HAMMERSHAIMB, E.: Das Verbum im Dialekt von Ras Schamra, 1941, in: ThLZ 68, 1943, Sp. 192–194.

SCHAEFFER, CL. F.-A., DOSSIN, G., VIROLLEAUD, C.: Les Dixième et Onzième Campagnes de Fouilles à Ras Shamra-Ugarit (Automne et Hiver 1938 à 1939), 1941, in: OLZ 46, 1943, Sp. 215–218.

SCHUBART, W.: Religion und Eros. Hrsg. von F. SEIFERT, 1941, in: HZ 167, 1943, S. 141 bis 144.

1944

Mythos und Sage in den Ras-Schamra-Texten, in: Beiträge zur Arabistik, Semitistik und Islamwissenschaft, hrsg. von R. HARTMANN und H. SCHEEL, 1944, S. 267–283 = Kl. Schr. II, S. 489–501.

Mythos und Sage in den Ras-Schamra-Texten, in: Der Orient in Deutscher Forschung. Vorträge der Berliner Orientalistentagung im Herbst 1942. Hrsg. von H. H. SCHAEDER, 1944, S. 117–125.

Wilhelm Gesenius 1786–1842, in: 250 Jahre Universität Halle. Streifzüge durch ihre Geschichte in Forschung und Lehre, 1944, S. 88–90 = Kl. Schr. II, S. 441–442.

[Die Ausgrabungen in Ras Schamra]: Die keilschriftalphabetischen Texte aus der zehnten und elften Ausgrabungskampagne, in: AfO 14, 1941/44, S. 371–375 = Kl. Schr. II, S. 356–364.

Die Wohnsitze der Götter von Ras Schamra, in: FF 20, 1944, S. 25–27 = Kl. Schr. II, S. 502–506.

Ein neuer Beleg für die Funktion der Sonnengöttin von Arinna als Oberhaupt des hethitischen Staates, in: FF 20, 1944, S. 242–244 = Kl. Schr. II, S. 507–513.

Ugaritisches (1–3), in: ZDMG 98, 1944, S. 84–104 = Kl. Schr. II, S. 514–528.

de Brauw, L.: Europe en de Stier, 1940, in: AfO 14, 1941/44, S. 79–80.

Goossens, G.: Hiérapolis de Syrie. Essai de monographie historique, 1943, in: OLZ 47, 1944, Sp. 108–111.

Hölscher, G.: Die Anfänge der hebräischen Geschichtsschreibung, 1942 (SAH 1941/42, Nr. 3), in: ThLZ 69, 1944, Sp. 110–113.

Jansen, H. L.: Die Politik Antiochos' des IV., 1943 (SNVAO 1942, No. 3), DLZ 65, 1944, Sp. 37–40.

Jirku, A.: Der Kampf um Syrien-Palästina, die Brücke zwischen Afrika und Asien, 1942, in: ThLZ 69, 1944, Sp. 22–23.

Lauha, A.: Zaphon. Der Norden und die Nordvölker im Alten Testament, 1943 (Annales Academiae Scientiarum Fennicae 49, 2), in: ZDMG 98, 1944, S. 158–160.

Nielsen, D.: Der Dreieinige Gott in religionshistorischer Bedeutung. Bd. I, 1922; Bd. II, 1, 1942, in: ZDMG 98, 1944, S. 160–163.

Noth, M.: Die Welt des Alten Testaments, 1940, in: AfO 14, 1941/44, S. 352–353.

1945

Christus-Monogramm und phönizisches „Hermes"-Emblem, in: ZDPV 68, 1945, S. 163 bis 182 = Kl. Schr. II, S. 542–557.

Zur textkritischen Auswertung der Mercatischen Hexapla-Fragmente, in: Festschrift E. Klostermann, 1945, S. 43–53 [photographisch vervielfältigt] = WO 1, [1947–1952] 1947, S. 93–97 = Kl. Schr. III, S. 9–13.

1947

Geschichtliches und Übergeschichtliches im Alten Testament, Berlin 1947, 54 S. (ThStKr 109, 1947, Heft 2).

„My God" in the Old Testament, in: The Evangelical Quarterly 19, 1947, S. 7–20.

Zerwan, Titan und Japetosthes. Eine Bemerkung zu Jacob Burckhardts „Weltgeschichtliche Betrachtungen", in: FF 21/23, 1947, S. 204–205 = Kl. Schr. III, S. 14–18.

The Beginnings of Phoenician Epigraphy according to a Letter written by Wilhelm Gesenius in 1835, in: PEQ 79, 1947, S. 68–86.

Die Geschichtswerke im Alten Testament (zu M. Noth: Überlieferungsgeschichtliche Studien I, 1943 [SKG 18, 2]), in: ThLZ 72, 1947, Sp. 71–76 = Kl. Schr. III, S. 19–27.

Kreter und Araber, in: ThLZ 72, 1947, Sp. 207–212 = Kl. Schr. III, S. 28–34.

Frey, H.: Das Buch der Kirche in der Weltwende. Die kleinen nachexilischen Propheten für Freunde und Verächter der Bibel ausgelegt, 1941, in: BL 1947, S. 20 = BLs 1957, S. 82.

Lettinga, J. P.: Overzichten van de Geschiedenis en de Opgravingen in het Nabije Oosten. I. rās eš-šamrā en mīnet el-beiḍā' 1942 (MEOL 6), in: BL 1947, S. 9–10 = BLs 1957, S. 71.

Martin, H. and Payne, E. A.: A Christian Year Book, 1947, in: Universitas 2, 1947, S. 1351.

Miller, A. und Schildenberger, J.: Die Bücher Tobias, Judith und Esther 1940/41 (HS 4, 3), in: BL 1947, S. 21 = BLs 1957, S. 83.

Rudolph, W.: Jeremia, 1947 (HAT I, 12), in: BL 1947, S. 22 = BLs 1957, S. 84.

Wutz, F.: Das Hohelied, 1940, in BL 1947, S. 23–24 = BLs 1957, S. 85–86.

Handbuch zum Alten Testament, hrsg. von O. Eissfeldt. Erste Reihe, Band 12: W. Rudolph, Jeremia. Tübingen 1947.

Geschichtsschreibung im Alten Testament. Ein kritischer Bericht über die neueste Literatur dazu. Berlin 1948, 48 S.

Von den Anfängen der phönizischen Epigraphik. Nach einem bisher unveröffentlichten Brief von Wilhelm Gesenius. Halle 1948, 18 S. (Hallische Monographien 5).

Literarische Schätze aus der Geniza der alten Synagoge in Alt-Kairo (zu P. KAHLE: The Cairo Geniza, 1947), in: ThLZ 73, 1948, Sp. 29–34.

Kultprophetische Genossenschaften und Hypostasierungen göttlicher Eigenschaften und Tätigkeiten im Alten Orient (zu A. HALDAR: Associations of Cult Prophets among the Ancient Semites, 1945 und H. RINGGREEN: Word and Wisdom. Studies in the Hypostatization of Divine Qualitics and Functions in the Ancient Near East, 1947), in: ThLZ 73, 1948, Sp. 151–155 = Kl. Schr. III, S. 48–54.

Zur Überlieferungsgeschichte der Prophetenbücher des Alten Testaments (zu S. MOWINCKEL: Prophecy and Tradition, 1946), in: ThLZ 73, 1948, Sp. 529–534 = Kl. Schr. III, S. 55–60.

„Mein Gott" im Alten Testament, in: ZAW 61, 1945/48, S. 3–16 = Kl. Schr. III, S. 35–47.

PAULYS Real-Encyclopädie der classischen Altertumswissenschaft. Neue Bearbeitung begonnen von G. WISSOWA. Zweite Reihe, Bd. 7: Tyros. 3) Stadt in Phoinikien Sp. 1876 bis 1908.

BEGRICH, J.: Studien zu Deuterojesaja, 1938 (BWANT 77), in: BL 1948, S. 31–32 = BLs 1957, S. 141–142.

CLEMEN, C.: Die phönikische Religion nach Philo von Byblos, 1939 (MVÄG 42, 3), in: BL 1948, S. 40 = BLs 1957, S. 150.

CORNELIUS, F.: Indogermanische Religionsgeschichte. Die Entwicklung der indogermanischen Einzelvölker, 1942, in: BiOr 5, 1948, S. 67–69.

DALMAN, G.: Arbeit und Sitte in Palästina. Bd. VII: Das Haus, Hühnerzucht, Taubenzucht, Bienenzucht, 1942, in: BL 1948, S. 13 = BLs 1957, S. 123.

GOOSSENS, G.: Hiérapolis de Syrie. Essai de monographie historique, 1943, in: BiOr 5 1948, S. 35–37.

HALDAR, A.: Associations of Cult Prophets among the Ancient Semites, 1945, in: Universitas 3, 1948, S. 1093–1094.

LETTINGA, J. P.: Overzichten van de Geschiedenis en de Opgravingen in het Nabije Oosten. I. ras eš-šamrā en minet el-beidā', 1942 (MEOL 6), in: ThLZ 73, 1948, Sp. 350.

ORTMANN, H.: Der Alte und der Neue Bund bei Jeremia, 1940, in: BL 1948, S. 47 = BLs 1957, S. 157.

RAD, G. V.: Deuteronomium-Studien, 1947 (FRLANT 58), in: DLZ 69, 1948, Sp. 125 bis 128.

RAVN, O. E.: Herodotus' Description of Babylon, 1942, in: ThLZ 73, 1948, Sp. 211–212.

RINGGREN, H.: Word and Wisdom. Studies in the Hypostatization of Divine Qualities and Functions in the Ancient Near East, 1947, in: Universitas 3, 1948, S. 1093–1094.

ROWLEY, H. H.: The Relevance of the Bible, 1944, in: ThLZ 73, 1948, Sp. 600–601.

DERS.: The Relevance of Apocalyptic. A Study of Jewish and Christian Apocalypses from Daniel to the Revelation, ²1947, in: ThLZ 73, 1948, Sp. 600–601 und Universitas 3, 1948, S. 327–329.

STEGMÜLLER, O.: Berliner Septuagintafragmente, 1939 (Berliner Klassikertexte aus den Staatlichen Museen zu Berlin, Heft 8), in: BL 1948, S. 22 = BLs 1957, S. 132.

STUMMER, F.: Geographie des Buches Judith, 1947, in: BL 1948, S. 36 = BLs 1957, S. 146.

WEISER, A.: Die Psalmen, Teil 1, 1939 (ATD 7, 1), in: BL 1948, S. 30 = BLs 1957, S. 140.

The Westminster Historical Atlas to the Bible, ed. by G. E. WRIGHT and F. V. FILSON, 1946, in: Universitas 3, 1948, S. 197–198.

ZIEGLER, J.: Die jüngeren griechischen Übersetzungen als Vorlagen der Vulgata in den Prophetischen Schriften (Beilage zum Personal- und Vorlesungsverzeichnis der Staatlichen Akademie zu Braunsberg [Ostpr.], 1943/44), in: BL 1948, S. 22–23 = BLs 1957, S. 132–133.

DERS.: Textkritische Notizen zu den jüngeren griechischen Übersetzungen des Buches Isaias (NGG N. F. 1, 1939, S. 75–102), in: BL 1948, S. 23 = BLs 1957, S. 133.

Hallische Monographien, hrsg. von O. EISSFELDT, Heft 1–20. Halle/S. 1948–1951.

Zeitschrift für die alttestamentliche Wissenschaft. Bd. 61 (N. F. 20) – 71 (N. F. 30) hrsg. von O. EISSFELDT und J. HEMPEL; Bd. 72 (N. F. 31) – 77 (N. F. 36) hrsg. von G. FOHRER, O. EISSFELDT und J. HEMPEL. Berlin 1948–1965.

1949

Gabelhürden im Ostjordanland, in: FF 25, 1949, S. 9–11 = Kl. Schr. III, S. 61–66.

Die Bedeutung der 1947 in Palästina aufgefundenen alten hebräischen und aramäischen Handschriften, in: FF 25, 1949, S. 196–200.

Nachtrag zu dem Aufsatz über die 1947 gefundenen hebräischen Handschriften, in: FF 25, 1949, S. 302.

Zahl und Art der in den Rollen enthaltenen Schriftwerke; ihre Entstehungszeit und ihre religionsgeschichtliche Einordnung (Der gegenwärtige Stand der Erforschung der in Palästina neu gefundenen hebräischen Handschriften 2), in: ThLZ 74, 1949, Sp. 95–98.

Varianten der Jesaja-Rolle (Der gegenwärtige Stand... 3), in: ThLZ 74, 1949, Sp. 221–226.

Ansetzung der Rollen nach paläographischen Kriterien (Der gegenwärtige Stand ... 4), in: ThLZ 74, 1949, Sp. 226–228.

Vorläufige Ergebnisse einer im Frühjahr 1949 vorgenommenen Untersuchung der Fundhöhle (Der gegenwärtige Stand .. 5), in: ThLZ 74, 1949, Sp. 228.

Neue Nachrichten über die im Frühjahr 1949 vorgenommene Untersuchung der Fundhöhle (Der gegenwärtige Stand... 6), in: ThLZ 74, 1949, Sp. 595–597.

Der Anlaß zur Entdeckung der Höhle und ihr ähnliche Vorgänge aus älterer Zeit (Der gegenwärtige Stand... 7), in: ThLZ 74, 1949, Sp. 597–600 = Kl. Schr. III, S. 71–74.

KÖHLER, L.: Kleine Lichter. 50 Bibelstellen erklärt, 1945, in: Universitas 4, 1949, S. 1097 bis 1099.

LETTINGA, J. P.: Oegarit (rās esj-sjamrā). Een nieuwe Phoinicische stad uit de Oudheid, 1948, in: BiOr 6, 1949, S. 109–110.

MENSCHING, G.: Geschichte der Religionswissenschaft, 1948, in: HZ 169, S. 610–611.

SEELIGMANN, I. L.: The Septuagint Version of Isaiah. A Discussion of its Problems, 1948 (MEOL 9), in: ThLZ 74, 1949, Sp. 475–480.

Handbuch zum Alten Testament, hrsg. von O. EISSFELDT. Erste Reihe, Band 20: W. RUDOLPH, Esra und Nehemia samt 3. Esra. Tübingen 1949.

1950

Die ältesten Traditionen Israels. Ein kritischer Bericht über C. A. SIMPSON's The Early Traditions of Israel. Berlin 1950, 100 S. (BZAW 71).

Zu den Urkunden über den Tribut Niqmads, Königs von Ugarit, an den hethitischen Groß-könig Schuppiluliuma, in: Festschrift A. Bertholet, 1950, S. 147—157 = Kl. Schr. III, S. 75—84.

nûaḥ „sich vertragen“, in: Festschrift für L. Köhler zu dessen 70. Geburtstag, 1950, S. 23 bis 26 = SThU 20, 1950, S. 71—74 = Kl. Schr. III, S. 124—128.

Jahwe Zebaoth, in: Miscellanea Academica Berolinensia II, 2, 1950, S. 128—150 = Kl. Schr. III, S. 103—123. Taf. V.

Schwerterschlagene bei Hesekiel, in: Studies in Old Testament Prophecy presented to Professor Theodore H. Robinson, 1950, S. 73—81 = Kl. Schr. III, S. 1—8.

Zur Deutung von Motiven auf den 1937 gefundenen phönizischen Elfenbeinarbeiten von Megiddo, in: FF 26, 1950, S. 1—4 = Kl. Schr. III, S. 85—93. Taf. II—III.

Ein Beleg für die Buchstabenfolge unseres Alphabets aus dem 14. Jahrhundert v. Chr., in: FF 26, 1950, S. 217—220 = Kl. Schr. III, S. 94—102. Taf. IV = Das Alphabet. Entstehung und Entwicklung der griechischen Schrift. Hrsg. von G. PFOHL, 1968, S. 221 bis 232.

Berichtigung [zu FF 26, 1950, S. 217—220], in: FF 26, 1950, S. 290.

Die Bestimmung der aramäisch geschriebenen Rolle als das apokryphe Lamech-Buch (Der gegenwärtige Stand... 8), in: ThLZ 75, 1950, Sp. 23—26.

Hauptmann Philippe Lippens' Bericht über die Wiederaufdeckung der Fundhöhle (Der gegenwärtige Stand ... 10), in: ThLZ 75, 1950, Sp. 145—152.

Die neueste Phase in der Entwicklung der Pentateuchkritik, in: ThR 18, 1950, S. 91—112. 179—215. 267—287.

Zur Neubesinnung auf die Biblische Theologie (zu TH. C. VRIEZEN: Hoofdlijnen der Theologie van het Oude Testament, 1949), in: ZAW 62, 1950, S. 312—313.

Ugaritisches (4), in: ZDMG 99, [1945—1949] 1950, S. 29—42 = Kl. Schr. II, S. 528—541.

Reallexikon für Antike und Christentum, hrsg. von TH. KLAUSER. Bd. I, Stuttgart 1950: [Mitarbeiter] Baal. A. Nichtchristlich, III. Katalog der Baale Sp. 1068—1101; Baalbek Sp. 1113—1118.

BEA, A.: Liber Ecclesiastae qui ab Hebraeis appelatur Qohelet, 1950, in: ThLZ 75, 1950, Sp. 603—604.

BENTZEN, A.: Introduction to the Old Testament. I, 1948; II, 1949, in: ThLZ 75, 1950, Sp. 541—545.

OBERMANN, J.: Ugaritic Mythology, 1948, in: Oriens 3, 1950, S. 132—135.

ROUX, R.: Le Problème des Argonautes. Recherches sur les aspects religieux de la légende, 1949, in: HZ 171, 1950, S. 185—186.

VOGT, J. (Hrsg.): Rom und Karthago. Ein Gemeinschaftswerk, 1943, in: BiOr 7, 1950, S. 29 bis 33.

WALLENSTEIN, M.: Hymns from the Judean Scrolls, 1950, in: ThLZ 75, 1950, Sp. 481 bis 482.

WAMBACQ, B. N.: L'épithète divine Jahvé Seba'ôt, 1947, in: ThLZ 75, 1950, Sp. 27—29.

Festschrift Alfred Bertholet zum 80. Geburtstag, gewidmet von Kollegen und Freunden, hrsg. durch W. BAUMGARTNER, O. EISSFELDT, K. ELLIGER, L. ROST. Tübingen 1950, VII, 578 S.

1951

El im ugaritischen Pantheon. Berlin 1951, 83 S. 1 Taf. (BAL 98, 4).

Variae Lectiones Rotulorum Manu Scriptorum Anno 1947 prope Mare Mortuum Repertorum ad Jes 1—66 et Hab 1—2 pertinentes praeparavit... Stuttgart 1951, 8 S.

The Prophetic Literature, in: The Old Testament and Modern Study, hrsg. von H. H. Rowley, 1951, S. 115–161.

Gott und Götter, in: Proceedings of the 7th Congress for the History of Religions. Amsterdam, 4th–9th September 1950, ed by C. J. Bleeker, G. W. J. Drewes, K. A. H. Hidding, 1951, S. 110–111.

Ernst Sellin, in: AfO 15, 1945/51, S. 189 = Kl. Schr. III, S. 129–130.

Otto Procksch, in: AfO 15, 1945/51, S. 189–190 = Kl. Schr. III, S. 131.

Ritus und Mythus im Alten Orient (zu Th. H. Gaster: Ritual, Myth and Drama in the Near East, 1950), in: BiOr 8, 1951, S. 56–63 = Kl. Schr. III, S. 168–183.

Ein gescheiterter Versuch der Wiedervereinigung Israels (II. Sam. 2, 12–3, 1), in: NC 3, 1951, S. 110–127 = Kl. Schr. III, S. 132–146

Von Joseph zu Josua (Zu H. H. Rowley: From Joseph to Joshua. Biblical Traditions in the Light of Archaeology, 1950), in: ThLZ 76, 1951, Sp. 281–285 = Kl. Schr. III, S. 151–158.

Israel und seine Geschichte (zu M. Noth: Geschichte Israels, 1950), in: ThLZ 76, 1951, Sp. 335–340 = Kl. Schr. III. S. 159–167.

Brønno, E.: Studien über hebräische Morphologie und Vokalismus. Auf Grundlage der Mercatischen Fragmente der zweiten Kolumne der Hexapla des Origenes, 1943 (AKM 28), in: ZDMG 100, [1950] 1951, S. 401–402.

Kötting, B.: Peregrinatio Religiosa. Wallfahrten in der Antike und das Pilgerwesen in der alten Kirche, 1950, in: HZ 172, 1951, S. 404–405.

[Nötscher, F.:] Alttestamentliche Studien. F. Nötscher zum 60. Geburtstag 19. Juli 1950. Dargeboten von H. Junker und J. Botterweck, 1950 (BBB 1), in: ZDMG 101, 1951, S. 378–382.

Schaeffer, Cl. F.-A.: Ugaritica II. Nouvelles études relatives aux découvertes de Ras Shamra, 1949 (BAH 47), in: ThLZ 76, 1951, Sp. 34–38.

Scharff, A. und Moortgat, A.: Ägypten und Vorderasien im Altertum, 1950, in: BL 1951, S. 28 = BLs 1957, S. 323 und ThLZ 76, 1951, Sp. 403–405.

Schrade, H.: Der verborgene Gott. Gottesbild und Gottesvorstellung in Israel und im Alten Orient, 1949, in: Gnomon 23, 1951, S. 97–99.

Young, E. J.: An Introduction to the Old Testament, 1949, in: BiOr 8, 1951, S. 34–35.

H. H. Rowley, Der Istanbuler Orientalistenkongreß und das Alte Testament, übers. von O. Eißfeldt, in: ThLZ 76, 1951, Sp. 699–702.

1952

Sanchunjaton von Berut und Ilumilku von Ugarit. Halle 1952, IX, 74 S. (Beiträge zur Religionsgeschichte des Altertums, Heft 5).

Taautos und Sanchunjaton. Berlin 1952, 70 S. (SAB 1952, Nr. 1).

Ein Lexikon der altpalästinischen und altorientalischen Geographie aus den Anfängen der Universität Wittenberg, in: 450 Jahre Martin-Luther-Universität Halle–Wittenberg, Bd. I, 1952, S. 239–253 = Kl. Schr. III, S. 184–199. Taf. VI–VII.

Noch einmal: Ein gescheiterter Versuch der Wiedervereinigung Israels, in: NC 4, 1952, S. 55–59 = Kl Schr. III, S. 147–150. Taf. VI.

Hans Schmidt zum 75. Geburtstag am 10. Mai 1952, in: ThLZ 77, 1952, Sp. 437–438.

Zeilenfüllung, in: VT 2, 1952, S. 87–92 = Kl. Schr. III, S. 205–209.

Die Menetekel-Inschrift und ihre Deutung, in: ZAW 63, [1951] 1952, S. 105–114 = Kl. Schr. III, S. 210–217.

[Becker, C. H.:] Carl Heinrich Becker. Ein Gedenkbuch, hrsg. von H. H. Schaeder, 1950, in: ZDMG 102, 1952, S. 371–372.

Beckerath, J. v.: Tanis und Theben, 1951 (Ägyptologische Forschungen, hrsg. von A. Scharff, 16), in: DLZ 73, 1952, Sp. 540–543.

Coche de la Ferté, E.: Essai de Classification de la Céramique Mycénienne d'Enkomi, (BAH 54), in: ThLZ 77, 1952, Sp. 84–85.

[Dhorme, E.:] Recueil Édouard Dhorme, 1951, in: ThLZ 77, 1952, Sp. 151–152.

Glueck, N.: Explorations in Eastern Palestine IV, 1951 (AASOR 25–28, 1945/49), in: ThLZ 77, 1952, Sp. 81–83.

Lindblom, J.: The Servant Songs in Deutero-Isaiah. A New Attempt to solve an old Problem, 1951 (LUA 47, 5), in: ThLZ 77, 1952, Sp. 725–728.

Nilsson, M. P.: Geschichte der Griechischen Religion. Bd. II: Die hellenistische und römische Zeit (Handbuch der Altertumswissenschaft 5, 2,2), 1950, in: HZ 173, 1952, S. 549–557.

Otten, H.: Mythen vom Gotte Kumarbi. Neue Fragmente, 1951 (VIOF 3), in: BL 1952, S. 18 = BLs 1957, S. 389.

Pritchard, J. B.: Ancient Near Eastern Texts relating to the Old Testament, 1950, in: ThLZ 77, 1952, Sp. 341–343.

Rowley, H. H.: Submission in Suffering and other Essays on Eastern Thought, 1951, in: ThLZ 77, 1952, Sp. 209–211.

Ders.: The Servant of the Lord and other Essays on the Old Testament, 1952, in: ThLZ 77, 1952, Sp. 728–730.

Schaeffer, Cl. F.-A.: Stratigraphie comparée et Chronologie de l'Asie occidentale (IIIe et IIe millénaires), 1948, in: ThLZ 77, 1952, Sp. 415–419.

Schmidtke, F.: Der Aufbau der babylonischen Chronologie, 1952, in: ThLZ 77, 1952, Sp. 348.

Schweizer, U.: Löwe und Sphinx im Alten Ägypten, 1948 (Ägyptologische Forschungen, hrsg. von A. Scharff, 15), in: DLZ 73, 1952, Sp. 540–543.

Starcky, J.: Palmyre, 1952, in: BiOr 9, 1952, S. 185–187.

Stevenson, W. B.: The Poem of Job. A Literary Study with a new translation, 1947; Critical Notes on the Hebrew Text of the Poem of Job, 1951, in: ThLZ 77, 1952, Sp. 212 bis 215.

de Vaux, R., Starcky, J. u. a.: Abraham, Père des Croyants. Mit einem Vorwort von Kardinal E. Tisserant, 1951 (Cahiers Sioniens, 5, 2), in: ThLZ 77, 1952, Sp. 31–32.

Wilcken, U.: Griechische Geschichte im Rahmen der Altertumsgeschichte, ^{7}1951, in: DLZ 73, 1952, Sp. 460–462.

C. C. Torrey: Ein griechisch transkribiertes und interpretiertes hebräisch-aramäisches Verzeichnis der Bücher des Alten Testaments aus dem 1. Jahrhundert n. Chr. Aus dem Englischen übersetzt und mit Anmerkungen versehen von O. Eissfeldt, in: ThLZ 77, 1952, Sp. 249–254.

Handbuch zum Alten Testament, hrsg. von O. Eissfeldt. Erste Reihe, Band 17: G. Hölscher, Das Buch Hiob. Tübingen ^{2}1952; Band 19: A. Bentzen, Daniel. Tübingen ^{2}1952.

1953

Der Gott Karmel. Berlin 1953, 48 S., 6 Taf., 1 Karte (SAB 1953, Nr. 1).

Psalm 80, in: Geschichte und Altes Testament. A. Alt zum 70. Geburtstag am 20. September 1953 in Dankbarkeit und Verehrung dargebracht, 1953, S. 65–78 (BHTh 16) = Kl. Schr. III, S. 221–232.

Gott und das Meer in der Bibel, in: Studia Orientalia Septuagenario Ioanni Pedersen, 1953 S. 76–84 = Kl. Schr. III, S. 256–264.

[Ras Schamra] Die keilschriftalphabetischen Texte der Kampagnen 1948–51, in: AfO 16, 1952/53, S. 116–122 = Kl. Schr. II, S. 365–374.

Hans Schmidt (10. Mai 1877 bis 20. Januar 1953), in: AfO 16, 1952/53, S. 398–399.

Ugarit und Alaschia (Akademieberichte: Leipzig, Phil.-hist. Klassensitzung 15. Dezember 1952), in: DLZ 74, 1953, Sp. 249–252 = Kl. Schr. III, S. 218–220.

Ein Lehrbuch der hebräischen Grammatik aus den Anfängen der Universität Wittenberg, in: Deutsches Pfarrerblatt 53, 1953, S. 488–489.

Albrecht Alt zum 70. Geburtstag am 20. September 1953, in: FF 27, 1953, S. 126–127 = Kl. Schr. III, S. 233–236.

Mose (zu E. Auerbach: Moses, 1953), in: OLZ 48, 1953, Sp. 490–505 = Kl. Schr. III, S. 240–255.

In memoriam Hans Schmidt (10. 5. 1877 bis 20. 1. 1953). Hans Schmidts Stelle in der theologischen Forschung der ersten Hälfte des 20. Jahrhunderts, in: ThLZ 78, 1953, Sp. 379–382 = Kl. Schr. III, S. 265–269.

Palästina und Israel (zu Th. C. Vriezen: Palestina en Israël, 1951), in: ZAW 64, [1952] 1953, S. 56–57.

Abel, F.-M.: Les livres des Maccabées, 1949, in: ThLZ 78, 1953, S. 334–337.

Altheim, F.: Aus Spätantike und Christentum, 1951, in: HZ 175, 1953, S. 624–625.

Anthes, R.: Die Maat des Echnaton von Amarna, 1952 (Suppl. to JAOS 14), in DLZ 74, 1953, Sp. 276–278.

Bardtke, H.: Die Handschriftenfunde am Toten Meer. Mit einer Einführung in die Text- und Kanonsgeschichte des Alten Testaments, 1952, in: BL 1953, S. 70 = BLs 1957, S. 523.

Bonkamp, B.: Die Psalmen nach dem hebräischen Grundtext übersetzt, 1949, in: ThLZ 78, 1953, Sp. 151–152.

Cross Jr., F. M. and Freedman D. N.: Early Hebrew Orthography. A Study of the Epigraphic Evidence, 1952, in: ThLZ 78, 1953, Sp. 507–508.

Jahrbuch für Kleinasiatische Forschung, Hrsg. von H. Th. Bossert und F. Steinherr. Bd. 1, [1950/51] 1951; Bd. 2, [1952/53] 1953, in: DLZ 74, 1953, Sp. 587–591.

Jalabert, L. et R. Mouterde: Inscriptions Grecques et Latines de la Syrie. III, 1. Région de l'Amanus 1950 (BAH 46), in: ThLZ 78, 1953, Sp. 215–216.

Kornfeld, W.: Studien zum Heiligkeitsgesetz (Lev 17–26), 1952, in: OLZ 48, 1953, Sp. 438–440.

Kukenheim, L.: Contributions à l'Histoire de la Grammaire Grecque, Latine et Hébraïque à l'Époque de la Renaissance, 1951, in: ThLZ 78, 1953, Sp. 423–424.

[Miller, A.:] Miscellanea Biblica et Orientalia R. P. A. Miller O. S. B., Secretario Pontificiae Commissionis Biblicae completis LXX annis oblata cura A. Metzinger, 1951, in: BiOr 10, 1953, S. 216–218.

Montgomery, J. A.: A critical and exegetical Commentary on the Book of Kings, ed. by H. S. Gehman, 1951, in: OLZ 48, 1953, Sp. 52–55.

Murtonen, A.: The Appearance of the Name Yhwh outside Israel, 1951 (Studia Orientalia 16, 3), in: OLZ 48, 1953, Sp. 313.

Needler, W.: Palestine ancient and modern. A Handbook and Guide to the Palestinian Collection of the Royal Ontario Museum of Archaeology, 1949, in: OLZ 48, 1953, Sp. 244–245.

Nielsen, D.: Der Dreieinige Gott in religionsgeschichtlicher Beleuchtung. Bd. II, 1, 1942, in: AfO 16, 1592/53, S. 106–109.

[Popper, W.:] Semitic and Oriental Studies. A Volume presented to W. Popper. Ed. by W. J. Fischel, 1951, in: BiOr 10, 1953, S. 165–166.
Reallexikon für Antike und Christentum, hrsg. von Th. Klauser u. a. Bd. 2, Liefg. 9–10, 1951/52, in: BL 1948, S. 10 = BLs 1957, S. 463.
Schaeffer, Cl. F.-A.: Enkomi-Alasia. Nouvelles Missions en Chypre 1946–1950, 1952, in: ThLZ 78, 1953, Sp. 415–417.
[Smith, W. R.:] Centenary of the Birth on 8th Nov. 1846 of the Rev. Prof. W. Robertson Smith, 1951, in: OLZ 48, 1953, Sp. 361.
Speiser, W.: Vorderasiatische Kunst, 1952, in: ThLZ 78, 1953, Sp. 522–523.
Vanderbuilt Studies in the Humanities. Vol. I. Ed. by R. C. Beatty, J. P. Hyatt and M. K. Spears 1951, in: ThLZ 78, 1953, Sp. 413–414.
Ziegler, J. (Ed.): Ezechiel, 1952 (Septuaginta. Vetus Testamentum Graecum auctoritate Societatis Litterarum Gottingensis editum, vol. XVI, 1), in: ThLZ 78, 1953, Sp. 274 bis 276.

Handbuch zum Alten Testament, hrsg. von O. Eissfeldt. Erste Reihe, Band 7: M. Noth, Das Buch Josua. Tübingen ²1953.
Orientalische Literaturzeitung, im Auftrage der Deutschen Akademie der Wissenschaften zu Berlin unter Mitwirkung von Prof. D. Dr. O. Eissfeldt ... hrsg. von Professor Dr. Richard Hartmann. Berlin 1953ff., seit 1962 hrsg. von Prof. Dr. Fritz Hintze.

1954

Neue Handschriftenfunde in Palästina, in: Deutsches Pfarrerblatt 54, 1954, S. 217–220. 246–248, 267–269.
Noch einmal: Gabelhürden im Ostjordanland, in: FF 28, 1954, S. 54–56 = Kl. Schr. III, S. 67–70. Taf. I.
Ugarit und Alalach, in: FF 28, 1954, S. 80–85 = Kl. Schr. III, S. 270–279.
Recht und Grenze archäologischer Betrachtung des Alten Testaments (zu C. H. Gordon: Introduction to Old Testament Times, 1953), in: OLZ 49, 1954, Sp. 101–108 = Kl. Schr. III, S. 280–287.
Partikularismus und Universalismus in der israelitisch-jüdischen Religionsgeschichte, in: ThLZ 79, 1954, Sp. 283–284 = Kl. Schr. III, S. 288.
Ein neues Handbuch der biblischen Archäologie (zu A. G. Barrois: Manuel d'Archéologie Biblique, I, 1939; II, 1953), in: ZAW 65, [1953] 1954, S. 82–85.
Das Alte Testament im Lichte der safatenischen Inschriften, in: ZDMG 104, 1954, S. 88 bis 118 = Kl. Schr. III, S. 289–317.
Reallexikon für Antike und Christentum. Bd. II, Stuttgart 1954: Christus I (Messias) Sp. 1250–1257.

Abel, F.-M.: Histoire de la Palestine depuis la conquête d'Alexandre jusqu'à l'invasion arabe. I. II, 1952, in: OLZ 49, 1954, Sp. 43–46.
Alt, A.: Kleine Schriften zur Geschichte des Volkes Israel. Bd. I. II, 1953, in: HZ 178, 1954, S. 399.
Bertholet, A.: Wörterbuch der Religionen, 1952, in: OLZ 49, 1954, Sp. 113–114.
Bruno, A.: Die Bücher Genesis–Exodus. Eine rhythmische Untersuchung, 1953; Jesaja. Eine rythmische Untersuchung, 1953; Jeremia. Eine rhythmische Untersuchung, 1954, in: ThLZ 79, 1954, Sp. 551–554.
Cornelius, F.: Geschichte des Alten Orients, 1950, in: ThLZ 79, 1954, Sp. 146.

DRIVER, G. R.: Aramaic Documents of the fifth Century B. C., 1954, in: OLZ 49, 1954, Sp. 525—529.

GROSS, H.: Weltherrschaft als religiöse Idee im Alten Testament, 1953 (BBB 6), in: ZDMG 104, 1954, S. 188—189.

HADAS, M.: The Third and Fourth Books of Maccabees, 1953, in: OLZ 49, 1954, Sp. 240 bis 241.

HOENIG, S. B.: The Great Sanhedrin, 1953, in: OLZ 49, 1954, Sp. 522—525.

JONGE, M. DE: The Testament of the twelve Patriarchs. A Study of their Text, Composition and Origin, 1953, in: ThLZ 79, 1954, Sp. 477—478.

KÖHLER, L.: Der hebräische Mensch. Eine Skizze. Mit einem Anhang: Die hebräische Rechtsgemeinde, 1953, in: OLZ 49, 1954, Sp. 329—331.

KUHL, C.: Die Entstehung des Alten Testaments, 1953, in: DLZ 75, 1954, Sp. 132—136.

LEIPOLDT, J. und MORENZ, S.: Heilige Schriften. Betrachtungen zur Religionsgeschichte der antiken Welt, 1953, in: BiOr 11, 1954, S. 141—142.

MAAG, V.: Text, Wortschatz und Begriffswelt des Buches Amos, 1951, in: OLZ 49, 1954, Sp. 138—139.

Mitteilungen des Instituts für Orientforschung, hrsg. von F. HINTZE. Bd. 1, Heft 1, 1953, in: BL 1954, S. 78 = BLs 1957, S. 615.

MOLIN, G.: Die Söhne des Lichtes, 1954, in: ThLZ 79, 1954, Sp. 605—606.

MORIARTY, F. L.: Foreword to the Old Testament Books, 1954, in: ThLZ 79, 1954, Sp. 555.

MOSCATI, S.: Geschichte und Kultur der semitischen Völker. Eine Einführung, 1953, in: Gnomon 26, 1954, S. 53—54.

MOWINCKEL, S.: Religion und Kultus, 1953, in: ThLZ 79, 1954, Sp. 475—476.

PARROT, A.: Découverts des mondes ensevelis, 1952, in: OLZ 49, 1954, Sp. 27—28.

ROSCH-PINNAH, E. [= E. E. ETTISCH]: Hebräisch ein uraltes Hieroglyphensystem, 1951; Astrologie im Hebräischen Alphabet, 1952, in: ZAW 65, [1953] 1954, S. 293.

SCHNEIDER, A. M.: Berichte über eine Reise nach Syrien und Jordanien, 1952 (NAWG 1952, Nr. 4), in: OLZ 49, 1954, Sp. 136—137.

SELLERS, O. R. and BARAMKI, D. C.: A Roman-Byzantine Burial Cave in Northern Palestine, 1953 (BASOR Suppl. Studies 15/16), in: OLZ 49, 1954, Sp. 235.

[STARR, J.:] The Joshua Starr Memorial Volume, 1953, in: OLZ 49, 1954, Sp. 423—424.

THIEBERGER, F.: Die Glaubensstufen des Judentums, 1952, in: DLZ 75, 1954, Sp. 391—394.

VERMÈS, G.: Les Manuscrits du Desert de Juda, 1953, in: ThLZ 79, 1954, Sp. 237—238.

VINCENT, L.-H. und STEVE, A. M.: Jérusalem de l'Ancient Testament. Part I: Archéologie de la Ville, 1954, in: ThLZ 79, 1954, Sp. 671—674.

WIESMANN, H.: Die Klagelieder, übersetzt und erklärt, 1954, in: ThLZ 79, 1954, Sp. 741 bis 742.

Handbuch zum Alten Testament, hrsg. von OTTO EISSFELDT. Erste Reihe, Band 14: TH. H. ROBINSON und F. HORST, Die Zwölf Kleinen Propheten. [TH. H. ROBINSON: Hosea bis Micha (Aus dem Englischen übersetzt vom Herausgeber); F. HORST: Nahum bis Maleachi]. Tübingen ^{2}1954; darin S. VII—IX: Das Zwölfprophetenbuch: Einleitung.

1955

Adrammelek und Demarus, in: Mélanges I. Lévy, 1955, S. 153—159 (AIPhHOS 13) = Kl. Schr. III, S. 335—339.

Zur Kompositionstechnik des Pseudo-Philonischen Liber Antiquitatum Biblicarum, in:

Interpretationes ad Vetus Testamentum pertinentes SIGMUNDO MOWINCKEL septuagenario missae, 1955, S. 53–71 (NTT 56) = Kl. Schr. III, S. 340–353.

Religionshistorie und Religionspolemik im Alten Testament, in: Wisdom in Israel and in the Ancient Near East, presented to Professor H. H. Rowley, 1955, S. 94–102 (SVT 3) = Kl. Schr. III, S. 359–366.

Das Lied Moses (5. Mose 32, 1–43) (Berichte der Akademien: Leipzig. Sitzung der phil.-hist. Klasse 24. Januar 1955), in: DLZ 76, 1955, Sp. 388.

Noch ein safatenisches Bild vom Grabmal des Hani', in: FF 29, 1955, S. 118–119 = Kl. Schr. III, S. 318–321.

Die Umrahmung des Mose-Liedes Dtn 32, 1–43 und des Mose-Gesetzes Dtn 1–30 in Dtn 31, 9–32, 47, in: WZ Halle 4, 1954/55, S. 411–417 = Kl. Schr. III, S. 322 bis 334.

Zwei verkannte militär-technische Termini im Alten Testament, in: VT 5, 1955, S. 232 bis 238 = Kl. Schr. III, S. 354–358.

Die Eroberung Palästinas durch Altisrael (zu Y. KAUFMANN: The Biblical Account of the Conquest of Palestine, transl. from the Hebrew manuscript by M. DAGUT, 1953), in: WO 2, [1954–59] 1955, S. 158–171 = Kl. Schr. III, S. 367–383.

Ugaritica et Palmyrenica, in: ZDMG 105, 1955, S. *37*–*38*.

ASHKENAZI, T.: A Bibliographical List of Writings 1922–1954, 1954, in: OLZ 50, 1955, Sp. 207.

BLOCH, J.: On the Apocalyptic in Judaism, 1952 (JBL Monograph Series 2), in: OLZ 50, 1955, Sp. 45–46.

BRATSIOTIS, P. I.: *'Ο Ἐκκλησιάστης*, 1951, in: ThLZ 80, 1955, Sp. 154–155.

BRUNO, A.: Die Psalmen. Eine rhythmische und textkritische Untersuchung, 1954, in: ThLZ 80, 1955, Sp. 153–154.

CASSUTO, U.: The Goddes Anath. Canaanite Epics of the Patriarchal Age. Text, Hebrew Translation, Commentary and Introduction, 1953, in: OLZ 50, 1955, Sp. 126–128.

DRIVER, G. R.: Semitic Writing from Pictograph to Alphabet, [2]1954, in: OLZ 50, 1955, Sp. 237–238.

Handbuch der Orientalistik. Hrsg. von B. SPULER. 1. Abt., Bd. 1: Ägyptologie. 2. Abschn.: H. BRUNNER u. a.: Literatur, 1952; Bd. 3: Semitistik. 1. Abschn.: C. BROCKELMANN u. a.: Kanaanäisch und Ugaritisch-Hebräisch, 1953; 2. u. 3. Abschn.: A. BAUMSTARCK † u. a.: Aramäisch und Syrisch. Arabisch und Äthiopisch, 1954; Bd. 6: Geschichte der islamischen Länder. 1. Abschn.: B. SPULER: Die Chalifenzeit, 1952; 2. Abschn.: DERS.: Die Mongolenzeit, 1954; 1. Erg.-Bd.: Papyrologie, Urkundenwesen u. a.: Heft 1: W. HINZ: Maße und Gewichte, 1955, in: DLZ 76, 1955, Sp. 810–812.

HARDING, G. L.: Four Tomb Groups from Jordan, 1953 (PEF Annual VI), in: OLZ, 50, 1955, Sp. 40–41.

[HEIM, K.:] Theologie als Glaubenswagnis. Festschrift zum 80. Geburtstag von Karl Heim, 1954, in: BL 1955, S. 9 = BLs 1957, S. 642.

Historia Mundi, hrsg. von F. VALIAVEC. Bd. 2: Grundlagen und Entfaltung der ältesten Hochkulturen, 1953, in: BL 1955, S. 25 = BLs 1957, S. 658 und DLZ 76, 1955, Sp. 403 bis 406.

IMSCHOOT, P. VAN: Théologie de l'Ancient Testament. Tome I: Dieu, 1954, in: ThLZ 80, 1955, Sp. 275–276.

INGHOLT, H.: Parthian Sculptures from Hatra, 1954, in: OLZ 50, 1955, Sp. 316–318.

The Interpreter's Bible. The Holy Scriptures in the King James and Revised Standard Version with General Articles and Introduction, Exegesis, Exposition for Each Book of the

Bible. Vol. 3: Kings, Chronicles, Ezra, Nehemiah, Ester, Job, 1954, in: JBL 74, 1955, S. 210–212.

JALABERT, L. et MOUTERDE, R.: Inscriptions Grecques et Latines de la Syrie. Tome III, Part 2: Antioche (suite), 1953, in: ThLZ 80, 1955, Sp. 419.

KAUFMANN, Y.: The Biblical Account of the Conquest of Palestine. Translated from the Hebrew manuscript by M. DAGUT, 1953, in: OLZ 50, 1955, Sp. 534–535.

KISCH, G.: Pseudo–Philo's Liber Antiquitatum Biblicarum, 1949, in: OLZ 50, 1955, Sp. 396–397.

NELIS, J. T.: Daniel 1954, (BOuT 11, 2), in: ZAW 66, [1954] 1955, S. 126–127.

NILSSON, M. P.: Geschichte der Griechischen Religion. Bd. I: Die Religion Griechenlands bis auf die griechische Weltherrschaft, ²1955 (Handbuch der Altertumswissenschaft 5, 2, 1), in: HZ 181, 1955, S. 676–677.

[NYBERG, H. S.:] Donum Natalicium H. S. Nyberg oblatum. Edendum curaverunt E. GREN, B. LEWIN, H. RINGGREN, S. WIKANDER, 1953, in: BiOr 12, 1955, S. 161–162.

ORLINSKY, H. M.: Ancient Israel, 1954, in: OLZ 50, 1955, Sp. 39–40.

PALACIOS, L.: Grammatica Aramaico-Biblica, ²1953, in: ZAW 66, [1954] 1955, S. 146–147.

PARROT, A.: Déluge et Arche de Noé, 1952, in: OLZ 50, 1955, Sp. 138–139.

DERS.: La Tour de Babel, 1953, in: OLZ 50, 1955, Sp. 139.

DERS.: Entdeckung begrabener Welten. Aus dem Französischen übersetzt von E. JENNI, 1954, in: Deutsches Pfarrerblatt 55, 1955, S. 69; OLZ 50, 1955, Sp. 380; ThLZ 80, 1955, Sp. 87.

PICARD, C.: Carthage, 1951, in: OLZ 50, 1955, Sp. 141–142

Reallexikon für Antike und Christentum Bd 2, Liefg. 11–16, 1952/54, in: BL 1955, S 11 = BLs 1957, S. 644.

[ROBERT, A.–TRICOT, A.:] Initiation Biblique. Introduction à l'étude des Saintes écritures, publ. sous la direction de A. ROBERT et A. TRICOT. 1954, in: ThLZ 80, 1955, Sp. 207–209.

SCHOEPS, H.-J.: Jüdische Geisteswelt. Zeugnisse aus zwei Jahrtausenden, 1953, in: HZ 180, 1955, S. 150.

SELMS, A. VAN: Marriage and Family Life in Ugaritic Literature, 1954 (Pretoria Oriental, Series 1), in: OLZ 50, 1955, Sp. 219–221.

STIER, F.: Das Buch Ijjob. Hebräisch und Deutsch. Übertragen, ausgelegt und mit Text- und Sacherläuterungen versehen, 1954, in: ThLZ 80, 1955, Sp. 277–278.

VAN'T LAND, F. A. W. und VAN DER WOUDE, A. S.: De Habakuk-rol van ‘Ain Fasha (Semietische Teksten met Vertaling I, hrsg. von J. H. HOSPERS und TH. C. VRIEZEN), 1954, in: ZAW 66, [1954] 1955, S. 135.

WOOLLEY, L.: Ein vergessenes Königreich. Die Ausgrabungen der zwei Hügel Atschana und al-Mina im türkischen Hatay, 1954, in: HZ 181, 1955, S. 190–191.

ZEITLIN, S.: The Second Book of Maccabees, ed. with Introduction and Commentary. English Translation by S. TEDESCHE, 1954, in: OLZ 50, 1955, Sp. 457–459.

Handbuch zum Alten Testament, hrsg. von OTTO EISSFELDT. Erste Reihe, Band 13: G. FOHRER, Ezechiel. Mit einem Beitrag von K. GALLING. Tübingen 1955; Band 21: W. RUDOLPH, Chronikbücher. Tübingen 1955.

1956

Einleitung in das Alte Testament unter Einschluß der Apokryphen und Pseudepigraphen sowie der apokryphen- und pseudepigraphenartigen Qumrān-Schriften. Entstehungsgeschichte des Alten Testaments. Zweite, völlig neubearbeitete Auflage. Tübingen 1956, XVI, 954 S.

Handschriftenfunde in der Wüste Juda, in: AMz Jahrbuch 1956, S. 271–283. 47 Abb.

Malkiel (König ist El) und Malkija (König ist Jahwe): Gottesglaube und Namennennung in Israel, in: Atti del' VIII Congresso Internazionale di Storia della Religioni [Rom 1955], 1956, S. 261–263 = Kl. Schr. III, S. 384–385.

Albrecht Alt †, in: Bibelblatt, hrsg. von der Privilegierten Württemberger Bibelanstalt Stuttgart, Nr. 82, 1956, S. 2–4 = Kl. Schr. III, S. 237–239.

Eine neue Konkordanz zum hebräischen Alten Testament (zu G. Lisowsky: Konkordanz zum Hebräischen Alten Testament. Liefg. 1–2, 1955/56), in: Deutsches Pfarrerblatt 56, 1956, S. 125–126.

El and Yahweh, in: JSSt 1, 1956, S. 25–37 = El und Jahwe, in: Kl. Schr. III, S. 386–397.

Art und Aufbau der Phönizischen Geschichte des Philo von Byblos, in: Syria 33, 1956, S. 88–98 = Kl. Schr. III, S. 398–406.

[Mitteilung an P. Katz über Änderung der Auffassung von der Priorität des Masoretischen Kanons, zitiert von P. Katz: The Old Testament Canon in Palestine and Alexandria, in: ZNW 47, 1956, S. 191–217 auf S. 203, Anm. 19a].

Albright, W. F.: Die Religion Israels im Lichte der archäologischen Ausgrabungen. Autorisierte Übersetzung mit Nachträgen des Verfassers von F. Cornelius, 1956, in: OLZ 51, 1956, Sp. 522–523.

Ders.: Recent Discoveries in Bible Lands, 1955, in: ThLZ 81, 1956, Sp. 611.

Altheim, F.: Gesicht vom Abend und Morgen. Von der Antike zum Mittelalter, 1954, in: HZ 182, 1956, S. 696.

Barthélemy, D. and Milik, J. T.: Discoveries in the Judaean Desert I: Qumran Cave I. With Contributions by R. de Vaux, G. M. Crowfoot, H. J. Plenderleith, G. L. Harding, 1955, in: OLZ 51, 1956, Sp. 55–56.

Birkeland, H.: The Evildoers in the Book of Psalms, 1955 (ANVAO 1955, No. 2), in: OLZ 51, 1956, Sp. 427–429.

Chary, Th.: Les Prophètes et le Culte à partir de l'exil, 1955, in: ThLZ 81, 1956, Sp. 432 bis 433.

Dentan, R. C.: The Idea of History in the Ancient Near East, 1955, in: OLZ 51, 1956, Sp. 400–401.

Dupont-Sommer, A.: The Jewish Sect of Qumran and the Essenes. New Studies in the Dead Sea Scrolls. Transl. from the French by R. D. Barnett, 1954, in: OLZ 51, 1956, Sp. 337.

Fronzaroli, P.: Leggenda di Aqhat, 1955, in OLZ 51, 1956, Sp. 136–137.

Gray, J.: The KRT Text in the Literature of Ras Shamra, 1955, in: ZDMG 106, 1956, S. 374.

Greenberg, M.: The Ḫab/piru, 1955, in: OLZ 51, 1956, Sp. 415–417.

Ingholt, H., Seyrig, H., Starcky, J.: Recueil des tessères de Palmyre, 1955, in: Gnomon 28, 1956, S. 467–468.

de Liagre Böhl, F. M. Th.: Opera Minora, 1953, in: ThLZ 81, 1956, Sp. 427.

Mendenhall, G. E.: Law and Covenant in Israel and the Ancient Near East, 1955, in: JSSt 1, 1956, S. 188–189.

Mendelsohn, I. (Hrsg.): Religions of the Ancient Near East. Sumero-Accadian Religious Texts and Ugaritic Epics. Edited with an Introduction, 1955, in: ThLZ 81, 1956, Sp. 428–429.

Palacios, L.: Grammatica Syriaca. 2. Auflage hrsg. von V. Camps, 1954, in: ZAW 67, [1955] 1956, S. 128.

Parrot, A.: Golgatha et Saint-Sépulcre, 1955 (CAB 6), in: OLZ 51, 1956, Sp. 241–242.

Pope, M. H.: El in the Ugaritic Texts, 1955 (SVT 2), in: JSSt 1, 1956, S. 402–403.

Pritchard, J. B.: The Ancient Near East in Pictures relating to the Old Testament, 1954, in: JSSt 1, 1956, S. 186–187.

Ricciotti, G.: Geschichte Israels. Bd. I, 1953, in: OLZ 51, 1956, Sp. 425–426.

Schilling, R.: La religion Romaine de Vénus depuis les origines jusqu'au temps d'Auguste, 1954, in: HZ 182, 1956, S. 444–445.

Schmökel, H.: Ur, Assur und Babylon. Drei Jahrtausende im Zweistromland, 1955, in: ZAW 67, [1955] 1956, S. 130–131.

Steinmann, J.: Les plus anciennes Traditions du Pentateuque, 1954, in: ThLZ 81, 1956, Sp. 671–672.

Sukenik, E. L. (Hrsg.): The Dead Sea Scrolls of the Hebrew University, 1955, in: OLZ 51, 1956, Sp. 523–524.

Sutcliffe, E. F.: Providence and Suffering in the Old and New Testament, 1955, in: JSSt 1, 1956, S. 404–405.

Vermès, G.: Les Manuscrits du Desert de Juda, [2]1954, in: ThLZ 81, 1956, Sp. 155.

Weltatlas, Großer Historischer. Hrsg. vom Bayer. Schulbuch-Verlag. Teil I: Vorgeschichte und Altertum. Erläuterungen zum I. Teil von H. Bengtson und V. Milojčić, 1954, in: OLZ 51, 1956, Sp. 488–489.

Wilson, J. A., Speiser, E. A., Güterbock, H. G. Mendelsohn, I., Ingalls, D. H. H. and Bodde, D.: Authority and Law in the Ancient Orient, 1954 (Suppl. to JAOS 17), in: OLZ 51, 1956, Sp. 36.

A. Alt: Utopien. Zum Druck besorgt von O. Eissfeldt, in: ThLZ 81, 1956, Sp. 521 bis 528.

1957

Textkritische Bemerkungen zu den in Eusebius' Praeparatio Evangelica enthaltenen Fragmenten des Philo Byblius, in: Festschrift Karl Mras, 1957, S. 94–99 (Wiener Studien 70) = Kl. Schr. III, S. 407–411.

Die Wandbilder der Synagoge von Dura-Europos, in: FF 31, 1957, S. 241–249 = Kl. Schr. III, S. 426–440, Taf. IX–XII.

Non dimittam te nisi benedixeris me, in: Mélanges Bibliques A. Robert, 1957, S. 77–81 = Kl. Schr. III, S. 412–416.

Silo und Jerusalem, in: Volumen du Congrès Strasbourg 1956, 1957, S. 138–147 (STV 4) = Kl. Schr. III, S. 417–425.

Psalm 76, in: ThLZ 82, 1956, Sp. 801–808 = Kl. Schr. III, S. 448–457.

Jahwes Verhältnis zu Eljon und Schaddaj nach Psalm 91, in: WO 2, [1954–1959] 1957, S. 343–348 = Kl. Schr. III, S. 441–447.

Die Religion in Geschichte und Gegenwart, hrsg. von K. Galling, Bd. 1, Tübingen [3]1957: Adonaj Sp. 97; Adonis Sp. 97–98; Ai Sp. 192; Anath Sp. 356; Aschera Sp. 637–638; Astarte Sp. 661; Atargatis Sp. 668; Baal Sp. 805–806; Baalat Sp. 806; Baudissin Sp. 919; Bethel Sp. 1095–1096.

W. F. Albright: Die Bibel im Licht der Altertumsforschung. Ein Bericht über die Arbeit eines Jahrhunderts. Stuttgart 1957, S. 7–8: Vorwort von O. Eissfeldt.

Albright, W. F.: L'Archéologie de la Palestine. Traduit de l'anglais par R. Alapetite, 1955, in: OLZ 52, 1957, Sp. 240.

Beaujeu, J.: La Religion Romaine à l'Apogée de l'Empire. I: La Politique Religieuse des Antonins (96–192), 1955, in: HZ 183, 1957, S. 439–440.

de Boer, P. A. H.: Second-Isaiah's Message, 1956 (OTS 11), in: OLZ 52, 1957, Sp. 139 bis 141.

Bruno, A.: Die Bücher Samuel. Eine rhythmische Untersuchung, 1955; Die Bücher der Könige. Eine rhythmische Untersuchung, 1955, in: ThLZ 82, 1957, Sp. 31.

Ders.: Die Bücher Josua, Richter, Ruth. Eine rhythmische Untersuchung, 1955; Das Hohe Lied. Das Buch Hiob. Eine rhythmische Untersuchung, 1956, in: ThLZ 82, 1957, Sp. 188.

[Christian, V.:] Festschrift für Prof. Dr. V. Christian... Hrsg. von K. Schubert, 1956, in: ThLZ 82, 1957, Sp. 569–571.

Driver, G. R.: Canaanite Myths and Legends, 1956 (Old Testament Studies 3), in: OLZ 52, 1957, Sp. 517–520.

Finet, A.: L'Accadien des Lettres de Mari, 1956, in: ZAW 68, [1956] 1957, S. 219.

Fritsch, C. T.: The Qumran Community. Its History and Scrolls, 1956, in: ThLZ 82, 1957, Sp. 266–267.

Fronzaroli, P.: La Fonetica Ugaritica, 1955, in: OLZ 52, 1957, Sp. 35.

Gordon, C. H.: Ugaritic Manual, 1955, in: OLZ 52, 1957, Sp. 237–240.

Handbuch der Orientalistik. Hrsg. von B. Spuler. 1. Abt. Bd. 2: Keilschriftforschung und Alte Geschichte. Abschn. 3: Schmölke, H.: Geschichte des Alten Vorderasien, 1957, in: DLZ 78, 1957, Sp. 780–783.

Dass.: 1. Abt., Bd. 4: Iranistik. Abschn. 3: Krause, W.: Tocharisch, 1955, in: DLZ 78, 1957, Sp. 783–784.

Historia Mundi, hrsg. von F. Valjavec. Bd. 3: Der Aufstieg Europas, 1954, in: DLZ 78, 1957, Sp. 132–137; Bd. 4: Römisches Weltreich und Christentum, 1956, in: DLZ 78, 1957, Sp. 1069–1073.

[Irwin, W. A.:] A Stubborn Faith. Papers on Old Testament and Related Subjects Presented to Honor W. A. Irwin, ed. by E. C. Hobbs, 1956, in: ZAW 69, 1957, S. 280 bis 281.

Israel. Nagels Reiseführer. Deutsche Ausgabe von Erna Stübel, 1953, in: OLZ 52, 1957, Sp. 42–43.

Jalabert, L. et Mouterde, R. avec la collaboration de C. Mondésert: Inscriptions Grecques et Latines de la Syrie. Tome IV: Laodicée Apamène, 1955 (BAH 61), in: ThLZ 82, 1957, Sp. 200.

Kahle, P.: Opera Minora. Festgabe zum 21. Januar 1956, 1956, in: ThLZ 82, 1957, Sp. 839 bis 841.

Kraeling, E. G.: Bible Atlas, 1956, in: ZAW 69, 1957, S. 277–278.

Kramer, S. N.: From the Tablets of Sumer. Twenty-Five Firsts in Man's recorded History, 1956, in: ThLZ 82, 1957, Sp. 105–106.

Kupper, J.-R.: Les nomades en Mésopotamie au temps des rois de Mari, 1957, in: ZAW 69, 1957, S. 278.

de Leeuw, V.: De Ebed Jahweh-Profetieen, 1956, in: ZAW 69, 1957, S. 278.

[Levi della Vida, G.:] Studi Orientalistici in Onore di G. Levi della Vida, Vol. I. II, 1956, in: BiOr 14, 1957, S. 68–70.

[Lévy, I.:] Mélanges Isidore Lévy, 1955 (AIPhHOS 13, 1953), in: OLZ 52, 1957, Sp. 111 bis 113.

Levy, G. R.: The Sword from the Rock. An Investigation into the Origins of Epic Literature and Development of the Hero, 1953, in: OLZ 52, 1957, Sp. 108–109.

Marks, J. H.: Der textkritische Wert des Psalterium Hieronymi juxta Hebraeos, 1956, in: ZAW 68, [1956] 1957, S. 266–267.

North, R.: Guide to Biblical Iran, 1956, in: ThLZ 82, 1957, Sp. 849–850.

Noth, M.: Geschichte Israels. 2., verb. Auflage 1954, in: ThLZ 82, 1957, Sp. 108–109.

Parrot, A.: Babylone et l'Ancien Testament, 1956 (CAB 8), in: OLZ 82, 1957, Sp. 431 bis 432.

Phillips, W.: Qataban and Sheba, 1955, in: OLZ 52, 1957, Sp. 52–53.

Schilling, O.: Das Buch Jesus Sirach übersetzt und erklärt, 1956 (Herders Bibelkommentar 7, 2), in: ThLZ 82, 1957, Sp. 423–424.

Segert, St.: Zur Habakuk-Rolle aus dem Funde vom Toten Meer I–IV (Sonder-Abdruck aus ArOr 21, 1953, S. 218–239; 22, 1954, S. 99–113. 444–459; 23, 1955, S. 178–183. 364–373. 575–619), in: OLZ 52, 1957, Sp. 525–526.

Staehelin, F.: Reden und Vorträge. Hrsg von W. Abt, 1956, in: OLZ 52, 1957, Sp. 342.

Studia Patristica. Vol. I, II. Papers presented to the Second International Conference on Patristic Studies held at Christ Church. Part I. II. Ed. by K. Aland and F. L. Cross, 1957 (TU 63. 64), in: ZAW 69, 1957, S. 261–262.

Vincent, L.-H. et Steve, A.-M.: Jérusalem de l'Ancien Testament. Recherches d'Archéologie et d'Histoire. II: Archéologie du Temple; III: Évolution historique de la ville, 1956, in ThLZ 82, Sp. 502–505.

Vriezen, Th. C.: Hoofdlijnen der Theologie van het Oude Testament, [2]1954, in: ZAW 68, [1956] 1957, S. 221–222.

Ders.: Het Nabije Oosten in de Branding, 1957, in: ThLZ 52, Sp. 581 und ZAW 69, 1957, S. 281.

1958

Die Genesis der Genesis. Vom Werden des ersten Buches der Bibel. Tübingen 1958, VII, 86 S.

Das Lied Moses Dt 32, 1–43 und das Lehrgedicht Asaphs Ps 78 samt einer Analyse der Umgebung des Mose-Liedes. Berlin 1958, 54 S. (BAL 104, 5).

Enno Littmann † am 4. Mai 1958. Gedenkrede... bei der Akademischen Gedächtnisfeier in Tübingen am 15. Juli 1958. Tübingen 1958, 28 S. (Tübinger Universitätsreden 5).

Psalm 78 (Berichte der Akademien: Leipzig. Sitzung der phil.-hist. Klasse 10. Februar 1958), in: DLZ 79, 1958, Sp. 564.

Biblos geneseōs, in: Gott und die Götter. Festgabe für E. Fascher, 1958, S. 31–40 = Kl. Schr. III, S. 458–470.

Des Matthäus Aurigallus hebräische Grammatik von 1523, in: WZ Halle 7, 1958, S. 885 bis 889 mit 6 Abb. = Kl. Schr. III, S. 200–204. Taf. VI. VIII.

Die Religion in Geschichte und Gegenwart. Bd. II, Tübingen [3]1958: Dura Europos I. Religionsgeschichtlich Sp. 287–288; El Sp. 413–414; Elohim Sp. 436.

Bruno, A.: Das Buch der Zwölf. Eine rhythmische und textkritische Untersuchung, 1957, in: ThLZ 83, 1958, Sp. 107–108.

Burrows, M.: Die Schriftrollen vom Toten Meer. Aus dem Amerikanischen übertragen von F. Cornelius, 1957, in: HZ 185, S. 94–95.

Charles-Picard, G.: Das wiederentdeckte Karthago. Mit einem Beitrag „Die Welt der Phönizier" von S. von Cles-Reden, 1957, in: HZ 185, 1958, S. 605–607.

Champdor, A.: Les Ruines de Palmyre, [6]1953, in: OLZ 53, 1958, Sp. 152–153.

Crahay, R.: La littérature oraculaire chez Hérodote, 1956, in: HZ 184, 1958, S. 671 bis 672.

Decamps de Mertzenfeld, C.: Inventaire commenté des Ivoires pheniciens et apparentés découverts dans le Proche-Orient. Texte et Album, 1954, in: OLZ 53, 1958, Sp. 209–210.

Del Medico, H. E.: L'énigme des manuscrits de la Mer Morte, 1957, in: OLZ 53, 1958, Sp. 554–557.
Driver, G. R.: Aramaic Documents of the Fifth Century B. C. Abridged and Revised Edition, 1957, in: OLZ 53, 1958, Sp. 559–560.
[Furlani, G.:] Scritti in onore di G. Furlani. I. II, 1957 (RStOr 32, 1957), in: BiOr 15, 1958, S. 178–180.
Gemser, B. u. a.: Studies on the Book of Genesis, 1958 (Oudtestamentische Studien 12), in: ThLZ 83, 1958, Sp. 621–622.
Gooding, D. W.: Recensions of the Septuagint Pentateuch, 1955, in: OLZ 53, 1958, Sp. 152.
Köbert, R.: Vocabularium Syriacum, 1956, in: OLZ 53, 1958, Sp. 52.
Kraeling, C. H. u. a.: The Synagogue (The Excavations at Dura-Europos, Final Report 8, 1, ed. by A. R. Bellinger u. a.), 1956, in: OLZ 53, 1958, Sp. 233–237.
Kuhn, K. G.: Phylakterien aus Höhle 4 von Qumran, 1957 (AAH 1957, Nr. 1), in: JSSt 3, 1958, S. 98.
Lambert, J. N.: Aspects de la civilisation à l'âge du fratiarcat. Étude d'histoire juridique et religieuse comparée, 1958, in: ZAW 70, 1958, S. 270.
Les manuscrits de la Mer Morte. Colloque de Strasbourg 25–27 mai 1955 (Travaux du centre d'études supérieures spécialisé d'histoire des religions de Strasbourg), 1957, in: ThLZ 83, 1958, Sp. 410.
Mayani, Z.: Les Hyksos et le Monde de la Bible, 1956, in: OLZ 53, 1958, Sp. 327–329.
Parrot, A.: Le Musée du Louvre et la Bible, 1957 (CAB 9), in: ThLZ 83, 1958, Sp. 342 bis 343.
Petit, P.: Libanius et la vie municipale à Antioche au IVe siècle après J.-C., 1955 (BAH 62), in: OLZ 53, 1958, Sp. 428–430.
[Robert, A.:] Mélanges bibliques rédigés en l'honneur de André Robert, 1957, in: ZAW 70, 1958, S. 148–149.
Robert, A. et A. Feuillet: Introduction à la Bible. Tome I: Introduction Générale. Ancien Testament par P. Auvray u. a., 1957, in: OLZ 53, 1958, Sp. 450–452.
Schaeffer, Cl. F.-A.: Reprise des Fouilles de Ras Shamra-Ugarit. Campagnes XII à XVII (1948–1953). Rapports préliminaires suives d'Études sur les Textes en cunéiformes par Ch. Virolleaud, 1955, in: OLZ 53, 1958, Sp. 445.
van Selms, A.: De Rol der Lofprijzingen. Een der Dode Zee-Rollen vertaald en toegelicht, 1957, in: OLZ 53, 1958, Sp. 454–456.
Vriezen, Th. C.: Der Nahe Osten in der Brandung. Aus dem Holländischen übertragen von A. Barkhuis, 1958, in: ZAW 70, 1958, S. 273.
Wellhausen, J.: Prolegomena to the History of Ancient Israel. With a reprint of the Article „Israel" from the „Encyclopaedia Brittanica". Preface by W. Robertson Smith, 1957, in: HZ 185, 1958, S. 661.
Yadin, Y.: The Scroll of the War of the Sons of Light against the Sons of Darkness. Edited with an introduction, emendations and a commentary, ^{2}1957, in: OLZ 53, 1958, Sp. 48–50.
Young, G. D.: Concordance of Ugaritic, 1956 (Analecta Orientalia 36), in: OLZ 53, 1958, Sp. 45–46.
Ziegler, J. (Ed.). Ieremias, Baruch, Threni, Epistula Ieremiae, 1957 (Septuaginta. Vetus Testamentum Graecum auctoritate Societatis Litterarum Gottingensis editum, vol. XV) in: ThLZ 83, 1958, Sp. 22–24.

Handbuch zum Alten Testament, hrsg. von Otto Eissfeldt. Erste Reihe, Band 12: W. Rudolph, Jeremia, Tübingen ^{2}1958.

Psalm 121, in: Stat crux dum volvitur orbis. Eine Festschrift für Landesbischof D. Hanns Lilje, 1959, S. 9–14 = Kl. Schr. III, S. 494–500.

Alexander von Humboldts „Kosmos" und der 104. Psalm, in: FF 33, 1959, S. 113–117 = Kl. Schr. III, S. 471–480.

René Dussaud (1868–1958), in: ZDMG 109, 1959, S. 1–8. 1 Abb. = Kl. Schr. III, S. 486–493.

Psalm 132, in: WdO 2, [1954–1959] 1959, S. 480–483 = Kl. Schr. III, S. 481–485.

Die Religion in Geschichte und Gegenwart. Bd. III, Tübingen [3]1959: Hadad Sp. 7–8; Jahu Sp. 515; Jahwe Sp. 515–516; Kanaan I. Geographisch (mit A. KUSCHKE) Sp. 1106 bis 1108; Kanaan III. Kanaanäische Religion Sp. 1111–1113.

Reallexikon für Antike und Christentum. Bd. IV, Stuttgart 1959: Dura Europos Sp. 358–370.

R. DE VAUX: Die hebräischen Patriarchen und die modernen Entdeckungen, 1959, S. III: Vorwort von OTTO EISSFELDT.

[BOTTÉRO, J. u. a.:] La Antiche Divinita Semitiche. Studi di J. BOTTÉRO, M. J. DAHOOD, W. CASKEL, 1958, in: BiOr 16, 1959, S. 100–101.

BARDTKE, H.: Die Handschriftenfunde am Toten Meer. II: Die Sekte von Qumran, 1958, in: ThLZ 84, 1959, Sp. 341–342.

BRUNO, A.: Sprüche, Prediger, Klagelieder, Esther, Daniel. Eine rhythmische und textkritische Untersuchung, 1958, in: ThLZ 84, 1959, Sp. 424–425.

CARRATELLI, G. P. e G. LEVI DELLA VIDA: Un Editto Bilingue Greco-Aramaico di Asóka, 1958, in: OLZ 54, 1959, Sp. 465–466.

DAWSON, CH.: Religion and culture, 1958, in: HZ 187, 1959, S. 669.

EHRLICH, E. L.: Geschichte Israels von den Anfängen bis zur Zerstörung des Tempels 70 n. Chr., 1958, in: HZ 187, 1959, S. 184.

EISSFELDT, O.: Die Genesis der Genesis. Vom Werden des ersten Buches der Bibel, 1985, in: ZAW 71, 1959, S. 255.

HOLLADAY, W. L.: The Root šûbh in the Old Testament, 1958, in: ThLZ 84, 1959, Sp. 911 bis 912.

HURWITZ, S.: Die Gestalt des sterbenden Messias, 1958, in: ZAW 71, 1959, S. 256.

KAPPLER, W. † und R. HANHART (Ed.): Maccabaeorum libri I–IV, fasc. II: Maccabaeorum liber II, 1959 (Septuaginta. Vetus Testamentum Graecum auctoritate Societatis litterarum Gottingensis editum, vol. IX, 2) in: ThLZ 84, 1959, Sp. 510–511.

KUHN, K. G.: Rückläufiges hebräisches Wörterbuch, 1958, in: OLZ 54, 1959, Sp. 277.

LICHT, J.: The Thanksgiving Scroll. A Scroll from the Wilderness of Judaea. Text, Introduction, Commentary and Glossary, 1957, in: OLZ 54, 1959, Sp. 281–284.

MADER, E.: Mambre. Die Ergebnisse der Ausgrabungen im heiligen Bezirk Râmet el-Halîl in Südpalästina 1926–1928, 1957, in: OLZ 54, 1959, Sp. 38–40.

NÖTSCHER, F.: Gotteswege und Menschenwege in der Bibel und Qumran, 1958 (BBB 15), in: OLZ 54, 1959, Sp. 597–598.

Orientalia et Biblica Lovaniensia I: L'Ancien Testament aux VIes Journées Bibliques de Louvain (11–13 septembre 1954), 1957, in: ThLZ 84, 1959, Sp. 179–181.

PARROT, A.: Samaria, die Hauptstadt des Reiches Israel. — Babylon und das Alte Testament. Aus dem Französischen übersetzt von M.-R. JUNG unter Mithilfe von E. JENNI, 1957, in: OLZ 54, 1959, Sp. 163–164.

RABIN, C. and YADIN, Y.: Aspects of the Dead Sea Scrolls, 1958 (Scripta Hierosolymitana 4), 1958, in: OLZ 54, 1959, Sp. 493–495.

REISEL, M.: The Mysterious Name of Y. H. W. H., 1957, in: OLZ 54, 1959, Sp. 164–165.
ROWLEY, H. H. (Ed.): Eleven Years of Bible Bibliography. The Book Lists of the Society for Old Testament Study 1946–1956, 1957, in: JSSt 4, 1959, S. 187–188.
SCHREINER, J.: Septuaginta-Massora des Buches der Richter. Eine textkritische Studie, 1957 (AnBibl 2), in: OLZ 54, 1959, Sp. 490–491.
SCHUBERT, K.: Die Gemeinde vom Toten Meer, 1958, in: OLZ 54, 1959, Sp. 392–393.
THOMAS, D. WINTON (Ed.): Documents from Old Testament Times, 1958, in: Gnomon 31, 1959, S. 543–544.
VRIEZEN, TH. C.: Der Nahe Osten in der Brandung. Aus dem Holländischen übertragen von A. BARKHUIS, 1958, in: ThLZ 84, 1959, Sp. 279.
WATTS, J. D. W.: Vision and Prophecy in Amos, 1955, in: ThLZ 84, 1959, Sp. 280.
Weltatlas, Großer Historischer. Hrsg. vom Bayer. Schulbuch-Verlag. Teil 3: Neuzeit, 1957, in: OLZ 54, 1959, Sp. 360–361.
WESTERMANNS Atlas zur Weltgeschichte. Vorzeit und Altertum, Mittelalter, Neuzeit, 1956, in: OLZ 54, 1959, Sp. 11–12.
WRIGHT, G. E.: Biblische Archäologie. Übersetzt aus dem Amerikanischen von CHR. VON MERTENS, 1959, in: OLZ 54, 1959, Sp. 592–593.

1960

Der Beutel der Lebendigen. Alttestamentliche Erzählungs- und Dichtungsmotive im Lichte neuer Nuzi-Texte. Berlin 1960, 40 S., 7 Taf. (BAL 105, 6).
Phönikische und griechische Kosmogonie, in: Travaux du centre d'études supérieures spécialisés d'histoire des religions du Strasbourg: Éléments orientaux dans la religion grecque ancienne, 1960, S. 1–15 = Kl. Schr. III, S. 501–512.
The Alphabetical Cuneiform Texts from Ras Shamra Published in „Le Palais Royal d'Ugarit" Vol. II, 1957, in: JSSt 5, 1960, S. 1–49 = Kl. Schr. II, S. 375–415.
Lade und Gesetzestafeln, in: ThZ 16, 1960, S. 281–284 = Kl. Schr. III, S. 526–529.
Daniels und seiner drei Gefährten Laufbahn im babylonischen, medischen und persischen Dienst, in: ZAW 72, 1960, S. 134–148 = Kl. Schr. III, S. 513–525.
Die Religion in Geschichte und Gegenwart. Bd. IV, Tübingen [3]1960: Literarkritische Schule Sp. 388–390; Menschenopfer 2. Im AT Sp. 868; Moloch Sp. 1089–1090.
AISTLEITNER, J.: Die mythologischen und kultischen Texte aus Ras Schamra übersetzt, 1959, in: OLZ 55, 1960, Sp. 145–146.
ALT, A.: Kleine Schriften zur Geschichte des Volkes Israel, Bd. 3, hrsg. von M. NOTH, 1959, in: HZ 191, 1960, S. 416.
BARAMKI, D. C.: The Archaeological Museum of the American University of Beirut, 1959, in: OLZ 55, 1960, Sp. 475.
BOWMAN, G.: Des Julian von Aecianum Kommentar zu den Propheten Osee, Joel und Amos, 1958 (AnBibl 9), in: OLZ 55, 1960, Sp. 273–274.
BRIGHT, J.: A History of Israel, 1959, in: JBL 79, 1960, S. 369–372.
BRUNO, A.: Ezechiel. Eine rhythmische und textkritische Untersuchung, 1959, in: ThLZ 85, 1960, Sp. 518.
EHRLICH, E. L.: Die Kultsymbolik im Alten Testament und im nachbiblischen Judentum, 1959, in: ThLZ 85, 1960, Sp. 35–36.
FUGMANN, E.: Hama. Fouilles et recherches de la fondation Carlsberg 1931–1938. II, 1: L'architecture des périodes pré-helléniques, 1958, in: OLZ 55, 1960, Sp. 503–505.
GORDON, C. H.: The World of the Old Testament, 1958, in: OLZ 55, 1960, Sp. 379.

Handbuch der Orientalistik. Hrsg. von B. SPULER. 1. Abt., Bd. 4: Iranistik. 1. Abschnitt: K. HOFFMANN u. a., Linguistik, 1958, in: DLZ 81, 1960, Sp. 308–311.

DASS.: 1. Abt., Bd. 2: Keilschriftforschung und Alte Geschichte. Abschn. 1 u. 2: Geschichte der Forschung, Sprache und Literatur. Liefg. 1: A. FALKENSTEIN, Das Sumerische 1959, in: DLZ 81, 1960, Sp. 311.

DASS.: 1. Abt., Bd. 6: Geschichte der islamischen Länder. 3. Abschn.: H. J. KISSLING u. a., Neuzeit, 1959, in: DLZ 81, 1960, Sp. 892–893.

KALLAI, Z.: The Northern Boundaries of Judah from the Settlement of the Tribes until the Beginning of the Hasmonaean Period [hebr., engl. summary], 1960, in: ZAW 72, 1960, S. 299–300.

Masada. Survey and Excavations 1955–1956 by the Hebrew University, Israel Exploration Society, Department of Antiquities, 1957, in: OLZ 55, 1960, Sp. 47–48.

MICHAUD, H.: Sur la Pierre et l'Argile. Inscriptions hébraïques et Ancien Testament, 1958 (CAB 10), in: OLZ 55, 1960, Sp. 269–270.

PORÙBČAN, ŠT.: Il Patto Nuovo in Is. 40–66, 1958 (AnBibl 8), in: OLZ 55, 1960, Sp. 381 bis 383.

PRITCHARD, J. B.: Hebrew Inscriptions and Stamps from Gibeon, 1959 (Museum Monographs), in: OLZ 55, 1960, Sp. 147–148.

RIIS, P. J. et POULSEN, V. avec le concours de E. HAMMERSHAIMB: Hama. Fouilles et recherches de la Fondation Carlsberg 1931–1938. IV, 2: Les verreries et poteries médiévales, 1957, in: OLZ 55, 1960, Sp. 501–503.

RINGBOM, L. I.: Paradisus Terrestris. Myt, Bild och Verklighet, 1958, in: ZAW 72, 1960, S. 97–98.

ROSENBACH, M.: Galliena Augusta. Einzelgötter und Allgott im gallienischen Panthon, 1958, in: HZ 190, 1960, S. 428.

[SCHOEPS, H.-J.:] Lebendiger Geist. Hans-Joachim Schoeps zum 50. Geburtstag von seinen Schülern dargebracht, hrsg. von H. DIWALD, 1959 (ZRGG Beiheft 4), in: HZ 191, 1960, S. 407–408.

TESTUZ, M.: Les Idées religieuses du Livre des Jubilés, 1960, in: DLZ 81, 1960, Sp. 890 bis 891.

VERGOTE, J.: Joseph en Égypte. Génèse Chap. 37–50 à la lumière des études égyptologiques récentes, 1959 (OBL 3), in: OLZ 55, 1960, Sp. 39–45.

WOLFF, H. W.: Dodekapropheton. Lfg. 1–2, 1957/58 (BK XIV, 1–2); KRAUS, H.-J. Psalmen . Lfg. 1–11, 1958–60 (BK XV, 1–11), in: OLZ 55, 1960, Sp. 493–495.

YADIN, Y., AHARONI, Y., AMIRAN, RUTH, DOTHAN, TRUDE, DUNAYEVSKY, I., PERROT, J.: Hazor I. An Account of the first Season of Excavations, 1955, in: OLZ 55, 1960, Sp. 149 bis 152.

THOMSEN, P.: Die Palästina-Literatur. Eine internationale Bibliographie. Band A. Die Literatur der Jahre 1878–1894. Durch den Druck geführt von O. EISSFELDT und L. ROST. Berlin 1960, V, 905 S.

1961

Die Genesis der Genesis. Vom Werdegang des ersten Buches der Bibel. 2., verb. u. verm. Aufl. Tübingen 1961, VIII, 103 S.

Toledot, in: Studien zum Neuen Testament und zur Patristik. E. Klostermann zum 90. Geburtstag dargebracht, 1961, S. 1–8 (TU 77) = Kl. Schr. IV, S. 1–7.

Psalm 46 (Arno Lehmann zu seinem 60. Geburtstag am 23. Mai 1961 als Manuskript überreicht), in: Kl. Schr. IV, S. 8–11.

Jeremias Drohorakel gegen Ägypten und gegen Babel, in: Verbannung und Heimkehr. W. Rudolph zum 70. Geburtstage, 1961, S. 31–37 = Kl. Schr. IV, S. 32–38.

Sinai-Erzählung und Bileam-Sprüche, in: HUCA 32, 1961, S. 179–190 = Kl. Schr. IV, S. 21–31.

Die älteste Erzählung vom Sinaibund, in: ZAW 73, 1961, S. 137–146 = Kl. Schr. IV, S. 12–20.

Die Religion in Geschichte und Gegenwart. Bd. V, Tübingen [3]1961: Philo Byblius Sp. 346 bis 347; Sanchunjaton Sp. 1361.

Ahlström, G. W.: Psalm 89. Eine Liturgie aus dem Ritual des leidenden Königs. Aus dem Schwedischen übersetzt von H.-K. Hacker und R. Zeitler, 1959, in: OLZ 56, 1961, Sp. 374–375.

The Annual of the Leeds University Oriental Society, ed. by J. Macdonald. Vol. 1, [1958 bis 1959] 1959, in: ThLZ 86, 1961, Sp. 585.

Brongers, H. A.: Oud-Oosters en Bijbels Recht, 1960, in: BiOr 18, 1961, S. 262–263.

Cramer, K.: Genesis 1–11: Urgeschichte? 1959 (SGV 224/225), in: ThLZ 86, 1961, Sp. 270.

Dupont-Sommer, A.: Les écrits esséniens découverts près de la mer Morte, 1959, in: OLZ 56, 1961, Sp. 51–53.

Ders.: Die Essenischen Schriften vom Toten Meer. Übers. von W. W. Müller, 1960, in: OLZ 56, 1961, Sp. 152.

Gabrieli, F. (Ed.): L'Antica Società Beduina, 1959 (Studi Semitici 2), in: ThLZ 86, 1961, Sp. 111–112.

Hanhart, R. (Ed.): Maccabaeorum libri I–IV, fasc. III: Maccabaeorum liber III, 1960 (Septuaginta. Vetus Testamentum Graecum auctoritate Societatis Litterarum Gottingensis editum, vol. IX, 3) in: ThLZ 86, 1961, Sp. 666.

Harding, G. L.: The Antiquities of Jordan, 1959, in: OLZ 56, 1961, Sp. 269–270.

Hitti, P. K.: Syria. A Short History, 1959, in: OLZ 56, 1961, Sp. 61–62.

Josephus, Flavius: De Bello Judaico (Der jüdische Krieg). Zweisprachige Ausgabe der sieben Bücher. I. Hrsg. und mit einer Einleitung sowie mit Anmerkungen versehen von O. Michel und O. Bauerfeind, 1960, in: OLZ 56, 1961, Sp. 505–506.

Melzer, J. (Hrsg.): Deutsch-jüdisches Schicksal in dieser Zeit. Wegweiser durch das Schrifttum der letzten 15 Jahre (1945–1960), 1960, in: OLZ 56, 1961, Sp. 264.

Meysels, Th. F.: Israel. Schwann Reiseführer, Sonderausgabe, 1959, in: OLZ 56, 1961, Sp. 605–606.

Montet, P.: L'Égypte et la Bible, 1959 (CAB 11), in: OLZ 56, 1961, Sp. 148–149.

Palache, J. L.: Sinai en Paran. Opera Minora, 1959; Semantic Notes on the Hebrew Lexicon. Transl. and ed. by R. J. Z. Werblowsky, 1959, in: OLZ 56, 1961, Sp. 255–256.

Perez Castro, F.: Séfer Abiša'. Edicion del Fragmento antiguo del Rollo Sagrado del Pentateuco Hebreo Samaritano de Nablus, 1959, in: OLZ 56, 1961, Sp. 377–378.

Rengstorf, K. H.: Ḫirbet Qumran und die Bibliothek vom Toten Meer, 1960, in: OLZ 56, 1961, Sp. 609–610.

Sacra Pagina. Miscellanea Biblica congressus internationalis catholici de re biblica. Ed. J. Coppens, A. Descamps, E. Masseaux, 1959 (BEThL 12–13), in: BiOr 18, 1961, S. 94–96.

Schilling, O.: Die Höhlenfunde vom Toten Meer: Ende, Anfang oder Übergang? (Vortrag), 1958, in: OLZ 56, 1961, Sp. 51.

Studia Biblica et Orientalia. Vol. 3: Oriens Antiquus, 1959 (AnBibl 12), in: OLZ 56, 1961, Sp. 12–14.

Textus. Annual of the Hebrew University Bible Project. Vol. 1, ed. by C. RABIN, 1960, in: ThLZ 86, 1961, Sp. 752—753.

WRIGHT, G. E.: Biblische Archäologie. Übersetzt aus dem Amerikanischen von CHR. v. MERTENS, 1958, in: DLZ 82, 1961, Sp. 926.

1962

Kleine Schriften I. Hrsg. von R. SELLHEIM und F. MAASS. Tübingen 1962, VII, 279 S.

Hexateuch-Synopse. Die Erzählung der fünf Bücher Mose und des Buches Josua mit dem Anfange des Richterbuches in ihre vier Quellen zerlegt und in deutscher Übersetzung dargeboten samt einer in Einleitung und Anmerkungen gegebenen Begründung. 2., unveränderte Auflage. Darmstadt 1962, XVI, 108, 285* S.

The Promises of Grace to David in Is. 55, 1—5, in: Israel's Prophetic Heritage. Essays in honor of J. Muilenburg, 1962, S. 196—207 = Die Gnadenverheißungen an David in Jes. 55, 1—5, in: Kl. Schr. IV, S. 44—52.

Der Beitrag der Württembergischen Bibelanstalt zur Förderung der biblischen Wissenschaft, in: Die Bibel in der Welt, Bd. 5 [1961—1962]. Stuttgart 1962, S. 76—84. 1 Taf.

Ba'al Ṣaphon von Ugarit und Amon von Ägypten, in: FF 36, 1962, S. 338—340 = Kl. Schr. IV, S. 53—57. Taf. I—IV.

Voraussage-Empfang, Offenbarungs-Gewißheit und Gebetskraft-Erfahrung bei Jeremia, in: NT 5, 1962, S. 77—81 = Kl. Schr. IV, S. 58—62.

Pergamente und Papyri aus Dura-Europos (zu C. B. WELLES, R. O. FINK and J. F. GILLIAM: The Parchments and Papyri, 1959 [The Excavations at Dura-Europos. Final Report 5, 1]), in: OLZ 57, 1962, Sp. 347—351.

Mesopotamische Elemente in den alphabetischen Texten von Ugarit, in: Syria 39, 1962, S. 36—41 = Kl. Schr. IV, S. 39—43.

Die Religion in Geschichte und Gegenwart. Bd. VI, Tübingen [3]1962: Sinai Sp. 44—45; Sodom und Gomorrha Sp. 114—115; Syrien II. Religionsgeschichtlich Sp. 576; Tanit Sp. 611—612; Ugarit, 3. Religiöse Texte Sp. 1103—1106; Wellhausen Sp. 1594—1595; Zebaoth Sp. 1875—1876; Zehnten, 1. Religionsgeschichtlich Sp. 1877—1878.

[ALBRIGHT, W. F.:] The Bible and the Ancient Near East. Essays in honor of William Foxwell Albright. Ed. by G. E. WRIGHT, 1961, in: BiOr 19, 1962, S. 129—132.

Annual of the Department of Antiquities of Jordan. Vol. 4. 5, 1960, in: OLZ 57, 1962, Sp. 499.

The Apocrypha of the Old Testament. Revised Standard Version, 1957, in: BiOr 19, 1962, S. 71.

BENOIT, P., MILIK, J. T., DE VAUX, R.: Discoveries in the Judaean Desert II: Les Grottes de Murabba'ât. 1. Texte; 2. Planches, 1961, in: OLZ 57, 1962, Sp. 389—392.

BERNHARDT, K.-H.: Das Problem der Altorientalischen Königsideologie im Alten Testament, 1961 (SVT VIII), in: BiOr 19, 1962, S. 62—64.

BRATSIOTIS, N. P.: Einleitung in die Monologe des Jeremia [griech.], 1959 (Diss. theol. Athen), in: OLZ 57, 1962, Sp. 607.

CARMIGNAC, J. et GUILBERT, P.: Les Textes de Qumran. I: La règle de la communauté. La règle de la guerre. Les hymnes, 1961, in: OLZ 57, 1962, Sp. 506—507.

GERLEMAN, G.: Ruth. Liefg. 1, 1960 (BK 18, 1), 1960, in: OLZ 57, 1962, Sp. 387—388.

GORDON, C. H.: New Horizons in Old Testament Literature, 1960, in: OLZ 57, 1962, Sp. 606—607.

Handbuch der Orientalistik. Hrsg. von B. SPULER. 1. Abt., Bd. 1: Ägyptologie. Abschn. 1: H. KEES u. a. Ägyptische Schrift und Sprache, 1959, in: DLZ 83, 1962, Sp. 7—8.

HARDING, G. L.: Auf biblischem Boden. Die Altertümer in Jordanien. Übers. aus dem Engl. von G. STEIGERWALD, 1961, in: OLZ 57, 1962, Sp. 496—497.

HITTI, P. K.: The Near East in History. A 5000 Year Story, 1961, in: OLZ 57, 1962, Sp. 507—508.

HORST, F.: Hiob. Liefg. 1, 1960 (BK 16, 1); Liefg. 2, 1960 (BK 16, 2), in: OLZ 57, 1962, Sp. 43—44. 387—388.

JACOB, E.: Ras Shamra-Ugarit et l'Ancien Testament, 1960 (CAB 12), in: ThLZ 87, 1962, Sp. 203.

JALABERT, L. † et MOUTERDE, R.: Inscriptions Grecques et Latines de la Syrie. Tome V: Émésène, 1959 (BAH 66), in: OLZ 57, 1962, Sp. 151—153.

JEAN, CH. F. et HOFTIJZER, J.: Dictionnaire des inscriptions sémitiques de l'ouest. Livraisons 1—2: א-כתב, 1960, in: ThLZ 87, 1962, Sp. 349.

[KAUFMANN, Y.:] Yehezkel Kaufmann Jubilee Volume, 1960, in: OLZ 57, 1962, Sp. 498—499.

KRAUS, H.-J.: Psalmen. Liefg. 12—13 und Einleitung, 1960 (BK 15, 12—13), in: OLZ 57, 1962, Sp. 43—44.

KUHN, K. G.: Konkordanz zu den Qumrantexten, 1960, in: OLZ 57, 1962, Sp. 47—48.

LIVERANI, M.: Storia di Ugarit nell'età degli archivi politici, 1962 (Studi Semitici 6), in: BiOr 19, 1962, S. 262—263.

MAIER, J.: Die Texte vom Toten Meer. Bd. 1: Übersetzung; Bd. 2: Anmerkungen, 1960, in: OLZ 57, 1962, Sp. 503—504.

MONTET, P.: Das alte Ägypten und die Bibel. Aus dem Französischen übers. von M. THURNEYSEN, 1960, in: OLZ 57, 1962, Sp. 267—268.

MURPHY, R. E.: Seven Books of Wisdom, 1960, in: OLZ 57, 1962, Sp. 606.

La Regalità Sacra. Contributi al Tema dell' VIII Congresso Internazionale di Storia della Religioni (Roma, Aprile 1955), 1959 (Studies in the History of Religions, Supplements to Numen 4), in: BiOr 19, 1962, S. 92—94.

SINCLAIR, L. A.: An archaeological Study of Gibeah (Tell el-Fûl); CLEVELAND, R. L.: The Excavation of the Conway High Place (Petra) and Soundings at Khirbet Ader, 1960 (AASOR 34—35, 1954—1956), in: OLZ 57, 1962, Sp. 502.

[SUKENIK, E. L.:] Essays on the Dead Sea Scrolls in Memory of E. L. Sukenik, ed. by C. RABIN und Y. YADIN, 1961, in: OLZ 57, 1962, Sp. 503.

VRIEZEN, TH. C.: De Literatuur van Oud-Israel, [2]1962, in: BiOr 19, 1962, S. 273.

WOLFF, H. W.: Dodekapropheton, Hosea. Liefg. 3, 1961 (BK 14, 3), 1961, in: OLZ 57, 1962, Sp. 387—388.

YADIN, Y., AHARONI, Y., AMIRAN, RUTH, DOTHAN, TRUDE, DUNAYEWSKY, I., PERROT, J.: Hazor II. An Account of the Second Season of Excavations 1956, 1960, in: OLZ 57, 1962, Sp. 155—156.

Internationale Zeitschriftenschau für Bibelwissenschaft und Grenzgebiete, hrsg. von F. STIER, Bd. 6, [1958/59] 1960, in: OLZ 57, 1962, Sp. 264—265.

VAN ZYL, A. H.: The Moabites, 1960 (Pret. Or. Ser. 3), in: OLZ 57, 1962, Sp. 384—386.

1963

Kleine Schriften II. Hrsg. von R. SELLHEIM und F. MAASS. Tübingen 1963, VIII, 557 S., 6 Taf.

„Gut Glück!" in semitischer Namengebung, in: JBL 82, 1963, S. 195—200 = Kl. Schr. IV, S. 73—78.

Jakobs Begegnung mit El und Moses Begegnung mit Jahwe, in: OLZ 58, 1963, Sp. 325 bis 331 = Kl. Schr. IV, S. 92–98.

Jahwe, der Gott der Väter, in: ThLZ 88, 1963, Sp. 481–490 = Kl. Schr. IV. S. 79–91.

„Juda" in II Könige 14, 28 und „Judäa" in Apostelgeschichte 2, 9, in: WZ Halle 12, 1963, S. 229–238 = Kl. Schr. IV, S. 99–120.

Ein Psalm aus Nord-Israel. Micha 7, 7–20, in: ZDMG 112, [1962] 1963, S. 259–268 = Kl. Schr. IV, S. 63–72.

Adam, A.: Antike Berichte über die Essener, 1961 (KIT 182), in: OLZ 58, 1963, Sp. 48–49.

Abr-Nahrain. An Annual under the Auspices of the Department of Semitic Studies University of Melbourne, edited by J. Bowman. Vol. 1, [1959/60] 1961, in: Der Islam 38, 1963, S. 205; Vol. 2, [1960/61] 1962, in: Der Islam 38, 1963, S. 312.

The Annual of Leeds University Oriental Society, ed. by J. Macdonald. Vol 2, [1959 bis 1961] 1961, in: ThLZ 88, 1963, Sp. 339.

Aistleitner, J.: Wörterbuch der ugaritischen Sprache. Hrsg. von O. Eissfeldt, 1963 (BAL 106, 3), in: ZAW 75, 1963, S. 248.

Allegro, J. M.: The Treasure of the Copper Scroll, 1960, in: OLZ 58, 1963, Sp. 160–161.

[Alleman, H. C.:] Biblical studies in memory of H. C. Alleman. Ed. by J. M. Myres, O. Reimherr, H. N. Bream, 1960, in: OLZ 58, 1963, Sp. 39–40.

La Sainte Bible. Traduite en français sous la direction de l'École Biblique de Jerusalem: Les Livres de Chroniques, trad. par H. Cazelles, 1961; Les Livres de Maccabées, trad. de F.-M. Abel (†), 1961, in: OLZ 58, 1963, Sp. 164–165.

Blank, S. H.: Jeremiah. Man and Prophet, 1961, in: OLZ 58, 1963, Sp. 573.

Brongers, H. A.: De Jozefgeschiedenis bij Joden, Christenen en Mohammedanen, 1962, in: BiOr 20, 1963, S. 189–190.

Downey, G.: A History of Antioch in Syria from Seleucus to the Arab Conquest, 1961, in: OLZ 58, 1963, Sp. 262–264.

[Driver, G. R.:] Studies in Honor of G. R. Driver dedicated on the occasion of his seventieth birthday by pupils and scholars, 1962 (= JSSt 7, 2), in: BiOr 20, 1963, S. 243–245.

Fuks, A. and I. Halpem (Ed.): Studies in History, 1961 (Scripta Hierosolymitana 7), in: HZ 196, 1963, S. 178–179.

Gray, J.: Archaeology and the Old Testament World, 1962, in: ZDMG 113, 1963, S. 253.

Holm-Nielsen, S.: Hodayot. Psalms from Qumran, 1960 (Acta Theologica Danica 2), in: OLZ 58, 1963, Sp. 49–50.

Malamat, A. (Ed.): The Kingdoms of Israel and Judah [hebr., engl. summary], 1961, in: OLZ 58, 1963, Sp. 258–259.

Michaeli, F.: Textes de la Bible et de l'Ancien Orient, 1961 (CAB 13), in: OLZ 58, 1963, Sp. 257–258.

Mulder, M. J.: Ba'al in het Oude Testament, 1962 (Academisch Proefschrift Vrije Universiteit Amsterdam), in: BiOr 20, 1963, S. 59.

Napier, B. D.: Song of the Vineyard. A theological Introduction to the Old Testament, 1962, in: OLZ 58, 1963, Sp. 467–469.

Neher, A. et R. : Histoire Biblique du Peuple d'Israël. I. II, 1962, in: OLZ 58, 1963, Sp. 469–471.

Nilsson, M. P.: Geschichte der Griechischen Religion. II: Die Hellenistische und Römische Zeit. 2., durchgesehene und ergänzte Auflage, 1961 (Handbuch der Altertumswissenschaft 5, 2, 2), in: HZ 196, 1963, S. 478–479.

Ders.: Die Entstehung und religiöse Bedeutung des griechischen Kalenders, 1962 (Scripta Minora Regiae Societatis Humaniorum Litterarum Lundensis 1960/61), in: HZ 196, 1963, S. 793.

Nötscher, F.: Vom Alten zum Neuen Testament. Gesammelte Aufsätze, 1962 (BBB 17), in: OLZ 58, 1963, Sp. 356–357.

Noth, M.: Die Ursprünge des alten Israel im Lichte neuer Quellen, 1961 (Veröffentlichungen der Arbeitsgemeinschaft für Forschung des Landes Nordrhein-Westfalen 94), in: ZDMG 113, 1963, S. 210.

Oxford Bible Atlas. Edited by H. G. May, 1962, in: ZDMG 113, 1963, S. 253–254.

Pritchard, J. B.: Die Archäologie und das Alte Testament. Aus dem Amerikanischen übertragen von Margaretha von Reischach-Scheffel und H. P. Rüger, 1962, in: DLZ 1963, Sp. 148–150.

Ders.: The Water System of Gibeon, 1961 (Museum Monographs), in: OLZ 58, 1963, Sp. 474.

Propyläen-Weltgeschichte. Eine Universalgeschichte hrsg. von G. Mann und A. Heuss. Bd. 2: Hochkulturen des mittleren und östlichen Asien, 1962, in: HZ 197, 1963, S. 371 bis 377.

Rabin, C. (Ed.): Studies in the Bible, 1961 (Scripta Hierosolymitana 8), in: OLZ 58, 1963, Sp. 44–46.

[v. Rad, G.:] Studien zur Theologie der alttestamentlichen Überlieferungen. Hrsg. von R. Rendtorff und K. Koch, G. v. Rad zum 60. Geburtstag, 1961, in: OLZ 58, 1963, Sp. 162–163.

Renckens, H.: De Godsdienst van Israël, 1962, in: BiOr 20, 1963, S. 176–177.

[Steinmann, J.:] Code Sacerdotal I. Genèse–Exode. Texte français, introduction et commentaire par J. Steinmann, 1961, in: DLZ 84, 1963, Sp. 303.

Strack, H. L. und Billerbeck, P.: Kommentar zum Neuen Testament aus Talmud und Midrasch. 6. Bd.: Verzeichnis der Schriftgelehrten. Geographisches Register. Hrsg. von J. Jeremias in Verbindung mit K. Adolph, 1961, in: OLZ 58, 1963, Sp. 475–576.

Textus. Annual of the Hebrew University Bible Project. Vol. 2, ed. by C. Rabin, 1962, in: ThLZ 88, 1963, Sp. 422.

Widengren, G.: Iranisch-Semitische Kulturbegegnung in parthischer Zeit, 1960 (Veröffentlichungen der Arbeitsgemeinschaft für Forschung des Landes Nordrhein-Westfalen 70), in: OLZ 58, 1963, Sp. 169–170.

Wolff, H. W.: Dodekapropheton, Hosea. Liefg. 4, 1961 (BK 14, 4), in: OLZ 58, 1963, Sp. 363.

Internationale Zeitschriftenschau für Bibelwissenschaft und Grenzgebiete, hrsg. von F. Stier Bd. 7, [1960/61] 1961, in: OLZ 58, 1963, Sp. 50.

Ziegler, J. (Ed.): Sapientia Salomonis, 1962 (Septuaginta. Vetus Testamentum Graecum auctoritate Societatis Litterarum Gottingensis editum, vol. XII, 1) in ThLZ 88, 1963, Sp. 586–587.

Aistleitner, J.: Wörterbuch der ugaritischen Sprache. Hrsg. von O. Eissfeldt. Berlin 1963, VI. 362 S. (BAL 106,3),

Handbuch zum Alten Testament, hrsg. von O. Eissfeldt. Erste Reihe, Band 16: B. Gemser, Sprüche Salomos, 2. Aufl. Tübingen 1963.

1964

Einleitung in das Alte Testament unter Einschluß der Apokryphen und Pseudepigraphen sowie der apokryphen- und pseudepigraphenartigen Qumrān-Schriften. Entstehungsgeschichte des Alten Testaments. 3., neubearbeitete Auflage. Tübingen 1964, XVI, 1129 S.

Kanaanäisch-ugaritische Religion, in: Handbuch der Orientalistik, hrsg. von B. SPULER. I. Abt., VIII. Bd., 1. Abschn., Liefg. 1, 1964, S. 76–91.

Joseph Aistleitner (2. Mai 1883–9. September 1960), in: Sächsische Akademie der Wissenschaften zu Leipzig. Jahrbuch 1960–62, 1964, S. 401–403 = Kl. Schr. IV, S. 143–145.

Nachruf auf E. Klostermann (14. Februar 1870 bis 18. September 1963), in: Jahrbuch der Deutschen Akademie der Wissenschaften 1963, 1964, S. 227–229 = Kl. Schr. IV, S. 149–152.

Nachruf auf Karl Mras (6. Juni 1877 bis 7. Juli 1962), in: Jahrbuch der Deutschen Akademie der Wissenschaften 1963, 1964, S. 210–211 = Kl. Schr. IV, S. 146–148.

Amos und Jona in volkstümlicher Überlieferung, in: „... und fragten nach Jesus". Festschrift für Ernst Barnikol, 1964, S. 9–13 = Kl. Schr. IV, S. 137–142.

„Juda" und „Judäa" als Bezeichnungen nordsyrischer Bereiche, in: FF 38, 1964, S. 20–25 = Kl. Schr. IV, S. 121–131.

Achronische, anachronische und synchronische Elemente in der Genesis, in: JEOL 17, [1963] 1964, S. 148–164 = Kl. Schr. IV, S. 153–169.

5000 Jahre Orient. (zu F. K. KIENITZ: 5000 Jahre Orient. Ein Gang durch die Geschichte und Kultur des Nahen Ostens von den Pharaonen bis zur Gegenwart, 1962), in: OLZ 59, 1964, Sp. 533–535.

Artur Weiser zum 70. Geburtstag am 18. November 1963, in: ThLZ 89, 1964, Sp. 71–72.

Psalm 80 und Psalm 89, in: WO 3, [1964–66] 1964, S. 27–31 = Kl. Schr. IV, S. 132 bis 136.

AHARONI, Y.: Excavations at Ramat Raḥel. Seasons 1959 and 1960, 1962, in: OLZ 59, 1964, Sp. 475–476.

ANDERSON, H.: Historians of Israel (2). 1. and 2. Chronicles, Ezra, Nehemia, 1962, in: OLZ 59, 1964, Sp. 364.

BAILLET, M., MILIK, J. T. et R. DE VAUX: Discoveries in the Judaean Desert of Jordan. III: Les „Petites Grottes" de Qumran. Exploration de la falaise. Les Grottes 2 Q, 3 Q, 5 Q, 6 Q, 7 Q à 10 Q. Le rouleau de cuivre. 1. Textes. 2. Planches, 1962, in: OLZ 59, 1964, Sp. 267–270.

BARDTKE, H.: Die Handschriftenfunde in der Wüste Juda, 1962, in: ThLZ 89, 1964, Sp. 420–421.

BEWER, J. A.: The Literature of the Old Testament. Third Edition Completely revised by E. G. KRAELING, 1962, in: OLZ 59, 1964, Sp. 472–473.

CASSUTO, U.: The Documentary Hypothesis and the Composition of the Pentateuch. Transl. from the Hebrew by I. ABRAHAMS, 1961, in: BiOr 21, 1964, S. 72–73.

CROSS, Jr., F. M.: The Ancient Library of Qumran and modern Biblical Studies, 1961 (The Haskell Lectures 1956–1957), in: OLZ 59, 1964, Sp. 157.

DAHOOD, M. J.: Proverbs and Northwestern Semitic Philology, 1963, in: DLZ 85, 1964, Sp. 304.

[G. R. DRIVER:] Hebrew and Semitic Studies Presented to Godfrey Rolles Driver in Celebration of His Seventieth Birthday 20 August 1962. Edited by D. W. THOMAS and W. D. MCHARDY, 1963, in: ZDMG 114, 1964, S. 418–420.

Expedition. The Bulletin of the University Museum of the University of Pennsylvania, Vol. 5, Number 1. Philadelphia, 1962, in: OLZ 59, 1964, Sp. 540–541.

FREEDMAN, D. N. and WRIGHT, G. E.: The Biblical Archaeologist Reader, 1961, in: OLZ 59, 1964, Sp. 271.

[GLUECK, N.:] NELSON GLUECK. A Bibliography, 1962, in: OLZ 59, 1964, Sp. 270.

GORDON, C. H.: Geschichtliche Grundlagen des Alten Testaments. 2., überarb. deutsche

Auflage. Aus dem Engl. übertragen von H. Marfurt, 1961, in: OLZ 59, 1964, Sp. 367 bis 368.

Haussig, H. W. (Hrsg.): Wörterbuch der Mythologie. I. Abt.: Die alten Kulturvölker. 1.–4. Lieferung. o. J., in: OLZ 59, 1964, Sp. 149–151.

Horst, F.: Hiob. Liefg. 3 u. 4 (1. Hälfte), 1962/63 (BK 16, 3. 4), in: OLZ 59, 1964, Sp. 266–267.

Hvidberg, F. F.: Weeping and Laughter in the Old Testament. A Study of Canaanite Religion. Posthumous Edition in English, 1962, in: BiOr 21, 1964, S. 330–331.

Jean, Ch.-F. et Hoftijzer, J.: Dictionnaire des inscriptions sémitiques de l'ouest, Livraison 3: סכל-כתב, 1962, in: ThLZ 89, 1964, Sp. 910.

Jirku, A.: Kanaanäische Mythen und Epen aus Ras Schamra-Ugarit, 1962, in: OLZ 59, 1964, Sp. 474–475.

Kahle, P. E.: Die Kairoer Genisa. Untersuchungen zur Geschichte des hebräischen Bibeltextes und seiner Übersetzungen. Deutsche Ausgabe bearbeitet und durch den Druck geführt von R. Meyer, 1962, in: ThLZ 89, 1964, Sp. 262–263.

Matthiae, P.: Ars Syra. Contributi alla storia dell' arte figurativa siriana nelle età del Medio e Tardo Bronzo, 1962, in: OLZ 59, 1964, Sp. 573–574.

[Mouterde, R.:] Mélanges offerts au Père René Mouterde pour son 80e anniversaire. I. II, 1962/63 (MUB 37–38, 1960/61. 1962), in: OLZ 59, 1964. Sp. 359–363.

Parrot, A.: Der Louvre und die Bibel. Aus dem Französischen übers. von M.-R. Jung, 1961 (Bibel und Archäologie 5), in: ThLZ 89, 1964, Sp. 350–351.

Porteous, N. W.: Das Danielbuch übersetzt und erklärt. Aus dem Englischen von W. Beyerlin und R. Walz, 1962 (ATD 23), in: OLZ 59, 1964, Sp. 265–266.

Robinson, G.: Historians of Israel (1). 1. and 2. Samuel, 1. and 2. Kings, 1962, in: OLZ 59, 1964, Sp. 364.

Supplements to Vetus Testamentum. Vol. 9: Congress Volume Bonn 1962, 1963, in: ZAW 76, 1964, S. 117.

Thompson, R. J.: Penitence and Sacrifice in Early Israel outside the Levitical Law, 1963, in: ThLZ 89, 1964, Sp. 744–747.

Transactions. Glasgow University Oriental Society, ed. by C. J. Mullo Weir. Vol. 18, [1959–60] 1961, in: OLZ 59, 1964. Sp. 363–364.

Vriezen, Th. Ch.: De Godsdienst van Israel, 1963, in: BiOr 21, 1964, S. 209–210.

Ders.: Jahwe en zijn Stad (Rede), 1962 (Koninklijke Nederlandse Akademie van Wetenschappen), in: OLZ 59, 1964, Sp. 265.

Yadin, Y.: The Scroll of the war of the Sons of Light against the Sons of Darkness, ed. with Commentary and Introduction. Transl. from the Hebrew by Batya and Chaim Rabin, 1962, in: OLZ 59, 1964, Sp. 159–160.

Handbuch zum Alten Testament, hrsg. von O. Eissfeldt. Erste Reihe, Band 14: Th. H. Robinson und F. Horst, Die Zwölf Kleinen Propheten [Th. H. Robinson: Hosea bis Micha (Aus dem Englischen übersetzt vom Herausgeber); F. Horst: Nahum bis Maleachi]. Tübingen ^{3}1964; darin S. V: Vorwort des Herausgebers zur dritten Auflage; S. VII–IX: Das Zwölfprophetenbuch: Einleitung.

1965

The Old Testament. An Introduction, including the Apocrypha and Pseudepigrapha, and also the works of similiar type from Qumran. The History of the Formation of the Old Testament. Transl. by Peter R. Ackroyd. Oxford 1965, XXIX, 861 S.

Palestine in the Time of the Nineteenth Dynastie. b) The Exodus and the Wanderings, in: Cambridge Ancient History, Revised Ed., Vol. 2, Chapter XXVIb, Cambridge 1965, 32 S.

The Hebrew Kingdom, in: Cambridge Ancient History, Revised Edition. Vol. 2, Chapter XXXIV, Cambridge 1965, 80 S.

Stammessage und Menschheitserzählung in der Genesis. — Wahrheit und Dichtung in der Ruth-Erzählung. Berlin 1965, 28 S. (SAL 110, 4).

Neue keilalphabetische Texte aus Ras Schamra-Ugarit. Berlin 1965, 49 S. (SAB 1965, Heft 6).

Jakob-Lea und Jakob-Rahel, in: Gottes Wort und Gottes Land. Festschrift für Hans-Wilhelm Hertzberg, 1965, S. 50—55 = Kl. Schr. IV, S. 170—175.

'äh^{e}yäh '^{a}šär 'äh^{e}yäh und ᵓEl ʿôlām, in: FF 39, 1965, S. 298—300 = Kl. Schr. IV, S. 193 bis 198.

Baruchs Anteil an Jeremia 38, 28b—40, 6, in: Oriens Antiquus 4, 1965, S. 31—34 = Kl. Schr. IV, S. 176—180.

The Hebrew Scriptures (zu S. SANDMEL: The Hebrew Scriptures, 1963), in: OLZ 60, 1965, Sp. 445—454 = Kl. Schr. IV, S. 199—208.

Unheils- und Heilsweissagungen Jeremias als Vergeltung für ihm erwiesene Weh- und Wohltaten, in: WZ Halle 14, 1965, S. 181—186 = Kl. Schr. IV, S. 181—192.

Abr-Nahrain. An Annual Published by the Department of Semitic Studies University of Melbourne in Association with the Department of Semitic Studies University of Sydney, ed. by J. BOWMAN. Vol. 3, [1961—62] 1963, in: Der Islam 41, 1965, S. 307.

ALBRIGHT, W. F.: The Biblical Period from Abraham to Ezra, 1963 (Harper Torchbooks. The Cloister Library, TB 102), in: OLZ 60, 1965, Sp. 574.

The Annual of Leeds University Oriental Society, ed. by J. MACDONALD. Vol. 3, [1961—62] 1963, in: ThLZ 90, 1965, Sp. 103—104.

BARAMKI, D.: Phoenicia and the Phoenicians, 1961, in: OLZ 60, 1965, Sp. 369.

BEEK, M. A.: Atlas of Mesopotamia. A survey of the history and civilisation of Mesopotamia from the Stone Age to the Fall of Babylon. Transl. by D. R. WELSH, ed. by H. H. ROWLEY, 1962, in: OLZ 60, 1965, Sp. 41—42.

BISHOP, E. F. F.: Prophets of Palestine. The Local Background to the Preparation of the Way, 1962, in: OLZ 60, 1965, Sp. 165—166.

BRAUN, F.-M.: The Work of Père Lagrange. Adapted from the French by R. T. A. MURPHY, 1963, in: OLZ 60, 1965, Sp. 162—163.

BUIS, P. et LECLERCQ, J.: Le deutéronome, 1963 (Sources Bibliques), in: ThLZ 90, 1965, Sp. 177—179.

FOHRER, G.: Überlieferung und Geschichte des Exodus. Eine Analyse von Exodus 1—15, 1964 (BZAW 91), in: ZAW 77, 1965, S. 114—115.

FRANKEN, H. J.: Heilig-Land en heilige huisjes (Antrittsvorlesung Rijksuniversiteit Leiden), 1962, in: OLZ 60, 1965, Sp. 165.

FRANKEN, H. J. and FRANKEN-BATTERSHILL, C. A.: A Primer of Old Testament Archaeology, 1963, in: OLZ 60, 1965, Sp. 163—165.

GEMSER, B. u. a.: Studies on Psalms, 1963 (Oudtestamentische Studiën 13), in: ThLZ 90, 1965, Sp. 28—29.

HABEL, N. C.: Yahweh versus Baal: A Conflict of Religious Cultures. A Study in the Relevance of Ugaritic Materials for the Early Faith of Israel, 1964, in: ThLZ 90, 1965, Sp. 674—676.

HARDEN, D.: The Phoenicians, 1962 (Ancient Peoples and Places 26), in: OLZ 60, 1965, Sp. 166—167.

HEMPEL, J.: Geschichten und Geschichte im Alten Testament bis zur persischen Zeit, 1964, in: HZ 200, 1965, S. 678.

JOSEPHUS, FLAVIUS: De Bello Judaico (Der Jüdische Krieg). Griechisch und Deutsch. Hrsg. und mit einer Einleitung sowie mit Anmerkungen versehen von O. MICHEL und O. BAUERNFEIND. Band I: Buch I–III. Zweite überprüfte Auflage, 1962; Band II, 1: Buch IV–V, 1963, in: HZ 200, 1965, S. 688.

KÜLLING, S. R.: Zur Datierung der „Genesis-P-Stücke", namentlich des Kapitels Genesis XVII, 1964, in: BiOr 22, 1965, S. 48–49.

MOWINCKEL, S.: Tetrateuch–Pentateuch–Hexateuch. Die Berichte über die Landnahme in den drei altisraelitischen Geschichtswerken 1964 (BZAW 90), in: ThLZ 90, 1965, Sp. 750–752.

[NEUMANN, A. A.:] Studies and Essays in Honor of Abraham A. Neumann. Ed. by M. BEN-HORIN, B. D. WEINRYB, S. ZEITLIN, 1962, in: OLZ 60, 1965, Sp. 47–48.

NORTH, CH. R.: The Second Isaiah. Introduction, Translation and Commentary to Chapters XL–LV, 1964, in BiOr 22, 1965, S. 51–52.

SCHAEFFER, CL. F.-A.: Ugaritica IV. Découvertes des XVIIIe et XIXe campagnes, 1954 bis 1955, 1962 (Missions de Ras Shamra, Tome 15), in: OLZ 60, 1965, Sp. 574–577.

SIBINGA, J. S.: The Old Testament of Justin Martyr. I. The Pentateuch, 1963, in: ThLZ 90, 1965, Sp. 341.

[SILVER, H.:] In the Time of Harvest. Essays in Honor of Abba Hillel Silver on the Occasion of his 70th Birthday. Editor: D. J. SILVER, 1963, in: OLZ 60, 1965, Sp. 370–371.

Sources Orientales: La Naissance du Monde, 1959; Les Danses Sacrées, 1963, in: Der Islam 41, 1965, S. 306.

Textus. Annual of the Hebrew University Bible Project. Vol. 3, ed. by C. RABIN, 1963, in: ThLZ 90, 1965, Sp. 263–264.

WELLHAUSEN, J.: Die kleinen Propheten. Übers. und erklärt. 4., unveränderte Auflage, 1963, in: DLZ 86, 1965, S. 486–487.

WOLFF, H. W.: Dodekapropheton, Joel. Liefg. 5, 1963 (BK 14, 5), in: OLZ 60, 1965, Sp. 573–574.

Internationale Zeitschriftenschau für Bibelwissenschaft und Grenzgebiete, hrsg. von F. STIER. Bd. 8, [1961/62] 1962; 9, [1962/63] 1963, in: OLZ 60, 1965, Sp. 262–263.

AISTLEITNER, JOSEPH: Wörterbuch der ugaritischen Sprache, hrsg. von O. EISSFELDT. 2., verb. Auflage. Berlin 1965, XVI, 362 S. (SAL 106, 3).

1966

Kleine Schriften III. Hrsg. von R. SELLHEIM und F. MAASS. Tübingen 1966, VIII, 529 S., 12 Taf.

Franz Delitzsch und Wolf Graf Baudissin. Dokumente über die Anfänge ihrer Freundschaft aus dem Erlanger Wintersemester 1866/67. Berlin 1966, 28 S., XXI Taf. (SAL 112, 1).

Die Komposition der Sinai-Erzählung Exodus 19–34. Berlin 1966, 31 S. (SAL 113, 1).

Die israelitisch-jüdische Religion, in: Saeculum-Weltgeschichte, Band 2, 1966, S. 217–260. 621. Taf. XVII.

Wahrsagung im Alten Testament, in: Travaux du centre d'études supérieures spécialisé d'histoire des religions de Strasbourg: La divination en Mésopotamie ancienne et dans les régions voisines, 1966, S. 141–146 = Kl. Schr. IV, S. 271–275.

Gottesnamen in Personennamen als Symbole menschlicher Qualitäten, in: Festschrift Walter Baetke, 1966, S. 110–117 = Kl. Schr. IV, S. 276–284.

Israels Führer in der Zeit vom Auszug aus Ägypten bis zur Landnahme, in: Studia Biblica et Semitica Theodoro Christiano Vriezen dedicata, 1966, S. 62–70 = Kl. Schr. IV, S. 297–304.

Paul Kahle (21. Januar 1875 bis 24. September 1964), in: AfO 21, 1966, S. 256–257 = Kl. Schr. IV, S. 215–218.

Johannes Hempel (30. Juli 1891 bis 9. Dezember 1964), in: AfO 21, 1966, S. 268 = Kl. Schr. IV, S. 219–220.

Die Komposition der Sinai-Erzählung Ex 19–34, in: FF 40, 1966, S. 213–215 = Kl. Schr. IV, S. 231–237.

Deuteronomium und Hexateuch, in: MIOF 12, 1966, S. 17–39 = Kl. Schr. IV, S. 238–258.

Etymologische und archäologische Erklärung alttestamentlicher Wörter, in: Oriens Antiquus 5, 1966, S. 165–176 = Kl. Schr. IV, S. 285–296.

Erwägungen zur Pentateuchquellenfrage (zu Sigmund Mowinckel: Erwägungen zur Pentateuchquellenfrage, 1964), in: OLZ 61, 1966, Sp. 213–218 = Kl. Schr. IV, S. 259 bis 263.

Sohnespflichten im Alten Orient, in: Syria 43, 1966, S. 39–47 = Kl. Schr. IV, S. 264 bis 270.

Sechs Jahrzehnte alttestamentlicher Wissenschaft, in: SVT 15, 1966, S. 1–13.

Das Gesetz ist zwischeneingekommen. Ein Beitrag zur Analyse der Sinai-Erzählung Ex 19–34 (Erdmann Schott zum 65. Geburtstag), in: ThLZ 91, 1966, Sp. 1–6 = Kl. Schr. IV, S. 209–214.

Goethes Beurteilung des kultischen Dekalogs von Ex 34 im Lichte der Pentateuchkritik (Hans Urner zu seinem 65. Geburtstag am 21. Mai 1966), in: ZThK 63, 1966, S. 135–144 = Kl. Schr. IV, S. 221–230.

[Aistleitner, J.:] Die mythologischen und kultischen Texte aus Ras Schamra. Übers. von J. Aistleitner †, ²1964, in: DLZ 87, 1966, Sp. 105 und OLZ 61, 1966, Sp. 155.

Altmann, A. (Ed.): Biblical and Other Studies, 1963 (Philip W. Lown Institute of Advanced Judaic Studies. Brandeis University, Studies and Texts, Vol. 1), in: OLZ 61, 1966, Sp. 46–47.

The Annual of Leeds University Oriental Society, ed. by J. Macdonald. Vol. 4, [1962 bis 1963] 1964, in: ThLZ 91, 1966, Sp. 737–738.

Baramki, D.: Die Phönizier. Aus dem Englischen von Helga Seeden, 1965 (Urban Bücher. Die Wissenschaftliche Taschenbuchreihe 85), in: HZ 203, 1966, S. 223.

de Boer, P. A. H. (Ed.): Oudtestamentische Studiën. Bd. 14: כה 1940–65, 1965, in: ThLZ 91, 1966, Sp. 658.

Carmignac, J., Cothenet, E. et H. Lignée: Les Textes de Qumran trad. et annotés. II. Règle de la congrégation. Recueil des Bénédictions. Interprétations de Prophètes et de Psaumes. Document de Damas. Apocryphe de la Genèse. Fragments des Grottes 1 et 4, 1963, in: OLZ 61, 1966, Sp. 51.

Clairmont, C. W.: The Glass Vessels, 1963 (The Excavations at Dura-Europos. Final Report 4, 5), in: OLZ 61, 1966, Sp. 50–51.

Dulière, W. L.: De la Dyade à l'Unité par la Triade, 1965, in: HZ 202, 1966, S. 240.

Falk, Z. W.: Hebrew Law in Biblical Times. An Introduction, 1964, in: ZDMG 116, 1966, S. 375–376.

Galling, K.: Studien zur Geschichte Israels im persischen Zeitalter, 1964, in: OLZ 61, 1966, Sp. 155–156.

Kenyon, Kathleen M.: Amorites and Canaanites, 1966 (The Schweich Lectures 1963), in: BiOr 23, 1966, S. 306.

Kitagawa, J. M.: Gibt es ein Verstehen fremder Religionen? Mit einer Biographie Joachim Wachs und einer vollständigen Bibliographie seiner Werke, 1963 (Beihefte der ZRGG 6 = Joachim Wach-Vorlesungen 1), in: OLZ 61, 1966, Sp. 156.

Kroeber, R.: Der Prediger. Hebräisch und deutsch, 1963 (Schriften und Quellen der alten Welt 13), in: OLZ 61, 1966, Sp. 49–50.

Lamparter, H.: Prophet wider Willen. Der Prophet Jeremia, 1964 (Die Botschaft des Alten Testaments 20), in: BiOr 23, 1966, S. 75.

Leclant, J. u. a.: Les Sagesses du Proche-Orient ancien, 1963 (Travaux du centre d'études supérieures spécialisé d'histoire des religions de Strasbourg), in OLZ 61, 1966, Sp. 442 bis 443.

Mansoor, M.: The Dead Sea Scrolls. A College Textbook and a Study Guide, 1964, in: OLZ 61, 1966, Sp. 573.

Mesnil du Buisson, R.: Les Tessères et les Monnaies de Palmyre, 1962, in: OLZ 61, 1966, Sp. 472.

Pritchard, J. B.: The Bronze Age Cemetery at Gibeon, 1963, in: OLZ 61, 1966, Sp. 41 bis 42.

Ringgren, H.: Israelitische Religion, 1963 (Die Religionen der Menschheit, Bd. 26), in: OLZ 61, 1966, Sp. 47–48.

Rowley, H. H.: The Servant of the Lord and other Essays on the Old Testament, 2nd. rev. Ed., 1965, in: ThLZ 91, 1966, Sp. 504–505.

Ruppert, L.: Die Josephserzählung der Genesis. Ein Beitrag zur Theologie der Pentateuchquellen, 1965 (StANT 11), in: ThLZ 91, 1966, Sp. 820–821.

Seierstad, I. P.: Die Offenbarungserlebnisse der Propheten Amos, Jesaja und Jeremia, ^{2}1965 (ANVAO 46, 2), in: BiOr 23, 1966, S. 183.

1967

Syrien und Palästina vom Ausgang des 11. bis zum Ausgang des 6. Jahrhunderts v. Chr. Vom Aufkommen des Königtums in Israel bis zum Ende des jüdischen Exils, in: Fischer Weltgeschichte. Band 4: Die Altorientalischen Reiche. III. Die erste Hälfte des 1. Jahrtausends. Frankfurt/M., 1967, S. 135–203.

„Bist du Elia, so bin ich Isebel" (I Kön 19, 2), in: Hebräische Wortforschung. Festschrift W. Baumgartner, 1967, S. 65–70. 1 Abb. (SVT 16) = Kl. Schr. V, S. 34–38. Taf. I.

Rechtskundige und Richter in Esther 1, 13–22, in: Festschrift für W. Eilers, 1967, S. 164 bis 166 = Kl. Schr. V, S. 31–33.

Die Komposition von I Reg 16, 29 – II Reg 13, 25, in: Das ferne und nahe Wort. Festschrift L. Rost, 1967, S. 49–58 (BZAW 105) = Kl. Schr. V, S. 21–30.

מַרְזֵחַ und מַרְזְחָא „Kultmahlgenossenschaft" im spätjüdischen Schrifttum (Ethelbert Stauffer zu seinem Geburtstag am 8. Mai 1967 als Manuskript überreicht), in: Kl. Schr. V, S. 136–142.

Israels Religion und die Religionen seiner Umwelt, in: NZSTh 9, 1967, S. 8–27 = Kl. Schr. V, S. 1–20.

Abr-Nahrain. An Annual Published by the Department of Semitic Studies University of Melbourne in Association with the Department of Semitic Studies University of Sydney, ed. by J. Bowman. Vol. 4, [1963–64] 1965, in: Der Islam 43, 1967, S. 236.

The Annual of Leeds University Oriental Society, ed. by J. Macdonald. Vol. 5, [1963 bis 1965] 1966, in: ThLZ 92, 1967, Sp. 818–819.

AP-THOMAS, D. R.: A Primer of Old Testament Text Criticism. 2nd revised Edition, 1965, in: DLZ 62, 1967, Sp. 261–262.

BARON, S. W.: History and Jewish Historians. Essays and Adresses. Compiled with a Foreword by A. HERTZBERG and L. A. FELDMAN, 1964, in: OLZ 62, 1967, Sp. 364.

BELTZ, W.: Die Kaleb-Traditionen oder Ein Beitrag zur theoretischen Diskussion in der Religionswissenschaft, 1966 (Diss. Budapest = Theologische Studien, Neue Folge 1), in: BiOr 24, 1967, S. 105–106.

DE BOER, P. A. H. (Ed.): Volume du Congrès. Genève 1965, 1966 (SVT 15), in: OLZ 62, 1967, Sp. 573–574.

BEIGHT, J.: Geschichte Israels. Von den Anfängen bis zur Schwelle des Neuen Bundes. Übers. von URSULA SCHIERSE, 1966, in: ThLZ 92, 1967, Sp. 498.

GLUECK, N.: Deities and Dolphins. The Story of the Nabataeans, 1965, in: OLZ 62, 1967, Sp. 571–572.

GOSHEN-GOTTSTEIN, M. H.: Hebrew and Semitic Languages. An Outline Introduction, 1965, in: OLZ 62, 1967, Sp. 156.

HARDING, G. L.: Baalbek, a new guide, 1963,in: OLZ 62, 1967, Sp. 471–472.

HERBERT, G.: The Old Testament from within, 1965, in: BiOr 24, 1967, S. 205.

JEAN, CH. F. et HOFTIJZER, J.: Dictionnaire des inscriptions sémitiques de l'ouest. Livraison 4–5: סכל–תתר, 1965, in: ThLZ 92, 1967, Sp. 353.

JIRKU, A.: Von Jerusalem nach Ugarit. Gesammelte Schriften. Ergänzt durch neue Anmerkungen und Tafeln, 1966, in: OLZ 62, 1967, Sp. 570–571.

DERS.: Der Mythus der Kanaanäer, 1966, in: HZ 205, 1967, S. 182.

MAZAR, B., DOTHAN, TRUDE, DUNAYEVSKY, I.: En-Gedi. The First and Second Seasons of Excavations 1961–1962, 1966 ('Atiqot. English Series 5), in: OLZ 62, 1967, Sp. 569 bis 570.

MOUTERDE, R.: Regards sur Beyrouth phénicienne, hellénistique et romain, 1966, in: OLZ 62, 1967, Sp. 570.

MULDER, M. J.: Kanaanitische Goden in het Oude Testament, 1965 (Exegetica. Oud-en Nieuw-Testamentische Studien IV, 4/5), in: OLZ 62, 1967, Sp. 466–467.

PRITCHARD, J. B.: Vinery, Defenses and Soundings at Gibeon, 1964 (Museum Monographs) in: OLZ 62, 1967, Sp. 154–155.

RAHLFS, A. †: Septuaginta-Studien I–III. 2. Aufl., vermehrt um einen unveröffentlichten Aufsatz und eine Bibliographie mit einem Nachruf von W. BAUER, 1965, in: ThLZ 92, 1967, Sp. 108.

Recent Progress in Biblical Scholarship. Published by The Lincombe Lodge Research Library, 1965, in: OLZ 62, 1967, Sp. 468.

Saeculum Weltgeschichte. I.: Ursprung und Frühkulturen. Primäre Zentren der Hochkultur. Weltgeschichtliche Berührungszonen. Von E. BAYER, D. O. EDZARD u. a., 1965, in: OLZ 62, 1967, Sp. 552–555.

SCHARBERT, J.: Die Propheten Israels bis 700 v. Chr., 1965, in: BiOr 24, 1967, S. 77.

VAN SELMS, A.: Genesis. Deel I. II, 1967 (De Prediking van het Oude Testament), in: BiOr 24, 1967, S. 354.

Textus. Annual of the Hebrew University Bible Project. Vol. 4, ed. by S. TALMON, 1964, in: ThLZ 92, 1967, Sp. 25–26; Vol. 5, ed. by S. TALMON, 1966, in: ThLZ 92, 1967, Sp. 352–353.

[TRANTHAM, H.:] The Teacher's Yoke: Studies in Memory of Henry Trantham, hrsg. von J. B. ADAIR, E. J. VARDAMAN, J. L. GARETT, 1964, in: OLZ 62, 1967, Sp. 254–255.

WINNETT, F. V. and REED, W. L.: The Excavations at Dibon (Dhiban) in Moab. I: The

First Campaign, 1950–51. By F. V. WINNETT; II: The Second Campaign, 1952. By W. L. REED, 1964 (AASOR 36–37 [1957–58]) in: OLZ 62, 1967, Sp. 259.

Internationale Zeitschriftenschau für Bibelwissenschaft, hrsg. von F. STIER, Bd. 10, [1963/64] 1964, in: OLZ 62, 1967, Sp. 42; Bd. 11, [1964/65] 1965, in: OLZ 62, 1967, Sp. 574–575.

ZIEGLER, J. (Ed.): Sapientia Jesu filii Sirach, 1965 (Septuaginta. Vetus Testamentum Graecum auctoritate Societas Litterarum Gottingensis editum, vol. XII, 2), in ThLZ 92, 1967, Sp. 269.

AISTLEITNER, JOSEPH: Wörterbuch der ugaritischen Sprache, hrsg. von O. EISSFELDT. 3., durchgesehene und ergänzte Auflage. Berlin 1967, XVI, 364 S. (SAL 106, 3).

1968

Kleine Schriften IV. Hrsg. von R. SELLHEIM und F. MAASS. Tübingen 1968, VIII, 304 S., 4 Taf.

Eine Qumran-Textform des 91. Psalms, in: Bibel und Qumran. Beiträge zur Erforschung der Beziehungen zwischen Bibel- und Qumranwissenschaft Hans Bardtke zum 22. 9. 1966, 1968, S. 82–85 = Kl. Schr. V, S. 45–49.

Der kanaanäische El als Geber der den israelitischen Erzvätern geltenden Nachkommenschaft- und Landbesitzverheißungen, in: Studia Orientalia in memoriam Caroli Brockelmann 1968, S. 45–53 (WZ Halle 15, 2/3) = Kl. Schr. V, S. 50–62.

Neue Belege für das phönizische Hermes-Emblem, in: In Memoriam Paul Kahle, 1968, S. 124–128. Taf. 8–15 (BZAW 103) = Kl. Schr. V, S. 63–67. Taf. II–VI.

Renaming in the Old Testament, in: Words and Meanings. Essays presented to David Winton Thomas, ed. by P. R. ACKROYD and B. LINDARS, 1968, S. 69–79 = Umnennungen im Alten Testament, in: Kl. Schr. V, S. 68–76.

Die Lade Jahwes in Geschichtserzählung, Sage und Lied, in: Das Altertum 14, 1968, S. 131–145 = Kl. Schr. V, S. 77–93. Taf. VII.

Protektorat der Midianiter über ihre Nachbarn im letzten Viertel des 2. Jahrtausends v. Chr. (mit Beiträgen von W. F. ALBRIGHT), in: JBL 87, 1968, S. 383–393; erweiterte Fassung in: Kl. Schr. V, S. 94–105.

ANDERSON, G. W.: The History and Religion of Israel, 1966 (The New Clarendon Bible. Old Testament I), in: ThLZ 93, 1968, Sp. 97–98.

BARNETT, R. D.: Illustrations of Old Testament History, 1966, in: OLZ 63, 1968, Sp. 468.

BRONGERS, H. A.: I Koningen, 1967 (De Prediking van het Oute Testament), in: BiOr 25, 1968, S. 220–221.

VAN DAALEN, ALEIDA G.: Simson. Een onderzoek naar de plaats, de opbouw en de funktie van het Simsonverhaal in het kader van de Oudtestamentische geschiedschrijving, 1966 (Studia Semitica Neerlandica 8), in: BiOr 25, 1968, S. 66.

DAHOOD, M. J.: Ugaritic-Hebrew Philology. Marginal Notes on Recent Publications, 1965 (Biblica et Orientalia 17), in: OLZ 63, 1968, Sp. 39.

DAVICO, A. u. a.: Missione Archaeologica Italiana in Siria. Rapporto preliminare della Campagna 1964, 1965, in: OLZ 63, 1968, Sp. 465–466.

VAN DIJK, H. J.: Ezekiel's Prophecy on Tyre (Ez. 26, 1–28, 19). A New Approach, 1968 (Biblica et Orientalia 20), in: ThLZ 93, 1968, Sp. 733–734.

DRIVER, G. R.: The Judaean Scrolls. The Problem and a Solution, 1965, in: OLZ 63, 1968, Sp. 147–148.

DUBARLE, A. M.: Judith. Formes et sens des diverses traditions. I. Études; II. Textes, 1966 (Analecta Biblica 24), in: OLZ 63, 1968, Sp. 362–363.

Fauth, W.: Aphrodite Parakyptusa. Untersuchungen zum Erscheinungsbild der vorderasiatischen Dea Prospiciens, 1967 (AAMz 1966, Nr. 6), in: GGA 220, 1968, S. 302–309 = Kl. Schr. V, S. 106–112.

Gerleman, G.: Ruth. Das Hohelied. Liefg. 2 u. 3, 1963/1965 (BK XVIII, 2. 3), in: OLZ 63, 1968, Sp. 39–40.

Goshen-Gottstein, M. H.: The Book of Isaiah. Sample Edition with Introduction, 1965 (The Hebrew University Bible Project), in: OLZ 63, 1968, Sp. 361–362.

Hanhart, R. (Ed.): Esther, 1966 (Septuaginta. Vetus Testamentum Graecum auctoritate Academiae Litterarum Gottingensis editum, vol. VIII, 3), in: ThLZ 93, 1968, Sp. 97.

Hoftijzer, J.: Religio Aramaica. Godsdienstige Verschijnselen in Aramese Teksten, 1968 (MEOL 16), in: BiOr 25, 1968, S. 224.

Horaïn, C. M.: L'identité de Masada et de quelques autres forteresses de la Palestine Ancienne, 1966, in: OLZ 63, 1968, Sp. 150.

Hyatt, J. P. (Ed.): The Bible in Modern Scholarship. Papers Read at the 10th Meeting of the Society of Biblical Literature, December 28–30, 1964, 1965, in: OLZ 63, 1968, Sp. 467–468.

Kapelrud, A. S.: Die Ras-Schamra-Funde und das Alte Testament. Übersetzt von F. Cornelius, 1967, in: ThLZ 93, 1968, Sp. 337.

Livre d'Or de l'Institut d'Archéologie Reine Elisabeth de Belgique de l'Université Hébraïque de Jérusalem, édité à l'occasion du 40me anniversaire de l'Université Hébraïque de Jérusalem, 1965, in: OLZ 63, 1968, Sp. 152–153.

Mansel, A. M.: Die Ruinen von Side, 1963 (Deutsches Archäologisches Institut, Abt. Istanbul), in: OLZ 63, 1968, Sp. 466–467.

Rabin, C. and Yadin, Y.: Aspects of the Dead Sea Scrolls, ²1965 (Scripta Hierosolymitana 4), in: OLZ 63, 1968, Sp. 470.

Rabinowitz, I.: Towards a Valid Theory of Biblical Hebrew Literature (The Classical Tradition: Literary and Historical Studies in Honor of Harry Caplan, ed. by L. Wallach, 1966, S. 315–328), in: OLZ 63, 1968, S. 470.

Reallexikon der Assyriologie. Bd. III, 1–3, hrsg. von E. Ebeling †, durch den Druck geführt und ergänzt von E. Weidner, S. 1–232: F-Geschwulst, 1957–1964, in: ThLZ 93, 1968, Sp. 25; Bd. III, 4, hrsg. von W. v. Soden, S. 233–320: Geschwulst–Gewand, 1966, in: ThLZ 93, 1968, Sp. 575; Bd. III, 5, S. 321–400: Gewand–Girsu, 1968, in: ThLZ 93, 1968, Sp. 905–906.

Roth, W. H. W.: Numerical Sayings in the Old Testament. A Form-Critical Study, 1965 (SVT 13), in: OLZ 63, 1968, Sp. 469–470.

Sanders, J. A.: The Psalm Scroll of Qumran Cave 11 (11 Q Ps a), 1965 (Discoveries of the Judaean Desert of Jordan IV), in: OLZ 63, 1968, Sp. 148–149.

Scharbert, J.: Die Propheten Israels um 600 v. Chr., 1967, in: BiOr 25, 1968, S. 221.

Stadelmann, R.: Syrisch-palästinische Gottheiten in Ägypten, 1967 (Probleme der Ägyptologie, 5), in: OLZ 63, 1968, Sp. 551–552.

Vanel, A.: L'Iconographie du Dieu de l'Orage dans le Proche-Orient Ancien jusqu'au VIIe siècle avant J.-C., 1965 (Cahiers de la Revue Biblique 3), in: OLZ 63, 1968, Sp. 155 bis 156.

Wildberger, H.: Jesaja. Liefg. 1, 1965 (BK X, 1), in: OLZ 63, 1968, Sp. 470.

Wolff, H. W.: Dodekapropheton 1. Hosea, 2., verb. u. ergänzte Auflage, 1965 (BK XIV, 1), in: OLZ 63, 1968, Sp. 39.

Wort und Botschaft. Eine theologische und kritische Einführung in die Probleme des Alten Testaments. Hrsg. von J. Schreiner, 1967, in: Bibl 49, 1968, S. 287–292.

Yadin, Y.: The Ben Sira Scroll from Masada. With Introduction, Emendations and Commentary [hebr.], 1965, in: OLZ 63, 1968, Sp. 149–150.

Yadin, Y., Aharoni, Y., Amiran, Ruth, Dothan, Trude, Dunayewsky, I., and J. Perrot: Hazor III–IV. An account of the third and fourth Season of Excavations, 1957 bis 1958, in: OLZ 63, 1968, Sp. 252–253.

Internationale Zeitschriftenschau für Bibelwissenschaft und Grenzgebiete, hrsg. von F. Stier. Bd. 12, [1965/66] 1967, in: OLZ 63, Sp. 566–567.

1969

Bleiben im Hause Jahwes, in: Beiträge zur Alten Geschichte und deren Nachleben. Festschrift für Franz Altheim zum 6. 10. 1968, 1969, S. 76–81 = Kl. Schr. V, S. 113–117.

„Das Gesetz des Menschen" in 2. Sam 7, 19 (Der Stiftung Johanneum in Berlin zu ihrer Jahrhundertfeier am 2. November 1969 als Manuskript überreicht), in: Kl. Schr. V, S. 143–151.

Kultvereine in Ugarit, in: Ugaritica VI, 1969, S. 187–195 = Kl. Schr. V, S. 118–126.

Nachträge zu „Adrammelek und Demarus" und zu „Bist Du Elia, so bin ich Isebel (1 Kön. XIX 2)", in: BiOr 26, 1969, S. 182–184 = Kl. Schr. V, S. 39–42.

Neue Belege für nabatäische Kultgenossenschaften, in: MIOF 15, 1969, S. 216–227 = Kl. Schr. V, S. 127–135. Taf. VIII.

Abr-Nahrain. An Annual Published by the Department of Semitic [Middle Eastern] Studies University of Melbourne in Association with the Department of Semitic Studies University Sidney, ed. by J. Bowman. Vol. 5, [1964–65] 1966; 6, [1965–66] 1967, in: Der Islam 45, 1969, S. 231–232.

Aharoni, Y.: The Land of the Bible. A Historical Geography. Translated from the Hebrew by A. F. Rainey, 1967, in: OLZ 64, 1969, Sp. 464–465.

Brunet, G.: Les Lamentations contre Jérémie, 1968, in: OLZ 64, 1969, Sp. 818–819.

[Dahood, M. J.] Hebrew-Ugaritic Index to the Writings of Mitchell J. Dahood. A Bibliography with Indices of Scriptural Passages, Hebrew and Ugaritic Words and Grammatical Observations, compiled by E. R. Martinez, 1967, in: OLZ 64, 1969, Sp. 352–354.

Dothan, M. and D. N. Freedman: Ashdod I. The First Season of Excavations 1962, 1967 ('Atiqot, English Series 7), in: OLZ 64, 1969, Sp. 353–354.

Gröndahl, F.: Die Personennamen der Texte aus Ugarit, 1967, in: ThLZ 94, 1969, Sp. 655.

Hachmann, R. und A. Kuschke: Bericht über die Ergebnisse der Ausgrabungen in Kamid el-Loz (Libanon) in den Jahren 1963 und 1964, 1966 (Saarbrücker Beiträge zur Altertumskunde 3), in: OLZ 64, 1969, Sp. 354–355.

Henninger, J.: Über Lebensräume und Lebensformen der Frühsemiten, 1968 (Veröffentlichungen der Arbeitsgemeinschaft für Forschung des Landes Nordrhein-Westfalen 151), in: ThLZ 94, 1969, Sp. 733–734.

Kraeling, C. H.: The Christian Building (The Excavations at Dura-Europos. Final Report, 8, 2), 1967, in: OLZ 64, 1969, Sp. 575–576.

Labuschagne, C. J.: The Incomparability of Yahweh in the Old Testament, 1966 (Pretoria Oriental Series 5), in: OLZ 64, 1969, Sp. 152–153.

Lapp, P. W.: The Dhar Mirzbâneh Tombs. Three Intermediate Bronze Cemeteries in Jordan, 1966 (AASOR. Publications of the Jerusalem School 4), in: OLZ 64, 1969, Sp. 153 bis 155.

Malamat, A.: Prophecy in the Mari Documents and the Bible [Three articles reprinted from „Eretz-Israel"], 1967, in: OLZ 64, 1969, Sp. 352.

MALINA, B. J.: The Palestinian Manna Tradition. The Manna Tradition in the Palestinian Targums and its Relationship to the New Testament Writings, 1968 (AGSJU 7), in: BiOr 26, 1969, S. 406.

MOSCATI, S.: Die Phöniker von 1200 vor Christus bis zum Untergang Karthagos. Übers. von R. SCHOTT, 1966, in: OLZ 64, 1969, Sp. 44–46.

PHILONENKO, M., PICARD, J.-CL., ROSENSTIEHL, J.-M., SCHMIDT, F.: Pseudépigraphes de l'Ancien Testament et manuscrits de la Mer Morte. I, 1967, in: OLZ 64, 1969, Sp. 570 bis 571.

ROSENTHAL, F. (Ed.): An Aramaic Handbook. Bd. I, 1: Texts; I, 2: Glossary; II, 1: Texts; 2: Glossary, 1967 (Porta Linguarum Orientalium N. S. 10), in: OLZ 64, 1969, Sp. 356 bis 357.

SCHMIDT, W. H.: Die Schöpfungsgeschichte der Priesterschrift, [2]1967 (WMANT 17), in: OLZ 64, 1969, Sp. 572–575.

TEIXIDOR, J.: Inventaire des Inscriptions de Palmyre. Fasc. 11, 1965, in: OLZ 64, 1969, Sp. 576.

Textus. Annual of the Hebrew University Bible Project. Vol. 6, ed. by S. TALMON, 1968, in: ThLZ 94, 1969, Sp. 258.

WILDBERGER, H.: Jesaja. 2. Lfg., 1966 (BK X, 2) in: OLZ 64, 1969, Sp. 257; 3. Lfg., 1968 (BK X, 3), in: OLZ 64, 1969, Sp. 572.

WOLFF, H. W.: Dodekapropheton, Amos. Lfg. 6, 1967 (BK XIV, 6), in: OLZ 64, 1969, Sp. 355–356.

ZEITLIN, S.: The Rise and Fall of the Judaean State. A Political, Social and Religious History of the Second Commonwealth. I, [2]1964; II, 1967, in: OLZ 64, 1969, Sp. 461 bis 462.

Internationale Zeitschriftenschau für Bibelwissenschaft und Grenzgebiete, hrsg. von F. STIER. Bd. 13, [1966/67] 1967, in: OLZ 64, 1969, Sp. 466.

Biblia Hebraica Stuttgartensia. Editio funditus renovat ed. K. ELLIGER et W. RUDOLPH. Liefg. 1: Genesis, praep. O. EISSFELDT, 1969.

1970

Introduzione all'Antico Testamento. Vol. I: Aspetti letterari dell'Antico Testamento. Edizione italiana a cura di A. ORNELLA, 1970 (Biblioteca teol. 1). Brescia 1970, 311 S.

Adonis und Adonaj. Berlin 1970. 28 S. 15 Taf. (SAL 115, 4).

Gilgal or Shechem?, in: Proclamation and Presence. Old Testament Essays in Honor of Gwynne Henton Davies, 1970, S. 90–101 = Gilgal oder Sichem? (Dtn 11, 29–32; 27, 11–13; 27, 1–8; Jos 8, 30–35), in: Kl. Schr. V, S. 165–173.

Alles Ding währt seine Zeit, in: Archäologie und Altes Testament. Festschrift für Kurt Galling, 1970, S. 70–74 = Kl. Schr. V, S. 174–178.

Walter Baumgartner (24. November 1887 bis 31. Januar 1970), in: AfO 23, 1970, S. 215 bis 218 = Kl. Schr. V, S. 161–164.

Harold Henry Rowley (1890–1969). Ein Herold weltweiter Wissenschaft vom Alten Testament, in: ThLZ 95, 1970, Sp. 81–88 = Kl. Schr. V, S. 152–160.

Πληρῶσαι πᾶσαν δικαιοσύνην in Matthäus 3, 15, in: ZNW 61, 1970, S. 209–215 = Kl. Schr. V, 179–184.

Theologisches Wörterbuch zum Alten Testament, hrsg. von G. J. BOTTERWECK und H. RINGGREN. Band I, Liefg. 1, 1970: אדון Sp. 62–78.

ALBRIGHT, W. F.: New Horizons in Biblical Research, 1966, in: OLZ 65, 1970, Sp. 156.

DE BOER, P. A. H. (Ed.): Oudtestamentische Studiën. Bd. 15: The Priestly Code and seven other Studies, 1969, in: ThLZ 95, 1970, Sp. 180–183.

DELEKAT, L.: Phönizier in Amerika. Die Echtheit der bekanntgewordenen kanaanäischen (altsidonischen) Inschriften aus Paraiba in Brasilien nachgewiesen, 1969 (BBB 32), in: JSSt 15, 1970, S. 102–104.

FITZMYER, J. A.: The Aramaic Inscriptions of Sefire, 1967 (Biblica et Orientalia 19), in: OLZ 65, 1970, Sp. 366.

HARNISCH, W.: Verhängnis und Verheißung der Geschichte. Untersuchungen zum Zeit- und Geschichtsverständnis im 4. Buch Esra und in der syrischen Baruchapokalypse, 1969 (FRLANT 97), in: ThLZ 95, 1970, Sp. 183–184.

JOSEPHUS, FLAVIUS: De Bello Judaico – Der Jüdische Krieg. Griechisch und Deutsch. Hrsg. und mit einer Einleitung sowie mit Anmerkungen versehen von O. MICHEL und O. BAUERNFEIND. Bd. II, 2: Buch VI–VII; Bd. III: Ergänzungen und Register, 1969, in: HZ 210, 1970, S. 676–677.

KALLAI (–KLEINMANN), Z.: The Tribes of Israel. A Study in the Historical Geography of the Bible (hebr.), 1967, in: OLZ 65, 1970, Sp. 270.

KRAUS, H.-J.: Geschichte der historisch-kritischen Erforschung des Alten Testaments. 2. überarbeitete und erweiterte Auflage, 1969, in: Bibl 51, 1970, S. 426–431.

VAN LEUWEN, C.: Hosea, 1968 (De Prediking van het Oude Testament), in: BiOr 27, 1970, S. 243.

MARTIN-ACHARD, R.: Actualité d'Abraham, 1967, in: ThLZ 95, 1970, Sp. 579.

Reallexikon der Assyriologie und vorderasiatischen Archäologie, hrsg. von W. v. Soden. Bd. III, 6, S. 401–480: Girsu–Götterreisen; Bd. III, 7, S. 481–560: Götterreisen–Gott, 1969, in: ThLZ 95, 1970, Sp. 495–496.

OLDENBURG, U.: The Conflict between El and Ba'al in Canaanite Religion, 1969 (Supplementa ad Numen. Altera Series, Vol. 3), in: BiOr 27, 1970, S. 380–381.

SANDERS, J. A.: The Dead Sea Psalms Scroll, 1967, in: OLZ 65, 1970, Sp. 149–150.

Textus. Annual of the Hebrew University Bible Project. Vol. 7, ed. by S. TALMON, 1969, in: ThLZ 95, 1970, Sp. 741.

Transactions. Glasgow University Oriental Society. Vol. 20, [1963–1964] 1965, in: OLZ 65, 1970, Sp. 307–308; Vol. 21, [1965–1966] 1967, in: OLZ 65, 1970, Sp. 417–418.

WIJNGAARDS, J. N. M.: The Dramatization of Salvific History in the Deuteronomic Schools, 1969 (Oudtestamentische Studiën 16), in: ThLZ 95, 1970, Sp. 180–183.

1971

Kleine Schriften zum Alten Testament. Hrsg. von KARL-MARTIN BEYSE und HANS-JÜRGEN ZOBEL. Berlin 1971, 482 S.

Die Psalmen als Geschichtsquelle, in: Near Eastern Studies in Honor of William Foxwell Albright, edited by H. GOEDICKE, 1971, S. 97–112 = Kl. Schr. V, S. 190–204.

Ein neuer Interpretations-Vorschlag zu Lukian, De Syria Dea, § 6, in: Mélanges Dunand, II, 1970/71, S. 235–241 [MUSJ 46] = Kl. Schr. V, S. 185–189.

Hebräisches *āh* und ugaritisches *āy* als Steigerungs-Afformative, in: Hommages à André Dupont-Sommer, 1971, S. 185–189 = Kl. Schr. V, S. 212–215.

Noch ein Nachtrag zu „Bist du Elia, so bin ich Isebel (I Kön XIX 2)", in: BiOr 28, 1971, S. 127 = Kl. Schr. V, S. 43–44.

Zusatz zu: Johannes Lindblom, Erwägungen zur Herkunft der Josianischen Tempelurkunde, in: I. LINDBLOM, Erwägungen zur Herkunft der Josianischen Tempelurkunde.

1971, S. 76–79 (Scripta Minora Regiae Societatis Humaniorum Litterarum Lundensis 1970/71, Nr. 3).

Der Zugang nach Hamath, in: Oriens Antiquus 10, 1971, S. 269–276 = Kl. Schr. V, S. 205–211.

Abr-Nahrain. An Annual published by the Department of Middle Eastern Studies University of Melbourne, ed. by J. BOWMAN. Vol. 8, [1968–1969] 1969, in: Der Islam 48, 1971, S. 183–184; Vol. 9, [1969–70] 1970, in: Der Islam 48, 1971, S. 184.

ATALLAH, W.: Adonis dans la littérature et l'art grecs, 1966, in: OLZ 66, 1971, Sp. 41–43.

BARR, J.: Comparative Philology and the Text of the Old Testament, 1968, in: OLZ 66, 1971, Sp. 477–478.

BORÉE, W.: Die alten Ortsnamen Palästinas. 2. Aufl. ergänzt um Namensregister und Nachträge, 1968, in: ThLZ 96, 1971, Sp. 91.

DONNER, H.: Herrschergestalten in Israel, 1970 (Verständliche Wissenschaft 103), in: HZ 213, 1971, S. 132–133.

JIDEJIAN, NINA: Byblos through the Ages. With a Foreword by M. DUNAND, 1968, in: OLZ 66, 1971, Sp. 558–559.

KENYON, KATHLEEN M.: Jerusalem. Die heilige Stadt von David bis zu den Kreuzzügen. Ausgrabungen 1961–1967, übers. von J. REHORK, 1968 (Neue Entdeckungen der Archäologie, hrsg. von M. WHEELER), in: OLZ 66, 1971, Sp. 372–373.

MULDER, M. J.: Het meisje van Sodom. De targumim op Genesis 18: 20, 21 tussen bijbeltekst en haggada (Rede), 1970, in: ThLZ 96, 1971, Sp. 420–421.

ORLINSKY, H. M. and N. H. SNAITH: Studies on the Second Part of the Book of Isaiah. The So-Called „Servant of the Lord" and „Suffering Servant" in Second Isaiah, by H. M. ORLINSKY; Isaiah 40–66. A Study of the Teaching of the Second Isaiah and its Consequences, by H. N. SNATH, 1967 (SVT 14), in: OLZ 66, 1971, Sp. 156.

Reallexikon der Assyriologie und vorderasiatischen Archäologie, hrsg. von W. v. SODEN. Bd. III, 8, S. 561–640: Gott–Grenze, 1971, in: ThLZ 96, 1971, Sp. 737.

SNIJDERS, L. A.: Jesaja deel I, 1969 (De Prediking van het Oude Testament), in: BiOr 28, 1971, S. 82.

[SPEISER, E. A.:] Oriental and Biblical Studies. Collected Writings of E. A. Speiser, ed. and with an Introduction by J. J. FINKELSTEIN and M. GREENBERG, 1967, in: OLZ 66, 1971, Sp. 252–253.

WEIDMANN, H.: Die Patriarchen und ihre Religion im Licht der Forschung seit Julius Wellhausen, 1968 (FRLANT 94), in: OLZ 66, 1971, Sp. 152–153.

WEIPPERT, M.: Die Landnahme der israelitischen Stämme in der neueren wissenschaftlichen Diskussion. Ein kritischer Bericht, 1967 (FRLANT 92), in: OLZ 66, 1971, Sp. 151.

WOLFF, H. W.: Dodekapropheten, Amos. Lfg. 7, 1969 (BK XIV, 7), in: OLZ 66, 1971, Sp. 258–259.

Internationale Zeitschriftenschau für Bibelwissenschaft und Grenzgebiete, hrsg. von F. STIER. Bd. 14, [1967/68] 1968, in: OLZ 66, 1971, Sp. 563–564.

1972

Jahwes Königspräzidierung als Verklärung national-politischer Ansprüche Israels, in: Wort, Lied und Gottesspruch. Festschrift für Joseph Ziegler. Hrsg. von J. SCHREINER. Bd. II, 1972, S. 51–55 (Forschungen zur Bibel 2) = Kl. Schr. V, S. 216–221.

Abr-Nahrain. An Annual published by the Department of Middle Eastern Studies University of Melbourne, ed. by J. BOWMAN. Vol. 10, [1970–71] 1971, in: Der Islam 48, 1972, S. 360.

The Annual of Leeds University Oriental Society, ed. by J. MACDONALD. Vol. 6, [1966 bis 1968] 1969, in: ThLZ 97, 1972, Sp. 355.

BEEK, M. A. u. a.: The Witness of Tradition, 1972 (Oudtestamentische Studiën 17), in: ThLZ 97, 1972, Sp. 827–828.

CODY, A.: A History of Old Testament Priesthood, 1969 (An Bibl 35), in: OLZ 67, 1972, Sp. 555–557.

DYSON, ST. L.: The Commonware Pottery. The Brittle Ware, 1968 (The Excavations at Dura-Europos. Final Report 4, 1, 3), in: OLZ 67, 1972, Sp. 48–49.

GOSHEN-GOTTSTEIN, M. H.: Introduction to the Lexicography of Modern Hebrew [Hebr., engl. summ.], 1969 (The Modern Hebrew Dictionary. Introductory Volume), in: OLZ 67, 1972, Sp. 559–560.

JALABERT, L. †, et MOUTERDE, R. continuée par C. MONDÉSERT et J.-P. REY-COQUAIS: Inscriptions Grecques et Latines de la Syrie. Tome VI: Baalbek et Beqa', 1967 (BAH 78), in: OLZ 67, 1972, Sp. 460–461.

LAPP, P. W. (Ed.): The 1957 Excavations at Beth-Zur, 1968 (AASOR 38), in: OLZ 67, 1972, Sp. 358–359.

OLDENBURG, U.: The Conflict between El and Ba'al in Canaanite Religions, 1969 (Supplementa ad Numen. Altera Series, Vol. 3), in: ThLZ 97, 1972, Sp. 581.

Reallexikon der Assyriologie und vorderasiatischen Archäologie, hrsg. von W. v. SODEN. Bd. III, 9, S. 640–722. 1 Falttabelle: Grenze – Gyges und Nachtrag, 1971, in: ThLZ 97, 1972, Sp. 347.

SCHAEDER, H. H.: Studien zur orientalischen Religionsgeschichte. Hrsg. mit einem Nachwort von C. Kolpe, 1968, in: OLZ 67, 1972, Sp. 360.

SCHAEFFER, CL. F.-A.: Ugaritica V. Nouveaux textes accadiens, hourites et ugaritiques des archives et bibliothèques privées d'Ugarit, 1968 (Missions de Ras Schamra, Tome 16), in: OLZ 67, 1972, Sp. 356–358.

SCHALIT, A.: Namenwörterbuch zu Flavius Josephus, 1968 (A Complete Concordance to Flavius Josephus, ed. by K. H. Rengstorf, Suppl. I), in: OLZ 67, 1972, Sp. 558–559.

SWAUGER, J. L. (Ed.): The Excavations of Bethel (1934–1960), 1968 (AASOR 39), in: OLZ 67, 1972, Sp. 359–360.

VOIGT, W. (Hrsg.): XVII. Deutscher Orientalistentag vom 21.–27. Juli 1968 in Würzburg. Vorträge, Teil 1–3, 1969 (= ZDMG Suppl. 1), in: OLZ 67, 1972, Sp. 26–30.

YADIN, Y.: Masada. Der letzte Kampf um die Festung des Herodes. Übers. aus dem Hebräischen von M. PEARLMAN, aus dem Englischen von E. u. A. EGGEBRECHT, 1967, in: OLZ 67, 1972, Sp. 257.

Internationale Zeitschriftenschau für Bibelwissenschaft und Grenzgebiete, hrsg. von F. STIER. Bd 15, [1968/69] 1969, in: OLZ 67, 1972, Sp. 557.

THOMSEN, PETER: Die Palästina-Literatur. Eine internationale Bibliographie in systematischer Ordnung mit Autoren- und Sachregister. Band VII: Die Literatur der Jahre 1940 bis 1945. Die Drucklegung überwachten O. EISSFELDT und L. ROST. Berlin 1972. 783 S.

1973

OTTO EISSFELDT und KARL HEINRICH RENGSTORF (Hrsg.): Briefwechsel zwischen Franz Delitzsch und Wolf Wilhelm Graf Baudissin 1866–1890. Opladen 1973 XXX, 543 S., 10 Abb. (Abhandlungen der Rheinisch-Westfälischen Akademie der Wissenschaften Bd. 43).

Kleine Schriften V. Hrsg. von R. SELLHEIM und F. MAASS. Tübingen 1973, 287 S., 8 Taf.

Kultzelt und Tempel, in: Wort und Geschichte. Festschrift für K. Elliger zum 70. Geburtstag, hrsg. von H. Gese und H.-P. Rüger, 1973, S. 51–55.

Monopol-Ansprüche des Heiligtums von Silo, in: OLZ 68, 1973, Sp. 327–333.

Allegro, J. M. with the Collaboration of A. A. Anderson: Qumran Cave 4. I (4Q 158 bis 4Q 186), 1968 (Discoveries in the Judaean Desert of Jordan V), in: OLZ 68, 1973, Sp. 42 bis 43.

Brown, J. P.: The Lebanon and Phoenicia. Ancient Texts illustrating their Physical Geography and Native Industries. I. The Physical Setting and the Forest, 1969, in: OLZ 68, 1973, Sp. 259–260.

Cintas, P.: Manuel d'Archéologie Punique. I. Histoire et Archéologie Comparées, 1970, in: OLZ 68, 1973, Sp. 592–593.

Congress Volume, Rome 1968, 1969 (SVT 17), in: OLZ 68, 1973, Sp. 107–108.

Elliger, K.: Jesaja II. 1. Lfg., 1970 (BK XI, 1), in: OLZ 68, 1973, Sp. 476.

Fohrer, G.: Studien zur alttestamentlichen Theologie und Geschichte (1949–1966), 1969 (BZAW 115), in: OLZ 68, 1973, Sp. 45.

[Glueck, N.]. Essays in Honor of Nelson Glueck, Near Eastern Archaeology in the Twenthieth Century, ed. by J. A. Sanders, 1970, in: OLZ 68, 1973, Sp. 587–588.

Görg, M.: Das Zelt der Begegnung. Untersuchung zur Gestalt der sakralen Zelttraditionen Altisraels, 1967 (BBB 27), in: OLZ 68, 1973, Sp. 475–476.

Jalabert, L. †, et Mouterde, R. continuée par l'Institut Fernand-Courby (Université de Lyon): Inscriptions Grecques et Latines de la Syrie. Tome VII: Arados et Régions Voisines par J.-P. Rey-Coquais, 1970 (BAH 89), in: OLZ 68, 1973, Sp. 593.

Lapp, P.: Biblical Archaeology and History, 1969, in: OLZ 68, 1973, Sp. 39–40

Mesnil du Buisson, R. de: Études sur les dieux phéniciens hérités par l'empire romain, 1970 (Études préliminaires aux religions orientales dans l' empire romain, publ. par M. J. Vermaseren, 4), in: OLZ 68, 1973, Sp. 474–475.

Pritchard, J. B. (Ed.). The Ancient Near East. Supplementary Texts and Pictures Relating to the Old Testament, 1969, in: OLZ 68, 1973, Sp. 473–474.

Wildberger, H.: Jesaja. 4. Lfg., 1969 (BK X, 4), in: OLZ 68, 1973, Sp. 254.

Wolff, H. W.: Amos. Lfg. 8. 9, 1969 (BK XIV, 8. 9), in: OLZ 68, 1973, Sp. 260.

Diccionario Teologico del Antiguo Testamento, dirigido por G. J. Botterweck y H. Ringgren. Traducción española de A. de la Fuente y M. Herranz. Tomo I, Fasc. I, 1973: אָרוֹן Sp. 62–78.

1974

Franken, H. J.: Excavations at Tell deir 'Allā. I: A stratigraphical and analytical Study of Early Iron Age Pottery. With Contributions by J. Kalsbek, 1969 (Documenta et Monumenta Orientis Antiqui 60), in: OLZ 69, 1974, Sp. 263–264.

Paul, S. M.: Studies in the Book of the Covenant in the Light of Cuneiform and Biblical Law 1970 (SVT 18), in: OLZ 69, 1974, Sp. 463–466.

Redford, D. B.: A Study of the Biblical Story of Joseph (Genesis 37–50), 1970 (SVT 20), in: OLZ 69, 1974, Sp. 562–564.

Thompson, H. O.: Mekal. The God of Beth-Shan, 1970, in: OLZ 69, 1974, Sp. 470–471.

Thompson, R. J.: Moses and the Law in a Century of Criticism since Graf 1970 (SVT 19), in: OLZ 69, 1974, Sp. 264–266.

Wildberger, H.: Jesaja. Liefg. 5, 1970 (BK X, 5), in: OLZ 69, 1974, Sp. 365.

[ALBRIGHT, W. F.] Near Eastern Studies in Honor of William Foxwell Albright. Ed. by HANS GOEDICKE, 1971, in: OLZ 70, 1975, Sp. 261–262.

BRONNER, LEAH: The Stories of Elijah and Elisha as Polemics against Baal Worship, 1968 (Pretoria Oriental Series VI), in: OLZ 70, 1975, Sp. 265.

FITZMYER, JOSEPH A.: The Genesis Apocryphon of Qumran Cave I. A Commentary. 2nd rev. Ed. 1971 (Biblica et Orientalia 18 A), in: OLZ 70, 1975, Sp. 148–149.

NOTH, MARTIN: Aufsätze zur biblischen Landes- und Altertumskunde, hrsg. von H. W. WOLFF. I. II, 1971, in: OLZ 70, 1975, Sp. 265–270.

REZENSIONEN VON ZEITSCHRIFTEN

Andrews University Seminary Studies

1, 1963, in: OLZ 59, 1964, Sp. 421–422; 2, 1964, in: OLZ 60, 1965, Sp. 209; 3, 1965, in: OLZ 61, 1966, Sp. 201–202; 4, 1966, in: OLZ 63, 1968, Sp. 303–304. 412–413; 5, 1967, in: OLZ 63, 1968, Sp. 606–607; 6, 1968; in: OLZ 64, 1969, Sp. 609–610; 7, 1969, in: OLZ 66, 1971, Sp. 93–94; 8, 1970, in: OLZ 66, 1971, Sp. 511; 9, 1971, in: OLZ 67, 1972, Sp. 610–612; 10, 1972, in: OLZ 70, 1975, Sp. 96.

'Atiqot
5, 1969, in: OLZ 65, 1970, Sp. 603.

The Australian Journal of Biblical Archaeology 1, 1968 – 4, 1971, in: OLZ 68, 1973, Sp. 313–314.

Berytus

10, 1, 1950/51, in: OLZ 48, 1953, Sp. 90–91; 10, 2, 1953, in: OLZ 48, 1953, Sp. 478; 11, 1, 1954, in: OLZ 50, 1955, Sp. 85; 11, 2, 1955, in: OLZ 52, 1957, Sp. 87–88; 12, 1, 1956/57, in: OLZ 54, 1959, Sp. 218; 12, 2, 1958, in: OLZ 55, 1960, Sp. 550–551; 13, 1, 1959/60, in: OLZ 55, 1960, Sp. 551; 13, 2, 1959/60, in: OLZ 57, 1962, Sp. 211; 14, 1 1961, in: OLZ 58, 1963, Sp. 96; 14, 2, 1961/63, in: OLZ 61, 1966, Sp. 204; 15, 1964, in: OLZ 61, 1966, Sp. 287; 16, 1966, in: OLZ 63, 1968, Sp. 610; 17, 1967/68, in: OLZ 65, 1970, Sp. 604–605.

Biblische Zeitschrift, Neue Folge

1, 1, 1957, in: OLZ 52, 1957, Sp. 561–562; 1, 2, 1957, in: OLZ 54, 1959, Sp. 316–317; 2, 1, 1958, in: OLZ 54, 1959, Sp. 317; 2, 2, 1958, in: OLZ 55, 1960, Sp. 98–99; 3, 1, 1959, in: OLZ 55, 1960, Sp. 99; 3, 2, 1959, in: OLZ 55, 1960, Sp. 551–552; 4, 1, 1960, in: OLZ 56, 1961, Sp. 443–444; 4, 2, 1960, in: OLZ 56, 1961, Sp. 550–551; 5, 1, 1961, in: OLZ 57, 1962, Sp. 215; 5, 2, 1961, in: OLZ 57, 1962, Sp. 322; 6, 1, 1962, in: OLZ 58, 1963, Sp. 96–97; 6, 2, 1962, in: OLZ 58, 1963, Sp. 614–615; 7, 1, 1963, in: OLZ 58, 1963, Sp. 615–616; 7, 2, 1963, in: OLZ 59, 1964, Sp. 427; 8, 1–2, 1964, in: OLZ 60, 1965, Sp. 215–217; 9, 1–2, 1965, in: OLZ 61, 1966, Sp. 607–608; 10, 1–2, 1966, in: OLZ 62, 1967, Sp. 414–415; 11, 1–2, 1967, in: OLZ 63, 1968, Sp. 610–611; 12, 1–2, 1968, in: OLZ 64, 1969, Sp. 303–304; 13, 1969, in: OLZ 65, 1970, Sp. 605–606; 14, 1970, in: OLZ 66, 1971, Sp. 515–516; 15, 1971, in: OLZ 68, 1973, Sp. 102–103; 16, 1972, in: OLZ 70, 1975, Sp. 206–207.

Forschungen und Fortschritte

19, 1943, in: ZAW 62, 1950, S. 291—292; 20, 1944, in: ZAW 62, 1950, S. 292—293; 21/23, 1947, in: ZAW 62, 1950, S. 293—294; 24, 1948, in: ZAW 62, 1950, S. 294—295; 25, 1949, in: ZAW 62, 1950, S. 295—297. 308—309; ZAW 63, 1951, S. 126; 26, 1950, in: ZAW 63, 1951, S. 126—129; 27, 1953, in: ZAW 66, 1954, S. 298—299; 28, 1954, in: ZAW 66, 1954, S. 299; ZAW 67, 1955, S. 271—272; 29, 1955, in: ZAW 67, 1955, S. 113—114; 30, 1956, in: ZAW 68, 1956, S. 198. 274—275; 31, 1957, in: ZAW 69, 1957, S. 253; 70, 1958, S. 142; 32, 1958, in: ZAW 70, 1958, S. 275; 33, 1959, in: ZAW 71, 1959, S. 221; 72, 1960, S. 77; 34, 1960, in: ZAW 73, 1961, S. 235—236; 35, 1961, in ZAW 73, 1961, S. 311; 74, 1962, S. 79—80. 222; 36, 1962, in: ZAW 74, 1962, S. 222—223. 334—335; 75, 1963, S. 105—106. 236; 37, 1963, in: ZAW 75, 1963, S. 236. 328; 76, 1964, S. 82; 38, 1964, in: ZAW 76, 1964, S. 208—209; 77, 1965, S. 99; 39, 1965, in: ZAW 77, 1965, S. 340; 78, 1966, S. 92. 238; 40, 1966, in: ZAW 78, 1966, S. 368; 79, 1967, S. 96—97. 245; 41, 1967, in: ZAW 79, 1967, S. 245—246; 80, 1968, S. 109.

Hebrew Union College Annual

23, 1, 1950/51, in OLZ 48, 1953, Sp. 154—156; 23, 2, 1950/51, in: OLZ 48, 1953, Sp. 526—527; 24, 1952/53, in: OLZ 50, 1955, Sp. 134—135; 25, 1954, in: OLZ 50, 1955, Sp. 426—428; 26, 1955, in: OLZ 52, 1957, Sp. 461—463; 27, 1956, in: OLZ 53, 1958, Sp. 280—282; 28, 1957, in: OLZ 54, 1959, Sp. 428—430; 29, 1958, in: OLZ 54, 1959, Sp. 540—542; 30, 1959, in: OLZ 56, 1961, Sp. 554—555; 31, 1960, in: OLZ 57, 1962, Sp. 329—330; 32, 1961, in: OLZ 57, 1962, Sp. 330—332; 33, 1962, in: OLZ 58, 1963, Sp. 622—623; 34, 1963, in: OLZ 59, 1964, Sp. 521—522; 35, 1964, in: OLZ 60, 1965, Sp. 428—429; 36, 1965, in: OLZ 62, 1967, Sp. 418—419; 37, 1966, in: OLZ 63, 1968, Sp. 614; 38, 1967, in: OLZ 64, 1969, Sp. 216—218; 39, 1968, in: OLZ 65, 1970, Sp. 605 bis 606; 39/40, 1969—1970, in: OLZ 66, 1971, Sp. 517.

Jaarbericht „Ex Oriente Lux"

18, (1964) 1965, in: OLZ 62, 1967, Sp. 419—420; 19, (1965—66) 1967, in: OLZ 64, 1969, Sp. 417—419.

Journal of the American Oriental Society

73, 1—3, 1952, in: OLZ 49, 1954, Sp. 283—284; 73, 4, 1953, in: OLZ 50, 1955, Sp. 91; 74, 1—3, 1954, in: OLZ 50, 1955, Sp. 557—558; 74, 4, 1954, in: OLZ 51, 1956, Sp. 89—90; 75, 1—4, 1955, in: OLZ 52, 1957, Sp. 185—187; 76, 1—3, 1956, in: OLZ 52, 1957, Sp. 466 bis 468; 76, 4, 1956, in: OLZ 53, 1958, Sp. 283—284; 77, 1, 1957, in: OLZ 53, 1958, Sp. 284; 77, 2—4, 1957, in: OLZ 54, 1959, Sp. 438—440; 78, 1—4, 1958, in: OLZ 54, 1959, Sp. 546—548; 79, 1—4, 1959, in: OLZ 56, 1961, Sp. 207—209; 80, 1, 1960, in: OLZ 56, 1961, Sp. 556; 80, 2—4, 1960, in: OLZ 57, 1962, Sp. 438—440; 81, 1—2, 1961, in: OLZ 58, 1963, Sp. 99—101; 81, 3—4, 1961, in: OLZ 58, 1963, Sp. 418—420; 82, 1—2, 1962, in: OLZ 59, 1964, Sp. 100—102; 82, 3—4, 1962, in: OLZ 59, 1964, Sp. 618—620; 83, 1—4, 1963, in: OLZ 60, 1965, Sp. 433—436; 84, 1—4, 1964, in: OLZ 62, 1967, Sp. 94—97; 85, 1—4, 1965, in: OLZ 62, 1967, Sp. 515—518; 86, 1—4, 1966, in: OLZ 63, 1968, Sp. 422 bis 425; 87, 1—4, 1967, in: OLZ 64, 1969, Sp. 309—312; 88, 1—4, 1968, in: OLZ 65, 1970, Sp. 610—613; 89, 1—4, 1969, in: OLZ 66, 1971, Sp. 97—100; 90, 1—4, 1970, in: OLZ 67, 1972, Sp. 413—418; 91, 1—4, 1971, in: OLZ 68, 1973, Sp. 109—110. 215—218; 69, 1974, Sp. 307—308.

Journal of Near Eastern Studies

12, 1–4, 1953, in: OLZ 49, 1954, Sp. 373–375; 13, 1, 1954, in: OLZ 49, 1954, Sp. 375–376; 13, 2, 1954, in: OLZ 50, 1955, Sp. 93–94; 13, 3–4, 1954, in: OLZ 50, 1955, Sp. 558–559; 14, 1–2, 1955, in: OLZ 50, 1955, Sp. 559–561; 14, 3–4, 1955, in: OLZ 52, 1957, Sp. 279–280; 15, 1, 1956, in: OLZ 52, 1957, Sp. 280; 15, 2–4, 1956, in: OLZ 52, 1957, Sp. 468–469; 16, 1–2, 1957, in: OLZ 53, 1958, Sp. 496–497; 16, 3–4, 1957, in: OLZ 54, 1959, Sp. 440–441; 17, 1–4, 1958, in: OLZ 55, 1960, Sp. 215–217; 18, 1, 1959, in: OLZ 55, 1960, Sp. 217–218; 18, 2–4, 1959, in: OLZ 56, 1961, Sp. 210–212; 19, 1–3, 1960, in: OLZ 56, 1961, Sp. 640–643; 19, 4, 1960, in: OLZ 57, 1962, Sp. 443; 20, 1–3, 1961, in: OLZ 57, 1962, Sp. 641; 20, 4, 1961, in: OLZ 58, 1963, Sp. 101; 21, 1–4, 1962, in: OLZ 58, 1963, Sp. 421–423; 22, 1, 1963, in: OLZ 59, 1964, Sp. 104; 22, 2–4, 1963, in: OLZ 59, 1964, Sp. 620–622; 23, 1–4, 1964, in: OLZ 61, 1966, Sp. 104–106; 24, 1–4, 1965, in: OLZ 62, 1967, Sp. 423–426; 25, 1–4, 1966 in: OLZ 62, 1967, Sp. 518–520; 26, 1–4, 1967, in: OLZ 64, 1969, Sp. 103–105; 27, 1–4, 1968, in: OLZ 64, 1969, Sp. 419–421; 28, 1–4, 1969, in: OLZ 65, 1970, Sp. 613–616; 29, 1–4, 1970, in: OLZ 67, 1972, Sp. 202–204; 30, 1–4, 1971, in: OLZ 68, 1973, Sp. 316–318; 31, 1–2, 1972, in: OLZ 70, 1975, Sp. 311–312.

Levant

1, 1969, in: OLZ 65, 1970, Sp. 420.

Mélanges de la Faculté Orientale de l'Université Saint-Joseph, Beyrouth

39, 1963, in: OLZ 60, 1965, Sp. 627.

La Nouvelle Clio.

1–2, 1949/50, in: ZAW 63, 1951, S. 134–138; 3, 1951, in: ZAW 64, 1952, S. 77–78; 4, 1952, in: ZAW 64, 1952, S. 78. 270–271; 65, 1963, S. 94; 5, 1953, in: ZAW 65, 1953, S. 277–278; 66, 1954, S. 131–132; 6, 1954, in: ZAW 66, 1954, S. 132. 302–303; 67, 1955, S. 117; 7/8/9, 1955/56/57, in: ZAW 68, 1956, S. 205; 69, 1957, S. 257–258; 70, 1958, S. 276; 10, 3, 1958–60, in: ZAW 73, 1961, S. 240–241; 10–12, 4–6, 1958/62, in: ZAW 76, 1964, S. 348–349.

Oriens Antiquus

1, 1962, in: OLZ 61, 1966, Sp. 106–107; 2, 1963, in: OLZ 61, 1966, Sp. 107; 3, 1964, in: OLZ 61, 1966, Sp. 107; 4, 1965, in: OLZ 62, 1967, Sp. 305–307; 5, 1966, in: OLZ 63, 1968, Sp. 516–518; 6, 1967, in: OLZ 64, 1969, Sp. 314–315; 7, 1968, in: OLZ 67, 1972, Sp. 100–101; 8, 1969, in: OLZ 67, 1972, Sp. 101; 9, 1970, in: OLZ 67, 1972, Sp. 302–304; 10, 1971, in: OLZ 68, 1973, Sp. 621–622.

Palestine Exploration Quarterly

84, 1–2, 1952, in: OLZ 49, 1954, Sp. 469–470; 85, 1–2, 1953, in: OLZ 49, 1954, Sp. 470–471; 86, 1–2, 1954, in: OLZ 50, 1955, Sp. 276–277; 87, 1, 1955, in: OLZ 51, 1956, Sp. 279–280; 87, 2, 1955, in: OLZ 52, 1957, Sp. 377–378; 88, 1–2, 1956, in: OLZ 53, 1958, Sp. 405–407; 89, 1, 1957, in: OLZ 53, 1958, Sp. 407–408; 89, 2, 1957, in: OLZ 54, 1959, Sp. 332; 90, 1–2, 1958, in: OLZ 55, 1960, Sp. 329–330; 91, 1, 1959, in: OLZ 55, 1960, Sp. 330–331; 91, 2, 1959, in: OLZ 56, 1961, Sp. 324; 92, 1, 1960, in: OLZ 56, 1961,

Sp. 324–325; 92, 2, 1960, in: OLZ 56, 1961, Sp. 652–653; 93, 1–2, 1961, in: OLZ 57, 1962, Sp. 554–556; 94, 1–2, 1962, in: OLZ 58, 1963, Sp. 512–513; 95, 1, 1963, in: OLZ 59, 1964, Sp. 216; 95, 2, 1963, in: OLZ 59, 1964, Sp. 523–524; 96, 1–2, 1964, in: OLZ 61, 1966, Sp. 290–291; 97, 1–2, 1965, in: OLZ 61, 1966, Sp. 421–422; 98, 1, 1966, in: OLZ 62, 1967, Sp. 104–105; 98, 2, 1966, in: OLZ 63, 1968, Sp. 617–618; 99, 1–2, 1967, in OLZ 63, 1968, 1968, Sp. 618–620.

Revista Biblica

20, 1958, Nr. 90, in ZAW 71, 1959, S. 233.

Revue Biblique

61, 1, 1954, in: OLZ 49, 1954, Sp. 473; 61, 2, 1954, in: OLZ 50, 1955, Sp. 187–188; 61, 3–4, 1954, in: OLZ 50, 1955, Sp. 279–280; 62, 1, 1955, in: OLZ 50, 1955, Sp. 280; 62, 2, 1955, in: OLZ 51, 1956, Sp. 280; 62, 3–4, 1955, in: OLZ 52, 1957, Sp. 471–472; 63, 1–3, 1956, in: OLZ 53, 1958, Sp. 411–414; 63, 4, 1956, in: OLZ 53, 1958, Sp. 499 bis 500; 64, 1, 1957, in: OLZ 53, 1958, Sp. 500–501; 64, 2–4, 1957, in: OLZ 54, 1959, Sp. 553–554; 65, 1–4, 1958, in: OLZ 55, 1960, Sp. 431–432; 66, 1–2, 1959, in: OLZ 55, 1960, Sp. 433–434; 66, 3–4, 1959, in: OLZ 56, 1961, Sp. 328–329; 67, 1–4, 1960, in: OLZ 56, 1961, Sp. 653–655; 68, 1, 1961, in: OLZ 56, 1961, Sp. 655; 68, 2–3, 1961, in: OLZ 57, 1962, Sp. 647–648; 68, 4, 1961, in: OLZ 58, 1963, Sp. 514–515; 69, 1–4, 1962, in: OLZ 58, 1963, Sp. 515–517; 70, 1–2, 1963, in: OLZ 59, 1964, Sp. 304–305; 70, 3, 1963, in: OLZ 59, 1964, Sp. 624; 70, 4, 1963, in: OLZ 61, 1966, Sp. 293; 71, 1–4, 1964, in: OLZ 61, 1966, Sp. 293–295; 72, 1–4, 1965, in: OLZ 62, 1967, Sp. 311–313; 73, 1–4, 1966, in: OLZ 63, 1968, Sp. 521–524; 74, 1–4, 1967, in: OLZ 64, 1969, Sp. 514–517; 75, 1968, in: OLZ 65, 1970, Sp. 516–518; 76, 1969, in: OLZ 68, 1973, Sp. 110.

Revue de Qumran

1, 1–2, 1958/59, in: OLZ 54, 1959, Sp. 555–556; 1, 3–4, 1958/59, in: OLZ 55, 1960, Sp. 437–438; 2, 1–2, 1959/60, in: OLZ 56, 1961, Sp. 331–332; 2, 3, 1959/60, in: OLZ 56, 1961, Sp. 665; 2, 4, 1959/60, in: OLZ 57, 1962, Sp. 103–104; 3, 1–3, 1961/62, in: OLZ 57, 1962, Sp. 650–653; 3, 4, 1961/62, in: OLZ 58, 1963, Sp. 518–519; 4, 1, 1963/64, in: OLZ 59, 1964, Sp. 306–308; 4, 2–3, 1963/64, in: OLZ 60, 1965, S. 106–107; 4, 4, 1963/64, in: OLZ 62, 1967, S. 105–106; 5, 1–4, 1964/66, in: OLZ 63, 1968, Sp. 310 bis 313; 6, 1967/68, in: OLZ 65, 1970, Sp. 521–524.

Revue de Théologie et de Philosophie

148, 1948–151, 1949, in: ZAW 62, 1950, S. 311.

Semitica

1, 1948; 2, 1949; 3, 1950; 4, 1951/52, in: ZAW 66, 1954, S. 135–138; 5, 1955, in: ZAW 68, 1956, S. 209–210; 6, 1956, in: ZAW 70, 1958, S. 154–155; 7, 1957, in: ZAW 70, 1958, S. 155; 8, 1958, in: ZAW 71, 1959, S. 238; 9, 1959, in: ZAW 72, 1960, S. 287; 10, 1960, in: ZAW 74, 1962, S. 96; 11, 1961, in: ZAW 74, 1962, S. 342; 12, 1962, in: ZAW 75, 1963, S. 339–340; 13, 1963, in: ZAW 76, 1964, S. 351–352; 14, 1964, in: ZAW 77, 1965, S. 348–349; 15, 1965, in: ZAW 78, 1966, S. 375–376; 16, 1966, in: ZAW 80,

1968, S. 116; 17, 1967, in: ZAW 80, 1968, S. 416–417; 18, 1968, in: ZAW 81, 1969, S. 406; 19, 1970, in: ZAW 82, 1970, S. 470; 20, 1970, in: ZAW 83, 1971, S. 411; 21, 1971, in: ZAW 85, 1973, S. 111–112.

Verbum Domini

Indices voluminum 15, 1935–30, 1952, in: ZAW 67, 1955, S. 123.

Vetus Testamentum

1, 1951, in: ZAW 64, 1952, S. 89–95; 2, 1952, in: ZAW 64, 1952, S. 95–98. 276–277. 282–284; 3, 1953, in: ZAW 65, 1953, S. 100–102. 284–286; 4, 1954, in: ZAW 66, 1954, S. 140–143. 308–310; 5, 1955, in: ZAW 67, 1955, S. 124–126. 281–283; 6, 1956, in: ZAW 68, 1956, S. 214–216; 69, 1957, S. 266–269; 7, 1957, in: ZAW 69, 1957, S. 269 bis 271; 70, 1958, S. 158–161; 8, 1958, in: ZAW 70, 1958, S. 161–162. 276–278; 71, 1959, S. 244–246; 9, 1959, in: ZAW 71, 1959, S. 246–248; 72, 1960, S. 89–92; 10, 1960, in: ZAW 72, 1960, S. 192–193. 290–291; 73, 1961, S. 122–124; 11, 1961, in: ZAW 73, 1961, S. 245–246. 323–325; 74, 1962, S. 99–101; 12, 1962, in: ZAW 74, 1962, S. 233–234. 344–347; 75, 1963, S. 123–124; 13, 1963, in: ZAW 75, 1963, S. 244. 343–344; 76, 1964, S. 96–98; 14, 1964, in: ZAW 76, 1964, S. 221–222. 355–357; 77, 1965, S. 235–236; 15, 1965, in: ZAW 77, 1965, S. 236–237. 352–353; 78, 1966, S. 109–110; 16, 1966, in: ZAW 78, 1966, S. 253–254. 378–379; 79, 1967, S. 109–111; 17, 1967, in: ZAW 79, 1967, S. 257–258. 393–394; 80, 1968, S. 118. 262–263; 18, 1968, in: ZAW 80, 1968, S. 263–265. 420–422; 81, 1969, S. 121–122; 19, 1969, ZAW 81, 1969, S. 268. 407–408; 82, 1970, S.145–146. 310–311; 20, 1970, in: ZAW 82, 1970, S. 311–312. 473–476; 83, 1971, S. 121–122; 21, 1971, in: ZAW 83, 1971, S. 286–287. 414–416; 84, 1972, S. 115 bis 116. 269; 22,1972, in: ZAW 84, 1972, S. 269–270; 85, 1973, S. 113.

Wissenschaftliche Zeitschrift Halle

4, 2, 1954/55, in: ZAW 67, 1955, S. 126.

Wissenschaftliche Zeitschrift Leipzig

3, 1, 1953/54, in: OLZ 49, 1954, Sp. 477–478.

Der ergänzte Abdruck dieser Bibliographie, erschienen in Band V der „Kleinen Schriften" von Otto Eißfeldt, S. 222–287, erfolgt mit freundlicher Genehmigung des Verlages J. C. B. Mohr (Paul Siebeck) in Tübingen.

Zur Biobibliographie

Schriftenverzeichnisse. K.-M. Beyse, siehe vorliegendes Jahrbuch S. 329–391. — Derselbe, O. Eißfeldt: Kleine Schriften *5* (Tübingen 1973) S. 222–287. — K. v. Rabenau, H.-J. Zobel, in: ThLZ *82* (1957) Sp. 701–714. — R. Sellheim, in: Festschrift zum 60. Geburtstag. Halle/S. 1947. S. 213–231. — O. V., in: Wiss. Z. Martin-Luther-Univ. Halle–Wittenberg, ges.-sprachwiss. R. *6* (1956/57) S. 158. — Kürschners Dtsch. Gelehrten-Kal. *1* (Berlin–Leipzig 1925) Sp. 195, *2*, *3*, *4*, *5*, *6*$_1$, *7*, *8*, *9*, *10* und *11* (Berlin 1970) S. 575–576. — Jahrbuch Sächs. Akad. Wiss. Leipzig *1957–1959* (Berlin 1961) S. 272; *1960–1962* (Berlin

1964) S. 305—306; *1963—1965* (Berlin 1967) S. 239—240; *1966—1968* (Berlin 1970) S. 242 bis 244; *1969—1970* (Berlin 1972) S. 174—175.

Festschriften. Festschrift Otto Eißfeldt zum 60. Geburtstag am 1. Sept. 1947. Hrsg. v. J. Fück. Halle (S.) 1947. V, 233 S. — Von Ugarit nach Qumran. Beiträge zur alttestamentlichen und altorientalischen Forschung. O. Eißfeldt zum 1. September 1957 dargebracht von Freunden und Schülern. Hrsg. v. J. Hempel, L. Rost. Berlin 1958, 303 S. (= Beihefte zur Z. f. d. alttestamentl. Wiss. 77). — Festgabe f. Prof. D. Dr. Eißfeldt zum 1. September 1957, In: Wiss. Z. Martin-Luther-Univ. Halle—Wittenberg, ges.-sprachwiss. R. *6* (1956/57) S. 697—749 (= Gottes ist der Orient. Festschrift f. Prof. D. Dr. Eißfeldt DD zu seinem 70. Geburtstag am 1. September 1957 [Berlin 1959] 197 S.).

Gedenkschrift. In memoriam Otto Eißfeldt. Zwei Reden aus Anlaß des ersten Todestages. Halle/Saale 1974. 20 S. (= Hallesche Universitätsreden, Wissenschaftliche Beiträge 9 [T 2]).

Würdigungen. Zum 60. Geburtstag: R. Meyer, Festschrift O. Eißfeldt zum 60. Geburtstag. In: DLZ *71* (1950) Sp. 1—5. — Zum 70. Geburtstag: F. Maass, in: ThLZ *82* (1957) S. 701—702. — K. v. Rabenau, Das wissenschaftliche Werk von Prof. D. Dr. Otto Eißfeldt DD, in: Festgabe, Wiss. Z. Martin-Luther-Univ. Halle—Wittenberg, ges.-sprachwiss. R. *6* (1956/57) S. 697—698 (m. Bildn.). — L. Stern, ebda, S. 699. — A. Lehmann, Grußwort an O. Eißfeldt. In: Gottes ist der Orient. Berlin 1959. S. 5. — K. v. Rabenau, ebda, S. 7—8 (m. Bildn.). — L. Rost, in: FuF *31* (1957) S. 285—286. — H.-J. Zobel, Erforscher einer verschollenen Kultur. In: Neue Zeit *18* (1957) Nr. 237 vom 10. Okt. — Zum 75. Geburtstag: H.-J. Zobel, Einblick in die Tiefen des Menschentums. Zum 75. Geburtstag von Prof. D. Dr. Dr. Otto Eißfeldt — Sein Werk als bleibendes Denkmal. In: Der Neue Weg *17* (1962) Nr. 209 vom 8. Sept. — O. V., in: Hochschulwesen *10* (1962) S. 712. — O. V., in: FuF *36* (1962) S. 256. — Zum 80. Geburtstag: Der Rat der Theologischen Fakultät Halle, in: ThLZ *92* (1967) Sp. 717—718. — H.-J. Zobel: Sein Werk weltweit gewürdigt. In: Neue Zeit *23* (1967) Nr. 204 vom 2. Sept. — Derselbe: Gründlichkeit und Weite. In: Zeichen der Zeit *21* (1967) S. 398—400. — Derselbe: Ursprung und Verwurzelung des Erwählungsglaubens Israels. In: ThLZ *93* (1968) Sp. 1—12. — Vorsitzender und Hauptvorstand der CDU, in: Die Union *22* (1967) Nr. 204 vom 1. Sept. — O. V., in: Der Neue Weg *22* (1967) Nr. 204 vom 1. Sept. — O. V., in: LDZ *22* (Halle/S. 1967) Nr. 203 vom 1. Sept. — O. V., in: Mitteldeutsche Neueste Nachrichten *16* (1967) Nr. 204 vom 1. Sept. — O. V., in: Universitätszeitung Martin-Luther-Universität Halle—Wittenberg *11* (1967) Nr. 27 vom 31. Aug. — Zum 85. Geburtstag: R. Meyer, in: OLZ *67* (1972) Nr. 9/10. — G. Wallis, in: Der Neue Weg *27* (1972) Nr. 209 vom 2. Sept. — O. V., in: LDZ *27* (Halle/S. 1972) Nr. 209 vom 2. Sept. — O. V., in: Mitteldeutsche Neueste Nachrichten *21* (1972) Nr. 208 vom 1. Sept. — Allgemein: K.-M. Beyse, H.-J. Zobel (Hrsg.): Otto Eißfeldt, Kleine Schriften zum Alten Testament. Berlin 1971. S. 7—8. — R. Galling, in: Religion in Geschichte und Gegenwart, 3. Aufl. *2* (Tübingen 1958) Sp. 407—408. — R. Sellheim, F. Maass, Vorwort zu: Otto Eißfeldt, Kleine Schriften *2* (Tübingen 1963) S. V. — J. J. Stamm, Das Einleitungswerk von Otto Eißfeldt. In: Theol. Z. *21* (Basel 1965) S. 386—399.

Nachrufe. F. Hintze, in: OLZ *68* (1973) H. 7/8. — A. Johnson, in: The Times vom 7. Mai 1973. — F. Maass, in: Frankfurter Allgemeine Zeitung Nr. 142 vom 22. Juni 1973. — R. Meyer, siehe vorliegendes Jahrbuch S. 324—329 (m. Bildn.). — A. Parrot, in: Syria *50* (1973) S. 475—476. — B. Satlow, in: Standpunkt. Evangelische Monatsschrift *1* (1973) H. 8, S. 217—218. — W. v. Soden, in: Jahrbuch der Akademie der Wissenschaften und der Literatur in Mainz *1973* (Wiesbaden 1973) S. 75—76. — G. Wallis, in: Bibliotheca Orientalis *30* (Leiden 1973) S. 347—348 (m. Bildn.). — Derselbe, in: Standpunkt. Evangelische Monatsschrift *2* (1974) H. 6, S. 152. — Derselbe, in: Der Neue Weg *28* (1973) Nr. 113 vom 12. Mai. — Derselbe, in: Neue Zeit *29* (1973) Nr. 111 vom 12. Mai. — O. V., in: Jahrbuch der Akademie der Wissenschaften der DDR *1973* (Berlin 1974) S. 51.

Franz Lau

18. 2. 1907—6. 6. 1973

Als Franz Lau 1965 zum ordentlichen Mitglied der Sächsischen Akademie der Wissenschaften gewählt wurde, hat er das als hohe Auszeichnung empfunden, und mit Recht. Denn nur in ganz wenigen Akademien der Wissenschaften (z. B. der Britischen) gibt es eine besondere theologische Abteilung; im Normalfall kann ein Theologe Mitglied einer Akademie nicht als Theologe werden, gleich welcher Rang ihm als solchem zukommt, sondern nur auf dem Umwege über das, was er „außerdem" noch ist: Historiker, Philologe, Philosoph usw. So bedeutet die Aufnahme in eine Akademie für ihn nicht die ausdrückliche Anerkennung durch die Fachgenossen, wie bei den Angehörigen anderer Disziplinen, sondern die noch schwerer zu erreichende durch die der Akademie angehörigen prominenten Vertreter der Nachbardisziplinen, der Historiker, Philologen, Philosophen usw.

Als Historiker, genauer gesagt als Kirchenhistoriker, ist Franz Lau in die Sächsische Akademie der Wissenschaften aufgenommen worden, der Tradition entsprechend, nach der auch in früheren Generationen prominente Vertreter dieses Faches der Akademie angehörten. Vom Kirchenhistoriker Lau ist also hier zu reden. Von seiner Dissertation in Leipzig 1932 an hat Lau sich mit aller Kraft der Reformationsgeschichte, insbesondere der Lutherforschung zugewandt. „‚Äußerliche Ordnung' und ‚Weltlich Ding' in Luthers Theologie" hieß das Thema dieser Dissertation (Bibliogr. Nr. 2). Dort geht es, in ausführlicher, vom Lehrer Laus in Leipzig, Horst Stephan, angeregter Fragestellung, um die weltlichen Ordnungen, ihre Göttlichkeit ebenso wie ihre Natürlichkeit und Weltlichkeit — bereits hier befindet sich Lau im Vorfeld von „Luthers Lehre von den beiden Reichen", so derTitel seines Buches von 1952, das sich nun in vollem Umfang an der damaligen, wissenschaftlich wie auch kirchlich ziemlich erhitzten Debatte beteiligte. Diese Schrift hat eine erhebliche Reichweite erlangt (vgl. Bibliogr. Nr. 24, 25, 30). Mehrfach hat Lau sich darüber hinaus zum Gegenstand geäußert (vgl. Bibliogr. Nr. 29, 45, 66, 69, 90, 91, 119, 120, 125), und zwar bis in die letzten Monate seines Lebens hinein, sich dabei zu einem der maßgebenden Partner in der bis in die Grundvoraussetzungen der Theologie Luthers reichenden Debatte entwickelnd. Aber dennoch wäre es ungerecht, Franz Laus wissenschaftliche Leistung nur in diese Rubrik einzuordnen — mag seine Arbeit auch in einer zur Rubrizierung geneigten Zeit bei vielen dieses Vorzeichen tragen. Denn seine Leistung reicht viel weiter, nicht nur was die Behandlung von Einzelthemen aus der Reformationszeit anbetrifft, (wovon gleich zu reden sein wird, vgl. S. 395), sondern auch was deren Gesamterfassung angeht. Entscheidende Bedeutung kommt dabei seiner Gesamtdarstellung der Reformationsgeschichte bis 1532 im Handbuch „Die Kirche in ihrer Geschichte“ von 1964 (Bibliogr. Nr. 86) und der sie vorbereitend ergänzenden Lutherbiographie von 1959 (Bibliogr. Nr. 56) zu.

Es ist nicht zufällig, daß seine Gesamtdarstellung, abgesehen von einer zweiten deutschen Auflage (vgl. Bibliogr. Nr. 121) — weitere werden folgen — eine Übersetzung ins Englische gefunden hat (Bibliogr. Nr. 118), und seine Lutherbiographie neben einer zweiten deutschen Auflage — es muß wiederholt werden: weitere werden folgen — eine Übersetzung ins Englische (und zwar in den USA wie in England), gleich zwei Mal ins Polnische und sogar ins Japanische (vgl. Bibliogr. Nr. 76, 80, 95—97). Die Gesamtdarstellung der frühen Reformationsgeschichte mußte, weil in ein Lehrbuchschema gezwängt, in welchem Luther und Zwingli wie Calvin einer besonderen Behandlung vorbehalten waren, im wesentlichen unter Ausklammerung Luthers geschehen — eigentlich eine Unmöglichkeit, nur in einer Anmerkung (S. 15,1) macht sich der Schmerz Laus darüber Luft: „Der Darsteller der Reformationsgeschichte von 1517—1532 ohne Luther konnte seinen Auftrag überhaupt nur wegen seiner eigenen Konzeption der Reformationsgeschichte übernehmen“. Dennoch hat er es vermocht, in äußerst konzentrierter Darstellung auf 65 Seiten die Entwicklung der Reformation und auf 151 Seiten Kleinformat ein Gesamtbild des Lebens, Wirkens und der Auswirkung Luthers zu zeichnen. Wenn Lau diese Darstellung in 5 Hauptabschnitten aufbaut: Luthers Welt, Luthers Werden, Luthers Durchbruch, Luthers Reformation, Luthers Kirche, so ist das nicht nur eindrucksvoll, sondern läßt auch seine Auffassung vom Ablauf der deutschen Reformationsgeschichte erkennen: sie beginnt nach Lau im eigentlichen Sinne erst 1521. Vorbereitet ist diese Reformation durch die von Luther ausgelöste reformatorische Bewegung, die 1521 noch lange nicht voll entwickelt ist, so daß die Zeit von 1521 bis 1525 die des „Wildwuchses“ der Reformation (ein für Lau charakteristischer, überall sich findender Ausdruck) bedeutet. Erst von 1525 ab entsteht das reformatorische Kirchentum, das zwar „die Tendenz zu einer großzügigen Homogenität“ aufweist, dem es aber nicht gelingt, „zu einer in sich homogenen Gestalt zu kommen“ (Bibliogr. Nr. 86, S. 15,1). Das ist eine, wenn man es so bezeichnen will, lutherisch-kirchlich bestimmte, institutionell ausgerichtete Auffassung der Reformationsgeschichte. Richtig ist selbstverständlich, daß der kirchliche Auf- und Ausbau der lutherischen Reformation erst 1525 beginnt, ebenso kann kein Zweifel daran bestehen, daß ihm — auch bei Luther selbst — eine Zeit der inneren Klärung vorangeht, zu fragen ist jedoch, ob man wirklich (S. 16) sagen kann: „Aber die Reformation setzte doch erst mit dem Reichstag zu Worms ein (1521)“. Dieser Reichstag verhängte doch über Luther und alle seine Anhänger die Reichsacht und schaltete ihn vom weiteren Geschehen durch die Festsetzung auf der Wartburg fast völlig aus. Dadurch wurde die Verwirrung der sog. Wittenberger Unruhen beinahe zwangsläufig ausgelöst. Außerdem würde der 1. Reichstag zu Speyer mit der Grundsatzerklärung der „protestantischen“ Stände, streng institutionell gedacht, eher als Ausgangsdatum für das reformatorische Kirchentum geeignet sein.

Aber das sind Randbemerkungen, die nichts an der in den beiden genannten Darstellungen vollbrachten Leistung Laus mindern; seine eigenständige Konzeption ist sogar als Anstoß zu neuen Überlegungen von besonderem Wert. Sie sind — neben den schon genannten über Luthers Zwei-Reiche-Lehre — begleitet von einer Fülle von Einzeluntersuchungen, von denen hier nur auf die wichtigsten eingegangen werden kann.

Außer zu den Anfängen Luthers (vgl. Bibliogr. Nr. 6, 46) hat Lau mehrfach zum Streit um den Thesenanschlag Stellung genommen (Bibliogr. Nr. 79, 98—102, 110, 124). Auch zur Frage der zeitlichen Ansetzung des reformatorischen Durchbruchs und seines Charakters hat Lau sich geäußert, wenn auch nicht in einer besonderen Publikation, sondern in Verbindung mit denen zum Anschlag der Thesen. Denn bei der Diskussion, ob Luther seine 95 Thesen angeschlagen habe oder nicht und ob dies am 31. Oktober 1517 geschehen sei, geht es nur vordergründig um ein historisch-chronologisches Spezialproblem, sondern tatsächlich um Grundfragen der Lutherdeutung, zumal wenn man, wie Lau das — als einer von wenigen — tut, die Frage der Datierung des sog. Turmerlebnisses Luthers (d. h. der reformatorischen Wende) mit einbezieht. Lau hat sich in diesen seit 1957 andauernden (und bis heute nicht abgeschlossenen) Auseinandersetzungen nicht nur für den Thesenanschlag am 31. Oktober 1517, sondern — wenn auch mit einiger Vorsicht — für die „Spätdatierung", d. h. die Ansetzung des reformatorischen Durchbruchs im Frühjahr 1518, ausgesprochen; seine Stimme wurde um so mehr gehört (am ausführlichsten und wichtigsten in diesem Zusammenhang ist Bibliogr. Nr. 98), als er inzwischen längst zum Wortführer der reformationsgeschichtlichen Forschung in der DDR geworden war, worüber noch zu reden sein wird (vgl. u. S. 397).

Als nächste Gruppe wären die mehrfachen Stellungnahmen Laus zu Luthers Haltung im Bauernkrieg zu nennen (Bibliogr. Nr. 36, 52, 58, 111, 115, 122), sie sind ebenso Zeugnis dafür, daß Kirchengeschichte für Lau eine aktuelle Wissenschaft war, wie sein Aufsatz mit der Frage nach dem Verhältnis von Bert Brechts Weltanschauung zu Luthers Theologie (Bibliogr. Nr. 78, 84), die Auseinandersetzung mit dem Lutherbild Friedrich Heers (Bibliogr. Nr. 61, 75), den extravaganten Thesen Reinhold Weijenborgs (Bibliogr. Nr. 62, 93), aber auch den Auffassungen moderner Lutherforscher (vgl. die Diskussion mit Löfgren, Bibliogr. Nr. 77/85 und die Forschungsberichte Nr. 44, 94, 116).

Jeder Reformationshistoriker wird seine Aufmerksamkeit nicht nur auf den äußeren Ablauf des Lebens Luthers und der Reformation richten, sondern dabei stets nach den wirkenden Kräften fragen und sich mit den theologischen Fragen der Zeit auseinandersetzen. Das ist selbstverständlich. Nicht selbstverständlich aber ist für den Historiker die Grenzüberschreitung zu einer sozusagen „hauptamtlichen" Beschäftigung mit Luthers Theologie; sie ist in der Regel (wenigstens im deutschen Sprachgebiet und in Skandinavien) den Vertretern der Systematischen Theologie vorbehalten. Für Franz Lau galt diese Grenzziehung nicht (wahrscheinlich unter dem Einfluß von Horst Stephan). Schon die Dissertation und anschließend sein immer tieferes Vordringen in die Probleme der Zwei-Reiche-Lehre Luthers zeigen das mit aller Deutlichkeit. Bereits 1937 behandelt er (Bibliogr. Nr. 7) „Luthers Schmalkaldische Artikel als eine Einführung in seine Theologie", 1948 schließt sich die Untersuchung über „Erstes Gebot und Ehre Gottes als Mitte von Luthers Theologie" an (Bibliogr. Nr. 21). Ganz charakteristisch ist, daß Laus dem vielbehandelten Thema „Paulus—Augustin—Luther" gewidmete Schrift von 1961 (Bibliogr. Nr. 70) den Untertitel trägt: „Rechtfertigung nicht aus dem Werk, sondern aus dem Glauben". Der Aufsatz über „Luthers Worttheologie in katholischer Sicht" (Bibliogr. Nr. 73) schließt sich dem folgerichtig an; als Lau zum 75. Geburtstag von Paul Althaus das Grußwort im Luther-Jahrbuch zu schreiben hat (Bibliogr. Nr. 82), wählt er dafür die Überschrift „Die Theologie Martin Luthers", in Aufnahme des Titels von Althaus' zusammenfassendem Werk über dieses Thema, und weiß dem namhaften Vertreter der Systematischen Theologie gleichberechtigt zu begegnen.

In diesem Aufsatz steht (S. 14) der schöne Satz: „Wer sich mit Leidenschaft mit Luther beschäftigt, kann zum kritiklosen Panegyriker Luthers werden". Diese kritische Anmerkung zur Lutherforschung (vor allem vergangener Zeiten, aber auch manchmal der Gegenwart) hat für Franz Lau keine Gültigkeit. Das beweist nicht nur der Aufsatz, aus dem der Satz stammt, sondern sein gesamtes Werk. Das ist um so erstaunlicher, als Franz Lau prominentes und überzeugtes Glied einer überzeugt lutherischen Landeskirche war, deren Leitung über Jahre hindurch in seinen Händen gelegen hat (vgl. dazu u. S. 398). Vor einer Lutherpanegyrik hat ihn die Objektivierung bewahrt, die jeder streng wissenschaftlichen Tätigkeit innewohnt, ebenso aber auch die Tatsache, daß er — obwohl Lutherforscher im eigentlichen Sinn — immer die Gesamtzusammenhänge der Reformationsgeschichte wie die Männer und Kräfte im Blickfeld gehabt hat, die sie neben Luther bestimmten. Von seiner Behandlung der Reformationsgeschichte bis 1532 war bereits die Rede (vgl. o. S. 394). Aber daneben ist noch seine Gesamtdarstellung der lutherischen Reformation im zwar allgemein verständlich gehaltenen, aber äußerst instruktiven Sammelband „Reformation in Europa" zu nennen (in deutscher, englischer, holländischer und schwedischer Ausgabe erschienen, Bibliogr. Nr. 103—106), ebenso wie seine Untersuchung über „Die Reformation in den norddeutschen Städten" (Bibliogr. Nr. 43) und die (in französischer Sprache erschienene) über das religiöse Leben in den protestantischen Ländern deutscher Sprache bis zum Ende des 16. Jahrhunderts (Bibliogr. Nr. 81). Die von Lau mit einer Einleitung versehene Sammlung von Augenzeugenberichten zur Reformation (Bibliogr. Nr. 114, 126) wie seine in deutscher, englischer und ungarischer Sprache erschienene kurzgefaßte Gesamtwürdigung der Reformation (Bibliogr. Nr. 107—109) gehören ebenfalls hierher. Aber nicht nur diese Bemühungen um eine Gesamterfassung des Reformationsgeschehens haben Franz Lau die Fähigkeit zur Objektivierung Luthers gegeben, sondern auch seine Beschäftigung mit den führenden Gestalten der Reformation neben Luther. Zwei Veröffentlichungen Laus über Melanchthon liegen vor (Bibliogr. Nr. 68, 74), zwei über Calvin (Bibliogr. Nr. 59, 64), eine über Georg III. von Anhalt (Bibliogr. Nr. 34), der nicht zufällig den Beinamen „der Gottselige" trägt und eben nicht nur Landesfürst, sondern auch Theologe ganz eigener Prägung war. Lau war bezeichnenderweise in der Lage, einen Sammelband „Der Glaube der Reformatoren" zu veröffentlichen (Bibliogr. Nr. 84a), welcher neben Luther auch Zwingli und Calvin nicht nur ausführlich, sondern auch objektiv zu Wort kommen läßt. Seine Arbeiten über den Kryptocalvinismus, „die zweite Reformation in Kursachsen“ (Bibliogr. Nr. 87) wie über die Geschichte der Waldenser (Bibliogr. Nr. 39) runden das Bild ab.

Wenn vorhin die Rede davon war, Franz Lau sei Wortführer der reformationsgeschichtlichen Forschung in der DDR gewesen (vgl. o. S. 395), so ist dieser Satz sicher richtig verstanden worden, vor allem von denen, die neben ihm in der DDR auf diesem Feld arbeiten. Die Persönlichkeit Franz Laus wie seine Leistungen auf dem Gebiet der Reformationsgeschichte, die hier kurz zu skizzieren versucht wurden (wobei von den selbständigen Publikationen und Aufsätzen ausgegangen wurde, unter bewußter Vernachlässigung der Fülle der Rezensionen, vgl. Bibliogr. Nr. 127 bis 256, an denen das auch hätte demonstriert werden können), sind die Voraussetzung dafür gewesen; die Übertragung der Herausgabe des Luther-Jahrbuches stellte die äußere Bekundung dieses Faktums dar, ebenso wie seine Berufung in die Kommission zur Herausgabe der Werke Martin Luthers (der sog. „Weimarer Ausgabe“). Schon lange vor dem Ende des 2. Weltkriegs hatten Zeitschrift und Jahrbuch der Luther-Gesellschaft ihr Erscheinen einstellen müssen, erst nach langem Zögern rief man sie wieder ins Leben: mit dem Jahrgang 1953 die Zeitschrift „Luther“, mit dem Jahrgang 1957 das Luther-Jahrbuch. Bereits Mitte 1956 (vgl. Bibliogr. Nr. 47) legte Lau das Programm dafür vor, das er dann Schritt für Schritt in vollem Umfang verwirklicht hat: einen Aufsatzteil, der einen „wirklichen Querschnitt durch die gesamte Lutherforschung“ bedeutete, einen Besprechungsteil, der dem entsprach, und eine Bibliographie, welche alljährlich das weitgespannte internationale Schrifttum über Luther und seine Wirkungen nahezu vollständig verzeichnete. Fünfzehn Jahrgänge hat Franz Lau aufopferungsvoll betreut, der letzte, der unter seiner Herausgeberschaft erschien, war der von 1971. Hier war sein bisheriger Mitarbeiter Helmar Junghans noch als Mitherausgeber genannt, für den Jahrgang 1972 zeichnete dieser bereits allein verantwortlich. Anläßlich dieses Wechsels, der durch die zunehmende Krankheit Laus und seine Konzentration auf den Abschluß einer Darstellung des Wiederaufbaus der sächsischen Landeskirche nach 1945 erzwungen wurde, hat der Vorsitzende der Luthergesellschaft dem Jahrbuch eine Würdigung der Arbeit Laus vorangeschickt. „Unter Ihrer Verantwortung“, erklärt W. v. Loewenich hier, hat „ein neuer Abschnitt in der Geschichte des Lutherjahrbuches begonnen, der ihm nach menschlichem Ermessen auch in dem kommenden Jahrzehnt einen internationalen Rang sichern wird“, das Luther-Jahrbuch ist „Ihr Lutherjahrbuch“ (S. 5f).

Nicht nur hier, sondern auch noch in anderer Hinsicht ist Franz Lau Wortführer der Kirchenhistoriker in der DDR gewesen. Diese Eigenschaft wird äußerlich bezeichnet durch seine Ämter als Vorsitzender der Arbeitsgemeinschaft für Sächsische Kirchengeschichte und vor allem der Vereinigten Arbeitsgemeinschaften für Kirchengeschichte in den evangelischen Landeskirchen der DDR, ebenso wie durch seine Mitgliedschaft in der Historischen Kommission der Akademie, deren stellvertretenden Vorsitz er innehatte. Von entscheidender Bedeutung war es, daß Franz Lau in dieser Eigenschaft das Erscheinen eines „Jahrbuchs für deutsche Kirchengeschichte“ ermöglichte, das den Titel „Herbergen der Christenheit“ erhielt. 1957 erschien der erste Jahrgang, sechs weitere sind ihm bis 1971 gefolgt. Der Fortschritt der historischen ebenso wie der kirchenhistorischen Arbeit ist entscheidend von dem abhängig, was an der „Basis" geschieht. Je genauer das lokale Geschehen erforscht wird, je mehr man um die Motive und das Handeln der Männer unterrichtet ist, die in den Gemeinden und Provinzen die historische Entwicklung bestimmten, je mehr man darüber weiß, was an und in den Gemeinden im „Kleinen" geschah, desto sicherer wird man in der Erfassung dessen und im Urteil darüber, was im „Großen" vor sich ging. Mit dem „Jahrbuch" gab Franz Lau zahlreichen Erforschern dessen, was man etwas abschätzig Territorialgeschichte nennt, eine Möglichkeit zur Publikation ihrer Forschungsergebnisse, ein Verdienst, das gar nicht hoch genug eingeschätzt werden kann.

Brechen wir damit ab. Denn nur vom Kirchenhistoriker Franz Lau zu reden, würde eine ungerechte Verkürzung bedeuten. Franz Lau war eben nicht nur gelehrter Forscher auf diesem Sektor der Wissenschaft, sondern ebenso Theologe und Mann der Kirche. Sein Lebenslauf spricht eine deutliche Sprache, ebenso wie sein Schrifttum. Am 18. 2. 1907 ist Franz Lau in Leipzig geboren. Nach seinem Studium der Geschichte und Theologie von 1925—1930 in seiner Vaterstadt, das durch einen Aufenthalt in Wien ergänzt wurde, legte er in den Jahren 1930—1932 die beiden theologischen Examina ab und trat 1933 in den kirchlichen Dienst ein (zunächst als Vikar an der Luther-Kirche und dann als Pfarrer in Regis-Breitingen). Eigentlich strebte Lau die wissenschaftliche Laufbahn an. 1932 war er zum Licentiaten promoviert worden und hatte sich sogleich danach an die Vorbereitung seiner Habilitationsschrift gemacht. 1933 bereits hatte er das Material dafür beisammen — aber inzwischen hatte sich die Lage in Staat und Kirche grundlegend geändert. Eine Habilitation war für Lau, der sich dem Pfarrernotbund anschloß, jetzt unmöglich geworden; er blieb in seinem Pfarramt in Regis-Breitingen. Als der Kampf des nationalsozialistischen Staates gegen die Kirche zur Zeit der sog. Kirchenausschüsse vorübergehend durch eine Epoche scheinbarer Befriedung abgelöst wurde, konnte Lau 1936 Studiendirektor am Predigerseminar Lückendorf werden. Aber schon 1938 wurde ihm dieses Amt wieder genommen, Ende 1939 wurde er ins Pfarramt an der Martin-Luther-Kirche in Dresden aufgenommen. Seine Haltung in der Zeit des sich vorbereitenden Zusammenbruchs — er mahnte nach der Katastrophe von Dresden seine Amtsbrüder mit aller Kraft, ihre Gemeinden nicht durch die Flucht im Stich zu lassen — führte dazu, daß er, der jüngste Pfarrer der Stadt, 1945 zunächst zum Superintendenten von Dresden und dann zum Landessuperintendenten, d. h. zum geistlichen Leiter der sächsischen Kirche, gewählt wurde. Dieses Amt hat er bis 1947 innegehabt, danach ging er den Weg an die Universität, der ihm 1933 verwehrt worden war. Als Professor mit Lehrstuhl für Kirchengeschichte hat er hier bis zu seiner Emeritierung 1971 gewirkt. Trotz aller Inanspruchnahme durch sein Lehramt (von 1954—1956 kam das des Dekans, von 1956—1958 das des Prodekans dazu) hat Lau nicht aufgehört, sich als Prediger in den Dienst der Kirche zu stellen. „Der Theologieprofessor gehört auch als Prediger vor die Gemeinde und ist, wenn schon seine Universität eine staatliche Anstalt ist, eine Persönlichkeit der Kirche", hat Lau 1961 in seiner Darstellung der Geschichte der Leipziger Theologischen Fakultät erklärt (Bibliogr. Nr. 72, S. 37). Daß er 1957 zum Domdechanten des Hochstifts Meißen ernannt wurde, bedeutete die verdiente Anerkennung seiner kirchlichen Arbeit auch im akademischen Lehramt. Am bedeutungsvollsten für diese kirchliche Wirksamkeit war es, daß er 1952 (im gleichen Jahr, das ihm den Ehrendoktor der Theologischen Fakultät Rostock brachte) das ihm angetragene Amt des Präsidenten des Gustav-Adolf-Werkes übernahm, jener die ganze Welt umspannenden Hilfsorganisation für die evangelische Diaspora. Auch trotz der Teilung Deutschlands war die Zentralleitung in Leipzig geblieben, lediglich die von der BRD aus zweckmäßiger zu erledigenden Aufgaben waren von den westdeutschen Untergliederungen übernommen worden.

Über die Einzelheiten dieser Arbeit kann hier nicht berichtet werden, es sei nur darauf hingewiesen, einen wie umfangreichen Niederschlag die kirchliche Arbeit Franz Laus in seinem Schrifttum gefunden hat. Neben allgemeinen theologischen Fragen (vgl. Bibliogr. Nr. 10—16, 51, 77, 85) hat Lau die grundsätzlichen und aktuellen Probleme des kirchlichen Lebens behandelt (vgl. Bibliogr. Nr. 8, 18, 19, 38, 54, 67). Eine lange Reihe von Aufsätzen und Betrachtungen gilt der Arbeit des Gustav-Adolf-Werkes und den damit verbundenen Problemen der Diaspora und der Ökumene (vgl. Bibliogr. Nr. 27, 31, 33, 37, 42, 48, 53, 55, 57, 63, 65, 83, 88, 89, 92). Lau hat sich darüber hinaus um Hilfen für ein besseres Verständnis der Bibel (vgl. Bibliogr. Nr. 9, 22, 23) und der Kirchengeschichte (vgl. Bibliogr. 26, 35) durch die Gemeinde bemüht. Über Jahre hindurch hat er an der „Pfarramtspraxis“ mitgewirkt, die „Homiletischen Monatshefte“ wie die Zeitschrift „Die evangelische Diaspora“ hat er lange Zeit als Herausgeber betreut. Im zweibändigen „Handbuch zu Freikirchen und Sekten“ hat er als Mitglied des zuständigen kirchlichen Ausschusses umfangreiche Mitarbeit geleistet (vgl. Bibliogr. S. 416f.). Wenn man Umfang und Bedeutung des kirchlichen Schrifttums Laus würdigen will, so muß man berücksichtigen, daß diese Angaben sich sämtlich auf selbständige Schriften und Aufsätze beziehen. Erst wenn man die 124 Titel der Bibliographie aus der Abteilung „Predigten, Reden, Berichte, Würdigungen“ (S. 411ff.) hinzunimmt, die fast sämtlich in den kirchlichen Bereich gehören, erhält man einen einigermaßen vollständigen Eindruck davon, welches Ausmaß und welche Bedeutung die kirchliche Arbeit Franz Laus besessen hat.

Angesichts der doppelten Belastung (wieviel Kraft hat allein das Amt des Präsidenten des Gustav-Adolf-Werkes verzehrt!) kann es nicht wundernehmen, daß der Körper sich schließlich weigerte, dem Geist und dem Willen zu folgen. Lau hat sehr darunter gelitten, daß er in den letzten Jahren seines Lebens nicht mehr das tun konnte, was er wollte, bzw. wollen zu müssen meinte. Seine letzte größere Arbeit war eine umfangreiche Niederschrift, welche die bewegten Jahre des Neuaufbaus der sächsischen Landeskirche von 1945—1947 darstellen sollte. Niemand war dazu berufener als er, der an allem führend mitgewirkt hatte. 200 Schreibmaschinenseiten waren bereits gefüllt, als er den Rundbrief an seine Freunde zum Jahre 1972 versandte. Zur Vollendung der Arbeit haben die Kräfte nicht mehr gereicht. Am 6. Juni 1973 ist Franz Lau im Alter von 66 Jahren gestorben.

Notwendigerweise muß der Versuch einer Würdigung seines Werkes, wie er hier unternommen wurde, unvollkommen bleiben. Von vielen Arbeiten Laus konnte nur ganz summarisch gesprochen werden, eine ganze Reihe von ihnen blieb völlig unbehandelt. So sei besonders auf die nachfolgende Bibliographie verwiesen. Drei ihrer Abschnitte, sämtlich bisher unerwähnt, sind besonders geeignet, einen zusammenfassenden Eindruck zu vermitteln: der über die Lexikonartikel (S. 415f.) — er gibt einen Einblick in die imponierende Reichweite wie die Intensität der wissenschaftlichen Arbeit des Verstorbenen —, der über die von Franz Lau betreuten Dissertationen und Habilitationsschriften (26 in weniger als 25 Jahren akademischer Wirksamkeit!) — er vermittelt einen Eindruck davon, wie sehr Franz Lau als akademischer Lehrer gewirkt hat und welche Kräfte er auszulösen wußte — und der letzte (S. 418f.), welcher die Würdigungen zusammenstellt — die ganz erstaunliche Zahl der Würdigungen zum 60. Geburtstag gibt wenigstens eine ungefähre Vorstellung davon, welch außergewöhnliche Persönlichkeit wir mit Franz Lau verloren haben.

Kurt Aland

Bibliographie Franz Lau

Selbständig erschienene Werke und Aufsätze

1. Zur kritischen Würdigung von Mehls „Evangelischem Brevier" durch Pfarrer Lic. Hanne in Wormditt (Ostpreußen). Monatsschrift für Gottesdienst und kirchliche Kunst *37* (1932) 138—141.
2. „Äußerliche Ordnung" und „Weltlich Ding" in Luthers Theologie. Göttingen 1933. 165 S. = Studien zur systematischen Theologie 12. Leipzig, Theol. Diss. 1932.
3. Zur kirchen- und theologiegeschichtlichen Bedeutung von Luthers Deutscher Bibel. Neues Sächsisches Kirchenblatt *41* (1934) 577—586.
4. Das eigene und das fremde Werk der Kirche. Neues Sächsisches Kirchenblatt *43* (1936) 547—554.
5. Das Heil des Volkes und das Evangelium. Geistlicher und leiblicher Segen. Ein Beitrag zu dem Thema: Luther und die Gegenwart. Leipzig 1937. 22 S. = Theologia militans 13.
6. Luthers Heidelberger Disputation in ihrer Bedeutung für Theologie und Kirche heute. SKB NF *1* (1937) 355—357. 363f.
7. Luthers Schmalkaldische Artikel als eine Einführung in seine Theologie. ZThK NF *18* (1937) 289—307.
8. Die kirchlichen Ordnungen als Erziehungsmacht. Leipzig 1938. 48 S. = Theologia militans 17.
9. Der Weg zum Glauben an den auferstandenen Christus. In: Der Heiland. Ein Buch von Jesus, dem Grund des Glaubens, dem Fürsten des Lebens und dem Herrn der Welt. Berlin 1938, 44—55.
10. Der protestantische und der reformatorische Mensch. SKB NF *2* (1938) 349—352.
11. Lagarde und die theologischen Fragen der Gegenwart. Allg. Ev.-Luth. Kirchenzeitung *71* (1938) 650—653, 666—672, 690—697.
12. Noch einmal: Paul de Lagarde. Allg. Ev.-Luth. Kirchenzeitung *71* (1938) 812—816.
13. Abgrenzung und Liebesgebot. Die Christliche Welt *52* (1938) 95—100.
14. Weltanschauung, Religion und christlicher Glaube. Neue kirchliche Zeitschrift *49* (1938) 299—312.

15. Synthese? (Stellungnahme zu Kurt Leeses Entgegnung auf „Der protestantische und der reformatorische Mensch"). SKB NF *3* (1939) 34—38.
16. Abschluß eines Gesprächs (mit Kurt Leese). SKB NF *3* (1939) 101.
17. Das Wesen des Christentums. Zu Emanuel Hirschs so betiteltem Buch. SKB NF *3* (1939) 334f.
18. Die Rede von der Sünde als Not und Aufgabe unserer Verkündigung. PB *82* (1939/40) 442—455.
19. Die Gestalt der Kirche in der Geschichte. Ev.-Luth. Kirchenzeitung *2* (1948) 107—110.
20. Die 600 Jahre alte Universität Prag und die Kirchengeschichte. Ev.-Luth. Kirchenzeitung *2* (1948) 257—259.
21. Erstes Gebot und Ehre Gottes als Mitte von Luthers Theologie. ThLZ *73* (1948) 719—730.
22. Das Matthäus-Evangelium. Übers. und ausgelegt. Kassel 1949. 210 S. = Bibelhilfe für die Gemeinde. Neutestamentliche Reihe 1.
23. Dass. Berlin 1951. 223 S. (2. Aufl. 1956).
24. Luthers Lehre von den beiden Reichen. Glaube und Heimat *6* (1951) Nr. 41 vom 14. 10. 1951, 4.
25. Luthers Lehre von den beiden Reichen. Berlin 1952. 96 S. (Nachdruck Nr. 30).
26. Aus der Geschichte der Kirche. In: Christenlehre. Ein Buch für die evangelische Jugend. Hrsg. von der Bibelanstalt Altenburg. 5. Aufl. Berlin 1952, 411—439.
27. Gustav Adolf — Werk und Kirche. Die Diaspora *1* (1952) Nr. 3, 1—9.
28. Luther und Leipzig. Der Sonntag *7* (1952) 197f. = Die Kirche *7* (1952) Nr. 21 vom 18. 5. 1952, 4 (gekürzt). = Die Union *7* (1952) Nr. 83 vom 17. 5. 1952, 3.
29. Bemerkungen zu Luthers Lehre von den beiden Reichen. Ev.-Luth. Kirchenzeitung *6* (1952) 234—236.
30. Luthers Lehre von beiden Reichen. Berlin: Luth. Verlagshaus 1953. 96 S. = Luthertum 8 (Nachdruck von Nr. 25).
31. Diaspora und ökumenisches Bewußtsein in der Geschichte der christlichen Kirche. EvD *24* (1953) 109—121.
32. Angriff auf Luther. InB *2* (1953) 288—290.
33. Das Bild Gustav Adolfs im Wandel der Zeit. Trägt das Gustav-Adolf-Werk seinen Namen noch zu Recht? Nachrichten der Ev.-Luth. Kirche in Bayern *8* (1953) 365—368. 382—384.
34. Georg III. von Anhalt (1507—1553), erster evangelischer „Bischof" von Merseburg. Seine Theologie und seine Bedeutung für die Geschichte der Reformation in Deutschland. Albrecht Alt zum 70. Geburtstag gewidmet. Wissenschaftliche Zeitschrift der Karl-Marx-Universität Leipzig. Gesellschafts- und Sprachwissenschaftliche Reihe *3* (1953/54) 139—154. = Alt-Festschrift, 93—106.
35. Aus der Geschichte der Kirche. In: Christenlehre. 6. Aufl. Berlin 1954, 411—439 (Vgl. Nr. 26).
36. Die prophetische Apokalyptik Thomas Müntzers und Luthers Absage an die Bauernrevolution. In: Gedenkschrift für D. Werner Elert. Beiträge zur historischen und systematischen Theologie. Berlin 1955, 163—170.
37. Ordnung und Freiheit. Eine kirchengeschichtliche Vorbesinnung zum Problem der freien Werke in der Kirche. EvD *26* (1955) 5—16. (Wiederabdruck Nr. 48).
38. Die Stunde der Evangelischen Kirche in Deutschland. EvD *26* (1955) 171—176.
39. Die Geschichte der Waldenser von den Anfängen bis zur Gegenwart. EvD *26* (1955) 218—234.

40. Die Einheit der Evangelischen Kirche in Deutschland. Junge Kirche *16* (1955) 453 bis 457.
41. Die Bedeutung des Augsburger Religionsfriedens von 1555. Nachrichten der Ev.-Luth. Kirche in Bayern *10* (1955) 261—265.
42. Probleme, Methoden und Erfahrungen katholischer Diaspora-Arbeit. EvD *27* (1956) 171—180.
43. Die Reformation in den norddeutschen Städten. InB *5* (1956) 113—118.
44. Internationale Lutherforschung. InB *5* (1956) 299—301.
45. Leges charitatis. Drei Fragen an Johannes Heckel. KD *2* (1956) 76—89. (Wiederabdruck Nr. 120).
46. Luthers Eintritt ins Erfurter Augustinerkloster. Luther *27* (1956) 49—70.
47. Luther-Jahrbuch 1956. Luther *27* (1956) 91—93.
48. Ordnung und Freiheit. In: Um Diaspora-Dienst und Diaspora-Fragen. Bruno Geißler, dem Achtzigjährigen. Kassel 1957, 5—16. (Wiederabdruck von Nr. 37).
49. (Bearb.) Luther-Bibliographie 1940—1953. (Auswahl). LuJ *24* (1957) 141—155.
50. (Bearb.) Luther-Bibliographie 1954—1972. LuJ *24* (1957) — *39* (1972). (Ab 1965 mit Helmar Junghans).
51. Der Kampf um das Naturrecht und die evangelische Theologie. In: Solange es „heute" heißt. Festgabe für Rudolf Hermann zum 70. Geburtstag. Berlin 1957, 179—199.
52. Luther — Revolutionär oder Reaktionär? Luther *28* (1957) 109—133.
53. Gustav-Adolf-Werk gedenkt seines Gründers Christian Gottlob Leberecht Großmann. Evang. Nachrichtendienst Ost *10* (1957) Nr. 23 vom 6. 6. 1957, 12.—15. = Die Kirche *12* (1957) Nr. 26 vom 30. 6. 1957, 3.
54. Die Konditional- oder Eventualtaufe und die Frage nach ihrem Recht in der lutherischen Kirche. LuJ *25* (1958) 110—140.
55. Die Vollversammlung des Lutherischen Weltbundes in Minneapolis 1957 und die Probleme der Diaspora. EvD *29* (1958) 8—18.
56. Luther. Berlin 1959. 151 S. = Sammlung Göschen 1187. (2. Aufl. 1966, Übersetzungen Nr. 76, 80, 95—97, 126a).
57. Die Verwandlung des Armutsproblems. In: Ökumenische Diakonie. Hrsg. von Christian Berg. Berlin 1959, 144. (Aus RGG *1*, 627f. vgl. Nr. 383).
58. Der Bauernkrieg und das angebliche Ende der lutherischen Reformation als spontaner Volksbewegung. *26* (1959) 109—134. (Wiederabdruck Nr. 111).
59. Die Bedeutung Calvins für die evangelische Diaspora. EvD *30* (1959) 133—143.
60. Martin Luther und sein Werk. EvD *30* (1959) 144—148.
61. Der verängstigte Luther. Zur Auseinandersetzung mit Friedrich Heer. Rheinische Post *14* (1959) Nr. 23 vom 31. 10. 1959.
62. Père Reinoud und Luther. Bemerkungen zu Reinhold Weijenborgs Lutherstudien. LuJ *27* (1960) 64—122. = (gekürzt) In: Bekenntnis zur Kirche. Festgabe für Ernst Sommerlath zum 70. Geburtstag. Berlin 1960, 94—111.
63. Vom Dienst an den Glaubensgenossen in der Zerstreuung. Werden und Wirken des Gustav-Adolf-Werkes. In: Unsere evangelische Kirche heute. Hrsg. von Hans Heuer. Nürnberg 1960, 507—522.
64. Die Calvinjubiläen 1959 in ihrer gesamtkirchlichen Bedeutung. InB *9* (1960) 65—70.
65. Lutherische Diasporaarbeit. InB *9* (1960) 252f.
66. Die Königsherrschaft Jesu Christi und die lutherische Zweireichelehre. KD *6* (1960) 306—326. (Wiederabdruck Nr. 119).
67. Staatskirchen überholt. Ev.-Luth. Kirchenzeitung *14* (1960) 116.

68. Philipp Melanchthon 1495–1560. Der Sonntag *15* (1960) Nr. 22 vom 24. 4. 1960, 1f.
69. Die Königsherrschaft Jesu Christi und die lutherische Zweireichelehre. Nachrichten der Luther-Akademie *1960* (1960) 27–30.
70. Paulus – Augustin – Luther. Rechtfertigung nicht aus dem Werk, sondern aus dem Glauben. Berlin 1961, 35 S. = Aufsätze und Vorträge zur Theologie und Religionswissenschaft 17.
71. Aus der Geschichte des Domes und des Domkapitels zu Wurzen. Herbergen der Christenheit *4* (1961) 8–19.
72. Aus der Geschichte der Leipziger Theologischen Fakultät. Herbergen der Christenheit 4 (1961) 27–39.
73. Luthers Worttheologie in katholischer Sicht. LuJ *28* (1961) 110–116.
74. Melanchthon und die Ordnung der Kirche. In: Philipp Melanchthon. Forschungsbeiträge zur vierhundertsten Wiederkehr seines Todestages, dargeboten in Wittenberg 1960. Göttingen, Berlin 1961, 98–115.
75. Der verängstigte Luther. Zu Friedrich Heers Humanismusbuch „Die dritte Kraft". In: Reich Gottes und Wirklichkeit. Festgabe für Alfred Dedo Müller zum 70. Geburtstag. Berlin 1961, 376–380.
76. Luther. Übers. von Robert H. Fischer. Philadelphia 1962. 178 S. (Übers. von Nr. 56, Nachdruck Nr. 80).
77. Theologie der Schöpfung gleich Theologie überhaupt? Zur Auseinandersetzung mit Löfgrens Luther-Buch. LuJ *29* (1962) 44–51. (Wiederabdruck Nr. 85).
78. Bert Brecht und Luther. Ein Versuch der Interpretation des „Guten Menschen von Sezuan". LuJ *29* (1962) 92–109.
79. Zweifel um den 31. Oktober 1517? Lutherische Monatshefte *1* (1962) 459–463.
80. Luther. Übers. von Robert H. Fischer. London 1963. 178 S. (Übers. von Nr. 56, Nachdruck von Nr. 76).
81. La vie religieuse dans les pays protestants de langue allemande à la fin du XVIe siècle. In: Colloque d'histoire religieuse (Lyon, octobre 1963). Grenoble 1963, 101–120.
82. „Die Theologie Martin Luthers". Ein Dankeswort an Paul Althaus. LuJ *30* (1963) 9–16.
83. Zehn Jahre evangelische Diaspora. EvD *34* (1963) 1–10.
84. Bert Brecht und Luther. Versuch einer Interpretation des Guten Menschen von Sezuan. Nachrichten der Luther-Akademie *1962* (1963) 33f.
84a. (Hrsg.) Der Glaube der Reformatoren. Luther – Zwingli – Calvin. Bremen 1964. XVI, 481 S. = Klassiker des Protestantismus 2 = Sammlung Dieterich 267.
85. Theologie der Schöpfung gleich Theologie überhaupt? Zur Auseinandersetzung mit Löfgrens Luther-Buch. In: ... und fragten nach Jesus. Beiträge aus Theologie, Kirche und Geschichte. Festschrift für Ernst Barnikol zum 70. Geburtstag. Berlin 1964, 140–146. (Wiederabdruck von Nr. 77).
86. Reformationsgeschichte Deutschlands bis 1555. Göttingen 1964. 170 S. = Die Kirche in ihrer Geschichte 3 K. (Mitverfasser Ernst Bizer). (2. Aufl. Nr. 121, engl. Übersetzung Nr. 118).
87. Die zweite Reformation in Kursachsen. Neue Forschungen zum sogenannten sächsischen Kryptocalvinismus. In: Verantwortung. Untersuchungen über Fragen aus Theologie und Geschichte. Zum sechzigsten Geburtstag von Landesbischof D. Gottfried Noth DD. Berlin 1964, 137–154.
88. Die Verantwortung der Kirche für ihre Diaspora. Mecklenburgische Kirchenzeitung *19* (1965) Nr. 21 vom 31. 5. 1965, 1.

89. Lutherische Diaspora-Fürsorge. EvD *36* (1965/66) 133—140. (Vgl. Nr. 384).
90. Die lutherische Lehre von den beiden Reichen. Lutherische Rundschau *15* (1965) 463—485. (Wiederabdruck Nr. 125).
91. The Lutheran Doctrine of the Two Kingdoms. Lutheran World *12* (1965) 355—372.
92. Trägt das Gustav-Adolf-Werk seinen Namen noch zu Recht? EvD *37* (1966) 9—16.
93. Briefwechsel zwischen P. Dr. Reinhold Weijenborg OFM und dem Herausgeber des Luther-Jahrbuches. LuJ *33* (1966) 118—121.
94. Lutherforschung. Lutherische Monatshefte *5* (1966) 512—519.
95. Marcin Luther (Luther poln.). Übersetzt von Janusz Narzyński. Warszawa 1966. 103 S. (Übersetzung von Nr. 56).
96. Dass. mit einem Nachwort von Oskar Bartel und Janusz Narzyński: Marcin Luter w Polsce. Warszawa 1966. 234 S.
97. Rutta-Ron (Luther jap.). Übers. von Shigeru Watanabe. Tokyo 1966. 206 S. (Übersetzung von Nr. 56).
98. Die gegenwärtige Diskussion um Luthers Thesenanschlag. Sachstandsbericht und Versuch einer Weiterführung durch Neuinterpretation von Dokumenten. LuJ *34* (1967) 11—59. = In: Erbe und Verpflichtung. Reformationsgedenkbuch. Berlin 1967, 13—72.
99. The Posting of Luther's Theses — Legend or Fact? Übers. von Elmer E. Foelber. Concordia Theological Montly *38* (1967) 691—703.
100. L'affichage des thèses de Luther une legende ou un fait? Luther a-t-il, ou non, affiché les 95 thèses? Positions lutheriennes *15* (1967) 270—282.
101. A afixação das teses de Lutero. Lenda ou Fato? (Luthers Thesenanschlag — Legende oder Tatsache? portug.). Übers. von M. Siegle und M. Dreher. Estudos teológicos *7* (1967) 170—185.
102. Nowy spór o tezy (Der neue Streit um die Thesen. poln.). Übers. von Karol Karski. Jednota *11* (1967) Nr. 10, 9—12 (Übersetzung von Nr. 124).
103. Aus Liebe zur Wahrheit. Deutschland. In: Reformation in Europa. Berlin, Kassel 1967, 9—24. (Übersetzungen Nr. 104—106).
104. For Love of Truth. In: Illustrated History of the Reformation. Leipzig 1967, 9 bis 24.
105. Uit liefde tot de waarheid (Aus Liebe zur Wahrheit. niederl.). In: De Reformatie in Europa. Übers. von A. van der Worp. Baarn 1967, 9—24.
106. Av kärlek till sanningen (Aus Liebe zur Wahrheit, schwed.). In: Reformationen i Europa. Stockholm 1967, 9—24.
107. Die Sache der deutschen Nation, die Kultur des Abendlandes und das Evangelium von der Herrlichkeit und Gnade Gottes. Amtsblatt der Evang.-Luth. Kirche in Thüringen *20* (1967) 119—224. Auch als Sonderdruck.
108. Germany, Western Culture, and the Gospel of the Glory and Grace of God. The Lutheran Quartely *19* (1967) 119—224. (Übersetzung von Nr. 107).
109. A reformáció mint nemzeti, európai és evangéliumi ügy (Die Sache der deutschen Nation, die Kultur des Abendlandes und das Evangelium von der Herrlichkeit und Gnade Gottes. ung. gekürzt). Lelkipasztor *42* (1967) 710—720.
110. Der junge Luther und der Beginn der Reformation. Zu dem so überschriebenen Aufsatz von Erwin Iserloh in Heft 11/1967. Die Zeichen der Zeit *21* (1967) 452—457.
111. Der Bauernkrieg und das angebliche Ende der lutherischen Reformation als spontaner Volksbewegung. In: Wirkungen der deutschen Reformation bis 1555. Darmstadt 1967, 68—100. (Wiederabdruck von Nr. 58).
112. Luthers Wahrheit und Wahrhaftigkeit. Gustav-Adolf-Bote *13* (1967) Heft 4.

113. Die Bedeutung Luthers für Theologie und Kirche unserer Zeit. Der Sonntag *22* (1967) Nr. 49 vom 29. 8. 1967, 1f.
114. Einleitung zu: Junghans, Helmar: Die Reformation in Augenzeugenberichten. 1. und 2. Aufl. Düsseldorf 1967, 13—34. (Wiederabdruck Nr. 126).
115. Luther und Balthasar Hubmaier. In: Humanitas — Christianitas. Walther von Loewenich zum 65. Geburtstag. Witten 1968, 63—73.
116. Stand der Lutherforschung heute. In: Reformation 1517—1967. Wittenberger Vorträge. Berlin 1968, 35—63.
117. Sinn und Aufgabe eines Domkapitels ehedem und heute. Der Sonntag *23* (1968) Nr. 25 vom 19. 5. 1968, 3.
118. A History of the Reformation in Germany to 1555. Übers. von B. A. Hardy. London 1968. XII, 249 S. (engl. Übersetzung von Nr. 86).
119. Die Königsherrschaft Jesu Christi und die lutherische Zweireichelehre. In: Reich Gottes und Welt. Darmstadt 1969. 484—513. (Wiederabdruck von Nr. 66).
120. Leges charitatis. In: Reich Gottes und Welt. Darmstadt 1969, 528—547. (Wiederabdruck von Nr. 45).
121. Reformationsgeschichte Deutschlands bis 1555. 2. Aufl. Göttingen 1969. 170 S. (Vgl. Nr. 86).
122. Luthers vierte Bauernschrift. In: Antwort aus der Geschichte. Betrachtungen und Erwägungen zum geschichtlichen Bild der Kirche. Walter Dress zum 65. Geburtstag. Berlin 1969, 84—98.
123. Vision eines ganz neuen Christentums. Luther *40* (1969) 86—91.
124. Der neue Thesenstreit. In: Die Reformation, ihr geschichtlicher Ablauf und ihre Bedeutung für die Theologie der Gegenwart. Referate, gehalten auf einer Tagung der Theologischen Fakultät der Martin-Luther-Universität Halle—Wittenberg anläßlich des 450. Jahrestages der Reformation am 27. 10. 1967. Halle 1969, 48—54. = Wissenschaftliche Beiträge der Martin-Luther-Universität 1969/8.
125. Die lutherische Lehre von den beiden Reichen. In: Luther und die Obrigkeit, Darmstadt 1972, 370—396. (Wiederabdruck von Nr. 90).
126. Einleitung zu: Junghans Helmar: Die Reformation in Augenzeugenberichten. München 1973, 13—34. = dtv Nr. 887 (Nachdruck von Nr. 114, ohne Bildteil).
126a. Lutero (Luther port.). Übers. von Walter O. Schlupp. São Leopoldo: Editora Sinodal 1974. 115 S. (Übersetzung von Nr. 56).

Besprechungen (Auswahl)

127. Reymann, Heinz: Glaube und Wirtschaft bei Luther. Gütersloh 1934. Neues Sächsisches Kirchenblatt *43* (1936) 431f.
128. Meinhold, Peter: Die Genesisvorlesung Luthers und ihre Herausgeber. Stuttgart 1936. ZThK NF *17* (1936) 379f.
129. Zoellner, Wilhelm, und Wilhelm Stählin: Die Kirche Jesu Christi und das Wort Gottes. Berlin 1937. SKB NF *1* (1937) 241.
130. Moldaenke, Günther: Schriftverständnis und Schriftdeutung im Zeitalter der Reformation. Teil 1: Matthias Flacius Illyricus. Stuttgart 1936. Die Christliche Welt *51* (1937) 86f.
131. Seeberg, Erich: Wer ist Christus. Tübingen 1937. SKB NF *2* (1938) 14.
132. Niemeier, G(ottfried): Wirklichkeit und Wahrheit. Gütersloh 1937. SKB NF *2* (1938) 250f.

133. Arnold, Franz Xaver: Zur Frage des Naturrechts bei Martin Luther. München 1937. ThLBl *59* (1938) 38.
134. Asmussen, Hans: Gottesdienstlehre. Bd. 1: Die Lehre vom Gottesdienst. München 1937. ThLBl *59* (1938) 188f.
135. Wünsch, Georg: Evangelische Ethik des Politischen. Tübingen 1936. ThLBl *59* (1938) 330–333.
136. Köhler, Manfred: Melanchthon und der Islam. Leipzig 1938. ZThK NF *19* (1938) 340.
137. Schwab, Wilhelm: Die Vereinigte evangelisch-protestantische Landeskirche Badens. Heidelberg 1938. ZThK NF *19* (1938) 366f.
138. Leese, Kurt: Die Religion des protestantischen Menschen. Berlin 1938. ThLBl *60* (1939) 153–156.
139. Müller, Alfred Dedo: Der Kampf um das Reich. Frankfurt (Main) 1935. SKB NF *4* (1940) 8.
140. Diem, Harald: Luthers Lehre von den beiden Reichen, untersucht von seinem Verständnis der Bergpredigt aus. München 1938. ThLBl *63* (1942) 125f.
141. Doerne, Martin: Der Mensch im Urteil der Bibel. Potsdam 1939. ThLBl *64* (1943) 22f.
142. Dilschneider, Otto: Evangelische Offenbarung. Gütersloh 1939. ThLBl *64* (1943) 44f.
143. Fangen, Ronald: ... daß sie Alle Eines seien. Die ökumenische Botschaft der Gruppenbewegung. Basel 1938. ThLZ *73* (1948) 540f.
144. Schlink, Edmund: Theologie der lutherischen Bekenntnisschriften. 2. Aufl. München 1946. ThLZ *74* (1949) 552–554.
145. Lübeck, Konrad: Das Bonifatiusgrab zu Fulda. Fulda 1947. ThLZ *75* (1950) 741f.
146. Leeuw, G(erardus) van der: Die Bilanz des Christentums. Zürich 1947. ThLZ *76* (1951) 221f.
147. Preuß, Hans: Die Geschichte der Abendmahlsfrömmigkeit in Zeugnisen und Berichten. Gütersloh 1949. ThLZ *76* (1951) 235.
148. Solzbacher, Joseph: Kaspar Ulenberg. Eine Priestergestalt aus der Zeit der Gegenreformation in Köln. Münster 1948. ThLZ *76* (1951) 742f.
149. Delius, Walter: Die Evangelische Kirche und die Revolution 1848. Berlin 1948. ThLZ *77* (1952) 231f.
150. Schuster, Hermann: Das Werden der Kirche. Eine Geschichte der Kirche auf deutschem Boden. Berlin 1950. ThLZ *77* (1952) 356f.
151. Rich, Arthur: Die Anfänge der Theologie Huldrych Zwinglis. Zürich 1949. ThLZ *77* (1952) 618f.
152. Joachimsen, Paul: Die Reformation als Epoche der deutschen Geschichte. München 1951. ThLZ *78* (1953) 288–290.
153. Melanchthons Werke in Auswahl. Hrsg. von Robert Stupperich. Bd 1: Reformatorische Schriften. Bd 2, 1: Loci communes von 1521. Loci praecipui theologici von 1559 (Teil 1). Güthersloh 1951/52. ThLZ *78* (1953) 292.
154. Harnack, Adolf von: Das Wesen des Christentums. Neuauflage zum 50. Jahrestag des ersten Erscheinens mit einem Geleitwort von Rudolf Bultmann. Stuttgart, Berlin 1950. ThLZ *78* (1953) 343–345.
155. Kähler, Ernst: Karlstadt und Augustin. Der Kommentar des Andreas Bodenstein von Karlstadt zu Augustins Schrift De spiritu et litera. Halle 1952. ThLZ *78* (1953) 427f.
156. Kosmos und Ekklesia. Festschrift für Wilhelm Stählin zu seinem 70. Geburtstag. Hrsg. von Heinz-Dietrich Wendland. Kassel 1953. EvD *25* (1954) 128.
157. Helbig, Herbert: Die Reformation der Universität Leipzig im 16. Jahrhundert.

Gütersloh 1953. EvD *25* (1954) 183f.; Blätter für deutsche Landesgeschichte *92* (1956) 554f.
158. Hudak, Adalbert: Die Kirche unserer Väter. Weg und Ende des deutschen Luthertums in der Slowakei. Stuttgart 1953. EvD *25* (1954) 187.
159. Niesel, Wilhelm: Das Evangelium und die Kirchen. Ein Lehrbuch der Symbolik. Neukirchen 1953. EvD *25* (1954) 128. 187f.
160. Hutten, Kurt: Seher — Grübler — Enthusiasten. Sekten und religiöse Sondergemeinschaften der Gegenwart. 3. Aufl. Stuttgart 1953. EVD *25* (1954) 233. (Bespr. der 5. Aufl. Nr. 225).
161. Meißinger, Karl August: Der katholische Luther. München 1952. EvD *25* (1954) 235f.
162. Viele Glieder — ein Leib. Kleine Kirchen. Freikirchen u. ä. Gemeinschaften in Selbstdarstellungen. Hrsg. von Ulrich Kunz. Stuttgart 1953. EvD *26* (1955) 77.
163. Bring, Ragnar: Das Verhältnis von Glauben und Werken in der lutherischen Theologie. München 1955. EvD *26* (1955) 199f.
164. Hermelink, Heinrich: Das Christentum in der Menschheitsgeschichte von der Französischen Revolution bis zur Gegenwart. 3 Bde. Stuttgart, Tübingen 1951—1955. EvD *26* (1955) 200—202.
165. Gruner, Max: Revolutionäres Christentum. 50 Jahre Geschichte der Heilsarmee in Deutschland. Berlin, Bochum 1954. EvD *26* (1955) 255f.
166. Hillerdal, Gunnar: Gehorsam gegen Gott und Menschen. Göttingen 1955, KD *1* (1955) 243—246.
167. Die Lage der Protestanten in katholischen Ländern. Zollikon, Zürich 1953. ThLZ *80* (1955) 224f.
168. Richter, Friedrich: Martin Luther und Ignatius von Loyola. Repräsentanten zweier Geisteswelten. Stuttgart 1954. ThLZ *80* (1955) 544f.
169. Pfister, Rudolf: Um des Glaubens willen. Die evangelischen Flüchtlinge von Locarno und ihre Aufnahme zu Zürich im Jahre 1555. ThLZ *80* (1955) 614f.
170. Vischer, Melchior: Jan Hus. Aufruhr wider Papst und Reich. Frankfurt/Main 1955. ThLZ *80* (1955) 673f.
171. Vinay, Valdo: Facoltà Valdese di Teologìa 1855—1955. Torre Pellice 1955. ThLZ *80* (1955) 753f.
172. Daumiller, Oscar: Südbayerns evangelische Diaspora. München 1955. EvD *27* (1956) 67f.
173. Flückinger, Felix: Der Ursprung des christlichen Dogmas. Eine Auseinandersetzung mit Albert Schweitzer und Martin Werner. Zollikon, Zürich 1955. EvD *27* (1956) 127f.
174. Beckmann, Joachim: Quellen zur Geschichte des christlichen Gottesdienstes. Gütersloh 1956. EvD *27* (1956) 128.
175. Blanke, Fritz: Kirchen und Sekten. Führer durch die religiösen Gruppen der Gegenwart. Zürich 1955. EvD *27* (1956) 128.
176. Wittenberg, Martin: Friede im Reich. Geschichte und Probleme des Religionsfriedens von Augsburg 1555. München 1956. EvD *27* (1956) 194.
177. Loewenich, Walther von: Der moderne Katholizismus. Witten 1955. EvD *27* (1956) 255.
178. Hardt, Karl (Hrsg.): Bekenntnis zur katholischen Kirche. Würzburg 1955. ThLZ *81* (1956) 174f.
179. Buchheim, Karl: Geschichte der christlichen Parteien in Deutschland. München 1953. ThLZ *81* (1956) 739—741.

180. Niemöller, Wilhelm: Die evangelische Kirche im dritten Reich. Bielefeld 1956. EvD *28* (1957) 240.
181. Luthers Werke in Auswahl. Bd 5: Der junge Luther. Hrsg. von Erich Vogelsang. Bd. 6: Luthers Briefe. Hrsg. von Hans Rückert. 2. Aufl. Berlin 1955. ThLZ *82* (1957) 48f.
182. Aland, Kurt: Hilfsbuch zum Lutherstudium. Berlin 1957. LuJ *25* (1958) 181f.
183. Joest, Wilfried: Gesetz und Freiheit. Göttingen 1956. LuJ *25* (1958) 190f.
184. Heckel, Johannes: Lex charitatis. München 1953. LuJ *25* (1958) 193–195.
185. Bornkamm, Heinrich: Luther im Spiegel der deutschen Geistesgeschichte. Heidelberg 1955. LuJ *25* (1958) 203.
186. Senft, Christoph: Wahrhaftigkeit und Wahrheit. Die Theologie des 19. Jahrhunderts zwischen Orthodoxie und Aufklärung. Tübingen 1957. EvD *29* (1958) 117f.
187. Hiltebrandt, Philipp: Papsttum und Kirche. Stuttgart 1957. EvD *29* (1958) 161.
188. Hauß, Fritz und Hans Georg Zier: Die Kirchenordnungen von 1556 in der Kurpfalz und in der Markgrafschaft Baden-Durlach. Karlsruhe 1956. EvD *29* (1958) 162.
189. Eggenberger, Oswald: Evangelischer Glaube und Pfingstbewegung. Zollikon 1956. EvD *29* (1958) 162f.
190. Hammerschmidt, Ernst: Grundriß der Konfessionskunde. Innsbruck, Wien, München 1955. EvD *29* (1958) 163.
191. Steck, Karl Gerhard: Luther und die Schwärmer. Zollikon 1955. EvD *29* (1958) 203f.
192. Marie-Eugène vom Kinde Jesu: Ich will Gott schauen. Die geistliche Lehre der Hl. Theresia von Avila, des Hl. Johannes vom Kreuz, der Hl. Theresia vom Kinde Jesu. II. Vereinigung der Seele mit Gott. Basel 1954. ThLZ *83* (1958) 205–207.
193. Link, Wilhelm: Das Ringen Luthers um die Freiheit der Theologie von der Philosophie. München 1955. ThLZ *83* (1958) 688–691.
194. Fischer, Max und Arthur Frey: Zeitgenössische Betrachtungen zur Jesuitenfrage. Zollikon, Zürich 1953. ThLZ *83* (1958) 692f.
195. Fabian, Ekkehart: Die Entstehung des Schmalkaldischen Bundes und seiner Verfassung 1529–1531/33. Tübingen 1956. ThLZ *83* (1958) 774f.
196. Gerstenkorn, Hans Robert: Weltlich Regiment zwischen Gottesreich und Teufelsmacht. Bonn 1956. LuJ *26* (1959) 141–145.
197. Bizer, Ernst: Fides ex auditu. Neukirchen 1958. LuJ *26* (1959) 154–156.
198. Haikola, Lauri: Studien zu Luther und zum Luthertum. Uppsala, Wiesbaden 1958. LuJ *26* (1959) 156–159.
199. Lutero, Martino: Scritti Religiosi. Bari 1958. LuJ *26* (1959) 159f.
200. Westin, Gunnar: Der Weg der freien christlichen Gemeinden durch die Jahrhunderte. Kassel 1956. EvD *30* (1959) 124f.
201. Dankbaar, Willem F.: Calvin. Neukirchen 1959. EvD *30* (1599) 195f.
202. Søe, Niels Hansen: Christliche Ethik. 2. Aufl. München 1957. EvD *30* (1959) 255f.
203. Kümmel, Werner Georg: Das Neue Testament. Geschichte der Erforschung seiner Probleme. München 1958. EvD *30* (1959) 256f.
204. Brandes, Erwin, und Willy Grüninger: Ein evangelischer Erzieher. Festschrift für Reinhold Sautter. Stuttgart 1954. ThLZ *84* (1959) 46.
205. Rautenberg, Werner: Johannes Bugenhagen. Berlin 1958. ThLZ *84* (1959) 299f.
206. Schmidt, Martin: John Wesley . Bd 1. Zürich 1953. ThLZ *84* (1959) 301–303.
207. Rost, Hans: Das konfessionelle Schicksal Deutschlands. 400 Jahre Glaubensspaltung und Una Sancta. Friedberg 1955. ThLZ *84* (1959) 372–374.

208. Historia Mundi 7: Übergang zur Moderne. Bern 1957. ThLZ *84* (1959) 431–434.
209. Franz, Georg: Kulturkampf. Staat und katholische Kirche in Mitteleuropa von der Säkularisation bis zum Abschluß des Preußischen Kulturkampfes. München 1954. ThLZ *84* (1959) 527f.
210. Klein, Joseph: Skandalon. Um das Wesen des Katholizismus. Tübingen 1968. ThLZ *84* (1959) 602–604.
211. Melanchthons Werke. Bd 2, 2: Loci praecipui theologici von 1559 (Teil 2) und Definitiones. Bd 6: Bekenntnisse und kleine Lehrschriften. Gütersloh 1953/55. ThLZ *84* (1959) 921f.
212. Lutherforschung heute. Berlin 1958. LuJ *27* (1960) 123f.
213. Thulin, Oscar: Martin Luther. München 1958. LuJ *27* (1960) 124.
214. Festgabe für Joseph Lortz. Baden-Baden 1958. LuJ *27* (1960) 124–126.
215. Bizer, Ernst: Luther und der Papst. München 1958. LuJ *27* (1960) 126f.
216. Lohse, Bernhard: Ratio und Fides. Göttingen 1958. LuJ *27* (1960) 131–133.
217. Luther, Martin: Ausgewählte Werke. Hrsg. von H(ans) H(einrich) Borcherdt und Georg Merz. Bd 6: Bibelübersetzung, Schriftauslegung, Predigt. München 1958. LuJ *27* (1960) 136f.
218. Cadier, Jean: Calvin. Zollikon 1959. EvD *31* (1960) 66f.
219. Jedin, Hubert: Kleine Konziliengeschichte. Freiburg 1959. EvD *31* (1960) 68.
220. Kirche im Osten. Bd 1–3. Stuttgart 1958–1960. EvD *31* (1960) 195f.
221. Bucsay, Mihaly: Geschichte des Protestantismus in Ungarn. Stuttgart 1959. EvD *31* (1960) 196–198.
222. Ausgewählte Quellen zur Kirchengeschichte Ostmitteleuropas. Ulm 1959. EvD *31* (1960) 198f.
223. Tyler, Royall: Kaiser Karl V. Stuttgart 1959. EvD *31* (1960) 199f.
224. Sperl, Adolf: Melanchthon zwischen Humanismus und Reformation. München 1959. EvD *31* (1960) 200.
225. Hutten, Kurt: Seher, Grübler – Enthusiasten. 5. Aufl. Stuttgart 1958. ThLZ *85* (1960) 59–61. (Bespr. der 3. Aufl. Nr. 160).
226. Stadelmann, Rudolf: Das Zeitalter der Reformation. Darmstadt 1954. ThLZ *85* (1960) 678f.
227. Kohler, Erika: Martin Luther und der Festbrauch. Köln, Graz 1959. LuJ *28* (1961) 124–126.
228. Pohlmann, Hans: Hat Luther Paulus entdeckt? Berlin 1959. LuJ *28* (1961) 127–130.
229. Lackmann, Max: Katholische Einheit und Augsburger Konfession. Graz, Wien, Köln 1959. EvD *32* (1961) 55.
230. Albrecht, Dieter: Richelieu, Gustav Adolf und das Reich. München 1959. EvD *32* (1961) 55f.
231. Gebhardt, Bruno: Handbuch der deutschen Geschichte. 8. Aufl. Bd 4: Erdmann, Karl Dietrich: Die Zeit der Weltkriege. Stuttgart 1959. EvD *32* (1961) 124f.
232. Brunner, Peter: Nikolaus von Amsdorf als Bischof von Naumburg. Gütersloh 1961. EvD *32* (1961) 191.
233. Comenius, Amos: Pampaedia. Heidelberg 1960. EvD *32* (1961) 250–252.
234. Geißler, Heinrich: Comenius und die Sprache. Heidelberg 1959. EvD *32* (1961) 250 bis 252.
235. Kurzer Bericht, wie der ehrwürdige Herr, unser lieber Vater und Präzeptor Philipp Melanchthon sein Leben hie auf Erden geendet und ganz christlich beschlossen hat. München 1960. EvD *32* (1961) 252.

236. Kantzenbach, Friedrich Wilhelm: Evangelium und Dogma. Stuttgart 1959. Ev.-Luth. Kirchenzeitung *15* (1961) 291.
237. Mulert, Hermann: Konfessionskunde. 3. Aufl. Berlin 1956. ThLZ *86* (1961) 511–514.
238. Luther, Martin: Vorlesung über den Römerbrief 1515/1516. Lateinisch-deutsche Ausgabe. Hrsg. von Martin Hofmann. 2 Bde. Weimar 1960. LuJ *29* (1962) 117f.
239. Hermann, Rudolf: Gesammelte Studien zur Theologie Luthers und der Reformation. Göttingen 1960. LuJ *29* (1962) 119–121; DLZ *83* (1962) 590–592.
240. Forck, Gottfried: Die Königsherrschaft Jesu Christi bei Luther. Berlin 1959. LuJ *29* (1962) 121–124.
241. Rogge, Joachim: Johann Agricolas Lutherverständnis unter besonderer Berücksichtigung des Antinomismus. Berlin 1960. LuJ *29* (1962) 126–129.
242. Schwarz, Reinhard: Fides, Spes und Caritas beim jungen Luther, unter besonderer Berücksichtigung der mittelalterlichen Tradition. Berlin 1962. EvD *33* (1962) 240; LuJ *30* (1963) 137–140.
243. Fabian, Ekkehart: Dr. Gregor Brück 1557–1957. Tübingen 1957. Ders.: Die Schmalkaldischen Bundesabschiede 1530–1532. Tübingen 1958. Ders.: Die Schmalkaldischen Bundesabschiede 1533–1536. Tübingen 1958. ThLZ *87* (1962) 530f.
244. Hornig, Gottfried: Die Anfänge der historischen Theologie. Johann Salomo Semlers Schriftverständnis und seine Stellung zu Luther. Göttingen 1961. LuJ *30* (1963) 142–145.
245. Niesel, Wilhelm: Calvin Bibliographie 1901–1959. München 1961. EvD *34* (1963) 67.
246. Binder, Gerhart: Lebendige Zeitgeschichte 1890 bis 1945. München 1961. EvD *34* (1963) 69f.
247. Kupisch, Karl: Quellen zur Geschichte des deutschen Protestantismus. 1871 bis 1945. Göttingen, Berlin, Frankfurt 1960. EvD *34* (1963) 69f; DLZ *84* (1963) 713f.
248. Bucer, Martin: Deutsche Schriften 1: Frühschriften 1520–1524. Gütersloh, Paris 1960. DLZ *84* (1963) 711–713.
249. Bornkamm, Heinrich: Das Jahrhundert der Reformation. Göttingen 1960. Zeitschrift für Kirchengeschichte *74* (1963) 178f.
250. Grane, Leif: Contra Gabrielem. Luthers Auseinandersetzung mit Gabriel Biel in der Disputatio Contra Scholasticam Theologiam 1517. Kopenhagen 1962. LuJ *31* (1964) 129–135.
251. Wünsch, Georg: Luther und die Gegenwart. Stuttgart 1961. LuJ *31* (1964) 140–144.
252. Vajta, Vilmos, und Hans Weißgerber: Das Bekenntnis im Leben der Kirche. Berlin 1963. Lutherische Rundschau *15* (1965) 226–228.
253. Vajta, Vilmos: Kirche und Abendmahl. Berlin 1963. Lutherische Rundschau *15* (1965) 229f.
254. Althaus, Paul: Die Ethik Martin Luthers. Gütersloh 1965. LuJ *33* (1966) 123–125.
255. Schlesinger, Walter: Kirchengeschichte Sachsens im Mittelalter. 2 Bde. Köln, Graz 1962. ThLZ *93* (1968) 359–363.
256. Schilling, Werner: Gotteslästerung strafbar? Religionswissenschaftliche, theologische und juristische Studie zum Begriff der Gotteslästerung und zur Würdigung von Religionsschutznormen im Strafgesetz. München 1966. ThLZ *94* (1969) 307–309.

Außerdem zahlreiche Besprechungen und Anzeigen in SKB NF *2* (1938) bis *4* (1940), ThLBl *59* (1938), Pfarramts-Praxis *12* (1938) – *17* (1943), Christliche Welt *53* (1939), EvD *24* (1953) – *34* (1963) und Das historisch-politische Buch. Im LuJ befinden sich seit 1961 Sammelbesprechungen, ab 1963 unter dem Titel „Luther und die Welt der

Reformation": LuJ *28* (1961) 130–135; *29* (1962) 130–134; *30* (1963) 142–145; *31* (1964) 144–153; *32* (1965) 145–154; *33* (1966) 143–160; *34* (1967) 141–144; *35* (1968) 108–122; *36* (1969) 93–132; *37* (1970) 129–134.

Varia (Predigten, Reden, Berichte, Würdigungen u. ä.)

257. Wird unsere Kirche in der Bedrängnis durchhalten? Abendandacht über Luk. 12, 32. 10. November 1932 Universitätskirche St. Pauli in Leipzig. SKB *83* (1933) 77–81.
258. Wer schlägt falsch Alarm und wer greift falsch an? (Antwort auf Müller, Hans Michael: Greif an, Kirche — aber richtig!) Neues Sächsisches Kirchenblatt *43* (1936) 737–742.
259. Das Predigerseminar Lückendorf. SKB NF *1* (1937) 142f.
260. Werkplan und Konfirmandenlehrplan. PB *80* (1937/38) 180–183.
261. Die wahre Glückseligkeit (Andacht über Matth. 5, 1–12). SKB NF *3* (1939) 41.
262. Die große Herablassung (Andacht über Joh. 4, 5–15). SKB NF *3* (1939) 49.
263. Die konfessionelle Frage (Andacht über Joh. 4, 19–30). SKB NF *3* (1939) 57.
264. Das Selbstbestimmungsrecht Jesu (Andacht über Matth. 16, 21–26). SKB NF *3* (1939) 65.
265. Die Apostelgeschichte in der Predigt. SKB NF *3* (1939) 243f.
266. Wirkliches Leben! (Predigt über 1. Joh. 1, 1–4). HM 15 (1939/40) 16–20.
267. Die Offenbarung der Gottesliebe (Predigt über 1. Joh. 4, 9–11). HM *15* (1939/40) 25–28.
268. Die christliche Hoffnung — der Anker unserer Seele! (Predigt über Hebr. 6, 17–20). HM *15* (1939/40) 32–35.
269. Die Apostelgeschichte in der Predigt. PB *81* (1938/39) 467–474.
270. Wir lesen aus unserer Bibel im Monat Januar vor allem Matthäus Kap. 1–4 und 8 und 9. Neue Saat *14* (1939) 2.
270a. Die Vollmacht der Kirche. Bibelarbeit über Matth. 16, 13–20. Gabe und Auftrag *6* (1939) 121–124.
271. Die Geburt Christi — ein Weltereignis. Predigt über Matth. 12, 1–12. In: Haltet an am Gebet! Lesepredigten für Kirche und Haus. Hrsg. von Walter Schadeberg. Dresden 1940, 94–105.
272. Weg mit der Sorge auch heute! Predigt über Matth. 6, 25–34. In: Haltet an am Gebet! Lesepredigten für Kirche und Haus. Hrsg. von Walter Schadeberg. Dresden 1940, 499–508.
273. Warum verwehrt uns Christus die Liebe zur Welt? (Predigt über 1. Joh. 2, 12–17). HM *15* (1939/40) 75–79.
274. Glaubensmut und Glaubenskraft (Predigt über 1. Joh. 2, 4f). HM *15* (1939/40) 89 bis 92.
275. Ostern der große Protest gegen unser landläufiges Christentum. (Predigt über 1. Kor. 15, 12–20). HM *15* (1939/40) 101–104.
276. Wozu brauchen wir heute Predigten? (Predigt über Röm. 10, 1–15). HM *15* (1939/40) 151–155.
277. Ein Mann wird vertraut mit der Heiligen Schrift (Predigt über Apg. 8, 26–39). HM *15* (1939/40) 172–176.
278. Was ist nötig zum rechten Aufbau einer christlichen Gemeinde (Predigt über 1. Tim. 4, 4–11). HM *15* (1939/40) 254–257.
279. Der Krieg eine ernste Prüfung der deutschen evangelischen Christenheit (Predigt über 5. Mose 13. 4). HM *15* (1939/40) 276–279.

280. Vor Gottes Angesicht (Predigt über Offb. 7, 15). HM *15* (1939/40) 284f.
281. Fürchte dich nicht, glaube nur! (Predigt über Mark. 5, 36). HM *15* (1939/40) 285–288.
282. Predigt über 1. Chron. 29, 11 und Hebr. 13, 8. PB *82* (1939/40) 123–125.
283. Traurede über Jer. 1, 7. HM *15* (1939/40) 288–290.
284. Wir warten auf das Gottesreich (Luk. 17, 20–25). HM *16* (1940/41) 1–4.
285. Die Öffentlichkeitsaufgabe der Gemeinde Jesu im kommenden Jahr (Predigt über Matth. 5, 13–16). HM *16* (1940/41) 32–34.
286. Weg des Leidens und doch Weg des Heils (Predigt über Matth. 16, 21–28). HM *16* (1940/41) 73–77.
287. Von der Macht des Glaubens (Predigt über Matth. 8, 5–13). HM *16* (1940/41) 172–176.
288. Der Kampf um das Himmelreich (Predigt über Matth. 13, 44–50). HM *16* (1940/41) 219–222.
289. Einheit der Kirche – die Botschaft für die evangelische Christenheit (Predigt über Joh. 20, 20–26). HM *16* (1940/41) 249–251.
290. Alle Dinge in der Liebe (Traurede über 1. Kor. 16, 14). HM *16* (1940/41) 283f.
291. Die große Freudenbotschaft (Abendmahlsrede über Luk. 2, 10). HM *17* (1941/42) 15f.
292. Aus der Angst in den Frieden (Predigt über Joh. 16, 33). HM *17* (1941/42) 28–30.
293. Der Himmelsweg (Abendmahlsrede über Joh. 3, 24). HM *17* (1941/42) 42f.
294. Die besondere Aufgabe der Christenheit heute (Predigt über Mark. 10, 35–45). HM *17* (1941/42) 49–51.
295. Unter dem Schutz des Hirten (Abendmahlsrede über Ps. 23, 1). HM *17* (1941/42) 63f.
296. Unser Christusbekenntnis (Konfirmationspredigt über Matth. 10, 32f). HM *17* (1941/42) 83–85.
297. Jesu Stimme (Abendmahlsrede über Joh. 18, 37c). HM *17* (1941/42) 88f.
298. Die 1. Offenbarung des auferstandenen Herrn (Predigt über Joh. 20, 11–18). HM *17* (1941/42) 99–102.
299. Der Sieg des Herrn (Abendmahlsrede über Ps. 118, 16). HM *17* (1941/42) 110f.
300. Mein Erlöser lebt (Begräbnisrede über Hiob 19, 25). HM *17* (1941/42) 111–113.
301. Von der Gewalt Jesu (Predigt über Matth. 28, 16–20). HM *17* (1941/42) 133–136.
302. Das Mahl der Lobgesänge (Abendmahlsrede über Eph. 5, 19b). HM *17* (1941/42) 136f.
303. Wie stehst du zu Jesus (Predigt über Luk. 9, 18–26)? HM *17* (1941/42) 154–157.
304. Die Frucht (Abendmahlsrede über Eph. 5, 9b). HM *17* (1941/42) 159f.
305. Unser Leistungsanteil am Gottesreich (Predigt über Mark. 4, 26–29). HM *17* (1941/42) 169–171.
306. Der Herzen Furcht ist der Anfang der Erkenntnis (Abendmahlsrede über Spr. 1, 7). HM *17* (1941/42) 179f.
307. Der Herr Jesus ist kein Pharisäer (Predigt über Luk. 7, 36–50). HM *17* (1941/42) 193–196.
308. Der Herr gibt, daß er fortgehe (Abendmahlsrede über Spr. 16, 9) HM *17* (1941/42) 203f.
309. Alle eure Sorge werfet auf Ihn (Abendandacht bei einem Gemeindehelfertag über 1. Petr. 5, 7). HM *17* (1941/42) 204–206.
310. Meine Taufe (Taufrede über Apg. 2, 38a). HM *17* (1941/42) 206f.
311. Der Christ und sein Sonntag (Predigt über Matth. 12, 1–8). HM *17* (1941/42) 223 bis 226.

312. Der heilsame Sinn dieses Erntedankfestes (Predigt über Matth. 4, 4). HM *17* (1941/42) 228–232.
313. Die Hauptsache im Christentum (Abendmahlsrede über 1. Joh. 4, 10). HM *17* (1941/42) 232f.
314. Den Kindern das Himmelreich (Predigt über Mark. 10, 13–16). HM *17* (1941/42) 247–250.
315. Bestehet in der Freiheit (Abendmahlsrede über Gal. 5, 1). HM *17* (1941/42) 253f.
316. Todesernst und Todestrost (Predigt über Ps. 39, 5–8). HM *17* (1941/42) 270–274.
317. Die große Erneuerung (Abendmahlsrede über Offb. 21, 5). HM *17* (1941/42) 274f.
318. Gott hat uns also geliebt (Predigt über 1. Joh. 4, 11). HM *18* (1942/43) 11–13.
319. Gnade um Gnade (Taufrede über Joh. 1, 16). HM *18* (1942/43) 16f.
320. Das Evangelium ein helles Licht (Predigt über 2. Kor. 4, 3–6). HM *18* (1942/43) 34–37.
321. Die enge Pforte (Taufrede über Matth. 7, 13a). HM *18* (1942/43) 40.
322. Gottes Gebote – eine sehr gefährliche Angelegenheit (Predigt über Röm. 7, 7–16). HM *18* (1942/43) 51–54.
323. Die Bibel ein Judenbuch (Predigt über 2. Kor. 3, 12–18)? HM *18* (1942/43) 57–59.
324. Arbeiter für den Herrn (Taufrede über Matth. 9, 38). HM *18* (1942/43) 65f.
325. Mitten unter uns und an meiner Seite (Taufrede über Matth. 18, 20). HM *18* (1942/43) 86f.
326. Das Leiden Jesu und unser Lebenswandel (Predigt über 1. Petr. 1, 17–25). HM *18* (1942/43) 97–99.
327. Gegen die Ermüdung der Christenheit (Predigt über Hebr. 12, 1–6). HM *18* (1942/43) 99–102.
328. Des Herrn Werke verkündigen (Taufrede über Ps. 118, 17). HM *18* (1942/43) 112f.
329. Kämpferisches Leiden für das Evangelium (Predigt über 2. Petr. 2, 8–13). HM *18* (1942/43) 125–128.
330. Aktive Christen (Taufrede über 1. Petr. 3, 15). HM *18* (1942/43) 132f.
331. Textbearbeitung zu Röm. 8, 33–39. PB *85* (1942/43) 341f.
332. Textbearbeitung zu 2. Kor. 9, 6–11. PB *85* (1942/43) 344f.
333. Textbearbeitung zu 1. Kor. 3, 11–23. PB *85* (1942/43) 363f.
334. Grußwort des Landeskirchenamtes an die ev.-luth. Gemeinden Sachsens. Dresden Juli 1945. Abgedruckt in verschiedenen Gemeindeblättern, u. a. Der Kirchenbote. Kamenz, 1. 8. 1945, 1 (Mitverfasser Erich Kotte).
335. Aus dem Tode in das Leben gekommen. In: Ostergruß der Ev.-Luth. Landeskirche Sachsens an die Gemeinden in Stadt und Land. Dresden März 1946, 2.
336. Ans Werk! In: Ostergruß der Ev.-Luth. Landeskirche Sachsens an die Gemeinden in Stadt und Land. Dresden März 1946, 8.
337. Die Kirche in der Gegenwart. Der Sonntag *1* (1946) 1f.
338. Dr. h. c. Rudolf Friedrichs zum Gedächtnis. Der Sonntag *2* (1947) 120.
339. Die Kirchenwahl am 17. August. Der Sonntag *2* (1947) 143f.
340. Predigt über Matth. 28, 16–20. In: Gnade und Wahrheit. Hrsg. von Walter Dreß und Max Loeser. Göttingen 1949, 238–242.
341. Predigt über 2. Kor. 3, 12–18. In: Gnade und Wahrheit. Göttingen 1949, 105–109.
342. Fruchtbare Traurigkeit (Predigt über Joh. 16, 16–23a). HM *24* (1948/49) 222–225.
343. Luthers Evangelium ein verlöschend Licht? Predigt zum Reformationsfest (Joh. 12, 35f). HM *24* (1948/49) 2–5.
343a. Daß der Glaube nicht aufhöre (Taufansprache über Luk. 22, 32a). HM *24* (1948/49) 336–338.

344. Die auf den Herrn wartende Gemeinde (Predigt über Offb. 22, 12—20). HM *25* (1949/50) 12—16.
345. Von der weltweiten Christenheit (Predigt über Apg. 10, 42—48). HM *25* (1949/50) 284—287.
346. Die erste Offenbarung des auferstandenen Herrn (Predigt über Joh. 20, 11—18). HM *26* (1950/51) 185—187.
347. Die Kirche nach Jesu Willen und Sinn (Predigt über Joh. 15, 9—16). HM *26* (1950/51) 261—264.
348. Den Kindern das Himmelreich (Predigt über Mark. 10, 13—16). HM *26* (1950/51) 444—447.
349. Der neu gewählte Präsident grüßt alle Mitarbeiter und Mitarbeiterinnen im Gustav Adolf-Werk der Evang. Kirche in Deutschland. Die Diaspora *1* (1952) Nr. 1, 1—3.
350. Sondert euch ab! (Predigt über 2. Kor. 6, 14 — 7, 1). HM *27* (1951/52) 95—98.
351. Die große Erneuerung (Abendmahlsrede über Offb. 21, 5). HM *28* (1952/53) 32.
352. Präsidialrede 1953. EvD *24* (1953) Anhang 90. Hauptversammlung, 10—13.
353. Begrüßung bei der 91. Hauptversammlung 1954. EvD *25* (1954) 246.
354. Ansprache auf der Festversammlung bei der 91. Hauptversammlung 1954. EvD *25* (1954) 250f.
355. Präsidialrede bei der 91. Hauptversammlung 1954. EvD *25* (1954) 255—259.
356. Die Theologische Fakultät. Leipziger Volkszeitung *9* (1954) Nr. 158 vom 11. 7. 1954, 4.
357. Ansprache beim Empfang auf der 92. Hauptversammlung 1955. EvD *26* (1955) Anhang 92. Hauptversammlung, 1f.
358. Ansprache bei der 92. Hauptversammlung 1955. EvD *26* (1955) Anhang 92. Hauptversammlung, 3f.
359. Arbeitstagung für Sächsische Kirchengeschichte. Neue Zeit *12* (1956) Nr. 101 vom 29. 4. 1956, 8.
360. Diaspora als Kirche. Kirche als Diaspora. Gustav Adolf-Blatt *2* (1956) Juli, 4f.
361. Ansprache bei der Festsitzung 1957. EvD *28* (1957) 241—249, 251.
362. Ernst Wagner, dem Achtzigjährigen. EvD *29* (1958) 1—3.
363. Gustav Adolf-Werk und ökumenische Diakonie EvD *29* (1958) Anhang 93. Hauptversammlung, 5—7, 8f.
364. Ansprache auf der Festversammlung bei der 93. Hauptversammlung 1958. EvD *29* (1958) Anhang 93. Hauptversammlung, 13—21.
365. Präsidialrede bei der 93. Hauptversammlung 1958. EvD *29* (1958) Anhang 93. Hauptversammlung, 25—31.
366. Wiedereinweihung der Trinitatiskirche in Warschau. Die Zeichen der Zeit *12* (1958) 343f.
367. Ansprache bei der 94. Hauptversammlung 1960. EvD *31* (1960) Anhang 94. Hauptversammlung, 4f.
368. Martin Niemöller zur Vollendung des 70. Lebensjahres. EvD *32* (1961) 243—245.
369. Der Weg zum Theologiestudium. Der Sonntag *16* (1961) Nr. 20 vom 9. 4. 1961, 3. = Die Kirche *16* (1961) Nr. 15 vom 9. 4. 1961, 4. = Die Union *16* (1961) Nr. 74 vom 28. 3. 1961, 5.
370. Richard Schlier, dem Siebzigjährigen. EvD *33* (1962) Heft 2 (Mitverfasser Fritz Hauß).
371. Paul Wilhelm Gennrich, dem Sechzigjährigen. EvD *33* (1962) Heft 3/4 (Mitverfasser Fritz Hauß).
372. Grußtelegramm an die 95. Hauptversammlung. EvD *33* (1962) Anhang 95. Hauptversammlung, 4f.

373. Auslegung zu Matth. 14, 22—34. In: Verkündigt das angenehme Jahr des Herrn. 1. Advent bis Estomihi. Ausgew. von Günther Baumbach. Berlin 1962, 218. = Predigtgedanken aus Vergangenheit und Gegenwart C 1.
374. Der Vorsitzende des Gustav-Adolf-Werkes grüßt. Die Saat. Kirchenbote für das evangelische Österreich *9* (1962) 62.
375. Ausdruck der Verbundenheit. Zum Gustav-Adolf-Gedenken am 6. November. Neue Zeit *18* (1962) Nr. 259 vom 6. 11. 1962, 4.
376. Gustav Adolf-Werk. Wir lieben die Brüder. Weg und Wahrheit *16* (Frankfurt/Main 1962) 433f.
377. Gustav Adolf-Werk im Dienst an den verstreuten Brüdern. Glaube und Heimat *18* (1963) Nr. 24 vom 16. 6. 1963, 1.
378. Präsidialrede bei der Jahresversammlung des Gustav Adolf-Werkes in der DDR, Bautzen 1964. EvD *35* (1964) Anhang, 6—8.
379. Abschied von der Zeitschrift. Gruß dem Jahrbuch „Die evangelische Diaspora". EvD *36* (1966) 197—199.
380. Präsidialrede bei der Jahresversammlung des Gustav Adolf-Werkes in der DDR, Neustrelitz 1965. EvD *36* (1966) Anhang, 1—4.

Lexikonartikel

381. Neue Deutsche Biographie.
Bd 1. Berlin 1953: Johann Abicht (Sp. 19f); Johann Agricola (Sp. 100); Johann Friedrich Ahlfeld (Sp. 110f); Valentin Alberti (Sp. 142); Christoph Friedrich von Ammon (Sp. 253f). — Bd 2, 1955: Martin Behm (Sp. 8); Adam Bernd (Sp. 106); Johann Georg Böse (Sp. 408); Georg Apollo Buchwald (Sp. 710f). — Bd 3, 1957: Johann Heinrich Callenberg (Sp. 96); Jakob Carpov (Sp. 155f). — Bd 4, 1959: Paul Drews (Sp. 118f); Johann August Ernesti (Sp. 604f Mitverfasser: Karlheinz Blaschke). — Bd 6, 1964: Georg III., Fürst von Anhalt (Sp. 197); Johann Gerhard (Sp. 281). — Bd 7, 1966: Daniel Greser (Sp. 49f.). — Bd. 8, 1969: Nicolaus Hausmann (Sp. 126). — Bd 9, 1972: Daniel Hof(f)mann (Sp. 404); Samuel Huber (Sp. 698f). — Bd 10, 1974: Ägidius Hunnius (Sp. 67f); Nikolaus Hunnius (Sp. 68); Leonhard Hutter (S. 104f).
382. Evangelisches Kirchenlexikon. Kirchlich-theologisches Handwörterbuch. Göttingen 1956—1961.
Bd 1, 1956: Antitrinitarier (Sp. 155f); Augsburger Religionsfrieden (Sp. 258f); Chalcedon (Sp. 674—676); Christentum I: Umrisse der geschichtlichen Entwicklung in ihren Hauptperioden (Sp. 709—717), II: Dogmen- und theologiegeschichtlich (Sp. 754 bis 773); Dogmengeschichte (Sp. 957—960); Ernst I. (der Fromme) (Sp. 1133f); Georg der Bärtige, Herzog von Sachsen (Sp. 1500); Gustav (II.) Adolf von Schweden (Sp. 1732f); Gustav-Adolf-Werk (S. 1733—1735).
Bd 2, 1958: Johann I. von Sachsen (Sp. 355); Moritz, Herzog und Kurfürst von Sachsen (Sp. 1452f).
Bd 3, 1959: Reformation I: Allgemein Historisches (Sp. 487—493); II: Im Gebiet des dt. Reiches (bis 1555) (Sp. 493—500), XI: Weiterentwicklung der Reformation 1555 bis 1648 (Sp. 526—528); Speyer (Sp. 1083f); Worms (Sp. 1844—1846).
383. Die Religion in Geschichte und Gegenwart. Handwörterbuch für Theologie und Religionswissenschaft. 3. völlig neu bearb. Aufl. Tübingen 1957—1965.
Bd 1, 1957: Achelis, Hans (Sp. 83); Adiaphora (Sp. 93—96); Alberti, Valentin (Sp. 216); Alesius, Alexander (Sp. 225f); Alvelt, Augustinus von (Sp. 301); Ammon, Christoph Friedrich von (Sp. 326); Amsdorf(f), Nikolaus von (Sp. 333f); Armut III:

Ethisch (627f); Askese V: In katholischer und protestantischer Ethik (Sp. 644—648); Baier, Johann Wilhelm (Sp. 846); Beruf III: Christentum und Beruf (Sp. 1076—1081); Billigkeit (Sp. 1292f); Brieger, Theodor (Sp. 1416); Brück, Gregor (Sp. 1432); Buchwald, Georg (Sp. 1467); Calixt, Georg (Sp. 1586f); Calov, Abraham (Sp. 1587); Camerarius, Joachim (Sp. 1602); Chemnitz, Martin (Sp. 1647f); Curäus, Joachim (Sp. 1890).

Bd 2, 1958: Deutschmann, Johann (Sp. 156); Diaspora II: Evangelische Diaspora (Sp. 177—180); Eck, Johann (Sp. 302f); Eigengegesetzlichkeit der Lebensgebiete (Sp. 354—356); Emser, Hieronymus (Sp. 462); Evangelische Räte (Sp. 785—788); Friedrich der Weise (Sp. 1150); Gelübde V: Im Protestantismus (Sp. 1325); Georg III. der Gottselige von Anhalt (Sp. 1395f); Gerechtigkeit des Menschen, ethisch (Sp. 1410 bis 1412); Gerhard, Johann (Sp. 1412f); Gesetz VII: Gesetz, ethisch (Sp. 1531—1533); Gute Werke (Sp. 1915f).

Bd 3, 1959: Habermann (Avenarius), Johann (Sp. 7); Hausmann, Nikolaus (Sp. 98); Höe von Höenegg, Matthias (Sp. 389f); Hoffmann, Daniel (Sp. 413); Horn(ejus), Konrad (Sp. 452); Huber, Samuel (Sp. 463); Hülsemann, Johann (Sp. 467); Hütter (Hutterus), Leonhard (Sp. 468); Hunn(ius), Aegidius (Sp. 490f); Interim (Sp. 791); König, Johann Friedrich (Sp. 1699).

Bd. 4, 1960: Kromayer, Hieronymus (Sp. 80f); Kryptocalvinisten (Sp. 84); Leipzig, Theol. Fakultät (Sp. 308—310); Leyser, Polycarp d. Ä. (Sp. 340f); Los-von-Rom-Bewegung (Sp. 452—455); Major, Georg (Sp. 617); Martini, Cornelius (Sp. 780f); Martini, Jakob (Sp. 781); Martini, Matthias (Sp. 781); Meisner, Balthasar (Sp. 832); Meisner, Johannes (Sp. 832f); Miltitz, Karl von (Sp. 954); Minderheiten (Sp. 957 bis 960); Moller, Martin (Sp. 1089); Musäus, Johannes (Sp. 1183); Musäus, Peter (Sp. 1193); Myslenta, Cölestin (Sp. 1231); Olearius, 1. Johannes (Sp. 1625); Olearius, 2. Johannes (Sp. 1625); Olearius, 3. Johannes (Sp. 1625); Orthodoxie, altprotestantische (Sp. 1719—1730).

Bd. 5, 1961: Pfeffinger, Johann (Sp. 306); Pflug, Julius (Sp. 319); Prierias, Silvester Mazzolini (Sp. 568); Quenstedt, Johann Andreas (Sp. 735); Rathmann, Hermann (Sp. 770); Rendtorff, Franz (Sp. 1064); Roeber, Paul (Sp. 1136); Scharf, Johann (Sp. 1389f); Scherzer, Johann Adam (Sp. 1402); Schöpfungsordnung (Sp. 1492—1494); Schupp, Johann Balthasar (Sp. 1584f); Seckendorf, Veit Ludwig (Sp. 1629f); Sennert, Andreas (Sp. 1700).

Bd 6, 1962: Synkretistischer Streit (Sp. 568f); Weller, Jakob (Sp. 1593f); Zwei-Reiche-Lehre (Sp. 1945—1949).

384. The Encyclopedia of the Lutheran Church. Hrsg. von Julius Bodensick. 3 Bde. Minneapolis 1965. Bd 1: Lutheran Diaspora Care (Sp. 700—704). (dt. Übersetzung vgl. Nr. 89).

Herausgeber

Pfarramtspraxis. Kartothek für Pfarramtskunde. *12* (1938) Lfg 9—17 (1943) Lfg 1/2. (Geht auf in ‚Christentum und Leben', für das Übergangsheft *18* [1943] H. 5/6 Mithrsg., dann geht ‚Christentum und Leben' in den prakt.-theol. Teil von ‚Pfarramt und Theologie' auf. Für die 7 erschienen Hefte bis 1944/45 schreibt Lau 8 Meditationen über Predigttexte).

Homiletische Monatshefte, *15* (1939/40) — *18* (1943/43) Heft 6. (Einstellung des Erscheinens, ab *16* (1940/41) Heft 8 unter Mitwirkung von Maximilian Meichßner).

Die evangelische Diaspora. Zeitschrift des Gustav Adolf-Werkes, *24* (1953) — *36* (1965/66).

Herbergen der Christenheit. Jahrbuch für deutsche Kirchengeschichte, *1* (1953) — *8* (1971).

FRANZ LAU

18. 2. 1907—6. 6. 1973

OTTO EISSFELDT

1. 9. 1887—23. 4. 1973

JOHANN FÜCK
8. 7. 1894–24. 11. 1974

JOHANNES KÜHN
24. 1. 1887–24. 2. 1973

(Beiträge zur Kirchengeschichte Deutschlands bzw. zur deutschen Kirchengeschichte).

Luther-Jahrbuch, *24* (1957) — *38* (1971). (1971 zusammen mit Helmar Junghans).

Handbuch zu Freikirchen und Sekten. (Mitarb.: Franz Lau u. a.). Hannover 1966. Teil 1: Freikirchen und Sondergemeinschaften. — Teil 2: Sekten und Weltanschauungsgemeinschaften. In zwei Plastikordnern.

Erbe und Verpflichtung. Reformationsgedenkbuch. Berlin 1967. (Vorwort S. 7—12).

Das Hochstift Meißen. Aufsätze zur sächsischen Kirchengeschichte. Berlin 1973. = Herbergen der Christenheit Sonderbd (Zur Einführung S. 9—14). (Mithrsg.) Kerygma und Dogma. Zeitschrift für theologische Forschung und kirchliche Lehre, *1* (1955) — *6* (1960).

Betreute Dissertationen und Habilitationen

1. Nagel, Walter: Leben und Theologie des Juan de Valdés. 1948.
2. Beyreuther, Erich: Bartholomäus Ziegenbalg, Theologie und Sendungsbewußtsein. 1951.
3. Haase, Martin: Der Einfluß des Kulturkampfes auf die kirchlichen Verhältnisse im ehem. Kgr. Sachsen. 1951.
4. Beyreuther, Erich: August Hermann Francke und die Ökumene. Habil. 1953. = Leipzig, Hamburg—Bergstedt 1957.
5. Ludolphy, Ingetraut: Henrich Steffens, sein Verhältnis zu den Lutheranern und sein Anteil an Entstehung und Schicksal der altlutherischen Gemeinde in Breslau. 1956. = Berlin 1962.
6. Jentsch, Joachim: Die Christl. Gemeinschaft Hirt und Herde. Ein Beitrag zur Sektenkunde. 1956.
7. Zimmermann, Hildegard: Adam Bernd und der Melodianismus. Teil 1: Die „Breslauer Frühaufklärung“ und ihr Einfluß auf Bernds Entwicklung. 1958.
8. Grafe, Hugald: Die volkstümliche Predigt des Ludwig Harms. 1959. = Göttingen 1965. 2. Aufl. Göttingen 1974.
9. Meier, Kurt: Die Deutschen Christen — ihre organisatorische Entwicklung von der Sportpalastkundgebung (November 1933) bis zum Beginn des zweiten Weltkrieges. 1960. = (bearb. und gekürzt) Die Deutschen Christen. Das Bild einer Bewegung im Kirchenkampf des Dritten Reiches. Halle, Göttingen 1964. 2. Aufl. Halle 1965. Nachdruck Halle 1967. 3. Aufl. Göttingen 1967.
10. Ludolphy, Ingetraut: Die Natur bei Luther. Habil. 1960.
11. Herrmann, Johannes: Augsburg—Leipzig—Passau (Das Leipziger Interim nach Akten des Landeshauptarchivs Dresden 1547—1552). 1962.
12. Seifert, Siegfried: Untergang und Wiederauferstehung des Katholizismus in Sachsen. 1962. = Niedergang und Wiederaufstieg der katholischen Kirche in Sachsen. 1517 bis 1733. Leipzig 1964.
13. Schwabe, Wolfgang: Richard Wagner und die Kirche. 1963.
14. Junghans, Helmar: Ockham im Lichte der neueren Forschung. Dargestellt an Hand der Arbeiten des Franziskaners Philotheus Böhner und seiner Gesprächspartner. 1964. = Berlin 1968.
15. Sygusch, Kurt: Die Auseinandersetzung der Oberlausitz mit dem Hussitismus unter besonderer Berücksichtigung ihrer Kirchengeschichte. 1966. = Frankfurt/M 1968.
16. Petzold, Helmut: Der Protestantismus in Böhmisch-Zinnwald, sein Ende 1728, und die Wirksamkeit des Geisinger Pfarrer Imanuel Heinrich Kauderbach. 1966.
17. Hertrampf, Hans-Dieter: Der kursächsische Oberhofprediger Matthias Höe von Höenegg — seine Theologie, Polemik und Kirchenpolitik. 1967. = (überarb.) Höe von

Höenegg — sächsischer Oberhofprediger 1631—1645. Herbergen der Christenheit *7* (1969) 129—148.
18. Minor, Rüdiger: Die Bischöfliche Methodistenkirche in Sachsen. Ihre Geschichte und Gestalt im 19. Jahrhundert in der Beziehung zur Umwelt. 1968.
19. Ders.: Quellen zur Geschichte des Methodismus in Sachsen (1860—1900). Habil. 1968.
20. Rudolph, Wolfgang: Paul Drews — der Vater der evangelischen Kirchenkunde und religiösen Volkskunde. 1968.
21. Delius, Hans-Ulrich: Der Briefwechsel des Nikolaus von Amsdorf als evangelischer Bischof von Naumburg—Zeitz. 1542—1546. Habil. 1968.
22. Zschuppe, Christian: Der Bund der Kämpfer für Glaube und Wahrheit. 1971.
23. Mühlmann, Sieghard: Luther und das Corpus Iuris Canonici bis zum Jahre 1530. 1972. = (Teildruck) Zeitschrift der Savigny-Stiftung für Rechtsgeschichte. Kanonistische Abt. *89* (1972) 233—305.
24. Petzold, Klaus: Studien zu einer Biographie Valentin Ernst Löschers, 1972.
25. Bräuer, Siegfried: Die zeitgenössischen Dichtungen über Thomas Müntzer und den Thüringer Bauernaufstand. 1973.
26. Rochler, Wolfgang: Luther und die Mystik. 1973.

Zur Biobibliographie

1. Vierhundertfünfzig Jahre lutherische Reformation 1517—1967. Festschrift für Franz Lau zum 60. Geburtstag. Hrsg. von Helmar Junghans, Ingetraut Ludolphy und Kurt Meier. Berlin, Göttingen 1967.
2. Gennrich, Paul Wilhelm: Die neue Schau der Diaspora. Franz Lau's Beitrag zur Diasporawissenschaft. EvD *38* (1967) 1—36.
3. Watanabe, Shigeru: Gendai ni okeru Rutaa—Zoo I. Franz Lau Kyooju ni okeru Rutaa-Kaishaku o Chuushin to shite (Das Lutherbild in der Gegenwart I. Im Hinblick auf die Lutherinterpretation von Prof. Franz Lau). Komazawa Daigaku Bungakubu Kenkyuu Kiyoo. Tokyo März 1964. 69—84.
4. Dank an Lic. Lau. Der Sonntag *3* (1947) 4.
5. Professor Lau — Präsident des Gustav Adolf-Werkes. Evangelischer Nachrichtendienst Ost *5* (1952) Nr. 6 vom 9. 2. 1952, 3.
6. Neuer Präsident des Gustav Adolf-Werkes. Die Kirche *7* (1952) Nr. 7 vom 17. 2. 1952, 2 (mit Bild). = Der Sonntag *7* (1952) 58.
7. Wir sprachen mit: Franz Lau. Die Kirche *12* (1957) Nr. 45 vom 10. 11. 1957, 3 (mit Bild).
8. Wagner, Heinz: Franz Lau zum 60. Geburtstag am 18. 2. 1967. ThLZ *92* (1967) 139f.
9. Loewenich, Walther von; Mülhaupt, Erwin: Grußwort an Professor D. Franz Lau. LuJ *34* (1967) 5f.
10 Professor D. Franz Lau 60 Jahre. Evangelischer Nachrichtendienst in der DDR *20* (1967) Nr. 5 vom 1. 2. 1967, 5f.
11. Tabula Gratulatoria zum 60. Geburtstag des Gustav Adolf-Werkes der EKD, der geistlichen Leiter der Diasporakirchen des Ostens und des Westens. EvD *38* (1967).
12. (Zum 60. Geburtstag). Der Sonntag 22 (1967) Nr. 13 vom 19. 2. 1967, 4.
13. Moritz, Hans: Franz Lau zum 65. Geburtstag am 18. Februar 1972. ThLZ *97* (1972) 151—153.
14. Loewenich, Walther von: Grußwort an Professor D. Franz Lau. LuJ *39* (1972) 5f.
15. Prof. D. Lau 65 Jahre. Evangelischer Nachrichtendienst in der DDR *25* (1972) Nr. 8 vom 23. 2. 1972, 7. = Die Zeichen der Zeit *26* (1972) 194.

16. Bibliographie Franz Lau. Zusammengestellt von Ingemaren Brüschke. ThLZ *92* (1967) 141—154. *97* (1972) 153—155.
17. Bibliographie Franz Lau. Zusammengestellt von Günther Wartenberg. S. vorliegendes Jahrb. S. 400—419.
18. Aland, Kurt: Franz Lau 18. 2. 1907—6. 6. 1973. S. vorliegendes Jahrb. S. 393—400.
19. Wartenberg, (Günther): Kirchenhistoriker von Rang. Zum Tode des Leipziger Theologen Prof. D. Franz Lau. Die Union *28* (1973) Nr. 141 vom 15. 6. 1973, 3.
20. Prof. D. Franz Lau verstorben. Evangelischer Nachrichtendienst in der DDR *26* (1973) Nr. 25 vom 20. 6. 1973, 4f.
21. M.: Ein Forscher von Weltruf. In memoriam Professor D. Franz Lau. Neue Zeit 29 (1973) Nr. 152 vom 30. 6. 1973, 5.
22. Herzog, Heinrich: Professor D. Franz Lau †. Der Sonntag *28* (1973) Nr. 26 vom 1. 7. 1973, 2 (mit Bild).
23. Kretzschmar, G(ottfried): Zum Tode von D. Franz Lau. Standpunkt *1* (1973) 241f.
24. Zimmermann, Gottfried: In memoriam Professor D. Franz Lau — Leipzig. Sächsische Heimat (1973) 341f. (mit Bild).
25. Hofmann, Hans: Franz Lau. 18. Februar 1907 bis 6. Juni 1973. Die Zeichen der Zeit *27* (1973) 428f.
26. Gennrich, Paul Wilhelm: Franz Lau zum Gedenken. EvD *44* (1974) 100—104 (mit Bild).
27. Kähler, Ernst: Franz Lau. D. Martin Luthers Werke. Kritische Gesamtausgabe. 41. Band: Revisionsnachtrag. Weimar 1974, 9.
28. Schmidt, Martin: Franz Lau zum Gedenken. 18. Februar 1907 bis 6. Juni 1973. Luther *45* (1974) 87—91.
29. In memoriam D. Franz Lau. LuJ *42* (1975) 7—10.
30. Hofmann, Hans: In memoriam Franz Lau. Catholica *29* (1975) 43—55.
31. Junghans, Helmar: In memoriam Franz Lau. Herbergen der Christenheit *10* (1975/76)

Dankenswerter Weise konnten Vorarbeiten von Ingemaren Brüschke (vgl. oben Nr. 16) benutzt werden.

Dr. Günther Wartenberg

Johann Fück

8. 7. 1894—24. 11. 1974

Am 24. November 1974, wenige Monate nach Vollendung seines 80. Lebensjahres, verstarb in Halle an der Saale Johann Fück, emeritierter Professor mit Lehrstuhl für Semitische Philologie und Islamkunde an der Martin-Luther-Universität. Halle—Wittenberg und seit 1948 ordentliches Mitglied der Sächsischen Akademie der Wissenschaften zu Leipzig. Mit ihm hat die internationale Arabistik und Islamwissenschaft einen Gelehrten von hohem Rang verloren, der durch die Qualität seiner wissenschaftlichen Arbeiten, aber auch durch ein seiner Natur so sehr gemäßes Wirken im Stillen — sei es durch das persönliche Geprách, sei es durch eine ausgedehnte Korrespondenz — das Niveau seiner Fächer im zweiten Drittel unseres Jahrhunderts wesentlich mitgeprägt und mitbestimmt hat.

Johann Fück wurde am 8. Juli 1894 in Frankfurt am Main als Sohn eines Schreiners geboren, und es mag nicht zuletzt an seiner Herkunft aus einer Handwerkerfamilie gelegen haben, daß er in seinem wissenschaftlichen Tun ebenso wie in seinen Forderungen an die ihm zur Ausbildung anvertrauten jungen Menschen stets den Maßstab größter Solidität, Exaktheit und Zuverlässigkeit anlegte. Nachdem er im Februar 1913 an einem Gymnasium seiner Vaterstadt die Reifeprüfung abgelegt hatte, entschied er sich für ein Studium der klassischen Philologie, das ihm den Weg in das Lehramt an einer Schule und damit den erforderlichen Lebensunterhalt sichern sollte. Interesse und Neigung zogen ihn jedoch von Anfang an zur Orientalistik, für die er sprachlich schon insoweit vorbereitet war, als er auf der Schule neben dem Hebräischen begonnen hatte, sich auf autodidaktischem Wege in das Arabische einzuarbeiten. So ausgerüstet ging er zunächst von April 1913 bis März 1914 nach Halle. Hier trieb er bei Carl Brockelmann Hebräisch und Arabisch und las arabische Historiker, belegte bei Hans Bauer Übungen zur Mesa-Inschrift sowie zu hebräischen und phönizischen Inschriften, hörte bei Paul Kahle Syrisch und Einführungen in den Talmud und in das Studium der Targume und widmete sich bei Eugen Hultzsch dem Neupersischen. Im Mai 1914 schrieb er sich für ein Semester in Berlin ein. Zu seinen Lehrern zählten dort Eugen Mittwoch, der ihn in die südarabische Epigraphik und in die Peschitta einführte, während er bei Eduard Sachau Harīrīs Makaman und bei Jakob Barth den Dīwān des Imra'alquais las. Daneben trieb er Assyrisch bei Franz Delitzsch und Neupersisch bei Josef Marquart. Wahrscheinlich bewogen durch den Ausbruch des 1. Weltkrieges wechselte er im November 1914 an die Universität Frankfurt am Main über, wo er vier Semester lang bis zum Juni 1917 unter der Anleitung von Josef Horovitz orientalistischen Studien oblag. Zu den Gegenständen seiner Wahl gehörten hier Übungen zur Urgeschichte des Islams, altarabische Poesie und die islamische Traditionsliteratur ebenso wie die aramäischen Dialekte, die Elephantine-Papyri, das Hebräische und Saʿdīs Gulistān. Nebenher betrieb er auch Indogermanistik und Sanskrit bei Eduard Hermann.

Horovitz, ein ausgezeichneter Kenner des Korans und der vor- und frühislamischen Dichtung und Prosaliteratur, war es, der Fücks weiterem wissenschaftlichen Werdegang die entscheidende Richtung gab. So verfaßte er neben der Vorbereitung auf den Schuldienst, für den er zwischen 1918 und 1920 die erforderlichen Prüfungen ablegte und die Lehrbefähigung für die klassischen Sprachen und das Hebräische erwarb, eine Dissertation mit dem Titel „Muḥammad Ibn Isḥāq. Literarhistorische Untersuchungen", mit der er 1921 von der Philosophischen Fakultät der Universität Frankfurt am Main zum Dr. phil. promoviert wurde. Fücks Erstlingsarbeit, die infolge der ungünstigen Zeitverhältnisse erst 1925 in einem Steindruck erscheinen und daher nicht die Verbreitung finden konnte, die ihr zu wünschen gewesen wäre, ist nicht nur eine beachtliche Leistung, die unsere Kenntnis von Umfang und Entwicklung der arabischen Literatur in den ersten beiden Jahrhunderten des Islams erheblich gefördert hat; sie zeigt auch bereits deutlich all jene Merkmale, welche Fücks gesamtes wissenschaftliches Lebenswerk kennzeichnen: eine umfassende Beherrschung des Materials auf der Grundlage einer bewundernswerten Kenntnis der ʿarabīya; die Fähigkeit, die gewonnenen Ergebnisse übersichtlich und sachgerecht gegliedert in präzisen sprachlichen Formulierungen vorzulegen; eine deutliche Scheu davor, Lücken des Materials oder Stellen, wo es Fragen offen ließ, durch

Spekulationen, kühne Hypothesen oder Theorien auszufüllen, und dementsprechend eine beträchtliche Skepsis, wenn dies von anderen versucht wurde.

Fücks Habilitation erfolgte acht Jahre später, im Oktober 1929, an derselben Universität auf Grund einer Schrift mit dem Titel „Grundriß der arabischen Dichtung. I. Die heidnischen Dichter der nordarabischen Stämme (mit Ausnahme der Iljās ibn Muḍar-Stämme)", der jedoch — vielleicht wegen der schlechten wirtschaftlichen Verhältnisse jener Zeit — eine Publikation in gedruckter Form versagt blieb. In diesem „Grundriß" hat Fück alle ihm erreichbaren Nachrichten über die Lebensumstände und die Lebenszeit der in den zugänglichen gedruckten Quellen genannten vorislamischen Dichter zusammengetragen. Streng dem genealogischen Schema folgend werden die Biographien der Dichter in chronologischer Abfolge geboten, gefolgt von Quellennachweisen und einem Verzeichnis der überlieferten Verse. Am Schluß des Vorwortes hat der Verfasser die Hoffnung geäußert, durch das von ihm erarbeitete chronologische Fundament sei „der Weg zu form- und motivgeschichtlichen und stilkritischen Untersuchungen der einzelnen Dichtungsgattungen freigemacht, jener Weg, an dessen Ende uns das große Ziel einer wahren Geschichte der arabischen Dichtung winkt".

Der Erwerb der venia legendi bot Fück die Möglichkeit, endgültig aus dem Schuldienst auszuscheiden und sich neben der Tätigkeit als Privatdozent ganz seiner wissenschaftlichen Arbeit zu widmen. Ende 1930 folgte er einem Ruf an die Universität Dacca im heutigen Bangla Desh, wo er bis Ende 1935 als Professor für Arabische und Islamische Studien und Leiter des gleichnamigen Departments wirkte. Die Erfahrungen jener Jahre und die unmittelbare Begegnung mit einem vom Islam geprägten Teil des indischen Subkontinents haben Fück nachhaltig und dauerhaft beeindruckt. Ein jeder, der mit ihm zusammentraf, konnte dies alsbald feststellen, und mehrere seiner Aufsätze und Rezensionen zeugen von dem damals geweckten Interesse und Verständnis für die Probleme des zeitgenössischen Islams, in Indien wie anderswo. Nach Deutschland zurückgekehrt, wurde er zunächst außerordentlicher Professor an der Universität Frankfurt am Main, ehe er im Oktober 1938 das durch den plötzlichen Tod Hans Bauers vakant gewordene Ordinariat für Semitische Philologie und Islamkunde in Halle übernahm, das er seit April desselben Jahres bereits vertretungsweise wahrgenommen hatte, und mit ihm das Direktorat des Orientalischen Seminars, das er bis 1966 versah. Fück hat der Universität, an welcher er ein Vierteljahrhundert zuvor zu studieren begonnen hatte und an der er 1962 emeritiert wurde, die Treue bewahrt und nach 1945 mehrfach Berufungen an andere Universitäten abgelehnt, darunter eine für ihn besonders ehrenvolle auf den einmal von Heinrich Leberecht Fleischer und später von dem von ihm hochgeschätzten August Fischer innegehabten arabistischen Lehrstuhl der Universität Leipzig. Der Grund für seine Ablehnung war stets die Bibliothek der Deutschen Morgenländischen Gesellschaft in Halle, deren reiche Bestände für seine Arbeit unersetzlich waren und um deren Schutz und Sicherung er sich in den schweren Jahren des 2. Weltkrieges in aufopferungsvoller Weise bemüht hatte.

In den halleschen Jahren entstanden, freilich auf der Grundlage langjähriger Vorarbeiten, die beiden Werke, auf denen Fücks internationaler wissenschaftlicher Ruf beruht: die „Arabiya. Untersuchungen zur arabischen Sprach- und Stilgeschichte" erschien 1950 in den Abhandlungen der Sächsischen Akademie der Wissenschaften und wurde schon ein Jahr später in Kairo ins Arabische und 1955 ins Französische übersetzt; „Die arabischen Studien in Europa bis in den Anfang des 20. Jahrhunderts" erschienen 1955 und bildeten in ihrem zweiten Teil die Fortsetzung einer elf Jahre zuvor während des Krieges veröffentlichten Arbeit, die unter dem Titel „Die arabischen Studien in Europa vom 12. bis in den Anfang des 19. Jahrhunderts" in den „Beiträgen zur Arabistik, Semitistik und Islamwissenschaft", herausgegeben von Richard Hartmann und Helmuth Scheel, erschienen war. Beide Bücher sind zu den Standardwerken der Arabistik und Islamwissenschaft zu rechnen.

In der „Arabiya" hat Fück anhand eines reichen Tatsachenmaterials eine lebendige, durch viele Details verdeutlichte Geschichte des Arabischen vom Tode Mohammeds bis zur Seldschukenzeit und zum Mongolensturm geliefert. Indem er im allgemeinen den Fragestellungen der einheimischen Grammatiker, Philologen und Sprachkritiker hinsichtlich der an ein normiertes und als korrekt geltendes Arabisch zu stellenden Anforderungen folgte, gelang es ihm, ein faszinierendes Bild von einem zentralen Problem der arabischen Kulturgeschichte, oder noch genauer der Bildungs- und Geistesgeschichte, zu entwerfen. Daß dieses Buch eine seit langem spürbare Lücke füllte, zeigten das lebhafte Echo, welches es in der internationalen Kritik fand, und die Diskussion über die linguistische Interpretation eines Teils des von Fück zusammengetragenen Materials sowie seiner Ansichten über die Funktion der Desinentialflexion und die damit eng zusammenhängende Frage des Verhältnisses von Hochsprache, Umgangssprache und Dialekten im Verlaufe der arabischen Sprachgeschichte.

Fücks Sinn für historische Prozesse und Zusammenhänge zeigte sich auch in hervorragender Weise in den „Arabischen Studien in Europa bis in den Anfang des 20. Jahrhunderts". Verständnisvoll, ausgewogen und feinfühlig zeichnete er hier die Geschichte seiner Wissenschaft von den ersten, noch ganz im Dienste der Missionsbestrebungen des christlichen Europa stehenden Anfängen im Mittelalter über die allmähliche Loslösung von der Theologie und der Bibelwissenschaft bis hin zu der im Anfang dieses Jahrhunderts endgültig vollzogenen Verselbständigung als wissenschaftliche Disziplin, in welcher bereits eine immer stärkere Spezialisierung um sich zu greifen begann. Weit entfernt davon, eine bloße Aufzählung von Gelehrtennamen und Publikationen zu liefern, hat Fück es verstanden, in seiner Darstellung die politischen und gesellschaftlichen Faktoren und die geistigen Strömungen deutlich zu machen, von denen die arabischen Studien in den einzelnen Etappen ihrer Entwicklung beeinflußt worden sind.

Seit Mitte der zwanziger Jahre arbeitete Fück, veranlaßt durch Handschriftenfunde in Istanbuler Bibliotheken und wichtige neue Texteditionen, an einer neuen Edition des für die frühe arabische Literaturgeschichte bedeutsamen Fihrist des Ibn al-Nadīm, welche die veraltete und mangelhafte Flügelsche Ausgabe ersetzen sollte. Wenn eine solche kritische Neu-Edition leider auch weiterhin ein Desiderat der Arabistik bleibt — zunächst verzögerte die Auffindung der Chester-Beatty-Handschrift die weitere Arbeit erheblich, und dann machte der 2. Weltkrieg mit seinen

Folgen den Druck unmöglich —, so sind doch aus den umfangreichen Vorarbeiten einige wichtige Aufsätze geflossen, wie die Untersuchung „Eine arabische Literaturgeschichte aus dem 10. Jahrhundert n. Chr. (Der Fihrist des Ibn an-Nadīm)" (1930), die Übersicht über „Neue Materialien zum Fihrist" (1936), die kommentierte Teilübersetzung „The Arabic Literature on Alchemy according to an-Nadīm (A. D. 987). A Translation of the Tenth Discourse of the Book of the Catalogue (al-Fihrist) with Introduction and Commentary" (1951) und die Edition einer längeren Passage, die in dem Flügelschen Text fehlt, unter dem Titel „Some hitherto unpublished texts on the Muʿtazilite movement from Ibn al-Nadīm's Kitāb al-Fihrist" (1955).

Fücks ganz besonderes Interesse galt stets der Person Mohammeds und der von ihm begründeten Religion. In einem Aufsatz unternahm er den überzeugenden Versuch, „Die Originalität des arabischen Propheten" (1936) unter anderem durch den Hinweis auf seine Lehre vom zyklischen Geschichtsverlauf zu erweisen und trat damit entschieden all jenen entgegen, die christlichen und jüdischen Einflüssen in seiner Gedankenwelt das entscheidende Gewicht beimaßen. Von demselben Grundtenor getragen sind seine Ausführungen in dem Aufsatz „Muhammed — Persönlichkeit und Religionsstiftung" (1954). Die intime Vertrautheit mit der für die Entwicklung des Islams so bedeutsamen Traditionsliteratur zeigte sich in den „Beiträgen zur Überlieferungsgeschichte von Buḫārī's Traditionssammlung" (1938); und in dem Aufsatz „Die Rolle des Traditionalismus im Islam" (1939) gelang ihm eine treffliche Darstellung eines für das Verständnis der islamischen Geschichte von ihren Anfängen bis zum Aufkommen des Wahhabismus wichtigen Phänomens, das in der Forderung nach der unbedingten Vorbildlichkeit des Propheten Mohammed für die islamische Gemeinde im ganzen wie für die Lebensführung jedes einzelnen Muslims bestand. Studien dieser Art befähigten Fück schließlich dazu, in einem Beitrag zum „Handbuch der Orientalistik", herausgegeben von B. Spuler, „Die Religion des sunnitischen Islams" (1961) in souveräner Weise zusammenfassend darzustellen.

Wenn es der zur Verfügung stehende Platz auch nicht zuläßt, auf weitere der Erwähnung würdige Arbeiten Fücks einzugehen, und der Verfasser dieser Zeilen es deshalb bei einem Hinweis auf die folgende Bibliographie bewenden lassen muß, so darf doch nicht unerwähnt bleiben, daß Fück stets mit großer Bereitwilligkeit und Zuverlässigkeit an internationalen wissenschaftlichen Gemeinschaftsunternehmen mitgearbeitet hat, wenn die Aufforderung dazu an ihn erging. So lieferte er Beiträge für die ersten beiden Bände (1936 und 1943) der von A. J. Wensinck begonnenen „Concordance et Indices de la Tradition Musulmane" und schrieb zahlreiche Artikel für die Neuauflage der „Encyclopaedia of Islam", nachdem er für die erste Auflage bereits den Artikel über (Ibn) al-Nadīm geliefert hatte. Bemerkenswert waren ferner der Ernst, die Sorgfalt und der Zeitaufwand, mit welchem er sich der Rezension einmal übernommener Fachliteratur widmete.

Alle diese Leistungen ruhten auf einem Fundament, das in seiner Breite und Festigkeit nur durch eine über Jahrzehnte mit immensem Fleiß betriebene Sammelarbeit errichtet werden konnte. Von ihr und einer außergewöhnlichen Belesenheit in der klassischen arabischen Literatur zeugen viele Tausende eng und mit kleiner Schrift beschriebene Zettel mit aus den Quellen zusammengetragenen Materialien zum

Fihrist des Ibn al-Nadīm, zu altarabischen Dichtern, zu Orientalisten und ihren Publikationen und zu einem Wörterbuch des Korans, Sammlungen, die mit dem Nachlaß Fücks in die Obhut der Universität Halle übergegangen sind.

Für seine wissenschaftlichen Leistungen ist Fück mehrfach geehrt worden. 1956 verlieh ihm die Regierung der Deutschen Demokratischen Republik den Nationalpreis. Der schon erwähnten Wahl zum ordentlichen Mitglied der Sächsischen Akademie der Wissenschaften folgte 1949 diejenige zum korrespondierenden Mitglied der Association pour l'Avancement des Études Islamiques in Paris; 1961 wurde er korrespondierendes Mitglied der Akademie der Arabischen Sprache in Kairo und Ehrenmitglied des Instituto Peruano de Altos Estudios Islamicos in Lima; und 1965 schließlich wählte ihn die Deutsche Morgenländische Gesellschaft zu ihrem Ehrenmitglied.

Fück hat alle diesen Ehrungen und Auszeichnungen mit großer Bescheidenheit entgegengenommen. Als er 1957 eine Würdigung der Leistungen und der Persönlichkeit seines ersten akademischen Lehrers und eines seiner halleschen Amtsvorgänger und späteren Kollegen mit den Worten einleitete: „Es war nicht Brockelmanns Art, von sich und seiner Arbeit viel Aufhebens zu machen", traf er eine Feststellung, die auch für ihn selbst galt. Im Vordergrund zu stehen, oder sich gar dorthin zu drängen, widerstrebte seinem Wesen. Und so hat er auch in den letzten, von Krankheit und Einsamkeit überschatteten Jahren seines Lebens unverdrossen in der Stille weitergearbeitet. Seinen Kollegen, Freunden und Schülern wird Johann Fück als ein bedeutender Arabist und Islamwissenschaftler und als ein liebenswerter Mensch, in dem sich Herzenswärme, Güte, Humor und Lebensweisheit zu einem eindrucksvollen und harmonischen Ganzen verbanden, in dauernder Erinnerung bleiben. Sein wissenschaftliches Lebenswerk ist Vermächtnis und Verpflichtung zugleich.

Prof. Dr. sc. phil. Manfred Fleischhammer

Bibliographie Johann Fück

Selbständige Werke

1. Muhammad Ibn Isḥâq. Literarhistorische Untersuchungen. Frankfurt am Main 1925. II, 46 S.
2. Arabiya. Untersuchungen zur arabischen Sprach- und Stilgeschichte. Berlin 1950. (Abhandlungen der SAW Phil.-hist. Kl., Bd 45, H. 1) 148 S.
   [Rez.: Isis *42* (1951) 358 (G. S.); RSO *26* (1951) 174—176 (F. Gabrieli); ZDMG *102* (1952) 179—186 (H. Wehr); JRAS 1952, 203—204 (A. S. Tritton); BSLP *48* (1952) 110—113 (J. Cantineau); JNES *12* (1953) 71—72 (G. E. von Grunebaum); OLZ *48* (1953), 258—265 (H. H. Giesecke); BO *10* (1953) 144—150 (A. Spitaler); Orientalia N. S. *22* (1953) 307—311 (F. Rosenthal); BSOAS *15* (1953) 608—613 (A. Guillaume); Oriens *7* (1954) 131—134 (B. Lewin); IQ *2* (1955) 146—147 (C. Rabin).]

3. *al-'Arabīya. Dirāsāt fī l-luġa wa-l-lahağāt wa-l-asālīb. Min 'amal Yūhān Fik. Naqalahū ilā l-'arabīya wa-ḥaqqaqahū wa-fahrasa lahū Dr. 'Abd al-Ḥalīm al-Naǧǧār. Bi-taṣdīr Dr. Aḥmad Amīn Bey wa-taqdīm Dr. Muḥammad Yūsuf Mūsā.* al-Qāhira 1370/1951. XXVII, 290 S.
4. 'Arabīya. Recherches sur l'histoire de la langue et du style arabe. Traduction par Cl. Denizeau. Avec une préface de l'auteur et une introduction par J. Cantineau. Paris 1955. (Institut des Hautes Études Marocaines. Notes et Documents XVI) XII, 237 S. [Rez.: MUSJ *32* (1955) 210 (H. Fleisch); Arabica *3* (1956) 330 (D. S.); JSS *1* (1956) 411—413 (A. J. Arberry); AOH *6* (1956) 290 (J. Somogyi); Anthropos *52* (1957) 696 (J. H.); BSLP *53* (1957—1958) 205—208 (E. Benveniste); Rocznik Orientalistyczny *22* (1958) 119—121 (J. Bielawski); al-Mašriq *53* (1959) 597—598 (H. Fleisch).]
5. Die arabischen Studien in Europa bis in den Anfang des 20. Jahrhunderts. Leipzig 1955. VIII, 335 S.
[Rez.: AOH *6* (1956) 291—292 (J. Somogyi); JRAS 1956, 216—217 (A. S. Tritton); Orientalia N. S. *26* (1957) 91—92 (R. Köbert); WZKM *53* (1957) 330—331 (H. L. Gottschalk); MUSJ *34* (1957) 323—325 (H. Fleisch); Irénikon *30* (1957) 492—493 (D. A. T.); OLZ *52* (1957) 432—433 (W. Björkman); Lychnos 1957/58, 346—347 (B. Lewin); JNES *17* (1958) 91 (N. Abbott); RSO *33* (1958) 149—151 (F. Gabrieli); DLZ *80* (1959), 394—395 (H. Simon); Der Islam *34* (1959) 237—240 (O. Pritsak); BSOAS *23* (1960) 624 (ohne Verf.); Przegląd Orientalistyczny 1961, Nr. 1 (37), 75—79 (J. Reychman).]

Aufsätze und Beiträge

6. Hosea Kapitel 3. In: ZAW *39* (1921) S. 283—290.
7. Gestaltwandel und Bedeutungsfunktion im Hebräischen. In: ZDMG *82* (1928) S. LXX.
8. Eine arabische Literaturgeschichte aus dem 10. Jahrhundert n. Chr. (Der *Fihrist* des Ibn an-Nadīm). In: ZDMG *84* (1930) S. 111—124.
9. The Political Reorganisation of the Near East after the War. In: MHM *6* (1932/33) S. 70—76.
10. Zu *an-nasī'* (Koran 9, 37). In: OLZ *36* (1933) Sp. 280—283. [Zu: A. Moberg, *An-Nasī'* (Koran 9, 37) in der islamischen Tradition. Lund/Leipzig 1931.]
11. Some Remarks on Neo-Arabic Literature. In: Dacca University Journal *9* (1933) S. 113—123.
12. Zum Problem der koranischen Erzählungen. In: OLZ *37* (1934) Sp. 73—77. [Zu: D. Sidersky, Les Origines des Légendes Musulmanes dans le Coran et dans les Vies des Prophètes. Paris 1933.]
13. Eine Konkordanz der kanonischen Tradition des Islams. In: OLZ *37* (1934) Sp. 722 bis 727. [Zu: A. J. Wensinck, Concordance et Indices de la Tradition Musulmane. Livr. I, II. Leiden 1934.]
14. The rôle of Manicheism under the early Abbasids. In: MHM *7* (1934) S. 92—101.
15. Spuren des Zindīqtums in der islamischen Tradition. In: Studien zur Geschichte und Kultur des Nahen und Fernen Ostens, Paul Kahle zum 60. Geburtstag überreicht von Freunden und Schülern aus dem Kreise des Orientalischen Seminars der Universität Bonn. Hrsg. von W. Heffening und W. Kirfel. Leiden 1935. S. 95—100.
16. Neue Materialien zum Fihrist. In: ZDMG *90* (1936) S. 298—321.
17. Die Originalität des arabischen Propheten. In: ZDMG *90* (1936) S. 509—525; [Nachdruck] Der Koran, hrsg. von R. Paret. Darmstadt 1975. S. 167—182 (= Wege der Forschung, Bd CCCXXVI).
18. *Taṣghīr al-ǧam'*. In: ZDMG *90* (1936) S. 626—636.
19. Eine wichtige Handschrift der Traditionssammlung des Buḫārī. In: OLZ *41* (1938)

Sp. 401–407. [Zu: A. MINGANA, An important Manuscript of the Traditions of Bukhāri. Cambridge 1936.]
20. Das Wesen des Islam nach Muḥammad ʿAlī. In: OLZ *41* (1938) Sp. 491–507. [Zu: Maulānā Muḥammad ʿAlī, The Religion of Islām. A Comprehensive Discussion of the Sources, Principles and Practices of Islām. Lahore 1936.]
21. Beiträge zur Überlieferungsgeschichte von Buḫārī's Traditionssammlung. In: ZDMG *92* (1938) S. 60–87.
22. Die Rolle des Traditionalismus im Islam. In: ZDMG *93* (1939) S. 1–32.
23. Die sufische Dichtung in der Landessprache des Panjab. In: OLZ *43* (1940) Sp. 1–11. [Zu: L. RAMA KRISHNA, Pañjābi Ṣūfī Poets A. D. 1460–1900. With a Foreword by A. C. WOOLNER. London 1938.]
24. Die arabischen Studien in Europa vom 12. bis in den Anfang des 19. Jahrhunderts. In: Beiträge zur Arabistik, Semitistik und Islamwissenschaft, hrsg. von R. HARTMANN und H. SCHEEL. Leipzig 1944. S. 85–253. [Rez.: JRAS 1950, 73 (A. J. Arberry).]
25. Zur Frage der Koranübersetzung. In: OLZ 47 (1944), Sp. 165–168. [Zu: A. FISCHER, Der Wert der vorhandenen Koranübersetzungen und Sure 111. Leipzig 1937. (BVSAW Phil.-hist. Kl., Bd 89, H. 2).]
26. Gedanken zur Methodik des hebräischen Unterrichts. In: Festschrift Otto Eißfeldt. Zum 60. Geburtstage 1. September 1947 dargebracht von Freunden und Verehrern. Hrsg. von J. FÜCK. Halle a. d. Saale 1947. S. 125–140.
27. The Arabic Literature on Alchemy according to an-Nadīm (A. D. 987). A Translation of the Tenth Discourse of The Book of the Catalogue (*al-Fihrist*) with Introduction and commentary. In: Ambix *4*, February 1951, Nos. 3 and 4, S. 81–144. [Rez.: Oriens *5* (1952) 336 (H. Ritter).]
28. Muhammed — Persönlichkeit und Religionsstiftung. In: Saeculum *3* (1952) S. 70–93.
29. Sechs Ergänzungen zu Sachaus Ausgabe von al-Bīrūnī's „Chronologie orientalischer Völker". In: Documenta Islamica Inedita (Ricardo Hartmann viro de studiis islamicis meritissimo sacrum). Hrsg. von J. FÜCK. Berlin 1952. S. 69–98.
30. Arabische Musikkultur und Islam. In: OLZ *48* (1953) Sp. 20–27. [Zu A. KUTZ, Musikgeschichte und Tonsystematik. Studien zur Entwicklung der Musik in der Stadtkultur. Berlin 1943.]
31. Geschichte der semitischen Sprachwissenschaft. In: HdO, hrsg. von B. SPULER. 3. Bd: Semitistik, 1. Abschnitt, 3. Kapitel. Leiden 1953, S. 31–39.
32. Geschichte der Arabistik. In: HdO, hrsg. von B. SPULER. 3. Bd: Semitistik, 2. und 3. Abschnitt, 13. Kapitel. Leiden 1954. S. 341–349.
33. Muhammad Iqbal und der indomuslimische Modernismus. In: Westöstliche Abhandlungen. Rudolf Tschudi zum siebzigsten Geburtstag überreicht von Freunden und Schülern. Hrsg. von F. MEIER. Wiesbaden 1954. S. 356–365.
34. Some hitherto unpublished texts on the Muʿtazilite movement from Ibn al-Nadīm's *Kitāb al-Fihrist*. In: Professor Muhammad Shafiʿ presentation volume. Edited by Dr. S. M. ABDULLAH, on behalf of the Majlis-e-Armughān-e-ʿIlmi Lahore (Pakistan). Lahore 1955. S. 50–74.
35. Der Ahn des Azraqī. In: Studi Orientalistici in onore di Giorgio Levi Della Vida. Vol. *I*. Rom 1956. S. 336–340.
36. Zur arabischen Wörterbuchfrage. In: ZDMG *107* (1957) S. 340–347. [Zu: Th. Nöldekes Belegwörterbuch zur klassischen arabischen Sprache. Bearb. und hrsg. von J. KRAEMER. Berlin: 1. Lfg 1952, 2. Lfg 1954.]
37. Asiatic Society of Pakistan. In: BO *15* (1958) S. 264.

38. Bemerkungen zur altarabischen Metrik. In: ZDMG *111* (1961) S. 464–469.
39. Die Religion des sunnitischen Islams. In: HdO, 1. Abteilung, 8. Bd, 2. Abschnitt. Leiden/Köln 1961. S. 405–448.
40. Islam as an historical problem in European historiography since 1800. In: Historians of the Middle East. Edited by B. Lewis and P. M. Holt (Historical Writing on the Peoples of Asia). London 1962. S. 303–314.
41. *Waraqa min ta'rīḫ al-istišrāq fī Almāniyā: Yūhān Ya'qūb Rāyskab* [Johann Jakob Reiske] In: FwF 2. Jahr (1964), Nr. 4, S. 44–54.
42. Zum Verständnis des dem Muḥyiddīn al-Iṣfahānī zugeschriebenen Sendschreibens über Monotheismus und Trinität. In: BO *21* (1964) S. 15–16. [Zu: Muḥyī al-Dīn al-Iṣfahānī, Épître sur l'unité et la trinité. Traité sur l'intellect, Fragment sur l'âme. Texte arabe édité, traduit et annoté par M. Allard et G. Troupeau. Beyrouth 1962 (RILOB XX).]
43. *Mulāḥaẓāt 'alā l-auzān al-'arabīya al-qadīma (naqalahā ilā l-'arabīya Dr. Murād Kāmil).* In: *Maǧallat Maǧma' al-luġa al-'arabīya 18* (1964) S. 33–47.
44. Zum Kalender von Cordoba vom Jahre 961. In: OLZ *60* (1965) Sp. 333–337. [Zu: R. Dozy, Le Calendrier de Cordoue publié. Nouv. éd. accompagnée d'une traduction française annotée par Ch. Pellat. Leiden 1961.]
45. Muḥammad Ibn Sa'd al-'Aufī. In: Studia Orientalia in memoriam Caroli Brockelmann. Hrsg. von M. Fleischhammer. Halle (Saale) 1968 (Unveränderter Nachdruck 1970). (= WZ Halle, Gesellschafts- und Sprachwiss. Reihe, Jahrg. 17). S. 85–86.
46. *Waraqa min ta'rīḫ al-istišrāq fī Almāniyā: Kārl Brūkalmān* [Carl Brockelmann] (1968–1956). (Übers. von M. A. Ḥašīšū). In: FwF *8*. Jahr (1970) Nr. 15, S. 78–85.
47. Georg Wilhelm Freytag. In: Bonner Gelehrte. Beiträge zur Geschichte der Wissenschaften in Bonn 1818–1968. Veröffentlichungen zur 150-Jahr-Feier. Bd *7*: Sprachwissenschaft. Bonn 1970. S. 293–295.
48. Mitarbeit an der Enzyklopaedie des Islām. Artikel in Bd *III* (1936): al-Nadīm.
49. Mitarbeit an The Encyclopaedia of Islam, New Edition.
    Artikel in Bd *I* (1960): Abu l-Aswad al-Du'alī, 'Amr b. Luḥayy, 'Arabiyya: *II 3* Middle Arabic, al-'Askarī, 'Ātika, al-Azraḳī, al-Baṣīr, al-Basūs, Bis͟hr b. Abī K͟hāzim, al-Bustī, Buzurg b. S͟hahriyār.
    Artikel in Bd *II* (1965): Fidjār, G͟hanī b. A'ṣur, G͟haṭafān, Banū G͟hifār.
    Artikel in Bd *III* (1971): Ḥabīb b. Maslama, al-Ḥādira, al-Ḥalabī, Ḥammād al-Rāwiya, Ḥayṣa Bayṣa, Ḥumayd b. T͟hawr, Ḥumayd al-Arḳaṭ, Ibn al-Dumayna, Ibn Durayd, Ibn Ḥibbān, Ibn al-Iṭnāba al-K͟hazradjī, Ibn Ḳays al-Ruḳayyāt, Ibn K͟hallikān, Ibn Mādja, Ibn Manẓūr, Ibn Misdjaḥ, Ibn al-Nadīm, Ibn Sa'd, Ibn Suraydj, Ibn Ya'īs͟h, Ibn al-Ziba'rā, Ibrāhīm al-Mawṣilī, 'Imrān b. Ḥiṭṭān.
    Artikel in Bd *IV* (1973ff.): 'Īsā b. 'Umar, Isḥāḳ b. Ibrāhīm al-Mawṣilī, Iyād, Kalb b. Wabara.
50. Mitarbeit an A. J. Wensinck, Concordance et Indices de la Tradition Musulmane. Leiden: Bd *I* (1936), Bd *II* (1943).
51. Mitarbeit an Neue Deutsche Biographie, hrsg. von der Historischen Kommission der Bayerischen Akademie der Wissenschaften.
    Artikel in Bd *I* (1953): Gottlieb (Theophil) Siegfried Bayer.
    Artikel in Bd *IV* (1959): Johann Albert Bernhard Dorn, Johann Elichman(n).
    Artikel in Bd *V* (1961): August Fischer, Heinrich Leberecht Fleischer, Gustav Leberecht Flügel, Christian Martin Joachim Frähn, Siegmund Fraenkel, Georg Wilhelm Freytag.
    Artikel in Bd *VII* (1966): Martin Hartmann.
    Artikel in Bd *XI* (im Druck): Paul Kahle.

Nachrufe und Würdigungen

52. Wilhelm Printz, 9. August 1887 — 23. Februar 1941. In: ZDMG *95* (1941) S. 157—164.
53. August Fischer zum Gedächtnis. In: FuF *25* (1949) S. 119.
54. August Fischer (1865—1949). In: ZDMG *100* (1950) S. 1—18.
55. Carl Brockelmann zum 85. Geburtstag. In: FuF *27* (1953) S. 122—123.
56. Carl Brockelmann †. In: FuF *30* (1956) S. 255—256.
57. Carl Brockelmann (1868—1956). In: BO *14* (1957) S. 60.
58. Carl Brockelmann 1868—1956 (Ins Arabische übersetzt von Dr. Murād Kāmil). In: *al-Maǧalla* (Kairo) 1957, Nr. 6, S. 1—4.
59. Carl Brockelmann als Orientalist (mit Schriftenverzeichnis). In: WZ Halle, Gesellschafts- und Sprachwiss. Reihe, Jahrg. *7* (1957/58) S. 847—875.
60. Carl Brockelmann (1868—1956). In: ZDMG *108* (1958) S. 1—13.
61. Carl Brockelmann, 17. 9. 1868—6. 5. 1956. In: SAW Jahrb. *1954—1956* (Berlin 1958) S. 268—270.
62. Paul Ernt Kahle (1875—1964). In: ZDMG *116* (1966) S. 1—7.

Herausgebertätigkeit

63. Festschrift Otto Eißfeldt. Zum 60. Geburtstage 1. September 1947 dargebracht von Freunden und Verehrern. Halle an der Saale 1947. V, 233 S.
    [Rez.: ZAW *61* (1945/48) 264—265 (J. Hempel); Book List 1948, 6 (G. W. Anderson); Le Muséon *61* (1948) 110—112 (G. Ryckmans); Biblica *30* (1949) 536—539 (A. Bea); ThLZ *75* (1950) 418—420 (L. Rost); Revue Biblique *57* (1950) 278 (R. de Vaux).]
64. Documenta Islamica Inedita (Ricardo Hartmann viro de studiis islamicis meritissimo sacrum). Berlin 1952. 277 S.
    [Rez.: OLZ *48* (1953) 156—159 (F. Bajraktarević); Orientalia N. S. *22* (1953) 315—320 (R. Köbert); Bijdragen tot de taal-, land- en volkenkunde van Nederlandsch-Indië 1954, 103—104 J. Houben); Der Islam *31* (1954) 129—130 (B. Spuler).]

Rezensionen und Anzeigen

65. Aartun, K.: Zur Frage altarabischer Tempora. Oslo 1963. In: BO *22* (1965) S. 216 bis 217.
    Abd al-Malik, Butrus, siehe Nr. 122.
66. Abdullah, S. M. (Hrsg.): Professor Muḥammad Shafi' Presentation Volume. Lahore 1955. In: OLZ *53* (1958) Sp. 316—319.
67. Abel, M.: Die arabische Vorlage des Suaheli-Epos Chuo Cha Herkal. Textkritische Edition und Übersetzung. Ein Beitrag zur Kenntnis der legendären Maġāzī-Literatur. Berlin/Hamburg 1938. (Beihefte zur Ztschr. für Eingeborenen-Sprachen, 18. H.). In: OLZ *43* (1940) Sp. 108—11.
    Abū 'Ubaid, siehe Nr. 188
    Abū Yūsuf, siehe Nr. 79, 79a.
68. Ahmad, A.: Islamic Modernism in India and Pakistan 1857—1964. London—Bombay-Karachi 1967. In: OLZ *67* (1972) Sp. 60—61.
69. Ahrens, K.: Muhammed als Religionsstifter. Leipzig 1935. (Abh. für die Kunde des Morgenlandes XIX 4). In: OLZ *39* (1936) Sp. 236—237.
70. Ali, M. M.: A Manual of Hadith. Lahore 1955. In: BO *14* (1957) S. 103—104.
    'Alī, Maulānā Muḥammad, siehe Nr. 20.
    Allard, M., siehe Nr. 42.
    Anderson, G. L., siehe Nr. 105.

71. ANDRAE, T.: Mohammed. Sein Leben und sein Glaube. Göttingen 1932. In: OLZ *38* (1935) Sp. 42–45.

72. ANDRUS, J. R. und A. F. MOHAMMED: The Economy of Pakistan. London 1958. In: OLZ *56* (1961) Sp. 82–87.

73. ASAD, M. (Übers.): *Ṣaḥîḥ al-Bukhârî.* Being Traditions of the sayings and doings of the Prophet Muḥammad, narrated by his companions to those who followed them, and compiled under the title *Kitâb al-Jâmi' al-Ṣaḥîḥ* by Imâm Abû 'Abd Allâh Muḥammad Ibn Ismâ'îl al-Bukhârî. Translated from the Arabic with explanatory notes by MUHAMMAD ASAD (LEOPOLD WEISS) in 8 vols (40 instalments). Vol. V-1st instalment. Lahore 1938. In: Der Islam *25* (1938) S. 282–283.

al-Aš'arī, siehe Nr. 172. – ASHTON, J. F., siehe Nr. 159.

al-Baiḍāwī, siehe Nr. 77.

74. BALJON, J. M. S.: Modern Muslim Koran Interpretation (1880–1960). Leiden 1961 In: OLZ *59* (1964) Sp. 160–161.

al-Bāqillānī, siehe Nr. 148.

75. BĀQIR, M.: Lahore Past and Present (Being an account of Lahore compiled from original sources). Lahore 1952. (Panjab University Oriental Publications N. 34). In: Oriens *8* (1955) S. 369–370.

76. BARTHÉLEMY, A.: Dictionnaire Arabe-Français. Dialectes de Syrie: Alep, Damas, Liban, Jerusalem. Fasc. 1–5. Paris 1935–1954. In: ZDMG *107* (1957) S. 205–206.

al-Baṣrī, Abū l-Ḥusain Muḥammad Ibn 'Alī Ibn al-Ṭaiyib, siehe Nr. 117.

BEKIR, M., siehe Nr. 117.

77. BEESTON, A. F. L.: Baiḍāwī's Commentary on Sūrah 12 of the Qur'ān. Text, accompanied by an interpretative rendering and notes. Oxford/London 1963. In: OLZ *61* (1966) Sp. 385–387.

BEN DAOUD, B., siehe Nr. 139.

78. BEN SHEMESH, A.: Taxation in Islam. Vol. I: Yaḥyā Ben Ādam's *Kitāb al-Kharāj.* Edited, translated and provided with an Introduction and Notes. With a Foreword by S. D. GOITEIN. Leiden 1958. In: OLZ *55* (1960) Sp. 172–173.

79. DERS.: Taxation in Islam. Vol. II: Qudāma b. Ja'far's *Kitāb al-Kharāj*, Part seven and excerpts from Abū Yūsuf's *Kitāb al-Kharāj.* Translated and provided with introduction and notes. Leiden 1965. In: OLZ *64* (1969) Sp. 363–364.

79a. DERS.: Taxation in Islam. Vol. III: Abū Yūsuf's *Kitāb al-Kharāj.* Translated and provided with an introduction and notes. Leiden 1969. In: OLZ *70* (1975) Sp. 278–279.

BERCHER, L., siehe Nr. 112.

80. BERGSTRÄSSER, G.: Plan eines Apparatus Criticus zum Koran. München 1930. (SbBAW Philos.-hist. Abt., Jahrg. 1930, H. 7). In: OLZ *35* (1932) Sp. 328–329.

81. DERS.: Nichtkanonische Koranlesarten im *Muḥtasab* des ibn Ǧinnī. (SbBAW Philos.-hist. Abt., Jahrg. 1933, H. 2). In: OLZ *38* (1935), Sp. 446–447.

82. DERS. und O. PRETZL (Hrsg.): Šamsaddīn Muḥammad Ibn al-Ǧazarī, Das biographische Lexikon der Koranleser. Teil I hrsg. von G. BERGSTRÄSSER. Teil II hrsg. von G. BERGSTRÄSSER und O. PRETZL. Leipzig 1933 und 1935. (BI 8a und 8b). In: OLZ *40* (1937) Sp. 742–744.

83. DERS. (Hrsg.): Ibn Ḫālawaih's Sammlung nichtkanonischer Koranlesarten. Leipzig 1934. (BI 7). In: OLZ *38* (1935) Sp. 689–690.

DERS., siehe Nr. 155.

84. BLACK, M. und G. FOHRER (Hrsg.): In Memoriam Paul Kahle. Berlin 1968. In: OLZ *68* (1973) Sp. 264–267.

85. Blau, J.: The Emergence and Linguistic Background of Judeo-Arabic. A Study of the Origins of Middle Arabic. London 1965. In: MIO *17* (1971/72) S. 152–153.
86. Bonelli, L.: Il Corano. Nuova versione letterale Italiana con Prefazione e note critico-illustrative. Seconda Edizione riveduta e migliorata. Milano 1940. In: OLZ *46* (1943) Sp. 34–35.
87. Bousquet, G. H.: Introduction à l'Étude de l'Islam Indonésien. Paris 1938. (Extrait de la Revue des Études Islamiques 1938, II–III). In: OLZ *45* (1942) Sp. 65–66.
Ders., siehe Nr. 192.
Bowen, H., siehe Nr. 109, 110.
Bräu, H. H., siehe Nr. 108.
Bräunlich, E., siehe Nr. 98.
al-Buḫārī, siehe Nr. 73.
Canard, M., siehe Nr. 136.
88. Corsepius, H. G.: Der Islam im Angriff. Potsdam 1938. In: OLZ *43* (1940) Sp. 193 bis 194.
al-Dānī, siehe Nr. 163, 164.
89. Dietrich, A.: Zum Drogenhandel im islamischen Ägypten. Heidelberg 1954. (Veröffentlichungen aus der Heidelberger Papyrussammlung, N. F. Nr. 1). In: ZDMG *107* (1957) S. 205.
90. Diez, E.: Glaube und Welt des Islam. Stuttgart [1941]. (Sammlung Völkerglaube). In: OLZ *46* (1943) Sp. 81–83.
91. Diwald-Wilzer, S. (Hrsg.): Aḥmad Ibn Yaḥyā Ibn al-Murtaḍā, Die Klassen der Mu'taziliten. Wiesbaden 1961. (BI 21). In: OLZ *59* (1964) Sp. 371–374.
Dozy, R., siehe Nr. 44.
92. Drower, E. S.: *Šarḥ d Qabin d Šišlam Rba* (D. C. 38). Explanatory commentary of the marriage-ceremony of the Great Šišlam. Rom 1950. (Sacra Scriptura antiquitatibus orientalibus illustrata, Biblica et Orientalia N. 12). In: BO *13* (1956) S. 73.
93. Faris, N. A.: The Antiquities of South Arabia being a Translation from the Arabic with Linguistic, Geographic and Historic Notes of the Eighth Book of Al-Hamdānī's *al-Iklīl*. Princeton 1938. (POT III). In: ZDMG *96* (1942) S. 158–159.
Ders., siehe Nr. 122.
94. Farrukh, O. A.: *Ikhwân Us-Safa* (Brethren of Purity). An analytical and critical study of the epistles of Ikhwân Us-Safa. In Arabic. 2nd Ed. Beirut 1953. In: OLZ *49* (1954) Sp. 432–433.
95. al-Faruqi, I. R. A.: 'Urubah and Religion. A Study of the Fundamental Ideas of Arabism and of Islam as its Highest Moment of Consciousness. Vol. I. Amsterdam 1962. (On Arabism, Vol. I). In: BO *21* (1964) S. 361–363.
96. Fischel, W. J.: The Bible in Persian Translation. (Repr. from the Harvard Theological Review *45* (1952) 1–45). In: OLZ *49* (1954) Sp. 153–154.
97. Fischer, A.: Muḥammad und Aḥmad, die Namen des arabischen Propheten. Leipzig 1932. (BVSAW Phil.-hist. Kl., 84. Bd, H. 3). In: OLZ *37* (1934) Sp. 174–175.
98. Ders. und E. Bräunlich: Schawāhid-Indices. Indices der Reimwörter und der Dichter der in den arabischen Schawāhidkommentaren und in verwandten Werken erläuterten Belegverse. Leipzig: 1. Lfg 1934, 2. Lfg 1938. In: ZDMG *93* (1939) S. 138–140.
99. Ders.: Der „Koran" des Abu'l-'Alā' al-Ma'arrī. Leipzig 1942. (BVSAW Phil.-hist. Kl. 94. Bd, H. 2). In: OLZ *47* (1944) Sp. 115–119.
Ders., siehe Nr. 25.

100. FLEISCH, H.: L'Arabe Classique. Esquisse d'une Structure Linguistique. Beyrouth 1956. In: ZDMG *109* (1959) S. 209–211.
101. DERS.: Traité de Philologie Arabe. Vol. I: Préliminaires, phonétique, morphologie nominale. Beyrouth 1961. In: ZDMG *112* (1962) S. 392–394.
102. DERS.: L'Arabe Classique. Esquisse d'une Structure Linguistique. Nouvelle Edition, revue et augmentée. Beyrouth 1968. In: ZDMG *122* (1972) S. 322.
FOHRER, G., siehe Nr. 84.
103. FOSTER, F. H.: A Brief Doctrinal Commentary on the Arabic Koran. With constant reference to the Historical Sequence of the Suras. London [1932]. In: OLZ *37* (1934) Sp. 616–617.
104. FRANKL, TH.: Die Entwicklung des Menschen nach dem Koran. Prag 1930. In: OLZ *35* (1932) Sp. 195.
105. FRENZ, H. und G. L. ANDERSON (Hrsg.): Indiana University Conference on Oriental-Western Literary Relations. Chapel Hill 1955. (University of North Carolina. Studies in Comparative Literature No. 13). In: Der Islam *33* (1958) S. 168–169.
106. FYZEE, A. A. A.: A Modern Approach to Islam. London 1963. In: MIO *13* (1967) S. 295–296.
107. DERS.: Cases in the Muhammadan Law of India and Pakistan. London 1965. In: MIO *13* (1967) S. 473–474.
al-Ġazālī, siehe Nr. 221.
108. GEYER, R. (Hrsg.): Die *Mukâṭarah* von aṭ-Ṭayâlisî. Mit einer Beilage: Die alte Einteilung der arabischen Dichter und das ʿAmr-Buch des Ibn al-Jarrâḥ von H. H. BRÄU. Wien und Leipzig 1927. (Akademie der Wiss. in Wien, Phil.-hist. Kl., Sitzungsber., 203. Bd, 4. Abh.). In: Islamica *4* (1931) S. 602–610.
109. GIBB, H. A. R. and H. BOWEN: Islamic Society and the West. A Study of the Impact of Western Civilization on Moslem Culture in the Near East. Vol. I: Islamic Society in the eighteenth Century. Part 1. Oxford Univ. Press 1950. In: BO *10*(1953) S. 71 bis 73.
110. DIES.: Islamic Society and the West. Vol. I, Part 2. London 1957. In: BO *16* (1959) S. 32.
DERS., siehe Nr. 144.
111. GIESECKE, H. H. (Hrsg.): Richard Hartmann — Bibliographie. Festgabe zum 70. Geburtstag. Leipzig 1951. In: ThLZ *77* (1952) Sp. 154.
112. GOLDZIHER, I.: Études sur la tradition islamique extraites du Tome II des Muhammedanische Studien, trad. par L. BERCHER. Paris 1952. In: OLZ *48* (1953) Sp. 528–529.
113. GRUNEBAUM, G. E. VON: Kritik und Dichtkunst. Studien zur arabischen Literaturgeschichte. Wiesbaden 1955. In: Oriens *10* (1957) S. 363–364.
114. DERS.: Der Islam im Mittelalter. Zürich–Stuttgart 1963. (Die Bibliothek des Morgenlandes [I]). In: Oriens *18–19* (1967) S. 429–431.
al-Ǧurǧānī, siehe Nr. 170, 171.
115. EL-HAJJÉ, H.: Le Parler Arabe de Tripoli (Liban), avec une Préface de J. CANTINEAU. Paris 1954. In: ZDMG *107* (1957) 206–207.
al-Ḥākim al-Nīsābūrī, siehe Nr. 173.
al-Hamdānī, siehe Nr. 93.
116. HAMIDULLAH, M.: The Battlefields of the Prophet Muhammad with maps, illustrations and sketches. Woking 1953/1373. In: Oriens *8* (1955) S. 309.
117. DERS., M. BEKIR und H. HANAFI (Hrsg.): Abū l-Ḥusain Muḥammad b. ʿAlī b. aṭ-Ṭaiyib al-Baṣrī, *Kitāb al-Muʿtamad fī Uṣūl al-Fiqh*. Édition critique. I. II. Damas 1964/1965. In: OLZ *63* (1968) Sp. 267–268.
HANAFI, H., siehe Nr. 117.

118. HARTMANN, R.: Islam und Nationalismus. Berlin 1948. (ADAW Jahrg. 1945/46, Philos.-hist. Kl. Nr. 5). In: ThLZ *76* (1951) Sp. 225–226.
DERS., siehe Nr. 111.
119. HASCHMI, M. J.: Wer war Mohammed? „Mohammeds Geburtstag im Zeichen der Wiedergeburt des Islams“. Berlin 1936. In: OLZ *40* (1937) Sp. 690–691.
120. HENNINGER, J.: Spuren christlicher Glaubenswahrheiten im Koran. Schöneck/Beckenried (Schweiz) 1951. (Schriftenreihe der Neuen Ztschr. für Missionswiss. X). In: ThLZ *79* (1954) Sp. 473.
121. HIRSCHBERG, J. W.: Jüdische und christliche Lehren im vor- und frühislamischen Arabien. Ein Beitrag zur Entstehungsgeschichte des Islams. Kraków 1939. (Polska Akademia Umiejętności. Mémoires de la Commission Orientaliste No. 32). In: OLZ *44* (1941) Sp. 76–77.
122. HITTI, PH. K., N. A. FARIS und B. ABD-AL-MALIK: Descriptive Catalog of the Garrett Collection of Arabic Manuscripts in the Princeton University Library. Princeton 1938. (POT V). In: ZDMG *94* (1940) S. 109–110.
123. HOLLISTER, J. N.: The Shi'a of India. London 1953. In: OLZ *51* (1956) Sp. 245–248.
124. HUMBERT, P.: Observations sur le Vocabulaire Arabe du Châhnâmeh. Neuchâtel 1953. (Mémoires de l'Université de Neuchâtel, tome XXII). In: DLZ *75* (1954) Sp. 398 bis 400.
125. Ibn Abī Ḥātim ar-Rāzī: *Taqdimatu'l ma'rifa li Kitābi'l jarḥ wa't-ta'dīl.* Hyderabad – Deccan 1952. In: BO *14* (1957) S. 106–107.
Ibn 'Aqīl, siehe Nr. 145, 146.
Ibn al-Ǧarrāḥ, siehe Nr. 108.
Ibn al-Ǧazarī, siehe Nr. 82, 165.
Ibn Ǧinnī, siehe Nr. 81.
Ibn Ḫālawaih, siehe Nr. 83.
Ibn al-Murtaḍā, siehe Nr. 91.
Ibn Qutaiba, siehe Nr. 140.
Ibn Quzmān, siehe Nr. 200.
126. IKRAM, S. M. und P. SPEAR: The Cultural Heritage of Pakistan. London [1955]. In: OLZ *54* (1959) Sp. 177–180.
IQBĀL, MUḤAMMAD, siehe Nr. 149, 186.
127. JACOB, G. (Übers.): Unio mystica. Sehnsucht und Erfüllung. Hafisische Lieder in Nachbildungen. Hannover 1922. In: EMM NF 66 (1922), S. 388.
128. JAIRAZBHOY, Q. 'A.: Muhammad. “A Mercy to all the Nations”. With a Foreword by his Highness The Aga Khan. London 1937. In: OLZ 41 (1938), Sp. 369–370.
129. Journal of the Asiatic Society of Pakistan *I*, 1 (1956). In: OLZ *53* (1958) Sp. 93–94.
130. JURJI, E. J.: Illumination in Islamic Mysticism. A Translation, with an introduction and notes, based upon a critical edition of Abu-al-Mawāhib al-Shādhilī's traetise entitled *Qawānīn Ḥikam al-Ishrāq.* London/Princeton 1938. (POT IV). In: OLZ *43* (1940) Sp. 433–434.
131. JUYNBOLL, G. H. A.: The Authenticity of the Tradition Literature. Discussions in Modern Egypt. Leiden 1969. In: OLZ *68* (1973) Sp. 270–271.
KAHLE, P., siehe Nr. 84.
al-Kāšī, siehe Nr. 142.
132. KEILANI, I.: Abū Ḥayyān at-Tawḥīdī, Essayiste arabe du IVe s. de l'Hégire (Xe s.). Introduction à son oeuvre. Damas 1950. In: BO *11* (1954) S. 138.
133. KEILANI, I. (Hrsg.): Trois Épîtres d'Abū Ḥayyān at-Tawḥīdī. Damas 1951. In: BO *11* (1954) S. 138.

134. KHADDURI, M.: War and Peace in the Law of Islam. Baltimore 1955. DLZ *78* (1957) Sp. 540–541.
135. KÖBERT, R.: Textus et Paradigmata Syriaca. Rom 1952. In: OLZ *49* (1954) Sp. 339.
Koran, siehe Nr. 77, 86, 99, 103, 139, 152, 153, 162, 194, 195, 218.
KRAEMER, J., siehe Nr. 36.
136. KRATCHKOVSKY, I.: Avec les Manuscrits Arabes (Souvenirs sur les livres et les hommes), traduit du russe par M. CANARD. Alger 1954. (Publications de l'Institut d'Études Orientales de la Faculté des Lettres d'Alger No. XIV). In: Oriens *10* (1957) S. 386.
137. KRISS, R. und H. KRISS-HEINRICH: Volksglaube im Bereich des Islam. Bd I: Wallfahrtswesen und Heiligenverehr. Wiesbaden 1960. In: ThLZ *87* (1962) Sp. 830–832.
138. DIES.: Volksglaube im Bereich des Islam. Bd II: Amulette, Zauberformeln und Beschwörungen. Wiesbaden 1962. In: ThLZ *89* (1964) Sp. 739–740.
KRISS-HEINRICH, H., siehe Nr. 137, 138.
KUTZ, A., siehe Nr. 30.
139. LAÏMÈCHE, A. et B. BEN DAOUD: Le Coran (Lecture par Excellence). Paris [o. J.]. In: OLZ *36* (1933) Sp. 627.
140. LECOMTE, G.: Le Traité des Divergences du Ḥadīṯ d'Ibn Qutayba (mort en 276/889). Traduction annotée du *Kitāb Ta'wīl Muḫtalif al-Ḥadīṯ*. Damas 1962. In: OLZ *64* (1969) Sp. 472–475.
141. DERS.: Ibn Qutayba (mort en 276/889). L'homme, son oeuvre, ses idées. Damas 1965. In: OLZ *64* (1969) Sp. 472–475.
LITTMANN, E., siehe Nr. 157.
142. LUCKEY, P. (Übers.): Ǧamšīd b. Mas'ūd al-Kāšī, Der Lehrbrief über den Kreisumfang (*Ar-risāla al-muḥīṭīya*). Übersetzt und erläutert. Hrsg. von A. SIGGEL. Berlin 1953. (ADAW Kl. für Mathematik und allg. Naturwiss., Jahrg. 1950, Nr. 6). In: OLZ *49* (1954) Sp. 147–149.
al Ma'arrī, siehe Nr. 99
MAHMASSANI, S., siehe Nr. 220.
Maimonides, siehe Nr. 209.
143. MAKDISI, G.: *Kitāb at-tauwābīn* „Le livre des pénitents". Damas 1961. In: BO *22* (1965) S. 327.
144. DERS. (Hrsg.): Arabic and Islamic Studies in Honor of Hamilton A. R. Gibb. Leiden 1965. In: BO *24* (1967) S. 223–224.
145. DERS. (Hrsg.): The Notebooks of Ibn 'Aqīl, *Kitāb al-funūn*. Arabic text edited with introduction and critical notes. Part I of the Paris manuscript. Beirut 1970. In: MIO *17* (1971/72) S. 318–319.
146. DERS. (Hrsg.): The Notebooks of Ibn 'Aqīl, *Kitāb al-funūn*. Arabic text edited with introduction and critical notes. Part II of the Paris manuscript. Beirut 1971. In: MIO *17* (1971/72) S. 701.
147. [MASSIGNON, L.] Mélanges Louis Massignon. Tome I–III. Damas 1956–1957. In: Oriens *12* (1959) S. 285–290.
148. MCCARTHY, R. J. (Hrsg.): al-Bāqillānī, *Kitāb At-Tamhīd*. Beyrouth 1957. (Publications of Al-Hikma University of Baghdad, Kalām Series No. 1). In: BO *17* (1960) S. 211–212.
149. MEYEROVITCH, E. (Übers.): MOHAMMED IQBAL, Reconstruire la pensée religieuse de l'Islam. Traduction et notes. Préface de L. MASSIGNON. Paris 1955. In: Oriens *10* (1957) 385–386.
150. MEZ, A.: Die Renaissance des Islams. Heidelberg 1922. In: EMM NF *68* (1924) S. 222–223.

Mingana, A., siehe Nr. 19.
Moberg, A., siehe Nr. 10.
Mohammed, A. F., siehe Nr. 72.
151. Mohibbul Hasan Khan: History of Tipu Sultan. Calcutta/Dacca 1951. In: ZDMG *107* (1957) S. 208—209.
152. Mohtar-Katirjioglu, M.: La sagesse coranique éclairée par des versets choisis reflétant la philosophie morale, religieuse et sociale de l'Islam suivis d'un exposé synoptique des enseignements du Coran. Paris 1935. In: OLZ *39* (1936) Sp. 435—436.
153. Ders. (Muḫtār-Kātircioglū, M.): The Wisdom of the Qur'ān, set forth in Selected Verses conveying the Moral, Religious and Social Philosophy of Islam. Preceeded by an Introduction expounding the Teachings of the Qur'ān. The English Translation by J. Naish. London 1937. In: OLZ *41* (1938) Sp. 694.
Muḥyī al-Dīn al-Iṣfahānī, siehe Nr. 42.
154. Nader, A. N.: Le Système Philosophique des Mu'tazila (Premiers penseurs de l'Islam). Beyrouth 1956. RILOB III). In: BO *17* (1960) S. 211.
Naish, J., siehe Nr. 153.
155. Nöldeke, Th.: Geschichte des Qorāns. 2. Aufl. III. Teil von G. Bergsträsser und O. Pretzl. Lfg 3 und 4. Leipzig 1936 und 1938. In: OLZ *42* (1939) Sp. 235—238.
Ders., siehe Nr. 36.
156. Paret, R.: Die legendäre Maghāzī-Literatur. Arabische Dichtungen über die muslimischen Kriegszüge zu Mohammeds Zeit. Tübingen 1930. In: OLZ *35* (1932) Sp. 266 bis 267.
157. Ders. und A. Schall (Hrsg.): Ein Jahrhundert Orientalistik. Lebensbilder aus der Feder von Enno Littmann und Verzeichnis seiner Schriften. Zum 80. Geburtstag am 16. 9. 1955. Wiesbaden 1955. In: OLZ *53* (1958) Sp. 51.
158. Ders.: The Study of Arabic and Islam at German Universities. Wiesbaden 1968. In: OLZ *68* (1973) Sp. 274—275.
159. Pearson, J. D.: Index Islamicus 1906—1955. A catalogue of articles on Islamic subjects in periodicals and other collective publications, with the assistance of J. F. Ashton. Cambridge 1958. In: ZDMG *109* (1959) S. 427—428.
160. Ders.: Index Islamicus Supplement 1956—1960. Cambridge 1962. In: ZDMG *113* (1963) S. 270.
161. Ders.: Index Islamicus Supplement 1961—1965. Cambridge 1967. In: ZDMG *119* (1970) S. 358—359.
Pellat, Ch., siehe Nr. 44.
162. Pickthall, M.: The Meaning of the Glorious Koran. An explanatory translation. London 1930. In: OLZ *35* (1932) Sp. 329—331.
163. Pretzl, O. (Hrsg.): Abū 'Amr 'Uṯmān Ibn Sa'īd ad-Dānī, Das Lehrbuch der sieben Koranlesungen. Istanbul 1930. (BI 2). In: OLZ *36* (1933) Sp. 42—44.
164. Ders. (Hrsg.): Abū 'Amr 'Uṯmān Ibn Sa'īd ad-Dānī, Orthographie und Punktierung des Koran. Zwei Schriften. Istanbul 1932. (BI 3). In: OLZ *38* (1935) Sp. 45—47.
165. Ders.: Šamsaddīn Muḥammad Ibn al-Ǧazarī, Das Biographische Lexikon der Koranleser. Indices. Leipzig 1935. (BI 8c). In: OLZ *42* (1939) Sp. 169—170.
Ders., siehe Nr. 82, 155.
Qudāma Ibn Ǧa'far, siehe Nr. 79.
166. Qureshi, I. H.: The Muslim Community of the Indo-Pakistan Subcontinent (610—1947). A brief historical Analysis. Den Haag 1962. In: OLZ *61* (1966) Sp. 57—63.
167. Rabin, C.: Ancient West Arabian. A Study of the Dialects of the Western Highlands

of Arabia in the Sixth and Seventh Centuries A. D. London 1951. In: IQ *1* (1954) S. 60–62.

RAMA KRISHNA, L., siehe Nr. 23.

168. REINERT, B.: Die Lehre vom tawakkul in der klassischen Sufik. Berlin 1968. (Studien zur Sprache und Kultur des islamischen Orients. Beihefte zur Ztschr. „Der Islam"). In: BO *28* (1971) S. 99.

169. Revue de l'Institut des Manuscrits Arabes 3–6. In: OLZ *54* (1959) Sp. 554–555; *55* (1960) Sp. 435–436; *56* (1961) Sp. 329–331; *58* (1963) Sp. 518.

170. RITTER, H. (Hrsg.): 'Abdalqāhir al-Ǧurǧānī, *Asrār al-balāġa*. The Mysteries of Eloquence. Istanbul 1954. In: ZDMG *108* (1958) S. 403–404.

171. DERS.: Die Geheimnisse der Wortkunst (*Asrār al-balāġa*) des 'Abdalqāhir al-Curcānī. Aus dem Arabischen übersetzt. Wiesbaden 1959. (BI 19). In: OLZ *57* (1962) Sp. 508 bis 510.

172. DERS. (Hrsg.): Abu l-Ḥasan 'Alī ibn Ismā'īl al-Aš'arī. Die dogmatischen Lehren der Anhänger des Islam. 2. Aufl. Wiesbaden 1963. (BI 1). In: OLZ *61* (1966) Sp. 387.

173. DERS.: *Mishkat-ul-Masabih*. English Translation with explanatory Notes. Part I. Lahore 1960. In: JSS *7* (1962) S. 102–104.

174. ROBSON, J. (Hrsg.): al-Ḥākim al-Naisābūrī, Abū 'Abdallāh Muḥammad b. 'Abdallāh: An Introduction to the Science of Tradition being *Al-madkhal ilā ma'rifat al-Iklīl*. Edited with Introduction, Translation and Notes. London 1953. (Oriental Translation Fund N. S. Vol. XXXIX). In: OLZ *49* (1954) Sp. 59–60.

175. DERS.: *Mishkāt al-Maṣābīḥ*. English Translation with explanatory notes. 4 vols. Lahore 1963–1965. In: JSS *11* (1966) S. 271.

176. DERS.: *Mishkat Al-Masabih*. English Translation with Explanatory Notes. I–IV. Lahore 1963–1965. In: OLZ *63* (1968) Sp. 379–380.

177. ROSENTHAL, E. I. J.: Political Thought in Medieval Islam. An introductory outline. Cambridge 1958. In: DLZ *81* (1960) Sp. 113–114.

178. ROSENTHAL, F.: Four Essays on Art and Literature in Islam. Leiden 1971. (The L. A. Mayer Memorial Studies in Islamic Art and Archaeology. II). In: BO *30* (1973) S. 297.

179. ROUX, J.-P.: L'Islam en Asie. Préface de LOUIS MASSIGNON. Paris 1958. In: DLZ *80* (1959) Sp. 305–306.

180. RUTTER, O.: Triumphant Pilgrimage. An English Muslim's Journey from Sarawak to Mecca. London 1937. In: OLZ *42* (1939) Sp. 35–36.

181. RYCKMANS, G.: Les Religions Arabes Préislamiques. 2. éd. Louvain 1951. (Bibliothèque du Muséon 26). In: DLZ *73* (1952) Sp. 710–712.

182. SABBAGH, T.: La Métaphore dans le Coran. Préface de M. LOUIS MASSIGNON. Paris 1943. In: ZDMG *98* (1944) S. 410–414.

al-Šāḏilī, siehe Nr. 130.

Šafī', M., siehe Nr. 66.

al-Sam'ānī, siehe Nr. 211.

183. SCHACHT, J.: Das Konstantinopler Fragment des *Kitāb Iḫtilāf al-Fuqahā'* des Abū Ǧa'far Muḥammad ibn Ǧarīr aṭ-Ṭabarī. Leiden 1933. (Veröffentl. der De Goeje-Stiftung Nr. X). In: OLZ *38* (1935) Sp. 626–629.

184. DERS.: The Origins of Muhammadan Jurisprudence. Oxford 1950. In: BO *10* (1953) S. 196–198.

DERS., siehe Nr. 192.

SCHALL, A., siehe Nr. 157.

185. Schimmel, A.: Gabriel's Wing. A Study into the religious ideas of Sir Muhammad Iqbal. Leiden 1963. (Studies in the History of Religions. Supplement to NUMEN VI). In: Der Islam *41* (1965) S. 301–303.
186. Dies. (Übers.): Sir Muhammad Igbal, Die Botschaft des Ostens (als Antwort auf Goethes West-Östlichen Divan). Aus dem Persischen übertragen und eingeleitet. Wiesbaden 1963. In: OLZ *60* (1965) Sp. 389–390.
187. Schipperges, H.: Ideologie und Historiographie des Arabismus. Wiesbaden 1961. (Sudhoffs Archiv für Geschichte der Medizin und der Naturwissenschaften, Beiheft 1). In: ZDMG *113* (1963) S. 266.
188. Sellheim, R.: Die klassisch-arabischen Sprichwörtersammlungen insbesondere die des Abū 'Ubaid. 's-Gravenhage 1954. In: ZDMG *106* (1956) S. 400–401.
189. Sezgin, F.: Geschichte des arabischen Schrifttums. I. Qur'ānwissenschaften, Ḥadīṯ, Geschichte, Fiqh, Dogmatik, Mystik bis ca. 430 H. Leiden 1967. In: OLZ *68* (1973) Sp. 150–151.
190. Sharma, S. R.: Mughal Government and Administration. Bombay [1951]. In: OLZ *48* (1953) Sp. 463–465.
Sidersky, D., siehe Nr. 12.
Siggel, A., siehe Nr. 142.
191. Simon, H.: Ibn Khalduns Wissenschaft von der menschlichen Kultur. Leipzig 1959. (Beiträge zur Orientalistik I). In: DLZ *82* (1961) Sp. 215–218.
192. Snouck Hurgronje, C.: Selected Works, edited in English and French by G. H. Bousquet and J. Schacht. Leiden 1957. In: BO *15* (1958) S. 223.
Spear, P., siehe Nr. 126.
193. Spitaler, A.: Die Verszählung des Koran nach islamischer Überlieferung. Mit Vorwort von O. Pretzl. München 1935. (SbBAW Philos.-hist. Abt., Jahrg. 1935, H. 11). In: OLZ *40* (1937) Sp. 529.
194. Strothmann, R. (Hrsg.): Ismailitischer Koran-Kommentar. Lfg. 1–3. Göttingen 1944 und 1948. (AAWG Philol.-hist. Kl., 3. Folge, Nr. 31). In: ThLZ *76* (1951) Sp. 226.
195. Ders. (Hrsg.): Ismailitischer Kommentar zum Koran. Abschnitt 11–20. Arabische Handschrift Ambrosiana H. 76. Lfg 4. Göttingen 1955. (AAWG Philol.-hist. Kl., 3. Folge, Nr. 31). In: ThLZ *83* (1958) Sp. 181.
196. Studia Islamica 1–2. In: OLZ *49* (1954) Sp. 473–474; *50* (1955) Sp. 281–282.
al-Ṭabarī, siehe Nr. 183.
197. Tamer, A. (Hrsg.): 'Alī b. Muḥammad al-Walīd, *dā'ī muṭlaq* des Ismā'īliens du Yémen (m. 612/1215): *Tāǧ al-'aqā' id wa-ma'din al-fawā'id.* Beyrouth 1967. (RILOB I: Pensée arabe et musulmane, 37. Textes pour l'étude de la pensée ismā'īlienne). In: OLZ *66* (1971) Sp. 481–482.
al-Tauḥīdī, siehe Nr. 133.
al-Ṭayālisī, siehe Nr. 108.
198. Torrey, Ch. C.: The Jewish Foundation of Islam. New York 1933. In: OLZ *38* (1935), Sp. 443–446.
Troupeau, G., siehe Nr. 42.
199. Trummeter, F.: Ibn Sa'īd's Geschichte der vorislamischen Araber. Stuttgart 1928. In: Der Islam *19* (1930) S. 169–170.
200. Tuulio, O. J. (Hrsg.): Ibn Quzmān, Poète Hispano-Arabe Bilingue. Édition critique Partielle et Provisoire. Chansons X, XIX, XX, LXXIX, LXXXIV, LXXXVII, XC. Helsinki 1941. (Studia Orientalia edidit Societas Orientalis Fennica IX, 2). In: ZDMG *96* (1942), S. 561–563.

201. Vacca, V.: L'India Musulmana. Milano [o. J.]. (Manuali di Politica Internazionale 31). In: OLZ *46* (1943) Sp. 376–377.
202. Vahiduddin, S.: Indisch-moslemische Werterlebnisse als Parallele zu europäischen Kulturwandlungen. Mit einer Einführung: Gemeinsame Probleme der Kulturmenschheit im Lichte der psychologischen Typenforschung. Von Prof. E. R. Jaensch. Leipzig 1937. (Abhandl. und Monographien zur Philosophie des Wirklichen, Nr. 8). In: OLZ *42* (1939) Sp. 320–322.
203. Waardenburg, J. D. J.: L'Islam dans le miroir de l'Occident. L'approche compréhensive et la formation de l'image de la religion islamique chez quelques orientalistes occidentaux. 's-Gravenhage 1961. In: BO *20* (1963) S. 87–89.
204. Waheed, K. A.: A Bibliography of Iqbal. Karachi 1965. In: MIO *13* (1967) S. 474–475.
al-Walīd, ʿAlī Ibn Muḥammad, siehe Nr. 197.
205. Watt, W. M.: The Faith and Practice of al-Ghazālī. London [1953]. (Ethical and Religious Classics of East and West). In: ThLZ *79* (1954) Sp. 472–473.
206. Ders.: Muhammad at Mecca. Oxford 1953. In: BO *10* (1953) S. 244–245.
207. Ders.: Muhammad at Medina. London 1956. In: BO *15* (1958) S. 144.
208. Wehr, H.: Arabisches Wörterbuch für die Schriftsprache der Gegenwart. Leipzig 1952. In: ZDMG *104* (1954) S. 505–506.
209. Weil, G. (Hrsg.): Maimonides, Über die Lebensdauer. Ein unediertes Responsum, herausgegeben, übersetzt und erklärt. Basel 1953. In: DLZ *74* (1953) Sp. 405–408.
Weiss, L., siehe Nr. 73.
210. Weisweiler, M.: Verzeichnis der arabischen Handschriften der Universitätsbibliothek Tübingen II. Leipzig 1930. In: OLZ *35* (1932) Sp. 411–412.
211. Ders. (Hrsg.): ʿAbd-al-Karīm Ibn Muḥammad As-Samʿānī, Die Methodik des Diktatkollegs (*Adab al-Imlā'wa'l-istimlā'*). Leiden 1952. (Veröffentl. der De Goeje-Stiftung Nr. XIV). In: BO *10* (1953) S. 248–249.
212. Wensinck, A. J.: Concordance et Indices de la Tradition Musulmane. Les six livres, le *Musnad* d'al-Dārimī, le *Muwaṭṭa'* de Mālik, le *Musnad* de Aḥmad Ibn Ḥanbal. Vol. I. Leiden 1934–1936. In: OLZ *41* (1938) Sp. 628–629.
213. Ders.: La Pensée de Ghazzālī. Paris 1940. In: ZDMG 96 (1942), S. 563.
Ders., siehe Nr. 13.
214. Westermarck, E.: Pagan Survivals in Mohammedan Civilisation. London 1933. In: OLZ *38* (1935) Sp. 448.
215. Wiener Zeitschrift für die Kunde des Morgenlandes 51–54. In: OLZ *49* (1954) Sp. 476; *50* (1955) Sp. 282; *51* (1956) Sp. 283–284; *53* (1958) Sp. 586; *54* (1959) Sp. 93–96.
216. Wild, S.: Das *Kitāb al-ʿAin* und die arabische Lexikographie. Wiesbaden 1965. In: BO *23* (1966) S. 199–200.
Yaḥyā Ibn Ādam, siehe Nr. 78.
217. Young, T. C. (Hrsg.): Near Eastern Culture and Society. A Symposium on the Meeting of East and West. Princeton 1951. (Princeton Oriental Studies Vol. 15). In: OLZ *49* (1954) Sp. 242–243.
218. Yusuf Ali, A.: The Holy Quran. English Translation and Commentary (with Arabic Text). Part I–III. Lahore [o. J.]. In: OLZ *41* (1938) Sp. 168–170.
219. Zbinden, E.: Die Djinn des Islam und der altorientalische Geisterglaube. Stuttgart 1953. In: Der Islam *32* (1957) S. 105.
220. Ziadeh, F. J. (Übers.): S. Mahmassani, *Falsafat al-Tashrīʿ fī al-Islām*. The Philosophy of Jurisprudence in Islam. Leiden 1961. In: OLZ *61* (1966) Sp. 161–164.

221. Zolondek, L.: Book XX of Al-Ghazālī's *Iḥyā' 'Ulūm Al-Dīn*. Leiden 1963. In: BO *24* (1967) S. 227–228.

222. Zwemer, S.: Studies in Popular Islam. A Collection of Papers dealing with the Superstitions and Beliefs of the Common People. London 1939. In: OLZ *43* (1940) Sp. 317 bis 318.

Akademie-Vorträge

223. Arabiya. Untersuchungen zur arabischen Sprach- und Stilgeschichte (17. 1. 1949). Siehe SAW Jahrb. *1949–1953* (Berlin 1954) S. 39; siehe ferner oben Nr. 2.

224. Neue Materialien zur Geschichte der Mu'tazila (16. 3. 1953). Siehe SAW Jahrb. *1949–1953* (Berlin 1954) S. 43; vgl. DLZ *74* (1953) Sp. 377–378.

225. Die Sprachtheorie des Sibawaih (9. 5. 1955). Siehe SAW Jahrb. *1954–1956* (Berlin 1958) S. 88–89; vgl. DLZ *76* (1955) Sp. 710–711.

226. Eine Apologie des sunnitischen Islam aus dem X. Jahrhundert (9. 3. 1959). Siehe SAW Jahrb. *1957–1959* (Berlin 1961) S. 162; vgl. DLZ *80* (1959) Sp. 564.

227. Islam und Hellenismus (21. 1. 1963). Siehe SAW Jahrb. *1963–1965* (Berlin 1967) S. 110–111; vgl. DLZ *84* (1963) Sp. 520–521.

Nachbemerkung

Bei der Zusammenstellung der Bibliographie konnte ich Aufzeichnungen benutzen, welche die Ehefrau des Verstorbenen, Dr. Katharina Fück geb. Boll, bis zu ihrem Tode am 14. November 1970 gemacht hat. Die von mir eingefügten Besprechungen von Arbeiten J. Fücks erheben keinen Anspruch auf Vollständigkeit.

Mehrere Hinweise erhielt ich von Frau Dipl.-Phil. Gudrun Goeseke, Leiterin der Bibliothek der Deutschen Morgenländischen Gesellschaft in Halle (Saale); bei der Ermittlung einiger bibliographischer Daten waren mir die Herren Prof. Dr. Rudolf Sellheim (Frankfurt/Main) und Prof. Dr. Manfred Ullmann (Tübingen) behilflich. Ihnen allen sei an dieser Stelle dafür herzlich gedankt.

Prof. Dr. phil. Manfred Fleischhammer

Zur Biobibliographie

Schriftenverzeichnisse. M. Fleischhammer, siehe vorliegendes Jahrbuch S. 424–438. – Jahrbuch Sächsische Akademie der Wissenschaften zu Leipzig *1954–1956* (Berlin 1958) S. 201; *1957–1959* (Berlin 1961) S. 274–275; *1960–1962* (Berlin 1964) S. 310; *1963–1965* (Berlin 1967) S. 245; *1966–1968* (Berlin 1970) S. 248. – Kürschners Deutscher Gelehrten-Kalender *9* (Berlin 1961) S. 509; *10* (Berlin 1966) S. 620; *11* (Berlin 1970) S. 764.

Würdigungen. Zum 65. Geburtstag: o. V., in: Hochschulwesen *7* (1959) S. 291. – Zum 70. Geburtstag: o. V., in: Neues Deutschland *19* (1964) Nr. 186 vom 8. Juli. – o. V., in: Freiheit *19* (Halle/S. 1964) Nr. 159B vom 8. Juli, S. 2. – o. V., in: FuF *38* (1964) S. 192. – o. V., in: Hochschulwesen *12* (1964) S. 443. – o. V., in: Sächsisches Tageblatt *19* (Leipzig 1964) Nr. 156 vom 8. Juli. – M. R., in: Mitteldeutsche Neueste Nachrichten *13* (Halle/S. 1964) Nr. 160 vom 10. Juli, S. 5. – Zum 75. Geburtstag: o. V., in: Neues Deutschland *24* (1969) Nr. 186 vom 8. Juli. – o. V., in: Liberal-Demokratische Zeitung Halle/S., Nr. 159 vom 8. Juli 1969. – o. V., in: Mitteldeutsche Neueste Nachrichten *18* (Halle/S. 1969)

Nr. 158 vom 8. Juli, S. 6. — Zum 80. Geburtstag: W. Walther, in: Asien, Afrika, Lateinamerika *2* (1974) S. 857—858. — Allgemein: Frankfurter Gelehrten-Handbuch. Hrsg. von E. Bergmann. Frankfurt/Main 1930. S. 54.

Nachrufe. W. Ende, in: Der Islam *53* (1976) S. 193—195. — M. Fleischhammer, siehe vorliegendes Jahrbuch S. 419—424. — Ders., in: Universitätszeitung Halle/S. vom 9. Jan. 1975, S. 2. — W. Walther, in: WZ Univ. Halle, ges. u. sprachwiss. R. *25* (1976) H. 4, S. 135—137.

Johannes Kühn

24. 1. 1887—24. 2. 1973

Am 24. Februar 1973 starb in Heidelberg im 87. Lebensjahr der dortige emeritierte Ordinarius der Geschichte Dr. phil. Johannes Kühn, seit 1943 ordentliches Mitglied unserer Akademie, seit 1946 ordentliches Mitglied der Historischen Kommission bei der Bayerischen Akademie der Wissenschaften. Im Jahr 1928 war Kühn an die Technische Hochschule Dresden, 1947 an die Universität Leipzig berufen worden, 1949 folgte er einem Ruf nach Heidelberg.

Geboren am 24. Januar 1887 als Sohn eines Pfarrers in Bogschütz in Schlesien, studierte Johannes Kühn in Breslau, München und besonders in Leipzig, wo in den Jahren vor dem ersten Weltkrieg in dem gespannten Zusammenwirken der Gerhard Seeliger und Erich Brandenburg einerseits, Karl Lamprechts und seiner Schule andererseits ideale Voraussetzungen für ein fruchtbares Studium der Geschichte gegeben waren. Als Schüler Seeligers wurde Kühn im Jahre 1911 auf Grund einer grundlegenden und bis heute zu zitierenden agrarhistorischen Arbeit promoviert (Das Bauerngut der alten Grundherrschaft. Eine Studie zur Geschichte des Verfalls der Grundherrschaft und der Entwicklung der Agrarverfassung in Südwestdeutschland). Nach der vom Krieg erzwungenen Arbeitspause knüpfte Kühn 1920 an seine agrar- und sozialhistorische Dissertation an mit einem Aufsatz zur Kritik der deutschen Weistümer (Festschrift für G. Seeliger) 1920, um sich alsbald seinem Haupt-Arbeitsgebiet: der Reformationsgeschichte, zuzuwenden. Noch 1914 hatte er sich zu der vielumstrittenen Frage der Entstehung des Wormser Ediktes geäußert (Zeitschrift für Kirchengeschichte *35*) — nun, 1923, erschien seine großangelegte Leipziger Habilitationsschrift: „Toleranz und Offenbarung. Eine Untersuchung der Motive und Motivformen der Toleranz im offenbarungsgläubigen Protestantismus, zugleich ein Versuch zur neueren Religions- und Geistesgeschichte.“

Bald nach der Berufung nach Dresden übernahm Kühn innerhalb der „Jüngeren“ Reihe der Deutschen Reichstagsakten die Edition des 7. Bandes, welcher den „Protestations“-Reichstag zu Speyer 1529 betraf. Die ebenso große wie entsagungsvolle Leistung gedieh im Jahre 1935, also in bemerkenswert kurzer Zeit, mit fast 1 500 Seiten zum Druck. Zur Bändigung der übermäßigen Aktenmasse wählte Kühn, im Einklang mit dem damaligen Abteilungsleiter Erich Brandenburg, eine neue Methode: zunächst aktengesättigtes Referat des Fortgangs der Verhandlungen, sodann (ausgewählte) Aktenstücke als „Beilagen“. Diese Methode ist seither oft diskutiert, gerühmt, auch kritisiert worden. Sie ist, angesichts der Verschiedenheit der Überlieferung, gewiß nicht auf alle Reichstage (schon gar nicht auf die der älteren Zeiten) anwendbar, hat sich aber bei der verhältnismäßig geschlossenen Überlieferung für den Tag von 1529 glänzend bewährt. Pünktlich zum Jubiläum lieferte der Herausgeber, seiner Edition vorgreifend, „Die Geschichte des Speyrer Reichstags 1529" (Schriften des Vereins für Reformationsgeschichte Nr. 146), wo auf nur 247 Seiten ein in jedem Stadium der Verwicklungen klares Bild des religiösen und politischen Kräftespiels geboten wird — eine echte, über Rankes klassische Darstellung sachlich weit hinausführende, die Analyse nie vernachlässigende Geschichtserzählung, so „prall voller Gestalten, Ereignisse, Spannungen und Probleme, daß einem dabei allein der Sinn und Blick für Geschichte überhaupt aufgehen kann" (Grundmann).

Es entsprach den besonderen Bedingungen historischer Lehre an einer Technischen Hochschule, daß der Inhaber des Lehrstuhls neben der Forschung besonders die Lehre, außer dem Besonderen des Allgemeine stark zu betonen hatte. Es entsprach aber auch der Natur Johannes Kühns: der wahrlich bewährte Forscher war ein Denker und ein Lehrer des Allgemeinen in der Geschichte. Von Kühns geschichtsphilosophischen und welthistorischen Überlegungen ist das Meiste gesprochenes Wort geblieben. Von Gedrucktem sei genannt: „Thomas Morus und Rousseau. Die Geburt einer Gesellschaftslehre aus einem Menschenideal" (Hist. Vjschrift 1925), „Dantes Geschichtsphilosophie" (Deutsches Dantejahrbuch 26. 1946).

Für K. war die Geschichte eine Einheit. Der junge Dozent, der erfahrene Professor las über Deutsche Mystik des 14. Jahrhunderts ebenso wie über Lenin, über den Revolutionsbegriff. Überblick über das Ganze der Geschichte ließ ihm knappe, durchdachte Studien gelingen, von denen nur der Aufsatz über Wilsons Friedensvermittlung im Weltkrieg (auf Grund der damals soeben in vier Bänden erschienenen „Intimate papers" des Obersten House, Wilsons rechter Hand — und Gegners — auf der Pariser Friedenskonferenz; Zeitschrift für Politik *18*, 1928) gedruckt ist.

Die beste, von einem Nachruf nicht zu leistende Würdigung ist Kühn von seinem ehemaligen Schüler Herbert Grundmann zuteil geworden unter dem das Wesen des Verstorbenen genau treffenden Titel: „Geschichtsforschung im Großen und im Kleinen" (Ruperto-Carola *XIX* 41. 1967, mit Nennung weiterer Arbeiten).

Hermann Heimpel

Bibliographie Johannes Kühn

Das Bauerngut der alten Grundherrschaft. Eine Studie zur Geschichte des Verfalls der Grundherrschaft und der Entwicklung der Agrarverfassung in Südwestdeutschland. (Leipziger historische Abhandlungen 28) Leipzig 1912 [Phil. Diss. Leipzig 1912].

Zur Entstehung des Wormser Edikts. Untersuchung über die Reihenfolge der Entwürfe Aleanders nebst einem Exkurs. In: Zeitschrift für Kirchengeschichte *35* (1914) S. 372 bis 392, 529—547.

[Hrsg.] Luther und der Wormser Reichstag. Aktenstücke und Briefe zusammengestellt von Johannes Kühn. (Voigtländers Quellenbücher 73) Leipzig 1914.

Zur Kritik der Weistümer (Nach oberrheinischen Quellen). In: Festgabe Gerhard Seeliger zum 60. Geburtstage dargebracht. Leipzig 1920. S. 29—50.

Toleranz und Offenbarung. Eine Untersuchung der Motive und Motivformen der Toleranz im offenbarunggläubigen Protestantismus, zugleich ein Versuch zur neueren Religions- und Geistesgeschichte. Leipzig 1923 [Phil. Habil.-Schr. Leipzig].

Thomas Morus und Rousseau. Die Geburt einer Gesellschaftslehre aus einem Menschenideal. In: Historische Vierteljahrschrift *23* (1926) S. 161–187.

Wer trägt die Verantwortung an der Entstehung des politischen Protestantismus? In: Kultur- und Universalgeschichte. Walter Goetz zu seinem 60. Geburtstage dargebracht. Leipzig 1927. S. 215–228.

Landgraf Philipp von Hessen. Der politische Sinn der sogenannten Packschen Händel 1528–1928. In: Staat und Persönlichkeit. Erich Brandenburg zum 60. Geburtstag dargebracht. Leipzig 1928. S. 107–129.

Die Friedensvermittlung des Präsidenten Wilson im Weltkrieg. In: Zeitschrift für Politik. *18* (1928) S. 209–230.

Die Geschichte des Speyrer Reichstags 1529. (Schriften des Vereins für Reformationsgeschichte 146) Leipzig 1929, H. 1.

[Hrsg.] Deutsche Reichstagsakten unter Kaiser Karl V. Bd *7*: 1527–1529. (Deutsche Reichstagsakten. Jüngere Reihe) Gotha 1935.

Über den Sinn des gegenwärtigen Krieges. In: Zeitschrift für Geopolitik. *17* (1940) S. 57 bis 62, 105–112, 156–171.

Über den Sinn des gegenwärtigen Krieges. (Schriften zur Geopolitik 19) Heidelberg, Berlin, Magdeburg 1940.

Die eigentliche Grundlage der US-amerikanischen Außenpolitik. In: Zeitschrift für Geopolitik. *20* (1943) S. 129–137.

Dantes Geschichtsphilosophie. In: Deutsches Dante-Jahrbuch. *26* (1946) S. 97–114.

Die Wahrheit der Geschichte und die Gestalt der wahren Geschichte. Oberursel 1947.

Das Zeitalter der Kirche. Eine weltgeschichtliche Betrachtung. (Leipziger Universitätsreden 17) Leipzig 1949.

Utopie und Geschichtsphilosophie. In: Welt als Geschichte *11* (1951) S. 1–11.

Thesen zur gegenwärtigen Krise der Geschichte. In: Ruperto-Carola *3* (1951) Nr. 4, S. 21 bis 24.

Fünf entscheidende Punkte der Erde. In: Außenpolitik. Zeitschrift für internationale Fragen *3* (1952) S. 141–156.

Böhmen in Mitteleuropa – Böhmen in der Welt. Gedanken über böhmische Geschichte. In: Aussenpolitik. Zeitschrift für internationale Fragen *4* (1953) S. 753–769.

Das Geschichtsproblem der Toleranz. In: Autour de Michel Servet et de Sebastien Castellion, ed. B. Becker. Haarlem 1953.

Willy Andreas zum 70. Geburtstage. In: Ruperto-Carola *6* (1954) S. 65–66.

Das Zeitalter und der Zeitpunkt. In: Ottheinrich, Gedenkschrift, ed. Georg Poensgen, Sonderband der Ruperto-Carola *1956*.

Zum 80. Geburtstag Fedor Stepuns. In: Ruperto-Carola *16* (1964) S. 84–92.

[Bespr.] Carl Auer: Das Luthervolk; Georg von Below: Die Ursachen der Reformation. In: Historische Vierteljahresschrift *XX* (1920/21).

[Bespr.] Karl Heussi: Altertum, Mittelalter und Neuzeit in der Kirchengeschichte; A. Willburger: Die Konstanzer Bischöfe Hugo von Landenberg, Balthasar Marklin, Johann von Lupfen (1496–1537) und die Glaubensspaltung. In: Historische Vierteljahreshefte *XXI* (1922/23).

[Bespr.] O. Merx: Akten zur Geschichte des Bauernkrieges in Mitteldeutschland. In: Historische Vierteljahreshefte *XXII* (1924/25).

[Bespr.] H. Jedin: Geschichte des Konzils von Trient. In: HZ *174* (1949) S. 587–595.

Schriftenverzeichnisse. H. Lietzmann, siehe vorliegendes Jahrbuch S. 440–442. – Jahrbuch Sächsische Akademie der Wissenschaften zu Leipzig *1963–1965* (Berlin 1967) S. 249. – Kürschners Deutscher Gelehrten-Kalender *1* (Berlin–Leipzig 1925) Sp. 554; *2* (Berlin–Leipzig 1926) Sp. 1049; *3* (Berlin–Leipzig 1928/29) Sp. 1301; *4* (Berlin–Leipzig 1931) Sp. 1617; *5* (Berlin–Leipzig 1935) Sp. 754; 6_1 (Berlin 1940/41) Sp. 1018–1019; *8* (Berlin 1954) Sp. 1298–1299; *10* (Berlin 1966) S. 1320; *11* (Berlin 1970) S. 1627–1628. – Wer ist wer? Hrsg. W. Habel *15* (Berlin 1967) S. 1067.

Würdigungen. Zum 80. Geburtstag: H. Grundmann: Geschichtsforschung im Kleinen und im Großen. In: Ruperto-Carola *19* (1967) S. 77–81. – Allgemein: Degeners Wer ist's? *10* (Berlin 1935) S. 907; *16* (Berlin 1970) S. 706.

Nachrufe. H. Heimpel, siehe vorliegendes Jahrbuch S. 439–440. – Derselbe, 1 S. [in dem Akten der Akademie]. – R. Koselleck, in: Ruperto-Carola *25* (1973) S. 143–144.

Dr. phil. Hilda Lietzmann

Friedrich Zucker

30. 6. 1881–4. 4. 1973

Am 4. 4. 1973 verstarb kurz vor Vollendung seines 92. Lebensjahres auf seinem Alterssitz in Holstein das auswärtige, früher ordentliche Mitglied der philologisch-historischen Klasse Friedrich Zucker, emeritierter ordentlicher Professor für Klassische Philologie an der Universität Jena.

Der am 30. 6. 1881 in Fürth geborene Gelehrte studierte in München und Berlin Klassische Altertumswissenschaft und Philosophie, promovierte 1904 in München mit einer Arbeit über den Aristarch-Schüler Apollodor und habilitierte sich ebenda im Jahre 1911, um anschließend an der gleichen Universität zunächst als Dozent, dann 1917 verwundet aus dem Kriege zurückgekehrt, als außerordentlicher Professor zu wirken. 1918 wurde Fr. Zucker als beamteter außerordentlicher nach Tübingen und noch im selben Jahr als ordentlicher Professor an die Universität Jena berufen, der er bis zur Emeritierung an seinem 80. Geburtstag im Jahre 1961 die Treue gehalten hat. 1928 und dann wieder bei der Neueröffnung im Jahre 1945 wurde er zu ihrem Rektor gewählt.

Der junge Gelehrte geriet bald nach seiner Promotion in den Bann der damals gerade auch in Deutschland mächtig aufblühenden jungen Sonderdisziplin der Papyrologie. Mehrere Male nahm er an Papyrusgrabungen in Ägypten teil, um dann von 1907 bis 1911 eine solche Expedition selber zu leiten. Er hatte das Arabische inzwischen so gut gelernt, daß er sich mit seinen Arbeitern in der Landessprache unterhalten konnte, was ihm Freude machte und ihm zugleich die verehrende Liebe der einfachen Fellachen eintrug. Er hat bis zuletzt mit einer gewissen Wehmut an diese Zeit immer neuer Entdeckungen zurückgedacht und gern und mit Laune von den mancherlei Begebnissen geplaudert, die solche Unternehmungen damals mit sich brachten. Als Papyrologe erntete Fr. Zucker denn auch seine ersten publizistischen

Erfolge; sie waren so überzeugend, daß der Habilitandus der Verpflichtung enthoben wurde, eine besondere Habilitationsschrift vorzulegen — damals ein seltener Fall.

Nach einer längeren Pause, die der überaus gewissenhafte Hochschullehrer der gründlichen Ausarbeitung seiner Vorlesungen schuldig zu sein glaubte, begann dann mit der noch heute viel beachteten Rektoratsrede von 1928 über den Begriff des Gewissens und des sittlichen Bewußtseins im antiken Denken eine neue Phase im Wirken von Fr. Zucker. Zwar betrachtete er nach wie vor die Beschäftigung mit den Papyrus-Urkunden als sein „Spezialgebiet", aber seither nehmen wort- und begriffsgeschichtliche Studien einen immer breiteren Raum in seinen Publikationen ein und darüber hinaus nun auch Fragen der Literatur- und Philosophiegeschichte der Antike, über die er denn auch in unserer wie in der Berliner Akademie öfter vorgetragen hat; in beide Körperschaften wurde Fr. Zucker gleich nach ihrer Wiedereröffnung im Jahre 1948 gewählt; bis ins hohe Alter hat er wohl an keiner ihrer Sitzungen gefehlt.

Zur Mitarbeit in den beiden Akademien und den mannigfachen amtlichen Pflichten, die Fr. Zucker immer sehr ernst genommen hat und bei deren Wahrnehmung er zu keinerlei Kompromissen bereit war (er war ein entschiedener Gegner des Nationalsozialismus und hat auch alles von sich abgewehrt, was er als Eingriff in seine persönliche Freiheit empfand): zu diesen Belastungen kamen seit 1946 noch die verantwortungsschwere Redaktion von zwei führenden Fachzeitschriften sowie die Vertretung und Repräsentation der Altertumswissenschaft der DDR bei zahlreichen internationalen Kongressen. Durch Verleihung des Nationalpreises im Jahre 1952 hat unser Staat dem verdienten und auch im Ausland angesehenen Gelehrten auch öffentlich die Anerkennung ausgesprochen, auf die Fr. Zucker als Lehrer, Forscher und Organisator sowie als allgemein verehrter Senior seiner Wissenschaft in der DDR einen unbestrittenen Anspruch hatte. Für das Ansehen, das er im Kreis seiner Kollegen und Freunde genoß, legt die Fr. Zucker zu seinem 70. Geburtstag gewidmete Festschrift Zeugnis ab. Eine besondere Freude wird es ihm damals gewesen sein, daß an ihr auch die ersten Schüler beteiligt waren, die er zu selbständiger wissenschaftlicher Arbeit herangebildet hatte; die Tüchtigsten unter ihnen halten heute sein Erbe in Jena in treuen Händen.

Strenges Verantwortungsgefühl, echte humanistische Gesinnung, absolute Zuverlässigkeit und Treue in der Wissenschaft wie im Menschlichen, Gründlichkeit, Sauberkeit und unbestechliche Sachlichkeit, Vornehmheit des Denkens wie des Auftretens sind die Eigenschaften, die Fr. Zucker vor allem ausgezeichnet haben. In der Verbindung von papyrologischen Studien mit antiker Geistesgeschichte (wie man damals sagte) hat Fr. Zucker immer eine Sonderstellung in der Philologie eingenommen; auch darin, daß beide scheinbar so weit auseinanderliegenden Gebiete stets mit der gleichen kompromißlosen Werktreue gepflegt wurden, die jeder in seiner Generation nicht selten zur Schau gestellten Selbstgefälligkeit kategorisch den Zugang verwehrte. Aber alle strengen Züge, die das Wesen von Fr. Zucker geprägt haben, wurden ergänzt, gemildert und überhellt von einer liebenswürdigen und doch auch immer wieder Distanz wahrenden Charis im Verkehr von Mensch zu Mensch, die uns allen unvergeßlich bleiben wird.

Unsere Akademie, der Friedrich Zucker so manchen schönen Vortrag geschenkt und an deren Leben er bis in sein hohes Alter lebendigen Anteil genommen hat, darf stolz darauf sein, ihn über ein Vierteljahrhundert lang zu den ihrigen gezählt zu haben.

Werner Peek

Bibliographie Friedrich Zucker

1904

Spuren von Apollodoros *περὶ θεῶν* bei christlichen Schriftstellern der ersten fünf Jahrhunderte. Diss. phil. München. Nürnberg, 38 S.

1906

Euhemeros und seine *Ἱερὰ ἀναγραφή* bei den christlichen Schriftstellern. In: Philologus *64*, 465–472.

1907–1911

Berichte über die archäologischen Funde dieser Jahre in Ägypten. In: Archäol. Anzeiger im Jahrb. d. Kaiserlichen Deutsch. Archäol. Instituts *22*, 154–162. *23*, 192–207. *24*, 176–189. *25*. 244–256. *26*, 238–242.

1909

Bericht über die Ausgrabungen auf Elephantine in d. J. 1907–08, zusammen mit W. Honroth. In: Zeitschr. f. ägypt. Sprache u. Altertumskunde *46*, 43–61.

1910

Urkunde aus der Kanzlei eines römischen Statthalters von Ägypten in Originalausfertigung. In: Sitzungsber. d. Königlich Preuß. Akad. d. Wiss. 1910, 710–730.

Zu den Klageschriften mit Schlußbitte um Registrierung. In: Philologus *69*, 449–465.

[Rez.:] O. Crusius, Paroemiographica, 1910, in: Lit. Zentralbl., Sp. 1481–1482.

1911

Ἐπίτροπος χαρτηρᾶς Ἀλεξανδρείας In: Philologus *70*, 79–105.

Zusammen mit U. von Wilamowitz-Moellendorff: Zwei Edikte des Germanicus auf einem Papyrus des Berliner Museums. In: Sitzungsber. d. Königlich Preuß. Akad. d. Wiss. 1911, 794–821.

[Rez.]: G. Plaumann, Ptolemais in Oberägypten, 1910, in: Lit. Zentralbl., Sp. 1367 bis 1369.

[Rez.:] K. Fitzler, Steinbrüche und Bergwerke im ptolemäischen und römischen Ägypten, 1911, ebda Sp. 1547–1549.

[Rez.:] A. S. Hunt, Catalogue of the Greek Papyri in the John Rylands Library. I: Literary Texts, 1911, in: Byzant. Zeitschr. *20*, 555–557.

1912

Beiträge zur Kenntnis der Gerichtsorganisation im ptolemäischen und römischen Ägypten. In: Philologus, Suppl.-Bd *12*, H. 1, 132 S.

Von Debôd bis Bâb Kalabsche (Die griech. u. lat. Inschriften von Unternubien, mit Kommentar). In der Publikationsreihe: Les temples immergés de la Nubie. Le Caire, Imprimerie de l'Institut Français d'archéologie orientale. 169 S.

[Rez.:] A. Patin, Ästhetisch-kritische Studien zu Sophokles, 1911, in: Lit. Zentralbl., Sp. 967—969.

1912—1936
(ohne die Kriegsjahre)

Byzantinische Zeitschr. *21*, 3.—4. Heft bis einschl. Bd *36*: in der Abt. III „Bibliographie" Bearbeitung der Neuerscheinungen auf dem Gebiet der Papyruskunde.

1913

[Rez.:] Dikaiomata. Auszüge aus alexandrinischen Gesetzen u. Verordnungen, hrsg. von der Graeca Halensis 1913, in: Wochenschr. f. klass. Philologie, Sp. 1305—1312, 1340 bis 1347.

[Rez.:] O. Gradenwitz, F. Preisigke, W. Spiegelberg: Ein Erbstreit aus dem ptolemäischen Ägypten, 1912, in: DLZ, Sp. 119—123.

[Rez.:] Friedrich Nietzsche, Philologica, Bd II, hrsg. von O. Crusius, 1912, in: Lit. Zentralbl., Sp. 712—714.

1914

[Rez.:] A. Steiner, Der Fiskus der Ptolemäer. I: Seine Spezialbeamten u. sein öffentlichrechtlicher Charakter, 1913, in: DLZ, Sp. 1713—1715.

[Rez.:] E. Majer-Leonhard, *'Αγράμματοι*, 1913, in: Wochenschr. f. klass. Philol., Sp. 998 bis 999.

1917

[Rez.:] F. Paulus, Prosopographie der Beamten des *'Αρσινοΐτης νόμος*. Diss. Greifswald, 1914, in: Wochenschr. f. klass. Philol., Sp. 1100—1103.

[Rez.:] C. Wessely, Aus der Welt der Papyri, 1914, ebda Sp. 952—955.

1919

[Rez.:] V. Gardthausen, Die griechischen Handzeichen, 1918, in: Wochenschr. f. klass. Philol., Sp. 101—102.

[Rez.:] W. Schubart, Einführung in die Papyruskunde, 1918, ebda 289—294, 313—323.

1923

[Rez.:] M. Rostovtzeff, A large estate in Egypt in the third century B. C., 1922, in: HZ *129*, 69—78.

1924

[Rez.:] W. Schubart, Ägypten. Von Alexander d. Gr. bis auf Mohammed, 1922, in: DLZ, Sp. 434–438.

1925

[Rez.:] De Engstroemii CLE n° 306, in: Eranos *23*, 206–207.

[Rez.:] B. Olsson, Papyrusbriefe aus der frühesten Römerzeit, 1925, in: DLZ, Sp. 1800 bis 1802.

1926

Zusammen mit F. Schneider: Jenaer Papyrusurkunden u. spätmittelalterliche Urkunden nebst den ersten Universitätsordnungen und Statuten vom Jahre 1548. Beitr. Zucker S. 1–6. Jena u. Greiz. 23 S. (= Veröff. des Thüringischen Staatsarchivs 6).

Zusammen mit P. Viereck: Ägyptische Urkunden aus den Staatlichen Museen zu Berlin. Griechische Urkunden, Bd VII: Papyri, Ostraka und Wachstafeln aus Philadelphia im Fayûm. Berlin, 276 S.

1927

Zur Textherstellung und Erklärung von Philodems V. Buch *περὶ ποιημάτων* (mit einem Exkurs über *εἰρησιλογεῖν, εἰρησιλογία, εἰρησίλογος*). In: Philologus *82*, 241–267.

Griechische Urkunde oberägyptischer Herkunft aus einem Erbstreit v. J. 226 v. Chr. In: Festschr. A. Cartellieri zum 60. Geburtstag, 168–180.

1928

Syneidesis — Conscientia. Ein Versuch zur Geschichte des sittlichen Bewußtseins im griechischen und im griechisch-römischen Altertum. Rekoratsrede 1928, 26 S. (Jenaer akademische Reden, Heft 6).

[Rez.:] Aristophanes, Lysistrata, erklärt von U. von Wilamowitz-Moellendorff, 1927, in: DLZ, Sp. 1059–1064.

[Rez.:] A. Calderini, *Θησαυροί*, 1924; P. Collomp, Recherches sur la chancellerie et la diplomatique des Lagides, 1926, in: Gnomon *4*, 374–384.

1929

Plinius epist. VIII 24 — ein Denkmal antiker Humanität. In: Philologus *84*, 209–232.

Zur Landeskunde Ägyptens aus griechischen und römischen Quellen. In: Festschrift W. Judeich zum 70. Geburtstag, 131–141.

[Rez.:] P. Collart, Les papyrus Bouriant, 1926, in: DLZ, Sp. 800–805.

1929 „Zum Geleit": Thüringens Höhere Schule, Zeitschr. d. Thüringer Philologen-Verbandes V (1929) Nr. 4/5, abgedruckt unter dem Titel „Was Not tut" in: Mitteil. d. Deutsch. Altphilol.-Verbandes *III*, 23–26.

1930

Über Sprache und Stil frühbyzantinischer Urkunden. In: Byz. Zeitschr. *30*, 146–155 (Festschr. A. Heisenberg zum 60. Geburtstag).

1931

Γυμνασίαρχος κώμης. In: Aegyptus *11*, 485–496.

[Rez.:] Die griechischen Zauberpapyri, hrsg. u. übersetzt von K. Preisendanz, Bd I, 1928, in: Byz. Zeitschr. 31, 355–363.

[Rez.:] Papyri Jandanae, fasc. 5: J. Sprey, Literarische Stücke u. Verwandtes, 1931, in: Gnomon *7*, 509–510.

1932

[Rez.:] Papyri russischer und georgischer Sammlungen, III: Spätrömische und byzantinische Texte, bearb. von G. Zereteli und P. Jernstedt, 1930, in: Byz. Zeitschr. *32*, 85–89.

[Rez.:] K. Ohly, Stichometrische Untersuchungen, 1928, in: Gnomon *8*, 383–388.

[Rez.:] H. Glitsch, De Ptolemaei et Apollonii, Glauciae filiorum, chartis quaestiones linguisticae, 1929, ebda 577–584.

1933

Zur Textherstellung und Erklärung von Pap. Enteux. II. In: Aegyptus *13* (Festschr. U. Wilcken zum 70. Geburtstag), 213–224.

Klassisches Altertum und deutsche Bildung. Vortrag gehalten am 18. Juli 1933 vor der philol. Fachschaft der Univ. Jena. Jena 1934. 22 S. (= Student und Leben. NF H. 1).

[Rez.:] R. Maschke, Die Willenslehre im griechischen Recht, 1926, in: Gnomon *9*, 191–201.

[Rez.:] Mitteilungen a. d. Papyrussammlung d. Nationalbibliothek in Wien. 1: Griechische literarische Papyri I, bearb. von H. Gerstinger, H. Oellacher, K. Vogel, 1932: Papyri Osloenses, fasc. 2, ed. by S. Eitrem and L. Amundsen, 1931, ebda 648–657.

[Rez.:] E. R. Hardy, The large estates of Byzantine Egypt, 1931, in: J. of Egyptian Archaelogy *19*, 99–103.

1935

[Rez.:] T. Kalén, Berliner Leihgabe griechischer Papyri, 1932, in: Gnomon *11*, 368–374.

[Rez.:] G. Rosenberger, Griechische Privathilfe, 1934, in: DLZ, Sp. 320–322.

1936

Neues aus dem hellenistisch-römischen Aegypten. In: Geistige Arbeit *3*, Nr. 5, 3–5.

[Rez.:] F. Solmsen, Antiphonstudien, 1931, in: Gnomon *12*, 442–444 und in: Lysias, Isaios, Aischines. Darmstadt 1976 (= Wege der Forschung, Bd CXXVII).

[Rez.:] W. Schubart u. D. Schäfer, Spätptolemäische Papyri aus amtl. Büros des Herakleopolites 1933, ebda 476–485.

[Rez.:] Leiv. Amundsen, Ostraca Osloensia, 1933; Greek Ostraca in the University of Michigan Collection, 1935, ebda 668–670.

[Rez.:] K. Müller, Die Epigramme des Antiphilos von Byzanz, 1935, in: Geistige Arbeit *3*, Nr. 15 u. 16.

[Rez.:] Die griechischen Zauberpapyri, hrsg. u. übersetzt von K. Preisendanz. Bd II, 1931, ebda 399–403.

[Rez.:] E. Seidl, Der Eid im römisch-ägyptischen Provinzialrecht, 2. Teil, 1935, in: Byz. Zeitschr. *36*, 449–451.

1937

Antiphilos von Byzanz Epigr. 49 M samt anderen Nachträgen zu Karl Müllers Antiphilos-Ausgabe. In: Philologus *91*, 361–372.

Zu Papyri Osloenses III: Symbolae Osloenses *17*, 54–56.

Πέρσαι und *Πέρσαι τῆς ἐπιγονῆς*. In: Realencyclopädie der Klass. Altertumswissenschaft *XIX*, Sp. 910–926.

[Rez.:] J. D. Denniston, The Greek particles, 1934, in: Gnomon *13*, 65–73.

1938

Doppelinschrift spätptolemäischer Zeit aus der Garnison von Hermopolis Magna. In: Abhandlungen d. Preuß. Akad. d. Wissensch., Jg. 1937, phil.-hist. Kl. Nr. 6. Berlin 1938, 63 S.

[Rez.:] Steuerlisten römischer Zeit aus Theadelphia. Bearb. von H. Kortenbeutel, 1937; Tax rolls from Karanis in two volumes. Part I: Text. Ed. by H. C. Yontie with the assistance of V. B. Schuman and O. M. Pearl, 1936, in: Gnomon *14*, 377–389.

Βάθος ἐλευθέριον. In: Philologus *193* (Festgabe für Ed. Schwartz zum 80. Geburtstag) 31–60.

Nachträge zur „Doppelinschrift spätptolemäischer Zeit aus der Garnison von Hermopolis Magna". In: Aegyptus *18*, 279–284.

1939

Germanen im römischen und byzantinischen Ägypten. In: Geistige Arbeit *6*, Nr. 13, 3–6.

[Rez.:] C. Clemen, Lukians Schrift über die syrische Göttin, 1938, in: Orientalia *8*, 382–385.

[Rez.:] Victor Antoun Girgis, Prosopografia e Aphroditopolis, 1938, in: Gnomon *15*, 498–500.

1940

[Rez.:] J. Walz, Der lysianische Epitaphios, 1936, in: Gnomon *16*, 268–281 und in: Lysias, Isaios, Aischines. Darmstadt 1976 (= Wege der Forschung, Bd CXXVII).

[Rez.:] Mitteilungen aus der Papyrussammlung d. Nationalbibliothek in Wien. Griech. Literar. Papyri, bearb. von H. Oellacher, 1939, in: DLZ, Sp. 327–331.

[Rez.:] Joh. Hummel, Griech. Wirtschaftsrechnungen und Verwandtes, 1938, in: Gnomon *16*, 425–426.

[Rez.:] Isocrate, Discours. Tome 2. Par G. Mathieu et É. Brémond, 1938, ebda 502–507.

1941

ΕΙΔΟΣ und *ΕΙΔΥΛΛΙΟΝ*. In: Hermes *76*, S. 382–392.

[Rez.:] A. Svensson, Die Wiener Handschrift zu Xenophons Anabasis, 1940, in: Gnomon *17*, 286–287.

1942

Die Bevölkerungsverhältnisse Ägyptens in hellenistisch-römischer Zeit. In: Das neue Bild der Antike *I*, 369–388.

[Rez.:] Hans Wersdörfer, die *φιλοσοφία* des Isokrates im Spiegel seiner Terminologie, 1940, in: Gnomon *18*, 12–18.

[Rez.:] Joh. de Vries, Untersuchungen über Sperrung von Substantiv und Attribut in der Sprache der attischen Redner, 1938, ebda 148–155.

1943

Semitische Namen auf den neugefundenen Inschriftstelen von Minturnae. In: Hermes *78*, 200—204.

Plutarch, Perikles 12,6 — ein Zeugnis für die Organisation der Arbeiten an den Parthenonskulpturen? In: Jahrbuch d. Deutschen Archäol. Inst. *58*, Arch. Anz. 106—110.

Der Namenbestand der Familien *τῶν ἀπὸ γυμνασίου* in den Epikrisis- und Ephebenurkunden des römischen Ägyptens. Festschrift der Universität Pisa zur 100. Wiederkehr des Todestages von Ippolito Rosellini.

1944

Socia unanimans, in: Rheinisches Museum *92* (1943/44) 193—217.

Beeidete Übernahme der Vormundschaft über eine Minderjährige, in: Festschr. f. L. Wenger II, 149—167 (= Münchener Beiträge z. Papyrusforschung u. Rechtsgeschichte, H. 35).

[Rez.]: Sophokles, Antigone, übersetzt von K. Reinhardt, o. J., in: DLZ, Sp. 294—296.

1947—1948

Politische Publizistik im griech. Altertum. In. Forum *1*, H. 2. 15—17; H. 3, 5—7.

[Rez.:] G. Ottervik, Koordination inkonzinner Gleder in der attischen Prosa, 1943, in: DLZ, Sp. 53—57

1949

Πάροχος (parochus), in: Realencyclopädie der klassischen Altertumswissenschaft *XVIII*, Sp. 1670—1684.

[Rez.:] Elio Aristide, In gloria di Roma, con introduzione, traduzione e commento di L. A. Stella, 1940, in: Gnomon *21*, 57—60.

[Rez.:] Carl Wendel, Die griechische und römische Buchbeschreibung verglichen mit der des vorderen Orients (1949) = Hallische Monografien, hrsg. v. O. Eißfeld, Nr. 3, in: Zentralbl. f. Bibliothekswesen *63*, 472—476.

[Hrsg.] Otto, Walter: Beiträge zur Hierodulie im hellenistischen Ägypten. München 1949. 74 S. (= Abh. Bayr. Akad. Wiss., philos.-hist. Kl. NF, H. 29).

1950

Plotin und Lykopolis. In: Sitzungsber. d. Deutsch. Akad. d. Wiss., Kl. f. Sprachen etc., H. 1, 20 S.

Freundschaftsbewährung in der neuen attischen Komödie — ein Kapitel hellenistischer Ethik und Humanität. Berlin 1950. 38 S. (= Sitzungsber. Sächs. Akad. Wiss., phil.-hist. Kl. Bd 98, H. 1).

Das Menschenbild im antiken Griechentum, in: Schriften der Evangelischen Forschungsakademie Ilsenburg. Berlin-Spandau 1950. 27 S. (= Erkenntnis u. Glaube, H. 3).

Athen und Ägypten bis zum Beginn der hellenistischen Zeit. In: Aus Antike und Orient, 146—165 = Festschrift Wilhelm Schubart.

Phoker oder Phokäer im Elymerland? In: Würzburger Jahrbücher *4*, 335—338.

[Rez.:] Henrik Zilliacus, Vierzehn Berliner griechische Papyri. Urkunden und Briefe,

Helsingfors 1941, derselbe, Zur Sprache griechischer Familienbriefe des 3. Jh. n. Chr. Helsingfors 1943, in: Gnomon *22*, 143–150.

[Rez.:] Theocriti carmina, ed. K. Latte, 1948, in: DLZ *71*, Sp. 535–538

1952

Studien zur Namenkunde vorhellenistischer und hellenistischer Zeit. In: Sitzungsber. d. Deutsch. Akad. d. Wiss., Kl. f. Sprachen etc., Jg. 1951, H. 1, 32 S.

Simaristos, in: Rhein. Museum *95*, 338–342.

Papyri Jenenses, in: Aegyptus *32*, 73–84.

Aus einer Urkundenfolge hadrianisch-antoninischer Zeit, in: Wissenschaftl. Z. Univ. Jena *2* (1952/53) 45–51.

[Rez.:] Papiri greci e latini, vol. XII, fasc. 2 (p. 97–282) a cura di V. Bartoletti, Firenze 1951, in: Gnomon *24*, 238–240.

1953

'Ανηθοποίητος. Eine semasiologische Untersuchung aus der antiken Rhetorik und Ethik. In: Sitzungsber. d. Deutsch. Akad. d. Wiss., Kl. f. Sprachen etc., Jg. 1952, H. 4, 25 S.

Bedeutung und Stand der griechisch-römischen Papyrologie. In: Wissenschaftl. Annalen *2*, 253–270.

Selbstbehauptung und Versagen des Griechentums in Ägypten bis zum Ende der Ptolemäerzeit. In: Gymnasium *60*, 7–20.

Ein angebliches Zeugnis für eine syrisch-hellenistische Götterdreiheit. In: Archiv f. Papyrusforschung *15*, 61–70.

Nachruf auf Ulrich Wilcken. Ebda 1–3.

Lydda-Lunda. In: Rhein. Museum *96*, 96.

Verbundenheit von Erkenntnis und Wille im griechischen Sprachbewußtsein, beleuchtet durch Erscheinungen aus der Bedeutungsentwicklung von *ἄγνοια, ἀγνοεῖν, ἀγνόημα*. Studies presented to D. M. Robinson *II*, 1063–1072.

[Rez.:] H. C. Yontie und J. G. Winter, Papyri and Ostraca from Karanis. Second Series = Michigan Papyri vol. VIII (Ann Arbor 1951), in: DLZ *74*, 339–342.

[Rez.:] M. Hombert et Claire Préaux, Recherches sur le recensement dans l'Egypte romaine, 1952, In: Gnomon *25*. 462–463

1954

Grabgedicht aus Herakleopolis. In: J. of Egyptian Archaelogy *40*, 118–123. = Festschr. f. Sir Harold J. Bell.

Isokrates' Panathenaikos. Berlin 1954. 31 S. (= Sitzungsber. Sächs. Akad. Wiss., phil.-hist. Kl. Bd 101, H. 7) und in: Isokrates. Darmstadt 1975. S. 227–252 (= Wege der Forschung, Bd CCCLI).

Zu hellenistischen Dichtern. In: Philologus *98*, 94–100.

In memoriam Johannes Stroux. In: FuF *28*, 318–319.

[Anz.:] L'originalité de l'Égypte dans le monde grécoromain = Museum Helveticum 10 (1953) fasc. 3/4, in: Gnomon *26*, 341–343.

[Rez.:] Papiri greci e latini, vol. XIII, a cura di M. Norsa e V. Bartoletti, Nr. 1296–1370, Firenze 1953, ebda 391–394.

1955

Formen gesteigert affektischer Rede in Sprechversen der Tragödie. In: Indogermanische Forschungen *62*, 62–77 (= Festschr. z. 70. Geburtstag von Ferdinand Sommer 1944) und in: Sophokles Darmstadt 1967. S. 252–267 (= Wege der Forschung, Bd XCV).

Die Bedeutung Pindars für Goethes Leben und Dichtung. In: Das Altertum *1*, 171–186.

[Rez.:] Démosthène: Plaidoyers politiques, t. I, par O. Navarre et P. Orsini, Paris 1954; Démosthène: Plaidoyers civils, t. I, par L. Gernet, Paris 1954, in: Gnomon *27*, 275–279.

1956

Der Stil des Gorgias nach seiner inneren Form. Mit Erörterungen über vorisokratische Periodik. In: Sitzungsber. d. Deutsch. Akad. d. Wiss., Kl. f. Sprachen etc., Jg. 1956, Nr. 1, 19 S.

Literatur und literarische Papyruskunde. In: Der Hellenismus in der deutschen Forschung 1938–1948, Wiesbaden 1956, S. 1–9.

Priester und Tempel in Ägypten in den Zeiten nach der decianischen Christenverfolgung. In: Akten des VIII. Internationalen Kongresses für Papyrologie Wien 1955, 167–174.

[Anz.:] Antifonte. Prima orazione. Sesta orazione. Con introduzione e note di A. Barigazzi, Firenze 1955, in: Gnomon *28*, 555.

[Hrsg.:] Archiv für Papyrusforschung *16*, S. 1–142.

1957

Verantwortung in Sprache und Denken der Griechen und Römer. In: Das Institut f. griech.-röm. Altertumskunde = Deutsche Akad. d. Wiss., Schriften d. Sekt. f. Altertumswiss. *8*, 54–70.

Versuch unzulässiger Überführung eines Handwerkers in ein anders Handwerk. In: Studi in onore di A. Calderini e R. Paribeni *II*, 339–343.

Personennamen in J. Scherers „Papyrus de Philadelphie". In: Symbolae Rafaeli Taubenschlag dedicatae *II*, 171–174 (= Eos *48*, 2).

ΣΙΜΑΡΙΣΤ⟨ΕΙ⟩ΟΙ. In: Philologus *101*, 164–166.

Wirtschaftliches aus einer ptolomaischen Reitertruppe. In: Univ. Z. Jena.

Nachruf auf Johannes Stroux. In: Jahrb. Deutsche Akad. Wiss. *1954* (1957) 388–392.

Nachruf auf Leopold Wenger. Ebda 405–407.

Nachruf auf Michael Rostovtzeff und Gunnar Rudberg. Ebda *1955* (1957) 291–294, 295 bis 296.

[Rez.:] John Day u. Clinton Walker Keyes, Tax Documents from Theadelphia, 1956, in: Göttingische Gelehrte Anzeigen *211*, 59–72.

[Rez.:] A. Lesky, Geschichte der griechischen Literatur, Lfg 1, in: DLZ *78*, Sp. 784–787.

1958

Ägypten im römischen Reich. Berlin 1958. 23 S. (= Sitzungsber. Sächs. Akad. Wiss., phil. hist. Kl. Bd 104, H. 7).

[Rez.:] A. Lesky, Geschichte der griechischen Literatur, Lfg 2–3, 4–7, in: DLZ *79*, Sp. 301 bis 304, 604–609.

[Rez.:] U. Wilcken, Urkunden der Ptolomäerzeit II, (Schluß-) Lfg 3, 1957, in: Gnomon *30*, 114–119.
[Rez.:] Démosthène, Plaidoyers civils, tome 2, par Louis Gernet, 1957, ebda, 550–552.
Urkundenreferat, in: Archiv f. Papyrusforschung *16*, 214–267.
[Hrsg.:] Archiv f. Papyrusforschung *16*, 143–267.

1959

Kallimachos ep. 20. In: Maia N. S. *11*, 87–88.
Der Name Thearistis. Ebda 162.
Steuerlisten des 2. Jahrhunderts n. Chr. aus Theadelphia und Nachbardörfern. In: Philologische Vorträge, hrsg. v. J. Irmscher u. W. Steffen, Wrocław, S. 79–83.
[Rez.:] A. Lesky, Geschichte der griechischen Literatur, Lfg 8–9, in: DLZ *80*, Sp. 103 bis 107.

1960

Ein neugefundenes griechisches Drama. Berlin 1960. 21 S. (= Sitzungsber. Dtsch. Akad. Wiss., Kl. f. Sprachen etc., Nr. 5).
[Rez.:] Urkundenreferat über Vittorio Bartoletti, Papiri greci e Latini XIV, Firenze 1957. In: Archiv f. Papyrusforschung *17*, 110–114.
[Rez.:] Démosthène, Plaidoyers politiques, tome 2: Contre Midias. Texte et trad. par J. Humbert et L. Gernet, in: Gnomon *32*, 606–608.
[Hrsg.:] Archiv f. Papyrusforschung *17*, 1–114.

1961

Nachruf auf Karl Reinhardt. In: Jahrb. Dtsch. Akad. Wiss. *1960* (Berlin 1961) 126–127.
Nachruf auf Richard Harder. Ebda 124–125.
Nachruf auf Wilhelm Schubart. In: Gnomon *33*, 108–109.
Verfahrensweisen in der Einführung gewisser Einrichtungen des Augustus in Ägypten. In: Revue Internationale des Droits de l'Antiquité 3ᵉ sér. *8*, 155–164.
[Rez.:] Démosthène, Plaidoyers civils, tome III: Discours 49–56, tome 4: Discours 57–59. Texte et trad. par L. Gernet. Paris 1959 u. 1960, in: Gnomon *33*, 132–134, 768–771
[Hrsg. m. B. Altaner u. a.] Studien zum Neuen Testament und zur Patristik. Erich Klostermann zum 90. Geburtstag dargebracht. Berlin 1961. 378 S. (= Texte u. Untersuchungen zur Geschichte der altchristlichen Literatur, Bd 77).

1962

Αὐθέντης und Ableitungen. Berlin 1962. 26 S. (= Sitzungsber. Sächs. Akad. Wiss., phil.-hist. Kl. Bd 107, H. 4).
Urkundenreferat, in: Archiv f. Papyrusforschung *17*, 211–262.
Zum Gedenken an Ulrich Wilcken. In: FuF *36*, 378–379.
Nachruf auf Wilhelm Schubart. In: Jahrb. Deutsche Akad. Wiss. *1961* (Berlin 1962) 887 bis 888.
[Rez.:] Andocide, L'orazione De reditu. Introduzione e commento a cura di U. Albini, in: Atene e Roma N. S. *6*, 183–187.
[Hrsg.:] Archiv f. Papyrusforschung *17*, 115–264.

1963

Semantica, Rhetorica, Ethica. Berlin 1963, 196 S. (= Deutsch. Akad. d. Wiss., Schriften d. Sekt. f. Altertumswiss. 38).

Sprachliche Bemerkungen zu griechischen Grabepigrammen. In: Maia N. S. *15*, 395–396.

Der jüngere Plinius und die Familie des Konsulars Q. Corellius Rufus. In: Das Altertum *9*, 37–43.

Worte des Gedenkens an Erich Bethe. In: FuF *37*, 124–125.

Karl Barwick zum 80. Geburtstag. Ebda 158–159.

1964

Beobachtungen an den permanenten Klerosnamen. In: Festschrift F. Oertel. Bonn, 101 bis 106.

M. Tulli Ciceronis pro L. Flacco oratio. Fridericus Zucker recognovit. In: M. Tulli Ciceronis opera omnia quae exstant critico apparatu instructa consilio et auctoritate Collegi Ciceronianis studiis promovendis sumptibus Arnoldi Mondadori. Romae 1964. 84 S.

Papyri der Thebais: Bestand und Wert. In: Actes du Xe Congrès International de Papyrologues Varsovie-Cracovie 1961. Wrocław–Varsovie–Cracovie 1964, 161–163.

[Rez :] A. Lesky, Geschichte der griechischen Literatur. 2., neubearb. u. erw. Aufl. Bern, München 1963, in: DLZ *85*, Sp. 614–616.

1965

Menanders Dyskolos in seiner Umwelt. In: Menanders Dyskolos als Zeugnis seiner Epoche. Berlin, 7–10.

[Rez.:] Paul Cloché, Isocrate et son temps, Paris 1963, in: Anzeiger f. Altertumswiss. *18*, Sp. 153–159.

[Rez.:] G. Pasquali, Filologia e Storia. Nuova edizione con una premessa di Alessandro Ronconi, Firenze 1964 (= Biblioteca del saggiatore 2), in: DLZ *86*, Sp. 411–413.

[Hrsg.:] Menanders Dyskolos als Zeugnis seiner Epoche. Berlin, 254 S.

1966

Materialsammlungen, Hilfsmittel, Abhandlungen. In: Archiv f. Papyrusforschung *18*, 57 bis 67.

Urkundenreferat. Ebda 68–123.

Albin Lesky zum 70. Geburtstag am 7. Juli 1966. In: FuF *40*, 222–223.

Otto Regenbogen zum 75. Geburtstag. Ebda 61–62.

Nachruf auf Vincenzo Arangio-Ruiz. In: Archiv f. Papyrusforschung *18*, 123.

Nachruf auf André Bataille. Ebda.

Nachruf auf Victor Martin. Ebda.

[Hrsg.:] Archiv f. Papyrusforschung *18*, 123 S.

1967

Nachruf auf Karl Barwick. In: Jahrb. Sächs. Akad. Wiss. *1963–1965*, 338–339.

[Rez.:] H. Braunert, Die Binnenwanderung, in: Historia *16*, H. 3, 377–384.

1969

[Hrsg.:] Ulrich von Wilamowitz-Moellendorff: Kleine Schriften. Bd 3: Griechische Prosa. Berlin 1969. VII, 523 S.

Biobibliographie Friedrich Zucker

Schriftenverzeichnisse. Siehe vorliegendes Jahrbuch S. 444—454. — Jahrbuch Sächsische Akademie der Wissenschaften zu Leipzig *1954—1956* (Berlin 1958) S. 206; *1960—1962* (Berlin 1964) S. 325; *1963—1965* (Berlin 1967) S. 265; *1966—1968* (Berlin 1970) S. 270. — Kürschners Deutscher Gelehrten-Kalender *1* (Berlin—Leipzig 1925) Sp. 1187; *2* (Berlin—Leipzig 1926) Sp. 2243—2244; *3* (Berlin—Leipzig 1928/29) Sp. 2775; *4* (Berlin—Leipzig 1931) Sp. 3410; 6_2 (Berlin 1940/41) Sp. 1171; *7* (Berlin 1950) Sp. 2354; *8* (Berlin 1954) Sp. 2684; *9* (Berlin 1961) S. 2357; 10_2 (Berlin 1966) S. 2799; *11* (Berlin 1970) S. 3405—3406.

Festschrift. Friedrich Zucker zum 70. Geburtstag. Gesamtredaktion W. Müller. Berlin 1954. 389 S., mit Bildnis.

Würdigungen. Zum 80. Geburtstag: [F. Kühnert], in: Sozialistische Universität, Jena 1961, 4. Juli. — A. Lesky, in: FuF *35* (1961) S. 187—188. — W. Müller: Ethik und Humanität der Antike. In: Neue Zeit, Berlin 1961, Nr. 150 vom 1. Juli. — Neues Deutschland *16* (Berlin 1961) Nr. 178 vom 30. Juni. — Volkswacht vom 30. Juni 1961. — Thüringische Landes-Ztg vom 1. Juli 1961. — Zum 90. Geburtstag: [F. Uebel], in: Archiv für Papyrusforschung *21* (1971) S. 3. — Zum Goldenen Doktor-Jubiläum: W. Müller, in: Hochschulwesen *2* (1954) H. 8/9, S. 85—86. — Allgemein: Die Mitglieder der Deutschen Akademie der Wissenschaften zu Berlin 1700—1950. Bearbeitet von E. Amburger. Berlin 1950. S. 49, 52. — Wiss. Annalen *2* (1953) H. 4, S. 272. — Degeners Wer ist's? *8.* Ausg. (Leipzig 1922) S. 1757.

Nachrufe. W. Peek, siehe vorliegendes Jahrbuch S. 442—444. — E. G. Schmidt, in: Sozialistische Universität, Jena 1974, Nr. 10 vom 4. Juni. — Derselbe, in: Gnomon, München 1977 [in Vorbereitung]. — [F. Uebel], in: Archiv für Papyrusforschung *22/23* (1974) 2 S. — Jahrbuch Akademie der Wissenschaften der DDR *1973* (Berlin 1974) S. 53.

G. W.

Albert Egges van Giffen

14. 3. 1884—31. 5. 1973

Am 31. Mai 1973 verstarb in Zwolle, Niederlande, das auswärtige Mitglied der Sächsischen Akademie der Wissenschaften zu Leipzig, der emeritierte Professor für Prä- und Protohistorie und Direktor a. D. des Biologisch-Archäologischen Instituts der Rijksunversiteit in Groningen, der emeritierte Professor und Direktor a. D. des Instituts für Prä- und Protohistorie der Universität Amsterdam, der ehemalige Direktor des Rijksdienst voor het Oudheidkundig Bodemonderzoek Amersfoort und nachmalige Rijksadviseur voor de Bescherming en der Instandhouding van de Hunnebedden en van de gerestaureerde archaeologische Monumenten, Albert Egges van Giffen im 90. Lebensjahr. 65 Jahre lang hat er seinem Fach mit vorbildlicher Aktivität gedient, vor allem für die niederländische Archäologie viele wesentliche und neue Grundlagen geschaffen und sich damit in der gesamten Fachwelt einen überall beachteten Namen gemacht.

Van Giffen wurde am 14. März 1884 in Nordhorn, Prov. Groningen geboren, in Friesland, Groningen und Drenthe wuchs er auf. In Groningen oblag er auch den Studien der Biologie und erhielt bereits als Student 1908 den Auftrag, sich an den Wurtenabtragungen in Dorkwerd in der Nähe von Groningen zu beteiligen. Damit kam er erstmals mit archäologischen Forschungen in praktischen Kontakt und konnte bei den dort durchgeführten interdisziplinären Arbeiten die vielseitigsten wissenschaftlichen Beobachtungen und Erkenntnisse sammeln, sich dabei aber auch in den reichen Diskussionen Methoden und Ergebnisse der einzelnen Wissenschaftszweige aneignen. Es handelte sich dabei vor allem um zoologische, geologische und archäologische, dann aber auch botanische Spezialuntersuchungen im Wurten-(Terpen-)Gebiet Frieslands. Die im wesentlichen neue Art des „Teamwork“ sollte für alle seine kommenden Unternehmungen richtungsweisend werden, und die Kombination verschiedenster Wissensgebiete, vor allem aber auch der Naturwissenschaften, blieb für seine folgenden Unternehmungen kennzeichnend. 1913 promovierte van Giffen mit einer Arbeit über die Fauna der Terpen. Die systematische Abtragung der Wurten erfolgte zur Materialgewinnung für die Bodenverbesserung und bedeutete eine allmähliche und teilweise fast systematische Vernichtung der mit den archäologischen Zeugnissen der ältesten Kultur des Küstenlandes angereicherten Erdschichten. Um den Verlust der wichtigen wissenschaftlichen Erkenntnisquellen möglichst zu verhindern, scharte das Centraal Bureau voor de kennis van der provincie Groningen en omgelegen streken Wissenschaftler um sich und gründete ein Komitee zur intensiven Erforschung der Geschichte und Geografie der Provinz Groningen. Dieses schon lange bestehende Zentralbüro schuf finanzielle Möglichkeiten, um eine Arbeitsgruppe aus einem Geologen, einem Zoologen, einem Historiker, einem klassischen und prähistorischen Archäologen zu bilden und bei ihren Untersuchungen zu unterstützen.

Ab 1912 war van Giffen am Staatlichen Museum für Altertümer in Leiden tätig, kehrte aber schon 1916 zur Facharbeit an die Groninger Universität zurück und fand große Unterstützung durch die neugegründete Vereinigung zur Terpenforschung und die Prähistorische Gesellschaft von Drenthe. 1917 wurde er Konservator des Museums von Oudheden voor de Province en Stad Groningen und des Provincial-Museums von Oudheden in Drenthe und 1920 zum Direktor des neuen Biologisch-Archäologischen Instituts der Universität Groningen berufen. Nun begann eine neue gesicherte Etappe systematischer archäologischer Untersuchungen besonders in den nördlichen Provinzen mit vielen Ausgrabungen auch in Verbindung mit den entsprechenden Lokalmuseen, aber auch Feldforschungen in Irland, Ungarn, Norddeutschland und Frankreich. Durch seine fortgeschrittenen Grabungsmethoden setzte er schon in den zwanziger und dreißiger Jahren europäische Maßstäbe. Durch die den jeweiligen Objekten angepaßten Kombinationen von Vertikal- und Horizontalschnitten gelang die dokumentarische Sicherung des dreidimensionalen Aufbaus der alten Siedlungsstätten und Gräber (hier besonders durch die Quadrantenmethode). Die Wurten ließen darüber hinaus durch seine konsequent angewandte Methode eine sich über Jahrhunderte erstreckende ununterbrochene Siedlungsabfolge erkennen mit allen Einzelheiten des Hausbaus und anderer wichtiger Siedlungshinterlassenschaften. Für seine Arbeiten bedurfte es überdurchschnittlich erfahrener und exakt arbeitender Grabungstechniker, die er sich selbst mit heranzog. Auch hier leistete der Verstorbene im besten Sinne des Wortes Pionierarbeit. Neben guter Organisation bedurfte es ja eines großen Stabes geschulten technischen Personals, um die oft über 100 Arbeitskräfte ohne Erkenntnisverluste für die Forschung rationell einzusetzen.

Van Giffens erfolgreiche und energische Maßnahmen bei der Prüfung und Anwendung neuer Methoden waren fast sprichwörtlich für den allem Neuen gegenüber äußerst aufgeschlossenen Forscher. Das trifft für die Pollenanalyse ebenso zu wie etwa die Dendrochronologie oder die C_{14}-Methode. Als tüchtiger Organisator verstand er es, alle als richtig erkannten Möglichkeiten zur Förderung der Forschungen zu nutzen und in kurzer Zeit entsprechende Arbeitsstellen in die Untersuchungen einzubeziehen und auszubauen oder gar neue zu schaffen, wie eines der ersten C_{14}-Laboratorien der Welt mit dem bekannten Spezialisten Hessel de Vries, der leider schon 1959 verstarb. Alle interdisziplinären Bemühungen brachten der Urgeschichtsforschung Hilfsmittel von unschätzbarem Wert und die Möglichkeit zur vielseitigen Absicherung der Forschungsergebnisse.

In seinen Publikationen ab 1910 berührte er zunächst geologische und zoologische Probleme, ging dann zur archäologischen Feldforschung über und beachtete vor allem die Fragen von Umwelt und Lebensweise in den vergangenen Jahrtausenden unter steigender Hervorhebung wirtschaftlicher und gesellschaftlicher Zusammenhänge. So ist es nicht verwunderlich, daß er die Entwicklung der Landwirtschaft ebenso erforschte wie etwa römische Kastelle, mittelalterliche Kirchen und Klöster sowie Stadtkerne. Seine Aufmerksamkeit galt gleichfalls den Hünenbetten wie der Sachsenforschung. Seine bekanntesten Publikationen betreffen die Wurtenforschung — an erster Stelle seine weltbekannten Untersuchungen in Ezinge — und die Bauart der Einzelgräber. Bei allen Arbeiten wurden die verschiedensten Wissenschaftszweige miteinander in Beziehung gebracht, wie seine interdisziplinäre Aktivität wohl überhaupt das hervorstechendste Merkmal seiner gesamten wissenschaftlichen Tätigkeit gewesen sein dürfte.

Als Lehrer kamen ihm seine großen organisatorischen Fähigkeiten, sein umfassendes Wissen und seine vielseitigen Erfahrungen zugute. Als echte Forscherpersönlichkeit verband sich in ihm die Theorie mit weitreichendster Praxis. Er verlangte auch von seinen Schülern von Anfang an die Nutzanwendung alles Gelernten, das Einpassen in kollektive Arbeit und war in der Vermittlung von Methoden, Stoff und Erfahrungen im Gelände am wirksamsten. Hinzu kam sein ansteckender Enthusiasmus und ein trotz aller wissenschaftlich bedingten Skepsis fast unbegrenzter Optimismus. Es war immer wieder erstaunlich, wie er es bei einer ungewöhnlichen Breite der Wissensskala stets verstand, bei allen Problemen noch gründlich in die Tiefe zu gehen.

Die Anerkennung seiner großen Leistungen blieb nicht aus und zeigte sich auch in der Wahl in viele Akademien und andere geachtete wissenschaftliche Gesellschaften in den Niederlanden, in Belgien, England, Schottland, Deutschland, Dänemark und Finnland sowie in der Verleihung hoher Orden und Medaillen, nicht zuletzt aber in seiner Hinzuziehung zu vielen schwierigen Ausgrabungen im In- und vor allem auch im Ausland bis ins höchste Alter. So war er ein bewährter Ausgräber, anregender und wegweisender Lehrer, erfahrener Museumsmann, Organisator der Denkmalpflege und nicht zuletzt ein außerordentlich fruchtbarer Forscher auf den verschiedensten Spezialgebieten.

Der Verstorbene hatte besonders am Anfang seiner Laufbahn aufreibende Kämpfe zu bestehen, konnte sich aber auf Grund seiner Leistungen, seiner besseren Argumente und seiner persönlichen Lauterkeit durchsetzen. Seine universal ausgerichteten Forschungen förderten die Suche nach der historischen Wahrheit. Außerordentliches Pflichtgefühl, große Begeisterungsfähigkeit sowie beispielhafte Bescheidenheit, Liebenswürdigkeit und Hilfsbereitschaft waren die Kennzeichen einer weit verehrten wissenschaftlichen Persönlichkeit, die wesentlich an der Schaffung des heutigen Geschichtsbildes beteiligt war, und der wir auch in Zukunft verpflichtet bleiben.

Werner Coblenz

Bibliographie Albert Egges van Giffen

(Abkürzungen siehe S. 563ff.)

1910

Het dalingsvraagstuk der Alluviale Noordzeekusten, in verband met bestudeering der terpen. TGLV *25*, S. 258–294.

1913

Die Fauna der Wurten. 1. Teil. Proefschrift ter verkrijging van den graad van Doctor in de Plant- en Dierkunde, aan de Rijksuniversiteit te Groningen, op gezag van den Rector-Magnificus Prof. Dr. J. H. Kern, tegen de bedenkingen van de Faculteit der Wis- en Natuurkunde in het openbaar te verdedigen op 20 Juni 1913. Leiden. (weiterhin erschienen unter: Onderzoekingen, verricht in het Zoölogisch Laboratorium der R. U. te Groningen *3*; und: TNDV [2. Serie] *13*, 1914, S. 1–167).

De Buinerbrug en het steenen voetpad aldaar. OMROL *7*, S. 51–90.

Iets over terpen en den terphond. Hand. 14e Ned. Nat. en Gen. Congres, 27., 28. und 29. März zu Delft, S. 468–481.

1915

Archaeologie en Natuurwetenschap. Hand. 15e Ned. Nat. en Gen. Congres, 8., 9. und 10. April zu Amsterdam, S. 471–483.

De statistische methode in het huisdiervraagstuk. TNDV (2. Serie) *14*, S. XLVIII–LX.

Nota n. a. v. den brief van Curatoren der Rijks-Universiteit te Leiden, 3. September, No. 561[2].

1917

Statistisch-biologische gegevens voor de palethnographie. Hand. 16e Net. Nat. en Gen. Congres, Gravenhage 12., 13. und 14. April, S. 256–258.

Voorlooping verslag von het proefonderzoek der Wierde, „de Wierhuizen" te Jukwerd bij Appingedam in 1916. JVT *1*, S. 7–31. (und: Bijdragen tot de kennis van de provincie Groningen en omgelegen streken *2* [mit deutscher Zusammenfassung von J. F. van Bemmelen], 1918, S. 390–409).

1918

Opgraving te Wierhuizen 1917; Verkort verslag. JVT *2* (1917–1918) S. 4–22.

Begin van een onderzoek van „de zoogenaamde voormalige Romeinsche legerplaats" en aangelegen grafheuvelveld te Zeijen. NDV *36*, S. 135–175.

1919

Mededeeling omtrent onderzoek en restauratie van het groote hunebed te Havelte. NDV *37*, S. 109–139.

De Drentsche voorgeschiedenis. Volksuniversiteit Assen. Cursus, 24. März, 7. und 14. April zu Assen.

1920

Hoofdstukken uit de voorgeschiedenis, in 't bijzonder van Noord-Nederland. Volksuniversiteit Groningen. Cursus 1919–1920. Groningen.

Grafheuvels uit den vroegen bronstijd bij Zeijen. (Voorlooping bericht). NDV *38*, S. 122–146.

Iets over terpen. JVT *3–4* (1918–1920) S. 7–31.

Een Karolingisch grafveld bij Godlinze. JVT *3–4* (1918–1920) S. 39–96.

Het in April 1818 ontdekte en in Otcober 1920 teruggevonden urnenveld met „kringgreptumuli" bij den „Pottenberg" onder Wollinghuizen, gem. Vlagtwedde. Verslag GM, *1* S. 33–59.

1921

Bijdrage tot de kennis van enkele geologisch-archaeologische verkenmerken in verband met het vraagstuk der bodemdaling in historischen tijd. Versl. Geol.-Mijnb. Gen. *2* (1915–1919) S. 159–220.

Hunsow. NDV *39*, S. 90–116.

Uit de Drentsche kinderwereld. NDV *39*, S. 117–120.

1922

De heuvel aan den Korrel bij 't Zandt. JVT *5–6* (1920–1922) S. 7–14.

Proefonderzoek in de terp te Hatsum bij Dronrijp (mit einem Nachtrag von P. C. J. A. Boeles). JVT *5–6* (1920–1922) S. 21–34.

Het Biologisch-Archaeologisch Instituut en zijn Taak. Rede vom 17. Juni 1922 bei der offiziellen Eröffnung des Biologisch-Archaeologisch Instituut aan de Rijksuniversiteit te Groningen. Groningen.

Een dubbele grafheuvel uit den steen- en bronstijd te Harendermolen bij Groningen. Verslag GM, S. 41–57.

Een kringgrepurnenveld bij Wessingtange, Gem. Vlachtwedde. Verslag GM, S. 59–67.

1923

Merkwaardige grafheuvels bij den Annertol, Zd.-Zd.-W. van Zuidlaren, en bronstijdgraven in Drente in het algemeen. NDV *41*, S. 156–205.

1924

Voortgezette mededeelingen op oudheidkundig gebied in Drente. De begraafplaats bij het Kamperechje tusschen Valthe en Weerdinge; Een neolithische grafheuvel met houtbouw bij de Tumulihoeve op het Zeyerveld. NDV *42*, S. 133–172.

Postscript to: H. Martin, Ontgraving van een prae-historischen tumulus te Soesterberg. Versl. Gesch. en Kunst (1923) S. 14–32.

Het hunebed te Rijs in Gaasterland. De Vrije Fries *27*, S. 307–325.

Voortgezet onderzoek in de terp te Hatsum bij Dronrijp. JVT *7–8* (1922–1924) S. 9–47.

Bericht omtrent een bezoek aan den vliedberg bij Duivendijke op Schouwen, den 16 October 1922. JVT *7–8* (1922–1924) S. 48–49.

Ein neolithischer Grabhügel mit Holzkonstruktion in Harendermolen, Gem. Haren, Prov. Groningen, Niederlande. Praehistorische Zeitschrift *15*, S. 52–61.

De oudste cultuuroverblijfselen in ons land. NTG *68*, S. 688–700.

Het verstoorde hunebed op de Eeze bij Steenwijk. Overijsselsch Regt en Gesch. *41*, S. 56 bis 71.

Excursie van de Drentsch Praehistorische Vereeniging op Zondag 5 October 1924 naar Emmen. Noord Barge enz. n. p.

1925

Een kringgrepurnenveld te Oosterwolde. OJ (3. Serie) *5*, S. 152–157.

De hunebedden in Nederland. Text: 2 Bde. Utrecht 1925–1927. Atlas: 2 Bde. Utrecht 1925.

De Zuid-Limburgische voorhistorische vuursteenindustrie tusschen Ryckholt en St. Geertruid (mit einem Anhang von W. G. N. van der Sleen). TKNAG *42*, S. 481–507.

Twee lááт-neolithische grafheuvels ten Zuiden von Zeijen, Gem. Vries. NDV *43*, S. 53–70.

De ligging der archae[o]logica in het hoogveen. Hand. 20e Ned. Nat. en Gen. Congres, 14., 15. und 17. April zu Groningen, S. 265–268.

Tweedaagsche excursie op Zaterdag 23 en Zohdag 24 Mei 1925. Toelichting: terpen, hunebedden, tumuli, urnenvelden en z. g. n. Romeinsche legerplaats. Mensch en Maatschappij *1*, S. 98–104.

Over de oudste bevolkingselementen van ons land. Mensch en Maatschappij *1*, S. 257–279.

1926

Resumé van de in de laatste vereenigingsjaren verrichte werkzaamheden ten behoeve van de Terpenvereeniging. De proefgraving te Peeloo. JVT *9–10* (1924–1926) S. 9–35.

Eene excursie over een aan praehistorische overblijfselen rijk heideveld in Oost-Drenthe, tusschen Emmen, Weerdinge, Valthe en Odoorn. NVD *44*, S. 67–98.

Mededeeling over het proefonderzoek van het Klooster Smalle Ee te Smalle Ee, aan de Smalle Ee, in Smallingerland. De Vrije Fries *28*, S. 101–110.

1927

Het Weerdinger hunebed. NDV *45*, S. 65–68.

Een rijengrafveld te Looveen bij Wijster, Gem. Beilen. NDV *45*, S. 83–122.

Het oudste huisdier en de palethnologie. Verslag van de gewone vergadering der afd. Natuurkunde van de KAW *36*, S. 1287–1297.

The hunebeds in the Netherlands. Part I. Utrecht. (Englische Übersetzung der Originalpublikation von 1925).

1928

Naschrift over de prae- en proto-historische Hillegomsche duinvondsten. In: R. Oppenheim, Zwerftochten in Oer-Nederland II, In de achterduinen (Zuid-Holland). De Levende Natuur *32*, S. 78–83.

Programme détaillé de l'excursion des membres de l'Assemblé générale triennale de l'Institut International d'Anthropologie (IIIe Sess.) à Amsterdam, du 20 au 28 Septembre 1927, faite dans: a) les environs de Groningue, le 27 Septembre; b) la province de Drenthe, le 28 Septembre. n. p.

De woerd, genaamd de „Kerkenhof" te Ressen, Gem. Ressen-Bemmel (mit M. A. Evelein). JVT *11–12* (1926–1928) S. 16–29.

Mededeeling omtrent de systematische onderzoekingen, verricht in de jaren 1926 en 1927, ten behoeve van de Terpenvereeniging, in Friesland en Groningen (mit einem Anhang von W. Beijerinck). JVT *11–12* (1926–1928) S. 30–40.

„De Parel" bij Tzum. JVT *11–12* (1926–1928) S. 49–51.

Het runenhoutje van Westeremden (mit einem Beitrag von O. von Friesen). JVT *11–12* (1926–1928) S. 52–70.

Bouwstoffen voor de Groningsche oergeschiedenis. Opgravingsresultaten in 1927: Het onderzoek bij Wessinghouizen, Gem. Onstwedde; Het onderzoek te Onnen, Gem. Haren. Verslag GM, S. 49–86.

Les Pays-Bas, considérés comme région intermédiaire entre les phénomènes culturaux néolithiques et énéolithiques du nord-ouest et du centre de l'Europe. Revue anthropologique *38*, S. 226–238.

1929

Grafheuvels in Oosterwolde. Opgravingen in 1928. De Vrije Fries *29*, S. 37–60.

Het provincial museum van oudheden en geschiedkundige voorwerpen te Assen, 8 Nov· 1854—8 Nov. 1929. Drenthe, S. 81—89.
On the oldest domesticated animal and its significance for palethnology. Proceedings KAW *32*, S. 321—329 (Englische Übersetzung der Originalpublikation von 1927).

1930

Die Ergebnisse der Warfengrabung. Archaeologisches Institut des Deutschen Reiches, Bericht über die Jahrhundertfeier 21.—25. April 1929. Berlin, S. 322—338.
De hoofdverschijnselen op het gebied der voor-Romeinsche beschaving in Nederland. Mensch en Maatschappij *6*, S. 297—304.
Archaeologische verkenmerken. Openbare voordracht, gehouden bij de aanvaarding van het ambt van lector aan de Rijks-Universiteit te Groningen, den 18den October 1930. Groningen.
Die Bauart der Einzelgräber. Beitrag zur Kenntnis der älteren individuellen Grabhügelstrukturen in den Niederlanden. 2 Teile Mannus-Bibliothek *44—45*. Leipzig.

1931

Mededeeling omtrent het systematisch onderzoek, verricht in de jaren 1928, 1929 en 1930. JVT *13—15* (1928—1931) S. 16—46.
De terpsector in Ezinge. OJ (3. Serie) *11*, S. 182—183.

1932

Een germaansch tempel- of kapelrayon te Looveen bij Wijster, Gem. Beilen? NDV *50*, S. 51—63.

1933

De terp te Ezinge. De Ingenieur *48*, S. A. 313—A. 315.

1934

Voorloopig bericht over de opgraving op het Domplein te Utrecht in December van 1929. (mit Anhang von J. P. Otto und A. Pulle). (Opgravingen op hel Domplein te Utrecht-Wetenschappelijke Verslagen. I). [Haarlem S. 3—34].
Ein Beitrag zur Germanenfrage im mittel- und westeuropäischen Grenzgebiet. Zweites Nordisches Thing: Veröffentlichungen der Väterkunde *2*. Bremen, S. 42—52.
Oudheidkundige aanteekeningen over Drentsche vondsten: De Hooge Loo te Noordbarge, gem. Emmen; Een heidensche begraafplaats bij Garminge, gem. Westerbork; Kringgrepurnenveld en tumuli in het Elper Noordeveld, gem. Westerbork; De zgn. heidensche legerplaats te Peest, gem. Norg; De Germaansche nederzetting uit den Romeinschen keizertijd in de zgn. Vijzelkampen op den Hondsrug bij Eext, gem. Anloo. NDV *52*, S. 85—116.

1935

Oudheidkundige aanteekeningen over Drentsche vondsten (II): Het Ballooërveld, Ndl. van Balloo, gem. Rolde; Twee grafheuvels te Nieuw Roden, gem. Roden; Oude cultuursporen te Leggeloo, gem. Dwingeloo. NDV *53*, S. 67–122.

Bouwstoffen voor de Groningsche oergeschiedenis (II). Opgravingsresultaten in 1932: Het grafveld van Laudermarke. Verslag GM, S. 47–87.

1936

Néolithique et Enéolithique. Age du bronze. Age du fer, in: L'homme préhistorique et protohistorique aux Pays-Bas, von J. Butter und A. E. van Giffen. Report of XVI International Geological Congress, Washington 1933. Washington, S. 1219–1229.

Oudheidkundige aanteekeningen over Drentsche vondsten (III): De „Tweeënbarg" en het kringgrepurnenveld op de Oeren bij Wapse, gem. Diever; Het Doddenveld bij het tumulibosch Zd. W. van Balloo, gem. Rolde; Het zgn. Jodenbergje op het Zeijerveld bij Zeijen gem. Vries; De Galgenberg te Westerbork, gem. Westerbork; De zgn. Galgenberg en het kringgrepurnenveld in de boschwachterij Sleenerzand, gem. Sleen-Zweeloo; De steenkransheuvel en de Paaschberg te Weerdinge, gem. Emmen; De meest ndl. tumuli op het Noordsche Veld bij Zeijen, gem. Norg; Een versterking naar Romeinsch patroon op het Noordsche Veld bij Zeijen, gem. Vries; Inheemsche behuizingen uit den Romeinschen keizertijd bij Diphoorn, gem. Sleen; Een inheemsch-Karolingische hutkom te Schipborg, gem. Anloo; Het zgn. Klooster te Kamps onder Deurze, gem. Rolde. NDV *54*, S. 75–140.

Der Warf in Ezinge, Provinz Groningen, Holland, und seine westgermanischen Häuser. Germania *20*, S. 40–47. (und in: JVT *16–19* [1931–1935] S. 1–8).

Die Siedlungen in den Warfen Hollands, besonders in dem Dorfwarf Ezinge. FuF *12*, S. 189–191.

Vóór- en vroeghistorische verschijnselen in Nederland. Reisboek voor Nederland, uitg. Kon. Ned. Toeristenbond A. N. W. B. 2. Auflage, S. 3–13.

De opgravingen 1936 op het Domplein te Utrecht. De beide Romeinsche vestingen. Nieuwsblad van jet Noorden 15. und 16. Juni.

De Emmerschans, de excursie von de Drentsch Praehistorische Vereeniging op 4. Juli l.l. Drente 8, nr. 3 (Aug.), S. 1–4.

Nog eens de Emmerschans. Drente *8*, nr. 5 (Oct.), S. 4–5.

1937

Oudheidkundige aanteekeningen over Drentsche vondsten (IV): De Kerk te Eelde, gem. Eelde; De Germaansche nederzetting in de zgn. Vijzelkampen te Eext, gem. Anloo; Een grafveld uit den hunebeddentijd op den Ekel(en)berg bij Kerkenbosch, gem. Zuidwolde; Twee grafvelden uit den hunebedden- en bekertijd in het Kruidhaarsveld bij Sleen, gem. Sleen; Een urnenveldje bij Roden, gem. Roden; Omheinde, inheemsche nederzettingen met aanliggende tumuli, leemkuilen en rijengrafveld te Rhee en Zeijen, gem. Vries; De Hooge Loo te Noordbarge, gem. Emmen; Een láát-middeleeuwsche welput te Zuidvelde, gem. Norg. NDV *55*, S. 67–88.

Tumuli-opgravingen in Gelderland, 1935/1936: Opgravingen in en bij Bennekom, Gem.

Ede; De Bergsham en omgeving, Gem. Barneveld; De Galgenberg, de zgn. Reigersberg en de zgn. Kloosberg bij Hoog-Soeren, Gem. Apeldoorn. Gelre *40*, S. 3—18.

Bouwsteenen voor de Brabantsche oergeschiedenis. Opgravingen in de provincie Noord-Brabant 1935: De zgn. „Vijfberg" op de Rechte Heide, Gem. Goirle, ringwalheuvel en palissadetumuli in Midden-Brabant; Het kringgrepurnenveld te Wilreit, Gem. Bergeyk. 's-Hertogenbosch.

Naar aanleiding van den dupliek von den Heer J. Bastaans „Nog eens van 'n Groaverei" in het Maartnummer 1937 van dit Maandblad op de „afstraffing" van Prof. Dr. G. S. Overdiep in het Januarinummer von dit orgaan. Drente *9*, S. 3—6.

De oudste woningen en nederzettingen op klei- en zandgronden in Noordelijk Nederland. Bijdragen en Mededelingen „Het Nederlandsch Openlucht-Museum" te Arnhem, S. 391—397.

1938

Oudheidkundige aanteekeningen over Drentsche vondsten (V): Het zgn. Jodenbergje op het Zeijerveld, gem. Vries (II), Omheinde, inheemsche nederzettingen, tumuli enz. te Rhee en Zeijen, gem. Vries (II); Een kringgrepurnenveld te Meppen, gem. Zweeloo; Een kringgrepurnenveld in het Zuidveld te Elp, gem. Westerbork; Twee steenkransheuvels in het Odoorner Noorderveld, gem. Odoorn; Een steentijdheuvel bij Weerdinge, gem. Emmen; Een kistgrafheuvel uit den bronstijd bij Emmen, gem. Emmen; Tumuli te Bargeroosterveld, gem. Emmen; Het urnenveld en de vergraven palissadeheuvel te Peeloo, gem. Assen; Een veenbrug bij Valthe, gem. Odoorn. NUV *56*, S. 91—118.

Vorgeschichtliche Beziehungen zwischen den Niederlanden und Westfalen. Westfälische Forschungen *1*, S. 115—123.

Das Kreisgraben-Urnenfeld bei Vledder, Provinz Drente, Niederlande. Mannus *30*, S. 331 bis 384.

Der Kreisgrabenfriedhof bei Sleen, Provinz Drente, Niederlande. Mannus *30*, S. 546 bis 548.

Continental Bell- or Disc-Barrows in Holland, with special reference to Tumulus I at Rielsch Hoefke. Proceedings of the Prehistoric Society *4*, S. 258—271.

1939

Oudheidkundige aanteekeningen over Drentsche vondsten (VI): Het zgn. Eppiesbergje, gem. Odoorn; Een tweenperioden-heuvel en twee steenkransheuvels bij Eext, gem. Anloo; Tumuli I en II bij Zuidvelde, gem. Norg; De brandheuvels 5 en 6 bij Peeloo, gem. Assen; Tumulus 43 in het Hijkerveld, gem. Beilen; Een brandheuvel bij Anloo, gem. Anloo; De tumuli a en b te Doldersum, gem. Vledder; De palissandeheuvel bij Diphoorn, gem. Sleen; Een kringgrepurnenveld bij Sleen, gem. Sleen; Woningsporen te Den Hool en Sleen, gem. Sleen. NDV *57*, S. 119—140.

Bouwstoffen voor de Groninger oergeschiedenis (III). Opgravingsresultaten, I Vóór- en vroeghistorische objecten: Kringgrepurnenveld te Jipsinghuizen, Gem. Vlachtwedde; Kringgrepurnenveld te Achterholte, Gem. Onstwedde; Nederzettingsoverblijfselen op den Beukhorst te Laude, Gem. Vlachtwedde; „Heidensche legerplaats" (celtic field) te Zuidveld bij Sellingen, Gem. Vlachtwedde. II Historische objecten: De kapel te Bedum; De oude kerkheuvel te Friescheloo. Verslag GM, S. 71—103.

1940

Oudheidkundige aanteekeningen over Drentsche vondsten (VII): Een láát-middeleeuwsche steenoven en oude nederzettingsporen bij Erm, gem. Sleen; Nederzettingen, grafheuvels, leemkuilen en rijengrafveld tusschen Rhee en Zeyen, gem. Vries; De zgn. Romeinsche of heidensche legerplaats op het Noordsche Veld bij Zeyen, gem. Vries; Grafheuvels (I—IV) en kringgrafveld bij Aalden, gem. Zweeloo; Een drie-periodenheuvel, de zgn. Galgenberg, boschwachterij „Sleenerzand", gem. Sleen-Zweeloo; Twee palissadeheuvels bij Westervelde, gem. Norg. NDV *58*, S. 180—216.

Die Wurtenforschung in Holland. Probleme der Küstenforschung im südlichen Nordseegebiet *1*, S. 70—86.

Vorwort zu: Beckers Sr., H. J. & G. A. J. Beckers Jr., Voorgeschiedenis van Zuid-Limburg; 20 jaren archaeologisch onderzoek. Maastricht, S. 5—6.

Een systematisch onderzoek in een der Tuinster wierden te Leens (mit deutscher Zusammenfassung). JVT *20—24* (1926—1940) S. 26—115.

1941

Oudheidkundige aanteekeningen over Drentsche vondsten (VIII): Kringgrepurnenveld in het zgn. Overdijksveld onder Gasteren, gem. Anloo; Twee samengestelde palissadeheuvels (I en II) bij Nijlande, gem. Rolde; Twee brandheuvels (I en II) in „De achterste Holten", Z. W. van Vries, gem. Vries; Drie grafheuvels (I—III), W. van Vries, gem. Vries; Kringgrepurnenveld bij Holsloot, gem. Sleen; Twee samengestelde palissadeheuvels (I en II) Z. W. van Zuidvelde, gem. Norg; Woningsporen uit den Romeinschen keizertijd N. van Wachtum, gem. Dalen; Neolithische grafheuvel (I) N. O. van Anloo, gem. Anloo; Tweeperioden-heuvel N. van Eext, gem. Anloo; Tweeperioden-heuvel N. van Gasteren, gem. Anloo; Neolithische grafheuvel bij het „Hillig Meer" Z.W. van Eext, gem. Anloo. NDV *59*, S. 101—141.

De tijd van vorming van heidepodsolprofielen aan de hand van archaeologische waarnemingen. Besprekingen over het Heidepodsolprofiel, gehouden op de bijeenkomst der sectie Nederland van de Intern. Bodemkundige Vereeniging op 18 en 19 April 1941. Groningen, S. 12—23.

1942

Oudheidkundige aanteekeningen over Drentsche vondsten (IX): Een tumulus uit den ouderen bronstijd bij Eext, gem. Anloo; Een tweeperioden-heuvel uit steen- en bronstijd en een kringgrepurnenveld bij Eldersloo, gem. Rolde; Drie grafheuvels (I—III) uit steen-(I) en bronstijd (II—III), met neolithisch „vlakgraf" onder III, N. van Vredenheim, gem. Rolde; Twee tumuli, een tweeperioden-heuvel (I) uit steen- en bronstijd en een ringslootheuvel (II) uit den steentijd, bij Eext, gem. Anloo; Een steenkransheuvel uit den ouderen bronstijd, bij Drouwen, gem. Borger. NDV *60*, S. 97—119.

Die vor- und frühgeschichtliche Abteilung des Groninger Museums. Ostfriesland *5* (Folge 8), S. 16—20.

1943

Oudheidkundige aanteekeningen over Drentsche vondsten (X): Tumulus II, de zgn. Paaschberg, te Weerdinge, gem. Emmen; Een urnenveld te Drouwen, gem. Borger; Een drieperioden-heuvel, tumulus II, bij Anloo, gem. Anloo; De zgn. Eexter grafkelder,

FRIEDRICH ZUCKER
30. 6. 1881—4. 4. 1973

ALBERT EGGES VAN GIFFEN
14. 3. 1884—31. 4. 1973

MARTIN JAHN
20. 9. 1888—11. 9. 1974

WOLFGANG SCHADEWALDT
15. 3. 1900—10. 11. 1974

hunebed D XIII, te Eext, gem. Anloo; Het ndl. hunebed (D XXVIII) te Buinen, gem. Borger. NDV *61*, S. 93–139.

Opgravingen in Drente tot 1941, in: Drente, een handboek voor het kennen van het Drentsche leven in voorbije eeuwen, ed. J. Poortman. Meppel, S. 391–564 2. Aufl. Meppel 1944, S. 393–568.

1944

Oudheidkundige aanteekeningen over Drentsche vondsten (XI): De twee vernielde hunebedden, D VIe en D VIf, bij Tinaarloo, gem. Vries; Een grafheuvel, tumulus I, bij Tinaarloo, gem. Vries; Een steenkeldertje, D XIIIa, te Eext, gem. Anloo; Twee vernielde hunebedden, D XIIIb en c, te Eext, gem. Anloo; Een grafheuvel, tumulus III, te Eext, gem. Anloo; Een grafheuvel, tumulus IV, te Eext, gem. Anloo; Een kringgrepurnenveld Z. O. van den Eexter grafkelder, D XIII, gem. Anloo; De ndl. tumuli groep op het Eexter Veld, gem. Anloo; De zdl. tumuligroep op het Eexter Veld, gem. Anloo; De tumuligroep bij de zgn. heidensche legerplaats op het Emmerveld, gem. Emmen. NDV *62*, S. 93–144.

Grafheuvels te Zwaagdijk, gem. Wervershoof (N. H.) (mit einem Beitrag von J. van der Spek). West-Friesland's Oud en Nieuw *17*, S. 121–231.

1945

Oudheidkundige aanteekeningen over Drentsche vondsten (XII): Het kringgrepurnenveld en de grafheuvels O. Z. O. van Gasteren, gem. Anloo (mit Anhang von H. T. Waterbolk und J. Leutscher). NDV *63*, S. 69–134.

1946

Oudheidkundige aanteekeningen over Drentsche vondsten (XIII): Een vernield hunebed, D LIIa, het zgn. Pottiesbargien, in het (vroegere) Wapserveld bij Diever, gem . Diever; Een neolithische grafheuvel N. W. van Rhee, gem. Vries; Een twee- of drieperiodenheuvel, met primairen subtumulus der Veluwsche klokbekercultuur, bij de Eenerschans, gem. Norg; Vroeg-middeleeuwsche gebouwsporen bij Hooghalen, gem. Beilen. NDV *64*, S. 61–88.

Alphen, gem. Alphen (Prov. Noord-Brabant); Velzen, gem. Velzen (Prov. Noord-Holland); Zesperioden-castellum te Valkenburg – Praetorium Agrippinae (Prov. Zuid-Holland); Nieuwe opgravingen te Vechten-Fectio (Prov. Utrecht), mit W. Glasbergen. l'Antiquité Classique *15*, Archeologie, S. 308–312.

Onze hunebedden. De Vrije Pers *2* (11. October), S. 4–5.

1947

Oudheidkundige aanteekeningen over Drentsche vondsten (XIV): De aeneolithische ringslootheuvel tussen Odoorn en Exloo, gem. Odoorn, een bijenkorfgrafheuvel met ringsloot, intermediair en centraal standspoor (mit einem Anhang von H. T. Waterbolk). NDV *65*, S. 79–135.

Het Hunsow-vraagstuk en hat oudheidkundig bodemonderzoek, in: Het boek der Podagristen, Drenthe in vlugtige en losse omtrekken geschetst door drie Podagristen (D. H. van der Scheer, H. Boom, A. L. Lesturgeon). Assen, S. 119—131.

Oudheidkundige perspectieven, in het bijzonder ten aanzien van de vaderlandsche prae- en protohistorie. Rede, uitgesproken naar aanleiding van der aanvaarding van het ambt van buitengewoon hoogleeraar aan de Universiteit van Amsterdam op 3 Februari 1947. Groningen (und in: Een kwart eeuw oudheidkundig bodemonderzoek in Nederland, Gedenkboek A. E. van Giffen. Meppel, S. 497—544).

De opgravingen in Nederland in 1947; Rhenen, gem. Rhenen (Prov. Utrecht), mit W. Glasbergen. l'Antiquité Classique *16*, Archeologie, S. 366—368.

1948

Oudheidkundige aanteekeningen over Drentse vondsten (XV): Een rendierjagerstation tussen Eext en Gasselte, gem. Gasselte (mit einem Anhang von H. T. Waterbolk); De vóórhistorische begraafplaats op de „Hankenberg" of het „Hankenbossien" bij Erica (mit einem Anhang von H. W. Assies). NDV *66*, S. 85—135.

De Romeinsche castella in den dorpsheuvel te Valkenburg aan den Rijn (Z. H.), Praetorium Agrippinae, I: De opgravingen in 1941, JVT *25—28* (1940—1944) S. 1—316.

Thermen en Castella te Heerlen (L.). Een rapport en een werkhypothese. l'Antiquité Classique 17, Miscellanea Philologica, Historica et Archaeologica in honorem Huberti van de Weerd, S. 199—236.

Excursie naar de opgravingen te Vechten-Fectio (mit W. Glasbergen). Bulletin van de Nederlandse Oudheidkundige Bond (5. Serie), *1*, S. 99—101.

Recent bodemonderzoek in de stad Groningen, I. Stad en Lande *1*, S. 19—23.

1949

Oudheidkundige aantekeningen over Drentse vondsten (XVI): Het onderzoek op het Noordse Veld bij Zeijen, gem. Vries, opgravingen in 1944 (mit einem Beitrag von H. T. Waterbolk). NDV *67*, S. 93—148.

Drentsch Praehistorische Vereeniging. Excursies 1948. I, Langs de Hondsrug in Drenthe. NDV *67*.

Inheemse en Romeinse terpen. Opgravingen in de dorpswierde te Ezinge en de Romeinse terpen van Utrecht, Valkenburg Z. H. en Vechten. Akademie-dagen *2*, S. 91—119.

Een vierperioden-tumulus met primaire klokbekerheuvel bij Schaaik, gem. Oss, N. Br. Brabants Heem *1*, S. 67—74.

De opgravingen in Nederland in 1948 (mit W. Glasbergen). l'Antiquité Classique *18*, Archeologie, S. 418—423.

Bouwstoffen voor de Groninger oergeschiedenis (IV): Opgraving urnenveld Wedderveer, mit H. T. Waterbolk, mit einem Anhang von H. W. Assies. Verslag GM (1944—1948) S. 49—119.

Toepassing van het hout in het voorhistorische Europa, gedurende het palaeolithicum en mesolithicum (mit A. Bohmers), in: W. Boerhave Beekman, Hout in alle tijden. Deventer, S. 235—266.

Toepassing van het hout in het voorhistorische Europa, gedurende het neolithicum, de bronstijd, de oude ijzertijd, de Romeinse en de Volksverhuizingstijd (mit A. Bohmers), in: W. Boerhave Beekman, Hout in alle tijden. Deventer, S. 267—328.

Opgravingen op het Domplein 1949. Nieuw Utrechts Dagblad 3. Febr., 11. März, 6. und 23. Mai.

1950

Nieuwe methoden en resultaten op het gebied van prae- en protohistorie. Vorderingen der wetenschap sinds 1940, ed. L. D. Eerland a. o., Zevende interfacultaire leergang: Academiejaar 1949—1950. Groningen, S. 43—72.

De opgravingen in Nederland in 1949 (mit W. Glasbergen). l'Antiquité Classique *19*, Archeologie, S. 427—436.

Oudheidkundige aanteekeningen over Drentse vondsten (XVII): De nederzettingsoverblijfselen in het Bolleveen en de versterking, de zgn. „legerplaats", aan het Witteveen op het Noordse Veld, beide bij Zeijen, gem. Vries (mit einem Beitrag von H. T. Waterbolk). NDV *68*, S. 89—124.

De jongste opgravingen op het terrein der oude abdij (mit W. Glasbergen). Tien eeuwen Egmond, ontstaan, bloei en ondergang van de regale abdij van Egmond, ed. A. Beekman. Heemstede, S. 145—157.

Van rendierjager tot ontginner, teogelicht met sprekende voorwerpen uit de voorhistorische verzameling van het Prov. Museum van Drenthe (mit W. Glasbergen). Assen.

Inheemse en Romeinse terpen. Opgravingen in de dorpswierde te Ezinge en de Romeinse terpen van Utrecht, Valkenburg (Z. H.) en Vechten. (mit englischer Zusammenfassung von J. Gerritsen). JVT *29—32* (1944—1948) S. 1—66.

De St. Walburgskerk. Nieuwsblad van het Noorden 27. Mai und 10., 11. und 12. Aug.

De excursie van de Drentsch Praehistorische Vereeniging op Zaterdag 16. September 1950, naar een boerderij te Bruntinge, het rijengrafveld te Aalden, kerk en „spijker" te Roden, de kerk te Vries. NDV *70*.

1951

Oudheidkundige aantekeningen over Drentse vondsten (XVIII): De Havelterberg en omgeving bij Havelte, gem. Havelte. Opgravingen in 1918, 1943, 1944 en 1946 (mit Beiträgen von H. T. Waterbolk und T. H. van Andel). NDV *69*, S. 97—162.

Biologisch-Archaeologisch Instituut (mit W. Glasbergen). Jaarboek der Rijksuniversiteit te Groningen, S. 83—87.

De excursie van de Drentsch Praehistorische Vereeniging in 1951, bezoek aan de grafkelder bij Eext, het urnenveld te Drouwen, de tumulus bij Schoonloo en het Asser Museum. Assen.

1952

Oudheidkundige aantekeningen over Drentse vondsten (XIX): Het Bolleveen bij Zeijen, gem. Vries (Postscript). NDV *70*, S. 89—108.

Praehistorische opgravingen in Nederland; Romeinse castella in Nederland; Rijengrafvelden in Drente en Groningen; Opgravingen van kerken in Nederland (mit W. Glasbergen). l'Antiquité Classique *21*, Archeologie, S. 127—130, 138—139, 423, 427.

Das Bauernhaus der Vor- und Frühgeschichte Nordwesteuropas. Niederschrift über die Tagung des Arbeitskreises für deutsche Hausforschung in Cloppenburg i. Oldbg. vom 22. bis 24. 8. 1952, S. 27—33.

Three Roman Frontier Forts in Holland at Utrecht, Valkenburg and Vechten. The Congress of Roman Frontier Studies 1949, ed. E. Birley. Durham, S. 31—40.

Vorwort zu: (V.) Gordon Childe, Van vuursteen tot wereldrijk (holländige Übersetzung von R. van Amerongen). Amsterdam, S. 5—9.

Praehistorie, Winkler Prins Encyclopaedie, 6. Aufl., *15*, S. 599.

1953

De vóórhistorische vuursteenexploitatie bij Rijckholt in Nederlands Limburg. Mélanges en hommage au professeur Hamal-Nandrin. Merksplas, S. 97–102.

Onderzoek van drie bronstijdgrafheuvels bij Grootebroek, gem. Grootebroek (Voorlopig verslag), mit Beiträgen von J. H. van Voorthuysen, H. W. Assies, H. T. Waterbolk und Hl. de Vries. West-Friesland's Oud en Nieuw *20*, S. 34–40.

1954

Oudheidkundige aantekeningen over Drentse vondsten (XX): En grafheuvelonderzoek op de Emelange bij Wijster, gem. Beilen (mit Beiträgen von W. van Zeist und H. W. Assies). NDV *72*, S. 159–200.

Archäologische Bemerkungen zur Frage der Niveauveränderungen im niederländischen Küstengebiet. Geologie en Mijnbouw, N. S. *16*, S. 226–231.

Onderzoekingen naar de oudste bewoning van Amstelland en Amsterdam (mit J. D. van der Waals). Amstelodamum *41*, S. 61–62.

Opgravingen bij het Ouderkerksplein (mit J. D. van der Waals). Amstelodamum *41*, ibid., S. 97–99.

Praehistorische huisvormen op de zandgronden. Kunsthistorisch Jaarboek *5*, (Festschrift Prof. Dr. A. W. Byvanck, S. 11–41).

Praehistorische opgravingen in Nederland in 1952 en 1953. (mit W. Glasbergen). l'Antiquité Classique *23*, Archeologie, S. 172–174.

Een meerperioden-heuvel, tumulus I, te Bennekom, gem. Ede (mit einem Beitrag von W. van Zeist). Gelre *44*, S. 9–26.

Zwei merkwürdige Funde. I: römische Schminkdose oder römisches Arzneikästchen; II: ein eiserner Bandspangenhelm von Groningen. Palaeohistoria *3*, S. 226–244.

De geschiedenis van het Provinciaal Museum van Drenthe. Drenthe *25*, Nr. 7, S. 3–7.

1955

Culture et Milieu: Relations entre phénomèmes culturels et naturels. Actes du Congrès Panafricain de Préhistoire, IIe Session – Alger – 1952. Paris, S. 779–782.

Die frühgeschichtlichen Marschensiedlungen, die „Terpen" oder Warfen. Jahrbuch der Männer vom Morgenstern *36*, S. 7–19.

De Romeinse Castella in de dorpsheuvel te Valkenburg aan de Rijn (Z. H.) (Praetorium Agrippinae). II: De opgravingen in 1942–1943 en 1946–1950 (mit 11 Nachträgen). JVT *33–37* (1948–1953) S. 1–209.

1956

Voorlopige mededeling over de opgravingen bij en in de Oude Kerk, 9 mei – october 1955 (mit H. Miedema). Amstelodamum *43*, S. 68–75.

Zur Frage der Einheitlichkeit der Hünenbetten. Das Riesen-Großsteinlanggrab bei Emmen, Prov. Drenthe. Zur Ur- und Frühgeschichte Nordwestdeutschlands, Festschrift zum 70. Geburtstage von K. H. Jacob-Friesen, ed. P. Zylmann. Hildesheim, S. 97–122.

1957

Een trechterbeker van de hunebeddencultuur uit Lutjesaaksum. Een merkwaardige vondst op een uitzonderlijke vindplaats. Groningse Volksalmanak, S. 148—151.

1958

Prähistorische Hausformen auf Sandböden in den Niederlanden. Germania *36*, S. 35—71 (Holländische Ausgabe s. 1954).

De ontwikkeling van de archaeologie in Nederland. Pro-Historia *3*, Nr. 7, S. 4—7.

1960

Herdenking van Vere Gordon Childe (14. april 1892—19. october 1957). Jaarboek KNAW *1959—1960*, S. 373—382.

Restauratie en onderzoek I en II. Groninger Universiteitsblad (Oud-Hoogleraar aan het woord) *11*, S. 29—32, 51—52.

In memoriam Ferdinand Lieftinck. NDV *78*, S. 12—14.

Drenthe: Borger; Coevorden; Emmen; Emmerschans; Odoorn; Oosterhesselen; Sleen Noord-Brabant: Veldhoven. Nieuws-Bulletin KNOB *13*, Archeologisch Nieuws, col. resp. 240, 240—241, 241—243 und 263—264, 243, 244, 244 und 264, 245, 250.

1961

Een vlakgraf van de trechterbekercultuur, gesneden door enn standkuil van hunebed D 32 te Odoorn (Dr.). Helinium *1*, S. 39—43. (und in: Voorlopige Monumentenraad. Verslag over het jaar 1959. 's-Gravenhage 1960, S. 8—11).

Een gereconstrueerd hunebed. NDV *79*, S. 189—198 (siehe auch 1967 und 1968).

De Groninger St. Walburg. Historische Avonden. Vierde bundel geschiedkundige opstellen, uitg. door ‚Het Historisch Genootschap te Groningen', t. g. v. zijn 75-jarig bestaan. Groningen (und: TG *74*, S. 330—343).

De Groninger Sint Walburg. Groningen (siehe auch 1970 und 1973).

Nederzettingssporen van de vroege klokbekercultuur bij Oostwoud (N. H.). In het voetspoor van A. E. van Giffen. Groningen, S. 66—71. 2. Aufl. Groningen 1966, S. 66—71.

Settlement Traces of the Early Bell Beaker Culture at Oostwoud (N. H.). Helinium *1*, S. 223 bis 228.

Drenthe: Ballo; Coevorden; Een; Emmen; Emmerschans; Oosterhesselen. Nieuws-Bulletin KNOB *14*, Archeologisch Nieuws, S. 33—34, 34, 34—35, 2—3, 35, 3 und 35.

1962

Grafheuvels uit de midden-bronstijd met nederzettingssporen van de klokbekercultuur by Oostwoud. West-Frieslands Oud en Nieuw *29*, S. 199—209.

Restauratie en na-onderzoek van het langgraf (D 43) te Emmen. Helinium *2*, S. 104—114.

Bijdrage tot de geschiedenis van de stad Groningen binnen de Diepen. *I* (mit H. Praamstra). Groningse Volksalmanak, S. 68—154.

Opgravingen in de Martinikerk. Beenderen van Wessel Gansfort gevonden? Nieuwsblad van het Noorden (7. Juni).

Drenthe: Borger, Nieuws-Bulletin KNOB *15*, Archeologisch Nieuws, 20—21.

1963

Totenritus bei den Normannen oder Vikingern an der Wolga im zehnten Jahrhundert. Numen *10*, S. 228–235.

Het bouwoffer uit de oudste hoeve te Ezinge (Gr.). Helinium *3*, S. 246–253.

Ausgrabungsergebnisse in römischen Limeskastellen am Rhein in Holland, bes. zu Valkenburg Z. H. (Praetorium Agrippinae). V Congressus Intern. Limitis Romani Studiosorum, Acta et Dissertationes Archaeologica *3*. Zagreb, S. 133–143.

Rapport over het VIe Congrès Intern. des Sciences Préhist. et Protohist., Rom 20. Aug. bis 3. Sept. 1962, mit G. H. R. von Koenigswald. KNAW *15*, S. 28–30.

1964

De vroegste faze van de TRB-cultuur in Nederland (mit W. Glasbergen). Helinium *4*, S. 40–48.

Herinnering aan dr. ir. Johan van Veen — De ouderdom onzer dijken. TKNAG *81*, S. 271 bis 286.

Mededeling omtrent een oudheidkundig bodemonderzoek in de kerk te Termunten, prov. Groningen. Opus Musivum, Een bundel studies, aangeboden aan Prof. dr. M. D. Ozinga. Assen, S. 31–58.

Dodenbestel bij de Noormannen of Vikingen aan de Wolga in de 10e eeuw. Palaeohistoria *10*, S. 317–323. (Deutsche Ausgabe s. 1963).

Een archaeologisch verkenmerk. Een niet geschreven oorkonde uit het bodemarchief te Grijpskerk (zeer voorlopige mededeling), mit einem Beitrag von W. Groenman-van Waateringe). Groningse Volksalmanak, S. 187–197.

Erinnerung an Heinrich Schütte und seine wissenschaftliche Auswirkung (anläßlich seines 100. Geburtstages am 28. Dezember 1963). Oldenburger Jahrbuch *63*, S. 121–124.

1965

Hilversum, grafheuvels 5 en 6 bij de Erfgooiersstraat (mit J. A. Bakker). Nieuws-Bulletin KNOB *18*, Archeologisch Nieuws, S. 120–121.

1966

Bijdrage tot de geschiedenis van de stad Groningen binnen de Diepen II (mit H. Praamstra). Groningse Volksalmanak, S. 1–243.

1967

Herinneringen bij $12^1/_2$ jaar archeologische werkgroep. West-Frieslands Oud en Nieuw *34*, S. 185–188.

De Papeloze Kerk. Het gereconstrueerde rijkshunebed D 49 bij Schoonoord, gem. Sleen. Geïll. gids (mit einem Beitrag von W. Groenman-van Waateringe). Groningen. (siehe auch 1961 und 1969).

1968

Das St. Bernardus Kloster in Aduard (vorläufige Mitteilung). Bonner Jahrbücher *168*, S. 307–314.

1969

De excursie naar Toterfout en Halve Mijl (mit W. Glasbergen). Akademiedagen *20*, S. 103 bis 107.

De Papeloze Kerk. Het gereconstrueerde rijkshunebed D 49 bij Schoonoord, gem. Sleen. Geïll. gids (mit einem Beitrag von W. Groenman-van Waateringe) 3. vermehrte Auflage Groningen.
(siehe auch 1961 und 1967).

1970

De Groninger Sint Walburg. 3. erweiterte Auflage. Groningen.
(siehe auch 1961 und 1973).

1971

Neues zu dem Limeskastellum unterhalb Valkenburg am Rhein. Roman Frontier Studies. Tel Aviv, S. 88–93.

Een grafheuvel te Putten (Gld.), mit M. Addink-Samplonius und W. Glasbergen. Helinium *11*, S. 105–123.

1973

De Groninger St. Walburg en haar ondergrond (mit H. Praamstra), mit einem Beitrag von D. Bouvy, Verhandelingen KNAW N. R. *78*, afd. Letterkunde. Amsterdam.
(siehe auch 1961 und 1970).

Nogmaals Ezinge. Archeologie en Historie, opgedragen aan Prof. dr. H. Brunsting bij zijn 70ste verjaardag. Bussum, S. 49–85.

Ohne Jahr

Het hunebed van Westervelde. Norg en omgeving. Geill. gids met wandelkaart, uitg. VVV Nord e. o.

Een trechterbeker van de hunnebeddencultuur uit Lutjesaaksum. Groningse Volksalmanak [o. J.] S. 148–151, 16 Taf.

[Nachr.] Vere Gordon Childe. D. Sc., D. Litt., F. B. A. (14. April 1892–19. Oktober 1957). Jaarboek der Koninklijke Nederlandse Akademie van Wetenschappen 1959–1960. S. 1–10.

[Nach Unterlagen von Herrn Prof. H. T. Waterbolk, Rijksuniversiteit Groningen]

Zur Biobibliographie

Schriftenverzeichnisse. W. Coblenz, siehe vorliegendes Jahrbuch S. 457–472. — Jahrbuch Sächsische Akademie der Wissenschaften zu Leipzig *1957–1959* (Berlin 1961) Beilage S. 10; *1966–1968* (Berlin 1970) S. 249; *1969–1970* (Berlin 1972) S. 177. — The Int. Who's who *20* (London 1956) S. 325; *22* (London 1958); *26* (London 1962/63) S. 365. — M. Brerma, in: Palaeohistoria *15* (1973) S. 15–34.

Festschriften. Een kwart eeuw oudheidkundig bodemonderzoek in Nederland. Gedenkboek A. E. van Giffen. Meppel 1947. — In het voetspoor van A. E. van Giffen. 1e dr. Groningen

1961; 2e dr. Groningen 1966. — Palaeohistorica *14*, 1966 [1967]. Neolithic Studies in Atlantic Europe. Proc. of the second Atlantic Colloquium, Groningen 6—11 April 1964, presented to A. E. van Giffen for his 80th Birthday.

Würdigungen. Zum 80. Geburtstag: FuF *38* (1964) S. 63. — Allgemein: Das Große Duden-Lexikon in 8 Bänden. *3* (Mannheim 1965) S. 561.

Nachrufe. H. Brunsting, in: Westerheem *22* (1973) S. 93—95. — A. T. Clason, in: Lutra *15* (1973) S. 13—15. — W. Coblenz, siehe vorliegendes Jahrbuch S. 455—457. — W. A. van Es, in: Die Kunde NF *24* (1973) S. 3—7. — W. Haarnagel, in: Praehist. Zschr. *49* (1974) S. 6—9. — S. J. de Laet, in: Helinium *13* (1973) S. 3—9. — H. T. Waterbolk, in: Palaeohistoria *15* (1973) S. 7—12.

Werner Coblenz

Martin Jahn

20. 9. 1888—11. 9. 1974

Kurz vor Vollendung seines 86. Lebensjahres verstarb am 11. September 1974 in Buckenhof bei Erlangen das auswärtige Mitglied der Sächsischen Akademie der Wissenschaften zu Leipzig, der em. Professor mit Lehrstuhl für Vor- und Frühgeschichte an der Martin-Luther-Universität Halle—Wittenberg und em. Direktor des Landesmuseums für Vorgeschichte in Halle, Martin Jahn. Bereits am 1. 8. 1948 war er zum ordentlichen Mitglied unserer Akademie gewählt worden und kam auf Grund seiner großen Verdienste in gleiche Funktionen beim Deutschen Archäologischen Institut in Berlin—Dahlem und der Deutschen Akademie der Naturforscher Leopoldina in Halle.

Martin Jahn war am 20. 9. 1888 als Sohn eines Lehrers in Berlin geboren worden und erwählte sich nach Ablegung des Abiturs den Beruf eines Prähistorikers, ohne zu wissen, ob und welche Aussichten sich für diese junge Wissenschaft vor dem 1. Weltkriege bieten könnten. Bei seinem Berliner akademischen Lehrer Gusta Kossinna erlebte er die kämpferischen Jahre um eine wissenschaftliche Methode an der Quelle mit und wurde sein spezielles Interesse auf die Erforschung der germanischen Frühgeschichte gelenkt. So nimmt es kaum wunder, daß seine Dissertation von 1912 über „Die Bewaffnung der Germanen in der älteren Eisenzeit“ handelte.

Die Berufspraxis begann Jahn 1912 am seinerzeitigen Museum für Kunstgewerbe und Altertümer in Breslau, wo er in Hans Seger einen tatkräftigen Förderer aller seiner wissenschaftlichen Belange fand, sich vor allem aber intensiv in die Museumsarbeit und die damals noch in den Anfängen steckende Bodendenkmalpflege einarbeiten konnte. Jahn ist ein Großteil der Erfolge aller seinerzeitigen archäologischen Entdeckungs- und Schutzmaßnahmen in Schlesien zu verdanken, von der einfachen Fundbergung bis zur systematischen Forschungsgrabung, von den Vorarbeiten zur Landesaufnahme, dem Schutz der Denkmale im Gelände bis zur wissenschaftlichen Auswertung und zur Popularisierung aller mit der Ur- und Frühgeschichtsforschung zusammenhängenden kulturellen Aufgaben.

Seine Erfolge, bedingt durch einen überdurchschnittlichen persönlichen Einsatz bei ständiger Nutzung der neuesten wissenschaftlichen Erkenntnisse, zu denen er selbst ebenfalls viel mit beitrug, und die organisatorischen Fähigkeiten ließen es als selbstverständlich erscheinen, daß er 1930 zum Direktor des neugegründeten Landesamtes für Vorgeschichte in Schlesien ernannt wurde. In seinem neuen Wirkungsgebiet hatte er sich materialbedingt verstärkt mit der Lausitzer Kultur und der Geschichte der Kelten befaßt, daneben mit der Skythenfrage, und sich 1929 habilitiert. Die Berufung auf den Breslauer Lehrstuhl für Vor- und Frühgeschichte erfolgte nach dem Abgang Segers 1934. Eine für das Fach große Zahl von Studenten der „Breslauer Schule“, mit der die Namen Segers und auch Jahns verbunden werden muß, legt Zeugnis von der Ausstrahlungskraft beider ab. Zur praktischen Tätigkeit und Lehre traten dann auch bei Jahn schon bald die Verpflichtungen zur Herausgabe wissenschaftlicher Reihen im Arbeitsgebiet, vor allem aber seit 1926 der archäologischen Monatsschrift und des wichtigsten deutschsprachigen Informationsorganes „Nachrichtenblatt für deutsche Vorzeit“, das bis 1943 fortgeführt werden konnte und trotz aller Beeinflussungsversuche in den letzten zehn Jahren seines Erscheinens das streng wissenschaftliche Anliegen nicht verriet.

Als Folge des 2. Weltkrieges verließ Martin Jahn kurz vor Jahresende 1945 Breslau, um sich im folgenden Jahre in Halle niederzulassen, wo inzwischen am 3. März 1946 das Landesmuseum für Vorgeschichte wieder eröffnet worden war. Er übernahm nicht nur die genannte Institution, sondern auch den Lehrstuhl an der Universität und hielt dort bereits im Oktober 1946 die erste Fachvorlesung nach dem Kriege. Es ist erstaunlich, wie er sich trotz des schweren Schicksalschlages — er hatte u. a. sein gesamtes in vielen Jahrzehnten zusammengetragenes wissenschaftliches Material mit dem Krieg verloren und stand vor einem völligen Neubeginn — und seines vorgeschrittenen Alters in die neuen Aufgaben einarbeitete und auch wissenschaftsorganisatorisch noch Hervorragendes leistete. Wohl kam ihm dabei zugute, daß beide Institutionen im gleichen Hause arbeiteten, das noch dazu den Krieg heil überstanden hatte, über eine intakte Bibliothek, ein Fundarchiv und außerordentlich reiche Bestände verfügte und auch personell gegenüber den anderen kleinen Forschungszentren der späteren DDR nicht zu übersehende Vorteile besaß, aber nur sein großer Einsatz und die nicht erlahmende Pflichterfüllung sicherten beiden Forschungseinrichtungen den weiteren Aufstieg. Schon 1949 rief er die Jahresschrift für mitteldeutsche Vorgeschichte wieder ins Leben und gab noch die Bände 33—42 heraus. Daneben förderte er seit 1949 die große Grabung auf dem Taubenberg bei Wahlitz und in deren Auswirkung die 1950 gegründete Arbeitsgemeinschaft zur Erforschung der Frühgeschichte der Landwirtschaft, die im großen Umfang die interdisziplinäre Zusammenarbeit zwischen den Historikern und den Naturwissenschaftlern der verschiedensten Sparten pflegte und von der ab 1953 eine Publikationsreihe ausging. Es ist selbstverständlich, daß Martin Jahn mit der Gründung der Sektion für Vor- und Frühgeschichte bei der seinerzeitigen Deutschen Akademie der Wissenschaften zu Berlin im April 1952 zu deren Mitglied berufen wurde und daß er auch in den wissenschaftlichen Beiräten am seinerzeitigen Staatssekretariat für Hoch- und Fachschulwesen die Ur- und Frühgeschichte maßgeblich vertrat, nachdem bereits 1948 von ihm ein Lehrheft für den Geschichtsunterricht der Oberschulen über allgemeine Vorgeschichte erschienen war.

Bis 1958 wirkte Martin Jahn an Universität und Landesmuseum in Halle. In diese Zeit fällt auch seine Tätigkeit in der Sächsischen Akademie der Wissenschaften zu Leipzig, in deren Publikationen er methodischen Problemen und allgemeineren Themen, wie der Frage des Handels und des ältesten Bergbaues in Europa, nachging. Außerordentlich verdienstvoll sind seine Bemühungen um die Bibliographie zur Vor- und Frühgeschichte Mitteldeutschlands, deren Abschluß er noch erleben konnte und zu deren Verfassern er für die ehemaligen Länder Sachsen-Anhalt und Thüringen Walther Schulz, für das Land Sachsen Georg Bierbaum gewinnen konnte. Beide nutzten ihre jahrzehntelangen Vorarbeiten und brachten sie in das große Werk ein, das trotz aller Unterschiedlichkeit im Aufbau und in der Auswertung für viele Jahrzehnte keine gleichwertige Nachfolge erleben dürfte. Wenn Jahn das Erscheinen der letzten Teile des Gesamtwerkes auch nach seiner Übersiedlung nach Buckenhof auf Grund vieler bearbeitungstechnischen Verzögerungen dann nicht mehr so intensiv fördern konnte, wie es ihm wohl vorgeschwebt hatte, so blieb er dieser Bibliographie doch bis zuletzt mit großem Interesse verbunden und konnte sich aus der Ferne noch der Vollendung freuen.

Mit Martin Jahn ging einer der letzten aus der Gründergeneration des Faches von uns, dessen Sachlichkeit und ruhige Güte ihn bis ins hohe Alter auszeichneten, dessen Leistung und Pflichterfüllung Vorbild sein sollten, und der durch Lob und Ermutigung so manchen Jüngeren vor Enttäuschungen bewahrte und stets anspornte.

Im Kreise der Fachkollegen und der Akademie hat er sich durch vorbildlichen Dienst an der Wissenschaft einen Ehrenplatz gesichert.

Werner Coblenz

Bibliographie Martin Jahn

(Abkürzungen siehe S. 563ff.)

A. Monographien

1914

Die Bewaffnung der Germanen in der älteren Eisenzeit. Diss.-Teildr.

1916

Die Bewaffnung der Germanen in der älteren Eisenzeit. Mannus-Bibl. *16.*

1921

Der Reitersporn, seine Entstehung und früheste Entwicklung. Ebda *21.*

1931

Die Kelten in Schlesien. Quellenschriften zur ostdeutschen Vor- und Frühgeschichte. Bd 1.

1948

Allgemeine Vorgeschichte, Teil 1. Sonderheft der Lehrhefte für den Geschichtsunterricht in der Oberschule. Verlag Volk und Wissen.

1952

Die Abgrenzung von Kulturgruppen und Völkern in der Vorgeschichte. Ber. Verhandl. Sächs. Akad. Wiss. Bd *99*, H. 3.

1956

Gab es in der vorgeschichtlichen Zeit bereits einen Handel? Abhandl. Sächs. Akad. Wiss. Bd *48*, H. 4.

1960

Der älteste Bergbau in Europa. Abhandl. Sächs. Akad. Wiss. Bd *52*, H. 2.

B. Zeitschriftenaufsätze

1912

Über märkische Latènefunde aus Wohnstätten. Mannus *4*, 138.

1913

Über Beziehungen zwischen Ostgermanen, Westgermanen und Kelten während der Spät-Latènezeit. Mannus *5*, 75–95.

Zusammenhänge der keltischen, römischen und germanischen Bewaffnung um Chr. Geburt. Mannus *6*, 143.

Fundberichte. Schlesien *7*, H. 4 u. 7.

1914

Die Bewaffnung der Germanen zur Römerzeit. Korrespondenzbl. Gesamtver. d. deutschen Geschichts- und Altertumsvereine, Protokolle der Hauptversammlung in Breslau *1913*, 140–146.

Die Bewaffnung der Germanen zur Römerzeit. Zschr. hist. Waffenkunde *6*, H. 8, 259–264 (mit Abbildungen und Berichtigung im nächst. Heft).

Die Bewaffnung der Germanen zur Römerzeit. Deutsche Geschichtsbl. *15*, 43.

Die Kriegführung der Germanen zur Römerzeit. Mannus *6*, 391 f.

1916

Die schlesischen verzierten Waffen der Eisenzeit. Schlesiens Vorzeit NF *7*, 93–112.

1918

Der Spätlatènefund von Tschiläsen, Kr. Guhrau. Mannus *10*, 15–24.

Die oberschlesischen Funde aus der römischen Kaiserzeit, Teil 1. Prähist. Zschr. *10*, 80–149.

1920

Die germanische Besiedlung Oberschlesiens in vorgeschichtlicher Zeit. Mannus *11/12*, 416–418.
Die Entstehung und Entwicklung der ältesten Sporenformen. Zschr. hist. Waffenkunde *8*, H. 10–11, 305–310.

1921

Aus Oberschlesiens Vorzeit. In: Aus Oberschlesiens Vergangenheit, hsg. v. Verein für Geschichte Schlesiens, 4–7.
Nachgedruckt in: Schullesebuch „Wägen und Wirken", Heimatausgabe für Schlesien Teil 4, *1926*, 51–54.
Die älteste germanische Besiedlung Oberschlesiens. Wochenschrift „Schlesische Heimat" *1921*, H. 9, 193–195.

1922

Zur Herkunft der schlesischen Wandalen. In: 25 Jahre Siedlungsarchäologie. Mannus-Bibl. *22*, 78–94.
Neue Funde aus der Kaiserzeit. Altschlesien *1*, 13–25.
Literatur zur Vorgeschichte Schlesiens 1900–1921. Mannus *14*, 275–299.
Literatur zur Vorgeschichte der Provinz Posen 1900–1920. Mannus *14*, 300–313.
Die oberschlesischen Funde aus der römischen Kaiserzeit. Teil 2 u. Schluß. Prähist. Zschr. *13/14*, 127–149.

1923

Zur Chronologie der „Lausitzer Kultur" auf Grund neuerer Grabungen in Schlesien. Mannus-Erg.-Bd *3*, 28–42.

1924

Funde aus dem vierten Jahrhundert n. Chr. Altschlesien *1*, 86–103.
Eine altgermanische Siedlung in Beuthen O. S. Monatsschrift „Heimat und Arbeit". Oktober-Heft, 19–21.
Die Besiedlung Schlesiens zur Völkerwanderungszeit. Schlesische Monatshefte *1*, 275–280.
Die Gliederung der wandalischen Kultur in Schlesien. Schlesiens Vorzeit NF *8*, 20–36.
Eine neue ostgermanische Kulturgruppe aus der späten Kaiserzeit. Petermanns Mitteilungen *1924*, 178.

1925

Schlesien zur Völkerwanderungszeit. Mannus-Erg.-Bd *4*, 147–156.
Gemeinsam mit O. Richter: Eine neue keltische Schwertform aus Süddeutschland. Mannus *17*, 92–104.
Die erste Ausgrabung des Megalithgrabes von Wulfen in Anhalt. Mannus *17*, 110f.
Neue spätkaiserzeitliche Funde aus der Lausitz. In: Götze-Festschrift, 190–198.
Die germanischen Kulturen in der Lausitz. Erg -Bd z. d. Bautzener Geschichtsheften *1925*, H. 2, 56–58.

1926

Hermann Schmidt †. Mannus *18*, 166.
Die Vorgeschichte des Friedländischen. Heimatkunde des Bezirks Friedland in Böhmen Bd *3*, H. 1. (auch in: Sudeta *2*, 1—21.)
Germanische Gräber aus der Kaiserzeit. Altschlesien *1*, 233—248.

1927

Vorzeitliche Beziehungen zwischen Ostbrandenburg und Schlesien. In: Lienau-Festschrift, 28—30.
Verzeichnis der Vertrauensmänner für kulturgeschichtliche Bodenaltertümer in Preußen. Nachrichtenbl. f. Deutsche Vorzeit *3*, 2—4.
Ausstellung von Neuerwerbungen im Breslauer Altertumsmuseum. Altschles. Bl. *1927*, 21—23.
Vorschläge zur Organisation und zum Ausbau der Denkmalpflege. Nachrichtenbl. f. Deutsche Vorzeit *3*, 34—36.
Hans Seger 20 Jahre Vorsitzender des Schlesischen Altertumsvereins. Altschles. Bl. *1927*, 33f.
Römische Medaillons als germanische Schmuckstücke. Bl. f. Münzfreunde *62*, 65f.
Ein neues Denkmalschutzgesetz. Nachrichtenbl. f. Deutsche Vorzeit *3*, 81—83.
gemeinsam mit O. Richter: Die keltischen Schwerter mit Knollenknauf. Mannus *19*, 266, 270.
Skythen in Ostdeutschland. Petermanns Mitteilungen, 363.

1928

Der Burgwall von Ober-Poppschütz. In: Heimatkalender für die Kreise Grünberg und Freystadt, 33f.
Die Skythen in Schlesien. Schlesiens Vorzeit *9*, 11—25.
Gustaf Kossinna zum 70. Geburtstage. Nachrichtenbl. f. Deutsche Vorzeit *4*, 129f.
Völkerwanderungen vor der Völkerwanderungszeit in Schlesien. Mannus-Erg. Bd *6* (Kossinna-Festschrift) 271—277.

1929

Breslau in der Vorzeit. In: Komm ich zeige Dir Breslau H 2, 64—70.
Vorgeschichte an deutschen Hochschulen. Nachrichtenbl. f. Deutsche Vorzeit *5*, 17—19 u. 102—104.
Robert Beltz zum 75. Geburtstage. Nachrichtenbl. f. Deutsche Vorzeit *5*, 33f.
Eine frühgeschichtliche Mahlsteinindustrie am Siling (Zobten). Altschlesien *2*, 282—289.
Hans Seger zum 65. Geburtstage. Altschles. Bl. *1929*, 69f.
Die ersten Reste des Eiszeitmenschen aus Niederschlesien. Der Wanderer im Riesengebirge, Jg *49*, H. 12, 178f.

1930

Die wichtigsten Grabungsergebnisse auf dem Gebiete der deutschen Vorgeschichte. FuF *6*, 2f.
Die ersten Reste des Eiszeitmenschen aus Niederschlesien. FuF *6*, 89f.

Vorgeschichte an deutschen Hochschulen. Nachrichtenbl. f. Deutsche Vorzeit *6*, 2f. u. 102f.

Das neue Denkmalschutzgesetz in Mecklenburg-Schwerin. Nachrichtenbl. f. Deutsche Vorzeit *6*, 33—39.

Wie ist die Vorgeschichtswissenschaft an den deutschen Universitäten vertreten? Nachrichtenbl. f. Deutsche Vorzeit *6*, 150—153.

Ein frühkaiserzeitlicher Prunksporn von der Donaugrenze. In: Schumacher-Festschrift, 282—284.

1931

Westdeutsche Gegenstücke zum Siling-Zobtenproblem. Altschles. Bl. *1931*, 2—5.

Vereinheitlichung der Berichterstattung über vor- und frühgeschichtliche Funde. Nachrichtenbl. f. Deutsche Vorzeit *7*, 33—35.

Vorgeschichte an deutschen Hochschulen. Nachrichtenbl. f. Deutsche Vorzeit *7*, 4—6 u. 116f

Der germanische Goldfund von Namiest in Mähren. Sudeta *7*, 39—51.

Gustaf Kossinna †. Nachrichtenbl. f. Deutsche Vorzeit *7*, 225f.

Die neuen Bestimmungen über Schatzfunde in England Nachrichtenbl. f. Deutsche Vorzeit *7*, 150f.

1932

Der Stand der schlesischen Bodendenkmalpflege. Niederschlesien, Mitteilungsbl. f. d. Mitglieder d. provinziellen Körperschaften *2*, Nr. 5, 6f.

Für das Landesamt für Vorgeschichtliche Denkmalpflege. Niederschlesien *3*, H. 1/2, 10.

Vorgeschichte an deutschen Hochschulen. Nachrichtenbl. f. Deutsche Vorzeit *8*, 6—8, 71—73 u. 198—200.

Der Wanderweg der Kimbern, Teutonen und Wandalen. Mannus *24*, 150—157.

Die Vorgeschichte des schlesischen Sudetengebietes. Altschlesien *4*, 1—13.

Neue skythische und keltische Funde. Altschlesien *4*, 112—126.

1933

Die älteste Münze aus Oberschlesien. Der Oberschlesier *15*, 349f. (auch in: Aus Oberschlesiens Urzeit *20*, 61f.)

Das Landesamt für vorgeschichtliche Denkmalpflege 1932. Altschles. Bl. *1933*, 41. (auch in: Niederschlesien *4*, H. 1—2, 15.)

Vorgeschichte an deutschen Hochschulen. Nachrichtenbl. f. Deutsche Vorzeit *9*, 66 u. 182—185.

Vorgeschichte der Sudetenländer. Schles. Jahrb. *4*, 13—20.

1934

Hans Seger zum 70. Geburtstage. Altschlesien *5*, 1—4.

Hans Seger zum 70. Geburtstage. Nachrichtenbl. f. Deutsche Vorzeit *10*, 51—53.

Vorgeschichte an deutschen Hochschulen. Nachrichtenbl. f. Deutsche Vorzeit *10*, 73—75 u. 227—229.

Noch einmal Wandalismus. Nachrichtenbl. f. Deutsche Vorzeit *10*, 156f.

1935

Arthur Meiner zum 70. Geburtstage. Nachrichtenbl. f. Deutsche Vorzeit *11*, 52.
Vorgeschichte an deutschen Hochschulen. Nachrichtenbl. f. Deutsche Vorzeit *11*, 76f., 109 u. 252—254.

1936

Mitteilungen des Herausgebers der Seger-Festschrift. Altschles. *6*, 430.
Michael Martin Lienau †. Nachrichtenbl. f. Deutsche Vorzeit *12*, 129.

1937

Der Siling, der heilige Berg der Wandalen. Erläuterung einer Wandtafel. Flemmings Verlag Breslau—Deutsch Lissa.
Die Heimat der Wandalen und Norwegen. Acta Archaeologica, Kopenhagen *8*, 149—167.
Rätische Kultureinflüsse in Böhmen und Schlesien während des letzten vorchristlichen Jahrhunderts. Sudeta *13*, 103—111.
Der Burgwall von Poppschütz. Altschlesien *7*, 93—112.

1938

Der Fund von Sacrau. Die Hohe Straße *1*, 60—65.
Die Wandalen und Burgunden in Brandenburg. Brandenburgisches Jahrb. *12*, 82—92.

1939

Robert Beltz zum 85. Geburtstage. Nachrichtenbl. f. Deutsche Vorzeit *15*, 69f.
Johannes Deichmüller zum 85. Geburtstage. Nachrichtenbl. f. Deutsche Vorzeit *15*, 70f.
Georg Florschütz zum 80. Geburtstage. Nachrichtenbl. f. Deutsche Vorzeit *15*, 71.
Georg Wilke †. Nachrichtenbl. f. Deutsche Vorzeit *15*, 72f.

1940

Ein germanisches Klappergefäß aus Schlesien und seine Gegenstücke aus der wandalischen Urheimat in Nordjütland. Altschles. Bl. *1940*, 45—48.
Ein schlesischer Fund wirft die germanische Zeitfolge um? Altschles. Bl. *1940*, 156—163.
Die Wandalen. In: Vorgeschichte der deutschen Stämme, hrsg. v. H. Reinerth, Bd *3*, 943—1032.

1941

Die deutsche Vorgeschichtsforschung in einer Sackgasse? Nachrichtenbl. f. Deutsche Vorzeit *17*, 73—82.
Die ersten Germanen in Südböhmen. Altböhmen und Altmähren *1*, 64—93.
Die Schicksale einer hannöverschen Altertümersammlung des frühen 19. Jahrhunderts. Nachr. Niedersachs. Urgesch. *5*, 73—101.
Walter Klammt †. Nachrichtenbl. f. Deutsche Vorzeit *17*, 269f.
Otto von Hagen 80 Jahre. Nachrichtenbl. f. Deutsche Vorzeit *17*, 264f.
Zum 100. Geburtstag von Hugo Jentsch. Nachrichtenbl. f. Deutsche Vorzeit *17*, 266.

1942

Die Entdeckung des Urnenfriedhofes von Wohlau-Ost vor 125 Jahren. Altschles. Bl. *1942*, 114–123.

Ludwig Schmidt 80 Jahre. Nachrichtenbl. f. Deutsche Vorzeit *18*, 105f.

Student Karl Nolte †. Nachrichtenbl. f. Deutsche Vorzeit *18*, 117f.

Herbert Mertin †. Nachrichtenbl. f. Deutsche Vorzeit *18*, 205.

Ein bedeutsamer Fund aus Brünn-Obersess. Zschr. Mährisch. Landesmus. NF *2*, 99–107.

Warum Bodendenkmalpflege? In: L. Zotz, Die erste Tagung des Forschungsrates für Vorgeschichte. Altböhmen und Altmähren *2*, 141f.

1944

Ein eigenartiger Vorschlag zur Umbenennung des Zobtens aus dem Jahre 1816. Altschles. Bl. *1944*, 8–10.

Germanische Grabfunde von Grünchen, Kr. Lissa. Posener Jahrb. f. Vorgesch. *1*, 28–44.

1949

Die ältesten Wirtschaftsstufen des Menschen. Jschr. f. mitteldt. Vorgesch. *33*, 86–97.

Zwei neue Faustkeile aus Mitteldeutschland. Jschr. f. mitteldt. Vorgesch. *33*, 108f.

Die kulturgeschichtliche Bedeutung der alt- und mittelsteinzeitlichen Menschenfunde von Weimar-Ehringsdorf und Bottendorf, Kr. Querfurt. Jschr. f. mitteldt. Vorgesch. *33*, 110–119.

1950

Wie der menschliche Erfindergeist einmal auf einen Holzweg geriet. Vorgesch. Museumsarbeit u. Bodendenkmalpflege *1*, 7–9.

Ur- und Frühgeschichte im Geschichtsunterricht. Geschichte in der Schule *3*, H. 3, 9–10.

Ein neues Altsteinzeitgerät aus Hundisburg. Jschr. f. mitteldt. Vorgeschichte *34*, 7f.

Ein kultureller Mittelpunkt bei Halle/S. während der frühen Bronzezeit. Jschr. f. mitteldt. Vorgesch. *34*, 81–89.

1951

Ein neues Altsteinzeitgerät aus Barleben. Jschr. f. mitteldt. Vorgesch. *35*, 5–7.

Kulturverbindungen zwischen Mitteldeutschland und Vorderasien zu Beginn des 2. Jahrtausends v. d. Z. Jschr. f. mitteldt. Vorgeschichte *35*, 57–70.

1952

Die Entwicklung der Vorgeschichtswissenschaft in Halle. In: „450 Jahre Martin-Luther-Universität Halle–Wittenberg“ Bd *2*, 283–289.

Zum Fernhandel vom Ostalpengebiet nach Skandinavien in der frührömischen Kaiserzeit. Jschr. f. mitteldt. Vorgesch. *36*, 93–101.

1959

Bibliographie zur Vor- und Frühgeschichte Mitteldeutschlands. Jahrb. Sächs. Akad. Wiss. *1954–1956*, 137–142.

Der älteste Bergbau in Europa. DLZ *1959*, 368f.

1961

Früheisenzeitliche Gesichtsurnen und Szenen-Darstellungen im Weichselmündungsgebiet. DLZ *1961*, 757f.

Bibliographie zur Vor- und Frühgeschichte Mitteldeutschlands. Jahrb. Sächs. Akad. Wiss. *1957–1959*, 219–221.

1963

Friedrich Behn 80 Jahre. FuF *37*, 61f.

C. Besprechungen

1911

F. Knorr, Friedhöfe der älteren Eisenzeit in Schleswig-Holst. In: Mannus *3*, 165f. u. *4*, 161.

G. Schwantes, Die Gräber der ältesten Eisenzeit im östlichen Hannover. In: Mannus *3*, 165f. u. 4, 161.

J. Wiedener-Stern, Das gallische Gräberfeld bei Münsingen. In: Mannus *3*, 166.

P. Kropp, Latènezeitlichen Funde an der keltisch-germanischen Völkergrenze zwischen Saale und Weißer Elster. In: Mannus *3*, 302f.

1915

O. Almgren, Die ältere Eisenzeit Gotlands, H. 1. In: Mannus *7*, 201–204.

1920

G. Kossinna, Die deutsche Ostmark ein Urheimatboden der Germanen. In: Mannus 11/12, 226f.; Korrespondenzbl. d. Gesamtver. d. deutschen Geschichts- u. Altertumsvereine 1919, 264–266; Petermanns Mitteilungen *1920*, 101f.

J. Kostrzewski, Die ostgermanische Kultur der Spätlatènezeit. In: Mannus *11/12*, 419–427.

1921

W. LaBaume, Vorgeschichte von Westpreußen. In: Mannus *13*, 127f.

A. Götze, Die vor- und frühgeschichtlichen Denkmäler des Kreises Ostprignitz, des Kreises Westprignitz und der Kreise Lebus und Stadt Frankfurt a. O. In: Mannus *13*, 234–237.

N. Aberg, Ostpreußen in der Völkerwanderungszeit. In: Mannus *13*, 340–343.

1922

G. Kossinna, Die Herkunft der Germanen. 2. Aufl. In: Mannus *14*, 314f; Korrespondenzbl. d. Gesamtver. d. deutschen Geschichts- u. Altertumsvereine *1922*, 39f.

G. Girke, Die Tracht der Germanen. In: Mannus *14*, 315f.

1923

H. Hahne, 25 Jahre Siedlungsarchäologie. In: Mannus *15*, 324f.

1924

O. Almgren-B. Nerman, Die ältere Eisenzeit Gotlands H. 2. In: Mannus *16*, 182f.
P. Vouga, La Tène. In: Mannus *16*, 184–187; Altschles. *1*, 114.
K. Schirmacher, Unsere Ostmark. In: Mannus *16*, 187.
A. Plettke, Ursprung und Ausbreitung der Angeln und Sachsen. In: Altschles. *1*, 113.
Blätter für deutsche Vorgeschichte, Danzig, 1. In: Altschles. *1*, 114.
K. Tackenberg, Neue schlesische Funde der frühgermanischen Zeit. In: Altschles. *1*, 114.
C. Diculescu, Die Wandalen und die Goten in Ungarn und Rumänien. In: Mannus *16*, 382f.
A. Kiekebusch, Die Ausgrabung des bronzezeitlichen Dorfes Buch bei Berlin. In: Wiener Prähist. Zschr. *11*, 151f.

1925

G. Behrens, Denkmäler des Wangionengebietes. In: Mannus *17*, 135f.
O. Almgren, Studien über nordeuropäische Fibelformen. In: Mannus *17*, 136f.
C. Diculescu, Die Gepiden. In: Mannus *17*, 137f.
L. Franz, W. Hesch, O. Menghin, H. Mitscha-Märheim, Die prähistorische Sammlung des Niederösterreichischen Landesmuseums. In: Mannus *17*, 138f.
F. Geschwendt, Breslau. In: Mannus *17*, 141f.
W. Schulz, Die germanische Familie. In: Mannus *17*, 256.
L. Schmidt, Geschichte der germanischen Frühzeit. In: Mannus *17*, 255f.
Götze-Festschrift. In: Mannus *17*, 392–395.
N. Aberg, Förhistorik nordisk Ornamentik. In: Mannus *17*, 396.

1926

O. Kunkel, Der Mäander in den vor- und frühgeschichtlichen Kulturen Europas. In: Mannus *18*, 241f.
O. Kunkel, Oberhessens vorgeschichtliche Altertümer. In: Mannus *18*, 365f.
Aus Pommerns Urgeschichte: In: Mannus *18*, 365f.
K. Tackenberg, Die Wandalen in Niederschlesien. In: Altschles. *1*, 279.
W. Schulz, Die germanische Familie in der Vorzeit; Staat und Gesellschaft in germanischer Vorzeit. In: Altschles. *1*, 282.
W. Schulz, Kartographische Darstellungen zur altgermanischen Religionsgeschichte. In: Altschles. *1*, 282.
H. von Buttel-Reepen, Über Fensterurnen. In: Altschles. *1*, 282f.
Festschrift zur 25-Jahr-Feier der Gesellschaft für Vorgeschichte und Geschichte der Oberlausitz zu Bautzen. In: Bautzener Geschichtshefte *4*, H. 5, 192–194.

1927

Schriften zur Vor- und Frühgeschichte Schlesiens aus dem Jahre 1926. In: Altschles. Bl. *1927*, 5–7.

Fundberichte aus Schwaben NF 3. In: Mannus *19*, 312f.
O. Menghin, Einführung in die Urgeschichte Böhmens und Mährens. In: Altschles. *2*, 69f.
H. Preidel, Germanen in Böhmen im Spiegel der Bodenfunde. In: Altschles. *2*, 70.
Literaturbericht zur Vorzeit und Frühgeschichte von Ost-, Mittel- und Norddeutschland. In: Jahresber. f. Deutsche Geschichte *1* (Leipzig 1925) 219—223.

1928

St. Bolin, Fynden av romerska mynt i det fria Germanien. In: Mannus *20*, 440—442.
Literaturbericht zur Vorzeit und Frühgeschichte von Ost-, Mittel- und Norddeutschland. In: Jahresber. f. Deutsche Geschichte *2* (Leipzig 1926) 244—251.
F. Wagner, Die Römer in Bayern. In: Altschles. *2*, 165.
F. Hertlein, Die Geschichte der Besetzung des römischen Württemberg. In: Altschles. *2*, 165f.
L. Franz und J. Weninger, Die Funde aus den prähistorischen Pfahlbauten im Mondsee. In: Altschles. *2*, 166.

1929

W. Matthes, Urgeschichte des Kreises Ostprignitz. In: Mannus *21*, 186f.
B. von Richthofen, Gehört Ostdeutschland zur Urheimat der Polen. In: Altschles. *2*, 310f.
F. Sprater, Die Urgeschichte der Pfalz. In: Altschles. *2*, 311.
G. Behrens, Bodenurkunden aus Rheinhessen. In: Altschles. *2*, 311f.
Schriften zur Vor- und Frühgeschichte Schlesiens aus dem Jahre 1928. In: Altschles. Bl. *1929*, 105—107.
Literaturbericht zur Vorzeit und Frühgeschichte von Ost-, Mittel- und Norddeutschland. In: Jahresber. f. Deutsche Geschichte *3* (Leipzig 1927) 188—195.

1930

E. Petersen, Die frühgermanische Kultur in Ostdeutschland und Polen. In: Altschles. *3*, 107f.
H. Hahne, Totenehre im alten Norden. In: Altschles. *3*, 109.
Burgenland, Jg *1—2*. In: Altschles. *3*, 109.
J. Schranil, Die Vorgeschichte Böhmens und Mährens. In: Jahrb. f. Kultur u. Gesch. d. Slaven NF *6*, 204—206.
Literaturbericht zur Vorzeit und Frühgeschichte von Ost-, Mittel- und Norddeutschland. In: Jahresber. f. Deutsche Geschichte *4* (Leipzig 1928) 164—170.

1931

Literaturbericht zur Vorzeit und Frühgeschichte von Ost-, Mittel- und Norddeutschland. In: Jahresber. f. Deutsche Geschichte *5* (Leipzig 1929), 187—195.

G. Rodenwaldt, Neue deutsche Ausgrabungen. In: Altschles. *3*, 314.
Fundberichte aus Schwaben NF 4—5. In: Altschles. *3*, 316f.
F. Sprater, Die Pfalz unter den Römern. In: Altschles. *3*, 323f.
J. Hagen, Römerstraßen der Rheinprovinz. In: Altschles. *3*, 324.

E. Beninger, Prähistorische, germanische und mittelalterliche Funde von Carnuntum und Umgebung. In: Altschles. *3*, 324f.
R. Pittioni, La Tène in Niederösterreich. In: Altschles. *3*, 324f.

1932

E. Beninger, Der westgotisch-alanische Zug nach Mitteleuropa. In: Sudeta *1932*, 73f.
E. Beninger, Der Wandalenfund von Czeke. In: Sudeta *1932*, 73f.
H. Obermaier, Urgeschichte der Menschheit. In: Altschles. *4*, 210f.
J. Steinhausen, Archäologische Karte der Rheinprovinz I, 1. In: Altschles. *4*, 217.
W. Matthes, Die nördlichen Elbgermanen in spätrömischer Zeit. In: Altschles. *4*, 221.
W. Matthes, Die Germanen in der Prignitz zur Zeit der Völkerwanderung. In: Altschles. *4*, 221.
Literaturbericht zur Vorzeit und Frühgeschichte von Ost-, Mittel- und Norddeutschland. In: Jahresber. f. Deutsche Geschichte *6* (Leipzig 1930) 154—159.
E. Sprockhoff, Zur Handelsgeschichte der germanischen Bronzezeit. In: Vierteljahrschr. f. Sozial- u. Wirtschaftsgesch. *25*, 365—367.

1933

W. Frenzel, Vorgeschichte der Lausitzen, In: Petermanns Mitteilungen *1933*, 101.
E. Wahle, Deutsche Vorzeit. In: Vierteljahrschr. f. Sozial- u. Wirtschaftsgesch. *26*, 373 bis 375.

1934

Literaturbericht zur Vorzeit und Frühgeschichte von Ost-, Mittel- und Norddeutschland. In: Jahresber. f. Deutsche Geschichte *7* (Leipzig 1931) 157—162.
O. Menghin, Geist und Blut. In: Altschles. *4*, 320.
P. Steiner, Vorzeitburgen des Hochwalds. In: Altschles. *4*, 325.
Max zu Löwenstein, Der Gallische Krieg von C. I. Caesar. In: Altschles. *4*, 325.
W. Petzsch, Die vorgeschichtlichen Münzfunde Pommerns. In: Vierteljahrschr. f. Sozial- u. Wirtschaftsgesch. *27*, 171.
Literaturbericht zur Vorzeit und Frühgeschichte von Ost-, Mittel- und Norddeutschland. In: Jahresber. f. Deutsche Geschichte *8* (Leipzig 1932) 168—173.

1935

L. Schmidt, Geschichte der germanischen Frühzeit. 2. Aufl. In: DLZ *1935*, 1396—1398.
F. Pfützenreiter, Die vor- und frühgeschichtliche Besiedlung des Kreises Fraustadt. In: Petermanns Mitteilungen *1935*, 382.
W. Bohm, Die ältere Bronzezeit in der Mark Brandenburg. In: DLZ *1935*, 2008—2011.
G. Ekholm, Romerska vinskopor och Kärl ar Hemmoortype i skandinaviska fynd. In: Vierteljahrschr. f. Sozial- u. Wirtschaftsgesch. *28*, 287.
R. Stampfuß, Gustaf Kossinna, ein Leben für die deutsche Vorgeschichte. In: HZ *153*, 167.
W. La Baume, Urgeschichte der Ostgermanen. In: Jahrb. f. Kultur u. Gesch. d. Slaven *1935*, 622f.
B. von Klingspor, Der Gang der ältesten Besiedlung Schwedens. In: Vierteljahrschr. f. Sozial- u. Wirtschaftsgesch. *28*, 413f.

1936

H. Groß, Erzbergbau — Hüttentechnik, Metallhandel und metallverarbeitende Gewerbe auf deutschem Boden. In: Vierteljahrschr. f. Sozial- u. Wirtschaftsgesch. *29*, 63.
G. Kossinna, Die Deutsche Vorgeschichte. 7. Aufl. In: Mannus *28*, 131f.
B. von Richthofen, Vorgeschichte der Menschheit. In: Altschles. *6*, 405.
G. Kossinna, Altgermanische Kulturhöhe. 5. Aufl. In: Altschles. *6*, 411.
R. Stampfuß, Gustaf Kossinna, ein Leben für die deutsche Vorgeschichte. In: Altschles. *6*, 411.
E. Beninger, Die Germanenzeit in Niederösterreich. In: Altschles. *6*, 422f.
E. Beninger, H. Freising, Die germanischen Bodenfunde in Mähren. In: Altschles. *6*, 422f.
A. Götze, Das Oderbruch in vorgeschichtlicher Zeit. In: Altschles. *6*, 424.
E. Petersen, Schlesien von der Eiszeit bis ins Mittelalter. In: DLZ *1936*, 1110—1112.
L. Schmidt, Geschichte der deutschen Stämme. Die Ostgermanen. 2. Aufl. In: Mannus *28*, 284—286.
G. Ekholm, Zur Geschichte des römisch-germanischen Handels. In: Mannus *28*, 290.
G. Ekholm, Forntid och fornforskning i Skandinavien. In: Mannus *28*, 293f.
M. Stenberger, Öland under äldre järnaldern. In: Mannus *28*, 398—401.
A. von Scheltema, Die Kunst unserer Vorzeit. In: DLZ *1936*, 1969—1971.

1937

E. Sturms, Die ältere Bronzezeit im Ostbaltikum. In: DLZ *1937*, 890f.
W. Czajka und Th. Mann, Lebensraum und Reiche der Germanen. Farbige Wandkarte. In: Altschles. *7*, 123.
H. Schlenger, Deutsche Stämme. In: Altschles. *7*, 123.
G. Wilke, Die Heilkunde in der europäischen Vorzeit. In: Altschles. *7*, 127.
K. Tackenberg, Die Kultur der frühen Eisenzeit in Mittel- und Westhannover. In: Altschles. *7*, 131f.
L. Lindner, Die Jagd der Vorzeit. In: DLZ *1937*, 1892f.
C. Engel und W. La Baume, Kulturen und Völker der Frühzeit im Preußenlande. In: Prähist. Zschr. *27*, 271—273.
K. Kersten, Zur älteren nordischen Bronzezeit. In: Prähist. Zschr. *27*, 289—292.
P. Paulsen, Studien zur Wikinger-Kultur. In: Prähist. Zschr. *27*, 307f.
G. Schwantes, Vorgeschichte Schleswig-Holsteins. Lieferung 1—5. In: Prähist. Zschr. *27*, 312—315.

1938

B. von Richthofen, Die Vor- und Frühgeschichtsforschung im neuen Deutschland. In: DLZ *1938*, 61f.
F. Tischler, Fuhlsbüttel, ein Beitrag zur Sachsenfrage. In: DLZ *1938*, 61f.
H. Jankuhn, Die Wehranlagen der Wikingerzeit zwischen Schlei und Treene. In: DLZ *1938*, 1353—1356.
A. van Scheltema, Der Osebergfund. 2. Auflage. In: DLZ *1938*, 1538f.
H. Gummel, Forschungsgeschichte in Deutschland. In: Altschles. *7*, 284.
E. Petersen, Germanen in Schlesien. In: Altschles. *7*, 286.
F. Geschwendt, Der Siling. In: Altschles. *7*, 286.
G. Hoffmann, Vor- und Frühgeschichte. In: Altschles. *7*, 286.

R. Pittioni, Allgemeine Urgeschichte und Urgeschichte Österreichs. In: Altschles. *7*, 288.
R. Pittioni, Österreichs Urzeit im Bilde. In: Altschles. *7*, 288.
L. Franz, Kelten und Germanen in Böhmen. In: Altschles. *7*, 289f.
H. Preidel, Germanen in Böhmens Frühzeit. In: Altschles. *7*, 289f.
C. Streit, Saazer Latènefunde. In: Altschles. *7*, 290.
J. Glott, Das bronzezeitliche Dorf von Triebschitz bei Brüx. In: Altschles. *7*, 290.
A. Rust, Das altsteinzeitliche Rentierjägerlager Meiendorf. In: Altschles. *7*, 293.
E. Forssander, Der ostskandinavische Norden während der ältesten Metallzeit Europas. In: Altschles. *7*, 293.
W. Wegewitz, Die langobardische Kultur im Gau Moswidi. In: Altschles. *7*, 296.
J. Barchewitz, Von der Wirtschaftstätigkeit der Frau in der vorgeschichtlichen Zeit. In: Altschles. *7*, 303f.
Chr. Albrecht, Das Römerlager in Oberaden. In: Prähist. Zschr. *28–29*, 422f.
K. Langenheim, Die Tonware der Riesensteingräber in Schleswig-Holstein. In: Prähist. Zschr. *28–29*, 442–445.
Marburger Studien. In: Prähist. Zschr. *28–29*, 445–447.

1939

R. von Uslar, Westgermanische Bodenfunde. In: DLZ *1939*, 240–243.
L. Schmidt, Geschichte der deutschen Stämme II, 1 (Westgermanen). In: DLZ *1939*, 313 bis 316.
A. Rust, Vor 20000 Jahren. In: DLZ *1939*, 491f.
W. Witter, Die Ausbeutung der mitteldeutschen Erzlagerstätten in der frühen Metallzeit. In: Vierteljahrschr. f. Sozial- u. Wirtschaftsgesch. *32*, 63f.
W. Witter, Die Kenntnis von Kupfer und Bronze in der alten Welt. In: Vierteljahrschr. f. Sozial- u. Wirtschaftsgesch. *32*, 157.
W. Buttler, Der donauländische und der westliche Kulturkreis der jüngeren Steinzeit. In: Altschles. *8*, 198f.
E. Sprockhoff, Die nordische Megalithkultur. In: Altschles. *8*, 198f.
G. Dorka, Urgeschichte des Weizackerkreises Pyritz. In: Altschles. *8*, 207.
Gothiskandza. In: Altschles. *8*, 207.
H. Hoffmann, Die Gräber der jüngeren Bronzezeit in Holstein. In: Altschles. *8*, 210f.
J. Filip, Die Urnenfelder und die Anfänge der Eisenzeit in Böhmen. In: Altschles. *8*, 211.
R. Stampfuß, Das Hügelgräberfeld Rheinberg, Kr. Mörs. In: Altschles. *8*, 212.
W. D. Asmus, Tonwarengruppen und Stammesgrenzen in Mecklenburg. In: Altschles. *8*, 212f.
G. Körner, Der Urnenfriedhof von Rebenstorf. In: Altschles. *8*, 213f.
M. Ohlhaver, Der germanische Schmied und sein Werkzeug. In: Altschles. *8*, 216f.
G. Merschberger, Die Rechtsstellung der germanischen Frau. In: Altschles. *8*, 217.

1940

E. Petersen, Der ostelbische Raum als germanisches Kraftfeld. In: DLZ *1940*, 196–199.
P. Paulsen, Axt und Kreuz bei den Nordgermanen. In: DLZ *1940*, 293–295.
P. Weiershausen, Vorgeschichtliche Eisenhütten Deutschlands. In: Vierteljahrschr. f. Sozial- u. Wirtschaftsgesch. *33*, 233f.
H. Ohlhaver, Der germanische Schmied und sein Werkzeug. In: Vierteljahrschr. f. Sozial- u. Wirtschaftsgesch. *33*, 234.

L. Franz, Vorgeschichte und Zeitgeschehen. In: Altschles. *9*, 189f.

G. Kossinna, Das Weichselland ein uralter Heimatboden der Germanen. 3. Aufl. In: Altschles. *9*, 190.

Herausgeber A. Funkenberg, Tracht und Schmuck im nordischen Raum. Bd 1—2. In: Altschles. *9*, 191f.

Th. Voigt, Die Germanen des 1. und 2. Jahrhunderts im Mittelelbegebiet. In: Altschles. *9*, 196f.

K. Ziegel, Die Thüringe der späten Völkerwanderungszeit im Gebiet östlich der Saale. In: Altschles. *9*, 196f.

H. von Petrikovits und R. Stampfuß, Das germanische Brandgräberfeld Keppeln, Kr. Kleve. In: Altschles. *9*, 198f.

F. Rademacher, Fränkische Goldscheibenfibeln. In: Altschles. *9*, 199.

L. Zotz und B. von Richthofen, Ist Böhmen-Mähren die Urheimat der Tschechen? In: Altschles. *9*, 199f.

1941

F. Holste, Die Bronzezeit im nordmainischen Hessen. In: DLZ *1941*, 220—222.

L. Schmidt und H. Zeiß, Geschichte der deutschen Stämme II, 1 (Westgermanen), 2. Aufl. in: DLZ *1941*, 458f.

W. A. von Jenny, Die Kunst der Germanen im frühen Mittelalter. In: DLZ *1941*, 845f.

H. Bohne-Fischer, Ostpreußens Lebensraum in der Steinzeit. In: DLZ *1941*, 944f.

K.-H. Jacob-Friesen, Einführung in Niedersachsens Urgeschichte. 3. Aufl. In: Altschles. *10*, 163f.

H. Reinerth und G. Merschberger, Handbuch der vorgeschichtlichen Sammlungen Deutschlands I. In: Altschles. *10*, 164.

W. Kimmig, Die Urnenfelderkultur in Baden. In: Altschles. *10*, 165.

1942

W. Dehn, Katalog Kreuznach. In: DLZ *1942*, 265—267.

L. Schmidt, Die Ostgermanen. 2. Aufl. In: DLZ *1942*, 949—950.

H. Urbanek, Die frühen Flachgräberfelder Ostpreußens. In: DLZ *1942*, 1109f.

J. Hoffmann, Die spätheidnische Kultur des Memellandes. In: DLZ *1942*, 1110f.

G. Schwantes, Vorgeschichte Schleswig-Holsteins. In: Prähist. Zschr. *32—33*, 422f.

W. Witter, Über den Stand der Metallforschung im Dienste der Vorgeschichtswissenschaft In: Vierteljahrschr. f. Sozial- u. Wirtschaftsgesch. *36*, 113.

1944

J. Werner, Der Fund von Ittenheim. In: DLZ *1944*, 82f.

H. Zeiß, Das Heilsbild in der germanischen Kunst des frühen Mittelalters. In: DLZ *1944*, 235—237.

1948

L. Schmidt, Geschichte der Wandalen. 2. Aufl. In: DLZ *1948*, 324f.

1949

H. Weinert, Menschen der Vorzeit. 2. Aufl. In: Urania *12*, Umschlagseite XI.

1951

Chr. Pescheck, Lehrbuch der Urgeschichtsforschung. In: Philosophia Naturalis *3*.

1953

E. Wahle, Zur ethnischen Deutung frühgeschichtlicher Kulturprovinzen. 2. Aufl. In: DLZ *1953*, 218—220.

Historia Mundi Bd *1*. In: DLZ *1953*, 666—672.

1954

F. Kern, Der Beginn der Weltgeschichte. In: DLZ *1954*, 218—220.

1957

K. Tackenberg, Der Neandertaler und seine Umwelt. In: DLZ *1957*, 712—714.

1958

E. Schwarz, Germanische Stammeskunde. In: DLZ *1958*, 888—890.

Herausgebertätigkeit

Nachrichtenblatt für Deutsche Vorzeit Bde 2—19 (1926—1943) mit einleitenden Bemerkungen des Herausgebers zu den Bänden *2*, *3*, *4*, *5* und *6*.

Altschlesien Bd *5* (1934) Seger-Festschrift (I).

Altschlesien Bd *8* (1939) Seger-Festschrift (II).

Quellenschriften zur ostdeutschen Vor- und Frühgeschichte Bde 3—6 (1937—1940) gemeinsam mit Hans Seger.

Mitteilungen des Instituts für Vor- und Frühgeschichte der Universität Breslau (1940ff.), mit Einführung in diese Schriftenreihe, siehe: Nachrichten aus Niedersachsens Urgeschichte *14* (1940) 28f.

Jahresschrift für mitteldeutsche Vorgeschichte Bde 33—42 (1949 bis 1958) mit Einführung des Herausgebers in Bd *33*, 1—4.

Arbeiten aus dem Institut für Vor- und Frühgeschichte der Martin-Luther-Universität Halle—Wittenberg (Wissenschaftliche Zeitschrift der Martin-Luther-Universität Halle—Wittenberg), Hefte 1—8 (1952 bis 1958) mit Einführung des Herausgebers in Heft 1, 81.

Veröffentlichungen des Landesmuseums für Vorgeschichte in Halle Bde *14—16* (1955 bis 1957) mit Einführung des Herausgebers in Bd *14*.

[Nach Jahresschrift für mitteldeutsche Vorgeschichte *47* (Halle 1963) S. 9—19. Zusammengestellt von Fr. Schlette]

Bibliographie zur Vor- und Frühgeschichte Mitteldeutschlands. Abhandlungen der Sächsischen Akademie der Wissenschaften zu Leipzig. Philologisch-historische Klasse Bd *47*, Heft 1, Bd *48*, Heft 2, Bd *50*, Heft 1a und 1b, Bd *55*, Heft 1a und 1b. Berlin 1955, 1957, 1959, 1962, 1969 und 1970 mit Einführung des Herausgebers in Bd *1*, *5*.

Bd *1* Walther Schulz, Bibliographie zur Vor- und Frühgeschichte von Sachsen-Anhalt und Thüringen.

Bd *2* Georg und Liddy Bierbaum, Bibliographie zur Vor- und Frühgeschichte Land Sachsen.

Germanische Grabfunde von Grünchen, Kreis Lissa. In: Bonner Hefte zur Vorgeschichte *10*, 1975 (Manuskript von 1943).

Zur Biobibliographie

Schriftenverzeichnisse. W. Coblenz, siehe vorliegendes Jahrbuch S. 474—489. — Kürschners Deutscher Gelehrten-Kalender *1* (Berlin—Leipzig 1925) Sp. 441; *2* (Berlin—Leipzig 1926) Sp. 833; *3* (Berlin—Leipzig 1928/29) Sp. 1037; *5* (Berlin—Leipzig 1935) Sp. 606; 6_1 (Berlin 1940/41) Sp. 811; *7* (Berlin 1950) Sp. 907; *9* (Berlin 1961) S. 869—870; *10* (Berlin 1966) S. 1052; *11* (Berlin 1970) S. 1301. — Jahrbuch Sächsische Akademie der Wissenscahften zu Leipzig *1954—1956* (Berlin 1958) S. 202; *1957—1959* (Berlin 1961) S. 277; *1960—1962* (Berlin 1964) S. 314; *1963—1965* (Berlin 1967) S. 249. — Jahresschrift für mitteldeutsche Vorgeschichte *47* (Halle 1963) S. 9—19.

Festschrift. Festgabe zum 60. Geburtstag von Martin Jahn. Strena Praehistorica. Hrsg. von K. Schwarz. Halle 1948, 253 S., 6 Ktn, 17 Taf. und 86 Abb.

Würdigungen. Zum 65. Geburtstag: Hochschulwesen 1953/54, H. 1, S. 44—45. (Zum 70. Geburtstag: Die Union vom 20. Sept. 1958. — Allgemein: S. Fornaçon, in: Jahrbuch für schlesische Kirche und Kirchengeschichte NF *35* (1956) S. 31—43. — Degeners Wer ist's? *10* (Berlin 1935) S. 749. — Zum 75. Geburtstag: F. Schlette, in: FuF *37* (1963) S. 285 286. — Derselbe, in: Jahresschrift für mitteldeutsche Vorgeschichte *47* (1963) S. 7—8.

Nachrufe. W. Coblenz, siehe vorliegendes Jahrbuch S. 472—474. — B. Hänsel, in: Erlanger Tageblatt Nr. 217 vom 18. Sept. 1974. — G. Smolla, in: Frankfurter Allgemeine Ztg Nr. 215 vom 17. Sept. 1974, S. 24. — Fr. Schlette, in: Zschr. f. Archäologie *9* (1975) S. 143—144. — J. Preuß, in: Ausgrabungen und Funde *20* (1975) S. 59—60.

Werner Coblenz

Wolfgang Schadewaldt

15. 3. 1900—10. 11. 1974

Wenige Monate vor Vollendung seines 75. Lebensjahres verstarb am 10. November 1974 in Tübingen der em. o. Professor für Klassische Philologie und für Geschichte der Fortwirkung des Griechentums an der Eberhard-Karls-Universität Tübingen. Seit 1935 Mitglied der Sächischen Akademie der Wissenschaften zu Leipzig, die er auch in der Internationalen Thesaurus-Kommission und innerhalb der Union Académique Internationale im Redaktionsbeirat des Mittellateinischen Wörterbuchs vertreten hat, war er außerdem Mitglied der Akademie der Wissenschaften der DDR und der Heidelberger Akademie der Wissenschaften sowie korrespondierendes Mitglied der Österreichischen Akademie der Wissenschaften in Wien. Wolfgang Schadewaldt war Träger des Ordens Pour le mérite für Wissenschaften und Künste und anderer hoher Auszeichnungen.

Nach seiner Habilitation in Berlin (1927) erhielt Schadewaldt ein Jahr später einen Ruf als o. Professor nach Königsberg. Die weiteren Stationen seines akademischen Lebens waren Freiburg im Breisgau (1929), Leipzig (1934), Berlin (1941) und schließlich Tübingen (1950). Neben Ulrich von Wilamowitz-Moellendorf hat vor allem Werner Jaeger einen nachhaltigen Einfluß auf ihn ausgeübt.

Das Hauptarbeitsgebiet Schadewaldts war die frühgriechische Dichtung und ihr Fortwirken in der neueren Literatur. Besonders hervorzuheben sind seine Untersuchungen über die griechische Tragödie, über Homer und über Goethes Verhältnis zur Antike. Der griechischen Tragödie galt bereits seine erste größere und in vieler Hinsicht grundlegende Arbeit „Monolog und Selbstgespräch. Untersuchungen zur Formgeschichte der griechischen Tragödie" (1926). Bis zum Ende seines Lebens hat ihn immer wieder vor allem Sophokles stark beschäftigt; ein schönes und eindrucksvolles Beispiel dafür, wie er Sophokles einem tieferen Verständnis auch über den engeren Kreis der Fachkollegen hinaus zu erschließen versucht hat, ist sein Potsdamer Vortrag „Sophokles und das Leid" (1944). Die Homerforschung hat Schadewaldt ganz wesentlich durch seine 1938 in den Abhandlungen der Sächsischen Akademie veröffentlichten „Iliasstudien" beeinflußt, denen die beiden wichtigen Aufsätze „Homerische Szenen" (I 1935, II 1936) vorangingen. Als fruchtbar erwies sich besonders der Begriff der homerischen Szene als einer relativ selbständigen Einheit, wodurch es möglich wurde, die Ilias wieder als ein einheitliches Werk des Dichters zu begreifen. Auch die frühgriechische Lyrik (Pindar, Sappho), die griechische Geschichtsschreibung und Philosophie und selbst die lateinischen Dichter Terenz und Vergil hat Schadewaldt in den weitgespannten Kreis seiner Forschung einbezogen. Besondere Erwähnung verdient seine Sappho-Monographie („Sappho — Welt und Dichtung", 1950). Kleinere Arbeiten sind vereinigt in den umfangreichen Sammelwerken „Von Homers Welt und Werk" (1944, in 4. Auflage 1965) und „Hellas und Hesperien" (1960, in 2. Auflage 1970). Die Nachwirkung der Antike auf die deutsche Klassik, vor allem auf Goethe und Hölderlin, ist das dritte wichtige Forschungsgebiet Schadewaldts gewesen, mit dem er erstmalig 1932 in der Freiburger Universitätsrede „Goethe und das Erleben des antiken Geistes" hervorgetreten ist. Zu der von Ernst Grumach herausgegebenen Sammlung „Goethe und die Antike" (1949) schrieb er das Nachwort („Goethes Beschäftigung mit der Antike"). Die Ergebnisse seiner Untersuchungen über Goethes Verhältnis zur Antike legte er in seinen „Goethestudien" (1963) vor. Mit seiner Denkschrift „Das Goethe-Wörterbuch" (1949) hat er das Projekt des Goethe-Wörterbuchs angeregt, das seit 1966 gemeinsam von der Akademie der Wissenschaften der DDR, der Akademie der Wissenschaften zu Göttingen und der Heidelberger Akademie der Wissenschaften herausgegeben wird; bis zu seinem Tod hat Schadewaldt zur wissenschaftlichen Leitung dieses großen Gemeinschaftsunternehmens gehört. Neben Goethe trat Hölderlin. Hier sind vor allem die Aufsätze „Hölderlin und Homer" (I 1950, II 1953) zu nennen. Aber auch über den Einfluß der Antike auf Schiller, Heinrich von Kleist, Richard Wagner und nicht zuletzt Carl Orff hat Schadewaldt gearbeitet.

Eine wichtige, vielleicht sogar die wichtigste Aufgabe sah Schadewaldt darin, die Wirkung und Wirkungsmöglichkeiten der griechischen Antike auf die Folgezeit bis in die unmittelbare Gegenwart hinein deutlich zu machen, ihre Aktualität auch für die heutige Zeit zu erweisen. Dieses Anliegen bestimmte von Anfang an weitgehend seine wissenschaftliche Arbeit und trat in den beiden letzten Jahrzehnten seines Lebens immer mehr in den Vordergrund. Bereits 1930 suchte er in dem Aufsatz „Begriff und Wesen der antiken Klassik“ aufzuzeigen, worin die Bedeutung der Antike für alle Zeiten begründet liegt. Immer wieder wies er auf die „Modellfunktion“ der antiken Klassik hin, auf ihren „Modellcharakter“, der es ermögliche, die verschiedenen Lösungsmöglichkeiten allgemein gültiger Probleme an den antiken Lösungsversuchen abzulesen. Das Bestreben, die Gegenwartsnähe der Antike zu demonstrieren, die Meisterwerke der griechischen Literatur für unsere Zeit lebendig und fruchtbar zu machen, lag auch der ausgedehnten Übersetzungstätigkeit Schadewaldts zugrunde. Seine Übersetzungen von Dramen des Aischylos, Sophokles, Euripides und Menander sind durch zahlreiche Bühnenaufführungen und Rundfunksendungen in einer breiten Öffentlichkeit wirksam geworden. In seiner Prosaübersetzung der homerischen Odyssee (1958, überarbeitete Neuausgabe 1960) suchte er durch ein „dokumentierendes“ Übersetzen die herbe Sachlichkeit Homers dem heutigen Leser nahezubringen. Das Übersetzen stand am Anfang und Ende seines Lebenswerkes: Das Thema seines öffentlichen Habilitationsvortrags an der Berliner Universität lautete: „Das Problem des Übersetzens“, und als letzte, noch unveröffentlichte Arbeit konnte er eine Übersetzung der homerischen Ilias in rhythmischer Prosa zum Abschluß bringen. Seine Verdienste als Übersetzer sind anerkannt und gewürdigt worden: er war Mitglied der Deutschen Akademie für Sprache und Dichtung in Darmstadt — 1965 erhielt er den Übersetzerpreis dieser Akademie — und des PEN-Clubs.

In einer Zeit, die sich mehr an den drängenden Aufgaben der Gegenwart und der Zukunft als an den großen Leistungen der Vergangenheit orientiert, in einer Zeit, in der jahrhundertealte Normen und Wertvorstellungen ihre Allgemeingültigkeit weitgehend verloren haben, hat Wolfgang Schadewaldt durch sein Lebenswerk in eindrucksvoller Weise den doppelten Auftrag der Klassischen Philologie auf seine Art zu erfüllen versucht: die griechisch-römische Antike zu erforschen und sie gleichzeitig für die Gegenwart lebendig und fruchtbar zu machen.

Friedmar Kühnert

Bibliographie Wolfgang Schadewaldt

Die folgende Bibliographie beruht in der Hauptsache — bis zu den Veröffentlichungen des Jahres 1970 — auf dem ‚Verzeichnis der Schriften Wolfgang Schadewaldts‘ von Reinhard Thurow in: HuH^2 II 829—844. Die neueren Veröffentlichungen wurden mit Hilfe von Maria Schadewaldt und Ingeborg Schudoma hinzugefügt.

Wiederabgedruckte (gegebenenfalls auch neubearbeitete) Aufsätze sind im Anschluß an die Erstveröffentlichung aufgeführt. Neuauflagen und Neuausgaben von selbständigen Publikationen sowie Übersetzungen in andere Sprachen sind jeweils gesondert unter den

Erscheinungsjahren verzeichnet und dort mit einem Sternchen (*) gekennzeichnet; bei der Erstveröffentlichung ist auf die späteren Erscheinungsjahre hingewiesen.

In dem Band ,Goethestudien' (1963) sind die bis dahin erschienenen Arbeiten zu Goethe vereinigt. Die Sammelwerke ,Von Homers Welt und Werk' (^{1}1944–^{4}1965) und ,Hellas und Hesperien' (^{1}1965, ^{2}1970) sind mit zwei Sternchen (**) gekennzeichnet und werden in abgekürzter Form angeführt: HWW und HuH (wobei die hochgestellten Zahlen die Auflagen angeben).

Ein Verzeichnis der Sprechplatten sowie der Bühnenaufführungen und der Rundfunksendungen von Wolfgang Schadewaldts Übersetzungen findet sich, bis zum Jahr 1970 reichend, in ,Antikes Drama auf dem Theater heute' (1970) 45–48 und in HuH^{2} II 845 bis 848.

1924

Zu einer Londoner Kanne aus Cerveteri. Mitteilungen des Deutschen Archäologischen Instituts, Römische Abteilung, Band 38/39, 1923/24, 490f., Tafel 4.2 — HuH^{1} 211–213 — HuH^{2} I 367–369.

1926

Monolog und Selbstgespräch. Untersuchungen zur Formgeschichte der griechischen Tragödie. Neue Philologische Untersuchungen 2, Berlin 1926, 270 S. — Berlin–Zürich–Dublin ^{2}1966.

1927

Das Problem des Übersetzens. Öffentlicher Habilitationsvortrag an der Universität Berlin 1927. Die Antike 3, 1927, 287–303 — HuH^{1} 523–537 — Das Problem des Übersetzens. Wege der Forschung 8, hrsg. von H. J. Störig, Darmstadt 1963, 249–267. ^{2}1969 — HuH^{2} II 608–622.

1928

Der Aufbau des Pindarischen Epinikion. Schriften der Königsberger Gelehrten Gesellschaft, Geisteswissenschaftliche Klasse, 5, 1928, Heft 3. — Gleichzeitig: Halle 1928, 85 S. — Tübingen ^{2}1966.

Besprechung: Eugen Petersen, Die attische Tragödie als Bild- und Bühnenkunst, Bonn 1915. In: Franz Studniczka, Eugen Petersen, Biographisches Jahrbuch für Altertumskunde 48, 1928, 137–140.

Zum «Phrixos» des Euripides. Hermes 63, 1928, 1–14 — HuH^{1} 305–316 — HuH^{2} I 505 bis 515.

1929

Die Geschichtschreibung des Thukydides. Ein Versuch. Berlin 1929, VIII. 100 S. Nach einem Vortrag, gehalten auf der Fachtagung der Klassischen Altertumswissenschaft zu Weimar im Mai 1928.

Sophokles, Aias und Antigone. Neue Wege zur Antike 8, 1929, 61–109.

1930

Begriff und Wesen der antiken Klassik. Die Antike 6, 1930, 265 — 283. — Das Problem des Klassischen und die Antike. Acht Vorträge, hrsg. von W. Jaeger, Leipzig–Berlin 1931, 15 bis 32. Darmstadt ^{2}1961.

1931

Bemerkungen zur «Hecyra» des Terenz. Hermes 66, 1931, 1—29 — HuH¹ 472—494 — HuH² I 722—744 — Die römische Komödie: Plautus und Terenz. Wege der Forschung 236, hrsg. v. E. Lefèvre, Darmstadt 1973, 279—311.

Sinn und Werden der Vergilischen Dichtung. Vortrag, gehalten als Festrede zum zweitausendsten Geburtstag Vergils auf dem Meersburger Ferienkurs am 11. August 1930, in neuer Fassung wiederholt in Freiburg bei einer Veranstaltung der Freunde des humanistischen Gymnasiums. In: Das Erbe der Alten, Zweite Reihe, Heft 20, Aus Roms Zeitwende. Von Wesen und Wirken des Augusteischen Geistes, Leipzig 1931, 67—95 — HuH¹ 498—519 — Wege zu Vergil. Drei Jahrzehnte Begegnungen in Dichtung und Wissenschaft. Wege der Forschung 19, hrsg. von H. Oppermann, Darmstadt 1963, 43—68 — HuH² I 701—722.

1932

Goethe und das Erlebnis des antiken Geistes. Freiburger Universitätsreden 8, 1932, Freiburg 1932, 21 S. — Goethestudien, 1963, 9—21.

Besprechung: Ernst Howald, Die griechische Tragödie, München—Berlin 1930. Gnomon 8, 1932, 1—13 — HuH¹ 335—346 — HuH² I 237—248.

Der Kommos in Aischylos' «Choephoren». Hermes 67, 1932, 312—354 — HuH¹ 106—141 — HuH² I 249—284.

Das Vermächtnis des Perikles. Aus dem zweiten Buche der Historien des Thukydides. Die Antike 8, 1932, 23—34.

1933

Lebenszeit und Greisenalter im frühen Griechentum. Die Antike 9, 1933, 282—302 — HuH¹ 41—59 — HuH² I 109—127.

Lukrez 1, 50. Hermes 68, 1933, 460—464 — HuH¹ 494—498 — HuH² I 744—749.

Die Neuausgabe von Bismarcks «Erinnerung und Gedanke» und die Klassische Philologie. Neue Jahrbücher für Wissenschaft und Jugendbildung 9, 1933, 244—251 — HuH¹ 850—858 — HuH² II 581—589.

1934

Die Anfänge der Geschichtschreibung bei den Griechen. Die Antike 10, 1934, 144—168 — HuH¹ 395—416 — Auszug: Herodot als erster Historiker, in: Herodot. Eine Auswahl aus der neueren Forschung. Wege der Forschung 26, hrsg. von W. Marg, Darmstadt 1962 (gleichzeitig München 1962), 109—121. ²1965 — HuH² I 559—580.

Einzelner und Staat im politischen Denken der Griechen. Vergangenheit und Gegenwart. Zeitschrift für Geschichtsunterricht und politische Erziehung 24, 1934, 16—32.

Besprechung: Leonhard Illig, Zur Form der pindarischen Erzählung, Berlin 1932. Deutsche Literaturzeitung 55, 1934, 1407—1412 — HuH¹ 94—98 — HuH² I 181—186.

Die «Niobe» des Aischylos. Sitzungsberichte der Heidelberger Akademie der Wissenschaften, Phil.-hist. Klasse, 1933/34, 3. Abh., Heidelberg 1934, 32 S. — HuH¹ 141—166 — HuH² I 284—308.

1935

Homerische Szenen I. Hektor und Andromache. Die Antike 11, 1935, 149—170 — HWW¹ 135—161 — HWW²⁻⁴ 207—233.

Sophokles und Athen. Antrittsrede, gehalten an der Universität Leipzig im Januar 1935. Wissenschaft und Gegenwart 7, Frankfurt am Main 1935, 23 S. — HuH¹ 215—230 — HuH² I 370—385.

1936

Aischylos' «Achilleis». Hermes 71, 1936, 25–69 – HuH¹ 166–211 – HuH² I 308–354.

Homerische Szenen II. Die Entscheidung des Achilleus. Die Antike 12, 1936, 173–201 – HWW¹ 162–195 – HWW²⁻⁴ 234–267.

Für Alfred Körte. Widmungsepigramm zu seinem siebzigsten Geburtstag am 5. September 1936. Hermes, Zeitschrift für Klassische Philologie. Alfred Körte-Festheft, Berlin 1936 (Hermes 71, 1936, Heft 3, ohne Widmung). – HuH¹ 1044 – HuH² II 689.

Alfred Körte zum 70. Geburtstag am 5. September 1936. Forschungen und Fortschritte 12, 1936 315 – HuH² II 690f.

Humanismus und Einzelforschung in England. In: Die Antike 14, 1938, 78–80 (Besprechung zu E. R. Dodds, Humanism and Technique in Greek Studies, Oxford 1936).

Sapphoverse auf einer Tonscherbe. In: Die Antike 14, 1938, 77–78.

Zu Sappho. Hermes 71 (Alfred Körte-Festheft), 1936, 363–373 – HuH¹ 66–77 – HuH² I 134–145.

1937

Besprechung: Karl Reinhardt, Sophokles, Frankfurt am Main 1933. Deutsche Literaturzeitung 58, 1937, 997–1002 – HuH¹ 296–300 – HuH² 496–500.

Besprechung: T. B. L. Webster, An Introduction to Sophocles, Oxford 1936. Gnomon 13, 1937, 586–590 – HuH¹ 301–305 – HuH² I 500–504.

1938

Homer und die Homerische Frage. Die Antike 14, 1938, 1–21 – Auch selbständig: Berlin 1938, 21 S. – HWW¹⁻⁴ 9–35.

Iliasstudien. Abhandlungen der Sächsischen Akademie der Wissenschaften, Phil.-hist. Klasse, Band 43, Nr. 6, Leipzig 1938, 182 S. – Leipzig ²1943 – Darmstadt ³1966.

Der Schild des Achilleus. Neue Jahrbücher für Antike und deutsche Bildung 1, 1938, 65–82 – HWW¹ 280–302 – HWW²⁻⁴ 352–374.

1940

Otfrid Becker †. Gnomon 16, 1940, 237–239.

1941

Homerforschung einst und jetzt. Frankfurter Zeitung, 8. Juli 1941.

Winckelmann und Homer. Vortrag, gehalten zur Winckelmann-Feier des Archäologischen Instituts in der Universität Leipzig am 7. Dezember 1940. Leipziger Universitätsreden 6, Leipzig 1941, 69 S. – HuH¹ 600–636 – HuH² II 37–73.

1942

Ital Gelzer †. Gnomon 18, 1942, 61–63.

Homer und sein Jahrhundert. Nach einem im April 1941 auf der Fachtagung der Klassischen Altertumswissenschaft zu Berlin gehaltenen Vortrag. In: Das Neue Bild der Antike, hrsg. von H. Berve, 1. Band: Hellas, Leipzig 1942, 51–90 – HWW¹⁻⁴ 87–129.

Legende von Homer dem fahrenden Sänger. Ein altgriechisches Volksbuch, übersetzt und erläutert von Wolfgang Schadewaldt. Leipzig 1942, 102 S. — Sonderausgabe 1947 — Neuausgabe 1959.

1943

* *Iliasstudien* (1938). 2., verb. Aufl., Leipzig 1943. (Durch Brand im Verlagshaus zerstört.)

Die Rede des Perikles für die Gefallenen. Aus dem Geschichtswerk des Thukydides verdeutscht. In: Hellas. Bilder zur Kultur des Griechentums, hrsg. von H. von Schoenebeck und W. Kraiker, Burg b. M. 1943, 103—107 — Überarbeitet: HuH^{1} 417—423 — HuH^{2} I 593—600 — In: Konrad Gaiser, Das Staatsmodell des Thukydides. Zur Rede des Perikles für die Gefallenen, Heidelberg 1975 (Heidelberger Texte / Didaktische Reihe 8) 7—16.

1944

Die Gestalt des homerischen Sängers. In: HWW$^{1-4}$ 54—86.

Hektors Tod. In: HWW^{1} 196—279 — HWW$^{2-4}$ 268—351.

** *Von Homers Welt und Werk.* Aufsätze und Auslegungen zur homerischen Frage. Leipzig 1944, 354 S., 23 Abb. auf 12 Tafeln — Stuttgart ^{2}1951, ^{3}1959, ^{4}1965 — Neugriechische Übersetzung in Vorbereitung.

Neues zur Ilias. In: HWW$^{1-4}$ 36—53.

Sophokles und das Leid. Potsdamer Vorträge 4, Potsdam 1944, 30 S. — 2., verb. Aufl. 1944, 31 S. — ^{3}1947, ^{4}1948 — HuH^{1} 231—247 — Gottheit und Mensch in der Tragödie des Sophokles. Vorträge von Hans Diller, Wolfgang Schadewaldt, Albin Lesky. Darmstadt 1963, 31—57 — HuH^{2} I 385—401.

1946

Die Heimkehr des Odysseus. In: Taschenbuch für junge Menschen, hrsg. von P. Suhrkamp, Berlin-Frankfurt 1946, 177—224 — HWW$^{3-4}$ 375—412.

1947

Legende von Homer dem fahrenden Sänger. Deutsch von Wolfgang Schadewaldt. Zeichnungen von Graf Heinrich Luckner. Potsdam 1947, 62 S. (Sonderausgabe: Übersetzung und Nachwort; erstmals 1942.)

* *Sophokles und das Leid* (^{2}1944). 3., unv. Aufl., Potsdam 1947.

1948

* *Sophokles und das Leid* (^{3}1947). 4., unv. Aufl., Potsdam 1948.

1949

Goethe und die Antike. Eine Sammlung, hrsg. von Ernst Grumach. Mit einem Nachwort von Wolfgang Schadewaldt. Berlin 1949, 2 Bände. — Das Nachwort «Goethes Beschäftigung mit der Antike» (II, 971—1050) in: Goethestudien, 1963, 23—126.

Goethe und Homer. Trivium 7, 1949, 200—232 — Goethestudien, 1963, 127—157.

Das Goethe-Wörterbuch. Eine Denkschrift. In: Goethe. Neue Folge des Jahrbuchs der Goethe-Gesellschaft 11, 1949, 293—305 — Auch selbständig als Privatdruck, 1955 — Auszug: Zu Goethes Sprache, in: Goethestudien, 1963, 397—404.

Nachruf auf Ludwig Deubner. In: Jahrbuch der Deutschen Akademie der Wissenschaften zu Berlin 1946–1949, Berlin 1950, 139–141 (mit einem Verzeichnis der Arbeiten Ludwig Deubners, S. 142–144). – HuH¹ 1023–1025 – HuH² II 736–739.

Odysseus-Abenteuer. Aus einer gesprächsweisen homerischen Improvisation über Irrfahrer-Angelegenheiten. In: Martin Heideggers Einfluß auf die Wissenschaften. Aus Anlaß seines sechzigsten Geburtstags verfaßt von C. Astrada u. a., Bern 1949, 94–121 – HuH¹ 982–1008 – HuH² II 753–779.

Ulrich von Wilamowitz-Moellendorff zum 100. Geburtstag. Ansprache, gehalten anläßlich der Gedenkfeier der Berliner Akademie der Wissenschaften zum hundertsten Geburtstag Ulrich von Wilamowitz-Moellendorffs. Gymnasium 56, 1949, 80f. – HuH¹ 1021f. – HuH² II 698f.

1950

Hölderlin und Homer. Erster Teil. In: Hölderlin-Jahrbuch 1950, Tübingen 1950, 2–27 – HuH¹ 681–705 – HuH² II 189–213.

Sappho – Welt und Dichtung. Dasein in der Liebe. Potsdam 1950, 191 S. – Auszug: Sappho. Das Schönste (fr. 27a D.), in: Ars Interpretandi 2. Antike Lyrik, hrsg. von W. Eisenhut, Darmstadt 1970, 73–81 – 2. Auflage in Vorbereitung.

Die Wandlung des Homerbildes in der Gegenwart. In: Vom Geist der deutschen Wissenschaft. Ansprachen und Vorträge, Wiesbaden 1950, 29–36 – Universitas 7, 1952, 233–240 – HuH¹ 9–16 – HuH² I 9–16 – *Sammelband 1975.*

1951

** *Von Homers Welt und Werk.* Aufsätze und Auslegungen zur homerischen Frage (1944). 2., verm. Aufl. Stuttgart 1951, 450 S., 28 Abb. auf 16 Tafeln.

Lear und Ödipus. In: Das neue Forum, hrsg. von E. Vietta und G. R. Sellner, Heft 1, Darmstadt 1951, 5–12 – HuH¹ 570–578 – HuH² II 28–36.

Carl Orff, Trionfo di Afrodite. Concerto scenico. Deutsche Übertragung von Wolfgang Schadewaldt. Mainz 1951 (Textbuch) – Neuausgabe 1965, in: Carl Orff, Trionfi. Trittico teatrale, Mainz 1965.

1952

Das Bild der exzentrischen Bahn bei Hölderlin. Vortrag, gehalten zur Jahresversammlung der Friedrich Hölderlin Gesellschaft am 7. Juni 1952 in Tübingen. In: Hölderlin-Jahrbuch 1952, Tübingen 1952, 1–16 – HuH¹ 666–680 – HuH² II 175–189.

Einblick in die Erfindung der Ilias. Ilias und Memnonis. In: Varia Variorum. Festgabe für Karl Reinhardt, dargebracht von Freunden und Schülern zum 14. Februar 1951, Münster – Köln 1952, 13–48 – HWW²⁻⁴ 155–202.

Eudoxos von Knidos und die Lehre vom unbewegten Beweger. In: Satura. Früchte aus der antiken Welt. Otto Weinreich zum 13. März 1951 dargebracht, Baden-Baden 1952, 103–129 – HuH¹ 451–471 – HuH² I 635–655.

Zu einem Florentiner Papyrusbruchstück aus dem «Alkmeon in Psophis» des Euripides. Hermes 80, 1952, 46–66 – HuH¹ 316–334 – Euripides. Wege der Forschung 89, hrsg. von E.-R. Schwinge, Darmstadt 1968, 153–176 – HuH² I 516–534.

Die homerische Gleichniswelt und die kretisch-mykenische Kunst. Zur homerischen Naturanschauung. In: Hermeneia. Festschrift, Otto Regenbogen zum 60. Geburtstag am 14. Februar 1951 dargebracht von Schülern und Freunden, Heidelberg 1952, 9–27, Tafel I–VIII – Gymnasium 60, 1953, 193–209, Tafel IX–XVI – HWW²⁻⁴ 130–154.

1953

Das Drama der Antike in heutiger Sicht. Universitas 8, 1953, 591—599 — HuH[1] 99—106 — Antike und Gegenwart, 1966, 7—15 — HuH[2] I 187—194 — *Sammelband 1975.*

Hölderlin und Homer. Zweiter Teil. In: Hölderlin-Jahrbuch 1953, Tübingen 1953, 1—53 — HuH[1] 705—752 — HuH[2] II 214—261.

Carl Orff, Carmina Burana. Lieder aus der Benediktbeurer Handschrift. Weltliche Gesänge für Soli und Chor mit Begleitung von Instrumenten und mit Bildern. Deutsche Übertragung von Wolfgang Schadewaldt. Mainz 1953 (Mit einem Nachwort von W. Schadewaldt) — Neuausgabe 1965, in: Carl Orff, Trionfi. Trittico teatrale, Mainz 1965.

Carl Orff, Trionfi. Zur Idee des Werks. In: Programm der deutschen konzertanten Erstaufführung unter Leitung von Eugen Jochum, München, Saal der Residenz, 5. März 1953 — HuH[1] 858—860 — HuH[2] II 418—420.

Sappho: «An die Tochter Kleis». In: Studies Presented to David Moore Robinson on his Seventieth Birthday, ed. by G. E. Mylonas and D. Raymond, Washington University, Saint Louis, Missouri, II, 1953, 499—506 — HuH[1] 77—85 — HuH[2] I 145—153.

1954

Bernhard Schweitzer zum sechzigsten Geburtstag am 3. Oktober 1952. In: Neue Beiträge zur Klassischen Altertumswissenschaft. Festschrift zum sechzigsten Geburtstag von Bernhard Schweitzer, hrsg. von R. Lullies, Stuttgart 1954, 5f.

Winckelmann als Exzerptor und Selbstdarsteller. Mit Beiträgen von Walther Rehm. In: Neue Beiträge zur Klassischen Altertumswissenschaft. Festschrift zum sechzigsten Geburtstag von Bernhard Schweitzer, hrsg. von R. Lullies, Stuttgart 1954, 391—409, Tafel 89—91 — HuH[1] 637—657 — HuH[2] II 74—95.

1955

Zur Entstehung der Elfenszene im 2. Teil des «Faust». Deutsche Vierteljahrsschrift für Literaturwissenschaft und Geistesgeschichte 29, 1955, 227—236 — Goethestudien, 1963, 251 bis 262.

Furcht und Mitleid? Zur Deutung des Aristotelischen Tragödiensatzes. Hermes 83, 1955, 129—171 — HuH[1] 346—388 — Antike und Gegenwart, 1966, 16—60 — HuH[2] I 194 bis 236.

Die homerische Frage gestern und heute. Stuttgarter Zeitung, 11. Jahrgang, Nr. 278, 3. Dezember 1955 — HuH[1] 16—21 — HuH[2] I 16—21 — *Sammelband 1975.*

Sophokles, König Ödipus. Deutsch von Wolfgang Schadewaldt. Berlin—Frankfurt am Main 1955, 95 S.

Zur Tragik Schillers. Ansprache bei der Schiller-Gedenkfeier im großen Sendesaal des Frankfurter Funkhauses am 9. Mai 1955. In: Schiller. Reden im Gedenkjahr 1955. Im Auftrag der Deutschen Schiller-Gesellschaft hrsg. von Bernhard Zeller. Veröffentlichungen der Deutschen Schiller-Gesellschaft 21, Stuttgart 1955, 303—316 — HuH[1] 832—842 — HuH[2] II 133—144.

Von der Wirkung des Trauerspiels. In: Sonderheft. Shakespeare-Tage 1955. Schauspielhaus Bochum, 5—8 — Prisma. Blätter des Schauspielhauses Bochum 1954/55, Heft 11, 126 bis 129 — HuH[1] 388—391 — *Sammelband 1975.*

1956

Faust und Helena. Zu Goethes Auffassung vom Schönen und der Realität des Realen im zweiten Teil des „Faust". In: Deutsche Vierteljahrsschrift für Literaturwissenschaft und Geistesgeschichte 30, 1956, 1–40 – Goethestudien, 1963, 165–205 – Goethe im XX. Jahrhundert. Spiegelungen und Deutungen, hrsg. v. Hans Mayer, Hamburg 1967, 241–279.

Furcht und Mitleid? Zu Lessings Deutung des Aristotelischen Tragödiensatzes. Deutsche Vierteljahrsschrift für Literaturwissenschaft und Geistesgeschichte 30, 1956, 137–140 – Gotthold Ephraim Lessing. Wege der Forschung 211, hrsg. von G. und S. Bauer, Darmstadt 1968, 336–342.

Gedanken zu Ziel und Gestaltung des Unterrichts in den alten Sprachen auf der Oberstufe unserer altsprachlichen Gymnasien. Gymnasium 63, 1956, 298–318 – HuH^1 950–972 – HuH^2 II 544–566.

Das Gedenken an die Toten in der Antike. Universitas 11, 1956, 33–39 – HuH^1 60–66 – HuH^2 I 127–134.

Goethes Begriff der Realität. In: Goethe. Neue Folge des Jahrbuchs der Goethe-Gesellschaft 18, 1956, 44–88 – Goethestudien, 1963, 207–249.

Griechische Sternsagen. Fischer Bücherei 129, Frankfurt am Main 1956, 191 S. – Dänische und japanische Übersetzung 1963 – Auszug in: Wort und Sinn. Lesebuch für den Deutschunterricht, III, Paderborn 1964, 87–99 – Neuausgabe 1970.

Hektor in der Ilias. In: Festschrift, Albin Lesky zum 7. Juli 1956 dargebracht von Freunden und Schülern. Wiener Studien 69, 1956, 5–25 – HuH^1 21–38 – HuH^2 I 21–38.

Die Hölderlinische «Antigone» des Sophokles von Carl Orff. In: Programmheft der Württembergischen Staatstheater Stuttgart, 1956 – HuH^1 861 f. – Antike und Gegenwart, 1966, 175 f.

Friedrich Hölderlin. In: Die großen Deutschen. Deutsche Biographie in vier Bänden, hrsg. von H. Heimpel, Th. Heuß und B. Reifenberg, Berlin 1956, II, 354–361 – Unter dem Titel «Hölderlins Weg zu den Göttern» in: Hölderlin-Jahrbuch 9, 1955/56, Tübingen 1957, 174–182 – HuH^1 658–665 – Hölderlin. Beiträge zu seinem Verständnis in unserm Jahrhundert, hrsg. von A. Kelletat. Schriften der Hölderlin-Gesellschaft 3, Tübingen 1961, 333–341 – HuH^2 II 167–175.

Der «König Ödipus» des Sophokles in neuer Deutung. Schweizer Monatshefte 36, 1956, 21–31 – HuH^1 277–287 – HuH^2 I 466–476.

Zum Osterspiel von Carl Orff. In: Programmheft der Württembergischen Staatstheater Stuttgart 8, 1956/57, 118 f. – Neufassung 1960.

Sinn und Wert der humanistischen Bildung im Leben unserer Zeit. Vortrag, gehalten zur Kundgebung der drei humanistischen Gymnasien in Stuttgart am 17. September 1955. In: Mitteilungen des Vereins der ehemaligen Schüler des Eberhard-Ludwigs-Gymnasiums, Sonderheft April 1956, 4–8 – Selbständig: Göttingen–Berlin–Frankfurt am Main 1956, 18. S. – ^{2}1957 – HuH^1 934–941 – Texte zur Antike, hrsg. von O. Leggewie, H. Lenzen, J. R. Zinken. Herder-Bücherei 290, Freiburg–Basel–Wien 1967, 29–36 – HuH^2 II 528–535.

Zum zweiten Stasimon des «König Ödipus». Studi Italiani di Filologia Classica 27/28, dedicato alla memoria di Giorgio Pasquali, 1956, 489–497 – HuH^1 287–294 – HuH^2 I 476 bis 483.

1957

Die Anforderungen der Technik an die Geisteswissenschaften. Vortrag, gehalten auf der Jahresversammlung des Stifterverbandes für die Deutsche Wissenschaft am 25. April 1957 in Wiesbaden. In: Forschung und Wirtschaft. Partner im Fortschritt, hrsg. vom Stifterverband für die Deutsche Wissenschaft 6, 1957, Heft 3 (Nicht im Handel) — Auszug: Stuttgarter Zeitung, 4. Mai 1957 — Auszug: Neue Ruhr Zeitung. Mensch und Werk an Rhein und Ruhr, Ausgabe Nr. 7 im Oktober 1959 — Selbständig, mit einem Anhang «Zum Bildungsproblem» (Bandaufnahme einer im Rahmen der Jahrestagung bei der Aussprache über «Fachliche und allgemeine Bildung» frei gehaltenen Diskussionsrede): Göttingen—Berlin—Frankfurt am Main 1957, 48 S. — HuH[1] 867—891 — Ohne den Anhang: Humanismus. Wege der Forschung 17, hrsg. von H. Oppermann, Darmstadt 1970, 468—492 — HuH[2] II 461—484.

Antike Tragödie auf der modernen Bühne. Zur Geschichte der Rezeption der griechischen Tragödie auf der heutigen Bühne. Sitzungsberichte der Heidelberger Akademie der Wissenschaften, Jahresheft 1955/56, Heidelberg 1957, 37—64, mit 16 Abb. auf 8 Tafeln — Zugleich selbständig (unveränderte Seitenzählung): Heidelberg 1957, mit 16 Abb. auf 8 Tafeln — Auszug: Programmheft. Carl Orff, Oedipus der Tyrann. Ein Trauerspiel des Sophokles von Friedrich Hölderlin. Uraufführung 11. Dezember 1959 im Großen Haus der Württembergischen Staatstheater Stuttgart, S. 26—30 — HuH[1] 543—570 — Antike und Gegenwart, 1966, 67—96 — Sophokles. Wege der Forschung 95, hrsg. von H. Diller, Darmstadt 1967, 500—536 — HuH[2] II 622—650.

Daten zur «Elektra» des Sophokles. In: Programmheft des Landestheaters Württemberg-Hohenzollern 1957/58, 6. Folge, S. 42—45 — HuH[1] 294—296.

Die Metaphysik der Hundemarke. In: Erziehung zur Menschlichkeit. Die Bildung im Umbruch der Zeit. Festschrift für Eduard Spranger zum 75. Geburtstag am 27. Juli 1957, hrsg. von H. W. Bähr, Th. Litt, N. Louvaris und H. Wenke, Tübingen 1957, 587—597 — HuH[1] 1009—1017 — HuH[2] II 814—822.

* *Sinn und Wert der humanistischen Bildung im Leben unserer Zeit* (1956). 2., unv. Aufl., Göttingen—Berlin—Frankfurt am Main 1957.

Sinnvolles Verstehen geistiger Zusammenhänge. In: Eduard Spranger. Bildnis eines geistigen Menschen unserer Zeit. Festschrift zum 75. Geburtstag, Heidelberg 1957, 476—480 — Unter dem Titel «Begegnung mit Eduard Spranger»: HuH[1] 1018—1020 — HuH[2] II 692 bis 695.

Sophokles, Tragödien. Deutsch von Friedrich Hölderlin, hrsg. und eingeleitet von Wolfgang Schadewaldt. Fischer Bücherei 162, Frankfurt am Main 1957, 261 S. — Einleitung und Anhang (S. 9—95, 258—261) «Hölderlins Übersetzung des Sophokles»: HuH[1] 767—824 — Ohne den Anhang: Antike und Gegenwart, 1966, 113—174 — Gekürzt, ohne den Anhang: Über Hölderlin. Aufsätze von Theodor W. Adorno u. a., hrsg. von J. Schmidt, Frankfurt am Main 1970, 237—293 — HuH[2] II 275—332.

Das Weltmodell der Griechen. Ein Vortrag. Die Neue Rundschau 68, 1957, 187—213 — HuH[1] 426—450 — Humanismus. Wege der Forschung 17, hrsg. von H. Oppermann, Darmstadt 1970, 322—352 — HuH[2] I 601—625 — *Sammelband 1975.*

Der «Zerbrochene Krug» von Heinrich von Kleist und Sophokles' «König Ödipus». Schweizer Monatshefte 37, 1957, 311—318 — HuH[1] 843—850 — Heinrich von Kleist. Aufsätze und Essays. Wege der Forschung 147, hrsg. von W. Müller-Seidel, Darmstadt 1967, 317—325 — HuH[2] II 333—340.

1958

Gruß an den großen Humanisten Werner Jaeger. Stuttgarter Zeitung, 14. Jahrgang, Nr. 149, 3. Juli 1958.

Richard Harder †. Gnomon 30, 1958, 73—76 — HuH[1] 1026—1029 — HuH[2] II 739—743.

Heimweh nach Hellas heute? Rheinische Post, 13. Jahrgang, Nr. 15, 18. Januar 1958 — Gymnasium 66, 1959, 1—5 — HuH[1] 929—933 — HuH[2] II 442—447.

Einleitung für die Schallplatte: Homer, Odyssee. Aus dem ersten und fünften Gesang. Zürich—Stuttgart 1958 — HuH^{1} 38—41 — HuH^{2} I 106—108.

Homer, Die Odyssee. Übersetzt in deutsche Prosa von Wolfgang Schadewaldt. Rowohlts Klassiker 29/30, Hamburg 1958, 331 S. — Neuausgabe 1966.

Die «Lysistrata» des Aristophanes in neuer Übersetzung. In: Programmheft der Schwetzinger Festspiele 1958, 21—25 — HuH^{1} 392—394 — HuH^{2} I 543—546.

Der Mensch und die Technik. Natur — Technik — Kunst. Vortrag, gehalten zur Zehnjahresversammlung des Bundesverbandes Steine und Erden zu München am 10. Oktober 1958. Bauverlag Wiesbaden—Berlin 1958, 32 S. (Nicht im Handel) — Auszug: Physikalische Blätter 15, 1959, 337—341 — Auszug: Programmheft. 32. Bühnentechnische Tagung vom 27.—30. Juli 1959 im Nationaltheater Mannheim, S. 66—78 — Natur—Technik—Kunst. Drei Beiträge zum Selbstverständnis der Technik in unserer Zeit, Göttingen—Berlin—Frankfurt am Main 1960, 7—31 — HuH^{1} 892—906 — HuH^{2} II 497—512.

Einleitung für die Schallplatte: Platon, Apologie des Sokrates. Zürich—Stuttgart 1958 — HuH^{1} 424—426 — HuH^{2} I 632—634.

Der Prolog der Odyssee. In: Festschrift für Werner Jaeger. Harvard Studies in Classical Philology 63, 1958, 15—32 — HuH^{2} I 42—58.

Karl Reinhardt und die Klassische Philologie. Schweizer Monatshefte 38, 1958, 737—744 — HuH^{1} 1030—1037 — HuH^{2} II 700—707.

Die Wiedergewinnung antiker Literatur auf dem Wege der nachdichtenden Übersetzung. Deutsche Universitätszeitung 13, 1958, 741—744 — Antike und Gegenwart, 1966, 61—66 — HuH^{1} 538—542.

1959

Zu Aischylos' «Persern». In: Blätter der Städtischen Bühne Heidelberg 1958/59, Heft 9 — HuH^{1} 213—215.

Die Gegenwärtigkeit der Antike in unserer Zeit. Zur Tagung des Deutschen Altphilologen-Verbandes in Stuttgart. Stuttgarter Zeitung, 15. Jahrgang, Nr. 111, 16. Mai 1959 — überarbeitet: Rheinische Post, 15. Jahrgang, Nr. 61, 12. 1960 — HuH^{1} 922—928— HuH^{2} II 436—442.

Goethes Knabenmärchen «Der neue Paris». Eine Deutung. Die Neue Rundschau 70, 1959, 599 bis 616 — Goethestudien, 1963, 263—282.

** *Von Homers Welt und Werk.* Aufsätze und Auslegungen zur homerischen Frage (^{2}1951) 3., erw. Aufl. Stuttgart 1959, 499 S., 28 Abb. auf 16 Tafeln.

Homer und der Becher von Ischia. In: HWW$^{3-4}$ 413—416.

Das humanistische Bildungsideal und die Forderungen unserer Zeit. Vortrag, gehalten im Süddeutschen Rundfunk im Mai 1959 in der Sendereihe «Probleme einer Schulreform». In: Probleme einer Schulreform. Kröners Taschenausgaben 301, Stuttgart 1959, 167—179 — HuH^{1} 942—950 — HuH^{2} II 536—543 — *Sammelband 1975.*

Kleiderdinge. Zur Analyse der Odyssee. Hermes 87 (Hermes Klingnerianus), 1959, 13—26 — HuH^{2} I 79—93.

* *Legende von Homer dem fahrenden Sänger.* Übertragen und erläutert von Wolfgang Schadewaldt. Lebendige Antike. Zürich—Stuttgart 1959, 80 S. (Neuausgabe, erstmals 1942).

Neue Kriterien zur Odyssee-Analyse. Die Wiedererkennung des Odysseus und der Penelope. Sitzungsberichte der Heidelberger Akademie der Wissenschaften, Phil.-hist. Klasse, 1959, 2. Abh., 28 S. — ^{2}1966 — HuH^{2} I 58—77.

Pindars Zehnte Nemeische Ode. In: Martin Heidegger zum siebzigsten Geburtstag. 26. September 1959, Pfullingen 1959, 252—263 — HuH^{1} 85—94 — HuH^{2} I 172—181.

Professor Hommel 60 Jahre alt. Schwäbisches Tagblatt, 15. Jahrgang, Nr. 112, 19. Mai 1959 — HuH^{2} II 695f.

Schillers Griechentum. In: Schiller. Reden im Gedenkjahr 1959. Im Auftrag der Deutschen Schiller-Gesellschaft hrsg. von Bernhard Zeller. Veröffentlichungen der Deutschen Schiller-Gesellschaft 24, Stuttgart 1959, 258—270 — Neufassung 1960.

1960

Anacreonteum Latine versum. Übersetzung eines Anacreonteum (23 Bergk) ins Lateinische; Lösung einer im Juli 1921 im Berliner Proseminar von Eduard Fraenkel gestellten Aufgabe. In: HuH^{1} 520 — Viva Camena. Latina huius aetatis carmina collecta et edita ab Iosepho Eberle, in aedibus Artemidos Turici et Stuttgardiae MCMLXI, p. 129— HuH^{2} I 797.

Antrittsrede, gehalten im Sommer 1943 in der Preußischen Akademie der Wissenschaften zu Berlin (Jahrbuch der Preußischen Akademie der Wissenschaften 1943, 172—174; ungedruckt). In: HuH^{1} 1038—1040 — HuH^{2} II 782—784.

Antrittsrede, gehalten im Sommer 1958 in der Heidelberger Akademie der Wissenschaften. In: Jahreshefte der Heidelberger Akademie der Wissenschaften 1958/59, Heidelberg 1960, 3—5 — HuH^{1} 1040—1043 — HuH^{2} II 784—787.

Die Begriffe «Natur» und «Technik» bei den Griechen. Zwei Rundfunkvorträge zum Selbstverständnis der Technik in unserer Zeit, gehalten am 25. Juni und 10. Juli 1959 im Norddeutschen Rundfunk Hannover. In: Natur—Technik—Kunst. Drei Beiträge zum Selbstverständnis der Technik in unserer Zeit, Göttingen—Berlin—Frankfurt am Main 1960, 35—53 — HuH^{1} 907—919 — HuH^{2} II 512—524.

Daten zu den «Sieben gegen Theben» des Aischylos. In: Programmheft des Landestheaters Württemberg-Hohenzollern 1959/60, 16. Folge, S. 122—124.

Der Eid des Hippokrates, griechisch und deutsch gesprochen von Wolfgang Schadewaldt. Deutsche Grammophon Gesellschaft 1960 — HuH^{2} I 655f. (Übersetzung).

Einleitung zur «Antigone» des Sophokles von Hölderlin in der Vertonung von Carl Orff. Drei Sendungen des Bayerischen Rundfunks im April 1959. In: HuH^{1} 247—277 — HuH^{2} I 434—465.

Die Empedokles-Tragödie Hölderlins. Vortrag, gehalten bei einer Morgenfeier im Staatstheater Stuttgart als Einführung zur Neueinstudierung der Tragödie am 18. November 1956. In: Hölderlin-Jahrbuch II, 1958/60, Tübingen 1960, 40—54 — HuH^{1} 753—766 — Auszug: Blätter des Deutschen Theaters in Göttingen, Spielzeit 1965/66, 16. Jahr, Heft 266, S. 134—137 — Antike und Gegenwart, 1966, 97—112 — HuH^{2} 261—275.

Fausts Ende und die «Achilleis». In: Weltbewohner und Weimaraner. Ernst Beutler zugedacht, hrsg. von B. Reifenberg und E. Staiger, Zürich 1960, 243—260 — Goethestudien, 1963, 283—300.

Zu Werner Heisenbergs Darstellung der Entwicklung der modernen Atomphysik. Beitrag zu einem Rundfunkgespräch. Sendung des Bayerischen Rundfunks am 22. Mai 1958. In: Natur—Technik—Kunst. Drei Beiträge zum Selbstverständnis der Technik in unserer Zeit, Göttingen—Berlin—Frankfurt am Main 1960, 57—61 — HuH^{1} 920—922 — HuH^{2} II 525—527.

Der Helioszorn in der Odyssee. In: Studi in onore di Luigi Castiglioni, Florenz 1960, (II) 861—876 — HuH^{2} I 93—104.

** *Hellas und Hesperien.* Gesammelte Schriften zur Antike und zur neueren Literatur. Zum sechzigsten Geburtstag von Wolfgang Schadewaldt am 15. März 1960 unter Mitarbeit von Klaus Bartels hrsg. von Ernst Zinn. Zürich—Stuttgart 1960, 1072 S., 6 Tafeln — Neuausgabe 1970.

Der «König Ödipus» des Sophokles. Daten zur Erläuterung der Tragödie. In: Programmheft. Städtische Bühnen Bielefeld. Stadttheater. 1959/60, 12. Folge.

Mein Gedicht. Hölderlin, Der Spaziergang. Die Zeit, 15. Jahrgang, Nr. 21, 20. Mai 1960, S. 6.

Von der Mündlichkeit des Worts. Stuttgarter Zeitung, 16. Jahrgang, Nr. 167, 23. Juli 1960, S. 49 — Neufassung 1968.

Natur — Technik — Kunst. Drei Beiträge zum Selbstverständnis der Technik in unserer Zeit. Göttingen—Berlin—Frankfurt am Main 1960, 64 S. Drei Vorträge, gehalten 1958. 1959.

Shakespeare und die griechische Tragödie. Sophokles' «Elektra» und «Hamlet». Vortrag, gehalten auf der Jahresversammlung der Deutschen Shakespeare-Gesellschaft zu Bochum am 26. April 1955. In: Jahrbuch der Deutschen Shakespeare-Gesellschaft 96, 1960, Heidelberg 1960, 7—34 — HuH^1 578—599 — HuH^2 II 7—27.

Die Situation der Klassischen Philologie heute. Vortrag, gehalten im Südfunk Stuttgart am 19. Dezember 1957 in der Sendereihe «Die Situation der Wissenschaften heute». In: HuH^1 973—981 — HuH^2 II 589—598.

Zur Uraufführung des «Oedipus, der Tyrann» nach Hölderlin von Carl Orff. Einführung zur gleichzeitigen Übertragung der Uraufführung im Südfunk Stuttgart am 11. Dezember 1959. In: HuH^1 866f. — Antike und Gegenwart, 1966, 177f.

Der Wandel des Griechenlandbildes in den letzten fünfzig Jahren. In: Begegnung mit Griechenland. Aus dem Programm der Griechischen Woche. Radio Bremen. Bremer Beiträge 1, Bremen 1960, 38—53 — Neufassung 1966.

Der Weg Schillers zu den Griechen. In: Die Gegenwart der Griechen im neueren Denken. Festschrift für Hans-Georg Gadamer zum 60. Geburtstag, hrsg. von D. Henrich, W. Schulz, K.-H. Volkmann-Schluck, Tübingen 1960, 225—232 — Jahrbuch der Deutschen Schiller-Gesellschaft 4, 1960, Stuttgart 1960, 90—97 — HuH^1 825—831 — HuH^2 II 127—133.

Das Werk Carl Orffs und sein neues Osterspiel. Ein Rundfunkvortrag. In: HuH^1 862—865 — HuH^2 II 420—423 (Zusammenfassung aus: Zum Osterspiel von Carl Orff, 1956, und: Das Werk Carl Orffs und sein neues Osterspiel. Sendung des Süddeutschen Rundfunks am 8. April 1956; ungedruckt).

Das Wort der Dichtung. Mythos und Logos. Vortrag, gehalten anläßlich der Vortragsreihe Wort und Wirklichkeit, veranstaltet von der Bayerischen Akademie der Schönen Künste vom 11. bis 15. Juli 1960 in der Aula der Universität München. In: Wort und Wirklichkeit. Sechste Folge des Jahrbuchs Gestalt und Gedanke, München 1960, 90—128 — Auszug: Süddeutsche Zeitung, Nr. 170, 16./17. Juli 1960 — Norwegische Übersetzung 1961 — Goethestudien, 1963, 405—432 — Sprache und Wirklichkeit. Essays. Deutscher Taschenbuch Verlag 432, München 1957, 91—121 — HuH^2 I 750—772.

1961

* *Diktningens ord.* Mythos og Logos. In: Litteraturforståelse. Diktning og Kritikk, Oslo 1961, 169—190 (Das Wort der Dichtung, 1960).

Epigramma invectivum, pluteo auditorii inscriptum. In: Viva Camena. Latina huius aetatis carmina collecta et edita ab Iosepho Eberle, in aedibus Artemidos Turici et Stuttgardiae MCMLXI, p. 129 (Collegium philosophicum) — HuH^2 II 823.

Die erste Begegnung des Odysseus und der Penelope. Odyssee 18, 158ff. Sitzungsberichte der Heidelberger Akademie der Wissenschaften, Jahresheft 1960/61, 3—5 — HuH^2 I 77f.

Des Geistes Wirksamkeit. Zum Tode von Werner Jaeger. Frankfurter Allgemeine Zeitung, Nr. 256, 3. November 1961, S. 28.

Goethe, Plutarch und Sophokles. Deutsche Vierteljahrsschrift für Literaturwissenschaft und Geistesgeschichte 35, 1961, 64—68 — Goethestudien, 1963, 159—163.

Lob Berlins. In: Berlin in Vergangenheit und Gegenwart. Tübinger Vorträge, hrsg. von H. Rothfels, Tübingen 1961, 75—92 — Neufassung 1963.

Das Reh. In: Dauer im Wandel. Festschrift zum 70. Geburtstag von Carl J. Burckhardt, hrsg. von H. Rinn und M. Rychner, München 1961, 375—380 — Schlesien. Eine Vierteljahresschrift für Kunst, Wissenschaft und Volkstum 10, 1965, 222—225 — Neufassung 1968.

Die Wappnung des Eteokles. Zu Aischylos' «Sieben gegen Theben». In: Eranion. Festschrift für Hildebrecht Hommel, dargebracht von seinen Tübinger Freunden und Kollegen. Unter Mitwirkung von E. Zinn hrsg. von J. Kroymann, Tübingen 1961, 105—116 — HuH² I 357—367.

1962

Gedenkrede auf Werner Jaeger. 30. Juli 1888 — 19. Oktober 1961. Gehalten an der Freien Universität Berlin am 12. Juli 1962. Schweizer Monatshefte 42, 1962, 755—769 — Selbständige Ausgabe 1963 (HuH² II 707—722).

Gerhart Hauptmann und die Griechen. Zum «Bogen des Odysseus». In: Jahrhundertfeier für Gerhart Hauptmann, 15.—21. November 1962, Köln 1962, 25—28 — HuH² II 406—410.

Humanität und Technik. Vortrag, gehalten auf dem Deutschen Ingenieurtag am 25. Mai 1962 in Karlsruhe. In: VDI—Zeitschrift 104, 1062, 1327—1331 — Englische Übersetzung 1963 — Technik und Humanität sind einander zugeordnet, Berichte und Informationen des Österreichischen Forschungsinstituts für Wirtschaft und Politik 19, 1964, 15f. — Spanische Übersetzung 1966.

Beiträge in: Pindar, Siegeslieder. Deutsche Übertragungen, zusammengestellt von U. Hölscher. Mit einem Nachwort von B. Snell. Fischer Bücherei, Exempla Classica 52, Frankfurt am Main—Hamburg 1962. (Der Band enthält neben Nem. 10 [1959] folgende neu übersetzte Oden: Ol. 7, 11, 13; Pyth. 12; Nem. 1, 2, 6, 7, 11.)

Das Religiös-Humane als Grundlage der geschichtlichen Objektivität bei Herodot. In: Geschichte und Gegenwartsbewußtsein. Festschrift für Hans Rothfels zum 70. Geburtstag, hrsg. von W. Besson und F. Frhr . Hiller von Gaertringen, Göttingen 1962, 217—229 — Herodot. Eine Auswahl aus der neueren Forschung. Wege der Forschung 26, hrsg. von W. Marg, Darmstadt 1962, 185—201. ²1965 — HuH² I 580—593.

Richard Wagner und die Griechen. Erster Teil. In: Lohengrin-Programmheft der Bayreuther Festspiele 1962, 2—32. — Richard Wagner und das neue Bayreuth, hrsg. von Wieland Wagner. List Bücher 237, München 1962, 149—174 — HuH² II 343—365 — Die in den Jahren 1962—1964 anläßlich der Bayreuther Festspiele gehaltenen drei Vorträge wurden auf Einladung Herbert von Karajans in den Jahren 1967, 1968 und 1969 anläßlich der Osterfestspiele in Salzburg wiederholt.

1963

Das Problem der Übersetzung antiker Dichtung. Referat und Schlußwort. In: Artemis-Symposion, Zürich—Stuttgart 1963, 22—41 und 51—55 (Nicht im Handel).

Zu den Begriffen Augenblick · Moment · Stunde. In: Goethestudien, 1963, 433—446.

Berlin und die Berliner. In: Berliner Geist. Fünf Vorträge der Bayerischen Akademie der Schönen Künste, Berlin 1963, 11—49 — HuH² II 787—808 — (Neufassung des Vortrags von 1961).

Gedenkrede auf Werner Jaeger. 1888–1961. Mit einem Verzeichnis der Schriften Werner Jaegers. Berlin 1963, 39 S. — HuH^2 II 707–722.

Goethes «Achilleis». Rekonstruktion der Dichtung. In: Goethestudien, 1963, 301–395.

Goethestudien. Natur und Altertum. Zürich–Stuttgart 1963, 531 S.

Der Gott von Delphi und die Humanitätsidee. Vortrag, gehalten anläßlich der Entgegennahme des Reuchlin-Preises der Stadt Pforzheim zu Pforzheim am 16. November 1963. Selbstverlag der Stadt Pforzheim 1963, 28 S. (Nicht im Handel) — Gekürzt: Stuttgarter Zeitung, 20. Jahrgang, Nr. 145, 27. Juni 1964 — Überarbeitung 1965 — Griechische Übersetzung 1965 — Neuausgabe 1965.

* *Graeske Stjernesagn,* på Dansk ved Otto Foss. Kopenhagen 1963, 172 S. (Griechische Sternsagen, 1956).

Beiträge in: Griechische Lyrik. Von den Anfängen bis zu Pindar. Griechisch und deutsch, hrsg. von G. Wirth. Rowohlts Klassiker 140/142, Hamburg 1963 (Drei Übertragungen aus: Sappho — Welt und Dichtung, 1950).

Griechische Sternsagen, erläutert von Yoshima Sugiyama, Tokio 1963, 57 S. (Abschnitt 2 und 4 der deutschen Ausgabe von 1956 mit japanischen Erläuterungen).

* *Griechische Sternsagen* (1956). In das Japanische übersetzt von T. Kawara, Tokio 1963, 230 S.

Beiträge in: Liebesdichtung der Griechen und Römer. Zweisprachig, ausgewählt und zum Teil neu übertragen von H. Gasse. Sammlung Dieterich 141, Leipzig 1963 (Sieben Übertragungen aus: Sappho — Welt und Dichtung, 1950).

Menander. Das Schiedsgericht — Der Menschenfeind. Mit einem Nachwort von Wolfgang Schadewaldt. Fischer Bücherei, Exempla Classica 72, Frankfurt–Hamburg 1963 (Darin: Das Schiedsgericht. Für die Bühne übersetzt und ergänzt von Wolfgang Schadewaldt / Der Menschenfeind. Deutsche Bühnenbearbeitung von Otto Vicenzi) — Das Nachwort, S. 136–150, auch in: HuH^2 I 549–558.

Der Mensch in der technischen Welt. Festvortrag anläßlich des 30. Fortbildungskurses für Ärzte in Regensburg am 23. Mai 1963, o. J. (1963), 3–22 (Nicht im Handel) — Bayerisches Ärzteblatt 18, 1963, 749–756 — Universitas 19, 1964, 449–467.

Sophokles, Die Tragödien. Mit einem Nachwort von Wolfgang Schadewaldt. Fischer Bücherei, Exempla Classica 81, Frankfurt am Main — Hamburg 1963 (Darin: Übersetzung des «König Ödipus» und der «Elektra»; weitere Übersetzungen von E. Buschor, K. Reinhardt, E. Staiger) — Das Nachwort, S. 375–390, wiederholt in der Sophokles-Ausgabe von 1968.

* *Technology and Man.* In: German Opinion on Problems of Today, ed. by W. Leifer. 2, 1963 «Man and Technology», München 1963, 51–59 (Humanität und Technik, 1962).

Übersetzung als geistige Aufgabe. Vortrag, gehalten am 23. Oktober 1958. In: Theater im Gespräch. Ein Forum der Dramaturgie. Aus den Tagungen 1953–1960 der Deutschen Dramaturgischen Gesellschaft ausgewählt und hrsg. von F. Schultze, München–Wien 1963, 325–339.

Richard Wagner und die Griechen. Zweiter Teil. Die Ringdichtung und Aischylos' «Prometheus». In: Meistersinger-Programmheft der Bayreuther Festspiele 1963, 24–44 — HuH^2 II 365–386.

1964

Beitrag in: Abschied von der Antike? Eine Enquête über die Rolle des griechisch-lateinischen Geisteserbes in der Bildungsgesellschaft von morgen. Wort und Wahrheit 19, 1964, 128f.

Beiträge in: Griechische Lyrik in deutschen Übertragungen. Eine Auswahl mit Anmerkungen und

Nachwort von W. Marg. Reclams Universal-Bibliothek 1921/1923, Stuttgart 1964 (Drei Übertragungen, nach verschiedenen Vorlagen).

Griechisches Theater. Deutsch von Wolfgang Schadewaldt. Frankfurt am Main 1964, 535 S. (Aischylos: Die Perser, Die Sieben gegen Theben. Sophokles: Antigone, König Ödipus, Elektra. Aristophanes: Die Vögel, Lysistrata. Menander: Das Schiedsgericht) — Daten zu Aischylos' «Persern», S. 503—505, auch in: Programmheft des Landestheaters Württemberg-Hohenzollern 1965/66, 11. Folge; HuH^2 I 354—357; Wege zu Aischylos. Wege der Forschung 87, hrsg. v. H. Hommel, Darmstadt 1974, Bd. II, 3—6 —Daten zu Aristophanes' «Vögeln» auch in HuH^2 I 540—543.

Einführung zu: Otto Rieth, Die Kunst Menanders in den «Adelphen» des Terenz. Mit einem Nachwort hrsg. von Konrad Gaiser, Hildesheim 1964.

Der Umfang des Begriffs der Literatur in der Antike. Vortrag, gehalten am 19. Oktober 1963 anläßlich der Herbsttagung in Darmstadt. In: Jahrbuch der Deutschen Akademie für Sprache und Dichtung Darmstadt 1963, Heidelberg—Darmstadt 1964, 98—115 — Auszug: Stuttgarter Zeitung, 20. Jahrgang, Nr. 62, 14. März 1964 — HuH^2 I 782—796 — Literatur und Dichtung. Versuch einer Begriffsbestimmung, hrsg. v. Horst Rüdiger, Stuttgart 1973, 12—25.

Richard Wagner und die Griechen. Dritter Teil: Nachlese. In: Meistersinger-Programmheft der Bayreuther Festspiele 1964, 3—30 — HuH^2 II 386—405.

1965

Gedenkworte für Werner Jaeger. In: Orden Pour le Mérite für Wissenschaften und Künste Reden und Gedenkworte 6, 1963/64, Heidelberg 1965, 31—43 — HuH^2 II 722—730.

Gedenkworte für Eduard Spranger. In: Orden Pour le Mérite für Wissenschaften und Künste. Reden und Gedenkworte 6, 1963/64, Heidelberg 1965, 121—131 — HuH^2 II 730—736.

Der Gott von Delphi und die Humanitätsidee. Vortrag, gehalten am 14. Mai 1964 in Athen. In: Wissenschaftliches Jahrbuch der Philosophischen Fakultät der Universität Athen (EEAth.) 15, 1964/65, Athen 1965, 59—75 — Auch selbständig, mit griechischer Zusammenfassung, in: Schriftenreihe der Griechischen Humanistischen Gesellschaft, 1. Reihe, Heft 26, Athen 1965, 23 S. (Neufassung des Vortrags von 1963) — *Sammelband 1975*.

Der Gott von Delphi und die Humanitätsidee. Vortrag, gehalten am 14. Mai 1964 in Athen. Griechische Übersetzung und Anmerkungen von Konstantin J. Merentitis. Athen 1965, 38 S.

* *Der Gott von Delphi und die Humanitätsidee*. Vortrag, gehalten anläßlich der Entgegennahme des Reuchlin-Preises der Stadt Pforzheim am 16. November 1963. Opuscula aus Wissenschaft und Dichtung 23, Pfullingen 1965, 31 S. (Neuausgabe in der überarbeiteten Fassung von 1964) — HuH^2 I 669—685.

** *Von Homers Welt und Werk*. Aufsätze und Auslegungen zur homerischen Frage (^{3}1959). 4., verb. Aufl. Stuttgart 1965.

* *Técnica y cultura desde la antiqua Grecia*. Revista de Occidente 3, 1965, 1—19 (Die Welt der modernen Technik ..., 1965).

Die Übersetzung im Zeitalter der Kommunikation. Festansprache anläßlich der Eröffnung der Frankfurter Buchmesse 1965. Börsenblatt für den Deutschen Buchhandel 21, 1965, 2267 bis 2270 — Der Übersetzer. Diskussionsbeiträge und Informationen 2, 1965, Nr. 12, 7. Dezember 1965, S. 1—3 — Staatsanzeiger für Baden-Württemberg, Nr. 100, 18. Dezember 1965, S. 5 — Nachrichten für Dokumentation 17, 1966, 1—5 — Integritas. Geistige Wandlung und menschliche Wirklichkeit, hrsg. von D. Stolte und R. Wisser. Karl Holzamer gewidmet. Tübingen 1966, 550—559 — HuH^2 II 680—688.

Die Welt der modernen Technik und die altgriechische Kulturidee. Vortrag gehalten in Athen 1964. Hellenika. Blätter der Vereinigung der Deutsch-Griechischen Gesellschaften 1965, Heft 1, 14–21 – Spanische Übersetzung 1965 – Schweizer Monatshefte 47, 1967, 969–983 – Französische Übersetzung 1967 – HuH² II 485–497.

1966

Antike und Gegenwart. Über die Tragödie. Deutscher Taschenbuch Verlag 342, München 1966, 178 S. (Acht Aufsätze, ausgewählt aus HuH¹).

* *Der Aufbau des Pindarischen Epinikion* (1928). 2., unv. Aufl. Tübingen 1966 – Lizenzausgabe: Darmstadt ²1966.

Einführung zu: Zeitalter der Menschheit. Eine Weltkulturgeschichte. Klassisches Griechenland, von Cecil M. Bowra und der Redaktion der Time-Life-Bücher, Time-Life International (Nederland) N. V. Deutsche Ausgabe 1966, 7 – HuH² II 447f.

Experimentelle Philologie. In: Donum natalicium, Albin Lesky zum 7. Juli 1966 dargebracht von Freunden und Schülern. Wiener Studien 79, 1966, 66–80 – HuH² I 483–496.

* *Homer, Die Odyssee.* Deutsch von Wolfgang Schadewaldt. Bibliothek der Alten Welt. Zürich–Stuttgart 1966, 448 S. (Überarbeitete Neuausgabe der Fassung von 1958) – Das «Verzeichnis der dem Dichter B angehörenden Partien der Odyssee» auch in: HuH² I 105 – Lizenzausgabe 1970.

Humanidad y técnica. Folia Humanistica 4, 1966, 455–465 (Humanität und Technik, 1962).

* *Iliasstudien* (²1943). 3., unv. Aufl. Darmstadt 1966.

Lebensgang. In: Forscher und Gelehrte, hrsg. von W. Ernst Böhm in Zusammenarbeit mit G. Paehlke, Stuttgart 1966, 101f. – Überarbeitet: HuH² II 780f.

* *Monolog und Selbstgespräch.* Untersuchungen zur Formgeschichte der griechischen Tragödie (1926). 2., unv. Aufl., Berlin–Zürich–Dublin 1966.

* *Neue Kriterien zur Odyssee-Analyse.* Die Wiedererkennung des Odysseus und der Penelope (1959). 2., unv. Aufl., Heidelberg 1966.

Der Tod des Empedokles. Zur gegenwärtigen Bühnen-Einrichtung. In: Blätter des Deutschen Theaters in Göttingen, Spielzeit 1965/66, 16. Jahr, Heft 266, S. 137–140.

Die Vergegenwärtigung des antiken Dramas auf der heutigen Bühne. In: Theaterblätter. Bühnen der Landeshauptstadt Kiel. Spielzeit 1965–1966. Die Vögel, S. 265–267.

Wandel des Griechenbildes. Hellenika. Zeitschrift für deutsch-griechische kulturelle und wirtschaftliche Zusammenarbeit, 1966, Heft 3, 9–16. – Texte zur Antike, hrsg. von O. Leggewie, H. Lenzen, J. R. Zinken. Herder-Bücherei 290, Freiburg–Basel–Wien 1967, 191–202 – HuH² II 448–460 (Neufassung des Rundfunkvortrags von 1960).

Aus der Werkstatt meines Übersetzens. Dargetan an der Anrufung des Eros in Sophokles' «Antigone». Schweizer Monatshefte 46, 1966, 851–859 – HuH² II 671–680.

Zwei Pindar-Oden. In: Jahresring 66/67, Stuttgart 1966, 241–254 – HuH² I 162–172 (Pyth. 1 und 3).

1967

Aristophanes, «Die Acharner oder Der private Sonderfrieden». Deutsche Übersetzung und Bühnenbearbeitung. Bühnenmanuskript des Suhrkamp-Verlags, Frankfurt a. M. 1967.

Einblick in die Werkstätte meiner Arbeit. Für Dino Larese. In: Dino Larese, Wolfgang Schadewaldt. Eine Lebensskizze. Amriswiler Bücherei, Amriswil 1967, 22–40 – HuH² II 598–606.

Gedenkworte für Thomas Stearns Eliot. In: Orden Pour le Mérite für Wissenschaften und

Künste. Reden und Gedenkworte 7, 1965/66, Heidelberg 1967, 45–54 – HuH² II 411–417.

* *Le monde de la technique moderne et l'idée de culture dans la philolophie de la Grèce antique.* In: The Living Heritage of Greek Antiquity – L'Héritage vivant de l'Antiquité grecque, Paris 1967, 39–55 (Die Welt der modernen Technik ..., 1965).

Sophokles – Ferment unserer Bühne. Siegfried Melchinger über griechische Tragödie und moderne Theaterpraxis. Frankfurter Allgemeine Zeitung, Nr. 88, 15. April 1967 (Besprechung zu S. Melchinger, Sophokles. Friedrichs Dramatiker des Welttheaters 12, Hannover 1966).

Richard Wagner und die Griechen. Theater heute, 1967, Heft 10, 2–4.

Erwin Wolff †. Gnomon 39, 1967, 319f. – HuH² II 747–749.

Zwei Oden Pindars auf Theron von Akragas, Sieger zu Olympia mit dem Viergespann. In: Soldatentum und Kultur. Festschrift zum siebzigsten Geburtstag von Hans Speidel, hrsg. von M. Horst, Berlin 1967, 91–102 – HuH² I 153–161 (Ol. 3 und 2; Fr. 129, 131a, 130, 131b, 133 Snell).

1968

Brünnhilde und Prometheus. Zu Richard Wagners «Walküre». In: Programmheft. Osterfestspiele Salzburg 1968, S. 63–66.

Von der Mündlichkeit des Wortes. In: Sprache und Politik. Festgabe für Dolf Sternberger zum sechzigsten Geburtstag, hrsg. von C.-J. Friedrich und B. Reifenberg, Heidelberg 1968, 68–77 – HuH² I 772–781.

Carl Orff und die griechische Tragödie. «Antigone» «Ödipus» «Prometheus». In: Prometheus. Mythos–Drama–Musik. Beiträge zu Carl Orffs Musikdrama nach Aischylos, hrsg. von F. Willnauer, Tübingen 1968, 27–42 – HuH² II 423–435.

Das Reh. In: Jahresring 68/69, Stuttgart 1968, 160–165 – HuH² II 809–813 (Überarbeitete Fassung, erstmals 1961).

Bernhard Schweitzer †. 3. 10. 1892 – 12. 7. 1966. In: Jahrbuch der Heidelberger Akademie der Wissenschaften 1966/67, Heidelberg 1968, 100–104 – HuH² II 743–747.

Sophokles, «Ödipus auf Kolonos», Deutsche Übersetzung. Bühnenmanuskript des Suhrkamp-Verlags, Frankfurt a. M. 1968.

Sophokles, Tragödien, Hrsg. und mit einem Nachwort versehen von Wolfgang Schadewaldt. Bibliothek der Alten Welt. Zürich–Stuttgart 1968, 458 S. (Der Band enthält «Aias», «Antigone», «König Ödipus», «Elektra» in der Übersetzung von W. Schadewaldt sowie Trachinierinnen», «Philoktet», «Ödipus auf Kolonos» in der Übersetzung von Ernst Buschor) – Das Nachwort «Leben und Werk», S. 415–454, auch in: HuH² I 402–434.

Winckelmann und Rilke. Zwei Beschreibungen des Apollon. Opuscula aus Wissenschaft und Dichtung 33, Pfullingen 1968, 32 S., 8 Tafeln. – HuH² II 95–116.

1969

Antikes und Modernes in Schillers «Braut von Messina». In: Jahrbuch der Deutschen Schiller-Gesellschaft 13, 1969, Stuttgart 1969, 286–307 – HuH² II 144–166.

Der harte Kern des Epos. Rheinische Post, 24. Jahrgang, Nr. 182, 9. August 1969 – Unter dem Titel «Zur Geschichtlichkeit des Troischen Krieges»: HuH² I 38–42.

Sophokles, Elektra. Übersetzung und Nachwort von Wolfgang Schadewaldt. Reclams Universal-Bibliothek 711, Stuttgart 1969, 80 S.

Die Acharner des Aristophanes. Daten zur Übersetzung und Bühnenbearbeitung. In: HuH² I 535–540.

Antikes Drama auf dem Theater heute. Übersetzung Inszenierung. Pfullingen 1970, 48 S., 17 Abb. – HuH² II 650–671.

Aristophanes, Die Vögel. Bühnenfassung. Übersetzung und Bearbeitung. Insel-Bücherei 946, Frankfurt am Main 1970, 112 S.

Das Bild des griechischen Menschen. Vortrag, gehalten im Bayerischen Rundfunk am 9. April 1968. In: Schweizer Monatshefte 49, 1969/70, 1079–1092 – HuH² I 657–669.

Das Bühnenwerk Carl Orffs. Grundzüge und Gruppierung. In: Carl Orff. Das Bühnenwerk. Mit einem Vorwort von Wolfgang Schadewaldt und einem chronologischen Werkverzeichnis. Katalog der Ausstellung anläßlich des fünfundsiebzigsten Geburtstags. 10. Juni bis 31. Juli 1970. Bayerische Staatsbibliothek München.

Das Bühnenwerk Carl Orffs. Vortrag, gehalten bei dem Festakt zum fünfundsiebzigsten Geburtstag Carl Orffs in der Bayerischen Staatsbibliothek zu München am 10. Juni 1970. In: Schönere Heimat, hrsg. vom Bayerischen Landesverein für Heimatpflege, 59, 1970, Heft 3, 561–566.

Daten zu den «Fröschen» des Aristophanes. In: HuH² I 546–549.

* *Griechische Sternsagen.* Mit Zeichnungen von Celestino Piatti. Ungekürzte, neubearbeitete Ausgabe. Deutscher Taschenbuch Verlag 660, München 1970, 171 S. (Erste Ausgabe 1956).

** *Hellas und Hesperien.* Gesammelte Schriften zur Antike und zur neueren Literatur in zwei Bänden. Zum siebzigsten Geburtstag von Wolfgang Schadewaldt am 15. März 1970. Zweite, neugestaltete und vermehrte Ausgabe unter Mitarbeit von Klaus Bartels herausgegeben von Reinhard Thurow und Ernst Zinn. Erster Band: Zur Antike, Zweiter Band: Antike und Gegenwart, Zürich und Stuttgart 1970, 827 u. 864 S.

* *Homer: Die Odyssee.* Deutsch von Wolfgang Schadewaldt, Zürich 1970 (Lizenzausgabe nach der Ausgabe 1966, für den Buchclub Ex Libris Zürich).

Lehren, Lernen, Erziehen, Bilden. Vortrag, gehalten am Dies Universitatis der Universität Tübingen am 13. November 1953. In: HuH² II 566–581.

Nachruf auf Kurd von Hardt. Nach dem Kallimachos. In: HuH² II 752.

Novo Ennio meo. Josef Eberle, dem Dichter in deutscher, schwäbischer und lateinischer Sprache zum fünfundsiebzigsten Geburtstag. 8. September 1966. In: HuH² II 697.

Carl Orff. Der Komponist – Der Dichter. In: Catulli Carmina – Carmina Burana. Programm. Bühnen der Stadt Köln, 1970.

Die «Philologische Methode». Ein Apophthegma von Ulrich von Wilamowitz-Moellendorff, niedergeschrieben für Wolfgang Schmidt auf dessen Wunsch als Diskussionsbeitrag für eine Aussprache über die Philologische Interpretation. In: HuH² II 606f.

Platon und Kratylos. Ein Hinweis. In: HuH² I 626–632. – Philomathes. Studies and Essays in the Humanities in Memory of Philip Merlan, Den Haag 1971, 3–11.

Sappho. In: Die Großen der Weltgeschichte, hrsg. v. K. Fassmann unter Mitwirkung v. M. Bill, Bd I, Zürich und München 1970, 307–321.

Wallenstein. Aus einem Brief an Hansgünther Heyme, geschrieben am 29. Januar 1970. In: Thema. Theater-Magazin. Bühnen der Stadt Köln, 1970, Heft 8, S. 8.

Ein Wort des Gedenkens für Adolf Deißmann. Zur Feier der Einweihung einer Gedächtnistafel am Geburtshaus Adolf Deißmanns in Langenscheid/Nassau am 24. Juni 1962. In: HuH² II 749–751.

Wort und Sache im Denken Goethes. In: Silvae. Festschrift für Ernst Zinn zum 60. Geburtstag, dargebracht von Kollegen, Schülern und Mitarbeitern. Hrsg. von M. v. Albrecht und E. Heck, Tübingen 1970, 199–209 – Schweizer Monatshefte 49, 1969/70, 926–934 – HuH² II 117–126.

1971

Anmut und Schicksal. Zwei Olympische Oden Pindars. In: Sprache und Bekenntnis. Festschrift für Hermann Kunisch. Sonderband des Literaturwissenschaftlichen Jahrbuchs, Berlin 1971, 1–5 (Übersetzung und Erläuterung der 14. und 12. Ol. Ode).

Aristophanes: Die Frösche. Bühnenfassung. Übersetzung und Bearbeitung. Insel-Bücherei 962, Frankfurt am Main 1971, 111 S.

Das Bild des griechischen Menschen. Zwei Vorträge (1. Der griechische Mann, 2. Bild der griechischen Frau). In: Jenseits von Resignation und Illusion. Festschrift des Lessing-Gymnasiums, Frankfurt am Main 1971, 84–103 (Vorträge, gehalten im Bayerischen Rundfunk 1968 und 1970, der erste Vortrag schon 1970 gedruckt).

* *Die Geschichtschreibung des Thukydides.* Ein Versuch (1929). Mit Nachwort 1971, Dublin/Zürich ²1971, 104 S.

Homer und das Dichterische. Überlebte oder bleibende Antike? Geistige Begegnungen in der Böttcherstraße. 50. Vortrag, Bremen 1971, 23 S. (nicht im Handel).

Sophokles: Die Frauen von Trachis. Übersetzung und Bühnenbearbeitung mit Daten. Bühnenmanuskript Suhrkamp Verlag, Frankfurt am Main 1971, 45 S.

1972

* *Aristophanes: Lysistrate.* Bühnenfassung. Übersetzung und Bearbeitung. Insel-Bücherei 967, Frankfurt am Main 1972, 91 S. (gegenüber der Fassung von 1964 verändert).

Diagramme zum Aufbau von Ilias und Odyssee. Beigegeben dem Aufsatz von H. Vester, Homer im Unterricht, in: Griechisch in der Schule, Frankfurt am Main 1972, 152–155.

Einführung zu: Griechenland/Von Homer bis zum Hellenismus, von C. M. Bowra und der Redaktion der Time-Life-Bücher, Reinbek bei Hamburg 1972, 6.

Euripides: Die Bakchen. Übersetzt und für die Bühne bearbeitet. Bühnenmanuskript Suhrkamp Verlag, Frankfurt am Main 1972, 50 S. (mit Einfügung von Fragmenten aus dem ‚Christos Paschon').

Mond und Sterne in Goethes Lyrik. Ein Beitrag zu Goethes erlebtem Platonismus. In: Goethe und die Tradition, hrsg. v. Hans Reiss, Frankfurt am Main 1972, 58–83.

Pindars Olympische Oden. Übersetzt und herausgegeben, Frankfurt am Main 1972, 128 S.

Sophokles: Philoktet. Übersetzung und Daten. Bühnenmanuskript Suhrkamp Verlag, Frankfurt am Main 1972, 60 S.

Zwei Lebensgedichte Goethes. In: Festschrift für Hermann Heimpel, Band III, Göttingen 1972, 708–712.

1973

Humanitas Romana. In: Aufstieg und Niedergang der römischen Welt. Geschichte und Kultur Roms im Spiegel der neueren Forschung, Band I 4, hrsg. v. Hildegard Temporini, Berlin 1973, 43–62. – Frühere Fassung schon in: HuH² I 685–700.

Lebenswerte des Griechischen. In: Der Lebenswert des Griechischen. Veröffentlichungen der Katholischen Akademie der Erzdiözese Freiburg 29, Karlsruhe 1973, 65–75. – Verkürzt schon in: Stuttgarter Zeitung. Die Brücke zur Welt, 4. 3. 1972.

Die Odyssee als Dichtung. In: Fortwirkende Antike. Dialog Schule und Wissenschaft. Klassische Sprachen und Literaturen Band VI, München 1973, 33—59. — Schon früher spanisch: La ‚Odisea' como poesia. In: Estudios de literatura griega. Cuadernos de la ‚Fundación Pastor' 18, Madrid 1971, 11—49.

* *Sophokles: König Ödipus.* Übertragen und herausgegeben. Mit einem Nachwort, drei Aufsätzen, Wirkungsgeschichte und Literaturnachweisen. Insel Taschenbuch 15, Frankfurt am Main 1973, 142 S. — Enthält die Übersetzung (1955, 1964, 1968) und die Aufsätze: Der „König Ödipus" des Sophokles in neuer Deutung (1956), Shakespeares „König Lear" und Sophokles' „König Ödipus" (1951), Der „Zerbrochene Krug" von Heinrich von Kleist und Sophokles' „König Ödipus" (1957).

Theia, die Mutter des Sonnengotts: Die Fünfte Isthmische Ode Pindars. In: Wirklichkeit und Reflexion. Festschrift für Walter Schulz, Pfullingen 1973, 355—363.

Übersetzungen in: Carl Orff, De Temporum fine Comoedia. Das Spiel vom Ende der Zeiten. Vigilia, Mainz 1973 (Textbuch).

1974

Die Zehnte Pythische Ode Pindars, in: Festschrift für Friedrich Beißner, hrsg. v. U. Gaier u. W. Volke, Bebenhausen 1974, 364—370. — Übersetzung und Erläuterung im besonderen Bezug auf Hölderlins späte Hymne ‚Die Wanderung'.

Sophokles: Antigone. Übertragen und herausgegeben. Mit einem Nachwort, einem Aufsatz, Wirkungsgeschichte und Literaturhinweisen. Insel Taschenbuch 70, Frankfurt am Main 1974, 152 S. — Enthält die Übersetzung (1964) und den Aufsatz: Einleitung zur „Antigone" des Sophokles von Hölderlin in der Vertonung von Carl Orff (1960).

Ursprung und frühe Entwicklung der attischen Tragödie. Eine morphologische Struktur-Betrachtung des Aischylos. In: Wege zu Aischylos. Wege der Forschung 87, hrsg. v. H. Hommel, Darmstadt 1974, Bd. I, 104—147.

1975

* *Der Gott von Delphi und die Humanitätsidee.* Bibliothek Suhrkamp, Frankfurt am Main 1975, 154 S. — Enthält: Der Gott von Delphi und die Humanitätsidee (1965), Das Welt-Modell der Griechen (1957), Die Wandlung des Homer-Bildes in der Gegenwart (1950), Die homerische Frage gestern und heute (1955), Das Drama der Antike in heutiger Sicht (1953), Von der Wirkung des Trauerspiels (1955), Das humanistische Bildungsideal und die Forderungen unserer Zeit (1959).

Homer: Ilias. Wolfgang Schadewaldts neue Übertragung. Mit Ergänzungsband: Der Aufbau der Ilias. Strukturen und Konzeptionen, Frankfurt am Main 1975, 432 und 100 S. Neue Übertragung von Wolfgang Schadewaldt mit antiken Vasenbildern. Insel Taschenbuch 153, Frankfurt am Main 1975, 432 S.

Unveröffentlicht

Pindars sämtliche Oden. — Druck in Vorbereitung.

Sophokles' sämtliche Tragödien. — Druck in Vorbereitung.

Aristophanes auf der modernen Bühne.

Nachschriften der Vorlesungen. — Im Besitz des Philologischen Seminars der Universität Tübingen. Publikationen in Vorbereitung.

Herausgeber- und Redaktionstätigkeit

Hermes. Zeitschrift für Klassische Philologie. Band 67, 1932 und Band 68, 1933: In Gemeinschaft mit Helmut Berve und Eduard Fraenkel hrsg. von Alfred Körte und Wolfgang Schadewaldt. — Band 69, 1934 bis Band 79, 1944: Hrsg. von Helmut Berve, Alfred Körte, Wolfgang Schadewaldt.

Die Antike. Zeitschrift für Kunst und Kultur des klassischen Altertums, hrsg. von Wolfgang Schadewaldt, Bernhard Schweitzer, Johannes Stroux. Band 13, 1937 (ab Heft 2, S. 77ff.) bis Band 20, 1944 (erschienen: 2 Hefte, bis S. 174).

Tübinger Beiträge zur Altertumswissenschaft. Heft 38, 1953 bis Heft 43, 1965: Hrsg. von Wolfgang Schadewaldt, Joseph Vogt, Otto Weinreich. — Heft 44, 1968ff.: Hrsg. von Wolfgang Schadewaldt, Joseph Vogt, Otto Weinreich, Ernst Zinn.

Edition der „Collectanea zu meinem Leben", besorgt von Wolfgang Schadewaldt und Ingeborg Schudoma. In: *Johann Joachim Winckelmann*, Briefe. In Verbindung mit H. Diepolder hrsg. von W. Rehm. Vierter Band. Dokumente zur Lebensgeschichte, Berlin 1957, 154—163 und 478—480.

Goethe-Wörterbuch. Hrsg. von der Deutschen Akademie der Wissenschaften zu Berlin (ab Lieferung 7: Akademie der Wissenschaften der DDR), der Akademie der Wissenschaften zu Göttingen und der Heidelberger Akademie der Wissenschaften. Wissenschaftliche Leitung Werner Hartke (Lieferung 2—8), Wolfgang Schadewaldt, Werner Simon (Lieferung 1—7), Wilhelm Wißman (Lieferung 1—2) und Hans Neumann (Lieferung 8). Erster Band 1. Lieferung (mit Einführung Sp. III—XVI von Wolfgang Schadewaldt), Stuttgart—Berlin—Köln—Mainz 1966; 8. Lieferung (bis Stichwort ‚aufsteigen') 1974, Sp. 897 bis 1024.

Zur Biobibliographie

Schriftenverzeichnisse. In: Dino Larese, Wolfgang Schadewaldt. Eine Lebensskizze, Amriswil 1967, 42—61. — Reinhard Thurow, Verzeichnis der Schriften Wolfgang Schadewaldts. In: Hellas und Hesperien ²1970, II 829—848.

Festschriften: Synusia. Festgabe für Wolfgang Schadewaldt zum 15. März 1965. Im Namen seiner Tübinger Schüler hrsg. von Hellmut Flashar und Konrad Gaiser, Pfullingen 1965, 465 S. — Das Altertum und jedes neue Gute. Für Wolfgang Schadewaldt zum 15. März 1970. hrsg. von Konrad Gaiser, Stuttgart 1970, 560 S.

Würdigungen. Hildebrecht Hommel, Wolfgang Schadewaldt zum 60. Geburtstag, in: Stuttgarter Zeitung 14. 3. 1960. — Olof Gigon, Arbeitsmethoden der klassischen Altertumswissenschaft heute. Zu Wolfgang Schadewaldts 65. Geburtstag, in: Universitas 20, 1965, 633—642. — Horst Rüdiger, Wolfgang Schadewaldt als Übersetzer, in: Jahrbuch 1965 der Deutschen Akademie für Sprache und Dichtung Darmstadt, Heidelberg—Darmstadt 1966, 36—48. — Ernst Zinn, Lebendiges Griechenland. Gruß an Wolfgang Schadewaldt, in: Stuttgarter Zeitung 12. 3. 1965. — Dino Larese, Wolfgang Schadewaldt. Eine Lebensskizze. Amriswiler Bücherei, Amriswil 1967, 7—21. — Klaus Bartels, Wolfgang Schadewaldt. Zum siebzigsten Geburtstag, in: Neue Zürcher Zeitung 14. 3. 1970. — Konrad Gaiser, 3000 Jahre Gegenwart, in: Stuttgarter Zeitung 13. 3. 1970; Antike dem Heute erschlossen, in: Schwäbisches Tagblatt 14. 3. 1970. — Albin Lesky, Wolfgang Schadewaldt 70, in: Die Presse 16. 3. 1970. — Clara Menck, Eine Art Freiheit, in: Frankfurter Allgemeine Zeitung 13. 3. 1970.

Autobiographisches. Einblick in die Werkstätte meiner Arbeit. In: Dino Larese, Wolfgang Schadewaldt. Eine Lebensskizze. Amriswiler Bücherei, Amriswil 1967, 22–40 = HuH² II 598–606.

Nachrufe. Hellmut Flashar, in: Rheinischer Merkur 22. 11. 1974; Gnomon. Kritische Zeitschrift für die gesamte klassische Altertumswissenschaft 47, 1975, S. 731–736. – Günther Fleckenstein, in: Göttingen. Informationszeitschrift für die Universitäts- und Kongreßstadt, Dezember 1974. – Konrad Gaiser, in: Stuttgarter Zeitung 12. 11. 1974; Nationalzeitung Basel 20. 11. 1974; Attempto. Nachrichten für die Freunde der Universität Tübingen 53/54, 1975, S. 134–136; Jahrbuch der Heidelberger Akademie der Wissenschaften 1975, S. 92–98. – Hildebrecht Hommel, in: Frankfurter Allgemeine Zeitung 13. 11. 1974. – Bernhard Kytzler, in: Die Welt 14. 11. 1974. – Walter Marg, in: Mainzer Allgemeine Zeitung 14. 11. 1974. – Albin Lesky, in: Die Presse 14. 11. 1974; Almanach für das Jahr 1974 der Österreichischen Akademie der Wissenschaften 124 (Wien 1975) S. 514 bis 523; ‚Reden und Gedenkworte' des Ordens Pour le Mérite für Wissenschaften und Künste, 12, 1974/75, S. 115–123. – Horst Rüdiger, in: Neue Zürcher Zeitung 16. 11. 1974; Jahrbuch Deutsche Akademie für Sprache und Dichtung Darmstadt 1974 (Heidelberg 1975) S. 135–138. – Albert von Schirnding, in: Süddeutsche Zeitung 13. 11. 1974. – Wilfried Stroh, in: Rhein-Neckar-Zeitung 15. 3. 1975. – Hans Tümmler, in: Goethe-Jahrbuth 93 (Weimar 1976) S. 247–248.

Prof. Dr. Konrad Gaiser

VERZEICHNIS DER MIT DER AKADAMIE IM SCHRIFTENTAUSCH STEHENDEN KÖRPERSCHAFTEN

I. Periodica

Ägypten — VAR

Kahira

Institut Français d'Archéologie Orientale
Annuaire de l'Egyptologie
Bibliothèque d'études
Bulletin
Conférences

National Information and Documentation Centre
Journal of Chemistry

Argentinien

Buenos Aires

Comisión Nacional de la Energia Atómica
Informe

Sociedad Cientifica Argentina
Anales

La Plata

Universidad Nacional. Facultad de Ciencias Fisicomatemáticas
Revista

Mendoza

Universidad Nacional de Cuyo. Instituto de Lingüistica
Anales

Australien

Adelaide

Barr Smith Library
The Australian Journal of Experimental Biology and Medical Science

Brisbane

Royal Society of Queensland
Proceedings

Canberra

National Library of Australia
Australian Bibliography and Bibliographical Services
Historical Records of Australia

Melbourne

Royal Society of Victoria
Proceedings

Sydney

Royal Society of New South Wales
Journal and Proceedings Monographs

Belgien

Anvers	*Académie Royale d'Archéologie de Belgique* Revue belge d'archéologie et d'histoire de l'art
Bruxelles	*Algemene Bibliotheek van de Vrije Uniteit* Tijdschrift Studiereeks
	Institut Royal des Sciences naturelles de Belgique Memoirs
	Société des Bollandistes Analecta Bollandiana
	Union Académique Internationale Compte rendu Archivum latinitatis medii aevi
	Université libre. Bibliothèque Centrale Revue de l'Université
Gent	*Koninklijke Vlaamse Aacdemie voor Taal- en Letterkunde* Jaarboek Verslagen en Mededeelingen Leonard Willemsfonds
Liège	*L'Institut Archéologique Liègeois* Bulletin Chronique archéologique du pays de Liège
Louvain	*Bibliothèque de l'Université* Recherches sociologiques Travaux de la Faculté de Philosophie et Lettres Travaux de l'Institut de Psychologie
	Universiteitsbibliotheek Bibliographia Academia Jaarboek Studia Hellenistica Ancient Society Humanistica Lovaniensia
Meise	*Jardin Botanique de l'Etat* Bulletin
Namur	*Rédaction „Les études classiques"* Les Etudes classiques

Brasilien

Rio de Janeiro	*Academia Brasileira de Ciências* Anais
	Centro Brasileiro de Pesquisas Fisicas Summa brasiliensis mathematicae

VR Bulgarien

Sofija	*Bulgarska Akademija na Naukite* Doklady Izvestija na Instituta za bulgarski ezik Izvestija na Instituta za literatura Izvestija na Zoologičeskija institut
	Universitetska Biblioteka Godišhik na Sofijskija Universitet. Filologičeski fakultet Juridičeski fakultet Naučni trudove na Visši medicinski institut
	Upravlenie za geoložki proučvanija Godišnik. Otdel A Priloženie kum Godišnik

Bundesrepublik Deutschland

Aachen	*Aachener Geschichtsverein* Zeitschrift
	Bibliothek der Rheinisch-Westfälischen Technischen Universität Jahrbuch
Augsburg	*Historischer Verein für Schwaben* Zeitschrift
	Naturwissenschaftlicher Verein für Schwaben Bericht
Bad Godesberg	*Deutsche Forschungsgemeinschaft* Bericht über ihre Tätigkeit Deutscher Forschungsdienst Deutscher Forschungsdienst. Angewandte Wissenschaft
Beuron	*Bibliothek der Erzabtei Beuron* Erbe und Auftrag Texte und Arbeiten Vetus lat. inst. Bericht

Braunschweig

Braunschweigische wissenschaftliche Gesellschaft
Abhandlungen

Bibliothek der Technischen Universität
Geographische Rundschau

Bremen

Überseemuseum
Veröffentlichungen

Die Wittheit zu Bremen
Abhandlungen des Naturwissenschaftlichen Vereins
Abhandlungen und Vorträge
Bremisches Jahrbuch
Deutsche geographische Blätter
Jahrbuch der Bremischen Wissenschaft
Veröffentlichungen aus dem Staatsarchiv

Bremerhaven

Institut für Meeresforschung
Veröffentlichungen

Düsseldorf

Düsseldorfer Geschichtsverein
Düsseldorfer Jahrbuch

Rheinisch-Westfälische Akademie der Wissenschaften
Vorträge. Reihe Geisteswissenschaften
Reihe Natur-, Ingenieur- u. Gesellschaftswissenschaften
Wissenschaftliche Abhandlungen

Erlangen

Institut für Fränkische Landesforschung
Jahrbuch
Schriften. Historische Reihe

Physikalisch-Medizinische Sozietät
Sitzungsberichte

Frankfurt a. M.

Philosophisch-Theologische Hochschule St. Georgen
Theologie und Philosophie

Physikalischer Verein
Jahresbericht

Senckenbergische Naturforschende Gesellschaft
Abhandlungen
Aufsätze und Reden
Natur und Museum
Senckenbergiana biologica
Senckenbergiana lethaea

Frobenius-Inst. an der J.-W.-Goethe-Univ.
Paideuma

Freiburg i. Br.	*Naturforschende Gesellschaft* Berichte
Gießen	*Bibliothek der Justus-Liebig-Universität* Hessische Blätter für Volkskunde Mitteilungen des Oberhessischen Geschichtsvereins Nachrichten der Gießener Hochschulgesellschaft Schriften der Justus-Liebig-Universität
	Oberhessische Gesellschaft für Natur- und Heilkunde Bericht. Medizinische Abteilung Naturwissenschaftliche Abteilung
Göttingen	*Akademie der Wissenschaften* Abhandlungen. Math.-phys. Klasse Phil.-hist. Klasse Göttingische gelehrte Anzeigen Jahrbuch Nachrichten. Math.-phys. Klasse Phil.-hist. Klasse
Hamburg	*Geographische Gesellschaft* Mitteilungen
	Gesellschaft für Natur- und Völkerkunde Ostasiens Mitteilungen Nachrichten
	Mathematische Gesellschaft Mitteilungen
	Staats- und Universitätsbibliothek Abhandlungen aus dem Gebiete der Auslandskunde Hamburger Universitätsreden
Hannover	*Akademie für Raumforschung und Landesplanung* Raumforschung und Raumordnung
	Bibliothek der Technischen Universität.
	Technische Informationsbibliothek Berichte der Naturhistorischen Gesellschaft Jahrbuch
	Niedersächsische Landesbibliothek Niedersächsisches Jahrbuch für Landesgeschichte Studia Leibnitiana
Heidelberg	*Heidelberger Akademie der Wissenschaften* Abhandlungen. Phil.-hist. Klasse Jahrbuch

Sitzungsberichte. Math.-nat. Klasse
Phil.-hist. Klasse

Naturhistorisch-medizinischer Verein
Verhandlungen

Universitätsbibliothek
Heidelberger Jahrbücher
Veröffentlichungen aus der Heidelberger Papyrussammlung

Karlsruhe

Bibliothek der Technischen Universität
Monographien

Kiel

Naturwissenschaftlicher Verein für Schleswig-Holstein
Schriften

Universitätsbibliothek
Schriften des Geographischen Instituts der Universität Kiel

Leverkusen

Farbenwerke Bayer
Therapeutische Berichte

Lübeck

Stadtbibliothek
Ausblick
Mitteilungen der Geographischen Gesellschaft und des Naturhistorischen Museums

Verein für Lübeckische Geschichte und Altertumskunde
Zeitschrift

Verein „Natur und Heimat"
Berichte

Mainz

Akademie der Wissenschaften und der Literatur
Abhandlungen der Klasse Literatur
Abhandlungen der geistes- u. sozialwiss. Klasse
Abhandlungen der mathematisch-naturwiss. Klasse
Jahrbuch
Karl-A.-Forster-Lectures
Mikrofauna des Meeresbodens

Mainzer Altertumsverein
Mainzer Zeitschrift

Marburg

Deutsche Morgenländische Gesellschaft
Zeitschrift

Gesellschaft zur Beförderung der gesamten Naturwissenschaften
Sitzungsberichte

Universitätsbibliothek
Alma Mater Philippina

München

Bayerische Akademie der Wissenschaften
Abhandlungen. Math.-nat. Klasse
Phil.-hist. Klasse
Jahrbuch
Sitzungsberichte. Math.-nat. Klasse
Phil.-hist. Klasse
Sonderschriften

Deutsches Museum
Abhandlungen und Berichte
Verwaltungsbericht

Max-Planck-Gesellschaft
Jahrbuch
Mitteilungen

Technische Universität. Universitätsbibliothek
Veröffentlichungen der Deutschen Geodätischen Kommission. Reihe E
Berichte über die Vollsitzungen der Deutschen Geodätischen Kommission

Zentralinstitut für Kunstgeschichte
Jahresbericht

Münster

Universitätsbibliothek
Westfälische geographische Studien

Nürnberg

Germanisches Nationalmuseum
Anzeiger
Bilder aus deutscher Vergangenheit

Offenbach

Bibliothek des Deutschen Wetterdienstes
Agrarmeteorologische Bibliographie
Berichte
Deutsches meteorologisches Jahrbuch

Remagen

Bundesanstalt für Landeskunde und Raumforschung
Berichte zur deutschen Landeskunde

Speyer

Pfälzische Landesbibliothek
Mitteilungen der Pollichia
Mitteilungen des Historischen Vereins der Pfalz
Pfälzische Bibliographie

Stuttgart

Kommission für geschichtliche Landeskunde in Baden-Württemberg
Württembergische Geschichtsliteratur
Zeitschrift für Württembergische Landesgeschichte

Museum für Länder- und Völkerkunde. Linden-Museum
Tribus. Jahrbuch

Staatliches Museum für Naturkunde
Stuttgarter Beiträge zur Naturkunde

Verband der deutschen Höhlen- und Karstforscher
Bibliographie zur Karst- und Höhlenkunde in Deutschland
Jahreshefte für Karst- und Höhlenkunde
Mitteilungen

Württembergische Landesbibliothek
Jahrbücher für Statistik und Landeskunde von Baden-Württemberg

Ulm

Verein für Kunst und Altertum in Ulm und Oberschwaben
Mitteilungen

Wiesbaden

Hessisches Landesamt für Bodenforschung
Abhandlungen
Hessisches Lagerstättenarchiv
Mitteilungen der Reichsstelle für Bodenforschung
Notizblatt

Nassauischer Verein für Naturkunde
Jahrbücher

Verein für Nassauische Altertumskunde und Geschichtsforschung
Nassauische Annalen

Wolfenbüttel

Braunschweigischer Geschichtsverein
Braunschweigisches Jahrbuch

Würzburg

Freunde Mainfränkischer Kunst und Geschichte
Mainfränkische Hefte
Mainfränkisches Jahrbuch für Geschichte und Kunst

Physikalisch-medizinische Gesellschaft
Berichte

Ceylon

Colombo

National Museum
Administration Report of the Director
Ceylon Journal of Science. Biological Series
Spolia Zeylanica. Geology, Zoology, Anthropology

Peradeniya

The Librarian, University of Ceylon
The Ceylon Journal of Medical Science

China

Pei-ching *Kuo li t'u shu kuan (National Library)*
Chinese Literature
Chinese Medical Journal
Scientia Sinica

Città del Vaticano
(vgl. auch Italien, Roma)

Archivum Historicum Societatis Jesu
Archivum

Biblioteca Apostolica Vaticana
Studi e Testi

Pontifico Istituto Biblico
Biblica
Orientalia

Specola Vaticana
Annual Report
Miscellanea astronomica
Richerche astronomiche
Richerche spettroscopiche

Dänemark

Århus *Statsbiblioteket*
Acta Jutlandica

København *Det Kongelige Danske Videnskaberne Selskab*
Arkeologisk-kunsthistoriske meddelelser
Arkeologisk-kunsthistoriske skrifter
Biologiske meddelelser
Biologiske skrifter
Historisk-filologiske meddelelser
Historisk-filologiske skrifter
Matematisk-fysiske meddelelser
Matematisk-fysiske skrifter
Oversigt over selskabets virksomhed

Universitets Matematiske Institut
Mathematica Scandinavica

Odense *Odense Universitetsbibliotek*
Mediaeval Scandinavia

Deutsche Demokratische Republik

Berlin

Akademie der Landwirtschaftswissenschaften der DDR
Albrecht-Thaer-Archiv

Akademie der Wissenschaften der DDR zu Berlin
Abhandlungen. Klasse f. Mathematik, Physik u. Technik
Klasse f. Chemie, Geologie u. Biologie
Klasse f. Bergbau, Hüttenwesen u. Montangeologie
Klasse f. Medizin
Klasse f. Sprachen, Literatur u. Kunst
Klasse f. Philosophie, Geschichte, Staats-, Rechts- u. Wirtschaftswissenschaften
Jahrbuch
Journal für Hirnforschung
Monatsberichte
Sitzungsberichte. Klasse f. Mathematik, Physik u. Technik
Klasse f. Chemie, Geologie u. Biologie
Klasse f. Bergbau, Hüttenwesen u. Montangeologie
Klasse f. Medizin
Klasse f. Sprachen, Literatur u. Kunst
Klasse f. Philosophie, Geschichte, Staats-, Rechts- u. Wirtschaftswissenschaften
Vorträge und Schriften

Deutsches Amt für Meßwesen und Warenprüfung
Wissenschaftliche Abhandlungen

Geologischer Dienst. Staatliche Geologische Kommission
Geologie. Nebst Beihefte
Zeitschrift für angewandte Geologie

Universitätsbibliothek der Humboldt-Universität
Wissenschaftliche Zeitschrift der Humboldt-Universität
Gesellschafts- u. sprachwiss. Reihe
Mathematisch-naturwiss. Reihe

Dresden

Bibliothek der Hochschule für Verkehrswesen
Wissenschaftliche Zeitschrift der Hochschule für Verkehrswesen

Bibliothek der Technischen Universität
Wissenschaftliche Zeitschrift der Technischen Universität

Erfurt

Medizinische Akademie
Personal- und Vorlesungsverzeichnis

Wissenschaftliche Bibliothek der Stadt Erfurt
Aus der Vergangenheit der Stadt Erfurt
Beiträge zur Geschichte der Stadt Erfurt

Greifswald	*Universitätsbibliothek* Wissenschaftliche Zeitschrift der Ernst-Moritz-Arndt-Universität Gesellschafts- u. sprachwiss. Reihe Mathematisch-naturwiss. Reihe
Halle (Saale)	*Deutsche Akademie der Naturforscher „Leopoldina“* Nova Acta Leopoldina
	Landesmuseum für Vorgeschichte Jahresschrift für mitteldeutsche Vorgeschichte Vorgeschichtliche Museumsarbeit und Bodendenkmalpflege
	Universitäts- und Landesbibliothek Hallische Monographien Schriften zum Bibliotheks- und Büchereiwesen in Sachsen-Anhalt Wissenschaftliche Zeitschrift der Martin-Luther-Universität Gesellschafts- u. sprachwiss. Reihe Mathematisch-naturwiss. Reihe
Ilmenau	*Technische Hochschule* Wissenschaftliche Zeitschrift
Jena	*Universitätsbibliothek* Wissenschaftliche Zeitschrift der Friedrich-Schiller-Universität Gesellschafts- u. sprachwiss. Reihe Mathematisch-naturwiss. Reihe
Leipzig	*Deutsches Institut für Länderkunde* Geographische Berichte
	Universitätsbibliothek Wissenschaftliche Zeitschrift der Karl-Marx-Universität Gesellschafts- u. sprachwiss. Reihe Mathematisch-naturwiss. Reihe
Magdeburg	*Technische Hochschule Otto von Guericke* Wissenschaftliche Zeitschrift
Potsdam	*Geodätisches Institut* Veröffentlichungen
Rostock	*Universitätsbibliothek* Archiv der Freunde der Naturgeschichte in Mecklenburg Wissenschaftliche Zeitschrift der Universität Rostock Gesellschafts- u. sprachwiss. Reihe Mathematisch-naturwiss. Reihe
Tautenburg (Kr. Jena)	*Karl-Schwarzschild-Observatorium* Mitteilungen
Weimar	*Zentralbibliothek der deutschen Klassik* Wissenschaftliche Zeitschrift der Hochschule für Architektur und Bauwesen

England

Cambridge	*Cambridge Philosophical Society* Proceedings
Edinburgh	*Royal Society of Edinburgh* Proceedings Transcations Year Book
Leeds	*Leeds Philosophical and Literary Society* Proceedings. Scientific section Literary and historical section
Liverpool	*Biological Society* Annual Report Proceedings and Transactions
London	*India Office Library* Report
	London Mathematical Society Journal
	Polski Uniwersytet na Obczyznie Zeszyty Naukowe. Wyzd. Humanist.
	University of London. Library Annual Report of the Library Committee Bulletin of the Institute of Historical Research Bulletin of the School of Oriental and African Studies Institute of Historical Research. Annual Report Slavonic and East European Review Theses and Dissertations accepted for Highet Degrees
	The Warburg Institute Journal of the Warburg and Courtauld Institutes
Manchester	*Manchester Literary and Philosophical Society* Memoirs and Proceedings
	University of Manchester. University Library Journal of Semitic Studies Rhodes-Livingstone Journal

Finnland

Helsinki	*Geologinen Tutkimuslaitos* Bulletin de la Commission Geologique de Finlande

Societas pro Fauna et Flora Fennica
Acta
Acta botanica fennica
Acta zoologica fennica
Memoranda

Societas Biologica Fennica „Vanamo"
Annales Botanici Fennici
Annales Zoologici Fennici

Suomen Tiedeseura (Societas Scientiarum Fennica)
Acta. Nova series. B. C.
Arsbok
Bidrag till kännedom af Finlands natur och folk
Commentationes biologicae
Commentationes humanorum litterarum
Commentationes physico-mathematicae

Suomalainen Tiedeakatemia (Academia Scientiarum Fennica)
Ergebnisse der magnetischen Beobachtungen
Folklore Fellows Communications
Sitzungsberichte
Toimituksia (Annales). Sarja A 1.2.3.4.5.6. B

Suomalais-Ugrilainen Seura, Kirjasto
(Société Finno-Ougrienne, Bibliothèque)
Finnisch-ugrische Forschungen

Turku

Academia Aboeniora
Acta. Humaniora
Mathematica et physica

Turun Yliopiston Kirjasto (Bibliotheca Universitatia Turkuensis)
Temenos
Turun Yliopiston Julkaisuja (Annales Universitatis Turkuensis). Ser. A I. II. B

Frankreich

Besançon

Université de Besançon
Annales littéraires
Annales scientifiques

Cherbourg

Société Nationale des Sciences Naturelles et Mathématiques
Mémoires

Clermont-Ferrand

Académie des Sciences, Belles-Lettres et Arts
Bulletin historique et scientifique de l'Auvergne

Lyon	*Bibliothèque de l'Université* Annales de l'Université. Droit Lettres Sciences. Sect. A. B. C. Fascicule spécial
	Société Linnéenne de Lyon Bulletin mensuel
Marseille	*Bibliothèque de l'Université d'Aix-Marseille* Annales de la faculté des sciences Comptes rendus des travaux
Montpellier	*Académie des Sciences et Lettres* Bulletin
Paris	*Ecole Normale Supérieure* Annales scientifiques
	Institut de France. Académie des Inscriptions et Belles-Lettres Compres rendus Mémoires Mémoires présentés par divers savants
	Institut de France. Académie des Sciences Annuaire Bulletin bibliographique des publications périodiques Comptes rendus hebdomadaires. Sér. A. B. C. D.
	Institut de France. Académie des sciences morales et politiques Notices biographiques et bibliographiques Revue des travaux
	Muséum National d'Histoire Naturelle Annuaire Bulletin
	Société Mathématique de France Bulletin Gazette des mathématiques
	Union Géodésique et Géophysique Internationale Chronique
Rennes	*Socété Scientifique de Bretagne* Bulletin
Saint-Martin-d'Hères	*Institut de mathématiques pures* Annales de l'Institut Fourier
Sèvres	*Bureau International des Poids et Mesures* Travaux et mémoires

Toulouse — *Académie des Sciences, Inscriptions et Belles-Lettres*
Mémoires

Université de Toulouse. Bibliothèque
Annales publ. de la Faculté des Lettres et des Sciences.
Séries Homo
    Littératures
    Pallas
    Travaux de l'Institut d'Art Préhistorique
    Via Domitia

Ghana

Accra — *Ghana Academy of Sciences*
Ghana Journal of Science
Proceedings

Griechenland

Athenai — *Ecole Française d'Athènes*
Bulletin de Correspondence hellénique

Thessaloniki — *Institute for Balkan Studies*
Balkan Studies

Indien

Allahabad — *Allahabad Mathematical Society*
Indian Journal of Mathematics

Indonesien

Djakarta — *Perpustakaan-Museum Pusat*
Madjalah ilmu sastra Indonesia

Irland

Dublin — *Royal Dublin Society*
Economic Proceedings
Report
Scientific Proceedings. Ser. A. B.

Royal Irish Academy
Proceedings. Sect. A. B. C.
    Minutes

Island

Reykjavik — *Visindafélag islendinga (Societas Scientiarum Islandica)*
The Eruption of Hekla
Rit
Science in Iceland

Israel

Jeruschalajim — *Jewish National and University Library*
Atiquot
Israel Exploration Journal
Israel Journal of Botany
Israel Journal of Chemistry
Israel Journal of Earth-Sciences
Israel Journal of Experimental Medicine
Israel Journal of Mathematics
Israel Journal of Technology
Israel Journal of Zoology
Kirjath Sepher
Scripta Hierosolymithana

Italien

Acireale — *Accademia di Scienze, Lettere e Belle Arti*
Memorie e Rendiconti. Classe di lettere

Bologna — *Accademia delle Scienze dell'Istituto*
Annuario
Atti. Classe di scienze fisiche
Memorie
Rendiconti

Istituto di Geologie e Paleontologia dell'Universitè
Giornale di geologia

Brescia — *Ateneo*
Commentari. Nebst Supplemento
Guardare il firmamento dalla specola cidnea
Quaderni della Biblioteca comunale

Cremona — *Biblioteca Governativa e Libreria Civica*
Annali

Ferrara — *Accademia delle Scienze. Biblioteca Ariόstea*
Atti

Firenze — *Accademia della Crusca*
Studi di filologia italiana
Studi di grammatica italiana

Accademia Toscana di Scienze e Lettere „La Colombaria“
Atti e Memorie

Biblioteca Nazionale Centrale
Bibliografia nazionale italiana

Genova	*Museo Civico di Storia Naturale „Giacomo Doria“* Annali Doriana
Lecce	*Università degli Studi di Lecce, Biblioteca Centrale* Annali, Fac. di Lettere e Filosofia Fac. di Magistero Quaderni della Biblioteca Centrale
Mantova	*Accademia Virgiliana di Scienze, Lettere ed Arti* Atti e Memorie
Milano	*Istituto Lombardo, Accademia di scienze e Lettere* Memorie. Classe di lettere e scienze morali e storiche Rendiconti. Classe di lettere e scienze morali e storiche Classe di scienze matematiche e naturali. A. B. Parte generale e atti ufficiali
	Società Storica Lombarda Archivio
Modena	*Accademia di Scienze, Lettere ed Arti* Atti e Memorie
	Istituto di Fisica dell'Università Atti del Seminario Matematico e Fisico
Napoli	*Società Nazionale di Scienze, Lettere ed Arti* Annuario Atti dell'Accademia di Scienze Fisiche e Matematiche Atti dell'Accademia di Scienze Morali e Politiche Atti dell'Accademia Pontaniana Memorie dell'Accademia di Archeologia, Lettere e Belle Arti Rendiconti dell'Accademia di Archeologia, Lettere e Belle Arti Rendiconti dell'Accademia di Scienze Fisiche e Matematiche Rendiconti ed Atti dell'Accademia di Scienze, Mediche e Chirurgiche
	Stazione Zoologica di Napoli Activity Report Pubblicazioni
	Università di Napoli. Biblioteca della Facoltà di lettere e filosofia Annali
Padova	*Seminario Matematico* Rendiconti

Pisa	*Scuola Normale Superiore* Annali. Classe di lettere, storia e filosofia Classe di scienze fisiche e matematiche
	Società Toscana di Scienze Naturali Atti. Ser. A. B.
Roma	*Accademia Nazionale dei Lincei* Atti. Rendiconti. Classe di scienze fisiche, matematiche e naturali Classe di scienze morali, storiche e filologiche Rendiconti della adunanze solenni
	American Academy in Rome Memoirs Papers and Monographs
	Servizio Geologico d'Italia Bolletino
	Istituto di Geologia e Paleontologia dell'Università Geologica Romana
Sassari	*Università, Istituto giuridico* Studi Sassaresi
Torino	*Accademia delle Scienze* Annuario Atti. Classe di scienze fisiche, matematiche e naturali Classe di scienze morali, storiche e filologiche Memorie. Classe di scienze fisiche, matematiche e naturali Classe di scienze morali, storiche e filologiche
Trento	*Museo di Storia Naturale della Venezia Tridentina* Memorie Studi Trentini
Trieste	*Museo Civico di Storia Naturale* Atti
	Università degli Studi Pubblicazioni Rendiconti dell'Istituto di matematiche
Venezia	*Istituto Veneto di Scienze, Lettere ed Arti* Atti. Classe di scienze matematiche e naturali Classe di scienze morali e lettere Parte generale e atti ufficiali
	Museo Civico di Storia Naturale Bolletino

Japan

Fukuoka

Kyusyu University. Faculty of Science. Mathematical Institute
Memoirs. Ser. A. Mathematics

Hiroshima

Hiroshima Universtiy. Library
Journal of Science. A—II. Physics and Chemistry
B—I. Zoology
B—II. Botany
C. Geology and Mineralogy
Memoirs of the Facolty of Engineering

The Hiroshima Institute of Technology
Research Bulletin

Laboratory for Amphibian Biology
Scientific Report

Okayama

Okayama University. Faculty of Science. Department of Mathematics
Mathematical Journal

Osaka

Kansai University. Faculty of Engineering
Technology Reports

Osaka City University. Faculty of Science
Annual Reports of Scientific Works
Osaka Journal of Mathematics

Osaka City University. Institute of Polytechnics
Journal of Biology

Sapporo

Hokkaido University. Faculty of Science
Journal of the Faculty of Science. Physics

Sendai

Tohoku University. Faculty of Letters. Laboratory of Psychology
Tohoku psychologica folia

Tokyo

Mathematical Society of Japan
Journal

Nippon Telegraph and Telephone Public Corporation. The Electrical Communication laboratory
Review

Tokyo University of Education. Mathematical Institute
Science Reports. Sect. A.

University of Tokyo. Faculty of Science
Journal of the Faculty of Science. Sect. 1.2.3.4.5.

Beograd

Društvo matematičara, fizičara i Astronoma N. R. S.
Mathematica balkanica
Vesnik

Matematički institut
Matematički vesnik
Publications

Observatoire Astronomique
Bulletin
Publications

Srpska akademija nauka i umetnosti
Balcanica
Bulletin de l'Académie Serbe des sciences
Glas
Glasnik
Glasnik Etnografskog instituta
Godišnjak
Godišnjak našeg neba
Gradja
Istoriski časopis
Istoriski institut. Posebna izdanja
Jugoslovenske zemlje u XX veku
Jugoslovenske zemlje u XIX veku
Juznoslovenski Filolog
Muzička izdanja
Naš jezik
Naučno-popularni spisi. Odeljenje društvenih nauka
Posebna izdanja
Predavanja
Pregled razmene izdanja
Spomenik
Srpski etnografski zbornik
Starinar
Zbornik radova
Zbornik radova Vizantološkog instituta
Zbornik za istočnjačku istorisku i književnu gradju
Zbornik za istoriju, jezik i književnost srpskog naroda

Institut de Geographie „Jovan Cvijić"
Posebna izdanja
Zbornik radova

Ljubljana

Conseil des Académies de la RPF Yougoslavie
Bulletin scientifique. Sect. A. B.

Geografsko društvo
Geografski vestnik

Nuklearni institut „Jozef Stefan“
Report

Slovenska akademija znanosti in umetnosti
Acta carsologica
Arheološki vestnik
Biblioteka
Dela. Classis I—V
Fontes rerum Slovenicarum
Geografski zbornik
Gradivo za narodopisje slovencev
Letopis
Linguistica
Razprave. Classis I—IV

Pula

Arheološki muzej Istre
Jadranski zbornik
Kulturhistorische Denkmäler in Istrien

Sarajevo

Biološki institut
Godišnjak

Zemalski muzej
Glasnik. Arheologija
Etnologija
Istorija

Skopje

Faculté de Philosophie de l'Université
Annuaire (Godišen zbornik)

Faculté des Sciences de l'Université.
Institut de Mathématique
Annuaire de la faculté des sciences (Godišen zbornik)

Filozofski fakultet na universitetot.
Seminar za klasična filologija
Ziva antika

Makedonische Akademie der Wissenschaften und Künste
Pristapni
Letopis

Narodna i Univerzitetska Biblioteka „Kliment Ohridski“
Macedonian Review
Razgledi

Split

Institut za nacionalnu arheologiju
Starohrvatska prosvjeta

Zagreb

Društvo matematičara i fizičara SRH
Glasnik matematičko-fizički i astronomski

Hrvatsko-prirodoslovno društvo
Biološki glasnik

Institut za geologiju i rudarstvu
Geološki Vjesnik

Jugoslovenska akademija znanosti i umetnosti
Bulletin international
Djela
Gradja za povijest književnosti hrvatske
Hrvatska bibliografija
Ljetopis
Prirodoslovna istraživanja
Rad
Zbornik Historijskog instituta

Staroslavenski Institut
Slovo

Kanada

Montréal — *Université. Départment d'Etudes Slaves*
Études slaves et est-européennes
Canadian Slavic Studies

Winnipeg — *Ukrainska Vil'na Akademija Nauk*
Literatura
Litopys
Onomastica
Slavica Canadiana
Ucrainica Canadiana

Kolumbien

Bogotá — *Academia Columbiana de Ciencias Exactas, Fisicas y Naturales*
Revista

Kongo

Kinshasa — *Université Lovanium*
Publications

Luxemburg

Luxemburg — *Institut Grand-Ducal de Luxembourg. Section des Sciences Naturelles, Physiques et Mathématiques*
Archives
Publications

Marokko

Rabat — *Société des Sciences Naturelles et Physiques du Maroc*
Bulletin
Cahiers de la Faculté des Sciences
Comptes rendus des séances mensuelles
Travaux de l'Institut scientifique Chérifien
Travaux de la section de pédologie

Mexico

Méjico — *Academia Nacional de Ciencias*
Memorias y Revista

Biblioteca Instituto de Geología. Ciudad Universitaria
Anales
Boletín
Paleontologia mexicana

Escuela Nacional de Ciencias Biologicas
Anales

Neuseeland

Auckland — *The University of Auckland*
Univ. of Auckland. Dpt. of Mathematics.
Report Series

Wellington — *Royal Society of New Zealand*
Bulletin
Journal
Procendings
Transactions

Niederlande

Amsterdam — *Koninklijke Nederlandse Akademie van Wetenschappen*
Jaarboek
Mededeelingen. Afd. Letterkunde
Proceedings. Ser. A. B. C.
Verhandelingen. Afd. Letterkunde
Afd. Natuurkunde

Wiskundig Genootschaap
Niew Archief voor Wiskunde
Wiskundige opgaven met de oplossingen

Harlem — *Hollandsche Matschæppij der Wetenschappen*
Archives nèerlandaises de zoologie
Physica

Musèe Teyler
Archives

Teyler's Tweede Genootschap
Verhandelingen

Den Haag — *Koninklijke Bibliotheek*
Bibliografie
Mnemosyne
Neophilologus

Leiden — *Koninklijke Nederlandsche Botanische Vereeniging*
Recueil des travaux botaniques néerlandais

Maatschappij der Nederlandsche Letterkunde
Jaarboek

Utrecht — *Physiologisch Laboratorium der Rijksuniversiteit*
Onderzoekingen

Provinciaal Utrechts Genootschap van Kunsten en Wetenschappen
Bijdragen en mededeelingen van het Historisch Genootschap
Verslag van het verhandelde

Norwegen

Bergen — *Universitetsbiblioteket*
Årbok. Hist.-antikv. rekke
Matemat.-naturvidenskapelig serie
Medisinsk rekke
Årsmelding
Skrifter. Bergens Historiske Forening

Oslo — *Norske Meteorologiska Institutt*
Norsk Meteorologisk Årbok

Norske Videnskaps-Akademi
Årbok
Astrophysica Norvegica
Avhandlingar. Hist.-filos. klasse
Matem.-naturvid. klasse
Geofysiske Publikasjoner
Hvalradets Skrifter
Physica Norvegica
Skrifter. Hist.-filos. klasse
Matem.-naturvid. klasse

Universitetets Oldsaksamling
Årbok
Norske Oldfunn
Skrifter
Viking

Universitetsbiblioteket
Archiv for Matematik og Naturvidenskab
Nytt Magasin for Botanikk
Nytt Magasin for Zoologi

Tromsö

Tromsö Museum
Acta Borealia. A. Scientia
B. Humaniora
Astarte

Trondheim

Det Kongelige Norske Videnskabers Selskabet
Årbok
Forhandlinger
Skrifter

Österreich

Graz

Historischer Verein für Steiermark
Blätter für Heimatkunde
Mitteilungen des Steiermärkischen Landesarchives
Zeitschrift. Nebst Sonderband

Naturwissenschaftlicher Verein für Steiermark
Mitteilungen

Steiermärkische Landesbibliothek
Jahrbuch der naturwissenschaftlichen Abteilung am Joanneum

Innsbruck

Universitätsbibliothek
Acta philologica Aenipontana
Innsbrucker Beiträge zur Kulturwissenschaft
Nebst Sonderheft

Wien

Bibliothek der Technischen Hochschule
Abhandlungen der Geographischen Gesellschaft
Mitteilungen der Geographischen Gesellschaft

Geologische Gesellschaft
Mitteilungen

Naturhistorisches Museum
Annalen

Österreichische Akademie der Wissenschaften
Almanach
Anzeiger. Math.-nat. Klasse
Phil.-hist. Klasse
Archiv für österreichische Geschichte
Denkschriften. Math.-nat. Klasse
Phil.-hist. Klasse
Fontes
Mitteilungen der prähistorischen Kommission

Monatshefte für Chemie
Sitzungsberichte. Math.-nat. Klasse
Phil.-hist. Klasse

Österreichische Mathematische Gesellschaft
Internationale mathematische Nachrichten

Universitätssternwarte
Annalen
Mitteilungen

Zoologisch-botanische Gesellschaft
Abhandlungen
Koleoptereologische Rundschau
Verhandlungen

Perú

Lima

Academia Nacional de Ciencias Exactas, Físicas y Naturales
Actas

VR Polen

Kraków

Polska Akademia Nauk. Oddzial w Krakowie
Acta archaeologica Carpathica
Acta biologica Cracoviensia. Ser. Botanica et Zoologica
Folia quaternaria
Mały Atlas gwar polskich
Materialy Komisji nauk historycznych
Prace Komisji archeologicznej
Prace Komisji filologii klasycznej
Prace Komisji nauk historycznych
Rocznik
Ruch literacki
Słownik staropolski

Polskie Towarzystwo Geologiczne
Regionalna Geologia Polski
Rocznik

Łódź	*Łódzkie Towarzystwo Naukowe* Biuletyn peryglacjalny
Lublin	*Biblioteka Uniwersytecka. Katolickiego Uniwersytetu Lubelskiego* Roczniki filozoficzne Roczniki teologiczno-kanoniczne
	Uniwersytet Marii Curie-Skłodowskiej Annales. A. AA. B. C. D. DD. E. F. G. H.
Poznań	*Polskie Towarzystwo Antropologiczne* Przeglad antropologiczny
	Polskie Towarzystwo Historyczne. Oddzial w Poznaniu Studia i materiały do dziejów Wielkopolski i Pomorza
	Poznańskie Towarzystwo przyjaciol nauk Bulletin. Ser. B. C. Prace Komisji filologicznej Prace Komisji historii sztuki Prace Komisji jezykoznawczej Roczniki historyczne Zabytki jezyka i literatury polskiej
Szczecin	*Muzeum Pomorza zachodniego* Materiały zachodnio-pomorskie
Toruń	*Towarzystwo Naukowe* Fontes Prace wydziału filologiczno-filosoficznego Roczniki Sprawozdania Studia. Sect. C. E. F. G. Zapiski
Warszawa	*Centralny Instytut Informacij Naukowo-Technicznej i Ekonomicznej* Obzor pol'skoj techniceskoj literatury
	Panstwowe Muzeum Archeologiczne Archeologia Materjały starozytne Materjały wczesnośredniowieczne Sprawozdania Wiadomości archeologiczne
	Polska Akademia nauk Archiwum budowy maszyn Bulletin de l'Académie Polonaise des sciences. Sér. des sciences biologiques Sér. des sciences chimiques Sér. des sciences géol. et géogr.

Sér. des sciences mathématiques
Sér. des sciences techniques
Nauka polska
Review of the Polish Academy of Sciences
Sprawozdania z prac naukowych
Zeszyty problemowe nauki polskiej

Polska Akademia nauk. Instytut matematyczny
Annales polonici mathematici
Roczniki. Ser. I. II.
Studia mathematica

Polska Akademia nauk. Instytut zoologiczny
Acta ornithologica
Annales zoologici
Fragmenta faunistica

Polskie Towarzystwo Botaniczne
Acta Societatis botanicorum Poloniae

Polskie Towarzystwo Matematyczne
Roczniki. Ser. A. Prace matematyczne
Ser. B. Wiadomości matematyczne

Warsaw Technical University. Institut of Mathematics
Demonstratio Mathematica

Wrocław

Muzeum Etnograficzne. Oddział Muzeum Śląskiego
Roczniki etnografii śląskiej
Zeszyty etnograficzne

Polski Zwiazek Entomologiczny
Polskie pismo entomologiczhe. Ser. B

Wrocławskie Towarzvstwo Naukowe
Comptes rendus
Prace. Ser. A. B
Sprawozdania. Ser. A. B

Portugal

Lisboa

Academia das Ciências de Lisboa
Anuário
Boletim

Faculdade das Ciências
Portugaliae physica

Oeiras

Instituto Gulbenkian de Ciência
Arquivo. A. Estudo matemáticos e fisico-matemáticos

SR Rumänien

București

Academia Republicii Socialiste România
Analele
Communicările
Publicaţiile Comisiei pentru studiul şi formarea limbii şi poporului romîn
Revue des études sud-est européennes
Revue roumaine de biologie.
Sér. botanique
Sér. de zoologie
Revue roumaine de chimie
Revue roumaine de géologie, géophysique et géographie
Revue roumaine de mathématiques pures et appliquées
Revue roumaine des sciences sociales
Sér. de philosophie et logique
Sér. de psychologie
Sér. des sciences économiques
Sér. des sciences juridiques
Revue roumaine des sciences techniques
Sér. de mécanique appliquée
Sér. de métallurgie
Sér. électrotechnique et énergétique
Revue roumaine d'historie
Progresele ştiintei
Studii. Revista de istorie
Studii clasice
Studii şi cercetări istorie veche
Studii şi materiale de istorie moderna
Viitorul social

Institutul Geologic
Anuarul Comitetului geologic
Dări de seama ale şedinţelor

Societatea de Ştiinţe Matematice si Fizice din R. P. R.
Bulletin mathématique
Gazeta matematica. Ser. A

Schweden

Göteborg

Chalmers Tekniska Högskolas Biblioteket
Handlingar

Kungl. Vetenskaps- och Vitterhetssamhället
Handlingar. Ser. A. B. Bihang

Universitetsbibliotek
Arsredogörelse
Etnologiska studier
Fynd
Göteborgs Universitets Årsskrift
Göteborger germanistische Forschungen
Gothenburg Studies in English
Romanica Gothoborgensia
Studia graeca et latina Gothoburgensia

Lund

Kungl. Fysiografiska Sällskapets
Förhandlingar

Kungl. Humanistiska Vetenskapssamfundet
Årsberättelse
Scripta minora
Skrifter

Slaviska Institutet vid Lunds Universitet
Årsbok
Meddelanden (Sprakliga bidrag)

Universitetsbiblioteket
Lunds Universitets Årsskrift
Mémoires du Musée historique
Scando-Slavica

Stockholm

Geologiska Föreningens
Förhandlingar
Sveriges Geologiska Undersökning

Humanistiska Biblioteket. Stockholms Högskola
The Ethnographical Museum of Sweden. Monograph Studies
Stockholm Studies in History of Literature

Kungl. Svenska Vetenskapsakademien
Arkiv för Astronomi
Arkiv för Botanik
Arkiv för Geofysik
Arkiv för Kemi
Arkiv för Matematik
Arkiv för Mineralogi och Geologi
Arkiv för Zoologi
Arsbok
Avhandlingar i naturskyddsärenden
Bidrag till Kungl. Svenska Vetenskapsakademiens historia
Chemica scripta
Handlingar
Levnadsteckningar
Physica scripta

Skrifter i naturskyddsärenden
Stockholms Observatoriums Annaler
Zoologica scripta

Kungl. Tekniska Högskolas Bibliotek
Forskning och Framsteg

Kungl. Vitterhets Historie och Antikvitets Akademien
Antikvariskt arkiv
Arsbok
Filologiskt arkiv
Fornvännen
Handlingar. Antikvariska serien
Filosofisk-filologiska serien
Historiska serien
Historiskt arkiv
Samlinggarnas tillväxt

Nordiska Museet
Fataburen
Handlingar

Uppsala

Kungl. Humanistiska Vetenskapssamfundet
Skrifter

Kungl. Vetenskaps-Societeten
Årsbok
Nova Acta

Universitetsbiblioteket
Acta Universitatis Upsaliensis
Acta Societatis linguisticae Upsaliensis
Historia litterarum
Historia religionum
Archives des traditions populaires Suédoises
Bulletin of the Geological Institute
Eranos
Uppsala Läkareförenings förhandlingar (Acta Societatis medicorum Upsaliensis)
Zoon

Schweiz

Aarau

Historische Gesellschaft des Kantons Aargau
Argovia

Basel

Historische und antiquarische Gesellschaft
Basler Bibliographie
Basler Zeitschrift für Geschichte und Altertumskunde

	Naturforschende Gesellschaft Verhandlungen
	Öffentliche Bibliothek der Universität Akademische Vorträge Jahresverzeichnis der Schweizerischen Hochschulschriften
Bern	*Allgemeine Geschichtsforschende Gesellschaft der Schweiz* Bibliographie der Schweizergeschichte Schweizerische Zeitschrift für Geschichte
	Naturforschende Gesellschaft Mitteilungen
	Schweizerische Geisteswissenschaftliche Gesellschaft Jahresbericht
	Schweizerische Naturforschende Gesellschaft Denkschriften Verhandlungen
Chur	*Naturforschende Gesellschaft Graubündens* Jahresbericht
Genf	*Institut National Genevois* Bulletin
	Musée d'Art et d'Histoire Genava
	Société Physique et d'Histoire Naturelle Archives des sciences Mémoires
Neuchâtel	*Bibliothèque Centrale de l'Université* Mémoires Publications du Séminaire de Géométrie Recueil des travaux Travaux de l'Institut botanique
Schaffhausen	*Naturforschende Gesellschaft* Mitteilungen Neujahrsblatt
Zürich	*Geographisch-ethnographische Gesellschaft* Geographica Helvetica
	Naturforschende Gesellschaft Neujahrsblatt Vierteljahrsschrift. Nebst Beihefte
	Schweizerisches Landesmuseum Jahresbericht

Senegal

Dakar	*Université de Dakar. Bibliothèque Universitaire* Annales de la Faculté des Scéances

Sowjetunion

Alma Ata	*Akademija nauk Kazachskoj SSR* Astrofizika Izvestija Astrofizičeskogo instituta Trudy Astrofizičeskogo instituta
Char'kov	*Char'kovskij gosudarstvennyj universitet im. A. M. Gor'kogo* Cirkuljar Astronomičeskoj observatorii
Dušanbe	*Akademija nauk Tadžikskoj SSR* Bjulleten' Instituta astrofiziki Bjulleten' Kommissii po kometan i meteoram Astronomičeskogo Soveta SSSR Trudy Instituta astrofiziki
Erevan	*Akademija nauk Armjanskoj SSR* Soobščenija Bjurakanskoj observatorii
Kazań	*Kazanskij gosudarstvennyj universitet* Bjulleteń Astronomičeskoj observatorii Izvestija Astronomičeskoj Engel'gardtovskoj observatorii
Kiev	*Akademieja nauk Ukrainskoj SSR* Avtomatičeskaja svarka Doslidžennja i materialy u ukrainśkoimovy Geologičnyj žurnal Izvestija Glavnoj astronomičeskoj observatorii Leksykografija i kul'tura reči Prykladna mechanika Trudy Poltavskoj gravimetričeskoj observatorii Visnyk
	Astronomičeskaja observatorija Gosudarstvennogo universiteta im. T. G. Ševčenko Cirkuljar Publikacii Kievskoj astronomičeskoj observatorii
Kišinev	*Akademija nauk Moldavskoj SSR* Monographien
Leningrad	*Akademija nauk SSSR* Astronomičeskij cirkuljar izdavaemyj Bjuro astronomičeskich soobščenij Akademii nauk SSSR Astronomičeskij ežegodnik Astronomičeskij žurnal

Bjulleteń Instituta teoretičeskoj astronomii
Bjulleteń Vsesojuznogo astronomo-geodezičeskogo obščestva
Efemeridy malych planet
Geobotanika
Izvestija Glavnoj astronomičeskoj observatorii v Pulkove
Izvestija Krymskoj astrofizičeskoj observatorii
Novosti sistematiki nizših rastenij
Novosti vysših rastenij
Peremennye zvezdy. Bjulleteń izdavaemyj Astronomičeskim Sovetom Akademii nauk SSSR
Sbornik Muzeja antropologii i etnografii
Soobščenija Gosudarstvennogo astronomičeskogo instituta im K. P. Šternberga
Trudy Gel'mintologičeskoj laboratorii
Trudy Glavnoj astronomičeskoj observatorii v Pulkove
Trudy i materialy naučnych kongressov i soveščanij, opublikovannye za rubežom
Trudy Laboratorii geologii dokembrija
Trudy Paleontologičeskogo instituta

Lvov
Astronomičeskaja observatorija L'vovskogo gosudarstvennogo universiteta im. Ivana Franko
Cirkuljar

Moskva
Gosudarstvennaja publičnaja naučno-techničeskaja biblioteka SSSR
Uspechi chimii

Riga
Akademija nauk Latvijskoj SSR. Astrofizičeskaja laboratorija
Trudy laboratorii
Zvaigžnota debess

Tartu
Akademija nauk Estonskoj SSR. Institut fiziki i astronomii
Soobščenija Tartuskoj astronomičeskoj observatorii
Trudy

Taškent
Akademija nauk Uzbekskoj SSR. Taškentskaja astronomičeskaja observatorija
Cirkuljar
Trudy

Tbilisi
Akademija nauk Gruzinskoj SSR
Bjulleteń

Spanien

Barcelona
Real Academia de Ciencias y Artes
Nomina de personal académico

Madrid
Ciudad Universitaria. Facultad de Filosofía y Letras
Cuadernos de Filología Clásica

Tschechoslowakische Sozialistische Republik

Bratislava	*Geologický ústav D. Štúra* Geologické prace Sbornik geologických vied
	Slovenská Aakdémia vied Acta geologica et geographica Ekonomický časopis Filozofia Jazykovedny časopis Slovenskie odborné nažvoslovie
	Slovenska Akademia Vied. Geofyzikalny Ustav Bulletin of the Slovak Seismographic Stations Contributions
Brno	*Brněnská základna Československé Akademie ved* Folia Geobotanica Práce
	Czechoslovak Academy of Sciences. Botanical Institute Folia geobotanica et phytotaxonomica bohemoslavica
	J. E. Purkyně-Universita Ročenka Sbornik praci Filosofické fakulty. Řada C. E. F. G. Spisy Filosofické fakulty Spisy Lékarské fakulty Spisy Prirodovědecké fakulty
	Vysoká Škola Zemědělská Sbornik. Ř. A—C. Spisy
Olomouc	*University Palackého. Přirodovedecké fakulty* Acta
Praha	*Československá Akademie věd* Collection of Czechoslovak Chemical Communications Rózpravy matemat. a přirod. věd Rózpravy společenských véd Rózpravy techn. věd Slovanský přehled Věstnik
	Československá Akademie věd. Historický ústav Československý časopis historický
	Československá Akademie věd. Geofysikálni ústav Bulletin čsl. seismických stanic Geofysikálni sbornik Special Papers Výsledky geomagnetikých

Československá Akademie věd. Matematický ústav
Aplikace matematiky
Časopis pro pestováni matematiky
Czechoslovak Mathematical Journal

Československá Akademie věd. Orientalni ústav
Archiv orientalni

Československá Akademie věd. Kabinet pro studia recká, římska a latinska
Byzantinoslavica

Československá Akademie věd. Ústav jazyku a literatur
Slavia

Československá Akademie věd. Ústav technické fysiky
Czechoslovak Journal of Physics

Československá Akademie věd. Ústředni ústav matematický
Časopis pro pěstováni matematiky a fysiki

Československá entomologická společnost
Acta entomologica bohemoslovaca

Ministerstvo vnitra
Archivni časopis
Sbornik archivnich praci

Národni muzeum. Entomologické odděleni
Acta faunistica entomologica
Sbornik (Acta entomologica)

Statni Židovské Muzeum v Praze
Judaica Bohemiae

Ústredni ústav geologický
Rozpravy
Sbornik

VR Ungarn

Budapest

Bányaszati Kutatóintézet
Mitteilungen

Budapesti Müszaki Egyetem Központi Könyvtara
Periodica polytechnica
Elfogadot doktori diszertaciok jegy zeke

Magyar Tudományos Akadémia
Acta biologica
Acta geologica
Acta historica
Acta historiae artium
Acta linguistica
Acta mathematica
Acta medica
Acta morphologica

Acta orientalia
Acta physiologica
Acta veterinaria
Könyvtárának kiadványai
Magyar tudomány
Studia historica

Magyar Tudományos Akadémia. Matematikai Kutató Intézetének
Studia scientiarum mathematicarum Hungaricae

Debrecen *Institutum mathematicum Universitatis Debreceniensis*
Publicationes mathematicae

Szeged *Institutum Bolyainum Universitatis*
Acta scientiarum Mathematicarum

Uruguay

Montevideo *Instituto de Matemática y Estadística. Facultad de Ingeniería y Agrimensura*
Publicaciones

Museo Nacional de Historia Natural
Anales
Communicaciones antropológicas
Communicaciones botánicas
Communicaciones paleontológicas
Communicaciones zoológicas
Flora del Uruguay

Sociedad de Biología
Archivos

Venezuela

Caracas *Universidad Central de Venezuela. Facultad de Ciencias. Escuela de Biología*
Acta biologica venezuelica

Vereinigte Staaten von Amerika

Albany *New York State Museum and Science Service*
Bulletin
Educational Leaflet
Map and Chart Series

Ann Arbor *University of Michigan*
Mathematical Reviews

Berkeley *University of California*
Lick Observatory Bulletin
Pacific Journal of Mathematics

Bloomington

Goethe-Linck-Observatory
Publications
Reprint

Indiana University Library
Indiana University Publications. Social Science Series
Journal of Mathematics and Mechanics

Boston

American Academy of Arts and Sciences
Daedalus
Memoirs
Records

Cambridge

Astronomical Observatory of Harvard College
Annual Report

Massachusetts Institute of Technology
Studies in applied mathematics

Museum of Comparative Zoology at Harvard College
Annual Report
Breviora
Bulletin

Chicago

Chicago Natural History Museum
Publication. Report Series

Columbus

Ohio State University
Chemical Abstracts

Eugene

University of Oregon. Library
Comparative Literature

Gainesville

University of Florida. Florida State Museum
Bulletin of the Biological Science
Report of the Director

Indianapolis

Indiana Academy of Science
Proceedings

Lawrence

Kansas Academy of Science
Transactions

University of Kansas. Library
Library Series
Science Bulletin

Lexington

University of Kentucky Library
Kentucky Geological Survey. Bulletin
Report of Investigations
Special Publication

Los Angeles

University of Southern California. Allan Hancock Foundation
Allan Hancock Monographs in marine biology

Madison	*Wisconsin Academy of Sciences, Arts and Letters* Transactions
Minneapolis	*University of Minnesota. Library* Agricultural Experiment Station. Bulletin
New Haven	*Connecticut Academy of Arts and Sciences* Memoirs Transactions
	Kline Science Library Bulletin of the Bingham Oceanographie Collection
New York	*American Academy of Arts and Letters* Proceedings
	American Council of Learned Societies Annual Report
	American Museum of Natural History Annual Report Anthropological Papers Bulletin Natural History Novitates Science Guide
	Rockefeller Institute for Medical Research Journal of General Physiology
Philadelphia	*Academy of Natural Sciences* Proceedings
	American Philosophical Society Memoirs Proceedings Transactions Year Book
Reston	*U. S. Geological Survey* Abstracts of North American Geology Bulletins New Publications
San Francisco	*California Academy of Sciences* Annual Report Occasional Papers Proceedings
Seattle	*University of Washington. Library* Annual Northwest Traffic Engineering Conference. Proceedings Publications in Anthropology Publications of Oceanography Trend in Engineering

Urbana — *University of Illinois. Library*
Agricultural Experiment Station.
  Bulletin
  Circular
Biological Notes
Engineering Experiment Station.
  Bulletin
Engineering Outlook
Illinois Natural History Survey.
  Bulletin
  Circular
  Publications

Washington — *Library of Congress*
American Bibliography of Russian and East European Studies
World list of Future International Meetings

National Bureau of Standards
Journal of Research. Sect. B.

Smithsonian Institution
Annual Report of the Board of Regents
Bureau of American Ethnology.
  Annual Report
  Bulletin
Contributions from the U. S. National Herbarium
Proceedings of the U. S. National Museum Smithsonian Contributions to Astrophysics Smithsonian Miscellaneous Collections

Westberlin

Deutsches Archäologisches Institut
Archäologische Bibliographie
Archäologischer Anzeiger
Jahrbuch

Staatsbibliothek. Preußischer Kulturbesitz
Monographien

Technische Universität
Humanismus und Technik

II. Einzelwerke

Es gingen ein an Einzelwerken im Jahr 1973 194 Bände
1974 278 Bände

Dr. Dietmar Debes

ANHANG

ZUM 200. GEBURTSTAG VON CHRISTIAN FRIEDRICH SCHWÄGRICHEN*

Als Sohn eines Kaufmanns am 16. September 1775 in Leipzig geboren, ist Christian Friedrich Schwägrichen seiner Heimatstadt und ihrer Universität zeitlebens treu geblieben: als Student (1799 Magister, später Dr. med.), als Professor für Naturgeschichte (seit 1802), als Professor der Botanik (seit 1806), als Beisitzer der Medizinischen Fakultät (seit 1819), als Direktor des Botanischen Gartens (bis 1837). In der Geschichte unserer am 1. Juli 1846 als Kgl. Sächsische Gesellschaft der Wissenschaften gegründeten Akademie steht er mit am Beginn. Nachdem er sich 1852 aus der akademischen Öffentlichkeit zurückgezogen hatte, ist er in Leipzig im 78. Lebensjahr am 2. Mai 1853 gestorben. Als Botaniker war Schwägrichen ein bekannter Moosforscher.

Den knappen biographischen Abriß von E. Wunschmann[1]) möchte ich im folgenden um einige Tatsachen ergänzen, die mir bei der Beschäftigung mit Christian Ludwig Brehm und mit Lorenz Oken bekannt geworden sind. Dabei beziehe ich Schwägrichens jüngeren Kollegen Gustav Kunze ein, der unserer Akademie gleichfalls angehört hat. Auch Kunze war fest in Leipzig verwurzelt. Er wurde dort am 4. Oktober 1793 geboren. Von Haus aus Mediziner (22. VI. 1819 Dr. med.), aber sehr früh auch entomologisch und botanisch interessiert, warf er sich schon vor seiner Promotion mit besonderem Eifer auf das Studium der Pilze, später mit noch größerer Begeisterung auf das der Farne, die in der Pflanzenwelt fortan seine Lieblinge geblieben sind und denen er hervorragende wissenschaftliche Werke gewidmet hat („einer der besten Kenner dieser Pflanzenformen auf dem Continent"). Seit 1822 ao. Professor der Medizin, bald darauf Kustos der Gehlerschen Bibliothek (bis 1848), 1835 ao. Professor der Botanik, seit 1837 Schwägrichens Nachfolger in der Leitung des Botanischen Gartens, 1845 o. Professor der Botanik, starb er, nachdem er eben noch im Botanischen Garten gearbeitet, plötzlich infolge eines Schlaganfalls am 30. April 1851, nur 57 Jahre alt. Kunze war ein hochbegabter, ungemein vielseitiger und

* Siehe Jahrb. Sächs. Akad. Wiss. *1960—1962* (Berlin 1964) S. 94; ebda *1971—1972* (Berlin 1974) S. 51.

[1]) ADB *33* (1891) S. 175. — Ein Bildnis von Schwägrichen scheint nicht zu existieren: Zaunick (s. Anm. 18) S. 39 Anm. 3. — Nachzutragen ist: Verzeichnis der Bibliothek des Herrn Dr. Christian Friedrich Schwägrichen. Leipzig 1852: Breitkopf & Härtel. 96 S.; dass., 2. Abt., 276 S. (Bibliographie zur Gesch. der Stadt Leipzig. Sonderband *IV*: Das Buch. Weimar 1967, S. 308).

kenntnisreicher Mann, „schon als Schüler und Student mit Naturforschern correspondirend“[2]).

Einer unter diesen Naturforschern, dessen Zuneigung Kunze bereits als Leipziger Medizinstudent erwarb, war der Pfarrer und Ornithologe Christian Ludwig Brehm in Unterrenthendorf (1787—1864). Aus der Verbindung Brehms mit dem sechs Jahre jüngeren Kunze überliefert dessen Nachlaß 24 Briefe[3]), die von Brehm zwischen 1814 und 1818 an Kunze geschrieben worden sind. Rudolf Möller hat vor Jahren auf sie aufmerksam gemacht und einen Einblick gegeben[4]). Veröffentlicht sind sie noch nicht, doch hoffe ich, sie zusammen mit anderen Brehm-Briefen demnächst in den Abhandlungen der Sächsischen Akademie herausgeben zu können. Brehm arbeitete damals an seinem ersten großen Werk, den „Beiträgen zur Vögelkunde“[5]), und studierte bereits systematisch die Anatomie der Vögel. Dazu benötigte er „den Tiedemann“, den er sich bandweise von Kunze auslieh[6]). Auch anderes Schrifttum suchte er durch Kunzes Vermittlung zu erhalten. Im Mai 1816 richtete er erstmalig an Kunze die Bitte, ihm die Mageninhalte von Vögeln auf Insekten und Pflanzen zu bestimmen[7]), während er seinerseits für Kunze auf Vögeln lebende

---

[2]) E. Wunschmann, ADB *17* (1883) S. 400ff. (402). — Ein ansprechendes Bildnis von Kunze bei Zaunick (s. Anm. 18) S. 55 (nach dem Original im Botanischen Institut der Karl-Marx-Universität Leipzig).

[3]) Univ.-Bibl. Leipzig: Ms 0352^{I}. Dem Leiter der Handschriftenabteilung, Herrn Dr. Dietmar Debes, danke ich für die Erlaubnis zur Benutzung und Veröffentlichung dieser Briefe.

[4]) Rudolf Möller, Christian Ludwig Brehms Briefe an Kunze, in: Der Falke *11* (1964) H. 3, S. 81—83.

[5]) Christian Ludwig Brehm, Beiträge zur Vögelkunde. 3 Bde. Bd *III* gemeinsam mit Wilhelm Schilling. Neustadt a. d. Orla 1820—1822. Nachdruck Leipzig 1971.

[6]) Brehm, 17. X. 1814: „Mit Tiedemanns Zoologie 2^{t} Band bin ich bald zu Ende; und habe mich besonders auch über das Capitel von den Respirationsorganen sehr gefreut“. — 29. XI. 1814: „Auf den neuen Band von Tiedemanns Zoologie bin ich sehr neugierig...“ — 8. III. 1815: „Schon lange hätte ich Ihnen Tiedemanns III zurückgeschickt, wenn ich nicht...“. — 7. VI. 1816: „Sie würden mir nämlich eine große Freude machen, wenn Sie mir die ornithologischen Theile von Tiedemanns Zoologie auf einige Zeit überlassen könnten; ich brauche sie um manches darinnen zu studieren...“. — 25. IV. 1818: „Ihren Tiedemann habe ich leider noch immer und brauche ihn aber auch gewiß nothwendiger wie Sie, weil Sie das Meiste darinnen schon wissen; ich aber es erst lernen muß ...“. — Über Brehms schon früh einsetzendes anatomisches Interesse s. auch Gerhard Buchda, Ludwig Brehm und Gottlieb Richter. (Zur Morphologie der Vögel.), in: FuF **39** (1965) H. 6, S. 165ff.

[7]) Brehm, 8. V. 1816: „Nun habe ich aber eine recht große Bitte an Sie, deren Gewährung ich von Ihrer Freundschaft hoffe. Ich sehe nämlich immer mehr ein, daß ein Mensch nicht ein Mal einen Zweig der Naturgeschichte ganz erforschen kann; deswegen nehme ich meine Zuflucht zu Ihrer großen Kenntnis der Insekten und bitte Sie freundschaftlich, mir so viel als möglich ist, die Namen der Insekten zu bestimmen, wie auch die der Pflanzen, welche die Vögel in Ihrem [!] Magen haben. Ich sende Ihnen also nicht bloß Schmarotzerinsekten, sondern auch den Inhalt des Magens verschiedener Vögel, und werde damit fortfahren, wenn Sie mir meine Bitte gewähren. Ich verkenne die Schwierigkeiten nicht, welche

Schmarotzerinsekten sammelte, weshalb er ihn wiederholt um die Zusendung von Gläschen und Stöpselchen ersuchte, weil er solche in Renthendorf nicht beschaffen konnte. Und es gab noch andere Dinge, die der hilfsbereite Kunze aus der sächsischen Handelsmetropole für seinen Vogelfreund im abgelegenen Renthendorf besorgte.

In Brehms Briefen an Kunze wird nun auch Schwägrichen genannt, und zwar erfahren wir, daß Brehm mit ihm und anderen Vögel tauschen wollte. Schwägrichen war also auch Ornithologe. Nach Peter Thomsen und Erwin Stresemann besaßen mehrere Mitglieder der 1818 gegründeten Naturforschenden Gesellschaft zu Leipzig „große Sammlungen, so vor allem Christian Friedrich Schwägrichen, der das Kabinett des Kassierers der Gesellschaft, Christoph Heinrich Ploss, übernommen hatte, in den Besitz Pallasscher Vögel aus Sibirien und Kamtschatka gelangt war und auch lebende Vögel hielt“[8]). Als der niederländische Ornithologe und Direktor des Reichsmuseums in Leiden, Coenraad Jacob Temminck, der Vorgänger des aus Altenburg stammenden Hermann Schlegel, Anfang August 1818 Johann Friedrich Naumann in Ziebigk besucht hatte, reiste er nach Leipzig, „wo ihm die Freude zuteil ward, bei Professor C. F. Schwägrichen eine ganze Anzahl von Vögeln aus Sibirien und Kamtschatka zu sehen, die der Leipziger Kupferstecher C. G. H. Geißler (dem sie einst von Pallas als Vorlagen für Tafeln zur Zoographia Rosso-Asiatica anvertraut worden waren) an Schwägrichen verkauft hatte, darunter *Cinclus pallasii*“[9]). Naumann trat im Oktober 1819 der Leipziger Naturforschenden Gesellschaft bei, vor allem wohl, um zu den Leipziger Ornithologen Kontakt zu gewinnen und aus ihren Sammlungen seine Kenntnisse zu erweitern[10]). „Von Schwägrichen tauschte Naumann einige Seevögel ein, die, aus den Gewässern von Kamtschatka‘ stammten und dort (wahrscheinlich vor allem 1790 auf Unalaschka) von Dr. Carl Heinrich Merck für Pallas gesammelt worden waren. Einen davon erkannte er als neue Art und beschrieb ihn in Okens ‚Isis‘ 1821 als *Mormon corniculata* [jetzt: *Fratercula corniculata*]“[11]).

Wir haben zeitlich etwas vorgegriffen und müssen zum Jahre 1814 zurückkehren.

---

hiermit verbunden sind, so wenig, als die große Mühe, welche es kosten wird; doch bin ich fest überzeugt, daß durch solche Untersuchungen die Naturgeschichte wahrhaft gefördert wird“.

[8]) Peter Thomsen u. Erwin Stresemann, Johann Friedrich Naumann der Altmeister der deutschen Vogelkunde. Sein Leben und seine Werke (Lebensdarst. dt. Naturforscher 6). Leipzig 1957, S. 99. — Über die Sammlung des Bankiers Christoph Heinrich Ploß (24. III. 1757—11. I. 1838) schrieb Chr. L. Brehm 1822, sie sei „die schönste Privatvögelsammlung“, die er kenne (Beitr. zur Vögelkunde III, 2, S. 894). Die Sammlung kam über Professor Schwägrichen an die Universität Leipzig, „um nun den Grundstock zu deren ornithologischer Sammlung zu bilden“: Richard Heyder, Die Vögel des Landes Sachsen. Leipzig 1952, S. 5 (mit Gewährsmann).

[9]) Erwin Stresemann, Die Entwicklung der Ornithologie von Aristoteles bis zur Gegenwart, Berlin 1951, S. 125f.

[10]) Thomsen-Stresemann, a. a. O., S. 99.

[11]) Thomsen-Stresemann, a. a. O., S. 99. — Der Titel von Naumanns Veröffentlichung in der „Isis“ 1821 unter dem Jahr 1821 des Verzeichnisses seiner Schriften, a. a. O., S. 199.

Gleich in dem ersten der Briefe an Kunze schreibt Brehm (20. VI. 1814): „Sollten Sie mir die Vögel, welche der Herr Professor Schwägrichen und Andere vertauschen wollen unter Bemerkung des Preises gefälligst anzeigen wollen: würden Sie mich sehr verbinden". Auch am 2. IX. 1814 geht es um den Vogeltausch: „Das Verzeichniß meiner Doupletten erfolgt Ihrem Wunsche gemäß, nur habe ich das der gewöhnlichen Singvögel, weil diese Arten zu leicht zu haben sind, nicht beygefügt; wünscht es aber der Herr D. Schwägrichen, dem ich mich gehorsamst zu empfehlen bitte: so soll es bald folgen". Am 16. X. 1814 meldet Brehm: „Die Kiste habe ich schon gestern nach Triptis geschickt ... In der Kiste befinden sich für Herrn D. Schwägrichen *Columba oenas* [Hohltaube], W. ohne Augen; für Herrn Dr. Kaulfuß ...". Doch scheinen die Tauschgeschäfte mit Schwägrichen nicht recht in Gang gekommen zu sein; denn enttäuscht fragt Brehm am 27. IV. 1815: „Haben Sie von Schwägrichen noch immer nichts erhalten? Dieser Herr scheint sehr saumseelig".

Von Kunze erbat Brehm auch spezielle Beobachtungen an Schwägrichens Vögeln. „Haben Sie die Güte, die Schneeammer beym Herrn D. Schwägrichen anzusehen, damit Sie mir sagen können, ob der meinige Männchen oder Weibchen ist" (2. IX. 1814). Bald darauf möchte Brehm einen solchen Vogel erwerben (29. XI. 1814): „Von Herrn D. Schwägrichen werden Sie schon noch einen *emberiza nivalis* [jetzt: *Plectrophenax nivalis nivalis* (L.)] (ein Männchen) oder sonst einen seltenen Vogel für mich erhalten", und nochmals am 8. III. 1815: „Haben Sie einen Schneeammer erhalten können?"

Schwägrichen besaß einen lebenden Adler. Über ihn will Brehm ganz Genaues wissen: „Tiedemann behauptet, die Vögel, die Eule ausgenommen, sähen blos von der Seite; daß ist nach meinen Beobachtungen falsch; machen Sie doch Versuche mit Schwägrichens Adler, indem Sie ihm Etwas gerade vor den Schnabel halten; er wird es gewiß sehen. Ueberhaupt bitte ich Sie inständig, dieses Thier ja recht sorgfältig zu beobachten, die Beobachtungen zu merken und mir mit der Zeit mitzutheilen. Sollte er mit Tode abgehen und auch an Schwanz und Flügeln zerstoßen seyn: so schicken Sie mir von ihm, wenn er etwa skelettirt wird, wenigstens die Haut, damit ich sie nur sehe, und vielleicht durch andere Adler vollständig machen kann" (29. XI. 1814). Einige Monate später fragt Brehm wieder nach dem Tier: „Was macht D. Schwägrichens Adler?" (8. III. 1815). Über diesen Steinadler lesen wir dann noch in Brehms „Beiträgen zur Vögelkunde" (I, 1 S. 67f.). Brehm schreibt an dieser Stelle, es handele sich bei der verbreiteten Angabe, die Eier des Steinadlers seien weiß und rot gefleckt, wohl um einen Irrtum. „Der Herr Professor, Doktor Schwägrichen in Leipzig, hat ein Steinadlerweibchen schon lange Zeit lebendig. Dieses legt fast alle Frühjahre ein Ei und frißt es; doch bleiben gewöhnlich einige Stückchen Schale übrig, und diese sind immer weiß, ohne Spuren eines röthlichen Flecks".

Zusammen mit seinem Jugendfreund Ludwig Reichenbach und einigen Männern, die nicht vom Gelehrtenstande, gründete Kunze 1818 die Naturforschende Gesellschaft zu Leipzig, die ihn zu ihrem Sekretär ernannte, als Reichenbach 1820 nach Dresden übersiedelte[12]). Ungefähr zur selben Zeit rief in Altenburg ein ähnlicher

---

[12]) E. Wunschmann, ADB *17*, S. 402.

Kreis Naturbeflissener die Naturforschende Gesellschaft des Osterlandes ins Leben[13]). Beide Gesellschaften knüpften bald freundschaftliche Beziehungen zueinander an, vereinbarten in einer bestimmten Beziehung auch Zusammenarbeit. Wieder ist von Schwägrichen die Rede. „Aus Dank ernannte die Naturforschende Gesellschaft des Osterlandes Schwägrichen, den Direktor der Leipziger Vereinigung und Reichenbach zum Mitgliede“[14]).

Noch einmal treten Schwägrichen und Kunze, die beiden Leipziger Botaniker, gemeinsam hervor: bei der Gründung der Gesellschaft Deutscher Naturforscher und Ärzte in Leipzig im Herbst 1822[15]). Der Initiator dieser heute noch blühenden Gesellschaft war bekanntlich Lorenz Oken in Jena, der in seiner „Isis“ bereits 1820 und ausführlich 1821 zu einer „Versammlung der deutschen Naturforscher“ aufgerufen[16]), schließlich definitiv für den 18. September 1822 nach Leipzig eingeladen hatte[17]). Schon für die Vorbereitungen war Schwägrichen dort sein Verbindungsmann. Dies beweisen zwei Briefe von Oken an Schwägrichen vom 2. und vom 25. Juni 1821[18]) wie auch die Ankündigungen und die Einladung Okens in der „Isis“.

Aus dem ersten Brief geht hervor, daß die Gründungsversammlung bereits für 1821 geplant und daß Schwägrichen bereits tätig geworden war. Da sich jedoch herausstellte, daß an dem ins Auge gefaßten Septembertag Schwägrichen nicht und Oken wahrscheinlich nicht von ihren Reisen zurücksein und demzufolge die beiden Hauptorganisatoren fehlen würden, hielt Oken es für zweckmäßig, erst für 1822 einzuladen. „Bey dem 19ten Sept. habe ich vorzüglich auf Ihre Rückkunft Rücksicht genommen. Da Sie nun nicht da sind, so weiß ich in der That nicht, ob es rathsam ist, die Versammlung in diesem Jahre [1821] zu halten. Ihr Eifer, welchen Sie sogleich an den Tag gelegt haben, indem Sie die Sache der naturf. Ges. vorgelegt und schon den Stadtrath um einen Saal gebeten haben, beweist, wie nothwendig Sie bey der Versammlung sind. Da auch ich eine Reise nach den Niederlanden machen muß, und zu zweifeln anfange, ob ich am 12. Sept. wieder da bin, so stelle ich es Ihnen ganz anheim, ob alles das, was Sie indessen gethan nicht für 1822 in Anspruch zu

---

[13]) Rudolf Möller, Die „Naturforschende Gesellschaft des Osterlandes zu Altenburg“ Teil I — Die Zeit von 1817—1836, in: Abh. u. Ber. des Naturkundl. Museums „Mauritianum“. Altenburg *7* (1972) S. 71—126, bes. S. 85f.

[14]) Möller, a. a. O., S. 86.

[15]) Zum Folgenden: Max Pfannenstiel, Kleines Quellenbuch zur Geschichte der Gesellschaft Deutscher Naturforscher und Ärzte. Gedächtnisschr. für die hundertste Tagung der Gesellschaft [Wiesbaden 1958] . Im Auftrage des Vorstandes der Gesellschaft verfaßt. Berlin, Göttingen, Heidelberg 1958.

[16]) A. a. O., S. 5 und S. 24—27.

[17]) A. a. O., S. 6 u. S. 33—34.

[18]) Rudolph Zaunick, Zwei Briefe Lorenz Okens vom Juni 1821 an den Leipziger Biologen Friedrich Schwägrichen. Im Rahmen der Dokumente zur Vorgeschichte der Leipziger Gründungstagung der Gesellschaft Deutscher Naturforscher und Ärzte, in: Nova Acta Leopoldina NF *29* Nr. 171 (1964) S. 37—56. Der erste Brief (früher im Stadtmuseum Jena) gehört zu den Kriegsverlusten, der zweite (1928 in der Autographensammlung von Dr. Kurt Loewenfeld, Hamburg) ist verschollen. Näheres bei Zaunick, S. 44, 46.

nehmen sey und ob man daher nicht dieses Jahr [1822] als den Termin für die erste Versammlung zu bestimmen habe ... Ich habe deshalb mit Kieser öfters hier geredet. Er meint zwar, daß die Zusammenkunft dieses Jahr [1821] gut wäre, damit nur einmal ein Anfang gemacht würde, glaubt aber, daß sie erst im nächsten Jahr [1822] vollständig gelingen könnte. — Ich überlasse Ihnen hiermit die ganze Entscheidung, die darauf gegründet seyn wird, daß Sie und wahrscheinlich auch ich nicht dabey seyn werden, und darauf, daß schon Schritte in Leipzig beim Stadtrath geschehen sind. Man könnte, denke ich, diese Schritte, den guten Willen der Leipziger bekannt machen und die Versammlung auf den Septbr. 1822 vestsetzen. Wie Sie mir hierüber schreiben, so werde ich thun .... Mit Hochachtung. Glück auf die Reise Ihr Oken"[19]. Was Schwägrichen geantwortet, ist nicht bekannt, doch dürfte er entschieden haben, die Versammlung um ein Jahr hinauszuschieben. Jedenfalls ist dann von Oken in der „Isis" 1821, H. 7 und nochmals H. 8 die Verlegung auf 1822 und in der „Isis" 1822 H. 6, H. 7 und H. 8 der 18. September 1822 als fester Termin bekannt gemacht worden[20]). Wiederholt nennt Oken in seinen Ankündigungen wie zuletzt in der endgültigen Einladung Professor Schwägrichen als diejenige Person, bei der sich melden könne, wer an der Leipziger Versammlung teilzunehmen gedenke[21]) und wer in einem Privatquartier zu wohnen wünsche[22]).

Derartige von Anfang an gezeigte Einsatzfreudigkeit macht begreiflich, daß Oken seinen zweiten Brief vom 25. Juni 1821 mit dem Satze beginnt: „Wenn sich jedermann unserer Wissenschaften mit solchem Interesse annimmt, wie dasjenige ist, welches sich in Ihrem Briefe so freundlich ausgesprochen hat, so ist an dem guten Fortgang der Sache, die wir vorhaben, nicht zu zweifeln". Bemerkenswert ist auch der sich anschließende Satz: „Von der Liberalität der sächsischen Regierung war ich von jeher überzeugt und die großen Kenntnisse S. M. des Königes in den Naturwissenschaften, besonders in der Botanik sind allgemein bekannt; ich dachte aber, es wäre der Form wegen gut höheren Ortes davon zu reden, was Sie also thun und am besten thun können". Oken fügt hinzu: „Für ihre [!] Bemühungen, denen Sie sich unterziehen wollen, wird Ihnen gewiß jedermann dankbar seyn. Den Lohn aber gibt die Wissenschaft."[23])

---

[19]) Zaunick, a. a. O., S. 45.

[20]) Zaunick, a. a. O., S. 50, 52.

[21]) Zaunick, a. a. O., S. 52; Pfannenstiel, Quellenbuch S. 33 oben. — „Das Melden kann bey Schwägrichen in Leipzig oder bey der Redaction der Isis geschehen" (so 1821 H. 12).

[22]) Zaunick, a. a. O., S. 50, 52; auch Pfannenstiel, Quellenbuch, S. 33 unter 3. — Isis 1821 H. 7, H. 8; 1822 H. 5, H. 6, H. 7, H. 8. — „Wer ..., beliebe dieses dem Herrn Professor Schwägrichen noch vor Ende Augusts anzuzeigen"; „wer .., beliebe sich deshalb noch vor Ende Augusts an Professor Schwägrichen zu wenden"; „Das Melden kann nach Bequemlichkeit bey Prof. Schwägrichen in Leipzig oder bey der Redaction der Isis geschehen". — Über das Tagungsbüro bemerkte Oken: „Sobald man ankommt, meldet man sich bey dem Ornithologen Heinr. Ploß (Grimmaische Gasse Nr. 593), um Auskunft über Wohnung, Versammlungsort und -zeit zu erhalten". Zaunick, a. a. O., S. 51, 53; Pfannenstiel, Quellenbuch, S. 34 unter 9.

[23]) Zaunick, a. a. O., S. 46.

TOPOGRAPHIAE BOTANICAE ET ENTOMOLOGICAE LIPSIENSIS

SPECIMEN PRIMVM

QVOD

AMPLISSIMI PHILOSOPHORVM ORDINIS

AVCTORITATE

DIE XIII. APRILI ANNI MDCCXCIX

DEFENDET

CHRISTIANVS FRIDERICVS SCHWAEGRICHEN

PHIL. D. ET A. A. L. L. MAG. MEDIC. BACC. SOCIETATIS

LINNAEAN. LIPSIENSIS SODALIS

LIPSIENSIS

SOCIO

IOANNO CHRISTIANO AVGVSTO CLARO

MED. BACC. SOCIET. LINNAEAN. LIPSIENSIS SODALI

COBVRGO-FRANCO

LIPSIAE

IMPRESSIT CAROLVS TAVCHNITZ

Titelblatt des ersten Teiles der von Schwägrichen verfaßten Leipziger botanisch-entomologischen Topographie von 1799

VIRO PRAENOBILISSIMO

CHRISTIANO GODOFREDO
SCHWAEGRICHEN

MERCATORI CLARISSIMO

PATRI

SVMMO PIETATIS ET AMORIS CVLTV

PROSEQVENDO

HOC LIBELLVM SACRVM ESSE VOLVIT

AVCTOR

Widmung von Schwägrichen an seinen Vater

Als der 18. September 1822 herangekommen war, fanden die ersten abendlichen Zusammenkünfte im Hause der Leipziger Naturforschenden Gesellschaft in der Grimmaischen Gasse Nr. 7 und 8 statt[24]). In einer kurzen Zeitungsnotiz über die Gründungsversammlung ist festgehalten: „Für diesmal war der verdiente Botaniker und Naturforscher, Professor Schwägrichen, ihr Geschäftsführer, Dr. Kunze Sekretär“[25]). In einem späteren Bericht heißt es dann: „Unter der Leitung dieser beyden Männer wurden sofort folgende Statuten der Gesellschaft entworfen, die auch bisher, mit Ausnahme der Einführung von Aufnahmskarten und der Eintheilung der Gesellschaft in mehrere wissenschaftliche Sectionen, unverändert beybehalten worden sind“[26]). Die Statuten, 20 Paragraphen umfassend, sind datiert und tragen die Unterschriften:[27])

Leipzig,
am 1ten Oktober 1822. Im Auftrage der Gesellschaft
der Geschäftsführer D. Friedrich Schwägrichen ord.
Prof. der Naturg[eschichte]
der Sekretär D. Gustav Kunze, außerord. Prof. d. Med.

Max Pfannenstiel hat seinem „Kleinen Quellenbuch“ ein Verzeichnis der [ersten] 100 Versammlungen der Gesellschaft [1822—1958] sowie ihrer Geschäftsführer und Sekretäre beigegeben und dazu bemerkt: „Wie viele gute Namen sind in diesen Verzeichnissen aufgeführt, Namen voll Stolz und Glanz, denen wir Nachfahren tiefen Dank schulden“[28]. Die Sächsische Akademie der Wissenschaften darf sich darüber freuen, daß zwei ihrer frühesten Mitglieder die Liste der Geschäftsführer und der Sekretäre dieser Versammlungen eröffnen.

Briefe Brehms an Schwägrichen (und umgekehrt) habe ich nicht ausfindig machen können; ebensowenig weiß ich, ob sich beide Männer einmal persönlich begegnet sind. Auch besitze ich keinen Anhaltspunkt dafür, weshalb die Briefe Brehms an

---

[24]) „Kurze Geschichte der zehn ersten Versammlungen“ der Gesellschaft Deutscher Naturforscher und Ärzte, von Frhr. von Jacquin und J. J. Littrow dargeboten im Rahmen des Berichts über die 10. Versammlung in Wien 1832, bei Pfannenstiel, Quellenbuch, S. 6, 38. Benutzt wurde auch ein kleines Zimmer in der Pleißenburg (Pfannenstiel, a. a. O., S. 5).

[25]) Allgemeine Zeitung vom 10. XII. 1822, Nr. 206, Beilage (S. 821/22) bei Pfannenstiel, Quellenbuch, S. 34f.

[26]) „Kurze Geschichte“ (s. Anm. 24) S. 39.

[27]) Ich folge hier dem Abdruck bei Zaunick, J. S. C. Schweiggers „Vorschläge zum Besten der Leopoldinisch-Carolinischen Akademie der Naturforscher“ (Oktober 1818) mit dem Plan zur Abhaltung von „akademischen Versammlungen der Naturforscher Deutschlands“, in: Nova Acta Leopoldina NF *29* Nr. 171 (1964) S. 7ff. (33), wo die Statuten mit Datum und Unterschriften [nach dem Originaldruck berichtigt] aus Schweiggers Bericht in seinem „Journal für Chemie und Physik“ *37* (1823) S. 455—458 übernommen sind. In der „Kurzen Geschichte“ (s. Anm. 24) bei Pfannenstiel, Quellenbuch, S. 40 anderes Datum und die Unterschriften ohne Beifügung der Professuren.

[28]) Pfannenstiel, Quellenbuch, S. 13. Die Verzeichnisse S. 160ff.

Kunze 1818 abbrechen. Doch ist kaum zu bezweifeln, daß sich die beiden alten Freunde begrüßt haben, als sie die 14. Versammlung Deutscher Naturforscher und Ärzte in Jena (18. September 1836) besuchten, wo Kieser als Geschäftsführer und Zenker als Sekretär amtierten[29]. Sowohl Brehm wie Kunze trugen etwas vor. Brehm lieferte in der Zoologischen Sektion am 20. September „Einige Bemerkungen über die große Verwandtschaft der Subspecies der Vögel", die er durch 44 Vogelbälge (darunter mehrere seiner berühmten Suiten) erläuterte[30], und in der 3. allgemeinen Sitzung am 23. September sprach er über „Das Betragen der männlichen Raubvögel gegen ihr brütendes Weibchen und die Jungen"[31]. Kunze berichtete über pflanzenschädliche Insekten und über Holzinsekten in Italien bei Mailand[32].

Gerhard Buchda

---

[29]) Ausführlicher Bericht in der „Isis" 1837 H. V–VII, Sp. 321 ff. – Amtlicher Bericht von Kieser. Weimar 1837. – Siehe auch Georg Uschmann, Geschichte der Zoologie und der zoologischen Anstalten in Jena 1779/1919. Jena 1959, S. 14 ff.; ders., Jena 1836. – Die 14. Versammlung Deutscher Naturforscher und Ärzte, in: Nova Acta Leopoldina NF *29* Nr. 171 (1964) S. 163–178.

[30]) „Isis" 1837 Sp. 498 f. – Zoologisch-anatomisch-physiologische Abteilung. 1. Sitzung am 20. September unter dem Vorsitz von Lichtenstein und Huschke.

[31]) „Isis", a. a. O., Sp. 367–373 unter 15.

[32]) „Isis", a. a. O., Sp. 526 unter Nr. 4 und Nr. 8.

ABKÜRZUNGEN

AAB, ADAW	Abhandlungen der Deutschen (Preußischen) Akademie der Wissenschaften zu Berlin. Berlin
AAH	Abhandlungen der Heidelberger Akademie der Wissenschaften. Heidelberg
AAMz	Abhandlungen der Akademie der Wissenschaften und der Literatur zu Mainz. Wiesbaden
AASOR	The Annual of the American Schools of Oriental Research. (New Heaven) Cambridge
AAWG	Abhandlungen der Akademie der Wissenschaften zu Göttingen. Göttingen
Abh.	Abhandlungen
AdW	Akademie der Wissenschaften der DDR
AfO	Archiv für Orientforschung. (Berlin) Graz
AGSJU	Arbeiten zur Geschichte des Spätjudentums und Urchristentums. Leiden
AIPhHOS	Annuaire de l'Institut de Philologie et d'Histoire Orientales et Slaves. Brüssel
AKM	Abhandlungen für die Kunde des Morgenlandes
Alman.	Almanach
AMz	Akademie der Wissenschaften und der Literatur zu Mainz, Jahrbuch. Wiesbaden
AnBibl	Analecta Biblica. Rom
Ann.	Annotation
Anz.	Anzeige
ANVAO	Avhandlinger utgitt av det Norske Videnskaps-Akademi i Oslo. Oslo
AO	Der Alte Orient. Leipzig
AOH	Acta Orientalia Academiae Scientiarum Hungaricae
Arch.	Archiv
ArOr	Archiv Orientalní. Prag
ARW	Archiv für Religionswissenschaft. Leipzig, Berlin
ATA	Alttestamentliche Abhandlungen. Münster
ATD	Das Alte Testament Deutsch. Göttingen
BAH	Bibliothèque Archéologique et Historique. Paris
BAL, BVSAW	Berichte über die Verhandlungen der Sächsischen Akademie der Wissenschaften zu Leipzig. Berlin
BASOR	Bulletin of the American Schools of Oriental Research. New Heaven/Baltimore

BBB	Bonner Biblische Beiträge. Bonn
Bd	Band, Bände
Beil.	Beilage
Beitr.	Beiträge zur Geschichte der deutschen Sprache und Literatur
Ber.	Berichte
Bespr.	Besprechung
BEThL	Bibliotheca Ephemeridum Theologicarum Lovaniensium. Paris, Gembloux
BFChTh	Beiträge zur Förderung Christlicher Theologie. Gütersloh
BHTh	Beiträge zur Historischen Theologie. Tübingen
BI	Bibliotheca Islamica
Bibl	Biblica. Rom
BK	Biblischer Kommentar. Neukirchen
BL	The Society for Old Testament Study: Book List. Oxford
BLs	Eleven Years of Bible Bibliography. The Book Lists of the Society for Old Testament Study 1946–1957, ed. by H. H. Rowley. Indian Hills 1957
BO, BiOr	Bibliotheca Orientalis. Leiden
BOuT	De Boeken van het Oude Testament. Roermond en Maaseik
BrP	Beiträge zur romanischen Philologie
BSLP	Bulletin de la Société de Linguistique de Paris
BSOAS	Bulletin of the School of Oriental and African Studies
BWA(N)T	Beiträge zur Wissenschaft vom Alten (und Neuen) Testament. (Leipzig) Stuttgart
BZAW	Beihefte zur Zeitschrift für die alttestamentliche Wissenschaft. (Gießen) Berlin
CAB	Cahiers d'Archéologie Biblique. Neuchâtel
CVA	Corpus Vasorum Antiquorum
ders., dies.	derselbe, dieselbe
DLZ	Deutsche Literaturzeitung. Berlin 1880ff.
DMG	Deutsche Morgenländische Gesellschaft
dtsch.	deutsch (e, er)
dtv	Deutscher Taschenbuch-Verlag
EH	Exegetisches Handbuch zum Alten Testament. Münster
Einltg.	Einleitung
EMM	Evangelisches Missionsmagazin
ET	The Expository Times. Edinburgh
EvD	Die evangelische Diaspora
Festschr.	Festschrift
Forsch.	Forschung(en)
FRLANT	Forschungen zur Religion und Literatur des Alten und Neuen Testaments. Tübingen
FuF, FF	Forschungen und Fortschritte. Berlin
FwF	Fikrun wa Fann

Gelre	Bijdragen en Medede(e)lingen van „Gelre“, Vereeiniging tot Beoefening van Geldersche Geschiedenis, Oudheidkunde en Recht
GGA	Göttingische Gelehrte Anzeigen. Göttingen
Hand. Ned. Nat. en Gen. Congres	Handelingen van het Nederlandsch Natuuren Geneeskundig Congres
HAT	Handbuch zum Alten Testament. Tübingen
Hdb.	Handbuch
HdO	Handbuch der Orientalistik
HM	Homiletische Monatshefte
HS	Die Heilige Schrift des Alten Testaments, hrsg. von F. Feldmann und H. Herkenne. Bonn
HSAT	Die Heilige Schrift des Alten Testaments, hrsg. von E. Kautzsch (4. Aufl. von A. Bertholet). Tübingen
HUCA	Hebrew Union College Annual. Cincinnati
HZ	Historische Zeitschrift
IF	Indogermanische Forschungen
InB	Informationsblatt für die Gemeinden in den niederdeutschen lutherischen Landeskirchen
Interv.	Interview
IQ	The Islamic Quaterly
JAOS	Journal of the American Oriental Society, Baltimore
Jb., Jahrb.	Jahrbuch
JBL	Journal of Biblical Literature. Philadelphia
JEOL	Jaarbericht van het Vooraziatisch-Egyptisch Gezelschap „Ex oriente lux“. Leiden
JNES	Journal of Near Eastern Studies
JPOS	Journal of the Palestine Oriental Society. Jerusalem
JRAS	Journal of the Royal Asiatic Society
JSS(t)	Journal of Semitic Studies. Manchester
JVT	Jaarverslag van de Vers(e)niging voor Terpenonderzoek
Kal.	Kalender
KAT	Kommentar zum Alten Testament. (Leipzig) Gütersloh
KD	Kerygma und Dogma
Kl. Schr.	Kleine Schriften
KlT	Kleine Texte für theologische und philosophische Vorlesungen. (Bonn) Berlin
KMU	Karl-Marx-Universität Leipzig
K(N)AW	Koninklijke (Nederlandse) Akademie van Wetenschappen
(K)NOB	(Koninklijke) Nederlandse Oudheidkundige Bond
Ktn	Karten
KZ	Kartell-Zeitung. Organ des Eisenacher Kartells Akademisch-Theologischer Vereine. Berlin
Lfg	Lieferung
LUA	Lunds Universitets Arsskrift. Lund
LuJ	Luther-Jahrbuch

MAOG	Mitteilungen der Altorientalischen Gesellschaft
Mbl.	Monatsblätter
MEOL	Mededeelingen en Verhandlingen van het Vooraziatisch-Egyptisch Gezelschap „Ex oriente lux". Leiden
MGH	Monumenta Germaniae Historica
MHM	The Muslim Hall Magazine. Dacca
MIO(F)	Mitteilungen des Instituts für Orientforschung. Berlin
Mitt.	Mitteilungen
MUSJ	Mélanges de l'Université Saint-Joseph. Beyrouth
MVÄG	Mitteilungen der Vorderasiatisch (-Ägyptischen) Gesellschaft. (Berlin) Leipzig
Nachr.	Nachruf
NAWG	Nachrichten von der Akademie der Wissenschaften zu Göttingen. Göttingen
NC	La Nouvelle Clio. Brüssel
NDV	Nieuwe Drents(ch)e Volksalmanak
NF	Neue Folge
NGG	Nachrichten von der Gesellschaft der Wissenschaften zu Göttingen. Göttingen
Nieuws-Bulletin KNOB	Nieuws-Bulletin van de Koninklijke Nederlandse Oudheidkundige Bond
NPT	Nationalpreisträger
NS	Neue Serie o. ä.
NT	Novum Testamentum
NTG	Nederlandsch Tijdschrift voor Geneeskunde
NTT	Norsk Teologisk Tidsskrift. Oslo
NZSTh	Neue Zeitschrift für Systematische Theologie. Berlin
OBL	Orientalia et Biblica Lovaniensia. Louvain
OJ	Oudheidkundig Jaarboek
OLZ	Orientalische Literaturzeitung. (Leipzig) Berlin
OMROL	Oudheidkundig Medede(e)lingen van het Rijksmuseum van Oudheden te Leinen
Onoma	Onoma. Louvain 1950ff.
OTS	Oudtestamentische Studien. Leiden
o. V.	ohne Verfasser
Overijsselsch Regt en Gesch.	Verslagen en Mededeelingen van de Vereeinigin tot Beoefining van Geschiedenis
PB	Pastoralblätter
PEF Annual	Palestine Exploration Fund, Annual. London
PEQ	Palestine Exploration Quarterly. London
PGM	Petermanns Geographische Mitteilungen
PH	Pädagogische Hochschule
PJb	Palästina-Jahrbuch. Berlin
Pl.	Plan, Pläne
POT	Princeton Oriental Texts
Prbl	Protestantenblatt. Wochenschrift für den deutschen Protestantismus. Berlin, Bremen

Pret. Or. Ser.	Pretoria Oriental Series. Pretoria
PrJb	Preußische Jahrbücher. Berlin
PrM	Protestantische Monatshefte. Leipzig
Psych.	Zeitschrift für Psychatrie, Neurologie und medizinische Psychologie
Rdsch.	Rundschau
Rev.	Revue
RGG	Religion in Geschichte und Gegenwart. Handwörterbuch für Theologie und Religionswissenschaften. Tübingen
RGW	Rat für gegenseitige Wirtschaftshilfe
RILOB	Recherches publiées sous la direction de l'Institut de Lettres Orientales de Beyrouth
RSO, RStOr	Rivista degli Studi Orientali. Rom
RuS	O. Eißfeldt, Ras Schamra und Sanchunjaton. Halle 1939
SAB	Sitzungsberichte der Deutschen (Preußischen) Akademie der Wissenschaften zu Berlin. Berlin
SAH	Sitzungsberichte der Heidelberger Akademie der Wissenschaften. Heidelberg
SAL	Sitzungsberichte der Sächsischen Akademie der Wissenschaften zu Leipzig. Berlin
SAT	Die Schriften des Alten Testaments. Göttingen
SbBAW	Sitzungsberichte der Bayerischen Akademie der Wissenschaften. München
SGV	Sammlung gemeinverständlicher Vorträge aus dem Gebiet der Theologie und Religionsgeschichte. Tübingen
SKB	Sächsisches Kirchenblatt
SKG	Schriften der Königsberger Gelehrten Gesellschaft. Halle
SNVAO	Skrifter utgitt av det Norske Videnskaps-Akademi i Oslo. Oslo
StANT	Studien zum Alten und Neuen Testament. München
SThU	Schweizerische Theologische Umschau. Bern
StOr	Studia Orientalia. Helsinki
SVT	Supplements to Vetus Testamentum. Leiden
T.	Teil
TB	Harper Torchbook
TH	Technische Hochschule
ThBl	Theologische Blätter. Leipzig
TG	Tijdschrift voor Geschiedenis
TGLV	Tijdschrift voor Geschiedenis, Land- en Volkenkunde
ThLB	Theologisches Literaturblatt
ThLZ	Theologische Literaturzeitung. (Leipzig) Berlin
ThR	Theologische Rundschau. Tübingen
ThStKr	Theologische Studien und Kritiken. Gotha
ThZ	Theologische Zeitschrift. Basel
TKNAG	Tijdschrift van het Koninklijk Nederlands(ch) Aardrijkskundig Genootschap
TNDV	Tijdschrift der Nederlandsche Dierkundige Vereeniging

TU	Texte und Untersuchungen zur Geschichte der altchristlichen Literatur. Berlin
TU	Technische Universität
Versl. Geol.-Mijnb. Gen.	Verslagen der Geologische Sectie van het Geologisch-Mijnbouwkundig Genootschap voor Nederland en Koloniën
Versl. Gesch. en Kunst	Verslag van de Provinciale Commissie van Toezicht op de Bewaring en Instandhouding van Voorwerpen van Waarde uit het oogpunt van Geschiedenis en Kunst
VerslagGM	Verslag van de(n) toestand van het Museum van Oudheden voor Provincie en Stadt Groningen
VIOF	Veröffentlichungen des Instituts für Orientforschung. Berlin
Vortr.	Vortrag
Vorw.	Vorwort
VT	Vetus Testamentum. Leiden
WHO	World Health Organisation
Wiss. Ann.	Wissenschaftliche Annalen. Berlin 1952ff.
WMANT	Wissenschaftliche Monographien zum Alten und Neuen Testament. Neukirchen
WO	Die Welt des Orients. Göttingen
Wschr.	Wochenschrift
WTA	Wissenschaftlich-technische Assistenten
WZ	Wissenschaftliche Zeitschrift
WZKM	Wiener Zeitschrift für die Kunde des Morgenlandes
Z., Zschr.	Zeitschrift
ZAA	Zeitschrift für Anglistik und Amerikanistik
ZAW	Zeitschrift für die alttestamentliche Wissenschaft. (Gießen) Berlin
Zbl.	Zentralblatt
ZDMG	Zeitschrift der Deutschen Morgenländischen Gesellschaft. (Leipzig) Wiesbaden
ZDPV	Zeitschrift des Deutschen Palästina-Vereins. (Leipzig) Wiesbaden
ZfdPh.	Zeitschrift für deutsche Philologie
ZfG	Zeitschrift für Geschichtswissenschaft
ZfS, ZfSl.	Zeitschrift für Slawistik
ZMR	Zeitschrift für Missionskunde und Religionswissenschaft. Berlin
ZNW	Zeitschrift für die neutestamentliche Wissenschaft. (Gießen) Berlin
ZRGG	Zeitschrift für Religions- und Geistesgeschichte. Köln
ZrP	Zeitschrift für romanische Philologie
ZS	Zeitschrift für Semistik und verwandte Gebiete. Leipzig
ZSS	Zeitschrift der Savigny-Stiftung für Rechtsgeschichte
Ztg	Zeitung
ZThK	Zeitschrift für Theologie und Kirche. Tübingen

NAMENREGISTER

(Mitglieder der Akademie und der mit ihnen im Zusammenhang stehenden Personen)

ZUWAHLEN

In der Plenarsitzung vom 15. Oktober 1976 wurden gewählt:

Auswärtiges Mitglied
der mathematisch-naturwissenschaftlichen Klasse

Skrjabin, Georgi Konstantinowitsch, Dr. rer. nat., Professor, Generalsekretär der Akademie der Wissenschaften der UdSSR, Direktor des Akademie-Instituts für Biochemie und Physiologie der Mikroorganismen in Puschtschino, Staatspreisträger der UdSSR, korresp. Mitglied der Akademie der Wissenschaften der UdSSR, Moskau V-71, Leninsky Prospekt 14.

Auswärtige Mitglieder
der philologisch-historischen Klasse

Angelov, Dimiter Simeonov, Dr. phil., o. Professor für byzantinische Geschichte an der Universität Sofia, Nationalpreisträger der VR Bulgarien, korresp. Mitglied der Bulgarischen Akademie der Wissenschaften in Sofia und Direktor des Archäologischen Instituts dieser Akademie, Vizepräsident der Association Internationale des études byzantines und Vorsitzender des Bulgarischen Nationalkomitees für Byzantinistik, geb. 2. II. 1917, Sofia 11, bulv. Chr. cabakčiev 79 „G".

Jarceva, Victoria N., Dr. phil., Professor, korresp. Mitglied der Akademie der Wissenschaften der UdSSR und Direktorin des Instituts für Sprachwissenschaft dieser Akademie in Moskau, Moskau, ul. Marxa i Engelsa 1/14.

Telegdi, Zsigmond, Dr. phil., o. Professor für Sprachwissenschaft an der Universität Budapest, Budapest V, Pesti-Barnabas út. 1.

Turner, Eric Gardner, M. A., Dr. phil. h. c. (Brüssel), o. Professor für Papyrologie an der Universität London, Präsident der Union Académique Internationale und der Internationale Association of Papyrologists, o. Mitglied der British Academy in London, auswärtiges Mitglied der Académie Royale des Sciences, des Lettres et des Beaux-Arts de Belgique in Brüssel und der Accademia delle Scienze di Napoli, Ehrenmitglied der Societas Scientiarium Fennica-Finska Vetenskaps Societeten-Suomen Tiedeseura in Helsinki, geb. 26. II. 1911.

Kolibri 360 GmbH
Ferdinand-Jühlke-Straße 7
99095 Erfurt, Deutschland
produktsicherheit@kolibri360.de